Springer Mineralogy

More information about this series at http://www.springer.com/series/13488

Nikita V. Chukanov • Marina F. Vigasina

Vibrational (Infrared and Raman) Spectra of Minerals and Related Compounds

Volume 2

Nikita V. Chukanov
Institute of Problems
of Chemical Physics
Russian Academy of Sciences
Chernogolovka, Russia

Marina F. Vigasina
Geological Faculty
Moscow State University
Moscow, Russia

Additional material to this book can be downloaded from http://extras.springer.com.

ISSN 2366-1585   ISSN 2366-1593   (electronic)
Springer Mineralogy
ISBN 978-3-030-26805-3   ISBN 978-3-030-26803-9   (eBook)
https://doi.org/10.1007/978-3-030-26803-9

© Springer Nature Switzerland AG 2020

This work is subject to copyright. All rights are reserved by the Publisher, whether the whole or part of the material is concerned, specifically the rights of translation, reprinting, reuse of illustrations, recitation, broadcasting, reproduction on microfilms or in any other physical way, and transmission or information storage and retrieval, electronic adaptation, computer software, or by similar or dissimilar methodology now known or hereafter developed.

The use of general descriptive names, registered names, trademarks, service marks, etc. in this publication does not imply, even in the absence of a specific statement, that such names are exempt from the relevant protective laws and regulations and therefore free for general use.

The publisher, the authors, and the editors are safe to assume that the advice and information in this book are believed to be true and accurate at the date of publication. Neither the publisher nor the authors or the editors give a warranty, expressed or implied, with respect to the material contained herein or for any errors or omissions that may have been made. The publisher remains neutral with regard to jurisdictional claims in published maps and institutional affiliations.

This Springer imprint is published by the registered company Springer Nature Switzerland AG.
The registered company address is: Gewerbestrasse 11, 6330 Cham, Switzerland

Alexandr Dmitrievich Chervonnyi
1948–2017

This book is dedicated to the memory of the outstanding scientist, a specialist in the field of inorganic materials and chemistry of rare-earth elements Dr. Alexandr Dmitrievich Chervonnyi. Him belongs the idea to publish this book.

Preface

This volume is the third and final part of the series of reference books on vibrational spectra of minerals. Unlike the two previous parts (Chukanov 2014; Chukanov and Chervonnyi 2016), this book contains not only infrared (IR) spectra of minerals but also data on their Raman spectra.

In Chap. 1, numerous examples of the application of IR spectroscopy to the analysis of crystal-chemical features of minerals are considered. In particular, spectral bands that characterize different local situations around OH^- and BO_3^{3-} groups in vesuvianite-group minerals are revealed. The effect of symmetry on the parameters of IR spectra of vesuvianite-group minerals is discussed. By means of IR and Raman spectroscopic methods, it is shown that the clathrate mineral melanophlogite is not a single species but a mineral group including minerals with different combinations of small molecules (CO_2, CH_4, H_2S, N_2, H_2O, C_2H_6) entrapped in structural cages. Based on numerous IR spectra of nakauriite samples from different localities, it is demonstrated that this mineral does not contain sulfate groups, and its tentative simplified formula $(Mg_3Cu^{2+})(OH)_6(CO_3)\cdot 4H_2O$ is suggested. A close crystal chemical relationship between nepskoeite and shabynite is demonstrated based on their IR spectra, compositional, and X-ray diffraction data. Contrary to the formula $Mg_4Cl(OH)_7\cdot 6H_2O$ accepted for nepskoeite, this mineral is a borate with the tentative simplified formula $Mg_5(BO_3)(Cl, OH)_2(OH)_5\cdot nH_2O$ ($n > 4$). Consequently, shabynite may be a product of nepskoeite dehydration. Based on IR spectroscopic data, it is also shown that some nominally boron-free lead carbonate minerals (molybdophyllite, hydrocerussite, plumbonacrite, somersetite) often contain minor BO_3^{3-} admixture which is overlooked in structural and chemical analyses.

Chapter 2 contains IR spectra of 1024 minerals and related compounds which were not included in the preceding reference books of this series (Chukanov 2014; Chukanov and Chervonnyi 2016). Most spectra are accompanied by the information about the origin of reference samples, methods of their identification, and analytical data.

In Chap. 3, possibilities, advantages, and shortcomings of Raman spectroscopy as a method of investigation and identification of minerals are discussed. Numerous examples illustrate capabilities of Raman spectroscopy in identification of minerals and analysis of their crystal chemical features, orientation, and polarization effects, selection rules, as well as difficulties encountered in the study of microscopic inclusions in minerals and minerals that are unstable under laser beam.

Chapter 4 contains data on 2104 Raman spectra of more than 2000 mineral species taken from various periodicals. The data are accompanied by some experimental details and information on the reference samples used.

A supplementary chapter provides comments on published IR spectra which are erroneous, dubious, or of poor quality. This chapter is provided by a separate list of references.

This work was carried out with assistance of numerous colleagues. The working partnership with Prof. I.V. Pekov, Dr. A.D. Chervonnyi, and Dr. S.A. Vozchikova was the most important.

Reference samples and valuable analytical data were kindly granted by A.V. Kasatkin, S. Jančev, E. Jonssen, R. Hochleitner, E.V. Galuskin, S. Weiss, N.V. Sorokhtina, Ł. Kruszewski, and many other mineralogists, as well as mineral collectors, of which the contribution of R. Kristiansen, G. Möhn, W. Schüller, B. Ternes, G. Blass, B. Dünkel, S. Möckel, and C. Schäfer was the most important. Collaboration with the crystallographers N.V. Zubkova, R.K. Rastsvetaeva, S.M. Aksenov, D.I. Pushcharovsky, T.L. Panikorovskii, O.I. Siidra, S.N. Britvin, M.G. Krzhizhanovskaya, D.A. Ksenofontov, S.V. Krivovichev, and I. Grey, as well as with specialists in different areas of geosciences and analytical methods (J. Göttlicher, K.V. Van, D.A. Varlamov, V.N. Ermolaeva, D.I. Belakovskiy, Yu.S. Polekhovsky, P. Voudouris, A. Magganas, A. Katerinopoulos, N.V. Shchipalkina, V.O. Yapaskurt, L.A. Pautov, V.S. Rusakov, R. Scholz, A.R. Kampf, S. Encheva, P. Petrov, Ya.V. Bychkova, N.N. Koshlyakova, P. Yu. Plechov, C.L.A. de Oliveira, I.S. Lykova, and T.S. Larikova) was especially fruitful. All of them are kindly appreciated.

This work was partly supported by the Russian Foundation for Basic Research, grant no. 18-29-12007. A part of analytical data on reference samples was obtained in accordance with the Russian Government task, registration no. 0089-2016-0001.

Chernogolovka, Russia	Nikita V. Chukanov
Moscow, Russia	Marina F. Vigasina

Reference

Chukanov NV, Chervonnyi AD (2016) Infrared spectroscopy of minerals and related compounds. Springer, Cham. (1109 pp)

Contents

3 Some Aspects of the Use of Raman Spectroscopy in Mineralogical Studies 721
  3.1 General Principles of Raman Spectroscopy 721
  3.2 Specific Features and Possibilities of Raman Spectroscopy: Practical Recommendations 722
    3.2.1 Advantages and Disadvantages of the Method 722
    3.2.2 Spectral Band Assignment 727
    3.2.3 Effect of Structural Disorder on Raman Spectra of Minerals 728
    3.2.4 Selection Rules 730
    3.2.5 The Longitudinal-Transverse Splitting 732
    3.2.6 Orientation and Polarization Effects; Analysis of Water and OH Groups 733
    3.2.7 Effect of Luminescence 736
    3.2.8 Destructive Effect of Laser Radiation 736

4 Raman Spectra of Minerals 741

References 1257

Index 1351

Contents for Volume 1

1 **Some Examples of the Use of IR Spectroscopy in Mineralogical Studies** 1
   1.1 Characteristic Bands in IR Spectra of Vesuvianite-Group Minerals 1
      1.1.1 O–H-Stretching Vibrations 1
      1.1.2 B–O-Stretching Vibrations 3
      1.1.3 Stretching and Bending Vibrations of SiO_4^{4-} and $Si_2O_7^{6-}$ Groups 5
   1.2 Problem of Melanophlogite 5
   1.3 Problem of Nakauriite 9
   1.4 Relationship Between Nepskoeite and Shabynite 13
   1.5 Orthoborate Groups in Lead Carbonate Minerals 14

2 **IR Spectra of Minerals and Related Compounds, and Reference Samples Data** 19
   2.1 Borates, Including Arsenatoborates and Carbonatoborates .. 21
   2.2 Carbonates 78
   2.3 Organic Compounds and Salts of Organic Acids 98
   2.4 Nitrides and Nitrates 111
   2.5 Oxides and Hydroxides 117
   2.6 Fluorides and Fluorochlorides 226
   2.7 Silicates 236
   2.8 Phosphides and Phosphates 350
   2.9 Sulfides, Sulfites, Sulfates, Carbonato-Sulfates, Phosphato-Sulfates, and Tellurato-Sulfates 470
   2.10 Chlorides and Hydroxychlorides 537
   2.11 Vanadates and Vanadium Oxides 555
   2.12 Chromates 590
   2.13 Germanates 597
   2.14 Arsenides, Arsenites, Arsenates, and Sulfato-Arsenates ... 597
   2.15 Selenides, Selenites, and Selenates 639
   2.16 Bromides 664
   2.17 Molybdates 665
   2.18 Tellurides, Tellurites, and Tellurates 680

2.19 Iodides, Iodites, and Iodates 699
2.20 Xenates 704
2.21 Tungstates and W-Bearing Oxides 706

Some Aspects of the Use of Raman Spectroscopy in Mineralogical Studies

3.1 General Principles of Raman Spectroscopy

Raman scattering is the process of inelastic light scattering that occurs on fluctuations in the polarizability of molecules which are excited to higher vibrational or rotational energy levels. This phenomenon was discovered in 1928 by C.V. Raman and K.S. Krishnan (for liquids) and L.I. Mandelshtam and G.S. Landsberg (for crystals).

In Raman scattering, a monochromatic line of exciting laser radiation after interaction with a substance is accompanied by an additional set of spectral components. The newly appeared lines are located on the scale of electromagnetic waves symmetrically with respect to the line of exciting radiation and are separated from it by the frequencies of atomic vibrations. The totality of newly emerging spectral components is the Raman spectrum of a substance, which is its diagnostic feature. The phenomenon of Raman scattering is characteristic of substances that are in a gaseous, liquid, or solid state consisting of molecules or molecular complexes with an internal structure, or atoms combined into crystalline structures. Monatomic gas particles that do not interact with each other (for example, inert gases) do not have Raman spectra.

The source of secondary radiation (Raman scattering) is a variable in time electric dipole moment that occurs in the medium as a result of the interaction of particles of a substance with the electric component of external electromagnetic radiation of the visible or close to visible range. The magnitude of the dipole moment depends on the magnitude of the external field and on the polarizability of the substance:

$$p = \hat{a}E$$

where p is the induced electric dipole moment vector, E is the vector of external electric field strength, and $\hat{a}$ is the polarizability.

The polarizability of the substance depends on the structure of the molecules or crystals forming it, on the types of bonds in the substance, as well as on the nature of the motions of the atoms in the molecules or in the crystals. Polarizability is a variable in time characteristic of the substance, which is modulated by the movements of the particles of the substance itself and the electrical component of the external electromagnetic field, causing the appearance of a variable electric dipole moment in the substance. In accordance with the basic rule of electrodynamics, a system with a variable electric dipole moment in time can be a source of electromagnetic radiation with a frequency of change in the dipole moment.

In experiments on Raman scattering using monochromatic radiation with a frequency of Ω, spectral components with frequencies Ω, $(\Omega + \omega_i)$ and $(\Omega - \omega_i)$ are recorded in the spectrum of scattered radiation, where ω_i are the frequencies of vibrational and rotational motions of particles of a substance. The spectral region in which the components are located with frequencies greater than the frequency of laser radiation $(\Omega + \omega_i)$ is commonly called the anti-Stokes region, and the region with lower frequencies $(\Omega - \omega_i)$ is called the Stokes region.

Under normal conditions, the intensity of the strongest lines in the Stokes region of the Raman spectra is usually 10^{-6}–10^{-8} of the intensity of the exciting line (Reshetnyak and Bukanov 1991). The intensity of the anti-Stokes component is even less by several decimal exponents of magnitude and decreases rapidly with increasing magnitude of the detuning from the laser line. For this reason, the bulk of the experiments are carried out in the Stokes spectral region. As a source of spectral information, mainly vibrational spectra are used.

Polarizability is anisotropic and is described by a second rank tensor, which can be written as a symmetric matrix:

$$\hat{a} = \begin{vmatrix} a_{xx} & a_{xy} & a_{xz} \\ a_{yx} & a_{yy} & a_{yz} \\ a_{zx} & a_{zy} & a_{zz} \end{vmatrix}$$

The component of the polarizability tensor a_{ij} determines the magnitude of the dipole moment arising in the medium along the i axis under the action of an electromagnetic field with the direction of the polarization vector of the electric field along the j axis. This means that for different orientations of the polarization vector of the laser radiation and the polarization vector of the detected scattered radiation in the Raman spectrum, scattering will be recorded on different components of the polarizability tensor. In a general case, Raman scattering occurs at different vibrations, and the recorded scattering lines in the Raman spectra have different frequencies and intensities. The intensity of the scattering line in the case of nonpolar normal vibrations is determined by the following formula:

$$I \sim \left[\sum f_i \alpha_{ij} e_j\right]^2,$$

$$i,j = x, y, z$$

where f_i and e_j are components of the unit vectors of the dipole moment polarization and laser radiation, respectively, and α_{ij} is the change of the polarizability tensor component at a given kind of normal vibrations. Not all types of vibrations can be detected as lines in the Raman spectra. For molecules and crystals with an inversion center, there is an alternative prohibition rule which is very important for experimental practice. According to this rule, for compounds with an inversion center, bands of antisymmetric (with respect to the inversion center) vibrations are forbidden in the Raman spectra, and symmetrical ones are forbidden in the IR spectra. The alternative prohibition rule relates simultaneously to Raman spectroscopy and IR absorption spectroscopy and indicates the complementary nature of these methods of molecular spectroscopy.

The theory of Raman spectroscopy is described in more detail in numerous publications (Brandmüller and Moser 1962; Anderson 1973; Sushchinsky 1981; Banwell 1983; Zhizhin et al. 1984; Nakamoto 2009).

3.2 Specific Features and Possibilities of Raman Spectroscopy: Practical Recommendations

3.2.1 Advantages and Disadvantages of the Method

The most important properties of Raman spectroscopy are that this method is nondestructive and local. The ability of laser radiation to penetrate inside transparent minerals makes Raman spectroscopy indispensable for diagnosing mineral phases in inclusions (see, e.g., Figs. 3.1 and 3.2). In this case, the minimum dimensions of the

3.2 Specific Features and Possibilities of Raman Spectroscopy: Practical Recommendations

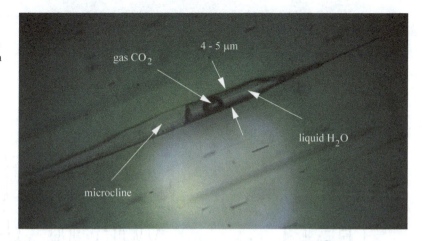

Fig. 3.1 A three-phase inclusion in aquamarine. The transverse size of the inclusion is about 4–5 μm

investigated phases are limited by the diameter of the focal spot of laser radiation. When using high-quality optical elements, laser radiation can be focused into a spot with a diameter of up to several microns. In this case, phase diagnostics can be carried out in situ, without damage the inclusions and disbalance the phase equilibrium. The highest quality Raman spectra are obtained for solid-state phases because of their high density (Fig. 3.3).

Figure 3.2 illustrates both the large diagnostic capabilities of Raman spectroscopy and the difficulties encountered in the study of inclusions. Laser radiation, before it reaches an inclusion, goes some distance in the host mineral causing Raman scattering in the latter. As a result, the resulting spectrum contains the scattering lines of both the studied inclusion and the host mineral. In the cases when the host mineral spectrum is rich in its own scattering lines, the diagnosis of the substances of microscopic inclusions can be significantly complicated. It should be taken into account that in the spectra of microscopic inclusions usually only most intense scattering lines can be observed, which may have low intensities against the background of the more powerful spectrum of the host mineral. Obtaining spectra of inclusions located as close as possible to the surface of the host mineral reduces the laser beam path through the latter. The depth at which diagnostics of inclusions is possible is limited by the focal length of the lens used.

In some cases, in multiphase inclusions, it is possible to diagnose not only solid, but also liquid (Fig. 3.4) and gaseous (Figs. 3.5 and 3.6) phases.

In the region of stretching vibrations of water, a strong and narrow scattering line with a frequency of 3610 cm^{-1} is recorded, which refers to H_2O molecules located in the channels of the aquamarine structure. Due to the small diameter of the channels (about 0.5 nm), water molecules exist in a constrained state with hydrogen atoms attached to the channel walls and do not form strong hydrogen bonds. This is reflected in the small half-width of the scattering line of about 4 cm^{-1}. The broad band at 3420 cm^{-1} with a shoulder at 3250 cm^{-1} corresponds to O–H-stretching vibrations of water molecules forming rather strong hydrogen bonds and belonging to the liquid phase of the inclusion.

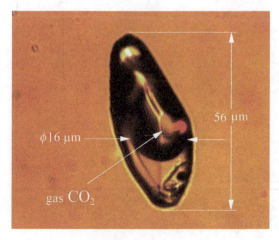

Fig. 3.2 A multiphase inclusion in topaz

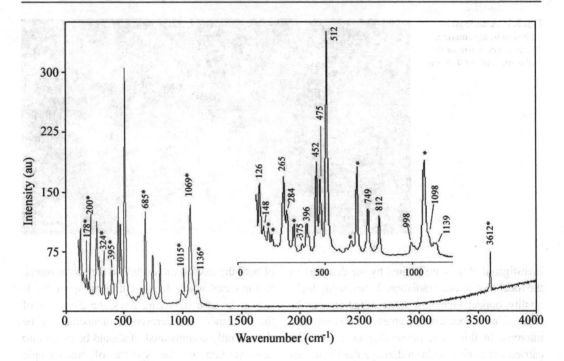

Fig. 3.3 Raman spectrum of a microscopic inclusion of microcline in aquamarine. Bands marked with an asterisk belong to the aquamarine matrix. The laser emission wavelength is 532 nm, the spectral resolution is about 6 cm^{-1}, and the laser power is 30 mW

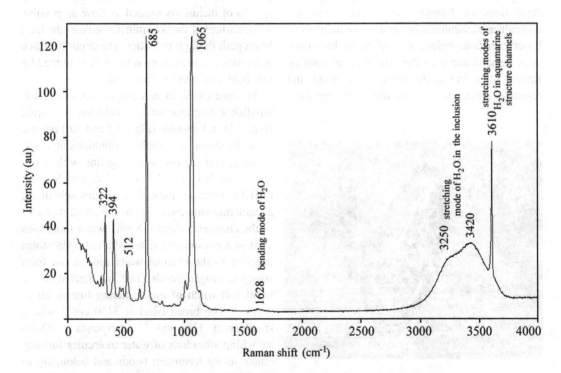

Fig. 3.4 Raman spectrum of the liquid phase of the inclusion in aquamarine

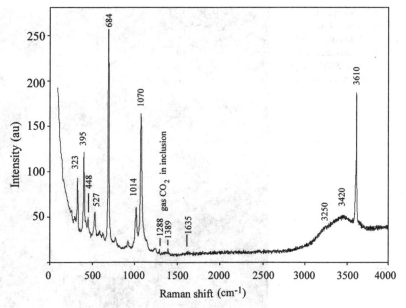

Fig. 3.5 Raman spectrum showing weak bands of gaseous CO_2 (gas bubble in the liquid) in the inclusion in aquamarine. All strong narrow bands correspond to aquamarine matrix

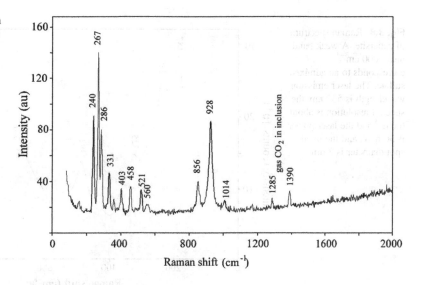

Fig. 3.6 Raman spectrum showing bands of gaseous CO_2 in a multiphase inclusion in topaz (see Fig. 3.2)

To obtain the maximum possible intensity of the Raman signal, the focal volume of the laser beam must be immersed in the substance of the host mineral and pointed at the object under investigation. In this case, a defocused luminous spot forms on the surface of the host mineral at the entry point of the laser beam, which can cover the working field. In such a situation, it is not possible to determine not only the diameter of the focal spot on the phase being diagnosed, but it is also generally difficult to understand whether the laser beam is focused on the inclusion. In this case, one can judge the success of the experiment only from the results of a comparison of the Raman spectra obtained in the pure region of the host mineral and the spectra obtained from inclusions. Reliable confirmation of the result in this situation is the reproducibility of spectral data.

Another difficulty in working with inclusions is a significant loss of laser power when passing

Fig. 3.7 Sword-like microcrystals of the rare mineral katiarsite KTiO(AsO$_4$) on arsmirandite crystal crust. SEM image (in secondary electrons)

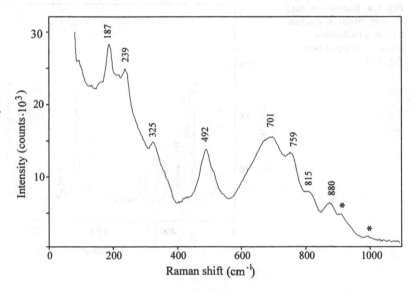

Fig. 3.8 Raman spectrum of catiarsite. A weak band near 1000 cm^{-1} corresponds to an admixed sulfate. The laser emission wavelength is 532 nm, the spectral resolution is about 6 cm^{-1} and the laser power is 30 mW, and the focal spot diameter is 3 µm

through the boundary of the inclusion. Uneven boundary can cause strong scattering of the laser beam, reducing the effective exciting power. Moreover, the total internal reflection conditions for a laser beam can be realized on the surface of the inclusion capsule. In this case, the laser beam will not penetrate into the inclusion, and obtaining a Raman spectrum will not be possible.

The "nondestructiveness" and locality of the Raman spectroscopy method were the reason for its widespread use in the study of unique minerals represented by single finds or microscopic monomineral aggregates. If the studied mineral is represented by individual prismatic crystals or thin needles (Fig. 3.7), then to obtain the Raman spectrum (Fig. 3.8), the sample area was chosen whose linear dimensions are larger than the diameter of the laser beam focal spot. Otherwise, there will be a loss of power of the exciting radiation and the Raman scattering signal.

The intensity of the scattered radiation depends on the number of scattering centers in the focal volume of the laser beam. Therefore, *ceteris paribus*, the best quality of the Raman spectrum will be obtained in the area of the sample where the mineral aggregate has the highest concentration of the substance. For example, with needle-like or finely prismatic microcrystal forms, the Raman spectrum should be recorded at the common base of needle growth. The interpretation of the Raman spectra obtained on the microaggregates of minerals and the identification of the scattering lines related to the mineral of interest require special attention and analysis. When working with microscopic aggregates of minerals, one should take into account the possible presence of mineral impurities, the removal of which is impossible due to the small size.

3.2.2 Spectral Band Assignment

Raman spectra primarily reflect the features of the anionic part of the mineral, as well as some polyatomic cations like NH_4^+ or UO_2^{2-}. The frequencies of symmetric stretching vibrations of some complex anions occurring in the structures of minerals change mainly in the following ranges (cm^{-1}):

Nesosilicates	820–980	Sulfates	970–1020
Carbonates	1050–1100	Arsenates	800–900
Molybdates	780–880	Tungstates	850–920
Orthophosphates	930–990	Orthovanadates	820–880

Raman shifts of the stretching vibration bands increase with increasing of polymerization of coordination polyhedra; e.g., for Al-poor framework silicates they are typically in the range 1040–1130 cm^{-1}.

Thus, scattering lines observed in these ranges can be used for a preliminary assignment of a mineral to one or more class of compounds as a step preceding a more precise specification. When conducting diagnostic studies, it is necessary to take into account that the approximate proportion of the intensities of the Raman lines in the spectra of complex anions is as follows: MoO_4^{2-} ($\approx WO_4^{2-}$): SO_4^{2-} : PO_4^{3-} : CO_3^{2-} : SiO_4^{4-} = 10 : 6 : 3 : 1.5 : 1 (for excitation radiation with a wavelength between 488 and 515 nm). This feature can be illustrated by the spectra of cancrinite $Na_6Ca_2[AlSiO_4]_6(CO_3)_2 \cdot 2H_2O$ (Fig. 3.9) and vishnevite $Na_8[AlSiO_4]_6(SO_4) \cdot 2H_2O$ (Fig. 3.10), structurally related tectosilicates with additional

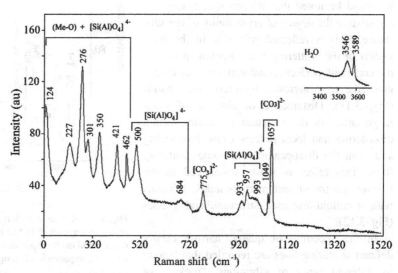

Fig. 3.9 Raman spectrum of cancrinite

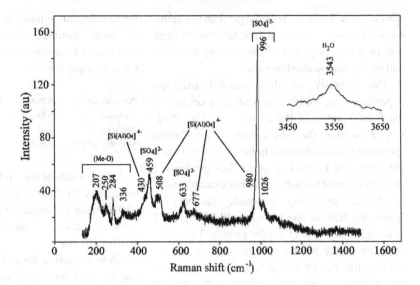

Fig. 3.10 Raman spectrum of vishnevite

anions. The additional anions CO_3^{2-} and SiO_4^{4-} play subordinate role in the chemical composition of these minerals. However, the strongest lines in the Raman spectra are the scattering lines corresponding to the internal fully symmetric stretching vibrations of just additional anions. This feature may cause difficulty in determining chemical class of a mineral.

3.2.3 Effect of Structural Disorder on Raman Spectra of Minerals

It should be noted that Raman spectroscopy is sensitive to the degree of crystallinity of the substance. This is reflected primarily in the half-widths of the scattering lines. Raman spectra of minerals with perfect crystal structures are distinguished by narrow well-resolved bands (Fig. 3.11). Disturbance or absence of long-range order in the structure of matter, cation disordering and local defects cause broadening and even the disappearance of some scattering lines. This effect is most pronounced in the Raman spectra of metamict minerals, minerals with a colloid-dispersed structure, and glasses (Fig. 3.12).

In the spectrum of quartz, narrow clearly defined scattering lines are recorded that belong to different types of vibrations: "rocking of tetrahedra" (210 cm^{-1}), "twisting of tetrahedra" (353 cm^{-1}), O–Si–O bending (467 cm^{-1}), and Si–O stretching (1086 cm^{-1}) (Ranieri et al. 2009). The absence of a long-range order in obsidian, which is a SiO_2-rich glass, results in the absence of specific lines corresponding to any symmetry elements. The broad bands at

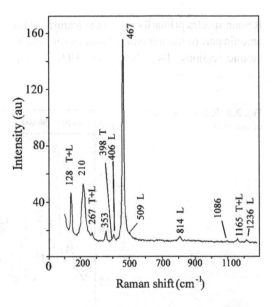

Fig. 3.11 Raman spectrum of quartz powder. The laser emission wavelength is 532 nm, the spectral resolution is 6 cm^{-1}, and the laser power is 30 mW. The letters L and T denote components of longitudinal-transverse splitting (see below)

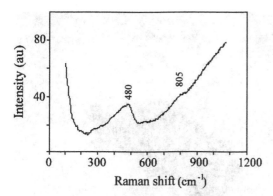

Fig. 3.12 Raman spectrum of obsidian (volcanic glass). The laser emission wavelength is 532 nm, the spectral resolution is 6 cm^{-1}, and the laser power is 30 mW

480 and 805 cm^{-1} correspond to the totality of vibrations in areas with different local structures.

The sensitivity of Raman scattering to the disordered distribution of atoms between crystallographic positions is clearly reflected in the spectra of feldspars. For example, in the structure of the disordered oligoclase (Na,Ca)[AlSi$_3$O$_8$], Al atoms are statistically distributed between different tetrahedral sites, whereas in an ordered variety they are concentrated mainly in only one of independent crystallographic positions. Raman spectra of the two oligoclase varieties differ markedly from each other: in the spectrum of an ordered oligoclase variety, a greater number of scattering lines (Fig. 3.13, upper curve) are recorded, which have a smaller half-width than in the spectrum of a disordered oligoclase (Fig. 3.13, lower curve).

In the study of black microscopic inclusions (Fig. 3.14) in aquamarine, it was found [based on the assignment by Sharma et al. (2001)] that they consist of crystalline graphite, which was diagnosed by the relatively narrow line at 1574 cm^{-1}, and X-ray amorphous carbon showing a broad band at 1336 cm^{-1} (Fig. 3.15). The Raman spectrum made it possible to suppose that the substance in the inclusion is compressed, since the frequencies of the recorded scattering lines differ from the values of the frequencies characteristic of the same substances under normal conditions (i.e., 1360 and 1582 cm^{-1} for amorphous carbon and graphite, respectively).

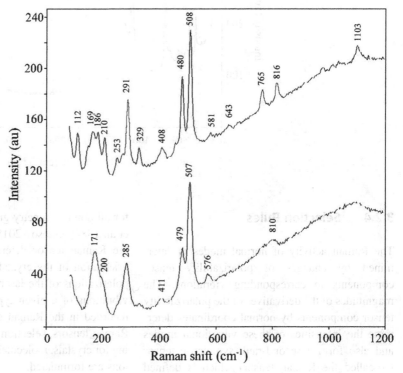

Fig. 3.13 Raman spectra of the ordered (upper curve) and disordered (lower curve) oligoclase varieties. The laser emission wavelength is 532 nm, the spectral resolution is 6 cm^{-1}, and the laser power is 14 mW

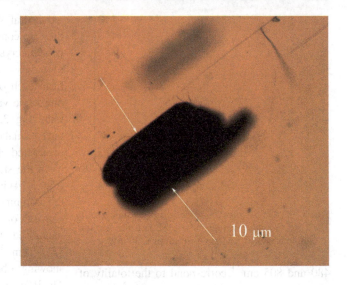

Fig. 3.14 Inclusion of carbonaceous matter in aquamarine

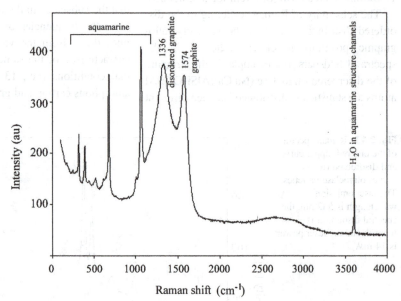

Fig. 3.15 Raman spectrum of a carbonaceous inclusion in aquamarine. The laser emission wavelength is 532 nm, the spectral resolution is 6 cm^{-1}, the laser power is 30 mW, and signal accumulation time is 120 s

3.2.4 Selection Rules

The Raman activity of normal modes is determined by changes of polarizability tensor components in corresponding vibrations. The magnitudes of the derivatives of the polarizability tensor components by normal coordinates determine the intensities of these vibrational modes and also form a second-rank symmetric tensor (so-called the Raman tensor) which is defined for all point symmetry groups (see, e.g., Zhizhin et al. 1984; Kolesov 2018). Nonzero elements of the Raman tensor determine at which relative orientation of the crystallographic axes and the polarizations of the laser and scattered radiation vibrations of a given type of symmetry will be recorded in the Raman spectrum. Based on the Raman tensors, selection rules in Raman scattering for crystals, molecules, molecular groups, and ions are formulated.

For objects with C_{2v} symmetry, to which H_2O and H_2S molecules belong, vibrations of the symmetry types A_1 (symmetric stretching and bending vibrations) and B_1 (antisymmetric stretching vibrations) are permitted by the selection rules. Corresponding Raman bands of water molecules in the gas phase are observed at 3657, 1595, and 3756 cm^{-1}, respectively (Halonen and Carrington Jr 1988).

For isolated undistorted planar trigonal AB_3 ions (symmetry group D_{3h}) like CO_3^{2-}, NO_3^-, and BO_3^{3-}, symmetric stretching vibrations ν_1 with the symmetry A_1' as well as stretching (ν_3) and in-plane bending (ν_4) doubly degenerate vibrations of type E' are permitted in Raman spectra by the selection rules. Out-of-plane bending ν_2 vibrations with the symmetry A_2'' are prohibited.

For isolated tetrahedral AB_4 ions with the T_d symmetry (MoO_4^{2-}, AsO_4^{3-}, PO_4^{3-}, VO_4^{3-}, SO_4^{2-}, SiO_4^{4-}, etc.) in the Raman spectra the stretching ν_1 mode (with the A_1 symmetry), bending doubly degenerate ν_2 mode (with the E symmetry), and triply degenerate ν_3 stretching and ν_4 bending modes (with the F_2 symmetry) are allowed by the selection rules.

In real structures of minerals, the symmetry of tetrahedral ions decreases, sometimes to D_{2h} (flattened tetrahedron) or even to C_s, which leads to the removal of degeneracy and splitting of degenerate vibrations into separate components. To identify the scattering lines, one can use the known regularity established by different authors in numerous experimental studies: in most cases, the scattering lines corresponding to fully symmetric stretching vibrations have a smaller width and higher peak intensity than the scattering lines corresponding to degenerate vibrations. This empirical regularity is explained by the greater polarizability of bonds with fully symmetric stretching vibrations (Kolesov 2018).

Symmetry types of the vibrations of isolated complex ions may differ from that in crystals. For example, in calcite $CaCO_3$ having D_{3d} symmetry, CO_3^{2-} ions are located in positions on the third-order axis (D_3 positional symmetry) and do not change their symmetry compared to the free state. In this case, the same selection rules are valid as for an isolated CO_3^{2-} ion. On the other hand, in aragonite (orthorhombic $CaCO_3$ polymorph with the D_{2h} symmetry), the crystal structure of which does not have axes of the third order, the positional symmetry of the CO_3^{2-} ion decreases to C_s, and according to the selection rules for the D_{2h} group, the out-of-plane bending mode (ν_2), which is classified as A_g, is active in the Raman spectrum of aragonite: corresponding band is observed at 853 cm^{-1} (Frech et al. 1980).

In vivianite $Fe_3^{2+}[PO_4]_2 \cdot 8H_2O$, which is monoclinic with the symmetry C_{2h}, there are no axes of the third order, and the symmetry of the phosphate ion also decreases as compared with isolated PO_4^{3-}. As a result, all vibrations are nondegenerate and are classified according to symmetry types as A_g, A_u, B_g, and B_u. In the Raman spectrum of vivianite, only A_g and B_g bands appear in accordance with the "alternative prohibition" rule applied to symmetry groups with an inversion center:

PO_4^{3-} in aqueous solution (Nakamoto 2009), cm^{-1}	PO_4^{3-} in vivianite (Piriou and Poullen 1984), cm^{-1}
A_1 (ν_1) 938	A_g (ν_1) 951
E (ν_2) 420	A_g (ν_2) 458; B_g (ν_2) 425
F_2 (ν_3) 1017	A_g (ν_3) 1053, 990; B_g (ν_3) 1018
F_2 (ν_4) 567	A_g (ν_4) 572, 539

For isolated regular octahedral AB_6 groups, ν_1 symmetric stretching mode with the symmetry A_{1g}, doubly degenerate ν_2 stretching mode with the symmetry E_g and triply degenerate ν_5 bending mode with the symmetry F_{2g} are permitted in the Raman spectra by the selection rules.

In accordance with the "alternative prohibition rule," in IR spectra of isolated regular octahedral AB_6 groups, only the ν_3 stretching band and the ν_4 bending band (both having the F_{1u} symmetry) are observed. Anionic groups of this type [Si(OH)$_6$, Al(OH)$_6$, Fe^{3+}(OH)$_6$, Mn^{4+}(OH)$_6$, etc.] occur in the structures of ettringite-group minerals. In ettringite $Ca_6[Al(OH)_6]_2(SO_4)_3 \cdot 26H_2O$, the symmetry of which is C_{3v}, with triad axes and reflection planes being the only symmetry elements. As a result, the symmetry of the [Al(OH)$_6$]$^{3-}$ group is also lowered. The ν_5 modes which were forbidden in the Raman spectrum according to the "alternative prohibition rule" for the free anion become nondegenerate and active. Vibrational modes of ettringite are classified according to the symmetry types A_1, A_2, and E. In accordance with the selection rules, only bands of A_1 and E vibrations appear in the Raman spectra (presumably, the bands at ~550 and ~345 cm^{-1}: see Deb et al. 2003; Renaudin et al. 2007). Vibrations of type A_1 can be separately recorded under the conditions when polarizations of laser and scattered radiation and the C_3 crystallographic axis of a single crystal are parallel to each other.

The C_6 symmetry group of thaumasite Ca$_3$[Si(OH)$_6$](SO$_4$)(CO$_3$)·12H$_2$O contains axes of the second, third, and sixth orders. In accordance with this set of symmetry elements, the "alternative prohibition rule" becomes inapplicable, and the degeneration is removed from the triple degenerate modes. This leads to an increase in the number of possible scattering lines in the spectrum. In accordance with the table of group characters, in the spectra of structures with such symmetry, the existence of vibrational modes of the A, B, E_1, and E_2 types is possible. In accordance with the selection rules and Raman scattering tensor, vibrations of symmetry types A, E_1, and E_2 are active in the Raman spectra. The study of the polarized Raman spectra of a thaumasite single crystal showed that the fully symmetric vibrations of the [Si(OH)$_6$]$^{2-}$ anion have a frequency of about 660 cm^{-1} (Kononov et al. 1990).

An example of the manifestation of the "alternative prohibition rule" is the absence of first-order Raman spectra in some minerals with inversion centers. Such minerals, for example, are halite NaCl and sylvite KCl having a cubic (O_h) symmetry. All atoms forming the structures of these minerals are located at the centers of inversion, and any displacements from their equilibrium positions violate the symmetry. As a result, the bands corresponding to all kinds of vibrations are forbidden in the first-order Raman spectra. However, with a large signal accumulation time, it is possible to record weak bands of the second order Raman spectra (Fig. 3.16). The selection rules for two-phonon spectra are determined using the tables

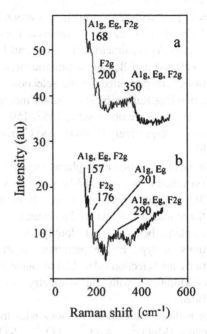

Fig. 3.16 Second-order Raman spectra of halite (a) and sylvite (b). The laser emission wavelength is 532 nm, the spectral resolution is 6 cm^{-1}, the laser power is 30 mW, and signal accumulation time is 200 and 25 s, respectively

of the characters of irreducible representations of the point group of the mineral under investigation. Thus, an analysis of the types of symmetry of two-phonon vibrations in crystals with a point group O_h shows that among the possible combination modes in this point group there are vibrations with symmetry types A_{1g}, E_g and F_{2g}, which are allowed in the Raman spectra. Consequently, second order Raman spectra can also be used for diagnostic purposes.

3.2.5 The Longitudinal-Transverse Splitting

The special feature of the Raman scattering method, which makes it difficult to interpret the spectra, includes the appearance of the longitudinal-transverse (LO-TO) splitting of lines in the spectra in crystals without an inversion center. In such crystals, some vibrations that are active in the Raman spectra are accompanied by changes in the dipole moment. As a result, vibrations of atoms lead to changes in the macroscopic electric dipole moment in the crystal. The resulting additional electromagnetic field in turn affects the atoms. In the Raman spectra, such an interaction can result in the appearance of additional scattering lines. The *LO-TO* splitting theoretically exists in all cases when scattering occurs on dipole-active vibrations. A weak splitting results in changes of the shapes of some lines and appearance of additional shoulders. However, in crystals with the ionic character of bonds the magnitude of the *LO-TO* splitting can reach considerable values. For example, in the Raman spectrum of LiH, it is almost 500 cm^{-1}. In Raman spectra of molecular crystals, the splitting value only in some cases reaches 15 cm^{-1} (Zhizhin et al. 1984), but in most molecular crystals, the *LO-TO* splitting is not observed. The prediction of the *LO-TO* splitting in the spectra of the Raman spectra goes beyond the framework of factor group analysis.

In the case when a mineral without an inversion center has several dipole-active vibrations, several additional scattering lines may appear in the spectrum owing to splitting into *LO-TO* components (see, e.g., Raman spectrum of quartz in Fig. 3.11). As a result, the total number of lines in the spectrum may exceed the number of normal vibrations expected according to group theory. It should be noted that in the infrared absorption spectra, the frequencies of the longitudinal vibrations are not recorded, since only transverse vibrations are excited.

3.2.6 Orientation and Polarization Effects; Analysis of Water and OH Groups

A specific feature of Raman scattering is its tensor character. As a result, Raman spectra of single crystals depend on their orientation and the direction of polarization of the vector of the electrical component of the electromagnetic wave of laser radiation. Spectra obtained in different experimental geometries may differ from each other by the number of recorded scattering lines and their intensity (Fig. 3.17).

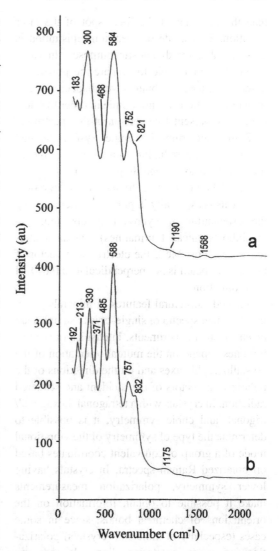

Fig. 3.17 Raman spectra of laachite (Ca,Mn)$_2$Zr$_2$Nb$_2$TiFeO$_{14}$ (monoclinic, point group C_{2h}) obtained with the polarization of the laser beam parallel (upper curve) and perpendicular (lower curve) to the *a* axis of the crystal. The laser emission wavelength is 532 nm, the spectral resolution is 2 cm^{-1}, the laser power is 6 mW, and the focal spot diameter is about 15 μm

With a random orientation of the single crystal, the scattering line intensities are also random. This uncertainty does not apply to spectra of powdery samples with chaotic orientation of microcrystals. A reproducible total spectrum averaged over all possible spatial orientations of the microcrystals can be obtained only if the size of the microcrystals in the powder is much less than the diameter of the focal spot of the laser radiation. This mode of spectrum registration is most suitable for diagnostic purposes. In cases where it is impossible to prepare the powder, it is recommended to obtain several spectra at different orientations of the sample in order to select the most representative version of the spectrum.

Raman spectrum of a single crystal, obtained using polarized radiation, makes it possible to draw conclusions regarding the directions of chemical bonds relative to the crystallographic axes. This is especially important for determining the orientation of hydroxyl groups (e.g., in amphiboles, micas, tourmalines). Raman scattering is only possible if the electric field vector of an incident beam is not perpendicular to the O–H bond direction.

To study structural features of minerals, spectra of Raman spectra of single-crystal samples are taken. In such experiments, intensities of scattering lines depend on the mutual orientation of the crystallographic axes and on the directions of the polarization vectors of the incident and scattered radiation. In crystals with a tetragonal, hexagonal/trigonal, and cubic symmetry, it is possible to determine the type of symmetry of the vibrational mode of a group of equivalent coordinates based on polarized Raman spectra. In crystals having lower symmetry, polarization measurements make it possible to obtain information on the orientations of chemical bonds, since in some cases (especially in molecular crystals) polarization of some scattering lines depends on vibrations of single bonds (Kolesov 2018).

The polarizability of a bond in the longitudinal direction is much greater than in the transverse directions. Therefore, the scattering line corresponding to stretching vibrations of this bond is most intense when the polarization of the exciting laser radiation (and in the ideal case of the scattered light) coincides with the direction of this bond. This regularity can be illustrated by the example of micas. In phlogopite $K(Mg,Fe^{2+})_3[AlSi_3O_{10}](OH,F)_2$ having monoclinic symmetry, OH groups coordinated by divalent octahedral cations are oriented almost parallel to the crystallographic c axis, perpendicular to the cleavage plane. The Raman scattering line at 3709 cm^{-1} corresponding to stretching vibrations of hydroxyl groups has a maximum intensity when the polarization of laser radiation is perpendicular to the cleavage plane of the mineral sample, i.e., parallel to the c axis. A weak additional line at about 3666 cm^{-1} is observed in Raman spectra of phlogopite samples containing trivalent impurity ions, Fe^{3+} or Al^{3+}. Since the orientation of hydroxyl groups coordinated by trivalent cations deviates from the direction perpendicular to the cleavage plane, a weak scattering line can be recorded in the Raman spectra with polarization of laser radiation parallel to the cleavage plane. In muscovite $KAl_2[AlSi_3O_{10}](OH)_2$, which is a dioctahedral mica, hydroxyl groups are almost parallel to the cleavage plane, and the band of stretching vibrations of OH groups (at 3628 cm^{-1}) has a maximum intensity in spectra excited by laser radiation with polarization parallel to the cleavage plane (Tlili et al. 1989).

Based on polarized spectra of tourmaline group minerals, it was found that the OH groups are mainly oriented along the threefold c crystallographic axis (Gasharova et al. 1997; Berryman et al. 2016). Polarized Raman spectra of the orthorhombic mineral magnesiocarpholite $MgAl_2Si_2O_6(OH)_4$ revealed the presence of three OH groups, one of which is oriented almost perpendicular to the c axis (Fuchs et al. 2001).

In some cases, the orientation of complex anionic groups can be determined from polarized Raman spectra. Based on the data obtained for a columnar thaumasite crystal, Kononov et al. (1990) have confirmed that almost flat CO_3^{2-} group (Edge and Taylor 1971) is oriented perpendicular to the C_6 axis of the crystal.

Raman spectroscopy has a low sensitivity in determination of water in minerals. The H_2O molecule has a weak polarizability and, as a result, a weak response to excitation radiation. Bands of O–H-stretching vibrations are usually observed in the range from 3000 to 3800 cm^{-1}, but bands of acidic OH groups and very strong hydrogen bonds may have Raman shifts below 3000 cm^{-1}. In most cases, bands of H–O–H bending vibrations of H_2O molecules are observed in the range 1600–1700 cm^{-1}, but with a low water content, these bands are recorded with difficulty and only with a successful selection of

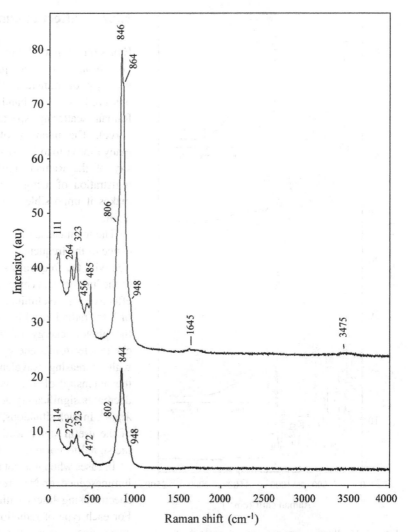

Fig. 3.18 Raman spectra of martyite $Zn_3(V_2O_7)(OH)_2 \cdot 2H_2O$ obtained at the laser emission wavelength of 532 nm, the spectral resolution of 2 cm^{-1}, the laser power of 4 and 13 mW, and the signal accumulation time of 200 and 50 s (for the upper and lower curves, respectively)

experimental conditions. An increase in the accumulation time at a given laser power or an increase in the laser power may lead to local overheating and dehydration of the sample (see Fig. 3.18).

In the structure of fluorapophyllite-(K), water molecules occupying a single crystallographic position are asymmetric: the positional symmetry of H_2O decreases to $C1$, and the hydrogen atoms belonging to the same molecule are nonequivalent. A significant difference in the interactions of the two hydrogen atoms with their nearest environment leads to the fact that stretching vibrations of OH bonds in the water molecule are practically independent (Ryskin and Stavitskaya 1990). The broad band with a maximum of about 3013 cm^{-1} and the narrow band at 3564 cm^{-1} refer to the stretching vibrations of OH bonds, which form strong and very weak hydrogen bonds, respectively (Fig. 3.19).

In the structure of hydroxylapophyllite-(K), both asymmetric H_2O molecules and OH groups are present. The broad band at about 2990 cm^{-1} and the narrow band at 3569 cm^{-1} correspond to a strong and a weak hydrogen bonds formed by H_2O, respectively. Another narrow band (with a

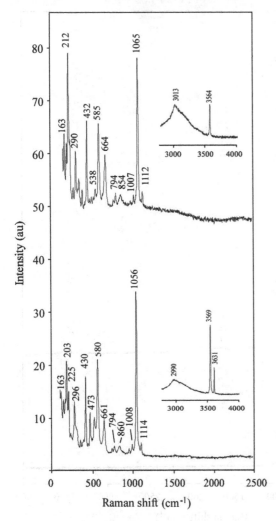

Fig. 3.19 Raman spectra of fluorapophyllite-(K) $KCa_4[Si_8O_{20}]F \cdot 8H_2O$ (upper curve) and hydroxylapophyllite-(K) $KCa_4[Si_8O_{20}]OH) \cdot 8H_2O$ (lower curve) obtained using 532 nm laser radiation

half-width of about 3 cm^{-1}) is observed at 3631 cm^{-1} and corresponds to stretching vibrations of the OH group.

In some cases when the number of hydrogen-containing groups in the mineral is insignificant, the Raman spectroscopy method does not allow one to unambiguously distinguish between water and hydroxyl groups. In such cases, it is more appropriate to use infrared absorption spectroscopy.

3.2.7 Effect of Luminescence

Emission of photoluminescence excited by the laser beam is a serious problem of Raman spectroscopy of minerals. Usually, minescence is observed as broad bands superimposed on the Raman scattering spectrum (Fig. 3.20, upper curve). The intensity of luminescence can be many times (up to 10^3–10^4) greater than the intensity of the Raman signal, which prevents the registration of a high-quality spectrum or even makes it impossible to obtain Raman spectrum at all.

The main cause of luminescence is the coincidence of the frequency of the exciting laser radiation with the frequencies of electronic transitions of the luminescent center in the mineral. The most effective way to eliminate luminescence is the use of laser radiation with a longer wavelength λ_{exc}, the photon energy of which is insufficient to excite electronic energy levels. Unfortunately, with increasing wavelength of laser radiation, the intensity of the useful Raman signal I_R decreases significantly according to the law $I_R \sim \lambda_{exc}^{-4}$. In such situations, long-time accumulation of the useful signal leads to an improvement in the signal-to-noise ratio (Fig. 3.20, lower curve).

In cases when it is not possible to eliminate the luminescence, one can resort to recording several spectra using lasers with different wavelengths. For each type of radiation, the spectral luminescence lines appear in a specific spectral range. The bands that will be present with a constant frequency in the spectra obtained at different wavelengths of the exciting radiation should be referred to the lines of Raman scattering.

3.2.8 Destructive Effect of Laser Radiation

Local temperature increase due to strong absorption of laser radiation may result in alteration or decomposition of the sample. In the Raman microprobe analysis this problem is especially

3.2 Specific Features and Possibilities of Raman Spectroscopy: Practical Recommendations

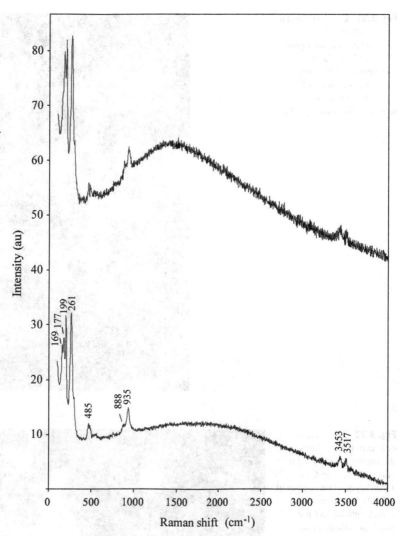

Fig. 3.20 Raman spectra of an unoriented sample of romanorlovite $K_8Cu_6Cl_{17}(OH)_3$ obtained at the laser emission wavelength of 532 nm, the spectral resolution of 6 cm^{-1}, the laser power of 3 and 1.5 mW, and the signal accumulation time of 4 and 17 min (for the upper and lower curves, respectively)

significant. Therefore, the selection of conditions for a nondestructive experiment plays an extremely important role in experiments on Raman scattering.

Special attention should be paid to proper selection of laser power when working with highly colored or opaque minerals, which include most sulfides and sulfosalts. Dark coloring of minerals causes a strong absorption of exciting radiation, which results in attenuation of scattered signal. To enhance the intensity of the Raman signal, an increase in the laser pump power is required. However, it should be borne in mind that increasing the power of laser radiation leads to an increase in the energy absorbed by the mineral. Since most minerals have relatively low thermal conductivity, the absorbed energy causes a local increase in temperature in the focal volume of the laser beam. Too high laser power can result in local thermal destruction of the mineral structure, leading to the formation of cavities on the sample surface (Figs. 3.21 and 3.22) or cavities with a destroyed structure inside the sample. However, in most cases optimizing the experimental conditions (reducing the laser excitation power and increasing the signal accumulation

Fig. 3.21 Caverns formed on the surface of melanarsite at a laser power of 13 mW and laser wavelength of 532 nm, focal spot diameter of ~15 μm, and signal accumulation time of 10 s

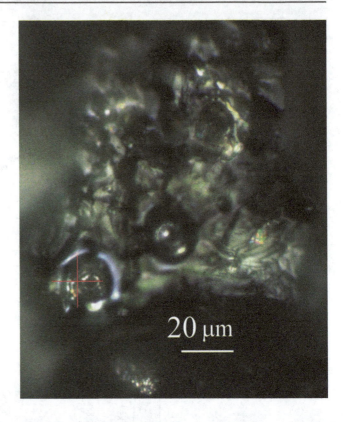

Fig. 3.22 The thermal destruction zone formed on the surface of a crystal of vorontsovite $(Hg_5Cu)TlAs_4S_{12}$ at a laser power of 1.5 mW and laser wavelength of 532 nm, focal spot diameter of ~15 μm, and signal accumulation time of 1 h. Field width 100 μm. It was not possible to obtain Raman spectrum of the mineral due to its thermal instability

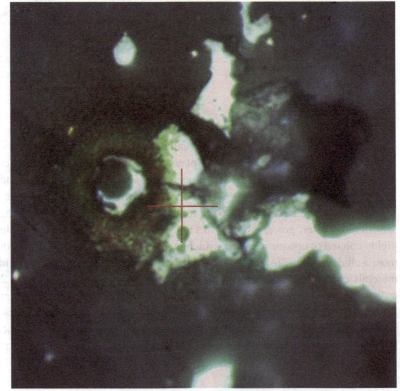

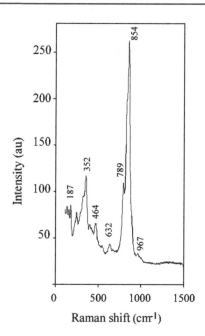

Fig. 3.23 Raman spectrum of melanarsite $K_3Cu_7Fe^{3+}O_4(AsO_4)_4$ obtained at the laser emission wavelength of 532 nm, spectral resolution of 6 cm^{-1}, laser power of 4 mW, focal spot diameter of ~15 μm, and signal accumulation time of 100 s

time) makes it possible to obtain scattering spectra even on thermally unstable samples (Fig. 3.23).

In some cases, it is possible to reduce the degree of overheating by placing microscopic mineral samples on a metal substrate, which leads to acceleration of the heat sink. The correct sequence of actions when working with an unknown colored opaque or especially valuable sample is the use of low-power laser source at the initial stage of research. A gradual increase in power with constant monitoring of the state of the sample will prevent its damage or destruction. To improve the quality of the Raman spectra at low power of the exciting radiation, an increase in the time of accumulation of the Raman signal may play a positive role.

Additional information on the practical application of Raman spectroscopy can be obtained from numerous books and review articles (Reshetnyak and Bukanov 1991; Nasdala et al. 2004; Larkin 2011; Vandenabeele 2013; Kolesov 2018).

Raman Spectra of Minerals

This chapter provides data on 2104 Raman spectra of minerals and their synthetic counterparts taken from various periodicals. The overwhelming majority of these spectra were obtained on arbitrarily oriented samples. As a result, absolute values band intensities are not very informative. For this reason, we do not show the spectra figures, but only give lists of Raman shifts with indication of the strongest and weakest bands.

Data on the Raman spectra are listed in alphabetical order of mineral names and are accompanied by brief descriptions of the conditions under which the spectra were taken. In most cases, comments are made regarding the quality of the spectrum and/or methods of identification of the reference sample.

Abellaite $NaPb_2(CO_3)_2(OH)$

Origin: Synthetic.
Experimental details: Raman scattering measurements have been performed on a powdered sample using 488 and 568.2 nm laser radiations. The laser radiation power at the sample was 200 and 100 mW, respectively.
Raman shifts (cm^{-1}): 3500w, 1750w, 1730w, 1392s, 1350sh, 1068, 1057s, 1052sh, 1036w, 868, 695sh, 681s, 285 (broad), 202, 125w, 98 (broad), 86 (broad), 52s, 37.
Source: Brooker et al. (1983).
Comments: The band 1036w is a satellite band arising from the isotopic moieties ($C^{18}O^{16}O_2^{2-}$). The sample identification was done and the purity of the substance was proved by powder X-ray diffraction data.

Abelsonite $NiC_{31}H_{32}N_4$

Origin: Green River Formation, Utah, USA (type locality).
Experimental details: Raman scattering measurements have been performed on an arbitrarily oriented sample using 532 nm laser radiation. The nominal laser radiation power was 90 mW.
Raman shifts (cm^{-1}): ~1230, ~1170, ~1150, ~1120, ~760.
Source: Liu et al. (2015e).
Comments: The wave numbers were estimated by us based on spectral curve analysis of the published spectrum. The sample was characterized by powder X-ray diffraction data and chemical analysis.

Abhurite $Sn^{2+}_{21}O_6(OH)_{14}Cl_{16}$

Origin: Synthetic.
Experimental details: Raman scattering measurements have been performed on suspension in water using 785.0 nm laser radiation. The laser radiation power at the sample was 200 mW.
Raman shifts (cm^{-1}): 256, 206s, 170, 156, 147.
Source: Chen and Grandbois (2013).
Comments: Spectral analysis and Raman shifts calculation were based on semi-quantitative indirect hard modeling (IHM) analysis.

Acanthite Ag_2S

Origin: Synthetic.
Experimental details: Raman scattering measurements have been performed on an unoriented crystal using 488.0 nm laser radiation. The laser radiation power is not indicated. The spectrum was measured at 30 K.
Raman shifts (cm^{-1}): 222s, 200.
Source: Milekhin et al. (2011).
Comments: For the Raman spectrum of acanthite see also Martina et al. (2012).

Acetamide solution CH_3CONH_2

Origin: Synthetic.
Experimental details: Raman scattering measurements have been performed on 0.1 M and 0.4 M aqueous solutions of acetamide. The wavelengths of laser excitation lines were 220, 240, 250, 365, and 560 nm. The laser radiation power is not indicated.
Raman shifts (cm^{-1}): 1662, 1616, 1457, 1404, 1360, 1131, 1005, 871.
Source: Dudik et al. (1985).
Comments: The line at 871 cm^{-1} is strong in experiments with the laser radiation wave lengths of 365 and 560 nm. The lines at 1457, 1616, and 1662 cm^{-1} are strong in experiments with the laser radiation wave length of 220 nm. For the Raman spectrum of acetamide solution see also Spinner (1959).

Actinolite $Ca_2(Mg_{4.5-2.5}Fe^{2+}_{0.5-2.5})Si_8O_{22}(OH)_2$

Origin: Košino, Macedonia.
Experimental details: Raman scattering measurements have been performed on a powdered sample using 514.5 and 532.0 nm laser radiations. The laser radiation power is not indicated.
Raman shifts (cm^{-1}): 1056s, 1026, 954, 926, 899, 744w, 672s, 527, 517w, 477w, 433w, 413w, 391, 368, 346.
Source: Jovanovski et al. (2009).
Comments: The identification of the sample was done by results of electron microprobe analysis; the purity of the substance was proved by powder X-ray diffraction data. For the Raman spectra of actinolite see also Gopal et al. (2004), Makreski et al. (2006a), Petry et al. (2006), Apopei and Buzgar (2010), Apopei et al. (2011), Andò and Garzanti (2014), and Leissner et al. (2015).

Actinolite $Ca_2(Mg_{4.5-2.5}Fe^{2+}_{0.5-2.5})Si_8O_{22}(OH)_2$

Origin: Tyrol, Austria.
Experimental details: Raman scattering measurements have been performed on an arbitrarily oriented sample using 532 nm laser radiation. The nominal laser radiation power was 100 mW.
Raman shifts (cm^{-1}): 3675, 3661, 1059, 1027s, 949, 929, 892w, 744, 670s, 577w, 522, 482w, 415, 392s, 369s, 294, 247w, 226s.
Source: Apopei et al. (2011).
Comments: No independent analytical data are given for the sample used. For the Raman spectra of actinolite see also Gopal et al. (2004), Makreski et al. (2006a), Petry et al. (2006), Jovanovski et al. (2009), Apopei and Buzgar (2010), Andò and Garzanti (2014), and Leissner et al. (2015).

Adachiite $CaFe^{2+}_3Al_6(Si_5AlO_{18})(BO_3)_3(OH)_3(OH)$

Origin: No data.
Experimental details: A single crystal was used. The wavelengths of laser excitation lines were 488.0 and 514.5 nm. Laser radiation power at the sample was 14 mW. Polarized spectra were collected in $y(zz)y$, $y(zx)y$, and $y(xx)y$ scattering geometries.
Raman shifts (cm^{-1}): 3679, 3625, 3570, 3565s, 3486.
Source: Watenphul et al. (2016a).
Comments: The sample was identified by electron microprobe analysis, and boron was determined by LA-ICP-MS measurement. The Raman shifts are given for the scattering geometry $y(zz)y$, in which the Raman intensities are most strong.

Adamite $Zn_2(AsO_4)(OH)$

Origin: Lavrion mining District, Attikí (Attika, Attica) Prefecture, Greece.
Experimental details: Raman scattering measurements have been performed on a powdered sample using 632.8 nm laser radiation. The laser radiation power is not indicated.
Raman shifts (cm^{-1}): 3552s, 890s, 846s, 820, 539, 496, 458, 422, 380, 325.
Source: Makreski et al. (2013a).
Comments: No independent analytical data are provided for the sample used. For the Raman spectrum of adamite see also Yang et al. (2001).

Adelite $CaMg(AsO_4)(OH)$

Origin: No data
Experimental details: Raman scattering measurements have been performed on an arbitrarily oriented sample using 633 nm He-Ne laser radiation. The Raman shifts have been determined for the maxima of individual peaks obtained as a result of the spectral curve analysis.
Raman shifts (cm^{-1}): 3550, 890, 846, 821, 803, 540, 497, 479, 421, 376, 322, 277, 253, 232, 211, 172, 134.
Source: Martens et al. (2003c).
Comments: No independent analytical data are given for the sample used. Intensities of Raman bands are not indicated.

Admontite $MgB_6O_{10} \cdot 7H_2O$

Origin: Synthetic.
Experimental details: Raman scattering measurements have been performed on a powdery sample. The wavelengths of the laser excitation line and laser radiation power are not indicated.
Raman shifts (cm^{-1}): 1092w, 963, 881, 637s, 523w, 500w, 428s, 412, 394, 320.
Source: Derun et al. (2015)
Comments: The sample was characterized by powder X-ray diffraction data. For the Raman spectrum of admontite see also Kipcak et al. (2014).

Adolfpateraite $K(UO_2)(SO_4)(OH)(H_2O)$

Origin: Svornost mine, Jáchymov, Krušné Hory (Ore Mts.), Bohemia, Czech Republic (type locality).
Experimental details: Raman scattering measurements have been performed on an arbitrarily oriented crystal using 780 nm laser radiation. The laser radiation power at the sample was from 4 to 8 mW.
Raman shifts (cm^{-1}): 1169w, 1149w, 1116w, 1063w, 1029w, 993s, 935s, 900s, 843w, 638w, 597w, 456, 442, 399w, 350w, 320, 270, 264, 219, 1198, 169w, 130w, 109w.
Source: Plášil et al. (2012b).
Comments: The sample was characterized by powder X-ray diffraction data and electron microprobe analyses. The crystal structure is solved.

Aegirine $NaFe^{3+}Si_2O_6$

Origin: Kangerdlnarsuk, Greenland (sample 1) and Brewig, Norway (sample 2).
Experimental details: Raman scattering measurements have been performed on unoriented crystals. The wavelength of the laser excitation line was 532 nm. The nominal laser radiation power was 100 mW.
Raman shifts (cm^{-1}): 1130w, 1044s, 953s, 866, 758, 678, 544s, 499, 465w, 385, 342s, 309, 294, 273 (sample 1); 1132w, 1041s, 971s, 952, 866, 757, 678, 558, 544s, 496, 466w, 385, 343s, 295, 267 (sample 2).
Source: Buzatu and Buzgar (2010).
Comments: No independent analytical data are provided for the samples used. For the Raman spectra of aegirite; see also Andò and Garzanti (2014) and Zhou et al. (2014).

Aegirine Li analogue $LiFe^{3+}(Si_2O_6)$

Origin: Synthetic.
Experimental details: Raman scattering measurements have been performed on an arbitrarily oriented sample using 1064 nm laser radiation. The nominal laser radiation power was 120 mW.
Raman shifts (cm^{-1}): 1084s, 1038s, 1012s, 976, 932, 857, 776w, 684, 575, 553, 519, 493, 387, 363, 347, 325, 313, 294, 284, 263w, 237, 214, 196, 175, 133w, 97.
Source: Zhang et al. (2002a).
Comments: The sample used was prepared from a stoichiometric mixture of finely ground Li_2CO_3, Fe_2O_3 and SiO_2 by solid-state ceramic sintering techniques at 1223 K and ambient pressure and characterized by neutron powder diffraction and Mössbauer measurements. Monoclinic, space group $C2/c$, $a = 9.6641(2)$, $b = 8.6612(3)$, $c = 5.2924(2)$ Å, $\beta = 110.12(1)°$.

4 Raman Spectra of Minerals

Aerinite $(Ca,Na)_6(Fe^{3+},Fe^{2+},Mg,Al)_4(Al,Mg)_6Si_{12}O_{36}(OH)_{12}(CO_3)\cdot 12H_2O$

Origin: Estopiñán dam, Estopiñándel Castillo, Huesca, Aragón, Spain.

Experimental details: Raman scattering measurements have been performed on an arbitrarily oriented sample using 633 nm laser radiation. The laser radiation power is not indicated. The Raman shifts have been determined for the maxima of individual peaks obtained as a result of the spectral curve analysis.

Raman shifts (cm^{-1}): 1236, 1072, 1049, 1013, 974, 933, 909, 885w, 546, 512, 465w, 392, 365s, 331s, 300, 281, 245, 222.

Source: Frost et al. (2015y).

Comments: The identification of the sample used was done only by means of scanning electron microscopy. In the cited paper, the band at 1236 cm^{-1} is assigned to CO_3^{2-} asymmetric stretching mode or to a Si–O vibrations. In both cases the position of this band would be anomalous.

Aeschynite-(Y) $(Y,Ln,Ca,Th)(Ti,Nb)_2(O,OH)_6$

Origin: A granitic pegmatite situated in the Aust-Agder province, southern Norway.

Experimental details: Raman scattering measurements have been performed on an arbitrarily oriented sample using 514.5 nm laser radiation. The nominal laser radiation power was 100 mW.

Raman shifts (cm^{-1}): 830, 665, 609, 396, 355, 256, 232, 159, 99.

Source: Tomašić et al. (2004).

Comments: The identification of the sample was done by powder X-ray diffraction. The chemical composition of the sample was determined by ICP measurement. The Raman spectrum was obtained on a sample which was regained crystal structure after heating up to 1000 °C. The spectrum is of poor quality because the crystallization was not completed.

Afmite $Al_3(OH)_4(H_2O)_3(PO_4)(PO_3OH)\cdot H_2O$

Origin: Fumade, Tarn, France (type locality).

Experimental details: Raman scattering measurements have been performed on an oriented crystal using 514 nm laser radiation. The laser beam was incident on the (001) crystal face. The radiation power on the sample was 5 mW.

Raman shifts (cm^{-1}): No data: only a figure of the Raman spectrum of afmite is given in the cited paper.

Source: Kampf et al. (2011b).

Comments: The sample was characterized by electron microprobe analysis and powder X-ray diffraction. The crystal structure is solved. For the Raman spectrum of afmite see also Sanchez-Moral et al. (2011).

Afwillite $Ca_3[SiO_4][SiO_2(OH)_2]\cdot 2H_2O$

Origin: Synthetic.

Experimental details: Raman scattering measurements have been performed on an arbitrarily oriented sample using 488 nm laser radiation. The radiation power on the sample was 150 mW.

Raman shifts (cm^{-1}): 3973, 856s, 819, 785, 550, 410.

Source: Stodolski et al. (1985).

Comments: The sample was characterized by powder X-ray diffraction.

Agakhanovite-(Y) $YCa\square_2KBe_3Si_{12}O_{30}$

Origin: Heftetjern pegmatite, Tørdal, Southern Norway (type locality).
Experimental details: No data.
Raman shifts (cm^{-1}): 3730w, 3670w, ~3560–3400 (broad), 1120, 1000w, 480s, 360w, 290w, 140.
Source: Hawthorne et al. (2014).
Comments: The sample was characterized by powder X-ray diffraction data and electron microprobe analysis. The crystal structure is solved.

Agardite-(Ce) $CeCu^{2+}_6(AsO_4)_3(OH)_6 \cdot 3H_2O$

Origin: Synthetic.
Experimental details: Raman scattering measurements have been performed on arbitrarily oriented crystals using a 633 nm He-Ne laser. The Raman shifts have been determined for the maxima of individual peaks obtained as a result of the spectral curve analysis. The laser radiation power at the sample was 1 mW.
Raman shifts (cm^{-1}): 885, 868s, 834, 809, 527w, 493w, 475, 461w, 427w, 393w, 319w, 283w, 235, 197w, 170w, 158w, 136w.
Source: Frost et al. (2004f).
Comments: The sample was characterized by powder X-ray diffraction data and electron microprobe analysis, but no analytical data are given in the cited paper.

Agardite-(La) $LaCu^{2+}_6(AsO_4)_3(OH)_6 \cdot 3H_2O$

Origin: Synthetic.
Experimental details: Raman scattering measurements have been performed on arbitrarily oriented crystals using a 633 nm He-Ne laser. The Raman shifts have been determined for the maxima of individual peaks obtained as a result of the spectral curve analysis. The laser radiation power at the sample was 1 mW.
Raman shifts (cm^{-1}): 888, 867s, 832, 803, 524w, 491w, 473, 470, 425w, 391, 317w, 274w, 234, 196w, 165, 136w.
Source: Frost et al. (2004f).
Comments: No independent analytical data are given in the cited paper.

Agardite-(Y) $YCu^{2+}_6(AsO_4)_3(OH)_6 \cdot 3H_2O$

Origin: Synthetic.
Experimental details: Raman scattering measurements have been performed on arbitrarily oriented crystals using a 633 nm He-Ne laser. The Raman shifts have been determined for the maxima of individual peaks obtained as a result of the spectral curve analysis. The laser radiation power at the sample was 1 mW.
Raman shifts (cm^{-1}): 915, 870s, 833, 798w, 557w, 539w, 514w, 486, 468w, 434w, 294w, 267w, 239, 192, 176w, 167w, 142.
Source: Frost et al. (2004f).
Comments: The sample was characterized by powder X-ray diffraction data and electron microprobe analysis, but no independent analytical data are given in the cited paper. For the Raman spectrum of agardite-(Y) see also Morrison et al. (2013).

Agricolaite $K_4(UO_2)(CO_3)_3$

Origin: Synthetic.

Experimental details: Raman scattering measurements have been performed on an arbitrarily oriented polycrystalline sample using 514.5 nm Ar^+ laser radiation. The laser radiation power is not indicated.

Raman shifts (cm^{-1}): 1569w, 1543w, 1354w, 1320, 1065sh, 1055, 1046sh, 879w, 863w, 806s, 785w, 725, 719w, 693w, 307, 288, 262s, 251w, 249w, 241, 191w, 176sh, 165, 154sh, 132, 123, 119sh, 106, 96w, 91w, 82, 74w, 63, 48w.

Source: Anderson et al. (1980).

Comments: The compound was synthesized hydrothermally and characterized by powder X-ray diffraction data.

Ahlfeldite $Ni(SeO_3) \cdot 2H_2O$

Origin: Pacajake Mine, Bolivia.

Experimental details: Raman scattering measurements have been performed on arbitrarily oriented crystals using a 633 nm He-Ne laser. The Raman shifts have been determined for the maxima of individual peaks obtained as a result of the spectral curve analysis. The laser radiation power is not indicated.

Raman shifts (cm^{-1}): 3466, 3385s, 3329, 3251, 2185, 2130, 861, 832s, 803s, 751, 719, 595, 532, 508, 430s, 421, 348, 267, 218, 177s, 149s.

Source: Frost and Keeffe (2009f).

Comments: No independent analytical data are provided for the sample used.

Ahrensite γ-$Fe_2(SiO_4)$

Origin: Tissint Martian meteorite (type locality).

Experimental detail: Raman scattering measurements have been performed on an arbitrarily oriented sample using 514.3 nm Ar^+ laser radiation. The laser radiation power is not indicated.

Raman shifts (cm^{-1}): 843, 795, 672, 213.

Source: Ma et al. (2016a).

Comments: The sample was characterized by powder X-ray diffraction data and electron microprobe analyses.

Aikinite $CuPbBiS_3$

Origin: Karrantza Valley, westerner area of the Basque Co., Spain.

Experimental details: Raman scattering measurements have been performed on an arbitrarily oriented single crystal using 633 nm laser radiation with the laser radiation power at the sample of 50 mW and 785 nm laser radiation with the nominal laser radiation power of 150 mW, using filters of 1% and 10%.

Raman shifts (cm^{-1}): 326s, 227s.

Source: Goienaga et al. (2011).

Comments: The sample identification was done by means of electron fluorescence analysis.

Ajoite $K_3Cu^{2+}{}_{20}Al_3Si_{29}O_{76}(OH)_{16} \cdot 8H_2O$

Origin: New Cornelia Mine in the Ajo District of Pima County, Arizona, USA (type locality).
Experimental details: Raman scattering measurements have been performed on arbitrarily oriented crystals using a 633 nm He-Ne laser. The Raman shifts have been determined for the maxima of individual peaks obtained as a result of the spectral curve analysis. The laser radiation power at the sample is not indicated.
Raman shifts (cm^{-1}): 3619, 3589, 3574, 3553, 3394, 1155, 1139, 1069, 1048, 1015, 962, 796, 672, 630, 516, 484, 437, 411, 343, 325, 304.
Source: Frost and Xi (2012m).
Comments: No independent analytical data are given for the sample used.

Akaganeite $(Fe^{3+},Ni^{2+})_8(OH,O)_{16}Cl_{1.25} \cdot nH_2O$

Origin: No data.
Experimental details: Raman scattering measurements have been performed on an arbitrarily oriented powdery sample using 636.4 nm laser radiation. The laser radiation power on the sample was 0.34 mW.
Raman shifts (cm^{-1}): 717, 614w, 537, 415sh, 390s, 311s.
Source: Nieuwoudt et al. (2011).
Comments: The identification of the sample was done by powder X-ray diffraction data. For the Raman spectra of akaganeite see also Das and Hendry (2011) and Aramendia et al. (2014).

Åkermanite $Ca_2MgSi_2O_7$

Origin: Synthetic.
Experimental details: Raman scattering measurements have been performed on an arbitrarily oriented powdered sample in a 90°-scattering geometry using 488 nm laser radiation. The laser radiation power on the sample was from 300 up to 500 mW.
Raman shifts (cm^{-1}): 1067w, 1012w, 990sh, 948w, 927sh, 906s, 664s, 603, 515, 484, 448, 361sh, 317, 269w, 220, 210, 97.
Source: Sharma et al. (1988).
Comments: The identification of the sample synthesized from glass was performed by comparison with published Raman data.

Åkermanite Sr analogue $Sr_2Mg(Si_2O_7)$

Origin: Synthetic.
Experimental details: Raman scattering measurements have been performed using 632.8 nm laser radiation. The laser radiation power is not indicated.
Raman shifts (cm^{-1}): 984sh, 975w, 901s, 838w, 653, 590w, 568w, 475w, 450w, 315.
Source: Gabelica-Robert and Tarte (1979).
Comments: The sample synthesized by solid-state reaction was characterized by powder X-ray diffraction.

Akimotoite $MgSiO_3$

Origin: Synthetic.
Experimental details: Raman scattering measurements have been performed on an arbitrarily oriented powdered sample using 514.5 nm laser radiation. The nominal laser radiation power was 200 mW.
Raman shifts (cm^{-1}): 802s, 687, 622, 481, 412, 352, 291.
Source: Okada et al. (2008).
Comments: The sample was characterized by powder X-ray diffraction data and chemical analysis. For the Raman spectra of akimotoite see also Ferroir et al. (2008) and Chen and Xie (2015).

Aklimaite $Ca_4[Si_2O_5(OH)_2](OH)_4 \cdot 5H_2O$

Origin: Lakargi Mt., Upper Chegem caldera, Kabardino-Balkaria, Northern Caucasus, Russia (type locality).
Experimental details: Raman scattering measurements have been performed on an arbitrarily oriented sample using 532 nm laser radiation. The laser radiation power on the sample was 44 mW.
Raman shifts (cm^{-1}): 3611, 3593, 3549, 3535, 3261, 1666, 1575, 1490, 1380, 1344, 1087w, 999s, 960sh, 924sh, 908s, 838, 680s, 569, 543, 488, 444, 406w, 340s, 254s, 191s, 142.
Source: Zadov et al. (2013).
Comments: The wavenumbers are indicated for the maxima of individual bands obtained as a result of the spectral curve analysis. The sample was characterized by powder X-ray diffraction data and chemical analysis. The crystal structure is solved.

Alabandite MnS

Origin: Synthetic.
Experimental details: Raman scattering measurements have been performed on an arbitrarily oriented sample using 514.4 nm laser radiation. The laser radiation power on the sample was 2 mW.
Raman shifts (cm^{-1}): 580w, 285, 230, 160.
Source: Avril et al. (2013).
Comments: The cubic monosulfide alabandite does not have a first-order Raman spectrum due to its ideal rock salt structure. Through local symmetry breaking, the inactive or infrared-active vibrational modes become Raman active. As a result, the Raman peaks are broad and have a very weak intensity. For the Raman spectrum of alabandite see also Ma et al. (2012b).

Alacránite As_8S_9

Origin: No data.
Experimental details: Raman scattering measurements have been performed on an arbitrarily oriented crystal using 785 nm laser radiation. The nominal laser radiation power was 3 mW. A non-oriented crystal.
Raman shifts (cm^{-1}): 384, 361, 350sh, 340s, 329, 307, 240sh, 230s.
Source: Pagliai et al. (2011).

Alamosite $PbSiO_3$

Origin: Synthetic.
Experimental details: Raman scattering measurements have been performed on an arbitrarily oriented single crystal using 488 nm laser radiation. A 90°-scattering geometry was employed. The nominal laser radiation power was in the range from 50 to 200 mW.

Raman shifts (cm^{-1}): 1047w, 1002sh, 987, 955, 946, 924w, 897w, 870w, 780w, 735w, 683w, 647w, 590, 534w, 495w, 469w, 448, 388w, 371w, 355, 324w, 306sh, 272w, 218w, 171w, 146w, 109s, 90, 82, 70, 59s.
Source: Furukawa et al. (1979).
Comments: The hydrothermally synthesized sample was characterized by powder X-ray diffraction.

Alarsite Al(AsO$_4$)

Origin: Synthetic.
Experimental detail: Raman scattering measurements have been performed on an arbitrarily oriented crystal using 514.5 nm laser radiation. The laser radiation power at the sample was 400 mW.
Raman shifts (cm^{-1}): 1031w, 1000w, 985s, 939, 930sh, 630w, 613w, 573w, 450sh, 420s, 390, 371w, 347, 319, 311, 260w, 226, 198, 136, 123, 98.
Source: Dultz et al. (1975).
Comments: No independent analytical data are provided for the sample used.

Albertiniite Fe^{2+}(SO$_3$)·3H$_2$O

Origin: Monte Falò Pb-Zn mine, Coiromonte, Armeno municipality, Verbano Cusio Ossola province, Piedmont, Italy (type locality).
Experimental details: Raman scattering measurements have been performed on an arbitrarily oriented single crystal using 473.1 nm laser radiation. The laser radiation power is not indicated.
Raman shifts (cm^{-1}): 3350s, 3215s, 1660w, 970s, 950s, 910, 860w, 825w, 660w, 600w, 495sh, 482, 457, 438sh, 324w, 279w, 241w, 197w, 172, 123.
Source: Vignola et al. (2016).
Comments: The sample was characterized by powder X-ray diffraction data and electron microprobe analyses. The crystal structure is solved.

Albite Na(AlSi$_3$O$_8$)

Origin: Alinci, Macedonia.
Experimental detail: Raman scattering measurements have been performed on a powdered sample using 532 nm laser radiation. The laser radiation power at the sample was 7 mW.
Raman shifts (cm^{-1}): 1115w, 100w, 1034w, 1006w, 976w, 816w, 764w, 726w, 646w, 579w, 508s, 479, 458w, 408w, 329w, 291s, 270sh, 251w, 209w, 185w, 171w, 162w, 148w, 113w.
Source: Makreski et al. (2009).
Comments: The sample was characterized by powder X-ray diffraction. The assignment of Raman bands given in the cited paper is incorrect. For the correct assignment see Többens et al. (2005). For the Raman spectra of albite see also Frezzotti et al. (2012), Karwowski et al. (2013), and McKeown (2005).

Aleksite PbBi$_2$Te$_2$S$_2$

Origin: Panarechensk volcanic-tectonic formation, Kola Peninsula, Russia.
Experimental details: Raman scattering measurements have been performed on an arbitrarily oriented sample using 514.5 nm Ar$^+$ laser radiation. The nominal laser radiation power was 50 mW.
Raman shifts (cm^{-1}): 232–235w, 143–147s, 99–103s, (75–79w).

Source: Voloshin et al. (2015a).
Comments: The samples used were characterized by electron microprobe analyses. For the Raman spectrum of aleksite see also Gehring et al. (2015).

Alforsite OH-analogue $Ba_5(PO_4)_3(OH)$

Origin: Synthetic.
Experimental details: Micro-Raman scattering measurements have been performed on an arbitrarily oriented sample using 532 nm laser radiation. The nominal laser radiation power at the sample was 10 mW.
Raman shifts (cm^{-1}): 3608, 3583w, 1057w, 1029, 1007w, 934s.
Source: Yoder et al. (2012).
Comments: The sample was characterized by powder X-ray diffraction data and chemical analysis. It contains 2.95 wt% of a carbonate.

Allactite $Mn^{2+}_7(AsO_4)_2(OH)_8$

Origin: Nordmark, Sweden (type locality).
Experimental details: Raman scattering measurements have been performed on arbitrarily oriented crystals using a 633 nm He-Ne laser. The Raman shifts have been determined for the maxima of individual peaks obtained as a result of the spectral curve analysis. The laser radiation power at the sample is not indicated.
Raman shifts (cm^{-1}): 3561s, 3523, 3489s, 3446, 3395, 3292, 1011, 909, 883s, 859s, 834s, 827s, 808s, 779s, 743, 633, 470, 452, 422, 393, 377, 360s, 350, 331, 323, 298, 288, 271, 241, 197, 158.
Source: Frost and Weier (2006).
Comments: The sample was characterized by electron microprobe analysis.

Allanite (Ce) $CaCe(Al_2Fe^{2+})[Si_2O_7][SiO_4]O(OH)$

Origin: Brahmaputra River, Bangladesh.
Experimental details: Raman scattering measurements have been performed on an arbitrarily oriented sample using 633 nm laser radiation. A nearly 180°-scattering geometry was employed. The laser radiation power on the sample is not indicated.
Raman shifts (cm^{-1}): 1062, 972s, ~900–920sh, 689, ~630, ~550, 457s, 421s, ~380, ~270, ~220, ~180.
Source: Andò and Garzanti (2014).
Comments: No independent analytical data are given for the sample used.

Allanite-(Nd) $CaNd(Al_2Fe^{2+})[Si_2O_7][SiO_4]O(OH)$

Origin: Kracovice pegmatite, Moldanubian Zone, Czech Republic.
Experimental details: Raman scattering measurements have been performed on an arbitrarily oriented crystal using 532 and 633 nm laser radiations. The laser radiation power is not indicated.
Raman shifts (cm^{-1}): 1052, 995, 964, 926, 897, 873, 841, 691, 632, 595, 573, 571, 493, 457, 442, 428, 385, 354, 321, 294, 279, 256, 221, 199, 160, 126, 107.
Source: Čopjaková et al. (2015).

Comments: The sample was characterized by ICP and electron microprobe analyses. The Raman shifts are indicated for the maxima of individual peaks obtained as a result of the spectral curve analysis.

Allanpringite $Fe^{3+}_3(PO_4)_2(OH)_3 \cdot 5H_2O$

Origin: Grube Mark, near Essershausen, *ca.* 5 km SE of Weilburg/Lahn, Taunus, Hesse, Germany (type locality).
Experimental details: Raman scattering measurements have been performed on an arbitrarily oriented sample using 488 nm laser radiation. A 180°-scattering geometry was employed. The laser radiation power is not indicated.
Raman shifts (cm^{-1}): ~3567, 3412, ~3197, ~3060, ~3052, ~1652, ~1100sh, 1060, 1023, 1009, 987, 590, 555, 516, 491, 470, 424, 355, ~309, 292.
Source: Kolitsch et al. (2006).
Comments: The sample was characterized by powder X-ray diffraction data and chemical analyses. The crystal structure is solved.

Allendeite $Sc_4Zr_3O_{12}$

Origin: Allende meteorite (type locality).
Experimental details: Micro-Raman scattering measurements have been performed on an arbitrarily oriented crystal using 514.5 nm laser radiation. The laser radiation power on the sample was 5 mW.
Raman shifts (cm^{-1}): No data: only a figure of the Raman spectrum is given in the cited paper.
Source: Ma et al. (2014b).
Comments: The sample was characterized by powder X-ray diffraction data and chemical analyses. The strong Raman peaks in the 1300–1000 cm^{-1} region may be caused by *REE* luminescence.

Allophane $Al_2O_3(SiO_2)_{1.3-2.0} \cdot 2.5–3.0H_2O$

Origin: Reppia, NW Italy.
Experimental details: Raman scattering measurements have been performed on an arbitrarily oriented sample using 532 and 785 nm laser radiations. The nominal laser radiation power on the sample was 1.4 and 4 mW, respectively.
Raman shifts (cm^{-1}): 3420s (broad), 2942w, 1638w, 1357w, 1103, 982w, 858–859s.

Almandine $Fe^{2+}_3Al_2(SiO_4)_3$

Origin: An unknown locality in Mongolia.
Experimental details: Micro-Raman scattering measurements have been performed on a single crystal using 488 nm Ar^+ laser radiation. The nominal laser radiation power was 100 mW.
Raman shifts (cm^{-1}): 1038s, 930, 916s, 897, 863, 630, 596, 556, 581, 500s, 475, 370s, 342s, 323, 314, 256, 216s, 171s, 166s.
Source: Kolesov and Geiger (1998).
Comments: Measurements were made perpendicular to the {100} and {110} faces of a single crystal. The sample identification was done by powder X-ray diffraction data and electron microprobe analyses. For the Raman spectra of almandine see also Mingsheng et al. (1994), Makreski et al.

(2005b), Bersani et al. (2009), Ferrari et al. (2009), Jovanovski et al. (2009), Frezzotti et al. (2012), and Andò and Garzanti (2014).

Almarudite K($\square$,Na)$_2$(Mn,Fe,Mg)$_2$[(Be,Al)$_3$Si$_{12}$]O$_{30}$

Origin: Bellerberg volcano, Eifel area, Germany (type locality).
Experimental details: Micro-Raman scattering measurements have been performed on a single crystal along [0001], using 633 nm He-Ne laser radiation. The nominal laser radiation power was 17 mW.
Raman shifts (cm^{-1}): 1134s, 1053, 991, 931, 847, 778, 698w, 653, 640, 608, 563, 554, 493s, 460, 382, 346w, 311sh, 290s, 276sh, 236w, 160, 135s, 108sh, 84.
Source: Lengauer et al. (2009).
Comments: The Raman shifts were determined by us based on spectral curve analysis of the published spectrum.

Alstonite Ba Ca(CO$_3$)$_2$

Origin: Moore Hill, England.
Experimental details: Raman scattering measurements have been performed on an arbitrarily oriented polycrystalline sample using 488 and 514.5 nm laser radiations. The nominal laser radiation power at the sample was 200 mW.
Raman shifts (cm^{-1}): 1489, 1092s, 1067s, 709sh, 693, 284, 252w, 211w, 196, 170, 145s, 130, 102, 79s.
Source: Scheetz and White (1977).
Comments: No independent analytical data are provided for the sample used.

Altaite PbTe

Origin: Synthetic.
Experimental details: Raman scattering measurements have been performed on an arbitrarily oriented sample using 532 laser radiation. The laser radiation power at the sample was from 1 to 2 mW.
Raman shifts (cm^{-1}): Altaite is inactive in Raman due to halite-type structure.
Source: Vymazalová et al. (2014).

Althausite Mg$_4$(PO$_4$)$_2$(OH,O)(F,$\square$)

Origin: Sapucaia pegmatite mine, Minas Gerais, Brazil (?).
Experimental details: Raman scattering measurements have been performed on arbitrarily oriented crystals using a 633 nm He-Ne laser. The Raman shifts have been determined for the maxima of individual peaks obtained as a result of the spectral curve analysis. The laser radiation power is not indicated.
Raman shifts (cm^{-1}): 3688w, 3653w, 3535, 3523sh, 3511sh, 3500s, 3488, 3472sh, 3455w, 1320br, 1154, 1130s, 1114, 1078w, 1062w, 1049w, 1033s, 993s, 986sh, 964s, 950sh, 917w, 902w, 860w, 668, 638sh, 628sh, 612, 594, 580, 537, 510sh, 499, 488sh, 466, 458, 436sh, 426, 412.
Source: Frost et al. (2014p).
Comments: The sample identification was made only by SEM. The broad band at 1320 cm^{-1} may be due to impurities.

Alum-(K) $KAl(SO_4)_2 \cdot 12H_2O$

Origin: No data.
Experimental details: Raman scattering measurements have been performed on an arbitrarily sample using 488 nm laser radiation. The nominal laser radiation power at the sample was 220 mW. A 90°-scattering geometry was employed.
Raman shifts (cm^{-1}): 3396s, 3072, 1130, 1104w, 989s, 974s, 614, 455, 442w.
Source: Barashkov et al. (2004).
Comments: For the Raman spectra of alum-(K) see also Makreski et al. (2005a), Brooker and Eysel (1990), and Rao (1941).

Alum-(K) $KAl(SO_4)_2 \cdot 12H_2O$

Experimental details: Raman scattering measurements have been performed on a powdered sample using 1064 nm laser radiation. The laser radiation power is not indicated.
Raman shifts (cm^{-1}): 1086w, 991s, 872w, 841w, 818w, 804w, 757w, 714w, 684w, 603, 517w, 416w, 386w.
Source: Makreski et al. (2005a).
Comments: No independent analytical data are provided for the sample used. For the Raman spectra of alum-(K) see also Rao (1941), Brooker and Eysel (1990), Barashkov et al. (2004), and Frezzotti et al. (2012).

Aluminite $Al_2(SO_4)(OH)_4 \cdot 7H_2O$

Origin: Newhaven, East Sussex, England, UK.
Experimental details: Raman scattering measurements have been performed on arbitrarily oriented sample using a 633 nm He-Ne laser. The Raman shifts have been determined for the maxima of individual peaks obtained as a result of the spectral curve analysis. The laser radiation power is not identified
Raman shifts (cm^{-1}): 3588, 3569, 3439, 3378, 3294, 3157, 2930, 2875, 1136w, 1094, 1069w, 999, 93s, 990, 793w, 680, 642, 631, 607, 575, 490, 440, 333, 317, 285.
Source: Frost et al. (2015k).
Comments: The sample was characterized only by qualitative EDS analysis.

Aluminocerite-(Ce) $(Ce,REE,Ca)_9(Al,Fe^{3+})(SiO_4)_3[SiO_3(OH)]_4(OH)_3$

Origin: Ratti quarry, near Baveno, Italy (type locality).
Experimental details: Raman scattering measurements have been performed on arbitrarily oriented crystals using 514.5 nm Ar$^+$ laser radiation. The nominal laser radiation power at the sample was from 10 to 50 mW.
Raman shifts (cm^{-1}): 3646w, 1098, 1083, 1010w, 977w, 869, 816, 763, 647w, 579, 507s, 478s, 455, 412, 327, 288, 266, 250w, 205, 181, 168, 147w.
Sourse: Nestola et al. (2009).
Comments: The sample was identified by electron microprobe analyses and single-crystal X-ray diffraction.

Aluminocopiapite $(Al,Mg)Fe^{3+}_4(SO_4)_6(OH,O)_2 \cdot 20H_2O$

Origin: Synthetic.
Experimental details: Raman scattering measurements have been performed on an arbitrarily oriented sample using 532.0 nm laser radiation. The laser radiation power is not indicated.
Raman shifts (cm^{-1}): 3566, 3384, 3164, 2428, 1638, 1220, 1123, 1019, 989, 636, 614, 598, 555, 476, 452, 300, 270, 247.
Source: Kong et al. (2011b).
Comments: The sample was identified by powder X-ray diffraction data.

Alumohydrocalcite $CaAl_2(CO_3)_2(OH)_4 \cdot 3H_2O$

Origin: Synthetic.
Experimental details: Raman scattering measurements have been performed on an arbitrarily oriented sample using 514.5 nm Ar$^+$ laser radiation. The laser radiation power is not indicated.
Raman shifts (cm^{-1}): 1091sh, 729, 590.
Source: Jay et al. (2015).
Comments: Alumohydrocalcite was identified by the Raman spectrum using data from the RRUFF database as a reference.

Alunite $KAl_3(SO_4)_2(OH)_6$

Origin: Argillic zone hosted by volcanic rocks in Bulgaria.
Experimental details: Raman scattering measurements have been performed on an arbitrarily oriented sample using 514.5 nm Ar$^+$ laser radiation. The laser radiation power at the sample was 1 mW.
Raman shifts (cm^{-1}): 3508s, 3480s, 3068w, 1186, 1151, 1077, 1024s, 653, 560, 508, 484, 381, 345, 234, 161.
Source: Maubec et al. (2012).
Comments: The sample was characterized by powder X-ray diffraction data. For the Raman spectra of alunite see also Toumi and Tlili (2008) and Frezzotti et al. (2012).

Alunogen $Al_2(SO_4)_3(H_2O)_{12} \cdot 5H_2O$

Origin: Synthetic.
Experimental details: Raman scattering measurements have been performed on a powdered sample using 532 nm laser radiation. The nominal laser radiation power was 9 mW.
Raman shifts (cm^{-1}): 3246br, 3078sh, 1126w, 1086w, 992s, 612, 528w, 470, 339w, 309w.
Source: Wang and Zhou (2014).
Comments: The sample was characterized by powder X-ray diffraction data. For the Raman spectrum of alunogen see also Buzatu et al. (2016).

Alwilkinsite-(Y) $Y(UO_2)_3(SO_4)_2O(OH)_3(H_2O)_7 \cdot 7H_2O$

Origin: Blue Lizard mine, San Juan Co., Utah, USA (type locality).
Experimental details: Raman scattering measurements have been performed on an arbitrarily oriented single sample using 780 nm laser radiation. The laser radiation power is not indicated.

Raman shifts (cm^{-1}): 1325, 1265s, 1135w, 1080w, 1035, 1015, 990, 900w, 840s, 605, 555, 530, 465w, 455, 380, 320, 288, 268, 240, 200, 170, 145w, 135w, 108w, 90w, 72w, 60.

Comments: The broad bands at around 1600 cm^{-1} are the result of the fluorescence. The sample was characterized by powder X-ray diffraction data and electron microprobe analyses. The crystal structure is solved.

Amarantite $Fe^{3+}_2O(SO_4)_2 \cdot 7H_2O$

Origin: Caracoles, Sierra Gorda district, Chile (type locality).

Experimental details: Raman scattering measurements have been performed on arbitrarily oriented crystals using a 633 nm He-Ne laser. The Raman shifts have been determined for the maxima of individual peaks obtained as a result of the spectral curve analysis. The laser radiation power is not indicated.

Raman shifts (cm^{-1}): 3529s, 3480, 3401, 3340, 3227, 3089, 3025, 1648, 1577, 1441, 1233, 1195, 1131, 1098s, 1054, 1039, 1017s, 1006sh, 650w, 622w, 602w, 543w, 491, 451sh, 409, 399sh, 346, 309, 290, 255, 247, 229, 212, 205, 195s, 183sh, 176, 162, 149, 139, 129, 113, 107.

Source: Frost et al. (2013a, d).

Comments: The sample identification was done only by SEM.

Amblygonite $LiAl(PO_4)F$

Origin: Penig, Saxony, Germany.

Experimental details: Raman scattering measurements have been performed on an arbitrarily oriented sample using 488.0 nm laser radiation. The nominal laser radiation power at the sample was 200 mW.

Raman shifts (cm^{-1}): 3348 (broad), 1066, 1043s, 1008s, 645s, 604s, 423, 154.

Source: Rondeau et al. (2006).

Comments: The sample identification was done by powder X-ray diffraction data and by electron microprobe analysis. For the Raman spectra of amblygonite see also Dias et al. (2011) and Frezzotti et al. (2012).

Ambrinoite $[K,(NH_4)]_2(As,Sb)_6(Sb,As)_2S_{13} \cdot H_2O$

Origin: Oulx, Susa Valley, Torino, Piedmont, Italy (type locality).

Experimental details: Raman scattering measurements have been performed on an arbitrarily oriented sample using 628 nm laser radiation. The laser radiation power is not indicated.

Raman shifts (cm^{-1}): 3475w, 3150w, 1595w, 1423w, 393s, 371, 364, 352, 341, 324, 294, 216, 207.

Source: Biagioni et al. (2011a).

Comments: The sample was characterized by powder X-ray diffraction data and electron microprobe analysis. The crystal structure is solved.

Ameghinite $NaB_3O_3(OH)_4$

Origin: Tincalayu deposit, Salta province, Argentina (type locality).

Experimental details: Raman scattering measurements have been performed on an arbitrarily oriented sample using 633 nm laser radiation. The laser radiation power is not indicated. The Raman shifts

have been determined for the maxima of individual peaks obtained as a result of the spectral curve analysis.

Raman shifts (cm^{-1}): 3385, 3343, 3275, 3249, 3230, 3203s, 3191, 1724, 1604, 1531, 1403, 1371, 1320, 1281, 1245, 1213, 1087, 1061, 1027s, 1014, 887, 861, 786, 769, 755, 730, 711, 701, 656, 624, 620s, 615, 574, 530, 503, 464, 405, 367, 352, 224, 197, 145, 135s, 119.

Source: Frost and Xi (2012i).

Comments: The sample was characterized by powder X-ray diffraction data and chemical analysis.

Amesite $Mg_2Al(AlSiO_5)(OH)_4$

Origin: Artificial (a product of hydrothermal alteration of olivine in the presence of $AlCl_3$).

Experimental details: Micro-Raman scattering measurements have been performed on an arbitrarily oriented sample using 514 nm laser radiation. The laser radiation power is not indicated.

Raman shifts (cm^{-1}): ~1080w, ~1000w, ~845w, ~700s, ~540s, ~495, ~480, ~405s, ~345w, ~273, ~190.

Source: Andreani et al. (2013).

Comments: The sample was identified as amesite by comparison to the reference spectrum from RUFF database.

Ammoniojarosite $(NH_4)Fe^{3+}_3(SO_4)_2(OH)_6$

Origin: Synthetic.

Experimental details: Raman scattering measurements have been performed on an arbitrarily oriented sample using 488 nm laser radiation. The nominal laser radiation power was less than 5 mW.

Raman shifts (cm^{-1}): 3485sh, 3434, 3360sh, 3210, 1164, 1092, 1006s, 637sh, 623s, 565, 451, 423s, 342w, 301, 218s, 136.

Source: Chio et al. (2010).

Comments: The sample was characterized by powder X-ray diffraction data. For the Raman spectrum of ammoniojarosite see also Sasaki et al. (1998).

Grunerite $\square Fe^{2+}_2 Fe^{2+}_5 Si_8 O_{22}(OH)_2$

Origin: South Africa.

Experimental details: Raman scattering measurements have been performed on an arbitrarily oriented fibrous aggregate ("amosite") using 632 nm laser radiation. The nominal laser radiation power was 20 mW.

Raman shifts (cm^{-1}): 1093w, 1020s, 968, 904w, 659s, 528, 507w, 423w, 400w, 368w, 348, 307w, 289w, 252w, 216, 182s, 155s.

Source: Rinaudo et al. (2004).

Comments: The sample was characterized by powder X-ray diffraction data and electron microprobe analysis.

Analcime $Na(AlSi_2O_6) \cdot H_2O$

Origin: Aussig, Bohemia, Czech Republic.

Experimental details: Raman scattering measurements have been performed on an arbitrarily oriented sample using Nd-YAG laser radiation. The laser radiation power at the sample was 300 mW.

Raman shifts (cm^{-1}): 1104, 591w, 483s, 390, 298.
Source: Mozgawa (2001).
Comments: The sample was characterized by powder X-ray diffraction data. For the Raman spectrum of analcime see also Frost et al. (2014i).

Anapaite $Ca_2Fe^{2+}(PO_4)_2 \cdot 4H_2O$

Origin: Bellver de la Cerdanya, Lérida, Spain.
Experimental details: Raman scattering measurements have been performed on arbitrarily oriented crystals using a 633 nm He-Ne laser. The Raman shifts have been determined for the maxima of individual peaks obtained as a result of the spectral curve analysis. The laser radiation power is not indicated.
Raman shifts (cm^{-1}): 3248, 3176, 3101, 3022, 2882, 2777, 1658w, 1373w, 1071, 1039, 992w, 965sh, 943s, 808w, 777, 654, 622, 582, 546, 445, 432, 352, 335, 287, 266, 231sh, 281, 202, 186, 170, 155, 149, 136.
Source: Frost et al. (2013t).
Comments: The sample identification was done only by SEM.

Anatase TiO_2

Origin: Synthetic.
Experimental details: Raman scattering measurements have been performed on a powdery sample using 488 nm laser radiation. The laser radiation power is not indicated. A 90°-scattering geometry was employed. The Raman shifts are indicated for the maxima of individual peaks obtained as a result of the spectral curve analysis.
Raman shifts (cm^{-1}): 640s, 515, 398, 198w, 147s.
Source: Balachandran and Eror (1982).
Comments: The sample was characterized by ICP method and electron microprobe analysis. For the Raman spectra of anatase see also Zajzon et al. (2013), Andò and Garzanti (2014), and Martins et al. (2014).

Anatase TiO_2

Origin: Perkupa, Hungary.
Experimental details: Raman scattering measurements have been performed on an arbitrarily oriented sample using 633 nm laser radiation. The nominal laser radiation power at the sample was 13 mW.
Raman shifts (cm^{-1}): 470, 248s, 226, 182s, 155.
Source: Zajzon et al. (2013).
Comments: The sample was characterized by powder X-ray diffraction data and electron microprobe analysis. For the Raman spectra of anatase see also Balachandran and Eror (1982), Andò and Garzanti (2014), and Martins et al. (2014).

Ancylite-(Ce) $CeSr(CO_3)_2(OH) \cdot H_2O$

Origin: Bear Lodge, Wyoming, USA.
Experimental details: Raman scattering measurements have been performed on an arbitrarily oriented sample using 532 nm laser radiation. The laser radiation power is not indicated.

Raman shifts (cm⁻¹): 1106, 1086s, 1077s, 861w, 744w, 725, 705w, 466, 299, 252s, 223s, 193, 130w, 123w.
Source: Chakhmouradian et al. (2017).
Comments: The sample was characterized by electron microprobe analyses.

Andalusite Al_2SiO_5

Origin: No data.
Experimental details: No data.
Raman shifts (cm⁻¹): 1111, 1065, 952, 920s, 834, 719, 606, 553, 453, 361, 323, 293s, 278.
Source: Frezzotti et al. (2012).
Comments: The data are from the database www.ens-lyon.fr/LST/Raman. For the Raman spectra of andalusite see also Mernagh and Liu (1991) and Andò and Garzanti (2014).

Andersonite $Na_2Ca(UO_2)(CO_3)_3 \cdot 6H_2O$

Origin: Grants, New Mexico, USA.
Experimental details: Raman scattering measurements have been performed on arbitrarily oriented crystals using a 633 nm He-Ne laser. The Raman shifts have been determined for the maxima of individual peaks obtained as a result of the spectral curve analysis. The laser radiation power is not indicated.
Raman shifts (cm⁻¹): 3558, 3510, 2415, 1406, 1370, 1092, 1080, 928, 833, 831, 742, 696, 299, 284, 272, 242, 224, 182, 164.
Source: Frost et al. (2004b).
Comments: For the Raman spectra of andersonite see also Stefaniak et al. (2009) and Driscoll et al. (2014).

Andradite $Ca_3Fe^{3+}_2(SiO_4)_3$

Origin: No data.
Experimental details: Raman scattering measurements have been performed on an arbitrarily oriented sample using 4480 Å laser radiation. The nominal laser radiation power was 100–150 mW.
Raman shifts (cm⁻¹): 1000w, 870s, 840w, 816w, 726, 576, 573s, 444, 363s, 366w, 228, 168.
Source: Mingsheng et al. (1994).
Comments: The sample was characterized by electron microprobe analyses. For the Raman spectra of andradite see also Kolesov and Geiger (1998), Bersani et al. (2009), Katerinopoulou et al. (2009), and Andò and Garzanti (2014).

Andradite $Ca_3Fe^{3+}_2(SiO_4)_3$

Origin: Maronia area, western Thrace, Greece.
Experimental details: Raman scattering measurements have been performed on an arbitrarily oriented sample using 514.5 nm laser radiation. The laser radiation power at the sample was 2 mW.
Raman shifts (cm⁻¹): 750, 520, 350.
Source: Katerinopoulou et al. (2009).
Comments: A Cr-, Ti-, and Zr-rich variety was investigated. The empirical formula is $(Ca_{2.99}Mn_{0.03})(Fe^{3+}_{0.67}Cr_{0.54}Al_{0.33}Ti_{0.29}Zr_{0.15})(Si_{2.42}Ti_{0.24}Fe_{0.18}Al_{0.14})O_{12}(OH)_{0.11}$. The Mössbauer analysis

showed that the total Fe is ferric, preferentially located at the octahedral site. For the Raman spectra of andradite see also Mingsheng et al. (1994), Kolesov and Geiger (1998), Bersani et al. (2009), and Andò and Garzanti (2014).

Andychristyite $PbCu^{2+}Te^{6+}O_5(H_2O)$

Origin: Otto Mt., near Baker, California, USA (type locality).
Experimental details: Raman scattering measurements have been performed on an arbitrarily oriented sample using 780 nm laser radiation. The laser radiation power at the sample was from 4 to 8 mW.
Raman shifts (cm^{-1}): 3306w, 870, 708s, 665s, 625s, 552, 461, 402, 316, 291, 242, 214, 181.
Source: Kampf et al. (2016b).
Comments: The sample was characterized by powder X-ray diffraction data and electron microprobe analysis. The crystal structure is solved.

Angastonite $CaMgAl_2(PO_4)_2(OH)_4 \cdot 7H_2O$

Origin: Angaston, South Australia (type locality).
Experimental details: Raman scattering measurements have been performed using 785 nm laser radiation. The method of sample preparation and the laser radiation power are not indicated.
Raman shifts (cm^{-1}): 1159, 988, 630, 539, 502, 415.
Source: Mills et al. (2008).
Comments: The sample was characterized by powder X-ray diffraction data and electron microprobe analyses.

Anglesite $Pb(SO_4)$

Origin: Monte Poni, Sardinia, Italy.
Experimental details: Raman scattering measurements have been performed on an arbitrarily oriented sample using 532 nm Nd-YAG laser radiation. The nominal laser radiation power at the sample was 100 mW.
Raman shifts (cm^{-1}): 1157, 1055w, 978s, 646w, 612, 553w, 450s.
Source: Buzgar et al. (2009).
Comments: The Raman shifts are indicated for the maxima of individual peaks obtained as a result of the spectral curve analysis. Methods of the identification of the sample are not indicated. For the Raman spectra of anglesite see also Bouchard and Smith (2003), Jehlička et al. (2009b), and Petrov (2014).

Anhydrite $Ca(SO_4)$

Origin: Bleiberg, Carinthia, Austria.
Experimental details: Raman scattering measurements have been performed on an arbitrarily oriented crystal using 532 nm Nd-YAG laser radiation. The nominal laser radiation power at the sample was 100 mW.
Raman shifts (cm^{-1}): 1160, 1129s, 1017s, 678, 630, 503s, 420w, 235w.
Source: Buzgar et al. (2009).
Comments: For the Raman spectra of anhydrite see also Sarma et al. (1998), Makreski et al. (2005a), White (2009), and Ciobotă et al. (2012).

Anilite Cu_7S_4

Origin: Synthetic.
Experimental details: No data.
Raman shift (cm^{-1}): 470s.
Source: Palve et al. (2016).
Comments: The sample was characterized by powder X-ray diffraction data.

Ankerite $Ca(Fe^{2+},Mg)(CO_3)_2$

Origin: Brusson, Val d'Ayas, Valle d'Aosta, Italy.
Experimental details: Raman scattering measurements have been performed on an arbitrarily oriented sample using 633 nm Ar$^+$ laser radiation. The laser radiation power is not indicated.
Raman shifts (cm^{-1}): 1091, 716, 283.
Source: Andò and Garzanti (2014).
Comments: No independent analytical data are provided for the sample used.

Annabergite $Ni_3(AsO_4)_2 \cdot 8H_2O$

Origin: 132 North Deposit, Widgiemooltha District, Western Australia.
Experimental details: Raman scattering measurements have been performed on arbitrarily oriented crystals using a 633 nm He-Ne laser. The Raman shifts have been determined for the maxima of individual peaks obtained as a result of the spectral curve analysis. The laser radiation power is not indicated.
Raman shifts (cm^{-1}): 3419, 3209, 3185, 3010, 941, 854, 800, 676, 466, 442, 401, 350, 321, 286, 260, 242, 225, 203, 175, 160, 155, 119.
Source: Frost et al. (2003g).
Comments: No independent analytical data are provided for the sample used.

Annite $KFe^{2+}_3(AlSi_3O_{10})(OH)_2$

Origin: Sierra los Filahres, Spain.
Experimental details: Raman scattering measurements have been performed on a partially oriented sample using 488 or 514.5 nm laser radiations. The spectrum was recorded with the electric field polarized either parallel or perpendicular to the cleavage plane. The laser radiation power is not indicated.
Raman shifts (cm^{-1}): 3654, 1045w, 676s, 272w, 182s.
Source: Tlili et al. (1989).
Comments: The sample was characterized by electron microprobe analyses. For the Raman spectrum of annite see also Rancourt et al. (2001).

Annite Cl-analogue $KFe^{2+}_3(AlSi_3O_{10})(Cl,OH)_2$

Origin: Khlebodarovka, Azov Sea region, Ukraine.
Experimental details: Raman scattering measurements have been performed on an arbitrarily oriented crystal. The wavelength of laser excitation line and the laser radiation power are not indicated.
Raman shifts (cm^{-1}): 3647s, 989s, 811, 660s, 553w, 403, 353.

Source: Sharygin et al. (2014).
Comments: The Cl-analogue of annite from Khlebodarovka contains up to 7.3 wt% Cl.

Anorpiment As_2S_3

Origin: Palomo mine, Castrovirreyna province, Huancavelica department, Peru (type locality).
Experimental details: Raman scattering measurements have been performed on an arbitrarily oriented sample using 532 nm laser radiation. The laser radiation power is not indicated.
Raman shifts (cm^{-1}): 643w, 387sh, 375, 348sh, 334s, 324sh, 234w, 192, 187, 176, 168sh.
Source: Kampf et al. (2011a).
Comments: The sample was characterized by powder X-ray diffraction data and electron microprobe analyses. The crystal structure is solved. The Raman shifts were determined by us based on spectral curve analysis of the published spectrum.

Anorthite $Ca(Al_2Si_2O_8)$

Origin: Synthetic.
Experimental details: Raman scattering measurements have been performed on a powdered sample using 488.0 nm laser radiation. The nominal laser radiation power was 600 mW. A 90°-scattering geometry was employed.
Raman shifts (cm^{-1}): 1124w, 1072s, 1044sh, 998sh, 974, 949sh, 908sh, 756w, 741sh, 681, 620w, 590w, 553, 503s, 484sh, 427w, 400w, 369w, 316w, 281, 273, 253, 200sh, 182, 139, 88, 63.
Source: Matson et al. (1986).
Comments: No independent analytical data are provided for the sample used. For the Raman spectrum of anorthite see also Ling et al. (2011).

Antarcticite $CaCl_2 \cdot 6H_2O$

Origin: Synthetic.
Experimental details: Raman scattering measurements have been performed on arbitrarily oriented crystals using 532.2 nm laser radiation. The nominal laser radiation power was 120 mW.
Raman shifts (cm^{-1}): 3431/3430s, 3410, 3404/3405sh, 3386, 3242.
Source: Baumgartner and Bakker (2010).
Comments: The Raman spectrum of antarcticite was recorded at −190 °C.

Antarcticite $CaCl_2 \cdot 6H_2O$

Origin: Synthetic.
Experimental details: Raman scattering measurements have been performed at −190 °C on an arbitrarily oriented sample using 514.5 nm Ar$^+$ laser radiation. The laser radiation power is not indicated.
Raman shifts (cm^{-1}): 3430s, 3384sh, 3245, 1664, 1647.
Source: Uriarte et al. (2015).
Comments: The sample was characterized by powder X-ray diffraction data.

Anthophyllite $\square Mg_2Mg_5Si_8O_{22}(OH)_2$

Origin: Bresimo Mine, near Trento, Trentino Alto Adige, Italy.
Experimental details: Raman scattering measurements have been performed on arbitrarily oriented fibers using 632 nm laser radiation. The nominal laser radiation power was 20 mW.
Raman shifts (cm^{-1}): 1044, 928, 699w, 674s, 539, 503w, 433, 410, 387, 342w, 304, 265, 254, 222w, 188.
Source: Rinaudo et al. (2004).
Comments: The sample was characterized by powder X-ray diffraction data and electron microprobe analyses. For the Raman spectrum of anthophyllite see also Petry et al. (2006), Kloprogge et al. (2001a), and Leissner et al. (2015).

Anthophyllite $\square Mg_2Mg_5Si_8O_{22}(OH)_2$

Origin: Origätri, Finland.
Experimental details: Raman scattering measurements have been performed on arbitrarily oriented fibers, using 244 nm laser radiation. The laser radiation power at the sample was less than 5 mW.
Raman shifts (cm^{-1}): 3691s, 3666s, 1052s, 1003sh, 669s, 530s.
Source: Petry et al. (2006).
Comments: The sample was characterized by electron microprobe analysis. For the Raman spectrum of anthophyllite see also Rinaudo et al. (2004), Kloprogge et al. (2001a), and Leissner et al. (2015).

Anthraxolite
Origin: Perya, Novaya Zemlya Islands, Russia.
Experimental details: Raman scattering measurements have been performed on an arbitrarily oriented sample using 514.5 nm Ar$^+$ laser radiation. The nominal laser radiation power was 1.2 mW.
Raman shifts (cm^{-1}): The first-order spectrum: 1180–1200, 1330–1350, ~1500, 1580–1600, 1610–1620sh. The second-order spectrum: 2500, 2700, 2850, 2950, 3230.
Source: Golubev et al. (2016).
Comments: The Raman shifts are indicated for the maxima of individual peaks obtained as a result of the spectral curve analysis. The sample was characterized by powder X-ray diffraction data and electron microprobe analyses.

Antigorite $Mg_3Si_2O_5(OH)_4$

Origin: Piedmont Alps, Italy.
Experimental details: Raman scattering measurements have been performed on an arbitrarily oriented sample using 1064 nm laser radiation. The laser radiation power is not indicated. A 180°-scattering geometry was employed.
Raman shifts (cm^{-1}): 1044, 683s, 635w, 520w, 375s, 230s.
Source: Rinaudo et al. (2003).
Comments: The sample was characterized by powder X-ray diffraction data and electron microprobe analyses.

Antigorite $Mg_3Si_2O_5(OH)_4$

Origin: Escambray Massif, Central Cuba.
Experimental details: Raman scattering measurements have been performed on an arbitrarily oriented sample using 514.5 nm Ar⁺ laser radiation. The laser output power was from 200 to 600 mW. A 180°-scattering geometry was employed.
Raman shifts (cm^{-1}): 3774, 3760w, 3729s, 3709, 3687, 3658, 3606w, 685s, 377s, 233s.
Source: Auzende et al. (2004).
Comments: The Raman shifts are partly indicated for the maxima of individual peaks obtained as a result of the spectral curve analysis.

Antimonselite Sb_2Se_3

Origin: Synthetic.
Experimental details: Raman scattering measurements have been performed on a μm thick layer using 532 nm laser radiation. The laser radiation power is not indicated.
Raman shifts (cm^{-1}): 252, 189.
Source: Zhou et al. (2014).
Comments: The sample was characterized by powder X-ray diffraction data and X-ray photoelectron spectroscopy.

Antimony Sb

Origin: Artificial.
Experimental details: Raman scattering measurements have been performed on an arbitrarily oriented sample using 632.8 nm laser radiation. The laser radiation power at the sample was 0.97 mW.
Raman shifts (cm^{-1}): 145s, 105.
Source: Makreski et al. (2013b).
Comments: Antimony was obtained as a product of photo-induced decomposition of stibnite Sb_2S_3 and identified due to the resemblance to the spectrum of Sb from Degtyareva et al. (2007). According to Degtyareva et al. (2007) Sb-I (99.999% purity) at ambient pressure is characterized by the Raman shifts of 151s and 114 cm^{-1}.

Antlerite $Cu^{2+}_3(SO_4)(OH)_4$

Origin: Chuquicamata, Chile.
Experimental details: The method of the sample preparation is not indicated. Raman scattering measurements have been performed using 780 nm laser radiation. The laser radiation power was less than 1 mW.
Raman shifts (cm^{-1}): 3580, 3488, 1905, 1266, 1173, 1148, 1135, 1079s, 990, 985, 902, 786, 759, 651, 629, 606, 600, 485, 469, 440, 415s, 335, 330, 295, 265, 259, 169, 151, 146, 141, 131.
Source: Martens et al. (2003a).
Comments: The Raman shifts are indicated for the maxima of individual peaks obtained as a result of the spectral curve analysis. The sample was characterized by powder X-ray diffraction data and electron microprobe analyses. The band at 1905 cm^{-1} is attributed to the first overtone of symmetric stretching vibrations of $[SO_4]^{2-}$. For the Raman spectra of antlerite see also Bouchard and Smith (2003), Frost et al. (2004n), Apopei et al. (2014a), and Coccato et al. (2016).

Apachite $Cu^{2+}_9Si_{10}O_{29} \cdot 11H_2O$

Origin: Christmas mine, Christmas area, Banner District, Dripping Spring Mts., Gila Co., Arizona, USA (type locality).

Experimental details: Raman scattering measurements have been performed on arbitrarily oriented crystals using a 633 nm He-Ne laser. The Raman shifts have been determined for the maxima of individual peaks obtained as a result of the spectral curve analysis. The laser radiation power is not indicated.

Raman shifts (cm^{-1}): 3651sh, 3614s, 3579, 3491, 3374, 3215, 2997, 2938, 2894, 2746, 1668, 1610, 1536sh, 1364, 1336, 1287, 1264, 1096, 997, 967s, 939, 898, 837, 777, 673s, 663sh, 529, 512, 479, 449sh, 435s, 402, 349, 335, 314, 305, 289, 253, 238, 221, 207, 195sh, 186, 180sh, 151s, 139sh, 113, 106.

Source: Frost and Xi (2012k).

Comments: No independent analytical data are provided for the sample used. The very intense sharp Raman band at 3614 cm^{-1} is assigned by the authors of the cited article to stretching vibrations of OH groups. The authors suppose that the correct formula of apachite could be $Cu_9Si_{10}O_{23}(OH)_{12} \cdot 5H_2O$.

Aphthitalite $K_3Na(SO_4)_2$

Origin: Vesuvius volcano, Somma-Vesuvius complex, Naples province, Campania, Italy.

Experimental details: Raman scattering measurements have been performed on an arbitrarily oriented sample using 514.5 nm Ar$^+$ laser radiation in a 180°-scattering geometry. The laser radiation power is not indicated.

Raman shifts (cm^{-1}): 1206–1201, 1084, 996–985s, 626–627, 618–617, 452–451s.

Source: Hansteen and Burke (1994).

Comments: The sample was characterized by electron microprobe analyses. For the Raman spectrum of aphthitalite see also Jentzsch et al. (2013).

Apjohnite $Mn^{2+}Al_2(SO_4)_4 \cdot 22H_2O$

Origin: Coranda-Hondol ore deposit, Certej, Romania.

Experimental details: Method of sample preparation is not indicated. Raman scattering measurements have been performed using 532 nm laser radiation. The laser radiation power at the sample was 22.9 mW.

Raman shifts (cm^{-1}): 3379, 3299, 3237sh, 3007sh, 1630w, 1227sh, 1141w, 1116w, 1108w, 1086w, 1073sh, 996s, 978sh, 619w, 529w, 469w, 432sh, 311w.

Source: Apopei et al. (2014a).

Comments: The sample was characterized by powder X-ray diffraction data and electron microprobe analyses. The Raman shifts are indicated for the maxima of individual peaks obtained as a result of the spectral curve analysis. For the Raman spectra of apjohnite see also Reddy et al. (2006) and Locke et al. (2007).

Apuanite $(Fe^{2+}Fe^{3+}_2)(Fe^{3+}_2Sb^{3+}_4)O_{12}S$

Origin: Bucadella Vena, Apuan Alps, Italy (type locality).

Experimental details: Raman scattering measurements have been performed on an arbitrarily oriented sample using 633 nm laser radiation. The laser radiation power is not indicated.

Raman shifts (cm^{-1}): 713sh, 669s, 632sh, 574w, 548, 491, 437sh, 435, 396, 332, 292sh, 274sh, 242sh, 230s, 196, 174, 162sh, 143sh, 121, 106.
Source: Bahfenne (2011).
Comments: The sample was characterized by powder X-ray diffraction data and electron microprobe analyses. The Raman shifts are indicated for the maxima of individual peaks obtained as a result of the spectral curve analysis. For the Raman spectrum of apuanite see also Bahfenne et al. (2011a).

Aradite $BaCa_6[(SiO_4)(VO_4)](VO_4)_2F$

Origin: Hatrurim Complex, Negev desert, Israel (type locality).
Experimental details: Experimental details are not identified. The methods used for the investigations are analogous to those reported by Galuskin et al. (2015b).
Raman shifts (cm^{-1}): 989, 968, 942, 874sh, 859s, 835sh, 449w, 386sh, 366, 293w, 222w.
Source: Galuskin et al. (2015e).
Comments: The investigated sample has the crystal-chemical formula $BaCa_6[(SiO_4)_{1.2}(VO_4)_{0.5}(PO_4)_{0.1}(SO_4)_{0.2}][(VO_4)_{1.51}(PO_4)_{0.59}]F$ and is a not an end-member of a complex solid solution. For the Raman spectrum of aradite see also Galuskin et al. (2015b).

Aragonite $Ca(CO_3)$

Origin: No data.
Experimental details: Methods of sample preparation are not indicated. Raman scattering measurements have been performed using 1064 nm laser radiation. The nominal laser radiation power was 100 mW.
Raman shifts (cm^{-1}): 2905w, 2835w, 1904w, 1574w, 1462w, 1086s, 854w, 717sh, 704w.
Source: Edwards et al. (2005).
Comments: No independent analytical data are provided for the sample used. For the Raman spectra of aragonite see also Buzgar and Apopei (2009), Behrens et al. (1995), White (2009), Wehrmeister et al. (2010), Frezzotti et al. (2012), Kristova et al. (2014), Shatskiy et al. (2015), and Sánchez-Pastor et al. (2016).

Aragonite $Ca(CO_3)$

Origin: Spania Dolina, Slovakia.
Experimental details: Methods of sample preparation are not indicated. Raman scattering measurements have been performed using 532 nm Nd-YAG laser radiation. The nominal laser radiation power was 1050 mW.
Raman shifts (cm^{-1}): 1573, 1461, 1083s, 701, 250.
Source: Buzgar and Apopei (2009).
Comments: No independent analytical data are provided for the sample used. For the Raman spectra of aragonite see also Edwards et al. (2005), Behrens et al. (1995), White (2009), Wehrmeister et al. (2010), Frezzotti et al. (2012), Kristova et al. (2014), Shatskiy et al. (2015), and Sánchez-Pastor et al. (2016).

Arapovite-related silicate $(Ca_{0.5}Na_{0.5})_2NaUSi_8O_{20}$

Origin: Synthetic.
Experimental details: Raman scattering measurements have been performed on arbitrarily oriented single crystals using 532 nm laser radiation. The laser radiation power is not indicated.

Raman shifts (cm^{-1}): 1139s, 1081, 655, 468, 303, 202w.

Comments: The sample was characterized by powder X-ray diffraction data. The crystal structure is solved.

Aravaipaite Pb$_3$AlF$_9$·H$_2$O

Origin: Grand Reef mine, Aravaipa mining district, Arizona, USA (type locality).

Experimental details: Raman scattering measurements have been performed on arbitrarily oriented crystals using 532 nm laser radiation. The laser radiation power is not indicated.

Raman shifts (cm^{-1}): ~3370s, ~3250w, ~2935w, ~1640w, ~620, ~550, ~530, ~400, ~385, ~370, ~330w, ~317w, ~260sh, ~233s, ~175, ~170sh.

Source: Kampf et al. (2011c).

Comments: The sample identification was done by single-crystal X-ray diffraction. The crystal structure is solved.

Arcanite K$_2$(SO$_4$)

Origin: Synthetic.

Experimental details: Methods of sample preparation are not indicated. Raman scattering measurements have been performed using 532 nm laser radiation. The nominal laser radiation power was 100 mW.

Raman shifts (cm^{-1}): 1147, 1108, 1094, 985s, 623, 458.

Source: Buzgar et al. (2009).

Comments: The Raman shifts are indicated for the maxima of individual peaks obtained as a result of the spectral curve analysis. No independent analytical data are provided for the sample used. For the Raman spectrum of arcanite see also Martínez-Arkarazo et al. (2007).

Archerite H$_2$K(PO$_4$)

Origin: Petrogale cave, Madura, Western Australia.

Experimental details: Raman scattering measurements have been performed on arbitrarily oriented crystals using a 633 nm He-Ne laser. The Raman shifts have been determined for the maxima of individual peaks obtained as a result of the spectral curve analysis. The laser radiation power is not indicated.

Raman shifts (cm^{-1}): 3235, 3191, 3151, 3089, 3018, 1724sh, 1704, 1660, 1421, 983sh, 917s, 562sh, 533, 477sh, 461sh, 393s, 347, 328sh, 270w, 180, 144.

Source: Frost et al. (2012f).

Comments: No independent analytical data are provided for the sample used.

Ardealite Ca$_2$(PO$_3$OH)(SO$_4$)·4H$_2$O

Origin: Moorba cave, Jurien Bay, Western Australia, Australia.

Experimental details: Raman scattering measurements have been performed on arbitrarily oriented crystals using a 633 nm He-Ne laser. The Raman shifts have been determined for the maxima of individual peaks obtained as a result of the spectral curve analysis. The laser radiation power is not indicated.

Raman shifts (cm^{-1}): 1141, 1102, 1001s, 998sh, 862, 670, 613, 598, 528sh, 505, 448, 421, 363, 230, 198sh, 188, 155, 143.

Source: Frost et al. (2011g).
Comments: No independent analytical data are provided for the sample used.

Ardennite-(As) $Mn^{2+}_4Al_4(AlMg)(AsO_4)(SiO_4)_2(Si_3O_{10})(OH)_6$

Origin: Salm-Château, Vielsalm, Stavelot massif, Luxembourg province, Belgium (type locality).
Experimental details: Raman scattering measurements have been performed on arbitrarily oriented crystals using a 633 nm He-Ne laser. The Raman shifts have been determined for the maxima of individual peaks obtained as a result of the spectral curve analysis. The laser radiation power is not indicated.
Raman shifts (cm^{-1}): 3298, 3211sh, 3149, 3041, 1605s, 1394sh, 1287sh, 1255s, 1218s, 1197s, 947sh, 935, 920sh, 890sh, 877, 865sh, 785sh, 779, 721, 713, 625, 601, 561, 544sh, 519, 471sh, 460, 445, 430, 414, 396, 365sh, 352, 314, 301, 228, 183, 167, 144.
Source: Frost et al. (2014s).
Comments: No independent analytical data are provided for the sample used. The strong band at 1605 cm^{-1} corresponds to an impurity. The spectrum contains broad bands of unknown origin near 2100 cm^{-1}.

Arfvedsonite $NaNa_2(Fe^{2+}_4Fe^{3+})Si_8O_{22}(OH)_2$

Origin: Vodno, Macedonia.
Experimental details: Raman scattering measurements have been performed on a powdered sample using 514.5 nm Ar$^+$ laser radiation. The nominal laser radiation power was from 50 to 100 mW.
Raman shifts (cm^{-1}): 1055, 1020w, 974s, 921, 892, 815, 793, 772, 749, 725, 676s, 640w, 610w, 583, 539, 478, 435, 370, 335, 317w, 255sh, 215, 172, 149, 112.
Source: Makreski et al. (2006a).
Comments: The sample used is an intermediate member between arfvedsonite and magnesioarfvedsonite. The sample was characterized by powder X-ray diffraction data. For the Raman spectrum of arfvedsonite see also Jovanovski et al. (2009) and Leissner et al. (2015).

Argentojarosite $AgFe^{3+}_3(SO_4)_2(OH)_6$

Origin: Synthetic.
Experimental details: Raman scattering measurements have been performed on an arbitrarily oriented sample prepared as a disk 10 mm in diameter with KBr powder. The wavelength of laser excitation line was 514.5 nm. The laser radiation power at the sample was 38 mW.
Raman shifts (cm^{-1}): 1161w, 1107, 1012, 623, 574w, 449sh, 442s, 363w, 306, 228.
Source: Sasaki et al. (1998).
Comments: The sample was characterized by powder X-ray diffraction data. For the Raman spectrum of argentojarosite see also Frost et al. (2006r).

Argutite GeO_2

Origin: Synthetic.
Experimental details: Raman scattering measurements have been performed on an arbitrarily oriented sample in a 180°-scattering geometry using 514.5 nm Ar$^+$ laser radiation. The laser radiation power is not indicated.

Raman shifts (cm^{-1}): 873, 700s, 170w.
Source: Madon et al. (1991).
Comments: The procedure of verification of the structure of the rutile form GeO_2 has been described by Richet (1990).

Arisite (Ce) $NaCe_2(CO_3)_2[F_{2x}(CO_3)_{1-x}]F$

Origin: Aris phonolite, Namibia (type locality).
Experimental details: Raman scattering measurements have been performed on an arbitrarily oriented sample using 532 nm laser radiation. The laser radiation power is not indicated.
Raman shifts (cm^{-1}): 3449, 3255s, 2642, 2458s, 2068, 1799, 1596, 1455, 1072s, 704s, 396, 187s, 152s.
Source: Piilonen et al. (2010).
Comments: The sample was characterized by powder X-ray diffraction data, LA-ICP-MS, and electron microprobe analyses. The band at 1596 cm^{-1} indicates possible presence of H_2O molecules.

Armalcolite $(Mg,Fe^{2+})Ti_2O_5$

Origin: Skallevikshalsen, Lützow-Holm Complex, East Antarctica.
Experimental details: Raman scattering measurements have been performed on an arbitrarily oriented thin section using 533 nm laser radiation. The laser radiation power is not indicated.
Raman shifts (cm^{-1}): 759, 633s, 393, 201s, 169sh.
Source: Kawasaki et al. (2013).
Comments: The sample was characterized by electron microprobe analyses.

Arrojadite-(KFe) $(KNa)Fe^{2+}(CaNa_2)Fe^{2+}_{13}Al(PO_4)_{11}(PO_3OH)(OH)_2$

Origin: Rapid Creek, Richardson Mts., Yukon Territory, Canada.
Experimental details: Raman scattering measurements have been performed on arbitrarily oriented crystals using a 633 nm He-Ne laser. The Raman shifts have been determined for the maxima of individual peaks obtained as a result of the spectral curve analysis. The laser radiation power is not indicated.
Raman shifts (cm^{-1}): 3574, 3564s, 3553, 3530, 3515, 1714, 1580, 1444, 1187w, 1148w, 1123w, 1092w, 1066, 1024sh, 1005s, 991s, 975s, 951sh, 903, 852, 638, 615, 604sh, 583s, 580, 557s, 548, 540, 513, 479, 463, 449, 424s, 403, 349, 306, 275, 251, 239, 202, 185, 162, 140.
Source: Frost et al. (2013ag).
Comments: There are discrepancies between some frequencies given in the text and figures. The sample identification was done by XRD and qualitative EDS analysis.

Arsenbrackebuschite $Pb_2(Fe^{3+},Zn)(AsO_4)_2(OH,H_2O)$

Origin: No data.
Experimental details: No data.
Raman shifts (cm^{-1}): 972w, 846sh, ~820s, 730, 620, ~464sh, 420, 405sh, ~345, 308, ~240w, ~102–170.
Source: Costin et al. (2014).
Comments: No independent analytical data are provided for the sample used.

Arsendescloisite Sr-analogue $SrZn(AsO_4)(OH)$

Origin: Synthetic.
Experimental details: Raman scattering measurements have been performed on an arbitrarily oriented sample using 632.8 or 473 nm laser radiation. The nominal laser radiation power was 10 or 3 mW.
Raman shifts (cm^{-1}): ~3400sh, 3300, ~3240w, 818s, 804, 790, ~780w, ~450–300w.
Source: Đorđević et al. (2016).
Comments: The sample was characterized by single-crystal X-ray diffraction data. The crystal structure is solved.

Arsenolamprite As

Origin: Muiane pegmatite, Mozambique.
Experimental details: Raman scattering measurements have been performed on a melt inclusion in a morganite crystal using 514 or 488 nm Ar^+ laser radiation. The laser radiation power at the sample was 14 mW.
Raman shifts (cm^{-1}): 253s, 225w, 220.
Source: Thomas and Davidson (2010).

Arseniosiderite $Ca_2Fe^{3+}_3O_2(AsO_4)_3 \cdot 3H_2O$

Origin: Romanêche, near Mâcon, Saoneet-Loirse, France.
Experimental details: Raman scattering measurements have been performed on an arbitrarily oriented sample using 632 nm laser radiation. The laser radiation power is not indicated.
Raman shifts (cm^{-1}): 927s, 852s, 828, 772sh, 621, 535, 479w, 441w, 389s, 331w, 298sh, 250s, 227, 197.
Source: Gomez et al. (2010b).
Comments: The sample was characterized by powder X-ray diffraction data and electron microprobe analysis. For the Raman spectrum of arseniosiderite see also Filippi et al. (2007).

Arsenogorceixite $BaAl_3(AsO_4)(AsO_3OH)(OH)_6$

Origin: Michael mine, Weiler, near Lahr, Schwarzwald (Black Forest), Germany.
Experimental details: Raman scattering measurements have been performed on an arbitrarily oriented sample using 633 nm He-Ne laser radiation. The laser radiation power at the sample was 0.01 mW. The Raman shifts have been determined for the maxima of individual peaks obtained as a result of the spectral curve analysis.
Raman shifts (cm^{-1}): 3691, 3621, 3473sh, 3301, 2973sh, 2961sh, 2930, 2891, 2863, 2849, 2725, 1597, 1447, 1332, 1307, 1208, 1160, 1148, 1057, 1014, 972s, 873, 818, 814s, 805, 776, 764, 617, 600, 556, 510, 462, 441s, 407, 388, 340s, 318, 264, 244, 189, 167, 137.
Source: Frost et al. (2012g).
Comments: No independent analytical data are provided for the sample used. There are discrepancies between the spectrum and its description in the cited paper.

Arsenolite As_2O_3

Origin: Cobalt City, Ontario, Canada.
Experimental details: Raman scattering measurements have been performed on an arbitrarily oriented sample using 633 nm He-Ne laser radiation. The laser radiation power is not indicated.

Raman shifts (cm⁻¹): 781, 560, 469, 413w, 368s, 265s, 180.

Source: Kloprogge et al. (2006).

Comments: The sample was characterized by SEM/EDS. Raman shifts are indicated for the maxima of individual peaks obtained as a result of the spectral curve analysis. For the Raman spectrum of arsenolite see also Guńka et al. (2012).

Arsenopyrite FeAsS

Origin: Nistru mine, Maramures, Romania.

Experimental details: Raman scattering measurements have been performed on an arbitrarily oriented sample in a 180°-scattering geometry using 532 nm laser radiation. The nominal laser radiation power was 210 mW.

Raman shifts (cm⁻¹): 478, 453, 427, 392, 362, 333, 303, 280s, 253, 231, 217s, 200sh, 187, 180sh, 170, 127, 108sh, 94, 84, 73.

Source: Kharbish and Andráš (2014).

Comments: The Raman shifts are indicated for the maxima of individual peaks obtained as a result of the spectral curve analysis. The sample was characterized by powder X-ray diffraction data and electron microprobe analysis. For the Raman spectrum of arsenopyrite see also Mernagh and Trudu (1993).

Arsentsumebite $Pb_2Cu(AsO_4)(SO_4)(OH)$

Origin: Tsumeb mine, Tsumeb, Namibia (type locality).

Experimental details: Raman scattering measurements have been performed on arbitrarily oriented crystals using a 633 nm He-Ne laser. The Raman shifts have been determined for the maxima of individual peaks obtained as a result of the spectral curve analysis. The laser radiation power is not indicated.

Raman shifts (cm⁻¹): 3458, 3324, 2925, 2876, 2857, 1446, 1161, 1121w, 1070w, 972s, 906, 853, 814s, 620, 604, 464, 442, 412, 390, 340, 324, 308, 248, 241, 197, 188, 171, 145, 102.

Source: Frost et al. (2011n).

Comments: No independent analytical data are provided for the sample used. There are discrepancies between the spectrum and its description in the cited paper. Spectroscopic data show possible presence of impurities in the investigated material. For the Raman spectrum of arsentsumebite see also Costin et al. (2014).

Arsenuranylite $Ca(UO_2)_4(AsO_4)_2(OH)_4 \cdot 6H_2O$

Origin: Cherkasar deposit, Uzbekistan.

Experimental details: Raman scattering measurements have been performed on arbitrarily oriented crystals using a 633 nm He-Ne laser. The Raman shifts have been determined for the maxima of individual peaks obtained as a result of the spectral curve analysis. The laser radiation power is not indicated.

Raman shifts (cm⁻¹): 3489, 2929, 2872, 926, 883, 795s, 787s, 561, 558, 494, 493, 462, 422, 388, 344, 298, 259, 213, 170, 150.

Source: Frost et al. (2009d).

Comments: No independent analytical data are provided for the sample used.

Arthurite $CuFe^{3+}_2(AsO_4)_2(OH)_2 \cdot 4H_2O$

Origin: Majuba Hill, Pershing Co., Nevada, USA.
Experimental details: Raman scattering measurements have been performed on an arbitrarily oriented polished sample. The wavelength of laser excitation line and the laser radiation power are not indicated.
Raman shifts (cm^{-1}): 3496, 3307, 3232, 3162, 1050, 907, 850, 812, 784, 551w, 508s, 450, 426, 368, 343, 288, 259, 245s, 221, 186, 153, 138, 101, 70.
Source: Jambor et al. (2002).
Comments: The sample was characterized by electron microprobe analysis. For the Raman spectra of arthurite see also Frost et al. (2003b) and Palmer and Frost (2011).

Artinite $Mg_2(CO_3)(OH)_2 \cdot 3H_2O$

Origin: Aichi prefecture, Japan.
Experimental details: Raman scattering measurements have been performed on arbitrarily oriented crystals using a 633 nm He-Ne laser. The Raman shifts have been determined for the maxima of individual peaks obtained as a result of the spectral curve analysis. The laser radiation power is not indicated.
Raman shifts (cm^{-1}): 3593s, 3589sh, 3573sh, 3229, 3030, 2291, 1673, 1453, 1092s, 1060sh, 913w, 698, 469, 432, 376, 324, 273, 244, 209, 188.
Source: Frost et al. (2009a).
Comments: No independent analytical data are provided for the sample used. There are discrepancies between the pattern of the spectrum and its description. For the Raman spectrum of artinite see also Edwards et al. (2005).

Arzakite $Hg^{2+}_3S_2(Br,Cl)_2$

Origin: No data.
Experimental details: Micro-Raman scattering measurements have been performed on an arbitrarily oriented sample using 514.5 or 785 nm laser radiation. The laser radiation power is not indicated.
Raman shifts (cm^{-1}): ~585, ~480w, ~390, ~270s, ~215s.
Source: Potgieter-Vermaak et al. (2012).

Asbolane $Mn^{4+}(O,OH)_2(Co,Ni,Mg,Ca)_x(OH)_{2x} \cdot nH_2O$

Origin: Democratic Republic of Congo.
Experimental details: No data.
Raman shifts (cm^{-1}): 3484w, 1206w, 1084w, 950w, 592s, 539s, 489, 376.
Source: Burlet et al. (2014).
Comments: The sample was characterized by electron microprobe analyses. For the Raman spectra of asbolane see also Burlet and Vanbrabant (2015) and Roqué-Rosell et al. (2010).

Asbolane $Mn^{4+}(O,OH)_2(Co,Ni,Mg,Ca)_x(OH)_{2x} \cdot nH_2O$

Origin: Democratic Republic of Congo.
Experimental details: No data.

Raman shifts (cm^{-1}): 3484w, 1206w, 1084w, 950w, 592s, 539s, 489, 376.
Source: Burlet et al. (2014).
Comments: The sample was characterized by electron microprobe analyses. For the Raman spectra of asbolane see also Burlet and Vanbrabant (2015) and Roqué-Rosell et al. (2010).

Asbolane $Mn^{4+}(O,OH)_2(Co,Ni,Mg,Ca)_x(OH)_{2x} \cdot nH_2O$

Origin: A Cu-Co supergene deposit, Ruashi, Katanga, Democratic Republic of Congo.
Experimental details: Raman scattering measurements have been performed on an arbitrarily oriented polished sample using 532 nm laser radiation. The laser radiation power was 0.2 mW.
Raman shifts (cm^{-1}): 3455–3475w, 1580–1600w, 1300–900w (a triplet), 627sh, 596s, 553s, 497, 456sh, 374w.
Source: Burlet and Vanbrabant (2015).
Comments: The sample was characterized by powder X-ray diffraction data and electron microprobe analyses. Raman shifts are indicated for the maxima of individual peaks obtained as a result of the spectral curve analysis. For the Raman spectra of asbolane see also Burlet et al. (2014) and Roqué-Rosell et al. (2010).

Aspedamite $\square\square_{12}(Fe^{3+},Fe^{2+})_3Nb_4[Th(Nb,Fe^{3+})_{12}O_{42}][(H_2O),(OH)]_{12}$

Origin: Herrebøkasa quarry, Aspedammen, Østfold, southern Norway (type locality).
Experimental details: No data.
Raman shifts (cm^{-1}): 3465w, 3556sh, 1610w, 933, 865sh, 812, 666s, 448w, 359, 234s, 169s, 117.
Source: Cooper et al. (2012b).
Comments: The sample was characterized by powder X-ray diffraction data and electron microprobe analyses.

Asselbornite $Pb(UO_2)_4(BiO)_3(AsO_4)_2(OH)_7 \cdot 4H_2O$

Origin: Horní Halže, the Krušnéhory Mountains, Czech Republic (type locality).
Experimental details: Methods of sample preparation are not indicated. Raman scattering measurements have been performed using 785 nm laser radiation. The laser radiation power at the sample was 4 mW.
Raman shifts (cm^{-1}): 1039w, 999w, 962w, 874sh, 842sh, 797s, 673w, 599w, 503, 450w, 395w, 321, 266.
Source: Sejkora and Čejka (2007).
Comments: The sample was characterized by powder X-ray diffraction data and electron microprobe analyses. The crystal structure is solved.

Atacamite $Cu_2Cl(OH)_3$

Origin: Atacama, Chile.
Experimental details: Raman scattering measurements have been performed on an oriented crystal using 532.8 nm laser radiation. The laser radiation power at the sample was from 0.05 to 1 mW. The spectrum was obtained in the scattering geometry with polarization of the laser beam oriented at 45° with respect to the c axis.

Raman shifts (cm^{-1}): 3433, 3348, 975s, 909s, 846, 819s, 587, 513s, 353w, 177s, 148s, 135, 117, 105s.
Source: Bertolotti et al. (2012).
Comments: No independent analytical data are provided for the sample used. For the Raman spectra of atacamite see also Frost et al. (2002b), Bouchard and Smith (2003), and Christy et al. (2004).

Atelestite $Bi_2O(AsO_4)(OH)$

Origin: Synthetic.
Experimental details: Raman scattering measurements have been performed on arbitrarily oriented crystals using a 633 nm He-Ne laser. The Raman shifts have been determined for the maxima of individual peaks obtained as a result of the spectral curve analysis. The laser radiation power is not indicated.
Raman shifts (cm^{-1}): 3095w, 1082w, 887sh, 834s, 802, 782, 767, 623w, 480, 450, 395sh, 370sh, 352, 324, 310, 278s, 219, 200, 173, 118.
Source: Frost et al. (2011b).
Comments: The sample was characterized by powder X-ray diffraction data and electron microprobe analysis.

Atelisite (Y) $Y_4Si_3O_8(OH)_8$

Origin: Synthetic.
Experimental details: Raman scattering measurements have been performed on an oriented crystal using 514.5 nm Ar$^+$ laser radiation. The laser radiation power at the sample was 8 mW. The spectra obtained in the scattering geometries $X(ZZ)X$, $X(YY)X$, and $Y(YZ)X$ are similar.
Raman shifts (cm^{-1}): 3225w, 2964w, 2905w, 885s, 755sh, 709w, 490w.
Source: Malcherek et al. (2012).
Comments: The sample was characterized by powder X-ray diffraction data and electron microprobe analysis.

Athabascaite Cu_5Se_4

Origin: Synthetic.
Experimental details: Raman scattering measurements have been performed on a nanocrystallione aggregate. The wavelengths of laser radiation and laser radiation power are not indicated.
Raman shifts (cm^{-1}): 128.
Source: Ge and Li (2003).
Comments: The sample was characterized by powder X-ray diffraction data and qualitative electron microprobe analysis.

Atokite Pd_3Sn

Origin: Synthetic.
Experimental details: Raman scattering measurements have been performed on an arbitrarily oriented sample using 532.068 nm Nd-YAG laser radiation. The laser radiation power at the sample was between 1 and 2 mW.
Raman shifts (cm^{-1}): The obtained spectrum does not show characteristic bands.
Source: Vymazalová et al. (2014).

Augelite $Al_2(PO_4)(OH)_3$

Origin: Ehrenfriedersdorf, Saxony, Germany.
Experimental details: Micro-Raman scattering measurements have been performed on an arbitrarily oriented inclusion in quartz using 514.5 nm Ar^+ laser radiation. The nominal laser radiation power at the sample was between 100 and 500 mW.
Raman shifts (cm^{-1}): 3537s, 3469s, 3428, 1107, 635s, 367, 252, 227.
Source: Thomas et al. (1998).
Comments: The sample was characterized by electron microprobe analysis. For the Raman spectrum of augelite see also Frost and Weier (2004b).

Augite $(Ca,Mg,Fe)_2Si_2O_6$

Origin: Sasa, Macedonia.
Experimental details: Raman scattering measurements have been performed on a powdered sample using 514.5 nm Ar^+ laser radiation. The nominal laser radiation power at the sample was 50 or 100 mW.
Raman shifts (cm^{-1}): 1025sh, 1009s, 907w, 851w, 654s, 549, 523, 510sh, 492w, 372, 336, 301, 229, 186, 178, 146, 122, 116w.
Source: Makreski et al. (2006b).
Comments: The Raman shifts are indicated for the maxima of individual peaks obtained as a result of the spectral curve analysis. The sample was characterized by powder X-ray diffraction data and neutron activation analysis. For the Raman spectra of augite see also Buzatu and Buzgar (2010) and Andò and Garzanti (2014).

Augite $(Ca,Mg,Fe)_2Si_2O_6$

Origin: Techereu area, Apuseni Mts., Romania.
Experimental details: Methods of sample preparation are not indicated. Raman scattering measurements have been performed using 532 nm laser radiation. The nominal laser radiation power was 100 mW.
Raman shifts (cm^{-1}): 1102sh, 1043sh, 1006s, 928w, 863, 769, 707w, 667s, 555, 533, 392, 355, 327, 299sh, 226w.
Source: Buzatu and Buzgar (2010).
Comments: No independent analytical data are provided for the sample used. For the Raman spectra of augite see also Makreski et al. (2006b) and Andò and Garzanti (2014).

Aurichalcite $(Zn,Cu)_5(CO_3)_2(OH)_6$

Origin: No data.
Experimental details: Raman scattering measurements have been performed on an arbitrarily oriented sample using 632.8 nm laser radiation. The nominal laser radiation power was 30 mW.
Raman shifts (cm^{-1}): 3331, 1511, 1479, 1074s, 843, 750, 734, 709, 503, 463, 437w, 389, 354w, 234, 211, 175, 141.
Source: Bouchard and Smith (2003).
Comments: For the Raman spectra of aurichalcite see also Frost et al. (2007j), Buzgar and Apopei (2009), and Rotondo et al. (2012).

Aurostibite $AuSb_2$

Origin: Synthetic.
Experimental details: Raman scattering measurements have been performed at 380 K on an arbitrarily oriented crystal using 514.5, 501.5, 476.5, and 457.9 nm laser radiation. The nominal laser radiation power at the sample was ~200, ~190, ~200, and ~150 mW, respectively. The incident laser light was scattered off a [100] natural cleavage.
Raman shifts (cm^{-1}): 158w, 151s, 122sh, 114s.
Source: Freund et al. (1977).
Comments: No independent analytical data are provided for the sample used.

Austinite $CaZn(AsO_4)(OH)$

Origin: Gold Hill Mine, Tooele Co., Utah, USA.
Experimental details: Raman scattering measurements have been performed on an arbitrarily oriented sample using 514 nm laser radiation. The laser radiation power is not indicated.
Raman shifts (cm^{-1}): 3270, 828s, 814, 802, 779, 418, 403.
Source: Liu et al. (2015a).
Comments: The sample used was characterized by powder and single-crystal X-ray diffraction data. IR spectrum shows that the sample contains minor SO_4^{2-} substituting arsenate anions. For the Raman spectrum of austinite see also Martens et al. (2003c).

Autunite $Ca(UO_2)_2(PO_4)_2 \cdot 10-12H_2O$

Origin: Merrivale Quarry, Tavistock, Cornwall, UK.
Experimental details: Methods of sample preparation are not indicated. Raman scattering measurements have been performed using 785 nm laser radiation. The nominal laser radiation power at the source was ~370 mW.
Raman shifts (cm^{-1}): 1008s, 990, 900sh, ~400, ~270, 210.
Source: Driscoll et al. (2014).
Comments: The sample was characterized by electron microprobe analysis. For the Raman spectra of autunite see also Frost (2004d), Frost and Weier (2004c, d).

Avicennite Tl_2O_3

Origin: Synthetic.
Experimental details: Raman scattering measurements have been performed on finely ground powder pressed into pellets using 488 and 514.5 nm Ar$^+$ laser radiation. The laser radiation power is not indicated.
Raman shifts (cm^{-1}): A spectrum of Tl_2O_3, shows no distinct features above 400 cm^{-1} except an absorption edge near 400 cm^{-1}.
Source: White and Keramidas (1972).

Avogadrite KBF_4

Origin: Synthetic.
Experimental details: Methods of sample preparation are not described. Raman scattering measurements have been performed in N_2 atmosphere using 514 nm laser radiation. The laser radiation power is not indicated.

Raman shifts (cm^{-1}): 1097, 1043, 775s, 534, 360.
Source: Zavorotynska et al. (2011).
Comments: The sample was characterized by powder X-ray diffraction data. For the Raman spectra of avogadrite see also Bonadeo and Silberman (1970) and Bates and Quist (1974).

Awaruite Ni_3Fe

Origin: Synthetic.
Experimental details: Raman scattering measurements have been performed on an arbitrarily oriented sample using 785 nm laser radiation. The laser radiation power at the sample was 3 mW.
Raman shifts (cm^{-1}): 701, 566s.
Source: Abelló et al. (2014).
Comments: The sample was characterized by powder X-ray diffraction data.

Axinite-(Fe) $Ca_4Fe^{2+}{}_2Al_4[B_2Si_8O_{30}](OH)_2$

Origin: Drum valley, Tulare Co., California, USA.
Experimental details: Raman scattering measurements have been performed on arbitrarily oriented crystals using a 633 nm He-Ne laser. The Raman shifts have been determined for the maxima of individual peaks obtained as a result of the spectral curve analysis. The laser radiation power is not indicated.
Raman shifts (cm^{-1}): 3368, 1084, 1057, 1005sh, 993sh, 979s, 964s, 959sh, 931, 909, 898, 869, 813, 768, 714s, 678, 645, 619, 590, 574, 562, 547, 512, 485, 445, 422, 418, 390, 344, 319, 275s, 256, 212, 170, 140, 110.
Source: Frost et al. (2007b).
Comments: No independent analytical data are provided for the sample used. For the Raman spectrum of axinite (Fe) see also Andò and Garzanti (2014).

Azurite $Cu_3(CO_3)_2(OH)_2$

Origin: Namibia.
Experimental details: Raman scattering measurements have been performed on an arbitrarily oriented sample using 532 nm laser radiation. The nominal laser radiation power was 100 mW.
Raman shifts (cm^{-1}): 1579w, 1459w, 1425w, 1098s, 939, 835s, 766, 544, 404s, 339w, 285, 250s.
Source: Buzgar and Apopei (2009).
Comments: No independent analytical data are provided for the sample used. For the Raman spectra of azurite see also Frost et al. (2002d), Bouchard and Smith (2003), and Frezzotti et al. (2012).

Backite Pb_2AlTeO_6Cl

Origin: Grand Central mine, Tombstone Hills, Cochise Co., Arizona, USA (type locality).
Experimental details: Raman scattering measurements have been performed on an arbitrarily oriented grain using 785 nm laser radiation. The laser radiation power at the sample was ~1 mW.
Raman shifts (cm^{-1}): 967, 733s, 625, 425, 350, 120s.
Source: Tait et al. (2015).

Comments: The sample was characterized by powder X-ray diffraction data and electron microprobe analyses. The crystal structure is solved. The Raman band at 967 cm^{-1} is assigned to the combination mode ~(350+625) cm^{-1}.

Baddeleyite ZrO_2

Origin: Synthetic.
Experimental details: Raman scattering measurements have been performed on an arbitrarily oriented crystal in a thin section using 514.5 nm Ar$^+$ laser radiation. The nominal laser radiation power at the sample was from 30 to 50 mW. A 90°-scattering geometry was employed.
Raman shifts (cm^{-1}): 757w, 643, 622, 559, 538, 503, 476, 384s, 350s, 308, 264w, 225, 190s, 182sh, 105.
Source: Galuskina et al. (2013a).
Comments: The sample was characterized by SEM/EBSD and electron microprobe analysis. The Raman shifts were determined by us based on spectral curve analysis of the published spectrum. For the Raman spectrum of baddeleyite see also Zhang et al. (2010a).

Bafertisite $BaFe^{2+}_2Ti(Si_2O_7)O(OH,F)_2$

Origin: Gremyakha-Vyrmes alkaline complex, Kola Peninsula, Russia.
Experimental details: Raman scattering measurements have been performed on an arbitrarily oriented sample using 632 nm laser radiation. The laser radiation power is not indicated.
Raman shifts (cm^{-1}): 1184w, 1116, 1027, 988, 966, 917s, 812s, 688s (broad), 593s, 478, 418, 355, 336, 264, 230w, 178, 163, 135.
Source: Cámara et al. (2016c).
Comments: The sample was characterized by single-crystal X-ray diffraction data, electron microprobe analyses, IR, and Mössbauer spectra.

Baghdadite $Ca_6Zr_2(Si_2O_7)_2O_4$

Origin: Synthetic.
Experimental details: Raman scattering measurements have been performed on an arbitrarily oriented sample using 532 nm laser radiation. The nominal laser radiation power at the sample was 10 mW.
Raman shifts (cm^{-1}): 1034w, 1011, 975, 958, 944, 921s, 855, 669s, 624s, 568, 542, 521, 476w, 452, 433, 409, 398, 376, 357s, 321, 295, 280w, 262, 245, 214, 198, 171, 148, 127, 122, 106s, 97, 86.
Source: Dul et al. (2015).
Comments: The sample was characterized by powder X-ray diffraction data. The crystal structure is solved by the Rietveld method.

Bahianite $Al_5Sb^{5+}_3O_{14}(OH)_2$

Origin: Paramirim region, Bahia Province, Brazil (type locality).
Experimental details: Raman scattering measurements have been performed on arbitrarily oriented crystals using a 633 nm He-Ne laser. The Raman shifts have been determined for the maxima of individual peaks obtained as a result of the spectral curve analysis. The laser radiation power is not indicated.

Raman shifts (cm^{-1}): 3495, 3462, 3190, 2955, 2718, 2531, 2389, 2273, 2079, 1929, 1808, 1756s, 1489, 1438, 998, 975w, 956w, 952w, 883sh, 856sh, 843sh, 818s, 770, 682, 669sh, 589, 567, 534s, 498, 478s, 471sh, 412s, 405sh, 386, 376, 352, 319, 294, 258s, 220, 199, 165, 158sh, 146sh.
Source: Frost and Bahfenne (2010c).

Bairdite $Pb_2Cu^{2+}_4Te^{6+}_2O_{10}(OH)_2(SO_4)\cdot H_2O$

Origin: Otto Mt., near Baker, California, USA (type locality).
Experimental details: Raman scattering measurements have been performed on a partly oriented platelet (from the [100] face of a crystal) using 514.5 nm Ar$^+$ laser radiation. The laser radiation power at the sample was 5 mW.
Raman shifts (cm^{-1}): 977, 721s, 634, 558, 518, 378, 336, 238, 208.
Source: Kampf et al. (2013a).
Comments: The sample was characterized by powder X-ray diffraction data. The crystal structure is solved. Chemical data are questionable (total sum is 92.97%).

Balestraite $KLi_2V^{5+}Si_4O_{12}$

Origin: Cerchiara mine, Eastern Liguria, Italy.
Experimental details: Raman scattering measurements have been performed on an arbitrarily oriented crystal in a 180°-scattering geometry using 632.8 nm laser radiation. The laser radiation power is not indicated.
Raman shifts (cm^{-1}): 1136, 973, 888, 868s, 707, 539, 437, 309, 261.
Source: Lepore et al. (2015).
Comments: The sample was characterized by powder X-ray diffraction data and electron microprobe analysis. The crystal structure is solved.

Balkanite $Ag_5Cu_9HgS_8$

Origin: Röhrerbühel, near Kitzbühel, Tyrol, Eastern Alps, Austria.
Experimental details: Methods of sample preparation are not indicated. Raman scattering measurements have been performed using 515 nm laser radiation. The laser radiation power at the sample was 6 mW.
Raman shifts (cm^{-1}): 325, 306.
Source: Steiner et al. (2010).
Comments: The Raman shifts are indicated for the maxima of individual peaks obtained as a result of the spectral curve analysis. The sample was characterized by powder X-ray diffraction data and electron microprobe analysis. For the Raman spectrum of balkanite see also Biagioni and Bindi (2016).

Bambollaite Te analogue $Cu(Te,Se)_2$

Origin: Ozernyi district, Salla-Kuolayarvi, Kola Peninsula, Russia.
Experimental details: Raman scattering measurements have been performed on an arbitrarily oriented sample using 514.5 nm Ar$^+$ laser radiation. The nominal laser radiation power was 50 mW.
Raman shifts (cm^{-1}): 157, 134s.
Source: Voloshin et al. (2015b).
Comments: The sample was characterized by electron microprobe analyses.

Baotite $Ba_4(Ti,Nb,W)_8O_{16}(SiO_3)_4Cl$

Origin: Bayan Obo *REE*–Fe–Nb deposit, Inner Mongolia, China (type locality).
Experimental details: Methods of sample preparation are not indicated. Raman scattering measurements have been performed using 1064 nm laser radiation. The nominal laser radiation power was from 300 to 380 mW.
Raman shifts (cm^{-1}): 982, 777, 550, 450, 392, 344, 295, 241, 172.
Source: Yuran and Li (1998).
Comments: No independent analytical data are provided for the sample used.

Barahonaite-(Al) $(Ca,Cu,Na,Fe^{3+},Al)_{12}Al_2(AsO_4)_8(OH,Cl)_x \cdot nH_2O$

Origin: Sapucaia pegmatite mine, Minas Gerais, Brazil.
Experimental details: Raman scattering measurements have been performed on an arbitrarily oriented sample using 633 nm laser radiation. The laser radiation power is not indicated.
Raman shifts (cm^{-1}): 3549sh, 3413, 3217, 2993sh, 2702w, 1657w, 1605w, 1450, 1351w, 1304w, 1228w, 1154w, 1072w, 998w, 890sh, 863s, 828s, 802sh, 723w, 529, 506sh, 449, 399, 360, 325, 300, 233, 159.
Source: López et al. (2014e).
Comments: The Raman shifts are indicated for the maxima of individual peaks obtained as a result of the spectral curve analysis. No independent analytical data are provided for the sample used. For the Raman spectrum of barahonaite-(Al) see also Viñals et al. (2008).

Barahonaite-(Fe) $(Ca,Cu,Na,Fe^{3+},Al)_{12}Fe^{3+}_2(AsO_4)_8(OH,Cl)_x \cdot nH_2O$

Origin: Dolores prospect, near the village of Pastrana, Murcia Province, southeastern Spain (type locality).
Experimental details: Raman scattering measurements have been performed on an arbitrarily oriented sample using 514.5 nm Ar$^+$ laser radiation. The laser radiation power is not indicated.
Raman shifts (cm^{-1}): 908sh, 860s, 828s, 799sh, 517, 508sh, 427, 360w, 325w, 219w, 162, 138, 87w, 65w, 40w.
Source: Viñals et al. (2008).
Comments: The sample was characterized by powder X-ray diffraction data and electron microprobe analyses.

Bararite $(NH_4)_2SiF_6$

Origin: Synthetic.
Experimental details: Raman scattering measurements have been performed in 90°-scattering geometry using 514.5 nm Ar$^+$ laser radiation. The laser radiation power at the sample was ~1 W.
Raman shifts (cm^{-1}): (*XX*)—3236, 1706, 1428, 1406, 650s, 466, 406s, 180; (*YY*)—1706, 1430, 650, 466, 406s, 180; (*ZZ*)—3235, 1428, 1406, 650s, 406s.
Source: Trefler and Wilkinson (1969).
Comments: No independent analytical data are provided for the sample used. For the Raman spectrum of bararite see also Jenkins (1986).

Barberiite $(NH_4)BF_4$

Origin: Synthetic.
Experimental details: Raman scattering measurements have been performed on a powdered sample using Ar^+ laser radiation. The wavelengths of laser excitation line and the laser radiation power are not indicated.
Raman shifts (cm^{-1}): 3400–3300w (broad), 3250s, 767s, 529, 523, 355.
Source: Schutte and Van Rensburg (1971).
Comments: No independent analytical data are provided for the sample used.

Barbosalite $Fe^{2+}Fe^{3+}_2(PO_4)_2(OH)_2$

Origin: Sapucaia mine, Galileia, Minas Gerais (type locality).
Experimental details: Raman scattering measurements have been performed on arbitrarily oriented crystals using a 633 nm He-Ne laser. The Raman shifts have been determined for the maxima of individual peaks obtained as a result of the spectral curve analysis. The laser radiation power is not indicated.
Raman shifts (cm^{-1}): 2261, 2216, 2090, 1138, 1083sh, 1067, 1044, 1033sh, 1020s, 988w, 968w, 831w, 702, 606, 589, 575sh, 503, 475, 461, 439, 398sh, 381, 361sh, 346s, 312s, 291, 275, 256, 241, 198sh, 187, 179, 166, 151, 145, 133s, 125, 113.
Source: Frost et al. (2013q).
Comments: The sample was characterized by powder X-ray diffraction data and electron microprobe analysis.

Bariandite Al-free analogue $V_{10}O_{24}\cdot 9H_2O$

Origin: Synthetic.
Experimental details: Raman scattering measurements have been performed on a powdery sample with crystalline nanoparticles using 514.5 nm Ar^+ laser radiation. The laser radiation power at the sample was 0.2 mW.
Raman shifts (cm^{-1}): 1022, 908s, 518s, 429w, 409w, 270s.
Source: Menezes et al. (2009).
Comments: The sample was characterized by powder X-ray diffraction data.

Baricite $(Mg,Fe)_3(PO_4)_2\cdot 8H_2O$

Origin: Big Fish River, Rapid creek, Richardson Mts., Yukon, Canada (type locality).
Experimental details: Raman scattering measurements have been performed on arbitrarily oriented crystals using a 633 nm He-Ne laser. The Raman shifts have been determined for the maxima of individual peaks obtained as a result of the spectral curve analysis. The laser radiation power is not indicated.
Raman shifts (cm^{-1}): 3480, 3300, 3231, 3121, 3025, 1057, 953s, 859w, 632, 576, 545, 527, 461, 428, 390, 340, 314, 281, 243, 212, 201, 170, 140.
Source: Frost et al. (2002f).
Comments: No independent analytical data are provided for the sample used.

Barioferrite $BaFe^{3+}_{12}O_{19}$

Origin: Synthetic.
Experimental details: Polarized Raman spectra were collected at temperatures from 10 to 200 K on a single crystal in $z(xx)-z$ and $z(yx)-z$ scattering geometries using 610 nm laser radiation. The laser radiation power at the sample was ~1 mW.
Raman shifts (cm^{-1}): ~730sh, ~695, ~627, ~538, ~346 [for the $z(xx)-z$ configuration at 200 K].
Source: Chen et al. (2013b).
Comments: No independent analytical data are provided for the sample used. For the Raman spectrum of barioferrite see also Kreisel et al. (1998a, b, 1999), and Zhao et al. (2008b).

Barioperovskite $BaTiO_3$

Origin: Synthetic.
Experimental details: Raman scattering measurements have been performed on a powdery sample using 785 nm laser radiation. The nominal laser radiation power was 500 mW.
Raman shifts (cm^{-1}): 719, 517, 292s.
Source: Cernea (2005).
Comments: The sample was characterized by powder X-ray diffraction data. For the Raman spectrum of barioperovskite see also Ma and Rossman (2008).

Barnesite $Na_2V^{5+}_6O_{16} \cdot 3H_2O$

Origin: Cactus Rat Mine, Thompson district, Grand Co., Utah, USA.
Experimental details: Raman scattering measurements have been performed on arbitrarily oriented crystals using a 633 nm He-Ne laser. The Raman shifts have been determined for the maxima of individual peaks obtained as a result of the spectral curve analysis. The laser radiation power is not indicated.
Raman shifts (cm^{-1}): 3494, 3435, 3403, 3330, 3253.
Source: Frost et al. (2004e).
Raman shifts (cm^{-1}): 1010s, 761, 728, 683, 670, 620, 534, 492, 433, 413, 341, 287, 284, 260, 248, 217, 192, 153.
Source: Frost et al. (2005d).
Comments: No independent analytical data are provided in the cited papers.

Barringtonite $Mg(CO_3) \cdot 2H_2O$

Origin: Synthetic.
Experimental details: Raman scattering measurements have been performed on a powdered sample containing other carbonates using 785 nm laser radiation. The nominal laser radiation power was 70 mW.
Raman shifts (cm^{-1}): ~1095.
Source: Kristova et al. (2014).
Comments: The sample identification was done by powder X-ray diffraction data.

Barrydawsonite-(Y) $Na_{1.5}Y_{0.5}CaSi_3O_9H$

Origin: Merlot Claim, North Red Wine Pluton, Labrador, Canada (type locality).
Experimental details: Raman scattering measurements have been performed on a powdered sample using 633 nm laser radiation. The nominal laser radiation power was 50 mW.

Raman shifts (cm^{-1}): 1037s, 1004s, 968, 907, 686, 655s, 524, 506, 463, 445, 416, 362, 312, 272, 244, 207, 148, 107.
Source: Mitchell et al. (2015).
Comments: Unpolarized and polarized single-crystal spectra showed only minor differences in relative band intensities. The sample was characterized by powder X-ray diffraction data and electron microprobe analysis. The crystal structure is solved.

Bartelkeite $PbFe^{2+}Ge^{6+}(Ge^{4+}{}_2O_7)(OH)_2 \cdot H_2O$

Origin: Tsumeb mine, Tsumeb, Namibia (type locality).
Experimental details: Raman scattering measurements have been performed on an arbitrarily oriented crystal using 532 nm laser radiation. The nominal laser radiation power was 200 mW.
Raman shifts (cm^{-1}): 3490, 3293, 1558, 812, 758s, 549, 492, 393.
Source: Origlieri et al. (2012).
Comments: The sample was characterized by powder X-ray diffraction data and electron microprobe analysis. The crystal structure is solved.

Barylite $BaBe_2Si_2O_7$

Origin: Zomba, Malawi.
Experimental details: Raman scattering measurements have been performed on an arbitrarily oriented sample using 532 nm laser radiation. The nominal laser radiation power was 150 mW.
Experimental details: Experimental details are not indicated.
Raman shifts (cm^{-1}): 1014w, 996w, 982w, 958s, 937, 888, 685s, 627, 573w, 542w, 464, 447, 423, 410, 384, 337s, 262w, 234, 202, 191.
Source: Yang et al. (2013b).
Comments: The sample was characterized by powder X-ray diffraction data. The Raman shifts were determined by us based on spectral curve analysis of the published spectrum.

Baryte $Ba(SO_4)$

Origin: Dufton, England.
Experimental details: Raman scattering measurements have been performed on an arbitrarily oriented crystal using 532 nm laser radiation. The nominal laser radiation power was 100 mW.
Raman shifts (cm^{-1}): 1167, 1143, 1085, 989s, 648, 619, 461.
Source: Buzgar et al. (2009).
Comments: No independent analytical data are given for the sample used. For the Raman spectra of baryte see also Jehlička et al. (2009b), White (2009), Ciobotă et al. (2012), and Andò and Garzanti (2014).

Baryte $Ba(SO_4)$

Origin: No data.
Experimental details: Methods of sample preparation are not indicated. Raman scattering measurements have been performed using 785 nm laser radiation. The laser radiation power at source was 320 mW.
Raman shifts (cm^{-1}): 1166w, 1138w, 1104w, 1084w, 988s, 647w, 617w, 453.

Source: Jehlička et al. (2009b).
Comments: For the Raman spectra of baryte see also Buzgar et al. (2009), White (2009), Ciobotă et al. (2012), and Andò and Garzanti (2014).

Barytocalcite $BaCa(CO_3)_2$

Origin: Alston Moor, England, UK (type locality).
Experimental details: Raman scattering measurements have been performed on a powdered sample. The wavelengths of laser radiation and laser radiation power are not indicated.
Raman shifts (cm^{-1}): 1510w, 1085s, 715, 700, 688, 314w, 261, 225, 209, 164, 154sh, 127w, 107s, 86w, 70.
Source: Scheetz and White (1977).
Comments: No independent analytical data are provided for the sample used. For the Raman spectrum of barytocalcite see also Frost and Dickfos (2008).

Bassanite $Ca(SO_4) \cdot 0.5H_2O$

Origin: Artificial (obtained by dehydration at ~338 K of gypsum from an unknown salt core location in India).
Experimental details: Raman scattering measurements have been performed in 90°-scattering geometry using 514.5 nm Ar$^+$ laser radiation. The nominal laser radiation power at the sample was 80 mW.
Raman shifts (cm^{-1}): 3552, ~1621, 1026w.
Source: Sarma et al. (1998).
Comments: The new line at 1026 cm^{-1} appears in the spectrum of partially dehydrated gypsum. For the Raman spectrum of bassanite see also Apopei et al. (2015).

Bassoite $SrV^{4+}_3O_7 \cdot 4H_2O$

Origin: Molinello mine, Val Graveglia, eastern Liguria, Italy (type locality).
Experimental details: Methods of sample preparation are not indicated. Raman scattering measurements have been performed using unpolarized 785 nm laser radiation. The nominal laser radiation power at the sample was ~3 mW.
Raman shifts (cm^{-1}): 3534, 3407, 1645.
Source: Bindi et al. (2011a).
Comments: Raman spectrum was obtained only in the regions 1250–2000 and 3000–4000 cm^{-1}. The sample was characterized by powder X-ray diffraction data and electron microprobe analyses. The crystal structure is solved.

Bastnäsite-(Ce) $Ce(CO_3)F$

Origin: No data.
Experimental details: Raman scattering measurements have been performed using 488 nm laser radiation. The laser radiation power was 300 mW.
Raman shifts (cm^{-1}): 2621s, 2059, 2009, 1899, 1476s, 1447s, 1345, 1279w, 1098s, 835w, 732w, 670w, 600w, 353, 259.
Source: Hong et al. (1999).

Comments: For the Raman spectra of bastnäsite-(Ce) see also Yang et al. (2008a) and Frost and Dickfos (2007a).

Batiferrite Co-bearing $BaFe_{9.4}Ti_{1.3}Co_{1.3}O_{19}$

Origin: Synthetic.
Experimental details: Raman scattering measurements have been performed on a platy single crystal oriented perpendicular to the c-axis using 514.5 nm Ar^+ laser radiation. The laser radiation power at the sample was 5 mW.
Raman shifts (cm^{-1}): 732sh, 694s, 626, 418, 362.
Source: Kreisel et al. (1999).
Comments: The spectrum was obtained only in the region from 200 to 900 cm^{-1}. The sample was characterized by powder X-ray diffraction data and electron microprobe analysis.

Baumhauerite $Pb_{12}As_{16}S_{36}$

Origin: Lengenbach, Binntal, Switzerland (type locality).
Experimental details: Raman scattering measurements have been performed on an arbitrarily oriented sample using slightly defocused 632.8 nm He-Ne laser radiation. The laser radiation power is not indicated.
Raman shifts (cm^{-1}): 380, 362, 331, 291s, 254, 222, 192, 161, 137, 114w, 93w, 82.
Source: Kharbish (2016).
Comments: The Raman shifts have been determined for the maxima of individual peaks obtained as a result of the spectral curve analysis.

Bavenite $Ca_4Be_{2+x}Al_{2-x}Si_9O_{26-x}(OH)_{2+x}$ ($x = 0$ to 1)

Origin: An unknown locality in Siberia.
Experimental details: Raman scattering measurements have been performed on an arbitrarily oriented sample using 532 and 785 nm laser radiation. The laser radiation power at the sample was up to 35 mW for the 785 nm laser.
Raman shifts (cm^{-1}): 945, 674, 544, 505, 448, 401w, 395, 348, 330, 286, 259w, 232, 182w, 173, 140s, 108s.
Source: Jehlička et al. (2012).
Comments: No independent analytical data are provided for the sample used. The spectrum may correspond to bohseite, a mineral related to bavenite.
Comments: For the Raman spectrum of bavenite see also Jehlička and Vandenabeele (2015).

Bayerite $Al(OH)_3$

Origin: Synthetic.
Experimental details: Raman scattering measurements have been performed on an arbitrarily oriented sample using 514.5 nm Ar^+ laser radiation. The laser radiation power at the sample was 25 mW.
Raman shifts (cm^{-1}): 3425s, 3439, 3449sh, 3546s, 3552, 3654, 904, 868, 547s, 525, 447w, 438, 392, 362w, 325s, 299s, 250, 240, 204w, 147w, 141, 118, 108.
Source: Rodgers (1993).

Comments: The sample was characterized by powder X-ray diffraction data. For the Raman spectra of bayerite see also Rodgers et al. (1989) and Ruan et al. (2001).

Bayldonite $Cu_3PbO(AsO_3OH)_2(OH)_2$

Origin: Tsumeb mine, Tsumeb, Namibia.
Experimental details: Methods of sample preparation are not indicated. Micro-Raman scattering measurements have been performed using 632.8 nm He-Ne laser radiation. The laser radiation power at the sample was 0.97 mW.
Raman shifts (cm^{-1}): 838s, 804, 759w, 495, 428, 397w, 314, 230, 165, 110.
Source: Makreski et al. (2015a).
Comments: The sample was characterized by powder X-ray diffraction data and electron microprobe analyses. Raman spectrum of bayldonite was also published by Frost et al. (2014d) who neither presented nor assigned the IR bands below 700 cm^{-1}: see comment made by Makreski et al. (2015a). The bayldonite formula $(Cu,Zn)_3Pb(AsO_3OH)_2(OH)_2$ given by Frost et al. (2014d) is not charge-balanced. The correct formula should be $(Cu,Zn)_3Pb(AsO_3OH)_2O_2$ or $(Cu,Zn)_3Pb(AsO_3OH)_2O(OH)_2$.

Bayleyite $Mg_2(UO_2)(CO_3)_3 \cdot 18H_2O$

Origin: Barbora shaft, Jáchymov ore district, Czech Republic.
Experimental details: Raman scattering measurements have been performed on an arbitrarily oriented sample using 532 nm laser radiation. The nominal laser radiation power was 3 mW.
Raman shifts (cm^{-1}): 3560sh, 3425s (broad), 3260, 3150sh, 1642w, 1627w, 1619w, 1608w, 1586w, 1380w, 1067s, 825–832s, 752s, 718sh, 665sh, 253, 234, 191, 166, 114, 59w.
Source: Škácha et al. (2014b).

Baylissite NH_4-analogue $(NH_4)_2Mg(CO_3)_2 \cdot 4H_2O$

Origin: Synthetic.
Experimental details: No data.
Raman shifts (cm^{-1}): 3240, 3174, 2885, 1703w, 1440w, 1421w, 1098s, 686, 488.
Source: Fischer (2007).
Comments: The sample was characterized by single-crystal X-ray diffraction data and electron microprobe analysis.

Bazhenovite $Ca_8S_5(S_2O_3)(OH)_{12} \cdot 20H_2O$

Origin: Chelyabinsk coal basin, South Urals, Russia (type locality).
Experimental details: Raman scattering measurements have been performed on an arbitrarily oriented sample using 514.5 nm Ar^+ laser radiation. The laser radiation power at the sample was 3.5 mW.
Raman shifts (cm^{-1}): 3473, 3227, 2500, 1620w, 940, 507s, 466s, 218.
Source: Bindi et al. (2005).
Comments: The sample was characterized by single-crystal X-ray diffraction data. The crystal structure is solved. The Raman line at 2500 cm^{-1} is attributed by authors to H_2S in a condensed form which may be present in the sample.

Bazirite $BaZrSi_3O_9$

Origin: Synthetic.

Experimental details: Raman scattering measurements have been performed on a polycrystalline sample using 514.5 nm Ar^+ laser radiation. The laser radiation power is not indicated.

Raman shifts (cm^{-1}): 1148, 1092, 1078, 1067, 1046, 1005, 995, 957s, 948, 937, 639, 579s, 544, 529, 521sh, 507, 480, 464, 448w, 425w, 384s, 369, 356w, 348w, 342, 328, 259, 239, 201s, 184, 173w, 161, 152sh, 143, 133, 117, 109.

Source: Takahashi et al. (2008).

Comments: The sample was characterized by powder X-ray diffraction data. The Raman shifts were determined by us based on spectral curve analysis of the published spectrum.

Bazzite $Be_3(Sc,Fe^{3+},Mg)_2Si_6O_{18}Na_{0.32} \cdot nH_2O$

Origin: Furkabasistunnel, Switzerland.

Experimental details: Raman scattering measurements have been performed on an oriented single crystal using 488 nm laser radiation. The laser radiation power is not indicated. Polarized spectra were collected in 90°-scattering geometry, and the polarization conditions were: (ZZ), (ZY), (XY), and (ZZ+ZX).

Raman shifts (cm^{-1}): (ZZ): 1093, 1060, 672, 570–610 (broad), 392, 315; (ZY): 970, 917, 902sh, 775, 672, 653sh, 603, 556w, 485w, 451, 375, 243, 224, 131; (XY): 1180, 1163, 970, 909, 777, 731, 672, 653sh, 554–601 (broad), 439, 377, 390, 315, 266w; (ZZ+ZX): 3594s, 3535w.

Source: Hagemann et al. (1990).

Comments: The sample was characterized by powder X-ray diffraction data and electron microprobe analyses.

Beaverite-(Cu) $Pb(Fe^{3+}_2Cu)(SO_4)_2(OH)_6$

Origin: Synthetic.

Experimental details: Raman scattering measurements have been performed on arbitrarily oriented crystals using a 633 nm He-Ne laser. The Raman shifts have been determined for the maxima of individual peaks obtained as a result of the spectral curve analysis. The laser radiation power is not indicated.

Raman shifts (cm^{-1}): 3421, 3380, 3354, 1176, 1164, 1156, 1117, 1103s, 1081, 1076, 1018, 1010s, 999s, 645, 624, 619, 577, 560, 481, 456, 441, 433, 392, 356, 335, 328, 298, 278, 259, 242, 216, 202, 173.

Source: Frost et al. (2005m).

Comments: The sample was characterized by powder X-ray diffraction data. For the Raman spectrum of beaverite-(Cu) see also Hudson-Edwards et al. (2008).

Becquerelite $Ca(UO_2)_6O_4(OH)_6 \cdot 8H_2O$

Origin: No data.

Experimental details: Raman scattering measurements have been performed on arbitrarily oriented crystals using a 633 nm He-Ne laser. The Raman shifts have been determined for the maxima of individual peaks obtained as a result of the spectral curve analysis. The laser radiation power is not indicated.

Raman shifts (cm^{-1}): 3547, 3429, 3211, 879, 854, 838s, 831, 814, 546, 508, 455, 399, 353, 328, 303, 260, 238, 192, 156, 142, 111.
Source: Frost et al. (2007h).
Comments: No independent analytical data are provided for the sample used. For the Raman spectrum of becquerelite see also Amme et al. (2002).

Behierite Ta(BO$_4$)

Origin: Synthetic.
Experimental details: Methods of sample preparation are not indicated. Raman scattering measurements have been performed using 488 and 514.5 nm laser radiation. The laser radiation power is not indicated.
Raman shifts (cm^{-1}): 1004, 978s, 900, 848, ~700w, 568s, 461s, 284, 204, 191.
Source: Heyns et al. (1990).
Comments: No independent analytical data are provided for the sample used. For the Raman spectra of behierite see also Ross (1972) and Blasse and van den Heuvel (1973).

Behoite Be(OH)$_2$

Origin: Synthetic.
Experimental details: Raman scattering measurements have been performed on an arbitrarily oriented sample using 514.5 nm Ar$^+$ laser radiation. The laser radiation power is not indicated.
Raman shifts (cm^{-1}): 3501, (3489), 3449s, 1133w, 1056, 1031sh, 1000w, 845w, 769, 701, 682, 602, 549w, 459, 446, 400s, 364w, 349s, 280w, 134, 77.
Source: Lutz et al. (1998).
Comments: No independent analytical data are provided for the sample used.

Běhounekite U(SO$_4$)$_2$(H$_2$O)$_4$

Origin: Geschieber vein, Jáchymov (St Joachimsthal), Czech Republic (type locality).
Experimental details: Methods of sample preparation are not indicated. Raman scattering measurements have been performed using 480 (in the region of 4000–1900 cm^{-1}) and 785 (in the region of 1900–200 cm^{-1}) nm laser radiations. The laser radiation power is not indicated.
Raman shifts (cm^{-1}): 3370s, 3206, 1269, 1251, 1177, 1158, 1102, 1037, 1023s, 994, 638, 619, 598, 451, 438, 417, 268, 250, 198, 178, 125, 116, 98.
Source: Plášil et al. (2011b).
Comments: The Raman shifts are indicated for the maxima of individual peaks obtained as a result of the spectral curve analysis. The sample was characterized by powder X-ray diffraction data and electron microprobe analyses.

Belakovskiite Na$_7$(UO$_2$)(SO$_4$)$_4$(SO$_3$OH)(H$_2$O)$_3$

Origin: Synthetic.
Experimental details: Methods of sample preparation are not indicated. Raman scattering measurements have been performed using 780 nm laser radiation. The laser radiation power at the sample was from 2 mW to 6 mW.
Raman shifts (cm^{-1}): 1185, 1070, 1040, 1000, 985w, 903w, 865sh, 840s, 820w, 660, 650, 605, 590sh, 480, 450.

Source: Plášil et al. (2015c).
Comments: The sample was characterized by single-crystal X-ray diffraction data.

Bellidoite Cu_2Se

Origin: Synthetic.
Experimental details: Raman scattering measurements have been performed on an arbitrarily oriented sample using 514.12 nm Ar^+ laser radiation. The laser radiation power at the sample was 10 mW.
Raman shifts (cm^{-1}): No data: Cu_2Se exhibits very weak Raman features.
Source: Izquierdo-Roca et al. (2009).

Bendadaite $Fe^{2+}Fe^{3+}_2(AsO_4)_2(OH)_2 \cdot 4H_2O$

Origin: Lavra do Almerindo (Almerindo mine), Linópolis, Divino das Laranjeiras Co., Minas Gerais, Brazil.
Experimental details: Raman scattering measurements have been performed on an arbitrarily oriented sample using 676 nm Kr^+ laser radiation. The nominal laser radiation power is unknown.
Raman shifts (cm^{-1}): 3385, 3275, 1690w, ~800s.
Source: Kolitsch et al. (2010).
Comments: The sample was characterized by powder X-ray diffraction data and electron microprobe analyses. The crystal structure is solved.

Benitoite $BaTiSi_3O_9$

Origin: Synthetic.
Experimental details: Methods of sample preparation are not indicated. Raman scattering measurements have been performed using 632.8 nm laser radiation. The nominal laser radiation power was 50 mW.
Raman shifts (cm^{-1}): 1080, 951, 939, 930s, 917, 577s, 539s, 505, 480, 398, 374s, 350, 267s.
Source: Choisnet et al. (1975).
Comments: No independent analytical data are given for the sample used. For the Raman spectra of benitoite see also Gaft et al. (2004), Ma and Rossman (2008), and Takahashi et al. (2008).

Benstonite $Ba_6Ca_6Mg(CO_3)_{13}$

Origin: Minerva mine, Cave-in-Rock, Hardin Co., Illinois, USA.
Experimental details: Raman scattering measurements have been performed on an arbitrarily oriented sample using 488 and 514.5 nm laser radiation. The nominal laser radiation power was 200 mW.
Raman shifts (cm^{-1}): 1100s, 1096s, 1081s, 1074s, 714, 691, 251, 236, 206w, 171, 136, 96w.
Source: Scheetz and White (1977).
Comments: No independent analytical data are given for the sample used.

Beraunite $Fe^{2+}Fe^{3+}_5(PO_4)_4(OH)_5 \cdot 6H_2O$

Origin: Boca Rica pegmatite, Minas Gerais, Brazil.
Experimental details: Raman scattering measurements have been performed on arbitrarily oriented crystals using a 633 nm He-Ne laser. The Raman shifts have been determined for the maxima of

individual peaks obtained as a result of the spectral curve analysis. The laser radiation power at the sample is not indicated.

Raman shifts (cm^{-1}): 1174, 1155, 1133, 1116, 1098, 1084sh, 1069, 1058, 1051, 1034, 1011s, 990sh, 969sh, 703sh, 687s, 673, 661, 644, 601, 582, 567s, 546, 503, 491, 478, 468, 455, 437, 403, 398, 336, 322, 309sh, 300, 289, 280, 254, 238, 230, 225, 200, 191sh, 153, 1432, 118, 107.

Source: Frost et al. (2014al).

Comments: No independent analytical data are given for the sample used. In the cited paper incorrect formula of beraunite is given. Brown color of the sample indicates that it is not beraunite, but its oxydation product eleonorite, a mineral isostructural with beraunite.

Berdesinskiite $V^{3+}_2TiO_5$

Origin: Vihanti, Northern Finland Region, Finland.

Experimental details: Raman scattering measurements have been performed on an arbitrarily oriented sample using 633 nm laser radiation. The nominal laser radiation power was 2 or 20 mW.

Raman shifts (cm^{-1}): 978w, 898w. 768s, 720s, 647, 593, 512w, 485, 445, 411, 388, 341, 311, 257w, 210, 136w, 83.

Source: Voloshin et al. (2014).

Comments: The sample was characterized by powder X-ray diffraction data and electron microprobe analyses.

Bergenite $Ca_2Ba_4(UO_2)_9O_6(PO_4)_6 \cdot 16H_2O$

Origin: Mechelgrün, Vogtland, Saxony, Germany.

Experimental details: Raman scattering measurements have been performed on arbitrarily oriented crystals using a 633 nm He-Ne laser. The Raman shifts have been determined for the maxima of individual peaks obtained as a result of the spectral curve analysis. The laser radiation power is not indicated.

Raman shifts (cm^{-1}): 3607, 3459, 3295, 2944, 1602, 1330, 1152, 1107, 1059, 995s, 991w, 971w, 961, 948, 810, 798sh, 777, 592, 547, 515, 444, 432, 408, 396, 391, 270, 265, 256, 224, 205, 178, 145, 133, 111.

Source: Frost et al. (2007e).

Comments: The sample was characterized by electron microprobe analysis.

Berlinite $Al(PO_4)$

Origin: Synthetic.

Experimental details: Raman scattering measurements have been performed on a powdered sample compacted as a pellet using 1064 nm laser radiation. The nominal laser radiation power was 350 mW.

Raman shifts (cm^{-1}): 1230, 1111s, 1104sh, 1021, 725w, 650, 566, 524, 462s, 439, 418, 379, 335, 306, 258sh, 221, 196, 163, 149sh, 139w, 119, 107, 84.

Source: Pînzaru and Onac (2009).

Comments: No independent analytical data are provided for the sample used. For the Raman spectra of berlinite see also Thomas and Webster (1999), O'Neill et al. (2006), and Frezzotti et al. (2012).

Berlinite tetragonal polymorph $Al(PO_4)$

Origin: Synthetic.
Experimental details: Micro-Raman scattering measurements have been performed on an arbitrarily oriented sample using 488 nm Ar$^+$ laser radiation. The laser radiation power is not indicated.
Raman shifts (cm^{-1}): 1124s, 485, 391, 382, 279, 191.
Source: O'Neill et al. (2006).
Comments: The sample was characterized by powder X-ray diffraction data.

Bermanite $Mn^{2+}Mn^{3+}{}_2(PO_4)_2(OH)_2 \cdot 4H_2O$

Origin: El Criolo granitic pegmatite, Cerro Blanco pegmatite group, Córdoba province, Argentina.
Experimental details: Raman scattering measurements have been performed on arbitrarily oriented crystals using a 633 nm He-Ne laser. The Raman shifts have been determined for the maxima of individual peaks obtained as a result of the spectral curve analysis. The laser radiation power is not indicated.
Raman shifts (cm^{-1}): 3515, 3470, 3425, 3355, 3285, 3202, 3110, 3038, 2961, 1142, 1117, 1071w, 1012sh, 999s, 991s, 978sh, 900, 586, 577, 552sh, 505sh, 489s, 473s, 455, 441, 419, 400, 341, 307, 270, 256, 249sh, 217s, 2009sh, 189, 171, 156sh, 147, 127.
Source: Frost et al. (2013x).
Comments: The sample was characterized only by qualitative electron microprobe analysis.

Bernalite $Fe(OH)_3$

Origin: Synthetic.
Experimental details: Methods of sample preparation are not indicated. Raman scattering measurements have been performed using 752.6 nm laser radiation. The laser radiation power was from 0.01 to 10 mW.
Raman shifts (cm^{-1}): 398, 299.
Source: Lepot et al. (2006).
Comments: No independent analytical data are provided for the sample used.

Berndtite SnS_2

Origin: Synthetic.
Experimental details: Raman scattering measurements have been performed on an arbitrarily oriented sample using 532 nm laser radiation. The laser radiation power at the sample was 0.4 mW.
Raman shifts (cm^{-1}): 355, 315s, 205, 140w, 44, 25.
Source: Fontané et al. (2013).
Comments: For the Raman spectra of berndtite see also Smith et al. (1977), Jiang and Ozin (1997), and Utyuzh et al. (2010).

Berthierite $FeSb_2S_4$

Origin: Zlatá Baňa deposit, eastern Slovakia.
Experimental details: Raman scattering measurements have been performed on an arbitrarily oriented sample in a 180°-scattering geometry using 632 nm laser radiation. The laser output radiation power was 210 mW.

Raman shifts (cm^{-1}): 347, 334, 318sh, 297, 277sh, 264s, 251sh, 226, 183, 150sh, 131, 90, 76s, 60.
Source: Kharbish and Andráš (2014).
Comments: The sample was characterized by powder X-ray diffraction data and electron microprobe analyses.

Bertrandite $Be_4Si_2O_7(OH)_2$

Origin: Albany, Maine, USA.
Experimental details: Raman scattering measurements have been performed on arbitrarily oriented crystals using 488 nm Ar$^+$ laser radiation. The laser radiation power is not indicated.
Raman shifts (cm^{-1}): 3587, 3551, 1073w, 990, 947w, 928, 821, 772, 753, 725sh, 711, 694, 579, 538, 487w, 427w, 386w, 358, 352, 330, 301, 241, 231s, 206s, 183s.
Source: Hofmeister et al. (1987).
Comments: Methods of the sample identification are not indicated. For the Raman spectrum of bertrandite see also Jehlička et al. (2012).

Beryl $Be_3Al_2Si_6O_{18}$

Origin: Čanište, Republic of Macedonia.
Experimental details: Raman scattering measurements have been performed on a powdered sample using 532 nm laser radiation. The laser radiation power is not indicated.
Raman shifts (cm^{-1}): 3663w, 3595s, 1610w, 1128, 1111, 1070, 1005, 915, 770, 689s, 616, 587, 571, 516s, 400, 326, 286s, 248, 196s, 143.
Source: Makreski and Jovanovski (2009).
Comments: Raman shifts are indicated for the maxima of individual peaks obtained as a result of the spectral curve analysis. The sample was characterized by powder X-ray diffraction data and electron microprobe analysis. For the Raman spectra of beryl see also Hagemann et al. (1990), Kloprogge and Frost (2000a), Jasinevicius (2009), Jehlička et al. (2012), and Jehlička and Vandenabeele (2015).

Beryl $Be_3Al_2Si_6O_{18}$

Origin: Čanište, Republic of Macedonia.
Experimental details: Raman scattering measurements have been performed on a powdered sample using 532 nm laser radiation. The laser radiation power is not indicated.
Raman shifts (cm^{-1}): 3663w, 3595s, 1610w, 1128, 1111, 1070, 1005, 915, 770, 689s, 616, 587, 571, 516s, 400, 326, 286s, 248, 196s, 143.
Source: Makreski and Jovanovski (2009).
Comments: The Raman shifts are indicated for the maxima of individual peaks obtained as a result of the spectral curve analysis. The sample was characterized by powder X-ray diffraction data and electron microprobe analysis. For the Raman spectra of beryl see also Hagemann et al. (1990), Kloprogge and Frost (2000a), Jasinevicius (2009), Jehlička et al. (2012), and Jehlička and Vandenabeele (2015).

Beryl Cs-bearing $CsLiBe_2Al_2Si_6O_{18}$

Origin: Piława Górna, Lower Silesia, SW Poland.
Experimental details: Methods of sample preparation are not described. Raman scattering measurements have been performed using 532 nm laser radiation. The laser radiation power is not indicated.

Raman shifts (cm^{-1}): 1130w, 1100sh, 1069s, 1008, 686s, 531, 400, 323, 245w, 125.
Source: Pieczka et al. (2016b).
Comments: The sample was characterized by powder X-ray diffraction data and electron microprobe analyses.

Beryllonite NaBe(PO$_4$)

Origin: Ehrenfriedersdorf complex, Erzgebirge (Ore Mts.), Germany.
Experimental details: Micro-Raman scattering measurements have been performed on a microscopic inclusion in quartz using 514 and 488 nm laser radiations. The laser radiation power at the sample was 10 mW.
Raman shifts (cm^{-1}): 1056s, 1012s, 547, 432, 354.
Source: Thomas et al. (2011b).
Comments: The sample was characterized by ion microprobe analysis.
Comments: For the Raman spectra of beryllonite see also Tait et al. (2010) and Frost et al. (2012k).

Berzeliite (NaCa$_2$)Mg$_2$(AsO$_4$)$_3$

Origin: Synthetic.
Experimental details: Methods of sample preparation are not described. Raman scattering measurements have been performed using 514.5 nm Ar$^+$ laser radiation. The nominal laser radiation power was between 200 and 300 mW.
Raman shifts (cm^{-1}): 912sh, 891, 841s, 800sh, 506, 473, 461, 431sh, 332, 170, 127, 115.
Source: Khorari et al. (1995).
Comments: The sample was characterized by powder X-ray diffraction data.

Berzeliite polymorph alluaudite-type (NaCa$_2$)Mg$_2$(AsO$_4$)$_3$

Origin: Synthetic.
Experimental details: Methods of sample preparation are not described. Raman scattering measurements have been performed using 514.5 nm Ar$^+$ laser radiation. The nominal laser radiation power was between 200 and 300 mW.
Raman shifts (cm^{-1}): 891, 860s, 800sh, 540w, 469, 426, 402sh, 386, 348, 304w, 217sh, 200.
Source: Khorari et al. (1995).
Comments: The sample was characterized by powder X-ray diffraction data.

Betalomonosovite Na$_6$□$_4$Ti$_4$(Si$_2$O$_7$)$_2$[PO$_3$(OH)][PO$_2$(OH)$_2$]O$_2$(OF)

Origin: Lovozero alkaline massif, Kola Peninsula, Russia.
Experimental details: Methods of sample preparation are not described. Raman scattering measurements have been performed in a 180°-scattering geometry using 532 nm laser radiation. The laser radiation power is not indicated.
Raman shifts (cm^{-1}): 1100, 1030, 925s, 862sh, 804, 678, 587, 548, 493, 456, 414.
Source: Sokolova et al. (2015a).
Comments: The sample was characterized by single-crystal X-ray diffraction data and electron microprobe analysis. The crystal structure is solved.

Beudantite $PbFe^{3+}_3(AsO_4)(SO_4)(OH)_6$

Origin: Tsumeb mine, Tsumeb, Namibia.
Experimental details: Raman scattering measurements have been performed on an arbitrarily oriented sample using a 633 nm He-Ne laser. The Raman shifts have been determined for the maxima of individual peaks obtained as a result of the spectral curve analysis. The laser radiation power is not indicated.
Raman shifts (cm^{-1}): 3449, 3391, 3202, 3123, 3005, 1674w, 1319, 1144, 1107, 1081, 998s, 874sh, 851s, 829sh, 807sh, 622, 616, 578w, 560, 476s, 443, 434s, 410sh, 371w, 328, 311, 301sh, 293, 259sh, 248, 216sh, 202s, 143, 137s.
Source: Frost et al. (2011i).
Comments: No independent analytical data are provided for the sample used.

Beusite $Mn^{2+}Fe^{2+}_2(PO_4)_2$

Origin: Bull Moose Mine, Custer, South Dakota, USA.
Experimental details: Micro-Raman scattering measurements have been performed on an arbitrarily oriented crystal using 532 nm laser radiation. The laser radiation power at the sample was about 1–5 mW.
Raman shifts (cm^{-1}): 1080w, 1068, 1012sh, 999, 950s, 628, 589, 573, 442, 409sh, 322, 250w, 237, 200w, 155.
Source: Schneider et al. (2013).
Comments: The Raman shifts were determined by us based on spectral curve analysis of the published spectrum.

Beyerite $CaBi_2O_2(CO_3)_2$

Origin: Synthetic.
Experimental details: Methods of sample preparation are not described. Raman scattering measurements have been performed using 785 nm Ar$^+$ laser radiation. The laser radiation power is not indicated.
Source: Malik et al. (2016).
Raman shifts (cm^{-1}): 1302, 1069, 674w, 425, 207, 163s.
Comments: The sample was characterized by powder X-ray diffraction data.

Bianchite or Goslarite $Zn(SO_4) \cdot 6H_2O$ or $Zn(SO_4) \cdot 7H_2O$

Origin: Minei Hill open pit, Baia Sprie deposit, Romania.
Experimental details: Raman scattering measurements have been performed on white fine deposition material using 632 nm laser radiation. The laser radiation power at the sample was 53.6 mW.
Raman shifts (cm^{-1}): 1191, 1080, 1024s, 914, 626, 506, 427, 280, 222.
Source: Buzatu et al. (2012).
Comments: No independent analytical data are provided for the sample used. The Raman spectrum may correspond to bianchite or goslarite. These two minerals have very similar Raman spectra, which makes the identification difficult.

4 Raman Spectra of Minerals

Bikitaite $LiAlSi_2O_6 \cdot H_2O$

Origin: Bikita, Zimbabwe (type locality).
Experimental details: Raman scattering measurements have been performed on an arbitrarily oriented sample in 90°- and 180°-scattering geometries using 488 and 514.5 nm Ar^+ laser radiations. The laser radiation power is not indicated.
Raman shifts (cm^{-1}): 3588, 3477, 3411, 1641, 964, 504, 396, 255, 155, 104—the spectrum obtained at 5 K.
Source: Kolesov and Geiger (2002).
Comments: Raman spectra in the region below 1900 cm^{-1} were obtained at 5 K. No independent analytical data are provided for the sample used.

Billietite $Ba(UO_2)_6O_4(OH)_6 \cdot 8H_2O$

Origin: No data.
Experimental details: Raman scattering measurements have been performed on arbitrarily oriented crystals using a 633 nm He-Ne laser. The Raman shifts have been determined for the maxima of individual peaks obtained as a result of the spectral curve analysis. The laser radiation power is not indicated.
Raman shifts (cm^{-1}): 3568, 3487, 3398, 3238, 1604, 963, 873, 831, 830, 810, 800, 795, 737, 556, 528, 460, 452, 416, 363, 337, 290, 259, 244, 208, 200, 167, 158, 117, 109.
Source: Frost et al. (2007h).
Comments: No independent analytical data are provided for the sample used. For the Raman spectrum of billietite see also Qader (2011).

Biotite $K(Mg,Fe)_3[(Si,Al)_4O_{10}](OH)_2$

Origin: No data.
Experimental details: No data.
Raman shifts (cm^{-1}): 3680, 3658, 767, 717, 679s, 549, 178.
Source: Frezzotti et al. (2012).

Biphosphammite $(NH_4,K)H_2(PO_4)$

Origin: Synthetic.
Experimental details: Micro-Raman scattering measurements have been performed on an arbitrarily oriented sample using 488 nm Ar^+ laser radiation. The laser radiation power is not indicated.
Raman shifts (cm^{-1}): 1024sh, 925s, 551w, 479, 340, 179.
Source: O'Neill et al. (2006).
Comments: The sample was characterized by powder X-ray diffraction data. For the Raman spectrum of biphosphammite see also Frost et al. (2011r).

Birnessite $(Na,Ca,K)_{0.6}(Mn^{4+},Mn^{3+})_2O_4 \cdot 1.5H_2O$

Origin: Synthetic.
Experimental details: Raman scattering measurements have been performed on an arbitrarily oriented sample using 514.5 nm Ar^+ laser radiation. The laser radiation power density was 100 W/cm^2.
Raman shifts (cm^{-1}): 730, 656s, 575s, 506, 303, 296.

Source: Julien et al. (2004).
Comments: The sample was characterized by powder X-ray diffraction data. For the Raman spectra of birnessite see also Julien et al. (2003) and Roqué-Rosell et al. (2010).

Birnessite $(Na,Ca,K)_{0.6}(Mn^{4+},Mn^{3+})_2O_4 \cdot 1.5H_2O$

Origin: Moa Bay lateritic deposits, eastern Cuba.
Experimental details: Raman scattering measurements have been performed on an arbitrarily oriented sample using 785 nm laser radiation. The nominal laser radiation power was 30 mW.
Raman shifts (cm^{-1}): 655s, 573, 491w, 281.
Source: Roqué-Rosell et al. (2010).
Comments: The spectrum in the region from 200 to 900 cm^{-1} was obtained. The sample was characterized by powder X-ray diffraction data and chemical analysis. For the Raman spectra of birnessite see also Julien et al. (2003, 2004).

Bischofite $MgCl_2 \cdot 6H_2O$

Origin: No data.
Experimental details: Raman scattering measurements have been performed on an arbitrarily oriented sample using 532 nm Nd-YAG pulsed laser with 45 mJ/pulse total energy, up to 20 Hz lasing frequency, and 8 ns pulse width. 90°-scattering geometry was employed.
Raman shifts (cm^{-1}): 3507, 3350.
Source: Garcia et al. (2006).
Comments: No independent analytical data are given for the sample used.

Bismite Bi_2O_3

Origin: Synthetic.
Raman shifts (cm^{-1}): 446, 410w, 314, 282, 210, 184, 151, 139, 118s, 102, 93s, 83s, 67s, 59, 53s.
Source: Betsch and White (1978).
Comments: The sample was characterized by powder X-ray diffraction data. For the Raman spectra of bismite see also Narang et al. (1994) and Prekajski et al. (2010).

Bismoclite BiOCl

Origin: Synthetic.
Experimental details: Raman scattering measurements have been performed on a powdered sample using 568.2 nm laser radiation. The laser radiation power at the sample was 25 mW.
Raman shifts (cm^{-1}): 400, 202, 146s, 60.
Source: Davies (1973).
Comments: No independent analytical data are given for the sample used. For the Raman spectrum of bismoclite see also Rulmont (1972).

Bismuth Bi

Origin: Synthetic.
Experimental details: Raman scattering measurements have been performed on a film with thickness about 0.5–1 mμ using 532 nm laser radiation. The laser radiation power is not indicated. A 180°-scattering geometry was employed.

Raman shifts (cm^{-1}): 91, 65s.
Source: Russo et al. (2008).
Comments: The sample was characterized by powder X-ray diffraction data and electron microprobe analysis.

Bismuthinite Bi_2S_3

Origin: Panarechensk volcanic-tectonic formation, Kola Peninsula, Russia.
Experimental details: Raman scattering measurements have been performed on an arbitrarily oriented sample using 514.5 nm Ar$^+$ laser radiation. The nominal laser radiation power was 50 mW.
Raman shifts (cm^{-1}): 261s, 236s, 185, 169, 100, 83w, 70, 53s.
Source: Voloshin et al. (2015a).
Comments: The samples used were characterized by electron microprobe analyses. For the Raman spectra of bismuthinite see also Kharbish et al. (2009) and Efthimiopoulos et al. (2014).

Bismutite $Bi_2O_2(CO_3)$

Origin: Synthetic.
Experimental details: Raman scattering measurements have been performed on a polycrystalline sample using 514.5 and 647.1 nm laser radiations. The laser radiation power is not indicated. A 90°-scattering geometry was employed.
Raman shifts (cm^{-1}): 1690w, 1392, 1360, 1067s, 688w, 667, 519w, 445w, 410, 351, 312w, 277w, 203, 172, 162, 158s, 118w, 97w, 94w, 70s, 53s, 51s, 41s, 23.
Source: Taylor et al. (1984).
Comments: The sample was characterized by powder X-ray diffraction data.

Bismutocolumbite $BiNbO_4$

Origin: Synthetic.
Experimental details: Raman scattering measurements have been performed on a powdered sample using 514.5 nm Ar$^+$ laser radiation. The nominal laser radiation power was 200 mW.
Raman shifts (cm^{-1}): 883, 730, 624s, 537, 424, 382, 368, 336, 272s, 255, 220, 199, 153s, 139s, 110, 108, 93, 84s, 60s.
Source: Ayyub et al. (1986).
Comments: No independent analytical data are provided for the sample used. For the Raman spectra of bismutocolumbite see also Ayyub et al. (1987), Yu et al. (1990), and Lee et al. (2003).

Bismutoferrite $Fe^{3+}_2Bi(SiO_4)_2(OH)$

Origin: Jáchymov U deposit, Krušné Hory (Ore Mts.), Western Bohemia, Czech Republic.
Experimental details: Raman scattering measurements have been performed on arbitrarily oriented crystals using a 633 nm He-Ne laser. The Raman shifts have been determined for the maxima of individual peaks obtained as a result of the spectral curve analysis. The laser radiation power is not indicated.
Raman shifts (cm^{-1}): 3541, 1598, 1536, 1475, 1290, 1219, 1160, 1093, 1004, 695, 669, 501, 472, 440, 430s, 417, 386, 348, 333, 323, 306, 280, 273, 244, 223, 217, 196, 165s, 151s, 144s.
Source: Frost et al. (2010a).

Comments: The sample was characterized by powder X-ray diffraction data and electron microprobe analysis.

Bismutotantalite $BiTaO_4$

Origin: Synthetic.
Experimental details: Raman scattering measurements have been performed on a polycrystalline sample using 1084 nm Ar^+ laser radiation. The nominal laser radiation power was 130 mW.
Raman shifts (cm^{-1}): 625s, 539, 371, 340, 274, 257s, 222, 201, 157s, 143s, 112, 87, 64.
Source: Lee et al. (2003).
Comments: The sample was characterized by powder X-ray diffraction data.

Bitikleite $Ca_3(SbSn)(AlO_4)_3$

Origin: Upper Chegem volcanic structure, Kabardino-Balkaria, Northern Caucasus, Russia (type locality).
Experimental details: Raman scattering measurements have been performed on an arbitrarily oriented crystal using 514.5 nm Ar^+ laser radiation. The laser radiation output power was 40–60 mW. A 0°-scattering geometry was employed.
Raman shifts (cm^{-1}): 832sh, 799, 760, 738sh, 624sh, 600, 500s, 440w, 409w, 356w, 299s, 252, 218w, 190, 161w, 151, 120.
Source: Galuskina et al. (2010a).
Comments: The sample was characterized by electron micro-diffraction, powder X-ray diffraction data, and electron microprobe analyses.

Bixbyite $Mn^{3+}_2O_3$

Origin: An unknown locality in Zaire.
Experimental details: Methods of sample preparation are not described. Raman scattering measurements have been performed on an arbitrarily oriented sample using 514.5 nm Ar^+ laser radiation. The nominal laser radiation power was 12.5 mW. A 180°-scattering geometry was employed.
Raman shifts (cm^{-1}): 650sh, 630, 581s, 509.
Source: Bernard et al. (1993a).
Comments: The sample was characterized by powder X-ray diffraction data. The Raman spectrum of natural bixbyite differs from that of synthetic α-Mn_2O_3: the Raman shifts of the latter are 698, 645, 592, 481, 404, 314, and 192 cm^{-1} (Julien et al. 2004). Raman spectrum of presumed bixbyite published by Baioumy et al. (2013) is questionable.

Blatterite $Sb^{5+}_3Mn^{3+}_9Mn^{2+}_{35}(BO_3)_{16}O_{32}$

Origin: Bergslagen ore province, south central Sweden.
Experimental details: Raman scattering measurements have been performed on an arbitrarily oriented sample using 532.4 nm laser radiation. The laser radiation power is not indicated.
Raman shifts (cm^{-1}): 573s, 214w.
Source: Enholm (2016).
Comments: The sample was characterized by electron microprobe analyses.

4 Raman Spectra of Minerals

Blödite $Na_2Mg(SO_4)_2 \cdot 4H_2O$

Origin: Synthetic.
Experimental details: Methods of sample preparation are not described. Raman scattering measurements have been performed on a powdered sample using 532 nm laser radiation. The laser radiation power at the sample was 2 mW.
Raman shifts (cm^{-1}): ~3300, 1184, 1104, 1058, 995s, 631, 463, 449.
Source: Jentzsch et al. (2011).
Comments: No independent analytical data are provided for the sample used.

Bluebellite $Cu_6(IO_3)(OH)_{10}Cl$

Origin: Shallow D shaft, Blue Bell claims, Central Mojave Desert, California, USA (type locality).
Experimental details: Raman scattering measurements have been performed on the (001) face of a flat single crystal using 514.3 nm laser radiation. Laser beam was incident approximately perpendicular to the (001) face. The laser radiation power at the sample was 2 mW.
Raman shifts (cm^{-1}): ~3500, 1007, 680s, 641, 544, 502w, 254, 203, 172.
Source: Mills et al. (2014b).
Comments: The sample was characterized by powder X-ray diffraction data and electron microprobe analyses. The crystal structure is solved.

Bluelizardite $Na_7(UO_2)(SO_4)_4Cl(H_2O)_2$

Origin: Blue Lizard mine, San Juan Co., Utah, USA (type locality).
Experimental details: Methods of sample preparation are not described. Raman scattering measurements have been performed using 532 nm laser radiation. The nominal laser radiation power was 3 mW.
Raman shifts (cm^{-1}): 3606, 3576, 3475, 3422, 3343, 3219, 1216, 1189, 1156, 1143, 1090, 1061, 1050, 1012, 1003, 998, 986, 951, 854s, 848sh, 651, 641, 620, 619, 607, 592, 260, 252, 237, 208.
Source: Plášil et al. (2014a).
Comments: The Raman shifts are indicated for the maxima of individual peaks obtained as a result of the spectral curve analysis. The sample was characterized by powder X-ray diffraction data and electron microprobe analyses. The crystal structure is solved. Plášil et al. (2015d) indicate additional bands at 465 and 445 cm^{-1}.

Bobcookite $NaAl(UO_2)_2(SO_4)_4 \cdot 18H_2O$

Origin: Blue Lizard mine, San Juan Co., Utah, USA (type locality).
Experimental details: Methods of sample preparation are not described. Raman scattering measurements have been performed using 532 and 780 nm laser radiation. The laser radiation power is not indicated.
Raman shifts (cm^{-1}): 3610, 3565, 3500, 3445, 3380, 3315sh, 3270sh, 3195sh, 1640, 1210, 1145, 1110, 1035, 1010s, 990sh, 845s, 630, 600w, 470, 450, 330w, ~210.
Source: Kampf et al. (2015b).
Comments: Raman shifts in the range of stretching vibrations of water molecules are indicated for the maxima of individual peaks obtained as a result of the spectral curve analysis. The sample was

characterized by powder X-ray diffraction data and electron microprobe analyses. The crystal structure is solved.

Bobdownsite $Ca_9Mg(PO_3F)(PO_4)_6$

Origin: Big Fish River, Yukon, Canada (type locality).
Experimental details: Raman scattering measurements have been performed on an arbitrarily oriented crystal using 532 nm laser radiation. The laser radiation power is not indicated.
Raman shifts (cm^{-1}): 1088, 1028w, 989, 966s, 923, 626, 605w, 554, 483w, 433, 406, 282, 158w.
Source: Tait et al. (2011).
Comments: The sample was characterized by powder X-ray diffraction data and electron microprobe analysis. The crystal structure is solved. The Raman shifts were partly determined by us based on spectral curve analysis of the published spectrum.

Bobierrite $Mg_3(PO_4)_2 \cdot 8H_2O$

Origin: Zheleznyi mine (Iron mine), Kovdor massif, Kola Peninsula, Russia.
Experimental details: Raman scattering measurements have been performed on arbitrarily oriented crystals using a 633 nm He-Ne laser. The Raman shifts have been determined for the maxima of individual peaks obtained as a result of the spectral curve analysis. The laser radiation power is not indicated.
Raman shifts (cm^{-1}): 3498, 3263, 3212, 3096, 2895, 1072, 998, 951, 909, 842, 787, 717, 693, 668, 631, 583, 557, 542, 468, 435, 420, 364, 318, 290, 282, 262, 233, 215, 182, 170, 149, 136.
Source: Frost et al. (2002f).
Comments: No independent analytical data are provided for the sample used.

Bobshannonite $Na_2KBa(Mn,Na)_8(Nb,Ti)_4(Si_2O_7)_4O_4(OH)_4(O,F)_2$

Origin: Mont Saint-Hilaire, Québec, Canada (type locality).
Experimental details: Methods of sample preparation are not described. Raman scattering measurements have been performed using 532 nm laser radiation. The laser radiation power is not indicated. A 180°-scattering geometry was employed.
Raman shifts (cm^{-1}): ~3655, ~3610, 1038, 970, 901, 716, 680, 608, 580, 510, 410, 341, 310, 240, 207, 143.
Source: Sokolova et al. (2015b).
Comments: The Raman shifts are indicated for the maxima of individual peaks obtained as a result of the spectral curve analysis. The sample was characterized by powder X-ray diffraction data and electron microprobe analyses. The crystal structure is solved.

Bohdanowiczite $AgBiSe_2$

Origin: Synthetic.
Experimental details: Methods of sample preparation are not described. Raman scattering measurements have been performed using 532 nm Nd-YAG laser radiation. The laser radiation power is not indicated.
Raman shift (cm^{-1}): 171, 161.
Source: Rajaji et al. (2016).

Comments: The sample was characterized by powder X-ray diffraction data. Trigonal, $a = 8.412(6)$ Å, $c = 19.63(3)$ Å.

Böhmite γ-AlO(OH)

Origin: Synthetic.
Experimental details: Methods of sample preparation are not described. Raman scattering measurements have been performed using 1064 nm Nd-YAG laser radiation. The nominal laser radiation power at the sample was 200 mW. A 180°-scattering geometry was employed.
Raman shifts (cm^{-1}): 3371, 3220, 3085, 2989, 1072, 732, 674s, 495s, 451, 360s, 228.
Source: Ruan et al. (2001).
Comments: No independent analytical data are provided for the sample used. The Raman shifts have been determined for the maxima of individual peaks obtained as a result of the spectral curve analysis.

Boleite $KAg_9Pb_{26}Cu_{24}Cl_{62}(OH)_{48}$

Origin: Amelia Mine, Santa Rosalia, Baja, California, Mexico (type locality).
Experimental details: Raman scattering measurements have been performed on arbitrarily oriented crystals using a 633 nm He-Ne laser. The Raman shifts have been determined for the maxima of individual peaks obtained as a result of the spectral curve analysis. The laser radiation power is not indicated.
Raman shifts (cm^{-1}): 3448, 3408, 3371sh, 921, 817, 757, 731, 696, 478, 455, 386, 361, 300, 234, 161s, 146sh, 128sh.
Source: Frost et al. (2003j).
Comments: No independent analytical data are provided for the sample used. For the Raman spectrum of boleite see also Frost and Williams (2004).

Boltwoodite $(K,Na)(UO_2)(SiO_3OH) \cdot 1.5H_2O$

Origin: Kladská U deposit, Slavkovský les Mountains, western Bohemia, Czech Republic.
Experimental details: Methods of sample preparation are not described. Raman scattering measurements have been performed using 780 nm laser radiation. The laser radiation output power was 10 mW.
Raman shifts (cm^{-1}): 3387w, 3351w, 3313w, 1606w, 1327w, 958, 938, 847sh, 833sh, 804s, 542, 496, 483, 435, 423, 398, 321, 280, 262, 220, 180, 152, 136, 110, 93, 75, 60.
Source: Plášil et al. (2016a).
Comments: The sample was characterized by powder X-ray diffraction data and electron microprobe analyses. For the Raman spectrum of boltwoodite see also Frost et al. (2006e).

Bonaccordite $Ni_2Fe^{3+}O_2(BO_3)$

Origin: Synthetic.
Experimental details: Methods of sample preparation are not described. Raman scattering measurements have been performed using 514.5 nm Ar$^+$ laser radiation. The nominal laser radiation power was 20 mW. Polarized spectrum was collected in the (zz) geometry.
Raman shifts (cm^{-1}): 650s, 582s, 570sh, 544, 526, 495, 473sh, 423, 388w, 357w, 314, 285w, 257w.

Source: Leite et al. (2002).
Comments: No independent analytical data are provided for the sample used. The Raman shifts were determined by us based on spectral curve analysis of the published spectrum.

Bonattite $Cu(SO_4) \cdot 3H_2O$

Origin: Synthetic.
Experimental details: Raman scattering measurements have been performed on a powdered sample using 632.8 nm He-Ne laser radiation. The laser radiation power is not indicated.
Raman shifts (cm^{-1}): 1126, 1009s, 620, 587, 481, 429, 386, 250, 160, 123.
Source: Fu et al. (2012).
Comments: Bands of H_2O stretching vibrations are very weak and poor-resolved. No independent analytical data are provided for the sample used.

Bonazziite β-As_4S_4

Origin: Synthetic.
Experimental details: Raman scattering measurements have been performed on an arbitrarily oriented sample using 647.1 nm Kr$^+$ laser radiation. The laser radiation power at the sample was 60 mW.
Raman shifts (cm^{-1}): 388w, 376w, 362s, 352, 343, 332sh, 217, 211sh, 187s, 164, 144, 62, 56, 42w, 32w.
Source: Muniz-Miranda et al. (1996).
Comments: The photo-induced transformation from β-As_4S_4 to pararealgar takes place in the sample under the exposure to a more short-wave radiation.

Bonazziite β-As_4S_4

Origin: Khaidarkan deposit, Kyrgyzstan (type locality).
Experimental details: Raman scattering measurements have been performed on an arbitrarily oriented sample using 875 nm diode laser radiation. The laser radiation power was 3 mW.
Raman shifts (cm^{-1}): 362s, 352, 343, 217s, 187s, 164.
Source: Bindi et al. (2015b).
Comments: The sample was characterized by powder X-ray diffraction data and electron microprobe analyses.

Boracite $Mg_3B_7O_{13}Cl$

Origin: Lüneburg, Lower Saxony, Germany (type locality).
Experimental details: Raman scattering measurements have been performed on arbitrarily oriented crystals using a 633 nm He-Ne laser. The Raman shifts have been determined for the maxima of individual peaks obtained as a result of the spectral curve analysis. The laser radiation power is not indicated.
Raman shifts (cm^{-1}): 3581sh, 3494, 3431sh, 3405s, 3334sh, 3277, 3254w, 2903, 2727sh, 1617, 1583, 1348, 1143, 1136, 1121, 1009s, 671, 660sh, 621, 611sh, 582sh, 536w, 515sh, 494, 473sh, 415, 317, 211s, 182s, 163s, 147, 134s.
Source: Frost et al. (2012m).
Comments: No independent analytical data are provided for the sample used. The spectrum shows the presence of OH groups.

Borax $Na_2B_4O_5(OH)_4 \cdot 8H_2O$

Origin: Synthetic.

Experimental details: Raman scattering measurements have been performed on a polycrystalline sample using 514.5 nm Ar^+ laser radiation. The nominal laser radiation power was about 100 mW.

Raman shifts (cm^{-1}): 3575s, 3495s, 3447s, 3400s, 3357s, 3140s, 1640w, 957, 860, 776, 590w, 530, 474, 390, 361, 160, 120, 90, 78.

Source: Devi et al. (1994).

Comments: No independent analytical data are provided for the sample used. Polarized spectra of borax single crystals were collected in $x(yy)z$, $x(yx)z$, $x(zy)z$, and $x(zx)z$ scattering geometries too. For the Raman spectra of borax see also Krishnamurti (1955) and Kipcak et al. (2014).

Bornite Cu_5FeS_4

Origin: No data.

Experimental details: Raman scattering measurements have been performed on an arbitrarily oriented sample using 638 nm laser radiation. The laser radiation power is not indicated. A 180°-scattering geometry was employed.

Raman shifts (cm^{-1}): 784w, 579, 464w, 377, 266s, 201s.

Source: Lanteigne et al. (2012).

Bosiite $NaFe^{3+}_3(Al_4Mg_2)(Si_6O_{18})(BO_3)_3(OH)_3O$

Origin: No data.

Experimental details: Polarized $-y(zz)y$ Raman scattering measurements have been performed on a single crystal using 488.0 or 514.5 nm Ar^+ laser radiations. The laser radiation power at the sample was about 14 mW.

Raman shifts (cm^{-1}): No data: only a figure of the Raman spectrum is given in the cited paper.

Source: Watenphul et al. (2016b).

Comments: The sample was characterized by electron microprobe analysis.

Botallackite $Cu_2Cl(OH)_3$

Origin: No data.

Experimental details: Methods of sample preparation are not described. Raman scattering measurements have been performed using 632.8 and 514.5 nm Ar^+ laser radiation. The laser radiation output power was 30 mW at 632.8 nm and is not indicated at 514.5 nm.

Raman shifts (cm^{-1}): 3504, 3420, 897, 857, 678w, 503, 450s, 401s, 324, 279, 251, 175, 155, 115.

Source: Bouchard and Smith (2003).

Comments: The sample was characterized by powder X-ray diffraction data and electron microprobe analysis.

Botryogen $MgFe^{3+}(SO_4)_2(OH) \cdot 7H_2O$

Origin: Alcaparrosa mine, Antofagasta Province, Chile.

Experimental details: Raman scattering measurements have been performed on arbitrarily oriented crystals using a 633 nm He-Ne laser. The Raman shifts have been determined for the maxima of individual peaks obtained as a result of the spectral curve analysis. The laser radiation power is not indicated.

Raman shifts (cm^{-1}): 3576, 3441, 3330, 3256, 3107sh, 3186, 1626, 1221, 1202, 1178, 1076, 1041, 1017, 1002s, 607, 563, 499s, 464, 384, 353, 271, 241s, 208, 180, 146.
Source: Frost et al. (2011f).
Comments: The sample was characterized by powder X-ray diffraction data and electron microprobe analysis.

Bottinoite $NiSb^{5+}{}_2(OH)_{12}\cdot 6H_2O$

Origin: Bottino Mine, Italy (type locality).
Experimental details: Raman scattering measurements have been performed on arbitrarily oriented crystals using a 633 nm He-Ne laser. The Raman shifts have been determined for the maxima of individual peaks obtained as a result of the spectral curve analysis. The laser radiation power is not indicated.
Raman shifts (cm^{-1}): 3510sh, 3458, 3368, 3291, 3228sh, 3223sh, 1648, 1163w, 1111w, 1080w, 1045w, 735w, 630sh, 618, 599sh, 575sh, 516sh, 501, 316sh, 336, 317, 302, 254, 235, 207, 169, 146, 125, 114.
Source: Frost and Bahfenne (2010e).
Comments: No independent analytical data are provided for the sample used. For the Raman spectra of bottinoite see also Rintoul et al. (2011) and Bahfenne (2011).

Boulangerite $Pb_5Sb_4S_{11}$

Origin: Zlatá Baňa, Slanské Vrchy Mts., central Slovakia.
Experimental details: Raman scattering measurements have been performed on an arbitrarily oriented crystal using 632 nm Nd-YAG laser radiation. The laser radiation power radiation density at the sample was $8.5\cdot 10^{-3}$ mW/µm^2. A 180°-scattering geometry was employed.
Raman shifts (cm^{-1}): 355s, 335, 315, 236, 220, 206, 189, 146, 129sh, 100, 85, 74sh, 62.
Source: Kharbish and Jeleň (2016).
Comments: The Raman shifts are indicated for the maxima of individual peaks obtained as a result of the spectral curve analysis. The sample was characterized by electron microprobe analysis.

Bournonite $CuPbSbS_3$

Origin: Felsöbanya, Romania.
Experimental details: Raman scattering measurements have been performed on an oriented single crystal with the laser polarization parallel to the a-, b-, and c-axes using 785 nm laser radiation. The laser radiation power at the sample was 1.7 mW. A 180°-scattering geometry was employed.
Raman shifts (cm^{-1}): 339, 324s, 192, 275, 227w, 197, 181, 166.
Source: Kharbish et al. (2009).
Comments: Slightly varying band positions among polarized spectra are averaged. The sample was characterized by electron microprobe analysis.

Boussingaultite $(NH_4)_2Mg(SO_4)_2\cdot 6H_2O$

Origin: Larderello, Tuscany, Italy.
Experimental details: Micro-Raman scattering measurements have been performed on an arbitrarily oriented sample using 785 nm diode laser radiation. The laser radiation power is not indicated.

Raman shifts (cm^{-1}): 3380w, 3290w, 3080, 3040, 2918w, 2845w, 1705, 1678, 1460, 1436, 1133w, 1096w, 1063, 983s, 626, 616, 454, 360w, 310w, <222w.
Source: Culka et al. (2009).
Comments: No independent analytical data are provided for the sample used. Raman shifts are indicated for the maxima of individual peaks obtained as a result of the spectral curve analysis.
For the Raman spectrum of boussingaultite see also Jentzsch et al. (2013).

Boussingaultite $(NH_4)_2Mg_2(SO_4)_3 \cdot 6H_2O$

Origin: Synthetic.
Experimental details: Raman scattering measurements have been performed on an arbitrarily oriented sample using 253.7 nm of mercury radiation and exposition of the order of 16 h.
Raman shifts (cm^{-1}): 3396, 3331, 3281, 3060, 2830, 1469w, 1433, 1141, 1121, 1102, 1091, 1072, 1061, 979s, 622, 455, 265w, 220w, 198, 147, 130, 89w, 54.
Source: Shantakumari (1953).
Comments: No independent analytical data are provided for the sample used.

Bowieite Rh_2S_3

Origin: Svetly Bor complex, Urals, Russia.
Experimental details: Raman scattering measurements have been performed on an arbitrarily oriented grain using 532 nm Nd-YAG laser radiation. The nominal laser radiation power was 100 mW.
Raman shifts (cm^{-1}): 374, 308s, 287.
Source: Zaccarini et al. (2016).
Comments: For the Raman spectrum of bowieite see also Singh et al. (2014).

Braccoite $NaMn^{2+}_5[Si_5O_{14}(OH)](AsO_3)(OH)$

Origin: Valletta mine, Maira Valley, Piedmont, Italy (type locality).
Experimental details: Raman scattering measurements have been performed on an arbitrarily oriented crystal using 632.8 and 532 nm He-Ne and Nd-YAG laser radiations. The nominal laser radiation power was 20 and 80 mW, respectively. A 180°-scattering geometry was employed.
Raman shifts (cm^{-1}): 1040sh, 1017s, 932sh, 907s, 829s, 748, 706, 665s, 618, 563w, 525w, 451, 390, 360w, 291, 261w, 226.
Source: Cámara et al. (2015).
Comments: The sample was characterized by powder X-ray diffraction data and electron microprobe analyses. The crystal structure is solved.

Bracewellite $CrO(OH)$

Origin: Artificial.
Experimental details: Raman scattering measurements have been performed on a chromium coupon using 647 nm Kr$^+$ laser radiation. The nominal laser radiation power was 8 mW. A nearly 180°-scattering geometry was employed.
Raman shifts (cm^{-1}): ~825w, ~780w, ~620, ~565s, ~310s.
Source: Maslar et al. (2001).
Comments: The sample was characterized by X-ray emission spectrum.

Brackebuschite $Pb_2Mn^{3+}(VO_4)_2(OH)$

Origin: Sierra de Cordoba, Argentina (type locality).
Experimental details: No data.
Raman shifts (cm^{-1}): 3145w, 859s, 687w, 450w, 334, 158.
Source: Lafuente and Downs (2016).
Comments: The sample was characterized by single-crystal X-ray diffraction data and electron microprobe analysis. The crystal structure is solved. The Raman shifts were partly determined by us based on spectral curve analysis of the published spectrum.

Bradleyite $Na_3Mg(PO_4)(CO_3)$

Origin: Synthetic.
Experimental details: Methods of sample preparation are not described. Raman scattering measurements have been using 532 nm laser radiation. The nominal laser radiation power was 240 mW.
Raman shifts (cm^{-1}): 1079s, 1067w, 1051, 1033, 971s, 733w, 694w, 627w, 591, 484w, 430, 262, 218, 198, 161.
Source: Gao et al. (2015).
Comments: The sample was characterized by powder X-ray diffraction data. The Raman shifts were partly determined by us based on spectral curve analysis of the published spectrum.

Braggite PtS

Origin: No data.
Experimental details: Raman scattering measurements have been performed on an arbitrarily oriented sample using 18,794.59 cm^{-1} Ar^+ laser radiation. The laser radiation power at the sample was between 1 and 2 mW.
Raman shifts (cm^{-1}): 378, 360s, 330s, 111.
Source: Bakker (2014).
Comments: For the Raman spectra of braggite see also Mernagh and Hoatson (1995), Pikl et al. (1999), and Merkle et al. (1999).

Brandholzite $MgSb_2(OH)_{12} \cdot 6H_2O$

Origin: Krížnica mine, Pernek deposit, Malé Karpaty Mts., Slovak Republic.
Experimental details: Raman scattering measurements have been performed on arbitrarily oriented crystals using a 633 nm He-Ne laser. The Raman shifts have been determined for the maxima of individual peaks obtained as a result of the spectral curve analysis. The laser radiation power is not indicated.
Raman shifts (cm^{-1}): 3552, 3483sh, 3466, 3383sh, 3240, 3205sh, 1648w, 1597sh, 1189sh, 1160w, 1093, 1043, 730, 630, 618s, 604, 578sh, 526, 503, 340s, 318s, 303sh, 252w, 232w, 191w, 147w, 115w.
Source: Frost et al. (2009e).
Comments: The sample was characterized by powder X-ray diffraction data and electron microprobe analysis. For the Raman spectra of brandholzite see also Rintoul et al. (2011) and Bahfenne (2011).

Brannerite UTi_2O_6

Origin: El Cabril mine, near Cordoba, Sierra Albarrana region, southern Spain.
Experimental details: Raman scattering measurements have been performed on an arbitrarily oriented sample using 514.5 nm Ar^+ laser radiation. The laser radiation power is not indicated.
Raman shifts (cm^{-1}): 759s, 615, 523, 435, 375, 327, 265, 194, 161.
Source: Zhang et al. (2013).
Comments: The Raman shifts are indicated for the maxima of individual peaks obtained as a result of the spectral curve analysis. The sample was preliminarily annealed and characterized by powder X-ray diffraction data and electron microprobe analysis. For the Raman spectra of brannerite see also Frost and Reddy (2011a) and Charalambous et al. (2012).

Brannockite $KSn_2(Li_3Si_{12})O_{30}$

Origin: Golden Horn Batholith, Washington, USA.
Experimental details: Raman scattering measurements have been performed on arbitrarily oriented crystals using 632 nm He-Ne laser radiation. The laser radiation power was 1 mW.
Raman shifts (cm^{-1}): 1160, 1141, 1042, 992, 947w, 841, 774, 616, 485, 462s, 382, 365w, 343, 282s, 257w, 248w, 204w, 156s, 128, 103s, 93sh, 63.
Source: Raschke et al. (2016).
Comments: The sample was characterized by single-crystal X-ray diffraction data and electron microprobe analysis. The Raman shifts were partly determined by us based on spectral curve analysis of the published spectrum.

Brassite $Mg(AsO_3OH) \cdot 4H_2O$

Origin: Jáchymov U deposit, KrušnéHory (Ore Mts.), Western Bohemia, Czech Republic (type locality).
Experimental details: Raman scattering measurements have been performed on arbitrarily oriented crystals using a 633 nm He-Ne laser. The Raman shifts have been determined for the maxima of individual peaks obtained as a result of the spectral curve analysis. The laser radiation power is not indicated.
Raman shifts (cm^{-1}): 3511, 3450, 3387, 3314, 3035, 878sh, 876sh, 862, 809, 739sh, 699, 609, 448, 404, 387sh, 358sh, 298, 274sh, 242, 199, 181, 158sh, 149, 121, 108.
Source: Frost et al. (2010h).
Comments: The sample was characterized by powder X-ray diffraction data and electron microprobe analysis.

Braunite $Mn^{2+}Mn^{3+}_6O_8(SiO_4)$

Origin: Synthetic.
Experimental details: Raman scattering measurements have been performed on microscopic inclusions in in-glaze pigment of the nineteenth-century relief tiles using 632.8 nm He-Ne laser radiation. The nominal laser radiation power was 17 mW.
Raman shifts (cm^{-1}): 958, 686, 617, 513s, 471, 330w, 217, 121.
Source: Coutinho et al. (2016).
Comments: The sample was characterized by electron microprobe analysis.

Brazilianite $NaAl_3(PO_4)_2(OH)_4$

Origin: Córrego Frio mine, Linópolis, Divino das Laranjeiras, Doce valley, Minas Gerais, Brazil (type locality).
Experimental details: Raman scattering measurements have been performed on arbitrarily oriented crystals using a 633 nm He-Ne laser. The Raman shifts have been determined for the maxima of individual peaks obtained as a result of the spectral curve analysis. The laser radiation power is not indicated.
Raman shifts (cm^{-1}): 3543, 3519, 3472s, 3447, 3417s, 3355, 3291, 3249, 3157, 1579w, 1395, 1150w, 1117, 1074, 1037sh, 1019s, 988s, 973, 953, 723, 660, 636, 615, 599s, 563s, 534, 508sh, 466, 441, 414, 358, 319, 287, 276, 253, 244, 231, 220, 208, 172, 162w, 149, 113, 105.
Source: Frost and Xi (2012l).
Comments: No independent analytical data are provided for the sample used.

Bredigite $(Ca,Ba)Ca_{13}Mg_2(SiO_4)_8$

Origin: Synthetic.
Experimental details: Methods of sample preparation are not described. Raman scattering measurements have been performed using 514.5 nm Ar$^+$ laser radiation. The laser radiation power is not indicated.
Raman shifts (cm^{-1}): 991, 977, 950w, 927w, 907sh, 895sh, 884s, 872s, 857s, 847, 575, 554w, 543w, 526w, 514w, 502w, 424w, 406w, 384w, 375w, 298w, 257w, 240w, 211w, 194w, 149w, 125, 109sh, 68w.
Source: Xiong et al. (2016).
Comments: The sample was characterized by powder X-ray diffraction data and electron microprobe analysis.

Breithauptite NiSb

Origin: Synthetic.
Experimental details: Raman scattering measurements have been performed on an arbitrarily oriented sample using 514.5 nm Ar$^+$ laser radiation. The laser radiation power is not indicated.
Raman shifts (cm^{-1}): 650.
Source: Xie et al. (2011).
Comments: The sample was characterized by powder X-ray diffraction data.

Brewsterite-Sr $Sr(Al_2Si_6)O_{16} \cdot 5H_2O$

Origin: Strontian, Agryll, Scotland, UK (type locality).
Experimental details: Methods of sample preparation are not described. Raman scattering measurements have been performed on an arbitrarily oriented sample using 1.06 μm Nd-YAG laser radiation. The nominal laser radiation power was 300 mW.
Raman shifts (cm^{-1}): 1136w, 495s, 387, 236, 171.
Source: Mozgawa (2001).
Comments: The sample was characterized by powder X-ray diffraction data.

Brianyoungite $Zn_3(CO_3,SO_4)(OH)_4$

Origin: Esperanza Mine, Laurion district, Greece.
Experimental details: Raman scattering measurements have been performed on arbitrarily oriented crystals using a 633 nm He-Ne laser. The Raman shifts have been determined for the maxima of individual peaks obtained as a result of the spectral curve analysis. The laser radiation power is not indicated.
Raman shifts (cm^{-1}): 3669, 3631, 3615, 3564, 3571, 3554, 3531, 3518, 3400, 3297, 3193, 3076, 2973, 2938, 2910, 2880, 2851, 1550, 1457, 1440, 1388, 1367, 1298, 1163, 1127, 1086, 1056s, 1038sh, 984, 973s, 958sh, 736, 704, 638, 609, 528, 507sh, 475, 451, 433, 423, 378, 367, 347, 306, 271, 257w, 277, 216, 160sh, 153sh, 151s, 143s, 132sh, 125, 113, 108.
Source: Frost et al. (2015r).
Comments: The sample was characterized only by qualitative electron microprobe analysis.

Briartite Cu_2FeGeS_4

Origin: Synthetic.
Experimental details: Methods of sample preparation are not described. Raman scattering measurements have been performed using 488.0 or 514.5 nm Ar^+ laser radiation. The laser radiation power at the sample was 2 mW. A 180°-scattering geometry was employed.
Raman shifts (cm^{-1}): 437, 412w, 396, 378, 342s, 329sh, 304w, 294, 272, 250, 224, 162w, 141, 109.
Source: Rincón et al. (2015).
Comments: The sample was characterized by powder X-ray diffraction data. For the Raman spectrum of briartite see also Himmrich and Haeuseler (1991).

Britvinite $Pb_{14}Mg_9(Si_{10}O_{28})(BO_3)_4(CO_3)_2(OH)_{12}F_2$

Origin: Långban deposit, Bergslagen ore region, Filipstad district, Värmland, Sweden (type locality).
Experimental details: Raman scattering measurements have been performed on an arbitrarily oriented crystal using 633 nm laser radiation. The laser radiation power is not indicated. A 180°-scattering geometry was employeded.
Raman shifts (cm^{-1}): 3697s, 3543, 1718, 1696w, 1420, 1335, 1230, 1193, 1093, 1041s, 992, 960, 905, 873w, 842, 817, 802, 775, 740w, 717, 690, 667, 593, 489, 412, 303s, 277, 258, 217, 154.
Source: Kolitsch et al. (2012).
Comments: The sample was characterized by powder X-ray diffraction data and electron microprobe analysis. The Raman shifts were partly determined by us based on spectral curve analysis of the published spectrum.

Brizziite $NaSbO_3$

Origin: Le Cetine mine, Chiusdino, Siena province, Tuscany, Italy (type locality).
Experimental details: Raman scattering measurements have been performed on arbitrarily oriented crystals using a 633 nm He-Ne laser. The Raman shifts have been determined for the maxima of individual peaks obtained as a result of the spectral curve analysis. The laser radiation power is not indicated.
Raman shifts (cm^{-1}): 830w, 749w, 660s, 617, 508, 315, 307, 230, 204, 158.
Source: Frost and Bahfenne (2010a).

Comments: No independent analytical data are provided for the sample used. For the Raman spectrum of brizziite see also Bittarello et al. (2015).

Brochantite $Cu_4(SO_4)(OH)_6$

Origin: Chuquicamata, Chile.
Experimental details: Micro-Raman scattering measurements have been performed on an arbitrarily oriented sample using 780 nm Nd-YAG laser radiation. The laser radiation power at the sample was <1 mW.
Raman shifts (cm^{-1}): 3580s, 3501sh, 3489s, 1906, 1265w, 1173w, 1135w, 1078, 990s, 786, 770, 749, 629, 608, 600, 517, 501, 482, 467, 442, 415, 340, 330, 295, 265, 247, 238, 228, 213, 172, 149, 141, 124.
Source: Martens et al. (2003a).
Comments: The Raman shifts are indicated for the maxima of individual peaks obtained as a result of the spectral curve analysis. The sample was characterized by XRD and EDX, but these data are not provided in the cited paper. For the Raman spectra of brochantite see also Makreski et al. (2005a), Schmidt and Lutz (1993), Bouchard and Smith (2003), Frost et al. (2004a), Apopei et al. (2014a), and Coccato et al. (2016).

Brochantite $Cu_4(SO_4)(OH)_6$

Origin: Bučim, Macedonia.
Experimental details: Raman scattering measurements have been performed on a powdered sample using 514.5 nm Ar^+ laser radiation. The laser radiation power is not indicated.
Raman shifts (cm^{-1}): 3587sh, 3565s, 3536sh, 3508w, 3475, 3388s, 3358, 1160sh, 1134w, 1121sh, 1099sh, 1076, 973s, 904, 871w, 840w, 765w, 710w, 660w, 619, 609sh, 590w, 551w, 510, 480w, 431w, 415w, 394, 363sh, 323w, 292w, 237w, 190, 163w, 141w, 137w, 126w, 116w, 83.
Source: Makreski et al. (2005a).
Comments: No independent analytical data are provided for the sample used.

Bromargyrite AgBr

Origin: Synthetic.
Experimental details: Raman scattering measurements have been performed on a partly oriented crystal, in the $x(zz)y + x(zx)y$ scattering geometry, using 514.5 nm Ar^+ laser radiation. The nominal laser radiation power was 100 mW.
Raman shifts (cm^{-1}): 262, 176, 82s.
Source: Bottger and Damsgard (1971).
Comments: The sample was characterized by means of flame emission spectroscopy and mass-spectrometry.

Bromellite BeO

Origin: Muiane pegmatite, Mozambique.
Experimental details: Raman scattering measurements have been performed on inclusions in morganite using 488 and 514.5 nm Ar^+ laser radiation. The laser radiation power at the sample was 14 mW.

Raman shifts (cm^{-1}): 1097, 1081, 722, 684s, 678, 388.
Source: Thomas and Davidson (2010).
Comments: For the Raman spectrum of bromellite see also Devanarayanan et al. (1991).

Brookite TiO$_2$

Origin: Magnet Cove, Arkansas.
Experimental details: Raman scattering measurements have been performed on oriented crystals using 458, 515 and 633 nm laser radiation. The laser radiation power is not indicated. Polarized spectra were collected in the following scattering geometries: (*xx*), (*yy*), and (*zz*) for the A$_{1g}$ Raman mode, (*xy*) for B$_{1g}$, (*xz*) for B$_{2g}$, and (*yz*) for B$_{3g}$.
Raman shifts (cm^{-1}): A$_{1g}$: 640, 545, 492w, 412w, 324w, 246, 194w, 152s, 125; B$_{1g}$: 622, 449, 381w, 327, 283s, 212, 169; B$_{2g}$: 584, 460s, 391w, 366s, 325, 254w, 160s; B$_{3g}$: 500, 416w, 318s, 212w, 132.
Source: Iliev et al. (2013).
Comments: No independent analytical data are provided for the sample used. For the Raman spectra of brookite see also Yanqing et al. (2000), Zajzon et al. (2013), and Andò and Garzanti (2014).

Browneite MnS

Origin: Zakłodzie meteorite, Poland (type locality).
Experimental details: Micro-Raman scattering measurements have been performed on an arbitrarily oriented grain using 514.5 nm Ar$^+$ laser radiation. The laser radiation power at the sample was 1.5 mW.
Raman shifts (cm^{-1}): ~620, ~460, ~400sh, ~220w.
Source: Ma et al. (2012b).
Comments: The sample was characterized by powder X-ray diffraction data and electron microprobe analyses.

Brownleeite MnSi

Origin: Synthetic.
Experimental details: Raman scattering measurements have been performed on an oriented single crystal using 532 nm Nd-YAG laser radiation. The laser radiation power density was 2×10^5 W/cm^2. A 180°-scattering geometry was employed. Polarized spectra were collected in the z(yy)-z, z(xx)-z, z(xy)-z, and z(yx)-z scattering geometries. At different scattering geometries the shifts of the Raman lines do not exceed 2 cm^{-1}.
Raman shifts (cm^{-1}): ~310, ~190.
Source: Tite et al. (2010).
Comments: The sample was characterized by powder X-ray diffraction data and electron microprobe analysis.

Brownmillerite Ca$_2$Fe^{3+}AlO$_5$

Origin: Synthetic.
Experimental details: Raman scattering measurements have been performed on a polycrystalline sample using 514.5 nm Ar$^+$ laser radiation. The laser radiation power is not indicated.

Raman shifts (cm^{-1}): 707s, ~550, ~420, ~380, ~310, ~290, 256s.
Source: Dhankhar et al. (2016).
Comments: The sample was characterized by powder X-ray diffraction data. For the Raman spectrum of brownmillerite see also Martínez-Ramírez and Fernández-Carrasco (2011).

Brucite Mg(OH)$_2$

Origin: Mariana convergent plate margin, western Pacific Ocean.
Experimental details: Raman scattering measurements have been performed on an arbitrarily oriented crystal in a polished thin section using 514.5 nm Ar$^+$ laser radiation. The laser radiation power is not indicated.
Raman shifts (cm^{-1}): 722, 442s, 276.
Source: Sagatowska (2010).
Comments: The sample was characterized by electron microprobe analysis. The Raman shifts were determined by us based on spectral curve analysis of the published spectrum. For the Raman spectrum of brucite see also Lutz et al. (1994).

Brüggenite Ca(IO$_3$)$_2$·H$_2$O

Origin: Synthetic.
Experimental details: Raman scattering measurements have been performed on an arbitrarily oriented sample using 514.5 nm Ar$^+$ laser radiation. The laser radiation power is not indicated.
Raman shifts (cm^{-1}): 3470, 3376, 832sh, 811, 767, 754sh, 746sh, 382, 343, 333sh, 322sh, 251sh, 229.
Source: Alici et al. (1992).
Comments: The sample was characterized by powder X-ray diffraction data.

Brugnatellite Mg$_6$Fe^{3+}(CO$_3$)(OH)$_{13}$·4H$_2$O

Origin: Monte Ramazzo, Genoa, Liguria, Italy.
Experimental details: Raman scattering measurements have been performed on an arbitrarily oriented sample using 633 nm He-Ne laser radiation. The laser radiation power at the sample was 5 mW. Raman shifts are indicated for the maxima of individual peaks obtained as a result of the spectral curve analysis.
Raman shifts (cm^{-1}): 3922, 3696s, 3685s, 3656sh, 2933, 1591w, 1323w, 1102, 1087, 959w, 765w, 698sh, 690s, 664w, 644w, 621.
Source: Frost and Bahfenne (2009).
Comments: No independent analytical data are given for the sample used. Raman spectrum from Monte Ramazzo given as Supplementary Information does not coincide with the spectrum given in the cited paper.

Brushite Ca(PO$_3$OH)·2H$_2$O

Origin: Moorba Cave, Jurien Bay, Dandaragan Shire, Western Australia, Australia.
Experimental details: Raman scattering measurements have been performed on an arbitrarily oriented sample using 632 nm He-Ne laser radiation. The laser radiation power is not detected. Raman shifts are indicated for the maxima of individual peaks obtained as a result of the spectral curve analysis.

Raman shifts (cm^{-1}): 3533, 3472, 1055, 1000, 985s, 872, 858sh, 576w, 519s, 498, 411s, 276, 209, 140, 109.

Source: Frost et al. (2012h).

Comments: No independent analytical data are given for the sample used. For the Raman spectrum of brushite see also Xu et al. (1999).

Buchwaldite $NaCa(PO_4)$

Origin: Synthetic.

Experimental details: Raman scattering measurements have been performed on a powdered sample using 514.5 nm Ar$^+$ laser radiation. The laser radiation power is not indicated.

Raman shifts (cm^{-1}): ~1048, ~1026, ~1015, ~966s, ~588s, ~450, ~427.

Source: Suchanek et al. (1998).

Comments: The sample was characterized by powder X-ray diffraction data.

Bukovskýite $Fe^{3+}_2(AsO_4)(SO_4)(OH)\cdot 7H_2O$

Origin: Kaňk, near Kutná Hora, central Bohemia, Czech Republic (type locality).

Experimental details: Raman scattering measurements have been performed on microcrystalline aggregates using a 633 nm He-Ne laser. The Raman shifts have been determined for the maxima of individual peaks obtained as a result of the spectral curve analysis. The laser radiation power is not indicated.

Raman shifts (cm^{-1}): 3420w, 3219w, 3102sh, 1652w, 1179, 1131, 1090, 1054, 1010, 984s, 911, 886, 847s, 816s, 613, 552, 511, 464, 428s, 315, 263s, 196, 147.

Source: Loun et al. (2011).

Comments: The sample was characterized by powder X-ray diffraction data and electron microprobe analysis. For the Raman spectrum of bukovskýite see also Culka et al. (2016).

Bunnoite $Mn^{2+}_6AlSi_6O_{18}(OH)_3$

Origin: Kamo Mt., Kochi prefecture, Japan (type locality).

Experimental details: Methods of sample preparation are not described. Raman scattering measurements have been performed using 514.5 nm Ar$^+$ laser radiation. The nominal laser radiation power was 100 mW.

Raman shifts (cm^{-1}): 3546sh, 3472, 835, 718, 663, 651, 578, 553, 515s, 490s, 464, 451, 438, 385, 319s, 309s, 267w, 235.

Source: Nishio-Hamane et al. (2016a).

Comments: The sample was characterized by powder X-ray diffraction data and electron microprobe analysis. The crystal structure is solved. The Raman shifts were determined by us based on spectral curve analysis of the published spectrum.

Bunsenite NiO

Origin: Synthetic.

Experimental details: Raman scattering measurements have been performed on nano-scaled and slightly agglomerated particles using 532 nm laser radiation. The laser radiation power at the sample was 2.48 mW.

Raman shifts (cm^{-1}): 1525s, 1093, 497.
Source: Thema et al. (2016).
Comments: The sample was characterized by powder X-ray diffraction data and TEM/EDX. The strong band at 1525 cm^{-1} is attributed to the double magnon scattering, but this assignment is questionable.

Burangaite $NaFe^{2+}Al_5(PO_4)_4(OH)_6 \cdot 2H_2O$

Origin: Hålsjöberg, Sweden.
Experimental details: Methods of sample preparation are not described. Raman scattering measurements have been performed using 514.5 nm Ar$^+$ laser radiation. The nominal laser radiation power was 150 mW.
Raman shifts (cm^{-1}): 3618s, 3251s, 1141, 1045s, 1023, 988, 618, 601, 587, 400, 369, 354.
Source: Thomas et al. (1998).
Comments: No independent analytical data are provided for the sample used.

Burbankite $(Na,Ca)_3(Sr,Ba,Ce)_3(CO_3)_5$

Origin: Kalkfeld carbonatite complex, Namibia.
Experimental details: Raman scattering measurements have been performed on microscopic particles in fluid inclusions using a He-Ne laser with the laser radiation power of 1.8 mW or an Ar$^+$ laser with the power of 2 mW.
Raman shifts (cm^{-1}): 1078s, ~700w.
Source: Bühn et al. (1999).
Comments: The sample was characterized by synchrotron powder X-ray diffraction data and electron microprobe analysis. For the Raman spectrum of burbankite see also Bühn et al. (2002) and Chakhmouradian et al. (2017).

Burbankite $(Na,Ca)_3(Sr,Ba,Ce)_3(CO_3)_5$

Origin: Bear Lodge carbonatite, Wyoming, USA.
Experimental details: Methods of sample preparation are not described. Micro-Raman scattering measurements have been performed using 532 nm laser radiation. The laser radiation power is not indicated.
Raman shifts (cm^{-1}): 1085s, 1069w, (970) (broad), 881w, 872w, 738sh, 728sh, 717w, 703w, 286w, 233, 209w, 160, 143.
Source: Chakhmouradian et al. (2017).
Comments: No independent analytical data are provided for the sample used. For the Raman spectrum of burbankite see also Bühn et al. (1999, 2002).

Burckhardtite $Pb_2(Fe^{3+}Te^{6+})(AlSi_3O_8)O_6$

Origin: Moctezuma, Sonora, Mexico (type locality).
Experimental details: Raman scattering measurements have been performed on an arbitrarily oriented sample using 514.3 nm cobalt solid-state laser radiation. The laser radiation power at the sample was 0.6 mW.

Raman shifts (cm^{-1}): 897w, 833w, 690s, 661w, 646s, 619s, 554, 477, ~463sh, 505sh, 391, 322, 295, 202.
Source: Christy et al. (2014).
Comments: The sample was characterized by powder X-ray diffraction data and electron microprobe analyses. The crystal structure is solved.

Burgessite $Co_2(H_2O)_4[AsO_3(OH)]_2(H_2O)$

Origin: Keeley mine, South Larrain Township, Timiskaming District, Ontario, Canada (type locality).
Experimental details: Raman scattering measurements have been performed on a powdered sample using a 633 nm He-Ne laser. The Raman shifts have been determined for the maxima of individual peaks obtained as a result of the spectral curve analysis. The laser radiation power is not indicated.
Raman shifts (cm^{-1}): 3591w, 3395, 3328, 3204sh, 3185sh, 852s, 830s, 806s, 740s, 447, 383, 353sh, 322sh, 215, 162.
Source: Čejka et al. (2011a).
Comments: The sample was characterized by powder X-ray diffraction data and electron microprobe analysis.

Burkeite $Na_4(SO_4)(CO_3)$

Origin: Searles Lake, San Bernardino Co., California, USA (type locality).
Experimental details: Raman scattering measurements have been performed on arbitrarily oriented crystals using a 633 nm He-Ne laser. The Raman shifts have been determined for the maxima of individual peaks obtained as a result of the spectral curve analysis. The laser radiation power is not indicated.
Raman shifts (cm^{-1}): 3465w, 3331w, 1244w, 1132, 1102, 1065s, 1008sh, 994s, 981sh, 704, 645, 635sh, 622, 475, 453, 352, 149s, 112.
Source: López et al. (2014d).
Comments: The sample was characterized by powder X-ray diffraction data and electron microprobe analysis. For the Raman spectra of burkeite see also Korsakov et al. (2009) and Jentzsch et al. (2013).

Buseckite $(Fe,Zn,Mn)S$

Origin: Zakłodzie meteorite, Poland (type locality).
Experimental details: Raman scattering measurements have been performed on an arbitrarily oriented sample using 514.5 nm Ar$^+$ laser radiation. The laser radiation power at the sample was 1.2 mW.
Raman shifts (cm^{-1}): 322sh, 296.
Source: Ma et al. (2012a).
Comments: The sample was characterized by powder X-ray diffraction data and electron microprobe analyses.

Bustamite $(Ca,Mg,Fe)_2Si_2O_6$

Origin: Sasa, Macedonia.
Experimental details: Raman scattering measurements have been performed on a powdered sample using 514.5 nm Ar$^+$ laser radiation. The nominal laser radiation power was 50 or 100 mW.

Raman shifts (cm⁻¹): 1034, 1007, 972s, 869s, 839, 812, 735w, 714w, 644s, 574w, 548w, 511, 489, 446w, 428w, 404, 364s, 350sh, 310, 285sh, 260w, 232, 172, 154w, 138w, 125w, 116.
Source: Makreski et al. (2006b).
Comments: The Raman shifts are indicated for the maxima of individual peaks obtained as a result of the spectral curve analysis. The sample was characterized by powder X-ray diffraction data and neutron activation analysis.

Butlerite $Fe^{3+}(SO_4)(OH) \cdot 2H_2O$

Origin: Alcaparrosa mine, Cerritos Bayos, Calama, El Loa province, Antofagasta, Chile.
Experimental details: Raman scattering measurements have been performed on arbitrarily oriented crystals using a 633 nm He-Ne laser. The Raman shifts have been determined for the maxima of individual peaks obtained as a result of the spectral curve analysis. The laser radiation power is not indicated.
Raman shifts (cm⁻¹): 3469sh, 3310w, 3155sh, 3012sh, 1225sh, 1198, 1145sh, 1109s, 1088sh, 1024s, 617sh, 600, 543, 469, 450sh, 408, 374sh, 294sh, 247, 221, 181, 154sh.
Source: Čejka et al. (2011b).
Comments: The sample was characterized by powder X-ray diffraction data and electron microprobe analysis.

Buttgenbachite $Cu_{36}(NO_3)_2Cl_8(OH)_{62} \cdot nH_2O$

Origin: No data.
Experimental details: Raman scattering measurements have been performed on an arbitrarily oriented sample using 632.8 nm He-Ne or 514.5 nm Ar⁺ laser radiation. The nominal laser radiation power was ≤ 30 mW.
Raman shifts (cm⁻¹): 1054, 1041, 985, 843, 622, 595, 489, 451, 409s, 349, 314, 259, 236, 193, 184, 164, 145s, 128s.
Source: Bouchard and Smith (2003).
Comments: The sample was characterized by powder X-ray diffraction data.

Byströmite $MgSb^{5+}_2O_6$

Origin: Synthetic.
Experimental details: No data.
Raman shifts (cm⁻¹): 810w, 749s, 676, 625, 592w, 560w, 570s, 480w, 357, 331, 300, 248w, 231w.
Source: Husson et al. (1979).
Comments: The sample was characterized by powder X-ray diffraction data.

Cabalzarite $CaMg_2(AsO_4)_2 \cdot 2H_2O$

Origin: No data.
Experimental details: Methods of sample preparation are not described. Raman scattering measurements have been performed using 532 nm laser radiation. The laser radiation power is not indicated.
Raman shifts (cm⁻¹): ~800, ~750s, ~700sh, ~445.
Source: Carey et al. (2015).

Cabvinite $Th_2F_7(OH)\cdot 3H_2O$

Origin: Su Seinargiu, Sarroch, Cagliari, Sardinia, Italy (type locality).
Experimental details: Raman scattering measurements have been performed on an arbitrarily oriented sample using 532 nm laser radiation. The laser radiation power is not indicated.
Raman shifts (cm^{-1}): (3407), 3257s, 461, 342, 209s, 113s.
Source: Orlandi et al. (2017).
Comments: The Raman shifts have been partly determined for the maxima of individual peaks obtained as a result of the spectral curve analysis. The sample was characterized by powder X-ray diffraction data and electron microprobe analysis. The crystal structure is solved.

Cacoxenite $Fe^{3+}_{24}AlO_6(PO_4)_{17}(OH)_{12}\cdot 75H_2O$

Origin: No data.
Experimental details: Raman scattering measurements have been performed on arbitrarily oriented crystals using a 633 nm He-Ne laser. The laser radiation power is not indicated. The Raman shifts have been determined for the maxima of individual peaks obtained as a result of the spectral curve analysis.
Raman shifts (cm^{-1}): 3599, 3504, 3429, 3251s, 3085, 2947, 2667, 2309, 1231w, 1213w, 1153, 1118, 1081, 1041, 1026, 979, 961, 926w, 619, 606w, 573, 526w, 495w, 433w, 411w, 371w, 307w, 273, 262, 245, 217, 199w, 169w, 155.
Source: Frost et al. (2003c).
Comments: No independent analytical data are given for the sample used.

Cadmoindite $CdIn_2S_4$

Origin: Synthetic.
Experimental details: Raman scattering measurements have been performed on an arbitrarily oriented crystal using 647.1 nm Kr$^+$ laser radiation. The nominal laser radiation power was <5 mW. A 180°-scattering geometry was employed.
Raman shifts (cm^{-1}): 367s, 360, 315, 301w, 249, 232, 207, 188, 93, 70.
Source: Ursaki et al. (2002).
Comments: No independent analytical data are given for the sample used. For the Raman spectra of cadmoindite see also Unger et al. (1978), Kulikova et al. (1988), and Syrbu et al. (1996a, b).

Cadmoselite $CdSe$

Origin: Synthetic.
Experimental details: Raman scattering measurements have been performed on a powdered sample using 514.5 nm Ar$^+$ laser radiation. The laser radiation power is not indicated. A 180°-scattering geometry was employed.
Raman shifts (cm^{-1}): 201 for CdSe particles with 2.26 nm in diameter and 205 for CdSe particles with 3.52 nm in diameter.
Source: Nien et al. (2008).
Comments: The samples were characterized by TEM. Bulk CdSe exhibits a Raman peak at 209 cm^{-1} (Widulle et al. 1999).

Cafarsite $Ca_{5.9}Mn_{1.7}Fe_3Ti_3(AsO_3)_{12} \cdot 4\text{-}5H_2O$

Origin: Cervandone Mt., Val Devero, Piedmont, Italy.
Experimental details: Raman scattering measurements have been performed on arbitrarily oriented crystals using a 633 nm He-Ne laser. The Raman shifts have been determined for the maxima of individual peaks obtained as a result of the spectral curve analysis. The laser radiation power at the sample was 1 mW.
Raman shifts (cm^{-1}): 869, 757, 725, 328, 286, 196s.
Source: Frost and Bahfenne (2010d).
Comments: No independent analytical data are given for the sample used. For the Raman spectra of cafarsite see also Kloprogge and Frost (1999b) and Bahfenne (2011).

Cafetite $CaTi_2O_5 \cdot H_2O$

Origin: Khibiny massif, Kola Peninsula, Russia.
Experimental details: Methods of sample preparation are not indicated. Raman scattering measurements have been using 532 nm solid-state laser radiation. The laser radiation output power was 50 mW.
Raman shifts (cm^{-1}): 825, 798, 732w, 602, 482, 449, 419s, 358w, 329, 302, 292, 251s, 203, 190, 177, 152, 126, 110.
Source: Martins et al. (2014).
Comments: The sample was characterized by powder X-ray diffraction data and electron microprobe analyses.

Cahnite $Ca_2B(AsO_4)(OH)_4$

Origin: No data.
Experimental details: Raman scattering measurements have been performed on a powdered sample using 632.8 nm He-Ne laser radiation. The laser radiation power is not indicated.
Raman shifts (cm^{-1}): 844, 791, 759, 548, 538, 448, 428, 378, 395, 324, 290.
Source: Ross (1972).
Comments: No independent analytical data are given for the sample used.

Cairncrossite $Sr_2Ca_7(Si_4O_{10})_4(OH)_2 \cdot 15H_2O$

Origin: Wesselsmine, Kalahari Manganese Field, South Africa (type locality).
Experimental details: Methods of sample preparation are not indicated. Raman scattering measurements have been performed using 488 nm Ar^+ laser radiation. The nominal laser radiation power was 8 mW.
Raman shifts (cm^{-1}): 3670, 3650, 3550s, 1145, 1060, 1030w, 1000sh, 777, 700, 610s, 456, 438, 346w, 280, 183w, 130.
Source: Giester et al. (2016).
Comments: The sample was characterized by powder X-ray diffraction data and electron microprobe analyses. The crystal structure is solved. The Raman shifts were partly determined by us based on spectral curve analysis of the published spectrum.

Calaverite $AuTe_2$

Origin: Synthetic.

Experimental details: Raman scattering measurements have been performed at 50 K on an oriented crystal using 514.5 nm Ar^+ laser radiation. The nominal laser radiation power was <10 mW. Polarized spectra were collected from the (20-12) face, which is parallel to the b-axis and makes an angle of 7°05′ with the a-axis. A 180°-scattering geometry with laser beam polarizations (xx) and (yy) for A_g, and (xy) for B_g was employed.

Raman shifts (cm^{-1}): $A_g(xx)$: 172, 162, 152, 143w, 128, 119, 101s, 92, 88, 73w, 57w, 47, 42w; $A_g(yy)$: 163, 155w, 151, 144w, 134, 127, 119, 109w, 106w, 101w, 97, 72, 57, 47; $B_g(xy)$: 162, 154s, 142s, 133, 125s, 118, 108s, 101, 91, 61, 54, 48.

Source: van Loosdrecht et al. (1992).

Comments: The sample was characterized by powder X-ray diffraction data.

Calciborite CaB_2O_4

Origin: Synthetic.

Experimental details: Methods of sample preparation are not described. Raman scattering measurements have been performed using 514.5 nm Ar^+ laser radiation. The laser radiation power is not indicated.

Raman shifts (cm^{-1}): 1632, 1525s, ~1450, 1428, 1295, 1231, 1172, 1080, 1002, 811, 788, 738s, 683, 653, 549, 504, 389, 328, 230, 209, 179, 171.

Source: Rulmont and Almou (1989).

Comments: Raman frequencies are given for a sample with the isotopic composition $^{40}Ca^{11}B_2O_4$. The sample was characterized by powder X-ray diffraction data.

Calcio-olivine $Ca_2(SiO_4)$

Origin: Synthetic.

Experimental details: Raman scattering measurements have been performed on a powdered sample using 514.5 nm Ar^+ laser radiation. The laser radiation output power was 0.4 mW.

Raman shifts (cm^{-1}): 924, 885, 857, 838s, 813s, 570w, 558w, 525w, 410, 400, 306, 269, 261, 251, 242, 193, 183, 177, 151, 134w, 125w, 118w.

Source: Remy et al. (1997).

Comments: The sample was characterized by powder X-ray diffraction data. For the Raman spectrum of calico-olivine see also Piriou and McMillan (1983).

Calcioaravaipaite $PbCa_2AlF_9$

Origin: Grand Reef mine, Arizona, USA (type locality).

Experimental details: Raman scattering measurements have been performed on an arbitrarily oriented single crystal using 532 nm solid-state laser radiation. The laser radiation power is not indicated.

Raman shifts (cm^{-1}): 3580s, 3296, 2925w, 2425, 2344, 2203, 2073, 1939w, 1822, 1441w, ~652, 560s, 538s, 417w, 390w, 366w, 323, 277sh, 263, 234, 203, 191, 177, 167.

Source: Kampf et al. (2011c).

Comments: The sample identification was done by single-crystal X-ray diffraction data. The crystal structure is solved. The Raman shifts were determined by us based on spectral curve analysis of the published spectrum.

Calciolangbeinite $K_2Ca_2(SO_4)_3$

Origin: Artificial (component of clinker).
Experimental details: Raman scattering measurements have been performed on an arbitrarily oriented sample using 514 nm laser radiation. The nominal laser radiation power was 5 mW.
Raman shifts (cm^{-1}): 1203, 1147, 1107, 993, 630.
Source: Black and Brooker (2007).
Comments: Identification of this phase is tentative and, probably, erroneous. For the Raman spectrum of calciolangbeinite see also Gastaldi et al. (2008).

Calciolangbeinite $K_2Ca_2(SO_4)_3$

Origin: Synthetic.
Experimental details: Methods of sample preparation are not described. Micro-Raman scattering measurements have been performed using 632.8 nm laser radiation. The nominal laser radiation power was 20 mW.
Raman shifts (cm^{-1}): 1144w, 1118w, 1025s, 1019s, 1006s, 645, 618, 466, 454.
Source: Gastaldi et al. (2008).
Comments: The sample was characterized by powder X-ray diffraction data. For the Raman spectrum of calciolangbeinite see also Black and Brooker (2007).

Calciopetersite $CaCu_6(PO_4)_2(PO_3OH)(OH)_6 \cdot 3H_2O$

Origin: Domaš, near Olomouc, northern Moravia, Czech Republic (type locality).
Experimental details: Raman scattering measurements have been performed on arbitrarily oriented crystals using a 633 nm He-Ne laser. The Raman shifts have been determined for the maxima of individual peaks obtained as a result of the spectral curve analysis. The laser radiation power is not indicated.
Raman shifts (cm^{-1}): 3494, 3351w, 3301w, 3243w, 2931w, 2882sh, 1606, 1457sh, 1110, 1079, 1042, 947, 873, 577s, 475s, 394, 341, 208, 174, 144, 118.
Source: Čejka et al. (2011c).
Comments: The sample was characterized by powder X-ray diffraction data and electron microprobe analyses.

Calcite $Ca(CO_3)$

Origin: No data.
Experimental details: Methods of sample preparation are not described. Raman scattering measurements have been performed using 1064 nm Nd-YAG laser radiation. The nominal laser radiation power was 100 mW.
Raman shifts (cm^{-1}): 2906w, 2835w, 2707w, 1903w, 1749w, 1436w, 1086s, 713w, 283, 156.
Source: Edwards et al. (2005).

Comments: No independent analytical data are provided for the sample used. For the Raman spectra of calcite see also Rutt and Nicola (1974), Behrens et al. (1995), Buzgar and Apopei (2009), Wehrmeister et al. (2010), Ciobotă et al. (2012), Frezzotti et al. (2012), Schmid and Dariz (2015), Sánchez-Pastor et al. (2016), and Perrin et al. (2016).

Calcurmolite $(Ca_{1-x}Na_x)_2(UO_2)_3(MoO_4)_2(OH)_{6-x} \cdot nH_2O$

Origin: Sokh-Karasu area, Kadzharan Mo Deposit, Kafan District, Armenia (type locality).
Experimental details: Raman scattering measurements have been performed on a powdered sample using a 633 nm He-Ne laser. The Raman shifts have been determined for the maxima of individual peaks obtained as a result of the spectral curve analysis. The laser radiation power is not indicated.
Raman shifts (cm^{-1}): 930, 900sh, 868, 823, 794s, 700, 644, 495, 378, 354sh, 271, 206sh, 144w.
Source: Frost et al. (2008c).
Comments: No independent analytical data are provided for the sample used. Strong bands in the range from 970 to 1150 cm^{-1} in the IR spectrum of calcurmolite given in the cited paper are mainly due to an impurity. The assignment of these bands to MOH bending modes is erroneous. The IR band at 3694 cm^{-1} may correspond to the admixture of a clay mineral.

Calderite $Mn^{2+}_3Fe^{3+}_2(SiO_4)_3$

Experimental details: Only a calculated Raman spectrum of calderite is given in the cited paper.
Raman shifts (cm^{-1}): 1017, 897, 887, 884, 871, 840, 597, 580, 552, 493, 486, 450, 373, 355, 349, 344, 300, 293, 284, 217, 211, 186, 169, 158.
Source: Arlt et al. (1998).

Calderónite $Pb_2Fe^{3+}(VO_4)_2(OH)$

Origin: Karrantza Valley, western area of the Basque Co., Spain.
Experimental details: Raman scattering measurements have been performed on an arbitrarily oriented sample using 514.5 nm Ar$^+$ laser (with the laser radiation power at the sample of 20 mW) and 785 nm diode laser radiation (with the radiation output power of 150 mW).
Raman shifts (cm^{-1}): 977s, 684w, 336, 210w, 159w.
Source: Goienaga et al. (2011).
Comments: No independent analytical data are given for the sample used.

Caledonite $Cu_2Pb_5(SO_4)_3(CO_3)(OH)_6$

Origin: Hard Luck Claim, near Baker, San Bernardino Co., California, USA.
Experimental details: Raman scattering measurements have been performed on arbitrarily oriented crystals using a 633 nm He-Ne laser. The Raman shifts have been determined for the maxima of individual peaks obtained as a result of the spectral curve analysis. The laser radiation power at the sample was 1 mW.
Raman shifts (cm^{-1}): 3439, 3417, 3379, 1674, 1392, 1358, 1124, 1109, 1083, 1053, 977s, 950, 848, 825, 791, 722, 628, 605, 475, 456, 427, 344, 316, 278, 251, 229, 152.
Source: Frost et al. (2003e).
Comments: No independent analytical data are provided for the sample used.

Callaghanite $Cu_2Mg_2(CO_3)(OH)_6 \cdot 2H_2O$

Origin: Gabbs occurrence, Nye Co., Nevada, USA.
Experimental details: Raman scattering measurements have been performed on arbitrarily oriented crystals using a 633 nm He-Ne laser. The Raman shifts have been determined for the maxima of individual peaks obtained as a result of the spectral curve analysis. The laser radiation power at the sample was 1 mW.
Raman shifts (cm^{-1}): 3620, 3575, 3564s, 3511, 3502, 3375, 3350, 3040, 2906, 1398s, 1087s, 1013, 961, 944, 871, 840, 749, 707, 688, 517, 499s, 481, 459, 445, 395, 380, 350, 336, 283, 277, 252, 218, 211, 195, 160, 147, 141, 127, 121, 100.
Source: Čejka et al. (2013).
Comments: The sample was characterized by powder X-ray diffraction data and electron microprobe analyses.

Calomel HgCl

Origin: Synthetic.
Experimental details: Raman scattering measurements have been performed on a single crystal in the (zz) polarization using 514.5 and 632.8 nm He-Ne laser radiations. The laser radiation power is not indicated.
Raman shifts (cm^{-1}): 272, 164s, 36s.
Source: Radepont (2013).
Comments: For the Raman spectrum of calomel see also Markov and Roginskii (2011).

Calumetite $Cu(OH)_2 \cdot 2H_2O$

Origin: No data.
Experimental details: Methods of sample preparation are not indicated. Raman scattering measurements have been performed using 632.8 nm He-Ne laser radiation. The output laser radiation power was 30 mW.
Raman shifts (cm^{-1}): 3550, 3450, 1054, 1041s, 985, 843, 622, 595, 489, 451, 409, 347, 314, 259, 236, 193, 184, 164, 145, 139, 128.
Source: Bouchard and Smith (2003).
Comments: The sample was characterized by powder X-ray diffraction data. For the Raman spectrum of calumetite see also Bouchard-Abouchacra (2001).

Camaronesite $Fe^{3+}_2(PO_3OH)_2(SO_4)(H_2O)_4 \cdot 1\text{-}2H_2O$

Origin: Camarones valley, Arica province, Chile (type locality).
Experimental details: Raman scattering measurements have been performed on an arbitrarily oriented sample using 514.5 nm Ar^+ laser radiation. The laser radiation power at the sample was 5 mW.
Raman shifts (cm^{-1}): 3463, 3363, 3140, ~1610, 1080s, 1014s, 937, 526, 305, 254, 227.
Source: Kampf et al. (2013d).
Comments: The sample was characterized by powder X-ray diffraction data and electron microprobe analysis. The crystal structure is solved.

Camerolaite $Cu_6Al_3(OH)_{18}(H_2O)_2[Sb(OH)_6](SO_4)$

Origin: Cap Garonne, France (type locality).

Experimental details: Raman scattering measurements have been performed on an oriented single crystal with the laser beam orthogonal to the elongation direction of the crystal (*b* axis) using 532 nm Nd-YAG laser radiation. The nominal laser radiation power was from 5 to 30 mW.

Raman shifts (cm^{-1}): 3596, 3558, 3495, 3200–3120, 2330w, 1064, 1050w, 968, 614s, 526s, 447, 347w, 325, 272, 237.

Source: Mills et al. (2014a).

Comments: The Raman shifts are given for the holotype sample no. 477.067. The sample was characterized by powder X-ray diffraction data.

Canavesite $Mg_2(HBO_3)(CO_3)\cdot 5H_2O$

Origin: No data.

Experimental details: Methods of sample preparation are not described. Micro-Raman scattering measurements have been performed using 514.5 nm Ar-Kr laser radiation. The laser radiation power was from 2 to 5 mW.

Raman shifts (cm^{-1}): 3657s, 3484sh, 3392, 3293, 2910, 1458w, 1284, 1105s, 982, 768s, 635w, 548, 174s.

Source: Grice et al. (1986).

Comments: No independent analytical data are provided for the sample used. The Raman shifts were determined by us based on spectral curve analysis of the published spectrum.

Cancrinite $(Na,Ca,\square)_8(Al_6Si_6)O_{24}(CO_3,SO_4)_2\cdot 2H_2O$

Origin: Cava Satom, Cameroon.

Experimental details: Raman scattering measurements have been performed on a single crystal using 532 nm Nd-YAG laser radiation. The laser radiation power at the sample was 5 mW.

Raman shifts (cm^{-1}): 3647w, 3536w, 1057s, 1042, 1002, 981, 976, 960, 937, 816, 768, 685, 631, 499, 469, 460, 440, 418, 401, 364, 350, 338, 293s, 277, 231, 161, 115, 108.

Source: Gatta et al. (2012a).

Comments: The sample was characterized by single-crystal X-ray diffraction data and electron microprobe analyses. The crystal structure is solved. Raman shifts are given for the crystal in two different orientations (crystal rotated by 90°). For the Raman spectra of cancrinite see also Mozgawa (2001) and Lotti (2014).

Cancrinite SO$_4$-rich $(Na,Ca,\square)_8(Al_6Si_6)O_{24}(CO_3,SO_4)_2\cdot 2H_2O$

Origin: Cinder Lake, Manitoba, Canada.

Experimental details: Raman scattering measurements have been performed on an arbitrarily oriented sample using 633 nm He-Ne laser radiation. The laser radiation power is not indicated.

Raman shifts (cm^{-1}): 3591w, 3544, 1060s, 992s, 774w, 720w, 687w, 633w, 515, 465, 441s, 340, 298, 275s, 230, 198, 89s.

Source: Martins et al. (2016).

Comments: The sample was characterized by X-ray microdiffraction data and electron microprobe analysis.

Canfieldite Ag_8SnS_6

Origin: Synthetic.
Experimental details: Raman scattering measurements have been performed on an arbitrarily oriented sample using 514 nm YAG laser radiation. The laser radiation power is not indicated.
Raman shifts (cm^{-1}): 310, 221w, 76s.
Source: Cheng et al. (2016).
Comments: The sample was characterized by powder X-ray diffraction data and Hall measurements.

Cannonite $Bi_2O(SO_4)(OH)_2$

Origin: Alfenza, Crodo, Italy.
Experimental details: Raman scattering measurements have been performed on an oriented crystal with the polarization of the incident laser beam parallel to Y using 514.5 nm Ar$^+$ laser radiation. The laser radiation power is not indicated.
Raman shifts (cm^{-1}): 3439, 3376, 3190w 1114, 1059, 984s, 621w, 605w, 562, 467, 452s, 438s, 400w, 337, 318s, 279, 222s, 189, 147s, 121s, 101s, 80w, 62.
Source: Capitani et al. (2013).
Comments: The Raman shifts are indicated for the maxima of individual peaks obtained as a result of the spectral curve analysis. The sample was characterized by powder X-ray diffraction data and electron microprobe analysis. For the Raman spectra of cannonite see also Gama (2000) and Capitani et al. (2014).

Carbocernaite $(Sr,Ce,La)(Ca,Na)(CO_3)_2$

Origin: Bear Lodge carbonatite, Wyoming, USA.
Experimental details: Raman scattering measurements have been performed on an arbitrarily oriented crystal using 532 nm solid-state laser radiation. The laser radiation power is not indicated.
Raman shifts (cm^{-1}): Ca-Sr-rich variety: 1099s, 1077s, 979s (broad), 864w, 744w, 728sh, 716, 693, 269s, 215, 181, 127. Na-*REE*-rich variety: 1104s, 1097s, 1079s, 983 (broad) 874w, 860w, 749, 732sh, 724, 716sh, 704w, 677w, 257s, 188, 122.
Source: Chakhmouradian et al. (2017).
Comments: The samples were characterized by single-crystal X-ray diffraction data and electron microprobe analyses. The broad bands at ~980 cm^{-1} may be due to fluorescence.

Carbonatecyanotrichite $Cu_4Al_2(CO_3)(OH)_{12} \cdot 2H_2O$

Origin: Grandview mine, Coconino Co., Arizona, USA.
Experimental details: Raman scattering measurements have been performed on an oriented single crystal with the laser beam orthogonal to the elongation direction of the crystal (*b* axis) using 532 nm Nd-YAG laser radiation. The nominal laser radiation power was from 5 to 30 mW.
Raman shifts (cm^{-1}): 3657, 3583, 3400–3300, 2329, 1141w, 977s, 591, 524s, 441, 273w, 233w.
Source: Mills et al. (2014a).
Comments: The sample was characterized by powder X-ray diffraction data.

Carletonite $KNa_4Ca_4Si_8O_{18}(CO_3)_4(F,OH) \cdot H_2O$

Origin: Poudrette Quarry, Saint Hilaire Mt., Quebec, Canada (type locality).
Experimental details: Raman scattering measurements have been performed on arbitrarily oriented crystals using a 633 nm He-Ne laser. The Raman shifts have been determined for the maxima of individual peaks obtained as a result of the spectral curve analysis. The laser radiation power is not indicated.
Raman shifts (cm^{-1}): 3595, 3584, 3572, 3570, 3235w, 2905, 2630, 1753w, 1732w, 1662w, 1617w, 1548w, 1481, 1426, 1217, 1086s, 1075s, 1066s, 840, 782, 756w, 735w, 726, 706, 698, 685, 663w, 547, 513, 495, 430, 401, 356, 342, 325w, 316, 289, 235, 217, 194, 174, 157, 142.
Source: Frost et al. (2013ai).
Comments: No independent analytical data are provided for the sample used.

Carlfrancisite $Mn^{2+}_3(Mn^{2+},Mg,Fe^{3+},Al)_{42}(As^{3+}O_3)_2(As^{5+}O_4)_4[(Si,As^{5+})O_4]_6[(As^{5+},Si)O_4]_2(OH)_{42}$

Origin: Kombat mine, Otavi valley, Namibia (type locality).
Experimental details: Methods of sample preparation are not described. Raman scattering measurements have been performed using 532 nm laser radiation. The laser radiation power at the sample was in the range 5–12.5 mW. A 180°-scattering geometry was employed.
Raman shifts (cm^{-1}): 3660w, 3600, 3532. 3463, 1600, 897, 836sh, 819sh, 791s, 732, 616, 514, 400.
Source: Hawthorne et al. (2013).
Comments: The sample was characterized by powder and single-crystal X-ray diffraction data, and electron microprobe analyses. The Raman shifts were partly determined by us based on spectral curve analysis of the published spectrum.

Carlfriesite $CaTe^{6+}(Te^{4+})_2O_8$

Origin: Moctezuma mine, New Mexico, USA.
Experimental details: Raman scattering measurements have been performed on arbitrarily oriented crystals using a 633 nm He-Ne laser. The Raman shifts have been determined for the maxima of individual peaks obtained as a result of the spectral curve analysis. The laser radiation power is not indicated.
Raman shifts (cm^{-1}): The IR and Raman spectra of presumed carlfriesite presented in the cited paper are wrong. Actually, spectra of calcite are given. The bands of calcite are erroneously assigned to Te–O-stretching vibrations.
Source: Frost et al. (2009g).
Comments: No independent analytical data are provided for the sample used.

Carlinite Tl_2S

Origin: Synthetic.
Experimental details: Methods of sample preparation are not described. Raman scattering measurements have been performed using 532 nm Nd-YAG laser radiation. The laser radiation power at the sample was 2.5 mW.
Raman shifts (cm^{-1}): ~280, ~160 (at room temperature); 280, 192, 171s, 143 (at 12 K).
Source: Chia et al. (2015).
Comments: The sample was characterized by powder X-ray diffraction data and electron microprobe analysis.

Carlosturanite $(Mg,Fe^{2+},Ti)_{21}(Si,Al)_{12}O_{28}(OH)_{34} \cdot H_2O$

Origin: Val Varaita, Piedmont, northern Italy.
Experimental details: Raman scattering measurements have been performed on an arbitrarily oriented sample using 632.8 nm He-Ne laser radiation. The nominal laser radiation power was 20 mW.
Raman shifts (cm^{-1}): 788sh, 776s, 765sh, 706, 692, 671, 451, 363, 329.
Source: Belluso et al. (2007).
Comments: The sample was characterized by powder X-ray diffraction data and electron microprobe analyses.

Carlsbergite CrN

Origin: Synthetic.
Experimental details: Methods of sample preparation are not described. Raman scattering measurements have been performed on a powdered single-layer coating on silicon substrate using 633 nm He-Ne laser radiation. The nominal laser radiation power was 20 mW.
Raman shifts (cm^{-1}): 619w, 238w.
Source: Barshilia and Rajam (2004).
Comments: The sample was characterized by powder X-ray diffraction data.

Carminite $PbFe^{3+}_2(AsO_4)_2(OH)_2$

Origin: No data.
Experimental details: Raman scattering measurements have been performed on arbitrarily oriented crystals using a 633 nm He-Ne laser. The Raman shifts have been determined for the maxima of individual peaks obtained as a result of the spectral curve analysis. The laser radiation power was <1 mW.
Raman shifts (cm^{-1}): 3254, 3217, 849, 835, 822, 738, 543, 497, 467, 350, 324, 259, 210.
Source: Frost and Kloprogge (2003).
Comments: No independent analytical data are provided for the sample used.

Carnallite $KMgCl_3 \cdot 6H_2O$

Origin: A dolerite sill in eastern Siberia, Russia.
Experimental details: Micro-Raman scattering measurements have been performed on an arbitrarily oriented crystal using 514.5 nm Ar$^+$ laser radiation. The laser radiation power is not indicated.
Raman shifts (cm^{-1}): 3472sh, 3444s, 3425s, 3401, 1641s, 401, 321w, 217, 202, 126, 97, 71, 59.
Source: Grishina et al. (1992).
Comments: No independent analytical data are provided for the sample used. For the Raman spectrum of carnallite see also Weber et al. (2012).

Carnegieite $NaAlSiO_4$

Origin: Synthetic.
Experimental details: Raman scattering measurements have been performed on a powdered sample using 488 nm Ar$^+$ laser radiation. The output laser radiation power was 600 mW. A 90°-scattering geometry was employed.

Raman shifts (cm^{-1}): 1072, 982, 964, 949, 803w, 721w, 685, 637w, 487, 444, 433, 404, 379s, 347sh, 340sh, 313sh, 262w, 217, 154, 114.

Source: Matson et al. (1986).

Comments: The sample was characterized by powder X-ray diffraction data.

Carnotite $K_2(UO_2)_2(VO_4)_2 \cdot 3H_2O$

Origin: Synthetic.

Experimental details: Raman scattering measurements have been performed on an arbitrarily oriented sample using 6471 Å Kr laser radiation. The laser radiation power is not indicated.

Raman shifts (cm^{-1}): 975, 825w, 737s, (645w), 585w, 540w, (475w), 410w, 380, 360w, (310w), 275w, 250w, 230w.

Source: Baran and Botto (1976).

Comments: The sample was characterized by powder X-ray diffraction data. For the Raman spectra of carnotite see also Biwer et al. (1990) and Frost et al. (2005c).

Carpathite $C_{24}H_{12}$

Origin: Picacho Peak Area, San Benito Co., California, USA.

Experimental details: Methods of sample preparation are not described. Raman scattering measurements have been performed using 514.5 nm Ar$^+$ laser radiation. The nominal laser radiation power was 100 mW.

Raman shifts (cm^{-1}): 1627, 1615, 1594, 1449, 1437, 1393w, 1366s, 1350s, 1337, 1289w, 1220, 1044w, 1026, 994w, 949w, 660w, 483s, 450, 369, 304w, 238w.

Source: Echigo et al. (2007).

Comments: The sample was characterized by powder X-ray diffraction data and electron microprobe analysis. The Raman shifts were determined by us based on spectral curve analysis of the published spectrum. For the Raman spectrum of carpathite see also Zhao et al. (2013b).

Carpholite $Mn^{2+}Al_2Si_2O_6(OH)_4$

Origin: Vrpsko, Macedonia.

Experimental details: Raman scattering measurements have been performed on a powdered sample using 514.5 nm Ar$^+$ laser radiation. The nominal laser radiation power at the sample was from 50 to 100 mW.

Raman shifts (cm^{-1}): 1087, 1029, 987sh, 958, 919s, 883, 828, 777, 735s, 709, 678, 658, 634, 609, 583sh, 559, 503, 469, 441, 404, 371, 345s, 315, 292, 281, 260, 238, 209, 162sh.

Source: Makreski et al. (2006b).

Comments: The sample was characterized by powder X-ray diffraction data and neutron activation analysis. For the Raman spectrum of carpholite see also Jovanovski et al. (2009).

Carrboydite $(Ni_{1-x}Al_x)(SO_4)_{x/2}(OH)_2 \cdot nH_2O$ ($x < 0.5$, $n > 3x/2$)

Origin: Widgiemooltha, Western Australia.

Experimental details: Raman scattering measurements have been performed on arbitrarily oriented crystals using a 633 nm He-Ne laser. The Raman shifts have been determined for the maxima of individual peaks obtained as a result of the spectral curve analysis. The laser radiation power at the sample was <1 mW.

Raman shifts (cm^{-1}): 3614, 3445 (broad), 1125, 981s, 631, 613, 563, 552, 499, 457, 403, 318, 248, 227, 205.
Source: Frost et al. (2003h).
Comments: No independent analytical data are provided for the sample used. For the Raman spectrum of carrboydite see also Lin et al. (2006).

Carrollite $CuCo_2S_4$

Origin: Synthetic.
Experimental details: Methods of sample preparation are not described. Raman scattering measurements have been performed using 532 nm laser radiation. The laser radiation power is not indicated.
Raman shifts (cm^{-1}): 660s, 509, 474.
Source: Nie et al. (2016).
Comments: The sample was characterized by powder X-ray diffraction data and electron microprobe analysis.

Caryopilite $Mn^{2+}_3Si_2O_5(OH)_4$

Origin: Santa Cruz Formation, Brazil.
Experimental details: Methods of sample preparation are not described. Raman scattering measurements have been performed using 514 nm laser radiation. The laser radiation power is not indicated.
Raman shifts (cm^{-1}): No data: only a figure of the Raman spectrum is given in the cited book.
Source: Johnson (2015).
Comments: The sample was characterized by electron microprobe analyses.

Cassiterite SnO_2

Origin: No data.
Experimental details: Methods of sample preparation are not described. Raman scattering measurements have been performed using 633 nm He-Ne laser radiation. The nominal laser radiation power was 30 mW.
Raman shifts (cm^{-1}): 842, 776, 635s, 475.
Source: Bouchard and Smith (2003).
Comments: The sample was characterized by powder X-ray diffraction data. For the Raman spectra of cassiterite see also Andò and Garzanti (2014) and Evrard et al. (2015).

Castellaroite $Mn^{2+}_3(AsO_4)_2 \cdot 4H_2O$

Origin: Monte Nero mine, Rocchetta Vara, La Spezia, Liguria, Italy (type locality).
Experimental details: Raman scattering measurements have been performed on an arbitrarily oriented crystal using 532 nm Nd-YAG laser radiation. The laser radiation output power was 500 mW. A 180°-scattering geometry was employed.
Raman shifts (cm^{-1}): 3942w, 3758w, 3491w, 3241w, 3116w, 2925w, 1663w, 934w, 911sh, 863s, 847sh, 822s, 801s, 578w, 509w, 459sh, 426, 372, 341sh, 275w, 215w, 182w, 150, 143sh, 109.
Source: Kampf et al. (2016a).

Comments: The sample was characterized by powder X-ray diffraction data and electron microprobe analyses. The crystal structure is solved.

Caswellsilverite $NaCrS_2$

Origin: Synthetic.
Experimental details: Raman scattering measurements have been performed on an oriented single crystal using 514.5 nm Ar^+ laser radiation. The nominal laser radiation power was 50 mW. Polarized spectra were collected in the $y(xy)z$ and $y(xx)z$ scattering geometries.
Raman shifts (cm^{-1}): 317, 252.
Source: Unger et al. (1979).
Comments: No independent analytical data are provided for the sample used. In the $y(xy)z$ scattering geometry only a band at 252 cm^{-1} is observed.

Catalanoite $Na_2(HPO_4) \cdot 8H_2O$

Origin: Synthetic.
Experimental details: Raman scattering measurements have been performed at 40 °C on an arbitrarily oriented sample using 514.5 nm Ar^+ laser radiation. The nominal laser radiation power was 30 mW.
Raman shifts (cm^{-1}): 3412s, 3238w, 1089, 987s, 869s, 518, 407w, 388.
Source: Ghule et al. (2003).
Comments: The sample was characterized by thermogravimetric data.

Catapleiite $Na_2Zr(Si_3O_9) \cdot 2H_2O$

Origin: No data.
Experimental details: No data.
Raman shifts (cm^{-1}): 1005s, 928, 622, 563.
Source: Gaft et al. (2015).
Comments: The sample was characterized by powder X-ray diffraction data and electron microprobe analysis, but no independent analytical data are provided for the sample used.

Cattierite CoS_2

Origin: Synthetic.
Experimental details: Raman scattering measurements have been performed on a CoS_2 film using 632.8 nm laser radiation. The laser radiation power is not indicated.
Raman shifts (cm^{-1}): 392s, 316w, 290.
Source: Kinner et al. (2016).
Comments: The sample was characterized by powder X-ray diffraction data.

Cavansite $Ca(V^{4+}O)(Si_4O_{10}) \cdot 4H_2O$

Origin: Wagholi Quarry, Maharashtra, India.
Experimental details: Raman scattering measurements have been performed on arbitrarily oriented crystals using a 633 nm He-Ne laser. The Raman shifts have been determined for the maxima of individual peaks obtained as a result of the spectral curve analysis. The laser radiation power is not indicated.

Raman shifts (cm^{-1}): 3654, 3604, 3577w, 3546, 3504, 3429sh, 1109w, 1088w, 1072w, 1043w, 981s, 973sh, 954sh, 935w, 842w, 823w, 713w, 672, 587sh, 574, 542, 477, 437, 388w, 350, 307sh, 291, 251, 230, 194, 131, 113.
Source: Frost and Xi (2012h).
Comments: No independent analytical data are provided for the sample used. For the Raman spectrum of cavansite see also Prasad and Prasad (2007).

Cebaite (Ce) $Ba_3Ce_2(CO_3)_5F_2$

Origin: No data.
Experimental details: Raman scattering measurements have been performed using 488 nm laser radiation. The nominal laser radiation power was 300 mW.
Source: Hong et al. (1999).
Raman shifts (cm^{-1}): 1088, 911, 1516, 718, 625.

Čejkaite $Na_4(UO_2)(CO_3)_3$

Origin: Svornost mine, Jáchymov, Krušné Hory (Ore Mts.), Czech Republic (type locality).
Experimental details: Raman scattering measurements have been performed on arbitrarily oriented crystals using a 633 nm He-Ne laser. The Raman shifts have been determined for the maxima of individual peaks obtained as a result of the spectral curve analysis. The laser radiation power is not indicated.
Raman shifts (cm^{-1}): 1630, 1371sh, 1342, 1327sh, 1074s, 807, 805s, 734, 730, 703, 693, 419, 412, 347, 311s, 291, 281, 262, 194, 165, 143, 122.
Source: Čejka et al. (2010b).
Comments: The sample was characterized by powder X-ray diffraction data and electron microprobe analysis.

Celadonite $KMgFe^{3+}Si_4O_{10}(OH)_2$

Origin: Akaky River, Cyprus.
Experimental details: Raman scattering measurements have been performed on a powdery sample using 514.5 nm Ar$^+$ laser radiation. The laser radiation power at the sample was 0.9 mW.
Raman shifts (cm^{-1}): 3604w, 3583w, 3566s, 3538w, 1597, 1132, 1086, 1056w, 1017sh, 961, 797w, 769w, 701, 551s, 460w, 445w, 393, 318w, 273, 174s.
Source: Correia et al. (2007).
Comments: No independent analytical data are provided for the sample used. For the Raman spectrum of celadonite see also Ospitali et al. (2008).

Celestine $Sr(SO_4)$

Origin: Dufton, England.
Experimental details: Raman scattering measurements have been performed on an arbitrarily oriented crystal using 532 nm Nd-YAG laser radiation. The nominal laser radiation power was 100 mW.
Raman shifts (cm^{-1}): 1160, 1112w, 1003s, 641sh, 623, 461.
Source: Buzgar et al. (2009).

Comments: The Raman shifts are indicated for the maxima of individual peaks obtained as a result of the spectral curve analysis. The sample was characterized by EMPA. No independent analytical data are provided for the sample used. For the Raman spectra of celestine see also Kloprogge et al. (2001b), Andò and Garzanti (2014), and Culka et al. (2016a).

Celsian $Ba(Al_2Si_2O_8)$

Origin: No data.

Experimental details: Raman scattering measurements have been performed on an arbitrarily oriented sample using 532 nm laser radiation. The laser radiation power is not indicated.

Raman shifts (cm^{-1}): ~1050w, ~990, ~940, ~920sh, ~750w, ~715w, ~695w, ~550w, ~505s, ~470, ~405w, ~365w, ~305, ~250, ~195, ~165.

Source: Galuskina et al. (2016b).

Comments: No independent analytical data are provided for the sample used. For the Raman spectra of celsian see also Graham et al. (1992) and Colomban et al. (2000).

Cerianite-(Ce) CeO_2

Origin: Kerimasi volcano, Gregory Rift, northern Tanzania.

Experimental details: Raman scattering measurements have been performed on an arbitrarily oriented sample using 532 nm laser radiation. The laser radiation power is not indicated.

Raman shifts (cm^{-1}): 825–820w, 571, 449s, 184.

Source: Zaitsev et al. (2011).

Comments: The sample was characterized by powder X-ray diffraction data and electron microprobe analysis. For the Raman spectra of cerianite-(Ce) see also Nakajima et al. (1994), Wang et al. (1998b), and Hao (2008).

Černýite Cu_2CdSnS_4

Origin: Synthetic.

Experimental details: Raman scattering measurements have been performed on a thin film. Other experimental details are not described.

Raman shifts (cm^{-1}): 333, 304, 284.

Source: Guo et al. (2016).

Comments: The sample was characterized by powder X-ray diffraction data and electron microprobe analysis. For the IR spectrum of černýite see also Rincón et al. (2015).

Ceruleite $Cu_2Al_7(AsO_4)_4(OH)_{13} \cdot 11.5H_2O$

Origin: Emma Louisa gold mine, Guanaco district, Antofagasta, Chile.

Experimental details: Raman scattering measurements have been performed on arbitrarily oriented crystals using a 633 nm He-Ne laser. The Raman shifts have been determined for the maxima of individual peaks obtained as a result of the spectral curve analysis. The laser radiation power is not indicated.

Raman shifts (cm^{-1}): 3611sh, 3608s, 3597s, 3384, 3222, 3198sh, 3056, 1654w, 1042, 1003w, 951, 932, 903s, 870, 845, 827, 747w, 700w, 662w, 597s, 579sh, 515, 500sh, 464, 451sh, 430sh, 417, 400, 373, 335, 316, 299, 280, 262, 239, 231sh, 208, 195sh, 184sh, 176, 152, 132, 118, 111.

Source: Frost et al. (2013b).

Comments: No independent analytical data are provided for the sample used.

Cerussite $Pb(CO_3)$

Origin: No data.
Experimental details: Methods of sample preparation are not described. Raman scattering measurements have been performed using 633 nm He-Ne laser radiation. The nominal laser radiation power was 30 mW.
Raman shifts (cm^{-1}): 1477, 1370, 1052s, 246, 225, 176, 152.
Source: Bouchard and Smith (2003).
Comments: No independent analytical data are provided for the sample used. For the Raman spectra of cerussite see also Ciomartan et al. (1996), Frost et al. (2003e, f), and Frezzotti et al. (2012).

Cervantite $Sb^{3+}Sb^{5+}O_4$

Origin: Synthetic.
Experimental details: Methods of sample preparation are not described. Raman scattering measurements have been performed using 514.5 nm Ar$^+$ laser radiation. The laser radiation power is not indicated.
Raman shifts (cm^{-1}): 402w, 199, 96s.
Source: Jamal et al. (2013).
Comments: The sample was characterized by powder X-ray diffraction data. For the Raman spectra of cervantite see also Cody et al. (1979) and Makreski et al. (2013b).

Cesanite $Ca_2Na_3(SO_4)_3OH$

Origin: No data.
Experimental details: No data.
Raman shifts (cm^{-1}): 1104sh, 1004s, 647, 626s, 474, 448s.
Source: Frezzotti et al. (2012).
Comments: No independent analytical data are provided for the sample used.

Chabazite-Ca $Ca_2[Al_4Si_8O_{24}]\cdot 13H_2O$

Origin: Nidda, Germany.
Experimental details: Methods of sample preparation are not described. Raman scattering measurements have been performed using 532 nm Nd-YAG laser radiation. The laser radiation power at the sample was 300 mW.
Raman shifts (cm^{-1}): 1161w, 1082w, 808w, 697w, 465s, 402w, 357, 264, 204, 128.
Source: Mozgawa (2001).
Comments: The sample was characterized by powder X-ray diffraction data. For the Raman spectrum of chabazite-Ca see also Pechar and Rykl (1983).

Chalcanthite $Cu(SO_4)\cdot 5H_2O$

Origin: No data.
Experimental details: Raman scattering measurements have been performed on an arbitrarily oriented sample using 632.8 nm He-Ne laser radiation. The laser radiation power at the source was 30 mW.
Raman shifts (cm^{-1}): 3482, 3345, 3206, 1143, 1096, 986s, 612, 465, 426, 332w, 281, 202, 135, 124.
Source: Bouchard and Smith (2003).

Comments: The sample was characterized by powder X-ray diffraction data. For the Raman spectra of chalcanthite see also Berger (1976), Christy et al. (2004), Fu et al. (2012), and Bissengaliyeva et al. (2016).

Chalcoalumite $CuAl_4(SO_4)(OH)_{12} \cdot 3H_2O$

Origin: Červená vein, Jáchymov, Czech Republic.
Experimental details: Micro-Raman scattering measurements have been performed on an arbitrarily oriented sample using 532 nm laser radiation. The output laser radiation power was 2 mW.
Raman shifts (cm^{-1}): 3670, 3610, 3450s, 3270sh, 2940w, 2780w, 2650w, 1610w, 1455, 1135, 1110sh, 1005, 981s, 803w, (642sh), 594s, 494, 455, 415w, 220w, 175.
Source: Plášil et al. (2014d).
Comments: The sample was characterized by powder X-ray diffraction data and electron microprobe analyses.

Chalcocite Cu_2S

Origin: Synthetic.
Experimental details: No data.
Raman shifts (cm^{-1}): 465s, 257w.
Source: Kumar and Nagarajan (2011).
Comments: The sample was characterized by powder X-ray diffraction data and ICP analysis.

Chalcocyanite $Cu(SO_4)$

Origin: Synthetic.
Experimental details: Raman scattering measurements have been performed on a powdered sample using 532 nm Nd-YAG laser radiation. The nominal laser radiation power was 100 mW.
Raman shifts (cm^{-1}): 1205, 1101, 1045s, 1014s, 671w, 623, 514, 480sh, 448sh, 423, 347, 270, 250sh.
Source: Buzgar et al. (2009).
Comments: No independent analytical data are provided for the sample used. For the IR spectrum of chalcocyanite see also Fu et al. (2012).

Chalcomenite $Cu(Se^{4+}O_3) \cdot 2H_2O$

Origin: El Dragon Mine, Potosi, Bolivia.
Experimental details: Raman scattering measurements have been performed on arbitrarily oriented crystals using a 633 nm He-Ne laser. The Raman shifts have been determined for the maxima of individual peaks obtained as a result of the spectral curve analysis. The laser radiation power is not indicated.
Raman shifts (cm^{-1}): 3506w, 3184w, 2953w, 813s, 720, 690sh, 552, 472, 400w, 367w, 260, 218, 141, 128.
Source: Frost and Keeffe (2008b).
Comments: The sample was characterized by powder X-ray diffraction data and electron microprobe analysis, but no analytical data are provided in the cited paper.

Chalconatronite $Na_2Cu(CO_3)_2 \cdot 3H_2O$

Origin: Product of surface alterations of bronze.
Experimental details: Raman scattering measurements have been performed on a polycrystalline sample using 514.5 nm Ar^+ laser radiation. The laser radiation power is not indicated.
Raman shifts (cm^{-1}): 1600, 1329, 1073, 1053, 764, 698w, 327s, 261–268.
Source: Chiavari et al. (2016).
Comments: The sample was characterized by powder X-ray diffraction data and electron microprobe analysis.

Chalcophanite $ZnMn^{4+}_3O_7 \cdot 3H_2O$

Origin: Xiangguang Mn-Ag deposit, northern China.
Experimental details: Methods of sample preparation are not described. Raman scattering measurements have been performed using 632 nm laser radiation. The laser radiation power is not indicated.
Raman shifts (cm^{-1}): 651, 606, 502sh, 384w, 273w, 163.
Source: Fan et al. (2015).
Comments: The sample was characterized by LA-ICP-MS method and electron microprobe analysis. For the Raman spectrum of chalcophanite see also Kim and Stair (2004).

Chalcophyllite $Cu_{18}Al_2(AsO_4)_4(SO_4)_3(OH)_{24} \cdot 36H_2O$

Origin: Burrus Mine, USA.
Experimental details: Raman scattering measurements have been performed on arbitrarily oriented crystals using a 633 nm He-Ne laser. The Raman shifts have been determined for the maxima of individual peaks obtained as a result of the spectral curve analysis. The laser radiation power is not indicated.
Raman shifts (cm^{-1}): 3555w, 3390sh, 3129sh, 1636w, 981, 968sh, 867sh, 841s, 814sh, 499, 386, 221sh, 202, 142.
Source: Frost et al. (2010g).
Comments: No independent analytical data are provided for the sample used. The IR spectrum of presumed chalcophyllite from Burrus Mine given in the cited paper differs substantially from IR spectra of chalcophyllite published elsewhere (Moenke 1962; Chukanov 2014).

Chalcopyrite $CuFeS_2$

Origin: Mt. Morgan, Queensland, Australia.
Experimental details: Raman scattering measurements have been performed on an arbitrarily oriented sample using 514.5 nm Ar^+ laser radiation. The laser radiation power at the sample was in the range 1–10 mW. A 180°-scattering geometry was employed.
Raman shifts (cm^{-1}): 378, 352, 322, 293s.
Source: Mernagh and Trudu (1993).
Comments: The sample was characterized by electron microprobe analyses. For the Raman spectra of chalcopyrite see also Sasaki et al. (2009) and White (2009).

Chalcosiderite $CuFe^{3+}_6(PO_4)_4(OH)_8 \cdot 4H_2O$

Origin: Siglo XX mine, Andes Mts., Bustillo province, Bolivia.

Experimental details: Raman scattering measurements have been performed on an arbitrarily oriented sample using a 633 nm He-Ne laser. The Raman shifts have been determined for the maxima of individual peaks obtained as a result of the spectral curve analysis. The laser radiation power is not indicated.

Raman shifts (cm^{-1}): 3543sh, 3514s, 3501sh, 3482sh, 3480sh, 3384sh, 3306, 3200sh, 1194, 1168sh, 1159, 1102, 1062sh, 1042s, 1027sh, 990, 826w, 794sh, 768, 741sh, 636, 598sh, 580, 536, 484, 475sh, 420s, 415sh, 388, 351sh, 333, 293, 272, 264sh, 243sh, 235, 203, 182sh, 176, 153sh, 136s, 125s, 107.

Source: Frost et al. (2013af).

Comments: The sample was characterized by powder X-ray diffraction data and SEM/EDS, which may correspond to turquoise.

Chalcostibite $CuSbS_2$

Origin: Synthetic.
Experimental details: No data.
Raman shifts (cm^{-1}): 329s, 251, 152.
Source: Zhang et al. (2015).
Comments: The sample was characterized by powder X-ray diffraction data and electron microprobe analysis. For the Raman spectrum of chalcostibite see also Rath et al. (2015).

Challacolloite KPb_2Cl_5

Origin: Artificial (a phase in a fifteenth-century polychrome terracotta relief).

Experimental details: Raman scattering measurements have been performed on a powdered sample using 785 nm laser radiation. The laser radiation power at the sample was 2 mW.

Raman shifts (cm^{-1}): 202, 169, 158, 119s, 96, 85, 73.

Source: Bezur et al. (2015).

Comments: The sample was characterized by powder X-ray diffraction data and electron microprobe analysis. For the Raman spectra of oriented challacolloite crystals see Vtyurin et al. (2004).

Chambersite $Mn_3B_7O_{13}Cl$

Origin: No data.

Experimental details: Raman scattering measurements have been performed on arbitrarily oriented crystals using a 633 nm He-Ne laser. The Raman shifts have been determined for the maxima of individual peaks obtained as a result of the spectral curve analysis. The laser radiation power is not indicated.

Raman shifts (cm^{-1}): 1684, 1660, 1634sh, 1596m 1548sh, 1426sh, 1412sh, 1399, 1368sh, 1346, 1326sh, 1209, 1169sh, 1146sh, 1129, 1091sh, 1075, 1056, 1045, 1027sh, 989w, 963, 942, 920, 902sh, 871, 853, 836, 797w,7766, 755w, 721, 705w, 679sh, 660s, 642sh, 617w, 597, 559, 544sh, 523, 508, 402, 393sh, 371sh, 357s, 339sh, 302, 273, 259, 241, 229, 209, 185sh, 177sh, 161s, 143s, 116.

Source: Frost et al. (2014f).
Comments: No independent analytical data are provided for the sample used.

Chamosite $(Fe^{2+},Mg,Al,Fe^{3+})_6(Si,Al)_4O_{10}(OH,O)_8$

Origin: Turamdih, India.
Experimental details: Raman scattering measurements have been performed on an arbitrarily oriented polished section using 514.5 nm Ar^+ laser radiation. The laser radiation power behind the objective was in the range 0.03–0.8 mW. A nearly 180°-scattering geometry was employed.
Raman shifts (cm^{-1}): 3644s, 3625s, 3560s, 3434, ~1030, 665s, 615sh, 545s, 518sh, 428w, 361, 198, 127.
Source: Nasdala et al. (2006).
Comments: The sample was characterized by electron microprobe analysis.

Changbaiite $PbNb_2O_6$

Origin: Synthetic.
Experimental details: No data.
Raman shifts (cm^{-1}): 700sh, 676s, 550sh, 529w, 477, 453sh, 423w, 379sh, 361, 323, 3009w, 277, 247, 230w, 202s, 179s, 141s, 119, 89s, 61.
Source: Repelin et al. (1980).
Comments: No independent analytical data are provided for the sample used.

Changoite $Na_2Zn(SO_4)_2 \cdot 4H_2O$

Origin: Synthetic.
Experimental details:
Raman shifts (cm^{-1}): 3409, 3118, 1189, 1160, 1099, 1064, 985s, 628sh, 612, 473sh, 451.
Source: Jentzsch et al. (2013).

Chapmanite $Fe^{3+}_2Sb^{3+}(SiO_4)_2(OH)$

Origin: Boňenov, near Mariánské Lázně, western Bohemia, Czech Republic.
Experimental details: Raman scattering measurements have been performed on arbitrarily oriented crystals using a 633 nm He-Ne laser. The Raman shifts have been determined for the maxima of individual peaks obtained as a result of the spectral curve analysis. The laser radiation power is not indicated.
Raman shifts (cm^{-1}): 3563, 3555, 1590, 1317w, 1120, 1077, 1013, 903, 808, 773, 709, 553, 473w, 436sh, 422, 408, 391, 361sh, 350, 334sh, 313, 293w, 256, 219, 208, 186, 177, 147, 114, 107.
Source: Frost et al. (2010a).
Comments: The sample was characterized by powder X-ray diffraction data and electron microprobe analysis.

Charoite $(K,Sr,Ba,Mn)_{15-16}(Ca,Na)_{32}[Si_{70}(O,OH)_{180}](OH,F)_4 \cdot nH_2O$

Origin: Murun massif (Murunskii alkaline complex), Aldan Shield, southwest Yakutia, Siberia, Russia (type locality).
Experimental details: Methods of sample preparation are not described. Raman scattering measurements have been performed using 532 nm Nd-YAG laser radiation. The nominal laser radiation power was 100 mW.

Raman shifts (cm^{-1}): 2403, 2367, 1135, 1116sh, 1054, 675s, 638, 434w, 242w.
Source: Buzatu and Buzgar (2010).
Comments: No independent analytical data are provided for the sample used.

Chegemite $Ca_7(SiO_4)_3(OH)_2$

Origin: Upper Chegem volcanic structure, Northern Caucasus, Kabardino-Balkaria, Russia (type locality).
Experimental details: Raman scattering measurements have been performed on an arbitrarily oriented sample using 514.5 nm Ar$^+$ laser radiation. The laser radiation power at the sample was 20 mW. A 0°-scattering geometry was employed.
Raman shifts (cm^{-1}): 3563s, 3551sh, 3532, 3478s, 924s, 893, 845, 818s, 774, 766, 549, 526s, 403, 389s, 311, 293, 273, 226.
Source: Galuskin et al. (2009).
Comments: The Raman shifts have been determined for the maxima of individual peaks obtained as a result of the spectral curve analysis. The sample was characterized by powder X-ray diffraction data and electron microprobe analysis. The crystal structure is solved.

Chekhovichite $Bi^{3+}{}_2Te^{4+}{}_4O_{11}$

Origin: Synthetic.
Experimental details: Raman scattering measurements have been performed on an arbitrarily oriented sample using 1064 nm Nd-YAG laser radiation. The nominal laser radiation power at the sample was 87 mW.
Raman shifts (cm^{-1}): 761w, 726s, 674sh, 642s, 551w, 419w, 380, 298, 270, 224, 180.
Source: Durand (2006).
Comments: The sample was characterized by powder X-ray diffraction data.

Chenevixite $Cu(Fe^{3+},Al)(AsO_4)(OH)_2$

Origin: Manto Cuba Mine, San Pedro de Cachiyuyo district, Chañara lprovince, Atacama, Chile.
Experimental details: Raman scattering measurements have been performed on arbitrarily oriented crystals using a 633 nm He-Ne laser. The Raman shifts have been determined for the maxima of individual peaks obtained as a result of the spectral curve analysis. The laser radiation power is not indicated.
Raman shifts (cm^{-1}): 3501sh, 3405w, 3315sh, 2931w, 2870w, 1688, 1613, 1536, 1238, 1211, 1151, 1130, 883sh, 855s, 836sh, 807s, 495, 450sh, 435s, 408sh, 359, 350, 300.
Source: Frost et al. (2015f).
Comments: The sample was characterized by qualitative electron microprobe analysis that shows admixture of a silicate. The bands at 1688 and 1613 cm^{-1} indicate the presence of H_2O molecules.

Cheralite $CaTh(PO_4)_2$

Origin: Synthetic.
Experimental details: Raman scattering measurements have been performed on a polycrystalline pellet using 532 nm Nd-YAG laser radiation. The laser radiation power at the sample was 10 mW. A 180°-scattering geometry was employed.

Raman shifts (cm^{-1}): 1088w, 982s, 623w, 597w, 573, 537w, 54, 425, 399sh, 289w, 235w.
Source: Raison et al. (2008).
Comments: The sample was characterized by XRD. For the IR spectrum of cheralite see also Terra et al. (2008).

Chernikovite $(H_3O)(UO_2)(PO_4) \cdot 3H_2O$

Origin: Synthetic.
Experimental details: Raman scattering measurements have been performed on an arbitrarily oriented sample using 532 nm Nd-YAG laser radiation. The laser radiation power at the sample was about 1–4 mW. A 180°-scattering geometry was employed.
Raman shifts (cm^{-1}): 3425w, 3215sh, 3078w, 999s, 986, 842s, 458w, 402, 287, 193s, 110.
Source: Clavier et al. (2016).
Comments: The Raman shifts are indicated for the maxima of individual peaks obtained as a result of the spectral curve analysis. The sample was characterized by powder X-ray diffraction data and ICP-AES.

Chernovite-(Y) $Y(AsO_4)$

Origin: Synthetic.
Experimental details: Raman scattering measurements have been performed on a powdered sample using 514.5 nm Ar$^+$ laser radiation. The nominal laser radiation power was 100 mW.
Raman shifts (cm^{-1}): 888s, 880s, 835s, 395, 255w, 234w, 177w.
Source: Pradhan et al. (1987).
Comments: The sample was characterized by powder X-ray diffraction data.

Chervetite $Pb_2V^{5+}{}_2O_7$

Origin: Synthetic.
Experimental details: Methods of sample preparation are not described. Raman scattering measurements have been performed using 488 nm Ar$^+$ laser radiation. The laser radiation power is not indicated.
Raman shifts (cm^{-1}): 876s, 817s, 751w, 673w, 582, 371, 351s, 324, 258w, 232, 196w, 181w, 140w, 124, 112w, 89, 74.
Source: Schwendt and Joniaková (1975).
Comments: The sample was characterized by powder X-ray diffraction data and chemical analysis.

Chiavennite $CaMn^{2+}(BeOH)_2Si_5O_{13} \cdot 2H_2O$

Origin: Prata, Italy.
Experimental details: Raman scattering measurements have been performed on an arbitrarily oriented crystal using 532 nm Nd-YAG laser radiation. The laser radiation power is not indicated.
Raman shifts (cm^{-1}): 1068w, 996w, 960w, 919w, 682w, 559, 508, 460s, 428w, 360w, 356w, 352w, 347w, 342w, 299w, 257w, 234w, 324w, 204w, 195w, 148w, 115w, 95w.
Source: Jehlička et al. (2012).
Comments: No independent analytical data are provided for the sample used.

Chibaite $SiO_2 \cdot n(CH_4, C_2H_6, C_3H_8, C_4H_{10})$ ($n_{max} = 3/17$)

Origin: Arakawa, Chiba prefecture, Honshu Island, Japan (type locality).
Experimental details: Raman scattering measurements have been performed on an arbitrarily oriented sample using 514.5 nm Ar$^+$ laser radiation. The laser radiation power is not indicated.
Raman shifts (cm^{-1}): 3050w, 2960w, 2936, 2908s, 2900, 2866, 989, 873, 805, and a series of bands below 400 cm^{-1}.
Source: Likhacheva et al. (2016).
Comments: The sample was characterized by powder X-ray diffraction data. For the Raman spectrum of chibaite see also Momma et al. (2011).

Childrenite $Fe^{2+}Al(PO_4)(OH)_2 \cdot H_2O$

Origin: Ponte do Piauí mine, Piauí valley, Itinga, Minas Gerais, Brazil.
Experimental details: Raman scattering measurements have been performed on arbitrarily oriented crystals using a 633 nm He-Ne laser. The Raman shifts have been determined for the maxima of individual peaks obtained as a result of the spectral curve analysis. The laser radiation power is not indicated.
Raman shifts (cm^{-1}): 3471, 3420sh, 3333, 3199sh, 3043, 1724, 1673, 1573, 1183w, 1142, 1091w, 1011s, 978sh, 969s, 864w, 816w, 608sh, 595s, 562, 466s, 427sh, 405, 347sh, 310, 251, 228, 208, 188sh, 147sh, 138.
Source: Frost et al. (2013am).
Comments: The sample was characterized by powder X-ray diffraction data and electron microprobe analysis.

Chiolite $Na_5Al_3F_{14}$

Experimental details: Raman scattering measurements have been performed on an oriented single crystal using 514.5 nm Ar$^+$ laser radiation. The laser radiation power at the sample was between 0.5 and 1 W.
Raman shifts (cm^{-1}): A_{1g} ($xx + yy$): 530s, 356, 203, 110s; A_{1g} (zz): 530s, 414, 356, ~312w, 203w, 110; B_{1g} ($xx - yy$): 441w, 320w, ~250w (?); B_{2g} (xy): 390w, ~250w (?), 223w, 208w; E_g(xz, yz): 426w, 408, 360w, 248s, 28.
Source: Rocquet et al. (1985).
Comments: The sample was characterized by powder X-ray diffraction data. For the Raman spectrum of chiolite see also Carey et al. (2015).

Chloraluminite $AlCl_3 \cdot 6H_2O$

Origin: Synthetic.
Experimental details: Raman scattering measurements have been performed on a polycrystalline sample using 514.5 nm Ar$^+$ laser radiation. The laser radiation power is not indicated.
Raman shifts (cm^{-1}): 1150, 835, 710w, 615, 572, 530, 430, 315s, 180, 115s, 82w, 57s.
Source: Stefov et al. (1992).
Comments: No independent analytical data are provided for the sample used. The Raman shifts were partly determined by us based on spectral curve analysis of the published spectrum.

Chlorapatite $Ca_5(PO_4)_3Cl$

Origin: No data.
Experimental details: No data.
Raman shifts (cm^{-1}): 1127w, 1039, 963s, 581w, 430w.
Source: Frezzotti et al. (2012).
Comments: The methods of the identification of the sample are not indicated. For the Raman spectrum of chlorapatite see also Chen et al. (1995).

Chlorargyrite $AgCl$

Origin: Synthetic.
Experimental details: Raman scattering measurements have been performed on an arbitrarily oriented crystal using 568 nm laser radiation. The nominal laser radiation power was 100 mW.
Raman shifts (cm^{-1}): 386, 240, 149.
Source: Bottger and Damsgard (1971).
Comments: Second order Raman spectrum at 300 K is given.

Chloritoid $Fe^{2+}Al_2O(SiO_4)(OH)_2$

Origin: Tipam Formation, Bangladesh.
Experimental details: Raman scattering measurements have been performed on an arbitrarily oriented crystalin Canada Balsam using 532 nm diode laser radiation. The laser radiation power is not indicated. A nearly 180°-scattering geometry was employed.
Raman shifts (cm^{-1}): 880w, 596s, 160.
Source: Andò and Garzanti (2014).
Comments: No independent analytical data are provided for the sample used.

Chlorkyuygenite $Ca_{12}Al_{14}O_{32}[(H_2O)_4Cl_2]$

Origin: Upper Chegem caldera, Northern Caucasus, Kabardino-Balkaria, Russia (type locality).
Experimental details: Methods of sample preparation are not described. Raman scattering measurements have been performed using 652 nm laser radiation. The laser radiation power is not indicated.
Raman shifts (cm^{-1}): ~3400 (broad), ~3200 (broad), 907sh, 881, 776s, 705s, 511s, 321, 208, 161.
Source: Galuskin et al. (2015b).
Comments: The sample was characterized by powder X-ray diffraction data and electron microprobe analysis. The crystal structure is solved. No Raman bands of H_2O molecules are observed in the range from 1500 to 1700 cm^{-1}.

Chlormayenite $Ca_{12}Al_{14}O_{32}[\square_4Cl_2]$

Origin: Ettringer Bellerberg volcano, near Mayen, Eifel Mts., Germany (type locality).
Experimental details: Raman scattering measurements have been performed on an arbitrarily oriented crystal using 514.5 nm Ar^+ laser radiation. The nominal laser radiation power was in the range 30–50 mW. A 180°-scattering geometry was employed.
Raman shifts (cm^{-1}): 3669, 3644sh, 3570sh, 3400, 1094w, 991, 881, 816sh, 772s, 703, 512s, 323.

Source: Galuskin et al. (2012c).
Comments: The sample was characterized by powder X-ray diffraction data and electron microprobe analysis. The crystal structure is solved. For the Raman spectrum of chlormayenite see also Ma et al. (2011a).

Chlorocalcite $KCaCl_3$

Origin: A dolerite sill in eastern Siberia, Russia.
Experimental details: Micro-Raman scattering measurements have been performed on an arbitrarily oriented crystal using 514.5 nm Ar+ laser radiation. The laser radiation power is not indicated.
Raman shifts (cm^{-1}): 195, 140s, 128sh, 82, 67, 57.
Source: Grishina et al. (1992).
Comments: No independent analytical data are provided for the sample used.

Chloromagnesite $MgCl_2$

Origin: Synthetic.
Experimental details: Raman scattering measurements have been performed on a powdered sample using 514.5 nm Ar$^+$ laser radiation. The laser radiation power at the sample was 5 mW.
Raman shifts (cm^{-1}): 243s, 157w.
Source: Brambilla et al. (2004).
Comments: No independent analytical data are provided for the sample used.

Chloroxiphite $Pb_3CuO_2Cl_2(OH)_2$

Origin: Merehead Quarry, Shepton Mallet, Somerset, UK (type locality).
Experimental details: Micro-Raman scattering measurements have been performed on arbitrarily oriented crystals using a 633 nm He-Ne laser. The Raman shifts have been determined for the maxima of individual peaks obtained as a result of the spectral curve analysis. The laser radiation power was 0.1 mW.
Raman shifts (cm^{-1}): 3466, 3437, 3400, 3338, 875, 782, 692, 482, 469, 406, 350, 312, 286, 250, 226, 190, 179, 166, 145, 139s.
Source: Frost and Williams (2004).
Comments: No independent analytical data are provided for the sample used.

Chondrodite $Mg_5(SiO_4)_2F_2$

Origin: Sparta, Sussex Co., New Jersey, USA.
Experimental details: Raman scattering measurements have been performed on arbitrarily oriented crystals using a 633 nm He-Ne laser. The Raman shifts have been determined for the maxima of individual peaks obtained as a result of the spectral curve analysis. The laser radiation power is not indicated.
Raman shifts (cm^{-1}): 3576, 3570, 3561, 967, 931, 878w, 860s, 845s, 832s, (818w), 786w, 755, 607, 587s, 572, 547s, 430, 390.
Source: Frost et al. (2007k).
Comments: The sample was characterized by powder X-ray diffraction data and electron microprobe analysis. Raman shifts are given for chondrodite with the empirical formula

$Mg_{5.04}Fe^{2+}_{0.09}Ti_{0.02}Na_{0.10}(SiO_4)_2F_{1.38}(OH)_x$. For the Raman spectra of chondrodite see also Mernagh et al. (1999) and Lin et al. (1999).

Chromatite $CaCr^{6+}O_4$

Origin: Synthetic.
Experimental details: Raman scattering measurements have been performed on an arbitrarily oriented crystal. Other experimental details are not described.
Raman shifts (cm^{-1}): ~900, ~875s.
Source: Sánchez-Pastor et al. (2010).
Comments: The sample was characterized by electron microprobe analysis.

Chromite $Fe^{2+}Cr_2O_4$

Origin: Synthetic.
Experimental details: Raman scattering measurements have been performed on a powdered sample using 647.1 nm Kr$^+$ laser radiation. The nominal laser radiation power was 0.1 mW.
Raman shifts (cm^{-1}): 678s, 635sh, 531.
Source: Hosterman (2011).
Comments: The sample was characterized by powder X-ray diffraction data and electron microprobe analysis. For the Raman spectra of chromite see also Reddy and Frost (2005), Karwowski et al. (2013), Chen et al. (2008a), Sagatowska (2010), Lenaz and Lughi (2013), Andò and Garzanti (2014), and D'Ippolito et al. (2015).

Chromite $Fe^{2+}Cr_2O_4$

Origin: Morasko iron meteorite.
Experimental details: Raman scattering measurements have been performed on an arbitrarily oriented sample using 632.8 nm He-Ne laser radiation. The laser radiation power at the sample was 10 mW.
Raman shifts (cm^{-1}): 683s, 642sh, 604w, 513w, 446w.
Source: Karwowski et al. (2013).
Comments: The sample was characterized by powder X-ray diffraction data and electron microprobe analysis. Its empirical formula is $(Mg_{0.34}Mn_{0.04}Zn_{0.07}Al_{0.01}Fe^{2+}_{0.55}Fe_{0.01}Cr_{1.96})O_4$. For the Raman spectra of chromite see also Hosterman (2011), Reddy and Frost (2005), Chen et al. (2008a), Sagatowska (2010), Lenaz and Lughi (2013), Andò and Garzanti (2014), and D'Ippolito et al. (2015).

Chrysoberyl $BeAl_2O_4$

Origin: Colatine, Esperito Santo, Brazil.
Experimental details: Raman scattering measurements have been performed on an arbitrarily oriented single crystal using 488 nm Ar$^+$ laser radiation. The laser radiation power is not indicated.
Raman shifts (cm^{-1}): 1040w(?), 1020w, 931s, 816, 776, 747, 711, 679, 658, 639s, 567, 546, 518s, 501, 477s, 459, 447, 422, 397, 371w, 242w, 218w.
Source: Hofmeister et al. (1987).
Comments: No independent analytical data are given for the sample used. For the Raman spectra of chrysoberyl see also Jehlička et al. (2012), Beurlen et al. (2013), and Culka et al. (2016a).

Chrysocolla $(Cu_{2-x}Al_x)H_{2-x}Si_2O_5(OH)_4 \cdot nH_2O$

Origin: An unknown locality in Peru.
Experimental details: Micro-Raman scattering measurements have been performed on an arbitrarily oriented sample using 785 nm diode laser radiation. The laser radiation power at the sample was below 0.5 mW.
Raman shifts (cm^{-1}): ~1045w, ~945w, ~798w, 676s, ~490sh, 413s, 341, ~210.
Source: Bernardino et al. (2016).
Comments: No independent analytical data are given for the sample used. For the Raman spectra of chrysocolla see also Frost and Xi (2013a) and Coccato et al. (2016).

Chrysothallite $K_6Cu_6Tl^{3+}Cl_{17}(OH)_4 \cdot H_2O$

Origin: Second scoria cone of the Northern Breakthrough of the Great Tolbachik Fissure Eruption, Tolbachik volcano, Kamchatka, Russia (type locality).
Experimental details: Raman scattering measurements have been performed on an arbitrarily oriented sample using 532 nm diode laser radiation. The laser radiation power at the sample was ~0.1 mW.
Raman shifts (cm^{-1}): 3443s, 1580w, 949, 902sh, 465, 320, 295, 275s, 250sh, 206s.
Source: Pekov et al. (2015c).
Comments: The sample was characterized by powder X-ray diffraction data and electron microprobe analysis. The crystal structure is solved.

Chrysotile $Mg_3Si_2O_5(OH)_4$

Origin: Salem, Tamil Nadu, India.
Experimental details: Raman scattering measurements have been performed on a powdered sample using 1064 nm Nd-YAG laser radiation. The laser radiation power is not indicated.
Raman shifts (cm^{-1}): 1105, 692s, 622, 464, 438w, 390s, 348, 325, 304w, 232s, 180.
Source: Anbalagan et al. (2010).
Comments: The sample was characterized by powder X-ray diffraction data. For the Raman spectra of chrysotile see also Rinaudo et al. (2003), Auzende et al. (2004), and Petry et al. (2006).

Chukanovite $Fe_2(CO_3)(OH)_2$

Origin: Synthetic.
Experimental details: Micro-Raman scattering measurements have been performed on an arbitrarily oriented sample using 532 nm Nd-YAG laser radiation. The laser radiation power at the sample was 0.1 mW.
Raman shifts (cm^{-1}): 3454, 3321, 1510, 1434w, 1070s, 926w, 730, 389, 238w.
Source: Saheb et al. (2011).
Comments: The sample was characterized by powder X-ray diffraction data. For the Raman spectrum of chukanovite see also Rémazeilles and Refait (2009).

Chukhrovite (Ca) $Ca_3Ca_{1.5}Al_2(SO_4)F_{13} \cdot 12H_2O$

Origin: Val Cavallizza Pb-Zn-(Ag) mine, Cuasso al Monte, Varese province, Italy (type locality).
Experimental details: Methods of sample preparation are not described. Raman scattering measurements have been performed using 632.8 nm He-Ne laser radiation. The laser radiation power is not indicated.

Raman shifts (cm⁻¹): 3560, 3470, 3440, 3270, 1632, 1112, 977s, 553, 449, 395w, 345, 281, 211, 181, 140w.
Source: Vignola et al. (2012).
Comments: The sample was characterized by powder X-ray diffraction data and electron microprobe analysis. The crystal structure is solved. The Raman shifts were partly determined by us based on spectral curve analysis of the published spectrum.

Churchite-(Nd) (?) $Nd(PO_4) \cdot 2H_2O$

Origin: Costa Balzi Rossi, Magliolo, Liguria, Italy.
Experimental details: No data.
Raman shifts (cm⁻¹): 1316, 1181s, 974, 872w, 805, 742, 677, 633, 532s, 475, 424w, 249, 170w.
Source: Bracco et al. (2012).
Comments: The sample of presumed churchite-(Nd) was characterized only by electron microprobe analyses. The Raman shifts were determined by us based on spectral curve analysis of the published spectrum.

Churchite-(Y) $Y(PO_4) \cdot 2H_2O$

Origin: Grube Leonie, Auerbach, Oberphalz, Germany.
Experimental details: Raman scattering measurements have been performed on arbitrarily oriented crystals using a 633 nm He-Ne laser. The Raman shifts have been determined for the maxima of individual peaks obtained as a result of the spectral curve analysis. The laser radiation power is not indicated.
Raman shifts (cm⁻¹): 3327sh, 3205w, 3127w, 3065sh, 1067s, 1029w, 995sh, 984s, 981sh, 707w, 681, 662sh, 649, 565,497, 369sh, 362, 344sh, 307, 287, 269, 249, 210, 199, 188, 180, 162w, 145, 116sh, 109.
Source: Frost et al. (2014g).
Comments: No independent analytical data are provided for the sample used.

Cinnabar α-HgS

Origin: Synthetic.
Experimental details: Raman scattering measurements have been performed on a powdered sample using 532 nm laser radiation. The laser radiation power at the sample was from 50 to 520 µW.
Raman shifts (cm⁻¹): 343w, 282w, 251s, 40s.
Source: Radepont (2013).
Comments: For the Raman spectra of cinnabar see also Lepot et al. (2006) and Frost et al. (2010c).

Claringbullite $Cu^{2+}_4Cl(OH)(OH)_6$

Origin: Nchanga Open Pit, Chingola, Zambia.
Experimental details: Raman scattering measurements have been performed on an arbitrarily oriented sample using 633 nm He-Ne laser radiation. The laser radiation power is not indicated. The Raman shifts have been determined for the maxima of individual peaks obtained as a result of the spectral curve analysis.

Raman shifts (cm^{-1}): 3458sh, 3433s, 3351s, 3331sh, 3211, 970, 906, 815, 579, 511s, 447, 389, 356, 260, 231, 163, 147, 136, 119.
Source: Frost et al. (2003i).
Comments: No independent analytical data are provided for the sample used.

Claudetite As$_2$O$_3$

Origin: Jáchymov U deposit, KrušnéHory (Ore Mts.), Western Bohemia, Czech Republic.
Experimental details: Raman scattering measurements have been performed on an arbitrarily oriented crystal using 514.5 nm Ar$^+$ laser radiation. The laser radiation power at the sample was 100 mW.
Raman shifts (cm^{-1}): 814, 632, 626s, 541s, 459s, 356, 354, 284, 284, 259s, 248s, 218, 193, 175.
Source: Origlieri et al. (2009).
Comments: For the Raman spectrum of claudetite see also Guńka et al. (2012).

Clausthalite PbSe

Origin: Synthetic.
Experimental details: Raman scattering measurements have been performed on a nanocrystalline aggregate. No other experimental details are described.
Raman shifts (cm^{-1}): 791, 136s.
Source: Ge and Li (2003).
Comments: The sample was characterized by powder X-ray diffraction data. The band at 791 cm^{-1} may correspond to an impurity.

Clinoatacamite Cu$_2$Cl(OH)$_3$

Origin: No data.
Experimental details: Methods of sample preparation are not described. Raman scattering measurements have been performed using 632.8 nm He-Ne laser radiation. The output laser radiation power was 30 mW.
Raman shifts (cm^{-1}): 3442, 3355, 3310, 930, 911, 896, 842s, 820s, 804, 590, 511s, 450, 420, 364.
Source: Bouchard and Smith (2003).
Comments: The sample was characterized by powder X-ray diffraction data. For the Raman spectra of clinoatacamite see also Chu et al. (2011), Bertolotti et al. (2012), and Coccato et al. (2016).

Clinoatacamite Cu$_2$Cl(OH)$_3$

Origin: No data.
Experimental details: Methods of sample preparation are not described. Raman scattering measurements have been performed using 632.8 nm He-Ne laser radiation. The laser radiation power at the sample was in the range 0.05–1 mW.
Raman shifts (cm^{-1}): 3434, 3348, 3326, 3308, 971, 927, 893, 818, 581, 511s, 446, 361, 139.
Source: Bertolotti et al. (2012).
Comments: The sample was characterized by electron microprobe analysis, X-ray photoelectron spectroscopy, and IR spectroscopy. The band at 3326 cm^{-1} may be due to an impurity. For the Raman spectra of clinoatacamite see also Bouchard and Smith (2003), Chu et al. (2011), and Coccato et al. (2016).

Clinobisvanite $Bi(VO_4)$

Origin: Londonderry feldspar quarry, Coolgardie area, Western Australia.
Experimental details: Micro-Raman scattering measurements have been performed on arbitrarily oriented crystals using a 633 nm He-Ne laser. The Raman shifts have been determined for the maxima of individual peaks obtained as a result of the spectral curve analysis.
Raman shifts (cm^{-1}): 828s, 756, 712, 368, 329, 245, 211, 185, 167.
Source: Frost et al. (2006i).
Comments: No independent analytical data are given for the sample used.

Clinocervantite $Sb^{3+}Sb^{5+}O_4$

Origin: Synthetic.
Experimental details: Raman scattering measurements have been performed on packed powder using 488 nm Ar$^+$ laser radiation. The laser radiation power at the sample was ~300 mW. A 90°-scattering geometry was employed.
Raman shifts (cm^{-1}): 754w, 635w, 466, 439w, 405s, 283w, 212s, 195sh, 142, 94, 79s.
Source: Cody et al. (1979).
Comments: The sample was characterized by powder X-ray diffraction data.

Clinochalcomenite $Cu(Se^{4+}O_3) \cdot 2H_2O$

Origin: El Dragon Mine, Potosi, Bolivia.
Experimental details: Raman scattering measurements have been performed on arbitrarily oriented crystals using a 633 nm He-Ne laser. The Raman shifts have been determined for the maxima of individual peaks obtained as a result of the spectral curve analysis.
Raman shifts (cm^{-1}): 3507w, 3193w, 2909w, 967w, 817sh, 811s, 792sh, 749, 700, 552, 489, 378sh, 361, 349sh, 219, 180, 129.
Source: Frost and Keeffe (2008b).
Comments: Questionable data: the Raman spectrum of presumed clinochalcomenite is very close to that of chalcomenite. No independent analytical data are given for the sample used.

Clinochlore $Mg_5Al(AlSi_3O_{10})(OH)_8$

Origin: Synthetic.
Experimental details: Raman scattering measurements have been performed on a powdered sample using 514.5 nm Ar$^+$ laser radiation. The laser radiation power is not indicated. A 135°-scattering geometry was employed.
Raman shifts (cm^{-1}): 3679, 3647, 3605, 3477, 679s, 548, 358, 198, 104.
Source: Kleppe et al. (2003).
Comments: End-member clinochlore, $(Mg_5Al)(Si_3Al)O_{10}(OH)_8$, has been studied. Raman shifts for the range of the OH-stretching vibrations are indicated for the maxima of individual peaks obtained as a result of the spectral curve analysis.

Clinoclase $Cu_3(AsO_4)(OH)_3$

Origin: Tin Stope, Majuba Hill mine, Utah, USA.
Experimental details: Raman scattering measurements have been performed on arbitrarily oriented crystals using a 633 nm He-Ne laser. The Raman shifts have been determined for the maxima of individual peaks obtained as a result of the spectral curve analysis.
Raman shifts (cm^{-1}): 3559, 3339, 983, 850, 832s, 783, 607, 539, 508, 482, 460, 438, 348, 318, 308, 306, 295, 247, 231, 185, 171, 160, 136.
Source: Frost et al. (2002e).
Comments: The sample identification was done by PXRD, by SEM and by EMPA, but corresponding analytical data are not given in the cited paper. For the Raman spectrum of clinoclase see also Martens et al. (2003b).

Clinoenstatite $Mg_2Si_2O_6$

Origin: Synthetic.
Experimental details: Raman scattering measurements have been performed on an arbitrarily oriented crystal using 514.5 nm Ar^+ laser radiation. The laser radiation power at the sample was ~50 mW. A 180°-scattering geometry was employed.
Raman shifts (cm^{-1}): 1034s, 1012s, 927w, 848w, 755w, 689s, 666s, 583w, 577w, 523, 480w, 453w, 432, 418, 405, 388w, 371, 344s, 324w, 304w, 279w, 245, 233w, 206, 195, 158, 143.
Source: Lin (2004).
Comments: The sample was characterized by powder X-ray diffraction data and electron microprobe analysis.

Clinohedrite $CaZn(SiO_4) \cdot H_2O$

Origin: No data.
Experimental details: Raman scattering measurements have been performed on a powdered sample using 1064 nm Nd-YAG laser radiation. The laser radiation power is not indicated.
Raman shifts (cm^{-1}): 985sh, 951, 857s, 844s, 568sh, 550, 501, 465, 387s, 340, 308, 280w, 235s, 208s.
Source: Annen and Davis (1993).
Comments: The sample was characterized by powder X-ray diffraction data. The Raman shifts were determined by us based on spectral curve analysis of the published spectrum.

Clinometaborite HBO_2

Origin: Synthetic.
Experimental details: Methods of sample preparation are not described. Raman scattering measurements have been performed using 488 nm Ar^+ laser radiation. The laser radiation power is not indicated.
Raman shifts (cm^{-1}): 3130, 2930, 2725w, 1430w, 1400w, 1341w, 1330w, 1264w, 1227w, 1173w, 1135w, 1080w, 980w, 918w, 785s, 765sh, 710w, 680sh, 655, 628, 536, 522, 477, 432, 397, 379, 346sh, 338, 307sh, 293, 277, 226w, 198, 185sh, 178, 147, 131, 119, 108, 94, 77.
Source: Bertoluzza et al. (1980).
Comments: The sample was characterized by IR spectrum.

Clinoptilolite Na $Na_6(Si_{30}Al_6)O_{72} \cdot 20H_2O$

Origin: Dylagówka, Poland.
Experimental details: Methods of sample preparation are not described. Raman scattering measurements have been performed using 1064 nm Nd-YAG laser radiation. The laser radiation power at the sample was 300 mW.
Raman shifts (cm^{-1}): 1129w, 799w, 514s, 471s, 410s, 182.
Source: Mozgawa (2001).
Comments: The idealized formula $(Na,K)_6(Al_6Si_{30}O_{72}) \cdot 20H_2O$ is given for the sample described in the cited paper, but no chemical data are presented. The sample was characterized only by powder X-ray diffraction data.

Clinotobermorite-like mineral $Ca_4Si_6O_{17}(H_2O)_2 \cdot (Ca \cdot 3H_2O)(?)$

Origin: Artificial (an intermediate clinotobermorite-like phase formed during the thermal conversion of tobermorite 11 Å to tobermorite 10 Å).
Experimental details: Raman scattering measurements have been performed on an arbitrarily oriented sample using 632.8 nm He-Ne laser radiation. The laser radiation power is not indicated. A 180°-scattering geometry was employed.
Raman shifts (cm^{-1}): 1077, 1011s, 996, 682s, 619, 523, 484, 445, 410, 360, 302.
Source: Biagioni et al. (2012).
Comments: The sample was characterized by powder X-ray diffraction data.

Clinozoisite $Ca_2Al_3[Si_2O_7][SiO_4]O(OH)$

Origin: Beura, Verbania, Piemonte, Italy.
Experimental details: Methods of sample preparation are not described. Raman scattering measurements have been performed using 785 nm laser radiation. The laser radiation power is not indicated.
Raman shifts (cm^{-1}): 1092, 1050w, 985, 963, 919, 875, 832, 692, 605, 570s, 527, 513, 468sh, 452, 428s, 396, 353w, 328w, 305w, 276, 252, 233, 166w, 138.
Source: Andò and Garzanti (2014).
Comments: No independent analytical data are provided for the sample used. The Raman shifts were partly determined by us based on spectral curve analysis of the published spectrum.

Clintonite $CaAlMg_2(SiAl_3O_{10})(OH)_2$

Origin: Ilmeno-Vishnevogorsky Complex, South Urals, Russia.
Experimental details: Micro-Raman scattering measurements have been performed on an arbitrarily oriented sample using 632.8 nm He-Ne laser radiation. The nominal laser radiation power was 20 mW.
Raman shifts (cm^{-1}): 894, 828, 800, 550, 656s, 397, 346, 233, 184, 123.
Source: Korinevsky (2015).
Comments: The Raman shifts were partly determined by us based on spectral curve analysis of the published spectrum. For the Raman spectrum of clintonite see also Neuville et al. (2002).

Coalingite $Mg_{10}Fe^{3+}{}_2(CO_3)(OH)_{24} \cdot 2H_2O$

Origin: Union Carbide mine, San Benito Co., California, USA.

Experimental details: Raman scattering measurements have been performed on an arbitrarily oriented sample using a 633 nm He-Ne laser. The laser radiation power at the sample was below 5 mW. The Raman shifts have been determined for the maxima of individual peaks obtained as a result of the spectral curve analysis.

Raman shifts (cm^{-1}): 3632, 3596, 3585, 3228, 3030, 2807sh, 2261w, 1768sh, 1655s, 1555sh, 1420, 1093s, 1065sh, 928w, 797w, 702.

Source: Frost and Bahfenne (2009).

Comments: The spectrum is questionable. No independent analytical data are provided for the sample used. Data in the paper do not coincide with data in supplementary information. IR spectrum of the sample used shows significant admixture of serpentine.

Cobaltarthurite $CoFe^{3+}{}_2(AsO_4)_2(OH)_2 \cdot 4H_2O$

Origin: Dolores showing, Pastrana, about 10 km east of Mazarrón, the province of Murcia, Spain (type locality).

Experimental details: Raman scattering measurements have been performed on an arbitrarily oriented sample using Ar^+ laser radiation. The laser radiation power is not indicated.

Raman shifts (cm^{-1}): 3557, 3250, 3186, 1750–1550w (broad), 1042, 907, 846s, 816, 779, 555w, 509, 458w, 405, 348, 277, 260w, 240s, 231sh, 187, 151, 136, 98, 79, 35.

Source: Jambor et al. (2002).

Comments: The sample was characterized by powder X-ray diffraction data and electron microprobe analyses.

Cobaltaustinite $CaCo(AsO_4)(OH)$

Origin: No data.

Experimental details: Raman scattering measurements have been performed on arbitrarily oriented crystals using a 633 nm He-Ne laser. The laser radiation power is not indicated. The Raman shifts have been determined for the maxima of individual peaks obtained as a result of the spectral curve analysis.

Raman shifts (cm^{-1}): 3289, 3284, 918sh, 825s, 808, 795, 765, 469, 430, 387, 339, 327, 226, 214, 213, 168.

Source: Martens et al. (2003c).

Comments: No independent analytical data are provided for the sample used. For the Raman spectrum of cobaltaustinite see also Yang et al. (2007a).

Cobaltkoritnigite $Co(AsO_3OH) \cdot H_2O$

Origin: Richelsdorf District, Hessen, Germany.

Experimental details: Raman scattering measurements have been performed on arbitrarily oriented crystals using a 633 nm He-Ne laser. The laser radiation power is not indicated. The Raman shifts have been determined for the maxima of individual peaks obtained as a result of the spectral curve analysis.

Raman shifts (cm^{-1}): 3438w, 3165w, 2922, 2862sh, 1687sh, 1611, 1435, 1347, 1291, 1071sh, 1058sh, 1050, 1013sh, 1001s, 985sh, 973sh, 927, 907sh, 838s, 828s, 812sh, 726sh, 681, 637sh, 570w, 554w, 513w, 482, 461w, 430sh, 416sh, 401sh, 386, 367sh, 352sh, 301, 248w, 237sh, 205w, 190w, 166w, 152w, 140sh, 117w, 110w.
Source: Frost et al. (2014o).
Comments: No independent analytical data are provided for the sample used.

Cobaltomenite $Co(Se^{4+}O_3) \cdot 2H_2O$

Origin: El Dragon Mine, Potosi, Bolivia.
Experimental details: Raman scattering measurements have been performed on arbitrarily oriented crystals using a 633 nm He-Ne laser. The laser radiation power is not indicated. The Raman shifts have been determined for the maxima of individual peaks obtained as a result of the spectral curve analysis.
Raman shifts (cm^{-1}): 3450sh, 3209, 2962sh, 815sh, 813s, 716, 512, 443, 368w, 280w, 196w.
Source: Frost and Keeffe (2008b).
Comments: No independent analytical data are provided for the sample used. The Raman spectrum may correspond to Co-rich ahlfeldite.

Cobaltpentlandite Co_9S_8

Origin: Synthetic.
Experimental details: Raman scattering measurements have been performed on nanoparticles using 514.5 nm Ar$^+$ laser radiation. The nominal laser radiation power was 20 mW.
Raman shifts (cm^{-1}): ~650.
Source: Feng et al. (2015b).
Comments: The sample was characterized by powder X-ray diffraction data. For the Raman spectrum of cobaltpentlandite see also Yin et al. (2008).

Coccinite HgI_2

Origin: Synthetic.
Experimental details: Raman scattering measurements have been performed on an oriented crystal using 632.8 nm He-Ne laser radiation. The laser radiation power is not indicated. A 90°-scattering geometry was employed.
Raman shifts (cm^{-1}): 116, 32, 21.
Source: Nakashima et al. (1973).
Comments: No independent analytical data are provided for the sample used. For the Raman spectrum of coccinite see also Durig et al. (1969).

Cochromite $CoCr_2O_4$

Origin: Synthetic.
Experimental details: Raman scattering measurements have been performed on an anosized (about 20–25 nm) sample using 632 or 780 nm He-Ne or diode laser radiation. The laser radiation powers are not indicated.
Raman shifts (cm^{-1}): 684, 603, 514, 449, 195.

Source: Zákutná et al. (2014).
Comments: The sample was characterized by powder X-ray diffraction data and electron microprobe analysis.

Coconinoite $Fe^{3+}_2Al_2(UO_2)_2(PO_4)_4(SO_4)(OH)_2 \cdot 20H_2O$

Origin: Jomac mine, White Canyon, San Juan Co., Utah, USA.
Experimental details: Raman scattering measurements have been performed on arbitrarily oriented crystals using a 633 nm He-Ne laser. The laser radiation power is not indicated. The Raman shifts have been determined for the maxima of individual peaks obtained as a result of the spectral curve analysis.
Raman shifts (cm^{-1}): 1103w, 1085w, 1044sh, 1020s, 998s, 985s, 974sh, 847s, 837s, 826sh, 637w, 620w, 551w, 502sh, 492w, 447w, 409w, 377w, 320w, 229sh, 210sh, 199s, 181s, 147s, 110sh.
Source: Frost et al. (2011d).
Comments: No independent analytical data are provided for the sample used.

Coesite SiO_2

Origin: No data.
Experimental details: Raman scattering measurements have been performed on arbitrarily oriented crystals using 457.8, 488.0, and 514.5 nm Ar^+ laser radiation. The laser radiation power is not indicated. A 135°-scattering geometry was employed.
Raman shifts (cm^{-1}): 1164w, 1144w, 1065w, 1036w, 815w, 795w, 661w, 521s, 466, 427, 355, 326, 269s, 204, 176s, 151, 116s, 77s.
Source: Hemley (1987a, b).
Comments: No independent analytical data are provided for the sample used. For the Raman spectra of coesite see also Yang et al. (2007b), Palmeri et al. (2009), Miyahara et al. (2013), and Perraki and Faryad (2014).

Coffinite $U(SiO_4) \cdot nH_2O$

Origin: Synthetic.
Experimental details: Raman scattering measurements have been performed on a powdered sample. Kind of laser radiation is not indicated.
Raman shifts (cm^{-1}): 919s, 906sh, 591, 424.
Source: Mesbah et al. (2015).
Comments: The sample was characterized by powder X-ray diffraction data. For the Raman spectrum of coffinite see also Clavier et al. (2014).

Colemanite $CaB_3O_4(OH)_3 \cdot H_2O$

Origin: Death Valley, Inyo Co., California, USA (type locality).
Experimental details: Raman scattering measurements have been performed on arbitrarily oriented crystals using a 633 nm He-Ne laser. The laser radiation power is not indicated. The Raman shifts have been determined for the maxima of individual peaks obtained as a result of the spectral curve analysis.

Raman shifts (cm^{-1}): 3605s, 3534, 3389sh, 3300, 3182, 3069sh, 1603w, 1527sh, 1323, 1301sh, 1257, 1154, 1084, 1065, 1000sh, 988, 970, 907sh, 892, 876s, 846sh, 813, 788, 745, 709w, 684, 669, 611s, 565, 534, 505, 498, 479w, 455, 436w, 388, 350w, 325sh, 309, 267, 241, 223, 178, 167sh, 149, 129.
Source: Frost et al. (2013z).
Comments: No quantitative analytical data are provided for the sample used. For the Raman spectrum of colemanite see also Krishnamurti (1955).

Colimaite K_3VS_4

Origin: Colima volcano, State of Colima, Mexico (type locality).
Experimental details: Methods of sample preparation are not described. Raman scattering measurements have been performed using 514.5 nm Ar$^+$ laser radiation. The laser radiation power is not indicated.
Raman shifts (cm^{-1}): 990, 968, 879, 848, 689, 517, 482, 454, 401s, 387, 367w, 347, 338, 318, 297, 277s, 264s, 245s, 227, 203s, 192s, 180, 168.
Source: Ostrooumov et al. (2009).
Comments: The sample was characterized by powder X-ray diffraction data and electron microprobe analyses.

Colinowensite $BaCuSi_2O_6$

Origin: Synthetic.
Experimental details: No data.
Raman shifts (cm^{-1}): 1060w, 976, 579s, 506s, 454, 372w, 347w, 268, 231w, 173.
Source: Finger et al. (1989).
Comments: The sample was characterized by single-crystal X-ray diffraction data and electron microprobe analysis. The crystal structure is solved.

Coloradoite HgTe

Origin: Synthetic.
Experimental details: Raman scattering measurements have been performed at 90 K on a thin oriented sample with a (111) face using 514.5 nm Ar$^+$ laser radiation. The nominal laser radiation power was <100 mW. A 180°-scattering geometry was employed.
Raman shifts (cm^{-1}): 137s, 117, ~106sh.
Source: Ingale et al. (1989).
Comments: No independent analytical data are provided for the sample used.

Columbite-(Mg) $MgNb_2O_6$

Origin: Synthetic.
Experimental details: Raman scattering measurements have been performed on a powdered sample using 647.1 nm Kr$^+$ laser radiation. The nominal laser radiation power was 400 mW.
Raman shifts (cm^{-1}): 997, 906s, 850, 721w, 658, 616, 568, 533s, 497, 489, 460w, 451, 411s, 403sh, 389, 379w, 344, 324, 222, 205, 190, 170, 151, 140w, 125sh, 112, 88.
Source: Husson et al. (1977a).
Comments: The sample was characterized by powder X-ray diffraction data.

Columbite-(Mn) $Mn^{2+}Nb_2O_6$

Origin: Synthetic.
Experimental details: Raman scattering measurements have been performed on a powdered sample using 647.1 nm Kr^+ laser radiation. The nominal laser radiation power was 400 mW.
Raman shifts (cm^{-1}): 877s, 823, 707w, 634, 624, 606, 531s, 485, 438w, 399, 386w, 361, ~340sh, 314s, 297sh, 287, 274, 263, 248sh, 244s, 214, 206sh, 189sh, 179, 160w, 140s, 127, 113, 89.
Source: Husson et al. (1977a).
Comments: The sample was characterized by powder X-ray diffraction data. For the Raman spectrum of columbite-(Mn) see also Moreira et al. (2010a).

Comancheite $Hg^{2+}{}_{55}N^{3-}{}_{24}(NH_2,OH)_4(Cl,Br)_{34}$

Origin: Mariposa mine, Terlingua district, Brewster Co., Texas, USA (type locality).
Experimental details: Methods of sample preparation are not described. Raman scattering measurements have been performed using 785 nm diode laser radiation. The laser radiation power is not indicated. A 180°-scattering geometry was employed.
Raman shifts (cm^{-1}): 633sh, 577s, 545s, 470w, 440w, 315s, 268, 228s, 188s, 174s, ~140s.
Source: Cooper et al. (2013a).
Comments: The sample was characterized by single-crystal X-ray diffraction data. The crystal structure is solved.

Combeite $Na_{4.5}Ca_{3.5}Si_6O_{17.5}(OH)_{0.5}$

Origin: Synthetic.
Experimental details: Micro-Raman scattering measurements have been performed on an arbitrarily oriented sample using 514.5 nm Ar^+ laser radiation. The laser radiation power at the sample was 22 mW.
Raman shifts (cm^{-1}): 1039w, 986s, 903, 620, 588s, 532, 453w, 423, 346w, 279.
Source: Lin et al. (2015).
Comments: The sample was characterized by powder X-ray diffraction data.

Compreignacite $K_2(UO_2)_6O_4(OH)_6 \cdot 7H_2O$

Origin: West Wheal Owles, St. Just, Cornwall, UK.
Experimental details: Methods of sample preparation are not described. Raman scattering measurements have been performed using 785 nm diode laser radiation. The output laser radiation power was 380 mW.
Raman shifts (cm^{-1}): 834s, 785sh, 549, 460, 402, 329, 204.
Source: Driscoll et al. (2014).
Comments: The sample was characterized by powder X-ray diffraction data and electron microprobe analysis. For the Raman spectrum of compreignacite see also Frost et al. (2008g).

Conichalcite $CaCu(AsO_4)(OH)$

Origin: Lorena mine, Cloncurry, Queensland, Australia.
Experimental details: Raman scattering measurements have been performed on arbitrarily oriented crystals using a 633 nm He-Ne laser. The Raman shifts have been determined for the maxima of

individual peaks obtained as a result of the spectral curve analysis. The laser radiation power at the sample was 1 mW.
Raman shifts (cm^{-1}): 3233, 3158, 3086, 962w, 907, 832s, 821sh, 811sh, 781m 750, 534, 463, 446m 430, 389, 358, 335, 328, 303, 286, 274, 206, 180, 161, 121.
Source: Martens et al. (2003c).
Comments: No independent analytical data are given for the sample used. For the Raman spectra of conichalcite see also Reddy et al. (2005) and Đorđević et al. (2016).

Connellite $Cu_{36}(SO_4)(OH)_{62}Cl_8 \cdot 6H_2O$

Origin: Monte Fucinaia, central Western Italy.
Experimental details: Methods of sample preparation are not described. Raman scattering measurements have been performed on an arbitrarily oriented sample using 532 and 785 nm laser radiations. The laser radiation power at the sample was 4 and 1.4 mW, respectively.
Raman shifts (cm^{-1}): 3550w, 984s, 585, 524, 446, 404s, 350, 262w, 236w, 192, 184, 132m.
Source: Coccato et al. (2016).
Comments: The sample identification was done by powder X-ray diffraction. For the Raman spectrum of connellite see also Bouchard and Smith (2003).

Cooperite PtS

Origin: Synthetic.
Experimental details: Methods of sample preparation are not described. Raman scattering measurements have been performed using 514.5 nm Ar$^+$ laser radiation. The laser radiation output power was 500 mW.
Raman shifts (cm^{-1}): 328, 325sh (for pure PtS); 454sh, 430, 405sh, 382s, 358, 335, 317 (for a Pd-bearing sample).
Source: Pikl et al. (1999).
Comments: The samples were characterized by powder X-ray diffraction data and electron microprobe analysis. For the Raman spectrum of cooperite see also Mernagh and Hoatson (1995).

Copiapite $Fe^{2+}Fe^{3+}_4(SO_4)_6(OH)_2 \cdot 20H_2O$

Origin: Synthetic.
Experimental details: Methods of sample preparation are not described. Raman scattering measurements have been performed on an arbitrarily oriented sample using 532 nm Nd-YAG laser radiation. The laser radiation power is not indicated.
Raman shifts (cm^{-1}): 3527, 3179, 1644, 1224, 1138sh, 1115, 1026, 1005s, 996, 637, 614, 594, 554, 304, 270, 243.
Source: Kong et al. (2011b).
Comments: The sample was characterized by powder X-ray diffraction data and laser-induced breakdown spectroscopy. For the Raman spectra of copiapite see also Frost (2011c), Sobron and Alpers (2013), Rull et al. (2014), and Apopei et al. (2014a).

Copiapite $Fe^{2+}Fe^{3+}_4(SO_4)_6(OH)_2 \cdot 20H_2O$

Origin: Coranda-Hondol ore deposit, Certej, Romania.
Experimental details: Methods of sample preparation are not described. Raman scattering measurements have been performed on an arbitrarily oriented sample using 532 nm Nd-YAG laser radiation. The nominal laser radiation power was 7.4 mW.

Raman shifts (cm^{-1}): 3147, 1651w, 1247, 1143sh, 1113, 1031s, 999s, 748w, 637sh, 609, 558, 477s, 304sh, 274s, 246sh.
Source: Apopei et al. (2015).
Comments: The sample was characterized by powder X-ray diffraction data. For the Raman spectra of copiapite see also Frost (2011c), Kong et al. (2011b), Sobron and Alpers (2013), and Rull et al. (2014).

Coquandite $Sb^{3+}_{6+x}O_{8+x}(SO_4)(OH)_x \cdot H_2O_{(1-x)}$ ($x = 0.3$)

Origin: Pereta Mine, Italy (type locality).
Experimental details: Raman scattering measurements have been performed on arbitrarily oriented crystals using a 633 nm He-Ne laser. The laser radiation power is not indicated. The Raman shifts have been determined for the maxima of individual peaks obtained as a result of the spectral curve analysis.
Raman shifts (cm^{-1}): 3449w, 3318w, 3193w, 3122w, 2961w, 2900w, 2764w, 1588w, 1217sh, 1168w, 1151, 1100, 1072sh, 1020sh, 1007, 990s, 980, 970, 949sh, 787, 751, 600w, 638, 629sh, 610, 600, 508, 459, 429, 417, 375, 359, 317w, 291w, 270sh, 253, 229, 218sh, 216s, 203s, 178, 167s, 149, 129.
Source: Frost and Bahfenne (2010f).
Comments: No independent analytical data are provided for the sample used.

Coquimbite $Fe^{3+}_2(SO_4)_3 \cdot 9H_2O$

Origin: Baia Spriemining area, Romania.
Experimental details: Methods of sample preparation are not described. Raman scattering measurements have been performed using 532 nm Nd-YAG laser radiation. The nominal laser radiation power was 100 mW.
Raman shifts (cm^{-1}): 3388w, 1681w, 1202, 1167w, 1098, 1024, 882w, 604, 503, 457w, 285.
Source: Buzatu et al. (2016).
Comments: The sample was characterized by powder X-ray diffraction data. For the Raman spectra of coquimbite see also Apopei et al. (2012, 2014a), Sobron and Alpers (2013), Rull et al. (2014), and Frost et al. (2014b).

Corderoite $Hg_3S_2Cl_2$

Origin: Synthetic.
Experimental details: No data.
Raman shifts (cm^{-1}): 288sh, 280s, 132, 115, 63, 50s, 36, 24s.
Source: Radepont (2013).
Comments: The sample was characterized by powder X-ray diffraction data and electron microprobe analysis.

Cordierite $Mg_2Al_4Si_5O_{18}$

Origin: Northern part of the Strangways Metamorphic Complex, central Australia.
Experimental details: Raman scattering measurements have been performed on a polished section ∥ (010) of a single crystal using 632.8 nm He-Ne laser radiation. The output laser radiation power was <0.8 mW. A 180°-scattering geometry was employed.

Raman shifts (cm^{-1}): 1184, 971, 669, 577, 555, 366, 260s, 127 (for $E \| a$); 1010s, 971, 669, 577, 554s, 260 (for $E \| c$).
Source: Nasdala et al. (2006).
Comments: The sample was characterized by electron microprobe analysis. The empirical formula is $Mg_{1.6}Fe^{2+}_{0.4}Al_4Si_5O_{18}$. For the Raman spectrum of cordierite see also Majumdar and Mathew (2015). For the Raman spectrum of cordierite ("iolite") see also Culka et al. (2016a).

Cordylite (Ce) $(Na,Ca,\square)BaCe_2(CO_3)_4(F,O)$

Origin: No data.
Experimental details: Raman scattering measurements have been performed using 488 nm laser radiation. The nominal laser radiation power was 300 mW.
Raman shifts (cm^{-1}): 1538, 1088, 967, 720, 628.
Source: Hong et al. (1999).

Corkite $PbFe^{3+}_3(SO_4)(PO_4)(OH)_6$

Origin: Horn Silver mine, near Frisco, Beacer Co., Utah, USA.
Experimental details: Raman scattering measurements have been performed on arbitrarily oriented crystals using a 633 nm He-Ne laser. The laser radiation power is not indicated. The Raman shifts have been determined for the maxima of individual peaks obtained as a result of the spectral curve analysis.
Raman shifts (cm^{-1}): 3436, 3347, 3163sh, 1184sh, 1162, 1104s, 1050, 1003s, 996s, 983sh, 857, 821sh, 629sh, 620, 608sh, 572sh, 554, 466sh, 446, 430sh, 381, 274w, 217sh, 201, 144, 131sh, 104.
Source: Frost and Palmer (2011a).
Comments: No independent analytical data are provided for the sample used.

Cornetite $Cu_3(PO_4)(OH)_3$

Origin: Banská Bystrica, central Slovakia.
Experimental details: Raman scattering measurements have been performed on an oriented crystal, at $E \| b$ and $E \perp b$, using 632 nm He-Ne laser radiation. The nominal laser radiation power was 17 mW. A 180°-scattering geometry was employed.
Raman shifts (cm^{-1}): 3414, 1137w, 1112w, 1083, 1055, 1016, 971, 945sh, 861, 818, 801w, 750, 703w, 664w, 639w, 606, 539, 515, 477s, 446s, 412, 363, 297, 254w, 241, 214, 209, 174s, 131s, 109s, 86s (for $E \perp b$).
Source: Kharbish et al. (2014).
Comments: The sample was characterized by powder X-ray diffraction data and electron microprobe analyses. For the Raman spectrum of cornetite see also Frost et al. (2002g).

Cornubite $Cu_5(AsO_4)_2(OH)_4$

Origin: Daly mine, Flinders Ranges, South Australia.
Experimental details: Raman scattering measurements have been performed on arbitrarily oriented crystals using a 633 nm He-Ne laser. The laser radiation power was 1 mW. The Raman shifts have

been determined for the maxima of individual peaks obtained as a result of the spectral curve analysis.

Raman shifts (cm^{-1}): 3324, 3042sh, 962w, 815s, 780w, 525w, 496w, 440s, 398, 365w, 327w, 301, 259, 249, 211, 168, 151.

Source: Frost et al. (2002e).

Comments: No independent analytical data are provided for the sample used. For the Raman spectrum of cornubite see also Janeczek et al. (2016).

Cornwallite $Cu_5(AsO_4)_2(OH)_4$

Origin: Penberthy Croft mine, St Hilary, Cornwall, UK.

Experimental details: Raman scattering measurements have been performed on arbitrarily oriented crystals using a 633 nm He-Ne laser. The laser radiation power was 1 mW. The Raman shifts have been determined for the maxima of individual peaks obtained as a result of the spectral curve analysis.

Raman shifts (cm^{-1}): 3411w, 3350sh, 962w, 877sh, 859s, 806w, 763w, 606, 542, 512w, 454, 449s, 436s, 422, 363sh, 347s, 330sh, 311sh, 275, 279, 246, 203, 169s, 160, 137.

Source: Frost et al. (2002e).

Comments: No independent analytical data are provided for the sample used. For the Raman spectra of cornwallite see also Ciesielczuk et al. (2016) and Janeczek et al. (2016).

Coronadite $Pb(Mn^{4+}_6Mn^{3+}_2)O_{16}$

Origin: Imini, Morocco.

Experimental details: Methods of sample preparation are not described. Raman scattering measurements have been performed using 514.5 nm Ar$^+$ laser radiation. The nominal laser radiation power was 10 mW. A nearly 180°-scattering geometry was employed.

Raman shifts (cm^{-1}): 626s, 585s, 495, 388, 332.

Source: Julien et al. (2004).

Comments: The sample was characterized by powder X-ray diffraction data. The Raman shifts given by Julien et al. (2004) in Table 5 do not correspond to band positions in Fig. 5 of the cited paper. For the Raman spectrum of coronadite see also Fan et al. (2015).

Correianevesite $Fe^{2+}Mn^{2+}_2(PO_4)_2 \cdot 3H_2O$

Origin: Cigana mine, Conselheiro Pena, Rio Doce valley, Minas Gerais, Brazil (type locality).

Experimental details: Raman scattering measurements have been performed on arbitrarily oriented crystals using a 633 nm He-Ne laser. The laser radiation power is not indicated. The Raman shifts have been determined for the maxima of individual peaks obtained as a result of the spectral curve analysis.

Raman shifts (cm^{-1}): (3462), 3445s, (3400), 3265sh, 1641w, 1587sh, 1572w, 1553sh, 1193w, 1104w, (1093), 1064w, 1007s, 970+963s (unresolved doublet?), (951), 753, 588s, 569sh, (549), 531, 504sh, 482, 458sh, 420, (405), 373, 330w, 286sh, 260, 241w, 223w, 179, 164, 144.

Source: Frost et al. (2012j).

Comments: The sample was characterized by powder X-ray diffraction data and electron microprobe analyses. In the cited paper the mineral is described under the name "reddingite." Our investigations showed that it is correianevesite, a new mineral species of the reddingite group with ordered Fe^{2+} and Mn^{2+}.

Corundum Al_2O_3

Origin: Synthetic.
Experimental details: Raman scattering measurements have been performed on an arbitrarily oriented crystal using 780 nm laser radiation. The laser radiation power is not indicated.
Raman shifts (cm^{-1}): 751, 645, 578, 451w, 432, 418s, 378s.
Source: Jasinevicius (2009).
Comments: No independent analytical data are provided for the sample used. For the Raman spectra of corundum see also Shoval et al. (2001) and Andò and Garzanti (2014).

Cosalite $Pb_2Bi_2S_5$

Origin: An abandoned mine in the Karrantza valley, Basque Co., Spain.
Experimental details: Raman scattering measurements have been performed on an arbitrarily oriented sample using 514.5 or 785 nm Ar$^+$ and diode laser radiation. The laser radiation power at the sample was 20 mW.
Raman shifts (cm^{-1}): 439, 251s, 140s.
Source: Goienaga et al. (2011).
Comments: The sample was characterized by energy dispersive X-ray fluorescence. For the Raman spectrum of cosalite see also Fermo and Padeletti (2012).

Cotunnite $PbCl_2$

Origin: No data.
Experimental details: Methods of sample preparation are not described. Raman scattering measurements have been performed using 632 nm He-Ne laser radiation. The laser radiation output power was 30 mW.
Raman shifts (cm^{-1}): 202, 169, 158.
Source: Bouchard and Smith (2003).
Comments: The sample was characterized by powder X-ray diffraction data.

Coulsonite $Fe^{2+}V^{3+}_2O_4$

Origin: Vihanti deposit, Northern Finland Region, Finland.
Experimental details: Raman scattering measurements have been performed on an arbitrarily oriented sample using 633 nm laser radiation. The nominal laser radiation power was 2 or 20 mW.
Raman shifts (cm^{-1}): 1188w, 1144w, 1119w, 1062, 1020w, 994w, 962w, 937w, 908, 873w, 840w, 813w, 772w, 670s, 576, 526, 498, 469, 398w, 351w, 296w, 268w, 236w, 212w, 167, 131, 116.
Source: Voloshin et al. (2014).
Comments: The sample was characterized by powder X-ray diffraction data and electron microprobe analysis.

Covellite CuS

Origin: Guinaoang, NW Luzon, Philippines.
Experimental details: Raman scattering measurements have been performed on an arbitrarily oriented sample using 514.5 nm Ar$^+$ laser radiation. The laser radiation power at the sample was <10 mW. A 180°-scattering geometry was employed.

Raman shifts (cm^{-1}): 471s, 263, 139, 116.
Source: Mernagh and Trudu (1993).
Comments: The sample was characterized by powder X-ray diffraction data and electron microprobe analyses.
Comments: For the Raman spectra of covellite see also Bouchard and Smith (2003) and Kumar and Nagarajan (2011).

Crandallite $CaAl_3(PO_4)(PO_3OH)(OH)_6$

Origin: Synthetic.
Experimental details: Methods of sample preparation are not described. Raman scattering measurements have been performed on an arbitrarily oriented sample using 632.8 nm He-Ne laser radiation. The nominal laser radiation power was 10 mW.
Raman shifts (cm^{-1}): 3546sh, 3471, 3305, ~3150, ~1330w, 1228w, 1160, 1108s, 1035s, 982s, 858, 828, 720w, 693, 634, 615, 580, 555sh, 528, 462, 396s, 365, 258s, 223sh, 184s.
Source: Breitinger et al. (2006).
Comments: The sample was characterized by powder X-ray diffraction data. For the Raman spectra of crandallite see also Frost et al. (2011w) and Grey et al. (2011).

Cranswickite $Mg(SO_4) \cdot 4H_2O$

Origin: Calingasta, San Juan province, Argentina (type locality).
Experimental details: Methods of sample preparation are not described. Raman scattering measurements have been performed on an arbitrarily oriented sample using 632 nm He-Ne laser radiation. The laser radiation power is not indicated. A 180°-scattering geometry was employed.
Raman shifts (cm^{-1}): ~3430, ~3300, 1156, 1120, 1090w, 1002s, 617, 466.
Source: Peterson (2011).
Comments: The sample was characterized by powder X-ray diffraction data, electron microprobe analysis, and ICP-MS.

Creaseyite $Cu_2Pb_2Fe^{3+}_2Si_5O_{17} \cdot 6H_2O$

Origin: St. Anthony Mine, Tiger, Pinal Co., Arizona, USA.
Experimental details: Raman scattering measurements have been performed on arbitrarily oriented crystals using a 633 nm He-Ne laser. The laser radiation power is not indicated. The Raman shifts have been determined for the maxima of individual peaks obtained as a result of the spectral curve analysis.
Raman shifts (cm^{-1}): 3626s, 3525s, 3470sh, 3162, 2902, 2750sh, 1603, 1543, 1348w, 1071, 998s, 958sh, 920sh, 869s, 802, 712s, 672sh, 603, 511, 481sh, 443, 371, 351, 318, 295, 258sh, 237s, 211sh, 196s, 152sh, 139s, 126.
Source: Frost and Xi (2012g).
Comments: No independent analytical data are given for the sample used.

Crednerite $CuMnO_2$

Origin: Synthetic.
Experimental details: Raman scattering measurements have been performed on a thin film on quartz substrate using 532 nm diode laser radiation. The nominal laser radiation power was 13 mW.

Raman shifts (cm^{-1}): 688s, 381w, 314w.
Source: Chen et al. (2015b).
Comments: The sample was characterized by powder X-ray diffraction data.

Creedite $Ca_3Al_2(SO_4)(OH)_2F_8 \cdot 2H_2O$

Origin: Santa Eulalia mining district, Chihuahua province, Mexico.
Experimental details: Raman scattering measurements have been performed on arbitrarily oriented crystals using a 633 nm He-Ne laser. The laser radiation power is not indicated. The Raman shifts have been determined for the maxima of individual peaks obtained as a result of the spectral curve analysis.
Raman shifts (cm^{-1}): 3584, 3524s, 3458, 3382sh, 3349, 3248, 1673w, 1575w, 1503, 1499, 1234w, 1184, 1135, 1084w, 1033w, 1026w, 989sh, 986s, 983sh, 922w, 891, 819w, 765, 663, 629, 601, 596sh, 568w, 548, 483, 457sh, 440, 394, 371w, 348w, 322w, 311w, 286w, 278, 353w, 225, 217sh, 203, 190, 173, 149, 126.
Source: Frost et al. (2013al).
Comments: The sample was characterized by qualitative electron microprobe analysis.

Cristobalite SiO_2

Origin: No data.
Experimental details: Raman scattering measurements have been performed on a powdered sample using 476.5 nm laser radiation. The nominal laser radiation power was 400 mW.
Raman shifts (cm^{-1}): 1195, 1089, 1079, 795, 785, 485, 416, 380, 368, 287, 275, 230, 110.
Source: Etchepare et al. (1978).
Comments: The sample was characterized by powder X-ray diffraction data. For the Raman spectra of cristobalite see also Ling et al. (2011), Shoval et al. (2001), Ilieva et al. (2007), Wilson (2014), and Ferrero et al. (2016).

Cristobalite SiO_2

Origin: Lunar soil.
Experimental details: Raman scattering measurements have been performed on an arbitrarily oriented sample using 532 nm laser radiation. The nominal laser radiation power was in the range from 3 to 5 mW.
Raman shifts (cm^{-1}): 1075w, 781w, 411s, 229s.
Source: Ling et al. (2011).
Comments: No independent analytical data are given for the sample used. For the Raman spectra of cristobalite see also Etchepare et al. (1978), Shoval et al. (2001), Ilieva et al. (2007), Wilson (2014), and Ferrero et al. (2016).

Crocoite $Pb(CrO_4)$

Origin: Dundas, Tasmania, Australia.
Experimental details: Raman scattering measurements have been performed on an arbitrarily oriented crystal using 514.5 nm Ar$^+$ laser radiation. The laser radiation power at the sampling lens was 100 mW.
Raman shifts (cm^{-1}): 853sh, 840s, 825sh, 400, 377, 358s, 348, 338, 326, 179w, 135, 118w.

Source: Rodgers (1992).
Comments: The Raman shifts are indicated for the maxima of individual peaks obtained as a result of the spectral curve analysis. The sample identification was done by XRD, by SEM, and by electron probe analysis. For the Raman spectra of crocoite see also Frost (2004c) and Nasdala et al. (2004).

Cryptohalite $(NH_4)_2SiF_6$

Origin: Synthetic.
Experimental details: Raman scattering measurements in the region of N–H-stretching vibrations have been performed using 488 and 514.5 nm Ar^+ laser radiations. The laser radiation power at the sample was about 100 mW.
Raman shifts (cm^{-1}): 3235.
Source: Jenkins (1986).

Cryptomelane $K(Mn^{4+}{}_7Mn^{3+})O_{16}$

Origin: Synthetic.
Experimental details: No data.
Raman shifts (cm^{-1}): 743w, 626s, 574s, 508, 470, 386, 257w, 183.
Source: Santos et al. (2012).
Comments: The sample was characterized by powder X-ray diffraction data and ICP analysis. For the IR spectrum of cryptomelane see also Kim and Stair (2004).

Cubanite $CuFe_2S_3$

Origin: Synthetic.
Experimental details: Methods of sample preparation are not described. Raman scattering measurements have been performed on an arbitrarily oriented sample using 488 nm Ar^+ laser radiation. The nominal laser radiation power was below 10 mW.
Raman shifts (cm^{-1}): 469s, 374, 328w, 286s.
Source: Chandra et al. (2011a, b).
Comments: The sample was characterized by powder X-ray diffraction data and Mössbauer spectroscopy. For the Raman spectrum of cubanite see also Petrov (2014).

Cuboargyrite $AgSbS_2$

Origin: Synthetic.
Experimental details: Raman scattering measurements have been performed on a bulk polycrystalline sample using 1064 nm Nd-YAG laser radiation. The nominal laser radiation power was 50 mW.
Raman shifts (cm^{-1}): 324s, ~311w, 292, 282w, 240sh, 140, 125sh.
Source: Gutwirth et al. (2006).
Comments: The sample was characterized by powder X-ray diffraction data and electron microprobe analysis.

Cumengeite $Pb_{21}Cu_{20}Cl_{42}(OH)_{40} \cdot 6H_2O$

Origin: Beleo, Baja California, Mexico.
Experimental details: Raman scattering measurements have been performed on arbitrarily oriented crystals using a 633 nm He-Ne laser. The laser radiation power is not indicated. The Raman shifts

have been determined for the maxima of individual peaks obtained as a result of the spectral curve analysis.

Raman shifts (cm^{-1}): 3588sh, 3482s, 3413s, 3366sh, 3180, 1023sh, 984, 891w, 830sh, 797, 715sh, 676s, 500, 465, 376, 347, 307, 271, 243, 192, 154s.

Source: Frost et al. (2003j).

Comments: No independent analytical data are provided for the sample used. For the Raman spectra of cumengeite see also Bouchard and Smith (2003) and Frost and Williams (2004).

Cummingtonite $\square Mg_2Mg_5Si_8O_{22}(OH)_2$

Origin: Ca'Mondey, Montescheno, Piemonte, Italy.

Experimental details: Methods of sample preparation are not described. Raman scattering measurements have been performed using 514 and 785 nm laser radiations. The laser radiation power is not indicated.

Raman shifts (cm^{-1}): 3668, 3653, 1038, 671s, 384, 200, 186.

Source: Andò and Garzanti (2014).

Comments: The methods of the identification of the sample are not indicated. Raman spectrum of presumed cummingtonite was published by Mohanan (1993) and Kloprogge et al. (2001a), but chemical composition of this sample (Mohanan 1993) does not correspond to cummingtonite. For the Raman spectrum of cummingtonite see also Leissner et al. (2015).

Cuprite Cu_2O

Origin: Synthetic.

Experimental details: Raman scattering measurements have been performed on a crushed sample and a single crystal using 647.1 nm Kr$^+$ laser radiation. The nominal laser radiation power was 20 mW. A 90°-scattering geometry was employed.

Raman shifts (cm^{-1}): 640w, 485w, 420w, 300w, 220s, 204, 192, 190, 160w, 150, 125w, 106 (crushed sample); 640, 220, 204, 192, 150 (single crystal).

Source: Taylor and Weichman (1971).

Comments: No independent analytical data are provided for the sample used. For the Raman spectrum of cuprite see also Bouchard and Smith (2003).

Cuprocopiapite $Cu^{2+}Fe^{3+}_4(SO_4)_6(OH)_2 \cdot 20H_2O$

Origin: Rio Tinto Valley near Nerva, Spain.

Experimental details: No data.

Raman shifts (cm^{-1}): 997s, 978s, ~620, ~450.

Source: Chemtob et al. (2006).

Comments: The Raman spectrum is questionable: blue color is very unusual for cuprocopiapite. No independent analytical data are provided for the sample used.

Cuproiridsite $CuIr_2S_4$

Origin: Synthetic.

Experimental details: Methods of sample preparation are not described. Micro-Raman scattering measurements have been performed in a back-scattering geometry using 514.5 nm laser radiation. The laser radiation power is not indicated.

Raman shifts (cm^{-1}): ~402, ~375, ~327, ~300s.
Source: Zhang et al. (2014).
Comments: The sample was characterized by powder X-ray diffraction data. For the Raman spectrum of cuproiridsite see also Zhang et al. (2010b).

Cupromolybdite $Cu^{2+}_3O(Mo^{6+}O_4)_2$

Origin: Synthetic.
Experimental details: Raman scattering measurements have been performed on a single crystal in different scattering geometries using laser radiation with different wavelengths. The laser radiation power is not indicated.
Raman shifts (cm^{-1}): ~961, ~934, ~864w, ~842, ~813 [for the 532 nm laser radiation, in the $x(y, y+z)$-x scattering geometry].
Source: Sato et al. (2014).
Comments: The sample was characterized by X-ray diffraction.

Cuprorhodsite $CuRh_2S_4$

Origin: Synthetic.
Experimental details: Methods of sample preparation are not described. Raman scattering measurements have been performed using 514.5 nm Ar$^+$ laser radiation. The laser radiation power is not indicated.
Raman shifts (cm^{-1}): ~388, ~358, ~318, ~277s.
Source: Ito et al. (2003).
Comments: No independent analytical data are provided for the sample used.

Cuprorivaite $CaCuSi_4O_{10}$

Origin: No data in the cited paper.
Experimental details: No data in the cited paper.
Raman shifts (cm^{-1}): 1086s, 1013, 991, 788, 572, 473, 431s.
Source: Boschetti et al. (2008).
Comments: Methods of the sample identification are not indicated. For the Raman spectrum of cuprorivaite see also Pagès-Camagna et al. (1999).

Cuprosklodowskite $Cu(UO_2)_2(SiO_3OH)_2 \cdot 6H_2O$

Origin: Shinkolobwe mine, Shaba province, Democratic Republic of Congo (type locality).
Experimental details: Raman scattering measurements have been performed on arbitrarily oriented crystals using a 633 nm He-Ne laser. The laser radiation power is not indicated. The Raman shifts have been determined for the maxima of individual peaks obtained as a result of the spectral curve analysis.
Raman shifts (cm^{-1}): 3694w, 3571w, 3499, 3435s, 3282s, 2920, 1504, 1297, 1246, 1156, 974, 919w, 917, 901w, 847w, 812sh, 787s, 774w, 759w, 747, 535, 507w, 477w, 411w, 387, 301, 277, 267, 218, 206, 185, 165w, 134w, 114.
Source: Frost et al. (2006e).

Comments: No independent analytical data are provided for the sample used. For the Raman spectrum of cuprosklodowskite see also Driscoll et al. (2014).

Cuprospinel $Cu^{2+}Fe^{3+}_2O_4$

Origin: Synthetic.
Experimental details: Methods of sample preparation are not described. Micro-Raman scattering measurements have been performed using 532 nm Nd-YAG laser radiation. The laser radiation output power was 3 mW.
Raman shifts (cm^{-1}): 632, 656, 549, 462s, 346, 271, 211, 168s (for a sample annealed at 1200°C).
Source: Silva et al. (2014).
Comments: The sample was characterized by powder X-ray diffraction data. For the Raman spectrum of cuprospinel see also Li et al. (2015).

Cuprotungstite $Cu^{2+}_3(WO_4)_2(OH)_2$

Origin: Cordillera Mine, Peelwood, Australia (?).
Experimental details: Raman scattering measurements have been performed on arbitrarily oriented crystals using a 785 nm Nd-YAG laser. The laser radiation power at the sample was 1 mW. The Raman shifts have been determined for the maxima of individual peaks obtained as a result of the spectral curve analysis.
Raman shifts (cm^{-1}): 926s, 769w, 566, 498w, 415, 353sh, 329, 253, 212, 173w.
Source: Frost et al. (2004d).
Comments: Questionable data: qualitative electron microprobe analysis given in the cited paper shows a high content of Ca.

Curienite $Pb(UO_2)_2(VO_4)_2 \cdot 5H_2O$

Origin: Mounana Mine, Haut Ogoue, Gabon (type locality).
Experimental details: Raman scattering measurements have been performed on arbitrarily oriented crystals using a 633 nm He-Ne laser. The laser radiation power is not indicated. The Raman shifts have been determined for the maxima of individual peaks obtained as a result of the spectral curve analysis.
Raman shifts (cm^{-1}): 976sh, 959, 860w, 825, 741s, 655w, 569, 534, 465, 410, 374s, 362, 312, 288, 264, 234, 193sh.
Source: Frost et al. (2005c).
Comments: No independent analytical data are given for the sample used.

Curite $Pb_{3+x}[(UO_2)_4O_{4+x}(OH)_{3-x}]_2 \cdot 2H_2O$

Origin: Ranger U Mine, Northern Territory, Australia.
Experimental details: Raman scattering measurements have been performed on arbitrarily oriented crystals using a 633 nm He-Ne laser. The laser radiation power is not indicated. The Raman shifts have been determined for the maxima of individual peaks obtained as a result of the spectral curve analysis.
Raman shifts (cm^{-1}): 3437, 3297, 1530, 886w, 807w, 791, 772, 742w, 650, 561, 503, 415sh, 455, 393, 367, 340, 301sh, 273, 250, 225w, 212, 184, 163sh, 154sh, 144.

Source: Frost et al. (2007g).
Comments: No independent analytical data are given for the sample used. For the Raman spectrum of curite see also Frost et al. (2007h).

Cuspidine $Ca_8(Si_2O_7)_2F_4$

Origin: Anakitskii massif, Siberia.
Experimental details: Raman scattering measurements have been performed on an arbitrarily oriented crystal. No other experimental details are described.
Raman shifts (cm^{-1}): 910s, 654, 361, 300.
Source: Sharygin et al. (1996a).
Comments: The sample was characterized by powder X-ray diffraction data and electron microprobe analysis. For the Raman spectrum of cuspidine see also Sharygin et al. (1996b).

Cyanochroite $K_2Cu(SO_4)_2 \cdot 6H_2O$

Origin: Synthetic.
Experimental details: Methods of sample preparation are not described. Raman scattering measurements have been performed using 532 nm laser radiation. The laser radiation power at the sample was 20 mW. A 180°-scattering geometry was employed.
Raman shifts (cm^{-1}): 1159w, 1133w, 1081, 983s, 629, 609, 457, 438, 256, 187, 112, 66.
Source: Majzlan et al. (2015)
Comments: For the Raman spectrum of cyanochroite see also Jentzsch et al. (2013).

Cyanotrichite $Cu_4Al_2(SO_4)(OH)_{12}(H_2O)_2$

Origin: Cap Garonne mine, near La Pradet, Var, France.
Experimental details: Raman scattering measurements have been performed on a single crystal with the laser beam orthogonal to the elongation direction (*b* axis) using 532 nm Nd-YAG laser radiation. The laser radiation power was below 30 mW.
Raman shifts (cm^{-1}): 3590, 3300–3400, 2330w, 1142w, 377s, 609w, 592, 525s, 448, 274, 230.
Source: Mills et al. (2014a).
Comments: The sample was characterized by powder X-ray diffraction data.

Cymrite $Ba(Si,Al)_4(O,OH)_8 \cdot H_2O$

Origin: Synthetic.
Experimental details: Unpolarized micro-Raman scattering measurements have been performed on a microcrystalline aggregate using 514.5 nm Ar^+ laser radiation. The laser radiation power is not indicated.
Raman shifts (cm^{-1}): 3567sh, 3500, 1626w, (1555w), 1091w, 953w, 800w, 673w, 470w, 396, 297w, 104s.
Source: Graham et al. (1992).
Comments: The sample was characterized by powder X-ray diffraction data and electron microprobe analysis.

Cyrilovite $NaFe^{3+}_3(PO_4)_2(OH)_4 \cdot 2H_2O$

Origin: Sapucaia (Proberil) mine, Conselheiro Pena pegmatitedistrict, Minas Gerais, Brazil.

Experimental details: Raman scattering measurements have been performed on arbitrarily oriented crystals using a 633 nm He-Ne laser. The laser radiation power is not indicated. The Raman shifts have been determined for the maxima of individual peaks obtained as a result of the spectral curve analysis.

Raman shifts (cm^{-1}): 3452, 3328, 3244sh, 3194sh, 1184sh, 1177, 1136, 1105, 1087sh, 1065sh, 1055s, 1038sh, 1013sh, 992s, 974sh, 852, 811, 7762sh, 631s, 612s, 588sh, 541, 498sh, 482, 437, 411, 3065s, 279sh, 261, 216, 197sh, 165, 156, 148, 131sh, 117.

Source: Frost et al. (2013u).

Comments: No independent analytical data are given for the sample used.

Czochralskiite $Na_4Ca_3Mg(PO_4)_4$

Origin: Morasko IAB-MG iron meteorite, Poland (type locality).

Experimental details: Raman scattering measurements have been performed on an arbitrarily oriented sample using 632.8 nm He-Ne laser radiation. The laser radiation power is not indicated.

Raman shifts (cm^{-1}): 1119, 1067w, 1053, 1039, 1022, 1011, 986s, 974s, 966s, 606, 585, 578, 441.

Source: Karwowski et al. (2016).

Comments: The sample was characterized by powder X-ray diffraction data and electron microprobe analyses. The crystal structure is solved.

Dachiardite Na $Na_4(Si_{20}Al_4)O_{48} \cdot 13H_2O$

Origin: Elba Island, Italy.

Experimental details: Methods of sample preparation are not described. Raman scattering measurements have been performed using 1060 nm Nd-YAG laser radiation. The laser radiation power at the sample was 300 mW.

Raman shifts (cm^{-1}): 1090w, 714s, 558, 479, 409, 248, 182.

Source: Mozgawa (2001).

Comments: The sample was characterized by powder X-ray diffraction data.

Danburite $CaB_2Si_2O_8$

Origin: No data.

Experimental details: Polarized Raman scattering measurements have been performed on an oriented crystal in different scattering configurations using 514.5 nm Ar^+ laser radiation. The laser radiation power is not indicated.

Raman shifts (cm^{-1}): $z(xx)y$—1115, 1018, 983, 957, 925, 917, 783, 635, 619s, 572, 489, 450, 434, 357, 321, 255, 220, 190, 176, 139; $z(xy)x$—1185w, 1087w, 1037w, 963w, 885w, 760w, 645w, 619w, 592w, 556w, 472w, 465w, 450w, 344w, 288, 268w, 216w, 201, 150, 136w; $z(xz)x$—1160w, 1050w, 1006w, 983, 908, 880w, 783w, 684w, 637, 619, 581w, 471, 450w, 434w, 339, 305, 255, 236, 204, 193, 135, 126; $z(yz)x$—1185w, 1037w, 1013w, 980w, 932w, 786w, 725w, 634w, 619w, 562w, 418w, 404w, 315w, 278w, 213, 190w, 162, 148w.

Source: Best et al. (1994).

Comments: No independent analytical data are provided for the sample used. For the Raman spectra of danburite see also Kimata (1993) and Manara et al. (2009).

Darapskite $Na_3(SO_4)(NO_3) \cdot H_2O$

Origin: Synthetic.
Experimental details: Methods of sample preparation are not described. Micro-Raman scattering measurements have been performed using 532 nm laser radiation. The laser radiation power at the sample was 2 mW.
Raman shifts (cm^{-1}): 3479sh, 3451w, 1424, 1353, 1171, 1124, 1085, 1060s, 993s, 729, 706, 639, 618, 472w, 455.
Source: Jentzsch et al. (2013).
Comments: The sample was characterized by powder X-ray diffraction data. For the Raman spectra of darapskite see also Jentzsch et al. (2012b) and Linnow et al. (2013).

Darrellhenryite $Na(Al_2Li)Al_6(Si_6O_{18})(BO_3)_3(OH)_3O$

Origin: No data in the cited paper.
Experimental details: Raman scattering measurements have been performed on an oriented crystal in the $-y(zz)y$ scattering geometry using 514.5 or 488 nm Ar$^+$ laser radiation. The laser radiation power at the sample was 14 mW.
Raman shifts (cm^{-1}): 3593±4, 3562±4, 3494±8, 3465±11, 1085, 975, 750, 707s, 643, 534sh, 515, 407, 374s, 335w, 315, 268, 249, 223s.
Source: Watenphul et al. (2016b).
Comments: The sample was characterized by electron microprobe analysis and LA-ICP-MS. The Raman shifts were partly determined by us based on spectral curve analysis of the published spectrum.

Dashkovaite $Mg(HCOO)_2 \cdot 2H_2O$

Origin: Synthetic.
Experimental details: Raman scattering measurements have been performed on an arbitrarily oriented sample using 514.5 nm Ar$^+$ laser radiation. The laser radiation power is not indicated. 90°-scattering geometry was employed.
Raman shifts (cm^{-1}): 1405, 1396, 1385, 1376s, 1367.
Source: Stoilova and Koleva (2000).
Comments: The sample was characterized by powder X-ray diffraction data.

Datolite $CaB(SiO_4)(OH)$

Origin: Canossa, Réggionell'Emilia, Italy.
Experimental details: Raman scattering measurements have been performed on arbitrarily oriented crystals using a 633 nm He-Ne laser. The laser radiation power is not indicated. The Raman shifts have been determined for the maxima of individual peaks obtained as a result of the spectral curve analysis.

Raman shifts (cm^{-1}): 3498s, 2892w, 1355, 1243w, 1202w, 1172, 1154w, 1148, 1077s, 1002sh, 985, 956w, 917, 8872w, 864w, 828w, 765w, 730w, 708sh, 693s, 669, 654, 601sh, 593w, 559w, 491, 466w, 458w, 440w, 424w, 392, 378sh, 362, 332w.
Source: Frost et al. (2013ah).
Comments: The sample was characterized by powder X-ray diffraction data and electron microprobe analysis. For the Raman spectrum of datolite see also Goryainov et al. (2015).

Daubréelite $FeCr_2S_4$

Origin: Bustee, Pesyanoe, and Aubresmeteorites.
Experimental details: Raman scattering measurements have been performed on arbitrarily oriented samples using 514.5 nm Ar$^+$ laser radiation. The laser radiation power at the sample was 2 mW.
Raman shifts (cm^{-1}): 365s, 290w, 255s, 160.
Source: Avril et al. (2013).
Comments: The sample was characterized by powder X-ray diffraction data and electron microprobe analyses. For the Raman spectrum of daubréelite see also Lutz et al. (1989).

Davidite-(La) $La(Y,U)Fe_2(Ti,Fe,Cr,V)_{18}(O,OH,F)_{38}$

Origin: Billeroo Prospect, South Australia and Radium Hill Mine, Mingary, Olary, South Australia.
Experimental details: Raman scattering measurements have been performed on amorphous, metamict samples using a 633 nm He-Ne laser. The laser radiation power is not indicated. The Raman shifts have been determined for the maxima of individual peaks obtained as a result of the spectral curve analysis.
Raman shifts (cm^{-1}): 1597w, 1424sh, 1322, 1250w, 1074w, 698sh, 641, 514, 412sh, 394, 293, 206, 169sh, 151s (for the sample from the Billeroo Prospect);
1357sh, 1308, 1255sh, 1065w, 812w, 651s, 609s, 497, 450sh, 408, 380, 297sh, 291s, 223s, 150 (for the sample from the Radium Hill Mine).
Source: Frost and Reddy (2011b).
Comments: No independent analytical data are provided for the sample used.

Davidlloydite $Zn_3(AsO_4)_2 \cdot 4H_2O$

Origin: Tsumeb mine, Tsumeb, Otjikoto (Oshikoto) region, Namibia (type locality).
Experimental details: No data.
Raman shifts (cm^{-1}): 3360, 3295, 3170, 3110, 2950, 865s, 841sh, 550, 504, 454s, 420sh, 394w, 353, 305, 294, 258, 211, 200sh, 182w, 170, 163w, 121w.
Source: Hawthorne et al. (2012).
Comments: The sample was characterized by powder X-ray diffraction data and electron microprobe analysis. The crystal structure is solved. The Raman shifts were partly determined by us based on spectral curve analysis of the published spectrum.

Davisite $CaScAlSiO_6$

Origin: Allende meteorite, Chihuahua, Mexico (type locality).
Experimental details: Raman scattering measurements have been performed on a grain in polished section using 514.5 nm Ar$^+$ laser radiation. The laser radiation power is not indicated.

Raman shifts (cm^{-1}): ~855, ~815, ~660, ~540, ~405, ~385, ~330.
Source: Ma and Rossman (2009b), Ma et al. (2012c).

Dawsonite $NaAl(CO_3)(OH)_2$

Origin: Poudrette Quarry, Mont Saint-Hilaire, Québéc, Canada.
Experimental details: Raman scattering measurements have been performed on arbitrarily oriented crystals using a 633 nm He-Ne laser. The laser radiation power is not indicated. The Raman shifts have been determined for the maxima of individual peaks obtained as a result of the spectral curve analysis.
Raman shifts (cm^{-1}): 3467w, (3341w), (3295), 3283, 3250, (3218w), 1760w, 1731w, 1691w, 1506s, 1484, 1366w, 1099, 1091s, 1068, 936w, 898, 824, (820w), 747, 731, 590s, 519, 443, 389, (374w), 361, 261, 219, 191s, (188), 152, (141w), 134.
Source: Frost et al. (2015h).
Comments: The sample was characterized by qualitative electron microprobe analysis. For the Raman spectra of dawsonite see also Serna et al. (1985), Vard and Williams-Jones (1993), Sirbescu and Nabelek (2003), and Frost and Bouzaid (2007).

Decrespignyite-(Y) $Y_4Cu(CO_3)_4Cl(OH)_5 \cdot 2H_2O$

Origin: Paratoo copper mine, near Yunta, Olary district, South Australia (type locality).
Experimental details: Methods of sample preparation are not described. Micro-Raman scattering measurements have been performed using 514.5 nm Ar$^+$ laser radiation. The laser radiation output power was 0.3 mW.
Raman shifts (cm^{-1}): 3463, 3419, 1094s, 1075s, 1062s, 822, 751, 478, 415, 344, 201s.
Source: Wallwork et al. (2002).
Comments: The sample was characterized by powder X-ray diffraction data and electron microprobe analyses. For the Raman spectrum of decrespignyite-(Y) see also Frost and Palmer (2011g).

Delafossite $Cu^{1+}Fe^{3+}O_2$

Origin: Synthetic.
Experimental details: Methods of sample preparation are not described. Raman scattering measurements have been performed using 514.5 nm Ar$^+$ laser radiation. The laser radiation power is not indicated.
Raman shifts (cm^{-1}): 692, 351.
Source: Pavunny et al. (2010).
Comments: The sample was characterized by powder X-ray diffraction data and qualitative electron microprobe analysis. For the Raman spectra of delafossite see also Aktas et al. (2011) and Kučerová et al. (2013).

Delhayelite $K_7Na_3Ca_5Al_2Si_{14}O_{38}F_4Cl_2$

Origin: Yukspor Mt., Khibiny massif, Kola Peninsula, Russia.
Experimental details: Methods of sample preparation are not described. Raman scattering measurements have been performed using 514.5 nm Ar$^+$ laser radiation. The nominal laser radiation power was 50 mW. A 180°-scattering geometry was employed.

Raman shifts (cm^{-1}): 3147, 3098, 2980, 2824, 1144, 1081, 998w, 757w, 606s, 581, 495, 407, 353, 307w.
Source: Sharygin et al. (2013b).
Comments: The sample was characterized by powder X-ray diffraction data and electron microprobe analysis.

Deliensite $Fe^{2+}(UO_2)_2(SO_4)_2(OH)_2 \cdot 7H_2O$

Origin: Schweitzer dump, Jáchymov ore district, Western Bohemia, Czech Republic.
Experimental details: Methods of sample preparation are not described. Micro-Raman scattering measurements have been performed on an arbitrarily oriented sample using 532 nm laser radiation. The nominal laser radiation power was 3 mW.
Raman shifts (cm^{-1}): ~6300, ~3510, 1637sh, 1625w, 1157w, 1068sh, 1050s, 1024s, 1010sh, 980, 932, 900, 853sh, 838s, 824sh, 747w, 725w, 710w, 653w, 627w, 610w, 573w, 545w, 478, 453, 364, 235, 200, 152sh, 105sh, 90.
Source: Plášil et al. (2012a).
Comments: The sample was characterized by powder X-ray diffraction data and electron microprobe analyses. The crystal structure is solved. The Raman shifts were partly determined by us based on spectral curve analysis of the published spectrum.

Dellaite $Ca_6(Si_2O_7)(SiO_4)(OH)_2$

Origin: Birkhin complex, Eastern Siberia, Russia.
Experimental details: Raman scattering measurements have been performed on an arbitrarily oriented sample using 514.5 nm Ar$^+$ laser radiation. The nominal laser radiation power was in the range from 30 to 50 mW.
Raman shifts (cm^{-1}): 3592, 3573, 998, 963, 956sh, 930s, 892, 867s, 821w, 666, 652sh, 551sh, 528, 411, 397, 382sh, 354, 278, 248w, 127w.
Source: Armbruster et al. (2011).
Comments: The sample was characterized by powder X-ray diffraction data and electron microprobe analyses.

Delvauxite $CaFe^{3+}_4(PO_4)_2(OH)_8 \cdot 4\text{-}5H_2O$

Origin: Berneau, near Vise, Liège, Belgium (type locality).
Experimental details: Raman scattering measurements have been performed on an amorphous SO$_4$-rich sample using a 633 nm He-Ne laser. The laser radiation power is not indicated. The Raman shifts have been determined for the maxima of individual peaks obtained as a result of the spectral curve analysis.
Raman shifts (cm^{-1}): 3317w, 3193sh, 3029w, 2900w, 1607, 1385sh, 1345, 1085sh, 1006s, 630, 583sh, 485sh, 464s, 439sh, 276, 208, 148sh, 128, 108.
Source: Frost et al. (2012d).
Comments: No independent analytical data are provided for the sample used. The Raman spectrum of delvauxite is given also by Frost and Palmer (2011d), but bands at 1167, 1095, 799, 780 and 692 cm^{-1} in the IR spectrum of this sample correspond to quartz.

Demartinite K_2SiF_6

Origin: Synthetic.
Experimental details: Methods of sample preparation are not described. Raman scattering measurements have been performed using 1064 nm Nd-YAG laser radiation. The nominal laser radiation was in the range 80–200 mW.
Raman shifts (cm^{-1}): 655, 478w, 408.
Source: Rissom et al. (2008).
Comments: No independent analytical data are given for the sample used.

Demesmaekerite $Pb_2Cu_5(UO_2)_2(Se^{4+}O_3)_6(OH)_6 \cdot 2H_2O$

Origin: Musonoi mine, Shaba, Democratic Republic of Congo (type locality).
Experimental details: Raman scattering measurements have been performed on arbitrarily oriented crystals using a 633 nm He-Ne laser. The laser radiation power is not indicated. The Raman shifts have been determined for the maxima of individual peaks obtained as a result of the spectral curve analysis.
Raman shifts (cm^{-1}): 3382, 3319, 1493, 1458sh, 1366, 1095, 1062, 822, 756, 719, 597w, 535, 510, 450sh, 432sh, 351, 295, 269, 215, 178s, 151s, 114sh.
Source: Frost et al. (2008d).
Comments: No independent analytical data are given for the sample used.

Demicheleite-(Br) BiSBr

Origin: Synthetic.
Experimental details: Raman scattering measurements have been performed on an arbitrarily oriented crystal using 676.4 or 632.8 nm Kr$^+$ laser radiations with the laser radiation power of 150 mW, as well as a He-Ne laser with the laser radiation power of 40 mW.
Raman shifts (cm^{-1}): 287s, 250, 121, 92, 54, 42sh [$x(yx)y$, A_g modes]; 234, 46s [$x(zx)y$, B_g modes].
Source: Furman et al. (1976).
Comments: No independent analytical data are given for the sample used.

Demicheleite-(Cl) BiSCl

Origin: Synthetic.
Experimental details: Raman scattering measurements have been performed on an arbitrarily oriented crystal using 676.4 or 632.8 nm Kr$^+$ laser radiations with the laser radiation power of 150 mW, as well as a He-Ne laser with the laser radiation power of 40 mW.
Raman shifts (cm^{-1}): 287, 260, 107, 95, 54, 44[$x(yx)y$, A_g modes]; 240, 138, 47[$x(zx)y$, B_g modes].
Source: Furman et al. (1976).
Comments: No independent analytical data are given for the sample used.

Demicheleite-(I) BiSI

Origin: Synthetic.
Experimental details: Raman scattering measurements have been performed on an oriented crystal in the $-y(xy)x$ geometry using 1064 nm Nd-YAG laser radiation. The laser radiation power is not indicated.

Raman shifts (cm^{-1}): 290, 227, 119sh, 108, 85s, 55s, 50, 37s, 30.
Source: Teng et al. (1978).
Comments: No independent analytical data are given for the sample used. The Raman shifts were determined by us based on spectral curve analysis of the published spectrum.

Denningite $CaMn^{2+}Te^{4+}_4O_{10}$

Origin: Bambolla mine, Moctezuma, Sonora, Mexico (type locality).
Experimental details: Raman scattering measurements have been performed on arbitrarily oriented crystals using a 633 nm He-Ne laser. The laser radiation power is not indicated.
Raman shifts (cm^{-1}): 766sh, 734s, 674, 479w, 450w, 381, 349s, 237, 155.
Source: Frost et al. (2008h).
Comments: No independent analytical data are provided for the sample used.

Derriksite $Cu_4(UO_2)(Se^{4+}O_3)_2(OH)_6$

Origin: Musonoimine, Katanga, Democratic Republic of Congo (type locality).
Experimental details: Raman scattering measurements have been performed on arbitrarily oriented crystals using a 633 nm He-Ne laser. The laser radiation power is not indicated. The Raman shifts have been determined for the maxima of individual peaks obtained as a result of the spectral curve analysis.
Raman shifts (cm^{-1}): 3530, 3407, 3247sh, 2917, 1623w, 1433w, 971s, 943sh, 881, 788s, 282, 257sh, 206, 162w, 137, 117sh.
Source: Frost et al. (2014a).
Comments: No independent analytical data are provided for the sample used.

Desautelsite $Mg_6Mn^{3+}_2(CO_3)(OH)_{16} \cdot 4H_2O$

Origin: No data.
Experimental details: Raman scattering measurements have been performed on arbitrarily oriented crystals using a 633 nm He-Ne laser. The laser radiation power at the sample was below 1 mW. The Raman shifts have been determined for the maxima of individual peaks obtained as a result of the spectral curve analysis.
Raman shifts (cm^{-1}): 3646, 3608, 3509, 3409, 3325, 2882, 2836w, 2776, 1676w, 1638w, 1610w, 1579w, 1440w, 1407w, 1393, 1372w, 1349, 1342, 1303w, 1110w, 1086, 1062w, 1055, 1016w, 883w, 878w, 873, 560, 535, 506, 455, 436w, 422w, 313w, 281, 254.
Source: Frost and Erickson (2005).
Comments: No independent analytical data are provided for the sample used.

Dessauite-(Y) $Sr(Y,U,Mn)Fe_2(Ti,Fe,Cr,V)_{18}(O,OH)_{38}$

Origin: Provence-Alpes-Côte d'Azur, France.
Experimental details: Raman scattering measurements have been performed on an arbitrarily oriented crystal using 632.8 nm He-Ne laser radiation. The laser radiation power is not indicated. A 180°-scattering geometry was employed.
Raman shifts (cm^{-1}): 812, 638s, 604sh, 520, (485w), 430, 360w, 329w, 293, 240.
Source: Bittarello et al. (2014).

Comments: The sample was characterized by powder X-ray diffraction data and electron microprobe analysis.

Destinezite $Fe^{3+}_2(PO_4)(SO_4)(OH)\cdot 6H_2O$

Origin: Żdanów, Bardzkie Mts. (West Sudetes), Poland.
Experimental details: Raman scattering measurements have been performed on an arbitrarily oriented thin section of a crystal using 514.5 nm Ar$^+$ laser radiation. The laser radiation power is not indicated.
Raman shifts (cm^{-1}): 3472, 3224w, 1357w, 1111w, 1048s, 980s, 626, 605sh, 540sh, 460, 268, 200.
Source: Koszowska et al. (2005).
Comments: The sample was characterized by powder X-ray diffraction data and electron microprobe analyses. Raman shifts are given for a sample with the highest content of poorly ordered phase. The empirical formula of the sample used is $Fe_{2.09}Al_{0.1}(PO_4)_{1.08}(SO_4)_{0.89}(SiO_4)_{0.13}(OH)\cdot 4.01H_2O$. For the Raman spectrum of destinezite see also Frost and Palmer (2011f).

Devilline $CaCu_4(SO_4)_2(OH)_6\cdot 3H_2O$

Origin: Ozernyi district, Salla-Kuolayarvi, Kola Peninsula, Russia.
Experimental details: Raman scattering measurements have been performed on an arbitrarily oriented sample using 514.5 nm Ar$^+$ laser radiation. The nominal laser radiation power at the sample was 50 mW.
Raman shifts (cm^{-1}): 3594w, 3563w, 3493w, 1121s, 1041, 992s, 819, 599, 412, 241, 210.
Source: Voloshin et al. (2015b).
Comments: The sample was characterized by powder X-ray diffraction data and electron microprobe analysis. For the Raman spectrum of devilline see also Majzlan et al. (2015).

Devitoite $[Ba_6(PO_4)_2(CO_3)][Fe^{2+}_7(OH)_4Fe^{3+}_2O_2(SiO_3)_8]$

Origin: Esquire #8 claim, Big Creek, Fresno Co., California, USA (type locality).
Experimental details: Raman scattering measurements have been performed on arbitrarily oriented crystals using 514.5 nm Ar$^+$ laser radiation. The laser radiation power at the sample was 2.5 mW.
Raman shifts (cm^{-1}): 1072, 1053, 914s, 700sh, 660s, 463w, 243w.
Source: Kampf et al. (2010b).
Comments: The sample was characterized by powder X-ray diffraction data and electron microprobe analyses. The crystal structure is solved.

Devitrite $Na_2Ca_3Si_6O_{16}$

Origin: Synthetic.
Experimental details: Raman scattering measurements have been performed on a single crystal at an angle of 45° between the electric field vector of the exciting laser and the [100] direction. A 633 nm He-Ne laser was used. The laser radiation power at the sample was 1 mW.
Raman shifts (cm^{-1}): 1126, 1110, 1094, 1079, 1059, 1051, 1010, 1003, 977, 966, 955, 948, 915, 908w, 890w, 836w, 805w, 791, 715, 698, 673s, 659s, 607s, 548, 520, 509, 499, 486, 472, 455, 444, 432, 416, 408, 397, 387, 342, 327, 306, 294, 286, 274, 253, 245, 226, 208, 193, 170, 151, 143, 126, 117, 105.

Source: Kahlenberg et al. (2010).
Comments: Devitrite is a crystalline impurity phase in industrial soda-lime glasses. The sample was characterized by single-crystal X-ray diffraction data. The crystal structure is solved.

Dewindtite $H_2Pb_3(UO_2)_6O_4(PO_4)_4 \cdot 12H_2O$

Origin: Ranger U mine, Northern Territory, Australia.
Experimental details: Raman scattering measurements have been performed on arbitrarily oriented crystals using a 633 nm He-Ne laser. The laser radiation power is not indicated. The Raman shifts have been determined for the maxima of individual peaks obtained as a result of the spectral curve analysis.
Raman shifts (cm^{-1}): 3524, 3456, 3299, 3297, 1659, 1623, 1117, 1069, 1033, 1021, 994, 978, 8687, 857, 831, 818, 808, 795, 783, 617, 576, 536, 485, 445, 415, 390, 369, 274, 260, 251, 204, 170, 143, 115.
Source: Frost et al. (2006c).
Comments: The sample was characterized by chemical analysis.

Diaboleite $CuPb_2Cl_2(OH)_4$

Origin: Mannoth mine, Tiger, Arizona, USA.
Experimental details: Raman scattering measurements have been performed on arbitrarily oriented crystals using a 633 nm He-Ne laser. The laser radiation power is not indicated. The Raman shifts have been determined for the maxima of individual peaks obtained as a result of the spectral curve analysis.
Raman shifts (cm^{-1}): 3525sh, 3465sh, 3452s, 3436sh, 3405sh, 3340, 978w, 781, 672, 538, 468w, 437w, 365, 294w, 227, 175, 149w, 130w.
Source: Frost et al. (2003j).
Comments: No independent analytical data are provided for the sample used. In the cited paper, there are differences between Raman shifts given in Fig. 2b and in Table 1. Only Raman shifts from Fig. 2b are listed above. For the Raman spectrum of diaboleite see also Frost and Williams (2004).

Diadochite $Fe^{3+}{}_2(PO_4)(SO_4)(OH) \cdot 6H_2O$

Origin: Alum Cave Bluff, Great Smoky Mts., Sevier Co., Tennessee, USA.
Experimental details: Raman scattering measurements have been performed on an arbitrarily oriented sample using a 633 nm He-Ne laser. The laser radiation power is not indicated. The Raman shifts have been determined for the maxima of individual peaks obtained as a result of the spectral curve analysis.
Raman shifts (cm^{-1}): 3533sh, 3428, 3226sh, 2998, 1202sh, 1085, 1045sh, 1005s, 615, 565sh, 487, 448sh, 297sh, 263sh, 220, 142, 109.
Source: Frost and Palmer (2011f).
Comments: No independent analytical data are provided for the sample used.

Diamond C

Origin: Kangjinla mining district, Luobusa ophiolite, Southern Tibet.
Experimental details: No data.

Raman shifts (cm^{-1}): 1332.6s.
Source: Xu et al. (2015b).
Comments: For the Raman spectra of diamond see also Knight and White (1989), Hänni et al. (1997), Yang et al. (2007b), Jasinevicius (2009), and Frezzotti et al. (2012).

Diaspore AlO(OH)

Origin: Nevada, USA.
Experimental details: Raman scattering measurements have been performed on an arbitrarily oriented sample using 1064 nm Nd-YAG laser radiation. The nominal laser radiation power was 200 mW. A 180°-scattering geometry was employed.
Raman shifts (cm^{-1}): 3445, 3363sh, 3226sh, 3119sh, 2936sh, 1186, 1067, 1045, 1018, 956, 918, 837, 812, 790, 705s, 564, 609, 580, 552, 495, 466, 446s, 436, 394, 381, 364, 329, 287, 260s, 216, 207.
Source: Ruan et al. (2001).
Comments: The Raman shifts are indicated for the maxima of individual peaks obtained as a result of the spectral curve analysis. No independent analytical data are provided for the sample used. For the Raman spectrum of diaspore see also Shoval et al. (2003).

Dickinsonite-(KMnNa) K(NaMn)CaNa$_3$AlMn$_{13}$(PO$_4$)$_{12}$(OH)$_2$

Origin: Branchville, Fairfield Co., Connecticut, USA (type locality).
Experimental details: Raman scattering measurements have been performed in the region of O–H-stretching vibrations, on an arbitrarily oriented crystal, using 514.5 nm Ar$^+$ laser radiation. The laser radiation power at the sample was in the range from 2 to 5 mW.
Raman shifts (cm^{-1}): 3557, 3520.
Source: Cámara et al. (2006).
Comments: The sample was characterized by powder X-ray diffraction data, electron microprobe analysis, and LA-ICP-MS. The Raman shifts were determined by us based on spectral curve analysis of the published spectrum.

Dickite Al$_2$Si$_2$O$_5$(OH)$_4$

Origin: No data.
Experimental details: Raman scattering measurements have been performed on a powdered sample using 1064 nm Nd-YAG laser radiation. The nominal laser radiation power was 100 mW. A 180°-scattering geometry was employed.
Raman shifts (cm^{-1}): 1190, 1160, 1080, 1025s, 910, 790, 750, 710, 665s, 560, 515, 480, 380, 340, 275, 235s.
Source: Frost et al. (1993).
Comments: No independent analytical data are given for the sample used. For the Raman spectra of dickite see also Frost (1995), Johnston et al. (1998), and Shoval et al. (2001).

Digenite Cu$_{1.8}$S

Origin: Synthetic.
Experimental details: Methods of sample preparation are not described. Raman scattering measurements have been performed at 160 °C using 496.5 nm Ar$^+$ laser radiation. The nominal laser radiation power was 2000 mW.

Raman shifts (cm^{-1}): 469.
Source: Liu et al. (2002).
Comments: The sample was characterized by powder X-ray diffraction data.

Dimorphite As_4S_3

Origin: Synthetic.
Experimental details: Raman scattering measurements have been performed on an arbitrarily oriented sample using 676.4 nm Kr^+ laser radiation. The nominal laser radiation power was 10 mW.
Raman shifts (cm^{-1}): 374sh, 369, 365, 357, 349, 341, 274s, 224, 213, 206, 199, 186, 179, 175, 120w, 54, 51, 44w, 41, 38, 34, 29s, 17.
Source: Chattopadhyay et al. (1982).
Comments: The sample was characterized by powder X-ray diffraction data.

Diomignite $Li_2B_4O_7$

Origin: No data.
Experimental details: No data.
Raman shifts (cm^{-1}): 1352w, 1097w, 1028s, 554w, 446w, 390w.
Source: Thomas and Davidson (2010).
Comments: No independent analytical data are given for the sample used.

Diopside $CaMgSi_2O_6$

Origin: Zillerthel, Tirol, Austria.
Experimental details: Raman scattering measurements have been performed on an arbitrarily oriented crystal using 532 nm Nd-YAG laser radiation. The nominal laser radiation power was 100 mW.
Raman shifts (cm^{-1}): 1045, 1010s, 853, 665s, 558w, 531w, 507w, 389, 358, 323, 296sh, 248w, 230w.
Source: Buzatu and Buzgar (2010).
Comments: No independent analytical data are given for the sample used. For the Raman spectra of diopside see also Richet et al. (1998), Jasinevicius (2009), Frezzotti et al. (2012), and Andò and Garzanti (2014).

Dioptase $CuSiO_3 \cdot H_2O$

Origin: No data.
Experimental details: Methods of sample preparation are not described. Raman scattering measurements have been performed using 632.8 nm He-Ne laser radiation. The output laser radiation power was 30 mW.
Raman shifts (cm^{-1}): 3371w, 1025, 1006s, 960, 916, 743, 660s, 525, 452, 431, 400, 357s, 325, 294, 265, 240w, 225, 161, 140, 133.
Source: Bouchard and Smith (2003).
Comments: The sample was characterized by powder X-ray diffraction data. For the Raman spectrum of dioptase see also McKeown et al. (1995).

Dissakisite-(La) $CaLa(Al_2Mg)[Si_2O_7][SiO_4]O(OH)$

Origin: Hochwartperidotite, Ultenzone, Alps, Italy.
Experimental details: Methods of sample preparation are not described. Raman scattering measurements have been performed on an arbitrarily oriented sample using 514 nm laser radiation. The laser radiation power is not indicated.
Raman shifts (cm^{-1}): 3068, 1184, 1063s, 959s, 871s, 684, 566, 455s, 426s, 311, 226, 119.
Source: Tumiati et al. (2005).
Comments: The sample was characterized by powder X-ray diffraction data and electron microprobe analysis.

Dixenite $Cu^{1+}Fe^{3+}Mn^{2+}_{14}(As^{5+}O_4)(As^{3+}O_3)_5(SiO_4)_2(OH)_6$

Origin: Långban deposit, Bergslagen ore region, Filipstad district, Värmland, Sweden (type locality).
Experimental details: Raman scattering measurements have been performed on arbitrarily oriented crystals using a 633 nm He-Ne laser. The laser radiation power is not indicated. The Raman shifts have been determined for the maxima of individual peaks obtained as a result of the spectral curve analysis.
Raman shifts (cm^{-1}): See comments below.
Source: Bahfenne and Frost (2010d).
Comments: No independent analytical data are given for the sample used. The Raman shifts 3644sh, 3582sh, 3449, 3247sh, 1389sh, 1386, 1347s, 1338sh, 1214, 1057s, 1026, 988sh, 944, 861, 751, 688w, 526s, 505sh, 428, 385, 312, 300sh, 282, 212sh, 170sh, and 143 do not correspond to an arsenite. The strongest IR bands of this sample are observed at 1565 and 1361 cm^{-1}. This indicates that the investigated sample is not dixenite, but may be a carbonate or an organic compound. The assignment of the band at 1361 cm^{-1} to "SiO_4^{2-} antisymmetric stretching vibrations" (obviously, the authors meant SiO_4^{4-}) is erroneous. Bands of stretching vibrations of SiO_4^{4-} groups (in the range from 850 to 930 cm^{-1}) are not observed in the IR spectrum of presumed "dixenite."

Djerfisherite $K_6(Fe,Cu,Ni)_{25}S_{26}Cl$

Origin: Guli dunite complex, Polar Siberia, Russia.
Experimental details: Raman scattering measurements have been performed on an arbitrarily oriented sample using 632.8 nm He-Ne laser radiation. The laser radiation power is not indicated.
Raman shifts (cm^{-1}): 300.
Source: Zaccarini et al. (2007).
Comments: The sample was characterized by powder X-ray diffraction data and electron microprobe analysis.

Dmisokolovite $K_3Cu_5AlO_2(AsO_4)_4$

Origin: Arsenatnaya fumarole, Tolbachik volcano, Kamchatka, Russia (type locality).
Experimental details: Raman scattering measurements have been performed on an arbitrarily oriented sample using 532 nm diode laser radiation. The laser radiation power is not indicated. A 180°-scattering geometry was employed.

Raman shifts (cm^{-1}): 900sh, 852s, 839s, 819sh, 640, 552sh, 525sh, 500s, 440, 400, 345, 223, 198, 121s, 96.
Source: Pekov et al. (2015d).
Comments: The sample was characterized by powder X-ray diffraction data and electron microprobe analysis. The crystal structure is solved.

Dmisteinbergite Ca(Al$_2$Si$_2$O$_8$)

Origin: Carbonaceous chondrite Northwest Africa 2086.
Experimental details: Raman scattering measurements have been performed on an arbitrarily oriented grain in polished thin section using 532.2 nm Nd-YAG laser radiation. The nominal laser radiation power was 10 mW.
Raman shifts (cm^{-1}): 912s, 893, 801, 504, 442s, 327, 219.
Source: Fintor et al. (2013, 2014).
Comments: The sample was characterized by electron microprobe analyses. For the Raman spectra of dmisteinbergite see also Nestola et al. (2010).

Dolomite CaMg(CO$_3$)$_2$

Origin: Azcáratequarry, Eugui, Esteríbar, Spain.
Experimental details: Micro-Raman scattering measurements have been performed on an arbitrarily oriented polished sample using 514.5 nm Ar$^+$ laser radiation. The laser radiation power is not indicated.
Raman shifts (cm^{-1}): 1758w, 1443w, 1098s, 882w, 340w, 301, 177.
Source: Perrin et al. (2016).
Comments: For the Raman spectra of dolomite see also Edwards et al. (2005), Ciobotă et al. (2012), and Frezzotti et al. (2012).

Domerockite Cu$_4$(AsO$_4$)(AsO$_3$OH)(OH)$_3$·H$_2$O

Origin: Dome Rock mine, South Australia (type locality).
Experimental details: Raman scattering measurements have been performed on an arbitrarily oriented crystal using 632.8 nm He-Ne laser radiation. The laser radiation power is not indicated. A 180°-scattering geometry was employed.
Raman shifts (cm^{-1}): 3420, 3235w, 875sh, 850s, 822s, 808s, 650, 478, 445, 390, 360.
Source: Elliott et al. (2013).
Comments: The sample was characterized by powder X-ray diffraction data and electron microprobe analyses. The crystal structure is solved.

Donnayite-(Y) NaSr$_3$CaY(CO$_3$)$_6$·3H$_2$O

Origin: Poudrette quarry, Mont Sainte-Hilaire, Québec, Canada (type locality).
Experimental details: Raman scattering measurements have been performed on arbitrarily oriented crystals using a 633 nm He-Ne laser. The laser radiation power is not indicated. The Raman shifts have been determined for the maxima of individual peaks obtained as a result of the spectral curve analysis.

Raman shifts (cm^{-1}): 3760w, 3657sh, 3507sh, 3414, 3297, 3277, 3204sh, 1735w, 1694sh, 1583w, 1520sh, 1382w, 1093sh, 1077s, 1070sh, 1059sh, 728, 716, 694, 669sh, 549, 387, 427, 422, 387, 373, 363sh, 357, 338, 325sh, 287, 239sh, 225, 156s, 143sh.
Source: Frost et al. (2016a).
Comments: The sample was characterized by qualitative electron microprobe analysis.

Dorallcharite $TlFe^{3+}_3(SO_4)_2(OH)_6$

Origin: Crven Dol, Allchar, Macedonia (type locality).
Experimental details: Methods of sample preparation are not described. Raman scattering measurements have been performed in the range 100–1200 cm^{-1} using 632.8 nm He-Ne laser radiation. The laser radiation power is not indicated.
Raman shifts (cm^{-1}): 1165, 1089, 1004s, 621s, 564, 451, 420s, 339w, 300s, 218s, 206, 136s.
Source: Makreski et al. (2017).
Comments: The sample was characterized by powder X-ray diffraction data and electron microprobe analysis.

Dorfmanite $Na_2(PO_3OH) \cdot 2H_2O$

Origin: Synthetic.
Experimental details: Raman scattering measurements have been performed at 65°C on an arbitrarily oriented sample using 514.5 nm Ar$^+$ laser radiation. The nominal laser radiation power was 30 mW.
Raman shifts (cm^{-1}): 3391w, 3320w, 3080w, 3028w, 1149, 1072, 950s, 867, 567w, 541w, 513w, 446, 411, 395.
Source: Ghule et al. (2003).
Comments: The sample was characterized by TG data. For the Raman spectra of dorfmanite see also Ramakrishnan and Aruldhas (1987) and Frost et al. (2011l).

Dovyrenite $Ca_6Zr(Si_2O_7)_2(OH)_4$

Origin: Ioko-Dovyren massif, Northern Baikal region, Russia (type locality).
Experimental details: Raman scattering measurements have been performed from the natural face (100) of a crystal using 514.5 nm Ar$^+$ laser radiation. The laser radiation power at the sample was 20 mW. A 0°-scattering geometry was employed.
Raman shifts (cm^{-1}): 3638s, 3632s, 3593sh, 3585s, 3567sh, 1051s, 1002, 948s, 933sh, 836, 814sh, 759, 662s, 595s, 557sh, 518, 463, 437, 416sh, 393w, 372sh, 358, 333, 314, 296, 277, 260, 232.
Source: Galuskin et al. (2007b).
Comments: The Raman shifts are indicated for the maxima of individual peaks obtained as a result of the spectral curve analysis. The sample was characterized by powder X-ray diffraction data and electron microprobe analyses. The crystal structure is solved.

Downeyite SeO_2

Origin: Synthetic.
Experimental details: Raman scattering measurements have been performed on an arbitrarily oriented sample using 435.8 nm radiation (mercury arc). The radiation power is not indicated.
Raman shifts (cm^{-1}): 940w, 909, 889s, 862w, 706, 597s, 524, 364w, 299, 287, 254s, 199, 124.
Source: Venkateswaran (1936).

Comments: No independent analytical data are provided for the sample used. For the Raman spectra of downeyite see also Gerding (1941) and Stanila et al. (2000).

Doyleite Al(OH)$_3$

Origin: Mont Saint-Hilaire, Rouville RCM (Rouville Co.), Montérégie, Québec, Canada (type locality).
Experimental details: Raman scattering measurements have been performed on a powdered sample using 514.5 nm Ar$^+$ laser radiation. The nominal laser radiation power was 25 mW. A 180°-scattering geometry was employed.
Raman shifts (cm^{-1}): 3615sh, 3545, 1080(?), 936, 840(?), 806(?), 580, 392, 305, 279, 229, 208(?), 187(?), 158(?), 124, 117, 107(?).
Source: Rodgers (1993).
Comments: The sample was characterized by powder X-ray diffraction data. Doubtful bands are marked with a question mark.

Dravite NaMg$_3$Al$_6$(Si$_6$O$_{18}$)(BO$_3$)$_3$(OH)$_3$(OH)

Origin: Synthetic.
Experimental details: Raman scattering measurements have been performed on an oriented crystal (in the scattering geometry with the electrical field vector of the linearly polarized laser light parallel to the crystallographic c-axis) using 488 and 473 nm laser radiations. The nominal laser radiation power was 30 and 12 mW, respectively.
Raman shifts (cm^{-1}): 3776w, 3740, 3641sh, 3622, 3577, 3549sh, 3513sh, 1060 (broad), 1036, 980sh, 700, 676, 661sh, 635, 550, 493, 400sh, 370s, 313, 215s, 132.
Source: Berryman et al. (2016).
Comments: In the O–H stretching vibration range Raman shifts are indicated for the maxima of individual peaks obtained as a result of the spectral curve analysis. The sample was characterized by powder X-ray diffraction data and electron microprobe analysis. The Raman shifts were partly determined by us based on spectral curve analysis of the published spectrum. For the Raman spectra of dravite see also Gasharova et al. (1997), Andò and Garzanti (2014), and Watenphul et al. (2016a, b).

Dreyerite Bi(VO$_4$)

Origin: Hirschhorn, near Kaiserlautern, Germany (type locality).
Experimental details: Raman scattering measurements have been performed on arbitrarily oriented crystals using a 633 nm He-Ne laser. The laser radiation power is not indicated.
Raman shifts (cm^{-1}): 1164, 1137, 1104, 1082, 987, 836s, 790, 646, 617, 452s, 462sh, 408w, 365, 321sh, 301.
Source: Frost et al. (2006i).
Comments: No independent analytical data are provided for the sample used. The Raman shifts were partly determined by us based on spectral curve analysis of the published spectrum.

Drysdallite MoSe$_2$

Origin: Synthetic.
Experimental details: Raman scattering measurements have been performed on an oriented crystal using 514.5 nm Ar$^+$ laser radiation. The laser radiation power is not indicated.

Raman shifts (cm^{-1}): 286 (E^1_{2g}), 242 (A_{1g}), 168 (E_{1g}), 25 (E^2_{2g}).
Source: Sekine et al. (1980).
Comments: No independent analytical data are provided for the sample used. For the Raman spectrum of drysdallite see also Agnihotri and Sehgal (1972).

Dufrénoysite $Pb_2As_2S_5$

Origin: Lengenbach, Binntal, Switzerland (type locality).
Experimental details: Raman scattering measurements have been performed on an arbitrarily oriented sample using 632.81 nm He-Ne laser radiation. The laser radiation is not indicated.
Raman shifts (cm^{-1}): 376, 364s, 327s, 292s, 260s, 221s, 172, 144, 122, 102s, (74w).
Source: Kharbish (2016).
Comments: The sample was characterized by powder X-ray diffraction data and electron microprobe analysis. The Raman shifts have been determined for the maxima of individual peaks obtained as a result of the spectral curve analysis.

Duftite $PbCu(AsO_4)(OH)$

Origin: No data.
Experimental details: Raman scattering measurements have been performed on arbitrarily oriented crystals using a 633 nm He-Ne laser. The laser radiation power is not indicated. The Raman shifts have been determined for the maxima of individual peaks obtained as a result of the spectral curve analysis.
Raman shifts (cm^{-1}): 3280sh, 3240, 3192sh, 834s, 813, 792, 769sh, 549, 512, 454, 429, 403, 359, 340, 325, 301, 270, 229.
Source: Martens et al. (2003c).
Comments: No independent analytical data are provided for the sample used.

Dumontite $Pb_2(UO_2)_3O_2(PO_4)_2 \cdot 5H_2O$

Origin: Shinkolowbe, Congo (type locality).
Experimental details: Raman scattering measurements have been performed on arbitrarily oriented crystals using a 633 nm He-Ne laser. The laser radiation power is not indicated. The Raman shifts have been determined for the maxima of individual peaks obtained as a result of the spectral curve analysis.
Raman shifts (cm^{-1}): 3552sh, 3475, 3352, 3189sh, 1054, 1024, 982, 974sh, 815sh, 800, 780, 571, 551sh, 445sh, 440sh, 293, 271, 246, 175, 149.
Source: Frost and Čejka (2009a).
Comments: No independent analytical data are provided for the sample used.

Dumortierite $AlAl_6BSi_3O_{18}$

Origin: Dehesa, California, USA.
Experimental details: Raman scattering measurements have been performed on an oriented single crystal (with the incident laser light perpendicular to the c-axis of the crystal) using 514.5 nm Ar$^+$ laser radiation. The nominal laser radiation power was 205 mW.
Raman shifts (cm^{-1}): 1174, 1126w, 1113w, 1099w, 1002, 950s, 905sh, 843w, 808w, 793w, 780w, 750, 705sh, 660sh, 633, 548sh, 510s, 411s, 374, 283s, 228sh, 208s, 163, 147.

Source: Goreva et al. (2001).
Comments: The sample was characterized by powder X-ray diffraction data and electron microprobe analysis. The Raman shifts were determined by us based on spectral curve analysis of the published spectrum. For the Raman spectrum of dumortierite see also Pieczka et al. (2013).

Dundasite $PbAl_2(CO_3)_2(OH)_4 \cdot H_2O$

Origin: Synthetic.
Experimental details: Raman scattering measurements have been performed on an arbitrarily oriented sample using 514.5 nm Ar^+ laser radiation. The laser radiation output power was 50 mW.
Raman shifts (cm^{-1}): 1090s, 234, 193w, 170, 152w.
Source: Goienaga et al. (2011).
Comments: The sample was characterized by qualitative electron microprobe analysis.

Durangite $NaAl(AsO_4)F$

Origin: Barranca Sn mine, Coneto de Comonfort, Durango, Mexico (type locality).
Experimental details: Raman scattering measurements have been performed on an arbitrarily oriented crystal using 532 nm laser radiation. The laser radiation power is not indicated.
Raman shifts (cm^{-1}): 923sh, 912s, 827s, 718w, 610w, 539, 496s, 464, 430, 306w, 267s, 250, 202, 179sh, 160w, 130w.
Source: Downs et al. (2012).
Comments: The sample was characterized by powder X-ray diffraction data and electron microprobe analysis. The crystal structure is solved. The Raman shifts were determined by us based on spectral curve analysis of the published spectrum.

Dussertite $BaFe^{3+}_3(AsO_4)(AsO_3OH)(OH)_6$

Origin: Horní Slavkov, Slavkovský les Mts., Bohemia, Czech Republic.
Experimental details: Raman scattering measurements have been performed on arbitrarily oriented crystals using a 633 nm He-Ne laser. The laser radiation power is not indicated. The Raman shifts have been determined for the maxima of individual peaks obtained as a result of the spectral curve analysis.
Raman shifts (cm^{-1}): 3452s, 3439w, 3371s, 3242s, 3015w, 1250, 1220, 1176, 1115w, 902s, 870s, 859, 825w, 754w, 724w, 567, 561, 474s, 429s, 409, 372w, 306w, 275s, 247s, 188.
Source: Frost et al. (2011a).
Comments: The sample was characterized by powder X-ray diffraction data and electron microprobe analysis.

Dwornikite $Ni(SO_4) \cdot H_2O$

Origin: Artificial (product of Ni corrosion in concentrated sulfuric acid).
Experimental details: Raman scattering measurements have been performed on anodic corrosion film using 514.5 nm Ar^+ laser radiation. The nominal laser radiation power was 90 mW.
Raman shifts (cm^{-1}): 3300 (broad), 1600w, 1190, 1080, 1016s, 890, 596, 418.
Source: Melendres and Tani (1986).
Comments: The sample was characterized by powder X-ray diffraction data.

Dypingite $Mg_5(CO_3)_4(OH)_2 \cdot 5H_2O$

Origin: Shinshiro Shi, Aichi Prefecture, Japan.
Experimental details: Raman scattering measurements have been performed on arbitrarily oriented crystals using a 633 nm He-Ne laser. The laser radiation power is not indicated. The Raman shifts have been determined for the maxima of individual peaks obtained as a result of the spectral curve analysis.
Raman shifts (cm^{-1}): 3648s, 3644, 3519, 3427, 3399sh, 2934w, 2880w, 1767w, 1751w, 1713w, 1601, 1447, 1452, 1366sh, 1122s, 1119s, 1092, 761, 727, 559w, 434w, 311, 249, 227, 203.
Source: Frost et al. (2009a).
Comments: Questionable data: no independent analytical data are given for the samples used. The IR spectra of the presumed dypingite sample from Shinshiro Shi correspond to the mixture of serpentine and a carbonate mineral. The IR spectra of presumed dypingite from Clear Creek given in the cited paper are wrong. Actually, they are IR spectra of serpentine with minor admixture of a carbonate other than dypingite. IR bands of serpentine are erroneously assigned to vibrations of dypingite. For Raman spectra of carbonate mixtures containing dypingite see Kristova et al. (2014).

Dzhalindite $In(OH)_3$

Origin: Synthetic.
Experimental details: Methods of sample preparation are not described. Raman scattering measurements have been performed using Nd-YAG laser radiation. The wavelengths of laser excitation line and laser radiation power are not indicated.
Raman shifts (cm^{-1}): 302, 207.
Source: Yan et al. (2008).
Comments: The sample was characterized by powder X-ray diffraction data.

Dzhuluite $Ca_3(SbSn)(Fe^{3+}O_4)_3$

Origin: Upper Chegem Caldera, Northern Caucasus, Kabardino-Balkaria, Russia (type locality).
Experimental details: Raman scattering measurements have been performed on an arbitrarily oriented crystal using 514.5 nm Ar^+ laser radiation. The nominal laser radiation power was in the range from 30 to 50 mW.
Raman shifts (cm^{-1}): 812sh, 783sh, 756, 726sh, 612sh, 581, 487s, 326sh, 209sh, 285, 264sh, 235, 183, 161, 140, 112w.
Source: Galuskina et al. (2013b).
Comments: The sample was characterized by powder X-ray diffraction data and electron microprobe analyses. The Raman shifts have been determined for the maxima of individual peaks obtained as a result of the spectral curve analysis.

Dzierżanowskite $CaCu_2S_2$

Origin: Jabel Harmun, Judean Desert, Palestine Autonomy, Israel (type locality).
Experimental details: Micro-Raman scattering measurements have been performed on an arbitrarily oriented grain using 488 nm solid-state laser radiation. The laser radiation power at the sample was 120 mW.
Raman shifts (cm^{-1}): 300s, 103w, 86w.
Source: Galuskina et al. (2016a).

Comments: The sample was characterized by powder X-ray diffraction data and electron microprobe analyses.

Eakerite $Ca_2Sn^{4+}Al_2Si_6O_{18}(OH)_2 \cdot 2H_2O$

Origin: Foote Mineral Company mine, Kings Mt., North Carolina, USA.
Experimental details: Raman scattering measurements have been performed on an arbitrarily oriented crystal using 532.0 nm solid-state laser radiation. The laser radiation power is not indicated.
Raman shifts (cm^{-1}): 3528, 3405, 3317, 1664, 1085, 1055, 1025, 996w, 974, 948w, 923w, 795, 747, 569s, 548s, 477, 441s, 418s, 393, 350, 337, 276, 234.
Source: Uchida et al. (2007).
Comments: The sample was characterized by single-crystal X-ray diffraction data. The crystal structure is solved. The Raman shifts were partly determined by us based on spectral curve analysis of the published spectrum.

Eastonite $KAlMg_2(Si_2Al_2)O_{10}(OH)_2$

Origin: Hessdalen, Norway.
Experimental details: Raman scattering measurements have been performed on an oriented sample, using 514.5 nm Ar$^+$ laser radiation. The laser radiation power is not indicated. The spectra were recorded with the electric field polarized perpendicular to the cleavage plane.
Raman shifts (cm^{-1}): 3700s, 3663s, 1014, 670s, 652s, 400, 361, 274w, 194s, 100.
Source: Tlili et al. (1989).
Comments: The sample was characterized by electron microprobe analysis. For the Raman spectra of eastonite see also Tlili and Smith (2007) and Wang et al. (2015).

Eastonite $KAlMg_2(Si_2Al_2)O_{10}(OH)_2$

Origin: Easton, Pennsylvania, USA.
Experimental details: Raman scattering measurements have been performed on an arbitrarily oriented sample (on either loose grains without preparation or on a pressed pellet of sample powder) using 532 nm Nd-YAG laser radiation. The laser radiation power at the sample was 13 mW.
Raman shifts (cm^{-1}): 3718s, 3693, 3678, 1082, 1038, 681s, 456, 427, 391, 351, 279, 192s.
Source: Wang et al. (2015).
Comments: The sample was characterized by powder X-ray diffraction data and electron microprobe analysis. The Raman shifts were partly determined by us based on spectral curve analysis of the published spectrum. For the Raman spectra of eastonite see also Tlili and Smith (2007).

Ecandrewsite $ZnTiO_3$

Origin: Synthetic.
Experimental details: Raman scattering measurements have been performed on a powdered sample compressed into pellet. The wavelengths of laser excitation line and the laser radiation power are not indicated.
Raman shifts (cm^{-1}): 716s, 624, 490, 474, 395, 350s, 270, 234, 181, 141w.
Source: Bernert et al. (2015).

Comments: The sample was characterized by powder X-ray diffraction data. For the Raman spectrum of ecandrewsite see also Beigi et al. (2011).

Ecdemite $Pb_6As^{3+}_2O_7Cl_4$

Origin: Harstigen mine, Pajsberg, near Filipstad, Värmland, Sweden.
Experimental details: Raman scattering measurements have been performed on an arbitrarily oriented sample, using 514.5 nm Ar$^+$ laser radiation. The laser radiation power was in the range from 20 to 50 mW.
Raman shifts (cm^{-1}): 1122s, ~694, 340–310, 154s, 129s.
Source: Jonsson (2003).
Comments: The sample was characterized by powder X-ray diffraction data and electron microprobe analyses.

Eckhardite $(Ca,Pb)Cu^{2+}Te^{6+}O_5(H_2O)$

Origin: Otto Mt., near Baker, California, USA (type locality).
Experimental details: Raman scattering measurements have been performed on an arbitrarily oriented polished grain using 514.5 nm Ar$^+$ laser radiation. The laser radiation power at the sample was 2.5 mW.
Raman shifts (cm^{-1}): 3440, 729s, 692s, 562w, 312w, 274, 260.
Source: Kampf et al. (2013b).
Comments: The sample was characterized by powder X-ray diffraction data and electron microprobe analysis. The crystal structure is solved. The empirical formula of the sample used is $Ca_{0.962}Pb_{0.073}Cu^{2+}_{0.971}Mg_{0.005}Fe^{3+}_{0.002}Te^{6+}_{0.986}O_6H_{2.052}$.

Edgrewite $Ca_9(SiO_4)_4F_2$

Origin: Upper Chegem caldera, Northern Caucasus, Kabardino-Balkaria, Russia (type locality).
Experimental details: Raman scattering measurements have been performed on an arbitrarily oriented single crystal using 514.5 nm Ar$^+$ laser radiation. The output laser radiation power was in the range 30–50 mW. A 180°-scattering geometry was employed. The Raman shifts have been determined for the maxima of individual peaks obtained as a result of the spectral curve analysis.
Raman shifts (cm^{-1}): 3554, 3547, 3540, 921, 889, 839, 815s, 667w, 556, 527w, 423sh, 406, 394sh, 309, 269, 195w, 172sh, 163, 108w.
Source: Galuskin et al. (2012d).
Comments: The sample was characterized by powder X-ray diffraction data and electron microprobe analysis. The crystal structure is solved. The Raman shifts are given for the member of the edgrewite $Ca_9(SiO_4)_4F_2$–hydroxyledgrewite $Ca_9(SiO_4)_4(OH)_2$ series with the content of the edgrewite end-member minal more than 50%.

Edingtonite $Ba(Si_3Al_2)O_{10}\cdot 4H_2O$

Origin: Ice River, near Golden, British Columbia, Canada.
Experimental details: Raman scattering measurements have been performed on an oriented sample with the long axis of crystal normal to the polarization direction of the laser beam, using 514.5 nm Ar$^+$ laser radiation. A 180°-scattering geometry was employed. The laser radiation power at the sample was 10 mW.

Raman shifts (cm^{-1}): 3480s, 1644, 1096s, 1085s, 1077s, 1061s, 1049, 1021, 986s, 722s, 662, 531s, 480, 454s, 447s, 433, 427, 409, 395, 358s, 343s, 337, 330, 323, 311, 302, 284, 272s, 256, 167, 153, 141s.
Source: Wopenka et al. (1998).
Comments: No independent analytical data are provided for the sample used. For the Raman spectrum of edingtonite see also Mozgawa (2001).

Edoylerite $Hg^{2+}_3(Cr^{6+}O_4)S_2$

Origin: No data.
Experimental details: Raman scattering measurements have been performed on an arbitrarily oriented sample, using 785 nm Nd-YAG laser radiation. The laser radiation power at the sample was 1 mW. The Raman shifts have been determined for the maxima of individual peaks obtained as a result of the spectral curve analysis.
Raman shifts (cm^{-1}): 840s, 382, 368, 363, 340, 325, 269s.
Source: Frost (2004c).
Comments: No independent analytical data are provided for the sample used.

Effenbergerite $BaCuSi_4O_{10}$

Origin: Artificial.
Experimental details: Raman scattering measurements have been performed on an arbitrarily oriented sample, using 514 nm Ar$^+$ laser radiation. The laser radiation power at the sample was 1 mW.
Raman shifts (cm^{-1}): 1097s, 986s, 788, 588, 573w, 558w, 517,454sh, 423s, 380, 339w, 276w.
Source: Xia et al. (2014).
Comments: No independent analytical data are provided for the sample used.

Eitelite $Na_2Mg(CO_3)_2$

Origin: Synthetic.
Experimental details: Kind of sample preparation is not indicated. Micro-Raman scattering measurements have been performed in Ar atmosphere using 514.5 nm Ar$^+$ laser radiation. The nominal laser radiation power was 1 mW. A 180°-scattering geometry was employed.
Raman shifts (cm^{-1}): 1105s, 721w, 263, 208, 91s.
Source: Shatskiy et al. (2013).
Comments: The sample was characterized by single crystal X-ray diffraction data and energy-dispersive X-ray scan data. For the Raman spectrum of eitelite see also Sharygin et al. (2013c).

Ekanite $Ca_2ThSi_8O_{20}$

Origin: Moneragala, Okkampitiya area, Eastern Sri Lanka.
Experimental details: Raman scattering measurements have been performed on an annealed metamict sample using 473 nm Ar$^+$ laser radiation. The laser radiation power is not indicated.
Raman shifts (cm^{-1}): 1113s, 1009, 992, 747w, 657, 575, 433s, 395, 368, 350, 274, 189, 157, 133, 113.
Source: Nasdala et al. (2016).

Comments: The sample was characterized by electron microprobe analysis. The Raman shifts were determined by us based on spectral curve analysis of the published spectrum.

Ekplexite $(Nb,Mo)S_2 \cdot (Mg_{1-x}Al_x)(OH)_{2+x}$

Experimental details: Raman scattering measurements have been performed on an arbitrarily oriented sample using 532 nm diode laser radiation. The laser radiation power at the sample was 3 mW. A 180°-scattering geometry was employed.
Raman shifts (cm^{-1}): 3530, 3326, 707, 526s, 438, 387sh, 364s, 232sh, 198sh, 161, 120sh.
Source: Pekov et al. (2014a).
Comments: The sample was characterized by powder X-ray diffraction data and electron microprobe analyses. The empirical formula of the sample used is $(Nb_{0.45}Mo_{0.38}W_{0.10}V_{0.04})S_2(Mg_{0.60}Al_{0.37}Fe_{0.02})(OH)_{2.36}$.

Elbaite $Na(Al_{1.5}Li_{1.5})Al_6(Si_6O_{18})(BO_3)_3(OH)_3(OH)$

Origin: Granite pegmatite in an unknown locality in Southern California, USA.
Experimental details: Raman scattering measurements have been performed on an oriented sample, using 488 and 514.5 nm Ar^+ laser radiations in the range 150–1550 cm^{-1}. The laser radiation power is not indicated. A 180°-scattering geometry was employed. Polarized spectra were collected in the $z(xx)z$, $x(yy)x$, $x(zz)x$, $z(xy)z$ and $x(zy)x$ scattering geometries.
Raman shifts (cm^{-1}): $z(xx)z$: 1412, 1190, 1077, 760, 731, 693, 641, 407, 373, 222(A_1), 340(E); $x(yy)x$: 1412, 1105, 1059, 989, 850, 760, 717, 632, 508, 407, 373, 244, 222(A_1), 340 (E); $x(zz)x$: 1442, 1412, 1105, 1059, 989, 860, 760, 717, 637, 508, 407, 373, 244, 222 (A_1); $z(xy)z$: 1190, 1077, 731, 373 (E); $x(zy)x$: 717, 700, 373, 350, 286 (E).
Source: Gasharova et al. (1997).
Comments: The sample was characterized by powder X-ray diffraction data, electron microprobe analysis, wet chemical analysis for most elements, atom absorption spectroscopy for Li, flame photometry for Na and K, and thermal analysis for OH content. The empirical formula of the sample used is $(Na_{0.86}K_{0.09}Ca_{0.05})(Li_{0.99}Mg_{0.27}Mn_{0.23}Fe^{2+}_{0.01}Al_{1.41})Al_6B_{2.93}Si_6O_{27.26}(OH)_{3.64}F_{0.10}$. For the Raman spectra of elbaite see also Natkaniec-Nowak et al. (2009), Hoang et al. (2011), and Fantini et al. (2014).

Elbaite $Na(Al_{1.5}Li_{1.5})Al_6(Si_6O_{18})(BO_3)_3(OH)_3(OH)$

Origin: Lucyen mines, Vietnam.
Experimental details: Raman scattering measurements have been performed on a powdered sample, using 457 nm solid-state laser radiation. The laser radiation power at the sample was 1 mW.
Raman shifts (cm^{-1}): 3655, 3585, 3560, 3490, 1407w, 1070s, 1033sh, 981, 880w, 836, 808, 733s, 673, 587sh, 551, 517, 476, 447, 404sh, 377s, 335sh.
Source: Hoang et al. (2011).
Comments: The sample was characterized by powder X-ray diffraction data and energy-dispersive X-ray scan data. The Raman shifts were partly determined by us based on spectral curve analysis of the published spectrum. For the Raman spectra of elbaite see also Gasharova et al. (1997), Natkaniec-Nowak et al. (2009), and Fantini et al. (2014).

Elbrusite $Ca_3(U^{6+}_{0.5}Zr_{1.5})(Fe^{3+}O_4)_3$

Origin: Upper Chegem caldera, Kabardino-Balkaria, Northern Caucasus, Russia (type locality).
Experimental details: Raman scattering measurements have been performed on an arbitrarily oriented single crystal. A 0°-scattering geometry was employed. The wavelength of laser excitation line and the laser radiation power are not indicated. The Raman shifts have been determined for the maxima of individual peaks obtained as a result of the spectral curve analysis.
Raman shifts (cm^{-1}): 805sh, 730, 478, 273, 222s.
Source: Galuskina et al. (2010b).
Comments: The sample was characterized by powder X-ray diffraction data and electron microprobe analysis.

Eleonorite $Fe^{2+}Fe^{3+}_5(PO_4)_4(OH)_5 \cdot 6H_2O$

Origin: Boca Rica pegmatite, Minas Gerais, Brazil.
Experimental details: Raman scattering measurements have been performed on arbitrarily oriented crystals using a 633 nm He-Ne laser. The Raman shifts have been determined for the maxima of individual peaks obtained as a result of the spectral curve analysis. The laser radiation power at the sample is not indicated.
Raman shifts (cm^{-1}): 1174, 1155, 1133, 1116, 1098, 1084sh, 1069, 1058, 1051, 1034, 1011s, 990sh, 969sh, 703sh, 687s, 673, 661, 644, 601, 582, 567s, 546, 503, 491, 478, 468, 455, 437, 403, 398, 336, 322, 309sh, 300, 289, 280, 254, 238, 230, 225, 200, 191sh, 153, 1432, 118, 107.
Source: Frost et al. (2014al).
Comments: No independent analytical data are given for the sample used. In the cited paper the mineral is described with the name beraunite, but an incorrect formula of beraunite is given. Brown color of the sample indicates that it is not beraunite, but its oxydation product eleonorite, a mineral isostructural with beraunite (Chukanov et al. 2017a).

Elpasolite K_2NaAlF_6

Origin: Synthetic.
Experimental details: Raman scattering measurements have been performed on a powdered sample using 488 nm Ar$^+$ laser radiation. The laser radiation power is not indicated.
Raman shifts (cm^{-1}): 561, 330, 138.
Source: Morss (1974).
Comments: The sample was characterized by powder X-ray diffraction data. For the Raman spectrum of elpasolite see also Frezzotti et al. (2012).

Eltyubyuite $Ca_{12}Fe^{3+}_{10}Si_4O_{32}Cl_6$

Origin: Upper Chegem caldera, Kabardino-Balkaria, Northern Caucasus, Russia (type locality).
Experimental details: Raman scattering measurements have been performed on an arbitrarily oriented single crystal, using 514.5 nm Ar$^+$ laser radiation. The laser radiation output power was in the range from 30 to 50 mW. A 180°-scattering geometry was employed. The Raman shifts have been determined for the maxima of individual peaks obtained as a result of the spectral curve analysis.
Raman shifts (cm^{-1}): 3395w, 936w, 927w, 861, 846, 816s, 784, 699, 558w, 532w, 472sh, 468, 412, 327sh, 309, 260.
Source: Galuskin et al. (2013a).

Comments: The sample was characterized by single-crystal electron backscatter diffraction data and electron microprobe analyses. The wavenumbers were partly determined by us based on spectral curve analysis of the published spectrum. The empirical formula of the sample used is $Ca_{12.12}Mg_{0.04}Ti_{0.11}Fe^{3+}_{9.41}Al_{1.26}Si_{2.98}O_{31.89}Cl_{5.04}$. The band positions denoted by Galuskin et al. (2013a) as 448 cm^{-1} (twice) were determined by us at 468 and 412 cm^{-1}. For the Raman spectrum of eltyubuyite see also Gfeller et al. (2015).

Emmonsite $Fe^{3+}_2(Te^{4+}O_3)_3 \cdot 2H_2O$

Origin: Moctezuma mine, New Mexico.
Experimental details: Raman scattering measurements have been performed on arbitrarily oriented crystals using a 633 nm He-Ne laser. The laser radiation power is not indicated. The Raman shifts have been determined for the maxima of individual peaks obtained as a result of the spectral curve analysis.
Raman shifts (cm^{-1}): 788, 764, 688sh, 658s, 440s, 400s, 326, 275s, 227w, 187s.
Source: Frost et al. (2008i).
Comments: No independent analytical data are provided for the sample used.

Enargite Cu_3AsS_4

Origin: Butte, Montana, USA.
Experimental details: Raman scattering measurements have been performed on an arbitrarily oriented sample using 514.5 nm Ar$^+$ laser radiation. The laser radiation power at the sample was in the range 1–10 mW. A 180°-scattering geometry was employed.
Raman shifts (cm^{-1}): 382s, 337s, 298, 269, 170, 151, 133.
Source: Mernagh and Trudu (1993).
Comments: The sample was characterized by electron microprobe analysis. For the Raman spectrum of enargite see also Gow (2015).

Enargite Cu_3AsS_4

Origin: Quirivilca region, Peru.
Experimental details: Raman scattering measurements have been performed on an arbitrarily oriented sample using 632.8 nm He-Ne laser radiation. The maximum output laser radiation power was 100 mW.
Raman shifts (cm^{-1}): 724w, 679, 384, 338s, 265.
Source: Gow (2015).
Comments: The sample was characterized with scanning electron microscopy. For the Raman spectrum of enargite see also Mernagh and Trudu (1993).

Enstatite $Mg_2Si_2O_6$

Origin: Synthetic.
Experimental details: Raman scattering measurements have been performed on an oriented single crystal, using 488 nm Ar$^+$ laser radiation. The laser radiation power is not indicated. Polarized spectra were collected in the $E \parallel (100)$, $E \parallel (010)$ and $E \parallel (001)$ scattering geometries.
Raman shifts (cm^{-1}): 1035s, 1013s, 937, 938, 853, 687s, 665s, 581, 554, 541, 527, 446, 423s, 403s, 385, 344s, 303, 239s, 206, 198s, 134s.

Source: Stalder et al. (2009).
Comments: The sample was characterized by electron microprobe analysis. Raman peak positions do not exhibit a significant dependence on scattering geometry. For the Raman spectra of enstatite see also Lin (2004), Frezzotti et al. (2012), and Andò and Garzanti (2014).

Eosphorite $Mn^{2+}Al(PO_4)(OH)_2 \cdot H_2O$

Origin: Roberto mine, Divino das Laranjeiras, Minas Gerais, Brazil.
Experimental details: Raman scattering measurements have been performed on arbitrarily oriented crystals, using a 633 nm He-Ne laser. The laser radiation power is not indicated. The Raman shifts have been determined for the maxima of individual peaks obtained as a result of the spectral curve analysis.
Raman shifts (cm^{-1}): 3477sh, 3460s, 3313, 3191sh, 3063, 1747, 1651, 1193w, 1142w, 1091w, 1011w, 978sh, 969s, 864sh, 816w, 634sh, 618, 595, 580sh, 560, 473, 462sh, 399, 386sh, 338w, 307, 291, 254, 232, 208, 200, 175, 171, 152, 118.
Source: Frost et al. (2013am).
Comments: The sample was characterized by electron microprobe analysis. The empirical formula of the sample used is $(Mn_{0.72}Fe_{0.13}Ca_{0.01})Al_{1.04}(PO_4,HPO_4)_{1.07}(OH_{1.89}F_{0.02}) \cdot 0.94H_2O$.

Epidote $Ca_2(Al_2Fe^{3+})[Si_2O_7][SiO_4]O(OH)$

Origin: Dunje, Macedonia.
Experimental details: Raman scattering measurements have been performed on a powdered sample, using 514.5 nm Ar$^+$ laser radiation. The laser radiation power is not indicated.
Raman shifts (cm^{-1}): 1084, 1040w, 980w, 914s, 885, 864sh, 832w, 599s, 565s, 522w, 508, 452s, 430, 390w, 350, 328w, 314w, 292w, 276, 243, 230, 168, 134.
Source: Jovanovski et al. (2009).
Comments: The sample was characterized by powder X-ray diffraction data and electron microprobe analysis. For the Raman spectra of epidote see also Jovanovski et al. (2009) and Andò and Garzanti (2014).

Epistilbite $Ca_3[Si_{18}Al_6O_{48}] \cdot 16H_2O$

Origin: Berufjord, Iceland.
Experimental details: Kind of sample preparation is not indicated. The measurements have been performed using 1064 nm Nd-YAG laser radiation with the radiation power at the sample of 300 mW.
Raman shifts (cm^{-1}): 1631sh, 1620, 1117, 801w, 434s, 409s, 255, 179.
Source: Mozgawa (2001).
Comments: The sample was characterized by powder X-ray diffraction data. The Raman shifts were partly determined by us based on spectral curve analysis of the published spectrum.

Epsomite $Mg(SO_4) \cdot 7H_2O$

Origin: Synthetic.
Experimental details: Raman scattering measurements have been performed on a powdered sample using 532 nm Nd-YAG laser radiation. The laser radiation power at the sample was 15 mW.

Raman shifts (cm^{-1}): 3425, 3303, 1672, 1134, 1095, 1061, 984s, 612, 447, 369, 245, 154.
Source: Wang et al. (2006a).
Comments: The sample was characterized by powder X-ray diffraction data. For the Raman spectra of epsomite see also Genceli et al. (2007), Buzgar et al. (2009), Apopei et al. (2012, 2014a), and Jentzsch et al. (2013).

Epsomite Mg(SO$_4$)·7H$_2$O

Origin: Coranda-Hondol ore deposit, Certej, Romania.
Experimental details: Kind of sample preparation is not indicated. The measurements have been performed using 532 nm Nd-YAG laser radiation with the power at the sample of 53.6 mW.
Raman shifts (cm^{-1}): 3285, 3219sh, 1668w, 1136w, 1097w, 1062w, 985s, 615, 449, 371w, 246w.
Source: Apopei et al. (2014a).
Comments: The sample was characterized by single-crystal X-ray diffraction data and scanning electron microscopy. For the Raman spectra of epsomite see also Wang et al. (2006a), Genceli et al. (2007), Buzgar et al. (2009), Apopei et al. (2012), and Jentzsch et al. (2013).

Ericlaxmanite Cu$_4$O(AsO$_4$)$_2$

Origin: Arsenatnaya fumarole, Tolbachik volcano, Kamchatka, Russia (type locality).
Experimental details: Raman scattering measurements have been performed on an arbitrarily oriented sample, using 532 nm laser radiation. The laser radiation power at the sample was about 1 mW.
Raman shifts (cm^{-1}): 889sh, 863sh, 845s, 753, 664, 608, 531sh, 493, 440, 401, 329, 292, 229s, 181, 112.
Source: Pekov et al. (2014c).
Comments: The sample was characterized by powder X-ray diffraction data and electron microprobe analyses. The crystal structure is solved. The empirical formula of the sample used is (Cu$_{3.97}$Zn$_{0.06}$Fe$_{0.02}$)(As$_{1.94}$P$_{0.02}$V$_{0.01}$S$_{0.01}$)O$_9$.

Erikapohlite Cu$^{2+}_3$(Zn,Cu,Mg)$_4$Ca$_2$(AsO$_4$)$_6$·2H$_2$O

Origin: Tsumeb mine, Tsumeb, Namibia (type locality).
Experimental details: Raman scattering measurements have been performed on an arbitrarily oriented sample, using 514.5 nm Ar$^+$ laser radiation. The laser radiation power is not indicated.
Raman shifts (cm^{-1}): 894, 854s, 797s, 582w, 500, 408, 355, 306, 280, 234, 191, 122.
Source: Schlüter et al. (2013).
Comments: The sample was characterized by powder X-ray diffraction data and electron microprobe analyses. The empirical formula of the sample used is Cu$_3$(Zn$_{2.48}$Cu$_{0.93}$Mg$_{0.77}$Fe$_{0.01}$)Ca$_{2.04}$As$_{6.20}$O$_{24.71}$·1.29H$_2$O. The Raman shifts were determined by us based on spectral curve analysis of the published spectrum.

Eringaite Ca$_3$Sc$_2$(SiO$_4$)$_3$

Origin: Wiluy River, Sakha-Yakutia Republic, Russia (type locality).
Experimental details: Raman scattering measurements have been performed on an arbitrarily oriented sample using 514.5 nm Ar$^+$ laser radiation. The laser radiation power was in the range from 20 to 40 mW.

Raman shifts (cm^{-1}): 936s, 877s, 815sh, 742, 638, 602, 540sh, 511s, 484sh, 440w, 408, 359s, 336, 309, 255, 220w.

Source: Galuskina et al. (2010d).

Comments: The sample was characterized by single-crystal X-ray diffraction data and electron microprobe analyses. The empirical formula of the sample used is $(Ca_{3.00}Y_{0.01})(Sc_{0.63}Ti^{4+}_{0.66}Fe^{3+}_{0.25}Zr_{0.30}Mg_{0.08}Cr^{3+}_{0.06}Fe^{2+}_{0.01})(Si_{2.13}Al_{0.26}Fe^{3+}_{0.61})O_{12}$. The Raman shifts were determined by us based on spectral curve analysis of the published spectrum. For the Raman spectrum of eringaite see also Yun-Fang et al. (2013).

Eriochalcite $CuCl_2 \cdot 2H_2O$

Origin: Main Lode, Great Australia mine, Cloncurry, Queensland, Australia.

Experimental details: Raman scattering measurements have been performed on arbitrarily oriented crystals using a 633 nm He-Ne laser. The laser radiation power is not indicated. The Raman shifts have been determined for the maxima of individual peaks obtained as a result of the spectral curve analysis.

Raman shifts (cm^{-1}): 3462sh, 3367, 3176w, 1620, 690, 672sh, 405, 390sh, 234, 215s, 117.

Source: Frost et al. (2003i).

Comments: The sample was characterized by powder X-ray diffraction data. For the Raman spectrum of eriochalcite see also Christy et al. (2004).

Erionite-Ca $Ca_5[Si_{26}Al_{10}O_{72}] \cdot 30H_2O$

Origin: Oregon, USA.

Experimental details: Micro-Raman scattering measurements have been performed on an oriented fiber using 632.8 nm He-Ne laser radiation. The nominal laser radiation power was 20 mW. Spectra were recorded by placing fiber elongation axis at 0°, 45°, 90°, and 135° with respect to the cross hair of the microscope ocular lens.

Raman shifts (cm^{-1}): 816w, 790w, 569, 486s, 469sh, 346.

Source: Croce et al. (2013).

Comments: The Raman shifts are indicated for the fiber orientation with respect to the direction of the laser beam. No independent analytical data are provided for the sample used.

Erlichmanite OsS_2

Origin: Santa Elena Nappe, Costa Rica.

Experimental details: Raman scattering measurements have been performed on an arbitrarily oriented sample using 532.6 nm Nd-YAG laser radiation. The nominal laser radiation power was 100 mW.

Raman shifts (cm^{-1}): 396w, 342s.

Source: Zaccarini et al. (2010).

Comments: The sample was characterized by electron microprobe analysis. The Raman shifts were partly determined by us based on spectral curve analysis of the published spectrum. For the Raman spectrum of erlichmanite see also Bakker (2014).

Ernstburkeite $Mg(CH_3SO_3)_2 \cdot 12H_2O$

Origin: Synthetic.
Experimental details: No data.

Raman shifts (cm^{-1}): 3021, 2939, 1421, 1055s, 974, 776, 545, 348.
Source: Güner et al. (2013).
Comments: The sample was characterized by powder X-ray diffraction data.

Erythrite $Co_3(AsO_4)_2 \cdot 8H_2O$

Origin: Mt. Cobalt, Queensland, Australia.
Experimental details: Raman scattering measurements have been performed on arbitrarily oriented crystals using a 633 nm He-Ne laser. The laser radiation power is not indicated. The Raman shifts have been determined for the maxima of individual peaks obtained as a result of the spectral curve analysis.
Raman shifts (cm^{-1}): 3337sh, 3200, 3052, 902, 852, 792, 727w, 652w, 467, 457s, 439s, 391, 378, 301, 263, 249s, 234, 223, 209s, 188, 162, 147w, 126.
Source: Frost et al. (2003g).
Comments: No independent analytical data are provided for the sample used. For the Raman spectra of erythrite see also Frost et al. (2004i) and Kloprogge et al. (2006).

Erythrosiderite $K_2Fe^{3+}Cl_5 \cdot H_2O$

Origin: Synthetic.
Experimental details: Raman scattering measurements have been performed on an oriented sample. The laser beam was allowed to fall on a crystal face at an angle of 45°. The measurements have been performed using He-Ne laser radiation. The laser radiation power was 50 mW.
Raman shifts (cm^{-1}): 384, 299s, 224, 174, 129.
Source: Sharma and Pandya (1974).
Comments: No independent analytical data are provided for the sample used. For the Raman spectrum of erythrosiderite see also Piszczek et al. (2003).

Erythrosiderite $K_2Fe^{3+}Cl_5 \cdot H_2O$

Origin: Synthetic.
Experimental details: Raman scattering measurements have been performed on an arbitrarily oriented sample using 1064 nm Nd-YAG laser radiation. The maximum output laser radiation power was about 100 mW.
Raman shifts (cm^{-1}): 1807, 1591, 371, 298, 221, 180, 173, 128.
Source: Piszczek et al. (2003).
Comments: The water content in the sample used was established by thermogravimetric analysis. For the Raman spectrum of erythrosiderite see also Sharma and Pandya (1974).

Eskebornite $CuFeSe_2$

Origin: Synthetic.
Experimental details: Raman scattering measurements have been performed on a nanocrystalline sample, using 514.5 nm Ar$^+$ laser radiation. The nominal laser radiation power was 10 mW.
Raman shifts (cm^{-1}): 471, 348w, 288s.
Source: Wang et al. (2009a).
Comments: No independent analytical data are provided for the sample used.

Eskolaite Cr_2O_3

Origin: Synthetic.
Experimental details: Raman scattering measurements have been performed on a powdered sample using 647.1 nm laser radiation. A 180°-scattering geometry was employed.
Raman shifts (cm^{-1}): 613, 552s, 527, 397w, 350, 300.
Source: Maslar et al. (2001).
Comments: No independent analytical data are provided for the reference sample used. For the Raman spectra of eskolaite see also Bouchard and Smith (2003), Hosterman (2011), and Adar (2014).

Esperite $PbCa_2(ZnSiO_4)_3$

Origin: Franklin, New Jersey, USA (type locality).
Experimental details: Raman scattering measurements have been performed on an arbitrarily oriented sample using 780 nm solid-state laser radiation. The laser radiation power is not indicated.
Raman shifts (cm^{-1}): 958w, 931w, 900, 846s, 589, 538w, 504, 447s, 409, 368, 335, 297w, 278w, 250, 224, 215w, 200w, 181s, 160, 153.
Source: Tait et al. (2010).
Comments: The sample was characterized by single-crystal X-ray diffraction data and electron microprobe analyses. The crystal structure is solved. The empirical formula of the sample used is $Pb_{1.00}(Ca_{1.86}Fe^{2+}_{0.07}Mn_{0.04}Cr^{3+}_{0.02})(Zn_{1.00}Si_{1.00}O_4)_3$. The Raman shifts were partly determined by us based on spectral curve analysis of the published spectrum.

Ettringite $Ca_6Al_2(SO_4)_3(OH)_{12} \cdot 26H_2O$

Origin: Synthetic.
Experimental details: Kind of sample preparation is not indicated. Raman scattering measurements have been performed in the spectral regions from 200 to 1300 cm^{-1} and from 2800 to 4000 cm^{-1} using 514.5 nm Ar$^+$ laser radiation. The laser radiation power at the sample was 15 mW. A 180°-scattering geometry was employed.
Raman shifts (cm^{-1}): 3638, 3440s, 1118w, 1087w, 990s, 615w, 549w, 451, 416w 346w.
Source: Renaudin et al. (2007).
Comments: The sample was characterized by powder X-ray diffraction data. The Raman shifts were partly determined by us based on spectral curve analysis of the published spectrum. For the Raman spectra of ettringite see also Deb et al. (2003) and Frost et al. (2013i).

Euchroite $Cu_2(AsO_4)(OH) \cdot 3H_2O$

Origin: L'ubietová-Svätoduška, Banská Bystrica Co., Banská Bystrica region, Slovakia.
Experimental details: Raman scattering measurements have been performed on arbitrarily oriented crystals using a 633 nm He-Ne laser. The laser radiation power is not indicated. The Raman shifts have been determined for the maxima of individual peaks obtained as a result of the spectral curve analysis.
Raman shifts (cm^{-1}): 3537, 3470, 3278sh, 3116, 2924sh, 1634, 1032w, 976w, 848sh, 836s, 821sh, 768w, 474, 441, 385, 358, 294sh, 246s, 233s, 227sh, 210, 203sh, 171sh, 161s, 144s, 134sh, 112s.
Source: Frost et al. (2010e).

Comments: The sample was characterized by powder X-ray diffraction data and electron microprobe analyses. The empirical formula of the sample used is $Cu_{2.06}[(AsO_4)_{0.96}(PO_4)_{0.04}](OH)_{1.13} \cdot 3H_2O$. For the Raman spectrum of euchroite see also Frost and Bahfenne (2010b).

Euclase $BeAlSiO_4(OH)$

Origin: An unknown locality in Minas Gerais, Brazil.
Experimental details: Raman scattering measurements have been performed on a powdered sample using 488 nm Ar^+ laser radiation. The laser radiation power is not indicated.
Raman shifts (cm^{-1}): 3587, 3575, 1120, 1059s, 1023, 975, 908s, 880, 806w, 790w, 756w, 747w, 667w, 642w, 602w, 583, 574, 545, 518w, 461sh, 452, 444, 423w, 411w, 397s, 384, 355w, 341sh, 334, 309w, 290s, 276w, 259s, 237, 201, 180, 145.
Source: Hofmeister et al. (1987).
Comments: No independent analytical data are provided for the sample used. For the Raman spectra of euclase see also Jehlička et al. (2012) and Jehlička and Vandenabeele (2015).

Eucryptite-β $Li(AlSiO_4)$

Origin: Synthetic.
Experimental details: Raman scattering measurements have been performed on a single crystal at various crystal orientations using 1064 nm Nd-YAG laser radiation. The obtained data were averaged to produce a final spectrum. The nominal laser radiation power was 100 mW. A 180°-scattering geometry was employed.
Raman shifts (cm^{-1}): 1099, 1086, 1067(?), 1049, 1032s, 987w, 762w, 711w, 636w, 483s, 466sh, 352, 282, 233w, 187, 157, 142, 108.
Source: Zhang et al. (2003).
Comments: The sample was characterized by powder X-ray diffraction data.

Kentbrooksite $(Na,REE)_{15}(Ca,REE)_6Mn^{2+}_3Zr_3Nb(Si_{25}O_{73})OF_2 \cdot 2H_2O$

Origin: Sushina Hill Region, Purulia district, West Bengal, India.
Experimental details: Raman scattering measurements have been performed on an arbitrarily oriented sample, using 514.5 nm Ar^+ laser radiation. The laser radiation power at the sample was 4 mW.
Raman shifts (cm^{-1}): 1007s, 991, 975, 727s, 438, 433s, 429w, 354s, 333, 220, 209w, 199, 191s.
Source: Chakrabarty et al. (2011).
Comments: The sample was characterized by electron microprobe data which correspond to Ca-rich kentbrooksite. In the cited paper the mineral is erroneously described under the name "eudialyte." The Raman shifts were determined by us based on spectral curve analysis of the published spectrum.

Eugsterite $Na_4Ca(SO_4)_3 \cdot 2H_2O$

Origin: Efflorescence on the walls of the Manasija Monastery, Serbia.
Experimental details: Raman scattering measurements have been performed on an arbitrarily oriented sample, using 532 and 780 nm laser radiations.
Raman shifts (cm^{-1}): 1125, 1084.
Source: Matović et al. (2014).
Comments: The sample was characterized by powder X-ray diffraction data and EDS analysis.

Eulytine $Bi_4(SiO_4)_3$

Origin: Synthetic.
Experimental details: Raman scattering measurements have been performed on an arbitrarily oriented single crystal, using 488 and 514.5 nm Ar^+ laser radiations. The laser radiation power is not indicated.
Raman shifts (cm^{-1}): 991w, 930, 896, 888, 868s, 623w, 547w, 533w, 503w, 491w, 437, 393, 333, 314w, 283w, 276w, 249w, 202, 149, 131, 106, 100sh, 94s, 67.
Source: Beneventi et al. (1995).
Comments: No independent analytical data are provided for the sample used.

Euxenite-(Y) $(Y,Ca,Ce,U,Th)(Nb,Ta,Ti)_2O_6$

Origin: Billeroo Prospect, South Australia.
Experimental details: Raman scattering measurements have been performed on arbitrarily oriented crystals using a 633 nm He-Ne laser. The laser radiation power is not indicated. The Raman shifts have been determined for the maxima of individual peaks obtained as a result of the spectral curve analysis.
Raman shifts (cm^{-1}): 1624sh, 1520w, 1300w, 842, 805sh, 658sh, 624, 493sh, 410, 197sh, 161sh, 152s.
Source: Frost et al. (2011h).
Comments: No independent analytical data are provided for the sample used. For the Raman spectrum of euxenite-(Y) see also Gong et al. (1995).

Evansite $Al_3(PO_4)(OH)_6·8H_2O$

Origin: Porto, Northwest Portugal.
Experimental details: Kind of sample preparation is not indicated. Micro-Raman scattering measurements have been performed with 532 nm laser radiation. The nominal laser radiation power was 6 mW.
Raman shifts (cm^{-1}): 1645, 1364, 1048s, 1008s, 830w, 634, 564, 486, 367.
Source: Sanchez-Moral et al. (2011).
Comments: The sample was characterized by powder X-ray diffraction data, electron microprobe analyses, and inductively coupled plasma mass spectrometry.

Eveite $Mn^{2+}_2(AsO_4)(OH)$

Origin: Långban, Filipstad, Värmland, Sweden.
Experimental details: No data.
Raman shifts (cm^{-1}): 3564, 870, 827s, 809s, 510, 474w, 439, 222w.
Source: Yang et al. (2001).
Comments: The sample was characterized by single-crystal X-ray diffraction data. The crystal structure is solved. The Raman shifts were partly determined by us based on spectral curve analysis of the published spectrum.

Evenkite $C_{23}H_{48}$

Origin: Mernek, Slovakia.
Experimental details: Kind of sample preparation is not indicated. Raman scattering measurements have been performed using 514.5 nm Ar^+ laser radiation with the nominal radiation power of 10 mW and/or 1064 nm Nd-YAG laser radiation with the power of 350 mW. The spectrum was obtained with a beam perpendicular to the (111) crystal face.
Raman shifts (cm^{-1}): 2883s, 2848, 2735, 1464, 1441s, 1420, 1383w, 1370w, 1295, 1265w, 1171, 1133, 1123, 1103w, 1062, 890w, 103w.
Source: Jehlička et al. (2007a).
Comments: No independent analytical data are provided for the sample used. For the Raman spectrum of evenkite see also Jehlička et al. (2009a).

Ezcurrite $Na_2B_5O_7(OH)_3 \cdot 2H_2O$

Origin: Tincalayu Mine, Salardel Hombre Muerto, Salta, Argentina (type locality).
Experimental details: Raman scattering measurements have been performed on arbitrarily oriented crystals using a 633 nm He-Ne laser. The laser radiation power is not indicated. The Raman shifts have been determined for the maxima of individual peaks obtained as a result of the spectral curve analysis.
Raman shifts (cm^{-1}): 3652sh, 3596sh, 3576, 3547, 3509sh, 3431s, 3329sh, 3247sh, 3186, 3098sh, 1691sh, 1641w, 1591sh, 1343, 1333sh, 1318sh, 1193, 1163, 1129sh, 1060, 1048sh, 1037, 1025sh, 1015, 1000, 968, 957sh, 947, 857sh, 842, 803sh, 782sh, 761, 746sh, 726sh, 693w, 590sh, 575s, 550sh, 529, 488, 473sh, 460, 445sh, 385, 350, 317, 286, 273sh, 209sh, 190, 162, 141, 128sh, 115, 108.
Source: Frost et al. (2014j).
Comments: The sample was characterized by qualitative electron microprobe analysis which shows Al impurity.

Fabriesite $Na_3Al_3Si_3O_{12} \cdot 2H_2O$

Origin: Tawmaw, Myanmar.
Experimental details: Raman scattering measurements have been performed on an arbitrarily oriented sample using 532 nm laser radiation. The laser radiation power at the sample was 5 mW.
Raman shifts (cm^{-1}): 3500, 3200, 1011, 946, 903, 731, 682, 524, 490, 451s, 406, 372, 332, 314, 258w, 228w, 209.
Source: Ferraris et al. (2014).
Comments: The sample was characterized by electron microprobe analysis and electron backscatter diffraction. The empirical formula of the sample used is $(Na_{2.937}Ca_{0.030}K_{0.008}Mg_{0.007}Fe_{0.004}Ba_{0.002}Mn_{0.001})Al_{2.996}Si_{2.999}O_{12} \cdot 2H_{1.993}O$. The Raman shifts were partly determined by us based on spectral curve analysis of the published spectrum.

Fairfieldite $Ca_2Mn^{2+}(PO_4)_2 \cdot 2H_2O$

Origin: Cigana mine, Conselheiro Pena, Rio Doce valley, Minas Gerais, Brazil.
Experimental details: Raman scattering measurements have been performed on arbitrarily oriented crystals using a 633 nm He-Ne laser. The laser radiation power is not indicated. The Raman shifts

have been determined for the maxima of individual peaks obtained as a result of the spectral curve analysis.

Raman shifts (cm^{-1}): 3271sh, 3139sh, 3040, 2961sh, 1663w, 1632w, 1577w, 1491w, 1466w, 1099s, 1027, 955s, 945sh, 925s, 906, 768, 604s, 584, 552, 442sh, 422s, 369sh, 312, 287, 251, 235, 214, 203sh, 185, 176sh, 136w.

Source: Frost et al. (2013ad).

Comments: The sample was characterized by electron microprobe data. The empirical formula of the sample used is $Ca_2(Mn_{0.56}Mg_{0.33}Fe_{0.11})(PO_4)_2 \cdot 2H_2O$.

Falcondoite $Ni_4Si_6O_{15}(OH)_2 \cdot 6H_2O$

Origin: No data.

Experimental details: Raman scattering measurements have been performed on an arbitrarily oriented sample using 1064 nm laser radiation. The laser radiation power is not indicated.

Raman shifts (cm^{-1}): 823, 705sh, 673s, 640, 386s, 196s.

Source: Villanova-de-Benavent et al. (2012).

Comments: No independent analytical data are provided for the sample used.

Falottaite $MnC_2O_4 \cdot 3H_2O$

Origin: Synthetic.

Experimental details: Kind of sample preparation is not indicated. The measurements have been performed with using 1064 nm Nd-YAG laser radiation. The laser radiation power is not indicated.

Raman shifts (cm^{-1}): 1614w, 1478s, 1427, 913, 885, 575, 515, 480w.

Source: Mancilla et al. (2009a).

Comments: The sample was characterized by powder X-ray diffraction data.

Fangite Tl_3AsS_4

Origin: Allchar deposit, Macedonia.

Experimental details: Raman scattering measurements have been performed on an arbitrarily oriented single crystal using 632.8 nm He-Ne laser radiation. The laser radiation power at the sample was 1.9 mW. A 180°-scattering geometry was employed.

Raman shifts (cm^{-1}): 397, 379s, 367sh, 323, 308, 289sh, 275, 209, 191w, 170, 137, 105.

Source: Makreski et al. (2014).

Comments: The sample was characterized by energy-dispersive X-ray scan data.

Farringtonite $Mg_3(PO_4)_2$

Origin: Synthetic.

Experimental details: Raman scattering measurements have been performed on an arbitrarily oriented sample using 488 nm Ar$^+$ laser radiation. The laser radiation power is not indicated.

Raman shifts (cm^{-1}): 1151, 1112, 1093w, 1074, 1027s, 981s, 654, 638, 621, 576, 502, 475, 422, 355, 320, 271, 245w, 225w, 189, 177.

Source: O'Neill et al. (2006).

Comments: The sample was characterized by powder X-ray diffraction data. The Raman shifts were determined by us based on spectral curve analysis of the published spectrum.

Fassinaite $Pb_2(CO_3)(S_2O_3)$

Origin: Trentini mine, Mt. Naro, Vicenza province, Veneto, Italy (type locality).
Experimental details: Raman scattering measurements have been performed on arbitrarily oriented crystals using 632.8 nm He-Ne laser radiation. The laser radiation power at the sample was 0.8 mW.
Raman shifts (cm^{-1}): 1690w, 1444w, 1322w, 1137w, 1082w, 1061s, 983, 845w, 722w, 661, 637s, 629sh, 602, 549, 520s, 438s, 358, 342, 250w, 203w, 180, 75s.
Source: Bindi et al. (2011b).
Comments: The sample was characterized by single-crystal X-ray diffraction data, electron microprobe data, and electron microprobe analyses. The crystal structure is solved. The empirical formula of the sample used is $Pb_{2.01}(CO_3)(S_{1.82}O_3)$.

Faujasite-Na $(Na,Ca,Mg)_2(Si,Al)_{12}O_{24} \cdot 15H_2O$

Origin: Sasbach, Keiserstuhl, Germany (type locality).
Experimental details: Kind of sample preparation is not indicated. Raman scattering measurements have been performed using 1064 nm Nd-YAG laser radiation. The laser radiation power at the sample was 300 mW.
Raman shifts (cm^{-1}): 1096, 477s, 308, 177.
Source: Mozgawa (2001).
Comments: The sample was characterized by powder X-ray diffraction data.

Favreauite $PbBiCu_6O_4(SeO_3)_4(OH) \cdot H_2O$

Origin: El Dragón mine, Bolivia (type locality).
Experimental details: Raman scattering measurements have been performed on an arbitrarily oriented sample using 514.5 nm Ar$^+$ laser radiation. The laser radiation power at the sample was 2.3 mW.
Raman shifts (cm^{-1}): 3525, 1341w, 1240w, 1065w, 989w, 847s, 795sh, 764w, 542, 493, 392, 320w, 261w, 182.
Source: Mills et al. (2014c).
Comments: The sample was characterized by powder X-ray diffraction data and electron microprobe analyses. The crystal structure is solved. The empirical formula of the sample used is $Pb_{0.95}Ca_{0.17}Bi_{0.90}Cu_{5.81}Se_{4.10}O_{15.96}(OH)_{1.04} \cdot H_2O$.

Fayalite $Fe^{2+}_2(SiO_4)$

Origin: Synthetic.
Experimental details: Kind of sample preparation is not indicated. Raman scattering measurements have been performed using 532 nm Nd-YAG laser radiation. The laser radiation power is not indicated.
Raman shifts (cm^{-1}): 940, 838sh, 817s, 724, 643, 588, 508, 384, 315, 293, 197, 173w, 157s, 122.
Source: Mouri and Enami (2008).
Comments: The Raman shifts were determined by us based on spectral curve analysis of the published spectrum. For the Raman spectrum of fayalite see also Andò and Garzanti (2014).

Feitknechtite $Mn^{3+}O(OH)$

Origin: Synthetic.
Experimental details: Raman scattering measurements have been performed in the spectral region from 100 to 1200 cm^{-1}, on an arbitrarily oriented sample on the surface of catalyst, using 514.5 nm Ar$^+$ laser radiation. The laser radiation power at the sample was 0.235 mW.
Raman shifts (cm^{-1}): 635–633, 554–553, 495–492.
Source: Wang et al. (2014).
Comments: The sample was characterized by X-ray diffraction and X-ray photoelectron spectroscopy.

Felsőbányaite $Al_4(SO_4)(OH)_{10} \cdot 4H_2O$

Origin: RecoaroTerme, Vicenza, Italy.
Experimental details: Raman scattering measurements have been performed on a polycrystalline sample using 532 nm laser radiation. The laser radiation power is not indicated.
Raman shifts (cm^{-1}): 3590, 3560w, 1024w, 979s, 883w, 690w, 604, 510w, 451, 388w, 371w, 340w, 313w, 288w, 266w, 254w.
Source: Boscardin et al. (2009).
Comments: The sample was characterized by powder X-ray diffraction data and electron microprobe analyses. The Raman shifts were partly determined by us based on spectral curve analysis of the published spectrum.

Ferberite $Fe^{2+}(WO_4)$

Origin: Synthetic.
Experimental details: Raman scattering measurements have been performed on an arbitrarily oriented sample using 785 nm Nd-YAG laser radiation. The laser radiation power at the sample was 1 mW. The Raman shifts have been determined for the maxima of individual peaks obtained as a result of the spectral curve analysis.
Raman shifts (cm^{-1}): 933, 868, 811, 702, 615, 379, 281, 215.
Source: Frost et al. (2004d).
Comments: No independent analytical data are provided for the sample used.

Fergusonite-(Ce)-β $CeNbO_4$

Origin: Synthetic.
Experimental details: Raman scattering measurements have been performed on an arbitrarily oriented sample using 632.8 nm He-Ne laser radiation. The laser radiation power at the sample was 6 mW. A 180°-scattering geometry was employed. The Raman shifts have been determined for the maxima of individual peaks obtained as a result of the spectral curve analysis.
Raman shifts (cm^{-1}): 801s, 687w, 677w, (654w), 455sh, 397, 390, 347sh, 328sh, 311s, 269, 207, (200), 168, (157), 120sh, 103sh, 96s.
Source: Siqueira et al. (2010).
Comments: The sample was characterized by powder X-ray diffraction data. The Raman shifts of 103 and 96 cm^{-1} were determined by us based on spectral curve analysis of the published spectrum.

Fergusonite-(La)-β LaNbO$_4$

Origin: Synthetic.

Experimental details: Raman scattering measurements have been performed on an arbitrarily oriented sample using 632.8 nm He-Ne laser radiation. The laser radiation power at the sample was 6 mW. A 180°-scattering geometry was employed. The Raman shifts have been determined for the maxima of individual peaks obtained as a result of the spectral curve analysis.

Raman shifts (cm^{-1}): 807s, 667, (658), 628, 427, 400, (393), 347sh, 332sh, 327s, 287, 224, 201, 179, 170sh, (137), 126, 115, 110s, 89.

Source: Siqueira et al. (2010).

Comments: The sample was characterized by X-ray diffraction data. The Raman shifts of 110 and 89 cm^{-1} were determined by us based on spectral curve analysis of the published spectrum.

Fergusonite-(Nd)-β NdNbO$_4$

Origin: Synthetic.

Experimental details: Raman scattering measurements have been performed on an arbitrarily oriented sample using 632.8 nm He-Ne laser radiation. The laser radiation power at the sample was 6 mW. A 180°-scattering geometry was employed. The Raman shifts have been determined for the maxima of individual peaks obtained as a result of the spectral curve analysis.

Raman shifts (cm^{-1}): 808s, 673, 657, 633, 445, 419, 409, 359sh, 335sh, 331s, 303, 230, 212w, 186sh, 182, 130, 123sh, 120.

Source: Siqueira et al. (2010).

Comments: The sample was characterized by powder X-ray diffraction data.

Fergusonite-(Y)-β YNbO$_4$

Origin: Synthetic.

Experimental details: Raman scattering measurements have been performed on an arbitrarily oriented polycrystalline sample using 514.5 nm Ar$^+$ laser radiation. The laser radiation power at the sample was 200 mW.

Raman shifts (cm^{-1}): 810s, 677, 660, 627sh, 556, 464, 440w, 420w, 400sh, 344, 330s, 288sh, 237, 200sh, 182sh, 168, 130sh, 117sh, 136, 76, 70w, 55w.

Source: Pradhan and Choudhary (1987).

Comments: The sample was characterized by powder X-ray diffraction data.

Fergusonite-(Y) YNbO$_4$

Origin: No data.
Experimental details: No data.
Raman shifts (cm^{-1}): 779, 685w, 401w, 310, 208.
Source: Tomašić et al. (2008).

Comments: The sample was characterized by powder X-ray diffraction data. The Raman shifts were determined by us based on spectral curve analysis of the published spectrum. For the Raman spectrum of fergusonite-(Y) see also Gieré et al. (2009).

Fergusonite-(Y) $YNbO_4$

Origin: A granitic pegmatite situated in the Adamello massif, Italy.
Experimental details: Kind of sample preparation is not indicated. Raman scattering measurements have been performed using 488 nm Ar^+ laser radiation. The laser radiation power behind the objective was 8 mW.
Raman shifts (cm^{-1}): 817s, 698, 665, 444, 423, 379sh, 329s, 289, 230sh, 205sh, 143s.
Source: Gieré et al. (2009).
Comments: The sample was characterized by electron microprobe analyses. According to powder X-ray diffraction data, the mineral is metamict. It was identified as fergusonite-(Y) on the basis of its Raman spectrum. For the Raman spectrum of fergusonite-(Y) see also Tomašić et al. (2008).

Fermiite $Na_4(UO_2)(SO_4)_3 \cdot 3H_2O$

Origin: Blue Lizard mine, San Juan Co., Utah, USA (type locality).
Experimental details: Kind of sample preparation is not indicated. Raman scattering measurements have been performed using 532 nm laser radiation. The laser radiation power is not indicated.
Raman shifts (cm^{-1}): 3540, 3465, 3285, 1606, 1228, 1180, 1120, 1104, 1080, 1013s, 996sh, 992w, 860sh, 830s, 816sh, 638w, 616w, 583w, 506, 443, 384, 188, 163, 153, 132, 110, 55.
Source: Kampf et al. (2015c).
Comments: The sample was characterized by powder X-ray diffraction data and electron microprobe analyses. The crystal structure is solved. The empirical formula of the sample used is $Na_{3.88}(U_{1.05}O_2)(S_{0.99}O_4)_3 \cdot 3H_2O$.

Feroxyhyte $Fe^{3+}O(OH)$

Origin: Synthetic.
Experimental details: Raman scattering measurements have been performed on an arbitrarily oriented sample using 632.8 nm He-Ne laser radiation. The laser radiation power at the sample was 0.04 mW. A 180°-scattering geometry was employed. The Raman shifts have been determined for the maxima of individual peaks obtained as a result of the spectral curve analysis.
Raman shifts (cm^{-1}): 1442sh, 1330, 713sh, 663, 485sh, 401s, 292s, 222s.
Source: Müller et al. (2010).
Comments: The sample was characterized by powder X-ray diffraction data. For the Raman spectrum of feroxyhyte see also Nieuwoudt et al. (2011) and Chen et al. (2014c).

Ferriallanite-(Ce) $CaCe(Fe^{3+}AlFe^{2+})[Si_2O_7][SiO_4]O(OH)$.

Origin: Paleokerasia ophiolitic mélange formation, South Othris, Greece.
Experimental details: No data.
Raman shifts (cm^{-1}): 952w, 887w, 654s, 508s, 478s, 417w, ~377s, 212w, ~188w.
Source: Koutsovitis et al. (2013).
Comments: No independent analytical data are provided for the sample used.

4 Raman Spectra of Minerals

Ferricopiapite $Fe^{3+}_{0.67}Fe^{3+}_{4}(SO_4)_6(OH)_2 \cdot 20H_2O$

Origin: Baia Sprie mining area, Romania.
Experimental details: Kind of sample preparation is not indicated. Raman scattering measurements have been performed on an arbitrarily oriented sample using 532 nm Nd-YAG laser radiation. The nominal laser radiation power was 100 mW.
Raman shifts (cm^{-1}): 3135w, 1643, 1226, 1108, 1021s, 992s, 762w, 611, 480, 454sh, 304, 270s.
Source: Buzatu et al. (2016).
Comments: The sample was characterized by powder X-ray diffraction data. For the Raman spectra of ferricopiapite see also Ling and Wang (2010), Frost (2011c), Kong et al. (2011b), Apopei et al. (2012, 2014a), Sobron and Alpers (2013), Rull et al. (2014), and Wang and Zhou (2014).

Ferri-eckermannite $NaNa_2Mg_4Fe^{3+}Si_8O_{22}(OH)_2$

Experimental details: Raman scattering measurements have been performed in the spectral region 2600–3800 cm^{-1} on an oriented crystal with the polarization of incident light E_i parallel to the polarization of scattered light E_s and with the direction of crystal elongation perpendicular to E_i and parallel to E_s. 514.5 nm Ar$^+$ laser radiation was used. The laser radiation power is not indicated. A 180°-scattering geometry was employed.
Raman shifts (cm^{-1}): 3722sh, 3698.
Source: Leissner et al. (2015).
Comments: The sample was characterized by electron microprobe analysis and laser ablation inductively coupled-plasma mass spectrometry. The empirical formula of the sample used is $Na_{0.51}K_{0.49}Na_{2.00}(Mg_{0.45}Fe^{3+}_{0.20}Fe^{2+}_{0.14}Mn_{0.10}Li_{0.07}Al_{0.02}Ti_{0.02})_5Si_{8.00}(OH_{0.58}F_{0.34}O_{0.08})_2$. The Raman shifts were determined by us based on spectral curve analysis of the published spectrum.

Ferrierite-K $(K,Na)_5(Si_{31}Al_5)O_{72} \cdot 18H_2O$

Origin: Synthetic.
Experimental details: Kind of sample preparation is not indicated. Raman scattering measurements have been performed in the spectral region from 200 to 1400 cm^{-1} using 532 nm laser radiation. The laser radiation power is not indicated.
Raman shifts (cm^{-1}): 1163w, 1056, 1029, 833, 797, 576sh, 566, 455s, 432s, 389sh, 370w, 360w, 340, 319, 291sh, 228.
Source: Suzuki et al. (2009).
Comments: The sample was characterized by powder X-ray diffraction data. The Raman shifts were determined by us based on spectral curve analysis of the published spectrum.

Ferrierite-Na $(Na,K)_5(Si_{31}Al_5)O_{72} \cdot 18H_2O$

Origin: Synthetic.
Experimental details: Kind of sample preparation is not indicated. Raman scattering measurements have been performed in the spectral region from 200 to 1400 cm^{-1} using 532 nm laser radiation. The laser radiation power is not indicated.
Raman shifts (cm^{-1}): 1157, 1058, 823, 801, 572, 552, 494sh, 452s, 432s, 342, 316, 220.
Source: Suzuki et al. (2009).

Comments: The sample was characterized by powder X-ray diffraction data. The Raman shifts were determined by us based on spectral curve analysis of the published spectrum.

Ferrihydrite $Fe^{3+}_{10}O_{14}(OH)_2$

Origin: Synthetic.
Experimental details: Raman scattering measurements have been performed on an arbitrarily oriented sample using 632.8 nm He-Ne laser radiation. The laser radiation power was 0.04 mW.
Raman shifts (cm^{-1}): 1377w, 722, 676sh, 513, 358s.
Source: Müller et al. (2010).
Comments: The sample was characterized by powder X-ray diffraction data. For the Raman spectra of ferrihydrite see also Mazzetti and Thistlethwaite (2002) and Das and Hendry (2011).

Ferrihydrite $Fe^{3+}_{10}O_{14}(OH)_2$

Origin: Synthetic.
Experimental details: Raman scattering measurements have been performed on an arbitrarily oriented sample using 785 nm diode laser radiation. The laser radiation power at the sample was 0.3 mW. A 180°-scattering geometry was employed.
Raman shifts (cm^{-1}): 1045s, 707, 508w, 361w.
Source: Das and Hendry (2011).
Comments: The sample was characterized by X-ray diffraction data. For the Raman spectra of ferrihydrite see also Mazzetti and Thistlethwaite (2002) and Müller et al. (2010).

Ferri-kaersutite $NaCa_2(Mg_3Fe^{3+}Ti)(Si_6Al_2)O_{22}O_2$

Experimental details: Raman scattering measurements have been performed on an oriented crystal with the polarization of incident light E_i parallel to the polarization of scattered light E_s and with the direction of crystal elongation perpendicular to E_i and parallel to E_s. 514.5 nm Ar^+ laser radiation was used. The laser radiation power is not indicated. A 180°-scattering geometry was employed.
Raman shifts (cm^{-1}): 3699, 3670sh, 1072sh, 1013, 895, 780sh, 755s, 666, 527, 422, 378, 350sh, 294, 238, 184, 166, 134sh.
Source: Leissner et al. (2015).
Comments: The sample was characterized by powder X-ray diffraction data and electron microprobe data. The empirical formula of the sample used is $Na_{0.51}K_{0.36}\square_{0.13}(Ca_{0.89}Na_{0.07}Mg_{0.04})_2(Mg_{0.60}Fe^{3+}_{0.21}Ti_{0.13}Al_{0.06})_5(Al_{0.27}Si_{0.73})_8(O_{0.65}OH_{0.32}F_{0.03})_2$. The Raman shifts were determined by us based on spectral curve analysis of the published spectrum.

Ferrilotharmeyerite $CaZnFe^{3+}(AsO_4)_2(OH)\cdot H_2O$

Origin: Tsumeb mine, Tsumeb, Namibia (type locality).
Experimental details: No data in the cited paper.
Raman shifts (cm^{-1}): 3440w, 2973sh, 2636sh, 919, 880s, 830s, 814s, 765s, 730, 510, 487sh, 421, 370, 325sh, 230.
Source: Frost and Weier (2004e).

Comments: No independent analytical data are provided for the sample used. The Raman shifts have been determined for the maxima of individual peaks obtained as a result of the spectral curve analysis.

Ferrimolybdite $Fe^{3+}_2(Mo^{6+}O_4)_3 \cdot 7H_2O$

Origin: Vrchoslav, Krušné Hory (Ore Mts.), northern Bohemia, Czech Republic.
Experimental details: Raman scattering measurements have been performed on an arbitrarily oriented sample, using 633 nm He-Ne laser radiation. The laser radiation power is not indicated. The Raman shifts have been determined for the maxima of individual peaks obtained as a result of the spectral curve analysis.
Raman shifts (cm^{-1}): 1882w, 1611w, 991, 968s, 951sh, 935, 836sh, 822, 804sh, 784s, 771sh, 357s, 347, 327sh, 299, 258, 232, 206, 180, 156, 139sh, 115.
Source: Sejkora et al. (2014).
Comments: The sample was characterized by powder X-ray diffraction data and electron microprobe analysis. The empirical formula of the sample used is $Fe_{1.98}[(MoO_4)_{2.91}(SO_4)_{0.08}(PO_4)_{0.03}] \cdot 8H_2O$.

Ferrinatrite $Na_3Fe^{3+}(SO_4)_3 \cdot 3H_2O$

Origin: Lignitized wood from La Plaine-Chevrière, Oise, France.
Experimental details: Raman scattering measurements have been performed on an arbitrarily oriented sample using 532 nm laser radiation. The nominal laser radiation power was 2.5 mW. The Raman shifts have been determined for the maxima of individual peaks obtained as a result of the spectral curve analysis.
Raman shifts (cm^{-1}): 1614, 1250, 1235, 1220sh, 1202, 1123, 1011, 1002sh, 995s, 965s, 613, 603sh, 533sh, 502sh, 492, 460w, 438w, 267, 247s, 217, 197, 161, 138.
Source: Rouchon et al. (2012).
Comments: The sample was characterized by powder X-ray diffraction data and electron microprobe analysis.

Ferristrunzite $Fe^{3+}Fe^{3+}_2(PO_4)_2(OH)_3 \cdot 5H_2O$

Origin: No data.
Experimental details: Raman scattering measurements have been performed on an arbitrarily oriented sample, using 633 nm He-Ne laser radiation. The laser radiation power is not indicated. The Raman shifts have been determined for the maxima of individual peaks obtained as a result of the spectral curve analysis.
Raman shifts (cm^{-1}): 3465, 3367s, 3226, 3042, 1105sh, 1078s, 1022s, 1008sh, 985s, 634sh, 576, 533sh, 503, 454, 394, 354, 321, 302sh, 283s, 259s, 235, 203s, 184, 167s, 147sh.
Source: Frost et al. (2002c).
Comments: The sample was characterized by powder X-ray diffraction data. For the Raman spectrum of ferristrunzite see also Frost et al. (2004m).

Ferro-actinolite $\square Ca_2(Mg_{2.5-0.0}Fe^{2+}_{2.5-5.0})Si_8O_{22}(OH)_2$

Origin: No data.
Experimental details: No data.

Raman shifts (cm^{-1}): ~1065w, ~740w, ~670s, ~560, ~530, ~380, ~365, ~295w, ~215, ~175.
Source: Bersani et al. (2014).
Comments: The sample was characterized by electron microprobe analysis. The Raman shifts were determined by us based on spectral curve analysis of the published spectrum.

Ferro-glaucophane $Na_2(Fe^{2+}_3Al_2)Si_8O_{22}(OH)_2$

Origin: Vernè, Val Varaita, Sampeyre, Cuneo, Piemonte, Italy.
Experimental details: Raman scattering measurements have been performed on an arbitrarily oriented sample using 532 nm solid-state laser radiation. The laser radiation power is not indicated. A 180°-scattering geometry was employed.
Raman shifts (cm^{-1}): 1101, 1043, 984, 892, 831w, 909, 773, 733w, 667s, 606, 551, 537sh, 486, 765, 444w, 400sh, 385s, 334, 302w, 290w, 253, 207s, 175s, 157, 137w, 118s.
Source: Andò and Garzanti (2014).
Comments: In the cited paper the mineral is named Fe-glaucophane. No independent analytical data are provided for the sample used. The Raman shifts were partly determined by us based on spectral curve analysis of the published spectrum.

Ferro-hornblende $(Na,K)_{0-1}Ca_2(Mg,Fe^{2+},Fe^{3+},Al)_5(Si,Al)_8O_{22}(OH)_2$

Origin: Pelagon, Macedonia.
Experimental details: Raman scattering measurements have been performed on a powdered sample using 532 nm Nd-YAG laser radiation. The laser radiation power is not indicated.
Raman shifts (cm^{-1}): 1054w, 1018, 914, 863, 788w, 767w, 665s, 611s, 547, 517sh, 466, 426, 386sh, 330, 319sh, 251s, 226s, 177sh, 159, 115.
Source: Makreski et al. (2006a) and Jovanovski et al. (2009).
Comments: The sample was characterized by powder X-ray diffraction data and electron microprobe analyses. For the Raman spectrum of hornblende see also Andò and Garzanti (2014).

Ferrocarpholite $Fe^{2+}Al_2Si_2O_6(OH)_4$

Origin: Cole d'Esischie, Cuneo, Piemonte, Italy.
Experimental details: Raman scattering measurements have been performed on an arbitrarily oriented sample using 532 nm solid-state laser radiation. The laser radiation power is not indicated. A 180°-scattering geometry was employed.
Raman shifts (cm^{-1}): 1096, 1037, 1017w, 995w, 961, 928, 880, 866, 742s, 711, 664, 610s, 580sh, 562s, 530w, 517w, 498w, 488w, 475, 444, 406s, 350, 310, 282, 261, 238, 211s, 168.
Source: Andò and Garzanti (2014).
Comments: In the cited paper the mineral is named Fe-carpholite. No independent analytical data are provided for the sample used. The Raman shifts were partly determined by us based on spectral curve analysis of the published spectrum.

Ferroceladonite $KFe^{2+}Fe^{3+}Si_4O_{10}(OH)_2$

Origin: Mont Saint-Hilaire, Rouville RCM (Rouville Co.), Montérégie, Québec, Canada.
Experimental details: Kind of sample preparation is not indicated. 514.5 nm Ar$^+$ laser radiation was used. The laser radiation power was below 5 mW.

Raman shifts (cm^{-1}): 958, 696, 535, 453, 440, 395, 281, 234sh, 199s, 169sh.
Source: Ospitali et al. (2008).
Comments: The sample was characterized by electron microprobe analyses.

Ferrohögbomite (Fe,Mg,Zn,Al)$_3$(Al,Ti,Fe)$_8$O$_{15}$(OH) (for the 2N2S polysome)

Origin: Aktyuz area, Northern Tien Shan, Kyrgyzstan.
Experimental details: Kind of sample preparation is not indicated. 514.5 nm Ar$^+$ laser radiation was used. The laser radiation power is not indicated.
Raman shifts (cm^{-1}): 830, 776, 711, 525, 412, 257.
Source: Orozbaev et al. (2011).
Comments: Intensities of the Raman bands are not indicated. The sample was characterized by electron microprobe analysis. The empirical formula of the sample used is Mg$_{1.47}$Fe$^{2+}_{3.02}$Zn$_{0.04}$Fe$^{3+}_{1.76}$Al$_{15.13}$Ti$_{0.56}$O$_{30}$(OH)$_2$. For the Raman spectrum of ferrohögbomite see also Tsunogae and Santosh (2005).

Ferrokësterite Cu$_2$(Fe,Zn)SnS$_4$

Origin: Synthetic.
Experimental details: Raman scattering measurements have been performed on a polycrystalline thin film using 532 nm laser radiation. The laser radiation power was less than 1 mW.
Raman shifts (cm^{-1}): 378sh, 319s, 284, 256w.
Source: Khadka and Kim (2014).
Comments: The sample was characterized by powder X-ray diffraction data and electron microprobe analysis. The empirical formula of the sample used is Cu$_2$(Fe$_{77}$Zn$_{23}$)SnS$_4$. The Raman shifts were determined by us based on spectral curve analysis of the published spectrum.

Ferroselite FeSe$_2$

Origin: Synthetic.
Experimental details: Raman scattering measurements have been performed on an oriented single crystal using 514.5 nm Ar$^+$ laser radiation. The laser radiation power is not indicated. Polarized spectra were collected in different scattering geometries.
Raman shifts (cm^{-1}): 221, 183s.
Source: Lutz and Müller (1991).
Comments The Raman shifts are given for the scattering geometry $x(yy)$-x, in which the Raman intensities are most strong. No independent analytical data are provided for the sample used. For the Raman spectrum of ferroselite see also Wei et al. (2016).

Ferrosilite Fe$^{2+}_2$Si$_2$O$_6$

Origin: Synthetic.
Experimental details: Raman scattering measurements have been performed on an oriented single crystal using 488 nm Ar$^+$ laser radiation. The laser radiation power is not indicated. Spectra were recorded with the polarization of the laser radiation parallel to the (100) and (001) directions.
Raman shifts (cm^{-1}): 994s, 951, 888, 660s, 634, 532, 525s, 503, 396s, 349s, 319s, 301s, 247, 168s, 152, 129s.

Source: Stalder et al. (2009).
Comments: The sample was characterized by electron microprobe analysis. The average Raman shifts are given because the peak positions do not exhibit a significant dependence on orientation.

Ferrostrunzite $Fe^{2+}Fe^{3+}_2(PO_4)_2(OH)_2 \cdot 6H_2O$

Origin: Arnsberg, Sauerland, Germany.
Experimental details: Raman scattering measurements have been performed on an arbitrarily oriented sample using 633 nm He-Ne laser radiation. The laser radiation power is not indicated. The Raman shifts have been determined for the maxima of individual peaks obtained as a result of the spectral curve analysis.
Raman shifts (cm^{-1}): 3492sh, 3405s, 3134, 2943, 1113, 1058s, 1038sh, 1010s, 987sh, 967sh, 634, 568, 531sh, 509sh, 471, 434sh, 408, 396, 328, 297, 249s, 226, 202, 184s, 163s.
Source: Frost et al. (2002c).
Comments: The sample was characterized by powder X-ray diffraction data and electron microprobe analysis. For the Raman spectrum of ferrostrunzite see also Frost et al. (2004m).

Ferruccite $NaBF_4$

Origin: Synthetic.
Experimental details: Raman spectrum was obtained in the spectral region from 150 to 3000 cm^{-1} for a sample closed in quartz cell. 514 nm laser radiation was used. The laser radiation power is not indicated.
Raman shifts (cm^{-1}): 1279w, 1122w, 1060, 1040sh, 784s, 554, 532, 369w, 344.
Source: Zavorotynska et al. (2011).
Comments: The sample was characterized by powder X-ray diffraction data. The Raman shifts were partly determined by us based on spectral curve analysis of the published spectrum. For the Raman spectra of ferruccite see also Bonadeo and Silberman (1970) and Bates et al. (1971).

Fersmite $(Ca,Ce,Na)(Nb,Ta,Ti)_2(O,OH,F)_6$

Origin: Synthetic.
Experimental details: Raman scattering measurements have been performed on an oriented single-crystal fiber using 632.8 nm He-Ne laser radiation. The laser radiation power at the sample was 8 mW. A 180°-scattering geometry was employed. Polarized spectra were collected in the (xx), (xy), (zx), and (zy) scattering geometries.
Source: Moreira et al. (2010b).
Raman shifts (cm^{-1}): (xx): 906, 664, 540, 486, 379, 294s, 289, 242s, 227, 197, 140s, 64; (xy): 850s, 708, 598, 498, 431, 379s, 346, 293s, 262s, 208, 189, 164; (zx): 852, 637, 498s, 457, 380, 315, 2167, 249, 227, 198s, 130s, 111; (zy): 857, 736, 629, 462, 433, 365, 340, 270, 215s, 167, 139s, 65.
Comments: The sample was characterized by X-ray diffraction data.

Feruvite $CaFe^{2+}_3(Al_5Mg)(Si_6O_{18})(BO_3)_3(OH)_3(OH)$

Origin: No data.
Experimental details: Raman scattering measurements have been performed on a powdered sample in the spectral regions from 150 to 1600 cm^{-1} and from 3000 to 4000 cm^{-1} using 457 nm solid-state laser radiation. The laser radiation power at the sample was about 1 mW.

Raman shifts (cm^{-1}): 3624w, 3550, ~3500sh, 1324, 1095sh, 1054, 1003sh, 808sh, 768, 675sh, 573, 495, 420, 380, 356.
Source: Hoang et al. (2011).
Comments: The sample was characterized by powder X-ray diffraction data and electron microprobe analysis.

Fichtelite $C_{19}H_{34}$

Origin: Třeboň basin, Southern Bohemia, Czech Republic.
Experimental details: Raman scattering measurements have been performed on an arbitrarily oriented sample using 1064 nm Nd-YAG and 514.5 nm Ar$^+$ laser radiations. The nominal laser radiation power was 350 and 10 mW, respectively.
Raman shifts (cm^{-1}): 2997w, 2987, 2972, 2962w, 2947sh, 2937, 2923s, 2910, 2888, 2864s, 2846sh, 2842sh, 2834sh, 2757w, 2662sh, 1470sh, 1457sh, 1442s, 1381, 1361s, 1335, 1321, 1302, 1293w, 1275, 1264w, 1247s, 1227w, 1213, 1200, 1175, 1155, 1143, 1119w, 1104, 1091w, 1073, 1061sh, 1025, 997, 977, 950sh, 936, 913, 886w, 870, 852, 836, 815, 796, 770, 717s, 579w, 553, 490, 479, 450w, 438w, 399w, 380, 344.
Source: Jehlička and Edwards (2008).
Comments: The sample was characterized by powder X-ray diffraction data. For the Raman spectra of fichtelite see also Jehlička et al. (2005, 2009a).

Fiedlerite $Pb_3Cl_4F(OH)\cdot H_2O$

Origin: No data.
Experimental details: Kind of sample preparation is not indicated. Raman scattering measurements have been performed on an arbitrarily oriented sample using 632.8 and 514.5 nm laser radiations. The laser radiation power at the source was 30 mW and less than 30 mW, respectively.
Raman shifts (cm^{-1}): ~737, 600, 331s, 272s, 133s.
Source: Bouchard and Smith (2003).
Comments: No independent analytical data are provided for the sample used.

Finnemanite $Pb_5(As^{3+}O_3)_3Cl$

Origin: Långban deposit, Filipstad district, Värmland province, Sweden (type locality).
Experimental details: Raman scattering measurements have been performed on an arbitrarily oriented sample using 633 nm He-Ne laser radiation. The laser radiation power is not indicated. The Raman shifts have been determined for the maxima of individual peaks obtained as a result of the spectral curve analysis.
Raman shifts (cm^{-1}): 808w, 733s, 726sh, 640, 575, 450w, 372, 354sh, 244, 196sh, 174s, 128s, 113s.
Source: Bahfenne et al. (2011c).
Comments: No independent analytical data are provided for the sample used. For the Raman spectrum of finnemanite see also Bahfenne (2011).

Flamite $Ca_{8-x}(Na,K)_x(SiO_4)_{4-x}(PO_4)_x$

Origin: Hatrurim Basin, Negev Desert, Israel (type locality).
Experimental details: Raman scattering measurements have been performed in backscattered geometry, on an arbitrarily oriented sample, using 514.5 nm Ar$^+$ laser radiation. The laser radiation power at the sample was up to 17 mW.

Raman shifts (cm^{-1}): 1003w, 952s, (885sh), 863s, 850s, 714w, 666w, 589sh, 575w, 558sh, 538, 520, 500sh, 439sh, 430w, 396, 294sh, 260, 199sh, 188, 170, 125w, 106w.
Source: Sokol et al. (2015).
Comments: The sample was characterized by powder X-ray diffraction data and electron microprobe analyses. For the Raman spectrum of flamite see also Gfeller et al. (2015).

Flinteite K_2ZnCl_4

Origin: Tolbachik volcano, Kamchatka, Russia (type locality).
Experimental details: Raman scattering measurements have been performed on an arbitrarily oriented sample in the spectral region from 50 to 4000 cm^{-1} using 532 nm diode laser radiation. The laser radiation power at the sample was 13 mW.
Raman shifts (cm^{-1}): 294s, 192w, 140s, 113sh.
Source: Pekov et al. (2015e).
Comments: The sample was characterized by powder X-ray diffraction data and electron microprobe analyses. The crystal structure is solved. The empirical formula of the sample used is $(K_{1.91}Tl_{0.09})_{\Sigma 2.00}Zn_{1.04}Cl_{3.96}$.

Florencite-(La) $LaAl_3(PO_4)_2(OH)_6$

Origin: Igarapé Bahia mine, Serra dos Carajás, Pará, Brazil.
Experimental details: Raman scattering measurements have been performed on arbitrarily oriented crystals using a 633 nm He-Ne laser. The laser radiation power is not indicated. The Raman shifts have been determined for the maxima of individual peaks obtained as a result of the spectral curve analysis.
Raman shifts (cm^{-1}): 3649, 3440sh, 3158s, 2988sh, 2906sh, 1914s, 1713sh, 1655s, 1479s, 1221w, 1112s, 1064sh, 1021s, 987s, 846w, 783, 766w, 716, 699, 647sh, 614, 680sh, 536s, 524sh, 464, 404, 310, 270s, 255, 202, 194sh, 158w, 130w.
Source: Frost et al. (2013an).
Comments: The sample was characterized qualitative electron microprobe analysis.

Fluellite $Al_2(PO_4)F_2(OH)\cdot 7H_2O$

Origin: Krásno, near Horní Slavkov, western Bohemia, Czech Republic.
Experimental details: Raman scattering measurements have been performed on an arbitrarily oriented single crystal, using 633 nm He-Ne laser radiation. The laser radiation power is not indicated. Raman spectrum was obtained in the spectral region from 200 to 4000 cm^{-1}. The Raman shifts have been determined for the maxima of individual peaks obtained as a result of the spectral curve analysis.
Raman shifts (cm^{-1}): 3667, 3396s, 3314, 3124s, 1670w, 1583w, 1122, 1096sh, 1036s, 897w, 835sh, 646, 588, 557, 525s, 513, 459, 410, 342, 311, 295, 279, 251, 220, 208, 199, 191, 173, 151, 139, 123, 116, 108.
Source: Čejka et al. (2014a).
Comments: The sample was characterized by powder X-ray diffraction data and electron microprobe data. The empirical formula of the sample used is $Al_{1.98}(PO_4)_{1.07}F_{1.99}(OH)_{0.75}\cdot 7H_2O$.

Fluocerite-(Ce) CeF_3

Origin: Synthetic.
Experimental details: Raman scattering measurements have been performed on an oriented sample using 514.5 nm Ar$^+$ laser radiation. The laser radiation power is not indicated. Polarized spectra were collected in $x(zx)y$ and $x(yz)y$ scattering geometries.
Raman shifts (cm^{-1}): 293s, 290sh, 204, 141.
Source: Gerlinger and Schaack (1986).
Comments: The Raman shifts are given for the scattering geometry $x(zx)y$, in which the Raman intensities are most strong. The z axis is taken as the symmetry axis of the crystal and the x and y axes are equivalent. No independent analytical data are provided for the sample used. The Raman shifts were determined by us based on spectral curve analysis of the published spectrum. For the Raman spectrum of fluocerite-(Ce) see also Bauman and Porto (1967).

Fluocerite-(La) LaF_3

Origin: Synthetic.
Experimental details: Raman scattering measurements have been performed on an arbitrarily oriented 1% praseodymium-doped LaF$_3$ rod using 435.8 nm Hg line as excitation.
Raman shifts (cm^{-1}): 392, 365, 310, 292, 75.
Source: Caspers et al. (1964).
Comments: No independent analytical data are provided. For the Raman spectrum of fluocerite-(La) see also Bauman and Porto (1967).

Fluorapatite $Ca_5(PO_4)_3F$

Origin: No data.
Experimental details: Raman scattering measurements have been performed on a microcrystalline sample in the spectral region from 140 to 1200 cm^{-1} using He-Ne laser radiation. The laser radiation power is not indicated.
Raman shifts (cm^{-1}): 1084w, 1053w, 1041w, 966s, 609, 598, 595, 451w, 434, 268w, 184.
Source: Griffith (1970).
Comments: No independent analytical data are provided for the sample used. For the Raman spectra of fluorapatite see also Adams and Gardner (1974), Harlov et al. (2003), and Frezzotti et al. (2012).

Fluorapatite As-rich $Ca_5(PO_4,AsO_4)_3F$

Origin: Calvario Mt., Etna, Italy.
Experimental details: Raman scattering measurements have been performed on an arbitrarily oriented crystal using 633 and 532 nm laser radiations. The nominal laser radiation power was 20 and 10 mW, respectively.
Raman shifts (cm^{-1}): 1062sh, 1057sh, 1046, 1034sh, 964s, 877sh, 857s, 827w, 590, 580, 477sh, 430, 392, 372.
Source: Gianfagna et al. (2014).
Comments: The sample was characterized by powder X-ray diffraction data and electron microprobe analyses. The Raman shifts are indicated for the maxima of individual peaks obtained as a result of the spectral curve analysis.

Fluorapophyllite-(K) $KCa_4Si_8O_{20}F \cdot 8H_2O$

Origin: Międzyrzecze Górne, near Bielsko-Biała, Poland.
Experimental details: Raman scattering measurements have been performed on an oriented crystal using 514.5 nm Ar$^+$ laser radiation. The laser radiation power at the sample was less than 30 mW. Raman measurements were performed in different scattering geometries.
Raman shifts (cm^{-1}): -$z(y'y')z$: 3627w, 3559, 3357sh, 3104sh, 3010, 1117w, 1062, 856w, 794w, 665, 584, 542w, 522w, 486w, 433, 401w, 374w, 337w, 298, 231, 209, 166, 133, 123w; -$z(x'y')z$— 3559w, 3357sh, 3104sh, 3010, 1117w, 1010w, 973w, 840w, 768w, 630w, 463w, 371w, 267w, 226w, 217w, 185, 161, 131; -$y'(zz)y'$: 1063s, 794w, 584, 542, 486, 433, 401, 231, 195, 166; -$y'(x'z)y'$: 1091w, 1011w, 601w, 503w, 487w, 460w, 431w, 342w, 320w, 270w, 216w, 202w, 174w, 139w, 127w.
Source: Włodyka and Wrzalik (2004).
Comments: The sample was characterized by powder X-ray diffraction data and electron microprobe analyses. The empirical formula of the sample used is $(K_{0.96}Na_{0.03})Ca_{3.96}(Si_{7.95}Al_{0.04}P_{0.03})O_{19.98}(F_{0.83}OH_{0.17}) \cdot 8.02H_2O$. For the Raman spectra of fluorapophyllite-(K) see also Frost and Xi (2012o) and Goryainov et al. (2012).

Fluor-buergerite $NaFe^{3+}_3Al_6(Si_6O_{18})(BO_3)_3O_3F$

Origin: Mexiquitic, San Luis Potosi, Mexico.
Experimental details: Raman scattering measurements have been performed on an oriented crystal using 514.5 and 457.9 nm Ar$^+$ laser radiations. The laser radiation power at the sample was about 25 mW. Polarized spectra were collected in (zz) and (xy,z) scattering geometries.
Raman shifts (cm^{-1}): A_1 (zz): 3530, 1289, 1110w, 1060, 1010, 811w, 753, 706, 638s, 605sh, 541sh, 523, 475w, 406sh, 373s, 300, 260sh, 232s, 154; E (xy,z): 1290, 1260, 1107,1063, 1012, 982, 957, 757, 733, 707, 656s, 636, 586sh, 550sh, 522w, 475sh, 458, 400s, 375, 328, 300, 275, 265, 233, 212, 154.
Source: McKeown (2008).
Comments: The sample was characterized by single-crystal X-ray diffraction data and electron microprobe analysis. The crystal structure is solved. The empirical formula of the sample used is $NaFe^{3+}_3Al_6(Si_6O_{18})(BO_3)_3(O_{0.92}OH_{0.08})_{\Sigma3}F$. For the Raman spectra of fluor-buergerite see also Gasharova et al. (1997) and Watenphul et al. (2016a, b).

Fluorcalciobritholite $(Ca,REE)_5(SiO_4,PO_4)_3F$

Origin: Synthetic.
Experimental details: Raman scattering measurements have been performed on an arbitrarily oriented sample. Experimental details are not described.
Raman shifts (cm^{-1}): 1026sh, 1049w, 958s, 856, 603sh, 580, 557sh, 447sh, 428.
Source: Dacheux et al. (2010).
Comments: The sample was characterized by X-ray diffraction data and electron microprobe analysis. The empirical formula of the sample used is $Ca_9Nd_{0.5}Th_{0.5}(PO_4)_{4.5}(SiO_4)_{1.5}F_2$.

Fluorcalciomicrolite $(Ca,Na,\square)_2Ta_2O_6F$

Origin: Volta Grande pegmatite, Nazareno, Minas Gerais, Brazil (type locality).
Experimental details: Raman scattering measurements have been performed on an arbitrarily oriented sample 532 nm solid-state laser radiation. The laser radiation power is not indicated.

Raman shifts (cm^{-1}): 891, 792, 664, 530sh, 505s, 417, 331, 292, 239w, 187, 168.
Source: Andrade et al. (2013a).
Comments: The sample was characterized by powder X-ray diffraction data and electron microprobe analyses. The crystal structure is solved. The empirical formula of the sample used is $(Ca_{1.07}Na_{0.12}Vac_{0.12})(Ta_{1.84}Nb_{0.14}Sn_{0.02})[O_{5.93}(OH)_{0.07}][F_{0.79}(OH)_{0.21}]$.

Fluorcalcioroméite $(Ca,Na)_2Sb^{5+}_2O_6F$

Origin: Starlera mine, Ferrera, Grischun, Switzerland (type locality).
Experimental details: Raman scattering measurements have been performed on an arbitrarily oriented sample using 532 nm solid-state laser radiation. The laser radiation power is not indicated.
Raman shifts (cm^{-1}): 3686, 3630, 827, 790sh, 518s, 468sh, 302w.
Source: Atencio et al. (2013).
Comments: The sample was characterized by powder X-ray diffraction data and electron microprobe analyses. The crystal structure is solved. The empirical formula of the sample used is $(Ca_{1.16}Na_{0.56}\square_{0.22}Fe^{2+}_{0.03}Mn^{2+}_{0.03})(Sb^{5+}_{1.98}Al_{0.01}W_{0.01})O_6[F_{0.62}(OH)_{0.28}O_{0.06}\square_{0.04}]$.

Fluorcaphite $SrCaCa_3(PO_4)_3F$

Origin: Lovozero alkaline complex, Kola Peninsula, Russia.
Raman shifts (cm^{-1}): See comment below.
Source: Chakhmouradian et al. (2005).
Comments: Raman microspectroscopy cannot be used to distinguish between fluorcaphite and fluorapatite because their nonpolarized spectra are virtually identical. Polarized spectra in the range 50–350 cm^{-1} are sensitive to the local structural environment of A-site cations and can potentially be used for that purpose.

Fluor-elbaite $Na(Li_{1.5}Al_{1.5})Al_6(Si_6O_{18})(BO_3)_3(OH)_3F$

Origin: Paprok mine, Nuristan, Afghanistan.
Experimental details: Raman scattering measurements have been performed on an arbitrarily oriented cross-section of a single crystal using Ar$^+$ laser radiation. The laser radiation power is not indicated.
Raman shifts (cm^{-1}): 3658, 3594, 3489, 1421, 1064s, 731s, 550, 381, 344, 223.
Source: Natkaniec-Nowak et al. (2009).
Comments: The sample was characterized by electron microprobe analyses. Raman shifts are given for the central sample zone (zone I) which was identified as fluor-elbaite. The band positions denoted by Natkaniec-Nowak et al. (2009) as 3549 and 224 cm^{-1} were determined by us at 3594 and 344 cm^{-1}, respectively.

Fluor-schorl $NaFe^{2+}_3Al_6(Si_6O_{18})(BO_3)_3(OH)_3F$

Origin: Steinberg, Zschorlau, Erzgebirge (Saxonian Ore Mountains), Saxony, Germany.
Experimental details: Raman scattering measurements have been performed on an arbitrarily oriented sample using 532 nm Nd-YAG laser radiation. The nominal laser radiation power was 10 mW.
Raman shifts (cm^{-1}): 3563, 1084sh, 1048, 1025sh, 969w, 767s, 694, 665, 632, 537w, 483, 400sh, 364s, 313, 237s, 199.
Source: Ertl et al. (2015).

Comments: The sample was characterized by X-ray diffraction data and electron microprobe analyses. The crystal structure is solved. The empirical formula of the sample used is $(Na_{0.82}K_{0.01}Ca_{0.01}\square_{0.16})$ $(Fe^{2+}_{2.30}Al_{0.38}Mg_{0.23}Li_{0.03}Mn^{2+}_{0.02}Zn_{0.01})(Al_{5.80}Fe^{3+}_{0.10}Ti_{0.10})(Si_{5.81}Al_{0.19}O_{18})$ $(BO_3)_3(OH)_3[F_{0.66}(OH)_{0.34}]$. The Raman shifts were partly determined by us based on spectral curve analysis of the published spectrum.

Fluorite CaF_2

Origin: Synthetic.
Experimental details: Methods of sample preparation are not described. Raman scattering measurements have been performed using 632.8 nm He-Ne laser radiation. The nominal laser radiation power was 40 mW.
Raman shifts (cm^{-1}): 322.
Source: Gee et al. (1966).
Comments: No independent analytical data are provided for the sample used. For the Raman spectra of fluorite see also Tsuda et al. (1993), Dill and Weber (2010), Frezzotti et al. (2012), and Andò and Garzanti (2014).

Fluorkyuygenite $Ca_{12}Al_{14}O_{32}[(H_2O)_4F_2]$

Origin: Hatrurim Basin, Negev Desert, Israel (type locality).
Experimental details: Raman scattering measurements have been performed on an arbitrarily oriented sample using 532 nm solid-state laser radiation. The laser radiation power is not indicated.
Raman shifts (cm^{-1}): 3210sh, 3065, 1610w, 940sh, 876sh, 862, 772s, 695, 575, 517s, 392, 344, 320, 234.
Source: Galuskin et al. (2015c).
Comments: The sample was characterized by single-crystal X-ray diffraction data and electron microprobe analyses. The crystal structure is solved. The empirical formula of the sample used is $Ca_{12.034}(Al_{13.344}Fe^{3+}_{0.398}Si_{0.224})O_{32}[(H_2O)_{3.810}F_{1.894}(OH)_{0.296}]$. The Raman shifts were partly determined by us based on spectral curve analysis of the published spectrum.

Fluorlamprophyllite $Na_3(SrNa)Ti_3(Si_2O_7)_2O_2F_2$

Origin: Poços de Caldas alkaline massif, Morro doSerrote, Minas Gerais, Brazil (type locality)
Experimental details: Raman scattering measurements have been performed on an arbitrarily oriented sample using 532 nm solid-state laser radiation. The laser radiation power is not indicated.
Raman shifts (cm^{-1}): 3675w, 1070, 940sh, 893s, 862sh, 825sh, 690, 635w, 590, 462w, 419w, 375w, 346w, 280sh, 230sh, 174w.
Source: Andrade et al. (2017).
Comments: The Raman shifts were partly determined by us based on spectral curve analysis of the published spectrum.

Fluormayenite-related garnet

Origin: Afrikanda complex, Kola Peninsula, Russia.
Experimental details: Raman scattering measurements have been performed on an arbitrarily oriented sample using 633 nm He-Ne laser radiation. The laser radiation power is not indicated.

Raman shifts (cm^{-1}): 3630, 1100w, 955w, 863s, 815, 633w, 519, 385sh, 352, 240w.
Source: Chakhmouradian et al. (2008).
Comments: In the cited paper the mineral is described under the name hibscite. It contains from 4.2 to 6.0 wt% F. The sample was characterized by micro-X-ray diffraction data, electron microprobe analyses, and single-crystal X-ray diffraction. The crystal structure is solved. The compositional range of the sample used may be described as $Grs_{57-63}Kt_{21-27}Fgr_{8-11}Adr_{0-13}$, where Grs, Kt, and Adr are the designations for the grossular, katoite, and andradite, respectively, and Fgr is the hypothetical $Ca_3Al_2F_{12}$ end-member. The Raman shifts were partly determined by us based on spectral curve analysis of the published spectrum.

Fluormayenite $Ca_{12}Al_{14}O_{32}[\square_4F_2]$

Origin: Jabel Harmun, Judean Mts., Palestinian Autonomy, Israel (type locality).
Experimental details: Raman scattering measurements have been performed on an arbitrarily oriented sample using 532 nm solid-state laser radiation. The laser radiation power is not indicated.
Raman shifts (cm^{-1}): 3674w, 3572, 918sh, 890, 844, 776s, 709, 619w, 583, 524s, 390, 318s, 276, 250sh, 186.
Source: Galuskin et al. (2015c).
Comments: The sample was characterized by single-crystal X-ray diffraction data and electron microprobe analyses. The crystal structure is solved. The empirical formula of the sample used is $(Ca_{11.951}Na_{0.037})(Al_{13.675}Fe^{3+}_{0.270}Mg_{0.040}Si_{0.009}P_{0.005}S^{6+}_{0.013})O_{31.503}(HO)_{1.492}[\square_{4.581}F_{1.375}Cl_{0.044}]$. The Raman shifts were partly determined by us based on spectral curve analysis of the published spectrum.

Fluorocronite PbF_2

Origin: Synthetic.
Experimental details: Raman scattering measurements have been performed on an arbitrarily oriented sample using 632.8 nm He-Ne laser radiation. The laser radiation output power was 300 mW.
Raman shifts (cm^{-1}): 257.
Source: Krishnamurthy and Soots (1970).
Comments: No independent analytical data are provided for the sample used.

Fluoro-edenite $NaCa_2Mg_5(Si_7Al)O_{22}F_2$

Origin: Biancavilla area, Mount Etna, Sicily, Italy (type locality).
Experimental details: Raman scattering measurements have been performed on an oriented sample with the elongation axis at 45° with respect to N-S direction of the cross-hair of the ocular lens using 632.8 nm He-Ne laser radiation. The nominal laser radiation power was 20 mW.
Raman shifts (for the sample no. 3 in Table 1 of the cited paper, cm^{-1}): 1042, 1018, 926, 768, 747, 678s, 586w, 557, 520, 494w, 474w, 436, 408, 381, 365sh, 313, 300sh, 236s, 213w.
Source: Fornero et al. (2008).
Comments: The sample was characterized by electron microprobe analyses. The Raman shifts were partly determined by us based on spectral curve analysis of the published spectrum.

Fluorowardite $NaAl_3(PO_4)_2(OH)_2F_2 \cdot 2H_2O$

Origin: Silver Coin mine, Valmy, Nevada, USA (type locality).
Experimental details: Raman scattering measurements have been performed on an arbitrarily oriented sample using 514.5 nm Ar^+ laser radiation. The laser radiation power at the sample was about 5 mW.
Raman shifts (cm^{-1}): 3614w, 3542, 1049s, 1005s, 604s, 489, 431, 384w, 315, 272, 257, 216, 182s, 143.
Source: Kampf et al. (2014a).
Comments: The sample was characterized by powder X-ray diffraction data and electron microprobe analyses. The crystal structure is solved. The empirical formula of the sample used is $(Na_{0.87}Ca_{0.13}Mg_{0.04})(Al_{2.96}Fe^{3+}_{0.04})(P_{1.96}As_{0.03})O_{8.12}(OH)_{2.35}F_{1.53} \cdot 2H_2O$. The Raman shifts were partly determined by us based on spectral curve analysis of the published spectrum.

Fluor-uvite $CaMg_3(Al_5Mg)(Si_6O_{18})(BO_3)_3(OH)_3F$

Origin: No data in the cited paper.
Experimental details: Raman scattering measurements have been performed on an oriented crystal in the -y(zz)y scattering geometry using 514.5 or 488.0 nm Ar^+ laser radiations. The laser radiation power at the sample was 14 mW.
Raman shifts (cm^{-1}): 3770, 3740, 3665, 3642, 3572, 3554sh, 3522sh, 3488sh, 1040, 960w, 798w, 759w, 702, 670, 650, 630, 496, 459, 411, 372s, 316, 243s, 213s, 149w.
Source: Watenphul et al. (2016a).
Coments: The sample was characterized by electron microprobe analyses and LA-ICP-MS data. The empirical formula of the sample used is $(Ca_{0.63}Na_{0.26}\square_{0.09})(Mg_{2.92}Ti_{0.07})(Al_{5.51}Mg_{0.49})(Si_6O_{18})(BO_3)_3(OH)_3[F_{0.55}(OH)_{0.36}O_{0.09}]$.

Fluorwavellite $Al_3(PO_4)_2(OH)_2F \cdot 5H_2O$

Origin: Silver Coin mine, Valmy, Iron Point district, Humboldt Co., Nevada, USA (type locality).
Experimental details: Methods of sample preparation are not described. A 785 nm diode laser radiation was used. The laser radiation power is not indicated.
Raman shifts (cm^{-1}): 1147w, 1065sh, 1022s, 636, 589sh, 568sh, 550sh, 544s, 410s, 315, 286sh, 277, 224sh, 211.
Source: Kampf et al. (2017a).
Comments: The sample was characterized by powder X-ray diffraction data and electron microprobe analyses. The crystal structure is solved. The empirical formula of the sample used is $Al_{2.96}(PO_4)_2(OH)_{1.98}F_{1.02} \cdot 5H_2O$ (+0.12H for charge balance). The Raman shifts were partly determined by us based on spectral curve analysis of the published spectrum.

Foitite $\square(Fe^{2+}_2Al)Al_6(Si_6O_{18})(BO_3)_3(OH)_3(OH)$

Origin: No data.
Experimental details: Raman scattering measurements have been performed on an oriented crystal with the crystallographic c axis parallel to the z axis using 514.5 and 488.0 nm Ar^+ laser radiations. The laser radiation power at the sample was 14 mW. Raman spectrum was obtained in the spectral region from 15 to 4000 cm^{-1}. Polarized spectra were collected in the -y(zz)y, y(zx)y, and y(xx)y scattering geometries.

Raman shifts (cm^{-1}): 3726w, 3670w, 3644, 3631, 3570sh, 3551s, 3517, 3484s, 3641, 3630, 3479, 3351, 1054, 1020, 967, 777w, 743w, 696, 677, 630, 493, 459, 401, 367s, 313, 253sh, 236s, 205sh, 192sh, 158w.

Source: Watenphul et al. (2016a).

Comments: The sample was characterized by electron microprobe and LA-ICP-MS analyses. The Raman shifts are given for the scattering geometry -$y(zz)y$, in which the Raman intensities are most strong. The empirical formula of the sample used is ($\square_{0.61}$Na$_{0.35}$Ca$_{0.03}$)(Fe$_{1.28}$Al$_{1.03}$Mn$_{0.41}$Li$_{0.18}$Mg$_{0.11}$)Al$_6$(Si$_6$O$_{18}$)(BO$_3$)$_3$(OH)$_3$[(OH)$_{0.93}$F$_{0.07}$]. The Raman shifts were partly determined by us based on spectral curve analysis of the published spectrum.

Foordite Sn^{2+}Nb$_2$O$_6$

Origin: Synthetic.

Experimental details: Raman scattering measurements have been performed on an oriented crystal using 532 nm solid-state laser radiation. The laser radiation power is not indicated. Polarized spectra were collected in the $z(xx)$-z scattering geometry.

Raman shifts (cm^{-1}): 794, 665, 575, 432, 372, 348, 303, 257, 249, 224, 201, 185, 161, 135.

Source: Noureldine et al. (2014).

Comments: The sample was characterized by powder X-ray diffraction data. The Raman shifts were determined by us based on spectral curve analysis of the published spectrum. For the Raman spectrum of foordite see also Lee et al. (2015).

Forêtite Cu$_2$Al$_2$(AsO$_4$)(OH,O,H$_2$O)$_6$

Origin: Cap Garonne mine, France (type locality).

Experimental details: Raman scattering measurements have been performed on an arbitrarily oriented sample in the spectral region from 50 to 4000 cm^{-1} using 532 nm laser radiation. The laser radiation power is not indicated. A 180°-scattering geometry was employed.

Raman shifts (cm^{-1}): 3534sh, 3469sh, 3428s, 3343s, 2924, 2889, 2848, 1585, 1458, 848s, 816sh, 495, 446w, 371, 269, 218, 171, 140sh, 114s, 93s.

Source: Mills et al. (2012a).

Comments: The sample was characterized by powder X-ray diffraction data and electron microprobe analyses. The empirical formula of the sample used is Cu$_{1.94}$(Al$_{1.96}$Fe$_{0.04}$)(As$_{0.84}$S$_{0.09}$Si$_{0.04}$)O$_{10}$H$_{5.19}$.

Formanite-(Y) YTaO$_4$

Origin: Synthetic.

Experimental details: No data.

Raman shifts (cm^{-1}): 825s, 720w, 705, 670, 655w, 480w, 450w, 375w, 345s, 320s, 215s, 120.

Source: Blasse (1973).

Comments: The sample was characterized by powder X-ray diffraction data. For the Raman spectrum of formanite-(Y) see also Nazarov (2010).

Formicaite Ca(CHOO)$_2$

Origin: Synthetic.

Experimental details: Raman scattering measurements have been performed on an oriented crystal using 488.0 nm Ar$^+$ and 632.8 nm He-Ne laser radiations. The laser radiation power is not indicated.

Polarized spectra were collected in $x(zz)y$, $x(yx)y$, $x(yx)z$, $x(yz)y$, $x(zy)z$, $x(zx)y$, $x(xz)z$ scattering geometries with 488.8 nm laser excitation, and in $y(xx)z$, $y(xy)z$, $y(zy)z$, $y(zx)z$ scattering geometries with 632.8 nm laser excitation.

Raman shifts (cm^{-1}): 2185w, 2180w, 1406s, 1393s, 801w, 783, 169w, 118, 106, 72s.

Source: Krishnan and Ramanujam (1973).

Comments: The Raman shifts are given for the scattering geometry $x(yy)z$ with 488.0 nm Ar$^+$ laser radiation, in which the Raman intensities are most strong.

Fornacite $CuPb_2(CrO_4)(AsO_4)(OH)$

Origin: Whim Creek Copper mine, Pilbara, Western Australia.

Experimental details: Raman scattering measurements have been performed on an arbitrarily oriented sample using 785 nm Nd-YAG laser radiation. The laser radiation power at the sample was 1 mW.

Raman shifts (cm^{-1}): 916sh, 890sh, 872sh, 867, 847s, 830s, 790sh, 778sh, 400, 388, 381, 369, 354, 343, 332, 305, 159, 139, 122.

Source: Frost (2004c).

Comments: No independent analytical data are provided for the sample used.

Forsterite $Mg_2(SiO_4)$

Origin: Synthetic.

Experimental details: Raman scattering measurements have been performed on a powdered sample in the spectral region from 100 to 1200 cm^{-1} using 488 nm Ar$^+$ laser radiation. The laser radiation power at the sample was 20 mW. A 135°-scattering geometry was employed.

Raman shifts (cm^{-1}): 964, 919, 880sh, 855s, 824s, 608, 589, 544, 434, 374, 337, 329, 314w, 303, 225.

Source: Mohanan et al. (1993).

Comments: The sample was characterized by powder X-ray diffraction data. For the Raman spectra of forsterite see also Piriou and McMillan (1983), Mouri and Enami (2008), Frezzotti et al. (2012), Andò and Garzanti (2014), and Culka et al. (2016a, b).

Fougèrite $Fe^{2+}_4Fe^{3+}_2(OH)_{12}(CO_3) \cdot 3H_2O$

Origin: Fougères Forest, Fougères, Ille-et-Vilaine, Brittany, France (type locality).

Experimental details: Raman scattering measurements have been performed on a fine-crystalline sample using 514.5 nm Ar$^+$ laser radiation. The laser radiation power at the sample was less than 1 mW. A 180°-scattering geometry was employed.

Raman shifts (cm^{-1}): 518, 427.

Source: Trolard et al. (2007).

Comments: The sample was characterized by powder X-ray diffraction data, Mössbauer spectroscopy, and X-ray absorption spectroscopy at the FeK edge. For the Raman spectrum of fougèrite see also Bourrié and Trolard (2010).

Fraipontite $(Zn,Al)_3(Si,Al)_2O_5(OH)_4$

Origin: Blue Bell mine, USA.

Experimental details: Raman scattering measurements have been performed on arbitrarily oriented crystals using a 633 nm He-Ne laser. The laser radiation power is not indicated. The Raman shifts

have been determined for the maxima of individual peaks obtained as a result of the spectral curve analysis.

Raman shifts (cm^{-1}): 3825w, 3810w, 3781w, 3769w, 3747w, 3737w, 3669w, 3384w, 3367w, (1734), (1408), 1392sh, (1094s), 1089sh, (731), 675, (305), 290sh, (197), 186sh, 157sh, 144s, 115sh, 108.

Source: Theiss et al. (2015b).

Comments: The sample was characterized by qualitative energy-dispersive X-ray scan data. No independent quantitative analytical data are provided for the sample used. The bands at 1734, 1408, 1094, 731, 305, and 197 cm^{-1} correspond to admixed smithsonite.

Francevillite $Ba(UO_2)_2(VO_4)_2 \cdot 5H_2O$

Origin: Mounana Mine, Haut Ogoue, Gabon (type locality).

Experimental details: Raman scattering measurements have been performed on arbitrarily oriented crystals using a 633 nm He-Ne laser. The laser radiation power is not indicated. The Raman shifts have been determined for the maxima of individual peaks obtained as a result of the spectral curve analysis.

Raman shifts (cm^{-1}): 977, 965sh, 861, 829, 747s, 609, 526, 485, 470, 405, 370s, 304, 240s, 186, 163sh.

Source: Frost et al. (2005c).

Comments: No independent analytical data are provided for the sample used.

Francisite $Cu_3Bi(Se^{4+}O_3)_2O_2Cl$

Origin: Synthetic.

Experimental details: Raman scattering measurements have been performed on an oriented single crystal with laser light polarized in the xy plane of the crystal. The laser radiation power and the wavelength of laser radiation are not indicated.

Raman shifts (cm^{-1}): 583, 538, 484, 324, 173.

Source: Miller et al. (2012).

Comments: The sample was characterized by powder X-ray diffraction data.

Franckeite $Pb_{21.7}Sn_{9.3}Fe_{4.0}Sb_{8.1}S_{56.9}$

Origin: No data.

Experimental details: Raman scattering measurements have been performed on a pressed powdered sample using 532 nm laser radiation. The nominal laser radiation power was 0.2 mW.

Raman shifts (cm^{-1}): 650–400sh, 318, 253, 194, 145w, 66s.

Source: Molina-Mendoza et al. (2016).

Comments: The sample was characterized by selected area electron diffraction data, micro-X-ray photoemission, and scanning tunneling spectroscopy.

Françoisite-(Nd) $Nd(UO_2)_3O(OH)(PO_4)_2 \cdot 6H_2O$

Origin: Synthetic.

Experimental details: Raman scattering measurements have been performed on a powdered sample in the spectral region from 150 to 2000 cm^{-1} using 632.8 nm He-Ne laser radiation. The laser radiation power is not indicated.

Raman shifts (cm^{-1}): 998, 934w, 830s, 440–417, 202, 151.
Source: Armstrong et al. (2011).
Comments: The sample was characterized by elemental and thermal analyses, X-ray diffraction data, and inductively coupled optical emission spectroscopy. The empirical formula of the sample used is $Nd_{0.92}[(UO_2)_{3.11}O(OH)(PO_4)_{2.00}] \cdot 5.98H_2O$.

Franconite $NaNb_2O_5(OH) \cdot 3H_2O$

Origin: Poudrette (Demix) quarry, Mont Saint-Hilaire, Rouville RCM (Rouville Co.), Montérégie, Québec, Canada (type locality).
Experimental details: Raman scattering measurements have been performed on an arbitrarily oriented sample in the spectral region from 50 to 4000 cm^{-1} using 638 nm laser radiation. The laser radiation power is not indicated.
Raman shifts (cm^{-1}): 3416, 924, 879s, 661s, 583, 461, 391, 297, 212.
Source: Haring and McDonald (2014b).
Comments: It may be that the wavelength of 638 nm is a misprint, and the authors meant 632.8 nm. The sample was characterized by powder X-ray diffraction data, by energy-dispersive X–ray scan data, and by single crystal X-ray diffraction data. The crystal structure is solved. The empirical formula of the sample used is $(Na_{0.73}Ca_{0.13}\square_{0.14})_{\Sigma 1.00}(Nb_{1.96}Ti_{0.02}Si_{0.02}Al_{0.01})O_5(OH) \cdot 3H_2O$.

Frankdicksonite BaF_2

Origin: Synthetic.
Experimental details: The Raman spectrum was obtained at 15 K. Other experimental details are not indicated.
Raman shifts (cm^{-1}): 259sh, 244.
Source: Harrington et al. (1971).
Comments: No independent analytical data are provided for the sample used. The Raman shift of 259 cm^{-1} was determined by us based on spectral curve analysis of the published spectrum.

Franklinite $ZnFe^{3+}_2O_4$

Origin: Franklin or Sterling Hill mine, New Jersey, USA.
Experimental details: No data.
Raman shifts (cm^{-1}): 1206, 661sh, 597s, 493w, 347.
Source: Welsh (2008).
Comments: No independent analytical data are provided for the sample used.

Freboldite CoSe

Origin: Synthetic.
Experimental details: Raman scattering measurements have been performed on a powdered sample nucleated after about 50 h milling time using 514.5 nm Ar$^+$ laser radiation. The nominal laser radiation power was less than 5 mW. A 180°-scattering geometry was employed.
Raman shifts (cm^{-1}): 174.
Source: Campos et al. (2004a).
Comments: The sample was characterized by X-ray diffraction data and electron microprobe analysis.

Fredrikssonite $Mg_2Mn^{3+}O_2(BO_3)$

Origin: Långban deposit, Bergslagen ore region, Filipstad district, Värmland, Sweden (type locality).
Experimental details: Raman scattering measurements have been performed on arbitrarily oriented crystals using a 633 nm He-Ne laser. The laser radiation power is not indicated. The Raman shifts have been determined for the maxima of individual peaks obtained as a result of the spectral curve analysis.
Raman shifts (cm^{-1}): See comment below.
Source: Frost (2011b).
Comments: All Raman shifts (1750, 1530, 1435, 1086, 712, 282, and 155 cm^{-1}) ascribed by Frost (2011b) to fredrikssonite correspond to calcite. The correct Raman shifts of fredrikssonite are (RRUFF ID: R130112; cm^{-1}): 933w, 752, 701w, 666, 591s, 520w, 341sh, 317s.

Fresnoite $Ba_2TiO(Si_2O_7)$

Origin: No data.
Experimental details: No data.
Raman shifts (cm^{-1}): 994w, 960w, 928w, 904, 876, 860s, 666, 600, 542w, 477w, 377, 343, 318w, 272, 226, 207w.
Source: Gabelica-Robert and Tarte (1981).
Comments: The sample was characterized by powder X-ray diffraction data. For the Raman spectra of fresnoite see also Mayerhöfer and Dunken (2001) and Ma and Rossman (2008).

Friedrichbeckeite $K(\square Na)Mg_2(Be_2Mg)Si_{12}O_{30}$

Origin: Bellerberg volcano, Eifel paleovolcanic area, Rhineland-Palatinate (Rheinland-Pfalz), Germany (type locality).
Experimental details: Raman scattering measurements have been performed on an oriented crystal with direction of the laser beam along [0001] using 633 nm He-Ne laser radiation. The nominal laser radiation power was 17 mW.
Raman shifts (cm^{-1}): 1132s, 947, 837, ~770, 650, 577, 544, 488s, 461sh, 385, 292s, 162, 128, 69.
Source: Lengauer et al. (2009).
Comments: The sample was characterized by powder X-ray diffraction data, electron microprobe analyses, and laser-ablation inductively coupled plasma mass spectroscopy data. The crystal structure is solved. The empirical formula of the sample used is $K_{0.87}Na_{0.86}(Mg_{1.57}Mn_{0.28}Fe_{0.24})$ $(Be_{1.83}Mg_{1.17})[Si_{12}O_{30}]$. The Raman shifts were determined by us based on spectral curve analysis of the published spectrum.

Frohbergite $FeTe_2$

Origin: Synthetic.
Experimental details: Raman scattering measurements have been performed at 90 K on an oriented sample using 514.5 nm Ar^+ laser radiation. The laser radiation power is not indicated. A 180-°-scattering geometry was employed. Polarized spectra were collected in $y(xx)$-y, $y(xz)$-y, $z(xy)$-z, $x(zy)$-x, and $y(x,xz)$-y scattering geometries.
Raman shifts (cm^{-1}): 155, 138s, 125.
Source: Lutz and Müller (1991).

Comments: The Raman shifts are given in the scattering geometry $y(x,xz)$-y. The authors note that the origin of the band at 138 cm^{-1} is not quite clear. No independent analytical data are provided for the sample used.

Frolovite Ca[B(OH)$_4$]$_2$

Origin: Synthetic.
Experimental details: Methods of sample preparation are not described. Raman scattering measurements have been performed in the spectral region from 300 to 1800 cm^{-1} using 514.5 nm Ar$^+$ laser radiation. The nominal laser radiation power was 300 mW.
Raman shifts (cm^{-1}): 854w, 758s, 547, 390w.
Source: Jun et al. (1995).
Comments: No independent analytical data are provided for the sample used.

Frondelite Mn^{2+}Fe$^{3+}_4$(PO$_4$)$_3$(OH)$_5$

Origin: Cigana mine, Conselheiro Pena, Rio Doce valley, Minas Gerais, Brazil.
Experimental details: Raman scattering measurements have been performed on arbitrarily oriented crystals using a 633 nm He-Ne laser. The laser radiation power is not indicated. The Raman shifts have been determined for the maxima of individual peaks obtained as a result of the spectral curve analysis.
Raman shifts (cm^{-1}): 3581, 3315, 3144sh, 3029sh, 2886, 2747sh, 1597, 1532, 1416, 1352, 1164sh, 1112, 1071sh, 1027s, 1000, 966sh, 748, 635sh, 612s, 589sh, 572, 481sh, 455, 436sh, 379sh, 329sh, 291, 226s, 207sh, 189, 172sh, 151s, 137sh, 126sh.
Source: Frost et al. (2013y).
Comments: The sample was characterized by X-ray diffraction data and electron microprobe analysis. The empirical formula of the sample used is (Mn$_{0.68}$Fe$_{0.32}$)Fe$^{3+}_{3.72}$(PO$_4$)$_{3.17}$(OH)$_{4.99}$. For the Raman spectrum of frondelite see also Faulstich et al. (2013).

Fulgurite (a high-silicon glass) ~(Si,O,Fe)O$_{2-x}$

Origin: Greensboro, North Carolina, USA.
Experimental details: Raman scattering measurements have been performed using 514.5 nm Ar$^+$ laser radiation. The laser radiation power at the sample was 1 to 2 mW.
Raman shifts (cm^{-1}): 1188w, 1057, 930w, 796, 603, 488s, 440s.
Source: Carter et al. (2010).
Comments: The sample was characterized by inductively coupled plasma optical emission spectrometry data. It contains 81.3 wt% SiO$_2$, 8.32 wt% Al$_2$O$_3$, 8.48 wt% Fe$_2$O$_3$, and minor amounts of other components.

Gadolinite-(Nd) Nd$_2$Fe^{2+}Be$_2$O$_2$(SiO$_4$)$_2$

Origin: Malmkärra mine, ~3.5 km WSW of Norberg, Sweden (type locality).
Experimental details: Raman scattering measurements have been performed on a polished section using 633 nm He-Ne laser radiation. The laser radiation power at the sample was 10 mW.
Raman shifts (cm^{-1}): 3525w, 970, 897s, 707w, 677sh, 615w, 550w, 501w, 483w, 428, 411, 383, 363, 339, 306, 292, 279, 265, 225w, 203w, 143w, 104w.

Source: Škoda et al. (2017).
Comments: The sample was characterized by powder X-ray diffraction data and electron microprobe analyses. The crystal structure is solved.

Gahnite $ZnAl_2O_4$

Origin: Jemaa, Kaduna State, Nigeria.
Experimental details: Raman scattering measurements have been performed on an arbitrarily oriented single crystal using 473.1 nm Nd-YAG laser radiation. The laser radiation power is not indicated.
Raman shifts (cm^{-1}): 661s, 510w, 420.
Source: D'Ippolito et al. (2013).
Comments: The sample was characterized by single crystal X-ray diffraction data and electron microprobe analyses. The crystal structure is solved. The empirical formula of the sample used is $(Zn_{0.94}Fe^{2+}_{0.03}Al_{0.03})(Al_{1.96}Fe^{2+}_{0.03}Fe^{3+}_{0.01})O_4$. For the Raman spectrum of gahnite see also Faulstich et al. (2016).

Gaidonnayite $Na_2ZrSi_3O_9 \cdot 2H_2O$

Origin: Toongi rare metal deposit, New South Wales, Australia
Experimental details: Raman scattering measurements have been performed on an arbitrarily oriented sample using 532 nm laser radiation. The laser radiation power is not indicated.
Raman shifts (cm^{-1}): 1093sh, 1053, 963s, 926, 704w, 663w, 548s, 526sh, 404, 337sh, 316, 255sh, 199s, 152s.
Source: Spandler and Morris (2016).
Comments: The sample was characterized by X-ray fluorescence data and laser-ablation inductively coupled plasma mass spectroscopy. The Raman shifts were determined by us based on spectral curve analysis of the published spectrum.

Galaxite $Mn^{2+}Al_2O_4$

Origin: Synthetic.
Experimental details: Raman scattering measurements have been performed on an arbitrarily oriented single crystal in the spectral region from 100 to 900 cm^{-1} using 473.1 nm Nd-YAG laser radiation. The laser radiation power at the sample was less than 1 mW. A nearly 180°-scattering geometry was employed.
Raman shifts (cm^{-1}): 775, 700w, 644w, 510, 395s, 374sh, 202s.
Source: D'Ippolito et al. (2015).
Comments: The sample was characterized by electron microprobe analysis.

Galena PbS

Origin: No data.
Experimental details: Raman scattering measurements have been performed on an arbitrarily oriented sample using 632.8 nm laser radiation. The laser radiation power is not indicated.
Raman shifts (cm^{-1}): ~450, 205.
Source: Sherwin et al. (2005).

Comments: No independent analytical data are provided for the sample used. The first-order Raman scattering is forbidden in minerals with the halite structure. However, the peak at 205 cm^{-1} attributed to forbidden first order spectrum has been registered for galena. For the Raman spectrum of galena see also Frezzotti et al. (2012).

Galileiite NaFe$^{2+}_4$(PO$_4$)$_3$

Origin: Yanzhuang H6 chondrite, Yanzhuang village, Wenyuan Co., Guangdong province, China.
Experimental details: Raman scattering measurements have been performed on an arbitrarily oriented sample using 514.5 nm Ar$^+$ laser radiation. The nominal laser radiation power was 26.8 mW.
Raman shifts (cm^{-1}): 1129–1124, 982–980s, 599–596, 558–554w, 417–416, 305–304w, 154–151.
Source: Xie et al. (2014).
Comments: The sample was characterized by electron microprobe analyses. The empirical formula of the sample used is (Na$_{0.89}$K$_{0.01}$Ca$_{0.03}$Cr$_{0.05}$)(Fe$_{3.61}$Mn$_{0.29}$Mg$_{0.02}$Si$_{0.03}$)P$_{2.99}$O$_{12}$.

Gallite CuGaS$_2$

Origin: Synthetic.
Description: No data.
Experimental details: Raman scattering measurements have been performed on an arbitrarily oriented green-colored crystal with formula closest to pure CuGaS$_2$ using 514.5 nm Ar$^+$ laser radiation. The laser radiation output power was 30 mW.
Raman shifts (cm^{-1}): 384, 347w, 309s, 274w, 162w, 112w, 91w, 72w, 32w.
Source: Julien et al. (1999).
Comments: The sample was characterized by X-ray diffraction data. For the Raman spectrum of gallite see also Cha and Jung (2014).

Gallium sulfide Ga$_2$S$_3$

Origin: Synthetic.
Description: No data.
Experimental details: Raman scattering measurements have been performed on an arbitrarily oriented sample using 514.5 nm Ar$^+$ laser radiation. The laser radiation output power was 30 mW.
Raman shifts (cm^{-1}): 422, 386w, 329, 307, 233s, 147, 140, 114, 86, 72.
Source: Julien et al. (1999).
Comments: The sample was characterized by X-ray diffraction data.

Galloplumbogummite Pb(Ga,Al,Ge)$_3$(PO$_4$)$_2$(OH)$_6$

Origin: Tsumeb mine, Tsumeb, Namibia (type locality).
Experimental details: Raman scattering measurements have been performed on an oriented sample with three-fold crystallographic axis close to the direction of the polarization of the incident light using 514.5 and 488.0 nm Ar$^+$ laser radiations. The nominal laser radiation power is not indicated. A 180°-scattering geometry was employed.
Raman shifts (cm^{-1}): 3252, 1087s, 1007s, 979, 899, 619, 602, 573w, 566w, 559w, 538w, 492s, 356, 317, 272, 181, 83, 55.
Source: Schlüter et al. (2014).

Comments: The sample was characterized by powder X-ray diffraction data and electron microprobe analyses. The crystal structure is solved. The empirical formula of the sample used is $(Pb_{1.04}Ca_{0.05})(Ga_{1.41}Al_{1.35}Ge_{0.38}Fe_{0.02})(P_{1.91}S_{0.14})O_{8.44}(OH)_{5.56}$. The Raman shifts were determined by us based on spectral curve analysis of the published spectrum.

Galuskinite $Ca_7(SiO_4)_3(CO_3)$

Origin: Birkhin gabbro massif, Eastern Siberia, Russia (type locality).

Experimental details: Raman scattering measurements have been performed on an arbitrarily oriented single crystal in the spectral region from 100 to 4000 cm^{-1} using 514.5 nm Ar^+ laser radiation. The laser radiation output power was in the range 30–50 mW. A 180°-scattering geometry was employed.

Raman shifts (cm^{-1}): 1077s, 1007, 972w, 950, 928, 917, 898w, 889w, 882, 863s, 851s, 843, 704, 661, 570, 555, 525, 439, 404, 392, 367, 315w, 292w, 276, 257, 238, 220, 205, 184, 163, 140w, 124w, 114w.

Source: Lazic et al. (2011).

Comments: The sample was characterized by single-crystal X-ray diffraction data and electron microprobe analyses. The crystal structure is solved. The empirical formula of the sample used is $(Ca_{6.936}Na_{0.086})(Si_{2.983}P_{0.018}S_{0.004})O_{12}(CO_3)$. The Raman shifts were partly determined by us based on spectral curve analysis of the published spectrum.

Gamagarite $Ba_2Fe^{3+}(VO_4)_2OH$

Origin: Synthetic.

Experimental details: Raman scattering measurements have been performed on a polycrystalline sample using 514.5 nm Ar^+ laser radiation. The nominal laser radiation power was 90 mW. A 180°-scattering geometry was employed.

Raman shifts (cm^{-1}): 953, 944sh, 918, 904, 874, 853, 846, 827, 799, 728w, 629, ~480sh, ~455sh, 442s, 411sh, 361, 346, 311, 268, 247sh, 229.

Source: Sanjeewa et al. (2015).

Comments: The sample was characterized by powder X-ray diffraction data and electron microprobe analysis. The crystal structure is solved. The Raman shifts were determined by us based on spectral curve analysis of the published spectrum.

Gananite BiF_3

Origin: Synthetic

Experimental details: Raman scattering measurements have been performed on a powdered sample using 488.0 and 514.5 nm Ar^+ laser radiations. The nominal laser radiation power was 200 mW. A 90°-scattering geometry was employed.

Raman shifts (cm^{-1}): 334sh, 312s, 278sh, 271s, 265sh, 248s, 213, 203, 192, 177sh, 155, 142sh, 127s, 115s, 84, 70sh.

Source: Kavun et al. (2010).

Comments: No independent analytical data are provided for the sample used. The Raman shifts were partly determined by us based on spectral curve analysis of the published spectrum.

Ganomalite $Pb_9Ca_6(Si_2O_7)_4(SiO_4)O$

Origin: Jakobsberg, Bergslagen, Sweden.
Experimental details: Raman scattering measurements have been performed on an arbitrarily oriented sample using 514.5 nm Ar^+ laser radiation. The laser radiation power at the sample was 6 mW.
Raman shifts (cm^{-1}): 1068, 1042, 1000, 902, 886, 848s, 810, 727w, 673w, 564s, 551sh, 520, 486w, 455, 427w, 401w, 373, 354, 291, 242, 208.
Source: Kampf et al. (2016c).
Comments: The sample was characterized by powder X-ray diffraction data. The Raman shifts were partly determined by us based on spectral curve analysis of the published spectrum.

Ganterite $Ba_{0.5}(Na,K)_{0.5}Al_2(Si_{2.5}Al_{1.5})O_{10}(OH)_2$

Origin: Berisal Complex, Simplon Region, Switzerland (type locality).
Experimental details: Raman scattering measurements have been performed on an arbitrarily oriented polished thin section of a single crystal using 514.5 nm Ar^+ laser radiation. The nominal laser radiation power was 25 mW.
Raman shifts (cm^{-1}): 1092, 1025, 948, 699s, 595s, 488s, 405sh, 266s.
Source: Graeser et al. (2003).
Comments: The sample was characterized by powder X-ray diffraction data and electron microprobe analyses. The empirical formula of the sample used is $(Ba_{0.44}K_{0.28}Na_{0.27})(Al_{1.84}Mg_{0.09}Fe_{0.04}Ti_{0.04})[Si_{2.72}Al_{1.28}O_{10}](OH)_{1.89}$.

Garavellite $FeSbBiS_4$

Origin: Malé Karpaty Mts., Western Carpathians, Slovakia.
Experimental details: Raman scattering measurements have been performed on an arbitrarily oriented polished section using 532 nm Nd-YAG laser radiation. The nominal laser radiation power was 210 mW. A 180°-scattering geometry was employed. The Raman shifts have been determined for the maxima of individual peaks obtained as a result of the spectral curve analysis.
Raman shifts (cm^{-1}): 364w, 338, 322, 302, 271sh, 249sh, 233sh, 214s, 197sh, 182, 167, 151, 137sh, 120, 99sh, 78s, 61.
Source: Kharbish and Andráš (2014).
Comments: The sample was characterized by powder X-ray diffraction data and electron microprobe analyses.

Garnet $Mg_3(MgSi)Si_3O_{12}$

Origin: Synthetic.
Experimental details: Raman scattering measurements have been performed on a polycrystalline sample using Ar^+ laser radiation. The laser radiation power at the sample was in the range 5–50 mW.
Raman shifts (cm^{-1}): 1065, 1034w, 989w, 964sh, 931s, 889, 873sh, 852sh, 802, 724w, 648w, 602s, 559w, 535w, 507w, 498w, 481w, 458, 429w, 398w, 367, 354sh, 336w, 311, 293w, 275w, 261w, 238sh, 226, 205sh, 197, 181, 159, 138w.
Source: McMillan et al. (1989).
Comments: No independent analytical data are provided for the sample used.

Gartrellite $PbCuFe^{3+}(AsO_4)_2(OH) \cdot H_2O$

Origin: Anticline deposit, Ashburton Downs, Western Australia.
Experimental details: Raman scattering measurements have been performed on arbitrarily oriented crystals using a 633 nm He-Ne laser. The laser radiation power is not indicated. The Raman shifts have been determined for the maxima of individual peaks obtained as a result of the spectral curve analysis.
Raman shifts (cm^{-1}): 3404, 3229, 2999, 1161, 1099, 995, 869s, 842s, 812, 785, 618, 560, 499, 474s, 438s, 357, 331, 304, 238, 201, 164, 140.
Source: Frost and Weier (2004e).
Comments: No independent analytical data are provided for the sample used. Raman shifts are given for gartrellite with partly isomorphic substitution of arsenate by sulfate. For the Raman spectrum of gartrellite see also López et al. (2014c).

Garutiite (Ni,Fe,Ir)

Origin: Loma Peguera, Dominican Republic (type locality).
Experimental details: Methods of sample preparation are not described. Raman scattering measurements have been performed using 532.6 nm Nd-YAG laser radiation. The nominal laser radiation power was 100 mW.
Raman shifts (cm^{-1}): The obtained Raman spectrum shows no discernible absorption bands over the range of 150–2000 cm^{-1}.
Source: McDonald et al. (2010).

Gaspéite $Ni(CO_3)$

Origin: Synthetic.
Experimental details: Methods of sample preparation are not described. Raman scattering measurements have been performed using 488.0 and 514.5 nm Ar^+ laser radiations. The nominal laser radiation power was about 100 mW.
Raman shifts (cm^{-1}): 1731, 1428, 1089, 736, 343, 235.
Source: Rutt and Nicola (1974).
Comments: The sample was characterized by powder X-ray diffraction data.

Gaudefroyite $Ca_4Mn^{3+}_3(BO_3)_3(CO_3)O_3$

Origin: N'Chwaning II mine, Kalahari manganese fields, South Africa.
Experimental details: Raman scattering measurements have been performed on an arbitrarily oriented sample using 633 nm He-Ne laser radiation. The laser radiation power is not indicated. Raman spectrum was obtained in the spectral region from 200 to 4000 cm^{-1}. The Raman shifts have been determined for the maxima of individual peaks obtained as a result of the spectral curve analysis.
Raman shifts (cm^{-1}): 3385, 3249, 3206, 1595w, 1508w, 1491w, 1358sh, 1306sh, 1283, 1263sh, 1227sh, 1210sh, 1194, 1182sh, 1153sh, 1130w, 1112w, 1076, 1070sh, 950sh, 939, 928s, 914sh, 768w, 764w, 748w, 743w, 671s, 649s, 635sh, 584sh, 573, 534, 405w, 389w, 342s, 334sh, 297sh, 287s, 255sh, 244s, 228sh, 211w.
Source: Frost et al. (2014ae).
Comments: No independent analytical data are provided for the sample used.

Gaylussite $Na_2Ca(CO_3)_2 \cdot 5H_2O$

Origin: Teels Marsh, Esmeralda Co., Nevada, USA.
Experimental details: Raman scattering measurements have been performed on arbitrarily oriented crystals using a 633 nm He-Ne laser. The laser radiation power is not indicated. The Raman shifts have been determined for the maxima of individual peaks obtained as a result of the spectral curve analysis.
Raman shifts (cm^{-1}): 3344, 3250sh, 2948sh, 1070s, 719, 698sh, 663w, 518, 258, 222, 158.
Source: Frost and Dickfos (2007b).
Comments: No independent analytical data are provided for the sample used.

Gazeevite $BaCa_6(SiO_4)_2(SO_4)_2O$

Origin: Jabel Harmun, Judean Mts., Palestinian Autonomy, Israel.
Experimental details: Raman scattering measurements have been performed on an arbitrarily oriented sample using 532 nm Nd-YAG laser radiation. The laser radiation power at the sample was 50 mW.
Raman shifts (cm^{-1}): 1135, 1099, 997s, 963, 865, 638, 555w, 529w, 464, 409, 315w, 266sh, 213, 160w.
Source: Galuskin et al. (2016a).
Comments: The sample was characterized by single-crystal X-ray diffraction data and electron microprobe analyses. The crystal structure is solved. The empirical formula of the sample used is $(Ba_{0.85}K_{0.12}Sr_{0.02})(Ca_{5.99}Na_{0.02})[(SiO_4)_{1.82}(PO_4)_{0.14}(AlO_4)_{0.04}(TiO_4)_{0.01}][(SO_4)_{1.85}(PO_4)_{0.15}]O_{0.84}F_{0.11}$. The Raman shifts were partly determined by us based on spectral curve analysis of the published spectrum.

Gearksutite $CaAlF_4(OH) \cdot H_2O$

Origin: Valedo Ribeira, south of São Paulo and northeast of Paraná, Brazil.
Experimental details: No data.
Raman shifts (cm^{-1}): 463, 407, 381, 298w, 226, 209, 169, 147s, 93, 84, 63, 54, 11.
Source: Ronchi (2003).
Comments: No independent analytical data are provided for the sample used. The Raman shifts were determined by us based on spectral curve analysis of the published spectrum.

Geffroyite $(Cu,Fe,Ag)_9Se_8$

Origin: Moroshkovoe lake, Southern Sopchinskoe deposit, Monchegorsk ore district, Kola Peninsula, Russia.
Experimental details: Raman scattering measurements have been performed on an arbitrarily oriented sample using 514.5 nm Ar$^+$ laser radiation or 785 nm diode laser radiation. The nominal laser radiation power was 50 mW and 500 mW, respectively.
Raman shifts (cm^{-1}): 445, 365, 264s, 186, 90w, 76.
Source: Voloshin et al. (2015a).
Comments: The sample was characterized by electron microprobe analysis. The empirical formula of the sample used is $(Cu_{9.20}Ag_{0.44})(Se_{4.95}S_{3.05}Te_{0.10})$.

Gehlenite $Ca_2Al(SiAl)O_7$

Origin: Synthetic.

Experimental details: Raman scattering measurements have been performed on a powdered sample using 488.0 nm Ar^+ laser radiation. The laser radiation power at the sample was 500 mW. A 90°-scattering geometry was employed.

Raman shifts (cm^{-1}): 1005w, 998, 977, 914, 841sh, 796, 655sh, 626s, 528, 459, 425w, 303, 254sh, 240, 218, 180, 150w, 89.

Source: Sharma et al. (1983).

Comments: The sample was characterized by powder X-ray diffraction data. For the Raman spectra of gehlenite see also Burshtein et al. (1993) and Bouhifd et al. (2002).

Geikielite $MgTiO_3$

Origin: Synthetic.

Experimental details: Methods of sample preparation are not described. Raman scattering measurements have been performed using 514.5 nm Ar^+ laser radiation. The nominal laser radiation power was 80 mW.

Raman shifts (cm^{-1}): 714s, 641, 487sh, 485s, 397, 352s, 327s, 306, 281s, 224.

Source: Okada et al. (2008).

Comments: The sample was characterized by powder X-ray diffraction data and electron microprobe analysis. For the Raman spectrum of geikielite see also Reynard and Guyot (1994).

Geminite $Cu^{2+}(AsO_3OH)\cdot H_2O$

Origin: Jáchymov ore district, Krušné Hory (Czech Ore Mts.), western Bohemia, Czech Republic.

Experimental details: Raman scattering measurements have been performed on an arbitrarily oriented sample using 633 nm He-Ne laser radiation. The laser radiation power is not indicated. Raman spectrum was obtained in the spectral region from 200 to 4000 cm^{-1}. The Raman shifts have been determined for the maxima of individual peaks obtained as a result of the spectral curve analysis.

Raman shifts (cm^{-1}): 3521w, 3448, 3314, 3152sh, 2814, 2438, 2288, 1299w, 885s, 871sh, 853s, 843sh, 813, 743w, 496, 481sh, 451sh, 421, 345, 333, 310, 284w, 244w, 213, 182sh, 178s, 161s, 136.

Source: Sejkora et al. (2010).

Comments: The sample was characterized by powder X-ray diffraction data and electron microprobe analyses. The empirical formula of the sample used is $Cu_{1.00}[AsO_3(OH)_{0.96}F_{0.04}]\cdot H_2O$.

Gerhardtite $Cu_2(NO_3)(OH)_3$

Origin: Great Australia mine, Cloncurry, Queensland, Australia.

Experimental details: Raman scattering measurements have been performed on an arbitrarily oriented sample using 633 nm He-Ne laser radiation. The laser radiation power is not indicated. The Raman shifts have been determined for the maxima of individual peaks obtained as a result of the spectral curve analysis.

Raman shifts (cm^{-1}): 3556sh, 3546, 3477, 3417, 3391sh, 1438s, 1417s, 1339, 1324s, 1052, 1048s, 1031, 1024, 887w, 805w, 720w, 711, 668w, 503, 474sh, 458, 437sh, 423sh, 410, 336w, 279, 258, 213, 189s, 165s, 149sh, 132w.

Source: Frost et al. (2004h).
Comments: No independent quantitative analytical data are provided for the sample used.

Gerstleyite $Na_2(Sb,As)_8S_{13} \cdot 2H_2O$

Origin: Baker mine, Kramer district, Kern Co., California, USA (type locality).
Experimental details: Raman scattering measurements have been performed on arbitrarily oriented crystals using a 633 nm He-Ne laser. The laser radiation power is not indicated. The Raman shifts have been determined for the maxima of individual peaks obtained as a result of the spectral curve analysis.
Raman shifts (cm^{-1}): 449, 308s, 286s, 251s, 224sh, 188, 144s.
Source: Frost et al. (2010c).
Comments: No independent analytical data are provided for the sample used.

Geschieberite $K_2(UO_2)(SO_4)_2 \cdot 2H_2O$

Origin: Svornost mine, Jáchymov, Western Bohemia, Czech Republic (type locality).
Experimental details: Methods of sample preparation are not described. Raman scattering measurements have been performed using 532 nm diode laser radiation. The laser radiation power at the sample was 3 mW.
Raman shifts (cm^{-1}): 3595, 3506, 3280w, 1216, 1126, 1008s, 992, 984, 832s, 822sh, 652, 606w, 584w, 471, 454w, 386w, 270, 246, 230, 180, 154, 132sh, 100, 80.
Source: Plášil et al. (2015c).
Comments: The sample was characterized by powder X-ray diffraction data and electron microprobe analyses. The crystal structure is solved. The empirical formula of the sample used is $(K_{1.72}Mg_{0.29}Na_{0.04}Ca_{0.01})(U_{0.98}O_2)(S_{0.98}O_4)_2 \cdot 2H_2O$.

Ghiaraite $CaCl_2 \cdot 4H_2O$

Origin: Synthetic.
Experimental details: Methods of sample preparation are not described. Raman scattering measurements have been performed using 514.5 nm Ar^+ laser radiation. The laser radiation power is not indicated.
Raman shifts (cm^{-1}): 3511, 3460sh, 3435, 3400, 3364sh, 3242sh, 1657, 1645sh, 1625, 801, 763, 713, 698, 679, 657, 595, 573, 551, 523s, 435s, 405s, 374, 335, 309, 283, 261w, 253w, 232, 212, 204, 184, 173, 163w, 154, 134, 127, 118, 108.
Source: Uriarte et al. (2015).
Comments: The sample was characterized by powder X-ray diffraction data. The Raman shifts were partly determined by us based on spectral curve analysis of the published spectrum.

Gibbsite $Al(OH)_3$

Origin: Synthetic.
Experimental details: Methods of sample preparation are not described. Raman scattering measurements have been performed using 1064 nm Nd-YAG laser radiation. The nominal laser radiation power was 200 mW. The Raman shifts have been determined for the maxima of individual peaks obtained as a result of the spectral curve analysis.

Raman shifts (cm^{-1}): 3617s, 3522s, 3433s, 3364s, 1051, 1018, 979, 924, 892, 844, 816, 788, 751, 710, 617w, 602w, 569, 538s, 506, 444, 428, 412, 396s, 380, 371, 322s, 306, 290, 264, 255, 242.

Source: Ruan et al. (2001).

Comments: No independent analytical data are provided for the sample used. For the Raman spectra of gibbsite see also Rodgers (1992, 1993).

Gilalite $Cu_5Si_6O_{17} \cdot 7H_2O$

Origin: São José da Batalha, Brazil.

Experimental details: Raman scattering measurements have been performed on an arbitrarily oriented sample using a 633 nm He-Ne laser. The laser radiation power is not indicated. Raman spectrum was obtained in the spectral region from 1200 to 4000 cm^{-1}. The Raman shifts have been determined for the maxima of individual peaks obtained as a result of the spectral curve analysis.

Raman shifts (cm^{-1}): 3706, 3669s, 3631sh, 3609, 3529, 3478, 3423sh, 3386s, 3347, 3313sh, 3259sh, 3207sh, 3154sh, 3075w, 2999, 2946, 2905sh, 2859sh, 1131sh, 1096sh, 1057sh, 1008, 964sh, 931sh, 898sh, 831, 779, 755sh, 675, 621s, 561, 509sh, 484sh, 443s, 400, 338, 314sh, 250sh, 214s, 150s, 123sh.

Source: López et al. (2014b).

Comments: The sample was characterized by semiquantitative electron microprobe analysis.

Gillardite $Cu_3NiCl_2(OH)_6$

Origin: Artificial (a product of Brass corrosion in NaCl solution).

Experimental details: Methods of sample preparation are not described. Raman scattering measurements have been performed in the spectral region from 80 to 2000 cm^{-1} using 514 nm Ar$^+$-Kr$^+$ laser radiation. The laser radiation power is not indicated.

Raman shifts (cm^{-1}): 941, 900, 566sh, 511s, 458, 418, 371, 145, 127s.

Source: Babouri et al. (2015).

Comments: The sample was characterized by electron microprobe analysis. The Raman shifts were determined by us based on spectral curve analysis of the published spectrum.

Gillespite $BaFe^{2+}Si_4O_{10}$

Origin: Minade Las Pozos, Tecate, Baja California, Mexico.

Experimental details: Raman scattering measurements have been performed on an oriented sample using 514.5 nm Ar$^+$ laser radiation. The nominal laser radiation power was 170 mW. A 135-°-scattering geometry was employed.

Raman shifts (cm^{-1}): $x-z(yy)z(A_{1g})$: 1018w, 963, 758, 527w, 450, 401, 346, 306, 246, 100s, 41s; $x-z(yx)z(B_{2g})$: 992, 971w, 761w, 588w, 491, 379, 307w, 218w, 123s, 64; $x-z(yy)z$ (B_{1g}): 1092s, 1025w, 856w, 789w, 558, 427s, 380, 306, 144, 123, 65w; $x-z(x+z,y)z$ (E_g): 1145w, 1018sh, 1017w, 927w, 885, 758, 663w, 558w, 524, 522, 427sh, 378, 341, 331w, 307, 283s, 250, 136sh, 102, 90, 70, 39.

Source: McKeown and Bell (1998).

Comments: The sample was characterized by powder X-ray diffraction data.

Giniite $Fe^{2+}Fe^{3+}_4(PO_4)_4(OH)_2 \cdot 2H_2O$

Origin: Synthetic.
Experimental details: Raman scattering measurements have been performed at 198 K on arbitrarily oriented crystals using a 633 nm He-Ne laser. The laser radiation power is not indicated. The Raman shifts have been determined for the maxima of individual peaks obtained as a result of the spectral curve analysis.
Raman shifts (cm^{-1}): 3387, 3206, 2918w, 1184sh, 1148sh, 1128, 1040sh, 1023s, 999sh, 948, 766, 627, 618, 584, 487, 461, 446, 396, 346, 327, 234, 202.
Source: Frost et al. (2007n).
Comments: The sample was characterized by X-ray diffraction data and qualitative electron microprobe analysis.

Gismondine $Ca_2(Si_4Al_4)O_{16} \cdot 8H_2O$

Origin: Capo di Bove, Italy.
Raman shifts (cm^{-1}): See comment below.
Source: Mozgawa (2001).
Comments: Raman spectrum of presumed gismondine given in the cited paper corresponds to calcite with minor admixture of a silicate.

Glauberite $Na_2Ca(SO_4)_2$

Origin: Synthetic.
Experimental details: Raman scattering measurements have been performed on a powdered sample using 532 nm Nd-YAG laser radiation. The laser radiation power at the sample was 2 mW.
Raman shifts (cm^{-1}): 1167w, 1154, 1138, 1104, 998s, 647, 642, 6321, 621sh, 616, 484w, 469, 452.
Source: Jentzsch et al. (2012a).
Comments:: The sample was characterized by powder X-ray diffraction data. For the Raman spectrum of glauberite see also López et al. (2014f).

Glaucocerinite $(Zn_{1-x}Al_x)(SO_4)_{x/2}(OH)_2 \cdot nH_2O$ ($x < 0.5$, $n > 3x/2$)

Origin: No data.
Experimental details: Raman scattering measurements have been performed on an arbitrarily oriented sample using a 633 nm He-Ne laser. The laser radiation power is not indicated. The Raman shifts have been determined for the maxima of individual peaks obtained as a result of the spectral curve analysis.
Raman shifts (cm^{-1}): 3609s, 3520s, 3435sh, 3353s, 3304sh, 1083w, 982, 903, 831w, 712w, 605sh, 559s, 512sh, 437, 384, 319s, 310sh, 243w, 147s, 111.
Source: Frost et al. (2014ah).
Comments: No independent analytical data are provided for the sample used.

Glauconite $(K,Na)(Fe^{3+},Al,Mg)_2(Si,Al)_4O_{10}(OH)_2$

Origin: No data.
Experimental details: Raman scattering measurements have been performed on an arbitrarily oriented sample using 514.5 nm Ar^+, 632.8 He-Ne, and 780.0 diode laser radiations. The laser radiation output power of the He-Ne laser was less than 5 mW.

Raman shifts (cm^{-1}): ~1100w, ~955w, 697s, 591s, 451sh, 389, 264s, 194s.
Source: Ospitali et al. (2008).
Comments: The sample was characterized by energy-dispersive X–ray scan data. The Raman shifts are given for Ar$^+$ laser excitation.

Glaucophane $\square$Na$_2$(Mg$_3$Al$_2$)(Si$_8$O$_{22}$)(OH)$_2$

Origin: Sesia-Lanzo zone, Western Alps.
Experimental details: Raman scattering measurements have been performed on the oriented samples using 514.5 nm Ar$^+$ laser radiation. The laser radiation power is not indicated. A 180°-scattering geometry was employed. Spectra were collected in the scattering geometries with the laser beam perpendicular to (100), (010), and (110) faces of several crystals.
Raman shifts (cm^{-1}): 3658, 3645, 3630w, 1108w, 1063sh, 1045, 1016, 1010, 1000, 988s, 960sh, 895, 886, 790, 779, 742, 700, 684s, 670s, 611, 560, 525, 490w, 445, 411, 385s, 338, 310w, 256, 231, 210s, 181s, 160, 120.
Source: Gillet et al. (1989).
Comments: The sample was characterized by powder X-ray diffraction data and electron microprobe analyses. Raman shifts are given as a sum of spectra in all scattering geometries. For the Raman spectra of glaucophane see also Makreski et al. (2006a), Jovanovski et al. (2009), Apopei and Buzgar (2010), Andò and Garzanti (2014), and Leissner et al. (2015).

Glaukosphaerite CuNi(CO$_3$)(OH)$_2$

Origin: Carr Boyd Ni mine, Carr Boyd Rocks, Western Australia.
Experimental details: Raman scattering measurements have been performed on arbitrarily oriented crystals using a 633 nm He-Ne laser. The laser radiation power is not indicated. The Raman shifts have been determined for the maxima of individual peaks obtained as a result of the spectral curve analysis.
Raman shifts (cm^{-1}): 3481sh, 3382w, 3307w, 1639w, 1522sh, 1496, 1460sh, 1367, 1097, 1087sh, 1065, 751w, 719w, 532w, 432, 352, 272, 222, 189, 157.
Source: Frost (2006).
Comments: The sample was characterized by X-ray diffraction data and electron microprobe analysis. The empirical formula of the sample used is (Cu$_{1.1}$Ni$_{0.7}$Mg$_{0.06}$)(CO)$_3$(OH)$_2$.

Glushinskite Mg(C$_2$O$_4$)·2H$_2$O

Origin: No data.
Experimental details: Raman scattering measurements have been performed on arbitrarily oriented crystals using a 633 nm He-Ne laser. The laser radiation power is not indicated. The Raman shifts have been determined for the maxima of individual peaks obtained as a result of the spectral curve analysis.
Raman shifts (cm^{-1}): 3391, 3367, 3254, 1720w, 1660, 1636, 1612, 1471s, 1454, 915, 861, 657, 585, 527sh, 521s, 310s, 265s, 237s, 226s, 221.
Source: Frost (2004d).
Comments: No independent analytical data are provided for the sample used. For the Raman spectra of glushinskite see also Frost and Weier (2003), Frost et al. (2004a), and Baran (2014).

Gmelinite-Na $Na_4(Si_8Al_4)O_{24} \cdot 11H_2O$

Origin: Nova Scotia, Canada.
Experimental details: Methods of sample preparation are not described. Raman scattering measurements have been performed using 1064 nm Nd-YAG laser radiation. The laser radiation power at the sample was 300 mW.
Raman shifts (cm^{-1}): 1642w, 1118, 464s, 330, 181, 137s.
Source: Mozgawa (2001).
Comments: The sample was characterized by powder X-ray diffraction data. The Raman shifts were partly determined by us based on spectral curve analysis of the published spectrum.

Goethite $FeO(OH)$

Origin: No data.
Experimental details: Raman scattering measurements have been performed on a powdered sample using 632.8 nm He-Ne laser radiation. The nominal laser radiation power was less than 0.7 mW.
Raman shifts (cm^{-1}): 993w, 685w, 550, 479s, 385s, 299, 243w.
Source: De Faria et al. (1997).
Comments: No independent analytical data are provided for the sample used. For the Raman spectra of goethite see also Kustova et al. (1992), Bouchard and Smith (2003), Lepot et al. (2006), Müller et al. (2010), Nieuwoudt et al. (2011), Roqué-Rosell et al. (2010), Das and Hendry (2011), and Ciobotă et al. (2012).

Goldfieldite $Cu_{10}Te_4S_{13}$

Origin: Guinaoang, NW Luzon, Philippines.
Experimental details: Raman scattering measurements have been performed on an arbitrarily oriented sample using 514.5 nm Ar^+ laser radiation. The laser radiation power at the sample was 10 mW. A 180°-scattering geometry was employed.
Raman shifts (cm^{-1}): 354s, 324sh.
Source: Mernagh and Trudu (1993).
Comments: The sample was characterized by electron microprobe analyses.

Goldmanite $Ca_3V^{3+}_2(SiO_4)_3$

Origin: Pyrrhotite Gorge, Khibiny massif, Kola Peninsula, Russia.
Experimental details: Raman scattering measurements have been performed on an arbitrarily oriented sample in the spectral region from 80 to 4000 cm^{-1} using 633 nm He-Ne laser radiation. The nominal laser radiation power was 2 or 20 mW.
Raman shifts (cm^{-1}): 989, 932, 880s, 817s, 557s, 527, 495, 374s, 268, 241, 166.
Source: Voloshin et al. (2014).
Comments: The sample was characterized by electron microprobe analysis.

Gonnardite $(Na,Ca)_2(Si,Al)_5O_{10} \cdot 3H_2O$

Origin: Blackhead Quarry, Dunedin, New Zealand.
Experimental details: Raman scattering measurements have been performed on an oriented sample using 514.5 nm Ar^+ laser radiation. The nominal laser radiation power was 120 mW.

A 180°-scattering geometry was employed. Crystal fibers were oriented E–W in a horizontal plane, and perpendicular to the laser beam.

Raman shifts (cm^{-1}): 3587s, 3455s, 3253s, 1615, 1434, 1328, 1293, 1043w, 1003, 916, 890w, 839, 759w, 742w, 595w, 530s, 496, 441, 362w, 333, 314w, 266w, 232w, 159s.

Source: Graham et al. (2003).

Comments: The sample was characterized by powder X-ray diffraction data and electron microprobe analyses. The compositional ranges of the sample used correspond to the formula $Na_{4.78-6.27}Ca_{1.31-2.12}(Al_{8.41-8.79}Si_{11.03-11.84}O_{40})$. For the Raman spectrum of gonnardite see also Wopenka et al. (1998).

Goosecreekite $Ca(Si_6Al_2)O_{16} \cdot 5H_2O$

Origin: No data.
Experimental details: No data.
Raman shifts (cm^{-1}): 558w, 550w, 537, 513sh, 497s, 475sh, 450, 419, 414sh, 396sh, 376w.
Source: Lewis et al. (2006).
Comments: No independent analytical data are provided for the sample used. The Raman shifts were determined by us based on spectral curve analysis of the published spectrum.

Gorceixite $BaAl_3(PO_4)(PO_3OH)(OH)_6$

Origin: Ilmeny (Il'menskie) Mts., South Urals, Russia.
Experimental details: No data.
Raman shifts (cm^{-1}): 1086, 1019, 979s, 906, 816, 601, 496, 456, 341, 243s, 161.
Source: Dubinina and Valizer (2011).
Comments: The sample was characterized by electron microprobe analyses. The Raman shifts were determined by us based on spectral curve analysis of the published spectrum

Görgeyite $K_2Ca_5(SO_4)_6 \cdot H_2O$

Origin: Synthetic.
Experimental details: Raman scattering measurements have been performed on an arbitrarily oriented sample in the spectral region from 200 to 4000 cm^{-1} using 633 nm He-Ne laser radiation. The laser radiation power is not indicated.
Raman shifts (cm^{-1}): 3579, 3525s, 1215, 1187, 1175, 1164, 1161, 1137, 1115, 1108, 1085, 1078, 1067, 1013s, 1005s, 711, 661, 654, 631, 602, 595, 480, 457, 440, 433, 281.
Source: Kloprogge et al. (2004a).
Comments: The sample was characterized by powder X-ray diffraction and qualitative electron microprobe analysis.

Gormanite $Fe^{2+}_3Al_4(PO_4)_4(OH)_6 \cdot 2H_2O$

Origin: Yukon, Canada.
Experimental details: Raman scattering measurements have been performed on arbitrarily oriented crystals using a 633 nm He-Ne laser. The laser radiation power is not indicated. The Raman shifts have been determined for the maxima of individual peaks obtained as a result of the spectral curve analysis.

Raman shifts (cm^{-1}): 3615w, 3419w, 3404w, 3296w, 2893w, 1478w, 1466w, 1414w, 1382, 1365w, 1342, 1291w, 1247, 1150sh, 1123, 1095s, 1053sh, 996, 969w, 899, 505, 459, 436, 405, 380sh, 364sh, 349, 330w, 314, 195sh, 172w, 151sh.
Source: Frost et al. (2003c).
Comments: No independent analytical data are provided for the sample used.

Goryainovite Ca$_2$(PO$_4$)Cl

Origin: Synthetic.
Experimental details: Methods of sample preparation are not described. Raman scattering measurements have been performed in the spectral region from 80 to 1400 cm^{-1} using 457.9 nm Ar$^+$ laser radiation. The laser radiation power at the sample was 250 mW. A 90°-scattering geometry was employed.
Raman shifts (cm^{-1}): 1120, 1052, 950s, 606, 388.
Source: Capobianco et al. (1992).
Comments: No independent analytical data are provided for the sample used. Raman shifts are given for a sample doped with MnO$_4^{3-}$ (with Mn^{5+} concentration of 0.047 wt%). For the Raman spectrum of goryainovite see also Ivanyuk et al. (2017).

Goslarite Zn(SO$_4$)·7H$_2$O

Origin: Synthetic.
Experimental details: Raman scattering measurements have been performed on a powdered sample using 532 nm Nd-YAG laser radiation. The nominal laser radiation power was 100 mW.
Raman shifts (cm^{-1}): 1492w, 1192, 1084, 1024s, 913w, 671w, 626, 511, 423, 281, 223.
Source: Buzgar et al. (2009).
Comments: No independent analytical data are provided for the sample used. For the Raman spectra of goslarite see also Coleyshaw et al. (1994) and Buzatu et al. (2012).

Götzenite NaCa$_6$Ti(Si$_2$O$_7$)$_2$OF$_3$

Origin: Pian di Celle volcano, Umbria, Italy.
Experimental details: No data.
Raman shifts (cm^{-1}): 1031, 942, 902, 822s, 774, 664s, 587s, 560sh, 416, 371, 321, 278, 252, 218, 176sh.
Source: Sharygin et al. (1996b).
Comments: The sample was characterized by powder X-ray diffraction data and electron microprobe analyses. The Raman shifts were partly determined by us based on spectral curve analysis of the published spectrum.

Goudeyite Cu$_6$Al(AsO$_4$)$_3$(OH)$_6$·3H$_2$O

Origin: Synthetic.
Experimental details: Raman scattering measurements have been performed on an arbitrarily oriented sample using 785 nm Nd-YAG laser radiation. The laser radiation power at the sample was 1 mW. The Raman shifts have been determined for the maxima of individual peaks obtained as a result of the spectral curve analysis.

Raman shifts (cm^{-1}): 3480s, 3393sh, 3361s, 3046, 2921, 2863, 1695w, 1605w, 1434, 13280, 1001, 938, 930, 894sh, 873s, 835s, 814sh, 800, 740, 700, 560, 540, 516sh, 495s, 486sh, 470, 449, 435, 403, 395, 350, 346, 327, 293, 269, 257, 240, 226, 201, 196.
Source: Frost et al. (2006n).
Comments: The sample was characterized by powder X-ray diffraction data and qualitative EDX analysis.

Gowerite Ca[B$_5$O$_8$(OH)][B(OH)$_3$]·3H$_2$O

Origin: Synthetic.
Experimental details: Methods of sample preparation are not described. Raman scattering measurements have been performed using 514.5 nm Ar$^+$ laser radiation. The nominal laser radiation power was 300 mW.
Raman shifts (cm^{-1}): 952w, 895s, 862s, 813w, 764s, 482, 391w.
Source: Jun et al. (1995).
Comments: The sample was characterized by powder X-ray diffraction data.

Goyazite SrAl$_3$(PO$_4$)(PO$_3$OH)(OH)$_6$

Origin: Synthetic.
Experimental details: Raman scattering measurements have been performed on a powdered sample using 1064 nm Nd-YAG laser radiation. The nominal laser radiation power was ~50 mW.
Raman shifts (cm^{-1}): 3360w, 3150w, 3066, 2910w, 2830w, 1698, 1300w, 1213, 1183w, 1127w, 1102, 1068w, 1046w, 1032s, 986s, 930, ~895sh, 833w, 757, 710, 653, 609, 579, 553, 511, 464, 421, 374s, 321w, ~280sh, 256s, 230sh, 186s.
Source: Breitinger et al. (2006).
Comments: The sample was characterized by powder X-ray diffraction data. For the Raman spectrum of goyazite see also López et al. (2013c).

Graemite Cu^{2+}(Te^{4+}O$_3$)·H$_2$O

Origin: Cole Shaft, Arizona, USA.
Experimental details: Raman scattering measurements have been performed on an arbitrarily oriented sample using 633 nm He-Ne laser radiation. The laser radiation power is not indicated. The Raman shifts have been determined for the maxima of individual peaks obtained as a result of the spectral curve analysis.
Raman shifts (cm^{-1}): 3450sh, 3268, 2937, 793s, 768sh, 708, 676sh, 648sh, 527sh, 508sh, 507s, 471sh, 438s, 411sh, 380sh, 358sh, 314sh, 291, 257, 184, 146s.
Source: Frost and Keeffe (2009b).
Comments: No independent analytical data are provided for the sample used.

Graeserite Fe$^{3+}_4$Ti$_3$As^{3+}O$_{13}$(OH)

Origin: Monte Leone nappe, Binntal region, Western Alps, Switzerland.
Experimental details: Raman scattering measurements have been performed on an arbitrarily oriented sample using Ar$^+$ laser radiation. The nominal laser radiation power was 25 mW.
Raman shifts (cm^{-1}): 1451.8, 743.3, 591.5, 421.6, 297.5, 163.4.

Source: Krzemnicki and Reusser (1998).
Comments: The sample was characterized by powder X-ray diffraction data and electron microprobe analyses. The empirical formula of the sample used is $(Fe^{3+}_{2.91}Fe^{2+}_{0.38}Ti_{0.54}Pb_{0.15})Ti_3(As^{3+}_{0.94}Sb^{3+}_{0.07})O_{13}(OH)$.

Graftonite $(Fe^{2+},Mn^{2+},Ca)_3(PO_4)_2$

Origin: Sowie Góry Mts, SW Poland.
Experimental details: Raman scattering measurements have been performed on an arbitrarily oriented single crystal in the spectral region from 250 to 1300 cm^{-1} using 514.5 nm Ar$^+$ laser radiation. The laser radiation power is not indicated. A 180°-scattering geometry was employed.
Raman shifts (cm^{-1}): 1111w, 1018, 966s, 651, 593, 480w, 417, 317w, 285.
Source: Łodziński and Sitarz (2009).
Comments: The sample was characterized by electron microprobe data. For the Raman spectrum of graftonite see also Schneider et al. (2013).

Gramaccioliite-(Y) $(Pb,Sr)(Y,Mn)Fe^{3+}_2(Ti,Fe^{3+})_{18}O_{38}$

Origin: Sambuco, Stura valley, Cuneo province, Italy.
Experimental details: No data.
Raman shifts (cm^{-1}): 812, ~710w, 638, 430, 360w, 330w, 293, ~240w.
Source: Bittarello et al. (2014).
Comments: No independent analytical data are provided for the sample used.

Grandaite $Sr_2Al(AsO_4)_2(OH)$

Origin: Valletta mine, Maira Valley, Piedmont, Italy (type locality).
Experimental details: Raman scattering measurements have been performed on an arbitrarily oriented sample in the spectral region from 100 to 2500 cm^{-1} using 632.8 nm He-Ne laser radiation. The laser radiation power is not indicated. A 180°-scattering geometry was employed.
Raman shifts (cm^{-1}): 899, 857, 833sh, 790, 547, 526, 512, 499, 425, 418, 386sh, 382, 347, 308, 213, 162, 120.
Source: Cámara et al. (2014a).
Comments: The sample was characterized by powder and single-crystal X-ray diffraction data and by electron microprobe analyses. The crystal structure is solved. The empirical formula of the sample used is $(Sr_{1.41}Ca_{0.64}Ba_{0.05}Pb_{0.01})(Al_{0.68}Fe^{3+}_{0.14}Mn_{0.12}Mg_{0.13})[(As_{0.96}V_{0.01})_{\Sigma0.97}O_4]_2(OH)$.

Grandidierite $MgAl_3O_2(BO_3)(SiO_4)$

Origin: Kolonne area, Sri Lanka.
Experimental details: No data.
Raman shifts (cm^{-1}): 1047s, 993s, 982s, 952s, 868s, 717s, 687, 659s, 615w, 551, 492s, 427, 362, 343, 269, 244, 228.
Source: Schmetzer et al. (2003).
Comments: The sample was characterized by powder X-ray diffraction data and electron microprobe analyses.

Graphite C

Origin: No data.

Experimental details: Raman scattering measurements have been performed on an oriented single crystal and on a microcrystalline sample using 488.0 and 514.5 nm Ar$^+$ laser radiations. The laser radiation power is not indicated. A 90°-scattering geometry was employed.

Raman shifts (cm^{-1}): 1575s, 1355.

Source: Tuinstra and Koenig (1970).

Comments: No independent analytical data are provided for the sample used. Different orientations of the graphite single crystal with respect to the incident beam were used, but no changes in the spectrum were detected, and the only Raman line observed occurs at 1575 cm^{-1}. Polycrystalline graphite exhibits another band at 1355 cm^{-1}, which is attributed to a particle size effect. For the Raman spectra of graphite see also Mishra and Bernhardt (2009), Kaliwoda et al. (2011), and Ogawara and Akai (2014).

Gratonite $Pb_9As_4S_{15}$

Origin: Binntal, Switzerland.

Experimental details: Methods of sample preparation are not described. Raman scattering measurements have been performed in the spectral region from 50 to 600 cm^{-1} using 632.8 nm He-Ne laser radiation. The laser radiation power is not indicated. A 180°-scattering geometry was employed. The Raman shifts have been determined for the maxima of individual peaks obtained as a result of the spectral curve analysis.

Raman shifts (cm^{-1}): 370sh, 357s, 333, 312, 237, 186s, 169, 155s, 89sh, 75s.

Source: Kharbish (2016).

Comments: The sample was characterized by powder X-ray diffraction data and electron microprobe analyses. The empirical formula of the sample used is $(Pb_{8.94}Zn_{0.04}Cu_{0.02})(As_{3.99}Sb_{0.01})S_{15}$.

Greenockite CdS

Origin: Synthetic.

Experimental details: Methods of sample preparation are not described. Raman scattering measurements have been performed using 785 nm diode laser radiation. The nominal laser radiation power was in the range from 0.6 to 1 mW.

Raman shifts (cm^{-1}): 598, 560, 425w, 366sh, 345, 305, 252, 232, 210s.

Source: Rosi et al. (2016).

Comments: The sample was characterized by powder X-ray diffraction data. The Raman shifts were partly determined by us based on spectral curve analysis of the published spectrum. For the Raman spectra of greenockite see also Wang et al. (1993) and Chi et al. (2011).

Gregoryite $Na_2(CO_3)$

Origin: Oldoinyo Lengai volcano, northern Tanzania (type locality).

Experimental details: Raman scattering measurements have been performed on an arbitrarily oriented sample using 532 nm Nd-Gd laser radiation. The laser radiation power is not indicated.

Raman shifts (cm^{-1}): 1077–1078s, 1003–1005w, 952–954w, 704–710w, 630–635w.

Source: Golovin et al. (2017).

Comments: For the Raman spectrum of gregoryite see also Zaitsev et al. (2009).

Greigite $Fe^{2+}Fe^{3+}_2S_4$

Origin: Artificial (from an archaeological artifact).
Experimental details: Raman scattering measurements have been performed on an arbitrarily oriented sample using 632.8 nm He-Ne laser radiation. The nominal laser radiation power was 0.1 mW.
Raman shifts (cm^{-1}): 365s, 350s, 250w, 188w, 138w.
Source: Rémazeilles et al. (2010).
Comments: The sample was characterized by X-ray diffraction data and electron microprobe analysis. For the Raman spectra of greigite see also Bourdoiseau et al. (2011), Li et al. (2014), and Eder et al. (2014).

Griceite LiF

Origin: Synthetic.
Experimental details: Raman scattering measurements have been performed on an arbitrarily oriented sample using 532 nm laser radiation. The nominal laser radiation power was 6 mW.
Raman shifts (cm^{-1}): 191sh, 168s, 142s, 111w, 92.
Source: Alharbi et al. (2012).
Comments: The sample was characterized by powder X-ray diffraction data. The Raman shifts were determined by us based on spectral curve analysis of the published spectrum.

Grimaldiite CrO(OH)

Origin: Synthetic.
Experimental details: Raman scattering measurements have been performed on an arbitrarily oriented sample using 633 nm He-Ne laser radiation. The laser radiation power is not indicated. The Raman shifts have been determined for the maxima of individual peaks obtained as a result of the spectral curve analysis.
Raman shifts (cm^{-1}): 1634sh, 1593s, 1537sh, 1504sh, 1179sh, 1153, 981, 889sh, 823s, 630s, 558sh, 452.
Source: Yang et al. (2011c).
Comments: The sample was characterized by powder X-ray diffraction data. For the Raman spectrum of grimaldiite see also Maslar et al. (2001).

Grimaldiite CrO(OH).

Origin: Artificial (a product of Cr corrosion).
Experimental details: Raman scattering measurements have been performed on a polycrystalline sample using 647.1 nm Kr$^+$ laser radiation. The laser radiation power at the sample was less than 64 mW. A nearly 180°-scattering geometry was employed.
Raman shifts (cm^{-1}): 665sh, 535s, 475, 345w.
Source: Maslar et al. (2001).
Comments: The sample was characterized by X-ray diffraction data and EDS analysis. For the Raman spectrum of grimaldiite see also Yang et al. (2011c).

Grimselite $K_3Na(UO_2)(CO_3)_3 \cdot H_2O$

Origin: No data.
Experimental details: No data.
Raman shifts (cm^{-1}): 1063, 812s, 727, 723sh, 692, 686.
Source: Biswas et al. (2016).
Comments: The sample was characterized by electron microprobe analysis. The Raman shifts were determined by us based on spectral curve analysis of the published spectrum.

Grossite $CaAl_4O_7$

Origin: Synthetic.
Experimental details: Methods of sample preparation are not described. Raman scattering measurements have been performed using 532 nm Nd-YAG laser radiation. The maximum output laser radiation power was 100 mW. A 180°-scattering geometry was employed.
Raman shifts (cm^{-1}): 942, 909s, 837, 807, 793, 756, 714, 686, 660, 630, 568, 457, 412s, 398, 356, 331, 322, 282, 268, 252, 220, 210, 203, 185, 134.
Source: Hofmeister et al. (2004).
Comments: The sample was characterized by single-crystal X-ray diffraction data. The crystal structure is solved.

Grossular $Ca_3Al_2(SiO_4)_3$

Origin: Mengyin, Shandong Province or Gejiu, Yunnan Province, China.
Experimental details: Raman scattering measurements have been performed on an arbitrarily oriented single crystal in the spectral region from 50 to 1200 cm^{-1} using 488 nm Ar$^+$ laser radiation. The nominal laser radiation power was in the range from 100 to 150 mW.
Raman shifts (cm^{-1}): 1000w, 876s, 821, 777w, 623w, 540, 502w, 405w, 364s, 268, 232, 194w, 172, 135, 108s.
Source: Mingsheng et al. (1994).
Comments: The sample was characterized by electron microprobe data. For the Raman spectra of grossular see also Kolesov and Geiger (1998), Bersani et al. (2009), Makreski et al. (2011), and Andò and Garzanti (2014).

Groutite $Mn^{3+}O(OH)$

Origin: Cuyana Range, Minnesota, USA.
Experimental details: Raman scattering measurements have been performed on a powdered and pelletised sample in the spectral region from 10 to 1000 cm^{-1} using 514.5 nm Ar$^+$ laser radiation. The nominal laser radiation power was 10 mW. A ~180°-scattering geometry was employed.
Raman shifts (cm^{-1}): 648w, 615, 552s, 528s, 384s, 352, 278w, 253, 213, 142.
Source: Julien et al. (2004).
Comments: The sample was characterized by powder X-ray diffraction data. The Raman shifts were partly determined by us based on spectral curve analysis of the published spectrum. For the Raman spectrum of groutite see also Bernard et al. (1993a).

Grumiplucite $HgBi_2S_4$

Origin: Rudňany deposit, Slovakia.
Experimental details: Raman scattering measurements have been performed on an arbitrarily oriented sample in the spectral region from 30 to 3500 cm^{-1} using 532 nm diode laser radiation. The nominal laser radiation power was 0.5 mW.
Raman shifts (cm^{-1}): 310, 275s, 258, 221s, 162, 144w, 127w, 106w, 92s, 82.
Source: Števko et al. (2015).
Comments: The sample was characterized by powder X-ray diffraction data and electron microprobe analyses. The empirical formula of the sample used is $Hg_{0.99}Bi_{1.94}S_{4.08}$. The Raman shifts were partly determined by us based on spectral curve analysis of the published spectrum. For the Raman spectrum of grumiplucite see also Lecker (2013).

Grunerite $\square Fe^{2+}_2Fe^{2+}_5Si_8O_{22}(OH)_2$

Origin: Schneeberg, Tirol, Austria.
Experimental details: Methods of sample preparation are not described. Raman scattering measurements have been performed in the spectral region from 210 to 3400 cm^{-1} using 532 nm Nd-YAG laser radiation. The nominal laser radiation power was 100 mW.
Raman shifts (cm^{-1}): 1098, 1027, 999w, 971, 909w, 785w, 761, 747sh, 665s, 566w, 533, 415, 363, 315w, 289, 242w.
Source: Apopei and Buzgar (2010).
Comments: No independent analytical data are provided for the sample used. For the Raman spectra of grunerite see also Apopei et al. (2011) and Leissner et al. (2015).

Grunerite $\square Fe^{2+}_2Fe^{2+}_5Si_8O_{22}(OH)_2$

Origin: No data.
Experimental details: Raman scattering measurements have been performed on an oriented crystal in the spectral regions from 2600 to 3800 cm^{-1} using 514.5 nm Ar$^+$ laser radiation. The laser radiation power is not indicated. Polarized spectra were collected in the parallel-polarized scattering geometries (the polarization of incident light E_i was parallel to the polarization of scattered light E_s) with the crystal-elongation direction parrallel to $E_i \| E_s$ and crystal-elongation direction perpendicular to $E_i \| E_s$.
Raman shifts (cm^{-1}): 3651, 3635, 3617.
Source: Leissner et al. (2015).
Commebts: The sample was characterized by electron microprobe data. The Raman shifts are given for the scattering geometry with the crystal-elongation direction pendicular to $E_i \| E_s$, in which the Raman intensities are most strong. The empirical formula of the sample used is $(\square_{0.97}Na_{0.01}Ca_{0.02})(Fe_{0.89}Mn_{0.10}Ca_{0.01})_2(Fe_{0.61}Mg_{0.39})_5Si_{8.00}OH_{2.00}$. For the Raman spectra of grunerite see also Apopei and Buzgar (2010) and Apopei et al. (2011).

Guanacoite $Cu_2Mg_3(OH)_4(AsO_4)_2 \cdot 4H_2O$

Origin: El Guanaco Mine, near Taltal, Chile (type locality).
Experimental details: Raman scattering measurements have been performed on an arbitrarily oriented single crystal using 633 nm laser radiation. The laser radiation power is not indicated.

Raman shifts (cm^{-1}): 3561s, 3510w, ~2996sh, 1619w, ~1069w, ~996w, 865s, 837s, 738, ~740sh, ~685sh, 490, 439, 411sh, 390s, ~367sh, 352sh, 315s, ~295sh, 271, 221, 212, 188, 171, 149, 131.
Source: Witzke et al. (2006).
Comments: The sample was characterized by powder and single-crystal X-ray diffraction data and electron microprobe analyses. The crystal structure is solved. The empirical formula of the sample used is $Cu_{2.32}Mg_{2.64}(OH)_{4.13}(AsO_4)_{1.93} \cdot 4.15H_2O$.

Guanine $C_5H_3(NH_2)N_4O$

Origin: Synthetic.
Experimental details: Raman scattering measurements have been performed on a polycrystalline sample using 632.8 nm He-Ne laser radiation. The laser radiation power at the sample was 40 mW. A 180°-scattering geometry was employed.
Raman shifts (cm^{-1}): 1675, 1602w, 1551, 1479sh, 1468, 1421, 1390, 1361, 1265, 1234, 1186, 1159w, 937, 879w, 848, 775w, 710w, 693w, 649s, 603w, 562, 547, 495, 397, 357w, 340.
Source: Giese and McNaughton (2002).
Comments: No independent analytical data are provided for the sample used. For the Raman spectrum of guanine see also Mathlouthi et al. (1986).

Gudmundite FeSbS

Origin: No data.
Experimental details: Raman scattering measurements have been performed on an arbitrarily oriented sample using 532 nm Nd-YAG laser radiation. The nominal laser radiation power was about 21 mW. A 180°-scattering geometry was employed. The Raman shifts have been determined for the maxima of individual peaks obtained as a result of the spectral curve analysis.
Raman shifts (cm^{-1}): 462w, 362, 352s, 316sh, 305s, 288s, 279sh, 253w, 241, 230sh, 217sh, 210, 201sh, 161sh, 152s, 145sh, 136sh.
Source: Kharbish and Andráš (2014).
Comments: The sample was characterized by powder X-ray diffraction data and electron microprobe analysis. The empirical formula of the sample used is $Fe_{1.01}As_{0.91}S_{1.08}$.

Guilleminite $Ba(UO_2)_3(Se^{4+}O_3)_2O_2 \cdot 3H_2O$

Origin: Musonoi mine, Kolwezi, Katanga (Shaba), Democratic Republic of Congo.
Experimental details: Raman scattering measurements have been performed on arbitrarily oriented crystals in the spectral region from 100 to 1700 cm^{-1} using a 633 nm He-Ne laser. The laser radiation power is not indicated. The Raman shifts have been determined for the maxima of individual peaks obtained as a result of the spectral curve analysis.
Raman shifts (cm^{-1}): 1585w, 1514w, 831sh, 747s, 675sh, 544sh, 478, 419sh, 345sh, 245, 150.
Source: Frost et al. (2009c).
Comments: No independent analytical data are provided for the sample used.

Gunningite $Zn(SO_4) \cdot H_2O$

Origin: Coranda-Hondolopen pit, Certej Gold-Silver deposit, Certej, Romania.
Experimental details: Methods of sample preparation are not described. Raman scattering measurements have been performed using 532 nm Nd-YAG laser radiation. The laser radiation power at the sample was 7.37 mW.

Raman shifts (cm^{-1}): 3230w, 1492w, 1192, 1087, 1024s, 884w, 665w, 626, 503, 423, 307sh, 277w, 219w.
Source: Apopei et al. (2014a).
Comments: The sample was characterized by powder X-ray diffraction data. For the Raman spectrum of gunningite see also Buzatu et al. (2016).

Gurimite Ba$_3$(VO$_4$)$_2$

Origin: Synthetic.
Experimental details: Raman scattering measurements have been performed on a powdered sample. The laser radiation power and wavelength are not indicated. A 180°-scattering geometry was employed.
Raman shifts (cm^{-1}): 835s, 777, 414w, 378w, 326s, 169w, 132w, 104w.
Source: Azdouz et al. (2010).
Comments: The sample was characterized by powder X-ray diffraction data. The crystal structure is solved. For the Raman spectra of gurimite see also Baran et al. (1972), Grzechnik and McMillan (1997), and Galuskina et al. (2016b).

Gwihabaite (NH$_4$)(NO$_3$)

Origin: Synthetic
Experimental details: No data.
Raman shifts (cm^{-1}): 1458w, 1410w, 1283w, 1040s, 711, 190.
Source: Morillas et al. (2016).
Comments: No independent analytical data are provided for the sample used. The Raman shifts were partly determined by us based on spectral curve analysis of the published spectrum. For the Raman spectrum of gwihabaite see also Martínez-Arkarazo et al. (2007).

Gypsum Ca(SO$_4$)·2H$_2$O

Origin: Coranda-Hondol open pit, Certej Au-Agdeposit, Romania.
Experimental details: Methods of sample preparation are not described. Raman scattering measurements have been performed using 532 nm Nd-YAG laser radiation. The laser radiation power at the sample was 14.3 mW.
Raman shifts (cm^{-1}): 3398, 1136, 1106sh, 1010s, 671w, 623w, 576w, 495w, 416, 312w, 215w.
Source: Apopei et al. (2014a).
Comments: The sample was characterized by powder X-ray diffraction data. For the Raman spectra of gypsum see also Anbalagan et al. (2009), Buzgar et al. (2009), Jehlička et al. (2009b), White (2009), Ciobotă et al. (2012), Capitani et al. (2014), and Wang and Zhou (2014).

Gyrolite NaCa$_{16}$(Si$_{23}$Al)O$_{60}$(OH)$_8$·14H$_2$O

Origin: No data.
Experimental details: No data.
Raman shifts (cm^{-1}): 1057, 1035, 774w, 703w, 628, 598s, 572s, 456, 400, 351, 280s, 207, 169w, 145.
Source: De Ferri et al. (2012).

Comments: No independent analytical data are provided for the sample used. The Raman shifts were determined by us based on spectral curve analysis of the published spectrum.

Hafnon $Hf(SiO_4)$

Origin: Synthetic.
Experimental details: Raman scattering measurements have been performed on an arbitrarily oriented sample using 532 nm laser radiation. The laser radiation power is not indicated.
Raman shifts (cm^{-1}): 1021s, 985s, 935, 637w, 548w, 497w, 448, 402, 350, 268w, 213, 166w, 155, 147w.
Source: Grüneberger et al. (2016).
Comments: The sample was characterized by electron microprobe analysis. The empirical formula of the sample used is $Hf_{0.99}Zr_{0.01}SiO_4$. For the Raman spectra of hafnon see also Nicola and Rutt (1974) and Manoun et al. (2006).

Haidingerite $Ca(AsO_3OH) \cdot H_2O$

Origin: Jáchymov, Bohemia, Krušné Hory (Czech Ore Mts.), Czech Republic.
Experimental details: Raman scattering measurements have been performed on arbitrarily oriented crystals using a 633 nm He-Ne laser. The laser radiation power is not indicated. The Raman shifts have been determined for the maxima of individual peaks obtained as a result of the spectral curve analysis.
Raman shifts (cm^{-1}): 3574w, 3455sh, 3412, 2842w, 886sh, 855s, 838sh, 823sh, 745, 739sh, 660w, 433sh, 420, 376sh, 369, 338sh, 323, 299sh, 268w, 220, 180, 145, 123, 115sh.
Source: Frost et al. (2010h).
Comments: The sample was characterized by powder X-ray diffraction data and electron microprobe analyses.

Haiweeite $Ca(UO_2)_2(Si_5O_{12})(OH)_2 \cdot 6H_2O$

Origin: Teófilo Otoni, Minas Gerais, Brazil.
Experimental details: Raman scattering measurements have been performed on arbitrarily oriented crystals using a 633 nm He-Ne laser. The laser radiation power is not indicated. The Raman shifts have been determined for the maxima of individual peaks obtained as a result of the spectral curve analysis.
Raman shifts (cm^{-1}): 3606sh, 3498, 3375sh, 3273sh, 2923, 2875, 2851, 1170, 1115, 1108, 1087, 1019, 1015sh, 936, 919sh, 887w, 808s, 799s, 756w, 724w, 589, 473, 418w, 375, 317sh, 307, 283w, 264, 260, 236sh, 192s, 148sh, 108.
Source: Frost et al. (2006d).
Comments: No independent analytical data are provided for the sample used.

Hakite $Cu_6[Cu_4Hg_2]Sb_4Se_{13}$

Origin: Příbram uranium and base-metal district, Central Bohemia, Czech Republic.
Experimental details: Micro-Raman scattering measurements have been performed on a polished section using 532 nm diode laser radiation. The nominal laser radiation power was 0.5 mW.
Raman shifts (cm^{-1}): 261, 228s, 211sh, 185, 166, 82, 69.

Source: Škácha et al. (2017).
Comments: The sample was characterized by powder X-ray diffraction data and electron microprobe analyses.

Halite NaCl

Origin: Kłodawa salt mine, Central Poland.
Experimental details: Raman scattering measurements have been performed on an arbitrarily oriented sample using Ar^+ laser radiation. The laser radiation power is not indicated.
Raman shifts (cm^{-1}): 335.
Source: Wesełucha-Birczyńska et al. (2008).
Comments: No independent analytical data are provided for the sample used. Possibly, the band at 335 cm^{-1} is a combination band of acoustical and optical modes of NaCl.

Halloysite-10Å $Al_2Si_2O_5(OH)_4 \cdot 2H_2O$

Origin: A Neogene cryptokarst, southern Belgium.
Experimental details: Raman scattering measurements have been performed on an oriented single crystal using 633 nm He-Ne laser radiation. The laser radiation power is not indicated. The Raman shifts have been determined for the maxima of individual peaks obtained as a result of the spectral curve analysis.
Raman shifts (cm^{-1}): 3703, 3688, 3642, 3625s, 3598, 3556, 944sh, 910, 826sh, 794, 748, 710, 548sh, 503sh, 465s, 430s, 359sh, 332, 275, 245.
Source: Kloprogge and Frost (1999e).
Comments: The sample was characterized by powder X-ray diffraction data. Raman shifts are given for an unspecified scattering geometry.

Halotrichite $Fe^{2+}Al_2(SO_4)_4 \cdot 22H_2O$

Origin: Corral Hollow, California, USA.
Experimental details: Raman scattering measurements have been performed on an arbitrarily oriented sample in the spectral region from 100 to 4000 cm^{-1} using 633 nm He-Ne laser radiation. The laser radiation power at the sample was 1 mW. The Raman shifts have been determined for the maxima of individual peaks obtained as a result of the spectral curve analysis.
Raman shifts (cm^{-1}): 3548sh, 3426w, 3270sh, 1147, 1086, 1031w, 985s, 608, 467, 444, 365w, 276sh, 247, 215sh.
Source: Locke et al. (2007).
Comments: The sample was characterized by X-ray diffraction data and electron microprobe analysis. For the Raman spectrum of halotrichite see also Buzatu et al. (2016).

Hambergite $Be_2(BO_3)(OH)$

Origin: Ehrenfriedersdorf, complex, Erzgebirge (Ore Mts.), Germany.
Experimental details: Raman scattering measurements have been performed on a microscopic inclusions in beryl using 488.0 and 514.5 nm Ar^+ laser radiations. The laser radiation power at the sample was about 10 mW.
Raman shifts (cm^{-1}): 3469, 3403s, 988s, 270, 154s, 147, 123.
Source: Thomas et al. (2011b).

Comments: No independent analytical data are provided for the sample used. For the Raman spectrum of hambergite see also Thomas and Davidson (2010).

Hanjiangite $Ba_2Ca(V^{3+}Al)(AlSi_3O_{10})(OH)_2F(CO_3)_2$

Origin: Shiti Ba deposit, Dabashan region, China (type locality).
Experimental details: Raman scattering measurements have been performed on an arbitrarily oriented single-crystal thin chip using 514.5 nm Ar$^+$ laser radiation. The laser radiation power is not indicated.
Raman shifts (cm^{-1}): 3581s, 3540s, 2945, 2250, 1092s, 855w, 699, 405, 265, 193.
Source: Liu et al. (2012).
Comments: The sample was characterized by powder and single crystal X-ray diffraction data and by electron microprobe analyses. The crystal structure is solved. The empirical formula of the sample used is $(Ba_{1.98}Na_{0.06}K_{0.01})(Ca_{0.76}Mg_{0.12}Y_{0.06}Sr_{0.03}La_{0.01}Nd_{0.01})(V_{1.15}Al_{0.75}Cr_{0.20}Ti_{0.12})[(Si_{2.84}Al_{1.16})_{\Sigma 4.00}O_{10}][(OH)_{1.25}O_{0.77}](F_{0.82}Cl_{0.01})(CO_3)_{2.05}$.

Hanksite $KNa_{22}(SO_4)_9(CO_3)_2Cl$

Origin: Searles Lake, California, USA (type locality).
Experimental details: Raman scattering measurements have been performed on an oriented sample using 488.0 nm Ar$^+$ laser radiation. The laser radiation power is not indicated. Polarized spectra were collected in the scattering geometries with the laser radiation polarized approximately parallel to the c axis, and approximately parallel to the a axis.
Raman shifts (cm^{-1}): 1190w, 1166w, 1156, 1142, 1135, 1124, 1117, 1096w, 1083s, 993s, 979sh, 712w, 634, 625, 620, 474, 470, 459.
Source: Palaich et al. (2013).
Comments: The Raman shifts were partly determined by us based on spectral curve analysis of the published spectrum. The sample was characterized by powder X-ray diffraction data. For the Raman spectrum of hanksite see also Morillas et al. (2016).

Hannayite $(NH_4)_2Mg_3(PO_3OH)_4 \cdot 8H_2O$

Origin: Lava Cave, near Skipton, Victoria, Australia.
Experimental details: Raman scattering measurements have been performed on an arbitrarily oriented sample using 633 nm He-Ne laser radiation. The laser radiation power is not indicated. The Raman shifts have been determined for the maxima of individual peaks obtained as a result of the spectral curve analysis.
Raman shifts (cm^{-1}): 3496, 3460, 3384, 3314, 3219, 3185, 3125, 3090sh, 2983sh, 2872, 2649, 2466, 2228, 1947, 1756, 1708, 1661w, 1457w, 1429w, 1227w, 1172w, 1119w, 1070, 1011, 974s, 971s, 882, 802w, 757w, 596w, 556, 522, 513, 436, 415sh, 379, 375, 356, 269, 250, 247, 205, 193w.
Source: Frost et al. (2005j).
Comments: No independent analytical data are provided for the sample used.

Hannebachite $Ca(SO_3) \cdot 0.5H_2O$

Origin: Hannebacher Ley Volcano, Eifel, Germany (type locality).
Experimental details: Raman scattering measurements have been performed on arbitrarily oriented crystals using a 633 nm He-Ne laser. The laser radiation power is not indicated.

Raman shifts (cm^{-1}): 1092–1094, 1005s, 969, 655w, 520, 492, 444, 174.
Source: Frost and Keeffe (2009d).
Comments: No independent analytical data are provided for the sample used.

Hansesmarkite $Ca_2Mn_2Nb_6O_{19} \cdot 20H_2O$

Origin: Tvedalen, Larvik Plutonic Complex, Vestfold, southern Norway (type locality).
Experimental details: Methods of sample preparation are not described. Raman scattering measurements have been performed using 532 nm laser radiation. The laser radiation power is not indicated.
Raman shifts (cm^{-1}): 913s, 865, 841sh, 734w, 520w, 473sh, 302, 217.
Source: Friis et al. (2016).
Comments: The sample was characterized by powder X-ray diffraction data and electron microprobe analyses. The crystal structure is solved. The empirical formula of the sample used is $(Ca_{1.93}Na_{0.02}K_{0.01})(Mn_{1.79}Fe_{0.11})Nb_{6.00}O_{18.84} \cdot 20H_2O$.

Hardystonite $Ca_2ZnSi_2O_7$

Origin: Synthetic.
Experimental details: Raman scattering measurements have been performed on an oriented single crystal using 514.5 nm Ar$^+$ laser radiation. The laser radiation power is not indicated. Polarized spectrum was collected in the $\sim y(zz)\sim y$ scattering geometry.
Raman shifts (cm^{-1}): 1060, 1020w, 906s, 663s, 614s, 550w, 480, 445, 265, 240sh, 220, 145w, 115w, 100, 60w.
Source: Kaminskii et al. (2011).
Comments: No independent analytical data are provided for the sample used.

Harmotome $Ba_2(Si_{12}Al_4)O_{32} \cdot 12H_2O$

Origin: Mannbühl (Giro) quarry, Dannenfels, Kirchheimbolanden, Rhineland-Palatinate, Germany.
Experimental details: Raman scattering measurements have been performed on an arbitrarily oriented single crystal using 633 He-Ne laser radiation. The laser radiation power is not indicated. The Raman shifts have been determined for the maxima of individual peaks obtained as a result of the spectral curve analysis.
Raman shifts (cm^{-1}): 3615sh, 3526w, (3418w), 3287sh, 1707w, 1648w, 1102, 1020sh, 768w, 728w, 699w, 561w, 546sh, 534w, (515), 491s, 470sh, 428, 358w, 350sh, 335, 319, 289, 199, 183sh, 169sh.
Source: Frost et al. (2015q).
Comments: The sample was characterized by qualitative electron microprobe analysis. For the Raman spectrum of harmotome see also Mozgawa (2001).

Harmunite $CaFe_2O_4$

Origin: Jabel Harmun, West Bank, Palestinian Autonomy, Israel (type locality).
Experimental details: Raman scattering measurements have been performed on an arbitrarily oriented sample using 488 nm solid-state laser radiation. The laser radiation power at the sample was 44 mW. The Raman shifts have been determined for the maxima of individual peaks obtained as a result of the spectral curve analysis.

Raman shifts (cm^{-1}): 1228, 648s, 585s, 519, 453sh, 4354, 379sh, 364s, 298, 270, 206, 182, 161, 117.
Source: Galuskina et al. (2014).
Comments: The sample was characterized by powder X-ray diffraction data and electron microprobe analyses. The crystal structure is solved. The empirical formula of the sample used is $Ca_{1.013}(Fe^{3+}_{1.957}Al_{0.015}Cr_{0.011}Ti_{0.004}Mg_{0.003})O_4$.

Harmunite Mn^{4+}-bearing $Ca_{1-x}(Fe^{3+},Mn^{4+})_2O_4$

Origin: Bellerberg volcano, Eifel, Rhineland-Palatinate, Germany.
Experimental details: Raman scattering measurements have been performed on an arbitrarily oriented sample using 488 nm solid-state laser radiation. The laser radiation power at the sample was 44 mW.
Raman shifts (cm^{-1}): 1225, 617sh, 590s, 526, 468, 391w, 327w, 289w, 205w, 161, 115w.
Source: Galuskin et al. (2016b).
Comments: The sample was characterized electron microprobe data. The empirical formula of the sample used is $Ca_{0.862}(Fe^{3+}_{1.719}Mn^{4+}_{0.265}Ti^{4+}_{0.012}Mg_{0.008})O_4$.

Hartite $C_{20}H_{34}$

Origin: Castelnuovo di Valdarno, Italy.
Experimental details: Methods of sample preparation are not described. Raman scattering measurements have been performed in the spectral region from 100 to 3500 cm^{-1} using 1064 nm Nd-YAG laser radiation. The nominal laser radiation power was 350 mW.
Raman shifts (cm^{-1}): 3000, 2978, 2962, 2942s, 2921s, 2907sh, 2886sh, 2866, 2851, 2768w, 2733w, 1480, 1468sh, 1440s, 1387, 1370w, 1356, 1341, 1320w, 1310, 1287, 1264, 1249w, 1230, 1217sh, 1208s, 1180, 1154sh, 1144, 1114(?), 1096s, 1085w, 1075w, 1063, 1041, 1027w, 1013, 996, 976, 963sh, 946, 936, 926(?), 916w, 892, 879, 845, 814, 792s, 770, 745w, 729s, 693, 639, 598w, 558, 543, 526, 489, 452, 403, 389w, 345, 326w, 305.
Source: Jehlička and Edwards (2008).
Comments: The sample was characterized by powder X-ray diffraction data. For the Raman spectrum of hartite see also Jehlička et al. (2005).

Hashemite $Ba(CrO_4)$

Origin: Synthetic.
Experimental details: Raman scattering measurements have been performed on a powdered sample prepared as a pellet using 488 nm Ar$^+$ laser radiation. The nominal laser radiation power was 200 mW.
Raman shifts (cm^{-1}): 907, 900, 885w, 873, 864s, 429w, 412w, 404w, 396w, 361, 351, 135w, 112w, 66w.
Source: Scheuermann and Schutte (1973a).
Comments: The sample was characterized by powder X-ray diffraction data.

Hatchetine A paraffin wax related to evenkite.

Origin: Zastávka, near Brno, Bohemian Massif, Czech republic.
Experimental details: Raman scattering measurements have been performed on a compacted powder of amorphous sample using 514.5 nm Ar$^+$ laser radiation. The nominal laser radiation power was 10 mW.

Raman shifts (cm^{-1}): 2883s, 2847, 2725, 1463s, 1440s, 1418, 1389w, 1371w, 1347, 1295s, 1171.
Source: Jehlička et al. (2007a).
Comments: No independent analytical data are provided for the sample used.

Hatrurite Ca$_3$SiO$_5$

Origin: Synthetic.
Experimental details: Raman scattering measurements have been performed on an arbitrarily oriented sample using 363.8 nm laser radiation. The nominal laser radiation power was 150 mW.
Raman shifts (cm^{-1}): 953sh, 929sh, 914, 893, 880, 847s, 812, 741, 540, 521sh, 392, 352, 326, 317sh, 267, 240, 224, 180, 126, 103, 80, 56.
Source: Fujimori et al. (2005).
Comments: No independent analytical data are provided for the sample used. The Raman shifts were determined by us based on spectral curve analysis of the published spectrum.

Hauerite MnS$_2$

Origin: Synthetic.
Experimental details: Raman scattering measurements have been performed on a single crystal using 514.5 nm Ar$^+$ laser radiation. The laser radiation power is not indicated. A 180°-scattering geometry was employed. Raman spectrum was collected in the scattering geometry with the laser radiation direction normal to the (111) plane of the crystal.
Raman shifts (cm^{-1}): 743w, 655, 486s, 246, 223.
Source: Verble and Humphrey (1974).
Comments: No independent analytical data are provided for the sample used.

Hausmannite Mn^{2+}Mn$^{3+}_2$O$_4$

Origin: Synthetic.
Experimental details: Raman scattering measurements have been performed on a powdered sample prepared as a pellet using 514.5 nm Ar$^+$ and 647.1 nm Kr$^+$ laser radiations. The laser radiation power is not indicated.
Raman shifts (cm^{-1}): 668s, 479w, 371, 328, 298.
Source: Lutz et al. (1991).
Comments: No independent analytical data are provided for the sample used. For the Raman spectra of hausmannite see also Bernard et al. (1993a), Julien et al. (2004), and Mironova-Ulmane et al. (2009).

Hausmannite Mn^{2+}Mn$^{3+}_2$O$_4$

Origin: Synthetic.
Experimental details: Raman scattering measurements have been performed using 514.5 nm Ar$^+$ laser radiation. The nominal laser radiation power was 10 mW. A nearly 180°-scattering geometry was employed. Raman spectrum was obtained in the spectral region from 10 to 1200 cm^{-1}.
Raman shifts (cm^{-1}): 653s, 579w, 485w, 357, 310.
Source: Julien et al. (2004).

Comments: The sample was characterized by powder X-ray diffraction data. For the Raman spectra of hausmannite see also Lutz et al. (1991), Bernard et al. (1993a), and Mironova-Ulmane et al. (2009).

Haüyne $Na_3Ca(Si_3Al_3)O_{12}(SO)_4$

Origin: Sacrofano, Latium, Italy.
Experimental details: Raman scattering measurements have been performed on an arbitrarily oriented single crystal using 514.5 nm Ar$^+$ laser radiation. The nominal laser radiation power was 300 mW.
Raman shifts (cm^{-1}): 1152, 1090, 1027sh, 998sh, 986s, 977s, 643, 610, 545w, 433s.
Source: Ballirano (2012).
Comments: The sample was characterized by single-crystal X-ray diffraction data and electron microprobe analysis. The crystal structure is solved. The empirical formula of the sample used is $(Na_{4.4}K_{1.1}Ca_{2.1})[Si_6Al_6O_{24}](SO_4)_{1.6}(S_3)_{0.3}(CO_2)_{0.1}$. The Raman shifts were partly determined by us based on spectral curve analysis of the published spectrum. For the Raman spectrum of haüyne see also Caggiani et al. (2014).

Hawleyite CdS

Origin: No data.
Experimental details: Raman scattering measurements have been performed on a powdered sample using 514.5 nm Ar$^+$ laser radiation. The nominal laser radiation power was in the range 0.3–5 mW.
Raman shifts (cm^{-1}): 902, 633sh, 598, 295, 254sh.
Source: Rosi et al. (2016).
Comments: The Raman shifts were determined by us based on spectral curve analysis of the published spectrum. For the Raman spectrum of hawleyite see also Wang et al. (1993).

Hawthorneite $BaMgTi_3Cr_4Fe^{2+}{}_2Fe^{3+}{}_2O_{19}$

Origin: Synthetic.
Experimental details: Methods of sample preparation are not described. Raman scattering measurements have been performed using 632 nm He-Ne laser radiation. The nominal laser radiation power was in the range 4–8 mW.
Raman shifts (cm^{-1}): 680s, 516, 459sh, 352, 285w.
Source: Konzett et al. (2005).
Comments: The sample was characterized by electron microprobe analysis.

Haynesite $(UO_2)_3(Se^{4+}O_3)_2(OH)_2 \cdot 5H_2O$

Origin: Repete Mine, Blanding, San Juan Co., Utah, USA (type locality).
Experimental details: Raman scattering measurements have been performed on arbitrarily oriented crystals using a 633 nm He-Ne laser. The laser radiation power is not indicated. The Raman shifts have been determined for the maxima of individual peaks obtained as a result of the spectral curve analysis.
Raman shifts (cm^{-1}): 3476w, 862w, 812s, 800s, 741, 582w, 472, 437sh, 419s, 367, 342, 278, 257sh, 219, 157sh, 142.
Source: Frost et al. (2006q).
Comments: No independent analytical data are provided for the sample used.

Hazenite $KNaMg_2(PO_4)_2 \cdot 14H_2O$

Origin: Mono Lake, California, USA.
Experimental details: Raman scattering measurements have been performed on an arbitrarily oriented sample using 532 nm solid-state laser radiation. The laser radiation power is not indicated.
Raman shifts (cm^{-1}): 3900–2500, 2380, 1620w, 1100–988sh, 932s, 685w, 559, 430, 290, 234.
Source: Yang et al. (2011b).
Comments: The sample was characterized by powder X-ray diffraction data and electron microprobe analysis. The crystal structure is solved. The empirical formula of the sample used is $K_{0.97}(Na_{0.96}Ca_{0.02})Mg_{2.07}[(P_{0.98}S_{0.02})O_4]_2 \cdot 13.90H_2O$. The Raman shifts were partly determined by us based on spectral curve analysis of the published spectrum.

Heazlewoodite Ni_3S_2

Origin: Synthetic.
Experimental details: Raman scattering measurements have been performed on a sample prepared as a pellet using 514.5 nm Ar$^+$ laser radiation. The nominal laser radiation power was 40 mW.
Raman shifts (cm^{-1}): 351s, 324w, 305, 223, 201, 190.
Source: Cheng and Liu (2007).
Comments: The sample was characterized by powder X-ray diffraction data and electron microprobe analysis. For the Raman spectrum of heazlewoodite see also Lanteigne et al. (2012).

Hectorite $Na_{0.3}(Mg,Li)_3Si_4O_{10}(F,OH)_2 \cdot nH_2O$

Origin: No data.
Experimental details: No data.
Raman shifts (cm^{-1}): 1082, 944w, 892, 783, 683s, 560, 516w, 461, 379sh, 333, 282, 206sh, 184s.
Source: Wang et al. (1998a).
Comments: No independent analytical data are provided for the sample used. The Raman shifts were determined by us based on spectral curve analysis of the published spectrum.

Hedenbergite $CaFe^{2+}Si_2O_6$

Origin: Sasa, Macedonia.
Experimental details: Raman scattering measurements have been performed on a powdered sample using 514.5 nm Ar$^+$ laser radiation. The laser radiation power is not indicated.
Raman shifts (cm^{-1}): 1030sh, 1010s, 904w, 850w, 655s, 549, 523, 494w, 370, 330, 300, 230w, 182, 145, 121, 115.
Source: Jovanovski et al. (2009).
Comments: The sample was characterized by powder X-ray diffraction data and electron microprobe analyses. The Raman shifts were partly determined by us based on spectral curve analysis of the published spectrum. For the Raman spectra of hedenbergite see also Huang et al. (2000), Buzatu and Buzgar (2010), and Andò and Garzanti (2014).

Hedenbergite $CaFe^{2+}Si_2O_6$

Origin: Nordmarken, Sweden.
Experimental details: Raman scattering measurements have been performed on an arbitrarily oriented single crystal. Experimental details are not described.

Raman shifts (cm^{-1}): 1012s, 894w, 852, 663s, 553w, 524, 499sh, 381, 346w, 315, 231w.
Source: Buzatu and Buzgar (2010).
Comments: No independent analytical data are provided for the sample used. For the Raman spectra of hedenbergite see also Huang et al. (2000), Jovanovski et al. (2009), and Andò and Garzanti (2014).

Hedyphane $Ca_2Pb_3(AsO_4)_3Cl$

Origin: Puttapa mine, Beltana, Flinders Ranges, South Australia, Australia.
Experimental details: Raman scattering measurements have been performed on arbitrarily oriented crystals using a 633 nm He-Ne laser. The laser radiation power is not indicated. The Raman shifts have been determined for the maxima of individual peaks obtained as a result of the spectral curve analysis.
Raman shifts (cm^{-1}): 3659sh, 3563sh, 3458sh, 3409, 3389, 3352sh, 3304sh, 3258, 3235, 848sh, 834sh, 819s, 811sh, 787s, 770, 452, 432sh, 394, 372sh, 349, 334sh, 319s, 201sh, 192sh, 177sh, 162, 151sh, 141sh.
Source: Frost et al. (2007c).
Comments: No independent analytical data are provided for the sample used. Raman shifts in the OH stretching region indicate that there is isomorphic replacement of Cl for OH.

Heisenbergite $(UO_2)(OH)_2 \cdot H_2O$

Origin: Menzenschwand, Schwarzwald (Black Forest), Germany (type locality).
Experimental details: Raman scattering measurements have been performed on an arbitrarily oriented sample in the spectral region from 150 to 4000 cm^{-1} using 638 nm laser radiation. The laser radiation power is not indicated. The Raman shifts have been determined for the maxima of individual peaks obtained as a result of the spectral curve analysis.
Raman shifts (cm^{-1}): 1005w, 923w, 829s, 799sh, 742s, 538, 438, 389sh, 338w, 247w, 190w.
Source: Walenta and Theye (2012).
Comments: The sample was characterized by powder X-ray diffraction data and electron microprobe analyses. The empirical formula of the sample used is $U_{1.044}Pb_{0.020}Ba_{0.004}Ca_{0.008}H_{3.672}O_5$.

Heliophyllite $Pb_6As_2O_7Cl_4$

Origin: Karrantza valley, Basque Co., Spain.
Experimental details: Raman scattering measurements have been performed on an arbitrarily oriented sample using 514.5 nm Ar$^+$ and 785 nm diode laser radiations. The laser radiation power at the sample was 20 mW with Ar$^+$ excitation, and the nominal laser radiation power was 150 mW with diode excitation.
Raman shifts (cm^{-1}): 808, 746, 718, 160s.
Source: Goienaga et al. (2011).
Comments: No independent analytical data are provided for the sample used.

Hellyerite $Ni(CO_3) \cdot 6H_2O$

Origin: Synthetic.
Experimental details: Methods of sample preparation are not described. Raman scattering measurements have been performed using 532 nm laser radiation. The laser radiation power is not indicated.

Raman shifts (cm^{-1}): 1397, 1092s, 721w, 409w, 293, 257, 224sh, 157s.
Source: Bette et al. (2016).
Comments: The sample was characterized by powder X-ray diffraction data. The crystal structure is solved.

Hematite Fe_2O_3

Origin: No data.
Experimental details: Raman scattering measurements have been performed on a powdered sample using 632.8 nm He-Ne laser radiation. The nominal laser radiation power was below 0.7 mW.
Raman shifts (cm^{-1}): 1320, 612, 497w, 411s, 299sh, 293, 246w, 227.
Source: De Faria et al. (1997).
Comments: The sample was characterized by X-ray diffraction data. For the Raman spectra of hematite see also Bouchard and Smith (2003), Lepot et al. (2006), Müller et al. (2010), Sagatowska (2010), Das and Hendry (2011), Hosterman (2011), Nieuwoudt et al. (2011), Ciobotă et al. (2012), Andò and Garzanti (2014), and Apopei et al. (2014a).

Hematite Fe_2O_3

Origin: Synthetic.
Experimental details: Raman scattering measurements have been performed on a powdered sample using 632.8 nm He-Ne laser radiation. The nominal laser radiation power was below 0.4 mW.
Raman shifts (cm^{-1}): 1300, 1046, 824w, 657, 610, 494, 408, 292s, 243w, 223s.
Source: Müller et al. (2010).
Comments: The sample was characterized by powder X-ray diffraction data. For the Raman spectra of hematite see also De Faria et al. (1997), Bouchard and Smith (2003), Lepot et al. (2006), Sagatowska (2010), Das and Hendry (2011), Hosterman (2011), Nieuwoudt et al. (2011), Ciobotă et al. (2012), Andò and Garzanti (2014), and Apopei et al. (2014a).

Hemihedrite $ZnPb_{10}(CrO_4)_6(SiO_4)_2F_2$

Origin: Florence Pb-Ag mine, Pinal Co., Arizona, USA.
Experimental details: Raman scattering measurements have been performed on an arbitrarily oriented sample using 532 nm solid-state laser radiation. The laser radiation power is not indicated.
Raman shifts (cm^{-1}): 3389w, 858s, 841s, 825, 784, 370sh, 340.
Source: Lafuente et al. (2016).
Comments: The sample was characterized by single-crystal X-ray diffraction data and electron microprobe analyses. The crystal structure is solved. The empirical formula of the sample used is $Pb_{10.21}(Cu_{0.65}Zn_{0.34})(Cr_{5.93}P_{0.07}S_{0.04})(Si_{1.83}As_{0.10})O_{34}H_{1.62}$. The Raman shifts were partly determined by us based on spectral curve analysis of the published spectrum. For the Raman spectrum of hemihedrite see also Frost (2004c).

Hemimorphite $Zn_4(Si_2O_7)(OH)_2 \cdot H_2O$

Origin: Sasa, Macedonia.
Experimental details: Raman scattering measurements have been performed on a powdered sample in the spectral region from 100 to 1300 cm^{-1} using 514.5 nm Ar$^+$ laser radiation. The laser radiation power is not indicated.

Raman shifts (cm^{-1}): 980w, 930s, 678, 559w, 516w, 452, 402, 332, 304w, 282w, 218w, 169, 132s.
Source: Makreski et al. (2007).
Comments: The sample was characterized by powder X-ray diffraction data and electron microprobe analyses. For the Raman spectrum of hemimorphite see also Jovanovski et al. (2009).

Hemleyite $FeSiO_3$

Origin: Suizhou L6 chondrite (type locality).
Experimental details: Raman scattering measurements have been performed on an arbitrarily oriented sample. Experimental details are not described.
Raman shifts (cm^{-1}): 795s, 673, 611, 476, 403, 342.
Source: Bindi et al. (2017).
Comments: The sample was characterized by powder X-ray diffraction data and electron microprobe analyses. The crystal structure is solved. The empirical formula of the sample used is $(Fe^{2+}_{0.48}Mg_{0.37}Ca_{0.04}Na_{0.04}Mn^{2+}_{0.03}Al_{0.03}Cr^{3+}_{0.01})Si_{1.00}O_3$.

Hemusite $Cu^{1+}_4Cu^{2+}_2SnMoS_8$

Origin: Zijin Cu-Au mine, China.
Experimental details: No data.
Raman shifts (cm^{-1}): 826, 658, 413, 348, 294, 262.
Source: Liu et al. (2012).
Comments: The empirical formula of the sample used is $Cu_{6.03}Sn_{0.95}Fe_{0.14}Mo_{0.97}S_{8.0}$.

Hendricksite $KZn_3(Si_3Al)O_{10}(OH)_2$

Origin: Franklin Furnace, New Jersey, USA.
Experimental details: Raman scattering measurements have been performed on an oriented sample using 514.5 and 488 nm Ar$^+$ laser radiations. The laser radiation power is not indicated. Spectra were collected in scattering geometries with incident laser polarization parallel and perpendicular to the cleavage plane.
Source: Tlili et al. (1989).
Raman shifts (cm^{-1}): 1028, 677s, 644, 321w, 317w, 278w, 189s, 94w.
Comments: The sample was characterized by electron microprobe analyses. The Raman shifts are given as the sum of the both scattering geometries.

Henmilite $Ca_2Cu[B(OH)_4]_2(OH)_4$

Origin: Fuka mine, Okayama prefecture, Honshu Island, Japan (type locality).
Experimental details: Raman scattering measurements have been performed on arbitrarily oriented crystals using a 633 nm He-Ne laser. The laser radiation power is not indicated. The Raman shifts have been determined for the maxima of individual peaks obtained as a result of the spectral curve analysis.
Raman shifts (cm^{-1}): 3609s, 3593sh, 3559s, 3501, 3457sh, 3424s, 3396sh, 3328s, 3269s, 3195sh, 3101sh, 1270w, 1208w, 984w, 969sh, 951w, 922w, 902sh, 834w, 823w, 758s, 752sh, 745sh, 697sh, 598w, 562w, 547w, 534sh, 479sh, 469, 415w, 403w,m 365, 353sh, 335w, 290, 267sh, 255, 240, 225, 217sh, 197, 181sh, 172, 162, 148, 128.

Source: Frost and Xi (2013e).
Comments: No independent analytical data are provided for the sample used.

Henritermierite $Ca_3Mn^{3+}_2(SiO_4)_2(OH)_4$

Origin: N'Chwaning II mine, Kalahari Manganese Fields, South Africa.
Experimental details: Raman scattering measurements have been performed on an arbitrarily oriented single crystal using 532 nm Nd-YAG laser radiation. The nominal laser radiation power was 200 mW.
Raman shifts (cm^{-1}): 3428, 989w, 921, 882, 834w, 567s, 547w, 500s, 469w, 435, 373, 337, 322sh, 278, 257sh, 248, 168, 151.
Source: Friedrich et al. (2015).
Comments: The sample was characterized by single-crystal X-ray diffraction data and electron microprobe analyses. The crystal structure is solved. The empirical formula of the sample used is $(Ca_{2.98}Na_{0.01}Mg_{0.01})(Mn_{1.95}Fe_{0.01}Al_{0.04})[SiO_4]_{2.07}[O_4H_4]_{0.93}$. The Raman shifts were partly determined by us based on spectral curve analysis of the published spectrum.

Henryite $Cu_4Ag_3Te_4$

Origin: Pyrrhotite gorge, Lovchorr Mt., Khibinymassif, Kola Peninsula, Russia.
Experimental details: Raman scattering measurements have been performed on an arbitrarily oriented sample using 514.5 nm Ar$^+$ or 785 nm diode laser radiation. The nominal laser radiation power was 50 mW or 500 mW, respectively.
Raman shifts (cm^{-1}): 148s, 118s, 91w.
Source: Voloshin et al. (2015a).
Comments: The sample was characterized by electron microprobe analyses. The empirical formula of the sample used is $Cu_{3.89}(Ag_{2.75}Au_{0.03})Te_{4.00}$.

Herbertsmithite $Cu_3Zn(OH)_6Cl_2$

Origin: No data.
Experimental details: Raman scattering measurements have been performed on an oriented sample with the laser beam direction normal to the xy plane of a crystal using 532 nm laser radiation. The nominal laser radiation power was 1 mW. A nearly 180°-scattering geometry was employed. Polarized spectra were collected during rotation of the sample within the crystallographic xy plane with the axis of rotation along of the incident light.
Raman shifts (cm^{-1}): 943, 702, 501s, 402, 365, 230w, 148, 123s [for the (xx) scattering geometry].
Source: Wulferding et al. (2010).
Comments: The empirical formula of the sample used is $Zn_{0.8}Cu_{3.2}(OH)_6Cl_2$. For the Raman spectrum of herbertsmithite see also Chu et al. (2011).

Hercynite $Fe^{2+}Al_2O_4$

Origin: No data.
Experimental details: Raman scattering measurements have been performed on an arbitrarily oriented single crystal using 632.8 nm He-Ne laser radiation. The laser radiation power was less than 1 mW. A 180°-scattering geometry was employed.

Raman shifts (cm^{-1}): 748s, 699w, 617, 504, 400, 366sh, 189s.
Source: D'Ippolito et al. (2015).
Comments: The sample was characterized by electron microprobe analysis.

Herderite CaBe(PO$_4$)F

Origin: Ehrenfriedersdorf complex, Erzgebirge (Ore Mts.), Germany.
Experimental details: Raman scattering measurements have been performed on an arbitrarily oriented microscopic inclusion in quartz using 488 nm Ar$^+$ laser radiation. The nominal laser radiation power was 450 mW.
Raman shifts (cm^{-1}): 1005, 983s, 595, 584.
Source: Rickers et al. (2006).
Comments: No independent analytical data are provided for the sample used.

Herzenbergite SnS

Origin: Synthetic.
Experimental details: Methods of sample preparation are not described. Raman scattering measurements have been performed using 532 nm laser radiation. The laser radiation power at the sample was less than 0.4 mW.
Raman shifts (cm^{-1}): 220, 192s, 165, 95, 67w, 48, 39s.
Source: Fontané et al. (2013).
Comments: The sample was characterized by X-ray diffraction data. For the Raman spectrum of herzenbergite see also Chandrasekhar et al. (1977).

Hessite Ag$_2$Te

Origin: Synthetic.
Experimental details: Raman scattering measurements have been performed on a polycrystalline sample using 488, 515, and 633 nm laser radiations. The nominal laser radiation power was 0.3 mW.
Raman shifts (cm^{-1}): 138, 110, 80.
Source: Milenov et al. (2014).
Comments: The sample was characterized by powder X-ray diffraction data and electron microprobe analysis.

Hetaerolite ZnMn$^{3+}_2$O$_4$

Origin: Madjarovo deposit, Eastern Rhodopes, Bulgaria.
Experimental details: Raman scattering measurements have been performed on an arbitrarily oriented sample using 532 nm Nd-YAG laser radiation. The nominal laser radiation power was 38 mW.
Raman shifts (cm^{-1}): 684s, 637, 595sh, 574, 515w, 487w, 388s, 374sh, 329, 309sh.
Source: Vassileva et al. (2005).
Comments: The sample was characterized by X-ray diffraction data and electron microprobe analyses. For the Raman spectrum of hetaerolite see also Javed et al. (2013).

Heterogenite $Co^{3+}O(OH)$

Origin: Mindigi mine, Katanga copperbelt, Katanga province, Democratic Republic of Congo.
Experimental details: Methods of sample preparation are not described. Raman scattering measurements have been performed on an arbitrarily oriented sample using 532 nm laser radiation. The nominal laser radiation power was 2 mW. The Raman shifts have been determined for the maxima of individual peaks obtained as a result of the spectral curve analysis.
Raman shifts (cm^{-1}): 1202w, 1133w, 670, 626, 572, 495s.
Source: Burlet et al. (2011).
Comments: The sample was characterized electron microprobe analyses. For the Raman spectrum of heterogenite see also Burlet et al. (2014).

Heterosite $Fe^{3+}(PO_4)$

Origin: Synthetic.
Experimental details: Raman scattering measurements have been performed on a powdered sample using 532 nm laser radiation. The nominal laser radiation power was 3 mW. A 180°-scattering geometry was employed.
Raman shifts (cm^{-1}): 1123s, 1078s, 1064s, 962, 912, 660, 602, 587, 575w, 516w, 492, 400w, 339, 308, 246, 199w, 175s, 107.
Source: Burba and Frech (2004).
Comments: The sample was characterized by powder X-ray diffraction data and electron microprobe analyses. The Raman shifts were partly determined by us based on spectral curve analysis of the published spectrum.

Heulandite $(Ca,Na,K)_5(Si_{27}Al_9)O_{72} \cdot 26H_2O(?)$

Origin: Paterson, Passaic Co., New Jersey, USA.
Experimental details: Methods of sample preparation are not described. Raman scattering measurements have been performed using Nd-YAG laser radiation. The laser radiation power at the sample was 300 mW.
Raman shifts (cm^{-1}): 1138w, 799w, 611w, 483s, 404s, 147.
Source: Mozgawa (2001).
Comments: The sample was characterized by powder X-ray diffraction data.

Hexacelsian $Ba(Al_2Si_2O_8)$

Origin: Hatrurim complex, Negev Desert, Israel.
Experimental details: Raman scattering measurements have been performed on an arbitrarily oriented sample using 488 nm solid-state laser radiation. The laser radiation power at the sample was 44 mW. The Raman shifts have been determined for the maxima of individual peaks obtained as a result of the spectral curve analysis.
Raman shifts (cm^{-1}): 1119, 1087w, 961, 924, 890, 809w, 678, 594, 480w, 461w, 406s, 296w, 107s.
Source: Galuskina et al. (2016b).
Comments: The sample was characterized by electron microprobe analyses. The empirical formula of the sample used is $(Ba_{0.911}K_{0.059}Ca_{0.042}Na_{0.010})Al_{1.891}Fe^{3+}_{0.072}Si_{2.034}O_8$. For the Raman spectra of hexacelsian see also Colomban et al. (2000), Kremenović et al. (2003), and Dondur et al. (2005).

Hexaferrum (Fe,Os,Ru,Ir)

Origin: Synthetic.
Experimental details: Raman scattering measurements have been performed on a high-purity polycrystalline Fe sample at pressures from 15 to 152 GPa. A 35° incidence angle for the exciting radiation was employed. No characteristics of the laser radiation are indicated.
Raman shifts (cm^{-1}): 245sh, 210 (at 22 GPa); 260 (at 82 GPa); 300 (at 152 GPa).
Source: Merkel et al. (2000).
Comments: No independent analytical data are provided for the sample used.

Hexahydrite $Mg(SO_4) \cdot 6H_2O$

Origin: Synthetic.
Experimental details: Raman scattering measurements have been performed on a powdered sample in the spectral region from 50 to 4300 cm^{-1} using 532 nm Nd-YAG laser radiation. The laser radiation power at the sample was 15 mW.
Raman shifts (cm^{-1}): 3540sh, 3428, 3258sh, 1655, 1146w, 1085w, 984s, 610w, 466w, 445w, 364, 245, 223.
Source: Wang et al. (2006a).
Comments: The sample was characterized by powder X-ray diffraction data. The Raman shifts were partly determined by us based on spectral curve analysis of the published spectrum. For the Raman spectrum of hexahydrite see also Apopei et al. (2015).

Hexahydroborite $Ca[B(OH)_4]_2 \cdot 2H_2O$

Origin: Synthetic.
Experimental details: Raman scattering measurements have been performed using 514.5 nm Ar$^+$ laser radiation. The nominal laser radiation power was 300 mW.
Raman shifts (cm^{-1}): 859s, 755s, 389w.
Source: Jun et al. (1995).
Comments: The sample was characterized by powder X-ray diffraction data.

Hiärneite $(Ca,Mn^{2+},Na)_2(Zr,Mn^{3+})_5(Sb,Ti,Fe)_2O_{16}$

Origin: Långban deposit, Bergslagen ore region, Filipstad district, Värmland, Sweden (type locality).
Experimental details: Raman scattering measurements have been performed on an arbitrarily oriented sample using 514.5 nm Ar$^+$ laser radiation. The nominal laser radiation power was 200 mW.
Raman shifts (cm^{-1}): 672, 602, 515, 453, 426s, 398s, 388s, 378, 305, 268, 215sh, 195, 167, (159), 141.
Source: Holtstam (1997).
Comments: The sample was characterized by powder and single-crystal X-ray diffraction data and electron microprobe analyses. The empirical formula of the sample used is $Na_{0.17}Ca_{1.57}Mn^{2+}_{0.62}Zr_{4.19}Hf_{0.02}Sb^{5+}_{1.37}Ti_{0.59}Mn^{3+}_{0.36}Mg_{0.02}Fe_{0.09}O_{16}$. The Raman shifts were determined by us based on spectral curve analysis of the published spectrum.

Hibonite-(Fe) $(Fe,Mg)Al_{12}O_{19}$

Origin: Allende meteorite.
Experimental details: Raman scattering measurements have been performed on an arbitrarily oriented sample using 514.5 nm Ar$^+$ laser radiation. The laser radiation power is not indicated.
Raman shifts (cm^{-1}): 1067s, 1014, 748, 728, 490, 432.
Source: Ma (2010).
Comments: The sample was characterized by electron microprobe data and electron microprobe analyses. The empirical formula of the sample used is $(Fe^{2+}_{0.34}Mg_{0.27}Na_{0.12}Al_{0.11}Ca_{0.03})(Al_{11.77}Si_{0.23})O_{19}$.

Hibonite $(Ca,Ce)(Al,Ti,Mg)_{12}O_{19}$

Origin: Synthetic.
Experimental details: Raman scattering measurements have been performed on an arbitrarily oriented single crystal using 532 nm Nd-YAG laser radiation. The laser radiation output power was 100 mW. A 180°-scattering geometry was employed.
Raman shifts (cm^{-1}): 910s, 873, 837, 796, 774, 741, 684, 640, 625, 565, 530, 489, 458, 450, 399, 332s, 274, 251, 209, 194sh.
Source: Hofmeister et al. (2004).
Comments: The sample was characterized by single-crystal X-ray diffraction data and electron microprobe analysis. The crystal structure is solved.

Hidalgoite $PbAl_3(SO_4)(AsO_4)(OH)_6$

Origin: Gold Hill mine, Tooele Co., Utah, USA.
Experimental details: Raman scattering measurements have been performed on an arbitrarily oriented sample using 633 nm He-Ne laser radiation. The laser radiation power is not indicated. The Raman shifts have been determined for the maxima of individual peaks obtained as a result of the spectral curve analysis.
Raman shifts (cm^{-1}): 3477, 3351sh, 3185sh, 1730, 1093, 1014s, 998sh, 879s, 853sh, 649, 631sh, 595, 528s, 513sh, 480, 433, 351sh, 334, 265, 234sh, 210s, 157, 142sh, 107.
Source: Frost et al. (2011o).
Comments: No independent analytical data are provided for the sample used.

Hieratite K_2SiF_6

Origin: Synthetic.
Experimental details: Raman scattering measurements have been performed on an arbitrarily oriented sample using 1064 nm Nd-YAG laser radiation. The nominal laser radiation power was between 80 and 200 mW.
Raman shifts (cm^{-1}): 655s, 478w, 408.
Source: Rissom et al. (2008).
Comments: The sample was characterized by powder X-ray diffraction data.

Hilarionite $Fe^{3+}_2(SO_4)(AsO_4)(OH)6H_2O$

Origin: Hilarion Mine, Lavrion, Greece (type locality).
Experimental details: Raman scattering measurements have been performed on an arbitrarily oriented sample using 514 nm diode laser radiation. The laser radiation power is not indicated. The Raman shifts have been determined for the maxima of individual peaks obtained as a result of the spectral curve analysis.
Raman shifts (cm^{-1}): 1015s, 877sh, 843sh, 807, 585w, 495, 448sh, 390w, 365, 292, 191, 143sh, 123.
Source: Liu et al. (2017).
Comments: The sample was characterized by powder X-ray diffraction data and electron microprobe analysis.

Hingganite-(Y) $BeY(SiO_4)(OH)$

Origin: Oppach, Lusatian Mts., Germany.
Experimental details: Raman scattering measurements have been performed on an arbitrarily oriented sample using 488 nm Ar^+ laser radiation. The laser radiation power at the sample was 45 mW.
Raman shifts (cm^{-1}): 3540, 922, 724, 332w.
Source: Thomas and Davidson (2017).
Comments: The sample was characterized by electron microprobe analyses. Fluorescence lines are excluded.

Hinsdalite $PbAl_3(SO_4)(PO_4)(OH)_6$

Origin: Sylvester mine, Zeehan, Tasmania, Australia.
Experimental details: Raman scattering measurements have been performed on arbitrarily oriented crystals using a 633 nm He-Ne laser. The laser radiation power is not indicated. The Raman shifts have been determined for the maxima of individual peaks obtained as a result of the spectral curve analysis.
Raman shifts (cm^{-1}): 3603sh, 3472w, 3247w, 1021s, 1007s, 997sh, 982s, 930w, 614, 581sh, 506, 463, 374, 279sh, 250s, 187, 143, 108.
Source: Frost et al. (2011k).
Comments: No independent analytical data are provided for the sample used.

Hiortdahlite $(Na,Ca)_2Ca_4Zr(Mn,Ti,Fe)(Si_2O_7)_2(F,O)_4$

Origin: Langezundfiord, Norway.
Experimental details: No data.
Raman shifts (cm^{-1}): 1047, 947s, 796, 661s, 258.
Source: Sharygin et al. (1996a, b).
Comments: No independent analytical data are provided for the sample used.

Hisingerite $Fe_2Si_2O_5(OH)_4 \cdot 2H_2O$

Origin: McMurdo Dry Valleys, Antarctica.
Experimental details: No data.
Raman shifts (cm^{-1}): 1055, 1031, 983, 899w, 863w, 782, 736, 666s, 546s, 504, 450, 388, 288, 236, 189, 145w, 121w.

Source: Edwards et al. (2004).
Comments: No independent analytical data are provided for the sample used. The Raman shifts were determined by us based on spectral curve analysis of the published spectrum.

Hochelagaite $CaNb_4O_{11} \cdot 8H_2O$

Origin: Mont Saint-Hilaire, Québec, Canada (type locality).
Experimental details: Methods of sample preparation are not described. Raman scattering measurements have been performed using 532 nm laser radiation. The laser radiation power is not indicated.
Raman shifts (cm^{-1}): 925, 878s, 663s, 587, 477, 387, 325w, 300w, 234, 196sh.
Source: Haring and McDonald (2017).
Comments: No independent analytical data are provided for the sample used. The Raman shifts were partly determined by us based on spectral curve analysis of the published spectrum.

Hoelite $C_{14}H_8O_2$

Origin: Kladno, Czech Republic.
Experimental details: Raman scattering measurements have been performed on an arbitrarily oriented sample using 514.5 nm Ar^+ laser radiation. The nominal laser radiation power was 10 mW.
Raman shifts (cm^{-1}): 3076w, 1665s, 1597, 1583, 1439w, 1387w, 1317, 1306w, 1239w, 1214, 1176s, 1146, 1080w, 1030s, 976w, 818w, 790w, 767, 682, 520w, 495w, 475, 437, 362, 238s.
Source: Jehlička et al. (2007b).
Comments: No independent analytical data are provided for the sample used. For the Raman spectrum of hoelite see also Jehlička et al. (2009a).

Hoganite $Cu(CH_3COO)_2 \cdot H_2O$

Origin: Potosi Pit, Broken Hill, Yancowinna Co., New South Wales, Australia (type locality).
Experimental details: Raman scattering measurements have been performed on an arbitrarily oriented sample using 532 nm Nd-YAG laser radiation. The laser radiation power is not indicated. The Raman shifts have been determined for the maxima of individual peaks obtained as a result of the spectral curve analysis.
Raman shifts (cm^{-1}): 3478, 3024s, 2989, 2941s, 2922s, 2862, 2788, 1449, 1440, 1418, 1360, 948s, 938, 703, 684, 322s, 297s, 266, 252, 230, 212, 184.
Source: Musumeci and Frost (2007).
Comments: The sample was characterized by powder X-ray diffraction data and chemical analysis. The empirical formula of the sample used is $C_4H_{7.89}O_{5.07}Cu_{1.00}Fe_{0.01}$.

Hogarthite $(Na,K)_2CaTi_2Si_{10}O_{26} \cdot 8H_2O$

Origin: Poudrette (Demix) quarry, Mont Saint-Hilaire, Rouville RCM (Rouville Co.), Montérégie, Québec, Canada (type locality).
Experimental details: Raman scattering measurements have been performed with a laser beam perpendicular to the {010} cleavage of a single using 532 nm laser radiation. The laser radiation power is not indicated.

Raman shifts (cm^{-1}): 3607, 3411sh, 3239sh, 1608w, 1190, 1052, 942s, 902sh, 794w, 714w, 679, 548s, 448s, 295s, 258w, 225, 173, 135, 105.
Source: McDonald et al. (2015).
Comments: The sample was characterized by powder X-ray diffraction data and electron microprobe analyses. The crystal structure is solved. The empirical formula of the sample used is $(Na_{0.78}K_{0.62}\square_{0.51}Ca_{0.09})Ca(Ti_{1.85}Zr_{0.09}Nb_{0.06})Si_{10.09}O_{26} \cdot 8H_2O$.

Hohmannite $Fe^{3+}_2O(SO_4)_2 \cdot 8H_2O$

Origin: Sierra Gorda District, Antofagasta Province, Antofagasta Region, Chile.
Experimental details: Raman scattering measurements have been performed on an arbitrarily oriented single crystal using 632.8 nm He-Ne laser radiation. The nominal laser radiation power was less than 1 mW.
Raman shifts (cm^{-1}): 3495sh, 3438w, 3292sh, 3204w, 1204w, 1166w, 1125, 1098s, 1075w, 1058, 1031s, 1018, 659w, 628w, 605w, 580w, 496w, 470w, 400s, 334s, 258s, 245sh, 227sh, 198, 163, 144w, 129w, 73, 49.
Source: Ventruti et al. (2015).
Comments: The sample was characterized by single-crystal X-ray diffraction data. The crystal structure is solved.

Holdawayite $Mn^{2+}_6(CO_3)_2(OH)_7(Cl,OH)$

Origin: Udachnaya-East kimberlite, Yakutia, Russia.
Experimental details: Raman scattering measurements have been performed on an arbitrarily oriented sample using 514.5 nm Ar$^+$ laser radiation. The laser radiation power at the sample was 5 mW.
Raman shifts (cm^{-1}): 1087s, ~700.
Source: Mernagh et al. (2011).
Comments: No independent analytical data are provided for the sample used. The Raman shifts have been determined for a polymineral aggregate. The Raman shifts of holdawayite from the RRUFF Project database (sample R090029) are: 1087s, 900, 701, 222, 164, 151.

Holfertite $(UO_2)_{1.75}Ca_{0.25}TiO_4 \cdot 3H_2O$

Origin: Starvation Canyon, Thomas Range, Juab Co., Utah, USA (type locality).
Experimental details: Raman scattering measurements have been performed on an arbitrarily oriented sample using a 633 nm He-Ne laser. The laser radiation power is not indicated. The Raman shifts have been determined for the maxima of individual peaks obtained as a result of the spectral curve analysis.
Raman shifts (cm^{-1}): 3500w, 3406sh, 2925w, 2866sh, 1628w, 1555sh, 1128w, 828s, 749s, 641sh, 474, 389s, 328, 257, 201sh, 144.
Source: Frost (2011a).
Comments: No independent analytical data are provided for the sample used.

Hollandite $Ba(Mn^{4+}_6Mn^{3+}_2)O_{16}$

Origin: Jhabua district, Madhya Pradesh, India.
Experimental details: Methods of sample preparation are not described. Raman scattering measurements have been performed in the spectral region from 10 to 1200 cm^{-1} using 514.5 nm

Ar⁺ laser radiation. The nominal laser radiation power was 10 mW. A nearly 180°-scattering geometry was employed.
Raman shifts (cm⁻¹): 705w, 628, 586, 558w, 507, 395w.
Source: Julien et al. (2004).
Comments: The sample was characterized by powder X-ray diffraction data.

Hollingworthite RhAsS

Origin: No data.
Experimental details: Experimental details are not indicated. Raman scattering measurements have been performed using 532 nm Nd-YAG laser radiation. The nominal laser radiation power was 100 mW.
Raman shifts (cm⁻¹): 1093w, 986w, 557w, 381sh, 360s, 347sh, 283, 272sh, 256s, 214, 146, 77sh.
Source: Bakker (2014).
Comments: The sample was characterized by electron microprobe data. The Raman shifts were partly determined by us based on spectral curve analysis of the published spectrum.

Holmquistite $\square Li_2(Mg_3Al_2)Si_8O_{22}(OH)_2$

Origin: Martin Marietta quarry, Bessemer, North Carolina, USA.
Experimental details: Raman scattering measurements have been performed on arbitrarily oriented crystals using a 633 nm He-Ne laser. The laser radiation power is not indicated. The Raman shifts have been determined for the maxima of individual peaks obtained as a result of the spectral curve analysis.
Raman shifts (cm⁻¹): 3661s, 3646s, 3631s, 3614, 1127, 1102sh, 1085s, 1045sh, 1022s, 791, 753, 694sh, 679s, 613, 582sh, 565sh, 551, 530sh, 502, 471, 456, 423, 408sh, 390s, 343, 309w, 297.
Source: Kloprogge et al. (2001a).
Comments: The sample was characterized by powder X-ray diffraction data and qualitative electron-microprobe analysis. The analytical data are insufficient for the mineral identification. For the Raman spectrum of holmquistite see also Kloprogge et al. (2001c).

Honessite $(Ni_{1-x}Fe^{3+}{}_x)(SO_4)_{x/2}(OH)_2 \cdot nH_2O$ ($x < 0.5$, $n < 3x/2$)

Origin: Linden, Upper Mississippi Valley district, Iowa Co., Wisconsin, USA (type locality).
Experimental details: Raman scattering measurements have been performed on an arbitrarily oriented sample using 632.8 nm He-Ne laser radiation. A 180°-scattering geometry was employed.
Raman shifts (cm⁻¹): 3614w, 2988sh, 2956, 2244, 1061w, 973, 852w, 527s, 460, 167w.
Source: Bindi et al. (2015a).
Comments: The sample was characterized by powder X-ray diffraction data and electron microprobe analyses. The crystal structure is solved. The empirical formula of the sample used is $[(Ni^{2+}{}_{0.902} Ca^{2+}{}_{0.002})(Co^{3+}{}_{0.072}Fe^{3+}{}_{0.024})](OH)_{1.884}Cl_{0.012}(H_2O)_{0.004}(SO_4)_{0.100} \cdot 0.900H_2O$.

Hopeite $Zn_3(PO_4)_2 \cdot 4H_2O$

Origin: Kabwe (Broken Hill)mine, Kabwe district, Central province, Zambia.
Experimental details: Raman scattering measurements have been performed on an arbitrarily oriented sample using 633 nm He-Ne laser radiation. The laser radiation power at the sample was 1 mW. The

Raman shifts have been determined for the maxima of individual peaks obtained as a result of the spectral curve analysis.
Raman shifts (cm^{-1}): 3456w, 3247sh, 1150, 1059w, 1000sh, 995s, 940.
Source: Frost (2004a).
Comments: No independent analytical data are provided for the sample used. For the Raman spectrum of hopeite see also O'Neill et al. (2006).

Hopeite $Zn_3(PO_4)_2 \cdot 4H_2O$

Origin: Synthetic.
Experimental details: Raman scattering measurements have been performed on an arbitrarily oriented sample using 488 nm Ar$^+$ laser radiation. The laser radiation power is not indicated.
Raman shifts (cm^{-1}): 1149, 1059, 998s, 942, 598, 368sh, 315.
Source: O'Neill et al. (2006).
Comments: The sample was characterized by powder X-ray diffraction data, but no independent analytical data are provided for the sample used. For the Raman spectrum of hopeite see also Frost (2004a).

Hörnesite $Mg_3(AsO_4)_2 \cdot 8H_2O$

Origin: Allchar (Alšar) deposit, Rožde, Kavadarci, Republic of Macedonia.
Experimental details: Raman scattering measurements have been performed on a powdered sample using 632.8 nm He-Ne laser radiation. The laser radiation power is not indicated.
Raman shifts (cm^{-1}): 875w, 808s, 468w, 430, 403, 365, 301w, 271, 262sh, 242, 205, 180w, 158, 138.
Source: Makreski et al. (2015b).
Comments: The sample was characterized by powder X-ray diffraction data. For the Raman spectrum of hörnesite see also Frost et al. (2003g).

Hsianghualite $Li_2Ca_3Be_3(SiO_4)_3F_2$

Origin: Xianghualing (Hsianghualing) mine, Linwu Co., Hunan province, China (type locality).
Experimental details: Methods of sample preparation are not described. Raman scattering measurements have been performed using 1064 nm Nd-YAG laser radiation. The nominal laser radiation power was in the range from 300 to 380 mW.
Raman shifts (cm^{-1}): 930, 894.
Source: Yuran and Li (1998).
Comments: No independent analytical data are provided for the sample used.

Huanghoite-(Ce) $BaCe(CO_3)_2F$

Origin: Bayan Obo deposit, Baotou prefecture, Inner Mongolia Autonomous Region, China (type locality).
Experimental details: Raman scattering measurements have been performed using 488 nm laser radiation. The nominal laser radiation power was 300 mW. Raman spectrum was obtained in the spectral regions from 200 to 1200 cm^{-1} and from 1300 to 3200 cm^{-1}.

Raman shifts (cm^{-1}): 2746, 2718sh, 2183, 1840w, 1596, 1525, 1374sh, 1089s, 720, 695, 649, 354, 273, 224, 106.
Source: Hong et al. (1999).
Comments: No independent analytical data are provided for the sample used. Raman bands in the range from 1300 to 2800 cm^{-1} are mainly due to luminescence.

Huanzalaite Mg(WO$_4$)

Origin: Synthetic.
Experimental details: Raman scattering measurements have been performed on a powdered sample using 514.5 nm Ar$^+$ laser radiation. The laser radiation power at the sample was 2 mW. A 180°-scattering geometry was employed.
Raman shifts (cm^{-1}): 917s, 809, 713, 684w, 552, 518, 420, 405w, 385w, 352, 314w, 294, 277, 267sh, 215w, 185, 156, 97s.
Source: Ruiz-Fuertes et al. (2010).
Comments: The sample was characterized by powder X-ray diffraction data.

Hubeite Ca$_2$Mn^{2+}Fe^{3+}Si$_4$O$_{12}$(OH)·2H$_2$O

Origin: Fengjiashan Mine, Daye Co., Huangshi prefecture, Hubei province, China (type locality).
Experimental details: Methods of sample preparation are not described. Raman scattering measurements have been performed using 532 nm laser radiation. The nominal laser radiation power was 0.23 mW.
Raman shifts (cm^{-1}): 1083w, 1033, 966s, 912w, 899, 892, 864, 740, 664s, 575, 498, 489s, 468, 406sh, 400s, 366s, 292, 254, 195, 176, 145, 125, 117.
Source: Ferras et al. (2016).
Comments: No independent analytical data are provided for the sample used. The Raman shifts were partly determined by us based on spectral curve analysis of the published spectrum.

Hübnerite Mn^{2+}(WO$_4$)

Origin: Synthetic.
Experimental details: Raman scattering measurements have been performed on arbitrarily oriented crystals using a 685 nm Nd-YAG laser. The laser radiation power is not indicated. The Raman shifts have been determined for the maxima of individual peaks obtained as a result of the spectral curve analysis.
Raman shifts (cm^{-1}): 885, 775, 697, 672, 543, 509, 445, 395, 353, 325, 290, 255, 202, 174, 161, 126.
Source: Frost et al. (2004b).
Comments: No independent analytical data are provided for the sample used. For the Raman spectra of hübnerite see also Kloprogge et al. (2004b) and Almeida et al. (2012).

Hughesite Na$_3$AlV$_{10}$O$_{28}$·22H$_2$O

Origin: Sunday mine, Gypsum valley, San Miguel Co., Colorado, USA (type locality).
Experimental details: Raman scattering measurements have been performed on an oriented single crystal with the laser beam perpendicular to the (001) cleavage surface (orientation 1) and with the laser beam parallel to the cleavage surface (orientation 2) in the spectral region from 100 to

1500 cm^{-1} using 632 nm He-Ne laser radiation. The nominal laser radiation power was 1 mW and 0.5 mW, respectively.

Raman shifts (cm^{-1}): 1007s, 994sh, 972, 959sh, 945sh, 877w, 596, 471w, 363, 319, 270sh, 247, 235, 218w, 201sh, 192 (orientation 1); 999s, 972s, 854, 591, 469w, 362sh, 318, 260, 231w, 214w, 182 (orientation 2).

Source: Rakovan et al. (2011).

Comments: The sample was characterized by powder X-ray diffraction data and electron microprobe analyses. The crystal structure is solved. The empirical formula of the sample used is $Na_{2.99}Al_{1.05}(V_{10}O_{28}) \cdot 22H_2O$.

Humberstonite $K_3Na_7Mg_2(SO_4)_6(NO_3)_2 \cdot 6H_2O$

Origin: Artificial (a component of gypsum-based plaster).

Experimental details: Raman scattering measurements have been performed on an arbitrarily oriented sample using 785 nm diode laser radiation. The laser radiation power is not indicated.

Raman shifts (cm^{-1}): 1067, 1048, 1013, 723, 632, 183.

Source: Morillas et al. (2015).

Comments: The sample was characterized by powder X-ray diffraction data.

Humboldtine $Fe^{2+}(C_2O_4) \cdot 2H_2O$

Origin: Bohemia, Czech Republic.

Experimental details: Raman scattering measurements have been performed on an arbitrarily oriented sample using 633 nm He-Ne laser radiation. The laser radiation power is not indicated. The Raman shifts have been determined for the maxima of individual peaks obtained as a result of the spectral curve analysis.

Raman shifts (cm^{-1}): 3315, 1708w, 1555w, 1468s, 1450sh, 913, 856w, 582, 518, 293w, 246s, 203s.

Source: Frost (2004d).

Comments: No independent analytical data are provided for the sample used. For the Raman spectra of humboldtine see also Frost and Weier (2003), Echigo and Kimata (2008), and D'Antonio et al. (2009).

Humite $Mg_7(SiO_4)_3(F,OH)_2$

Origin: Monte Somma, Somma-Vesuvius complex, Naples, Italy.

Experimental details: Raman scattering measurements have been performed on an arbitrarily oriented sample using 633 nm He-Ne laser radiation. The laser radiation power is not indicated. The Raman shifts have been determined for the maxima of individual peaks obtained as a result of the spectral curve analysis.

Raman shifts (cm^{-1}): 3576, 3572, 3560, 966, 930, 876, 859, 844, 831, 784, 757, 747, 606, 587, 570, 549, 547, 539, 442, 428, 391.

Source: Frost et al. (2007k).

Comments: The sample was characterized by powder X-ray diffraction data and electron microprobe analysis. The empirical formula of the sample used is $Mg_{6.33}Fe^{2+}_{0.50}(SiO_4)_3(OH)_{1.66}$, which corresponds to the OH-analogue of humite.

Humite $Mg_7(SiO_4)_3(F,OH)_2$

Origin: Monte Somma, Somma-Vesuvius complex, Naples, Italy.
Experimental details: Raman scattering measurements have been performed on an arbitrarily oriented sample using 633 nm He-Ne laser radiation. The laser radiation power is not indicated. The Raman shifts have been determined for the maxima of individual peaks obtained as a result of the spectral curve analysis.
Raman shifts (cm^{-1}): 3576, 3572, 3560, 966, 930, 876, 859, 844, 831, 784, 757, 747, 606, 587, 570, 549, 547, 539, 442, 428, 391.
Source: Frost et al. (2007k).
Comments: The sample was characterized by powder X-ray diffraction data and electron microprobe analysis. The empirical formula of the sample used is $Mg_{6.33}Fe^{2+}_{0.50}(SiO_4)_3(OH)_{1.66}$, which corresponds to the OH-analogue of humite.

Hummerite $KMgV^{5+}_5O_{14} \cdot 8H_2O$

Origin: Hummer mine, Paradox valley, Montrose Co., Colorado, USA (type locality).
Experimental details: Raman scattering measurements have been performed on an arbitrarily oriented sample using 633 nm He-Ne laser radiation. The laser radiation power is not indicated. The Raman shifts have been determined for the maxima of individual peaks obtained as a result of the spectral curve analysis.
Raman shifts (cm^{-1}): 3599, 3526, 3416, 3404, 3296, 3230, 3223, 2929, 2902, 1621, 1600, 999s, 962s, 833w, 817sh, 590, 532, 442w, 360sh, 326s, 314sh, 254sh, 241, 227sh, 208, 183s, 146.
Source: Frost et al. (2004e, 2005a).
Comments: No independent analytical data are provided for the sample used.

Hungchaoite $MgB_4O_5(OH)_4 \cdot 7H_2O$

Origin: Synthetic.
Experimental details: Raman scattering measurements have been performed on an arbitrarily oriented sample in the spectral region from 300 to 1800 cm^{-1} using 514.5 nm Ar^+ laser radiation. The nominal laser radiation power was 300 mW.
Raman shifts (cm^{-1}): 1628w, 1352, 1045, 949s, 897w, 855s, 816, 787s, 716w, 583s, 556w, 519, 491s, 450, 408s, 371, 345, 316s.
Source: Li et al. (1995) and Jia et al. (2001).
Comments: The sample was characterized by powder X-ray diffraction data. The Raman shifts were partly determined by us based on spectral curve analysis of the published spectrum.

Huntite $CaMg_3(CO_3)_4$

Origin: No data.
Experimental details: Methods of sample preparation are not described. Raman scattering measurements have been performed in the spectral region from 50 to 1200 cm^{-1} using 1064 nm Nd-YAG laser radiation. The nominal laser radiation power was 100 mW.
Raman shifts (cm^{-1}): 2905w, 1761w, 1459w, 1123s, 878, 742w, 723, 705w, 386w, 364w, 316, 272, 253, 231w, 155, 118.
Source: Edwards et al. (2005).

Comments: The sample was characterized by powder X-ray diffraction data. For the Raman spectrum of huntite see also Scheetz and White (1977).

Hureaulite $Mn^{2+}_5(PO_3OH)_2(PO_4)_2 \cdot 4H_2O$

Origin: Cigana mine, Conselheiro Pena, Rio Doce valley, Minas Gerais, Brazil.
Experimental details: Raman scattering measurements have been performed on an arbitrarily oriented sample using 633 nm Nd-YAG laser radiation. The laser radiation power is not indicated. The Raman shifts have been determined for the maxima of individual peaks obtained as a result of the spectral curve analysis.
Raman shifts (cm^{-1}): 3424, 3322, 3185, 2973sh, 2818, 1648, 1571, 1109sh, 1083, 1047, 1024, 1007sh, 989, 950, 941sh, 778w, 726, 598sh, 582, 564sh, 543sh, 531, 455, 414, 398sh, 381, 304w, 267w, 237sh, 221w, 194w, 155w, 137w, 120w, 104w.
Source: Frost et al. (2013aj).
Comments: The sample was characterized by powder X-ray diffraction data and electron microprobe analyses. Powder X-ray diffraction data are not provided in the cited paper. The empirical formula of the sample used is $(Mn_{3.23}Fe_{1.04}Ca_{0.19}Mg_{0.13})(PO_4,HPO_4)_{4.13}(OH,H_2O)_x$, which indicates a significant deficit of (Mn,Fe,Ca,Mg)-cations.

Hurlbutite $CaBe_2(PO_4)_2$

Origin: Nanping no. 31 granitic pegmatite dyke, Fujian province, southeastern China.
Experimental details: Raman scattering measurements have been performed on an arbitrarily oriented sample. No other details are indicated.
Raman shifts (cm^{-1}): 1021, 587, 575, 550, 404, 132.
Source: Rao et al. (2011).
Comments: The sample was characterized by electron microprobe analyses.

Hydroboracite $CaMg[B_3O_4(OH)_3]_2 \cdot 3H_2O$

Origin: Kohnstein quarry, Thuringia, Germany.
Experimental details: Raman scattering measurements have been performed on arbitrarily oriented crystals using a 633 nm He-Ne laser. The laser radiation power is not indicated. The Raman shifts have been determined for the maxima of individual peaks obtained as a result of the spectral curve analysis.
Raman shifts (cm^{-1}): 3632s, 3563s, 3551sh, 3507s, 3384sh, 3371, 3255, 3138sh, 3076, 1685w, 1433w, 1394, 1379, 1318, 1268w, 1229, 1157, 1144sh, 1063sh, 1039s, 955w, 925, 910sh, 869w, 846, 825, 753, 730, 721sh, 696sh, 647, 612w, 582, 560s, 556sh, 526w, 491, 478sh, 459, 437, 421sh, 355sh, 342, 330sh, 304, 243, 229, 212, 168, 159sh, 135.
Source: Frost et al. (2014af).
Comments: No independent analytical data are provided for the sample used.

Hydrocalumite $Ca_4Al_2(OH)_{12}(Cl,CO_3,OH)_2 \cdot 4H_2O$

Origin: Synthetic.
Experimental details: Raman scattering measurements have been performed on a powdered sample using 1064 nm Nd-YAG laser radiation. The nominal laser radiation power was 200 mW. The

Raman shifts have been determined for the maxima of individual peaks obtained as a result of the spectral curve analysis.
Raman shifts (cm^{-1}): 1086, 1078sh, 712, 704sh, 531, 397, 359, 296sh, 281, 271sh.
Source: Frost et al. (2011j).
Comments: Questionable data: Raman spectra of samples with different Ca:Al ratios (from 2:1 to 4:1) are almost identical. According to powder X-ray diffraction data, the samples used may contain admixed calcite.

Hydrocerussite $Pb_3(CO_3)_2(OH)_2$

Origin: Synthetic.
Experimental details: Methods of sample preparation are not described. Raman scattering measurements have been performed in the spectral region from 50 to 3600 cm^{-1} using 1064 nm Nd-YAG laser radiation. The laser radiation power is not indicated.
Raman shifts (cm^{-1}): 1731w, 1467sh, 1365, 1050s, 862w, 837w, 707sh, 693sh, 679, 411, 321, 267w, 150sh, 113.
Source: Ciomartan et al. (1996).
Comments: The sample was characterized by powder X-ray diffraction data. For the Raman spectra of hydrocerussite see also Frost et al. (2003e) and Bouchard and Smith (2003).

Hydrocerussite $Pb_3(CO_3)_2(OH)_2$

Origin: No data.
Experimental details: Raman scattering measurements have been performed on an arbitrarily oriented sample using a 633 nm He-Ne laser. The laser radiation power is not indicated. The Raman shifts have been determined for the maxima of individual peaks obtained as a result of the spectral curve analysis.
Raman shifts (cm^{-1}): 3576, 3536w, 1736, 1705, 1679, 1479w, 1420w, 1378w, 1375w, 1365sh, 1053s, 1031sh, 887w, 866w, 837w, 737, 694w, 681, 671w, 417, 391, 376, 318, 221, 177w, 152.
Source: Frost et al. (2003e).
Comments: No independent analytical data are provided for the sample used. For the Raman spectra of hydrocerussite see also Ciomartan et al. (1996) and Bouchard and Smith (2003).

Hydrodelhayelite-related compound $KCa_2Na(Si_8O_{19})\cdot 5H_2O$

Origin: Synthetic.
Experimental details: Methods of sample preparation are not described. Raman scattering measurements have been performed using He-Ne laser radiation. The laser radiation power is not indicated.
Raman shifts (cm^{-1}): 3500 (broad), 3300sh, 1636w, 1174, 1125sh, 1103, 848w, 783, 697, 610s, 496sh, 469, 437, 376, 351w, 317, 286, 240w, 190, 156, 127, 113, 98sh, 80sh.
Source: Cadoni and Ferraris (2009).
Comments: The sample was characterized by single-crystal X-ray diffraction data and semiquantitative electron microprobe analyses. The crystal structure is solved. The Raman shifts were partly determined by us based on spectral curve analysis of the published spectrum

Hydrohalite $NaCl \cdot 2H_2O$

Origin: Synthetic.
Experimental details: Raman scattering measurements have been performed on an arbitrarily oriented sample at $-190\,°C$, in the spectral region from 2800 to 4000 cm^{-1}, using 532 nm Nd-YAG laser radiation. The laser radiation power at the sample was between 1 and 1.5 mW.
Raman shifts (cm^{-1}): 3536, 3432, 3418, 3402, 3321w, 3300w.
Source: Baumgartner and Bakker (2010).
Comments: No independent analytical data are provided for the sample used. For the Raman spectra of hydrohalite see also Bakker (2004), Sakurai et al. (2010), and Okotrub and Surovtsev (2013).

Hydrohonessite $(Ni_{1-x}Fe^{3+}{}_x)(SO_4)_{x/2}(OH)_2 \cdot nH_2O$ ($x < 0.5$, $n > 3x/2$)

Origin: Kambalda, Western Australia, Australia.
Experimental details: Raman scattering measurements have been performed on arbitrarily oriented crystals using a 633 nm He-Ne laser. The laser radiation power at the sample was less than 1 mW. The Raman shifts have been determined for the maxima of individual peaks obtained as a result of the spectral curve analysis.
Raman shifts (cm^{-1}): 3493s, 3405, 1638w, 1135, 1115sh, 1008s, 671, 619, 579, 493, 414, 318w, 209w, 182, 164.
Source: Frost et al. (2003h).
Comments: No independent analytical data are provided for the sample used.

Hydromagnesite $Mg_5(CO_3)_4(OH)_2 \cdot 4H_2O$

Origin: Salda Golulake, Turkey.
Experimental details: Methods of sample preparation are not described. Raman scattering measurements have been performed using 1064 nm Nd-YAG laser radiation. The nominal laser radiation power was 100 mW.
Raman shifts (cm^{-1}): 2904w, 1871w, 1521w, 1487w, 1451w, 1119s, 757, 727, 706, 669w, 653sh, 329, 247, 232, 202, 184, 147.
Source: Edwards et al. (2005).
Comments: No independent analytical data are provided for the sample used. For the Raman spectra of hydromagnesite see also Frost (2011d) and Kristova et al. (2014).

Hydroniumjarosite $(H_3O)Fe^{3+}{}_3(SO_4)_2(OH)_6$

Origin: Synthetic.
Experimental details: Raman scattering measurements have been performed on a powdered sample using 514.5 nm Ar^+ laser radiation. The laser radiation power at the sample was in the range from 0.22 to 1 mW.
Raman shifts (cm^{-1}): 3431, 3365, 2522, 2420, 1167, 1101s, 1019, 1013, 769, 633, 619, 565, 458, 420, 359, 281, 227.
Source: Murphy et al. (2009).
Comments: The sample was characterized by powder X-ray diffraction data and inductively coupled plasma–optical emission spectrometry. The empirical formula of the sample used is $(H_3O)Fe^{3+}{}_{2.93}(SO_4)_2(OH)_{6.79}(H_2O)_{0.2}$. For the Raman spectra of hydroniumjarosite see also Frost et al. (2006r), Plášil et al. (2014b), and Apopei et al. (2014a).

Hydroniumjarosite $(H_3O)Fe^{3+}_3(SO_4)_2(OH)_6$

Origin: Cerros Pintados, Pampa del Tamarugal, Iquique Province, Tarapacá Region, Chile.
Experimental details: Raman scattering measurements have been performed on an arbitrary oriented single crystal using 532 nm laser radiation. The nominal laser radiation power was 3 mW. Raman spectrum was obtained in the spectral region from 50 to 4000 cm^{-1}.
Raman shifts (cm^{-1}): 2989w, 2941w, 2878, 2328w, 2248, 2002w, 1742w, 1678w, 1622w, 1477w, 1449w, 1329w, 1164, 1103s, 1012s, 859w, 812w, 722w, 643sh, 620s, 569, 497w, 454sh, 424s, 367sh, 227s, 172w, 135, 62w.
Source: Plášil et al. (2014b).
Comments: The sample was characterized by single-crystal X-ray diffraction data and electron microprobe analyses. The crystal structure is solved. The empirical formula of the sample used is $(H_3O)^+_{0.77}Na_{0.20}K_{0.02}(Fe_{2.95}Al_{0.03})(OH)_{6.12}[(SO_4)_{1.97}(SiO_4)_{0.03}]$. The Raman shifts were partly determined by us based on spectral curve analysis of the published spectrum. For the Raman spectra of hydroniumjarosite see also Frost et al. (2006r), Murphy et al. (2009), and Apopei et al. (2014a).

Hydroromarchite $Sn^{2+}_3O_2(OH)_2$

Origin: Synthetic.
Experimental details: Raman scattering measurements have been performed on an arbitrarily oriented sample using 785 nm laser radiation. The nominal laser radiation power was 400 mW. The Raman shifts have been determined for the maxima of individual peaks obtained as a result of the spectral curve analysis.
Raman shifts (cm^{-1}): 264w, 226s, 184w, 53s.
Source: Chen and Grandbois (2013).
Comments: The sample was characterized by powder X-ray diffraction data.

Hydrotalcite-2H $Mg_6Al_2(CO_3)(OH)_{16} \cdot 4H_2O$

Origin: Kovdor massif, Kola Peninsula, Russia.
Experimental details: Raman scattering measurements have been performed on arbitrarily oriented crystals using a 633 nm He-Ne laser. The laser radiation power is not indicated. Raman spectrum was obtained in the spectral region from 200 to 4000 cm^{-1}. The Raman shifts have been determined for the maxima of individual peaks obtained as a result of the spectral curve analysis.
Raman shifts (cm^{-1}): 3573sh, 3487, 3371sh, 1237sh, 1223w, 1200w, 1101sh, 1064s, 1059sh, 1045sh, 973, 696w, 595sh, 558s, 484w, 408w, 151s, 111s.
Source: Frost et al. (2014y).
Comments: No independent analytical data are provided for the sample used. Based on the origin of the sample and its morphological features, one cannot exclude that it is quintinite.

Hydrotalcite $Mg_6Al_2(CO_3)(OH)_{16} \cdot 4H_2O$

Origin: Synthetic.
Experimental details: Raman scattering measurements have been performed on a powdered sample pressed as pellet using 1064 nm Nd-YAG laser radiation. The laser radiation power is not indicated. A 180°-scattering geometry was employed.

Raman shifts (cm^{-1}): 1403, 1355, 1134(?), 1110, 1053, 1044s, 982, 936, 712, 694, 626, 611, 461, 453.
Source: Kloprogge et al. (2002).
Comments: The sample was characterized by powder X-ray diffraction data. The empirical formula of the sample used is $Mg_{5.8}Al_{2.2}(OH)_{16}(CO_3)_{0.92}(NO_3)_{0.26} \cdot nH_2O$. For the Raman spectrum of hydrotalcite see also Palmer et al. (2011).

Hydrotungstite $WO_2(OH)_2 \cdot H_2O$

Origin: No data.
Experimental details: Raman scattering measurements have been performed on oriented and arbitrarily oriented samples in the spectral region from 150 to 1050 cm^{-1} using 488 nm Ar$^+$ laser radiation. The laser radiation power was in the range from 0.06 to 0.15 mW. A 180°-scattering geometry was employed.
Raman shifts (cm^{-1}): 956s, 677s, 661s, 580sh, 275w, 225w.
Source: Tarassov et al. (2002).
Comments: The Raman shifts were determined by us based on spectral curve analysis of the published spectrum.

Hydroxyapophyllite-(K) $KCa_4Si_8O_{20}(OH,F) \cdot 8H_2O$

Origin: Pune (Poonah) district, Maharashtra State, India.
Experimental details: Raman scattering measurements have been performed on an arbitrarily oriented sample using 633 nm He-Ne laser radiation. The laser radiation power is not indicated. The Raman shifts have been determined for the maxima of individual peaks obtained as a result of the spectral curve analysis.
Raman shifts (cm^{-1}): 3626w, 3614sh, 3557s, 3365sh, 3085sh, 3007s, 2893sh, 2813sh, 1705w, 1683sh, 1626w, 1523w, 1114sh, 1086sh, 1059s, 1043sh, 1007w, 970w, 846w, 791, 765, 663s, 633sh, 583s, 538, 511, 485w, 462w, 431s, 409sh, 373, 337, 325sh, 297, 266w, 228, 209s, 185, 161, 132, 123, 106.
Source: Frost and Xi (2012o).
Comments: No independent analytical data are provided for the sample used. For the Raman spectrum of hydroxyapophyllite-(K) see also Goryainov et al. (2012).

Hydroxycalciobetafite (?) $(Ca,U)_2(Ti,Nb)_2O_6(OH)$

Origin: Antanifotsy, Betafo district, Madagascar.
Experimental details: Methods of sample preparation are not described. Raman scattering measurements have been performed using a 633 nm He-Ne laser. The Raman shifts have been determined for the maxima of individual peaks obtained as a result of the spectral curve analysis. The laser radiation power is not indicated.
Raman shifts (cm^{-1}): 2184, 2075, 1956, 1584, 1456sh, 1327, 1161, 893, 810, 657, 601sh, 391, 325sh, 283, 218, 162, 149.
Source: Frost and Reddy (2010).
Comments: A metamict sample was used. For the Raman spectra of minerals and compounds related to the betafite group see also McMaster et al. (2013, 2014, 2015).

Hydroxycalciomicrolite $Ca_{1.5}Ta_2O_6(OH)$

Origin: Volta Grande pegmatite, Nazareno, Minas Gerais, Brazil (type locality).
Experimental details: Raman scattering measurements have been performed on an arbitrarily oriented sample using 532 nm solid-state laser radiation. The laser radiation power is not indicated.
Raman shifts (cm^{-1}): 3614, 3586, 1028w, 892w, 803, 665, 638sh, 564sh, 519s, 440sh, 419, 398sh, 342, 311s, 285, 243, 230, 211, 195.
Source: Andrade et al. (2016).
Comments: The sample was characterized by powder X-ray diffraction data and electron microprobe analyses. The crystal structure is solved. The empirical formula of the sample used is $(Ca_{1.48}Na_{0.06}Mn_{0.01})(Ta_{1.88}Nb_{0.11}Sn_{0.01})O_{6.00}[(OH)_{0.76}F_{0.20}O_{0.04}]$. The Raman shifts were partly determined by us based on spectral curve analysis of the published spectrum.

Hydroxycalciopyrochlore $(Ca,Na,U,\square)_2(Nb,Ti)_2O_6(OH)$

Origin: Bližná, Southwestern Czech Republic.
Experimental details: Raman scattering measurements have been performed on an arbitrarily oriented sample in the spectral region from 70 to 6300 cm^{-1} using 532.17 nm laser radiation. The nominallaser radiation power was 100 mW.
Raman shifts (cm^{-1}): 3850, 3670sh, 2065s, 770, 635sh, 240, 95.
Source: Drábek et al. (2017).
Comments: The sample was characterized by powder X-ray diffraction data and electron microprobe data. The empirical formula of the sample used is $(Ca_{0.48}Na_{0.02}Mg_{0.06}Mn_{0.01}Y_{0.06}REE_{0.27}Th_{0.27}U_{0.01})(Nb_{1.06}Ti_{0.79}Fe_{0.14}W_{0.01})(O_{4.96}OH_{1.04})(OH)_{0.81} \cdot H_2O$. The Raman shifts were determined by us based on spectral curve analysis of the published spectrum. The band at 2065 cm^{-1} may be due to fluorescence.

Hydroxycalcioroméite $(Ca,Sb^{3+})_2(Sb^{5+},Ti)_2O_6(OH)$

Origin: No data.
Experimental details: Raman scattering measurements have been performed on an arbitrarily oriented sample using laser radiation. The laser radiation power is not indicated. The Raman shifts have been determined for the maxima of individual peaks obtained as a result of the spectral curve analysis.
Raman shifts (cm^{-1}): 3542sh, 3471w, 3328sh, 3200sh, 2920w, 2860w, 1965w, 1599w, 1447w, 1353w, 984, 806, 725, 611, 565, 517s, 507sh, 487sh, 400sh, 356, 300, 179sh, 150sh, 129s.
Source: Bahfenne and Frost (2010b).
Comments: No independent analytical data are provided for the sample used. The Raman shifts were partly determined by us based on spectral curve analysis of the published spectrum.

Hydroxyferroroméite $(Fe^{2+}_{1.5}\square_{0.5})Sb^{5+}_2O_6(OH)$

Origin: Correc d'en Llinassos, Oms, Pyrénées-Orientales Department, France (type locality).
Experimental details: Raman scattering measurements have been performed on an arbitrarily oriented single crystal using 532 nm laser radiation. The nominal laser radiation power was 80 mW.
Raman shifts (cm^{-1}): 3634w, 3074w, 2936w, 1773w, 1706w, 1608, 709sh, 650s, 568sh, 466sh, 436, 358w, 271sh, 180.
Source: Mills et al. (2017b).

Comments: The sample was characterized by powder X-ray diffraction data, electron microprobe analyses, and X-ray photoelectron spectroscopy. The empirical formula of the sample used is $(Fe^{2+}_{1.07}Cu^{2+}_{0.50}Zn_{0.03}Sr_{0.03}Ca_{0.01}\square_{0.36})(Sb^{5+}_{1.88}Si_{0.09}Al_{0.02}As_{0.01})O_6[(OH)_{0.86}O_{0.14}]$.

Hydroxykenoelsmoreite $(\square,Pb)_2(W,Fe^{3+},Al)_2(O,OH)_6(OH)$

Origin: Masaka gold mine, Burundi (type locality).
Experimental details: Methods of sample preparation are not described. Raman scattering measurements have been performed using 782 nm diode laser radiation. The nominal laser radiation power was 1.15 mW.
Raman shifts (cm^{-1}): 3443, 2932w, 1610w, 929s, 853sh, 691, 476, 402, 298w, 225, 157.
Source: Mills et al. (2017a).
Comments: The sample was characterized by powder and single crystal X-ray diffraction data and electron microprobe analyses. The crystal structure is solved. The empirical formula of the sample used is $(\square_{1.668}Pb_{0.315}Ca_{0.009}Na_{0.005}K_{0.003}Ba_{0.001})(W^{6+}_{1.487}Fe^{3+}_{0.357}Al_{0.156})[O_{4.119}(OH)_{1.881}](OH)$.

Hydroxylapatite $Ca_5(PO_4)_3OH$

Origin: Tadano, Fukushima prefecture, Japan.
Experimental details: Raman scattering measurements have been performed on an oriented single crystal thin section cut nearly perpendicular to the c axis using 532 nm laser radiation. The laser radiation utput power was 11.8 mW.
Raman shifts (cm^{-1}): 3561, 3537, 1080w, 1059, 1035, 968s, 612, 586, 449, 433.
Source: Banno et al. (2016).
Comments: The sample was characterized by powder X-ray diffraction data and electron microprobe analyses. The empirical formula of the sample used is $Ca_{5.022}(P_{2.943}Si_{0.024}S_{0.011})O_{11.960}(OH_{0.546}F_{0.406}Cl_{0.048})$. The Raman shifts were partly determined by us based on spectral curve analysis of the published spectrum. For the Raman spectra of hydroxylapatite see also Penel et al. (1998), Koutsopoulos (2002), O'Neill et al. (2006), and Pasteris et al. (2012).

Hydroxylbastnäsite-(Ce) $Ce(CO_3)OH$

Origin: Trimouns, Luzenac, France.
Experimental details: Raman scattering measurements have been performed on an arbitrarily oriented sample using 532 nm laser radiation. The laser radiation power is not indicated.
Raman shifts (cm^{-1}): 3637, 3567, 3492, 3257, 1475, 1431sh, 1415, 1100s, 1088s, 1082s, 1007, 872, 851, 786, 728w, 694, 665w, 582, 342, 257, 239sh, 195, 177, 156, 137, 118, 70w.
Source: Yang et al. (2008a).
Comments: The sample was characterized by single-crystal X-ray diffraction data and electron microprobe analyses. The crystal structure is solved. The empirical formula of the sample used is $(Ce_{0.50}Nd_{0.24}La_{0.23}Y_{0.03})(CO_3)[(OH)_{0.65}F_{0.35}]$. The Raman shifts were determined by us based on spectral curve analysis of the published spectrum. For the Raman spectra of hydroxylbastnäsite-(Ce) see also Frost and Dickfos (2007a) and Michiba et al. (2013).

Hydroxylchondrodite $Mg_5(SiO_4)_2(OH)_2$

Origin: Synthetic.
Experimental details: Raman scattering measurements have been performed on an arbitrarily oriented single crystal using 514.5 nm Ar$^+$ laser radiation. The laser radiation power at the sample was about 60 mW.
Raman shifts (cm^{-1}): 3571sh, 3554s, 3515s, 3226, 2930, 955, 921, 848sh, 835s, 754w, 723w, 597, 570, 535w, 473w, 420w, 368w, 327w, 282w, 222w.
Source: Lin et al. (1999).
Comments: The sample was characterized by powder X-ray diffraction data. For the Raman spectrum of hydroxylchondrodite see also Mernagh et al. (1999).

Hydroxylclinohumite $Mg_9(SiO_4)_4(OH)_2$

Origin: Synthetic.
Experimental details: Raman scattering measurements have been performed on an arbitrarily oriented crystal in the spectral regions from 100 to 1600 cm^{-1} and from 2700 to 4000 cm^{-1} using 514.5 nm Ar$^+$ laser radiation. The laser radiation power at the sample was 5 mW.
Raman shifts (cm^{-1}): 3612w, 3580sh, 3564s, 3527s, 964, 856s, 847s, 838s, 826s, 765w, 737w, 716sh, 603, 583, 576w, 566w, 545w, 533sh, 499w, 468w, 456w, 430, 398w, 388sh, 378w, 364w, 342sh, 331w, 301w, 265w, 250w, 230w, 210w, 181w, 154w, 140w.
Source: Lin et al. (2000).
Comments: The sample was characterized by powder X-ray diffraction data and electron microprobe analyses. For the Raman spectrum of hydroxylclinohumite see also Hurai et al. (2014).

Hydroxylclinohumite $Mg_9(SiO_4)_4F_2$

Origin: Namibwuste, Namibia.
Experimental details: Raman scattering measurements have been performed on arbitrarily oriented crystals using a 633 nm He-Ne laser. The Raman shifts have been determined for the maxima of individual peaks obtained as a result of the spectral curve analysis.
Raman shifts (cm^{-1}): 3579, 3570, 3560, 3412, 3390, 967, 877sh, 862s, 846, 831, 808w, 785, 760, 744, 607, 587.
Source: Frost et al. (2007k).
Comments: The sample described as "clinohumite" was characterized by powder X-ray diffraction data and electron microprobe analysis. Actually, the empirical formula $Mg_{7.35}Fe^{2+}_{0.13}Ti_{0.08}Ca_{0.50}Al_{0.46}(SiO_4)_4Cl_{0.05}(OH)_{1.61}$ corresponds to cation-deficient hydroxylclinohumite.

Hydroxyledgrewite $Ca_9(SiO_4)_4(OH)_2$

Origin: Upper Chegem caldera, Northern Caucasus, Kabardino-Balkaria, Russia (type locality).
Experimental details: Raman scattering measurements have been performed on an arbitrarily oriented single crystal using 514.5 nm laser radiation. The nominal laser radiation power was in the range from 30 to 50 mW. A 180°-scattering geometry was employed.
Raman shifts (cm^{-1}): 3550, 3475w, 923, 890w, 840w, 821sh, 814, 559, 527w, 419sh, 404, 394sh, 324w, 295, 256, 166, 160sh.
Source: Galuskin et al. (2012d).

Comments: The sample was characterized by powder and single-crystal X-ray diffraction data and electron microprobe analyses. The crystal structure is solved. The sample used is a member of the solid-solution series $Ca_9(SiO_4)_4(F,OH)_2$ with F:OH = 63:37.

Hydroxylellestadite $Ca_5(SiO_4)_{1.5}(SO_4)_{1.5}OH$

Origin: Cioclovina Cave, Şureanu Mts., Romania.
Experimental details: Raman scattering measurements have been performed on an arbitrarily oriented sample using 514.5 nm Ar^+ laser radiation. The nominal laser radiation power was 100 mW. A 180°-scattering geometry was employed.
Raman shifts (cm^{-1}): 3564, 3517sh, 1144, 1122, 1066w, 1002s, 954s, 853s, 642sh, 625, 579, 530, 462, 431sh, 397sh, 312w.
Source: Onac et al. (2006).
Comments: The sample was characterized by single-crystal X-ray diffraction data and electron microprobe analyses. The crystal structure is solved. The empirical formula of the sample used is $Ca_{10.27}[(SiO_4)_{2.53}(SO_4)_{2.17}(PO_4)_{1.27}][(OH)_{1.66}F_{0.21}Cl_{0.16}]$. For the Raman spectrum of hydroxylellestadite see also Comodi et al. (1999).

Hydroxylherderite $CaBe(PO_4)(OH)$

Origin: Bennett pegmatite, Buckfield, Oxford Co., Maine, USA.
Experimental details: Raman scattering measurements have been performed on an arbitrarily oriented single crystal with polarizers and without polarizers using 458 nm laser radiation. The laser radiation power at the sample was 18 mW. Polarized spectra were collected in scattering geometries with three different polarized angles, 0, 45 and 90° from an arbitrary reference plane.
Raman shifts (cm^{-1}): 3620s, 3610sh, 3575sh, 3565s, 1142s, 1130sh, 1087w, 1005s, 985sh, 915, 910, 875sh, 770, 705sh, 680, 615sh, 598s, 590sh, 580, 532, 519, 448, 428w, 353w, 340w, 328w, 305w, 272w, 257w, 228w, 197w, 185w, 170w, 145w, 125w.
Source: Gatta et al. (2014).
Comments: The sample was characterized by single-crystal X-ray and neutron diffraction data, electron microprobe analyses and inductively coupled plasma-atomic emission spectroscopy. The crystal structure is solved. The empirical formula of the sample used is $(Ca_{1.01}Na_{0.01})(Be_{0.98}Li_{0.01})(P_{0.98}Si_{0.03})O_4[(OH)_{0.67}F_{0.33}]$. For the Raman spectrum of hydroxylherderite see also Frost et al. (2014a).

Hydrozincite $Zn_5(CO_3)_2(OH)_6$

Origin: No data.
Experimental details: Raman scattering measurements have been performed on an arbitrarily oriented sample using 632.8 nm He-Ne or 514.5 nm Ar^+ laser radiation. The nominal laser radiation power was 30 mW or less than 30 mW, respectively.
Raman shifts (cm^{-1}): 1544, 1371, 1061s, 732, 704, 389, 340w, 230, 152s, 139, 121, 81.
Source: Bouchard and Smith (2003).
Comments: No independent analytical data are provided for the sample used.

Hypersthene $(Mg,Fe)SiO_3$

Origin: Pietra Nera, Agrigento, Sicilia, Italy.
Experimental details: Raman scattering measurements have been performed on an arbitrarily oriented sample using 532 nm solid-state laser radiation. The laser radiation power is not indicated. A nearly 180°-scattering geometry was employed.
Raman shifts (cm^{-1}): 1006s, 681s, 661, 339, 233, 129.
Source: Andò and Garzanti (2014).
Comments: No independent analytical data are provided for the sample used.

Ianbruceite $Zn_2O[AsO_3(OH)](H_2O)_{3.53}$

Origin: Tsumeb mine, Otjikoto (Oshikoto) region, Namibia (type locality).
Experimental details: Raman scattering measurements have been performed on an arbitrarily oriented sample using 532 nm laser radiation. The nominal laser radiation power was 12.5 mW. A 180-°-scattering geometry was employed.
Raman shifts (cm^{-1}): 3600w, 3441sh, 3224w, 2740sh, 840s, 773sh, 534, 448, 420, 192.
Source: Cooper et al. (2012a).
Comments: The sample was characterized by powder X-ray diffraction data and electron microprobe analyses. The crystal structure is solved. The crystal-chemical formula of the sample used is $K_{0.02}(Zn_{1.93}Fe^{2+}_{0.03}Al_{0.02}Mn^{2+}_{0.01})(OH)_{0.96}(H_2O)(As^{5+}O_4)[As^{3+}(OH)_2O]_{0.04}(H_2O)_{1.96}$. The Raman shifts were partly determined by us based on spectral curve analysis of the published spectrum.

Iangreyite $Ca_2Al_7(PO_4)_2(PO_3OH)_2(OH,F)_{15} \cdot 8H_2O$

Origin: Silver Coin mine, Nevada, USA (type locality).
Experimental details: Raman scattering measurements have been performed on an arbitrarily oriented sample using 785 nm laser radiation. The laser radiation power is not indicated.
Raman shifts (cm^{-1}): 1345, 1200, 1095, 1078w, 1033, 1009s, 979, 923, 707, 622s, 510, 485w, 455, (377), 362, 338, 332, 271, 189s, 107, 114.
Source: Mills et al. (2011a).
Comments: The sample was characterized by powder X-ray diffraction data and electron microprobe analyses. The crystal structure is solved. The empirical formula of the sample used is $Ca_{1.42}K_{0.22}Na_{0.09}Ba_{0.03}Sr_{0.01}Al_{6.51}Mg_{0.09}Fe_{0.02}Cu_{0.01}Zn_{0.01}P_{3.81}F_{5.24}H_{30.21}O_{33.76}$. In the cited paper an erroneous figure of the Raman spectrum with displaced scale of Raman shifts is given. The Raman shifts listed above have been determined based on the analysis the correct spectral curve from the manuscript submitted to the Mineralogical Magazine, which was kindly provided by the authors.

Ice H_2O

Origin: Artificial.
Experimental details: Raman scattering measurements have been performed at -190 °C on an arbitrarily oriented sample in the spectral region from 2800 to 4000 cm^{-1} using 532.2 nm Nd-YAG laser radiation. The laser radiation power at the sample was in the range from 1 to 1.5 mW.
Raman shifts (cm^{-1}): 3370sh, 3321w, 3218, 3090s.

Source: Baumgartner and Bakker (2010).
Comments: For the Raman spectra of ice see also Giguére and Harvey (1956) and Garcia et al. (2006). The Raman shifts were partly determined by us based on spectral curve analysis of the published spectrum.

Idaite Cu_3FeS_4

Origin: No data.
Experimental details: No data.
Raman shifts (cm^{-1}): 466s, 403w, 356w, 265w, 209.
Source: Parker et al. (2008).
Comments: No independent analytical data are provided for the sample used. The Raman shifts were partly determined by us based on spectral curve analysis of the published spectrum.

Idrialite $C_{22}H_{14}$

Origin: Idrija mercury ore field, External Dinarides, Slovenia (type locality).
Experimental details: Raman scattering measurements have been performed on an arbitrarily oriented sample using 1064 nm Nd-YAG and 785 nm diode laser radiations. The laser radiation power is not indicated.
Raman shifts (cm^{-1}): 3049, 2905sh, 2891, 1617s, 1579s, 1445w, 1437sh, 1428sh, 1393s, 1375, 1367, 1352w, 1301, 1268, 1209, 1183, 1151, 1017s, 1009w, 911w, 960w, 825, 752s, 728sh, 710s, 679sh, 663, 642, 611, 599, 590sh, 579, 563sh, 552, 522, 498, 130 (?).
Source: Jehlička et al. (2006).
Comments: The sample was characterized by powder X-ray diffraction data.

Ikaite $Ca(CO_3)\cdot 6H_2O$

Origin: Ika fjord, Ivigtut, southern Greenland (type locality).
Experimental details: Raman scattering measurements have been performed at -80 °C on an arbitrarily oriented samplein the spectral region from 400 to 4000 cm^{-1} using 532 nm Nd-YAG laser radiation. The nominal laser radiation power was 8 mW.
Raman shifts (cm^{-1}): 3421, 3240, 3182, 1483w, 1072s, 873w, 722.
Source: Coleyshaw et al. (2003).
Comments: The sample was characterized by powder X-ray diffraction data. For the Raman spectra of ikaite see also Mikkelsen et al. (1999), Shahar et al. (2005), and Sánchez-Pastor et al. (2016).

Ikaite $Ca(CO_3)\cdot 6H_2O$

Origin: Synthetic.
Experimental details: Raman scattering measurements have been performed at 4°C on an arbitrarily oriented sample using 532 nm Nd-YAG laser radiation. The nominal laser radiation power was 10 mW. The Raman shifts have been determined for the maxima of individual peaks obtained as a result of the spectral curve analysis.
Raman shifts (cm^{-1}): 3432, 3336, 3257sh, 3165sh, 1066s, 715, 263, 214sh, 199s, 183, 156sh, 137, 116sh.
Source: Sánchez-Pastor et al. (2016).

Comments: The sample was characterized by powder X-ray diffraction analysis. For the Raman spectra of ikaite see also Mikkelsen et al. (1999), Coleyshaw et al. (2003), and Shahar et al. (2005).

Ilesite $Mn^{2+}(SO_4) \cdot 4H_2O$

Origin: Artificial (degradation product from black slag).
Experimental details: Raman scattering measurements have been performed on an arbitrarily oriented sample in the spectral region from 200 to 2000 cm^{-1} using 514.5 nm laser radiation. The laser radiation power is not indicated.
Raman shifts (cm^{-1}): 1024s, 622, 488, 427, 263, 207.
Source: Gómez-Nubla et al. (2013).
Comments: No independent analytical data are provided for the sample used.

Ilmenite $Fe^{2+}Ti^{4+}O_3$

Origin: Lunar basalt from Taurus–Littrow floor, 15 m northeast of a 10 m diameter crater with blocky ejecta.
Experimental details: Raman scattering measurements have been performed on an arbitrarily oriented sample using 532 nm laser radiation. The laser radiation power is not indicated.
Raman shifts (cm^{-1}): 679s, 449, 370, 330, 254, 227, 200w, 190w, 162w.
Source: Ling et al. (2011).
Comments: No independent analytical data are provided for the sample used. The Raman shifts were partly determined by us based on spectral curve analysis of the published spectrum. For the Raman spectrum of ilmenite see also Andò and Garzanti (2014).

Ilvaite $CaFe^{3+}Fe^{2+}_2O(Si_2O_7)(OH)$

Origin: Sasa, Macedonia.
Experimental details: Raman scattering measurements have been performed on a powdered sample in the spectral region from 100 to 1300 cm^{-1} using 514.5 nm Ar$^+$ laser radiation. The laser radiation power is not indicated.
Raman shifts (cm^{-1}): 1084s, 614s, 563w, 530w, 492w, 440w, 370w, 223s, 153w, 140.
Source: Makreski et al. (2007).
Comments: The sample was characterized by powder X-ray diffraction data and electron microprobe analyses. For the Raman spectrum of ilvaite see also Jovanovski et al. (2009).

Imogolite $Al_2SiO_3(OH)_4$

Origin: Synthetic.
Experimental details: Raman scattering measurements have been performed on a powdered sample using 1024 nm laser radiation. The nominal laser radiation power was 800 mW. A 180°-scattering geometry was employed.
Raman shifts (cm^{-1}): 959, 925, 866, 698w, 550sh, 514s, 400sh, 367s, 251, 118s, 107.
Source: Creton et al. (2008).
Comments: The sample was characterized by inductive coupled plasma atomic emission spectroscopy analysis. The wavelength of 1024 nm indicated by the authors may be a misprint. The Raman shifts were determined by us based on spectral curve analysis of the published spectrum.

Inderite $MgB_3O_3(OH)_5 \cdot 5H_2O$

Origin: Boron, Kern Co., California, USA.

Experimental details: Raman scattering measurements have been performed on arbitrarily oriented crystals using a 633 nm He-Ne laser. The laser radiation power is not indicated. The Raman shifts have been determined for the maxima of individual peaks obtained as a result of the spectral curve analysis.

Raman shifts (cm^{-1}): 3616s, 3479sh, 3429, 3365shg, 3292w, 3254, 3127, 3067, 2988w, 2931, 2843w, 2750w, 1397w, 1349, 1282sh, 1249, 1195, 1167sh, 1138sh, 1058w, 1006, 948s, 879, 811w, 743, 664, 637, 580, 551, 492s, 440, 420, 379, 347w, 313, 248w, 193, 166.

Source: Kloprogge and Frost (1999a).

Comments: No independent analytical data are provided for the sample used. For the Raman spectrum of inderite see also Frost et al. (2013f).

Indite $FeIn_2S_4$

Origin: Synthetic.

Experimental details: Methods of sample preparation are not described. Raman scattering measurements have been performed using 785 nm diode laser radiation. The laser radiation power is not indicated. The Raman shifts have been determined for the maxima of individual peaks obtained as a result of the spectral curve analysis.

Raman shifts (cm^{-1}): 370s, 334sh, 329sh, 313, 271sh, 253s, 231sh, 182, 167w, 96, 82.

Source: Guc et al. (2012).

Comments: The sample was characterized by electron microprobe analysis.

Inesite $Ca_2Mn^{2+}_7Si_{10}O_{28}(OH)_2 \cdot 5H_2O$

Origin: N'chwaning mine, Kalahari Manganese Fields, South Africa.

Experimental details: Raman scattering measurements have been performed on arbitrarily oriented crystals using a 633 nm He-Ne laser. The laser radiation power is not indicated. The Raman shifts have been determined for the maxima of individual peaks obtained as a result of the spectral curve analysis.

Raman shifts (cm^{-1}): 3661sh, 3642, 3612sh, 3496sh, 3420, 3362sh, 3300sh, 3246sh, 1856sh, 1825, 1775, 1730sh, 1671sh, 1653, 1608, 1546sh, 1383sh, 1365w, 1207sh, 1090sh, 1067sh, 1051sh, 1031s, 997sh, 958w, 933, 907sh, 764sh, 736sh, 716sh, 684sh, 653s, 631sh, 608sh, 467sh, 448sh, 428sh, 410, 374sh, 354, 301sh, 283, 248, 218, 156sh, 140s, 114s.

Source: Frost et al. (2014r).

Comments: No independent analytical data are provided for the sample used.

Innsbruckite $Mn_{33}(Si_2O_5)_{14}(OH)_{38}$

Origin: Staffelsee (Geier), Navis valley, Tyrol, Austria (type locality).

Experimental details: Methods of sample preparation are not described. Raman scattering measurements have been performed on an arbitrarily oriented sample in the spectral regions from 100 to 1250 cm^{-1} and from 3520 to 3700 cm^{-1} using 532 nm Nd-YAG laser radiation. The nominal laser radiation power was 30 mW. Raman spectra were collected inscattering geometries with a 90° sample rotation between the data collections.

Raman shifts (cm^{-1}): 1190w, 1032s, 1016w, 1010w, 998w, 792w, 777w, 717w, 693w, 671w, 649, 629, 609, 477w, 455w, 417, 405, 394, 360w, 352w, 336sh, 327w, 322w, 315, 312w, 305sh, 288, 277, 259, 252, 231, 221, 199sh, 193, 185, 178sh, 169, 160, 137, 131sh, 118w, 114w, 108w.
Source: Krüger et al. (2014).
Comments: The sample was characterized by single-crystal synchrotron radiation diffraction analysis and electron microprobe analyses. The crystal structure is solved. The empirical formula of the sample used is $Mn_{31.58}Fe_{0.19}Mg_{1.29}Si_{27.82}Al_{0.20}O_{108}H_{37.97}$.

Insizwaite $PtBi_2$

Origin: No data.
Experimental details: Methods of sample preparation are not described. Raman scattering measurements have been performed using 532.1 nm Nd-YAG laser radiation. The nominal laser radiation power was 100 mW.
Raman shifts (cm^{-1}): 126, 114s, 96.
Source: Bakker (2014).
Comments: No independent analytical data are provided for the sample used.

Inyoite $CaB_3O_3(OH)_5 \cdot 4H_2O$

Origin: Mount Blanco mine, Black Mountains, Death Valley, Inyo Co., California, USA.
Experimental details: Raman scattering measurements have been performed on an arbitrarily oriented sample using 633 nm He-Ne laser radiation. The laser radiation power is not indicated. The Raman shifts have been determined for the maxima of individual peaks obtained as a result of the spectral curve analysis.
Raman shifts (cm^{-1}): 3444w, 3389sh, 3153w, 2828w, 1689sh, 1656w, 1430w, 1376w, 1336sh, 1322w, 1254w, 1204w, 1177sh, 1062w, 1048sh, 1013w, 971sh, 957, 925sh, 910, 808w, 731, 615s, 596sh, 535, 521sh, 503, 474sh, 465, 408, 388, 352, 326w, 268sh, 258w, 206sh, 192sh, 182sh, 1721, 160sh.
Source: Frost et al. (2015i).
Comments: The sample was characterized by powder X-ray diffraction data and electron microprobe analysis.

Iodargyrite AgI

Origin: Synthetic.
Experimental details: Raman scattering measurements have been performed on an oriented single crystal with the c axis normal to the scattered plane using 457.9, 488.0, 546.1 nm Ar$^+$ and 568.2, 647.1 nm Kr$^+$ laser radiations. The laser radiation power is not indicated. A 180°-scattering geometry was employed.
Raman shifts (cm^{-1}): 104w, 85w, 37w, 17s.
Source: Hanson et al. (1975).
Comments: No independent analytical data are provided for the sample used. The Raman shifts are given for the $z(yy)$–z scattering geometry.

Iowaite $Mg_6Fe^{3+}_2(OH)_{16}Cl_2 \cdot 4H_2O$

Origin: Australia.

Experimental details: Raman scattering measurements have been performed on a powdered sample using 633 nm He-Ne laser radiation. The laser radiation power is not indicated. The Raman shifts have been determined for the maxima of individual peaks obtained as a result of the spectral curve analysis.

Raman shifts (cm^{-1}): 3720w, 3707, 3700w, 3691sh, 3685, 3674sh.

Source: Reddy et al. (2010).

Comments: The sample was characterized by electron paramagnetic resonance. Raman shifts are given for region from 3600 to 3740 cm^{-1}.

Iowaite $Mg_6Fe^{3+}_2(OH)_{16}Cl_2 \cdot 4H_2O$

Origin: Mount Keith, Western Australia.

Experimental details: Raman scattering measurements have been performed on an arbitrarily oriented sample using 633 nm He-Ne laser radiation. The laser radiation power is not indicated. The Raman shifts have been determined for the maxima of individual peaks obtained as a result of the spectral curve analysis.

Raman shifts (cm^{-1}): 708sh, 690, 620, 527s, 495sh, 456, 430w, 386, 348w, 312, 298w, 282, 231, 188, 153, 146w, 140w, 132w.

Source: Frost et al. (2010d).

Comments: No independent analytical data are provided for the sample used. For the Raman spectrum of iowaite see also Reddy et al. (2010).

Iranite $CuPb_{10}(CrO_4)_6(SiO_4)_2(OH)_2$

Origin: No data.

Experimental details: Raman scattering measurements have been performed on an arbitrarily oriented sample using 785 nm Nd-YAG laser radiation. The laser radiation power is not indicated. The Raman shifts have been determined for the maxima of individual peaks obtained as a result of the spectral curve analysis.

Raman shifts (cm^{-1}): 916sh, 891sh, 865s, 846s, 818s, 790, 535sh, 404, 389, 380sh, 369s, 354sh, 343, 333, 307, 240, 222, 196, 163, 139.

Source: Frost (2004c).

Comments: No independent analytical data are provided for the sample used.

Irarsite IrAsS

Origin: Santa Elena Nappe, Costa Rica.

Experimental details: Raman scattering measurements have been performed on an arbitrarily oriented sample using 532 nm Nd-YAG laser radiation. The laser radiation power is not indicated.

Raman shifts (cm^{-1}): 399, 375s, 353s, 289, 263s, 219, 211sh, 177, 145.

Source: Zaccarini et al. (2010).

Comments: The sample was characterized by electron microprobe analyses. The Raman shifts were partly determined by us based on spectral curve analysis of the published spectrum.

Iriginite $(UO_2)Mo^{6+}_2O_7 \cdot 3H_2O$

Origin: Hervey's Range deposit, 55 km W of Townsville, Queensland, Australia.

Experimental details: Raman scattering measurements have been performed on an arbitrarily oriented sample using 785 nm Nd-YAG laser radiation. The laser radiation power at the sample was 1 mW. The Raman shifts have been determined for the maxima of individual peaks obtained as a result of the spectral curve analysis.

Raman shifts (cm^{-1}): 965sh, 950s, 888, 826sh, 818, 693sh, 668, 487s, 457, 413, 373, 337, 301, 246s, 198, 164s.

Source: Frost et al. (2004c).

Comments: No independent quantitative analytical data are provided for the sample used.

Irinarassite $Ca_3Sn_2(SiAl_2)O_{12}$

Origin: Upper Chegem caldera, Northern Caucasus, Kabardino-Balkaria, Russia (type locality).

Experimental details: Raman scattering measurements have been performed on an arbitrarily oriented sample in the spectral region from 100 to 4000 cm^{-1} using 514.5 nm Ar$^+$ laser radiation. The laser radiation output power was in the range from 30 to 50 mW. A 0°-scattering geometry was employed. The Raman shifts have been determined for the maxima of individual peaks obtained as a result of the spectral curve analysis.

Raman shifts (cm^{-1}): 915w, 818, 787, 739, 578, 503s, 420, 316, 269sh, 250, 197, 156, 112w.

Source: Galuskina et al. (2013a).

Comments: The sample was characterized by single-crystal electron backscatter diffraction and electron microprobe analyses. The empirical formula of the sample used is $(Ca_{2.965}Fe^{2+}_{0.035})(Sn_{1.016}Zr_{0.410}Ti_{0.262}Sb^{5+}_{0.237}Fe^{2+}_{0.035}U^{6+}_{0.017}Sc_{0.014}Hf_{0.006}Nb_{0.004})(Al_{1.386}Fe^{3+}_{0.804}Si_{0.446}Ti^{4+}_{0.364})O_{12}$. The Raman shifts were partly determined by us based on spectral curve analysis of the published spectrum.

Iron Fe

Origin: Synthetic.

Experimental details: Raman scattering measurements have been performed on an arbitrarily oriented sample using 514.5 nm Ar$^+$ laser radiation. The laser radiation output power was 4–5 mW. A 180°-scattering geometry was employed.

Raman shifts (cm^{-1}): (187), 139w.

Source: Campos et al. (2004b).

Comments: The sample was characterized by powder X-ray diffraction data. The Raman shifts were determined by us based on spectral curve analysis of the published spectrum.

Iseite $Mn_2Mo_3O_8$

Origin: Synthetic.

Experimental details: Raman scattering measurements have been performed at 400 K on a powdered sample using 514.5 nm Ar$^+$ laser radiation. The laser radiation power is not indicated.

Raman shifts (cm^{-1}): 920s, 870, 815, 425w, 360, 325, 260w.

Source: Das et al. (2009).

Comments: The sample was characterized by powder X-ray diffraction data.

Isocubanite $CuFe_2S_3$

Origin: No data.
Experimental details: Raman scattering measurements have been performed on an arbitrarily oriented thin section of a sample using 532 nm laser radiation. The laser radiation power at the sample was about 20 mW.
Raman shifts (cm^{-1}): 440w, 386s, 350.
Source: White (2009).
Comments: No independent analytical data are provided for the sample used. For the Raman spectrum of isocubanite see also Chandra et al. (2011a, b).

Isokite $CaMg(PO_4)F$

Origin: Ehrenfriedersdorf, Erzgebirge (Ore Mts.), Germany.
Experimental details: Raman scattering measurements have been performed on an arbitrarily oriented sample using 514.5 nm Ar^+ laser radiation. The nominal laser radiation power was 150 mW.
Raman shifts (cm^{-1}): 1021, 955s, 607, 457, 425, 273.
Source: Thomas et al. (1998).

Isomertieite $Pd_{11}Sb_2As_2$

Origin: No data.
Experimental details: Methods of sample preparation are not described. Raman scattering measurements have been performed using 532 nm Nd-YAG laser radiation. The nominal laser radiation power was 100 mW.
Raman shifts (cm^{-1}): 136.
Source: Bakker (2014).
Comments: No independent analytical data are provided for the sample used.

Ivsite $Na_3H(SO_4)_2$

Origin: Synthetic.
Experimental details: Raman scattering measurements have been performed on a polycrystalline sample in the spectral region from 30 to 4000 cm^{-1} using 488 nm Ar^+ laser radiation. The laser radiation power is not indicated.
Raman shifts (cm^{-1}): 1198, 1162, 1112, 1004s, 973sh, 636s, 613s, 605s, 525w, 493w, 437s, 115, 80s.
Source: Damak et al. (1985).
Comments: The sample was characterized by powder X-ray diffraction data.

Iwateite $Na_2BaMn(PO_4)_2$

Origin: Tanohata mine, Iwate Prefecture, Japan (type locality).
Experimental details: Methods of sample preparation are not described. Raman scattering measurements have been performed using 514.5 nm Ar^+ laser radiation. The nominal laser radiation power was 100 mW.
Raman shifts (cm^{-1}): 1004w, 990w, 973s, 808w, 584w, 577w, 428w.

Source: Nishio-Hamane et al. (2014).
Comments: The sample was characterized by powder X-ray diffraction data and electron microprobe analyses. The empirical formula of the sample used is $Na_{2.026}(Ba_{0.993}Sr_{0.101})(Mn_{0.801}Mg_{0.164})P_{1.971}O_8$. The Raman shifts were partly determined by us based on spectral curve analysis of the published spectrum.

Iyoite $MnCuCl(OH)_3$

Origin: Sadamisaki Peninsula, Ehime Prefecture, Japan (type locality).
Experimental details: Methods of sample preparation are not described. Raman scattering measurements have been performed using 514.5 nm Ar^+ laser radiation. The nominal laser radiation power was 50 mW.
Raman shifts (cm^{-1}): 3558w, 3521s, 3513sh, 458, 438.
Source: Nishio-Hamane et al. (2016b).
Comments: The sample was characterized by powder X-ray diffraction data and electron microprobe analyses. The crystal structure is solved. The empirical formula of the sample used is $Mn_{1.085}Cu_{0.915}Cl_{1.058}(OH)_{2.942}$.

Jáchymovite $(UO_2)_8(SO_4)(OH)_{14} \cdot 13H_2O$

Origin: Jáchymov, Krušné Hory (Czech Ore Mts.), Bohemia, Czech Republic (type locality).
Experimental details: Raman scattering measurements have been performed on arbitrarily oriented crystals using a 633 nm He-Ne laser. The laser radiation power is not indicated. The Raman shifts have been determined for the maxima of individual peaks obtained as a result of the spectral curve analysis.
Raman shifts (cm^{-1}): 3504w, 3180w, 2257w, 1688w, 1614w, 1348w, 1125sh, 1094, 1068sh, 1015, 1010sh, 1003sh, 839sh, 828s, 807sh, 800sh, 667, 562, 542, 474sh, 454, 434, 405, 357sh, 337, 322sh, 276sh, 261, 252sh, 242sh, 208sh, 195, 172, 147, 114.
Source: Čejka et al. (2009a).
Comments: The sample was characterized by powder X-ray diffraction data and wet chemical analyses. The empirical formula of the sample used is $(UO_2)_{8.01}(SO_4)_{0.95}(OH)_{14.12} \cdot 13.06H_2O$.

Jacobsite $Mn^{2+}Fe^{3+}_2O_4$

Origin: Synthetic.
Experimental details: Raman scattering measurements have been performed on a powdered sample using 532 nm laser radiation. The laser radiation power is not indicated.
Raman shifts (cm^{-1}): 646s, 563sh, 456w, 340.
Source: Rafique et al. (2013).
Comments: The sample was characterized by powder X-ray diffraction data and electron microprobe analysis. For the Raman spectrum of jacobsitesee also Clark et al. (2007).

Jadeite $NaAlSi_2O_6$

Origin: Uru River area (?), north-central Myanmar.
Experimental details: Raman scattering measurements have been performed on an oriented thin section of a sample with the b axis parallel to the laser beam using 532 nm laser radiation. The nominal laser radiation power was 20 mW.

Raman shifts (cm^{-1}): 1309, 991, 700s, 575, 524, 434, 374, 292, 223, 203s, 144, 80.
Source: Leander et al. (2014).
Comments: The sample was characterized by electron microprobe analyses. For the Raman spectrum of jadeite see also Többens et al. (2005).

Jakobssonite α-CaAlF$_5$

Origin: Synthetic.
Experimental details: Raman scattering measurements have been performed on a pressed-disk sample in the spectral region from 280 to 800 cm^{-1} using 488 nm Ar$^+$ laser radiation. The laser radiation power is not indicated. A 90°-scattering geometry was employed.
Raman shifts (cm^{-1}): 588, 440.
Source: Kawamoto and Kono (1986) and Inoue et al. (1988).
Comments: The sample was characterized by powder X-ray diffraction data.

Jalpaite Ag$_3$CuS$_2$

Origin: No data.
Experimental details: Raman scattering measurements have been performed on an arbitrarily oriented sample using 514.5 nm Ar$^+$ laser radiation. The nominal laser radiation power was not higher than 0.3 mW.
Raman shifts (cm^{-1}): 258.
Source: De Caro et al. (2016).
Comments: The sample was characterized by electron microprobe analysis.

Jamborite Ni$^{2+}_{1-x}$Co$^{3+}_x$(OH)$_{2x}$(SO$_4$)$_x$·nH$_2$O [$x \leq 1/3$, $n \leq (1-x)$]

Origin: Rio Vesale, Sestola, Val Panaro, Modena province, Italy.
Experimental details: Raman scattering measurements have been performed on an arbitrarily oriented sample using 632.8 nm He-Ne laser radiation. The laser radiation power is not indicated.
Raman shifts (cm^{-1}): 3614w, 2988sh, 2956, 2244, 1061w, 973, 852w, 527s, 460, 167w.
Source: Bindi et al. (2015a).
Comments: The sample was characterized by powder X-ray diffraction data and electron microprobe analyses. The crystal structure is solved. The empirical formula of the sample used is (Ni$^{2+}_{0.902}$Co$^{3+}_{0.072}$Fe$^{3+}_{0.024}$Ca$_{0.002}$)(OH)$_{1.884}$Cl$_{0.012}$(SO$_4$)$_{0.100}$·0.904H$_2$O.

Jamesonite Pb$_4$FeSb$_6$S$_{14}$

Origin: Zlatá Baňa, Slanské Vrchy Mts., central Slovakia.
Experimental details: Raman scattering measurements have been performed on a polycrystalline sample in the spectral region from 10 to 600 cm^{-1} using 532 nm Nd-YAG laser radiation. The laser radiation power is not indicated. A 180°-scattering geometry was employed. The Raman shifts have been determined for the maxima of individual peaks obtained as a result of the spectral curve analysis.
Raman shifts (cm^{-1}): 344sh, 326s, 298, 277, 269sh, 251, 236, 225, 215sh, 199, 173, 163sh, 147, 128, 110, 92, 75, 59sh.
Source: Kharbish and Jeleň (2016).

Comments: The sample was characterized by electron microprobe analyses. The empirical formula of the sample used is $Pb_{4.01}Fe_{0.99}Sb_{6.01}S_{14.00}$.

Jarosite $KFe^{3+}_3(SO_4)_2(OH)_6$

Origin: No data.
Experimental details: Raman scattering measurements have been performed on an arbitrarily oriented thin section of a sample using 514.5 nm Ar$^+$ laser radiation. The laser radiation power at the sample was about 1 mW.
Raman shifts (cm^{-1}): 3415, 3395sh, 1156, 1102s, 1008s, 643sh, 626, 573, 551sh, 454sh, 434s, 354, 301, 223s, 140.
Source: Maubec et al. (2012).
Comments: The sample was characterized by powder X-ray diffraction data and electron microprobe analysis. The empirical formula of the sample used is $(K_{0.8}Na_{0.1})(Fe_{2.7}Al_{0.3})(SO_4)_{2.0}(OH)_{5.9}$. For the Raman spectra of jarosite see also Sasaki et al. (1998), Makreski et al. (2005a), Frost et al. (2006a), Murphy et al. (2009), Chio et al. (2010), Ciobotă et al. (2012), and Spratt et al. (2013).

Jeffbenite $Mg_3Al_2Si_3O_{12}$

Origin: An inclusion in an alluvial diamond, São Luizriver, Juina district, Mato Grosso, Brazil (type locality).
Experimental details: Raman scattering measurements have been performed on an arbitrarily oriented single crystal using 532 nm laser radiation. The nominal laser radiation power was in the range from 3 to 5 mW.
Raman shifts (cm^{-1}): 1056sh, 995, 926s, 865s, 635, 610sh, 542w, 499, 393w, 318s, 284, 233, 204.
Source: Nestola et al. (2016).
Comments: The sample was characterized by powder X-ray diffraction data and electron microprobe analyses. The crystal structure is solved. The empirical formula of the sample used is $(Mg_{0.82}Fe^{3+}_{0.12})(Al_{1.86}Cr_{0.16})(Mg_{1.80}Fe^{2+}_{0.15}Mn_{0.05}Ca_{0.01}Na_{0.01})(Si_{2.82}Al_{0.18})O_{12}$.

Jennite $Ca_9(Si_3O_9)_2(OH)_6 \cdot 8H_2O$

Origin: Maroldsweisach, Bavaria, Germany.
Experimental details: Raman scattering measurements have been performed on an oriented sample with longest axis corresponding to the [010] direction parallel and perpendicular with respect to laser polarization using 532 nm laser radiation. The nominal laser radiation power was 9 mW. The Raman shifts have been determined for the maxima of individual peaks obtained as a result of the spectral curve analysis.
Raman shifts (cm^{-1}): 3631, 3590sh, 3580s, 3534, 3489, 3464, ~3149, ~1640w, 1048, 1015sh, 1000s, 986s, 969s, 950sh, 906, 677sh, 658s, 632sh, 507sh, 492s, 479sh, 361sh, 335s, 312, 287, 270, 251, 204sh, 185s, 165sh, 142, 127, 113.
Source: Müller et al. (2015).
Comments: The sample was characterized by powder X-ray diffraction data and electron microprobe analyses. The empirical formula of the sample used is $Ca_{8.57-9.43}Si_{5.56-7.28}Al_{0.02-0.07}O_{18}(OH)_6 \cdot nH_2O$. The Raman shifts are given for total spectrum including all scattering geometries. For the Raman spectrum of jennite see also Kirkpatrick et al. (1997).

Ježekite $Na_8[(UO_2)(CO_3)_3](SO_4)_2 \cdot 3H_2O$

Origin: Jáchymov, Krušné Hory (Czech Ore Mts.), Bohemia, Czech Republic (type locality).
Experimental details: Methods of sample preparation are not described. Raman scattering measurements have been performed using 532 nm solid-state laser radiation. The nominal laser radiation power was 2.5 mW.
Raman shifts (cm^{-1}): 3680, 3380sh, 2740w, 1710w, 1656w, 1600w, 1550w, 1375w, 1355w, 1195w, 1130w, 1110sh, 1060s, 1050w, 996s, 896w, 825s, 731, 715, 688w, 629, 622, 458, 379sh, 277, 248, 188s, 161s, 85.
Source: Plášil et al. (2015a).
Comments: The sample was characterized by powder X-ray diffraction data and electron microprobe analyses. The crystal structure is solved. The empirical formula of the sample used is $Na_{7.88}(UO_2)(CO_3)_3(S_{1.01}O_4)_2 \cdot 3H_2O$.

Jixianite $(Pb,\square)_2(W,Fe^{3+})_2(O,OH)_7$

Origin: Yanhe Mine, Ji Co., Tianjin Municipality, China (type locality).
Experimental details: Methods of sample preparation are not described. Raman scattering measurements have been performed using 1064 nm Nd-YAG laser radiation. The laser radiation output power was in the range 300–380 mW.
Raman shifts (cm^{-1}): 907, 709, 433w, 371sh, 288, 174.
Source: Yuran and Li (1998).
Comments: No independent analytical data are provided for the sample used.

Joaquinite-(Ce) $NaBa_2Fe^{2+}Ti_2Ce_2(Si_4O_{12})_2O_2(OH) \cdot H_2O$

Origin: Benitoite Gem Mine, Southern San Benito Co., California (type locality).
Experimental details: Raman scattering measurements have been performed on an arbitrarily oriented sample using 633 nm He-Ne laser radiation. The laser radiation power is not indicated. The Raman shifts have been determined for the maxima of individual peaks obtained as a result of the spectral curve analysis.
Raman shifts (cm^{-1}): 3584sh, 3572sh, 3559sh, 3548w, 3509w, 3494sh, 3384sh, 3340sh, 3316w, 3242w, 1111, 1038, 1022, 991, 925, 902s, 891sh, 864w, 732s, 720sh, 686sh, 664s, 636s, 601s, 542, 492, 469, 440, 432, 377s, 358, 313, 299, 276, 250, 200, 180sh, 150.
Source: Frost and Pinto (2007).
Comments: The sample was characterized by powder X-ray diffraction data and electron microprobe analysis, however barium content is not indicated.

Joegoldsteinite $MnCr_2S_4$

Origin: Synthetic.
Experimental details: Raman scattering measurements have been performed on a polycrystalline sample hot-pressed as a pellet using 488.0 nm and 514.5 nm Ar^+ and 647.1 nm and 676.4 nm Kr^+ laser radiations. The laser radiation power is not indicated. A 180°-scattering geometry was employed.
Raman shifts (cm^{-1}): 378s, 282w, 251.
Source: Lutz et al. (1989).

Comments: The sample was characterized by powder X-ray diffraction data. The Raman shifts are given for the 647.1 nm radiation.

Joëlbruggerite $Pb_3Zn_3Sb^{5+}As_2O_{13}(OH)$

Origin: Black Pine mine, Montana, USA (type locality).
Experimental details: Methods of sample preparation are not described. Raman scattering measurements have been performed in the spectral region from 150 to 3500 cm^{-1} using 785 nm diode laser radiation. The laser radiation power is not indicated.
Raman shifts (cm^{-1}): 3030w, 3015w, 818, 680, 506, 475, 427, 383, 235, 200, 183, 168, 150.
Source: Mills et al. (2009).
Comments: The sample was characterized by powder X-ray diffraction data and electron microprobe analyses. The crystal structure is solved. The empirical formula of the sample used is $Pb_{3.112}(Zn_{2.689}Fe^{2+}_{0.185})(Sb^{5+}_{0.650}Te^{6+}_{0.451})(As_{1.551}P_{0.203}Si_{0.160})O_{13.335}(OH)_{0.665}$.

Johachidolite $CaAlB_3O_7$

Origin: An unknown locality in Myanmar.
Experimental details: Methods of sample preparation are not described. Raman scattering measurements have been performed using 514 nm laser radiation. The laser radiation power is not indicated.
Raman shifts (cm^{-1}): 1191, 1112, 684sh.
Source: Chadwick and Breeding (2008).
Comments: The sample was characterized by laser-ablation inductively coupled plasma mass spectroscopy analysis and by energy-dispersive X–ray fluorescence analysis.

Johannite $Cu(UO_2)_2(SO_4)_2(OH)_2 \cdot 8H_2O$

Origin: Saint Agnes, Cornwall, England.
Experimental details: Raman scattering measurements have been performed on arbitrarily oriented crystals using a 633 nm He-Ne laser. The laser radiation power is not indicated. The Raman shifts have been determined for the maxima of individual peaks obtained as a result of the spectral curve analysis.
Raman shifts (cm^{-1}): 3593, 3523w, 3387sh, 3234w, 1147, 1100, 1090, 1042, 975, 948sh, 812s, 788sh, 756sh, 539, 481, 384, 302s, 277s, 205s, 184sh.
Source: Frost et al. (2005e).
Comments: No independent analytical data are provided for the sample used. For the Raman spectrum of johannite see also Driscoll et al. (2014).

Johannite $Cu(UO_2)_2(SO_4)_2(OH)_2 \cdot 8H_2O$

Origin: Geevor mine, Pendeen, Cornwall, UK.
Experimental details: Methods of sample preparation are not described. Raman scattering measurements have been performed using 325 nm He-Cd, as well as 532 and 785 nm diode laser radiations. The laser radiation output power was 270, 380, and 370 mW, respectively.
Raman shifts (cm^{-1}): 1095, 1045, 836s, 448w, 352w, 244w, 203w.
Source: Driscoll et al. (2014).

Comments: The Raman shifts are given for 785 nm laser excitation. The sample was characterized by electron microprobe analysis. The proposed empirical formula of the sample used is $Cu_{1.4}(UO_2)_2(SO_4)_{1.8} \cdot nH_2O$. For the Raman spectrum of johannite see also Frost et al. (2005e).

Johnbaumite $Ca_5(AsO_4)_3(OH)$

Origin: Franklin or Sterling Hill, New Jersey, USA.
Experimental details: Methods of sample preparation are not described. Raman scattering measurements have been performed on an arbitrarily oriented sample using 633 nm laser radiation. The laser radiation power is not indicated.
Raman shifts (cm^{-1}): 960w, 888w, 865s, 840, 830.
Source: Crimmins (2012).
Comments: The sample was characterized by single-crystal X-ray diffraction data and electron microprobe analyses. The empirical formula of the sample used is $(Ca_{4.87}Pb_{0.07}Mn_{0.05}Sr_{0.01})[(As_{0.94}P_{0.06})O_4]_3[(OH)_{0.94}Cl_{0.06}]$. The Raman shifts were partly determined by us based on spectral curve analysis of the published spectrum.

Johninnesite $Na_2Mn^{2+}{}_9Mg_7(AsO_4)_2(Si_6O_{17})_2(OH)_8$

Origin: Schmorrasgrat deposit, Schams nappes, Val Ferrera, Graubünden, Switzerland.
Experimental details: Methods of sample preparation are not described. Raman scattering measurements have been performed using 514.5 nm Ar$^+$ laser radiation. The laser radiation power is not indicated.
Raman shifts (cm^{-1}): 1662w, 1616w, 1424w, 1322w, 1084w, 1055w, 1029, 1015, 939, 886w, 836s, 799s, 785s, 705, 667, 401, 347.
Source: Brugger and Berlepsch (1997).
Comments: The sample was characterized by powder X-ray diffraction data and electron microprobe analyses. The empirical formula of the sample used is $Na_{2.01}Mn^{2+}{}_{9.00}(Mg_{3.74}Mn_{2.66}Fe_{0.01})(As_{1.74}V_{0.03})Si_{12.58}O_{42}(OH)_8$. The Raman shifts were partly determined by us based on spectral curve analysis of the published spectrum.

Jordisite MoS_2

Origin: Zunyi Formation, southern China.
Experimental details: Experimental details are not indicated. Raman scattering measurements have been performed on an arbitrarily oriented sample. Raman spectrum was obtained in the spectral region from 150 to 1800 cm^{-1}.
Raman shifts (cm^{-1}): 600w, 438sh, 403 (h-MoS$_2$), 370 (h-MoS$_2$), 339w, 303w, 258sh, 216, 184sh, 148sh.
Source: Orberger et al. (2007).
Comments: The sample was characterized by a combination of methods including electron microprobe analysis, inductively coupled plasma mass spectroscopy, inductively coupled plasma atomic emission spectroscopy analysis, and proton-induced X-ray emission analysis.

Joteite $Ca_2CuAl(AsO_4)[AsO_3(OH)]_2(OH)_2 \cdot 5H_2O$

Origin: Jotemine, Tierra Amarilla, Copiapó Province, Atacama, Chile (type locality).
Experimental details: Raman scattering measurements have been performed on an arbitrarily oriented sample using 514.5 nm Ar^+ laser radiation. The laser radiation power at the sample was about 5 mW.
Raman shifts (cm^{-1}): 3429w, 3260w, 3068sh, 2930sh, 900sh, 861s, 849s, 822sh, 725, 521sh, 506s, 461s, 451sh, 414, 384, 349, 334, 317, 300, 283, 270, 259, 201, 162, 140, 119, 112sh.
Source: Kampf et al. (2013c).
Comments: The sample was characterized by powder X-ray diffraction data and electron microprobe analysis. The crystal structure is solved. The empirical formula of the sample used is $Ca_{1.98}Cu_{1.00}Al_{1.15}As_{2.87}H_{14.24}O_{19}$. The Raman shifts were partly determined by us based on spectral curve analysis of the published spectrum.

Kaersutite $NaCa_2(Mg_3AlTi^{4+})(Si_6Al_2)O_{22}O_2$

Origin: An unknown locality in Czech Republic.
Experimental details: Methods of sample preparation are not described. Raman scattering measurements have been performed using 532 nm Nd-YAG laser radiation. The nominal laser radiation power was 100 mW.
Raman shifts (cm^{-1}): 1183w, 1063sh, 1013w, 975w, 893w, 788sh, 764s, 666, 590s, 544sh, 514sh, 423sh, 363, 347sh, 331sh, 306sh, 292w, 249, 189w, 157, 144sh, 125w.
Source: Apopei et al. (2011).
Comments: No independent analytical data are provided for the sample used. For the Raman spectra of kaersutite see also Andò and Garzanti (2014) and Leissner et al. (2015).

Kainite $KMg(SO_4)Cl \cdot 3H_2O$

Origin: Synthetic.
Experimental details: Methods of sample preparation are not described. Raman scattering measurements have been performed using 514.5 nm or 785 nm laser radiation. The laser radiation power at the sample was no more than 1 mW.
Raman shifts (cm^{-1}): 1196w, 1040s, 1023s, 630w, 450w.
Source: Morillas et al. (2016).
Comments: The sample was characterized by electron microprobe analysis.

Kalgoorlieite As_2Te_3

Origin: Synthetic.
Experimental details: Raman scattering measurements have been performed on a powdered sample using 632.8 nm He-Ne laser radiation. The laser radiation output power was below 0.3 mW.
Raman shifts (cm^{-1}): 193, 171s, 142, 137sh, 128s, 123sh, 119sh, 99s, 91w, 67w, 49.
Source: Cuenca-Gotor et al. (2016).
Comments: The sample was characterized by powder X-ray diffraction data.

Kaliborite $KHMg_2B_{12}O_{16}(OH)_{10} \cdot 4H_2O$

Origin: Inder boron deposit, Atyrau region, Kazakhstan.
Experimental details: Raman scattering measurements have been performed on arbitrarily oriented crystals using a 633 nm He-Ne laser. The laser radiation power is not indicated. The Raman shifts have been determined for the maxima of individual peaks obtained as a result of the spectral curve analysis.
Raman shifts (cm^{-1}): 3603sh, 3597s, 3590sh, 3517sh, 3398sh, 3360sh, 3336, 3245sh, 3202, 3172sh, 3133, 3041sh, 2929, 1595w, 1448sh, 1444, 1309w, 1229w, 1144, 1084sh, 1065, 967sh, 944, 881, 847w, 793w, 775sh, 756, 670, 639, 630sh, 609, 567, 551, 526, 519sh, 485, 476sh, 454w, 426sh, 414sh, 402, 394sh, 387, 337, 327sh, 320, 310sh, 286w, 259, 252sh, 242, 218, 197, 184, 177sh, 1709sh, 151, 121.
Source: López et al. (2015a).
Comments: No independent analytical data are provided for the sample used.

Kalicinite $KH(CO_3)$

Origin: Synthetic.
Experimental details: Methods of sample preparation are not described. Raman scattering measurements have been performed on a powder sample using Ar^+ laser radiation. The laser radiation power at the sample was about 5 mW.
Raman shifts (cm^{-1}): 1037s, 678, 637.
Source: Kagi et al. (2003).
Comments: The sample was characterized by powder X-ray diffraction data. The crystal structure is solved by the Rietveld method. For the Raman spectrum of kalicinite see also Frezzotti et al. (2012).

Kalininite $ZnCr_2S_4$

Origin: Synthetic
Experimental details: Raman scattering measurements have been performed on a polycrystalline sample using 647.1 nm Kr^+ laser radiation. The laser radiation power is not indicated. A 180-°-scattering geometry was employed.
Raman shifts (cm^{-1}): 392s, 293, 260s.
Source: Lutz et al. (1989).
Comments: The sample was characterized by powder X-ray diffraction analysis. For the Raman spectrum of kalininite see also Kushwaha (2008).

Kalinite $KAl(SO_4)_2 \cdot 11H_2O$

Origin: No data.
Experimental details: Raman scattering measurements have been performed on arbitrarily oriented crystals using a 633 nm He-Ne laser. The laser radiation power is not indicated. The Raman shifts have been determined for the maxima of individual peaks obtained as a result of the spectral curve analysis.
Raman shifts (cm^{-1}): 3528sh, 3379, 1678, 1630, 1132, 1104sh, 990s, 975, 618, 536sh, 501sh, 454, 440sh, 327w.
Source: Frost and Kloprogge (2001).
Comments: No independent analytical data are provided for the sample used.

Kaliophilite $KAlSiO_4$

Origin: Artificial.
Experimental details: Raman scattering measurements have been performed on an arbitrarily oriented sample using 514.5 nm Kr^+-Ar^+ laser radiation. The laser radiation power at the sample was no more than 2 mW. A 180°-scattering geometry was employed.
Raman shifts (cm^{-1}): 402, 357.
Source: Jay and Cashion (2013).
Comments: The sample was characterized by electron microprobe analysis.

Kalsilite $KAlSiO_4$

Origin: San Venanzo, Terni province, Umbria, Italy.
Experimental details: No data.
Raman shifts (cm^{-1}): ~350s.
Source: Uchida et al. (2006).
Comments: The sample was characterized by single-crystal X-ray diffraction data. The crystal structure is solved. The empirical formula of the sample used is $(K_{0.92}Na_{0.07})(Al_{0.93}Fe^{3+}_{0.04}Si_{1.03})O_4$.

Kamacite (Fe,Ni)

Origin: Almahata Sitta meteorite.
Experimental details: Raman scattering measurements have been performed on an arbitrarily oriented sample using 532 nm Nd-YAG laser radiation. The nominal laser radiation power was 22.5 mW. A 180°-scattering geometry was employed.
Raman shifts (cm^{-1}): 275, 214, 175sh, 153sh.
Source: Kaliwoda et al. (2013).
Comments: The sample was characterized by electron microprobe analyses. It contains about 4 wt% Ni and 0.2 wt% Co. The Raman shifts were determined by us based on spectral curve analysis of the published spectrum.

Kamotoite-(Y) $Y_2O_4(UO_2)_4(CO_3)_3 \cdot 14H_2O$

Origin: Kamoto East open cut, Kolwezi, Shaba Cu belt, Democratic Republic of Congo (type locality).
Experimental details: Raman scattering measurements have been performed on arbitrarily oriented crystals using a 633 nm He-Ne laser. The laser radiation power is not indicated. The Raman shifts have been determined for the maxima of individual peaks obtained as a result of the spectral curve analysis.
Raman shifts (cm^{-1}): 3516w, 3361w, 1634sh, 1551w, 1338w, 1131, 1125sh, 815s, 810sh, 745, 584, 547sh, 418, 336.
Source: Frost et al. (2006k).
Comments: No independent analytical data are provided for the sample used.

Kampelite $Ba_6Mg_3Sc_8(PO_4)_{12}(OH)_6 \cdot 7H_2O$

Origin: Kovdor phoscorite-carbonatite complex, Kola Peninsula, Russia (type locality).
Experimental details: Methods of sample preparation are not described. Raman scattering measurements have been performed using 514.5 nm Ar$^+$ laser radiation. The laser radiation power is not indicated.
Raman shifts (cm^{-1}): 1604w, 1092, 975s, 932sh, 848, 715, 591, 456, (402), 297w, 172, 77s.
Source: Yakovenchuk et al. (2017).
Comments: The sample was characterized by powder X-ray diffraction data and electron microprobe analyses. The crystal structure is solved.

Kamphaugite-(Y) $CaY(CO_3)_2(OH)\cdot H_2O$

Origin: Goudini carbonatite, South Africa.
Experimental details: No data.
Raman shifts (cm^{-1}): 3473sh, 3383s, 3298s, 3208s, 3140s, 2953w, 2668, 2500, 2357w, 2206w, 1087s, 1041sh, 953w, 761w, 523w, 433w, 250.
Source: Verwoerd (2008).
Comments: The sample was characterized by powder X-ray diffraction data and electron microprobe analyses. The empirical formula of the sample used is $(Ca_{1.84}REE_x)(Y_{1.46}REE_{0.54-x})(CO_3)_4(OH)_{1.65}\cdot 2H_2O$. For the Raman spectrum of kamphaugite-(Y) see also Frost et al. (2015a).

Kangite $(Sc,Ti,Al,Zr,Mg,Ca,\square)_2O_3$

Origin: Allende meteorite. (2015a)
Experimental details: Raman scattering measurements have been performed on an arbitrarily oriented polished section of a sample using 514.5 nm Ar$^+$ laser radiation. The laser radiation power at the sample was 5 mW.
Raman shifts (cm^{-1}): No Raman shifts for kangite can be distinguished because of strong fluorescence features due to high concentrations of *REE*s in the sample.
Source: Ma et al. (2013).
Comments: The empirical formula of the sample used is $(Sc_{0.54}Al_{0.16}Y_{0.07}V_{0.03}Gd_{0.01}Dy_{0.01}Er_{0.01})(Ti_{0.66}Zr_{0.13})_{0.79}(Mg_{0.11}Ca_{0.06}Fe_{0.02})_{0.19}O_3$. In the cited paper a figure of the Raman spectrum of synthetic Sc_2O_3 is given.

Kaňkite $Fe^{3+}(AsO_4)\cdot 3.5H_2O$

Origin: No data.
Experimental details: Raman scattering measurements have been performed on an arbitrarily oriented sample using 633 nm He-Ne laser radiation. The laser radiation power at the sample was less than 0.1 mW in the hydroxyl stretching region. The laser radiation power in the other spectral regions was not indicated. The Raman shifts have been determined for the maxima of individual peaks obtained as a result of the spectral curve analysis.
Raman shifts (cm^{-1}): 3408w, 3221, 3112sh, 1629, 1469, 1065, 881s, 832sh, 808sh, 790, 733w, 564, 492s, 398w, 373w, 290w, 240, 228, 198, 179w.
Source: Frost and Kloprogge (2003).
Comments: No independent analytical data are provided for the sample used. For the Raman spectra of kaňkite see also Frost et al. (2015w) and Culka et al. (2016b).

Kanoite $MnMgSi_2O_6$

Origin: Artificial.
Experimental details: Raman scattering measurements have been performed on an arbitrarily oriented sample using 785 nm laser radiation. The laser radiation power at the sample was in the range 0.2–1 mW.
Raman shifts (cm^{-1}): 1010, 930w, 666, 563, 531, 393, 325, (231w).
Source: Tomasini et al. (2015).
Comments: The sample was characterized by powder X-ray diffraction data and electron microprobe analysis.

Kaolinite $Al_2Si_2O_5(OH)_4$

Origin: Washington County, Georgia, USA.
Experimental details: Methods of sample preparation are not described. Raman scattering measurements have been performed using 532 nm Nd-YAG laser radiation. The laser radiation power is not indicated.
Raman shifts (cm^{-1}): 3683s, 3657w, 3644w, 3616s, 912, 789, 745, 462, 430, 334, 274.
Source: Wang et al. (2015).
Comments: The sample was characterized by powder X-ray diffraction data and by electron microprobe analyses. For the Raman spectra of kaolinite see also Frost et al. (1993) and Frost (1995).

Kapellasite $Cu_3Zn(OH)_6Cl_2$

Origin: Sounion No. 19 mine, Kamariza, Lavrion, Greece (type locality).
Experimental details: Raman scattering measurements have been performed on an arbitrarily oriented single crystal using 633 nm laser radiation. The laser radiation power is not indicated. A 180-°-scattering geometry was employed.
Raman shifts (cm^{-1}): 3457, ~908w, 481, 409, 326, ~279sh, ~266sh, 247, ~232.
Source: Krause et al. (2006).
Comments: The sample was characterized by powder X-ray diffraction data and electron microprobe analyses. The crystal structure is solved. The empirical formula of the sample used is $(Cu_{3.24}Zn_{0.75})(OH)_{5.94}Cl_{2.0}$.

Kapundaite $CaNaFe^{3+}_4(PO_4)_4(OH)_3 \cdot 5H_2O$

Origin: Tom's quarry, Kapunda, Mt. Lofty Ranges, South Australia, Australia (type locality).
Experimental details: Raman scattering measurements have been performed on arbitrarily oriented crystals using 633 nm He-Ne and 785 nm laser radiations. The laser radiation powers are not indicated. The Raman shifts have been determined for the maxima of individual peaks obtained as a result of the spectral curve analysis.
Raman shifts (cm^{-1}): 3530w, 3449sh, 3311w, 3151sh, 2905w, 1687w, 1616w, 1549sh, 1443w, 1391w, 1203w, 1159sh, 1128sh, 1114s, 1089s, 1077sh, 1062sh, 1040sh, 1024, 1009sh, 988sh, 972, 963sh, 940, 912w, 675sh, 667sh, 658sh, 646, 633sh, 609, 588sh, 562sh, 547sh, 491sh, 475sh, 460sh, 448sh, 435, 412sh, 395s, 381sh, 361s, 295sh, 275s, 257s,244sh, 221sh, 186sh, 162s, 137sh, 113sh.
Source: Frost et al. (2014q).

Comments: The sample was characterized by qualitative electron microprobe analysis.

Karelianite V_2O_3

Origin: Merelani Hills gem zoisite deposit, Tanzania.
Experimental details: Raman scattering measurements have been performed on an arbitrarily oriented sample using 514.5 nm Ar$^+$ laser radiation. The laser radiation power at the sample was 1 mW.
Raman shifts (cm^{-1}): 506, 354, 219.
Source: Giuliani et al. (2008).
Comments: The sample was characterized by electron microprobe analyses. For the Raman spectrum of karelianite see also Voloshin et al. (2014).

Karelianite V_2O_3

Origin: Pyrrhotite gorge, Khibiny Mts., Kola Peninsula, Russia.
Experimental details: Raman scattering measurements have been performed on an arbitrarily oriented sample using 633 nm He-Ne laser radiation. The nominal laser radiation power was 2 mW.
Raman shifts (cm^{-1}): 972w, 712w, 644w, 576w, 503s, 305s, 227.
Source: Voloshin et al. (2014).
Comments: The sample was characterized by electron microprobe analyses. For the Raman spectrum of karelianite see also Giuliani et al. (2008).

Karrooite $MgTi_2O_5$

Origin: Synthetic.
Experimental details: Raman scattering measurements have been performed on an oriented single crystal with the b axis oriented parallel to the laser beam and the c-axis vertical using 785 nm Ti-sapphire laser radiation. The nominal laser radiation power was 50 mW.
Raman shifts (cm^{-1}): 1446, 1366, 1253, 1165, 1112, 913w, 790s, 632w, 522w, 499w, 422, 370, 329, 270s, 207, 165, 140, 124, 105, 88.
Source: Liermann et al. (2006).
Comments: No independent analytical data are provided for the sample used.

Karpenkoite $Co_3(V_2O_7)(OH)_2 \cdot 2H_2O$

Origin: Little Eva mine, Grand Co., Utah, USA (type locality).
Experimental details: Raman scattering measurements have been performed on a polycrystalline sample using 532 nm diode laser radiation. The laser radiation output power was 4 mW. A 180°-scattering geometry was employed.
Raman shifts (cm^{-1}): ~3500, 1670w, 823s, 474, 443, 312, 253.
Source: Kasatkin et al. (2015).
Comments: The sample was characterized by powder X-ray diffraction data and electron microprobe analyses. The empirical formula of the sample used is $(Co_{2.06}Zn_{0.72}Ni_{0.13}Mn_{0.09}Ca_{0.02}Cu_{0.02}Mg_{0.01})V_{1.98}O_7(OH)_2 \cdot 2H_2O$.

Kashinite Ir_2As_3

Origin: No data.

Experimental details: Methods of sample preparation are not described. Raman scattering measurements have been performed using 532 nm Nd-YAG laser radiation. The laser radiation power at the sample was in the range from 1 to 2 mW.

Raman shifts (cm^{-1}): 389, 367, 311, 290s, 200, 169, 152.

Source: Bakker (2014).

Comments: The sample was characterized by electron microprobe analysis. The Raman shifts were partly determined by us based on spectral curve analysis of the published spectrum. For the Raman spectrum of kashinite see also Zaccarini et al. (2016).

Kasolite $Pb(UO_2)(SiO_4) \cdot H_2O$

Origin: Sierra Albarrana, Córdoba, Spain.

Experimental details: Raman scattering measurements have been performed on an arbitrarily oriented sample using 632.8 nm He-Ne laser radiation. The nominal laser radiation power was 20 mW.

Raman shifts (cm^{-1}): 972w, 949w, 912w, ~800sh, 768s, 553, 424, 237, 217, 107.

Source: Bonales et al. (2015).

Comments: The sample was characterized by electron microprobe analyses. For the Raman spectra of kasolite see also Frost et al. (2006e) and Driscoll et al. (2014).

Kassite $CaTi_2O_4(OH)_2$

Origin: Prairie Lake carbonatite, Ontario, Canada.

Experimental details: Raman scattering measurements have been performed on an arbitrarily oriented sample using 532 nm solid-state laser radiation. The nominal laser radiation power was 50 mW.

Raman shifts (cm^{-1}): 3440sh, 3200–3194sh, 3161–3157, 3088sh, 2662w, 2596w, 2501w, 696–693s, 614sh, 472–463, 450–446, 398, 368, 336–332, 300–299s, 248sh, 193, 190sh, 165w, 147–146w, 124–123.

Source: Martins et al. (2014).

Comments: The sample was characterized by X-ray diffraction data and by electron microprobe analyses. The empirical formula of the sample used is $(Ca_{0.90}Ce_{0.03}Nd_{0.02}La_{0.01}Mn_{0.01})(Ti_{1.94}Fe_{0.04}Si_{0.01}Al_{0.01}Nb_{0.01})O_4(OH)_2$.

Katayamalite $KLi_3Ca_7Ti_2(SiO_3)_{12}(OH)_2$

Origin: Iwagi Island, Inland Sea, Ehime prefecture, Japan (type locality).

Experimental details: Methods of sample preparation are not described. Raman scattering measurements have been performed using 532 nm solid-state laser radiation. The laser radiation power is not indicated.

Raman shifts (cm^{-1}): ~3678, 1141, 1115w, 1019w, 982s, 960s, 915w, 904w, 700w, 668w, 570s, 514w, 495sh, 488, 454, 412, 376s, 297, 278, 257, 230w, 218, 195w.

Source: Andrade et al. (2013b).

Comments: The sample was characterized by single-crystal X-ray diffraction data and electron microprobe analyses. The crystal structure is solved. The empirical formula of the sample used is $(K_{0.89}Na_{0.12})Li_{3.21}(Ca_{6.87}Mn_{0.04}Ba_{0.02})(Ti_{1.79}Zr_{0.14}Fe_{0.04}Sn_{0.02})(SiO_3)_{12}(OH_{1.55}F_{0.45})$. The Raman shifts were determined by us based on spectral curve analysis of the published spectrum.

Katoite $Ca_3Al_2(OH)_{12}$

Origin: Synthetic.

Experimental details: Raman scattering measurements have been performed on an oriented single crystal from the (110) face of the crystal with parallel (A_{1g} and E_g) and cross (F_{2g}) polarizations of the incident and scattered light using 488 and 514.5 nm Ar$^+$ laser radiations. The laser radiation power is not indicated. A 180°-scattering geometry was employed.

Raman shifts (cm^{-1}): 3648s, 534, 327 ($A_{1g} + E_g$); 3653, 780w, 688sh, 535w, 388w, 331w, 231w, 163 (F_{2g}).

Source: Kolesov and Geiger (2005).

Comments: The sample was characterized by powder X-ray diffraction data.

Kawazulite Bi_2Te_2Se

Origin: Ozernyi district, Salla-Kuolayarvi, Kola Peninsula, Russia.

Experimental details: Methods of samples preparation are not described. Raman scattering measurements have been performed using 514.5 nm Ar$^+$ and 785 nm diode laser radiations. The nominal laser radiation power was 50 and 500 mW, respectively.

Raman shifts (cm^{-1}): 141–140, 103s, 61s.

Source: Voloshin et al. (2015a).

Comments: The samples were characterized by electron microprobe analyses. The empirical formulae of the samples used are $Bi_{2.052}Te_{1.797}Se_{1.151}$ and $(Bi_{1.986}Ni_{0.068})Te_{2.011}Se_{0.936}$. For the Raman spectra of kawazulite see also Akrap et al. (2012) and Gehring et al. (2013).

Kazanskyite $BaNa_3Ti_2Nb(Si_2O_7)_2O_2(OH)_2(H_2O)_4$

Origin: Kukisvumchorr Mt., Khibiny alkaline massif, Kola Peninsula, Russia (type locality).

Experimental details: Raman scattering measurements have been performed on an arbitrarily oriented sample using 532 nm laser radiation. The nominal laser radiation power was in the range from 5 to 12.5 mW.

Raman shifts (cm^{-1}): 3628w, 3545, 3462sh, 1071w, 1001w, 935, 885s, 822, 722w, 680, 580, 521w, 455, 411, 317, 214w, 151.

Source: Cámara et al. (2012b).

Comments: The sample was characterized by powder X-ray diffraction data and electron microprobe analyses. The crystal structure is solved. The empirical formula of the sample used is $(Na_{2.55}Mn_{0.31}Ca_{0.11}Fe^{2+}_{0.03})(Ba_{0.70}Sr_{0.28}K_{0.21}Ca_{0.03})(Ti_{2.09}Nb_{0.63}Mn_{0.26}Al_{0.02})Si_{4.05}O_{21.42}H_{9.45}F_{0.59}$. The Raman shifts were partly determined by us based on spectral curve analysis of the published spectrum.

Keilite FeS

Origin: Zakłodzie meteorite.

Experimental details: Raman scattering measurements have been performed in the spectral region from 100 to 4000 cm^{-1} on an arbitrarily oriented sample using 514.5 nm laser radiation. The laser radiation power at the sample was 1.2 mW.

Raman shifts (cm^{-1}): ~335w, ~280.

Source: Ma et al. (2012a).

Comments: The sample was characterized by electron backscatter diffraction and electron microprobe analysis. The empirical formula of the sample used is $(Fe_{0.43}Mn_{0.35}Mg_{0.16}Cr_{0.02}Ca_{0.02})S$. The Raman shifts were determined by us based on spectral curve analysis of the published spectrum.

Keiviite-(Yb) $Yb_2Si_2O_7$

Origin: Synthetic.
Experimental details: Raman scattering measurements have been performed on an oriented single crystal and on a powdered sample using 514.5 nm Ar^+ laser radiation. The nominal laser radiation power was 4 W.
Raman shifts (cm^{-1}): 952s, 923s, 663, 521, 484w, 424, 413, 370, 362, 277, 203w, 145s, 88w.
Source: Bretheau-Raynal et al. (1979).
Comments: The sample was characterized by electron microprobe analysis. The Raman shifts are given for a powdered sample.

Kemmlitzite $SrAl_3(AsO_4)(SO_4)(OH)_6$

Origin: Oschatz, Saxony, Germany (type locality).
Experimental details: Raman scattering measurements have been performed on arbitrarily oriented crystals using a 633 nm He-Ne laser. The laser radiation power is not indicated. The Raman shifts have been determined for the maxima of individual peaks obtained as a result of the spectral curve analysis.
Raman shifts (cm^{-1}): 3566sh, 3441s, 3374sh, 3269sh, 3047w, 1591w, 1524sh, 1356sh, 984s, 957sh, 825, 772sh, 690, 631sh, 564w, 482sh, 427, 388sh, 342s, 218, 208, 199, 149, 108w.
Source: Frost et al. (2012a).
Comments: No independent analytical data are provided for the sample used.

Kempite $Mn^{2+}_2Cl(OH)_3$

Origin: Artificial.
Experimental details: Raman scattering measurements have been performed on arbitrarily oriented crystals using 785 nm laser. The laser radiation power at the sample was less than 5 mW.
Raman shifts (cm^{-1}): 506, 478, 413w, 293s.
Source: Vallette Campos and Alvarado Aguayo (2015).
Comments: No independent analytical data are provided for the sample used.

Kentrolite $Pb_2Mn^{3+}_2O_2(Si_2O_7)$

Origin: Artificial.
Experimental details: Micro-Raman scattering measurements have been performed on an arbitrarily oriented sample using 532 nm laser radiation. The nominal laser radiation power was 300 mW.
Raman shifts (cm^{-1}): 950, 893, 593s, 540, 344, 305.
Source: Vieira Ferreira et al. (2014).
Comments: No independent analytical data are provided for the sample used.

Kenyaite $Na_2Si_{22}O_{41}(OH)_8 \cdot 6H_2O$

Origin: Synthetic.
Experimental details: Methods of sample preparation are not described. Raman scattering measurements have been performed using 1064 nm Nd-YAG laser radiation. The laser radiation power at the sample was 90 mW.
Raman shifts (cm^{-1}): 3200w, 1179, 1061, 1049, 819, 801w, 780w, 622, 493sh, 465s, 431sh, 398w, 376, 349, 316, 258, 242w, 204, 166, 162, 155w, 148w, 139w, 129w, 123w, 115w, 106.
Source: Huang et al. (1999a).
Comments: The sample was characterized by powder X-ray diffraction data.

Kerimasite $Ca_3Zr_2(SiFe^{3+}_2)O_{12}$

Origin: Kerimasi volcano, northern Tanzania (type locality).
Experimental details: Raman scattering measurements have been performed on an arbitrarily oriented sample using 633 nm He-Ne laser radiation. The laser radiation power is not indicated.
Raman shifts (cm^{-1}): 3537w, 3420sh, 3400w, 3380sh, 3240w, 875w, 830sh, 785sh, 732, 573w, 500s, 414w, 298s, 243, 152.
Source: Zaitsev et al. (2010).
Comments: The sample was characterized by powder X-ray diffraction data and electron microprobe analyses. The crystal structure is solved. The empirical formula of the sample used is $(Ca_{3.00}Mn_{0.01}Ce_{0.01}Nd_{0.01})(Zr_{1.72}Nb_{0.14}Ti_{0.08}Mg_{0.02}Y_{0.02})(Fe^{3+}_{1.23}Si_{0.86}Al_{0.82}Ti_{0.09})O_{12}$. For the Raman spectra of kerimasite see also Galuskina et al. (2013a) and Uher et al. (2015).

Kermesite Sb_2OS_2

Origin: Pernek, Slovak Republic.
Experimental details: Raman scattering measurements have been performed on an oriented crystal with the laser polarization parallel and perpendicular to the cleavage and elongation of kermesite. 632.8 nm He-Ne and 785 nm solid-state laser radiations were used. The nominal laser radiation power was 1.7 and 8.5 mW, respectively. A 180°-scattering geometry was employed.
Raman shifts (cm^{-1}): 334, 317s, 303w, 289w, 276w, 245, 237w, 231, 206w, 175w, 162w, 148, 130, 111w, 105, 97w, 84w, 72w, 64, 59, 48s.
Source: Kharbish et al. (2009).
Comments: The sample was characterized by electron microprobe analysis. The empirical formula of the sample used is $Sb_{1.9}S_{2.1}O$. The Raman shifts are given as the sum of the spectra of all scattering geometries with He-Ne laser excitation.

Keyite $Cu^{2+}_3Zn_4Cd_2(AsO_4)_6 \cdot 2H_2O$

Origin: Tsumeb, Namibia (?).
Experimental details: Raman scattering measurements have been performed on an arbitrarily oriented single crystal using 514.5 nm Ar$^+$ laser radiation. The laser radiation power is not indicated.
Raman shifts (cm^{-1}): 856, 803, 735, 590, 505, 411, 355, 307sh, 288, 239, 195, 134, 65.
Source: Schlüter et al. (2013).
Comments: No independent analytical data are provided for the sample used. The Raman shifts were determined by us based on spectral curve analysis of the published spectrum.

Khademite $Al(SO_4)F \cdot 5H_2O$

Origin: Kladno mine, Central Bohemia, Czech Republic (type locality).
Experimental details: Raman scattering measurements have been performed on arbitrarily oriented crystals using a 633 nm He-Ne laser. The laser radiation power is not indicated. Raman spectrum was obtained in the spectral region from 200 to 4000 cm^{-1}. The Raman shifts have been determined for the maxima of individual peaks obtained as a result of the spectral curve analysis.
Raman shifts (cm^{-1}): 3380, 3146sh, 2991sh, 1763sh, 1609w, 1449sh, 1132, 1104sh, 991s, 975sh, 618, 534sh, 505sh, 455, 324sh, 253sh, 226sh, 192, 150, 113sh.
Source: Frost et al. (2013m).
Comments: No independent analytical data are provided for the sample used.

Khatyrkite $CuAl_2$

Origin: Synthetic.
Experimental details: No data.
Raman shifts (cm^{-1}): 260, 103.
Source: Bahrami et al. (2014).
Comments: The sample was characterized by powder X-ray diffraction data and electron microprobe analyses.

Khesinite $Ca_4(Mg_3Fe^{3+}{}_9)O_4(Fe^{3+}{}_9Si_3)O_{36}$

Origin: Gurim anticline, near Arad city, Hatrurim Complex, Negev Desert, Israel (type locality).
Experimental details: Raman scattering measurements have been performed on an arbitrarily oriented sample using 532 nm solid-state laser radiation. The laser radiation power at the sample was in the range from 10 to 20 mW. The Raman shifts have been determined for the maxima of individual peaks obtained as a result of the spectral curve analysis.
Raman shifts (cm^{-1}): 1638, 1495sh, 1403, 1132w, 947, 814sh, 749sh, 696, 610, 522, 481sh, 336sh, 310s, 256sh, 208sh, 159, 121.
Source: Galuskina et al. (2017).
Comments: The sample was characterized by single-crystal X-ray diffraction data and electron microprobe analyses. The crystal structure is solved. The empirical formula of the sample used is $Ca_4(Fe^{3+}{}_{8.528}Mg_{1.635}Ca_{0.898}Ti^{4+}{}_{0.336}Ni^{2+}{}_{0.217}Mn^{2+}{}_{0.155}Cr^{3+}{}_{0.132}Fe^{2+}{}_{0.098})[(Fe^{3+}{}_{6.827}Al_{2.506}Si_{2.667})O_{40}]$.

Khvorovite $Pb_4Ca_2[Si_8B_2(Si,B)_2O_{28}]F$

Origin: Dara-I Pioz glacier, Dara-I Pioz alkaline massif, Tien Shan Mts., Tajikistan (type locality).
Experimental details: Raman scattering measurements have been performed on an arbitrarily oriented sample using 532 nm laser radiation. The laser radiation power is not indicated.
Raman shifts (cm^{-1}): 1100w, 1017s, 937, 783, 713, 642, 531s, 485, 425s, 296, 256s, 221w, 203w, 170, 162.
Source: Pautov et al. (2015).
Comments: The sample was characterized by powder X-ray diffraction data and electron microprobe analyses. The crystal structure is solved. The empirical formula of the sample used is $(Pb^{2+}{}_{2.76}Ba_{0.62}K_{0.56}Na_{0.16})(Ca_{1.86}Sr_{0.06}Y_{0.04}Na_{0.04})[Si_8B_2(Si_{1.46}B_{0.65})O_{28}](F_{0.71}O_{0.29})$.

Kiddcreekite Cu_6SnWS_8

Origin: Zijin Cu-Au mine, China.
Experimental details: No data.
Raman shifts (cm^{-1}): 860, 654, 430, 345, 291, 256.
Source: Liu et al. (2012).
Comments: No independent analytical data are provided for the sample used.

Kidwellite $NaFe^{3+}_{9+x}(PO_4)_6(OH)_{11} \cdot 3H_2O$ ($x \gg 0.33$)

Origin: Savannah River, Girard Barke Co., Georgia, USA.
Experimental details: Raman scattering measurements have been performed on arbitrarily oriented crystals using a 633 nm He-Ne laser. The laser radiation power is not indicated. Raman spectrum was obtained in the spectral region from 200 to 4000 cm^{-1}. The Raman shifts have been determined for the maxima of individual peaks obtained as a result of the spectral curve analysis.
Raman shifts (cm^{-1}): 3580, 3466, 3356, 3231, 3122, 1188, 1144w, 1129, 1082, 1063, 1050w, 1034, 1014s, 978s, 931w, 919, 875w, 653w, 644w, 631w, 588, 570, 557, 539, 500, 490, 467, 453, 444, 418w, 405, 333, 322, 285, 271, 253w, 223, 189, 181, 169, 144, 139, 118, 111.
Source: Frost et al. (2014k).
Comments: The sample was characterized by qualitative electron microprobe analysis.

Kieftite $CoSb_3$

Origin: Synthetic.
Experimental details: Raman scattering measurements have been performed on a polished plate cut perpendicular to the [101] direction of a single crystal using 476.5, 488, and 514.5 nm Ar$^+$ laser radiations. The laser radiation power is not indicated. A 135°-scattering geometry was employed.
Raman shifts (cm^{-1}): 188, 180, 171w, 154s.
Source: Nolas et al. (1996).
Comments: The sample was characterized by X-ray diffraction data. The Raman shifts are given for 514.5 nm laser radiation.

Kieserite $Mg(SO_4) \cdot H_2O$

Origin: Synthetic.
Experimental details: Raman scattering measurements have been performed on a powdered sample using 532 nm Nd-YAG laser radiation. The laser radiation power at the sample was 15 mW.
Raman shifts (cm^{-1}): 3297, 1509, 1215w, 1117, 1046s, 629, 481, 436, 272, 218.
Source: Wang et al. (2006a).
Comments: The sample was characterized by powder X-ray diffraction analysis. For the Raman spectrum of kieserite see also Wang et al. (2008).

Kilchoanite $Ca_6(SiO_4)(Si_3O_{10})$

Origin: Birkhin massif, Baikal Lake area, Siberia, Russia.
Experimental details: Raman scattering measurements have been performed on an arbitrarily oriented sample using 514.5 nm Ar$^+$ laser radiation. The nominal laser radiation power was in the range from 30 to 50 mW.

Raman shifts (cm^{-1}): 1012, 988, 967, 927, 911w, 871sh, 864s, 828s, 782w, 671s, 596w, 571w, 552w, 485w, 401, 358w, 333w, 306w, 259w, 246w, 208w, 195w, 188w, 163, 133, 121w, 112sh.
Source: Galuskin et al. (2012b).
Comments: The sample was characterized by single-crystal X-ray diffraction data and electron microprobe analyses. The crystal structure is solved.

Killalaite $Ca_{6.4}[H_{0.6}Si_2O_7]_2(OH)_2$

Origin: Upper Chegem caldera, Northern Caucasus, Kabardino-Balkaria, Russia.
Experimental details: Raman scattering measurements have been performed on an approximately oriented crystal using 514.5 nm Ar$^+$ laser radiation. The laser radiation output power was in the range from 30 to 50 mW. A 180°-scattering geometry was employed. Spectra were collected on a cross-section approximately perpendicular to z-axis (I scattering geometry) and on a cross-section approximately parallel to z-axis (II scattering geometry).
Raman shifts (cm^{-1}): 3562, 3523, 994, 943, 912, 677, 551, 433, 371, 302, 284, 266, 185, 129 (I scattering geometry); 3567, 3530, 1077, 999, 945, 912, 883, 678, 552, 435, 367, 284, 266, 185, 148, 107 (II scattering geometry).
Source: Armbruster et al. (2012).
Comments: The sample was characterized by single-crystal X-ray diffraction data and electron microprobe analyses. The crystal structure is solved. The Raman shifts were determined by us based on spectral curve analysis of the published spectrum.

Kimzeyite $Ca_3Zr_2(SiAl_2)O_{12}$

Origin: Wiluy River basin, Sakha-Yakutia, Russia.
Experimental details: Raman scattering measurements have been performed on an arbitrarily oriented single crystal using 514.5 nm Ar$^+$ laser radiation. The laser radiation power is not indicated.
Raman shifts (cm^{-1}): 937w, 879w, 785, 728, 499s, 303s, 248sh, 153.
Source: Galuskina et al. (2005).
Comments: The sample was characterized by electron microprobe analyses.

Kinoite $Ca_2Cu_2Si_3O_{10} \cdot 2H_2O$

Origin: Christmas mine, Gila Co., Arizona, USA.
Experimental details: Raman scattering measurements have been performed on arbitrarily oriented crystals using a 633 nm He-Ne laser. The laser radiation power is not indicated. The Raman shifts have been determined for the maxima of individual peaks obtained as a result of the spectral curve analysis.
Raman shifts (cm^{-1}): 3572sh, 3519, 3441sh, 3237w, 3022w, 1585w, 1186, 1052, 1000, 994sh, 975sh, 951s, 859sh, 847s, 765, 742, 642, 543sh, 531s, 486s, 456s, 422s, 400s, 352sh, 339, 324sh, 309, 301, 286, 266sh, 251, 233sh, 225, 194, 183, 163sh, 153s, 138s, 118sh, 110.
Source: Frost and Xi (2012c).
Comments: No independent analytical data are provided for the sample used.

Kinoshitalite $BaMg_3(Si_2Al_2O_{10})(OH)_2$

Origin: Hokkejino, Kyoto prefecture, Kinki region, Honshu Island, Japan.
Experimental details: Raman scattering measurements have been performed on an arbitrarily oriented sample using 780 nm laser radiation. The nominal laser radiation power was 600 mW.
Raman shifts (cm^{-1}): 653s, 210, 150, 140.
Source: Manuella et al. (2012).
Comments: The sample was characterized by X-ray diffraction data.

Kintoreite $PbFe^{3+}_3(PO_4)(PO_3OH)(OH)_6$

Origin: Broken Hill, New South Wales, Australia (type locality).
Experimental details: Raman scattering measurements have been performed on arbitrarily oriented crystals using a 633 nm He-Ne laser. The laser radiation power is not indicated. The Raman shifts have been determined for the maxima of individual peaks obtained as a result of the spectral curve analysis.
Raman shifts (cm^{-1}): 3435sh, 3391sh, 3225, 2968sh, 1413, 1229, 1140, 1110, 1075, 1021, 1003, 975, 851, 814, 625, 573sh, 562, 551, 477sh, 459sh, 440s, 422sh, 420sh, 372, 336, 307, 253, 219, 204.
Source: Frost et al. (2006p).
Comments: No independent analytical data are provided for the sample used.

Kipushite $Cu_6(PO_4)_2(OH)_6 \cdot H_2O$

Origin: Miedzianka (former Kupferberg), Sudety Mts., SW Poland.
Experimental details: Raman scattering measurements have been performed on an arbitrarily oriented sample using 532 nm Nd-YAG laser radiation. The nominal laser radiation power was 40 mW. The Raman shifts have been determined for the maxima of individual peaks obtained as a result of the spectral curve analysis.
Raman shifts (cm^{-1}): 3549, 3482, 3444sh, 3251sh, 1078sh, 1045w, 1021w, 975s, 942w, 875sh, 869s, 843sh, 814, 796sh, 639w, 617w, 552sh, 521w, 493sh, 464, 438w, 396, 369sh, 317w, 297w, 256w, 221w.
Source: Ciesielczuk et al. (2016).
Comments: The sample was characterized by powder X-ray diffraction data and electron microprobe analyses. The Raman shifts are given for As-bearing kipushite with 1.21 *apfu* P and 0.67 *apfu* As.

Kirschsteinite $CaFe^{2+}(SiO_4)$

Origin: Artificial.
Experimental details: Raman scattering measurements have been performed on an arbitrarily oriented polished sample cross-section using 785 nm laser radiation. The laser radiation power at the sample was 6.6 mW.
Raman shifts (cm^{-1}): 932, 901w, 849sh, 815s, 635w, 566w, 522w, 391, 291, 249.
Source: Kramar et al. (2015).
Comments: The sample was characterized by powder X-ray diffraction data and electron microprobe analyses. The Raman shifts were partly determined by us based on spectral curve analysis of the published spectrum.

Kladnoite $C_6H_4(CO)_2NH$

Origin: Kladno (Schöller) mine, Libušin, Kladno, Bohemia, Czech Republic (type locality).
Experimental details: Raman scattering measurements have been performed on an arbitrarily oriented sample using 514.5 nm Ar^+ laser radiation. The nominal laser radiation power was 10 mW.
Raman shifts (cm^{-1}): 3085, 3071w, 3068sh, 3046, 1755s, 1726, 1606, 1468w, 1386, 1373, 1305, 1165, 1142, 1091w, 1047w, 1012s, 901w, 809w, 805w, 795w, 743s, 666w, 641, 550, 531w, 350, 260, 200, 163s, 72.
Source: Jehlička et al. (2007b).
Comments: No independent analytical data are provided for the sample used. For the Raman spectra of kladnoite see also Moroz et al. (2004) and Jehlička et al. (2009a).

Klaprothite $Na_6(UO_2)(SO_4)_4 \cdot 4H_2O$

Origin: Blue Lizard mine, San Juan Co., Utah, USA (type locality).
Experimental details: Methods of sample preparation are not described. Raman scattering measurements have been performed using 532 nm laser radiation. The laser radiation power is not indicated.
Raman shifts (cm^{-1}): 3620, 3440, 1640, 1248, 1216w, 1175, 1161, 1070, 1036, 997s, 974, 944, 830s, 658, 624, 540w, 495w, 456, 430, 335w, 287w, 230sh, 215, 146w, 130w, 98w, 78w.
Source: Kampf et al. (2016g).
Comments: The sample was characterized by powder X-ray diffraction data and electron microprobe analyses. The crystal structure is solved. The empirical formula of the sample used is $Na_{6.01}(U_{1.03}O_2)(S_{0.993}O_4)_4(H_2O)_4$. The Raman shifts were determined by us based on spectral curve analysis of the published spectrum. For the Raman spectrum of klaprothite see also Plášil et al. (2015d).

Klebelsbergite $Sb^{3+}_4O_4(SO_4)(OH)_2$

Origin: Pereta mine, Grosseto province, Tuscany, Italy.
Experimental details: Raman scattering measurements have been performed on arbitrarily oriented crystals using a 633 nm He-Ne laser. The laser radiation power is not indicated. The Raman shifts have been determined for the maxima of individual peaks obtained as a result of the spectral curve analysis.
Raman shifts (cm^{-1}): 3457sh, 3435w, 3357w, 1142, 1139sh, 1089sh, 1074, 1029, 971s, 936sh, 723, 662, 627sh, 611sh, 604s, 581s, 489s, 481sh, 446, 435, 410, 326, 303, 266w, 238s, 225s, 205s, 194sh, 170s, 160sh, 142, 130, 116.
Source: Frost and Bahfenne (2011c).
Comments: No independent analytical data are provided for the sample used.

Klockmannite $Cu_{5.2}Se_6$

Origin: Synthetic.
Experimental details: Raman scattering measurements have been performed on a polycrystalline sample using 514.5 nm Ar^+ laser radiation. The laser radiation power is not indicated.
Raman shifts (cm^{-1}): 263s, 206w, 192w, 45sh, 43s, 17.
Source: Ishii et al. (1993).
Comments: The sample was characterized by powder X-ray diffraction data.

Knorringite $Mg_3Cr_2(SiO_4)_3$

Origin: Synthetic.
Experimental details: Raman scattering measurements have been performed on a polished surface of an arbitrarily oriented single crystal using 632 nm He-Ne laser radiation. The laser radiation power is not indicated.
Raman shifts (cm^{-1}): 936, 908, 866, 718, 551s, 368.
Source: Bykova et al. (2014).
Comments: The sample was characterized by powder X-ray diffraction data and electron microprobe analyses. The crystal structure is solved. The empirical formula of the sample used is $Mg_{3.21}Cr_{1.58}Si_{3.21}O_{12}$.

Kobokoboite $Al_6(PO_4)_4(OH)_6 \cdot 11H_2O$

Origin: No data.
Experimental details: No data.
Raman shifts (cm^{-1}): 1630, 1463, 1170sh, 1095, 1046, 980sh, 916, 764, 600, 514, 490sh.
Source: Sanchez-Moral et al. (2011).
Comments: No independent analytical data are provided for the sample used. The Raman shifts were determined by us based on spectral curve analysis of the published spectrum.

Koechlinite Bi_2MoO_6

Origin: Pittong, Victoria, Australia.
Experimental details: Raman scattering measurements have been performed on an arbitrarily oriented sample using 785 nm Nd-YAG laser radiation. The laser radiation power at the sample was 1 mW. The Raman shifts have been determined for the maxima of individual peaks obtained as a result of the spectral curve analysis.
Raman shifts (cm^{-1}): 843, 797, 773sh, 715, 401, 349, 321, 293sh, 281, 268sh, 228, 195, 154sh, 141.
Source: Frost et al. (2004c).
Comments: No independent analytical data are provided for the sample used.

Kojonenite $Pd_{7-x}SnTe_2$ ($0.3 \leq x \leq 0.8$)

Origin: Synthetic.
Experimental details: Raman scattering measurements have been performed on a polished surface of an arbitrarily oriented crystal using 532 nm Nd-YAG laser radiation. The laser radiation power at the sample was in the range from 1 to 2 mW.
Raman shifts (cm^{-1}): 197w.
Source: Vymazalová et al. (2014).
Comments: The sample was characterized by electron microprobe analysis.

Kokchetavite $K(AlSi_3O_8)$

Origin: Synthetic.
Experimental details: Raman scattering measurements have been performed on an arbitrarily oriented sample using 488 nm Ar$^+$ laser radiation. The nominal laser radiation power was 60 mW. A 180°-scattering geometry was employed.

Raman shifts (cm^{-1}): 835, 390s, 109s.
Source: Kanzaki et al. (2012).
Comments: The sample was characterized by powder X-ray diffraction data. For the Raman spectra of kokchetavite see also Hwang et al. (2004) and Ferrero et al. (2016).

Koktaite (NH$_4$)$_2$Ca(SO$_4$)$_2$·H$_2$O

Origin: Synthetic.
Experimental details: Raman scattering measurements have been performed on an arbitrarily oriented sample using 532 nm Nd-YAG laser radiation. The laser radiation power at the sample was 2 mW.
Raman shifts (cm^{-1}): 3353sh, 3149, 2852w, 1723w, 1677w, 1453, 1419, 1153, 1132, 1104w, 1087, 996s, 980s, 656, 642, 625, 616, 603, 487, 474, 437, 423.
Source: Jentzsch et al. (2012a).
Comments: The sample was characterized by powder X-ray diffraction data.

Kolskyite CaNa$_2$Ti$_4$(Si$_2$O$_7$)$_2$O$_4$(H$_2$O)$_7$

Origin: Kirovskii mine, Kukisvumchorr Mt., Khibiny alkaline massif, Kola Peninsula, Russia (type locality).
Experimental details: Raman scattering measurements have been performed on an arbitrarily oriented sample using 532 nm laser radiation. The laser radiation power is not indicated. A 180°-scattering geometry was employed.
Raman shifts (cm^{-1}): 1072w, 997w, 925s, 800w, 685s, 586, 505w, 435, 420, 397, 297, 214w, 151.
Source: Cámara et al. (2013).
Comments: The sample was characterized by powder X-ray diffraction data and electron microprobe analyses. The crystal structure is solved. The empirical formula of the sample used is (Na$_{1.93}$Mn$_{0.04}$Ca$_{0.03}$)(Ca$_{0.67}$Sr$_{0.45}$Ba$_{0.19}$K$_{0.15}$)(Ti$_{2.93}$Nb$_{0.46}$Mn$_{0.33}$Mg$_{0.17}$Fe$^{2+}_{0.10}$Zr$_{0.01}$)Si$_{4.00}$O$_{24.67}$H$_{13.60}$F$_{0.33}$. The Raman shifts were partly determined by us based on spectral curve analysis of the published spectrum.

Kolwezite (Cu,Co)$_2$(CO$_3$)(OH)$_2$

Origin: Mupine, Shaba province, Zaire.
Experimental details: Raman scattering measurements have been performed on an arbitrarily oriented sample using 633 nm He-Ne laser radiation. The laser radiation power is not indicated. The Raman shifts have been determined for the maxima of individual peaks obtained as a result of the spectral curve analysis.
Raman shifts (cm^{-1}): 3439sh, 3389, 3310sh, 3284sh, 1515sh, 1495, 1456sh, 1363, 1093s, 1059s, 757w, 718w, 530, 431, 346, 264, 176, 159.
Source: Frost (2006).
Comments: No independent analytical data are provided for the sample used.

Konyaite Na$_2$Mg(SO$_4$)$_2$·5H$_2$O

Origin: Synthetic.
Experimental details: Raman scattering measurements have been performed on an arbitrarily oriented sample using 532 nm Nd-YAG laser radiation. The laser radiation power at the sample was 2 mW.

Raman shifts (cm^{-1}): 3295, 3185sh, 1179w, 1144w, 1091, 1003s, 981s, 648w, 619sh, 610, 602sh, 471sh, 455sh, 448.

Source: Jentzsch et al. (2011).

Comments: No independent analytical data are provided for the sample used. For the Raman spectrum of konyaite see also Jentzsch et al. (2013). The Raman shifts were partly determined by us based on spectral curve analysis of the published spectrum.

Koritnigite $Zn(AsO_3OH) \cdot H_2O$

Origin: Jáchymov ore district, Krušné Hory Mts., Bohemia, Czech Republic.

Experimental details: Raman scattering measurements have been performed on an arbitrarily oriented sample using 633 nm He-Ne laser radiation. The laser radiation power is not indicated. The Raman shifts have been determined for the maxima of individual peaks obtained as a result of the spectral curve analysis.

Raman shifts (cm^{-1}): 3474w, 3297sh, 3182w, 3005w, 2770w, 2434sh, 2285w, 1755sh, 1597w, 1303w, 877s, 842s, 813s, 766, 330, 261, 284, 237, 219, 187, 172s, 154, 140, 120, 110.

Source: Frost et al. (2011q).

Comments: The sample was characterized by powder X-ray diffraction data and electron microprobe analysis. The empirical formula of the sample used is $(Zn_{0.79}Co_{0.14}Ni_{0.02})[AsO_3(OH)_{0.99}F_{0.01}]_{1.00} \cdot H_2O$. The Raman shifts were partly determined by us based on spectral curve analysis of the published spectrum. For the Raman spectrum of koritnigite see also Frost et al. (2014a).

Kornelite $Fe^{3+}_2(SO_4)_3 \cdot 7H_2O(?)$

Origin: Synthetic.

Experimental details: Raman scattering measurements have been performed on an arbitrarily oriented sample using 532 nm laser radiation. The laser radiation power is not indicated. The Raman shifts have been determined for the maxima of individual peaks obtained as a result of the spectral curve analysis.

Raman shifts (cm^{-1}): 3587w, 3352sh, 3123, 1696sh, 1658w, 1613sh, 1182, 1151sh, 1124w, 1078w, 1033s, 1021sh, 993w, 838w, 672w, 636, 597, 476, 452, 439, 269, 248, 209, 187.

Source: Ling and Wang (2010).

Comments: The sample was characterized by powder X-ray diffraction data. For the Raman spectra of kornelite see also Ling et al. (2009) and Kong et al. (2011a).

Kornerupine $(Mg,Fe^{2+},Al,\square))_{10}(Si,Al,B)_5O_{21}(OH,F)_2(?)$

Origin: Mautia Hill, Tanzania.

Experimental details: Raman scattering measurements have been performed on an arbitrarily oriented single crystal using 633 nm He-Ne laser radiation. The laser radiation power is not indicated. The Raman shifts have been determined for the maxima of individual peaks obtained as a result of the spectral curve analysis.

Raman shifts (cm^{-1}): 3619sh, 3612s, 3599sh, 3556sh, 3547s, 3538sh, 3521w, 3275w, 1084w, 1051, 1035sh, 995, 947sh, 923, 668, 648sh, 620, 586sh, 554w, 507sh, 487, 477sh, 459sh, 403, 394sh, 364w, 355w, 336, 324, 316sh, 298w, 261, 254sh, 236w, 224, 219sh, 191sh, 180, 150, 138sh, 123.

Source: Frost et al. (2015t).

Comments: The sample was characterized by qualitative electron microprobe analysis. For the Raman spectrum of kornerupine see also Wopenka et al. (1999).

Kosmochlor $NaCr^{3+}Si_2O_6$

Origin: Morasko iron meteorite.
Experimental details: Raman scattering measurements have been performed on an arbitrarily oriented sample using 632.8 nm He-Ne laser radiation. The laser radiation power at the sample was 10 mW.
Raman shifts (cm^{-1}): 1055, 1033s, 991, 952w, 863, 680, 565sh, 552s, 520, 413, 380w, 339s, 320w, 294, 252, 202, 142w.
Source: Karwowski et al. (2013).
Comments: The sample was characterized by electron microprobe analyses. The empirical formula of the sample used is $Na_{0.91}Ca_{0.07}Mg_{0.09}Fe_{0.02}Cr_{0.82}Al_{0.01}V_{0.01}Ti_{0.07}Si_{2.00}O_{6.00}$. The Raman shifts were partly determined by us based on spectral curve analysis of the published spectrum. For the Raman spectrum of kosmochlor see also Leander et al. (2014).

Kosnarite $KZr_2(PO_4)_3$

Origin: Jenipapo district, Itinga, Minas Gerais, Brazil.
Experimental details: Raman scattering measurements have been performed on arbitrarily oriented crystals using a 633 nm He-Ne laser. The laser radiation power is not indicated. The Raman shifts have been determined for the maxima of individual peaks obtained as a result of the spectral curve analysis.
Raman shifts (cm^{-1}): 1149w, 1116w, 1088, 1079sh, 1063, 1060, 1031sh, 1026s, 1022sh, 1017sh, 979sh, 638, 595, 561w, 437s, 421, 405, 387w, 318, 290s, 263, 188sh, 175, 156, 141sh, 122.
Source: Frost et al. (2012i).
Comments: The sample was characterized by qualitative electron microprobe analysis.

Kotoite $Mg_3(BO_3)_2$

Origin: Snezhnoye deposit, East Verkhoyan'ye, Sakha Yakutia, Russia.
Experimental details: Raman scattering measurements have been performed on an arbitrarily oriented polished sample using 514.5 nm Ar^+ laser radiation. The laser radiation power at the sample was less than 20 mW. A 0°-scattering geometry was employed.
Raman shifts (cm^{-1}): 1255w, 1088w, 920s, 866w, 847, 708, 796, 691, 553, 357s, 327, 310.
Source: Galuskina et al. (2008).
Comments: The sample was characterized by powder X-ray diffraction data and electron microprobe analyses. For the Raman spectra of kotoite see also Frost and Xi (2013c) and Kipcak et al. (2013).

Köttigite $Zn_3(AsO_4)_2 \cdot 8H_2O$

Origin: Ojuela Mine, Mapini, Durango, Mexico.
Experimental details: Raman scattering measurements have been performed on an arbitrarily oriented sample using 633 nm He-Ne laser radiation. The laser radiation power is not indicated. The Raman shifts have been determined for the maxima of individual peaks obtained as a result of the spectral curve analysis.

Raman shifts (cm^{-1}): 3458sh, 3215w, 868sh, 835s, 810sh, 790s, 547, 479sh, 451s, 432s, 371w, 332w, 286sh, 249s, 220s, 194sh, 142w.
Source: Frost et al. (2003g).
Comments: No independent analytical data are provided for the sample used.

Kotulskite Pd(Te,Bi)$_{2-x}$ ($x \approx 0.4$)

Origin: Synthetic.
Experimental details: Raman scattering measurements have been performed on an arbitrarily oriented sample using 532 nm Nd-YAG laser radiation. The laser radiation power at the sample was in the range from 1 to 2 mW.
Raman shifts (cm^{-1}): 97.
Source: Vymazalová et al. (2014).
Comments: The sample was characterized by electron microprobe analysis.

Kovdorskite Mg$_2$(PO$_4$)(OH)·3H$_2$O

Origin: Kovdor massif, Kola Peninsula, Russia (type locality).
Experimental details: Raman scattering measurements have been performed on an arbitrarily oriented sample using 532 nm solid-state laser radiation. The laser radiation power is not indicated.
Raman shifts (cm^{-1}): 3681, 3395, 3219, 2967, 1550, 1089w, 1056, 964s, 870w, 566sh, 536, 453, 410w, 375, 345, 320w, 303, 255w, 228, 201, 158s, 135.
Source: Morrison et al. (2012).
Comments: The sample was characterized by powder X-ray diffraction data and electron microprobe analyses. The crystal structure is solved. The empirical formula of the sample used is Mg$_{2.00}$PO$_{4.00}$(OH)·2.67H$_2$O. The Raman shifts were partly determined by us based on spectral curve analysis of the published spectrum. For the Raman spectrum of kovdorskite see also Frost et al. (2013a).

Kozyrevskite Cu$_4$O(AsO$_4$)$_2$

Origin: Arsenatnaya fumarole, Tolbachik volcano, Kamchatka, Russia (type locality).
Experimental details: Raman scattering measurements have been performed on an arbitrarily oriented sample using 532 nm diode laser radiation. The laser radiation power at the sample was about 3 mW. A 180°-scattering geometry was employed.
Raman shifts (cm^{-1}): 875s, 840sh, 826s, 757sh, 497w, 445, 394, 340, 216, 170sh, 137sh, 112.
Source: Pekov et al. (2014c).
Comments: The sample was characterized by powder X-ray diffraction data and electron microprobe analyses. The crystal structure is solved. The empirical formula of the sample used is (Cu$_{3.95}$Zn$_{0.07}$Fe$_{0.01}$)(As$_{1.83}$P$_{0.09}$S$_{0.03}$V$_{0.02}$Si$_{0.01}$)O$_9$.

Kremersite (NH$_4$)$_2$Fe^{3+}Cl$_5$·H$_2$O

Origin: Synthetic.
Experimental details: Raman scattering measurements have been performed on an arbitrarily oriented sample using He-Ne laser radiation. The nominal laser radiation power was 50 mW. A 135°-scattering geometry was employed.

Raman shifts (cm^{-1}): 359w, 296s, 211, 194, 130sh, 115w.
Source: Sharma and Pandya (1974).
Comments: No independent analytical data are provided for the sample used.

Krieselite $Al_2(GeO_4)F_2$

Origin: Tsumeb mine, Namibia (type locality).
Experimental details: Raman scattering measurements have been performed on an arbitrarily oriented sample using 632 nm He-Ne laser radiation. The laser radiation power is not indicated. A 180°-scattering geometry was employed.
Raman shifts (cm^{-1}): 862, ~802sh, 718w, 294s, 224.
Source: Schlüter et al. (2010).
Comments: The sample was characterized by powder X-ray diffraction data and electron microprobe analyses. The empirical formula of the sample used is $(Al_{1.860}Ga_{0.102}As^{3+}_{0.036}Zn_{0.020}Mg_{0.016}Fe^{3+}_{0.012}Na_{0.009}Sb^{3+}_{0.005}Ti_{0.003}Cu_{0.001})(Ge_{0.844}Al_{0.143}Si_{0.013})O_4(F_{1.103}OH_{0.897})$.

Kröhnkite $Na_2Cu(SO_4)_2 \cdot 2H_2O$

Origin: Synthetic.
Experimental details: Raman scattering measurements have been performed on a polycrystalline sample using 488 and 514.5 nm Ar$^+$ laser radiations. The laser radiation power is not indicated.
Raman shifts (cm^{-1}): 3280w, 3200w, 3140w, 3100w, 1660w, 1610w, 1590w, 1182w, 1175w, 1162, 1150, 1128w, 1090w, 1045s, 989s, 840w, 740, 715w, 655, 645, 615, 585w, 560, 464s, 444s, 430sh, 300s, 280w, 260, 210w, 175w, 145w, 100s, 80, 70, 55.
Source: Pillai et al. (1997).
Comments: No independent analytical data are provided for the sample used. For the Raman spectra of kröhnkite see also Frost et al. (2013v) and Majzlan et al. (2015).

Krotite $CaAl_2O_4$

Origin: Northwest Africa 1934 meteorite (type locality).
Experimental details: Raman scattering measurements have been performed on an arbitrarily oriented sample using 514.5 nm Ar$^+$ laser radiation. The laser radiation power is not indicated.
Raman shifts (cm^{-1}): 789, 686w, 647w, 543s, 520s, 456w, 404w, 312w, 174, 150, 141.
Source: Ma et al. (2011b).
Comments: The sample was characterized by powder X-ray diffraction data and electron microprobe analyses. The crystal structure is solved. The empirical formula of the sample used is $Ca_{1.02}Al_{1.99}O_4$. For the Raman spectrum of krotite see also Janáková et al. (2007).

Krut'aite $CuSe_2$

Origin: Synthetic.
Experimental details: Raman scattering measurements have been performed on an arbitrarily oriented sample using 514.5 nm Ar$^+$ and 632.8 nm He-Ne lasers radiations. The nominal laser radiation power was 70 and 50 mW, respectively. A 180°-scattering geometry was employed.
Raman shifts (cm^{-1}): 270sh, 260, 115w.
Source: Anastassakis (1973).
Comments: No independent analytical data are provided for the sample used.

Kryzhanovskite $(Fe^{3+}, Mn^{2+})_3(PO_4)_2(OH, H_2O)_3$

Origin: Hagendorf South pegmatite, Waidhaus, Upper Palatinate, Bavaria, Germany.

Experimental details: Raman scattering measurements have been performed on arbitrarily oriented crystals using a 633 nm He-Ne laser. The laser radiation power is not indicated. The Raman shifts have been determined for the maxima of individual peaks obtained as a result of the spectral curve analysis.

Raman shifts (cm^{-1}): 3562, 3531, 3432w, 1596w, 1530sh, 1379w, 1114sh, 1081sh, 1046s, 1020sh, 1001sh, 970s, 907sh, 829w, 622sh, 601s, 581sh, 547, 512sh, 477, 414, 363sh, 329, 283s, 251sh, 199s, 179sh, 147sh, 113.

Source: Frost et al. (2016d).

Comment The sample was characterized by qualitative electron microprobe analysis.

Ktenasite $(Cu, Zn)_5(SO_4)_2(OH)_6 \cdot 6H_2O$

Origin: No data.

Experimental details: Raman scattering measurements have been performed on arbitrarily oriented crystals using a 633 nm He-Ne laser. The laser radiation power is not indicated. The Raman shifts have been determined for the maxima of individual peaks obtained as a result of the spectral curve analysis.

Raman shifts (cm^{-1}): 994, 981, 973, 604, 475, 449.

Source: Frost et al. (2013a).

Comments: No independent analytical data are provided for the sample used.

Kuksite $Pb_3Zn_3TeO_6(PO_4)_2$

Origin: Blue Bellclaims, California, USA, and Black Pine mine, Montana, USA.

Experimental details: Methods of samples preparation are not described. Raman scattering measurements have been performed using 785 nm diode laser radiation. The laser radiation power is not indicated. A 180°-scattering geometry was employed.

Raman shifts (cm^{-1}): 3036w, 1006, 734, 529, 493, 416 (for a sample from Blue Bell claims); 1017, 731, 497, 476, 424 (for a sample from Black Pine mine).

Source: Mills et al. (2010).

Comments: The samples were characterized by powder X-ray data and electron microprobe analyses. The crystal structure is solved. The empirical formulae of the samples used are $(Pb_{2.89}Bi_{0.10})(Zn_{2.84}Cu_{0.20}Fe_{0.02})Te_{1.05}(P_{1.52}Si_{0.44}As_{0.02})O_{14}$ and $Pb_{2.93}(Zn_{2.74}Cu_{0.06}Fe_{0.01})(Te_{0.58}Sb_{0.33})(P_{1.44}As_{0.74}Si_{0.11})O_{14}$, respectively.

Kulanite $BaFe^{2+}_2Al_2(PO_4)_3(OH)_3$

Origin: Rapid Creek, Dawson Mining District, Yukon, Canada (type locality).

Experimental details: Raman scattering measurements have been performed on arbitrarily oriented crystals using a 633 nm He-Ne laser. The laser radiation power is not indicated. The Raman shifts have been determined for the maxima of individual peaks obtained as a result of the spectral curve analysis.

Raman shifts (cm^{-1}): 3533, 3513w, 3339sh, 3211, 3095w, 2960sh, 2754sh, 1303w, 1235w, 1182sh, 1146, 1110, 1076sh, 1039sh, 1022s, 1012sh, 1006sh, 967, 928w, 665w, 624sh, 616, 585, 569, 553, 525, 492, 456, 438, 418, 358w, 343, 317w, 279, 213, 196sh, 186sh, 161sh, 137, 122.
Source: Frost et al. (2013k).
Comments: No independent analytical data are provided for the sample used.

Kullerudite NiSe$_2$

Origin: Synthetic.
Experimental details: Raman scattering measurements have been performed on a polycrystalline thin film using 514.5 nm Ar$^+$ laser radiation. The laser radiation power density on the surface was of the order of 100 kW cm^{-2}. A 180°-scattering geometry was employed.
Raman shifts (cm^{-1}): 243, 214s, 170, 152.
Source: De las Heras and Agulló-Rueda (2000).
Comments: The sample was characterized by powder X-ray diffraction data. For the Raman spectrum of kullerudite see also Zhuo et al. (2015).

Kumdykolite Na(AlSi$_3$O$_8$)

Origin: Village of Staré, České Středohoří Mts., Czech Republic.
Experimental details: Raman scattering measurements have been performed on an arbitrarily oriented sample using 532 nm laser radiation. The nominal laser radiation power was 10 mW.
Raman shifts (cm^{-1}): 492s, 464, 407w, 284sh, 265, 222, 155s.
Source: Kotková et al. (2014).
Comments: The sample was characterized by electron microprobe analyses. For the Raman spectra of kumdykolite see also Hwang et al. (2009) and Ferrero et al. (2016).

Kumtyubeite Ca$_5$(SiO$_4$)$_2$F$_2$

Origin: Upper Chegem volcanic structure, Kabardino-Balkaria, Northern Caucasus, Russia (type locality).
Experimental details: Raman scattering measurements have been performed on an arbitrarily oriented single crystal using 514.5 nm Ar$^+$ laser radiation. The nominal laser radiation power was in the range from 20 to 40 mW. A 0°-scattering geometry was employed.
Raman shifts (cm^{-1}): 3561, 3553s, 3544, 925w, 901, 849, 822s, 547, 525w, 420, 397, 323, 299w, 281.
Source: Galuskina et al. (2009).
Comments: The sample was characterized by powder X-ray diffraction data and electron microprobe analyses. The crystal structure is solved. The empirical formula of the sample used is Ca$_5$(Si$_{1.99}$Ti$_{0.01}$)O$_8$(F$_{1.39}$OH$_{0.61}$).

Kuramite Cu$_3$SnS$_4$

Origin: Synthetic.
Experimental details: Raman scattering measurements have been performed on an arbitrarily oriented sample using 785 nm Ar$^+$ laser radiation. The laser radiation power is not indicated.
Raman shifts (cm^{-1}): 346s, 333s, 316s, 289.

Source: Gusain et al. (2015).
Comments: The sample was characterized by powder X-ray diffraction data and electron microprobe analyses.

Kuranakhite $PbMn^{4+}Te^{6+}O_6$

Origin: Moctezuma, Sonora, Mexico.
Experimental details: Raman scattering measurements have been performed on an arbitrarily oriented sample using 488 nm Ar$^+$ laser radiation. The laser radiation power is not indicated.
Raman shifts (cm^{-1}): 677, 619s, 508, 380, 320, 310.
Source: Grundler et al. (2008).
Comments: The sample was characterized by electron microprobe analyses. Raman spectrum of presumed kuranakhite published by Frost and Keeffe (2009e) without accompanying analytical data is questionable.

Kuratite $Ca_2(Fe^{2+}{}_5Ti)O_2[Si_4Al_2O_{18}]$

Origin: D'Orbigny angrite meteorite (type locality).
Experimental details: Raman scattering measurements have been performed on an arbitrarily oriented sample using 514.5 nm Ar$^+$ laser radiation. The laser radiation power is not indicated.
Raman shifts (cm^{-1}): 996, 856, 699s, 563, 500sh, 351.
Source: Hwang et al. (2016).
Comments: The sample was characterized by selected area electron diffraction and electron microprobe analysis. The empirical formula of the sample used is $(Ca_{3.88}Na_{0.02}REE^{3+}{}_{0.03}Mn_{0.03}Mg_{0.01}Ni_{0.02}Zn_{0.01}Sr_{0.01})(Fe^{2+}{}_{9.98}Ti_{2.00})(Si_{7.80}Al_{3.52}Fe^{3+}{}_{0.64}P_{0.05}S_{0.02})O_{39.98}F_{0.01}Cl_{0.01}$.

Kurnakovite $MgB_3O_3(OH)_5 \cdot 5H_2O$

Origin: No data.
Experimental details: Raman scattering measurements have been performed on an arbitrarily oriented sample using 514.5 nm Ar$^+$ laser radiation. The nominal laser radiation power was 300 mW.
Raman shifts (cm^{-1}): 944, 850s, 627, 466w, 422, 391.
Source: Jun et al. (1995).
Comments: The sample was characterized by powder X-ray diffraction data.

Kusachiite $Cu^{2+}Bi^{3+}{}_2O_4$

Origin: Synthetic.
Experimental details: Raman scattering measurements have been performed on a powder sample using 1064 nm laser radiation. The laser radiation power is not indicated.
Raman shifts (cm^{-1}): 914, 879, 625, 599, 560, 497, 482s, 451s, 415.
Source: Anandan et al. (2012).
Comments: The sample was characterized by powder X-ray diffraction data and electron microprobe analyses.

Kushiroite $CaAlAlSiO_6$

Origin: ALH 85085 CH chondrite.

Experimental details: Raman scattering measurements have been performed on an arbitrarily oriented sample using 488 and 514.5 nm Ar⁺ laser radiations. The nominal laser radiations power was in the range from 12 to 20 mW.

Raman shifts (cm^{-1}): 959, 675, 520w, 410sh, 369, 334.

Source: Kimura et al. (2009).

Comments: The sample was characterized by electron backscatter diffraction and electron microprobe analyses. The empirical formula of the sample used is $Ca_{1.008}(Mg_{0.094}Fe_{0.034}Al_{0.878})(Al_{0.921}Si_{1.079})O_6$. For the Raman spectrum of kushiroite see also Ma et al. (2009).

Kutnohorite $CaMn^{2+}(CO_3)_2$

Origin: No data.

Experimental details: Methods of sample preparation are not described. Raman scattering measurements have been performed using 514.5 nm Ar⁺ laser radiation. The laser radiation output power was in the range from 200 to 300 mW. A 180°-scattering geometry was employed.

Raman shifts (cm^{-1}): 1740w, 1420w, 1086s, 716, 284.

Source: Herman et al. (1987).

Comments: The sample was characterized by powder X-ray diffraction data and by quantitative chemical analysis.

Kuzminite $HgBr$

Origin: Synthetic.

Experimental details: Raman scattering measurements have been performed on an oriented single crystal using 514.5 nm Ar⁺ and 632.8 nm He-Ne laser radiations. The nominal laser radiation power was in the range from tens to hundreds mW. Spectra were collected in the (zz), (xz), and (yz) scattering geometries.

Raman shifts (cm^{-1}): 208, 128, 85, 35.

Source: Markov and Roginskii (2011).

Comments: No independent analytical data are provided for the sample used. The Raman shifts are given as the sum of the Raman shifts for different scattering geometries. For the Raman spectrum of kuzminite see also Ōsaka (1971).

Kyanite Al_2OSiO_4

Origin: Harts Range, Northern Territory, Australia.

Experimental details: Raman scattering measurements have been performed on an arbitrarily oriented single crystal using 514.5 nm Ar⁺ laser radiation. The laser radiation power at the sample was 60 mW.

Raman shifts (cm^{-1}): 998w, 952s, 900w, 669sh, 654, 632, 606, 562, 486s, 437, 419, 405, 386, 360, 325, 302.

Source: Mernagh and Liu (1991).

Comments: The sample was characterized by electron microprobe analysis. The Raman shifts were partly determined by us based on spectral curve analysis of the published spectrum. For the Raman spectra of kyanite see also Makreski et al. (2005b), Yang et al. (2007b), Andò and Garzanti (2014), and Culka et al. (2016a).

Kyawthuite $Bi^{3+}Sb^{5+}O_4$

Origin: Synthetic.
Experimental details: Raman scattering measurements have been performed on an arbitrarily oriented sample using 488 nm Ar$^+$ laser radiation. The nominal laser radiation power was less than 10 mW. A 180°-scattering geometry was employed. The Raman shifts for asymmetric peaks have been determined for the maxima of individual peaks obtained as a result of the spectral curve analysis.
Raman shifts (cm^{-1}): 783w, 730w, 636w, 603w, 452, 420w, 394, 387sh, 319w, 252, 158s, 137s, 133sh, 56s.
Source: Errandonea et al. (2016).
Comments: The sample was characterized by powder X-ray diffraction data and electron microprobe analyses. For the Raman spectrum of kyawthuite see also Loubbidi et al. (2014).

Laachite $(Ca,Mn)_2Zr_2Nb_2TiFeO_{14}$

Origin: Dellen (Zieglowski) pumice quarry, 1.5 km NE of Mendig, Laacher See volcano, Eifel region, Rhineland-Palatinate, Germany (typa locality).
Experimental details: Raman scattering measurements have been performed on an oriented single crystal using 532 nm diode laser radiation. The laser radiation power about 6 mW. Raman spectra were collected with the polarization of the laser beam parallel to the a axis of the crystal (A) and with the polarization of the laser beam lying in the plane (010), perpendicular to the a axis of the crystal (B) scattering geometries. The Raman shifts have been determined for the maxima of individual peaks obtained as a result of the spectral curve analysis.
Raman shifts (cm^{-1}): 1568w, 1190w, 821sh, 752, 584s, 468sh, 300s, 183w (A); 1175w, 832sh, 757, 588s, 485, 371sh, 330, 213sh, 192 (B).
Source: Chukanov et al. (2014a).
Comments: The sample was characterized by powder X-ray diffraction data and electron microprobe analysis. The crystal structure is solved. The empirical formula of the sample used is $(Ca_{0.66}Mn_{0.37}Th_{0.25}Y_{0.20}La_{0.11}Ce_{0.34}Nd_{0.11})(Zr_{1.36}Mn_{0.64})(Nb_{1.81}Ti_{1.19})(Fe_{0.69}Al_{0.17}Mn_{0.14})O_{14.00}$.

Lacroixite $NaAl(PO_4)F$

Origin: Ehrenfriedersdorf, Germany.
Experimental details: No data.
Raman shifts (cm^{-1}): 1001s, 623, 609.
Source: Frezzotti et al. (2012).
Comments: No independent analytical data are provided for the sample used.

Lafossaite TlCl

Origin: Synthetic.
Experimental details: Raman scattering measurements have been performed on an arbitrarily oriented polished crystals using 457.9 nm, 476.5 nm, and 514.5 nm Ar$^+$ laser radiations. The sample was

immersed in pumped liquid helium or in liquid nitrogen. The laser radiation power at the sample was below 200 mW.
Raman shifts (cm^{-1}): 340s, 237, 190, 172s, 147, 111, 98, 85sh, 70sh, 36, 26sh.
Source: Nanba et al. (1987).
Comments: No independent analytical data are provided for the sample used. In Iafossaite, the first order Raman scattering process is forbidden by the inversion symmetry (the alternative prohibition rule). However, the second order Raman scattering spectrum is allowed. Raman shifts are given for a sample at 77 K.

Laihunite $(Fe^{3+},Fe^{2+},\square)_2(SiO_4)$

Origin: Lau-Hi, China.
Experimental details: Experimental details are not indicated. Raman scattering measurements have been performed on an arbitrarily oriented polished sample.
Raman shifts (cm^{-1}): 896s, 785w, 592sh, 568s, 506w, 428w, 355sh, 312s.
Source: Kuebler et al. (2011).
Comments: The sample was characterized by electron microprobe analyses.

Lakargiite $CaZrO_3$

Origin: Upper Chegem caldera, Kabardino-Balkaria, Northern Caucasus, Russia.
Experimental details: Raman scattering measurements have been performed on an arbitrarily oriented polished sample using 514.5 nm Ar$^+$ laser radiation. The nominal laser radiation power was in the range from 40 to 60 mW. A 0°-scattering geometry was employed. Raman spectrum was obtained in the spectral region from 50 to 4000 cm^{-1}.
Raman shifts (cm^{-1}): ~800w, ~715w, ~525w, ~470, ~442s, ~355s, ~285s, ~262s, ~240w, ~215w, ~182w, ~152w.
Source: Galuskin et al. (2011b).
Comments: The sample was characterized by powder X-ray diffraction analysis and by electron microprobe analyses. Raman shifts are given for lakargiite with the following chemical composition: lakargiite CaZrO$_3$ 67%, megawite CaSnO$_3$ 27%, perovskite CaTiO$_3$ 2%, and others 4%.

Lamprophyllite $Na_3(SrNa)Ti_3(Si_2O_7)_2O_2(OH)_2$

Origin: Rasvumchorr Mt., Khibiny massif, Kola Peninsula, Russia.
Experimental details: Raman scattering measurements have been performed on arbitrarily oriented crystals using a 633 nm He-Ne laser. The laser radiation power is not indicated. The Raman shifts have been determined for the maxima of individual peaks obtained as a result of the spectral curve analysis.
Raman shifts (cm^{-1}): 1113w, 1072w, 1049w, 1028w, 1001w, 972w, 940s, 918s, 888sh, 861w, 852sh, 801w, 782sh, 707, 671, 595sh, 576, 538w, 516w, 459w, 445sh, 411w, 349, 319w, 294sh, 282sh, 270, 257sh, 227w, 208sh, 201, 177sh, 168sh, 151, 137sh, 114w.
Source: Frost et al. (2015ab).
Comments: The sample was characterized by qualitative electron microprobe analysis.

Lanarkite $Pb_2O(SO_4)$

Origin: Leadhills, Scotland, UK.
Experimental details: Raman scattering measurements have been performed on an arbitrarily oriented polished cross-section of the sample using 632.8 nm He-Ne laser radiation. The laser radiation power at the sample was in the range from 0.02 to 2 mW.
Raman shifts (cm^{-1}): 1070, 1055, 976s, 619w, 601w, 439w, 426w, 334, 284, 147s.
Source: Correia et al. (2007).
Comments: The sample was characterized by powder X-ray diffraction data and electron microprobe analysis.

Långbanite $Mn^{2+}_4Mn^{3+}_9Sb^{5+}O_{16}(SiO_4)_2$

Origin: Långban mine, Bergslagen ore district, Filipstad, Värmland, Sweden (type locality).
Experimental details: Raman scattering measurements have been performed on arbitrarily oriented crystals. Other experimental details are not indicated. The Raman shifts have been determined for the maxima of individual peaks obtained as a result of the spectral curve analysis.
Raman shifts (cm^{-1}): 3699sh, 3680w, 3476sh, 3076sh, 2636w, 2488w, 2240w, 1947w, 1718w, 1432w, 1200sh, 1130, 1094sh, 1034sh, 1012sh, 986s, 964s, 897sh, 872, 671, 646sh, 558sh, 542, 463w, 415sh, 386, 351, 330sh, 258w, 202.
Source: Bahfenne and Frost (2010a).
Comments: No independent analytical data are provided for the sample used.

Langbeinite $K_2Mg_2(SO_4)_3$

Origin: Synthetic.
Experimental details: Methods of sample preparation are not described. Raman scattering measurements have been performed on an arbitrarily oriented sample using 514 nm and 785 nm lasers radiations. The laser radiation power at the sample was less than 1 mW.
Raman shifts (cm^{-1}): 1245w, 1134w, 1123w, 1053s, 626w, 466w, 457w.
Source: Morillas et al. (2016).
Comments: The sample was characterized by electron microprobe analysis.

Langite $Cu_4(SO_4)(OH)_6 \cdot 2H_2O$

Origin: Cornwall, UK.
Experimental details: Raman scattering measurements have been performed on an arbitrarily oriented sample using 780 nm Nd-YAG laser radiation. The nominal laser radiation power was less than 1 mW. The Raman shifts have been determined for the maxima of individual peaks obtained as a result of the spectral curve analysis.
Raman shifts (cm^{-1}): 3587, 3564, 3405, 3372, 3260w, 1911, 1906, 1266, 1172, 1149, 1128, 1102, 1076, 982sh, 974s, 911w, 773w, 732, 621, 609, 596s, 507, 481, 449, 420, 391, 317, 273sh, 258sh, 241, 226sh, 1912, 183sh, 175, 167, 155, 147, 139, 130sh, 118sh.
Source: Martens et al. (2003a).
Comments: No independent analytical data are provided for the sample used.

Lanmuchangite $TlAl(SO_4)_2 \cdot 12H_2O$

Origin: Synthetic.
Experimental details: The stimulated Raman scattering measurements have been performed on an oriented sample using 532 nm and 1064 nm Nd-YAG laser radiations. The laser radiation power is not indicated. Raman spectra were collected in the scattering geometries with pumping and registration along [110] direction and polarization both emissions perpendicular to the [110] direction.
Raman shifts (cm^{-1}): 991s.
Source: Kaminskii et al. (2004).
Comments: No independent analytical data are provided for the sample used.

Lansfordite $Mg(CO_3) \cdot 5H_2O$

Origin: Synthetic.
Experimental details: Raman scattering measurements have been performed below 0°C on an arbitrarily oriented microcrystalline sample using 532 nm Nd-YAG laser radiation. The nominal laser radiation power was about 8 mW.
Raman shifts (cm^{-1}): 3264s, 1705w, 1514w, 1424w, 1098s, 774w, 698w, 225s.
Source: Coleyshaw et al. (2003).
Comments: No independent analytical data are provided for the sample used.

Lanthanite-(Nd) $Nd_2(CO_3)_3 \cdot 8H_2O$

Origin: Whitianga quarry, Coromandel Peninsula, New Zealand.
Experimental details: Raman scattering measurements have been performed on arbitrarily oriented crystal blades using 514.5 nm Ar$^+$ laser radiation. The nominal laser radiation power was 25 mW. A 180°-scattering geometry was employed.
Raman shifts (cm^{-1}): 3471s, 3280s, 3072s, 2865s, 1702w, 1636w, 1615w, 1612w, 1581, 1576w, 1559w, 1557w, 1513w, 1505w, 1487w, 1459w, 1454w, 1418w, 1394w, 1365, 1294w, 1292w, 1286w, 1173w, 1093s, 968w, 762w, 732w, 686w, 372w, 356w, 277w, 233.
Source: Graham et al. (2007).
Comments: The sample was characterized by powder X-ray diffraction data, electron microprobe analysis and laser-ablation inductively coupled plasma mass spectrometry. The empirical formula of the sample used is $(Nd_{0.63}La_{0.59}Ce_{0.35}Pr_{0.15}Sm_{0.10}Gd_{0.069}Y_{0.06}Eu_{0.03}Dy_{0.02}Ga_{0.01})(CO_3)_3 \cdot 8H_2O$.

Lapeyreite $Cu_3O[AsO_3(OH)]_2 \cdot H_2O$

Origin: Alpes-Maritimes Region, Nice, France (type locality).
Experimental details: Raman scattering measurements have been performed on an arbitrarily oriented fragment of the holotype sample using 532 nm He-Ne laser radiation. The laser radiation power is not indicated.
Raman shifts (cm^{-1}): 1047, 627, 405, 214s, 198sh, 188sh, 141, 100.
Source: Hatipoglu and Babalik (2012).

Larnite $Ca_2(SiO_4)$

Origin: No data.

Experimental details: Raman scattering measurements have been performed on an arbitrarily oriented sample using 514.5 nm Ar^+ laser radiation. The laser radiation power at the sample was 17 mW. A 180°-scattering geometry was employed.

Raman shifts (cm^{-1}): 1577w, 1112w, 1085w, 976, 949, 914, 897w, 871sh, 858s, 845sh, 669w, 564w, 554w, 536, 524sh, 516sh, 443, 368w, 300, 274sh, 252w, 241, 222w, 201w, 165, 146, 101w, 76w.

Source: Sokol et al. (2015).

Comments: The sample was characterized by powder X-ray diffraction data and electron microprobe analysis. For the Raman spectrum of larnite see also Piriou and McMillan (1983).

Laueite $Mn^{2+}Fe^{3+}_2(PO_4)_2(OH)_2 \cdot 8H_2O$

Origin: Cigana mine, Conselheiro Pena, Minas Gerais, Brazil.

Experimental details: Raman scattering measurements have been performed on an arbitrarily oriented sample using 633 nm He-Ne laser radiation. The laser radiation power at the sample was 0.1 mW. The Raman shifts have been determined for the maxima of individual peaks obtained as a result of the spectral curve analysis.

Raman shifts (cm^{-1}): 3515sh, 3478sh, 3430, 3379, 3297sh, 3080sh, 1692sh, 1613w, 1504sh, 1096sh, 1069, 1045s, 1021, 997sh, 980s, 864w, 731, 551, 542, 525, 472, 456, 404, 357, 335, 279, 265, 253s, 240, 226sh, 186, 172, 161s, 138, 115, 110.

Source: Frost et al. (2016b).

Comments: The empirical formula based on the semiquantitative chemical analyses of the sample used is $(Mn^{2+}_{0.85}Fe^{2+}_{0.10}Mg_{0.05})(Fe^{3+}_{1.90}Al_{0.10})(PO_4)_2(OH)_2 \cdot 8H_2O$.

Laumontite $CaAl_2Si_4O_{12} \cdot 4H_2O$

Origin: Grodziszcze, Poland.

Experimental details: Methods of sample preparation are not described. Raman scattering measurements have been performed using Nd-YAG laser radiation. The laser radiation power at the sample was 300 mW.

Raman shifts (cm^{-1}): 1023, 948, 817, 674, 593, 517s, 493s, 385, 327s, 201, 164.

Source: Mozgawa (2001).

Comments: The sample was characterized by powder X-ray diffraction data.

Laurentianite $[NbO(H_2O)]_3(Si_2O_7)_2[Na(H_2O)_2]_3$

Origin: Poudrette quarry, Mont Saint-Hilaire, Quebec, Canada (type locality).

Experimental details: Raman scattering measurements have been performed on an arbitrarily oriented single crystal using 532 laser radiation. The laser radiation power is not indicated. A 180°-scattering geometry was employed.

Raman shifts (cm^{-1}): 3421, 33267, 3024, 1103w, 1054w, 920s, 841s, 771w, 700, 628w, 597, 560w, 486w, 467w, 405, 344w, 309, 292, 241, 218w, 193sh, 176, 138sh, 122, 90.

Source: Haring et al. (2012).

Comments: The sample was characterized by powder X-ray diffraction data and electron microprobe analyses. The crystal structure is solved. The empirical formula of the sample used is $[(Nb_{0.99}Ti_{0.01})O(H_2O)]_3(Si_{2.00}O_7)_2[(Na_{0.86}\square_{0.10}K_{0.02}Ca_{0.01})(H_2O)_2]_3$.

Laurionite PbCl(OH)

Origin: Synthetic.
Experimental details: Raman scattering measurements have been performed on a polycrystalline and on an oriented single crystal using 514.5 nm Ar$^+$ laser radiation. The laser radiation power is not indicated. 90° and 180°-scattering geometries were employed in single-crystal experiments. Raman spectra were obtained in the spectral region from 50 to 4000 cm^{-1}. Polarized spectra were collected in the $c(xx)$–c, $c(yy)$–c, $a(zz)$–a, $a(yx)c$, $a(zx)c$, and $a(zy)c$ scattering geometries.
Raman shifts (cm^{-1}): 3517s, 665sh, 595, 505w, 446w, 327, 272, 175w, 123sh, 111s, 105s, 87sh, 51s.
Source: Lutz et al. (1995).
Comments: The sample was characterized by powder X-ray diffraction analysis. The Raman shifts are given for a polycrystalline sample.

Laurite RuS$_2$

Origin: Santa Elena Nappe, Costa Rica.
Experimental details: Raman scattering measurements have been performed on an arbitrarily oriented sample using 532.6 nm Nd-YAG laser radiation. The laser radiation power is not indicated.
Raman shifts (cm^{-1}): ~395w, 330s.
Source: Zaccarini et al. (2010).
Comments: The sample was characterized by electron microprobe analyses.

Laurite RuS$_2$

Origin: No data.
Experimental details: Methods of sample preparation are not described. Raman scattering measurements have been performed using 532.1 nm Nd-YAG laser radiation. The laser radiation power at the sample was in the range from 1 to 2 mW.
Raman shifts (cm^{-1}): ~395sh, 364–351s.
Source: Bakker (2014).
Comments: No independent analytical data are provided for the sample used.

Lausenite Fe$^{3+}_2$(SO$_4$)$_3 \cdot$5H$_2$O

Origin: Synthetic.
Experimental details: Methods of sample preparation are not described. Raman scattering measurements have been performed using 532 nm laser radiation. The laser radiation power is not indicated.
Raman shifts (cm^{-1}): 3425sh, 3323sh, 3195, 3057sh, 1653w, 1605w, 1189, 1119w, 1087w, 1052sh, 1036s, 1017, 799w, 652w, 631w, 614w, 599w, 494, 468, 457, 441sh, 416sh, 281, 253.
Source: Ling and Wang (2010).
Comments: The sample was characterized by powder X-ray diffraction data.

Lautarite $Ca(IO_3)_2$

Origin: Synthetic.
Experimental details: No data.
Raman shifts (cm^{-1}): 830s, 808w, 794s, 775, 759s, 737s, 427, 394, 362sh, 351, ~333sh, 326, ~315sh, 266w, 235.
Source: Alici et al. (1992).
Comments: The sample was characterized by powder X-ray diffraction data.

Lavendulan $NaCaCu_5(AsO_4)_4Cl \cdot 5H_2O$

Origin: Alice Mary Mine, Kundip, Western Australia, Australia.
Experimental details: Experimental details are not indicated. Raman scattering measurements have been performed on an arbitrarily oriented sample using 633 nm He-Ne laser radiation. The Raman shifts have been determined for the maxima of individual peaks obtained as a result of the spectral curve analysis.
Raman shifts (cm^{-1}): 1053w, 981w, 893sh, 878sh, 856s, 783, 614w, 543s, 406, 342w, 278w, 226, 176.
Source: Frost et al. (2007m).
Comments: No independent analytical data are provided for the sample used.

Lavinskyite $K(LiCu)Cu_6(Si_4O_{11})_2(OH)_4$

Origin: Wessels mine, Kalahari Manganese Fields, South Africa (type locality).
Experimental details: Methods of sample preparation are not described. Raman scattering measurements have been performed on an arbitrarily oriented sample using 532 nm solid-state laser radiation. The laser radiation power is not indicated.
Raman shifts (cm^{-1}): 3694, 3662s, 3630sh, 3390, 1090, 1043, 991, 919, 891, 685s, 580w, 562, 503, 445+424s, 401.
Source: Yang et al. (2014).
Comments: The sample was characterized by powder X-ray diffraction data and electron microprobe analyses. The crystal structure is solved. The empirical formula of the sample used is $(K_{0.99}Ba_{0.01})(Li_{1.04}Cu_{0.93}Na_{0.10})(Cu_{5.57}Mg_{0.43}Mn_{0.01})(Si_{4.00}O_{11})_2(OH)_4$.

Lawrencite $FeCl_2$

Origin: Synthetic.
Experimental details: Raman scattering measurements have been performed at 277 K using 647.1 nm Kr$^+$ laser radiation. The laser light was directed along the c face of the crystal with polarization in the plane of incidence. In this configuration the scattered light is largely depolarised and there were no polarization effects. The nominal laser radiation power was about 100 mW. A 90°-scattering geometry was employed.
Raman shifts (cm^{-1}): 246, 144s.
Source: Johnstone et al. (1978).
Comments: No independent analytical data are provided for the sample used.

Lawsonite $CaAl_2(Si_2O_7)(OH)_2 \cdot H_2O$

Origin: Tiburon Peninsula, California, USA (type locality).
Experimental details: Raman scattering measurements have been performed at different pressures, on an arbitrarily oriented single-crystal slice oriented parallel to (001), using 514.5 nm Ar^+ laser radiation. The laser radiation power is not indicated. A 180°-scattering geometry was employed.
Raman shifts (cm^{-1}): 3541, 961sh, 940s, 918sh, 800, 696, 565s, 462w, 434w, 364w, 330, 282, 280.
Source: Daniel et al. (2000).
Comments: The sample was characterized by powder X-ray diffraction data. Raman shifts are given for sample at ambient conditions (pressure 0.1 MPa).

Lazaridisite $Cd_3(SO_4)_3 \cdot 8H_2O$

Origin: Synthetic.
Experimental details: Raman scattering measurements have been performed on an arbitrarily oriented microcrystal using 632.8 nm He-Ne laser radiation. The laser radiation power is not indicated.
Raman shifts (cm^{-1}): 1117, 1004s, 625w, 460, 330w.
Source: Falgayrac et al. (2013).
Comments: The sample was characterized by powder X-ray diffraction data.

Lazulite $MgAl_2(PO_4)_2(OH)_2$

Origin: Gentil mine, Mendes Pimentel, east of Minas Gerais, Brazil.
Experimental details: Raman scattering measurements have been performed on an arbitrarily oriented sample using 633 nm He-Ne laser radiation. The laser radiation power is not indicated. The Raman shifts have been determined for the maxima of individual peaks obtained as a result of the spectral curve analysis.
Raman shifts (cm^{-1}): 3478w, 3402s, 3385sh, 3373sh, 3146w, 1684w, 1528sh, 1509w, 1271w, 1214w, 1139sh, 1137s, 1102, 1089sh, 1060s, 1019, 1004sh, 865w, 790w, 742, 714w, 669sh, 648, 633sh, 623, 613sh, 605sh, 580w, 568w, 527w, 479, 460w, 425, 414, 394sh, 378, 365sh, 347, 333sh, 322, 282, 254, 225, 197sh, 195sh, 190s, 173, 137.
Source: Frost et al. (2013p).
Comments: The sample was characterized by powder X-ray diffraction data and electron microprobe analyses. For the Raman spectrum of lazulite see also Frezzotti et al. (2012).

Lazurite $Na_3Ca(Si_3Al_3)O_{12}S$

Origin: Badakhshan, Afghanistan.
Experimental details: Raman scattering measurements have been performed on an arbitrarily oriented crystal using 514.5 nm Ar^+ laser radiation. The nominal laser radiation power was 10 mW.
Raman shifts (cm^{-1}): 1090, 970w, 801w, 636w, 582sh, 545s, 413w, 258.
Source: Caggiani et al. (2014).
Comments: The sample was characterized by powder X-ray diffraction analysis and by energy-dispersive X-ray scan analysis. For the Raman spectra of lazurite see also Ostroumov et al. (2002).

Lead Pb

Origin: Karrantza Valley, the westerner area of the Basque Co., Spain.
Experimental details: Raman scattering measurements have been performed on an arbitrarily oriented sample using 514.5 nm Ar$^+$ laser radiation. The laser radiation power at the sample was 20 mW.
Raman shifts (cm^{-1}): 153s.
Source: Goienaga et al. (2011).
Comments: No independent analytical data are provided for the sample used.

Leadhillite $Pb_4(SO_4)(CO_3)_2(OH)_2$

Origin: Hard Luck Claim, near Baker, San Bernardino Co., California, USA.
Experimental details: Raman scattering measurements have been performed on an arbitrarily oriented sample using 633 nm He-Ne laser radiation. The laser radiation power is not indicated. The Raman shifts have been determined for the maxima of individual peaks obtained as a result of the spectral curve analysis.
Raman shifts (cm^{-1}): 3481, 3386sh, 1719, 1704, 1674, 1375, 1097sh, 1054s, 1016sh, 964s, 856, 703w, 677w, 626w, 599w, 458sh, 428s, 360w, 307w, 262w, 220sh, 195sh, 173s.
Source: Frost et al. (2003e).
Comments: No independent analytical data are provided for the sample used.

Lechatelierite SiO_2

Origin: Synthetic.
Experimental details: Raman scattering measurements have been performed on an arbitrarily oriented sample using 632.8 nm He-Ne laser radiation. The laser radiation power at the sample was 3 mW.
Raman shifts (cm^{-1}): 1110w, 802, 603, 493, 441, ~300sh.
Source: Kowitz et al. (2013).
Comments: The sample was characterized by energy-dispersive X-ray scan analysis. Raman shifts are given for a sample subjected to a shock more than 36 GPa.

Leguernite $Bi_{12.67}O_{14}(SO_4)_5$

Origin: La Fossa crater, Vulcano, Aeolian Islands, Italy (type locality).
Experimental details: Raman scattering measurements have been performed on an arbitrarily oriented single crystal using 532 nm laser radiation. The nominal laser radiation power was 1.4 mW.
Raman shifts (cm^{-1}): 2430, 1145sh, 1019, 970s, 603sh, 473, 429, 279s, 243, 183, 150.
Source: Garavelli et al. (2014).
Comments: The sample was characterized by powder X-ray diffraction data and electron microprobe analysis. The crystal structure is solved. The empirical formula of the sample used is $(Bi_{12.40}Pb_{0.15})S_{5.08}O_{34}$.

Leightonite $K_2Ca_2Cu(SO_4)_4 \cdot 2H_2O$

Origin: Chuquicamata mine, Antofagasta region, Chile (type locality).
Experimental details: Raman scattering measurements have been performed on an arbitrarily oriented single crystal using 633 nm He-Ne laser radiation. The laser radiation power is not indicated. The

Raman shifts have been determined for the maxima of individual peaks obtained as a result of the spectral curve analysis.

Raman shifts (cm^{-1}): 3457sh, 3435, 3386sh, 3364sh, 3350s, 3329s, 3310sh, 3206w, 3088w, 2911w, 2856w, 1803w, 1748w, 1700sh, 1670w, 1446w, 1177w, 1163w, 1137w, 1120w, 1047, 990sh, 975sh, 912, 864w, 846sh, 823, 753w, 654, 640sh, 612, 601, 589sh, 513, 463, 446, 425sh, 413, 394, 361, 298, 266, 238w, 217, 155sh, 149, 137, 126sh, 120, 106.

Source: Frost et al. (2013j).

Comments: The sample was characterized by qualitative electron microprobe analysis.

Leiteite $ZnAs^{3+}_2O_4$

Origin: Synthetic.

Experimental details: Raman scattering measurements have been performed on an arbitrarily oriented sample and oriented crystal using 633 nm He-Ne laser radiation. The laser radiation power is not indicated. Polarized spectra were collected in the $c(bb)c$, $c(ba)c$, $c(aa)c$, $b(ac)b$, $b(aa)b$, $a(bb)a$, $a(bc)a$, and $a(cc)a$ scattering geometries. The Raman shifts have been determined for the maxima of individual peaks obtained as a result of the spectral curve analysis.

Raman shifts (cm^{-1}): 804, 763, 647w, 600, 566sh, 548, 457s, 3768w, 366w, 304w, 265sh, 254s, 217, 199w.

Source: Bahfenne et al. (2011b).

Comments: The sample was characterized by powder X-ray diffraction analysis and electron microprobe analysis. For the Raman spectra of leiteite see also Origlieri et al. (2009) and Frost and Bahfenne (2010d).

Lemanskiite $NaCaCu_5(AsO_4)_4Cl \cdot 5H_2O$

Origin: El Guanqco mine, Antofagsta, Chile (type locality).

Experimental details: Raman scattering measurements have been performed on an arbitrarily oriented sample using a 633 nm He-Ne laser. The Raman shifts have been determined for the maxima of individual peaks obtained as a result of the spectral curve analysis.

Raman shifts (cm^{-1}): 1369, 1264, 1165w, 910w, 878s, 853s, 800w, 775w, 545, 479w, 440w, 400w, 345, 280sh, 262w, 243w, 220, 172.

Source: Frost et al. (2007m).

Comments: No independent analytical data are provided for the sample used.

Lemoynite $Na_2CaZr_2Si_{10}O_{26} \cdot 5\text{-}6H_2O$

Origin: Poudrette quarry, Mont Saint-Hilaire, Montérégie, Quebec, Canada (type locality).

Experimental details: Methods of sample preparation are not described. Raman scattering measurements have been performed using 532 nm laser radiation. The laser radiation power is not indicated.

Raman shifts (cm^{-1}): 952–961, ~600–605s, 533–540s, 426–429s, (~360w), (~325w), (~280), (~250w).

Source: McDonald et al. (2015).

Comments: No independent analytical data are provided for the sample used.

Leogangite $Cu_{10}(AsO_4)_4(SO_4)(OH)_6 \cdot 8H_2O$

Origin: Monte Avanza Mine, Formi Avoltri, Udine Province, Fruili Venezia Giulia, Italy.

Experimental details: Raman scattering measurements have been performed on arbitrarily oriented crystals using a 633 nm He-Ne laser. The laser radiation power is not indicated. The Raman shifts have been determined for the maxima of individual peaks obtained as a result of the spectral curve analysis.

Raman shifts (cm^{-1}): 3619sh, 3496, 3316, 3181sh, 2929, 2884sh, 2854w, 1627w, 1576w, 1461sh, 1441w, 1070w, 996w, 904sh, 868s, 827sh, 628w, 606sh, 519, 498sh, 431, 416sh, 336sh, 324, 312sh, 263, 228sh, 205, 153, 139sh, 106.

Source: Frost et al. (2011t).

Comments: No independent analytical data are provided for the sample used.

Leószilárdite $Na_6Mg(UO_2)_2(CO_3)_6 \cdot 6H_2O$

Origin: Markey Mine, San Juan County, Utah, USA (type locality).

Experimental details: Raman scattering measurements have been performed on an arbitrarily oriented sample using 785 nm diode laser radiation. The nominal laser radiation power was 200 mW.

Raman shifts (cm^{-1}): 1535w, 1396w, 1328w, 1078, 1062, 1052, 824s, 742, 728, 705w, 695w, 345, 290, 254, 193sh, 172sh, 161s, 144s, 125s.

Source: Olds et al. (2016b).

Comments: The sample was characterized by powder X-ray diffraction data and electron microprobe analyses. The crystal structure is solved. The empirical formula of the sample used is $Na_{5.60}Mg_{0.90}U_2O_{28}C_6H_{12.60}$.

Lepidocrocite $Fe^{3+}O(OH)$

Origin: Synthetic.

Experimental details: Raman scattering measurements have been performed on an arbitrarily oriented sample using 785 nm diode laser radiation. The output laser radiation power was more than 300 mW. A 180°-scattering geometry was employed.

Raman shifts (cm^{-1}): 647w, 524s, 374s, 345s, ~315w, 284s, 249sh, 214, 140.

Source: Das and Hendry (2011).

Comments: The sample was characterized by powder X-ray diffraction data. For the Raman spectra of lepidocrocite see also De Faria et al. (1997) and Bouchard and Smith (2003).

Letovicite $(NH_4)_3H(SO_4)_2$

Origin: Synthetic.

Experimental details: Raman scattering measurements have been performed at different temperatures on an oriented crystal using 514.5 nm Ar$^+$ laser radiation. The nominal laser radiation power was 4 W. A 90°-scattering geometry was employed. Spectra were collected in (yy) scattering geometry.

Raman shifts (cm^{-1}): ~670sh, ~655, ~480s, ~275w, ~200w, ~90.

Source: Schwalowsky et al. (1996).

Comments: The sample was characterized by powder X-ray diffraction data and synchrotron diffraction analysis. The Raman shifts are given for a sample at 298 K.

Leucite K(AlSi$_2$O$_6$)

Origin: Swan City, Colorado, USA.
Experimental details: Raman scattering measurements have been performed on a powdered sample using 488 nm Ar$^+$ laser radiation. The nominal laser radiation power was 600 mW. A 90°-scattering geometry was employed.
Raman shifts (cm^{-1}): 1066, 984sh, 786w, 678w, 618w, 528sh, 498s, 432w, 394w, 338, 304, 272w, 266w, 216, 180, 152, 112, 76.
Source: Matson et al. (1986).
Comments: No independent analytical data are provided for the sample used. For the Raman spectrum of leucite see also Castriota et al. (2008).

Leucophosphite KFe$^{3+}_2$(PO$_4$)$_2$(OH)·2H$_2$O

Origin: Sapucaia mine, Conselheiro Pena pegmatite district, Brazil.
Experimental details: Raman scattering measurements have been performed on an arbitrarily oriented sample using 633 nm He-Ne laser radiation. The laser radiation power is not indicated. The Raman shifts have been determined for the maxima of individual peaks obtained as a result of the spectral curve analysis.
Raman shifts (cm^{-1}): 3535sh, 3456, 3355sh, 3225, 3171sh, 2892sh, 1632w, 1255w, 1177, 1135, 1104, 1087sh, 1058s, 1028sh, 1014sh, 994s, 973sh, 850w, 789w, 630s, 611s, 589sh, 550w, 497sh, 481, 436, 420, 407sh, 380sh, 336sh, 310sh, 303s, 282sh, 262sh, 226sh, 215, 204sh, 190sh, 164, 152, 142, 129sh, 117.
Source: Frost et al. (2013ac).
Comments: The sample was characterized by powder X-ray diffraction data and qualitative electron microprobe analysis.

Lévyne-Ca Ca$_3$(Si$_{12}$Al$_6$)O$_{36}$·18H$_2$O

Origin: Stolpen, Germany.
Experimental details: Methods of sample preparation are not described. Raman scattering measurements have been performed using Nd-YAG laser radiation. The laser radiation power at the sample was 300 mW.
Raman shifts (cm^{-1}): 917w, 709s, 430s, 290, 264w, 202s.
Source: Mozgawa (2001).
Comments: The sample was characterized by powder X-ray diffraction data.

Leydetite Fe(UO$_2$)(SO$_4$)$_2$·11H$_2$O

Origin: Mas d'Alary, Lodève, France (type locality).
Experimental details: Raman scattering measurements have been performed on an arbitrarily oriented sample using 532 nm laser radiation. The nominal laser radiation power was 2.5 mW.
Raman shifts (cm^{-1}): 3492, 3404, 3237, 3130, 1679, 1649, 1203, 1180, 1150, 1139, 1135, 1113, 1099, 1038s, 1030, 1023, 1015, 937, 930, 858, 851, 846, 843, 836, 828, 686s, 675, 666, 608, 538, 522, 504, 485, 464, 443, 420, 394, 373, 290, 260, 236, 223, 196, 182, 165, 138, 123, 116, 102, 89, 77, 65.
Source: Plášil et al. (2013a).

Comments: The sample was characterized by powder X-ray diffraction data and electron microprobe analyses. The crystal structure is solved. The empirical formula of the sample used is $(Fe_{0.93}Mg_{0.07}Al_{0.04}Cu_{0.01})(U_{1.01}O_2)(S_{1.96}Si_{0.02})O_8(H_2O)_{11}$.

Libethenite $Cu_2(PO_4)(OH)$

Origin: Banská Bystrica, central Slovakia.
Experimental details: Raman scattering measurements have been performed on an arbitrarily oriented sample using 632.8 nm He-Ne laser radiation. The nominal laser radiation power was 17 mW. A 180°-scattering geometry was employed. Polarized spectra were collected in parallel and perpendicular to the c-axis scattering geometries. The Raman shifts have been determined for the maxima of individual peaks obtained as a result of the spectral curve analysis.
Raman shifts (cm^{-1}): 3485–3475, 1130w, 1102w, 1075sh, 1052, 1022s, 1008sh, 979s, 944sh, 864w, 818w, 650w, 625, 590, 561, 461, 430sh, 392, 371sh, 319sh, 301s, 270w, 250w, 227s, 195s, 160s, 140sh, 113w, 92sh, 74s.
Source: Kharbish et al. (2014).
Comments: The sample was characterized by powder X-ray diffraction data and electron microprobe analysis. Raman shifts are given as sum of spectra at parallel and perpendicular to the c-axis scattering geometries. For the Raman spectra of libethenite see also Frost et al. (2002g), Bouchard and Smith (2003), Belik et al. (2007, 2011), and Majzlan et al. (2015).

Liebenbergite $Ni_2(SiO_4)$

Origin: Synthetic.
Experimental details: Raman scattering measurements have been performed at different pressures on an arbitrarily oriented sample using 514.5 nm Ar^+ laser radiation. The laser radiation power at the sample was about 60 mW. A 180°-scattering geometry was employed.
Raman shifts (cm^{-1}): 952, 889, 868sh, 831s, 819s, 593sh, 560, 521, 414, 344w, 298, 272w, 252w, 221w, 191w, 181w.
Source: Lin (2001).
Comments: The sample was characterized by powder X-ray diffraction data. The Raman shifts are given for a sample at ambient condirions.

Liebigite $Ca_2(UO_2)(CO_3)_3 \cdot 11H_2O$

Origin: Kroderen, Snarum, Norway.
Experimental details: Raman scattering measurements have been performed on arbitrarily oriented crystals using a 633 nm He-Ne laser. The laser radiation power is not indicated. The Raman shifts have been determined for the maxima of individual peaks obtained as a result of the spectral curve analysis.
Raman shifts (cm^{-1}): 3468s, 3258s, 1566, 1409, 1381, 1087s, 1073sh, 1007w, 838sh, 822s, 816sh, 758w, 747sh, 248.
Source: Frost et al. (2005g).
Comments: No independent quantitative analytical data are provided for the sample used.

Likasite $Cu_3(NO_3)(OH)_5 \cdot 2H_2O$

Origin: Great Australian Mine, Queensland, Australia.
Experimental details: Raman scattering measurements have been performed on an arbitrarily oriented sample using 633 nm He-Ne laser radiation. The laser radiation power is not indicated. The Raman shifts have been determined for the maxima of individual peaks obtained as a result of the spectral curve analysis.
Raman shifts (cm^{-1}): 3567sh, 3522s, 3452sh, 3338sh, 3281, 3040w, 1628w, 1394w, 1319w, 1050, 1049, 980w, 831, 763, 715w, 706w, 529, 514, 493, 459w, 377w, 341w, 233w, 210w, 190w, 175w, 165w, 140w.
Source: Frost et al. (2005h).
Comments: No independent analytical data are provided for the sample used.

Lime CaO

Origin: Synthetic.
Experimental details: Raman scattering measurements have been performed on a powdered sample using 632.8 nm He-Ne laser radiation. The laser radiation power at the sample was 10 mW.
Raman shifts (cm^{-1}): ~680w.
Source: Schmid and Dariz (2015).
Comments: No independent analytical data are provided for the sample used. The CaO phase has halite structure with cubic unit cell (Fm3m) and does not have the first-order Raman scattering. The given Raman shift belongs to the second-order Raman scattering.

Linarite $CuPb(SO_4)(OH)_2$

Origin: No data.
Experimental details: Raman scattering measurements have been performed on an arbitrarily oriented sample using 632 nm He-Ne laser radiation with output radiation power 30 mW, and 514.5 nm Ar$^+$ laser radiation with a low radiation power.
Raman shifts (cm^{-1}): 3471sh, 3448, 3220w, 1141, 1019, 968s, 818w, 632, 610, 594w, 513, 461, 436, 365, 345sh, 326w, 230w, 163.
Source: Bouchard and Smith (2003).
Comments: The sample was characterized by powder X-ray diffraction analysis. For the Raman spectra of linarite see also Buzgar et al. (2009) and Hrazdil et al. (2016).

Lindbergite $Mn(C_2O_4) \cdot 2H_2O$

Origin: Synthetic.
Experimental details: Raman scattering measurements have been performed on an arbitrarily oriented sample using 514.5 nm Ar$^+$ laser radiation. The nominal laser radiation power was 100 mW.
Raman shifts (cm^{-1}): 3326s, 1469s, 908, 579, 517, 240, 199.
Source: Echigo and Kimata (2008).
Comments: The sample was characterized by powder X-ray diffraction data.

Lindbergite $Mn(C_2O_4)\cdot 2H_2O$

Origin: Synthetic.
Experimental details: Raman scattering measurements have been performed on a powdered sample using 1064 nm Nd-YAG laser radiation. The laser radiation power is not indicated.
Raman shifts (cm^{-1}): 1625w, 1465s, 1410w, 909, 855w, 579, 517.
Source: Mancilla et al. (2009a).
Comments: No independent analytical data are provided for the sample used.

Lindgrenite $Cu_3(Mo^{6+}O_4)_2(OH)_2$

Origin: Broken Hill, NSW, Australia.
Experimental details: Raman scattering measurements have been performed on an arbitrarily oriented sample using 785 nm Nd-YAG laser radiation. The laser radiation power is not indicated. The Raman shifts have been determined for the maxima of individual peaks obtained as a result of the spectral curve analysis.
Raman shifts (cm^{-1}): 932s, 887w, 839w, 798w, 775w, 496w, 399, 349, 325, 313, 302s, 287w, 251, 217sh, 190w, 171, 158, 123w.
Source: Frost et al. (2004c).
Comments: The sample was characterized by qualitative electron microprobe analysis.

Lindsleyite $(Ba,Sr)(Zr,Ca)(Fe,Mg)_2(Ti,Cr,Fe)_{18}O_{38}$

Origin: Synthetic.
Experimental details: Raman scattering measurements have been performed on an arbitrarily oriented sample using 632 nm He-Ne laser radiation. The nominal laser radiation power was in the range from 2 to 4 mW.
Raman shifts (cm^{-1}): 1702s, 661s, 560, 433, 327.
Source: Konzett et al. (2005).
Comments: The sample was characterized by single-crystal X-ray diffraction data and electron microprobe analysis. The formula of the sample used is $Ba(Ti_{12}Cr_4Fe_2ZrMg_2)O_{38}$.

Lingunite $NaAlSi_3O_8$

Origin: Shocked Sixiangkou L6 chondrite.
Experimental details: Methods of sample preparation are not described. Raman scattering measurements have been performed using 514.5 nm Ar$^+$ laser radiation. The laser radiation power at the sample was about 16 mW. A 180°-scattering geometry was employed.
Raman shifts (cm^{-1}): 975, 844, 798sh, 767s, 717w, 625, 595w, 531w, 494, 430sh, 277, 213.
Source: Liu and El Gorsey (2007).
Comments: No independent analytical data are provided for the sample used.

Lingunite K-analogue $KAlSi_3O_8$

Origin: Synthetic.
Experimental details: Raman scattering measurements have been performed on an arbitrarily oriented single crystals using 514.5 nm Ar$^+$ laser radiation. The laser radiation power at the sample was about 10 mW. A 180°-scattering geometry was employed.

Raman shifts (cm^{-1}): 1726w, 1601w, 1580w, 1452w, 1043, 952w, 866sh, 838w, 761s, 721sh, 655w, 621, 539, 521sh, 405w, 380sh, 283, 214s.
Source: Liu et al. (2009).
Comments: The sample was characterized by X-ray diffraction data.

Linnaeite $Co^{2+}Co^{3+}_2S_4$

Origin: Synthetic.
Experimental details: Raman scattering measurements have been performed on a bulk sample and on an ultrathin sheet. Other experimental details are not described.
Raman shifts (cm^{-1}): ~400, ~340, ~235w.
Source: Liu et al. (2015d).
Comments: The samples were characterized by X-ray diffraction and SAED data.

Linzhiite $FeSi_2$

Origin: Synthetic.
Experimental details: Raman scattering measurements have been performed on the oriented single crystals in the forms as thin crystalline needles and plaquets in different scattering geometries, using 514.5 nm Ar$^+$ laser radiation. The laser radiation power at the sample was 100 mW. A 90°-scattering geometry was employed.
Raman shifts (cm^{-1}): ~340w, ~298w, ~270–275w, ~245s, and a series of peaks below 200 cm^{-1}.
Source: Guizzetti et al. (1997).

Liroconite $Cu_2Al(AsO_4)(OH)_4·4H_2O$

Origin: Cornwall deposit, UK.
Experimental details: Methods of sample preparation are not described. Raman scattering measurements have been performed using 632.8 nm He-Ne laser radiation. The nominallaser radiation power was 0.97 mW.
Raman shifts (cm^{-1}): 3580w, 3550w, 865s, 846sh, 567w, 418w, 376w, 316sh, 299, 182, 165, 109.
Source: Makreski et al. (2015a).
Comments: The sample was characterized by powder X-ray diffraction data and thermal analysis.

Liskeardite $(Al,Fe)_{32}(AsO_4)_{18}(OH)_{42}(H_2O)_{22}·52H_2O$

Origin: Penberthy Croft Mine, St. Hilary, Cornwall, England UK.
Experimental details: Raman scattering measurements have been performed on an arbitrarily oriented sample using 633 nm He-Ne laser radiation. The laser radiation power is not indicated. The Raman shifts have been determined for the maxima of individual peaks obtained as a result of the spectral curve analysis.
Raman shifts (cm^{-1}): 3618sh, 3577, 3504, 3446sh, 3289sh, 3077sh, 2930, 2762sh, 1687sh, 1611w, 1554w, 1532w, 1453w, 1138w, 1124w, 1111sh, 1007w, 987w, 931sh, 914sh, 893s, 867s, 843s, 813sh, 769sh, 750w, 723w, 651w, 624w, 579, 554sh, 528sh, 514s, 499sh, 485sh, 477s, 454, 431w, 406w, 386sh, 373, 343w, 336sh, 305, 285, 263sh, 245sh, 230sh, 217sh, 196sh, 182sh, 162s, 143, 126sh, 110w.
Source: Frost et al. (2015w).
Comments: The sample was characterized by qualitative electron microprobe analysis.

Litharge PbO

Origin: Synthetic.

Experimental details: Raman scattering measurements have been performed on a powdered sample using 1064 nm Nd-YAG laser radiation. The laser radiation power is not indicated.

Raman shifts (cm^{-1}): 381w, 339, 288w, 146s, 82w.

Source: Ciomartan et al. (1996).

Comments: The sample was characterized by powder X-ray diffraction data. For the Raman spectrum of litharge see also Bouchard and Smith (2003).

Lithiophilite LiMn^{2+}(PO$_4$)

Origin: Cigana pegmatite, Conselheiro Pena, Minas Gerais, Brazil.

Experimental details: Raman scattering measurements have been performed on arbitrarily oriented crystals using a 633 nm He-Ne laser. The laser radiation power is not indicated. The Raman shifts have been determined for the maxima of individual peaks obtained as a result of the spectral curve analysis.

Raman shifts (cm^{-1}): 1081w, 1068, 1018w, 1000, 955sh, 950s, 944sh, 627w, 591w, 575w, 443w, 424sh, 403sh, 317w, 288w, 247w, 235w, 199w, 154sh, 146w, 135sh, 105w.

Source: Frost et al. (2013ak).

Comments: The sample was characterized by electron microprobe analysis. The empirical formula of the sample used is Li$_{1.01}$(Mn$_{0.60}$Fe$_{0.41}$Mg$_{0.01}$Ca$_{0.01}$)(PO$_4$)$_{0.99}$.

Lithiophorite (Al,Li)(Mn^{4+},Mn^{3+})$_2$O$_2$(OH)$_2$

Origin: No data.

Experimental details: Raman scattering measurements have been performed on an arbitrarily oriented sample using 532 nm laser radiation. The laser radiation output power was 0.2 mW. The Raman shifts have been determined for the maxima of individual peaks obtained as a result of the spectral curve analysis.

Raman shifts (cm^{-1}): 3458–3465w, 1600w, (1183), (1058), (938), 621–629s, 575–579s, 541sh, 482–487, 460sh, 378–383w.

Source: Burlet et al. (2014), Burlet and Vanbrabant (2015).

Comments: The sample was characterized by powder X-ray diffraction data, energy-dispersive X-ray scan analysis, and flame emission analysis.

Lithiophosphate Li$_3$(PO$_4$)

Origin: Synthetic.

Experimental details: Methods of sample preparation are not described. Raman scattering measurements have been performed at different temperatures using 514.5 nm Ar$^+$ laser radiation. The laser radiation output power was 120 mW.

Raman shifts (cm^{-1}): 1061w, 1022, 942s, ~630s, 586s, 474, 442, 376, 352.

Source: Popović et al. (2003).

Comments: The sample was characterized by powder X-ray diffraction data. The Raman shifts are given for a sample at room temperature.

Lithiotantite $LiTa_3O_8$

Origin: Eastern Brazilian Pegmatite Province, Minas Gerais, Brazil.
Experimental details: Experimental details are not indicated. Raman scattering measurements have been performed on an arbitrarily oriented sample using 532 nm laser radiation. The laser radiation power is not indicated.
Raman shifts (cm^{-1}): 902, 826, 682s, 630w, 565s, 463w, 426sh, 416w, 370w, 342, 283, 259w, 229, 180.
Source: Menezes Filho et al. (2016).
Comments: The sample was characterized by electron microprobe analysis. The empirical formula of the sample used is $(Li_{0.96}Mn_{0.02}Fe_{0.01}Na_{0.01})(Ta_{2.18}Nb_{0.79}Sn_{0.03})O_{8.00}$.

Liversidgeite $Zn_6(PO_4)_4 \cdot 7H_2O$

Origin: Broken Hill, New South Wales, Australia (type locality).
Experimental details: Raman scattering measurements have been performed on an arbitrarily oriented single crystal using 632.8 nm He-Ne laser radiation. The nominal laser radiation power was 17 mW.
Raman shifts (cm^{-1}): 3220, 2895, 1645w, 1142w, 1050w, 1004, 986, 958s, 610, 584, 476, 464, 430, 244, 210.
Source: Elliott et al. (2010).
Comments: The sample was characterized by powder X-ray diffraction data and electron microprobe analyses. The crystal structure is solved. The empirical formula of the sample used is $Pb_{0.01}(Zn_{5.86}Mn_{0.06})(P_{4.01}As_{0.05}S_{0.04})O_{16.20} \cdot 6.8H_2O$.

Livingstonite $HgSb_4S_6(S)_2$

Origin: Huitzuco, Mexico.
Experimental details: Methods of sample preparation are not described. The Raman signal was excited by a 532 nm solid-state laser. The nominal laser radiation power was 0.5 mW.
Raman shifts (cm^{-1}): 308s, 284s, 238, 191, 157w, 125w, 106w, 75w.
Source: Števko et al. (2015).
Comments: The empirical formula of the sample used is $Hg_{1.01}(Sb_{3.89}As_{0.08})S_{8.01}$.

Lizardite $Mg_3Si_2O_5(OH)_4$

Origin: Monte Fico, Elba Island, Italy
Experimental details: Raman scattering measurements have been performed on a powdered sample using 1064 nm Nd-YAG laser radiation. The nominal laser radiation power was 120 mW. A 180°-scattering geometry was employed.
Raman shifts (cm^{-1}): 1096w, 690s, 630w, 510w, 388s, 350w, 233s.
Source: Rinaudo et al. (2003).
Comments: The sample was characterized by powder X-ray diffraction data and electron microprobe analyses. For the Raman spectra of lizardite see also Auzende et al. (2004) and Frezzotti et al. (2012).

Löllingite $FeAs_2$

Origin: Synthetic.
Experimental details: Raman scattering measurements have been performed on an oriented single crystal, using 676.4 nm Kr^+ laser radiation. The laser radiation power is not indicated. A 180°-scattering geometry was employed. Polarized spectra were collected in the $z(xxy)z$-, $y(xx)y$-, $x(xy)x$-, $y(xz)y$-, and $z(xy)z$- scattering geometries.
Source: Lutz and Müller (1991).
Raman shifts (cm^{-1}): 271s, 269sh, 241sh, 236s.
Comments: The Raman shifts are given for the scattering geometry $z(xxy)z$-. The notation $z(xxy)z$- means that the incident laser light is polarized parallel to x, scattered light is of all polarizations (x, y). No independent analytical data are provided for the sample used.

Lomonosovite $Na_5Ti_2(Si_2O_7)(PO_4)O_2$

Origin: Kirovskii apatite mine, Kukisvumchorr Mt., Khibiny Massif, Kola Peninsula, Russia.
Experimental details: Raman scattering measurements have been performed in the spectral regions from 100 to 550 and from 750 to ~3800 cm^{-1} on an arbitrarily oriented sample using 633 nm He-Ne laser radiation. The laser radiation power is not indicated. The Raman shifts have been determined for the maxima of individual peaks obtained as a result of the spectral curve analysis.
Raman shifts (cm^{-1}): 1084w, 1080w, 1070w, 999, 975, 939sh, 925sh, 909s, 882sh, 853, 838sh, 803, 789sh, ..., 534, 509sh, 499w, 457, 440sh, 427sh, 408s, 393sh, 368sh, 351, 319w, 302w, 284sh, 272, 223sh, 204, 173sh, 150sh, 145s, 112.
Source: Frost et al. (2015m).
Comments: The sample was characterized by qualitative electron microprobe analysis.

Lonecreekite $(NH_4)Fe^{3+}(SO_4)_2 \cdot 12H_2O$

Origin: No data.
Experimental details: Raman scattering measurements have been performed on an arbitrarily oriented sample using 633 nm He-Ne laser radiation. The laser radiation power is not indicated. The Raman shifts have been determined for the maxima of individual peaks obtained as a result of the spectral curve analysis.
Raman shifts (cm^{-1}): 1131, 1108w, 1099w, 991s, 701, 636, 615, 525, 463, 435, 307.
Source: Frost and Kloprogge (2001).
Comments: The sample was analyzed for chemical composition, and some substitution with Al^{3+} for Fe^{3+} was detected. Raman shifts are given for sample at 77 K because of the fluorescence at 298 K. For the Raman spectrum of lonecreekite see also Jentzsch et al. (2013).

Lonsdaleite C

Origin: Popigai crater, Siberia, Russia.
Experimental details: Raman scattering measurements have been performed on the arbitrarily oriented carbon platelets with the lonsdaleite fraction in in the range from 0.29 to 0.565 using 325 nm He-Cd laser radiation. The laser radiation power at the sample was 0.5 mW. A 180°-scattering geometry was employed. The Raman shifts have been determined for the maxima of individual peaks obtained as a result of the spectral curve analysis.

Raman shifts (cm^{-1}): 1303–1292, 1244–1219w.
Source: Goryainov et al. (2014).
Comments: The sample was characterized by powder X-ray diffraction data. The lonsdaleite/diamond molar ratio was estimated using the Rietveld method.

Lópezite $K_2Cr_2O_7$

Origin: Synthetic.
Experimental details: Raman scattering measurements have been performed on an arbitrarily oriented single crystal using 632.8 nm He-Ne laser radiation. The nominal laser radiation power was 6 mW.
Raman shifts (cm^{-1}): 950w, 938w, 935, 913w, 910w, 893, 744, 564, 553sh, 527, 385, 370sh, 357sh, 230, 220sh, 130w.
Source: Mathur et al. (1968).
Comments: No independent analytical data are provided for the sample used.

Lorándite $TlAsS_2$

Origin: Allchar, Republic of Macedonia.
Experimental details: Raman scattering measurements have been performed on an oriented sample using 632.8 nm He-Ne laser radiation. The nominal laser radiation power was 17 mW. A 180°-scattering geometry was employed. Polarized spectra were collected with the laser polarization parallel to the b- and c-axes scattering geometries.
Raman shifts (cm^{-1}): 398, 380s, 366sh, 325, 317sh, 311sh, 275, 263sh, 211, 203sh, 193, 172, 157sh, 135w.
Source: Kharbish (2011).
Comments: The sample was characterized by single-crystal X-ray diffraction and electron microprobe analyses. The Raman shifts are given for the scattering geometry with the laser polarization parallel to the b-axes. For the Raman spectra of lorándite see also Minceva-Sukarova et al. (2003) and Makreski et al. (2014).

Lorenzenite $Na_2Ti_2O_3(Si_2O_6)$

Origin: Synthetic.
Experimental details: Raman scattering measurements have been performed on a pellet of pressed powdered sample using 488 nm Ar^+ laser radiation. The laser radiation power at the sample was about 300 mW. A 180°-scattering geometry was employed.
Raman shifts (cm^{-1}): 1049w, 984, 963, 899w, 856w, 834w, 704, 637s, 579w, 538, 486, 451, 349, 305s, 274, 258sh, 233w, 215.
Source: Su et al. (2000).
Comments: The sample was characterized by powder X-ray diffraction data.

Löweite $Na_{12}Mg_7(SO_4)_{13} \cdot 15H_2O$

Origin: Synthetic.
Experimental details: Raman scattering measurements have been performed on an arbitrarily oriented sample using 532 nm Nd-YAG laser radiation. The laser radiation power at the sample was 2 mW.
Raman shifts (cm^{-1}): 1210, 1142w, 1117w, 1079, 1039s, 1001s, 980, 971sh, 641sh, 622, 606sh, 471sh, 462, 453sh.

Source: Jentzsch et al. (2011).
Comments: No independent analytical data are provided for the sample used.

Luddenite $Cu_2Pb_2Si_5O_{14} \cdot 14H_2O$

Origin: Artillery Peak, Mohave Co., Arizona, USA.
Experimental details: Raman scattering measurements have been performed on arbitrarily oriented crystals using a 633 nm He-Ne laser. The laser radiation power is not indicated. The Raman shifts have been determined for the maxima of individual peaks obtained as a result of the spectral curve analysis.
Raman shifts (cm^{-1}): 3329sh, 3317, 3284sh, 1658sh, 1603, 1557sh, 1482, 1455sh, 1368sh, 1346, 1301, 1276sh, 1160, 1148sh, 1122sh, 986sh, 978, 970sh, 831sh, 808, 801sh, 696sh, 676, 648sh, 501sh, 473sh, 464s, 449sh, 413w, 403, 394sh, 356, 344sh, 263, 213sh, 201, 174sh, 167.
Source: Frost et al. (2015u).
Comments: No independent analytical data are provided for the sample used.

Ludjibaite $Cu_3(PO_4)(OH)_3$

Origin: Banská Bystrica, central Slovakia.
Experimental details: Raman scattering measurements have been performed on an arbitrarily oriented sample using 632.8 nm He-Ne laser radiation. The nominal laser radiation power was 17 mW. A 180°-scattering geometry was employed. Polarized spectra were collected in parallel and perpendicular to the c-axis scattering geometries.
Raman shifts (cm^{-1}): 3470, 1115, 1072sh, 1046sh, 1019, 981s, 925sh, 855w, 815w, 784w, 736sh, 633, 586, 557sh, 449, 402sh, 387, 368sh, 301s, 263sh, 226, 190sh, 160s.
Source: Kharbish et al. (2014).
Comments: The sample was characterized by powder X-ray diffraction analysis and by electron microprobe analyses. For the Raman spectrum of ludjibaite see also Frost et al. (2002g).

Ludlamite $Fe^{2+}_3(PO_4)_2 \cdot 4H_2O$

Origin: Boa Vista mine, Galiléia, Minas Gerais, Brazil.
Experimental details: Raman scattering measurements have been performed on an arbitrarily oriented sample using 633 nm He-Ne laser radiation. The laser radiation power is not indicated. The Raman shifts have been determined for the maxima of individual peaks obtained as a result of the spectral curve analysis.
Raman shifts (cm^{-1}): 3190sh, 3137, 3013sh, 2896, 2730, 2605, 1160w, 1080, 1044, 992sh, 973sh, 950s, 916sh, 774w, 665sh, 634, 599, 564sh, 548, 494, 465, 371sh, 369, 345w, 302, 286, 266, 249, 244sh, 207sh, 199, 182sh, 172, 164sh, 145, 140sh, 103.
Source: Frost et al. (2013w).
Comments: The empirical formula of the sample used is $(Fe_{2.35}Mn_{0.25}Mg_{0.22})(PO_4)_{2.08} \cdot 4.0H_2O$.

Ludlockite $PbFe^{3+}_4As^{3+}_{10}O_{22}$

Origin: Tsumeb mine, Tsumeb, Namibia.
Experimental details: Raman scattering measurements have been performed on arbitrarily oriented crystals using a 633 nm He-Ne laser. The laser radiation power is not indicated. The Raman shifts

have been determined for the maxima of individual peaks obtained as a result of the spectral curve analysis.

Raman shifts (cm^{-1}): 798s, 756s, 743sh, 674sh, 666, 639, 611w, 579sh, 549s, 536sh, 524sh, 501sh, 486s, 470sh, 436w, 420, 408sh, 381sh, 368, 348, 332, 287, 266sh, 246s, 221, 204, 193sh.

Source: Bahfenne and Frost (2009).

Comments: No independent analytical data are provided for the sample used.

Ludwigite $Mg_2Fe^{3+}O_2(BO_3)$

Origin: No data.

Experimental details: Raman scattering measurements have been performed on an oriented sample using 514.5 nm Ar$^+$ laser radiation. The nominal laser radiation power was 20 mW. A nearly 180°-scattering geometry was employed. Polarized spectra were collected in the (zz) scattering geometry.

Raman shifts (cm^{-1}): 640s, 568, 480sh, 456w, 399s, 367sh, 298, 269, 233, 191, 160, 132w, 109.

Source: Leite et al. (2002).

Comments: No independent analytical data are provided for the sample used.

Lueshite $NaNbO_3$

Origin: Synthetic.

Experimental details: Raman scattering measurements have been performed on a powdered sample using 532 nm Nd-YAG laser radiation. The laser radiation power is not indicated.

Raman shifts (cm^{-1}): 868w, 599s, 428w, 238s.

Source: Wu et al. (2010b).

Comments: The sample was characterized by powder X-ray diffraction data. For the Raman spectrum of lueshite see also Fresno et al. (2016).

Lulzacite $Sr_2Fe^{2+}{}_3Al_4(PO_4)_4(OH)_{10}$

Origin: Saint-Aubindes-Châteauax, Loire-Atlantigue, France.

Experimental details: Methods of sample preparation are not described. Raman scattering measurements have been performed using 514.5 nm laser radiation. The laser radiation power is not indicated.

Raman shifts (cm^{-1}): 989s, 918, 842, 568, 505, 420, 290, 142.

Source: Moëlo et al. (2000).

Comments: The sample was characterized by powder X-ray diffraction analysis and by electron microprobe analyses. The empirical formula of the sample used is $(Sr_{0.96}Ba_{0.04})_2Fe^{2+}(Fe^{2+}{}_{0.63}Mg_{0.37})_2Al_4[(P_{0.98}V_{0.02})O_4]_4(OH)_{10}$.

Lüneburgite $Mg_3[B_2(OH)_6(PO_4)_2]\cdot 6H_2O$

Origin: Mejillones Peninsula, Antofagasta Province, Chile.

Experimental details: Raman scattering measurements have been performed on an arbitrarily oriented sample using 325 nm laser radiation. The laser radiation power at the sample was about 8 mW. The Raman shifts have been determined for the maxima of individual peaks obtained as a result of the spectral curve analysis.

Raman shifts (cm^{-1}): 3504sh, 3438sh, 3392s, 3272, 3207, 1087, 1032, 999sh, 877, 734, 590, 465.
Source: Korybska-Sadło et al. (2016).
Comments: The sample was characterized by powder X-ray diffraction data and electron microprobe analyses.

Luogufengite Fe_2O_3

Origin: Synthetic.
Experimental details: Raman scattering measurements have been performed on a powdered sample using 532 nm Nd-YAG laser radiation. The output laser radiation power was in the range from 0.2 to 30 mW.
Raman shifts (cm^{-1}): 1641, 1474, 1435, 1378, 1329, 1276, 1188, 829, 731, 704, 669, 643, 597, 579, 559, 488, 461, 439, 419, 397, 378, 362, 346, 330, 299, 267, 214, 195, 165, 146, 116.
Source: López-Sánchez et al. (2016).
Comments: The sample was characterized by powder X-ray diffraction data.

Lusernaite-(Y) $Y_4Al(CO_3)_2(OH,F)_{11} \cdot 6H_2O$

Origin: Luserna valley, Piedmont, Italy (type locality).
Experimental details: Raman scattering measurements have been performed on an arbitrarily oriented sample using 473.1 nm Nd-YAG and 632.8 nm He-Ne lasers radiations. The laser radiations powers are not indicated.
Raman shifts (cm^{-1}): 1096.
Source: Biagioni et al. (2013a).
Comments: The sample was characterized by powder X-ray diffraction data and electron microprobe analyses. The crystal structure is solved. The empirical formula of the sample used is $(Y_{3.41}Dy_{0.16}Er_{0.15}Yb_{0.09}Gd_{0.07}Ca_{0.05}Pb_{0.02}Sm_{0.01})Al_{1.06}(CO_3)_{2.00}(OH_{10.35}F_{0.65}) \cdot 6H_2O$. Due to the strong luminescence only one Raman band was registered in the spectrum, confirming the presence of CO_2^{3-} groups in the structure.

Macedonite $PbTiO_3$

Origin: Synthetic.
Experimental details: Raman scattering measurements have been performed on $PbTiO_3$ single crystals with the tetragonal c axis normal to the scattered plane, using a 633 nm He-Ne laser. Polarized spectra were collected in $x(zz)y$, $x(zx)y$, and $x(yx)y$ scattering geometry.
Raman shifts (cm^{-1}): 508s, 440, 290s, 220, 130, 89.
Source: Fontana et al. (1991).
Comments: The sample was identified by electron microprobe analysis; boron was determined by LA-ICP-MS. The Raman shifts are given for the scattering geometry $x(zx)y$.

Mackayite $Fe^{3+}Te^{4+}_2O_5(OH)$

Origin: An unknown locality in Nevada, USA (?).
Experimental details: Raman spectra of unoriented samples were obtained using a He-Ne laser with the wavelengths of laser excitation line of 633 nm. The Raman shifts have been determined for the maxima of individual peaks obtained as a result of the spectral curve analysis.

Raman shifts (cm^{-1}): 907w, 872w, 782w, 732, (644), 635s, (602w), 579, 513, (502w), 436, 424s, 379s, 349, 306, 177s, 150, 124.
Source: Frost and Dickfos (2009).
Comments: The IR spectra of presumed mackayitegiven in the cited paper are wrong: the strongest IR bands correspond to a sulfate. Possibly, the correct locality is Bambolla mine, Moctezuma, Sonora, Mexico.

Mackinawite (Fe,Ni)$_{1+x}$S ($x = 0$–0.07)

Origin: Synthetic (corrosion film formed after exposure of iron to saline H$_2$S saturated aceticsolution).
Experimental details: Raman spectrum of an unoriented sample was obtained at the wavelength of laser excitation line of 532.1 nm.
Raman shifts (cm^{-1}): (587), 474w, 385, 274s, 208.
Source: Genchev and Erbe (2016).

Macquartite Cu$_2$Pb$_7$(CrO$_4$)$_4$(SiO$_4$)$_2$(OH)$_2$

Origin: No data.
Experimental details: Raman spectra of crystals oriented to provide maximum intensity were obtained using a He-Ne laser with the wavelengths of laser excitation line of 785 nm. The Raman shifts have been determined for the maxima of individual peaks obtained as a result of the spectral curve analysis.
Raman shifts (cm^{-1}): 968w, 936w, 857, 840, 814s, 463, 439, 374, 349s, 340, 194, 152.
Source: Frost (2004c).
Comments: The sample was identified by electron microprobe analysis; boron was determined by LA-ICP-MS. The Raman shifts are given for the scattering geometry $y(zz)y$, in which the Raman intensities are most strong.

Magadiite Na$_2$Si$_{14}$O$_{29}$·11H$_2$O

Origin: Synthetic.
Experimental details: Raman scattering measurements have been performed on an unoriented sample using a 1064.1 nm Nd^{3+}:YAG laser. The laser radiation power at the sample was about 90 mW.
Raman shifts (cm^{-1}): 3266w, 1151w, 1131w, 1101w, 1085w, 1064, 1049w, 992s, 823w, 792w, 705w, 645w, 632w, 620, 587w, 488sh, 464s, 442sh, 398w, 373w, 338w.
Source: Huang et al. (1999b).
Comments: The sample was characterized by powder X-ray diffraction data.

Magbasite KBaFe^{3+}Mg$_7$Si$_8$O$_{22}$(OH)$_2$F$_6$

Origin: Eldor carbonatite complex, Quebec, Canada.
Experimental details: Polarized single-crystal Raman spectra were collected in the range from 3200 to 3800 cm^{-1} with the polarizer parallel and perpendicular to the length of the crystal using 460 and 532 nm laser radiations with a nominal output power of 50 mW.
Raman shifts (cm^{-1}): 3735w, 3719w, 3636s.
Source: Welch et al. (2014)

Comments: The sample was characterized by powder X-ray diffraction data and electron microprobe analyses. The Raman shifts are given for the radiation polarization parallel to the length of the crystal.

Maghemite Fe_2O_3

Origin: Synthetic.
Experimental details: Raman scattering measurements have been performed on a powdery sample using a 636.4 nm tuneable dye laser. The laser radiation power at the sample was 0.34 mW.
Raman shifts (cm^{-1}): 648s, 527, 377, 350, 309, 250w, 220w.
Source: Nieuwoudt et al. (2011).
Comments: The sample was characterized by powder X-ray diffraction data.

Magnesio-arfvedsonite $NaNa_2(Mg_4Fe^{3+})Si_8O_{22}(OH)_2$

Origin: No data.
Experimental details: Raman scattering measurements have been performed in the region of O–H-stretching vibrations, in backscattering geometry using a 514.5 nm Ar$^+$ laser.
Raman shifts (cm^{-1}): 3666.
Source: Leissner et al. (2015).
Comments: The sample was characterized by EMPA and ICP-MS. The empirical formula of the sample used is $(Na_{0.25}K_{0.13})(Na_{0.89}Ca_{0.11})_2(Mg_{0.70}Fe^{3+}_{0.26}Al_{0.01}Mn_{0.01})_5(Si_{0.99}Al_{0.01})_8O_{22}(OH)_{1.7}F_{0.3}$.

Magnesiocarpholite $MgAl_2Si_2O_6(OH)_4$

Origin: Monte Leoni, Monticiano-Roccastrada Unit, Northern Apennine, southern Tuscany, Italy.
Experimental details: Raman scattering measurements have been performed on single crystals in four different orientations, using a 514.5 nm Ar$^+$ laser. The laser emission power was 300 mW.
Raman shifts (cm^{-1}): 3633w, 3594s, 3571sh, 1098w, 1037w, 936, 879, 783s, 747, 688, 560, 445, 351, 278w, 263, 207w, 161w, 117w.
Source: Fuchs et al. (2001).
Comments: The sample was characterized electron microprobe analyses and Mössbauer spectroscopy. The Raman shifts are given for the scattering geometry with c axis vertical and most developed crystal face normal to the polarization direction.

Magnesiochloritoid $MgAl_2(SiO_4)O(OH)_2$

Origin: Synthetic
Experimental details: Raman scattering measurements have been performed on single crystals with different orientations using a 514.5 nm Ar$^+$ laser.
Raman shifts (cm^{-1}): 3455, 3076, 1096, 985, 909, 881, 847, 805, 738, 594, 551, 531, 513, 411–412.
Source: Koch-Müller et al. (2002).
Comments: Band intensities are not indicated in the cited paper.

Magnesiochromite $MgCr_2O_4$

Origin: Synthetic.
Experimental details: Raman scattering measurements have been performed on an unoriented sample using a 514.5 nm Ar$^+$ laser.
Raman shifts (cm^{-1}): 684s, 613, 543s, 447.
Source: Yong et al. (2012).
Comments: The sample was characterized by powder X-ray diffraction data. For the Raman spectra of magnesiochromite see also Hosterman (2011), Lenaz and Lughi (2013), Andò and Garzanti (2014), and D'Ippolito et al. (2015).

Magnesiocopiapite $MgFe^{3+}_4(SO_4)_6(OH)_2 \cdot 20H_2O$

Origin: Synthetic.
Experimental details: Raman scattering measurements have been performed on an unoriented sample using a 532 nm Nd-YAG laser.
Raman shifts (cm^{-1}): 3499, 3331, 3314, 3167, 1645, 1225, 1218, 1129, 1102, 1019, 1004, 995, 639, 613, 597, 557, 305, 270, 252, 227.
Source: Kong et al. (2011b).
Comments: The sample was characterized by powder X-ray diffraction data. Band intensities are not indicated in the cited paper. For the Raman spectra of magnesiocopiapite see also Frost (2011c) and Rull et al. (2014).

Magnesioferrite $MgFe^{3+}_2O_4$

Origin: Synthetic.
Experimental details: The spectrum of an unoriented sample was recorded using 632.8 nm line of a He-Ne laser.
Raman shifts (cm^{-11}): 707∗s, 661sh, 596∗w, 550w, 479∗, 377w, 332∗, 214∗.
Source: D'Ippolito et al. (2015).
Comments: The modes marked with an asterisk are provoked by the inversion. For the Raman spectra of magnesioferrite see also Lenaz and Lughi (2013) and Aramendia et al. (2014).

Magnesio-foitite $(Mg_2Al)Al_6(Si_6O_{18})(BO_3)_3(OH)_3(OH)$

Origin: Synthetic.
Experimental details: Raman scattering measurements have been performed on single crystals with the electrical field vector of the linearly polarized laser light parallel to the crystallographic c axis using a 488 or 473 nm laser. The laser radiation power at the sample was 30 or 12 mW. The Raman shifts in the region of O–H-stretching vibrations have been determined for the maxima of individual peaks obtained as a result of the spectral curve analysis.
Raman shifts (cm^{-1}): 3657w, 3619w, 3551, 3511, 3459, 688, ~370s, 311, 267, 228.
Source: Berryman et al. (2016).
Comments: The sample was characterized by powder and single-crystal X-ray diffraction data and electron microprobe analysis. For the Raman spectrum of magnesio-foitite see also Fantini et al. (2014).

Magnesiohögbomite-2N4S $(Mg,Fe^{2+})_{10}Al_{22}Ti^{4+}_2O_{46}(OH)_2$

Origin: Central Sør Rondane Mts., Queen Maud Land, East Antarctica (type locality).
Experimental details: Raman scattering measurements have been performed on an unoriented single crystal using a 532.1 nm Ar$^+$ laser.
Raman shifts (cm^{-1}): ~3400, 872s, 780, 709, 659, 536, 498, 479, 419, 342w, 302, 263w, 217, 142, 104w.
Source: Shimura et al. (2012).
Comments: The sample was characterized by powder X-ray diffraction data and electron microprobe analyses. The crystal structure is solved.

Magnesio-hornblende $Ca_2(Mg_4Al)(Si_7Al)O_{22}(OH)_2$

Origin: No data.
Experimental details: Raman scattering measurements have been performed in backscattering geometry using a 514.5 nm Ar$^+$ laser.
Raman shifts (cm^{-1}): 3672–3673. Only a figure of the Raman spectrum of magnesio-hornblende is given in the spectral range from 10 to 1200 cm^{-1}.
Source: Leissner et al. (2015).
Comments: The sample was characterized by electron microprobe analyses and ICP-MS. The empirical formula of the sample used is $(Na_{0.45}K_{0.04})(Ca_{0.87}Fe_{0.10}Mn_{0.02}Na_{0.01})_2(Mg_{0.52}Fe_{0.36}Al_{0.12})_5(Si_{0.86}Al_{0.14})_8O_{22}(OH)_{2.00}$.

Magnesiotaaffeite-2N'2S $Mg_3BeAl_8O_{16}$

Origin: Ratnapura district, Sri Lanka (type locality).
Experimental details: No data
Raman shifts (cm^{-1}): 809w, 758, 703, 662w, 489, 447, 435, 415s, 305s.
Source: Kiefert and Schmetzer (1998).
Comments: The sample was characterized by chemical and X-ray diffraction data.

Magnesiotaaffeite-6N'3S $Mg_2BeAl_6O_{12}$

Origin: Casey Bay, Antarctica.
Experimental details: No data
Raman shifts (cm^{-1}): 803, 713s, 660, 564w, 489, 443, 412s, 326s.
Source: Kiefert and Schmetzer (1998).
Comments: The sample was characterized by chemical and X-ray diffraction data.

Magnesite $Mg(CO_3)$

Origin: Brumado, Bahia, Brazil.
Experimental details: Raman scattering measurements have been performed on an arbitrarily oriented polished sample using 514.5 and 532 nm Ar$^+$ lasers.
Raman shifts (cm^{-1}): 1763w, 1446w, 1095s, 738w, 331s, 214.
Source: Perrin et al. (2016).

Comments: The sample was characterized by electron microprobe analysis. For the Raman spectra of magnesite see also Rutt and Nicola (1974), Edwards et al. (2005), Frezzotti et al. (2012), and Bernardino et al. (2016).

Magnetite $Fe^{2+}Fe^{3+}_2O_4$

Origin: Minas Gerais, Brazil.
Experimental details: Raman scattering measurements have been performed using a 632.8 nm He-Ne laser. Laser beam was focused on the sample to give a spot size of *ca.* 1 μm. The laser radiation power at the sample was 0.7 mW.
Raman shifts (cm^{-1}): 663s, 534, 513, 302.
Source: De Faria et al. (1997).
Comments: The sample was characterized by powder X-ray diffraction data. For the Raman spectra of magnetite see also Castriota et al. (2008), Nieuwoudt et al. (2011), Hosterman (2011), Saheb et al. (2011), Das and Hendry (2011), Andò and Garzanti (2014), and D'Ippolito et al. (2015).

Magnetoplumbite $PbFe^{3+}_{12}O_{19}$

Origin: Synthetic.
Experimental details: Raman scattering measurements have been performed on arbitrarily oriented nanoparticles using a 514.5 nm Ar$^+$ laser. The laser radiation powerat the sample was 20 mW.
Raman shifts (cm^{-1}): 1320 (broad), 680s, 608, 519, 403, (328w), 325s, (290), 206w, 176.
Source: Yang et al. (2007c).
Comments: The sample was characterized by powder X-ray diffraction data. For the Raman spectra of magnetoplumbite see also Kreisel et al. (1999), Konzett et al. (2005), and Zhukova et al. (2016).

Majorite $Mg_3(MgSi)(SiO_4)_3$

Origin: Synthetic.
Experimental details: Raman scattering measurements have been performed on a polycrystalline aggregate using a micro-Raman system with an argon ion laser. The laser radiation power at the sample was in the range from 5 to 50 mW.
Raman shifts (cm^{-1}): 1065, 989, 931s, 889, 802, 602s, 458, 367, 311, 226s, 159.
Source: De La Pierre and Belmonte (2016).

Makatite $Na_2Si_4O_8(OH)_2 \cdot 4H_2O$

Origin: Synthetic.
Experimental details: Raman scattering measurements have been performed on an aggregate of arbitrarily oriented particles using a 1064.1 nm Nd-YAG laser.
Raman shifts (cm^{-1}): 3368sh, 3122w (broad), 1244w, 1060w, 1025s, 986w, 945w, 919w, 793w, 567s, 482, 460, 414, 381w, 339w, 324, 292s, 266w, 226w, 219w, 194w, 167w, 151w, 122, 100.
Source: Huang et al. (1999a).
Comments: The sample was characterized by powder X-ray diffraction data.

Malachite $Cu_2(CO_3)(OH)_2$

Origin: Eisenzeche, Germany.
Experimental details: Raman scattering measurements have been performed on an arbitrarily oriented aggregate using a 532 nm Nd-YAG laser. The nominal laser radiation power was 100 mW.
Raman shifts (cm^{-1}): 3383, 3311w, 1639w, 1495s, 1462, 1369, 1098, 1059, 821w, 755, 722, 597w, 537, 435s, 355, 270s, 215s.
Source: Buzgar and Apopei (2009).
Comments: For the Raman spectra of malachite see also Frost et al. (2002g), Bouchard and Smith (2003), Frezzotti et al. (2012), Capitani et al. (2014), and Coccato et al. (2016).

Malayaite $CaSnO(SiO_4)$

Origin: Skarn approximately 4 km north of Ash Mountain, near Mc-Dame, northern British Columbia, Canada.
Experimental details: Raman scattering measurements have been performed on a sample with continuously variable polarization directions using a 488 nm Ar$^+$ laser.
Raman shifts (cm^{-1}): 895w, 837s, 802w, 595s, 520s, 510w, 450w, 412w, 340s, 328w, 305s, 295w, 280s, 250s, 227w, 197w, 176s, 142s, 109s, 75w.
Source: Groat et al. (1996).
Comments: The sample was characterized by single-crystal X-ray diffraction data and electron microprobe analyses. The crystal structure is solved. For the Raman spectrum of malayaite see also Heyns and Harden (1999).

Malladrite Na_2SiF_6

Origin: Synthetic.
Experimental details: Raman scattering measurements have been performed on an arbitrarily oriented sample using a 4358 Å laser.
Source: Begun and Rutenberg (1967).
Raman shifts (cm^{-1}): 592, 559s, 477, 300, 252.

Mallardite $Mn(SO_4)\cdot 7H_2O$

Origin: Synthetic.
Experimental details: Raman scattering measurements have been performed on an arbitrarily oriented single crystal using a 4358 Å laser.
Raman shifts (cm^{-1}): 3467, 3399, 1148, 1085, 1084, 993–994, 693, 603, 457, 330.
Source: Rao (1941).
Comments: Krishnamurti (1958) notes that actually this study does not reveal the existence of lines at 330 and 693 cm^{-1} and, based on the crystallization conditions, it is probable that the results refer to $MnSO_4\cdot 5H_2O$.

Mallestigite $Pb_3Sb(SO_4)(AsO_4)(OH)_6\cdot 3H_2O$

Origin: A waste dump from a Cu-Pb-Zn mine, 1 km NW of Mallestiger, Carinthia, Austria.
Experimental details: Raman scattering measurements have been performed on an arbitrarily oriented crystal using a 633 nm He-Ne laser. The Raman shifts have been determined for the maxima of individual peaks obtained as a result of the spectral curve analysis.

Raman shifts (cm^{-1}): 1261w, 1234w, 1158, 1151, 1062w, 978s, 865w, 827, 803, 641, 631w, 619w, 606, 460, 449s, 437s, 416w, 374, 340w.
Source: Frost et al. (2011p).
Comments: The IR spectrum of presumed mallestigite published in the cited paper corresponds to quartz.

Mandarinoite $Fe^{3+}_2(Se^{4+}O_3)_3 \cdot 6H_2O$

Origin: El Dragon Mine, Potosi, Bolivia (type locality).
Experimental details: Raman scattering measurements have been performed on an arbitrarily oriented crystal using a 633 nm He-Ne laser. The Raman shifts have been determined for the maxima of individual peaks obtained as a result of the spectral curve analysis.
Raman shifts (cm^{-1}): 3507, 3189s, (3046), 2926, (2796w), 1666w, 1563, 814s, 744, 723, 695w, 553, 474, 398w, 355w, 262, 212, 186, 129.
Source: Frost and Keeffe (2009a).
Comments: No independent analytical data are provided for the sample used.

Manganite $Mn^{3+}O(OH)$

Origin: Synthetic.
Experimental details: Raman scattering measurements have been performed on an arbitrarily oriented sample using a 632.8 nm He-Ne laser. In spite of the strong opacity of the mineral, it was possible to record a Raman spectrum by using a very weak laser beam intensity of 1 mW and a long duration of 500 s.
Raman shifts (cm^{-1}): 622, 558s, 530, 490w, 387, 357.
Source: Bouchard and Smith (2003).
Comments: For the Raman spectrum of manganite see also Bernard et al. (1993a).

Manganlotharmeyerite $CaMn^{3+}_2(AsO_4)_2(OH)_2$

Origin: Starlera Fe-Mn deposit, Middle Penninic domain, Eastern Swiss Alps.
Experimental details: Unpolarized Raman spectrum was obtained on an arbitrarily oriented single crystal using a 488 nm Ar$^+$ laser.
Raman shifts (cm^{-1}): 3000 (broad), 880, 830 (broad), 765, 475, 426, 365, 344.
Source: Brugger et al. (2002).
Comments: The sample was characterized by optical and structural data and electron microprobe analysis. The band at 475 cm^{-1} is absent in the figure of manganlotharmeyerite given in the cited paper. The strongest band in this figure has a maximum at ~520 cm^{-1}.

Manganochromite $Mn^{2+}Cr_2O_4$

Origin: Synthetic.
Experimental details: Raman scattering measurements have been performed on a powdery sample consisting of octahedral nanocrystals using a Raman microscope.
Raman shifts (cm^{-1}): 652, 555s.
Source: Tong et al. (2015).
Comments: The sample was characterized by powder X-ray diffraction data. For the Raman spectrum of manganochromite see also Chen et al. (2007c).

Manganolangbeinite $K_2Mn^{2+}_2(SO_4)_3$

Origin: Synthetic.
Experimental details: Unpolarized and polarized (with different scattering geometries) Raman scattering measurements have been performed on single crystals using a 488 nm Ar^+ laser at the power of 5 mW.
Raman shifts (cm^{-1}): 1224w, 1153w, 1138w, 1113w, 1107w, 1031, 1022sh, 651w, 645w, 628, 620, 604, 597sh, 590w, 473sh, 447, 436w, 426w.
Source: Kreske and Devarajan (1982).
Comments: The Raman shifts are given for the scattering geometry $z(yz)x$.

Manganosite MnO

Origin: Synthetic (Alfa Aesar).
Experimental details: Raman scattering measurements have been performed on an arbitrarily oriented sample using a 514.5 nm Ar^+ laser.
Raman shifts (cm^{-1}): 654, (591), 531s, 250w.
Source: Julien et al. (2004).
Comments: The sample was characterized by powder X-ray diffraction data. For the Raman spectrum of manganosite see also Mironova-Ulmane et al. (2009).

Manjiroite $Na(Mn^{4+}_7Mn^{3+})O_{16}$

Origin: Bahariya depression, Western Desert, Egypt.
Experimental details: Raman scattering measurements have been performed on an arbitrarily sample using a 532 nm Nd-YAG laser. The laser radiation power at the sample was between 20 and 200 µW.
Raman shifts (cm^{-1}): ~1300, 643, ~395, ~292.
Source: Ciobotă et al. (2012).
Comments: No independent analytical data are given for the sample used.

Marcasite FeS_2

Origin: A seafloor hydrothermal vent field.
Experimental details: Micro-Raman scattering measurements have been performed on an arbitrarily oriented sample using a 532 nm laser. The maximum laser radiation power was 20 mW.
Raman shifts (cm^{-1}): 386, 323s.
Source: White (2009).
Comments: For the Raman spectra of marcasite see also Lutz and Müller (1991), Mernagh and Trudu (1993), and Frezzotti et al. (2012).

Margarite $CaAl_2Si_2Al_2O_{10}(OH)_2$

Origin: Rekwika, Troms, Norway.
Experimental details: Micro-Raman scattering measurements have been performed on a single crystal using a 514.5 nm Ar^+ laser. Sample orientation is not indicated.
Raman shifts (cm^{-1}): 3635s, 917s, 711s, 676, 648, (553), 489, 393s, (348), 315, 271s, 248s, (225), 115s, 84.

Source: Tlili et al. (1989).
Comments: The sample was characterized by electron microprobe analyses. For the Raman spectra of margarite see also Graeser et al. (2003) and Wang et al. (2015).

Margarosanite $Ca_2PbSi_3O_9$

Origin: Franklin, New Jersey, USA.
Experimental details: Raman scattering measurements have been performed on an arbitrarily oriented sample using 437 nm laser radiation.
Raman shifts (cm^{-1}): 1014s, 966, 907, 659s, 581, 498, 443, 389, 262, 128.
Source: Gaft et al. (2013).
Comments: Weak peaks are not indicated.

Marićite $NaFe^{2+}(PO_4)$

Origin: Synthetic.
Experimental details: Raman scattering measurements have been performed on arbitrarily oriented powder using 532 nm Ar$^+$ laser radiation. The power at the laser beam was 10 mW.
Raman shifts (cm^{-1}): 1125w, 1080w, 1052w, 972s, 943sh.
Source: Burba and Frech (2006).
Comments: For the Raman spectrum of marićite see also Burba (2006).

Markascherite $Cu_3(MoO_4)(OH)_4$

Origin: Copper Creek, Pinal Co., Arizona, USA (type locality).
Experimental details: Micro-Raman scattering measurements have been performed on a randomly oriented crystal using 532 nm laser radiation. The laser radiation power at the sample was 200 mW.
Raman shifts (cm^{-1}): 3560sh, 3541sh, 3527s, 3510, 911s, 886sh, 864sh, 489w, 449w, 425, 402w, 329s.
Source: Yang et al. (2012).
Comments: The sample was characterized by powder and single-crystal X-ray diffraction data and electron microprobe analyses.

Marokite $CaMn^{3+}_2O_4$

Origin: Synthetic.
Experimental details: Raman scattering measurements have been performed on an arbitrarily oriented sample using 785 nm laser radiation. The laser radiation power at the sample was 3 mW.
Raman shifts (cm^{-1}): 720, 679, 633s, 614s, 577, 536, 517w, 475, 397s, 376s, 350, 299s, 284sh, 272, 241, 221w, 206, 192w, 177s, 123.
Source: Wang et al. (2003).
Comments: For the Raman spectrum of marokite see also Ivanov et al. (2014).

Marthozite $Cu^{2+}(UO_2)_3(Se^{4+}O_3)_2O_2 \cdot 8H_2O$

Origin: Synthetic.
Experimental details: Raman scattering measurements have been performed on arbitrarily oriented crystal using a 633 nm He-Ne laser. The Raman shifts have been determined for the maxima of individual peaks obtained as a result of the spectral curve analysis.

Raman shifts (cm⁻¹): 3180, 3271s, 3381, 3524, 1672, 1616, 1414w, 1358w, 1283w, 869, 812s, 797s, 739, 571w, 449, 424s, 360, 257, 199, 139.
Source: Frost et al. (2008e).
Comments: No independent analytical data are provided for the sample used. The Raman shifts have been determined for the maxima of individual peaks obtained as a result of the spectral curve analysis. The IR spectrum of presumed marthozite given in the cited paper is wrong and corresponds to malachite.

Martyite $Zn_3(V_2O_7)(OH)_2 \cdot 2H_2O$

Origin: Little Eva mine, Grand Co., Utah, USA (type locality).
Experimental details: Raman scattering measurements have been performed on an arbitrarily oriented polycrystalline aggregate using 532 nm laser radiation. The output power of the laser beam was about 4 mW.
Raman shifts (cm⁻¹): 3475, 1600w, 943sh, 864sh, 844s, 800sh, 483, 440, 319, 258, 111.
Source: Kasatkin et al. (2015).
Comments: The sample was characterized by powder X-ray diffraction data and electron microprobe analysis.

Maruyamaite $K(MgAl_2)(Al_5Mg)(BO_3)_3(Si_6O_{18})(OH)_3O$

Origin: Synthetic.
Experimental details: Raman scattering measurements have been performed using 514.5 nm Ar⁺ laser radiation with laser beam perpendicular and parallel to the c-axis.
Raman shifts (cm⁻¹): 3572, 1106sh, 1091, 977, 789s, 703s, 669, 538w, 500w, 367s, 242, 212s, 155w.
Source: Lussier et al. (2016).
Comments: The sample was characterized by powder X-ray diffraction data and electron microprobe analysis. The Raman shifts are given for the spectrum obtained with laser beam perpendicular to the c-axis.

Mascagnite $(NH_4)_2(SO_4)$

Origin: Synthetic (commercial reactant).
Experimental details: Raman scattering measurements have been performed on an arbitrarily oriented sample using 514.5 nm Ar⁺ laser radiation. The laser radiation power at the objective lens was about 100 mW.
Raman shifts (cm⁻¹): 975s, ~615w, ~610w, ~450w.
Source: Sakurai et al. (2010).
Comments: For the Raman spectrum of maskagnite see also Morillas et al. (2016).

Maskelynite A feldspar glass

Origin: Sixiangkou meteorite (L6 chondrite).
Experimental details: No data.
Raman shifts (cm⁻¹): 1090 (broad), 575sh, 487s (broad).
Source: Gillet et al. (2000).

Comments: The Raman spectrum contains numerous narrow peaks of admixed lingunite, the high-pressure hollandite-type phase $KAlSi_3O_8$.

Massicot PbO

Origin: Synthetic.
Experimental details: Raman scattering measurements have been performed on an arbitrarily oriented single sample using 1064 nm Nd-YAG laser radiation.
Raman shifts (cm^{-1}): 423w, 384, 341w, 289s, 248sh, 217w, 174sh, 143s, 88w, 73w.
Source: Ciomartan et al. (1996).
Comments: For the Raman spectra of massicot see also Madsen and Weaver (1998), Bouchard-Abouchacra (2001), Bouchard and Smith (2003), and Lepot et al. (2006).

Mathesiusite $K_5(UO_2)_4(SO_4)_4(VO_5)(H_2O)_4$

Origin: Jáchymov, Krušné Hory Mts. (Ore Mts.), Bohemia, Czech Republic (type locality).
Experimental details: Raman scattering measurements have been performed on the {110} face of a single crystal using depolarized 780 nm radiation of a frequency-stabilized single mode diode laser. The laser radiation power at the sample was between 4 and 8 mW.
Raman shifts (cm^{-1}): 1329w, 1210w, 1114w, 1007s, 982, 896, 888sh, 844sh, 830s, 742, 644, 619, 598, 557w, 480, 460s, 447sh, 370w, 276, 248.
Source: Plášil et al. (2014c).
Comments: The sample was characterized by powder X-ray diffraction data and electron microprobe analyses. The crystal structure is solved.

Mathiasite $(K,Ba,Sr)(Zr,Fe)(Mg,Fe)_2(Ti,Cr,Fe)_{18}O_{38}$

Origin: Synthetic.
Experimental details: Unpolarized micro-Raman spectrum was obtained on an arbitrarily oriented sample using 632 nm He-Ne laser radiation. The laser radiation power at the sample was between 4 and 8 mW.
Raman shifts (cm^{-1}): 649s, 550, 450, 337, 243w.
Source: Konzett et al. (2005).
Comments: The sample was characterized by electron microprobe analysis.

Matildite $AgBiS_2$

Origin: Synthetic.
Experimental details: Raman scattering measurements have been performed on an arbitrarily oriented sample consisting of nanocrystals using 514 nm laser radiation.
Raman shifts (cm^{-1}): ~140, ~120.
Source: Guin et al. (2016).
Comments: The sample was characterized by powder X-ray diffraction data.

Matioliite $NaMgAl_5(PO_4)_4(OH)_6 \cdot 2H_2O$

Origin: Gentil mine, Minas Gerais, Brazil.
Experimental details: Raman scattering measurements have been performed on arbitrarily oriented crystal using a 633 nm He-Ne laser. The Raman shifts have been determined for the maxima of individual peaks obtained as a result of the spectral curve analysis.
Raman shifts (cm^{-1}): 3643w, 3630s, 3399w, 3281s, (3253), 3232s, (3231), 3072, 2920w, 1751w, 1714w, 1610w, 1562w, 1230w, 1211, 1181, 1155w, 1104, 1068s, 1048s, 1025s, (1007), (994), 985, 965, 892w, 811,
Source: Scholz et al. (2013b).
Comments: The sample was characterized by electron microprobe analysis.

Matlockite PbClF

Origin: Synthetic.
Experimental details: Raman scattering measurements have been performed on an arbitrarily oriented sample using a 632.8 nm He-Ne laser. In spite of the strong opacity of the mineral, it was possible to record a Raman spectrum by using a very weak intensitylaser (1 mW) and a long duration of 500 s.
Raman shifts (cm^{-1}): 238, 227, 163s, 155, 132, 105, 84.
Source: Bouchard and Smith (2003).
Comments: For the Raman spectrum of matlockite see also Bouchard-Abouchacra (2001).

Mattagamite $CoTe_2$

Origin: Synthetic.
Experimental details: Raman scattering measurements have been performed on an arbitrarily oriented sample using 532 nm laser radiation.
Raman shifts (cm^{-1}): 121s.
Source: McKendry et al. (2016).
Comments: The sample was characterized by powder X-ray diffraction data and electron microprobe analysis.

Matteuccite $NaH(SO_4) \cdot H_2O$

Origin: Synthetic.
Experimental details: No data
Raman shifts (cm^{-1}): 2940w, 1656, 1549w, 1308, 1268w, 1241w, 1191, 1060s, 874s, 655, 612, 577, 435, 411, 278w, 224w, 188w, 140w, 87.
Source: Baran et al. (1999a).

Maxwellite $NaFe^{3+}(AsO_4)F$

Origin: No data.
Experimental details: Raman scattering measurements have been performed on arbitrarily oriented crystals using a 633 nm He-Ne laser. The laser radiation power is not indicated. The Raman shifts have been determined for the maxima of individual peaks obtained as a result of the spectral curve analysis.

Raman shifts (cm^{-1}): 954w, 814sh, 894sh, 871s, (849), 812, 753, 542, (523), (487), 455w, (373), 360sh, (346), 327, 309sh, (291), (274w), 259w, (164w), 151, (140w), 111w.
Source: Frost et al. (2014aa).
Comments: The sample was characterized by qualitative electron microprobe analysis. For the Raman spectrum of maxwellite see also Downs et al. (2012).

Mbobomkulite $(Ni,Cu)Al_4(NO_3,SO_4)_2(OH)_{12}\cdot 3H_2O$

Origin: Synthetic.
Experimental details: Raman scattering measurements of SO_4-free analogue of mbobomkulite have been performed on arbitrarily oriented sample using a 633 nm He-Ne laser. The Raman shifts have been determined for the maxima of individual peaks obtained as a result of the spectral curve analysis.
Raman shifts (cm^{-1}): 3647, 3576, 3544w, 3468s, 3447w, 3422w, 3250s, 2900, 1645w, 1413, 1342w, 1058, 1050, 1045, 713w, 676, 614, 570, 545w, 541, 494, 447w, 350, 340, 337w, 324w, 297w, 217w, 184w, 181w, 160w, 142w.
Source: Frost et al. (2005f).
Comments: No independent analytical data are provided for the sample used.

Mcallisterite $Mg_2[B_6O_7(OH)_6]_2\cdot 9H_2O$

Origin: Synthetic.
Experimental details: No data.
Raman shifts (cm^{-1}): 951, 637s, 526w, 488w, 410s, 321.
Source: Kipcak et al. (2014).
Comments: The sample was characterized by powder X-ray diffraction data and may contain admixture of admontite. For the Raman spectrum of mcallisterite see also Derun and Tugce (2014).

Mcalpineite $Cu_3Te^{6+}O_6$

Origin: Gambatesa mine, eastern Liguria, Italy.
Experimental details: Micro-Raman scattering measurements have been performed on earthy mcalpineite aggregate using 633 nm laser radiation.
Raman shifts (cm^{-1}): 740s, 690s.
Source: Carbone et al. (2013).
Comments: The sample was characterized by powder X-ray diffraction data and electron microprobe analyses.

Mcconnellite $Cu^{1+}CrO_2$

Origin: Synthetic.
Experimental details: Micro-Raman scattering measurements have been performed on a single crystal using 632 nm He-Ne laser radiation. The laser radiation power at the sample was 0.3 mW. Incident beam power with a 3 μm spot size (4000 W/cm^2) was used. Polarized spectra were collected in the $z(yy)z$ and $z(yx)z$ scattering geometries.
Raman shifts (cm^{-1}): 692s, 351w.
Source: Aktas et al. (2011).

Comments: The Raman shifts are given for the scattering geometry $z(yy)z$. For the Raman spectra of mcconnellite see also Shu et al. (2009) and Elkhouni et al. (2013).

Mcguinnessite $CuMg(CO_3)(OH)_2$

Origin: Red Mountain, Mendocino Co., California, USA.
Experimental details: Raman scattering measurements have been performed on arbitrarily oriented crystal using a 633 nm He-Ne laser. The Raman shifts have been determined for the maxima of individual peaks obtained as a result of the spectral curve analysis.
Raman shifts (cm^{-1}): 3594, 3522s, 3381, 3309, 1567w, 1540, 1494s, 1359, 1090s, 1060, 914w, 741w, 707w, 516, 433, 269, 166, 147.
Source: Frost (2006).

Megawite $CaSnO_3$

Origin: Upper Chegem caldera, Kabardino-Balkaria, Northern Caucasus, Russia (type locality).
Experimental details: Raman scattering measurements have been performed on an arbitrarily oriented single crystal using 514.5 nm Ar^+ laser radiation. The laser radiation power was in the range from 40 to 60 mW.
Raman shifts (cm^{-1}): 705w, 557 (broad), 474w, 443, 355s, 283, 262, 183, 159.
Source: Galuskin et al. (2011b).
Comments: The sample was characterized by electron data and electron microprobe analyses. The empirical formula of the sample used is $CaSn_{0.6}Zr_{0.3}O_3$. For the Raman spectrum of megawite see also Zheng et al. (2012).

Meisserite $Na_5(UO_2)(SO_4)_3(SO_3OH)(H_2O)$

Origin: Blue Lizard mine, San Juan Co., Utah, USA (type locality).
Experimental details: Raman scattering measurements have been performed on an arbitrarily oriented sample using 532 nm frequency-stabilized single mode diode laser radiation. The laser radiation power at the sample was 3 mW.
Raman shifts (cm^{-1}): 3497w, 3366w, 1239, 1213, 1186, 1153, 1139, 1102, 1068, 1045,1031s, 1019s, 990, 975, 847s, 633, 606, 589, 464, 448, 414, 241, 199, 171, 123, 96, 61.
Source: Plášil et al. (2013c).
Comments: The sample was characterized by powder X-ray diffraction data and electron microprobe analyses. The crystal structure is solved.

Meixnerite $Mg_6Al_2(OH)_{16}(OH)_2 \cdot 4H_2O$

Origin: Synthetic.
Experimental details: Micro-Raman scattering measurements have been performed on an arbitrarily oriented sample using 632.8 nm Ar^+ laser radiation. The output power of the laser was set to 50 mW.
Raman shifts (cm^{-1}): 3492w (broad), 2392w, 1808, 1512, 1368w, 1165w, 936s, 827s, 748s, 686s, 629, 572, 555, 491, 416, 394w, 342, 313w, 238s, 207w, 180, 157, 128, 101, 76w, 60w.
Source: Kagunya et al. (1998).
Comments: The sample was characterized by powder X-ray diffraction data.

Melanarsite $K_3Cu_7Fe^{3+}O_4(AsO_4)_4$

Origin: Arsenatnaya fumarole, Tolbachik volcano, Kamchatka Peninsula, Russia (type locality).
Experimental details: Raman scattering measurements have been performed on an arbitrarily oriented single crystal using 532 nm laser radiation. The laser output power was 4 mW.
Raman shifts (cm^{-1}): 854s, 789, 632w, 550w, 464, 403, 352s, 331, 241, 187, 164, 142.
Source: Pekov et al. (2016d).
Comments: The sample was characterized by powder X-ray diffraction data and electron microprobe analyses. The crystal structure is solved.

Melanophlogite $C_2H_{17}O_5 \cdot Si_{46}O_{92}$

Origin: Mt. Hamilton, California, USA.
Experimental details: Polarized (XX and XY) Raman spectra have been obtained on a single crystal using 514.5 or 488 nm Ar$^+$ laser radiation. The laser radiation power at the sample was 2 mW.
Raman shifts (cm^{-1}): 3050w, 2909, 2900, 2321, 1380, 1277w, 803, 590, 364w, 268s, 165.
Source: Kolesov and Geiger (2003).
Comments: The Raman shifts are given with the XX polarization. For the Raman spectrum of melanophlogite see also Tribaudino et al. (2008).

Melanterite $Fe(SO_4) \cdot 7H_2O$

Origin: Synthetic.
Experimental details: Raman scattering measurements have been performed on an arbitrarily oriented sample using 532 nm Nd-YAG laser radiation. The nominal laser radiation power was 100 mW.
Raman shifts (cm^{-1}): 3385, 1147, 1074, 992s, 612, 480, 457, 284, 241w, 214w.
Source: Buzgar et al. (2009).
Comments: For the Raman spectra of melanterite see also Chio et al. (2007), Sobron and Alpers (2013), Jentzsch et al. (2013), Wang and Zhou (2014), Apopei et al. (2015), Buzatu et al. (2016), and Kompanchenko et al. (2016).

Meliphanite $Ca_4(Na,Ca)_4Be_4AlSi_7O_{24}(F,O)_4$

Origin: Østskogen larvikittbrudd, Tvedalen, Larvik kommune, Vestfold fylke, Norway.
Experimental details: Raman scattering measurements have been performed on an arbitrarily oriented sample using a 633 nm He-Ne laser. The Raman shifts have been determined for the maxima of individual peaks obtained as a result of the spectral curve analysis.
Raman shifts (cm^{-1}): 3805w, 3693w, 3595w, 3503, 3412, 3330, 3304w, 3207w, 3155w, 1095w, 1050w, 1016s, 991, 968, 932, 893w, 870w, 774, 745w, 721, 666s, 636w, 625s, 611w, 555w, 534w, 510w, 472, 421w, 382, 285w, 258, 207, 180, 147s, 113.
Source: Frost et al. (2015o).

Mellite $Al_2C_6(COO)_6 \cdot 16H_2O$

Origin: Bílina, near Most, Czech Republic.
Experimental details: Raman scattering measurements have been performed on an arbitrarily oriented sample using 1064 nm Nd-YAG laser radiation. The nominal laser radiation power was 350 mW.

Raman shifts (cm^{-1}): 3250w (broad), 1552, 1467s, 1386w, 1343, 1224, 805 (broad), 772, 538w, 376sh, 353w, 325, 242, 202sh, 178s, 162w, 151w, 134, 117w.
Source: Jehlička and Edwards (2008).

Mellizinkalite $K_3Zn_2Cl_7$

Origin: Glavnaya Tenoritovaya fumarole, Tolbachik volcano, Kamchatka, Russia (type locality).
Experimental details: Raman scattering measurements have been performed on an arbitrarily oriented polycrystalline sample using 532 nm laser radiation. The laser radiation power at the sample was about 3–3.5 mW.
Raman shifts (cm^{-1}): 310, 274s, 264s, 188, 128s.
Source: Pekov et al. (2015f).
Comments: The sample was characterized by powder X-ray diffraction data and electron microprobe analyses. The crystal structure is solved.

Mendipite $Pb_3O_2Cl_2$

Origin: Synthetic.
Experimental details: Raman scattering measurements have been performed on an arbitrarily oriented sample using a 632.8 nm He-Ne laser.
Raman shifts (cm^{-1}): 3504, 732, 601w, 474w, 435, 330s, 273s, 134s.
Source: Bouchard and Smith (2003).
Comments: The sample was characterized by powder X-ray diffraction data. For the Raman spectrum of mendipite see also Frost and Williams (2004).

Mercallite $KH(SO_4)$

Origin: Synthetic.
Experimental details: Raman scattering measurements have been performed on an arbitrarily oriented single crystal using 514.5 nm Ar$^+$ laser radiation. The laser radiation power at the sample was 20 mW.
Raman shifts (cm^{-1}): 1174, 1126w, 1027s, 1001s, 872sh, 855, 581, 572, 448, 412.
Source: Ayta et al. (2010).
Comments: A Mn-doped crystal characterized by ESR was used.

Merelaniite $Mo_4Pb_4VSbS_{15}$

Origin: Merelani Tanzanite deposit, Lelatema Mts., Manyara Region, Tanzania (type locality).
Experimental details: Raman scattering measurements have been performed on a curved surface of a cylindrical whisker using 633 nm laser radiation. The laser radiation power at the sample was less than 3 mW.
Raman shifts (cm^{-1}): 780w, 570w, 450, 401s, 390, 379, 324, 245, 133.
Source: Jaszczak et al. (2016).
Comments: The sample was characterized by powder X-ray diffraction data and electron microprobe analyses.

Merenskyite $PdTe_2$

Origin: Synthetic.
Experimental details: Raman scattering measurements have been performed on an arbitrarily oriented sample using 532 nm laser radiation. Neutral filter was used to decrease the laser radiation power to prevent sample damage.
Raman shifts (cm^{-1}): 132s, (105).
Source: Bakker (2014).
Comments: For the Raman spectrum of merenskyite see also Vymazalová et al. (2014).

Meridianiite $Mg(SO_4) \cdot 11H_2O$

Origin: Synthetic.
Experimental details: Raman scattering measurements have been performed on an arbitrarily oriented sample using 513 nm Ar$^+$ laser radiation. The nominal laser radiation power was 20 mW.
Raman shifts (cm^{-1}): 3520sh, 3395s, 1116, 1071, 990s, 620, 155sh, 444, 233w, 190.
Source: Genceli et al. (2007).
Comments: The sample was characterized by single-crystal X-ray diffraction data. The crystal structure is solved. For the Raman spectra of meridianiite see also Genceli et al. (2009) and Sakurai et al. (2010).

Merrillite Na-free analogue $Ca_{9.5}Mg(PO_4)_7$

Origin: Synthetic.
Experimental details: Raman scattering measurements have been performed on an arbitrarily oriented sample using 532 nm Nd-YAG laser radiation. The nominal laser radiation power was 20 mW.
Raman shifts (cm^{-1}): 1071, 1010, 966s, 951s, 761, 656w, 618, 601, 548, 437, 406.
Source: Jolliff et al. (2006).
Comments: The empirical formula of the sample used is $Ca_{18.70}REE_{0.05}(Mg,Fe)_{2.00}Na_{0.17}P_{13.90}Si_{0.16}O_{56}$.

Merrillite $Ca_9NaMg(PO_4)_7$

Origin: Suizhou meteorite, Dayanpo, Hubei, China.
Experimental details: Micro-Raman scattering measurements have been performed on an arbitrarily oriented sample using 514 nm Ar$^+$ laser radiation.
Raman shifts (cm^{-1}): 1080, 1026w, 972s, 956s, 604, 550w, 445, 408, 178.
Source: Xie et al. (2002).
Comments: The empirical formula of the sample used is $Ca_{8.82}Na_{0.88}Mg_{0.91}Fe_{0.07}P_{7.14}O_{28}$. For the Raman spectra of merrillite see also Cooney et al. (1999) and Xie et al. (2015).

Merwinite $Ca_3Mg(SiO_4)_2$

Origin: Synthetic.
Experimental details: Raman scattering measurements have been performed on an arbitrarily oriented polycrystalline sample using 5145 or 4880 Å laser radiation.

Raman shifts (cm^{-1}): 1011, 991w, 980w, 939, 921, 911, 887s, 872s, 860, 845, 667w, 640w, 579, 540, 529, 424, 394, 374w, 330, 270, 227w, 203w, 194w, 156s, 143s, 126, 119, 78, 42.
Source: Piriou and McMillan (1983).
Comments: The sample was characterized by electron microprobe analysis. For the Raman spectrum of merwinite see also Zedgenizov et al. (2014).

Mesolite $Na_2Ca_2(Si_9Al_6)O_{30}\cdot 8H_2O$

Origin: Talisker, Isle of Skye, Scotland.
Experimental details: Micro-Raman scattering measurements have been performed on an arbitrarily oriented sample using 514.5 nm Ar$^+$ laser radiation. The laser radiation power at the sample was 10 mW.
Raman shifts (cm^{-1}): 3583s, 3510s, 3412s, 3329, 3242s, 1666, 1652, 1099s, 1073, 1049s, 1023, 1007, 990s, 955, 951, 762, 757s, 753, 735, 727, 717, 710, 708s, 673s, 668, 538s, 496, 441s, 409s, 381s, 374, 368, 346, 330s, 283s, 273, 255s, 225s, 204, 183s, 158s, 143.
Source: Wopenka et al. (1998).
Comments: Bands whose intensities are definitely dependent upon polarization are indicated in the cited paper. For the Raman spectra of mesolite see also Pechar (1983) and Mozgawa (2001).

Meta-ankoleite $K(UO_2)(PO_4)\cdot 3H_2O$

Origin: Synthetic.
Experimental details: Raman scattering measurements have been performed on an arbitrarily oriented sample using 532 nm Nd-YAG laser radiation. The laser radiation power at the sample was 1–4 mW.
Raman shifts (cm^{-1}): 3805, 3498, 3375, 3237, 3110, 2786, 1004s, 994s, 831s, 826s, 400, 291, 195, 173, 113, 108.
Source: Clavier et al. (2016).
Comments: For the Raman spectrum of meta-ankoleite see also Pham-Thi et al. (1985).

Meta-autunite $Ca(UO_2)_2(PO_4)_2\cdot 6H_2O$

Origin: Autun, France (type locality).
Experimental details: Raman scattering measurements have been performed on an arbitrarily oriented crystal using a 633 nm He-Ne laser. Power at the sample was measured as 1 mW. The Raman shifts have been determined for the maxima of individual peaks obtained as a result of the spectral curve analysis.
Raman shifts (cm^{-1}): 3508, 3456, 3244, 1093, 1033, 1018, 1007s, 989, 915, 890, 850, 833s, 818, 643, 507, 453, 387s, 263, 222, 190.
Source: Frost and Weier (2004d).
Comments: For the Raman spectra of meta-autunite see also Frost (2004b) and Stefaniak et al. (2009).

Metacinnabar β-HgS

Origin: No data.
Experimental details: No data.
Raman shifts (cm^{-1}): 339, 280w, 253s.
Source: Radepont (2013).

Metahewettite $CaV^{5+}_6O_{16} \cdot 3H_2O$

Origin: The Fish, Eureka Co., Nevada, USA.
Experimental details: Raman scattering measurements have been performed on arbitrarily oriented crystal using a 633 nm He-Ne laser. The Raman shifts have been determined for the maxima of individual peaks obtained as a result of the spectral curve analysis.
Raman shifts (cm^{-1}): 1013, 994s, 954s, 878, 692, 530, 470, 425, 404, 290, 280s, 240, 188, 154, 140s.
Source: Frost et al. (2005d).

Metakirchheimerite $Co(UO_2)_2(AsO_4)_2 \cdot 8H_2O$

Origin: Jáchymov, Krušné Hory Mts. (Ore Mts.), Bohemia, Czech Republic.
Experimental details: Raman scattering measurements have been performed on an arbitrarily oriented sample using 730 nm laser radiation. The laser radiation power at the sample was 10 mW.
Raman shifts (cm^{-1}): 908, 896sh, 883sh, 816s, 801sh, 449, 320w, 206sh, 191.
Source: Plášil et al. (2009).
Comments: The sample was characterized by powder X-ray diffraction data and electron microprobe analyses.

Metalodèvite $Zn(UO_2)_2(AsO_4)_2 \cdot 10H_2O$

Origin: Jánská vein, Březové Hory deposit, Příbram ore district, Czech Republic.
Experimental details: Raman scattering measurements have been performed on an arbitrarily oriented sample using 532.2 nm laser radiation. The laser radiation power was 5 mW.
Raman shifts (cm^{-1}): 3418, 994, 977, 892, 866, 819s, 522s, 469, 447, 399, 319, 303, 280w, 191.
Source: Plášil et al. (2010d).
Comments: The sample was characterized by powder X-ray diffraction data and electron microprobe analyses.

Metamunirite $NaV^{5+}O_3$

Origin: Synthetic.
Experimental details: Raman scattering measurements have been performed using 4880 Å Ar^+ laser radiation on the (110) cleavage plate, with incident light perpendicular to the plane, and polarized perpendicular to the scattering plane. The laser radiation power at the sample was 100 mW. No analyser was in the path of the scattered radiation but a polarization scrambler was used.
Raman shifts (cm^{-1}): 948s, 911, 887s, 737, 557, 431, 288, 257, 203, 168, 135, 124, 81w, 53w.
Source: Seetharaman et al. (1983).

Metarauchite $Ni(UO_2)_2(AsO_4)_2 \cdot 8H_2O$

Origin: Jáchymov, Krušné Hory Mts. (Ore Mts.), Bohemia, Czech Republic.
Experimental details: Raman scattering measurements have been performed on an arbitrarily oriented sample using 514.5 nm Ar^+ laser radiation. The laser radiation power was 10 mW. The Raman shifts have been partly determined for the maxima of individual peaks obtained as a result of the spectral curve analysis.

Raman shifts (cm^{-1}): 3265w (broad), 3079, 1124w, 911sh, 898s, 893w, 883w, 878w, 817s, 804sh, 785sh, 771sh, 682w, 534w, 445, 395, 361w, 330, 319, 248, 211, 204w.
Source: Plášil et al. (2010c).
Comments: The sample was characterized by powder X-ray diffraction data and electron microprobe analyses.

Metarossite $CaV^{5+}_2O_6 \cdot 2H_2O$

Origin: Blue Cap mine, San Juan Co., Utah, USA.
Experimental details: Raman scattering measurements have been performed on an arbitrarily oriented single crystal using 532 nm laser radiation. The laser radiation power was 150 mW.
Raman shifts (cm^{-1}): 3398, 3240, 3189, 2954, 2904.
Source: Kobsch et al. (2016).
Comments: The sample was characterized by the crystal structure refinement based on single-crystal X-ray diffraction data. For the Raman spectrum of metarossite see also Frost et al. (2005d).

Metastibnite Sb_2S_3

Origin: Synthetic.
Experimental details: Raman scattering measurements have been performed on an arbitrarily oriented sample using 488 nm Ar$^+$ laser radiation. The nominal laser radiation power was 25 mW.
Raman shifts (cm^{-1}): 290s, 170.
Source: Watanabe et al. (1983).
Comments: The sample used was prepared as amorphous film by thermal evaporation of bulk Sb-S alloy with corresponding composition.

Metastudtite $UO_4 \cdot 2H_2O$

Origin: Synthetic.
Experimental details: Raman scattering measurements have been performed on an arbitrarily oriented single crystal using 632.8 nm He-Ne laser radiation. The laser radiation power at the sample was 6 mW.
Raman shifts (cm^{-1}): 3446s, 3245w, 1716sh, 1621, 1111w, 927s, 909s, 791w, 558w, 474.
Source: Bastians et al. (2004).
Comments: The sample was characterized by powder X-ray diffraction data.

Metathénardite $Na_2(SO_4)$

Origin: Synthetic.
Experimental details: Raman scattering measurements have been performed at 523 K on a single crystal, in different scattering geometries, using 457.9 nm Ar$^+$ laser radiation. The laser radiation power at the sample was 150 mW.
Source: Choi and Lockwood (1989).
Raman shifts (cm^{-1}): 1175, 1100, 993.5s, 628–626s, 467sh, 464s.
Comments: The sample was characterized by powder X-ray diffraction data. For the Raman spectrum of metathénardite see also Murugan et al. (2000).

Metatorbernite $Cu(UO_2)_2(PO_4)_2 \cdot 8H_2O$

Origin: Synthetic.
Experimental details: Raman scattering measurements have been performed on an arbitrarily oriented sample using 532 nm laser radiation. The nominal laser radiation power was 10 mW.
Raman shifts (cm^{-1}): 1630w, 1514w, 1414w, 1014w, 997s, 986w, 905w, 831w, 827s, 806w, 627w, 602w, 560w, 541w, 464w, 443w, 429w, 412w, 406, 399w, 291w, 253w, 221w, 196, 145w, 125w, 113w, 99, 81w.
Source: Sánchez-Pastor et al. (2013).
Comments: The sample was characterized electron microprobe analysis. For the Raman spectra of metatorbernite see also Čejka Jr (1984), Frost (2004b), Frost and Weier (2004a), and Faulques et al. (2015a, b).

Metatyuyamunite $Ca(UO_2)_2(VO_4)_2 \cdot 3H_2O$

Origin: No data.
Experimental details: Raman scattering measurements have been performed on an arbitrarily oriented sample using 514.5 nm Ar^+ laser radiation.
Raman shifts (cm^{-1}): 962, 829, 747s, 646w, 582, 570, 532, 467, 411w, 369s, 310.
Source: Botto et al. (1989).
Comments: The sample was characterized by powder X-ray diffraction data and analysis of H_2O.

Metauranocircite-I $Ba(UO_2)_2(PO_4)_2 \cdot 6H_2O$

Origin: Synthetic.
Experimental details: Raman scattering measurements have been performed on an arbitrarily oriented sample using 532 nm laser radiation. The nominal laser radiation power was 10 mW.
Raman shifts (cm^{-1}): 1635, 1543, 1420, 1039, 1023, 1003, 989s, 960, 821, 815s, 809, 627, 566, 502, 425, 409, 398, 376, 222.
Source: Sánchez-Pastor et al. (2013).
Comments: The sample was characterized electron microprobe analysis.

Metauranospinite $Ca(UO_2)_2(AsO_4)_2 \cdot 8H_2O$

Origin: Příbram, Central Bohemia region, Czech Republic.
Experimental details: Raman scattering measurements have been performed on arbitrarily oriented crystals using a 633 nm He-Ne laser. The Raman shifts have been determined for the maxima of individual peaks obtained as a result of the spectral curve analysis.
Raman shifts (cm^{-1}): 3549w, 3428, 3238w, 2106w, 1891w, 1787w, 1617w, 1518w, 1366w, 907w, 896, 815s, 806, 458, 397, 321, 275, 196, 187s, 150, 111.
Source: Čejka et al. (2009b).
Comments: The sample was characterized by powder X-ray diffraction data and electron microprobe analysis.

Metavariscite Al(PO$_4$)·2H$_2$O

Origin: Mt Lucia, Utah, USA.
Experimental details: Raman scattering measurements have been performed at 77 K on anarbitrarily oriented sample using a 633 nm He-Ne laser. The Raman shifts have been determined for the maxima of individual peaks obtained as a result of the spectral curve analysis.
Raman shifts (cm^{-1}): 1889, 1628, 1362, 1249, 1150, 1081s, 1063, 1033, 1018s, 643, 585, 574, 553, 499, 460, 446, 427s, 400, 387, 380, 374, 367, 360, 353s, 347, 340, 329s, 302, 297, 290, 280, 273, 258, 253, 244, 239, 230s, 211, 201, 187s, 171, 152, 147, 133, 124.
Source: Frost et al. (2004l).

Metavivianite Fe^{2+}Fe$^{3+}_2$(PO$_4$)$_2$(OH)$_2$·6H$_2$O

Origin: Boa Vista pegmatite, near Galiléia, Minas Gerais, Brazil.
Experimental details: Raman scattering measurements have been performed on an arbitrarily oriented sample using 514.5 nm Ar$^+$ laser radiation. The laser radiation power at the sample was 10 mW.
Raman shifts (cm^{-1}): 3431, 3378w, 3299w, 3257sh, 3194sh, 1089w, 1022, 972s, 579, 506s, 461, 374w, 322sh, 289, 256, 236, 197, 166, 143.
Source: Chukanov et al. (2012b).
Comments: The sample was characterized by powder X-ray diffraction data, Mössbauer spectroscopy, electron microprobe analysis and gas-chromatographic determination of H$_2$O. The empirical formula of the sample used is (Fe$^{3+}_{1.64}$Fe$^{2+}_{1.23}$Mg$_{0.085}$Mn$_{0.06}$)$_{\Sigma 3.015}$(PO$_4$)$_{1.98}$(OH)$_{1.72}$·6.36H$_2$O. The crystal structure is solved. For the Raman spectrum of metavivianite see also Frost et al. (2004m).

Metazeunerite Cu(UO$_2$)$_2$(AsO$_4$)$_2$·8H$_2$O

Origin: Gilgai, New England, New South Wales, Australia, or Wheal Edward Bottalock, Cornwell, England (not specified in the cited paper).
Experimental details: Raman scattering measurements have been performed on arbitrarily oriented crystals using a 633 nm He-Ne laser. The Raman shifts have been determined for the maxima of individual peaks obtained as a result of the spectral curve analysis.
Raman shifts (cm^{-1}): 3371, 3238, 3136, 910, 888, 819s, 809, 793, 449, 398, 320, 276, 240, 218.
Source: Frost (2004b).
Comments: For the Raman spectra of metazeunerite see also Frost and Weier (2004c) and Frost et al. (2004k).

Meurigite-Na [Na(H$_2$O)$_{2.5}$][Fe$^{3+}_8$(PO$_4$)$_6$(OH)$_7$(H$_2$O)$_4$]

Origin: Silver Coin mine, Valmy, Iron Point district, Nevada, USA (type locality).
Experimental details: No data.
Raman shifts (cm^{-1}): 3270, 1125, ~1005s, ~960s, ~925, 876, 568, 490, 441s, 401, 281, 222, 170.
Source: Kampf et al. (2009a).
Comments: Maybe, an erroneous spectrum: wavenumbers of the strongest bands of phosphate groups in the range from 900 to 1100 cm^{-1} are anomalously high. The sample was characterized by powder X-ray diffraction data and electron microprobe analyses. The crystal structure is solved.

Meyerhofferite $CaB_3O_3(OH)_5 \cdot H_2O$

Origin: Bigadic deposits, Turkey.
Experimental details: Raman scattering measurements have been performed on arbitrarily oriented crystals using a 633 nm He-Ne laser. The Raman shifts have been determined for the maxima of individual peaks obtained as a result of the spectral curve analysis.
Raman shifts (cm^{-1}): 3608s, (3505w), 3483s, (3421w), 3400s, 3344w, 3287, 3232, (3092), 3031s, (2908w), 1621w, (1592w), 1551w, 1367w, 1201, 1135, 1110, 1046, 1002, 958, 944, 935, 880, 728, 698w, (627w), 609s, 592, 493, 474, 435, 398w, 381w, 366w, 336, 235w, 227, 204, 189, 164, 124, 118.
Source: Frost et al. (2013h).
Comments: The sample was characterized by powder X-ray diffraction data and qualitative electron microprobe analysis.

Meymacite monoclinic analogue $WO_3 \cdot 2H_2O$

Origin: Synthetic.
Experimental details: Raman scattering measurements have been performed on a powdery sample using Ar$^+$ laser radiation.
Raman shifts (cm^{-1}): 3530, 3370sh, 3160s, ~1600, 960s, 685s, 662s, 380, 268, 235, 210, 110.
Source: Daniel et al. (1987).

Miargyrite $AgSbS_2$

Origin: Synthetic.
Experimental details: Raman scattering measurements have been performed on an arbitrarily oriented sample using 532 nm Nd-YAG laser radiation.
Raman shifts (cm^{-1}): 447, 370w, 322w, 250s, 185s, 134w, 115.
Source: Minceva-Sukarova et al. (2003).
Comments: For the Raman spectrum of miargyrite see also Makreski et al. (2013b).

Microcline $K(AlSi_3O_8)$

Origin: Čanište, Macedonia.
Experimental details: Raman scattering measurements have been performed on an arbitrarily oriented sample using 514.5 nm Ar$^+$ laser radiation. The nominal laser radiation power was 50 mW.
Raman shifts (cm^{-1}): 1139sh, 1124, 1099, 1051w, 994, 814w, 750, 651, 585w, 513s, 475s, 454, 402, 372, 331w, 285, 266, 258, 199, 179, 155, 127, 108.
Source: Makreski et al. (2009).
Comments: The sample was characterized by electron microprobe analysis. For the Raman spectra of microcline see also Ciobotă et al. (2012) and Frezzotti et al. (2012).

Miersite AgI

Origin: Synthetic.
Experimental details: Raman scattering measurements have been performed at 573 K on a single crystal using 6328 Å He-Ne laser radiation. The laser radiation power at the sample was 10 mW. Polarized spectra were collected in the $y[z(x/y)]z$ and $y(zx)z$ scattering geometries.

Raman shifts (cm^{-1}): ~100 (broad), ~30s (broad).
Source: Delaney and Ushioda (1976).
Comments: The sample was characterized by powder X-ray diffraction data and electron microprobe analysis. The Raman shifts are given for the scattering geometry $y[z(x/y)]z$. In the scattering geometry $y(zx)z$ only a peak at ~30 cm^{-1} is observed.

Mikasaite $Fe^{3+}_2(SO_4)_3$

Origin: Synthetic.
Experimental details: Micro-Raman scattering measurements have been performed on an arbitrarily oriented sample using 532 nm laser radiation.
Raman shifts (cm^{-1}): 1123s, 1098s, 1078s, 1069, 1040, 677, 657, 628, 613, 600, 468, 461, 448, 295, 234, 178.
Source: Ling and Wang (2010).
Comments: The sample was characterized by powder X-ray diffraction data. For the Raman spectrum of mikasaite see also Apopei et al. (2015).

Milarite $KCa_2(Be_2AlSi_{12})O_{30} \cdot H_2O$

Origin: Giuv Tavetsch, CH.
Experimental details: Raman scattering measurements have been performed on an arbitrarily oriented single crystal using 785 nm laser radiation.
Raman shifts (cm^{-1}): 1126, 834, 538, 479s, 436w, 382w, 288w, 161w.
Source: Jehlička and Vandenabeele (2015).
Comments: For the Raman spectra of milarite see also Lengauer et al. (2009) and Jehlička et al. (2012).

Millerite β-NiS

Origin: Synthetic.
Experimental details: Raman scattering measurements have been performed on an arbitrarily oriented sample using 514.5 nm laser radiation. The laser radiation power was 15 mW.
Raman shifts (cm^{-1}): 372, 350, 301, 283, 246s, 222, 181, 174s, 142.
Source: Bishop et al. (2000).
Comments: The sample was characterized by powder X-ray diffraction.

Millosevichite $Al_2(SO_4)_3$

Origin: Synthetic.
Experimental details: Raman scattering measurements have been performed on an arbitrarily oriented sample using 633 nm He-Ne laser radiation. The laser radiation power is not indicated. The Raman shifts have been determined for the maxima of individual peaks obtained as a result of the spectral curve analysis.
Raman shifts (cm^{-1}): 1053, 1009, 990s, 979, 726, 630, 614, 572, 496, 459, 446.
Source: Kloprogge and Frost (1999c).

Mimetite $Pb_5(AsO_4)_3Cl$

Origin: Synthetic.
Experimental details: Raman scattering measurements have been performed on an arbitrarily oriented sample using Nd-YAG laser radiation. The laser radiation power at the sample was 300 mW.
Raman shifts (cm^{-1}): 1043, 1000w, 982w, 949, 920s, 812s, 791sh, 744sh, 553w, 546sh, 426sh, 411sh, 409, 391, 372, 338s, 314s.
Source: Bajda et al. (2011).
Comments: The empirical formula of the sample used is $Pb_5[(AsO_4)_{2.4}(PO_4)_{0.6}]Cl$. The bands in the ranges 900–1050 and 540–560 cm^{-1} correspond to phosphate groups. For the Raman spectra of mimetite see also Levitt and Condrate, Sr (1970), Adams and Gardner (1974), Bartholomäi and Klee (1978), Frost et al. (2007c), and Bajda (2010).

Minguzzite $K_3Fe^{3+}(C_2O_4)_3 \cdot 3H_2O$

Origin: Synthetic.
Experimental details: Raman scattering measurements have been performed on arbitrarily oriented crystals using 514.5 nm laser radiation.
Raman shifts (cm^{-1}): 1720, 1451s, 1252, 898, 782, 558s, 370, 257s, 136s.
Source: Narsimhulu et al. (2015).

Minium $Pb^{2+}_2Pb^{4+}O_4$

Origin: Synthetic.
Experimental details: Raman scattering measurements have been performed on an arbitrarily oriented sample using a 632.8 nm He-Ne laser. In spite of the strong opacity of the mineral, it was possible to record a Raman spectrum by using a very weak intensity laser (1 mW) and a long duration of 500 s.
Raman shifts (cm^{-1}): 549s, 480, 391, 313, 223, 150, 120s, 84, 70, 60, 51.
Source: Bouchard and Smith (2003).
Comments: For the Raman spectra of minium see also Bouchard-Abouchacra (2001), Burgio and Clark (2001), and Lepot et al. (2006).

Minnesotaite $Fe^{2+}_3Si_4O_{10}(OH)_2$

Origin: No data in the cited paper.
Experimental details: Raman scattering measurements have been performed on an arbitrarily oriented sample using 532 nm laser radiation. The laser radiation power at the sample was about 13 mW.
Raman shifts (cm^{-1}): 3654s, 3639s, 3625s, 660s, 545, 440, 407, 350, 251, 188.
Source: Wang et al. (2015).
Comments: The sample was characterized by powder X-ray diffraction data and electron microprobe analysis.

Minyulite $KAl_2(PO_4)_2F \cdot 4H_2O$

Origin: Minyulo Well, Australia (type locality).
Experimental details: Raman scattering measurements have been performed on an arbitrarily oriented crystal using a 633 nm He-Ne laser. The Raman shifts have been determined for the maxima of individual peaks obtained as a result of the spectral curve analysis.

Raman shifts (cm^{-1}): 3692w, 3669, 3661w, 3324 (broad), 3225 (broad), 1584w, (1190w), 1176w, 1155w, (1136w), (1105w), 1091, 1077, (1047), 1012s, (991w), 657, (628w), (606w), 592s, 575, (551w), (535w), (522w), (506), 494s, (481), 448, (420w), 407.
Source: Frost et al. (2014l).
Comments: The empirical formula of the sample used is $(K_{0.82}Ca_{0.05}Na_{0.03})Al_{2.04}Fe_{0.08}(PO_4)_2[F_{0.55}(OH)_{0.45}] \cdot 4H_2O$.

Mirabilite $Na_2(SO_4) \cdot 10H_2O$

Origin: No data in the cited paper.
Experimental details: No data.
Raman shifts (cm^{-1}): 3506s, 3340, 1129w, 989s, 627, 458.
Source: Frezzotti et al. (2012).
Comments: For the Raman spectrum of mirabilite see also Sakurai et al. (2010).

Misakiite $Cu_3Mn(OH)_6Cl_2$

Origin: Sadamisaki Peninsula, Ehime prefecture, Japan (type locality).
Experimental details: Raman scattering measurements have been performed on an arbitrarily oriented single sample using 514.5 nm Ar$^+$ laser radiation. The laser radiation power at the sample was 50 mW.
Raman shifts (cm^{-1}): 3552w, 3505, 3460s, 470, 397s, 321, 265.
Source: Nishio-Hamane et al. (2016b).
Comments: The sample was characterized by powder and single-crystal X-ray diffraction data, and electron microprobe analyses.

Mitscherlichite $K_2CuCl_4 \cdot 2H_2O$

Origin: Synthetic.
Experimental details: No data.
Raman shifts (cm^{-1}): 3340, 3275, 3250, 3235, 3170, 1630–1628, 685, 633, 550, 404, 395.
Source: Thomas et al. (1974).

Mixite $Cu_6Bi(AsO_4)_3(OH)_6 \cdot 3H_2O$

Origin: Smrkovec ore occurrence, the Slavkovský Les Mts., western Bohemia, Czech Republic.
Experimental details: Raman scattering measurements have been performed on arbitrarily oriented crystals using a 633 nm He-Ne laser. The Raman shifts have been determined for the maxima of individual peaks obtained as a result of the spectral curve analysis.
Raman shifts (cm^{-1}): 3470, 3392, 1588w, 1513w, 995w, (855), 850s, 805, 553, 529, (494), 472s, (460), 421, 390, 311, 284, 252, 232, 192, 169, 138, (112), 105s.
Source: Frost et al. (2009f).
Comments: The sample was characterized by X-ray diffraction data and electron microprobe analyses. For the Raman spectrum of mixite see also Frost et al. (2006m).

Moctezumite $Pb(UO_2)(Te^{4+}O_3)_2$

Origin: Moctezuma (Bambolla) mine, Sonora, Mexico.
Experimental details: Raman scattering measurements have been performed on arbitrarily oriented crystals using a 633 nm He-Ne laser. The Raman shifts have been determined for the maxima of individual peaks obtained as a result of the spectral curve analysis.
Raman shifts (cm^{-1}): 826, 758, 723s, 656s, 623, 511w, 455, 356, 308w, 252, 212, 142.
Source: Frost et al. (2009b).
Comments: IR spectrum of presumed moctezumite given in the cited paper corresponds to quartz.

Mogánite $SiO_2 \cdot nH_2O$

Origin: Mogán, Gran Canaria (Grand Canary), Las Palmas Province, Canary Islands, Spain (type locality).
Experimental details: Raman scattering measurements have been performed on an arbitrarily oriented sample using 514.5 nm Ar$^+$ laser radiation.
Raman shifts (cm^{-1}): 1177w, 1171w, 1084w, 1058w, (978w), (950w), 833w, 792w, 693w, 501s, 463, 449, 432w, 398w, 377w, 370w, 317w, 265w, 220s, 141, 129s.
Source: Kingma and Hemley (1994).
Comments: No independent analytical data are provided for the sample used.

Mohite Cu_2SnS_3

Origin: Synthetic.
Experimental details: Raman scattering measurements have been performed on an arbitrarily oriented sample using 532 or 785 nm laser radiation. The laser radiation power at the sample was below 0.4 mW.
Raman shifts (cm^{-1}): 374w, 352s, 314, 290s, 263sh, 248sh, 224w.
Source: Fontané et al. (2013).
Comments: No independent analytical data are provided for the sample used.

Mohrite $(NH_4)_2Fe^{2+}(SO_4)_2 \cdot 6H_2O$

Origin: Synthetic.
Experimental details: Micro-Raman scattering measurements have been performed on arbitrarily oriented individual particles using 532 nm Ar$^+$ laser radiation. The laser radiation power at the sample was 2 mW.
Raman shifts (cm^{-1}): 3355, 3290, 3101sh, (2913w), (2852w), 1705w, 1678w, 1430w, 1148sh, 1129, 1091, 1067, 980s, 622, 610, 457sh, 450.
Source: Jentzsch et al. (2013).
Comments: No independent analytical data are provided for the sample used.

Moissanite SiC

Origin: Synthetic.
Experimental details: Raman scattering measurements have been performed on an arbitrarily oriented sample using 632.8 or 488 nm laser radiation.

Raman shifts (cm^{-1}): 963, 783s, 762, 150.
Source: Andò and Garzanti (2014).
Comments: For the Raman spectra of moissanite see also Xu et al. (2008, 2015b) and Kompanchenko et al. (2016).

Mojaveite $Cu_6[Te^{6+}O_4(OH)_2](OH)_7Cl$

Origin: Blue Bell claims, near Baker, San Bernardino Co., California, USA (type locality).
Experimental details: Raman scattering measurements have been performed on a single crystal (probably on the (001) face) using 514.3 nm laser radiation. The laser radiation power at the sample was 2 mW.
Raman shifts (cm^{-1}): ~3500w, 1112w, 967w, 694s, 654, 624, 555, 510, 475sh, 414w, 286, 254, 233w, 203, 172.
Source: Mills et al. (2014b).
Comments: The sample was characterized by powder and single-crystal X-ray diffraction data and electron microprobe analyses.

Molybdenite MoS_2

Origin: Wolfram Camp, Qld., Australia.
Experimental details: Raman scattering measurements have been performed on an arbitrarily oriented sample using 514.5 nm Ar$^+$ laser radiation. The laser radiation power at the sample was between 1 and 10 mW.
Raman shifts (cm^{-1}): 451, 408s, 382s, 285.
Source: Mernagh and Trudu (1993).
Comments: For the Raman spectra of molybdenite see also Windom et al. (2011) and Štengl and Henych (2013).

Molybdite MoO_3

Origin: Synthetic.
Experimental details: Raman scattering measurements have been performed on a powdery sample using 632.8 nm He-Ne laser radiation. The nominal laser radiation power was 15 mW.
Raman shifts (cm^{-1}): 995s, 820s, 666, 285s, 158.
Source: Windom et al. (2011).
Comments: For the Raman spectra of molybdite see also Seguin et al. (1995), Nitta et al. (2006), and Camacho-López et al. (2011).

Molybdofornacite $CuPb_2(MoO_4)(AsO_4)(OH)$

Origin: No data.
Experimental details: Raman scattering measurements have been performed on an arbitrarily oriented sample using 785 nm Nd-YAG laser radiation. The laser radiation power at the sample was 1 mW.
Raman shifts (cm^{-1}): 1089, 1048, 1014, 855, 713, 666, 558, 387, 355, 320, 278, 226, 196, 159, 152.
Source: Frost (2004c).
Comments: The data are questionable: no information on the sample origin and no independent analytical data on the sample used are given. Band intensities are not indicated.

Molybdophyllite $Pb_8Mg_9[Si_{10}O_{28}(OH)_8O_2(CO_3)_3]\cdot H_2O$

Origin: Långban deposit, Bergslagen ore region, Filipstad district, Värmland, Sweden (type locality).
Experimental details: Raman scattering measurements have been performed on an single crystal using unpolarized 633 nm laser radiation.
Raman shifts (cm^{-1}): 3696s, ~3600, ~1050, ~998.
Source: Kolitsch et al. (2012).
Comments: The sample was characterized by single crystal X-ray diffraction data.

Molysite $FeCl_3$

Origin: Synthetic.
Experimental details: Raman scattering measurements have been performed on an arbitrarily oriented sample using a 632.8 nm He-Ne laser. Laser radiation power of 30 mW at the source was reduced considerably by various filters.
Raman shifts (cm^{-1}): 667, 598, 373,293s, 259.
Source: Bouchard and Smith (2003).
Comments: For the Raman spectrum of molysite see also Bouchard-Abouchacra (2001).

Monazite-(Ce) $Ce(PO_4)$

Origin: Synthetic.
Experimental details: Micro-Raman scattering measurements have been performed on an arbitrarily oriented sample using 632.8 He-Ne laser radiation. The laser radiation power behind the microscope objective was 8 mW.
Raman shifts (cm^{-1}): 1073, 1056, 992, 970s, 620, 572w, 467, 397, 220, 102, 88.
Source: Ruschel et al. (2012).
Comments: The sample was characterized by electron microprobe analysis. For the Raman spectra of monazite-(Ce) see also Begun et al. (1981), O'Neill et al. (2006), Silva et al. (2006), Andò and Garzanti (2014), and Heuser et al. (2014).

Monazite-(La) $La(PO_4)$

Origin: Synthetic.
Experimental details: Raman scattering measurements have been performed on an arbitrarily oriented sample using 514.5 nm laser radiation.
Raman shifts (cm^{-1}): 1073, 1065w, 1055s, 1025w, 991, 967s, 619, 589w, 572, 537w, 465, 414, 394, 271, 255w, 227, 220, 183w, 170, 151, 131w, 120w, 100, 90.
Source: Begun et al. (1981).
Comments: For the Raman spectra of monazite-(La) see also Silva et al. (2006), Frezzotti et al. (2012), and Heuser et al. (2014).

Monazite-(Nd) $Nd(PO_4)$

Origin: Synthetic.
Experimental details: Polarized Raman scattering measurements have been performed on a single crystal using 488 nm laser radiation. Scattering geometry is not indicated.

Raman shifts (cm^{-1}): 1079, 1061, 1033, 998, 977s, 625s, 601, 575, 539, 471s, 419, 398, 291, 264, 236, 228, 183, 175, 160, 154, 106, 89.
Source: Silva et al. (2006).
Comments: For the Raman spectra of monazite-(Nd) see also Begun et al. (1981) and Heuser et al. (2014).

Monazite-(Sm) Sm(PO$_4$)

Origin: Synthetic.
Experimental details: Polarized Raman scattering measurements have been performed on a single crystal using 514.5 nm Ar$^+$laser radiation. Scattering geometry is not indicated.
Raman shifts (cm^{-1}): 1084, 1065, 1035, 999, 983s, 629s, 603, 577, 542, 474s, 424, 404, 293, 265, 243, 231, 185, 177, 159, 155, 107, 88.
Source: Silva et al. (2006).
Comments: For the Raman spectra of monazite-(Sm) see also Begun et al. (1981) and Heuser et al. (2014).

Moncheite Pt(Te,Bi)$_2$

Origin: No data.
Experimental details: Raman scattering measurements have been performed on an arbitrarily oriented sample using 532.068 nm laser radiation. The laser radiation power at the sample was in the range from 1 to 2 mW.
Raman shifts (cm^{-1}): 155, 115s.
Source: Bakker (2014).
Comments: No independent analytical data are provided for the sample used. For the Raman spectrum of moncheite see also Mernagh and Hoatson (1995).

Monetite Ca(PO$_3$OH)

Origin: Synthetic (commercial reactant).
Experimental details: Raman scattering measurements have been performed on an arbitrarily oriented sample using 1064.1 nm Nd-YAG laser radiation. The laser radiation power at the sample was below 320 mW.
Raman shifts (cm^{-1}): 2814w, 2421w, 1617w, 1133, 1095, 988s, 901s, 779, 693, 591, 574, 562, 474w, 420, 395, 274w, 182w, 143w.
Source: Xu et al. (1999).
Comments: For the Raman spectrum of monetite see also Frost et al. (2013r).

Monipite MoNiP

Origin: Allende meteorite.
Experimental details: Micro-Raman scattering measurements have been performed on an arbitratily oriented grain in a polished section using 514.5 nm laser radiation.
Raman shifts (cm^{-1}): 430, 350s, 280.
Source: Ma et al. (2014a).
Comments: The sample was characterized by powder X-ray diffraction data and electron microprobe analyses.

Monohydrocalcite $Ca(CO_3) \cdot H_2O$

Origin: Sainte Guillaume vein, St.-Marie-aux-Mines, Haut Rhin, France.
Experimental details: Raman scattering measurements have been performed on an arbitrarily oriented sample using 514.5 nm Ar^+ laser adiation. The laser emitted power was about 8 mW.
Raman shifts (cm^{-1}): 3425w, 3326, 3224, 1069s, 876w, 723w, 699w, 208.
Source: Coleyshaw et al. (2003).
Comments: The sample was characterized by powder X-ray diffraction data.

Montebrasite $LiAl(PO_4)OH$

Origin: Minas Gerais, Brazil.
Experimental details: Raman scattering measurements have been performed on an arbitrarily oriented sample using 488 nm Ar^+ laser radiation. The nominal laser radiation power was 200 mW.
Raman shifts (cm^{-1}): 3329s, 1189w, 1108, 1057s, 1046s, 1012s, 799, 645, 627, 601w, 483w, 429, 298s, 278, and a series of weak bands below 270 cm^{-1}.
Source: Rondeau et al. (2006).
Comments: The sample was characterized by powder X-ray diffraction data and electron microprobe analysis. For the Raman spectrum of montebrasite see also Dias et al. (2011).

Monteponite CdO

Origin: Synthetic.
Experimental details: Raman scattering measurements have been performed on an arbitrarily oriented sample using 532 nm laser radiation. The laser radiation power was 2.48 mW.
Raman shifts (cm^{-1}): 938w, 390s, 330sh, 259.
Source: Thema et al. (2015).
Comments: For the Raman spectrum of monteponite see also Falgayrac et al. (2013).

Montgomeryite $Ca_4MgAl_4(PO_4)_6(OH)_4 \cdot 12H_2O$

Origin: Katies Bower, Chifley Cave, Jenolan Caves, New South Wales, Australia.
Experimental details: Raman scattering measurements have been performed on arbitrarily oriented crystals using a 633 nm He-Ne laser. The Raman shifts have been determined for the maxima of individual peaks obtained as a result of the spectral curve analysis.
Raman shifts (cm^{-1}): 1709, 1669w, 1606, (1582w), 1339, (1286), 1260, 1214, 1143, 1088, 1011s, 979s, 655w, (609), 591, 511s, 475, 457s, 391, 318, 292s, 268, (251), 202, 176, 161s, 146.
Source: Frost et al. (2012e).
Comments: Powder X-ray diffraction data indicate that the sample used contains minor admixture of variscite.

Monticellite $CaMg(SiO_4)$

Origin: Synthetic.
Experimental details: Raman scattering measurements have been performed on an arbitrarily oriented sample using 514.5 or 488.0 nm laser radiation.
Raman shifts (cm^{-1}): 950, 900, 852s, 818s.

Source: Piriou and McMillan (1983).
Comments: For the Raman spectrum of monticellite see also Mouri and Enami (2008).

Montmorillonite $(Na,Ca)_{0.3}(Al,Mg)_2Si_4O_{10}(OH)_2 \cdot nH_2O$

Origin: Bidahochi formation, Cheto district, Apache Co., Arizona, USA.
Experimental details: Raman scattering measurements have been performed on an arbitrarily oriented sample using 1064 nm laser radiation. The nominal laser radiation power was 0.2 mW.
Raman shifts (cm^{-1}): 1112, 1029, 917, 885, 840, 792, 709s, 705sh, 571w, 505sh, 433s, 330, 290, 262, 200s, 176, 81.
Source: Bishop and Murad (2004).
Comments: The sample was characterized by chemical analyses. For the Raman spectrum of montmorillonite see also Frost and Rintoul (1996).

Montroseite $(V^{3+},Fe^{2+},V^{4+})O(OH)$

Origin: Akouta mine, Niger.
Experimental details: Micro-Raman scattering measurements have been performed on an arbitrarily oriented single crystal using 514.5 nm Ar$^+$ laser radiation.
Source: Forbes and Dubessy (1988).
Raman shifts (cm^{-1}): 990s, 960, 925, 905w, 845, 780, 750w, 690, 560w, 515, 470, 400, 340w, 295sh, 280s, 144.
Comments: The sample was characterized by powder X-ray diffraction data and chemical analysis.

Montroydite HgO

Origin: Synthetic.
Experimental details: Raman scattering measurements have been performed on an arbitrarily oriented sample using 514.5 nm Ar$^+$ laser radiation.
Raman shifts (cm^{-1}): 568, 327s, 120w.
Source: Zhou et al. (1998).
Comments: The sample was characterized by powder X-ray diffraction data.

Moolooite $Cu(C_2O_4) \cdot nH_2O$

Origin: Synthetic.
Experimental details: Raman scattering measurements have been performed on an arbitrarily oriented sample using 514.5 nm Ar$^+$ laser radiation.
Raman shifts (cm^{-1}): 1620, 1513s, 1486s, 922, 845, 608, 584s, 558s, 300, 209s.
Source: D'Antonio et al. (2007).
Comments: The sample was characterized by powder X-ray diffraction data and TG analysis. For the Raman spectra of moolooite see also Frost and Weier (2003), Frost (2004d), Castro et al. (2008), and Romann et al. (2009).

Mopungite $NaSb^{5+}(OH)_6$

Origin: Pereta mine, Tuscany, Italy.
Experimental details: Micro-Raman scattering measurements have been performed on an arbitrarily oriented sample using 632.8 nm He-Ne laser radiation. The nominal laser radiation power was 20 mW. The Raman shifts have been partly determined for the maxima of individual peaks obtained as a result of the spectral curve analysis.
Raman shifts (cm^{-1}): 3423s, 3353, 3331, 3239, 3178s, 3134, (671), (648), 626s, (605), 362, 350, 204, 189.
Source: Bittarello et al. (2015).
Comments: The sample was characterized by single-crystal X-ray diffraction data and electron microprobe analysis. For the Raman spectra of mopungite see also Bahfenne (2011) and Rintoul et al. (2011).

Moraskoite $Na_2Mg(PO_4)F$

Origin: Morasko IAB-MG iron meteorite, Poland (type locality).
Experimental details: Raman scattering measurements have been performed on an arbitrarily oriented sample using 514.5 nm Ar$^+$ laser radiation. The laser radiation power at the sample was between 30 and 50 mW.
Raman shifts (cm^{-1}): 1114s, 1027s, 962s, 589, 438w, 336w, 308w, 279w, 262w, 244w, 193w, 184w, 147w, 131w.
Source: Karwowski et al. (2015).
Comments: The sample was characterized by single-crystal X-ray diffraction data and electron microprobe analyses.

Mordenite $(Na_2,Ca,K_2)_4(Al_8Si_{40})O_{96} \cdot 28H_2O$

Origin: Faröes Islands.
Experimental details: Raman scattering measurements have been performed on an arbitrarily oriented sample using Nd-YAG laser radiation. The nominal laser radiation power was 300 mW.
Raman shifts (cm^{-1}): 1087w, 1046, 977w, 713w, 534s, 448, 427, 297w, 229w, 160.
Source: Mozgawa (2001).
Comments: The sample was characterized by powder X-ray diffraction data.

Morenosite $Ni(SO_4) \cdot 7H_2O$

Origin: Synthetic.
Experimental details: Micro-Raman spectrum was recorded with a medium Hilger quartz spectrograph using the $\lambda = 2536.5$ Å resonance radiation of mercury.
Raman shifts (cm^{-1}): 3532w, 3437s, 3266, 1158w, 1139, 1095, 1057s, 985s, 642, 622w, 612s, 463s, 442s, 419w, 402, 251s, 233w, 207s, 152s, 131w, 111s, 88, 75s, 60.
Source: Krishnamurti (1958).

Moschelite HgI

Origin: Synthetic.
Experimental details: No data.
Raman shifts (cm^{-1}): 194, 113, 65, 31.
Source: Ōsaka (1971).
Comments: For the Raman spectrum of moschelite see also Cooney et al. (1968).

Mosesite $(Hg_2N)Cl$

Origin: No data.
Experimental details: Micro-Raman scattering measurements have been performed on an arbitrarily sample using 785 nm laser radiation.
Raman shifts (cm^{-1}): ~547, ~538.
Source: Cooper et al. (2013a).

Mottramite $PbCu(VO_4)(OH)$

Origin: Synthetic.
Experimental details: Raman scattering measurements have been performed on arbitrarily oriented crystals using a 633 nm He-Ne laser. The Raman shifts have been determined for the maxima of individual peaks obtained as a result of the spectral curve analysis.
Raman shifts (cm^{-1}): 3162w, (2927)w, (891w), (849), (841), 829s, (821), 802, (796), (747w), 716w, 612w, 500w, 451w, 411, 366, (354), 333, (307), (301), 293, 247w, 227, 202w, 172w, 151, (136), (129), 118.
Source: Frost et al. (2014ai).
Comments: The sample was characterized by qualitative electron microprobe analysis. For the Raman spectrum of mottramite see also Frost et al. (2001).

Mountkeithite $(Mg_{1-x}Fe^{3+}{}_x)(SO_4)_{x/2}(OH)_2 \cdot nH_2O$ ($x < 0.5, n > 3x/2$)

Origin: No data.
Experimental details: Raman scattering measurements have been performed on an arbitrarily oriented sample using a 633 nm He-Ne laser. The Raman shifts have been determined for the maxima of individual peaks obtained as a result of the spectral curve analysis.
Raman shifts (cm^{-1}): 3698, 3688, 3654, 2240w, 1937, 1679s, 1613s, 1525s, 1439, 1273s, 1122, 1109, 920, 691, 621w, 528, 468w, 390, 348w, 233, 202w.
Source: Frost et al. (2003h).
Comments: No independent analytical data are provided for the sample used.

Moydite-(Y) $YB(OH)_4(CO_3)$

Origin: Evans-Lougranitic pegmatite, near Wakefield, Quebec, Canada (type locality).
Experimental details: Micro-Raman scattering measurements have been performed on an arbitrarily oriented sample using 514.5 nm Ar$^+$ laser radiation. The laser radiation power at the sample was between 20 and 50 mW.
Raman shifts (cm^{-1}): 1610s, 1410, 1124s, 1106, 765, 700w.

Source: Grice et al. (1986).

Comments: The sample was characterized by powder X-ray diffraction data and electron microprobe analyses. There are discrepancies between Raman shifts and figure of the Raman spectrum given in the cited paper.

Mukhinite $Ca_2(Al_2V^{3+})[Si_2O_7][SiO_4]O(OH)$

Origin: Pyrrhotite gorge, Khibiny Mts., Kola Peninsula, Russia.

Experimental details: Micro-Raman scattering measurements have been performed on an arbitrarily oriented polished surface using 633 nm He-Ne laser radiation. The laser radiation power at the sample was betweem 2 and 20 mW.

Raman shifts (cm^{-1}): 1235w, 1213w, 1087w, 1052w, 965, 910, 849, 805, 741, 693, 623, 596s, 547, 518s, 485, 322, 284, 235, 155, 125, 85s.

Source: Voloshin et al. (2014).

Comments: The sample was characterized by electron microprobe analyses.

Mukhinite V-rich analogue $Ca_2(AlV^{3+}_2)[Si_2O_7][SiO_4]O(OH)$

Origin: Pyrrhotite gorge, Khibiny Mts., Kola Peninsula, Russia.

Experimental details: Micro-Raman scattering measurements have been performed on an arbitrarily oriented polished surface using 633 nm He-Ne laser radiation. The laser radiation power at the sample was betweem 2 and 20 mW.

Raman shifts (cm^{-1}): 1089s, 1012s, 929, 891, 816, 700, 668, 607, 552, 499, 389s, 357, 328, 229w, 136w.

Source: Voloshin et al. (2014).

Comments: The sample was characterized by electron microprobe analyses.

Mullite $Al_{4+2x}Si_{2-2x}O_{10-x}$ ($x \approx 0.4$)

Origin: NW of Ormsaig, Ross of Mull, Scotland, UK (type locality).

Experimental details: Micro-Raman scattering measurements have been performed on an arbitrarily oriented sample using 457 nm Ar$^+$ laser radiation.

Raman shifts (cm^{-1}): 1130, 1035, 960s, 870s, 720, 600s, 415, 340, 305.

Source: Bost et al. (2016).

Comments: For the Raman spectrum of mullite see also Shoval et al. (2001).

Muscovite $KAl_2(Si_3Al)O_{10}(OH)_2$

Origin: Rebra Valley, Rodnei (Rodna) Mts., Romania.

Experimental details: Raman scattering measurements have been performed on an arbitrarily oriented single sample using 532 nm laser radiation. The laser radiation power at the sample was 35 mW.

Raman shifts (cm^{-1}): 119, 957, 914, 755, 704s, 641, 412s, 266s, 219.

Source: Buzgar (2008).

Comments: According to the Raman Spectra Database, Siena (http://www.dst.unisi.it/geofluids/ raman/spectrum_frame.htm), Raman spectrum of muscovite contains also a band of O–H-stretching vibrations at 3627 cm^{-1}. For the Raman spectra of muscovite see also Haley et al. (1982), Tlili et al. (1989), Wada and Kamitakahara (1991), Graeser et al. (2003), and Frezzotti et al. (2012).

Nabiasite $BaMn_9(VO_4)_6(OH)_2$

Origin: Nabias hamlet, Central Pyrenees, France (type locality).
Experimental details: Methods of sample preparation are not described. Raman scattering measurements have been performed using 514.5 nm Ar^+ laser radiation. The laser radiation power is not indicated.
Raman shifts (cm^{-1}): 880sh, 867, 807s, 768.
Source: Brugger et al. (1999).
Comments: The sample was characterized by single-crystal X-ray diffraction data and electron microprobe analysis. The crystal structure is solved.

Nabimusaite $KCa_{12}(SiO_4)_4(SO_4)_2O_2F$

Origin: Jabel Harmun, Palestinian Autonomy, Israel (type locality).
Experimental details: Methods of sample preparation are not described. Micro-Raman scattering measurements have been performed using 488 nm solid-state laser radiation. The laser radiation power at the sample was 44 mW. The Raman shifts have been determined for the maxima of individual peaks obtained as a result of the spectral curve analysis.
Raman shifts (cm^{-1}): 1121, 993s, 948, 930sh, 885, 849sh, 831, 637, 563, 524, 463, 403, 129.
Source: Galuskin et al. (2015d).
Comments: The sample was characterized by single-crystal X-ray diffraction data and electron microprobe analyses. The crystal structure is solved.

Nacrite $Al_2Si_2O_5(OH)_4$

Origin: Commonwealth Scientific and Industrial Research Organization, Division of Soils, Glen Osmond, South Australia.
Experimental details: Raman scattering measurements have been performed on arbitrarily oriented crystals using a 1064 nm Nd-YAG laser. The nominal laser radiation power was 100 mW.
Raman shifts (cm^{-1}): 1175, 1160, 1085, 1020, 920s, 810, 720, 710, 660, 550, 520, 480s, 400, 340s, 275s, 250s.
Source: Frost et al. (1993).
Comments: No independent analytical data are provided for the sample used.

Nadorite $PbSb^{3+}O_2Cl$

Origin: Harstigen mine, Pajsberg, near Filipstad, Värmland, Sweden.
Experimental details: Raman scattering measurements have been performed on an arbitrarily oriented single crystal using 514.5 nm Ar^+ laser radiation. The nominal laser radiation power was in the range from 20 to 50 mW.
Raman shifts (cm^{-1}): 214, 165–117.
Source: Jonsson (2003).
Comments: The sample was characterized by powder X-ray diffraction data and electron microprobe analyses.

Nafertisite $Na_3Fe^{2+}{}_{10}Ti_2(Si_6O_{17})_2O_2(OH)_6F(H_2O)_2$

Origin: Kukisvumchorr, Khibiny alkaline massif, Kola peninsula, Russia (type locality).
Experimental details: Raman scattering measurements have been performed on an arbitrarily oriented crystal using 633 nm laser radiation. The laser radiation power is not indicated.
Raman shifts (cm^{-1}): 1030, 922, 688s, 658s, 569, 530, 490, 420w, 345, 297, 240, 212w, 184, 145.
Source: Cámara et al. (2014b).
Comments: The sample was characterized by single-crystal X-ray diffraction data, electron microprobe analyses, and Mössbauer spectroscopy. The crystal structure is solved.

Nagelschmidtite $Ca_7(SiO_4)_2(PO_4)_2$

Origin: Artificial (compoinent of slags produced from a solid radioactive waste).
Experimental details: Methods of sample preparation are not described. Raman scattering measurements have been performed using 532 nm laser radiation. The laser radiation power is not indicated.
Raman shifts (cm^{-1}): 1176, 1075, 1008s, 916, 843w, 702, 535w, 352, 217.
Source: Malinina and Stefanovsky (2014).
Comments: The sample was characterized by powder X-ray diffraction data and electron microprobe analysis.

Nahcolite $NaH(CO_3)$

Origin: Ryoke metamorphic rocks, Kasado Island, Yamaguchi prefecture, Japan.
Experimental details: Raman scattering measurements have been performed on a microscopic inclusion using 532 nm laser radiation. The nominal laser radiation power was 100 mW.
Raman shifts (cm^{-1}): 1268, 1046s, 687.
Source: Hoshino et al. (2006).
Comments: No independent analytical data are provided for the sample used.

Nahcolite $NaH(CO_3)$

Origin: No data in the cited paper.
Experimental details: No data in the cited paper.
Raman shifts (cm^{-1}): 1432w, 1271, 1048s, 688.
Source: Frezzotti et al. (2012).
Comments: No independent analytical data are provided for the sample used. For the Raman spectra of nahcolite see also Edwards et al. (2007), Kaminsky et al. (2009), and Frezzotti et al. (2012).

Nahpoite $Na_2(PO_3OH)$

Origin: Synthetic.
Experimental details: Raman scattering measurements have been performed at 84°C on an arbitrarily oriented sample using 514.5 nm Ar$^+$ laser radiation. The nominal laser radiation power was 30 mW.
Raman shifts (cm^{-1}): 1139, 1072, 1001w, 939s, 855, 583w, 571w, 556, 501w, 461, 397.
Source: Ghule et al. (2003).
Comments: The sample was characterized by TG data.

Namibite $Cu(BiO)_2(VO_4)(OH)$

Origin: Lodi No. 4 mine, Plumas Co., California, USA.
Experimental details: Raman scattering measurements have been performed on arbitrarily oriented crystals using a 633 nm He-Ne laser. The laser radiation power is not indicated.
Raman shifts (cm^{-1}): 899w, 846s, (842), (769w), 736, 677w, 563s, 410w, 370s, 328s, 288, 247, 212.
Source: Frost et al. (2006i)
Comments: No independent analytical data are provided for the sample used.

Nantokite $CuCl$

Origin: Synthetic.
Experimental details: No data.
Raman shifts (cm^{-1}): ~197, ~155, ~117s, ~58, ~40sh.
Source: Vardeny and Brafman (1979).
Comments: No independent analytical data are provided for the sample used. For the Raman spectra of nantokite see also Bouchard-Abouchacra (2001) and Frost et al. (2003i).

Naquite $FeSi$

Origin: Synthetic.
Experimental details: Raman scattering measurements have been performed on an oriented crystal at 340 K. The laser wavelength and the laser radiation power are not indicated.
Raman shifts (cm^{-1}): 436, (333w), 315s, (260), 219, (193), 180s.
Source: Nyhus et al. (1995).
Comments: No independent analytical data are provided for the sample used.

Narsarsukite $Na_2(Ti,Fe^{3+})Si_4(O,F)_{11}$

Origin: Synthetic.
Experimental details: Raman scattering measurements have been performed on pellets of pressed powder using 488 nm Ar$^+$ laser radiation. The laser radiation power at the sample was about 300 mW. A 180°-scattering geometry was employed.
Raman shifts (cm^{-1}): 1009, 907, 764s, 518w, 480w, 422w, 362.
Source: Su et al. (2000).
Comments: The sample was characterized by powder X-ray diffraction data.

Natalyite $NaV^{3+}Si_2O_6$

Origin: Synthetic.
Experimental details: Methods of sample preparation are not described. Micro-Raman scattering measurements have been performed in the backscattering configuration, using 514.5 nm Ar$^+$ laser radiation. The laser radiation power is not indicated.
Raman shifts (cm^{-1}): 1042, 1025w, 972s, 954s, ~910w, ~865, ~830, ~680, ~640w, ~550s, ~505, ~390, ~360, ~345sh, ~340s, and a series of bands below 340 cm^{-1}.
Source: Konstantinović et al. (2002).

Comments: No independent analytical data are provided for the sample used. For the Raman spectra of natalyite see also Popović et al. (2006) and Ullrich et al. (2009).

Natisite $Na_2TiO(SiO_4)$

Origin: Synthetic.
Experimental details: Raman scattering measurements have been performed on pellets of pressed powder using 488 nm Ar^+ laser radiation. The laser radiation power at the sample was about 300 mW. A 180°-scattering geometry was employed.
Raman shifts (cm^{-1}): 926w, 898, 869, 851s, 830s, 677w, 533w, 497w, 379sh, 360.
Source: Su et al. (2000).
Comments: The sample was characterized by powder X-ray diffraction data.

Natrite $Na_2(CO_3)$

Origin: Synthetic.
Experimental details: Methods of sample preparation are not described. Raman scattering measurements have been performed using 532 nm Nd-Yag laser radiation. The nominal laser radiation power was 100 mW.
Raman shifts (cm^{-1}): 1429w, 1080s, 702, 290.
Source: Buzgar and Apopei (2009).
Comments: No independent analytical data are provided for the sample used. For the Raman spectra of natrite see also Edwards et al. (2007), Frezzotti et al. (2012), and Shatskiy et al. (2013).

Natroalunite $NaAl_3(SO_4)_2(OH)_6$

Origin: No data.
Experimental details: Micro-Raman scattering measurements have been performed on a thin section using 514.5 nm Ar^+ laser radiation. The nominal laser radiation power at the sample was 20 or 50 mW.
Raman shifts (cm^{-1}): 1183, 1163w, 1024s, 652s, 572, 519, 482, 395, 345, 234s, 163.
Source: Maubec et al. (2012).
Comments: The sample was characterized by electron microprobe analysis. Its empirical formula is $Na_{0.6}K_{0.4}Al_{2.9}(SO_4)_{2.1}(OH)_{5.5}$.

Natrochalcite $NaCu_2(SO_4)_2(OH)\cdot H_2O$

Origin: Chuquicamata, Chile.
Experimental details: Raman scattering measurements have been performed on arbitrarily oriented crystals using a 633 nm He-Ne laser. The laser radiation power is not indicated. The Raman shifts have been determined for the maxima of individual peaks obtained as a result of the spectral curve analysis.
Raman shifts (cm^{-1}): 3196, 3156, 1046, 1023, 997s, 930, 709, 636s, 607, 466, 445, 429, 402, 212.
Source: Frost and Weier (2004e).
Comments: No independent analytical data are provided for the sample used.

Natrodufrénite NaFe^{2+}Fe$^{3+}_5$(PO$_4$)$_4$(OH)$_6$·2H$_2$O

Origin: Divino das Laranjeiras, eastern Minas Gerais, Brazil.
Experimental details: Raman scattering measurements have been performed on arbitrarily oriented crystals using a 633 nm He-Ne laser. The laser radiation power is not indicated. The Raman shifts have been determined for the maxima of individual peaks obtained as a result of the spectral curve analysis.
Raman shifts (cm^{-1}): 3573, 3187, (3119), 1634, 1335w, 1188, 1059, 1003, 961, 914, 814, 668, 619w, 582, 560, 507, 477s, (444), 425, 376, 356, 298, 217, 194, 163, 142, 120.
Source: López et al. (2013b).
Comments: The sample was characterized by powder X-ray diffraction data and electron microprobe analyses.

Natrojarosite NaFe$^{3+}_3$(SO$_4$)$_2$(OH)$_6$

Origin: The sample was obtained from the Dogan Paktunc of Mineral Resources Laboratories, CANMET, Ottawa, ON, Canada. The locality is not indicated.
Experimental details: Methods of sample preparation are not described. Raman scattering measurements have been performed using 785 nm solid-state laser radiation. The laser radiation output power was above 300 mW.
Raman shifts (cm^{-1}): 1105, 1008s, 618, 555, 356, 288, 221, 134.
Source: Das and Hendry (2011).
Comments: The sample was characterized by powder X-ray diffraction data. For the Raman spectra of natrojarosite see also Sasaki et al. (1998), Frost et al. (2006r), Murphy et al. (2009), and Chio et al. (2010).

Natrolemoynite Na$_4$Zr$_2$Si$_{10}$O$_{26}$·9H$_2$O

Origin: Poudrette (Demix) quarry, Mont Saint-Hilaire, Rouville RCM (Rouville Co.), Montérégie, Québec, Canada (type locality) (?).
Experimental details: Raman scattering measurements have been performed with laser radiation perpendicular to the {010} cleavage of a single crystal using 532 nm laser radiation. The laser radiation power is not indicated.
Raman shifts (cm^{-1}): 968s, ~610, 538, 428, ~290w.
Source: McDonald et al. (2015).
Comments: No independent analytical data are provided for the sample used.

Natrolite Na$_2$(Si$_3$Al$_2$)O$_{10}$·2H$_2$O

Origin: Chimney Rock Quarry, Bound Brook, New Jersey, USA (?).
Experimental details: Methods of sample preparation are not described. Raman scattering measurements have been performed on an arbitrarily oriented sample using 514.5 nm Ar$^+$ laser radiation. The laser radiation power at the sample was 10 mW.
Raman shifts (cm^{-1}): 3543s, 3476, 3329s, 3231, 1637, 1093, 1085, 1071, 1064, 1042s, 1019, 1009, 987, 977, 966, 727s, 718, 707s, 534s, 443s, 417, 393, 360s, 333s, 308s, 290, 276, 259, 241, 218s, 207, 186, 163s, 145s.
Source: Wopenka et al. (1998).

Comments: The sample was characterized by electron microprobe analyses. For the Raman spectra of natrolite see also Belitsky et al. (1992), Goryainov and Smirnov (2001), Mozgawa (2001), Jehlička et al. (2012), and Jehlička and Vandenabeele (2015).

Natron $Na_2(CO_3) \cdot 10H_2O$

Origin: Synthetic.
Experimental details: Methods of sample preparation are not described. Raman scattering measurements have been performed on an arbitrarily oriented sample using 785 or 514.5 nm laser radiation. The nominal laser radiation power was in the range from 5 to 100 mW.
Raman shifts (cm^{-1}): 1061–1070s, 335w, 224w, 185w.
Source: Edwards et al. (2007).
Comments: No independent analytical data are provided for the sample used.

Natroniobite $NaNbO_3$

Origin: Synthetic.
Experimental details: Methods of sample preparation are not described. Raman scattering measurements have been performed at 560 K in backscattering geometry using 514.5 nm Ar$^+$ laser radiation. The laser radiation power is not indicated.
Raman shifts (cm^{-1}): ~570s, ~240, ~205, ~160, ~120, ~50.
Source: Yuzyuk et al. (2005).
Comments: The sample was characterized by powder X-ray diffraction data. The Raman shifts are given for a sample at 560 K.

Natropalermoite $Na_2SrAl_4(PO_4)_4(OH)_4$

Origin: Palermo No. 1 mine, Groton, New Hampshire, USA (type locality).
Experimental details: Micro-Raman scattering measurements have been performed on an arbitrarily oriented crystal using 532 nm solid-state laser radiation. The laser radiation power is not indicated.
Raman shifts (cm^{-1}): ~3215, ~1145s, ~1079, ~1008s, ~935, ~640, ~622, ~595, ~524, ~431s, ~413, ~318, and a series of bands below 300 cm^{-1}.
Source: Schumer et al. (2016).
Comments: The sample was characterized by powder X-ray diffraction data and electron microprobe analyses. The crystal structure is solved.

Natrophilite $NaMn^{2+}(PO_4)$

Origin: Synthetic.
Experimental details: Raman scattering measurements have been performed on an aggregate of microtubes using 1064 nm Nd-YAG laser radiation. The laser radiation power is not indicated.
Raman shifts (cm^{-1}): 1054, 1010, 948, 614, 579, 454, 410.
Source: Shi et al. (2005).
Comments: The sample was characterized by powder X-ray diffraction data and energy dispersive X-ray spectral analysis.

Natrosilite $Na_2Si_2O_5$

Origin: Synthetic.
Experimental details: Methods of sample preparation are not described. Raman scattering measurements have been performed using 510.5 nm, 10 kHz pulsed copper vapor laser radiation with the laser radiation power at the sample of about 30 kW within a pulse.
Raman shifts (cm^{-1}): 1072s, ~1010w, ~955w, ~760w, 517s, ~470w, 384, 337, 270, 223, 149.
Source: You et al. (2001).
Comments: No independent analytical data are provided for the sample used. For the Raman spectrum of natrosilite see also Fleet and Henderson (1997).

Natrouranospinite $Na_2(UO_2)_2(AsO_4)_2 \cdot 5H_2O$

Origin: Měděnec deposit, the Krušné Hory (Ore Mts.), northern Bohemia, Czech Republic.
Experimental details: Raman scattering measurements have been performed on arbitrarily oriented crystals using a 633 nm He-Ne laser. The laser radiation power is not indicated. The Raman shifts have been determined for the maxima of individual peaks obtained as a result of the spectral curve analysis.
Raman shifts (cm^{-1}): 3493, (3450), 3404, 3260sh, 1136, 1008s, (904w), 893, 816s, (810), 671, 621, 579w, 494, 461, 415s, 400sh, 322, 267, (245), 209sh, (199), 188s, 156sh, 148, (140), 110.
Source: Čejka et al. (2009c).
Comments: The sample was characterized by powder X-ray diffraction data and electron microprobe analyses.

Natroxalate $Na_2(C_2O_4)$

Origin: Alluaiv Mt., Lovozero massif, Kola Peninsula, Russia.
Experimental details: Raman scattering measurements have been performed on arbitrarily oriented crystals using a 633 nm He-Ne laser. The laser radiation power is not indicated. The Raman shifts have been determined for the maxima of individual peaks obtained as a result of the spectral curve analysis.
Raman shifts (cm^{-1}): 1750, 1643, 1614, 1456s, 1358s, 884, 875, 567, 481, 221, 156, 117.
Source: Frost (2004d).
Comments: No independent analytical data are provided for the sample used. For the Raman spectra of natroxalate see also Frost and Weier (2003) and Frost et al. (2003k).

Natrozippeite $Na_5(UO_2)_8(SO_4)_4O_5(OH)_3 \cdot 12H_2O$

Origin: Mecsek Mountains, Hungary.
Experimental details: Raman scattering measurements have been performed on an arbitrarily oriented sample using 785 nm laser radiation. The laser radiation power is not indicated.
Raman shifts (cm^{-1}): 1239, 1137, 1099w, 1009, 825s, 665w, 461, 409, 273.
Source: Stefaniak et al. (2009).
Comments: The sample was characterized by electron microprobe analysis. For the Raman spectra of natrozippeite see also Frost et al. (2007f) and Driscoll et al. (2014).

Naumannite Ag_2Se

Origin: Synthetic.
Experimental details: No data.
Raman shifts (cm^{-1}): 141.
Source: Ge and Li (2003).
Comments: The sample was characterized by powder X-ray diffraction data and electron microprobe analysis.

Nealite $Pb_4Fe(AsO_3)_2Cl_4 \cdot 2H_2O$

Origin: Lavrion, Greece.
Experimental details: Raman scattering measurements have been performed on arbitrarily oriented crystals using a 633 nm He-Ne laser. The laser radiation power is not indicated. The Raman shifts have been determined for the maxima of individual peaks obtained as a result of the spectral curve analysis.
Raman shifts (cm^{-1}): 3451, 3431, 3387w, 3357sh, 3215m 2892w, 2849w, 1015w, 831w, 732, 808s, 632, 604sh, 548, 471, 418sh, 393, 371sh, 342sh, 320, 299sh, 245, 212, 194sh, 183sh, 160sh, 149s, 137, 129sh, 119.
Source: Frost and Bahfenne (2011a).
Comments: No independent analytical data are provided for the sample used. For the Raman spectra of nealite see also Bahfenne (2011).

Negevite NiP_2

Origin: Synthetic.
Experimental details: Raman scattering measurements have been performed on nanoparticles using 532 nm laser radiation. The laser radiation power is not indicated.
Raman shifts (cm^{-1}): 454, 586s.
Source: Zhuo et al. (2015).
Comments: The sample was characterized by powder X-ray diffraction data and electron microprobe analysis.

Neighborite $NaMgF_3$

Origin: Synthetic.
Experimental details: Raman scattering measurements have been performed in the geometry $y(zz)\text{-}y$. Characteristics of laser radiation are not indicated.
Raman shifts (cm^{-1}): ~230s, ~212w, ~187s, ~140, ~130sh, ~105.
Source: Oçafrain et al. (1996).
Comments: The sample was characterized by optical methods.

Nekoite $Ca_3Si_6O_{15} \cdot 7H_2O$

Origin: Iron Cap Mine, near Klondyke, Cochise Co., Arizona, USA.
Experimental details: Raman scattering measurements have been performed on arbitrarily oriented crystals using a 633 nm He-Ne laser. The laser radiation power is not indicated. The Raman shifts

have been determined for the maxima of individual peaks obtained as a result of the spectral curve analysis.
Raman shifts (cm^{-1}): (3567), 3502, 3380sh, 3071sh, 2810sh, (1647), 1623, 1567sh, 1132sh, 1092s, 1061s, 1023, 994sh, 974, 774, 661s, (560), 588s, 525, 437, 416sh, 398, 362sh, 345, 303, 287, (259), 240, 198, 180, 156, 136, 106w.
Source: Frost and Xi (2012n).
Comments: No independent analytical data are provided for the sample used.

Nenadkevichite (Na,□)$_8$Nb$_4$(Si$_4$O$_{12}$)$_2$(O,OH)$_4$·8H$_2$O

Origin: Synthetic.
Experimental details: Raman scattering measurements have been performed on an arbitrarily oriented sample using 632.8 nm He-Ne laser radiation. The output laser radiation power was 25 mW.
Raman shifts (cm^{-1}): 940w, 878w, 668s, ~500w, ~480w, 226s.
Source: Rocha et al. (1996).
Comments: A sample with the Ti:Nb molar ratio of 0.8 was used. The sample was characterized by powder X-ray diffraction data. For the Raman spectrum of nenadkevichite see also Rocha et al. (1996).

Nepheline NaAlSiO$_4$

Origin: Synthetic.
Experimental details: Raman scattering measurements have been performed on a powdered sample using 488 nm Ar$^+$ laser radiation. The nominal laser radiation power was 600 mW. A 90°-scattering geometry was employed.
Raman shifts (cm^{-1}): 1081, 984, 973, 690w, 616w, 497, 469, 427s, 399s, 331, 264, 214, 151, 138w, 123.
Source: Matson et al. (1986).
Comments: No independent analytical data are provided for the sample used.

Nesquehonite Mg(CO$_3$)·3H$_2$O

Origin: A natural sample. The locality is not indicated.
Experimental details: Raman scattering measurements have been performed on an arbitrarily oriented sample using 1064 nm Nd-YAG laser radiation. The nominal laser radiation power was 100 mW.
Raman shifts (cm^{-1}): 2905w, 1890w, 1708w, 1515w, 1428w, 1100s, 772w, 703w, 344w, 311w, 273w, 228, 199, 187, 119.
Source: Edwards et al. (2005).
Comments: The sample was characterized by powder X-ray diffraction data. For the Raman spectra of nesquehonite see also Coleyshaw et al. (2003), Kloprogge et al. (2003), and Kristova et al. (2014).

Nestolaite CaSeO$_3$·H$_2$O

Origin: Little Eva mine, Grand County, Utah, USA (type locality).
Experimental details: Raman scattering measurements have been performed on an arbitrarily oriented sample using 532 nm diode laser radiation. The laser radiation power at the sample was 5 mW.
Raman shifts (cm^{-1}): 1680w, 825s, 750s, 470w, 405w, 360w.

Source: Kasatkin et al. (2014).
Comments: The sample was characterized by powder X-ray diffraction data and electron microprobe analyses. The crystal structure is solved.

Newberyite $Mg(PO_3OH) \cdot 3H_2O$

Origin: Lava Cave, Skipton, Victoria, Australia (type locality).
Experimental details: Raman scattering measurements have been performed on arbitrarily oriented crystals using a 633 nm He-Ne laser. The laser radiation power is not indicated. The Raman shifts have been determined for the maxima of individual peaks obtained as a result of the spectral curve analysis.
Raman shifts (cm^{-1}): 3514, 3479, 3456, 3381, 3265, 3181, 2880, 1648, 1620, 1272, 1195, 1154, 984s, 967, 893, 555, 498, 400, 369, 327, 283, 266, 244, 219, 199, 180, 158, 139.
Source: Frost et al. (2005j).
Comments: No independent analytical data are provided for the sample used.

Nežilovite $PbZn_2Mn^{4+}_2Fe^{3+}_8O_{19}$

Origin: "Mixed series" metamorphic complex, near the Nežilovo village, 40 km SW of Veles, Pelagonian massif, Macedonia (type locality).
Experimental details: Raman scattering measurements have been performed on an arbitrarily oriented sample using 532 nm Nd-YAG laser radiation. The laser radiation power is not indicated.
Raman shifts (cm^{-1}): Raman shifts of presumed nežilovite given in the cited paper (*i. e.* 1099s, 725w, 483w, 340w, 300, 176) correspond to associated dolomite.
Source: Stamatovska et al. (2011).

Nickelaustinite $CaNi(AsO_4)(OH)$

Origin: No data.
Experimental details: Raman scattering measurements have been performed on arbitrarily oriented crystals using a 633 nm He-Ne laser. The laser radiation power is not indicated. The Raman shifts have been determined for the maxima of individual peaks obtained as a result of the spectral curve analysis.
Raman shifts (cm^{-1}): 3344, 3320, 917, 828s, 811, 799, 777, 495, 475s, 430s, 398, 369, 348s, 332, 216, 164, 147.
Source: Martens et al. (2003c).
Comments: No independent analytical data are provided for the sample used.

Nickelbischofite $NiCl_2 \cdot 6H_2O$

Origin: Synthetic.
Experimental details: Raman scattering measurements have been performed at 80 K on oriented single crystal using Kr$^+$ laser radiation. The laser radiation power is not indicated. Polarized spectra were collected in different scattering geometries.
Raman shifts (cm^{-1}): 3503, 3424, 3411, 3371, 3160, 1671, 1623, 869, 699, 685, 571, 543, 372, 362.
Source: Agulló-Rueda et al. (1987).

Comments: No independent analytical data are provided for the sample used. For the Raman spectrum of nickelbischofite see also Cariati et al. (1989).

Nickelboussingaultite $(NH_4)_2Ni(SO_4)_2 \cdot 6H_2O$

Origin: Cameron, Coconino Co., Arizona, USA.
Experimental details: Micro-Raman scattering measurements have been performed on a fine-grained sample using 785 nm diode laser radiation. The laser radiation power is not indicated.
Raman shifts (cm^{-1}): 3280w, 2940w, 1660w, 1460w, 1149w, 1093w, 1063w, 1027, 990s, 652w, 624w, 602w, 482w, 457, 440w, 341w, 312w, <240.
Source: Culka et al. (2009).
Comments: No independent analytical data are provided for the sample used. The Raman shifts have been partly determined for the maxima of individual peaks obtained as a result of the spectral curve analysis.

Nickelhexahydrite $Ni(SO_4) \cdot 6H_2O$

Origin: Synthetic.
Experimental details: Raman scattering measurements have been performed on an arbitrarily oriented crystal using 253.65 nm radiation of mercury. The radiation power is not indicated.
Raman shifts (cm^{-1}): 3441s, 3403s, 3302, 3250, 1131, 1088, 1050w, 987s, 971w, 639, 620, 596.
Source: Krishnamurti (1958).
Comments: The sample was characterized by optical methods.

Nickeline NiAs

Origin: Synthetic.
Experimental details: Raman scattering measurements have been performed on an arbitrarily oriented sample using 514.5 nm Ar^+ laser radiation at a power below 2 mW.
Raman shifts (cm^{-1}): 218, 154
Source: Watté et al. (1994).
Comments: The sample was characterized by powder X-ray diffraction data.

Nickelpicromerite $K_2Ni(SO_4)_2 \cdot 6H_2O$

Origin: Synthetic.
Experimental details: Raman scattering measurements have been performed on an arbitrarily oriented crystal using 253.65 nm radiation of mercury. The incident light was normal to the (110) face and the scattered light was taken parallel to the (110) face and roughly perpendicular to the (001) face. The radiation power is not indicated.
Raman shifts (cm^{-1}): 3310s, 3234, 3148, 3050–3540, 1230w, 1155, 1127, 1113, 1085s, 990s, 910w, 845w, 800w, 634, 611, 462, 448, (374), 320sh, 305w, 269w, 225, (184), 132sh, 115, (93w), 74sh, 62, 46.
Source: Ananthanarayanan (1961).
Comments: The sample was characterized by morphological features.

Nierite hexagonal polyborph β-Si_3N_4

Origin: Synthetic.
Experimental details: Raman scattering measurements have been performed on an arbitrarily oriented polycrystalline sample using 488 nm laser radiation. The laser radiation power at the sample was 50 mW.
Raman shifts (cm^{-1}): 1045, 937, 927, 863, 730w, 618w, 525s, 449w, 227, 206, 183.
Source: Muraki et al. (1997).
Comments: No independent analytical data are provided for the sample used. For the Raman spectrum of β-Si_3N_4 see also Honda et al. (1999).

Nifontovite $Ca_3[BO(OH)_2]_6 \cdot 2H_2O$

Origin: Fuka Mine, Okayama Prefecture, Japan.
Experimental details: Raman scattering measurements have been performed on an arbitrarily oriented sample using Ar$^+$ laser radiation. The laser radiation power at the sample was about 20 mW.
Raman shifts (cm^{-1}): 3650w, 3611s, 3575s, 3434s, 3393, 3311, 3217sh, 3177s, 1607w, 1235w, 1208w, 1186w, 1154w, 1122w, 1030w, 991w, 932w, 918w, 896w, 831w, 806w, 749sh, 720, 653s, 633sh, 574w, 558w, 543w, 441w, 420, 391w, 371w, 348w, 322w, 306, 275w, 262sh, 225w, 213w, 190w, 171w, 140w, 123.
Source: Bermanec et al. (2010).
Comments: The sample was characterized by powder X-ray diffraction and thermal data.

Nimite $(Ni,Mg,Al)_6(Si,Al)_4O_{10}(OH)_8$

Origin: Bon Accord, Barberton, South Africa.
Experimental details: No data.
Raman shifts (cm^{-1}): 3645, 678, 547, 279, 196.
Source: Villanova-de-Benavent et al. (2014).
Comments: The sample was characterized by powder X-ray diffraction data and electron microprobe analyses.

Niningerite MgS

Origin: Meteorite Sahara 97158.
Experimental details: Raman scattering measurements have been performed on an arbitrarily oriented sample using 514.5 nm Ar$^+$ laser radiation. The nominal laser radiation power was 2 mW.
Raman shifts (cm^{-1}): ~280s, ~220.
Source: Avril et al. (2013).
Comments: Fe,Mn-bearing sample with the formula $(Mg_{0.73}Fe_{0.16}Mn_{0.11})S$ was used.

Nioboholtite $(Nb_{0.6}\square_{0.4})Al_6BSi_3O_{18}$

Origin: Szklary pegmatite, Lower Silesia, Poland (type locality).
Experimental details: Micro-Raman scattering measurements have been performed on an arbitrarily oriented sample using 514.5 nm Ar$^+$ laser radiation. The laser radiation power at the sample was about 5.5 mW.

Raman shifts (cm^{-1}): ~1110, ~1045, ~965sh, ~945s, ~895sh, ~710w, ~612, ~555s, ~507s, ~460, ~420, ~380, ~270w, ~205w.
Source: Pieczka et al. (2013).
Comments: The sample was characterized by powder X-ray diffraction data and electron microprobe analyses.

Niter K(NO$_3$)

Origin: Synthetic.
Experimental details: Methods of sample preparation are not described. Raman scattering measurements have been performed using 1064 nm Nd-YAG laser radiation. The nominal laser radiation power was between 80 and 200 mW.
Raman shifts (cm^{-1}): 1359, 1345, 1050s.
Source: Rissom et al. (2008).
Comments: No independent analytical data are provided for the sample used.

Nitratine Na(NO$_3$)

Origin: Dolomite Cave of Pozalagua, Karrantza, Basque Co., northern Spain.
Experimental details: Raman scattering measurements have been performed on an arbitrarily oriented sample using 785 nm diode laser radiation. The laser radiation power is not indicated.
Raman shifts (cm^{-1}): See comment below.
Source: Martínez-Arkarazo et al. (2007).
Comments: In the cited paper Raman spectra of nitratine-bearing polymineral samples are given. Raman shifts of pure nitratine are: 1066s, 725, 185s (see RRUFF ID R050394, for an unoriented sample).

Nitrobarite Ba(NO$_3$)$_2$

Origin: Synthetic.
Experimental details: Methods of sample preparation are not described. Raman scattering measurements have been performed using 785 nm diode laser radiation. The laser radiation power is not indicated.
Raman shifts (cm^{-1}): 1631w, 1402w, 1045s, 1025w, 729.
Source: Maguregui et al. (2008).
Comments: No independent analytical data are provided for the sample used.

Nitrocalcite Ca(NO$_3$)$_2$·4H$_2$O

Origin: Dolomite Cave of Pozalagua, Karrantza, Basque Co., northern Spain.
Experimental details: Raman scattering measurements have been performed on an arbitrarily oriented sample using 785 nm diode laser radiation. The laser radiation power is not indicated.
Raman shifts (cm^{-1}): See comment below.
Source: Martínez-Arkarazo et al. (2007).
Comments: In the cited paper Raman spectra of nitrocalcite-bearing polymineral samples are given. Raman shifts of pure nitratine are: 1067s, 740, 282w, ~195 (RRUFF ID R120047, for an unoriented sample, with 780 nm radiation).

Nitromagnesite $Mg(NO_3)_2 \cdot 6H_2O$

Origin: Synthetic.
Experimental details: Methods of sample preparation are not described. Raman scattering measurements have been performed using 785 or 514 nm laser radiation. The nominal laser radiation power was 350 or 50 mW.
Raman shifts (cm^{-1}): 1059s.
Source: Morillas et al. (2016).
Comments: No independent analytical data are provided for the sample used.

Nobleite $CaB_6O_9(OH)_2 \cdot 3H_2O$

Origin: Synthetic.
Experimental details: Raman scattering measurements have been performed on a sample held in a pyrex tube using 514.5 nm Ar$^+$ laser radiation. The nominal laser radiation power was 300 mW.
Raman shifts (cm^{-1}): 996w, 956w, 855s, 745, 636, 572w, 460, 383.
Source: Jun et al. (1995).
Comments: The sample was characterized by chemical analyses.

Noelbensonite $BaMn^{3+}_2Si_2O_7(OH)_2 \cdot H_2O$

Origin: Postmasburg manganese field, Northern Cape Province, South Africa.
Experimental details: Methods of sample preparation are not described. Raman scattering measurements have been performed using 514.5 nm Ar$^+$ laser radiation. The laser radiation power is not indicated.
Raman shifts (cm^{-1}): 925s, 904, 650, 572, 523s, 436, 386, 331w, 304w, 245, 185, 160.
Source: Costin et al. (2015).
Comments: The sample was characterized by electron backscatter diffraction and electron microprobe analyses.

Nolanite $(V^{3+},Fe^{3+},Fe^{2+})_{10}O_{14}(OH)_2$

Origin: Vihanti, Northern Ostrobothnia region, Finland.
Experimental details: Methods of sample preparation are not described. Raman scattering measurements have been performed using 633 nm He-Ne laser radiation. The nominal laser radiation power was 2 or 20 mW.
Raman shifts (cm^{-1}): 587, 504, 478, 323w, 293w, 221s, 85s.
Source: Voloshin et al. (2014).
Comments: The sample was characterized by powder X-ray diffraction data and electron microprobe analysis.

Nontronite $Na_{0.3}Fe^{3+}_2(Si,Al)_4O_{10}(OH)_2 \cdot nH_2O$

Origin: No data.
Experimental details: Methods of sample preparation are not described. Raman scattering measurements have been performed using 632.8 nm laser radiation. The laser radiation power is not indicated.

Raman shifts (cm^{-1}): ~3575s, ~3410, ~890, ~810w, ~760w, ~680, ~570w, ~520, ~420, ~360, ~290s, ~240s.
Source: Wang et al. (1998a).
Comments: No independent analytical data are provided for the sample used.

Norbergite Mg$_3$(SiO$_4$)F$_2$

Origin: Franklin Limestone quarry, Franklin, Sussex Co., New Jersey, USA.
Experimental details: Raman scattering measurements have been performed on arbitrarily oriented crystals using a 633 nm He-Ne laser. The laser radiation power is not indicated. The Raman shifts have been determined for the maxima of individual peaks obtained as a result of the spectral curve analysis.
Raman shifts (cm^{-1}): 3583, 3488, 979, 955, 900, (884w), 855s, 614, 572, 555s, 435s, 382.
Source: Frost et al. (2007k).
Comments: The sample was characterized by electron microprobe analysis. The empirical formula shows deficit of Mg+Fe.

Nordenskiöldine CaSn(BO$_3$)$_2$

Origin: Gejiu tin deposit, Yunnan, China.
Experimental details: Methods of sample preparation are not described. Raman scattering measurements have been performed using 458 nm Ar$^+$ laser radiation. The nominal laser radiation power was 120 mW.
Raman shifts (cm^{-1}): 1453w, 1205s, 944w, 847w, 764, 743, 449s, 389, 277, 214w.
Source: Li et al. (1994).
Comments: The sample was characterized by powder X-ray diffraction data and electron microprobe analysis.

Nordstrandite Al(OH)$_3$

Origin: Gunong Kapor, Bau mining district, West Sarawak, Borneo.
Experimental details: Raman scattering measurements have been performed on grains placed on glass slide using 514.5 nm Ar$^+$ laser radiation. The laser radiation power at the sample was 25 mW.
Raman shifts (cm^{-1}): 3623, 3567s, 3523w, 3490, 3349, 1093, 985, 959w, 898, 657, (633w), 596, 542s, 507, 492, 464w, 437w, 412, 390, 378, (355w), (344w), 305s, 286, (267w), 252, 228w, 177, 119s.
Source: Rodgers (1993).
Comments: The sample was identified by Raman spectrum only. No independent analytical data are provided for the sample used.

Normandite Na$_2$Ca$_2$(Mn,Fe)$_2$(Ti,Nb,Zr)$_2$(Si$_2$O$_7$)$_2$O$_2$F$_2$

Origin: Partomchorr Mt., Khibiny Massif, Kola Peninsula, Russia.
Experimental details: Raman scattering measurements have been performed on arbitrarily oriented crystals using a 633 nm He-Ne laser. The laser radiation power is not indicated. The Raman shifts have been determined for the maxima of individual peaks obtained as a result of the spectral curve analysis.

Raman shifts (cm^{-1}): 813s, 803, 782, 748s, 724, 667sh, 656, (649), 641sh, 520sh, 513s, (505), 477, 454, 412w, 404sh, 382w, 371w, 361w, (352w), 285, 267s, (259), 198, 179, 151s, 141, 125, 109.
Source: Frost et al. (2015p).
Comments: The sample was characterized by qualitative electron microprobe analysis.

Norsethite BaMg(CO$_3$)$_2$

Origin: Synthetic.
Experimental details: Methods of sample preparation are not described. Raman scattering measurements have been performed using 488 or 514.5 nm Ar$^+$ laser radiation. The nominal laser radiation power was 200 mW.
Raman shifts (cm^{-1}): 1115s, 695, 260, 130.
Source: Scheetz and White (1977).
Comments: The sample was characterized by powder X-ray diffraction data. For the Raman spectra of norsethite see also Schmidt et al. (2013) and Effenberger et al. (2014).

Northupite Na$_3$Mg(CO$_3$)$_2$Cl

Origin: Searles Lake, California, USA (type locality).
Experimental details: Raman scattering measurements have been performed on arbitrarily oriented crystals using a 633 nm He-Ne laser. The laser radiation power is not indicated. The Raman shifts have been determined for the maxima of individual peaks obtained as a result of the spectral curve analysis.
Raman shifts (cm^{-1}): 1554, 1115s, 1107sh, 714w, 250, 213, 180, 148.
Source: Frost and Dickfos (2007a).
Comments: No independent analytical data are provided for the sample used.

Nosean Na$_8$(Si$_6$Al$_6$)O$_{24}$(SO$_4$)·H$_2$O

Origin: Laacher See (Laach Lake) volcano, Eifel, Germany.
Experimental details: Micro-Raman scattering measurements have been performed on an arbitrarily oriented sample using 532 laser radiation. The laser radiation power is not indicated.
Raman shifts (cm^{-1}): ~1060, ~980s, ~630, ~605w, ~530, ~435.
Source: Hettmann et al. (2012).
Comments: No independent analytical data are provided for the sample used.

Nováčekite-II Mg(UO$_2$)$_2$(AsO$_4$)$_2$·10H$_2$O

Origin: Wheal Edward, St. Just, Cornwall, UK.
Experimental details: Methods of sample preparation are not indicated. Raman scattering measurements have been performed using 785 nm laser radiation. The nominal laser radiation power at the source was ~370 mW.
Raman shifts (cm^{-1}): The strongest band is observed at 817 cm^{-1}. Other Raman shifts are not indicated.
Source: Driscoll et al. (2014).
Comments: The sample was characterized by electron microprobe analysis.

Novgorodovaite $Ca_2(C_2O_4)Cl_2 \cdot 2H_2O$

Origin: Synthetic.
Experimental details: Methods of sample preparation are not described. Raman scattering measurements have been performed using 532 nm solid-state laser radiation. The nominal laser radiation power was 10 mW.
Raman shifts (cm^{-1}): 3350sh, 3330s, 1717w, 1630w, 1477s, 1402w, 904s, 859s, 730w, 673w, 600w, 503s, 472s.
Source: Piro et al. (2017).
Comments: The sample was characterized by powder X-ray diffraction data.

Nsutite $Mn^{2+}_{x}Mn^{4+}_{1-x}O_{2-2x}(OH)_{2x}$

Origin: No data.
Experimental details: Methods of sample preparation are not described. Raman scattering measurements have been performed using 514.5 nm Ar$^+$ laser radiation. The nominal laser radiation power was 10 mW.
Raman shifts (cm^{-1}): 732, 634, 572, 515, 458, 382, 280.
Source: Julien et al. (2004).
Comments: No independent analytical data are provided for the sample used. For the Raman spectrum of nsutite see also Julien et al. (2003).

Nullaginite $Ni_2(CO_3)(OH)_2$

Origin: Otway Prospect, Nullagine district, Western Australia, Australia (type locality).
Experimental details: Raman scattering measurements have been performed on arbitrarily oriented crystals using a 633 nm He-Ne laser. The laser radiation power is not indicated. The Raman shifts have been determined for the maxima of individual peaks obtained as a result of the spectral curve analysis.
Raman shifts (cm^{-1}): 3506w, 1734w, (1441w), 1426, (1092), 1089s, 742, 528, 342s, 232
Source: Frost (2006).
Comments: Raman spectrum of presumed nullaginite was published also by Frost et al. (2008l). However IR spectra of presumed nullaginite published in this paper are wrong: the strongest bands correspond to a serpentine-type silicate.

Nyerereite $Na_2Ca(CO_3)_2$

Origin: Oldoinyo Lengai volcano, Tanzania (type locality).
Experimental details: Methods of sample preparation are not described. Raman scattering measurements have been performed using 514.5 nm Ar$^+$ laser radiation. The nominal laser radiation power was 20 mW.
Raman shifts (cm^{-1}): 1086s, 1078sh, 1001w, 723–725sh, 709, 682–684sh.
Source: Golovin et al. (2015).
Comments: The sample was characterized by powder X-ray diffraction data and electron microprobe analyses. For the Raman spectra of nyerereite see also Kaminsky et al. (2009), Zaitsev et al. (2009), Golovin et al. (2014), and Shatskiy et al. (2015).

Offretite $KCaMg(Si_{13}Al_5)O_{36} \cdot 15H_2O$

Origin: No data.
Experimental details: Raman scattering measurements have been performed on an arbitrarily oriented sample using 632.8 nm He-Ne laser radiation. The nominal laser radiation power was 20 mW.
Raman shifts (cm^{-1}): 788w, 480s, 330.
Source: Croce et al. (2013).
Comments: No independent analytical data are provided for the sample used. For the Raman spectrum of offretite see also Mozgawa (2001).

Okenite $Ca_{10}Si_{18}O_{46} \cdot 18H_2O$

Origin: Pune (Poonah) district, Maharashtra, India.
Experimental details: Raman scattering measurements have been performed on arbitrarily oriented crystals using a 633 nm He-Ne laser. The laser radiation power is not indicated. The Raman shifts have been determined for the maxima of individual peaks obtained as a result of the spectral curve analysis.
Raman shifts (cm^{-1}): (3607), 3531, 3417, (3284), 3029w, 1631w, 1180, 1125sh, 1090s, (1075), 1048w, 1024, (1014), 973, 943, 801, 668, 651, 617, (603), (581), 569, 515, 496sh, 445, 423, 403, 385, 352, 302, 254, 228, 211sh, 190, 155sh, 133.
Source: Frost and Xi (2012n).
Comments: No independent analytical data are provided for the sample used.

Oldhamite CaS

Origin: Synthetic.
Experimental details: Raman scattering measurements have been performed on an arbitrarily oriented sample using 514.5 nm Ar$^+$ laser radiation. The laser radiation power at the sample was 20 mW.
Raman shifts (cm^{-1}): ~485, ~350s, ~285s, ~215, 185, 160.
Source: Avril et al. (2013).
Comments: The sample was characterized by powder X-ray diffraction data and electron microprobe analysis.

Olgite $(Ba,Sr)(Na,Sr,REE)_2Na(PO_4)_2$

Origin: Synthetic.
Experimental details: Methods of sample preparation are not described. Raman scattering measurements have been performed using 1064 nm Nd-YAG laser radiation. The nominal laser radiation power was 50 mW.
Raman shifts (cm^{-1}): 1100w, 1001w, 952s, 608, 570, 423, 86.
Source: Huang et al. (2007).
Comments: The Raman spectrum was obtained for a Ce^{3+}-doped sample. The sample was characterized by powder X-ray diffraction data.

Olivenite $Cu_2(AsO_4)(OH)$

Origin: Synthetic.
Experimental details: Methods of sample preparation are not described. Raman scattering measurements have been performed using 532 nm Ar^+ laser radiation. The laser radiation power at the sample was 20 mW. A 180°-scattering geometry was employed.
Raman shifts (cm^{-1}): 854s, 818, 514, 498, 424w, 345w, 311, 284, 220, 184w, 157, 116, 95, 82, 67.
Source: Majzlan et al. (2015).
Comments: The sample was characterized by powder X-ray diffraction data. For the Raman spectra of olivenite see also Yang et al. (2001), Frost et al. (2002e, 2009i), and Martens et al. (2003b).

Olivine P-rich variety $(Fe,Mg)_{2-x}(SiO_4,PO_4)$

Origin: Prehistoric slag from Goldbichl, Igls, Tyrol, Austria.
Experimental details: Raman scattering measurements have been performed on an arbitrarily oriented sample using 633 nm He-Ne laser radiation. The nominal laser radiation power was 170 mW.
Raman shifts (cm^{-1}): 1096, 1049, 975, 936s, 832+822s (unresolved doublet?), 725, 680, 631, 582, 561, 510, 457, 405, 368, 331, 291, 228, 163, 113.
Source: Schneider et al. (2013).
Comments: The sample was characterized by electron microprobe analyses. The contents of P and Fe are from 0.36 to 0.54 and from 0.77 to 1.08 atoms per formula unit, respectively.

Olmiite $CaMn[SiO_3(OH)](OH)$

Origin: N'Chwaning II mine, Kalaharimanganese fields, South Africa (type locality).
Experimental details: Raman scattering measurements have been performed on arbitrarily oriented crystals using a 633 nm He-Ne laser. The laser radiation power is not indicated. The Raman shifts have been determined for the maxima of individual peaks obtained as a result of the spectral curve analysis.
Raman shifts (cm^{-1}): (3550), 3543, 3511, 3467sh, 2807, (064w), 953, 914w, 853s, 811, 799sh, 782sh, 726w, 513, 484, 436, 420, 400, 378w, 335.
Source: Frost et al. (2013o).
Comments: The sample was characterized by electron microprobe analysis.

Olshanskyite $Ca_2[B_3O_3(OH)_6]OH \cdot 3H_2O$

Origin: Fuka mine, Okayama prefecture, Japan.
Experimental details: Raman scattering measurements have been performed on arbitrarily oriented crystals using a 633 nm He-Ne laser. The laser radiation power is not indicated. The Raman shifts have been determined for the maxima of individual peaks obtained as a result of the spectral curve analysis.
Raman shifts (cm^{-1}): 3621, 3536, 3482s, (3440), 3373, 3240, 3100sh, 2919, 1365w, 1206w, 1141w, 1069, 1025sh, (1014), 1003, 989, (976), 961, (919), 850sh, (696), 679s, (658), 579w, 516w, 463, 448w, 388w, 345, (335), 330, 322, 315, (303).
Source: Frost et al. (2014ak).
Comments: No independent analytical data are provided for the sample used.

Omongwaite $Na_2Ca_5(SO_4)_6 \cdot 3H_2O$

Origin: A recent salt lake deposit at Omongwa pan, Namibia (type locality).
Experimental details: No data.
Raman shifts (cm^{-1}): 3527w, 1143, 1013s, 665w, 637, 608w, 476, 436w,
Source: Mees et al. (2008).
Comments: The sample was characterized by powder X-ray diffraction data and electron microprobe analyses. For the Raman spectra of omongwaite see also Tong et al. (2011).

Omphacite $(Ca,Na)(Mg,Fe,Al)Si_2O_6$

Origin: Uru River area (?), north-central Myanmar.
Experimental details: Raman scattering measurements have been performed on an oriented grain in a polished sample using 532 nm laser radiation, with laser beam parallel to the b axis. The nominal laser radiation power was 20 mW.
Raman shifts (cm^{-1}): 1016, 684s, 567, 382, 144, 76.
Source: Leander et al. (2014).
Comments: The sample was characterized by electron microprobe analysis. For the Raman spectra of omphacite see also Buzatu and Buzgar (2010), and Andò and Garzanti (2014).

Onoratoite $Sb_8O_{11}Cl_2$

Origin: Synthetic.
Experimental details: Methods of sample preparation are not described. Raman scattering measurements have been performed using 514.5 nm Ar^+ laser radiation. The nominal laser radiation power was 20 mW.
Raman shifts (cm^{-1}): ~477s, ~456s, ~430, ~372, ~325, ~225sh, ~205, ~195, and a series of weak bands below 190 cm^{-1}.
Source: Orman et al. (2008), Orman (2010).
Comments: The sample was characterized by powder X-ray diffraction data and electron microprobe analysis.

Opal-A $SiO_2 \cdot nH_2O$

Origin: No data.
Experimental details: Methods of sample preparation are not described. Raman scattering measurements have been performed using 1064 nm laser radiation. The laser radiation power is not indicated.
Raman shifts (cm^{-1}): ~3460, 410s.
Source: Kiefert and Karampelas (2011).
Comments: No independent analytical data are provided for the sample used.

Opal-CT $SiO_2 \cdot nH_2O$
Origin: No data.
Experimental details: Methods of sample preparation are not described. Raman scattering measurements have been performed using 1064 nm laser radiation. The laser radiation power is not indicated.

Raman shifts (cm^{-1}): ~3450, 335s.
Source: Kiefert and Karampelas (2011).
Comments: No independent analytical data are provided for the sample used. For the Raman spectra of opal-CT see also Ilieva et al. (2007) and Wilson (2014).

Ophirite $Ca_2Mg_4[Zn_2Mn^{3+}_2(H_2O)_2(Fe^{3+}W_9O_{34})_2]\cdot 46H_2O$

Origin: Ophir Hill Consolidated mine, Ophir district, Oquirrh Mts., Tooele Co., Utah, USA (type locality).
Experimental details: Methods of sample preparation are not described. Raman scattering measurements have been performed using 514.5 nm Ar$^+$ laser radiation. The nominal laser radiation power at the sample was between 0.5 and 5 mW.
Raman shifts (cm^{-1}): ~955s, ~920sh, ~850, ~335w, ~220w.
Source: Kampf et al. (2014b).
Comments: The sample was characterized by powder X-ray diffraction data and electron microprobe analyses. The crystal structure is solved.

Oppenheimerite $Na_4(UO_2)(SO_4)_3\cdot 3H_2O$

Origin: Blue Lizard mine, San Juan Co., Utah, USA (type locality).
Experimental details: Methods of sample preparation are not described. Raman scattering measurements have been performed using 532 nm laser radiation. The laser radiation power is not indicated.
Raman shifts (cm^{-1}): 3526, 3400, 3218, ~1600w, 1215w, 1156w, 1060w, 1013, 1002, 986, 970s, 841s, 825sh, 651w, 603w, 459, 378sh, 345sh, 207, 188, 163, 153, 132, 110, 55.
Source: Kampf et al. (2015c).
Comments: The sample was characterized by powder X-ray diffraction data and electron microprobe analyses. The crystal structure is solved.

Ordoñezite $ZnSb^{5+}_2O_6$

Origin: Synthetic.
Experimental details: No data.
Raman shifts (cm^{-1}): 830w, 798w, 748s, 730, 670s, 615, 570sh, 538s, 480w, 364w, 331sh, 323, 300sh, 292, 248sh, 261, 220w.
Source: Husson et al. (1979).
Comments: No independent analytical data are provided for the sample used.

Orpiment As_2S_3

Origin: Synthetic.
Experimental details: Methods of sample preparation are not described. Raman scattering measurements have been performed using 532 nm Nd-YAG laser radiation. The laser radiation power is not indicated.
Raman shifts (cm^{-1}): 379w, 364sh, 352s, 308s, 291, 200w, 178w, 153, 143w, 135w.
Source: Minceva-Sukarova et al. (2003).

Comments: No independent analytical data are provided for the sample used. For the Raman spectra of orpiment see also Forneris (1969), Trentelman et al. (1996), Burgio and Clark (2001), Frost et al. (2010c), and Kampf et al. (2011a).

Orschallite $Ca_3(S^{4+}O_3)_2(SO_4) \cdot 12H_2O$

Origin: Hannebacher Ley volcano, Eifel, Germany (type locality).
Experimental details: Raman scattering measurements have been performed on arbitrarily oriented crystals using a 633 nm He-Ne laser. The laser radiation power is not indicated. The Raman shifts have been determined for the maxima of individual peaks obtained as a result of the spectral curve analysis.
Raman shifts (cm^{-1}): 3383, 1215w, 1096, (1011), 1005s, 984, 971sh, (657), 651, 532sh, 521, 492, 441, 244, 194, 173, 145, 119.
Source: Frost and Keeffe (2009d).
Comments: No independent analytical data are provided for the sample used.

Orthoclase $K(AlSi_3O_8)$

Origin: Bahariya depression, Western Desert, Egypt.
Experimental details: Methods of sample preparation are not described. Raman scattering measurements have been performed using 532 nm Nd-YAG laser radiation. The laser radiation power at the sample was between 20 and 200 mW.
Raman shifts (cm^{-1}): ~1125, ~750w, ~650, ~640, ~590, 515s, 475s, 453, ~403, 282.
Source: Ciobotă et al. (2012).
Comments: No independent analytical data are provided for the sample used. For the Raman spectra of orthoclase see also Frezzotti et al. (2012) and Culka et al. (2016a).

Orthojoaquinite-(Ce) $NaBa_2Fe^{2+}Ce_2Ti_2(SiO_3)_8O_2(O,OH) \cdot H_2O$

Origin: Benitoite Gem Mine, San Benito Co., California, USA (type locality).
Experimental details: Raman scattering measurements have been performed on arbitrarily oriented crystals using a 633 nm He-Ne laser. The laser radiation power is not indicated. The Raman shifts have been determined for the maxima of individual peaks obtained as a result of the spectral curve analysis.
Raman shifts (cm^{-1}): 3574sh, 3558, (3515), 3506, 3444sh, 3397, 3344+3317 (unresolved doublet?), 3230w, 3192w, 1112, 1033, 982, 956, 933, 896s, 732, 687sh, 668, 636, 600, 526, 494, 511, 375, 358, 304, 267, (202), 191, (165), 145, 112.
Source: Frost and Pinto (2007).
Comments: No independent analytical data are provided for the sample used.

Osakaite $Zn_4(SO_4)(OH)_6 \cdot 5H_2O$

Origin: Block 14 Opencut, Broken Hill, New South Wales, Australia.
Experimental details: Raman scattering measurements have been performed on an arbitrarily oriented single crystal using 632.8 He-Ne laser radiation. The nominal laser radiation power was 17 mW. A 180°-scattering geometry was employed.

Raman shifts (cm^{-1}): 3633, 3550, 3510, 3455, 3330, 3257sh, 3243, 3175, ~1670w, ~1620w, 1160, 1112, 1051, 1024, 1011, 964s, 636, 604, 508, 456, 430, 398.
Source: Elliott (2010).
Comments: The sample was characterized by powder X-ray diffraction data and electron microprobe analyses. The crystal structure is solved.

Osbornite TiN

Origin: Synthetic.
Experimental details: Raman scattering measurements have been performed on a coating deposited on silicon (111) substrate using He-Ne laser radiation. The nominal laser radiation power was 20 mW.
Raman shifts (cm^{-1}): 563, 242.
Source: Barshilia and Rajam (2004).
Comments: The sample was characterized by powder X-ray diffraction data. For the Raman spectra of osbornite see also Spengler et al. (1978).

Oskarssonite AlF$_3$

Origin: Synthetic.
Experimental details: Raman scattering measurements have been performed on a powder sample using 1064 nm Nd-YAG or 514.5 nm Ar$^+$ laser radiation. The nominal laser radiation power was up to 400 mW or 10–25 mW, respectively.
Raman shifts (cm^{-1}): 478w, 382w, 157s, 96.
Source: Groß et al. (2007).
Comments: For the Raman spectrum of oskarssonite see also Daniel et al. (1990).

Osumilite KFe$_2$(Al$_5$Si$_{10}$)O$_{30}$

Origin: Rundvågshetta, Lützow-Holm Complex, East Antarctica.
Experimental details: Raman scattering measurements have been performed on a grain included within a garnet porphyroblast using 532 nm laser radiation. The nominal laser radiation power was 10 mW.
Raman shifts (cm^{-1}): 1103, 920, 549, 477s, 383w, 350w, 281.
Source: Kawasaki et al. (2011).
Comments: The sample was characterized by electron microprobe analyses.

Otavite Cd(CO$_3$)

Origin: Synthetic.
Experimental details: Methods of sample preparation are not described. Raman scattering measurements have been performed using 488 and 514.5 nm Ar$^+$ laser radiations. The nominal laser radiation power was in the range from 100 to 500 mW.
Raman shifts (cm^{-1}): 1718, 1388, 1084s, 712, 271, 158.
Source: Rutt and Nicola (1974).
Comments: The sample was characterized by powder X-ray diffraction data. For the Raman spectra of otavite see also Minch et al. (2010) and Falgayrac et al. (2013).

Ottemannite Sn_2S_3

Origin: Synthetic.
Experimental details: Methods of sample preparation are not described. Raman scattering measurements have been performed using 514.5 nm Ar^+ laser radiation. The laser radiation power at the sample was below 0.4 mW.
Raman shifts (cm^{-1}): 310s, 303sh, 247w, 231w, 67w, 56, 48.
Source: Fontané et al. (2013)
Comments: No independent analytical data are provided for the sample used. For the Raman spectrum of ottemannite see also Price et al. (1999).

Ottensite $Na_3(Sb_2O_3)_3(SbS_3)\cdot 3H_2O$

Origin: Pereta mine, Tuscany, Italy.
Experimental details: Raman scattering measurements have been performed on an arbitrarily oriented crystal using 632.8 nm He-Ne laser radiation. The nominal laser radiation power was 20 mW.
Raman shifts (cm^{-1}): 766w, 615w, 538, 479w, 355s, 344sh, 299, 254w, 225w, 153.
Source: Bittarello et al. (2015).
Comments: The sample was characterized by powder X-ray diffraction data and electron microprobe analyses.

Ottohahnite $Na_6(UO_2)_2(SO_4)_5(H_2O)_7 \cdot 1.5H_2O$

Origin: Blue Lizard mine, San Juan Co., Utah, USA (type locality).
Experimental details: Methods of sample preparation are not described. Raman scattering measurements have been performed using 532 nm laser radiation. The laser radiation power is not indicated.
Raman shifts (cm^{-1}): ~3595, ~3490w, ~1630w, ~1220, ~1200, ~1155, ~1050, ~1010, ~980, ~930, ~840s, ~655, ~203s, and a series of weak bands in the range from 300 to 640 cm^{-1}.
Source: Kampf et al. (2016g).
Comments: The sample was characterized by powder X-ray diffraction data and electron microprobe analyses. The crystal structure is solved.

Otwayite $Ni_2(CO_3)(OH)_2 \cdot H_2O$

Origin: Mt. Grey, Tasmania, Australia.
Experimental details: Raman scattering measurements have been performed on arbitrarily oriented crystals using a 633 nm He-Ne laser. The laser radiation power is not indicated. The Raman shifts have been determined for the maxima of individual peaks obtained as a result of the spectral curve analysis.
Raman shifts (cm^{-1}): 3612, (3610), 3579, 3538, 3470sh, (3288), (2989), 2935, 2879sh, 1690, 1600, 1353w, 1073s, (1068), 981s, 937, (835), 708, 703, 617, (545), 529, (469), 445, 395, 308sh, 232, 216, 194sh, 177sh.
Source: Frost et al. (2006o).
Comments: No independent analytical data are provided for the sample used.

Oxammite $(NH_4)_2(C_2O_4) \cdot H_2O$

Origin: No data.
Experimental details: Raman scattering measurements have been performed on arbitrarily oriented crystals using a 633 nm He-Ne laser. The laser radiation power is not indicated. The Raman shifts have been determined for the maxima of individual peaks obtained as a result of the spectral curve analysis.
Raman shifts (cm^{-1}): 3235, 3030, 2995, 2900, 2879, 2161, 1902, 1737, 1695s, 1605, 1473, 1451, 1447, 1430, 1417, 1312, 892s, 866, 815, 642, 489s, 438, 278, 224, 210, 198, 181, 160.
Source: Frost and Weier (2003).
Comments: No independent analytical data are provided for the sample used. For the Raman spectra of oxammite see also Frost et al. (2003k), Frost (2004d), and Frost and Weier (2004a).

Oxycalcioroméite $Ca_2Sb^{5+}_2O_7$

Origin: Bucadella Vena mine, Apuan Alps, Tuscany, Italy (type locality).
Experimental details: Methods of sample preparation are not described. Micro-Raman scattering measurements have been performed using 532 nm laser radiation. The laser radiation power is not indicated.
Raman shifts (cm^{-1}): 913w, 777, 666, 540sh, 509s, 426, 295, 199w.
Source: Biagioni et al. (2013c).
Comments: The sample was characterized by powder X-ray diffraction data and electron microprobe analyses. The crystal structure is solved.

Oxy-dravite $Na(Al_2Mg)(Al_5Mg)(Si_6O_{18})(BO_3)_3(OH)_3O$

Origin: No data available.
Experimental details: Raman scattering measurements have been performed on an oriented crystal with the crystallographic c axis parallel to the Cartesian coordinate z axis using 514.5 and 488.0 nm Ar^+ laser radiations. The laser radiation power at the sample was 14 mW. Raman spectrum was obtained in the spectral region from 15 to 4000 cm^{-1}. Polarized spectrum was collected in the $-y(zz)y$ scattering geometry.
Raman shifts (cm^{-1}): 3776w, 3738w, 3674w, 3642w, 3567+3529 (unresolved doublet?), (3480w), ~700, 368s, 312, 242, 217, and a series of relatively weak bands in the range from 400 to 700 cm^{-1}.
Source: Watenphul et al. (2016a, b).
Comments: The sample was characterized by electron microprobe and LA-ICP-MS analyses.

Oxykinoshitalite $BaMg_2Ti^{4+}O_2(Si_2Al_2)O_{10}$

Origin: S. Demetrio High, Hyblean plateau, Sicily, Italy.
Experimental details: Micro-Raman scattering measurements have been performed on a grain in thin section using 632.8 nm He-Ne laser radiation. The laser radiation power at the sample was 6 mW.
Raman shifts (cm^{-1}): 996, 880, 725s, 135, 122.
Source: Manuella et al. (2012).
Comments: The sample was characterized by electron microprobe analyses.

Oxynatromicrolite $(Na,Ca,U)_2(Ta,Nb)_2O_6(O,F)$

Origin: Guanpo, Henan province, China (type locality).
Experimental details: No data.
Raman shifts (cm^{-1}): ~3450sh, ~770, ~650s.
Source: Guang et al. (2016).
Comments: The sample was characterized by powder X-ray diffraction data and electron microprobe analyses, and differential thermal analysis.

Oxyplumboroméite $Pb_2Sb^{5+}_2O_7$

Origin: Synthetic.
Experimental details: Micro-Raman scattering measurements have been performed on an arbitrarily oriented sample using 532.0 nm Nd-YAG laser radiation. The laser radiation power at the sample was between 0.8 and 4 mW.
Raman shifts (cm^{-1}): 807w, 513s, 423, 355, 298, 230s, 190, 107s.
Source: Rosi et al. (2009).
Comments: The sample was characterized by powder X-ray diffraction data. For the Raman spectrum of oxyplumboroméite see also Kendix et al. (2008).

Oyelite $Ca_{10}B_2Si_8O_{29} \cdot 12H_2O$

Origin: N'Chwaning II mine, Manganese Fields, Kalahari desert, Republic of South Africa.
Experimental details: Methods of sample preparation are not described. Non-polarized micro-Raman spectrum has been obtained in a nearly backscattered geometry using 632.8 nm He-Ne laser radiation. The laser radiation power is not indicated.
Raman shifts (cm^{-1}): ~1025, ~1005, ~990, ~910, ~860, ~715s, ~685s, ~460, ~350, ~320.
Source: Biagioni et al. (2012).
Comments: The sample was characterized by powder X-ray diffraction data.

Ozokerite

Origin: Boryslav, Poland.
Experimental details: Kind of sample preparation is not indicated. Raman scattering measurements have been performed using 514.5 nm Ar$^+$ laser radiation with the nominal radiation power of 10 mW (micro-Raman measurements) and/or 1064 nm Nd-YAG laser radiation with the power of 350 mW.
Raman shifts (cm^{-1}): 2977, 2961w, 2930w, 2923s, 2917s, 2901w, 2880s, 2855, 2848w, 2836, 2733w, 2724, 1498w, 1486w, 1466, 1442, 1420w, 1416w, 1387w, 1369w, 1349w, 1321w, 1297w, 1169w, 1154w, 1133s.
Source: Jehlička et al. (2007a).
Comments: No independent analytical data are provided for the sample used.

Pabstite $BaSnSi_3O_9$

Origin: Synthetic.
Experimental details: Raman scattering measurements have been performed on a powdery sample using 514.5 nm Ar$^+$ laser radiation. The laser radiation power is not indicated.

Raman shifts (cm^{-1}): 994w, ~876w, ~817w, 574s.
Source: Takahashi et al. (2008).
Comments: The sample was characterized by powder X-ray diffraction data.

Padmaite PdBiSe

Origin: Southern Sopchinskoe deposit, Monchegorsk district, Kola Peninsula, Russia.
Experimental details: Raman scattering measurements have been performed on an arbitrarily oriented sample using 514.5 nm Ar$^+$ laser radiation. The nominal laser radiation power was 50 mW.
Raman shifts (cm^{-1}): 235s, 177w, 154, 115, 91, 82, 61s, 52sh.
Source: Voloshin et al. (2015a).
Comments: The sample was characterized electron microprobe analyses.

Pakhomovskyite $Co_3(PO_4)_2 \cdot 8H_2O$

Origin: Synthetic.
Experimental details: Methods of sample preparation are not described. Raman scattering measurements have been performed using 514.5 nm Ar$^+$ laser radiation. The laser radiation power is not indicated.
Raman shifts (cm^{-1}): 1043, 1023, 956s, 894, 560, 462, 370w, 260, 249, 202, 171.
Source: Shao et al. (2016).
Comments: The sample was characterized by powder X-ray diffraction data.

Palermoite $Li_2SrAl_4(PO_4)_4(OH)_4$

Origin: Palermo No. 1 mine, Groton, New Hampshire, USA (type locality).
Experimental details: Raman scattering measurements have been performed on an arbitrarily oriented crystal using 532 nm solid-state laser radiation. The laser radiation power is not indicated.
Raman shifts (cm^{-1}): 1077, 1026s, 1005s, 982sh, 657, 637, 603s, 534, 512, 345, 417, 367w, 328, 282, 253, 238.
Source: Schumer et al. (2016).
Comments: No independent analytical data are provided for the sample used.

Palladinite PdO

Origin: Synthetic.
Experimental details: Methods of sample preparation are not described. Raman scattering measurements have been performed using 18794.59 cm^{-1} laser radiation. The laser radiation power at the sample was between 1 and 2 mW.
Raman shifts (cm^{-1}): 723w, 650s, 445.
Source: Bakker (2014).
Comments: No independent analytical data are provided for the sample used. For the Raman spectrum of palladinite see also McBride et al. (1991).

Palladosilicide Pd_2Si

Origin: Synthetic.
Experimental details: Raman scattering measurements have been performed on a thin film using 514.5 nm Ar^+ laser radiation. The nominal laser radiation power was $\leq$ 10 mW.
Raman shifts (cm^{-1}): 115s, 90.
Source: Nemanich (1986).

Palygorskite $(Mg,Al)_2Si_4O_{10}(OH) \cdot 4H_2O$

Origin: Glasgow, Virginia, USA.
Experimental details: Methods of sample preparation are not described. Raman scattering measurements have been performed on randomlyoriented crystals in back-scattering geometry, using 1064 nm Nd-YAG laser radiation. The nominal laser radiation power was 900 mW.
Raman shifts (cm^{-1}): 1109, 988, 971, 809, 774, 680, 638, 556, 512, 488, 473, 437, 406, 354, 327, 268, 205, 183, 167, 130 (A_g modes); 1211, 1160, 1077, 986, 904, 800, 704, 658, 597, 540, 512, 456, 435, 410, 397, 353, 327, 243, 216, 205, 157, 139 (B_g modes). The strongest peaks are observed in the ranges 250–300 and 650–720 cm^{-1}.
Source: McKeown et al. (2002).
Comments: The empirical formula of the sample used is $(Mg_{2.00}Al_{1.96}Fe_{0.06})Si_{7.94}O_{21} \cdot nH_2O$.

Panguite $(Ti,Al,Sc,Mg,Zr,Ca)_{1.8}O_3$

Origin: Allende meteorite (type locality).
Experimental details: Raman scattering measurements have been performed on an arbitrarily oriented grain in a polished section using 514.5 nm Ar^+ laser radiation. The laser radiation power at the sample was ~5 mW.
Raman shifts (cm^{-1}): 405, 380s.
Source: Ma et al. (2012c).
Comments: The sample was characterized by electron backscatter diffraction data and electron microprobe analyses.

Panichiite $(NH_4)_2SnCl_6$

Origin: Synthetic.
Experimental details: No data.
Raman shifts (cm^{-1}): ~430s, ~330s, ~290, ~265, ~210, ~193, ~148, ~122.
Source: Podsiadlo et al. (2015).
Comments: The sample was characterized by X-ray diffraction data. The crystal structure is solved.

Papagoite $CaCuAlSi_2O_6(OH)_3$

Origin: Cornelia mine, Ajo, Pima Co., Arizona, USA (type locality).
Experimental details: Raman scattering measurements have been performed on arbitrarily oriented crystals using a 633 nm He-Ne laser. The laser radiation power is not indicated. The Raman shifts have been determined for the maxima of individual peaks obtained as a result of the spectral curve analysis.

Raman shifts (cm^{-1}): 3614sh, 3573s, (3567), 3545s, (3533), (3490), 3453, (3368), (1079), 1053, 986, 942w, 867, 830sh, 812, 755, (644w), 630, 573, 536, 472sh, 460s, 438sh, 419s, 382, 298, 279, 264, 251, 243sh, 236sh, 205, (199), 185sh, 178, 170sh, 163, 156w, 150w, 147sh, 136, 131sh, 119, 113sh, 107.
Source: Frost and Xi (2013d).
Comments: No independent analytical data are provided for the sample used.

Parabutlerite $Fe^{3+}(SO_4)(OH) \cdot 2H_2O$

Origin: Alcaparrosa mine, Cerritos Bayos, Calama, El Loa province, Antofagasta, Chile (type locality).
Experimental details: Methods of sample preparation are not described. Polarized Raman spectra have been obtained using 633 nm He-Ne laser radiation. The laser radiation power is not indicated. The Raman shifts have been determined for the maxima of individual peaks obtained as a result of the spectral curve analysis.
Raman shifts (cm^{-1}): 3504sh, 3316, 3200sh, (3133), 1202, 1164, 1109s, (1095), 1044, 1026s, 1014, 990sh, 655sh, 614, 550, 468s, 406, 368, 263sh, 237, 214, 186, 155sh.
Source: Čejka et al. (2011b).
Comments: The sample was characterized by powder X-ray diffraction data and electron microprobe analyses.

Paracoquimbite $Fe^{3+}_2(SO_4)_3 \cdot 9H_2O$

Origin: Synthetic.
Experimental details: Methods of sample preparation are not described. Raman scattering measurements have been performed using 532 nm laser radiation. The laser radiation power is not indicated.
Raman shifts (cm^{-1}): 3577, 3412, 3245, 3046, 1682, 1620, 1200, 1170, 1112, 1093, 1037, 1025s, 1012, 877, 675, 628, 602, 514, 502, 478, 286, 211.
Source: Ling and Wang (2010).
Comments: The sample was characterized by powder X-ray diffraction data.

Paragonite $NaAl_2(Si_3Al)O_{10}(OH)_2$

Origin: Rebra II Formation, Rodnei Mts., Eastern Carpathians, Romania.
Experimental details: Methods of sample preparation are not described. Raman scattering measurements have been performed using 532 nm laser radiation. The laser radiation power at the sample was 35 mW.
Raman shifts (cm^{-1}): 911, 704, 612, 442s, 266, 217w.
Source: Buzgar (2008).
Comments: No independent analytical data are provided for the sample used. For the Raman spectra of paragonite see also Tlili et al. (1989), Graeser et al. (2003), and Frezzotti et al. (2012).

Paraguanajuatite Bi_2Se_3

Origin: Synthetic.
Experimental details: Raman scattering measurements have been performed on a crystal with the trigonal c axis parallel to the laser beam direction using 514.5 nm Ar$^+$ laser and 647.1 nm Kr$^+$ laser

radiations. The laser radiation power is not indicated. A 180°-scattering geometry was employed. Polarized spectra were collected in z(xx)–z and z(xy)–z scattering geometries.
Raman shifts (cm^{-1}): z(xx)–z (A$_{1g}$): 175, 72; z(xy)–z (E$_g$): 132.
Source: Richter et al. (1977).

Parahopeite $Zn_3(PO_4)_2 \cdot 4H_2O$

Origin: Reaphook Hill, Martins Well, Flinders Ranges, South Australia, Australia.
Experimental details: Raman scattering measurements have been performed on arbitrarily oriented crystals in the spectral range from 700 to 4000 cm^{-1} using a 633 nm He-Ne laser. The laser radiation power at the sample was 1 mW. The Raman shifts have been determined for the maxima of individual peaks obtained as a result of the spectral curve analysis.
Raman shifts (cm^{-1}): 3439sh, 3293, 3163, 3027, 1053, 1033, 1003s, 959.
Source: Frost (2004a).
Comments: No independent analytical data are provided for the sample used.

Paramontroseite VO_2

Origin: Product of heating of karelianite from Pyrrhotite gorge, Khibiny massif by the laser beam.
Experimental details: Methods of sample preparation are not described. Raman scattering measurements have been performed using 633 nm He-Ne laser radiation. The nominal laser radiation power was 2 or 20 mW.
Raman shifts (cm^{-1}): 991, 897w, 688, 405, 281, 222, 190, 139s, 98.
Source: Voloshin et al. (2014).
Comments: The sample was characterized by powder X-ray diffraction data and electron microprobe analysis.

Paranatrolite $Na_2(Si_3Al_2)O_{10} \cdot 3H_2O$

Origin: Poudrette (Demix) quarry, Mont Saint-Hilaire, Rouville RCM (Rouville Co.), Montérégie, Québec, Canada.
Experimental details: No data.
Raman shifts (cm^{-1}): ~3465w, ~1095w, ~527s, ~430, ~330, ~135s.
Source: Belitsky et al. (1992).
Comments: No independent analytical data are provided for the sample used.

Paraotwayite $Ni(OH)_{2-x}(SO_4,CO_3)_{0.5x}$

Origin: Otway deposit, Western Australia, Australia (type locality).
Experimental details: Raman scattering measurements have been performed on arbitrarily oriented crystals using a 633 nm He-Ne laser. The laser radiation power is not indicated. The Raman shifts have been determined for the maxima of individual peaks obtained as a result of the spectral curve analysis.
Raman shifts (cm^{-1}): 3606sh, 3590, 3568, (3566), 3532, 2909w, 2852w, 1115, 987s, (977), 642w, 606, 487, 473, 451, 416, 297w, (260), (238), (218), (196), (176).
Source: Frost et al. (2006o).
Comments: No independent analytical data are provided for the sample used.

Parapierrotite $TlSb_5S_8$

Origin: Synthetic.
Experimental details: Raman scattering measurements have been performed on a single crystal with the laser polarization parallel to the b- and c-axes using 632.8 nm He-Ne laser radiation. The laser radiation power at the sample was 1.7 mW.
Raman shifts (cm^{-1}): 334, 321, 310 (very strong for $E\|b$), 293, 275w, 260w, 242, 227(very weak for $E\|b$), 204w, 178w, 162w, 145w, 127, 115w, 106w, 94w, 94, 81, 62, 51 (strong for $E\|c$), 41.
Source: Kharbish (2011).
Comments: The sample was characterized by powder X-ray diffraction data. For the Raman spectra of parapierrotite see also Makreski et al. (2013b, 2014).

Pararealgar As_4S_4

Origin: Synthetic.
Experimental details: Raman scattering measurements have been performed on a polycrystalline sample using 647.1 nm laser radiation. The laser radiation power is not indicated.
Raman shifts (cm^{-1}): 383w, 370w, 364, 346, 334, 340sh, 322w, 316w, 275, 236s, 230s, 204, 198, 191w, 175, 172, 167sh, 158, 152, 142, 135w, 118, 52sh, 45, 32.
Source: Muniz-Miranda et al. (1996).
Comments: For the Raman spectra of pararealgar see also Trentelman et al. (1996) and Burgio and Clark (2001).

Pararobertsite $Ca_2Mn^{3+}{}_3(PO_4)_3O_2 \cdot 3H_2O$

Origin: Tip Top mine, Custer, South Dakota, USA (type locality).
Experimental details: Raman scattering measurements have been performed on an arbitrarily oriented crystal using 780 nm laser radiation. The laser radiation power is not indicated.
Raman shifts (cm^{-1}): ~1197w, ~1112w, ~1042, ~1035, ~983, ~965, ~953sh, ~896w, ~623s, ~547w, ~498w, ~461, ~403, ~365w, ~309, ~277, ~258.
Source: Andrade et al. (2012).
Comments: The sample was characterized by powder X-ray diffraction data.

Parascholzite $CaZn_2(PO_4)_2 \cdot 2H_2O$

Origin: No data.
Experimental details: Raman scattering measurements have been performed on arbitrarily oriented crystals using a 633 nm He-Ne laser. The laser radiation power at the sample was 1 mW.
Raman shifts (cm^{-1}): 1170, 1115, 1086, 999, 925s, 553s, 409, 302, 286, 271, 236.
Source: Scholz et al. (2013a).
Comments: No independent analytical data are provided for the sample used. For the Raman spectrum of parascholzite see also Frost (2004a).

Parascorodite $Fe^{3+}(AsO_4) \cdot 2H_2O$

Origin: Kaňk (near Kutná Hora) or Lehnschafter gallery in Mikulov (near Teplice), both Czech Republic.

Experimental details: Methods of sample preparation are not described. Micro-Raman scattering measurements have been performed using 514.5 nm Ar⁺ laser radiation. The nominal laser radiation power was between 1 and 2 mW.

Raman shifts (cm^{-1}): 892, 859, 815s, 800sh, 782s, 492, 458sh, 437s, 413sh, 382w, 349sh, 337, 292sh, 280, 269sh, 245, 231w, 181s, 164sh.

Source: Culka et al. (2016b).

Comments: The sample was characterized by powder X-ray diffraction data. Additionally, in the cited paper Raman spectra of parascorodite obtained with 785 and 532 nm lasers are given.

Parasibirskite $Ca_2B_2O_5 \cdot H_2O$

Origin: Synthetic.

Experimental details: Raman scattering measurements have been performed on a powder sample was first pressed into a tablet using 532 nm Nd-YAG laser radiation. The nominal laser radiation power was 40 mW. Backscattered spectra were collected from focal spot of diameter of 2 μm.

Raman shifts (cm^{-1}): 3354, 3309, 1278, 1250, 1153, 1086, 908, 712, 489, 295, 281, 263, 252, 216, 175, 147, 135.

Source: Goryainov et al. (2017).

Comments: The sample was characterized by powder X-ray diffraction data.

Parasymplesite $Fe^{2+}_3(AsO_4)_2 \cdot 8H_2O$

Origin: No data.

Experimental details: Raman scattering measurements have been performed on arbitrarily oriented crystals using a 633 nm He-Ne laser. The laser radiation power is not indicated. The Raman shifts have been determined for the maxima of individual peaks obtained as a result of the spectral curve analysis.

Raman shifts (cm^{-1}): 3460, 3215, 860, 835, 810, 780, 545, 480, 450, 432, 371, 332, 286, 249, 220, 194, 142.

Source: Frost et al. (2003g).

Comments: In the cited paper the mineral is named "parasymplesite/symplesite." Band intensities are not indicated. No independent analytical data are provided for the sample used.

Paratacamite $Cu^{2+}_3(Cu,Zn)(OH)_6Cl_2$

Origin: Widgiemooltha, Western Australia.

Experimental details: Raman scattering measurements have been performed on arbitrarily oriented crystals using a 633 nm He-Ne laser. The laser radiation power is not indicated. The Raman shifts have been determined for the maxima of individual peaks obtained as a result of the spectral curve analysis.

Raman shifts (cm^{-1}): 3508, 3446, 3395, 3364, 3341, 3232, 942, 890, 732, 513, 501, 474, 404, 367, 277, 243, 148, 124.

Source: Frost et al. (2002b).

Comments: No independent analytical data are provided for the sample used. For the Raman spectrum of paratacamite see also Chu et al. (2011).

Paratellurite α-TeO$_2$

Origin: Synthetic.
Experimental details: Raman scattering measurements have been performed in a back-scattering geometry on an arbitrarily oriented sample using 514.5 nm Ar$^+$ laser radiation. The nominal laser radiation power was 150 mW.
Raman shifts (cm^{-1}): 786, 769, 649s, 642, 592, 575, 415, 392, 379, 330, 315, 297, 281, 259, 235, 218, 210, 179, 174, 157, 152s, 121s, 82, 62.
Source: Mirgorodsky et al. (2000).
Comments: For the Raman spectra of paratellurite see also Bürger et al. (1992), Berthereau (1995), and Noguera et al. (2003).

Paratooite-(La) (La,Ca,Na,Sr)$_6$Cu(CO$_3$)$_8$

Origin: Paratoo Cu mine, Olary district, South Australia (type locality).
Experimental details: Raman scattering measurements have been performed on an arbitrarily oriented sample using 785 nm laser radiation. The laser radiation power is not indicated.
Raman shifts (cm^{-1}): ~1440w, 1095s, 1075s, ~220s, and a series of weak bands in the range from 300 to 900 cm^{-1}.
Source: Pring et al. (2006).
Comments: The sample was characterized by powder X-ray diffraction data and electron microprobe analyses.

Paravauxite Fe^{2+}Al$_2$(PO$_4$)$_2$(OH)$_2$·8H$_2$O

Origin: Siglo XX mine, Bustillo province, Potosí department, Bolivia.
Experimental details: Raman scattering measurements have been performed on arbitrarily oriented crystals using a 633 nm He-Ne laser. The laser radiation power is not indicated. The Raman shifts have been determined for the maxima of individual peaks obtained as a result of the spectral curve analysis.
Raman shifts (cm^{-1}): 3648w, 3505s, 3421, 3315, (3215), 3086, 1639w, 1582sh, 1490sh, 1148, 1115sh, 1058sh, 1020s, 643, 609, 570, 537sh, 420sh, 393s, 253, 319, 299, 227sh, 214, 196w, 164, 148, 110.
Source: Frost et al. (2013n).
Comments: The sample was characterized by electron microprobe analysis.

Pargasite NaCa$_2$(Mg$_4$Al)(Si$_6$Al$_2$)O$_{22}$(OH)$_2$

Origin: Edenville, New York, USA.
Experimental details: Methods of sample preparation are not described. Raman scattering measurements have been performed using 532 nm Nd-YAG laser radiation. The nominal laser radiation power was 100 mW.
Raman shifts (cm^{-1}): 1095w, 1045, 1009, 971w, 956w, 924, 910, 885, 758s, 663s, 581sh, 547, 526, 514, 475w, 415w, 322, 292s, 226s.
Source: Apopei and Buzgar (2010).

Comments: No independent analytical data are provided for the sample used. For the Raman spectra of pargasite see also Apopei et al. (2011), Frezzotti et al. (2012), Andò and Garzanti (2014), and Leissner et al. (2015).

Parisite-(Ce) $CaCe_2(CO_3)_3F_2$

Origin: Snowbird mine, Fish Creek, Alberton, Mineral Co., Montana, USA.
Experimental details: Raman scattering measurements have been performed on arbitrarily oriented crystals using a 633 nm He-Ne laser. The laser radiation power is not indicated. The Raman shifts have been determined for the maxima of individual peaks obtained as a result of the spectral curve analysis.
Raman shifts (cm^{-1}): 3661sh, 3517sh, 3316, 3180, 1420w, 1088s, 742, 682w, 601, 263, 152.
Source: Frost and Dickfos (2007a).
Comments: No independent analytical data are provided for the sample used. For the Raman spectrum of parisite-(Ce) see also Hong et al. (1999).

Parisite-(La) $CaLa_2(CO_3)_3F_2$

Origin: Mula mine, Tapera village, Novo Horizonte Co., Bahia, Brazil (type locality).
Experimental details: Raman scattering measurements have been performed on an arbitrarily oriented crystal using 532 nm solid-state laser radiation. The nominal laser radiation power was 150 mW.
Raman shifts (cm^{-1}): 1428, 1331w, 1098s, 1091s, 1081s, 970, 871, 737, 600, 453, 394, 350, 331, 262s, 162.
Source: Menezes Filho et al. (2017).
Comments: The sample was characterized by powder X-ray diffraction data and electron microprobe analyses. All peaks above 1500 cm^{-1} are due to fluorescence.

Parnauite $Cu_9(AsO_4)_2(SO_4)(OH)_{10} \cdot 7H_2O$

Origin: Cap Garonne, Var, France.
Experimental details: Raman scattering measurements have been performed on a surfaces nearly perpendicular to the {010} cleavage of a single crystal using 532 nm laser radiation. The laser radiation power is not indicated.
Raman shifts (cm^{-1}): 3544sh, 3457, 3365sh, 2926, 2880, 2848, 1608w, 1442, 1320w, 1118sh, 1039, 975, 849s, 814sh, 688sh, 604sh, 493s, 447sh, 378, 295, 268, 218sh, 168sh, 110s.
Source: Mills et al. (2013).
Comments: The sample was characterized by single-crystal X-ray diffraction data and electron microprobe analyses. The crystal structure is solved. For the Raman spectra of parnauite see also Frost et al. (2009j) and Frost and Keeffe (2011).

Parsonsite $Pb_2(UO_2)(PO_4)_2$

Origin: Ranger U Mine, Northern Territory, Australia.
Experimental details: Raman scattering measurements have been performed on arbitrarily oriented crystals using a 633 nm He-Ne laser. The laser radiation power is not indicated. The Raman shifts have been determined for the maxima of individual peaks obtained as a result of the spectral curve analysis.

Raman shifts (cm^{-1}): 3404w, 3329, 1590, 1078, (1074), 1024, (998), (987), 967, (943), (872), (831), 807s, 796sh, 609, 595, (591), 582, 560, 540, 465, 439, 406, 394, 281, 255, 227sh, 206, 188sh, 171, 155, 136sh, 111sh.
Source: Frost et al. (2006f).
Comments: The sample was characterized by chemical analyses.

Parthéite $Ca_2(Si_4Al_4)O_{15}(OH)_2 \cdot 4H_2O$

Origin: Denezhkin Kamen' Mt., Middle Urals, Russia.
Experimental details: Raman scattering measurements have been performed on an arbitrarily oriented single crystal using 532 and 633 nm laser radiations. The nominal laser radiation power was 100 and 17 mW, respectively.
Raman shifts (cm^{-1}): 3574w, 3476sh, 3417, 3384, 3308w, 3256, ~1100w, ~1048, ~978, ~955, ~930, ~770, ~740, ~714, ~640, ~550, ~500s, ~450, ~420, ~400, ~360, ~310s, ~260w, ~250w, ~235w, ~220, ~200, ~180, ~160s, ~150, ~140, ~113.
Source: Lazic et al. (2012).
Comments: The sample was characterized by single-crystal X-ray diffraction data. The crystal structure is solved.

Partzite $Cu_2Sb^{5+}_2O_7$

Origin: Blind Spring Hill district, near Benton, Mono Co., California, USA.
Experimental details: Raman scattering measurements have been performed on arbitrarily oriented crystals using a 633 nm He-Ne laser. The laser radiation power is not indicated. The Raman shifts have been determined for the maxima of individual peaks obtained as a result of the spectral curve analysis.
Raman shifts (cm^{-1}): 3622w, 3586, 3563, 3485w, 3407, 3376, 3266, 2947sh, (1455w), (1396w), 1126, 1096, 1074, 982sh, 971s, (938), 907w, 837w, 777w, 730w, 675w, 620, 607, 594, 520s, 479, 449, 418, 387, 362, 316, 258sh, 241, 195.
Source: Bahfenne and Frost (2010c).
Comments: Questionable data. In particular, the strongest band at 971 cm^{-1} may correspond to an impurity. No independent analytical data are provided for the sample used.

Pašavaite $Pd_3Pb_2Te_2$

Origin: Synthetic.
Experimental details: Methods of sample preparation are not described. Raman scattering measurements have been performed using 532 nm Nd-YAG laser radiation. The nominal laser radiation power was 100 mW.
Raman shifts (cm^{-1}): 159s, 119.
Source: Bakker (2014).
Comments: No independent analytical data are provided for the sample used. For the Raman spectrum of pašavaite see also Vymazalová et al. (2014).

Pascoite $Ca_3V^{5+}_{10}O_{28} \cdot 17H_2O$

Origin: Vanadium Queen mine, San Juan Co., Utah, USA.
Experimental details: Raman scattering measurements have been performed on arbitrarily oriented crystals using a 633 nm He-Ne laser. The laser radiation power is not indicated. The Raman shifts have been determined for the maxima of individual peaks obtained as a result of the spectral curve analysis.
Raman shifts (cm^{-1}): 3570w, 3466, 3405, 3254, 3125, (1668), 1644, 1514w, 993s, 961s, 841, 621s, 588s, 540, 459, 360sh, 337, 320sh, 292, 238, 193, 159.
Source: Frost and Palmer (2011c).
Comments: No independent analytical data are provided for the sample used. For the Raman spectra of pascoite see also Frost et al. (2004e, 2005d).

Patrónite VS_4

Origin: Synthetic.
Experimental details: No data.
Raman shifts (cm^{-1}): ~990, ~690, ~400, ~280, ~190, ~140s, ~100.
Source: Kozlova et al. (2015).
Comments: The sample was characterized by powder X-ray diffraction data and qualitative electron microprobe analysis.

Pattersonite $PbFe_3(PO_4)_2(OH)_5 \cdot H_2O$

Origin: Grube Vereinigung, near Eisenbach, Taunus, Hesse, Germany (type locality).
Experimental details: Raman scattering measurements have been performed on an arbitrarily oriented sample using 633 nm laser radiation. The laser radiation power is not indicated. A 180°-scattering geometry was employed.
Raman shifts (cm^{-1}): 3547, 3526s, 3291 (broad), ~1610 (broad), 1084s, 1046s, 996, 973s, 927, 636, 571s, 523, 504, 460s, 407, ~377, 360, 325s, ~296, ~273, 259.
Source: Kolitsch et al. (2008).
Comments: The sample was characterized by powder X-ray diffraction data and electron microprobe analyses. The crystal structure is solved.

Pauflerite $VO(SO_4)$

Origin: Synthetic.
Experimental details: Methods of sample preparation are not described. Raman scattering measurements have been performed using 488 and 514.5 nm Ar$^+$ laser radiations. The laser radiation power is not indicated. A 90°-scattering geometry was employed.
Raman shifts (cm^{-1}): 1125, 1112, 1095, 1075, 1002, 925s, 654, 596w, 488w, 425w, 395, 361w, 335w, 311w, 285, 269, 253w, 231, 184w, 167w, 136w, 96.
Source: Boghosian et al. (1995).
Comments: The sample was characterized by single-crystal X-ray diffraction data. The crystal structure is solved.

Paulingite-K K,Ca,Na,Ba,$\square$)$_{10}$(Si,Al)$_{42}$O$_{84}\cdot$34H$_2$O

Origin: Vinařická Hora Hill, near Kladno, Czech Republic.
Experimental details: Raman scattering measurements have been performed on an arbitrarily oriented sample using 532 nm laser radiation. The laser radiation power is not indicated.
Raman shifts (cm^{-1}): 3554s (broad), 3433sh, 3265sh, 2945w, 2330w, 1640w, 1110, 993, 937, 774, 557sh, 496s, 474, 422w.
Source: Gatta et al. (2015b).
Comments: The sample was characterized by single-crystal X-ray diffraction data and electron microprobe analyses. The crystal structure is solved.

Paulmooreite Pb$_2$As$^{3+}_2$O$_5$

Origin: Långban, Filipstad district, Värmland province, Sweden (type locality).
Experimental details: Raman scattering measurements have been performed from the a and c faces an oriented single crystal the using 633 nm He-Ne laser radiation. The laser radiation power is not indicated.
Raman shifts (cm^{-1}): 751, 732, 658w, 562w, 501, 433, 410, 367, 344w, 312, 270, 210, 186s, 138s, 106w (for the spectrum collected from the $a$face of the crystal).
Source: Bahfenne et al. (2012).
Comments: For the Raman spectrum of paulmooreite see also Bahfenne (2011).

Pauloabibite NaNbO$_3$

Origin: Synthetic.
Experimental details: Methods of sample preparation are not described. Raman scattering measurements have been performed using 647.1 nm Kr$^+$ laser radiation. The nominal laser radiation power is not indicated.
Raman shifts (cm^{-1}): 790, 630sh, 590s, 505sh, 387, 295 (broad), 255w, 733s, 673w, 487, 476sh, 287, 257, 212, 202, 165.
Source: Baran et al. (1986).
Comments: No independent analytical data are provided for the sample used.

Paulscherrerite (UO$_2$)(OH)$_2$

Origin: No. 2 Workings, Radium Ridge, Northern Flinders Ranges, South Australia.
Experimental details: Raman scattering measurements have been performed on an arbitrarily oriented sample using 488 nm Ar$^+$ laser radiation. The laser radiation power is not indicated. The Raman shifts have been determined for the maxima of individual peaks obtained as a result of the spectral curve analysis.
Raman shifts (cm^{-1}): 3340, ~3290, 864sh, (850), 843s, 831s, 557w, 505w, 460w, 360w.
Source: Brugger et al. (2011).
Comments: The sample was characterized by powder X-ray diffraction data and electron microprobe analyses. For the Raman spectra of paulscherrerite see also Hoekstra and Siegel (1973), and Walenta and Theye (2012).

Pavlovskyite $Ca_8(SiO_4)_2(Si_3O_{10})$

Origin: Lakargi Mt., Upper Chegem caldera, North Caucasus (cotype locality).
Experimental details: Raman scattering measurements have been performed on an arbitrarily oriented sample using 514.5 nm Ar^+ laser radiation. The nominal laser radiation power at the sample was from 30 to 50 mW. A 180°-scattering geometry was employed.
Raman shifts (cm^{-1}): 1088w, 993s, 974, 925w, 908, 892, 858s, 847s, 821s, 669s, 569w, 546, 484w, 428w, 400, 358, 329, 305, 262w, 215w, 176w, 118s.
Source: Galuskin et al. (2012b).
Comments: The sample was characterized by single-crystal X-ray diffraction data and electron microprobe analyses. The crystal structure is solved.

Peatite-(Y) $Li_4Na_{12}(Y,Na,Ca,REE)_{12}(PO_4)_{12}(CO_3)_4(F,OH)_8$

Origin: Poudrette (Demix) quarry, Mont Saint-Hilaire, Rouville RCM (Rouville Co.), Montérégie, Québec, Canada (type locality).
Experimental details: Raman scattering measurements have been performed from a face of single crystal using 532 nm laser radiation. The laser radiation power is not indicated.
Raman shifts (cm^{-1}): 3659w, 1117sh, 1088s, 1042sh, 1000s, 623, 568, 494, 405sh, 370s, 260, 183, 139w.
Source: McDonald et al. (2013).
Comments: The sample was characterized by powder X-ray diffraction data and electron microprobe analyses.

Pecoraite $Ni_3Si_2O_5(OH)_4$

Origin: Nullagine region, Western Australia.
Experimental details: Raman scattering measurements have been performed on an arbitrarily oriented sample using a 633 nm He-Ne laser. The laser radiation power is not indicated. The Raman shifts have been determined for the maxima of individual peaks obtained as a result of the spectral curve analysis.
Raman shifts (cm^{-1}): (3638), (3613), 3586, 3535, 3460sh, 3271sh, 2934w, 2896w, 2854w, 1593, 1384, 1075s, 979s, 930w, 821, 761, 616, 451, 397, 235, 194, 151.
Source: Frost et al. (2008j).
Comments: No independent analytical data are provided for the sample used.

Pectolite $NaCa_2Si_3O_8(OH)$

Origin: No data.
Experimental details: Raman scattering measurements have been performed on arbitrarily oriented crystals using a 633 nm He-Ne laser. The laser radiation power is not indicated. The Raman shifts have been determined for the maxima of individual peaks obtained as a result of the spectral curve analysis.
Raman shifts (cm^{-1}): (2896), 2879w, 2851w, 2809sh, 1615, 1444w, 1413sh, 1388, (1047), 1026s, 998, 974s, (953), 936sh, 911, 706w, 687, 667sh, 653s, (642), (532w), 518sh, 508+500w (unresolved doublet?), 463w, (432), 415, 378, (370), 358, 325sh, 317, 276, 259sh, 225, (217), 203, 186, 152, 143, 134sh, 111.

Source: Frost et al. (2015n).
Comments: The sample was characterized by qualitative electron microprobe analysis. For the Raman spectra of pectolite see also Mitchell et al. (2015) and Origlieri et al. (2017).

Peisleyite $Na_3Al_{16}(PO_4)_{10}(SO_4)_2(OH)_{17} \cdot 20H_2O$

Origin: Tom's phosphate quarry, near the town of Kapunda, South Australia (type locality).
Experimental details: Raman scattering measurements have been performed on arbitrarily oriented crystals using a 633 nm He-Ne laser. The laser radiation power is not indicated. The Raman shifts have been determined for the maxima of individual peaks obtained as a result of the spectral curve analysis.
Raman shifts (cm^{-1}): 3505s, 3429, 3272, 2888, 1358w, 1289w, 1247w, 1144, 1023s, 989w, 634, 547, 412s, 314w, 276, 207, 188w, 159w.
Source: Frost et al. (2004j).
Comments: The sample was characterized TG/DTG data.

Péligotite $Na_6(UO_2)(SO_4)_4 \cdot 4H_2O$

Origin: Blue Lizard mine, San Juan Co., Utah, USA (type locality).
Experimental details: Methods of sample preparation are not described. Raman scattering measurements have been performed using 532 nm laser radiation. The laser radiation power is not indicated.
Raman shifts (cm^{-1}): ~3460, ~1210, ~1165, ~1080, ~1045, ~1000s, ~980, ~960, ~945, ~830s, ~655, ~620.
Source: Kampf et al. (2016g).
Comments: The sample was characterized by powder X-ray diffraction data and electron microprobe analyses. The crystal structure is solved.

Penkvilksite-2*O* $Na_2TiSi_4O_{11} \cdot 2H_2O$

Origin: Synthetic.
Experimental details: Methods of sample preparation are not described. Raman scattering measurements have been performed using 562 nm He-Ne laser radiation. The laser radiation power is not indicated.
Raman shifts (cm^{-1}): 3527w, 3471sh, 3410, 3354sh. In the spectral range below 3200 cm^{-1}, only a figure of the Raman spectrum of Penkvilksite-2*O* is given by Cadoni and Ferraris (2008).
Source: Cadoni and Ferraris (2008).
Comments: The sample was characterized by single-crystal X-ray diffraction data. The crystal structure is solved. For the Raman spectrum of penkvilksite see also Frost and Xi (2013b).

Penroseite $(Ni,Co,Cu)Se_2$

Origin: Synthetic.
Experimental details: Methods of sample preparation are not described. Raman scattering measurements have been performed by means of back scattering technique. Characteristics of the laser radiation are not indicated.
Raman shifts (cm^{-1}): 216.5.

Source: Yang et al. (2001).
Comments: The sample was characterized by powder X-ray diffraction data.

Pentagonite $CaV^{4+}OSi_4O_{10} \cdot 4H_2O$

Origin: Pune (Poonah) district, Maharashtra, India.
Experimental details: Raman scattering measurements have been performed on arbitrarily oriented crystals using a 633 nm He-Ne laser. The laser radiation power is not indicated. The Raman shifts have been determined for the maxima of individual peaks obtained as a result of the spectral curve analysis.
Raman shifts (cm^{-1}): 3640sh, (3580), 3532, (3499), 1634w, 1612w, 1191w, 1153w, 1089, 1047w, 971s, 765w, 651, 559, 524, 494w, 479w, 398, 344sh, 324, 305, 288, 261w, 230, 206, 158sh, 140, 123sh.
Source: Frost and Xi (2012h).
Comments: No independent analytical data are provided for the sample used.

Pentahydrite $Mg(SO_4) \cdot 5H_2O$

Origin: Synthetic.
Experimental details: Raman scattering measurements have been performed on randomly oriented fine grains using 532 nm Nd-YAG laser radiation. The nominal laser radiation power was 15 mW.
Raman shifts (cm^{-1}): 3553sh, 3494, 3391, 3343, 3289, 1650w, 1159, 1106w, 1005s, 602, 447w, 371, 241, 206, 165, 119.
Source: Wang et al. (2006a).
Comments: The sample was characterized by powder X-ray diffraction data. For the Raman spectra of pentahydrite see also Ling et al. (2009) and Frezzotti et al. (2012).

Pentahydroborite $CaB_2O(OH)_6 \cdot 2H_2O$

Origin: Fuka mine, Okayama prefecture, Japan.
Experimental details: No data.
Raman shifts (cm^{-1}): 3595w, 3499, 3445s, 3399, 3371sh, 3324s, 3196s, 3041s (broad), 2938w, 1610w, 1446w (broad), 1305, 1249, 1223, 1157, 1032, 981, 957, 918, 842, 781w, 725s, 678w, 611s, 584, 562, 508, 492, 472, 416sh, 401, 355, 330, 313sh, 272, 251, 243sh, 214.
Source: Bermanec et al. (2010).
Comments: The sample was characterized by powder X-ray diffraction data and thermal analysis.

Pentlandite $(Ni,Fe)_9S_8$

Origin: Kambalda, Western Australia, Australia.
Experimental details: Methods of sample preparation are not described. Raman scattering measurements have been performed on an arbitrarily oriented sample using 514.5 nm Ar$^+$ laser radiation. The laser radiation power at the sample was between 1 and 10 mW. A 180°-scattering geometry was employed.
Raman shifts (cm^{-1}): 370.
Source: Mernagh and Trudu (1993).
Comments: No independent analytical data are provided for the sample used.

4 Raman Spectra of Minerals

Peretaite $CaSb^{3+}_4O_4(SO_4)_2(OH)_2 \cdot 2H_2O$

Origin: Pereta mine, Scansano, Grosseto province, Tuscany, Italy (type locality).
Experimental details: Raman scattering measurements have been performed on arbitrarily oriented crystals using a 633 nm He-Ne laser. The laser radiation power is not indicated. The Raman shifts have been determined for the maxima of individual peaks obtained as a result of the spectral curve analysis.
Raman shifts (cm^{-1}): 3334, 1152, 1142, 1115w, 1092w, 1060, 980s, 710wm, 650w, 610sh, 595s, 589sh, 482w, 434, 417, 373w, 337, 229s, (219), 215sh, 196, 175w, 156, 137sh.
Source: Frost et al. (2010f).
Comments: No independent analytical data are provided for the sample used.

Perhamite $Ca_3Al_{77}Si_3P_4O_{23.5}(OH)_{14.1} \cdot 8H_2O$

Origin: Dunton Gem Quarry, Newry, Oxford Co., Maine, USA (type locality).
Experimental details: Raman scattering measurements have been performed on arbitrarily oriented crystals using a 633 nm He-Ne laser. The laser radiation power is not indicated. The Raman shifts have been determined for the maxima of individual peaks obtained as a result of the spectral curve analysis.
Raman shifts (cm^{-1}): 1884, 1354, 1245, 1153, 1131, 1096, 1059w, 1032, 1005s, 996s, 929w, 708, 636, 615, 586w, 554, 520s, 506s, 468, 442w, 385, 375, 363w, 334, 276, 267, 207, 191, 170w, 148w, 129w.
Source: Frost et al. (2007l).
Comments: No independent analytical data are provided for the sample used.

Perite $PbBiO_2Cl$

Origin: Homeward Bound mine, Mannahill, South Australia.
Experimental details: Raman scattering measurements have been performed on arbitrarily oriented crystals using a 633 nm He-Ne laser. The laser radiation power is not indicated. The Raman shifts have been determined for the maxima of individual peaks obtained as a result of the spectral curve analysis.
Raman shifts (cm^{-1}): 506, 484, 389, 367, 295, 253, 180, 157.
Source: Frost and Williams (2004).
Comments: The sample was characterized by X-ray diffraction and chemical analysis using ICP-AES techniques, but no analytical data are provided in the cited paper.

Permingeatite Cu_3SbSe_4

Origin: Příbram, Central Bohemia region, Czech Republic.
Experimental details: Raman scattering measurements have been performed on grains mounted in a polished section in backscattering geometry using 514.5 nm Ar^+ laser radiation. The laser radiation power at the sample was ~1 mW.
Raman shifts (cm^{-1}): 374w, 366w, 357w, 318w, 276w, 251w, 137, 229s, 214, 205, 193, 184s, 176, 167, 159w, 140w, 127, 78, 75, 63, 59, 51w, 45, 40, 26s, 21s, 17, 14.
Source: Škácha et al. (2014a).

Comments: The sample was characterized by powder X-ray diffraction data and electron microprobe analyses.

Perovskite $CaTiO_3$

Origin: Rocca Castellaccio, Ciappanico, Malenco Valley, Sondrio province, Lombardy, Italy.
Experimental details: Methods of sample preparation are not described. Raman scattering measurements have been performed on an arbitrarily oriented sample using 632.8 nm He-Ne or 488 nm Ar^+ laser radiation. The laser radiation power is not indicated.
Raman shifts (cm^{-1}): 638 (broad), 471, 248s, 227s, 182s.
Source: Andò and Garzanti (2014).
Comments: No independent analytical data are provided for the sample used. For the Raman spectra of perovskite see also Ma et al. (2013), Zajzon et al. (2013), and Martins et al. (2014).

Pertsevite-(OH) $Mg_2(BO_3)(OH)$

Origin: Snezhnoe boron deposit, Dogdo River basin, Saha Republic (Yakutia), Russia (type locality).
Experimental details: Raman scattering measurements have been performed on a polished section using 514.5 nm Ar^+ laser radiation. The laser radiation power at the sample was 20 mW. A $0°$-scattering geometry was employed.
Raman shifts (cm^{-1}): 3560 (broad), 919s, 862s, 738, 681, 602, 545.
Source: Galuskina et al. (2008).
Comments: The sample was characterized by single-crystal X-ray diffraction data and electron microprobe analyses. The crystal structure is solved.

Petalite $LiAlSi_4O_{10}$

Origin: Laghman province, Nuristan, Afghanistan.
Experimental details: Unpolarized Raman scattering measurements have been performed on an arbitrarily oriented sample using 514.5 nm Ar^+ laser radiation. The laser radiation power is not indicated.
Raman shifts (cm^{-1}): 1138, 1060w, 790w, 490s, 467w, 383s, 357s, 280, 143, 113, 85, 60w.
Source: Kaminskii et al. (2015).
Comments: The sample was characterized by X-ray diffraction data.

Petersite-(Ce) $Cu_6Ce(PO_4)_3(OH)_6 \cdot 3H_2O$

Origin: Yavapai County, Arizona, USA (type locality).
Experimental details: Raman scattering measurements have been performed on an arbitrarily oriented crystal using 532 nm solid-state laser radiation. The nominal laser radiation power was 60 mW.
Raman shifts (cm^{-1}): 3499, 3411, 3292, 3072, 2934, 2873, 2862, 1095, 1084, 1043s, 945, 580s, 528, 472s, 393.
Source: Morrison et al. (2016).
Comments: The sample was characterized by powder X-ray diffraction data and electron microprobe analyses. The crystal structure is solved.

Petitjeanite $Bi_3O(PO_4)_2(OH)$

Origin: Cetoraz, near Pacov, Czech Republic.
Experimental details: Methods of sample preparation are not described. Raman scattering measurements have been performed using 633 nm laser radiation. The laser radiation power is not indicated.
Raman shifts (cm^{-1}): 1038, 952, 584w, 548w, 405, 298, 260, 222s, 177s, 102s.
Source: Losertová et al. (2014).
Comments: The sample was characterized by electron microprobe analyses.

Petrukite $(Cu,Fe,Zn,Ag)_3(Sn,In)S_4$

Origin: Synthetic.
Experimental details: Raman scattering measurements have been performed on a powdery sample using 644 nm laser radiation. The laser radiation power at the sample was 0.02 mW.
Raman shifts (cm^{-1}): 990w, 940w, 637, 597sh, 346, 337, 317s, 292, 280, 261sh, 150, 130w, 99, 93.
Source: Dzhagan et al. (2014).
Comments: The sample was characterized by powder X-ray diffraction data. For the Raman spectrum of petrukite see also Fernandes et al. (2010).

Petterdite $PbCr_2(CO_3)_2(OH)_4 \cdot H_2O$

Origin: Red Lead mine, Zeehan–Dundas mining field, Tasmania, Australia (type locality).
Experimental details: Raman scattering measurements have been performed on a powder sample using 514.5 nm Ar$^+$ laser radiation. The laser radiation output power was 0.3 mW.
Raman shifts (cm^{-1}): 3540, 3470, 3282w, 2948w, 2924, 2854w, 2072w, 1641w, 1516s, 1493sh, 1394sh, 1343s, 1122w, 1089w, 956w, 881sh, 852w, 830w, 812w, 744w, 650sh, 626w, 592w, 541, 504s, 433.
Source: Birch et al. (2000).
Comments: The sample was characterized by powder X-ray diffraction data, as well as by electron microprobe and HCN analyses.

Petzite Ag_3AuTe_2

Origin: Coranda-Hondol open pit, Certej Au-Ag deposit, Romania.
Experimental details: Methods of sample preparation are not described. Raman scattering measurements have been performed using 632.8 He-Ne laser radiation. The laser radiation power is not indicated.
Raman shifts (cm^{-1}): ~320, 174, 163s.
Source: Apopei et al. (2014b).
Comments: The sample was characterized by electron microprobe analyses.

Pezzottaite $CsLiBe_2Al_2Si_6O_{18}$

Origin: Piława Górna, Lower Silesia, SW Poland.
Experimental details: Raman scattering measurements have been performed on a powder sample using 532 nm laser radiation. The laser radiation power is not indicated.

Raman shifts (cm^{-1}): 1106s, 1067s, 1024, 1007, 689s, 542, 461, 406s, 329, 250, 118s.
Source: Pieczka et al. (2016b).
Comments: The sample was characterized by powder X-ray diffraction data and electron microprobe analyses. Pezzottaite differs from beryl by intensive Raman bands at 118 and 1106 cm^{-1}. For the Raman spectrum of pezzottaite see also Lambruschi et al. (2014).

Pharmacolite Ca(AsO$_3$OH)·2H$_2$O

Origin: Jáchymov, Krušné Hory (Ore Mts.), Czech Republic.
Experimental details: Raman scattering measurements have been performed on arbitrarily oriented crystals using a 633 nm He-Ne laser. The laser radiation power is not indicated. The Raman shifts have been determined for the maxima of individual peaks obtained as a result of the spectral curve analysis.
Raman shifts (cm^{-1}): 3525, 3435, 3266sh, (3239), 3186, 1652w, 1179w, 885w, 865s, (858), 844, 706s, 676sh, 545w, 448, 397, 371, 357, 337sh, (309), 305, 286s, 195, (187), 176sh, 155, 133sh, 124.
Source: Frost et al. (2010b).
Comments: The sample was characterized by powder X-ray diffraction data and electron microprobe analyses.

Pharmacosiderite KFe$^{3+}_4$(AsO$_4$)$_3$(OH)$_4$·6-7H$_2$O

Origin: Mokrsko-west Au deposit, Příbram district, Central Bohemia region, Czech Republic.
Experimental details: Methods of sample preparation are not described. Raman scattering measurements have been performed using 532.2 nm diode laser radiation. The laser radiation power at the sample was 0.5 mW.
Raman shifts (cm^{-1}): 886, 830sh, 803s, 475s, 383, 336, 290w, 279, 244w, 179, 137w.
Source: Filippi et al. (2007).
Comments: The sample was characterized by powder X-ray diffraction data and electron microprobe analyses. For the Raman spectra of pharmacosiderite see also Frost and Kloprogge (2003) and Bossy et al. (2010).

Pharmazincite KZn(AsO$_4$)

Origin: Arsenatnaya fumarole, Tolbachik volcano, Kamchatka, Russia (type locality).
Experimental details: Raman scattering measurements have been performed on an arbitrarily oriented crystal using 532 nm laser radiation. The laser radiation output power was 30 mW.
Raman shifts (cm^{-1}): 853s, 513sh, 453w, 430w, 406w, 343, 323, 291.
Source: Pekov et al. (2016a).
Comments: The sample was characterized by powder X-ray diffraction data and electron microprobe analyses. The crystal structure is solved.

Phenakite Be$_2$(SiO$_4$)

Origin: San Miguel de Piracicaba, Minas Gerais, Brazil.
Experimental details: Methods of sample preparation are not described. Raman scattering measurements have been performed using 785 nm diode laser radiation. The maximum output powder of 300 mW was filtered to diminish the power at the sample.

Raman shifts (cm^{-1}): 1021w, 952, 938, 918, 879s, 786, 775w, 761w, 728w, 702w, 686w, 666w, 616w, 601w, 527, 463w, 446, 384, 347w, 283w, 233w, 223.
Source: Jehlička et al. (2012).
Comments: No independent analytical data are provided for the sample used. For the Raman spectra of phenakite see also Hofmeister et al. (1987), Annen and Davis (1993), Pilati et al. (1998), and Jehlička and Vandenabeele (2015).

Philipsbornite $PbAl_3(AsO_4)(AsO_3OH)(OH)_6$

Origin: Red Lead Mine, Dundas mineral field, Zeehan district, West Coast municipality, Tasmania, Australia (type locality).
Experimental details: Raman scattering measurements have been performed on arbitrarily oriented crystals using a 633 nm He-Ne laser. The laser radiation power is not indicated. The Raman shifts have been determined for the maxima of individual peaks obtained as a result of the spectral curve analysis.
Raman shifts (cm^{-1}): (865), (857), 846+831s (unresolved doublet?), (820), 399, 376, 357s, 347, 336, 325, 189w, 180w, 134, 115w.
Source: Frost et al. (2013s).
Comments: No independent analytical data are provided for the sample used.

Philipsburgite $(Cu,Zn)_6(AsO_4,PO_4)_2(OH)_6 \cdot H_2O$

Origin: Miedzianka (former Kupferberg), Sudety Mts., SW Poland.
Experimental details: Raman scattering measurements have been performed on an arbitrarily oriented sample using 532 nm Nd-YAG laser radiation. The nominal laser radiation power was 40 mW. The Raman shifts have been determined for the maxima of individual peaks obtained as a result of the spectral curve analysis.
Raman shifts (cm^{-1}): 3550w, 3489, 3429w, 3215sh, 1060w, (994w), 970, (946w), 865s, 847sh, 809s, 791sh, 667w, 564sh, (491w), 474, 396, 368sh, 347sh, 317, 307sh, 249w, 218w.
Source: Ciesielczuk et al. (2016).
Comments: The sample was characterized by powder X-ray diffraction data and electron microprobe analyses.

Phillipsite-K $K_6(Si_{10}Al_6)O_{32} \cdot 12H_2O$

Origin: Capo di Bove, Rome province, Latium, Italy (type locality).
Experimental details: Methods of sample preparation are not described. Raman scattering measurements have been performed using Nd-YAG laser radiation. The laser radiation power at the sample was 300 mW.
Raman shifts (cm^{-1}): 815, 743, 472s, 424sh, 187w.
Source: Mozgawa (2001).
Comments: The sample was characterized by powder X-ray diffraction data.

Philolithite $Pb_{12}O_6Mn_7(SO_4)(CO_3)_4Cl_4(OH)_{12}$

Origin: Långban, Värmland, Sweden (type locality).
Experimental details: Methods of sample preparation are not described. Raman scattering measurements have been performed using 514.5 nm Ar$^+$ laser radiation. The nominal laser radiation power was 20 mW.

Raman shifts (cm^{-1}): A group of bands around 3400 cm^{-1}, 1122, 1111, 1073, 1011, 420.
Source: Kampf et al. (1998).
Comments: The sample was characterized by powder X-ray diffraction data and electron microprobe analyses. The crystal structure is solved.

Phlogopite KMg$_3$(AlSi$_3$O$_{10}$)(OH)$_2$

Origin: Arendal Fe Mines, Aust-Agder, Norway.
Experimental details: Micro-Raman scattering measurements have been performed on a single crystal using a 514.5 nm Ar$^+$ laser. Sample orientation is not indicated.
Raman shifts (cm^{-1}): (1024), (1000), 675s, 650, 273, 190s, 97.
Source: Tlili et al. (1989).
Comments: The sample was characterized by powder X-ray diffraction data and electron microprobe analyses. For the Raman spectra of phlogopite see also McKeown et al. (1999), Tlili and Smith (2007), and Frezzotti et al. (2012).

Phoenicochroite Pb$_2$(CrO$_4$)O

Origin: Synthetic.
Experimental details: Methods of sample preparation are not described. Raman scattering measurements have been performed using 647.1 nm Kr$^+$ laser radiation. The laser radiation power is not indicated.
Raman shifts (cm^{-1}): 849, 838, 826s, 382, 356w, 343, 333w, 324.
Source: Roncaglia et al. (1985).
Comments: The sample was characterized by powder X-ray diffraction data. For the Raman spectrum of phoenicochroite see also Frost (2004c).

Phosgenite Pb$_2$(CO$_3$)Cl$_2$

Origin: No data.
Experimental details: Raman scattering measurements have been performed on an arbitrarily oriented sample using 632.8 nm He-Ne or 514.5 nm Ar$^+$ laser radiation. The nominal laser radiation power was $\leq$30 mW.
Raman shifts (cm^{-1}): 1063s, 668, 281, 252, 181w, 154, 129, 87s, 81, 53, 47s.
Source: Bouchard and Smith (2003).
Comments: The sample was characterized by powder X-ray diffraction data. For the Raman spectra of phosgenite see also Frost et al. (2003j) and Frost and Williams (2004).

Phosphammite (NH$_4$)$_2$(PO$_3$OH)

Origin: Synthetic.
Experimental details: Raman scattering measurements have been performed at 260 K on crystals in a sealed glass cell using 514.5 nm Ar$^+$ laser radiation. The nominal laser radiation power was between 0.1 and 0.2 mW.
Raman shifts (cm^{-1}): 3200sh, 3048s, 2805s, 2203, 1948, 1743, 1720, 1696, 1441w, 1404, 1094w, 1062w, 1052w, 997w, 949s, 900, 856w, 565, 557, 522, 510w, 400, 380.
Source: Hadrich et al. (2000).
Comments: No independent analytical data are provided for the sample used.

Phosphohedyphane $Ca_2Pb_3(PO_4)_3Cl$

Origin: Root (Bonanza Hill) mine, Goodsprings district, Spring Mts., Clark Co., Nevada, USA.

Experimental details: Raman scattering measurements have been performed on arbitrarily oriented crystals using a 633 nm He-Ne laser. The laser radiation power is not indicated. The Raman shifts have been determined for the maxima of individual peaks obtained as a result of the spectral curve analysis.

Raman shifts (cm^{-1}): (3421), 3395, 3344, 1226w, 1188w, (1084), 1073, (1030), (980), 975s, (966), 933s, 835, 812w, 595, 577sh, 557, 437s, (421), 400, 208, 148, 113sh, 106.

Source: Frost et al. (2014w).

Comments: No independent analytical data are provided for the sample used.

Phosphophyllite $Zn_2Fe^{2+}(PO_4)_2 \cdot 4H_2O$

Origin: Hagendorf South pegmatite, Bavaria, Germany.

Experimental details: Raman scattering measurements have been performed on arbitrarily oriented crystals using a 633 nm He-Ne laser. The laser radiation power is not indicated. The Raman shifts have been determined for the maxima of individual peaks obtained as a result of the spectral curve analysis.

Raman shifts (cm^{-1}): 3567s, (3561), 3362sh, 3258, 3146, (3034), 1603, (1571), 1135, 1073, 995s, 939, 744w, 633, 592, 571sh, 505, 415, 322, (297), 269, 199, 181sh, 142, 130sh, 119.

Source: Scholz et al. (2013a).

Comments: The sample was characterized by qualitative electron microprobe analyses. No independent quantitative analytical data are provided.

Phosphosiderite $Fe^{3+}(PO_4) \cdot 2H_2O$

Origin: Synthetic.

Experimental details: Raman scattering measurements have been performed on a powdered sample using 514.5 nm Ar^+ laser radiation. The laser radiation power is not indicated.

Raman shifts (cm^{-1}): 1032sh, 1000sh, 988s, 570w, 485sh, 447, 330sh, 302, 258, 200, 126w, 70.

Source: Zaghib and Julien (2005).

Comments: The sample was characterized by powder X-ray diffraction data. For the Raman spectrum of phosphosiderite see also Frost et al. (2004l).

Phosphuranylite $KCa(H_3O)_3(UO_2)_7(PO_4)_4O_4 \cdot 8H_2O$

Origin: Bedford, Westchester Co., New York, USA.

Experimental details: Raman scattering measurements have been performed on arbitrarily oriented crystals using 647.1 nm Kr^+ laser radiation. The laser radiation power is not indicated.

Raman shifts (cm^{-1}): 1034, 981, 827s, 398, 264, 237.

Source: Faulques et al. (2015a, b).

Comments: No independent analytical data are provided for the sample used. For the Raman spectra of phosphuranylite see also Frost et al. (2008a) and Driscoll et al. (2014).

Phurcalite $Ca_2(UO_2)_3O_2(PO_4)_2 \cdot 7H_2O$

Origin: Posey mine, Red Canyon, White Canyon district, San Juan Co., Utah, USA.
Experimental details: Raman scattering measurements have been performed on arbitrarily oriented crystals using a 633 nm He-Ne laser. The laser radiation power is not indicated. The Raman shifts have been determined for the maxima of individual peaks obtained as a result of the spectral curve analysis.
Raman shifts (cm^{-1}): 3613w, 3538w, 3421, (3238), 1769w, 1615w, 1155w, 1118sh, 1108, 1059, (1009), 1004s, 995sh, 969, 960, 950sh, 864w, (819), 810, (800), 546, (434), 431, 408, 391sh.
Source: Čejka et al. (2014b).
Comments: The sample was characterized by powder X-ray diffraction data and qualitative electron microprobe analysis.

Pickeringite $MgAl_2(SO_4)_4 \cdot 22H_2O$

Origin: San Bernadino Co., California, USA.
Experimental details: Raman scattering measurements have been performed on arbitrarily oriented crystals using a 633 nm He-Ne laser. The laser radiation power at the sample was 1 mW. The Raman shifts have been determined for the maxima of individual peaks obtained as a result of the spectral curve analysis.
Raman shifts (cm^{-1}): 3449, 3279, 3082sh, 1145w, 1114w, 1071w, (990), 986s, 975, 621, 530, 468, 424, 344w, 315, 221.
Source: Locke et al. (2007).
Comments: The sample was characterized by powder X-ray diffraction data and electron microprobe analyses.

Picromerite $K_2Mg(SO_4)_2 \cdot 6H_2O$

Origin: Synthetic.
Experimental details: Raman scattering measurements have been performed on an arbitrarily oriented crystal using 253.65 nm radiation of mercury. The incident light was normal to the (110) face and the scattered light was taken parallel to the (110) face and roughly perpendicular to the (001) face. The radiation power is not indicated.
Raman shifts (cm^{-1}): 3344, 3308s, 3250, 3150, 1234w, 1156, 1129s, 1111, 1082s, 990s, 835w, 792w, 767w, 632, 614, 462s, 448s, 372, 320w, 305w, 269w, 225, 177, 136, 115, 93w, 74, 62, 46.
Source: Ananthanarayanan (1961).
Comments: The sample was characterized by morphological features. For the Raman spectrum of picromerite see also Bouchard and Smith (2003).

Picropharmacolite $Ca_4Mg(AsO_3OH)_2(AsO_4)_2 \cdot 11H_2O$

Origin: Synthetic.
Experimental details: Raman scattering measurements have been performed on arbitrarily oriented crystals using a 633 nm He-Ne laser. The laser radiation power was below 0.1 mW. The Raman shifts have been determined for the maxima of individual peaks obtained as a result of the spectral curve analysis.
Raman shifts (cm^{-1}): 3448, 3212, (2922), 980, 866s, 750, 530, 460, 397, 325, 230.

Source: Frost and Kloprogge (2003).
Comments: No independent analytical data are provided for the sample used.

Pieczkaite $Mn_5(PO_4)_3Cl$

Origin: Cross Lake pegmatite field, Manitoba, Canada (type locality).
Experimental details: Methods of sample preparation are not described. Raman scattering measurements have been performed using 532 nm laser radiation. The nominal laser radiation power was between 5 and 12.5 mW.
Raman shifts (cm^{-1}): 1095s, 1000sh, 960sh, 795w, 560s, 480sh.
Source: Tait et al. (2015).
Comments: The sample was characterized by powder X-ray diffraction data and electron microprobe analyses. The crystal structure is solved.

Piemontite $Ca_2(Al_2Mn^{3+})[Si_2O_7][SiO_4]O(OH)$

Origin: Prabornaz (Praborna) mine, Saint-Marcel, Aosta Valley, Italy.
Experimental details: Methods of sample preparation are not described. Raman scattering measurements have been performed on an arbitrarily oriented sample using 632.8 nm He-Ne or 488 nm Ar$^+$ laser radiation. The laser radiation power is not indicated.
Raman shifts (cm^{-1}): 916s, 886s, 601, 565, 453s, 350s, 244, 172.
Source: Andò and Garzanti (2014).
Comments: No independent analytical data are provided for the sample used.

Pigeonite $(Mg,Fe,Ca)_2Si_2O_6$

Origin: Synthetic.
Experimental details: Raman scattering measurements have been performed on an arbitrarily oriented crystal in a polished thin section using 632.8 nm He-Ne laser radiation. The laser radiation power at the sample was ≤ 1 mW.
Raman shifts (cm^{-1}): ~1035, ~1010, 685s, ~665, ~415, ~400, ~385w, 341s, ~300w, ~240.
Source: Tribaudino et al. (2011).
Comments: The Raman shifts are given for a Fe-free sample with the diopside to enstatite ratio of 15:85. In the cited paper Raman spectra of natural pigeonite samples with different Fe:Mg ratios are given.

Pilsenite Bi_4Te_3

Origin: Synthetic.
Experimental details: Raman scattering measurements have been performed on a thin film grown on Si (111) substrate using 532 nm Nd-YAG laser radiation. The material grew along its c-axis. The nominal laser radiation power was 2 mW. A 180°-scattering geometry was employed.
Raman shifts (cm^{-1}): 204, 183w, 115, 101sh, 88s, 57, 37.
Source: Xu et al. (2015a).
Comments: The sample was characterized by X-ray diffraction data.

Pilsenite Bi_4Te_3

Origin: Panarechensk volcanic-tectonic formation, Kola Peninsula, Russia.
Experimental details: Raman scattering measurements have been performed on an arbitrarily oriented sample using 514.5 nm Ar^+ laser radiation. The nominal laser radiation power was 50 mW.
Raman shifts (cm^{-1}): 204w, 171–174, 131sh–133, 97–106s, 81.
Source: Voloshin et al. (2015a).
Comments: The samples used were characterized by electron microprobe analyses.

"Pimelite" $Ni_3Si_4O_{10}(OH)_2 \cdot 4H_2O$

Origin: Falcondo mine, Bonao, La Vega province, Dominican Republic.
Experimental details: Raman scattering measurements have been performed on an arbitrarily oriented sample using 1064 nm laser radiation. The laser radiation power is not indicated.
Raman shifts (cm^{-1}): 822, 735w, 675s, 640, 386, 362, 188s.
Source: Villanova-de-Benavent et al. (2012).
Comments: No independent analytical data are provided for the sample used.

Pinakiolite $(Mg,Mn)_2(Mn^{3+},Sb^{5+})O_2(BO_3)$

Origin: Långban deposit, Bergslagen ore region, Filipstad district, Värmland, Sweden (type locality).
Experimental details: Raman scattering measurements have been performed on arbitrarily oriented crystals using a 633 nm He-Ne laser. The laser radiation power is not indicated. The Raman shifts have been determined for the maxima of individual peaks obtained as a result of the spectral curve analysis.
Raman shifts (cm^{-1}): See comment below.
Source: Frost (2011b).
Comments: All Raman spectra of presumed orthoborates (azoproite, fredrikssonite, pinakiolite and takéuchiite) given in the cited paper are almost identical and correspond to calcite. In particular, for "pinakiolite" the following Raman shifts have been determined: 1748w, 1435w, 1086s, 712, 283, 154. The correct Raman shifts of pinakiolite are (RRUFF (2007), sample R050636; cm^{-1}): 686s, 644s, 550sh, 445, 391w, 352w, 322, 280, 200w, 153.

Pinnoite $MgB_2O(OH)_6$

Origin: Inder boron deposit, Atyrau region, Kazakhstan.
Experimental details: Raman scattering measurements have been performed on arbitrarily oriented crystals using a 633 nm He-Ne laser. The laser radiation power is not indicated. The Raman shifts have been determined for the maxima of individual peaks obtained as a result of the spectral curve analysis.
Raman shifts (cm^{-1}): 3579s, (3569), 3554s, (3415w), 3399s, 3290, 3179, (3085w), 1320, 1299, 1260, 1186, 1157, 1140sh, 1049, 1020, 945, 900s, 875, (799w), 745, 630, 605, (594w), 578, 538sh, 524, 508, 491w, 480, 468sh, 403sh, 388, 375, 357, (338), 288, 273, 260w, 230, 209sh, 193, 180, 172w, 143, 126.
Source: Frost and Xi (2014).
Comments: No independent analytical data are provided for the sample used.

Pirssonite $Na_2Ca(CO_3)_2 \cdot 2H_2O$

Origin: Green River formation, Sweetwater Co., Wyoming, USA.
Experimental details: Raman scattering measurements have been performed on arbitrarily oriented crystals using a 633 nm He-Ne laser. The laser radiation power is not indicated. The Raman shifts have been determined for the maxima of individual peaks obtained as a result of the spectral curve analysis.
Raman shifts (cm^{-1}): 3502, 3065, 1070s, 717, 659, 298, 253, (199), 126.
Source: Frost and Dickfos (2007b).
Comments: No independent analytical data are provided for the sample used.

Pitticite $[Fe,AsO_4,SO_4,H_2O]$ (?)

Origin: Synthetic.
Experimental details: Raman scattering measurements have been performed on arbitrarily oriented crystals using a 633 nm He-Ne laser. The laser radiation power is not indicated. The Raman shifts have been determined for the maxima of individual peaks obtained as a result of the spectral curve analysis.
Raman shifts (cm^{-1}): 3490, 3327, 3186, (3060), (2723), (1182w), 1096, 916sh, 845+837s (unresolved doublet?), 617w, (504), 457, 428, 401sh, 372w, 349sh, 335, 322, 309, 297sh, 278, 260, 236, 221sh, 207w, 194w, 181w, 166w, 131sh, 118+111 (unresolved doublet?).
Source: Frost et al. (2012l).
Comments: No independent analytical data are provided for the sample used.

Plancheite $Cu_8(Si_4O_{11})_2(OH)_4 \cdot H_2O$

Origin: Tsumeb mine, Tsumeb, Namibia.
Experimental details: Raman scattering measurements have been performed on arbitrarily oriented crystals using a 633 nm He-Ne laser. The laser radiation power is not indicated. The Raman shifts have been determined for the maxima of individual peaks obtained as a result of the spectral curve analysis.
Raman shifts (cm^{-1}): See comment below.
Source: Frost and Xi (2012d).
Comments: No independent analytical data are provided for the sample used. The Raman and IR spectra of presumed plancheite given in the cited paper are wrong and correspond to a carbonate. The correct Raman shifts of plancheite are (RRUFF (2007), sample R070233; cm^{-1}): 780, 674s, 553, 499, 441s, 400, 336, 328sh, 316, 262, (248), (239).

Plášilite $Na(UO_2)(SO_4)(OH) \cdot 2H_2O$

Origin: Blue Lizard mine, San Juan Co., Utah, USA (type locality).
Experimental details: Raman scattering measurements have been performed on an arbitrarily oriented crystal using 532 nm diode-pumped solid-state laser radiation. The nominal laser radiation power was 3 mW.
Raman shifts (cm^{-1}): 3600, 3520, 3385w, 1180, 1069w, 1035, 997, 986.5, 905w, 838s, 824, 645w, 603, 480, 445, 349, 243, 210, 186, 170.
Source: Kampf et al. (2015a).

Comments: The sample was characterized by powder X-ray diffraction data and electron microprobe analyses. The crystal structure is solved.

Platarsite PtAsS

Origin: Munni complex, west Pilbara block, Western Australia.
Experimental details: Raman scattering measurements have been performed on an arbitrarily oriented grain in a polished section using 514.5 nm Ar⁺ laser radiation. The nominal laser radiation power was between 1 and 10 mW. A 180°-scattering geometry was employed.
Raman shifts (cm^{-1}): 427, 350s, 291, 283, 253, 214.
Source: Mernagh and Hoatson (1995).
Comments: The sample was characterized by electron microprobe analyses.

Platinum Pt

Origin: Synthetic.
Experimental details: Methods of sample preparation are not described. Raman scattering measurements have been performed using 514.5 nm Ar⁺ laser radiation. The nominal laser radiation power was 100 mW.
Raman shifts (cm^{-1}): ~220, ~185, ~170, ~155.
Source: Vermaak (2005).

Plattnerite PbO_2

Origin: Synthetic.
Experimental details: Raman scattering measurements have been performed on a pressed disc using 632.8 nm laser radiation. The laser radiation power at the sample was 0.27 mW.
Raman shifts (cm^{-1}): 424w, 515s, 653w.
Source: Burgio et al. (2001).
Comments: Plattnerite slowly decomposes under the laser beam. For the Raman spectrum of plattnerite see also Inguanta et al. (2008).

Plavnoite $K_{0.8}Mn_{0.6}[(UO_2)_2O_2(SO_4)] \cdot 3.5H_2O$

Origin: Jáchymov, Krušné Hory (Ore Mts.), Bohemia, Czech Republic (type locality).
Experimental details: Methods of sample preparation are not described. Raman scattering measurements have been performed using 532 nm diode laser radiation. The laser radiation power at the sample was about 2 mW.
Raman shifts (cm^{-1}): 3533, 3385w, 1630w, 1106, 1027, 817s, 502, 475, 435, 377w, 348w, 292, 267, 229, 164s, 129w, 106w.
Source: Plášil et al. (2017).
Comments: The sample was characterized by powder X-ray diffraction data and electron microprobe analyses. The crystal structure is solved.

Plimerite $ZnFe^{3+}_4(PO_4)_3(OH)_5$

Origin: Huber open pit, Huber stock, Krásno, Horní Slavkov, Karlovy Vary region, Bohemia, Czech Republic (type locality).
Experimental details: Methods of sample preparation are not described. Raman scattering measurements have been performed using 532.2 nm laser radiation. The nominal laser radiation power was 5 mW.
Source: Sejkora et al. (2011).
Raman shifts (cm^{-1}): ~3590, 3228 (broad), 1600w, (1164), 1118s, (1098w), (1079), 1051s, 1014s, 993, 964, 930sh, 774w.

Plombièrite $Ca_4Si_6O_{16}(OH)_2(H_2O)_2 \cdot (Ca \cdot 5H_2O)$

Origin: Crestmore quarry, north of Riverside, Riverside Co., California, USA.
Experimental details: Methods of sample preparation are not described. Raman scattering measurements have been performed in nearly backscattered geometry using 632.8 nm He-Ne laser radiation. The laser radiation power at the sample was 1.5 mW.
Raman shifts (cm^{-1}): 1057sh, 1025, 996, 680sh, 664s.
Source: Biagioni et al. (2013b).
Comments: The sample was characterized by powder X-ray diffraction data.

Plumbogummite $PbAl_3(PO_4)(PO_3OH)(OH)_6$

Origin: Guochengmine, Yangshuo, Guangxi province, China.
Experimental details: Raman scattering measurements have been performed on arbitrarily oriented crystals using a 633 nm He-Ne laser. The laser radiation power is not indicated. The Raman shifts have been determined for the maxima of individual peaks obtained as a result of the spectral curve analysis.
Raman shifts (cm^{-1}): 3602sh, 3479, (3372), 3249, (3121), (1182w), 1106s, (1057), 1023s, (1002), 980s, (971w), (826w), (634w), 613, 579sh, 507, (494), 464, 388sh, 368, 281sh, 251s, 187, (163w), 145.
Source: Frost et al. (2013l).
Comments: The sample was characterized by powder X-ray diffraction data and electron microprobe analyses.

Plumbojarosite $Pb_{0.5}Fe^{3+}_3(SO_4)_2(OH)_6$

Origin: Synthetic.
Experimental details: Raman scattering measurements have been performed on a sample diluted with KBr powder and compressed to form a disk. A 514.5 nm Ar^+ laser was used. The laser radiation power at the sample was 38 mW.
Raman shifts (cm^{-1}): 1169w, 1120sh, 1108, 1015sh, 1002, 623, 583w, 452sh, 440s, 341w, 221.
Source: Sasaki et al. (1998).
Comments: The sample was characterized by powder X-ray diffraction data and chemical analyses. For the Raman spectra of plumbojarosite see also Frost et al. (2006r) and Spratt et al. (2013).

Plumbophyllite $Pb_2Si_4O_{10} \cdot H_2O$

Origin: Blue Bell mine, Soda Mts, Silver Lake District, San Bernardino Co., California, USA (type locality).
Experimental details: Raman scattering measurements have been performed on arbitrarily oriented crystals using a 633 nm He-Ne laser. The laser radiation power is not indicated. The Raman shifts have been determined for the maxima of individual peaks obtained as a result of the spectral curve analysis.
Raman shifts (cm^{-1}): 3567sh, 3494+3470 (unresolved doublet?), (3443), 3215w, 1153w, 1137w, 1095w, (1039), 1027s, 972, 956, 926, (657), 643s, (634), 506w, (500w), 485w, 438w, 409w, (398w), 381, (368w), 349sh, 332, 309, 253, 203sh, 182, (155), 147s, 112.
Source: Frost et al. (2014t).
Comments: The sample was characterized by qualitative electron microprobe analysis.

Plumbotsumite $Pb_{13}(CO_3)_6(Si_{10}O_{27}) \cdot 3H_2O$

Origin: St. Anthony deposit, Mammoth district, Pinal Co., Arizona, USA.
Experimental details: Raman scattering measurements have been performed on arbitrarily oriented crystals using a 633 nm He-Ne laser. The laser radiation power is not indicated. The Raman shifts have been determined for the maxima of individual peaks obtained as a result of the spectral curve analysis.
Raman shifts (cm^{-1}): 4620w, 3546, 3510w, 2950w, 2720w, 2674sh, 1744w, 1732sh, 1716w, (1709), 1685w, 1479, 1424, 1379, 1333sh, 1084w, 1060sh, 1055s, 1047, 844sh, 839, 772w, 729w, 697, 683, 673sh, 636w, 609, 581sh, 481, 458, 432w, 396, 346w, 288sh, 280w, 246, 227, 179, 154+143s (unresolved doublet?), 107+103 (unresolved doublet?).
Source: López et al. (2013a).
Comments: The sample was characterized by qualitative electron microprobe analysis.

Poitevinite $Cu(SO_4) \cdot H_2O$

Origin: Synthetic.
Experimental details: Methods of sample preparation are not described. Raman scattering measurements have been performed using 632.8 nm He-Ne laser radiation. The laser radiation power is not indicated.
Raman shifts (cm^{-1}): 1204, 1097, 1043.5s, 1014s, 669, 620.5, 607, 515, 419, 345, 268, 244, 207.5, 130w, 105w.
Source: Fu et al. (2012).

Pokrovskite $Mg_2(CO_3)(OH)_2$

Origin: Lytton, Sonoma Co., California, USA.
Experimental details: Raman scattering measurements have been performed on arbitrarily oriented crystals using a 633 nm He-Ne laser. The laser radiation power is not indicated. The Raman shifts have been determined for the maxima of individual peaks obtained as a result of the spectral curve analysis.
Raman shifts (cm^{-1}): 3556w, 3444, 1582, 1452, 1386, 1088s, 929, 734, 703w, 521w, 446w, 402, 282, 172, 143.

Source: Frost (2006).
Comments: No independent analytical data are provided for the sample used.

Poldervaartite $Ca(Ca,Mn)(SiO_3OH)(OH)$

Origin: N'Chwaning IImine, Kalahari manganese fields, South Africa (type locality).
Experimental details: Raman scattering measurements have been performed on arbitrarily oriented crystals using a 633 nm He-Ne laser. The laser radiation power is not indicated. The Raman shifts have been determined for the maxima of individual peaks obtained as a result of the spectral curve analysis.
Raman shifts (cm^{-1}): 3547, 3521sh, 3509, (3502w), 3487, 952, (943w), 917, (907), 900, (865), 852s, 807, 792sh, 528sh, 513, (498w), 485, 473sh, 245w, 239w, 213, 203, 163sh, 153s, (148), (145), (116), 109, 104.
Source: Frost et al. (2015g).
Comments: The sample was characterized by qualitative electron microprobe analysis which may correspond to olmiite.

Pollucite $Cs(Si_2Al)O_6 \cdot nH_2O$

Origin: Auburn, Androscoggin Co., Maine, USA.
Experimental details: Methods of sample preparation are not described. Raman scattering measurements have been performed using Nd-YAG laser radiation. The laser radiation power at the sample was 300 mW.
Raman shifts (cm^{-1}): 1109, 478s, 393, 299.
Source: Mozgawa (2001).
Comments: The sample was characterized by powder X-ray diffraction data.

Polycrase-(Y) $Y(Ti,Nb)_2(O,OH)_6$

Origin: No data.
Experimental details: Methods of sample preparation are not described. Raman scattering measurements have been performed using 514.5 nm Ar$^+$ laser radiation. The nominal laser radiation power was 100 mW.
Raman shifts (cm^{-1}): ~1480w, ~1420w, ~1130w, ~1080w, ~845, ~695, ~530w, ~400s, ~280s, ~225s.
Source: Tomašić et al. (2004).
Comments: The Raman shifts are given for an initially metamict sample heated at 1000°C to regain its crystal structure. The sample was characterized by powder X-ray diffraction data and chemical analyses.

Polydymite $Ni^{2+}Ni^{3+}_2S_4$

Origin: Synthetic.
Experimental details: Raman scattering measurements have been performed on grains in a Ni-based composite after exposure to an H$_2$S-containing fue 1514 nm laser radiation was used. The nominal laser radiation power was 40 mW.
Raman shifts (cm^{-1}): 379, 337, 287s, 223.

Source: Cheng and Liu (2007).
Comments: The sample was characterized by powder X-ray diffraction data and electron microprobe analyses.

Polyhalite $K_2Ca_2Mg(SO_4)_4 \cdot 2H_2O$

Origin: Synthetic.
Experimental details: Raman scattering measurements have been performed on an arbitrarily oriented crystal using 532 nm Nd-YAG laser radiation. The laser radiation power at the sample was 2 mW.
Raman shifts (cm^{-1}): 3437, 3288w, 1181w, 1165, 1144w, 1130, 1094w, 1069, 1014s, 987s, 652w, 641w, 626, 620sh, 477sh, 464, 448, 436.
Source: Jentzsch et al. (2012a).
Comments: The sample was characterized by powder X-ray diffraction data. For the Raman spectrum of polyhalite see also Jentzsch et al. (2013).

Popovite $Cu_5O_2(AsO_4)_2$

Origin: Arsenatnaya fumarole, Tolbachik volcano, Kamchatka, Russia (type locality).
Experimental details: Raman scattering measurements have been performed on an arbitrarily oriented crystal using 532 nm laser radiation. The laser radiation power at the sample was ~9 mW.
Raman shifts (cm^{-1}): 846s, 810s, 642, 547, 489, 478w, 427, 377w, 344, 281, 258w, 212, 128, 97s.
Source: Pekov et al. (2015b).
Comments: The sample was characterized by powder X-ray diffraction data and electron microprobe analyses. The crystal structure is solved.

Portlandite $Ca(OH)_2$

Origin: Synthetic.
Experimental details: Methods of sample preparation are not described. Raman scattering measurements have been performed using 632.8 nm He-Ne laser radiation. The nominal laser radiation power was 10 mW.
Raman shifts (cm^{-1}): 3620s, ~680 (broad), 356s, 252.
Source: Schmid and Dariz (2015).
Comments: No independent analytical data are provided for the sample used. For the Raman spectrum of portlandite see also Lutz et al. (1994).

Posnjakite $Cu_4(SO_4)(OH)_6 \cdot H_2O$

Origin: Ozernyi district, Sallo-Kuolajarvi, Kola Peninsula, Russia.
Experimental details: Raman scattering measurements have been performed on an arbitrarily oriented sample using 514.5 nm Ar$^+$ laser radiation. The nominal laser radiation power was 50 mW.
Raman shifts (cm^{-1}): 3587s, 3566s, 3400, 971s, 620sh, 607, 594, 508, 479, 449w, 418w, 387, 316w, 241w, 194w, 136, 88s.
Source: Voloshin et al. (2015b).
Comments: The sample was characterized by electron microprobe analyses. For the Raman spectra of posnjakite see also Martens et al. (2003a), Frost et al. (2004n), and Lepot et al. (2006).

Potarite PdHg

Origin: Munni Munni layered intrusion, West Pilbara Block, Western Australia.

Experimental details: Raman scattering measurements have been performed on arbitrarily oriented grains in a polished section using 514.5 nm Ar$^+$ laser radiation. The laser radiation power at the sample was between 1 and 10 mW. A 180°-scattering geometry was employed.

Raman shifts (cm^{-1}): 362, 340s, 285, 254.

Source: Mernagh and Hoatson (1995).

Comments: The sample was characterized by electron microprobe analyses.

Pottsite $(Pb_3Bi)Bi(VO_4)_4 \cdot H_2O$

Origin: Las Tapias, Cordoba province, Argentina.

Experimental details: Raman scattering measurements have been performed on arbitrarily oriented crystals using a 633 nm He-Ne laser. The laser radiation power is not indicated. The Raman shifts have been determined for the maxima of individual peaks obtained as a result of the spectral curve analysis.

Raman shifts (cm^{-1}): 2912w, 885, 874, 707, 643, 465, 413s, 404, 370, 348, 331, 264w, 204, 185.

Source: Frost et al. (2006i).

Comments: No independent analytical data are provided for the sample used.

Poubaite $PbBi_2(Se,Te,S)_4$

Origin: Ozernyi district, Salla-Kuolajarvi, Kola Peninsula, Russia.

Experimental details: Raman scattering measurements have been performed on an arbitrarily oriented sample using 514.5 nm Ar$^+$ laser radiation. The nominal laser radiation power was 50 mW.

Raman shifts (cm^{-1}): 139, 102s, 59.

Source: Voloshin et al. (2015a).

Comments: The samples used were characterized by electron microprobe analyses.

Poudretteite $KNa_2(B_3Si_{12})O_{30}$

Origin: Mogok valley, Shan State, Myanmar.

Experimental details: Raman scattering measurements have been performed on a fragment of a single crystal with the beamdirection parallel to the c-axis using 325 nm laser excitation. The laser radiation power is not indicated.

Raman shifts (cm^{-1}): 1799w, 1660w, 1556w, 1176s, 1147, 1045w, 1011, 928, 908w, 849w, 788, 696, 662, 594w, 552s, 490s, 429w.

Source: Smith et al. (2003).

Comments: The sample was characterized by powder X-ray diffraction data and electron microprobe analyses.

Poughite $Fe^{3+}{}_2(Te^{4+}O_3)_2(SO_4) \cdot 3H_2O$

Origin: Wendy Pit, El Indio gold mine, Coquimbo, Chile.

Experimental details: Raman scattering measurements have been performed on arbitrarily oriented crystals using a 633 nm He-Ne laser. The laser radiation power is not indicated. The Raman shifts

have been determined for the maxima of individual peaks obtained as a result of the spectral curve analysis.

Raman shifts (cm^{-1}): 3509, 3481, (3477w), 2330, 2231, 2155, 1926w, 1793, 1706, 1582, 1335, 1187w, 1152w, 1078, 1026s, (1022), (655), 653s, 561, 509, 485, 382, 347, 236s, (234), 163, (161).
Source: Frost and Keeffe (2008c).
Comments: No independent analytical data are provided for the sample used.

Povondraite $NaFe^{3+}_3(Fe^{3+}_4Mg_2)(Si_6O_{18})(BO_3)_3(OH)_3O$

Origin: No data.
Experimental details: Raman scattering measurements have been performed using 488 or 514.5 nm Ar$^+$ laser radiation. The laser radiation power at the sample was 14 mW. Polarized Raman spectra were collected from raw crystal surfaces in the spectral range 15–4000 cm^{-1} in -$y(zz)y$, -$y(zx)y$, and -$y(xx)y$ scattering geometries.
Raman shifts (cm^{-1}): ~3555, ~990, ~800w, ~670w, ~625w, ~545, ~460, ~430, ~400w, ~315, ~280s, ~230s, ~165, ~140w.
Source: Watenphul et al. (2016a, b).
Comments: The Raman shifts are given for the scattering geometry -$y(zz)y$.

Powellite $Ca(MoO_4)$

Origin: Dundas, Tasmania, Australia.
Experimental details: Raman scattering measurements have been performed on crystals oriented to provide maximum intensity using a 785 nm Nd-YAG laser. The laser radiation power at the sample was 1 mW. The Raman shifts have been determined for the maxima of individual peaks obtained as a result of the spectral curve analysis.
Raman shifts (cm^{-1}): 879, 847, 794, 513, 456, 403, 392, 324, 267.
Source: Frost et al. (2004c).
Comments: No independent analytical data are provided for the sample used.

Prehnite $Ca_2Al(Si_3Al)O_{10}(OH)_2$

Origin: Valtournanche, Aosta valley, Italy.
Experimental details: Methods of sample preparation are not described. Raman scattering measurements have been performed on an arbitrarily oriented sample using 632.8 nm He-Ne or 488 nm Ar$^+$ laser radiation. The laser radiation power is not indicated.
Raman shifts (cm^{-1}): ~1110w, ~1080w, 991, ~950, 931, ~640, ~600, 541, 520s, 388s, 318.
Source: Andò and Garzanti (2014).
Comments: No independent analytical data are provided for the sample used.

Preiswerkite $KFe^{2+}_3(AlSi_3O_{10})(OH)_2$

Origin: Liset, Selje, Sognog Fjordane, Norway.
Experimental details: Raman scattering measurements have been performed on oriented samples using 514.5 nm Ar$^+$ laser radiation. The laser radiation power is not indicated. The spectra were recorded with the electric field polarized perpendicular to the cleavage plane.

Raman shifts (cm^{-1}): 3628s, 3620, 916, 648s, 292s, 216s (Sample 8); 916, 679, 648s, 488, (400), (379), 330, 292s, (280), 216s, 156, 108 (Sample 9).
Source: Tlili et al. (1989).
Comments: The samples were characterized by electron microprobe analyses. For the Raman spectra of preiswerkite see also Tlili and Smith (2007) and Orozbaev et al. (2011).

Pretulite Sc(PO$_4$)

Origin: Saint-Aubin-des-Châteaux, Armorican Massif, France.
Experimental details: Raman scattering measurements have been performed on an arbitrarily oriented crystal using 514.5 nm Ar$^+$ laser radiation. The laser radiation power at the sample was 3 mW.
Raman shifts (cm^{-1}): 1079s, 1043, 1024s, 595w, 475, 326, 234, 186w.
Source: Moëlo et al. (2002).
Comments: The sample was characterized by single-crystal X-ray diffraction data and electron microprobe analyses. The crystal structure is solved. For the Raman spectrum of pretulite see also Giarola et al. (2011).

Příbramite CuSbSe$_2$

Origin: Synthetic.
Experimental details: Raman scattering measurements have been performed on a polycrystalline film using 532 nm laser radiation. The laser radiation power is not indicated.
Raman shifts (cm^{-1}): 226s, 200, 153, 114.
Source: Xue et al. (2015).
Comments: The sample was characterized by powder X-ray diffraction data.

Priceite Ca$_2$B$_5$O$_7$(OH)$_5$·H$_2$O

Origin: 20-Mule-Team Canyon, Furnace Creek district, California, USA.
Experimental details: Raman scattering measurements have been performed on arbitrarily oriented crystals using a 633 nm He-Ne laser. The laser radiation power is not indicated. The Raman shifts have been determined for the maxima of individual peaks obtained as a result of the spectral curve analysis.
Raman shifts (cm^{-1}): 3669, 3579sh, 3555s, 3510sh, 3496, 3468sh, 3404, 3385sh, 3221w, 1211, 1169w, 1127, 1100w, 1071, 1019, 991, 974sh, 956, 927w, 894, 842, 826, 736, (697), 689s, 674s, (660), 634, 602, 563sh, 545s, 511w, 481sh, 471, 450w, 433w, 409w, 387w, 368, 353sh, 306, 287sh, 266sh, 253, 231, 217w, 195, (183), 173, 148sh, 138, 129sh, 109.
Source: Frost et al. (2015l).
Comments: The sample was characterized by qualitative electron microprobe analysis.

Priderite Mg-analogue K$_2$(Ti$_7$Mg)O$_{16}$

Priderite Al-analogue K$_2$(Ti$_6$Al$_2$)O$_{16}$

Origin: Synthetic.
Experimental details: Raman scattering measurements have been performed on oriented single crystals using 488 and 514.5 nm Ar$^+$ laser radiations. The laser radiation power is not indicated.

Raman shifts (cm^{-1}): 840, 702, 690, 635, 610, 550, 505, 497, 460, 455, 370, 350, 344, 331, 260, 193, 144, 123 (for the priderite Al analogue); 840, 710, 700, 640, 623, 580, 511, 463, 461, 380, 375, 360, 350, 285, 210, 152, 131 (for the priderite Mg analogue).
Source: Ohsaka and Fujiki (1982).
Comments: The empirical formulae of the samples used are $K_{1.6}(Ti_{7.2}Mg_{0.8})O_{16}$ and $K_{1.6}(Ti_{6.4}Al_{1.6})O_{16}$. The intensities are given for the α_{zz} polarization.

Priderite Cr-analogue $(K,Ba)_{2-x}(Ti_6Cr_2)O_{16}$

Origin: Gföhl granulite, Bohemian massif, Czech Republic.
Experimental details: Raman scattering measurements have been performed on an arbitrarily oriented grain using 532 nm laser radiation. The laser radiation power at the sample was 1.4 mW.
Raman shifts (cm^{-1}): 800sh, 685s, 620, 545w, 350, 277, 195w, 150.
Source: Naemura et al. (2015).
Comments: The sample was characterized by electron microprobe analyses.

Prismatine $(Mg,Al,Fe)_6Al_4(Si,Al)_4(B,Si,Al)(O,OH,F)_{22}$

Origin: Madagascar.
Experimental details: Methods of sample preparation are not described. Micro-Raman scattering measurements have been performed on a single crystal using 514.5 nm Ar$^+$ laser radiation with the polarizationdirection parallel to the crystal elongation. The laser radiation power at the sample was 5 W. A 180°-scattering geometry was employed.
Raman shifts (cm^{-1}): 3615w, 3556, 1120w, 1028s, 973s, 864, 803, 760w, 610, 537, 459, 384, 286, 260s.
Source: Wopenka et al. (1999).
Comments: The sample contains 0.86 boron atoms per formula unit.

Proustite Ag_3AsS_3

Origin: Jáchymov, Krušné Hory (Ore Mts.), Bohemia, Czech Republic.
Experimental details: Raman scattering measurements have been performed on a polished crystal using 632.8 nm He-Ne laser radiation. The laser radiation power at the sample was 1.7 mW. A 180°-scattering geometry was employed.
Raman shifts (cm^{-1}): 364s, 337, 315sh, 274w.
Source: Kharbish et al. (2009).
Comments: The sample was characterized by electron microprobe analyses. For the Raman spectra of proustite see also Byer et al. (1973) and Makreski et al. (2004).

Pseudoboleite $Pb_{31}Cu_{24}Cl_{62}(OH)_{48}$

Origin: Siklverton Barrier Range, New South Wales, Australia.
Experimental details: Raman scattering measurements have been performed on arbitrarily oriented crystals using a 633 nm He-Ne laser. The laser radiation power is not indicated. The Raman shifts have been determined for the maxima of individual peaks obtained as a result of the spectral curve analysis.

Raman shifts (cm^{-1}): 3467, 3434, 3350, 3330, 973w, 908w, 817w, ~680s, 584, 512, 449, (388), (236), (179), 148s, 137s.
Source: Frost and Williams (2004).
Comments: No independent analytical data are provided for the sample used.

Pseudobrookite (Fe$^{3+}_2$Ti)O$_5$

Origin: Marion Island, Hawaiian archipelago.
Experimental details: Methods of sample preparation are not described. Raman scattering measurements have been performed using 514.5 nm Ar$^+$ laser radiation. The nominal laser radiation power was between 2.5 and 40 mW.
Raman shifts (cm^{-1}): 1334, 660s, 507w, 411w, 340, 226s, 200s.
Source: Prinsloo et al. (2011).
Comments: For the Raman spectra of pseudobrookite see also Bersani et al. (2000) and Wang et al. (2016).

Pseudocotunnite K$_2$PbCl$_4$ (?)

Origin: Synthetic.
Experimental details: Methods of sample preparation are not described. Raman scattering measurements have been performed using 488 nm Ar$^+$ laser radiation. The laser radiation power is not indicated.
Raman shifts (cm^{-1}): ~230.
Source: Oyamada (1974).

Pseudojohannite Cu$_3$(OH)$_2$[(UO$_2$)$_4$O$_4$(SO$_4$)$_2$]·12H$_2$O

Origin: Rovnost shaft, Jáchymov uranium deposit, Krušné Hory (Ore Mts.), Western Bohemia, Czech Republic (type locality).
Experimental details: Raman scattering measurements have been performed on arbitrarily oriented crystals using a 633 nm He-Ne laser. The laser radiation power is not indicated. The Raman shifts have been determined for the maxima of individual peaks obtained as a result of the spectral curve analysis.
Raman shifts (cm^{-1}): 3483w, (3353), (3226), 1625w, 1554w, 1333w, 1100, 1017, 810+805s (unresolved doublet?), 755sh, 539sh, 496, 465, 423s, 279, 210, 162+151 (unresolved doublet?).
Source: Frost et al. (2009h).
Comments: The sample was characterized by electron microprobe analyses.

Pseudolaueite Mn^{2+}Fe$^{3+}_2$(PO$_4$)$_2$(OH)$_2$·7-8H$_2$O

Origin: Hagendorf South pegmatite, Bavaria, Germany.
Experimental details: Raman scattering measurements have been performed on arbitrarily oriented crystals using a 633 nm He-Ne laser. The laser radiation power is not indicated. The Raman shifts have been determined for the maxima of individual peaks obtained as a result of the spectral curve analysis.

Raman shifts (cm^{-1}): 3593sh, 3485, 3376, 3209w, 3110sh, 1123s, 1066w, 1046, (1034), 1000+993s (unresolved doublet?), 976, 843, 643, 626sh, (565), 501, (485), 471s, 456sh, 435, 408, 373w, 332, 303, 286, 271, 249, 223, 215, 201s, 189, 183, 164.
Source: Frost et al. (2015ac).
Comments: The sample was characterized by qualitative electron microprobe analysis.

Pseudomalachite $Cu_5(PO_4)_2(OH)_4$

Origin: Banská Bystrica, central Slovakia.
Experimental details: Raman scattering measurements have been performed on an oriented crystal using 632.8 nm He-Ne laser radiation. The laser radiation power at the sample was 17 mW. A 180°-scattering geometry was employed. Polarized spectra were collected with $E\|b$ and $E\perp b$.
Raman shifts (cm^{-1}): 3414, 1137w, 1112w, 1083, 1055, 996sh, 971, 945sh, 861w, 801w, 750, (703w), (664w), 750, 703w, (664w), (639w), 606, 539sh, 515, 477s, 446s, 412, 297, 254w, 241, 214, 209, 174s, 131s, 109s, 86s (for $E\perp b$).
Source: Kharbish et al. (2014).
Comments: The sample was characterized by powder X-ray diffraction data and electron microprobe analyses. For the Raman spectra of pseudomalachite see also Frost et al. (2002a, g), Bouchard and Smith (2003), Majzlan et al. (2015), and Ciesielczuk et al. (2016).

Pseudowollastonite $CaSiO_3$

Origin: Synthetic.
Experimental details: Methods of sample preparation are not described. Raman scattering measurements have been using 488 or 514.5 nm Ar$^+$ laser radiation. The nominal laser radiation power was 500 mW.
Raman shifts (cm^{-1}): 1075, 989s, 932, 714w, 580s, 558, 511, 428w, 373s, 341, 327, 315, 301, 217w, 193, 167.
Source: Richet et al. (1998).
Comments: The sample was characterized by powder X-ray diffraction data.

Pucherite $Bi(VO_4)$

Origin: Pucher shaft, Schneeberg, Germany (type locality).
Experimental details: Raman scattering measurements have been performed on arbitrarily oriented crystals using a 633 nm He-Ne laser. The laser radiation power is not indicated. The Raman shifts have been determined for the maxima of individual peaks obtained as a result of the spectral curve analysis.
Raman shifts (cm^{-1}): 1002, (881), 872s, 710w, 693w, 647w, 415, 372s, 346, 333s, 256, 224, 196, 188.
Source: Frost et al. (2006i).
Comments: No independent analytical data are provided for the sample used.

Pumpellyite-(Al) $Ca_2Al_3(Si_2O_7)(SiO_4)O(OH)\cdot H_2O$

Origin: New Caledonia.
Experimental details: Methods of sample preparation are not described. Raman scattering measurements have been performed using 633 nm He-Ne or 325 nm He-Cd laser radiation. The laser radiation power at the sample was 5 mW.

Raman shifts (cm^{-1}): 1092w, 1015, 1002, 984, 957w, 922, 865, 816, 764w, 696s, 646w, 610, 589, 535, 508s, 480, 459, 429s, 361, 322, 293, 227, 208, 187, 172, 154, 135.
Source: Krenn et al. (2004).
Comments: The sample was characterized by electron microprobe analyses.

Pyrargyrite Ag$_3$SbS$_3$

Origin: Jáchymov, Krušné Hory (Ore Mts.), Bohemia, Czech Republic.
Experimental details: Raman scattering measurements have been performed on a polished crystal using 632.8 nm He-Ne laser radiation. The laser radiation power at the sample was 1.7 mW. A 180°-scattering geometry was employed.
Raman shifts (cm^{-1}): 323s, 300, 274sh, 252sh.
Source: Kharbish et al. (2009).
Comments: The sample was characterized by electron microprobe analyses. For the Raman spectra of pyrargyrite see also Byer et al. (1973), Nilges et al. (2002), and Makreski et al. (2004).

Pyrite FeS$_2$

Origin: Guinaoang, NW Luzon, Philippines (isotropic variety) and Pine Creek, Northern Territory, Australia (anisotropic variety).
Experimental details: Methods of sample preparation are not described. Raman scattering measurements have been performed using 514.5 nm Ar$^+$ laser radiation. The laser radiation power at the sample was between 1 and 10 mW. A 180°-scattering geometry was employed.
Raman shifts (cm^{-1}): 446, 387s, 353s (for the isotropic variety); 428, 377s, 342s (for the anisotropic variety).
Source: Mernagh and Trudu (1993).
Comments: The samples were characterized by electron microprobe analyses. For the Raman spectra of pyrite see also Kleppe and Jephcoat (2004), White (2009), Frezzotti et al. (2012), and Andò and Garzanti (2014).

Pyroaurite Mg$_6$Fe$^{3+}_2$(CO$_3$)(OH)$_{16}$·4H$_2$O

Origin: Synthetic.
Experimental details: Raman scattering measurements have been performed on an arbitrarily oriented sample using 532.12 nm Nd-YAG laser radiation. The laser radiation power at the sample was about 40 mW.
Raman shifts (cm^{-1}): See comments below.
Source: Rozov et al. (2010).
Comments: Raman spectra of the members of the hydrotalcite-pyroaurite series containing less than 1 Fe atom per formula unit contain bands at ~3500, ~1380, ~1060 and ~545s cm^{-1}. In the spectrum of the sample with the approximate formula Mg$_6$(AlFe^{3+})(CO$_3$)(OH)$_{16}$·4H$_2$O, the only band at ~545 cm^{-1} is observed. Measurements with greater Fe contents were precluded by fluorescence.

Pyrochroite Mn^{2+}(OH)$_2$

Origin: A sediment-hosted Mn deposit, Mesoarchean Mozaan Group, Pongola Supergroup, South Africa.

Experimental details: Methods of sample preparation are not described. Raman scattering measurements have been performed using 532 nm laser radiation. The laser radiation power is not indicated.
Raman shifts (cm^{-1}): 635, 330.
Source: Ossa et al. (2016).
Comments: The sample was characterized by powder X-ray diffraction data and electron microprobe analyses.

Pyrolusite MnO_2

Origin: Kisenge Mine, Zaire.
Experimental details: Methods of sample preparation are not described. Raman scattering measurements have been performed using 244 nm laser radiation. The laser radiation power is not indicated. A 180°-scattering geometry was employed.
Raman shifts (cm^{-1}): 666w, 610, ~550s, ~525s, 482, 384.
Source: Kim and Stair (2004).
Comments: No independent analytical data are provided for the sample used. For the Raman spectrum of pyrolusite see also Julien et al. (2004).

Pyromorphite $Pb_5(PO_4)_3Cl$

Origin: Synthetic.
Experimental details: Raman scattering measurements have been performed on an arbitrarily oriented sample using Nd-YAG laser radiation. The laser radiation power at the sample was 300 mW.
Source: Bajda et al. (2011).
Raman shifts (cm^{-1}): 1047w, 1012w, 984w, 945s, 920s, 813s, 781w, 764sh, 577w, 553, (548), (434w), 411, 392, 334, 324sh.
Comments: The empirical formula of the sample used is $Pb_5[(PO_4)_{2.4}(AsO_4)_{0.6}]Cl$. The bands in the ranges 750–820 and 320–350 cm^{-1} correspond to arsenate groups. For the Raman spectra of pyromorphite see also Levitt and Condrate Sr (1970), Bartholomäi and Klee (1978), Botto et al. (1997), Bouchard and Smith (2003), and Coccato et al. (2016).

Pyromorphite As-rich $Pb_5(PO_4,AsO_4)_3Cl$

Origin: Bunker Hill Mine, Kellogg, Idaho, USA.
Experimental details: Raman scattering measurements have been performed on arbitrarily oriented crystals using a 633 nm He-Ne laser. The Raman shifts have been determined for the maxima of individual peaks obtained as a result of the spectral curve analysis. The laser radiation power is not indicated.
Raman shifts (cm^{-1}): 3444, 3378, 3325, 3291, 3256, 1014, 979, 944, 943, 920, 917, 825, 818, 815s, 776, 768, 573, 549, 433, 414, 409, 391, 388, 377, 344, 339, 318, 206, 186, 177, 152, 111, 105.
Source: Frost et al. (2007c).
Comments: No independent analytical data are provided for the sample used.

4 Raman Spectra of Minerals

Pyrope $Mg_3Al_2(SiO_4)_3$

Origin: Synthetic.
Experimental details: Raman scattering measurements have been performed on a polycrystalline grain using Ar^+ laser radiation. The laser radiation power at the sample was between 5 and 50 mW.
Raman shifts (cm^{-1}): 1064, 928s, 910sh, 868, ~690w, 648, 562, 512w, 492w, 382sh, 364, 340w, 320w, 218w, 210.
Source: McMillan et al. (1989).
Comments: The sample was characterized by ^{27}Al MAS NMR spectroscopy. For the Raman spectra of pyrope see also Mingsheng et al. (1994), Kolesov and Geiger (1998), Bersani et al. (2009), Frezzotti et al. (2012), Andò and Garzanti (2014), Gilg and Gast (2015), and Du et al. (2017).

Pyrophanite $Mn^{2+}TiO_3$

Origin: Perkupa evaporite mine, Bódva valley, inner Western Carpathians, Hungary.
Experimental details: Raman scattering measurements have been performed on an arbitrarily oriented sample using 632.8 nm He-Ne laser radiation. The laser radiation power at the sample was 12 mW.
Raman shifts (cm^{-1}): 684s, ~600w, 466, 360, 334, 263, 235w, 202, 164.
Source: Zajzon et al. (2013).
Comments: The sample was characterized by electron microprobe analyses.

Pyrophyllite $Al_2Si_4O_{10}(OH)_2$

Origin: York Spring, Pennsylvania, USA.
Experimental details: No data.
Raman shifts (cm^{-1}): 3670s, 707s, 360, 261s, 193s.
Source: Wang et al. (2015)
Comments: Only the strongest Raman bands are indicated in the cited paper. The sample was characterized by powder X-ray diffraction data and electron microprobe analyses. For the Raman spectrum of pyrophyllite see also Wada and Kamitakahara (1991).

Pyrosmalite-(Fe) $Fe^{2+}_8Si_6O_{15}(OH)_{10}$

Origin: Cannington mine, McKinlay Shire, Queensland, Australia.
Experimental details: No data.
Raman shifts (cm^{-1}): 1023s, 820, 767w, 740w, 663w, 614s, 468, 365w, 325, 193.
Source: Dong and Pollard (1997).
Comments: The sample was characterized by quqlitative electron microprobe analysis.

Pyrosmalite-(Mn) $Mn^{2+}_8Si_6O_{15}(OH,Cl)_{10}$

Origin: A Zn-Pb-Ag sulfide deposit art Dugald River, NW Queensland, Australia.
Experimental details: Micro-Raman scattering measurements have been performed on an inclusion in quartz using 514.5 nm Ar^+ laser radiation. The laser radiation power is not indicated.
Raman shifts (cm^{-1}): 615s, 696w, 803, 1020.
Source: Xu (1998).
Comments: No independent analytical data are provided for the sample used.

Pyroxmangite $Mn^{2+}SiO_3$

Origin: No data.
Experimental details: Raman scattering measurements have been performed on an arbitrarily oriented single crystal using 532.2 nm laser radiation. The laser radiation power is not indicated.
Raman shifts (cm^{-1}): ~995s, ~915w, ~810w, ~660s, ~535, ~390, ~307.
Source: Wang et al. (2001a).
Comments: No independent analytical data are provided for the sample used. Another sample shows a doublet 965+995 cm^{-1}.

Pyrrhotite Fe_7S_8

Origin: A dolerite sill, Siberian Precambrian platform, eastern Siberia, Russia.
Experimental details: Micro-Raman scattering measurements have been performed on arbitrarily oriented inclusions in halite using 514.5 nm Ar$^+$ laser radiation. The laser radiation power is not indicated.
Raman shifts (cm^{-1}): 372–378s, 339–342.
Source: Grishina et al. (1992).
Comments: The samples were characterized by electron microprobe analyses. For the Raman spectrum of pyrrhotite see also Lanteigne et al. (2012).

Qandilite $(Mg,Fe^{3+})_2(Ti,Fe^{3+},Al)O_4$

Origin: Synthetic.
Experimental details: Methods of sample preparation are not described. Raman scattering measurements have been performed using 514.5 nm Ar$^+$ laser radiation. The nominal laser radiation power was 20 mW.
Raman shifts (cm^{-1}): 730s, 605w, 517, 387, 325, 272sh, 235sh, 141w.
Source: De Lima (2016).
Comments: A Fe-free sample was used. The sample was characterized by powder X-ray diffraction data.

Qingheiite $Na_2MnMgAl(PO_4)_3$

Origin: Santa Ana Pegmatite, Totoral pegmatitic field, Coronel Pringles department, San Luis, Argentina.
Experimental details: Raman scattering measurements have been performed on arbitrarily oriented crystals using a 633 nm He-Ne laser. The laser radiation power is not indicated. The Raman shifts have been determined for the maxima of individual peaks obtained as a result of the spectral curve analysis.
Raman shifts (cm^{-1}): 1136+1140w (unresolved doublet?), 1130w, 1106w, 1083, 1068w, 1058w, 1047w, 1021, 980s, 964sh, 945sh, 690w, 644w, 606, 572, 504w, 472w, 453w, 420w, 369w, 308w, 280w, 229w, 217w, 167sh, 152w, 143sh.
Source: Frost et al. (2013d).
Comments: The sample was characterized by electron microprobe analyses.

Qingsongite BN

Origin: Synthetic.
Experimental details: Raman scattering measurements have been performed on a powdery sample using 244 nm Ar$^+$ laser radiation. The laser radiation power is not indicated.
Raman shifts (cm^{-1}): 1304, 1055s.
Source: Reich et al. (2005).
Comments: The sample was produced from commercial hexagonal BN by nucleation under high pressure (4.2 GPa) and temperature (1800–1900 K) using a MgBN catalyst system.

Quadridavyne $[(Na,K)_6Cl_2][Ca_2Cl_2][(Si_6Al_6O_{24})]$

Origin: Monte Somma caldera, Somma-Vesuvius complex, Napoli, Campania, Italy (type locality).
Experimental details: No data.
Raman shifts (cm^{-1}): See comments below.
Source: Binon et al. (2004).
Comments: In the cited paper only ranges of Raman bands are indicated (2400–3700, 1050–1100, and 980–990 cm^{-1}). The precise Raman shifts of quadridavyne are (RRUFF R141084, cm^{-1}): 1047s, 985w, 966w, ~757w, 657, 471s, 425, 286 (λ = 532 nm).

Quartz SiO_2

Origin: Spruce Claim, King Co., Washington, USA.
Experimental details: Raman scattering measurements have been performed on an arbitrarily oriented crystal using 514.5 nm Ar$^+$ laser radiation. The nominal laser radiation power was 150 mW.
Raman shifts (cm^{-1}): 1231w, 1160w, 1083w, 808w, 697w, 463s, 401w, 354, 263w, 205, 128.
Source: Jasinevicius (2009).
Comments: The sample was characterized by powder X-ray diffraction data. For the Raman spectra of quartz see also Hemley (1987a, b), Lepot et al. (2006), Ling et al. (2011), Ciobotă et al. (2012), Frezzotti et al. (2012), and Karwowski et al. (2013).

Quenstedtite $Fe^{3+}_2(SO_4)_3 \cdot 11H_2O$

Origin: Allan Hills 77005 martian meteorite.
Experimental details: Raman scattering measurements have been performed on an arbitrarily oriented sample using 532.3 nm Nd-YAG laser radiation. The nominal laser radiation power was ≤6 mW.
Raman shifts (cm^{-1}): 1130, 1107, 1024s, 985, 614w, 600, 479s, 308, 275s, 247, 157.
Source: Kuebler (2013b).
Comments: No independent analytical data are provided for the sample used.

Quetzalcoatlite $Cu^{2+}_3Zn_6Te^{6+}_2O_{12}(OH)_6(Ag,Pb,\square)Cl$

Origin: Vlue Bell mine, Soda Mts., Calofornia, USA (?).
Experimental details: Raman scattering measurements have been performed on arbitrarily oriented crystals using a 633 nm He-Ne laser. The laser radiation power is not indicated. The Raman shifts have been determined for the maxima of individual peaks obtained as a result of the spectral curve analysis.

Raman shifts (cm^{-1}): 754sh, 719+693s (unresolved doublet?), 606, 602sh, 477, 403sh, (364w), 319, (248), 197, 141s, 108s.
Source: Frost and Dickfos (2009).
Comments: No independent analytical data are provided for the sample used.

Quintinite $Mg_4Al_2(OH)_{12}(CO_3)\cdot 3H_2O$

Origin: No data.
Experimental details: Raman scattering measurements have been performed on arbitrarily oriented crystals using a 633 nm He-Ne laser. The laser radiation power is not indicated. The Raman shifts have been determined for the maxima of individual peaks obtained as a result of the spectral curve analysis.
Raman shifts (cm^{-1}): 3545sh, 3465, 1586w, 1406sh, 1346s, 1061s, 1045sh, 974, 950, 859sh, 722s, 696w, 669, 627sh, 558, 483w, 179sh, 156s.
Source Frost and Dickfos (2007b).
Comments: Questionable data: in Figure 2 of the cited paper the band at 1061 cm^{-1} (a band of symmetric C–O-stretching vibrations that should be strong for a carbonate) is absent. No independent analytical data are provided for the sample used.

Quintinite $Mg_4Al_2(OH)_{12}(CO_3)\cdot 3H_2O$

Origin: Jacupiranga mine, Cajati, São Paulo, Brazil.
Experimental details: Raman scattering measurements have been performed on arbitrarily oriented crystals using a 633 nm He-Ne laser. The laser radiation power is not indicated. The Raman shifts have been determined for the maxima of individual peaks obtained as a result of the spectral curve analysis.
Raman shifts (cm^{-1}): 3485, 3334sh, 3078sh, 1698w, 1440w, 1346sh, 1062s, 1046sh, 973, 833w, 698, (684), 559s, 484w, 401w, 367w, 308w, 183sh, 155.
Source: Theiss et al. (2015a).
Comments: The sample was characterized by qualitative electron microprobe analysis.

Rabejacite $Ca_2[(UO_2)_4O_4(SO_4)_2](H_2O)_8$

Origin: Ranger No. 1deposit, Jabiru, Northern Territory, Australia.
Experimental details: Raman scattering measurements have been performed on arbitrarily oriented crystals using a 633 nm He-Ne laser. The laser radiation power is not indicated. The Raman shifts have been determined for the maxima of individual peaks obtained as a result of the spectral curve analysis.
Raman shifts (cm^{-1}): 3547+3465 (unresolved doublet), 1175w, 1123sh, 1086s, 1010s, 848, 832, ~672, ~620, ~492, ~415, ~245w, ~198, ~181.
Source: Frost et al. (2004g).
Comments: No independent analytical data are provided for the sample used.

Raguinite $TlFeS_2$

Origin: Crven Dol, Allchar, Roszdan, Republic of Macedonia (type locality).
Experimental details: Raman scattering measurements have been performed on an arbitrarily oriented single crystal using 632.8 nm He-Ne laser radiation. The laser radiation power at the sample was 1.9 mW. A 180°-scattering geometry was employed.

Raman shifts (cm^{-1}): 395, 377s, 367, 321, 306, 275, 206w, 190, 166w, 137, 126w, 102w.
Source: Makreski et al. (2014).
Comments: The sample was characterized by electron microprobe analyses.

Rajite $CuTe^{4+}_2O_5$

Origin: Lone Pine mine, Catron Co., New Mexico, USA.
Experimental details: Raman scattering measurements have been performed on arbitrarily oriented crystals using a 633 nm He-Ne laser. The laser radiation power is not indicated. The Raman shifts have been determined for the maxima of individual peaks obtained as a result of the spectral curve analysis.
Raman shifts (cm^{-1}): (775), 754s, 731s, 661sh, 652s, 603, 540, 459w, 430, 393w, 318, 299, 267w, 237, 204, 187s, 162, 146s, 127, or 740, 676, 592, 438s, 370, 347, 212.
Source: Frost et al. (2008h).
Comments: Questionable data. No independent analytical data are provided for the sample used. Two widely different Raman spectra are provided for rajite from New Mexico in the cited paper.

Ramanite-(Cs) $CsB_5O_6(OH)_4 \cdot 2H_2O$

Origin: Island of Elba, Italy (type locality).
Experimental details: Raman scattering measurements have been performed on an arbitrarily oriented inclusion in quartz using 488 nm Ar$^+$ laser radiation. The laser radiation power at the sample was about 30 mW.
Raman shifts (cm^{-1}): 907, 768, 548s, 462, 293, 98.
Source: Thomas et al. (2008).
Comments: The sample was characterized by electron microprobe analyses. The peak at 462 cm^{-1} is influenced by a strong Raman band of the quartz matrix. For the Raman spectrum of ramanite-(Cs) see also Frezzotti et al. (2012).

Ramanite-(Rb) $RbB_5O_6(OH)_4 \cdot 2H_2O$

Origin: Island of Elba, Italy (type locality).
Experimental details: Raman scattering measurements have been performed on an arbitrarily oriented inclusion in quartz using 488 nm Ar$^+$ laser radiation. The laser radiation power at the sample was about 30 mW.
Raman shifts (cm^{-1}): 914, 765, 554s, 508, 101.
Source: Thomas et al. (2008).
Comments: The sample was characterized by electron microprobe analyses. For the Raman spectrum of ramanite-(Rb) see also Frezzotti et al. (2012).

Rambergite MnS

Origin: Synthetic.
Experimental details: Methods of sample preparation are not described. Raman scattering measurements have been performed using 488 nm Ar$^+$ laser radiation. The laser radiation output power was 50 mW.
Raman shifts (cm^{-1}): 473s, 288w, 221.

Source: Fernandez et al. (2015).
Comments: The sample was characterized by powder X-ray diffraction data.

Rameauite $K_2CaO_8(UO_2)_6 \cdot 9H_2O$

Origin: Margnac mine, Compreignac, Haute-Vienne, Limousin, France.
Experimental details: Methods of sample preparation are not described. Raman scattering measurements have been performed using 633 nm He-Ne laser radiation. The laser radiation power at the sample was about 4.5 mW.
Raman shifts (cm^{-1}): 3450w (broad), 1635w, 813s, 791, 732sh, 578, 453, 363sh, 331, 298, 274, 260sh, 215w, 188, 139, 115w, 78.
Source: Plášil et al. (2016b).
Comments: The sample was characterized by single-crystal X-ray diffraction data. The crystal structure is solved.

Ramikite-(Y) $Li_4(Na,Ca)_{12}(Y,Ca,REE)_6Zr_6(PO_4)_{12}(CO_3)_4O_4[(OH),F]_4$

Origin: Poudrette (Demix) quarry, Mont Saint-Hilaire, Rouville RCM (Rouville Co.), Montérégie, Québec, Canada (type locality).
Experimental details: Raman scattering measurements have been performed from a face of single crystal using 532 nm laser radiation. The laser radiation power is not indicated.
Raman shifts (cm^{-1}): 3659w, 1117sh, 1088s, 1042sh, 1000s, 623, 568, 494, 405sh, 370s, 260, 183, 139w.
Source: McDonald et al. (2013).
Comments: The sample was characterized by powder X-ray diffraction data and electron microprobe analyses.

Ramsdellite MnO_2

Origin: An unknown locality in New Mexico, USA.
Experimental details: Methods of sample preparation are not described. Raman scattering measurements have been performed on an arbitrarily oriented sample using 514.5 nm Ar^+ laser radiation. The nominal laser radiation power was 12.5 mW. 180°-scattering geometry was employed.
Raman shifts (cm^{-1}): 775, 650s, 576s, 523s, 490, 392.
Source: Bernard et al. (1993a).
Comments: The sample was characterized by powder X-ray diffraction data. For the Raman spectra of ramsdellite see also Julien et al. (2004) and Kim and Stair (2004).

Ranciéite $(Ca,Mn^{2+})_{0.2}(Mn^{4+},Mn^{3+})O_2 \cdot 0.6H_2O$

Origin: Xiangguang Mn-Ag deposit, northern China.
Experimental details: Methods of sample preparation are not described. Raman scattering measurements have been performed using 632 nm laser radiation. The laser radiation power is not indicated.
Raman shifts (cm^{-1}): 645s, 370w, 304w.
Source: Fan et al. (2015).

Comments: The sample was characterized by powder X-ray diffraction data and electron microprobe analyses.

Rankamaite $(Na,K)_3(Ta,Nb,Al)_{11}(O,OH)_{31}$

Origin: Urubu pegmatite, Itinga, Minas Gerais, Brazil.
Experimental details: Raman scattering measurements have been performed on an arbitrarily oriented sample using 532 nm Nd-YAG laser radiation. The laser radiation power is not indicated. A 180°-scattering geometry was employed.
Raman shifts (cm^{-1}): 948w, 876w, 805w, 633s, 328, 275, 239.
Source: Atencio et al. (2011).
Comments: The sample was characterized by powder X-ray diffraction data and electron microprobe analyses. The crystal structure is solved.

Rankinite $Ca_3Si_2O_7$

Origin: Upper Chegem caldera, Kabardino-Balkaria, Northern Caucasus, Russia.
Experimental details: Methods of sample preparation are not described. Raman scattering measurements have been performed using 514.5 nm Ar$^+$ laser radiation. The laser radiation output power was between 30 and 50 mW. A 180°-scattering geometry was employed.
Raman shifts (cm^{-1}): 1048w, 1007, 971, 960, 947w, 891s, 671, 552, 507w, 472w, 450w, 347, 319w, 275w, 245w, 212w, 187, 140w.
Source: Galuskin et al. (2011c).
Comments: The sample was characterized by powder X-ray diffraction data and electron microprobe analyses.

Rapidcreekite $Ca_2(SO_4)(CO_3) \cdot 4H_2O$

Origin: Bahariya depression, Western Desert, Egypt.
Experimental details: Raman scattering measurements have been performed on an arbitrarily oriented sample using 532 nm Nd-YAG laser radiation. The laser radiation power at the sample was between 20 and 200 mW.
Raman shifts (cm^{-1}): 1129, 1080s, 1003s, 664, 482, 411.
Source: Ciobotă et al. (2012).
Comments: No independent analytical data are provided for the sample used.

Raspite $Pb(WO_4)$

Origin: Broken Hill, Yancowinna Co., New South Wales, Australia.
Experimental details: Raman scattering measurements have been performed on a powdered sample using 632.8 nm He-Ne laser radiation. The laser radiation power at the sample was 6 mW. A 180°-scattering geometry was employed.
Raman shifts (cm^{-1}): 870s, 747, 667w, 645w, 523w, 494w, 395, 300, 282w, 205w, 196sh, 184w.
Source: Bastians et al. (2004).
Comments: No independent analytical data are provided for the sample used. For the Raman spectra of raspite see also Frost et al. (2004d) and Andrade et al. (2014).

Rasvumite KFe_2S_3

Origin: Miller Range 03346 nakhlite meteorite.
Experimental details: Raman scattering measurements have been performed on an arbitrarily oriented sample in a thin section using 532 nm Nd-YAG laser radiation. The laser radiation power is not indicated.
Raman shifts (cm^{-1}): 474, 221, 154.
Source: Ling and Wang (2015).
Comments: The sample was characterized by optical reflectance.

Ravatite $C_{14}H_{10}$

Origin: Synthetic.
Experimental details: Raman scattering measurements have been performed on a single crystal. No other data are provided.
Raman shifts (cm^{-1}): 3088, 3082, 3071, 3055, 3047, 3033, 3024, 3015, 3003, 1684w, 1660w, 1622, 1613, 1600, 1591, 1570, 1523, 1500w, 1491w, 1481w, 1456w, 1440s, 1429, 1418, 1404, 1377, 1363, 1348s, 1336, 1318, 1303, 1295, 1280, 1244, 1224, 1200, 1170, 1164, 1153, 1140, 1092, 1072w, 1035s, 1000, 972, 968, 950, 944, 875, 860, 829, 817, 791, 760, 753, 734, 713, 710s, 616, 547, 536, 498, 442, 428, 410s, 397, 282, 248s, 234.
Source: Godec and Colombo (1976).
Comments: For the Raman spectra of ravatite see also Witt and Mecke (1967), and Nasdala and Pekov (1993).

Raygrantite $Pb_{10}Zn(SO_4)_6(SiO_4)_2(OH)_2$

Origin: Big Horn Mts., Maricopa Co., Arizona, USA (type locality).
Experimental details: Raman scattering measurements have been performed on an arbitrarily oriented crystal using 532 nm solid-state laser radiation. The laser radiation power is not indicated.
Raman shifts (cm^{-1}): 3515, 1075, 1011, 971s, 964s, 907, 876, 838, 832, 613, 597, 463, 452, 437, 325, 250, 231.
Source: Yang et al. (2016a).
Comments: The sample was characterized by powder X-ray diffraction data and electron microprobe analyses. The crystal structure is solved.

Realgar AsS

Origin: Synthetic.
Experimental details: Raman scattering measurements have been performed on a polycrystalline sample using 647.1 nm laser radiation. The laser radiation power is not indicated.
Raman shifts (cm^{-1}): 376w, 370w, 355s, 345, 341sh, 330w, 222s, 214w, 210w, 196s, 184s, 173w, 167w, 144, 67w, 61, 57, 52, 48, 41w, 28.
Source: Muniz-Miranda et al. (1996). For the Raman spectra of realgar see also Forneris (1969), Trentelman et al. (1996), Burgio and Clark (2001), and Frost et al. (2010c).

Rebulite $Tl_5Sb_5As_8S_{22}$

Origin: Crven Dol, Allchar, Roszdan, Republic of Macedonia (type locality).
Experimental details: Raman scattering measurements have been performed on an arbitrarily oriented single crystal using 632.8 nm He-Ne laser radiation. The laser radiation power at the sample was 1.9 mW. A 180°-scattering geometry was employed.
Raman shifts (cm^{-1}): 395, 377s, 321, 306, 275, 206w, 190, 166w, 137, 126w, 102w.
Source: Makreski et al. (2014).
Comments: The sample was characterized by electron microprobe analyses.

Reedmergnerite $NaBSi_3O_8$

Origin: No data.
Experimental details: Raman scattering measurements have been performed using 514.5 nm Ar$^+$ laser radiation. The nominal laser radiation power was 10 W.
Raman shifts (cm^{-1}): 584s, 540, 517, 505, 464, 314, 261, 237, 224, 162, 142, 129.
Source: Kimata (1993).
Comments: No independent analytical data are provided for the sample used. For the Raman spectra of Reedmergnerite see also Manara et al. (2009).

Reevesite $Ni_6Fe^{3+}{}_2(CO_3)(OH)_{16}\cdot 4H_2O$

Origin: Synthetic.
Experimental details: Raman scattering measurements have been performed on arbitrarily oriented crystals using a 633 nm He-Ne laser. The laser radiation power is not indicated. The Raman shifts have been determined for the maxima of individual peaks obtained as a result of the spectral curve analysis.
Raman shifts (cm^{-1}): 3598, 3451sh, (3250), 1382, 1163w, 1074s, 832w, 676+621 (broad, unresolved doublet), 550sh, 526, 423, 308, 162sh, 145,
Source: Frost et al. (2010d).
Comments: The sample was characterized by powder X-ray diffraction data. For the Raman spectrum of reevesite see also Frost et al. (2003h).

Reichenbachite $Cu_5(PO_4)_2(OH)_4$

Origin: Banská Bystrica, central Slovakia.
Experimental details: Raman scattering measurements have been performed on an arbitrarily oriented sample using 632.8 nm He-Ne laser radiation. The laser radiation power at the sample was 17 mW. A 180°-scattering geometry was employed.
Raman shifts (cm^{-1}): 3428, 3380, 1120sh, 1083, 1055, 1027w, 998, 971, (951sh), 863, 804, 752, 700w, (636sh), 607, (572sh), 540, 515, 480s, 453, 412sh, 365, 298, 255, 214, 188sh, 175s, 135, 110s, 89.
Source: Kharbish et al. (2014).
Comments: The sample was characterized by powder X-ray diffraction data and electron microprobe analyses. For the Raman spectra of reichenbachite see also Frost et al. (2002a, 2003a).

Reinerite $Zn_3(AsO_3)_2$

Origin: Tsumeb mine, Tsumeb, Otavi district, Oshikoto, Namibia (type locality).
Experimental details: Raman scattering measurements have been performed on arbitrarily oriented crystals using a 633 nm He-Ne laser. The laser radiation power is not indicated.
Raman shifts (cm^{-1}): 804, 772, 752s, 722, 658, 297w, 279w, 240w, 219w, 188w, 176w, 152w, 141, 135s.
Source: Frost and Bahfenne (2010d).
Comments: No independent analytical data are provided for the sample used. For the Raman spectrum of reinerite see also Bahfenne (2011).

Reinhardbraunsite $Ca_5(SiO_4)_2(OH)_2$

Origin: Upper Chegem volcanic structure, Northern Caucasus, Kabardino-Balkaria, Russia.
Experimental details: Raman scattering measurements have been performed on an arbitrarily oriented grain using 514.5 nm Ar$^+$ laser radiation. The laser radiation power at the sample was below 20 mW.
Raman shifts (cm^{-1}): 3562, 3532w, 3480, 924, (834sh), 821s, 554, 421, 409sh, 310w, 280w, 253w.
Source: Galuskin et al. (2009).
Comments: The sample was characterized by powder X-ray diffraction data and electron microprobe analyses.

Rengeite $Sr_4Ti_4ZrO_8(Si_2O_7)_2$

Origin: Itoigawa region, central Japan.
Experimental details: Raman scattering measurements have been performed using 532 nm laser radiation. The laser radiation power is not indicated.
Raman shifts (cm^{-1}): 660, 265, 239.
Source: Ogawara et al. (2010).

Retgersite $Ni(SO_4)\cdot 6H_2O$

Origin: Synthetic.
Experimental details: Raman scattering measurements have been performed on an arbitrarily oriented crystal using 514.5 nm Ar$^+$ laser radiation. The laser radiation power is not indicated.
Raman shifts (cm^{-1}): (1090w), 983s, 616.
Source: Petrova et al. (2012).
Comments: No independent analytical data are provided for the sample used. For the Raman spectra of retgersite see also Krishnamurti (1958), Jain et al. (1974), Cancela et al. (1983), and Aramendia et al. (2014)

Reyerite $Na_2Ca_{14}Al_2Si_{22}O_{58}(OH)_8\cdot 6H_2O$

Origin: No data.
Experimental details: Methods of sample preparation are not described. Raman scattering measurements have been performed using 488 nm Ar$^+$ laser radiation. The nominal laser radiation power was about 25 mW.

Raman shifts (cm^{-1}): 1172, 1095w, 1078w, 1050, 1021, 907w, 752, 613s, 569s, 352, 300sh, 280s, 202, 169w.
Source: De Ferri et al. (2012).
Comments: No independent analytical data are provided for the sample used. The Raman shifts were determined by us based on spectral curve analysis of the published spectrum.

Rhabdophane-(Ce) $Ce(PO_4)·H_2O$

Origin: Synthetic.
Experimental details: Methods of sample preparation are not described. Raman scattering measurements have been performed using 514.5 nm Ar$^+$ laser radiation. The nominal laser radiation power was between 5 and 100 mW.
Raman shifts (cm^{-1}): 1088, 1057, 977s, 642, 624, 571, 469s, 417.
Source: Assaaoudi et al. (2001).
Comments: The sample was characterized by powder X-ray diffraction data.

Rhabdophane-(Nd) $Nd(PO_4)·H_2O$

Origin: Synthetic.
Experimental details: Methods of sample preparation are not described. Raman scattering measurements have been performed using 514.5 nm Ar$^+$ laser radiation. The nominal laser radiation power was between 5 and 100 mW.
Raman shifts (cm^{-1}): 1094, 1057, 1033, 983s, 630, 582. 546, 470s, 419.
Source: Assaaoudi et al. (2001).
Comments: The sample was characterized by powder X-ray diffraction data.

Rheniite ReS_2

Origin: Synthetic.
Experimental details: Methods of sample preparation are not described. Raman scattering measurements have been performed using 633 nm laser radiation. The laser radiation power at the sample was 1 mW.
Raman shifts (cm^{-1}): 438w, 419w, 407, 378w, 369, 349w, 325, 321, 311w, 308w, 284w, 278w, 237, 217s, 164, 153s, 146, 140w.
Source: Feng et al. (2015c).
Comments: For the Raman spectrum of rheniite see also Tongay et al. (2014).

Rhodizite $KBe_4Al_4(B_{11}Be)O_{28}$

Origin: Antsongombato pegmatite, Central Madagascar.
Experimental details: Raman spectra were obtained on the dodecahedral and tetrahedral faces. Characteristics of the laser radiation are not indicated.
Raman shifts (cm^{-1}): 857, 803, 651, 544, 470s, 430s, 294.
Source: Laurs et al. (2002).
Comments: The sample was characterized by electron microprobe analyses. It contains zones corresponding to rhodizite and londonite. For the Raman spectrum of rhodizite see also Frost et al. (2014a).

Rhodochrosite $Mn(CO_3)$

Origin: Kohlenbachvalley, Eiserfeld, Siegerland, North Rhine-Westphalia, Germany.
Experimental details: Methods of sample preparation are not described. Raman scattering measurements have been performed using 532 nm Nd-YAG laser radiation. The nominal laser radiation power was 100 mW.
Raman shifts (cm^{-1}): 1752w, 1439w, 1094s, 726, 293.
Source: Buzgar and Apopei (2009).
Comments: No independent analytical data are provided for the sample used. For the Raman spectra of rhodochrosite see also Rutt and Nicola (1974), Frezzotti et al. (2012), and Capitani et al. (2014).

Rhodonite $Mn^{2+}SiO_3$

Origin: Sverdlovsk region, Urals, Russia.
Experimental details: Methods of sample preparation are not described. Raman scattering measurements have been performed using 532 nm Nd-YAG laser radiation. The nominal laser radiation power was 100 mW.
Raman shifts (cm^{-1}): 1038, 996s, 973s, 939w, 910w, 878, 714w, 667s, 557w, 510, 417, 385w, 347sh, 327, 265, 250.
Source: Buzatu and Buzgar (2010).
Comments: No independent analytical data are provided for the sample used. For the Raman spectra of rhodonite see also Mills et al. (2005), Makreski et al. (2006b), and Can et al. (2011).

Rhomboclase $(H_5O_2)Fe^{3+}(SO_4)_2 \cdot 2H_2O$

Origin: Coranda-Hondol open pit, Certej Au-Ag deposit, Romania.
Experimental details: Methods of sample preparation are not described. Raman scattering measurements have been performed using 532 nm Nd-YAG laser radiation. The nominal laser radiation power was 100 mW.
Raman shifts (cm^{-1}): 2775w, 2661w, 1456w, 1181, 1081sh, 1028sh, 1014s, 763w, 650sh, 622sh, 603, 472sh, 454, 381, 265sh, 242.
Source: Apopei et al. (2015).
Comments: The sample was characterized by powder X-ray diffraction data. For the Raman spectrum of rhomboclase see also Ling and Wang (2010).

Rhönite $Ca_4[Mg_8Fe^{3+}_2Ti_2]O_4[Si_6Al_6O_{36}]$

Origin: Eifel, Germany.
Experimental details: Raman scattering measurements have been performed on a arbitrarily oriented grains using 785 nm laser radiation. The laser radiation power is not indicated.
Raman shifts (cm^{-1}): 980, 840, 705, 655s, 535s, 470.
Source: Treiman (2008).
Comments: No independent analytical data are provided for the sample used.

Richelsdorfite $Ca_2Cu_5Sb^{5+}(AsO_4)_4(OH)_6Cl\cdot 6H_2O$

Origin: Wilhelm mine, Bauhaus, Richelsdorf District, Hesse, Germany (type locality).
Experimental details: Raman scattering measurements have been performed on arbitrarily oriented crystals using a 633 nm He-Ne laser. The laser radiation power is not indicated. The Raman shifts have been determined for the maxima of individual peaks obtained as a result of the spectral curve analysis.
Raman shifts (cm^{-1}): 1564w, 1376w, 1082w, 988w, 910sh, 849s, (835), (792), 546, 498sh, 415w, 344, 268sh, 185sh, 144s.
Source: Frost et al. (2011c).
Comments: No independent analytical data are provided for the sample used.

Richterite $Na(NaCa)Mg_5Si_8O_{22}(OH)_2$

Origin: No data.
Experimental details: Raman scattering measurements have been performed in the range from 3500 to 3800 cm^{-1} in backscattering geometry, using a 514.5 nm Ar^+ laser. The laser radiation power is not indicated.
Source: Leissner et al. (2015).
Raman shifts (cm^{-1}): 3730, 3712sh.
Comments: The sample was characterized by EMPA and ICP-MS.

Riebeckite $\square Na_2(Fe^{2+}_3Fe^{3+}_2)Si_8O_{22}(OH)_2$

Origin: Iacobdeal, Dobrogea, Romania.
Experimental details: Methods of sample preparation are not described. Raman scattering measurements have been performed using 532 nm Nd-YAG laser radiation. The nominal laser radiation power was 100 mW.
Raman shifts (cm^{-1}): 1084, 980sh, 966s, 885, 666s, 576, 537s, 431w, 363, 325, 244w, 222sh, 198s, 171s, 140.
Source: Apopei et al. (2011).
Comments: No independent analytical data are provided for the sample used. For the Raman spectra of riebeckite see also Apopei and Buzgar (2010).

Riebeckite (Crocidolite) $\square Na_2(Fe^{2+}_3Fe^{3+}_2)Si_8O_{22}(OH)_2$

Origin: No data.
Experimental details: Raman scattering measurements have been performed on an unoriented fibrous aggregate using 632.8 nm He-Ne laser radiation. The laser radiation power was 20 mW.
Raman shifts (cm^{-1}): 1082s, 1030, 967s, 889, 771w, 733w, 664s, 577s, 537, 506sh, 470sh, 428, 374, 360sh, 331, 300, 272, 246, 211, 195, 162s.
Source: Rinaudo et al. (2004).
Comments: The sample was characterized by powder X-ray diffraction data and electron microprobe analyses. For the Raman spectra of crocidolite see also Petry et al. (2006) and Croce et al. (2013).

Rimkorolgite $BaMg_5(PO_4)_4 \cdot 8H_2O$

Origin: Zheleznyi (Iron) mine, Kovdor massif, Kola Peninsula, Russia (type locality).

Experimental details: Raman scattering measurements have been performed on arbitrarily oriented crystals using a 633 nm He-Ne laser. The laser radiation power is not indicated. The Raman shifts have been determined for the maxima of individual peaks obtained as a result of the spectral curve analysis.

Raman shifts (cm^{-1}): (3444), 3272 (broad), (2991), 2913, 2859, 1492sh, 1480, 1467, 1455, 1436, 1236, 1135, 1105, 1073sh, 1052, (1035), 1016, (992), 975+964s (unresolved doublet?), (951), 930, 653sh, 622sh, 599, 570, 511, (485), 472, 439, 426, 373, 279, 262+252 (unresolved doublet?), 222, 195w, 159+146 (unresolved doublet?), 109.

Source: Frost et al. (2014h).

Comments: The sample was characterized by qualitative electron microprobe analysis.

Ringwoodite $Mg_2(SiO_4)$

Origin: Grove Mountains 052049 meteorite.

Experimental details: Raman scattering measurements have been performed on arbitrarily oriented grains in polished sections using 514.5 nm Ar^+ laser radiation. The nominal laser radiation power was 20 mW.

Raman shifts (cm^{-1}): 841–849, 783–796, 285–296.

Source: Feng et al. (2011).

Comments: The samples were characterized by electron microprobe analyses. As the fayalite content increases from 27.8 to 81.6 mol %, the bands at 783–796 and 285–296 cm^{-1} shift towards lower frequencies, whereas the band at 841–849 cm^{-1} does not show significant correlation with the fayalite content. For the Raman spectrum of ringwoodite see also Akaogi et al. (1984).

Rinkite $TiNa_2Ca_4REE(Si_2O_7)_2OF_3$

Origin: Khibiny massif, Kola Peninsula, Russia.

Experimental details: Raman scattering measurements have been performed on an annealed metamict sample using 632.8 nm laser radiation. The laser radiation power is not indicated.

Raman shifts (cm^{-1}): ~960, ~460.

Source: Zubko et al. (2013).

Riomarinaite $Bi(SO_4)(OH) \cdot H_2O$

Origin: No data.

Experimental details: No data.

Raman shifts (cm^{-1}): ~1190w, ~1160w, ~1095w, ~1002, ~960s, ~630, ~525, ~402w, ~198s.

Source: Capitani et al. (2014).

Comments: No independent analytical data are provided for the sample used.

Robertsite $Ca_2Mn^{3+}_3O_2(PO_4)_3 \cdot 3H_2O$

Origin: Tip Top mine, Custer Co., South Dakota, USA (type locality).

Experimental details: Raman scattering measurements have been performed on an arbitrarily oriented crystal using 532 nm laser radiation. The laser radiation power is not indicated.

Raman shifts (cm^{-1}): ~1187s, 1036, 947, 625s, 552, 497, 385, 289s.
Source: Andrade et al. (2012).
Comments: The sample was characterized by single-crystal X-ray diffraction data. The crystal structure is solved.

Robinsonite $Pb_4Sb_6S_{13}$

Origin: Zlatá Baňa, Slanské Vrchy Mts., central Slovakia.
Experimental details: Raman scattering measurements have been performed on a polycrystalline sample in the spectral region from 10 to 600 cm^{-1} using 532 nm Nd-YAG laser radiation. The laser radiation power is not indicated. A 180°-scattering geometry was employed. The Raman shifts have been determined for the maxima of individual peaks obtained as a result of the spectral curve analysis.
Raman shifts (cm^{-1}): 337sh, 328, 314, 308, 249sh, 228sh, 209s, 188sh, 168, 149, 133, 111sh, 101, 91sh, 75w.
Source: Kharbish and Jeleň (2016).
Comments: The sample was characterized by electron microprobe analyses. The empirical formula of the sample used is $Pb_{4.01}Sb_{5.99}S_{13.00}$.

Rockbridgeite $Fe^{2+}Fe^{3+}_4(PO_4)_3(OH)_5$

Origin: Galileia region, Minas Gerais, Brazil.
Experimental details: Raman scattering measurements have been performed on a radiated aggregate using 514.5 nm Ar$^+$ laser radiation. The laser radiation power at the sample was 1 mW.
Raman shifts (cm^{-1}): 1186, 1137, 1061s, 981, 937, 638, 616, 576, 463, 382, 333s, 299, 241.
Source: Faulstich et al. (2013).
Comments: The sample was characterized by electron microprobe analyses.

Rodalquilarite $H_3Fe^{3+}_2(Te^{4+}O_3)_4Cl$

Origin: Grand Central Mines, Tombstone, Cochise Co., Arizona, USA.
Experimental details: Raman scattering measurements have been performed on arbitrarily oriented crystals using a 633 nm He-Ne laser. The laser radiation power is not indicated. The Raman shifts have been determined for the maxima of individual peaks obtained as a result of the spectral curve analysis.
Raman shifts (cm^{-1}): (2998), 2870+2796 (unresolved doublet?), 2341, 781, 756sh, 725, 660sh, 641s, 612s, (599), 473, 449, 412, 400sh, 345s, 321s, (312), 233, 191, 179, 142, 110.
Source: Frost and Keeffe (2009i).
Comments: No independent analytical data are provided for the sample used.

Rodolicoite $Fe^{3+}(PO_4)$

Origin: Synthetic.
Experimental details: Raman scattering measurements have been performed on a polycrystalline sample using 514.5 nm Ar$^+$ laser radiation. The nominal laser radiation power was 150 mW. A 180°-scattering geometry was employed.
Raman shifts (cm^{-1}): 1018s, 436, 415, 390, 336, 280, 199, 161.

Source: Murli et al. (1997).
Comments: The sample was characterized by powder X-ray diffraction data. For the Raman spectrum of rodolicoite see also Bhalerao et al. (2012).

Rokühnite $FeCl_2 \cdot 2H_2O$

Origin: Synthetic.
Experimental details: Raman scattering measurements have been performed at 10 K on a single crystal using 514.5 nm Ar^+ laser radiation, with the incident beam parallel to the c-axis. The laser radiation power is not indicated.
Raman shifts (cm^{-1}): 747w, 717w, 661sh, 640.5s, 594s, 552w, 501, 375, 202s, 196s, 146, 141w, 33.7, 30.6s.
Source: Graf and Schaack (1976).
Comments: For the Raman spectrum of rokühnite see also Graf (1978).

Romanèchite $(Ba,H_2O)_2(Mn^{4+},Mn^{3+})_5O_{10}$

Origin: Bahariya depression, Western Desert, Egypt.
Experimental details: Raman scattering measurements have been performed on an arbitrarily sample using a 532 nm Nd-YAG laser. The laser radiation power at the sample was 20 to 200 µW.
Source: Ciobotă et al. (2012).
Raman shifts (cm^{-1}): ~1300, ~1100w, 644sh, 583s, ~500sh, ~390.
Comments: No independent analytical data are given for the sample used. For the Raman spectra of romanèchite see also Julien et al. (2003, 2004).

Romanorlovite $K_8Cu_6Cl_{17}(OH)_3$

Origin: Second scoria cone, Northern Breakthrough of the Great Tolbachik Fissure Eruption, Tolbachik, Kamchatka, Russia (type locality).
Experimental details: Raman scattering measurements have been performed on a polycrystalline sample using 532 nm laser radiation. The laser radiation output was 3 mW. A 180°-scattering geometry was employed.
Raman shifts (cm^{-1}): 3512w, 3440w, 931w, 879w, 548, 477, 264s, 178s.
Source: Pekov et al. (2016b).
Comments: The sample was characterized by powder X-ray diffraction data and electron microprobe analyses. The crystal structure is solved.

Romarchite SnO

Origin: Synthetic.
Experimental details: Raman scattering measurements have been performed on a arbitrarily oriented particles using 785 nm laser radiation. The laser radiation power at the sample was ~200 mW. A 180°-scattering geometry was employed.
Raman shifts (cm^{-1}): 210, 240.
Source: Chen and Grandbois (2013).
Comments: The sample was characterized by powder X-ray diffraction data.

Römerite $Fe^{2+}Fe^{3+}_2(SO_4)_4 \cdot 14H_2O$

Origin: Medvedza lens, Košice-Bankov magnesite deposit, Slovak Republic.
Experimental details: Raman scattering measurements have been performed on arbitrarily oriented crystals using a 633 nm He-Ne laser. The laser radiation power is not indicated. The Raman shifts have been determined for the maxima of individual peaks obtained as a result of the spectral curve analysis.
Raman shifts (cm^{-1}): 3465sh, 3340, 3235, 3029, 1642w, 1164, 1117, 1058, 1035s, 1012s, 999, 733w, 650, 608, 472, 447, 399, 278+264 (unresolved doublet?), 231, 173, 145.
Source: Frost et al. (2011f).
Comments: The sample was characterized by powder X-ray diffraction data and electron microprobe analyses.

Rondorfite $Ca_8Mg(SiO_4)_4Cl_2$

Origin: Upper Chegem caldera, Kabardino-Balkaria, Northern Caucasus, Russia.
Experimental details: Methods of sample preparation are not described. Raman scattering measurements have been performed using 514.5 nm Ar^+ laser radiation. The laser radiation output power was between 30 and 50 mW. A 180°-scattering geometry was employed.
Raman shifts (cm^{-1}): ~1000w, ~975, ~950w, ~920w, ~862s, ~820, ~570, ~520, ~415w, ~385, ~330w, ~263.
Source: Galuskin et al. (2013a).
Comments: The sample was characterized by electron microprobe analyses.

Rongibbsite $Pb_2(Si_4Al)O_{11}(OH)$

Origin: Big Horn Mts., Maricopa Co., Arizona, USA (type locality).
Experimental details: Raman scattering measurements have been performed on an arbitratily oriented sample using 532 nm laser radiation. The laser radiation power is not indicated.
Raman shifts (cm^{-1}): 3525s, 3430s, 962, 630sh, 602, 488, 453, 422, 372, 283, 258, 196s.
Source: Yang et al. (2013a).
Comments: The sample was characterized by single-crystal X-ray diffraction data and electron microprobe analyses. The crystal structure is solved.

Ronneburgite $K_2MnV_4O_{12}$

Origin: Ronneburg, Thuringia, Germany (type locality).
Experimental details: Micro-Raman scattering measurements have been performed on an arbitrarily oriented sample using 676 nm Kr^+ laser radiation. The laser radiation power was 1.5 mW.
Raman shifts (cm^{-1}): 952s, 911, 878, 830, 658w, 461, 350, 336, 261.
Source: Witzke et al. (2001).
Comments: The sample was characterized by powder X-ray diffraction data and electron microprobe analyses. The crystal structure is solved.

Rooseveltite $Bi(AsO_4)$

Origin: Synthetic.
Experimental details: Raman scattering measurements have been performed on a polycrystalline sample using 514.5 nm Ar^+ laser radiation.
Raman shifts (cm^{-1}): 841s, 795, 768, 425, 413, 386, 346s, 333sh, 276, 221.
Source: Roncaglia et al. (1993).
Comments: The sample was btained by slow addition of diluted arsenic acid to a diluted stoichiometric $Bi(NO_3)_3 \cdot 5H_2O$ solution and subsequent heating of the precipitated material at 600 °C during 12 h. The purity was checked by chemical analysis and powder X-ray diffractometry.

Roquesite $CuInS_2$

Origin: Synthetic.
Experimental details: Raman scattering measurements have been performed on a polycrystalline sample using 514.5 nm Ar^+ laser radiation. The laser radiation power is not indicated. A 180°-scattering geometry was employed.
Raman shifts (cm^{-1}): 340sh, 298, 240sh.
Source: Dutková et al. (2016).
Comments: The sample was characterized by powder X-ray diffraction data. For the Raman spectrum of roquesite see also Ho et al. (2012).

Rosasite $CuZn(CO_3)(OH)_2$

Origin: No data.
Experimental details: Raman scattering measurements have been performed on an arbitrarily oriented sample using a 632.8 nm He-Ne laser 30 mW. The laser radiation power at source reduced considerably by various filters.
Raman shifts (cm^{-1}): 3470, 3422w, 3232, 1540, 1514, 1453, 1086s, 1060, 843, 833w, 702, 508, 482w, 409, 390w, 332, 308w, 231, 208w, 193s, 146s, 126.
Source: Bouchard and Smith (2003).
Comments: No independent analytical data are provided for the sample used. For the Raman spectra of rosasite see also Frost (2006) and Rotondo et al. (2012).

Roselite $Ca_2Co(AsO_4)_2 \cdot 2H_2O$

Origin: Bou Azzer, Morocco.
Experimental details: Raman scattering measurements have been performed on arbitrarily oriented crystals using a 633 nm He-Ne laser. The laser radiation power is not indicated. The Raman shifts have been determined for the maxima of individual peaks obtained as a result of the spectral curve analysis.
Raman shifts (cm^{-1}): 3450, 3208, (3121), 3042, 1688, 1611, 1118, 976, 909, 864, (800), 798, 719, 659, 653, 540, 463, 440, 399, 373, 338, 307, 264, 243, 211, 197, 179, 155, 117.
Source: Frost (2009a).
Comments: No independent analytical data are provided for the sample used. Intensities of the bands are not indicated.

Rosiaite $PbSb_2O_6$

Origin: Synthetic.
Experimental details: Methods of sample preparation are not described. Diffusion Raman scattering measurements have been performed using 488 nm Ar^+ laser radiation. The nominal laser radiation power was 600 mW.
Raman shifts (cm^{-1}): 670s, 510, 498w, 318, 278w, 211.
Source: Vandenborre et al. (1980).
Comments: The sample was characterized by powder X-ray diffraction data.

Rostite $Al(SO_4)(OH) \cdot 5H_2O$

Origin: Le Cetine mine, Rosia, Chiusdino, Siena, Italy.
Experimental details: Raman scattering measurements have been performed on arbitrarily oriented crystals using a 633 nm He-Ne laser. The laser radiation power is not indicated. The Raman shifts have been determined for the maxima of individual peaks obtained as a result of the spectral curve analysis.
Raman shifts (cm^{-1}): 3295sh, (3222), 3155, 3082, (2948), (2764), 1692, 1605w, 1390w, 1312, 1227, (1145w), 1131, (1093w), 1083, (1070w), (998), 991s, (986), 939w, 874w, 854sh, 632, 620, 590, 570, 530w, 504, 434sh, 420, 340w, 319sh, 307sh, 295, 281sh, 216, 203sh, 169.
Source: Frost et al. (2015x).
Comments: The sample was characterized by qualitative electron microprobe analysis.

Rouaite $Cu_2(NO_3)(OH)_3$

Origin: Synthetic.
Experimental details: Methods of sample preparation are not described. Raman scattering measurements have been performed using 785 nm laser radiation. The laser radiation was 3 mW.
Raman shifts (cm^{-1}): 1423, 1321, 1047s, 714, 500.5, 456, 408, 331, 255.
Source: Nytko et al. (2008).
Comments: The sample was characterized by powder X-ray diffraction data. For the Raman spectrum of rouaite see also Nytko (2008).

Roumaite $(Nb,Ti)(Ca,Na,\square)_3(Ca,REE)_4(Si_2O_7)_2(OH)F_3$

Origin: Rouma Island, Los Archipelago, Guinea (type locality).
Experimental details: Raman scattering measurements have been performed on an arbitrarily oriented sample using 632.8 nm He-Ne laser radiation. The laser radiation power is not indicated. A nearly 180°-scattering geometry was employed.
Raman shifts (cm^{-1}): 1582w (broad).
Source: Biagioni et al. (2010).
Comments: The Raman spectrum shows important contributions of fluorescence effects related to the presence of *REE*, which does not allow an accurate study of the region above 3000 cm^{-1}. No data on the Raman spectrum below 1582 cm^{-1} are provided.

Rowleyite [Na(NH$_4$,K)$_9$Cl$_4$][(V^{5+},V^{4+})$_2$(P,As)O$_8$]$_6$·n[H$_2$O,Na,NH$_4$,K,Cl]

Origin: Rowley mine, about 100 km SW of Phoenix, Arizona, USA (type locality).
Experimental details: Methods of sample preparation are not described. Raman scattering measurements have been performed using 532 nm laser radiation. The nominal laser radiation power was 5 mW.
Raman shifts (cm^{-1}): ~1340, 1065w, 1002, 980, 825s, 683w, 565w, 460w, 325s, 280s, 180.
Source: Kampf et al. (2017b).
Comments: The sample was characterized by powder X-ray diffraction data and electron microprobe analyses. The crystal structure is solved.

Roxbyite Cu$_9$S$_5$

Origin: Synthetic.
Experimental details: No data.
Raman shifts (cm^{-1}): 466s, 259w.
Source: Kumar and Nagarajan (2011).
Comments: The sample was characterized by powder X-ray diffraction data.

Rozenite Fe^{2+}(SO$_4$)·4H$_2$O

Origin: Coranda-Hondol open pit, Certej Au-Ag deposit, Romania.
Experimental details: Methods of sample preparation are not described. Raman scattering measurements have been performed using 532 nm Nd-YAG laser radiation. The nominal laser radiation power was 100 mW.
Raman shifts (cm^{-1}): 3388, 3329sh, 3261sh, 1594w, 1176sh, 1149w, 1098sh, 1073w, 992s, 658sh, 612w, 480, 461sh, 383w, 284w, 239sh.
Source: Apopei et al. (2015).
Comments: The sample was characterized by powder X-ray diffraction data and electron microprobe analyses. For the Raman spectra of rozenite see also Chio et al. (2007), Buzatu et al. (2012, 2016), Jentzsch et al. (2013), Aramendia et al. (2014), and Kompanchenko et al. (2016).

Rruffite Ca$_2$Cu(AsO$_4$)$_2$·2H$_2$O

Origin: Maria Catalina mine, Tierra Amarilla, Chile (type locality).
Experimental details: Raman scattering measurements have been performed on an arbitrarily oriented single crystal using 532 nm solid-state laser radiation. The laser radiation power is not indicated.
Raman shifts (cm^{-1}): 3335w (broad), 3147w (broad), 866w, 839s, 803w, 715w, 485w, 451w, 426w, 335, 294.
Source: Yang et al. (2011a).
Comments: The sample was characterized by powder X-ray diffraction data and electron microprobe analyses. The crystal structure is solved.

Rucklidgeite $PbBi_2Te_4$

Origin: Ozernyi district, Salla-Kuolajarvi, Kola Peninsula, Russia.
Experimental details: Raman scattering measurements have been performed on an arbitrarily oriented sample using 514.5 nm Ar$^+$ laser radiation. The nominal laser radiation power was 50 mW.
Raman shifts (cm^{-1}): 127, 102s, 57.
Source: Voloshin et al. (2015a).
Comments: The sample was characterized by electron microprobe analyses.

Rudashevskyite $(Fe,Zn)S$

Origin: Indarch meteorite (an EH4 enstatite chondrite).
Experimental details: Raman scattering measurements have been performed on arbitrarily oriented grains using 514.5 nm Ar$^+$ laser radiation. The laser radiation power at the sample was 1.2 mW.
Raman shifts (cm^{-1}): ~460w, ~317s.
Source: Ma et al. (2012a).
Comments: No independent analytical data are provided for the sample used.

Ruizite $Ca_2Mn^{3+}_2Si_4O_{11}(OH)_4 \cdot 2H_2O$

Origin: Wessels mine, Hotazel, Kalahari Manganese Field, Northern Cape Province, South Africa.
Experimental details: Raman scattering measurements have been performed on an arbitrarily oriented sample using 532 nm laser radiation. The nominal laser radiation power was 150 mW.
Raman shifts (cm^{-1}): ~3578, ~3355w, ~3235w, ~2942, 924s, 727w, 639, 571s, 506w, 477, 433, 411, 364, 288, 256, 220, 180.
Source: Fendrich et al. (2016).
Comments: The sample was characterized by powder X-ray diffraction data and electron microprobe analyses. The crystal structure is solved.

Rusinovite $Ca_{10}(Si_2O_7)_3Cl_2$

Origin: Upper Chegem caldera, Kabardino-Balkaria, northern Caucasus, Russia (type locality).
Experimental details: Methods of sample preparation are not described. Raman scattering measurements have been performed using 514.5 nm Ar$^+$ laser radiation. The laser radiation output power was between 30 and 50 mW. A 180°-scattering geometry was employed.
Raman shifts (cm^{-1}): 1074w, 1036, 994w, 968w, 900s, 869, 830w, 652, 635, 568w, 549w, 528w, 430w, 409w, 365, 325w, 295w, 282w, 232w, 125w.
Source: Galuskin et al. (2011c).
Comments: The sample was characterized by powder X-ray diffraction data and electron microprobe analyses. The crystal structure is solved.

Russellite Bi_2WO_6

Origin: No data.
Experimental details: Raman scattering measurements have been performed on arbitrarily oriented crystals using a 785 nm Nd-YAG laser. The laser radiation power at the sample was 1 mW.
Raman shifts (cm^{-1}): 844, 795, 716, 667, 405, 349, 324, 284, 263

Source: Frost et al. (2004d).
Comments: No independent analytical data are provided for the sample used.

Rustumite $Ca_{10}(Si_2O_7)_2(SiO_4)(OH)_2Cl_2$

Origin: Upper Chegem Caldera, Northern Caucasus, Russia.
Experimental details: Methods of sample preparation are not described. Raman scattering measurements have been performed using 514.5 nm Ar^+ laser radiation. The laser radiation output power was between 30 and 50 mW.
Raman shifts (cm^{-1}): 3632w, 3585, 1056w, 1004, 914s, 868w, 812, 648, 556w, 538w, 381, 335w.
Source: Gfeller et al. (2013).
Comments: The sample was characterized by single-crystal X-ray diffraction data and electron microprobe analyses. The crystal structure is solved.

Rutherfordine $(UO_2)(CO_3)$

Origin: Sierra Albarrana, Córdoba, Spain.
Experimental details: Methods of sample preparation are not described. Raman scattering measurements have been performed using 632.8 nm He-Ne laser radiation. The laser radiation output power was 20 mW.
Raman shifts (cm^{-1}): 1120, 889s, 833, 789w, 220w, 162, 142.
Source: Bonales et al. (2015).
Comments: The sample was characterized by electron microprobe analyses. For the Raman spectra of rutherfordine see also Frost and Čejka (2009b) and Bonales et al. (2016).

Rutile TiO_2

Origin: Santa Benedetta, Canavese, Italy.
Experimental details: Methods of sample preparation are not described. Raman scattering measurements have been performed on an arbitrarily oriented sample using 632.8 nm He-Ne or 488 nm Ar^+ laser radiation. The laser radiation power is not indicated.
Raman shifts (cm^{-1}): 611s, 441s, 242, 142w.
Source: Andò and Garzanti (2014).
Comments: No independent analytical data are provided for the sample used. For the Raman spectrum of rutile see also Balachandran and Eror (1982).

Rynersonite $CaTa_2O_6$

Origin: Synthetic.
Experimental details: Raman scattering measurements have been performed on a single crystal fiber using 633 nm He-Ne laser radiation. The laser radiation power at the sample was 6 mW. A 180°-scattering geometry was employed.
Raman shifts (cm^{-1}): 690s, 654, 632w, 556w, 475, 342, 285, 243, 236, 175, 167, 151, 88s (with laser beam parallel to the fiber).
Source: Almeida et al. (2014).

Sabugalite $HAl(UO_2)_4(PO_4)_4 \cdot 16H_2O$

Origin: No data.
Experimental details: Raman scattering measurements have been performed at different temperatures on arbitrarily oriented crystals using a 633 nm He-Ne laser. The laser radiation power is not indicated. The Raman shifts have been determined for the maxima of individual peaks obtained as a result of the spectral curve analysis.
Raman shifts (cm^{-1}): 1008, 984, 970, 848, 826s, 806 (for the spectrum obtained at 40°C).
Source: Frost et al. (2005k).
Comments: The sample was characterized by powder X-ray diffraction data. For the Raman spectra of sabugalite see also Frost and Weier (2004c).

Sahlinite $Pb_{14}O_9(AsO_4)_2Cl_4$

Origin: Långban, near Pajsberg and Filipstad, Värmland, Sweden (type locality).
Experimental details: Micro-Raman scattering measurements have been performed on an arbitrarily oriented sample using 514.5 nm Ar$^+$ laser radiation. The nominal laser radiation power was 20 mW.
Raman shifts (cm^{-1}): 3526w, 819s, 806s.
Source: Jonsson (2003).
Comments: The sample was characterized by powder X-ray diffraction data and electron microprobe analyses.

Sailaufite $(Ca,Na,\square)_2Mn^{3+}_3O_2(AsO_4)_2(CO_3) \cdot 3H_2O$

Origin: Hartkoppe hill, Ober–Sailauf, Spessart Mts., Germany (type locality).
Experimental details: Raman scattering measurements have been performed on an arbitrarily oriented single crystal using 633 nm He-Ne laser radiation. The laser radiation power is not indicated.
Raman shifts (cm^{-1}): See comment below.
Source: Wildner et al. (2003).
Comments: The authors of the cited paper write: "Bands or band components observed in the IR- or Raman spectra around 730, 880, 1120, and 1420 cm^{-1} can be assigned to ... vibrational modes of the two different CO$_3$ groups." However these bands could be assigned only to a single CO$_3$ group. Moreover, Raman spectrum of sailaufite is not given by Wildner et al. (2003). The bands at 730, 880, and 1420 cm^{-1} can correspond to admixed Mn-bearing dolomite that is present in association with sailaufite.

Sakhaite $Ca_{48}Mg_{16}Al(SiO_3OH)_4(CO_3)_{16}(BO_3)_{28} \cdot (H_2O)_3(HCl)_3$

Origin: Titovskoe, Sakha (Yakutia) Republic, Russia (type locality).
Experimental details: Raman scattering measurements have been performed on arbitrarily oriented crystals using a 633 nm He-Ne laser. The laser radiation power is not indicated. The Raman shifts have been determined for the maxima of individual peaks obtained as a result of the spectral curve analysis.
Raman shifts (cm^{-1}): 3546, 3391, 2897, 1727, 1703w, 1560w, 1524w, 1479, 1349w, 1312w, 1218, 1167sh, 1134+1123s (unresolved doublet?), 968s, 950sh, 855w, 737sh, 725, 651, 627, 396w, 310, 211, 156, (132).
Source: Frost and Xi (2012j).
Comments: No independent analytical data are provided for the sample used.

Salammoniac NH_4Cl

Origin: Burning coal wastepile materials, Douro coalfield, Portugal.
Experimental details: No data.
Raman shifts (cm^{-1}): ~3120s, ~3050s, ~2807, (~2500 broad), (~2020 broad), ~1760, ~1705, ~1500, ~1400.
Source: Ribeiro et al. (2010).
Comments: The sample was characterized by powder X-ray diffraction dataю

Saléeite $Mg(UO_2)_2(PO_4)_2 \cdot 10H_2O$

Origin: East Alligator River, Northern Territory, Australia.
Experimental details: Raman scattering measurements have been performed on arbitrarily oriented crystals using 647.1 nm Kr$^+$ and 785 diode laser radiations. The laser radiation power is not indicated.
Raman shifts (cm^{-1}): 999s, 837s, 405, 284w, 194.
Source: Faulques et al. (2015a, b).
Comments: No independent analytical data are provided for the sample used. For the Raman spectra of saléeite see also Frost (2004b) and Frost and Weier (2004c).

Samarskite-(Y) $(Y,Ce,U,Fe,Nb)(Nb,Ta,Ti)O_4$

Origin: Beinmyr pegmatite, Landås, Iveland, Aust-Agder, Norway.
Experimental details: Raman scattering measurements have been performed on a metamictsample using 514.5 nm Ar$^+$ laser radiation. The laser radiation power at the sample was 20 mW.
Raman shifts (cm^{-1}): ~785s (broad), ~620 (broad), ~230sh (for a metamict sample); ~795, ~670s, ~535, ~417, ~360s, ~335s, ~230s, ~190s, ~115 (for a sample recrystallised in air at 1000°C.)
Source: Tomašić et al. (2010).
Comments: The sample was characterized by electron microprobe analyses.

Sampleite $NaCaCu_5(PO_4)_4Cl \cdot 5H_2O$

Origin: Northparkes mine, Goonumbla, New South Wales, Australia.
Experimental details: Raman scattering measurements have been performed on arbitrarily oriented crystals using a 633 nm He-Ne laser. The laser radiation power is not indicated. The Raman shifts have been determined for the maxima of individual peaks obtained as a result of the spectral curve analysis.
Raman shifts (cm^{-1}): 1269, 1152, 1088, (1016), 997s, 962s, 924sh, 643s, 604, 591, 557, 455s, 356, 282, 224, 190, 172.
Source: Frost et al. (2007m).
Comments: No independent analytical data are provided for the sample used.

Sanderite $Mg(SO_4) \cdot 2H_2O$

Origin: Synthetic.
Experimental details: Raman scattering measurements have been performed on a powdered sample using 532 nm Nd-YAG laser radiation. The laser radiation power at the sample was 15 mW.

Raman shifts (cm^{-1}): 3539, 3446s, 1647, 1164, 1034s, 630w, 597w, 492w, 447, 266.
Source: Wang et al. (2006a).
Comments: The sample was characterized by powder X-ray diffraction data. For the Raman spectra of sanderite see also Frezzotti et al. (2012) and Brotton and Kaiser (2013).

Sanguite KCuCl$_3$

Origin: Glavnaya Tenoritovaya fumarole, Second scoria cone, Tolbachik volcano, Kamchatka, Russia (type locality).
Experimental details: Raman scattering measurements have been performed on an arbitrarily oriented sample using 532 nm laser radiation. The laser radiation power at the sample was 3 mW.
Raman shifts (cm^{-1}): 547, 296sh, 272, 192s, 137, 117.
Source: Pekov et al. (2015a).
Comments: The sample was characterized by powder X-ray diffraction data and electron microprobe analyses. The crystal structure is solved. For the Raman spectrum of sanguite see also Choi et al. (2005).

Sanidine K(AlSi$_3$O$_8$)

Origin: Zvegor, Republic of Macedonia.
Experimental details: No data.
Raman shifts (cm^{-1}): 1117w, 1040w, 515s, 473, 450w, 406w, 379w, 338w, 283, 264w, 225w, 198w, 160, 122w, 108w.
Source: Makreski et al. (2009).
Comments: The sample was characterized by powder X-ray diffraction data and electron microprobe analysis. For the Raman spectra of sanidine see also Matson et al. (1986), Edwards et al. (2004), and Frezzotti et al. (2012).

Sanjuanite Al$_2$(PO$_4$)(SO$_4$)(OH)·9H$_2$O

Origin: Chica de Zonda, San Juan province, Argentina (type locality).
Experimental details: Raman scattering measurements have been performed on arbitrarily oriented crystals using a 633 nm He-Ne laser. The laser radiation power is not indicated. The Raman shifts have been determined for the maxima of individual peaks obtained as a result of the spectral curve analysis.
Raman shifts (cm^{-1}): 3575, 3509, 3406, (3330), 3152+3090 (unresolved doublet?), 1457w, 1438, 1305w, 1148w, 1102, 1037, 984s, 609, 523w, 466, 430, 400w, 365sh, 351, 337sh, 218w, 197+184 (unresolved doublet?), (152), 142, 108.
Source: Frost and Palmer (2011h).
Comments: No independent analytical data are provided for the sample used.

Sanmartinite Zn(WO$_4$)

Origin: Synthetic.
Experimental details: Raman scattering measurements have been performed on arbitrarily oriented crystals using a 633 nm He-Ne laser. The laser radiation power is not indicated.
Raman shifts (cm^{-1}): 904s, ~180, 705, 675, ~540, ~510w, ~340, 273, ~180.

Source: Kloprogge et al. (2004b).
Comments: The sample was characterized by powder X-ray diffraction data. For the Raman spectrum of sanmartinite see also Errandonea et al. (2008).

Santabarbaraite $Fe^{3+}_3(PO_4)_2(OH)_3 \cdot 5H_2O$

Origin: Santa Barbara mine, Tuscany, Italy (type locality).
Experimental details: Raman scattering measurements have been performed on arbitrarily oriented crystals using a 633 nm He-Ne laser. The laser radiation power is not indicated. The Raman shifts have been determined for the maxima of individual peaks obtained as a result of the spectral curve analysis.
Raman shifts (cm^{-1}): (3541), 3435, 3266sh, 1634, (1549), 1095sh, 1007s, 630sh, 592, 561sh, 478, 431sh, (318), 272, 221sh, (197), (159), 145, 111.
Source: Frost et al. (2016c).
Comments: The sample was characterized by qualitative electron microprobe analysis.

Santarosaite CuB_2O_4

Origin: Santa Rosa mine, Atacama desert, Chile (type locality).
Experimental details: Methods of sample preparation are not described. Raman scattering measurements have been performed using 514.5 nm Ar$^+$ laser radiation. The laser radiation power is not indicated.
Raman shifts (cm^{-1}): ~1110, ~1010, ~872s, ~775, ~745w, ~702, ~472, ~320, ~170w, (~70), (~60), (~30).
Source: Schlüter et al. (2008).
Comments: The sample was characterized by powder X-ray diffraction data, electron microprobe analyses andelectron energy loss spectroscopy.

Santite $KB_5O_6(OH)_4 \cdot 2H_2O$

Origin: Synthetic.
Experimental details: No data.
Raman shifts (cm^{-1}): 918s, 780–785w, 765–766, 556–557s, 510, 457, 369w, 296–299.
Source: Asensio et al. (2016).
Comments: The sample was characterized by powder X-ray diffraction data.

Saponite $(Ca,Na)_{0.3}(Mg,Fe)_3(Si,Al)_4O_{10}(OH)_2 \cdot 4H_2O$

Origin: Synthetic.
Experimental details: Raman scattering measurements have been performed on a powdery sample using 1064 nm Nd-YAG laser radiation. The laser radiation power is not indicated.
Raman shifts (cm^{-1}): 1082, 1051w, 998, 918w, 778, 683s, (660), 464-550, 432, 360s, 340sh, 288, 265w, 229w, 202.
Source: Kloprogge and Frost (2000c).
Comments: The Raman shifts are given for a Na-saturated sample. For the Raman spectrum of saponite see also Wang et al. (1999).

Sarcopside $Fe^{2+}_3(PO_4)_2$

Origin: Sowie Góry Mts., Lower Silesia, southwestern Poland.
Experimental details: Raman scattering measurements have been performed on an arbitrarily oriented crystal using 514.5 nm Ar^+ laser radiation. The laser radiation power is not indicated.
Raman shifts (cm^{-1}): 1109w, 1078, 1039, 1021w, 974s, 930s, 624, 606w, 553, 478, 400, 284w.
Source: Łodziński and Sitarz (2009).
Comments: The sample was characterized by electron microprobe analyses. For the Raman spectrum of sarcopside see also Schneider et al. (2013).

Sarkinite $Mn^{2+}_2(AsO_4)(OH)$

Origin: Långban deposit, Bergslagen ore region, Filipstad district, Värmland, Sweden.
Experimental details: Raman scattering measurements have been performed on an arbitrarily oriented sample using 632.8 nm He-Ne laser radiation. The laser radiation power is not indicated.
Raman shifts (cm^{-1}): 3550, 3535, 3528, 3519, 888, 839s, 826sh, 475, 380, 325.
Source: Makreski et al. (2013a).
Comments: No independent analytical data are provided for the sample used. For the Raman spectrum of sarkinite see also Hålenius and Westlund (1998).

Sarmientite $Fe^{3+}_2(AsO_4)(SO_4)(OH)\cdot 5H_2O$

Origin: Santa Elena mine, San Juan province, Argentina (type locality).
Experimental details: Methods of sample preparation are not described. Raman scattering measurements have been performed using Ar^+ laser radiation. The laser radiation power is not indicated.
Raman shifts (cm^{-1}): 3482w, 3336sh, 3184w, 1614w, 1130sh, 1118w, 1081, 998s, 889, 868, 818s, 638w, (590), 570, 477, 444, 405, 370, 322, 294, 259, 202, 191.
Source: Colombo et al. (2014).
Comments: The sample was characterized by powder X-ray diffraction data and electron microprobe analyses. The crystal structure is solved.

Sartorite $PbAs_2S_4$

Origin: Binntal, Switzerland.
Experimental details: Methods of sample preparation are not described. Raman scattering measurements have been performed in the spectral region from 50 to 600 cm^{-1} using 632.8 nm He-Ne laser radiation. The laser radiation power is not indicated. A 180°-scattering geometry was employed. The Raman shifts have been determined for the maxima of individual peaks obtained as a result of the spectral curve analysis.
Raman shifts (cm^{-1}): 375sh, 363s, 352sh, 336, 318sh, 300s, 281sh, 259sh, 229, 204w, 178, 167sh, 123w, 101w, 91, 85w, 75sh.
Source: Kharbish (2016).
Comments: The sample was characterized by powder X-ray diffraction data and electron microprobe analyses.

Sassolite $B(OH)_3$

Origin: Synthetic.
Experimental details: No data.
Raman shifts (cm^{-1}): ~1160w, 880s, 499, ~205.
Source: Peretyazhko et al. (2000).
Comments: No independent analytical data are provided for the sample used. For the Raman spectra of sassolite see also Thomas (2002), Michel et al. (2007), Thomas and Davidson (2010), and Frezzotti et al. (2012).

Scacchite $MnCl_2$

Origin: Synthetic.
Experimental details: Raman scattering measurements have been performed on a single crystal with laser beam directed along the c axis of the crystal and the scattered light at approximately 90° with respect to the incident beam. 488 and 514.5 nm Ar$^+$/Kr$^+$ laser radiations were used. The laser radiation power is not indicated.
Raman shifts (cm^{-1}): 234s, 144.
Source: Piseri and Pollini (1984).
Comments: No independent analytical data are provided for the sample used.

Schafarzikite $Fe^{2+}(Sb^{3+})_2O_4$

Origin: Pernek, Malé Karpaty Mts., Slovak Republic (type locality).
Experimental details: Raman scattering measurements have been performed on an arbitrarily oriented single crystal using 633 nm laser radiation. The laser radiation power is not indicated.
Raman shifts (cm^{-1}): (709w), 668s, 617sh, 558w, 526, (479w), 465, 403w, 353, (345w), 295, 249w, 219, 186w, 159, 132w, 119, 107.
Source: Bahfenne (2011).
Comments: No independent analytical data are provided for the sample used. For the Raman spectra of schafarzikite see also Sejkora et al. (2007) and Kharbish (2012).

Scheelite $Ca(WO_4)$

Origin: Lodrino, Riviera, Ticino (Tessin), Switzerland.
Experimental details: Methods of sample preparation are not described. Raman scattering measurements have been performed on an arbitrarily oriented sample using 632.8 nm He-Ne or 488 nm Ar$^+$ laser radiation. The laser radiation power is not indicated.
Raman shifts (cm^{-1}): 913s, 841w, 799w, 399, 332, 210.
Source: Andò and Garzanti (2014).
Comments: No independent analytical data are provided for the sample used. For the Raman spectra of scheelite see also Frost et al. (2004d) and Kloprogge et al. (2004b).

Schiavinatoite $Nb(BO_4)$

Origin: Synthetic.
Experimental details: Methods of sample preparation are not described. Raman scattering measurements have been performed using 488 and 514.5 nm Ar$^+$ laser radiations. The laser radiation power is not indicated.

Raman shifts (cm^{-1}): 986, 956s, 880, 815, 710w, 544s, 433, ~310w, 252, 235s.
Source: Heyns et al. (1990).
Comments: No independent analytical data are provided for the sample used.

Schlossmacherite $(H_3O)Al_3(SO_4)_2(OH)_6$

Origin: Emma Luisa Au, Guanaco district, Antofagasta, Chile (type locality).
Experimental details: Raman scattering measurements have been performed on arbitrarily oriented crystals using a 633 nm He-Ne laser. The laser radiation power at the sample was 0.1 mW. The Raman shifts have been determined for the maxima of individual peaks obtained as a result of the spectral curve analysis.
Raman shifts (cm^{-1}): 3537, 3449w, 3410, (3382), (3363), 2918, 1868, (2850), 1651sh, 1590w, 1458 +1442 (unresolved doublet?), 1304w, 1139w, 1082w, 1031w, 1000w, 938sh, 915, 864s, 819+809s (unresolved doublet?), 601w, 513, 459sh, 437s, 392, 358, 338sh, 312, 297, 263, 224, 203, 184sh, 147.
Source: Frost et al. (2012c).
Comments: No independent analytical data are provided for the sample used.

Schmiederite $Cu_2Pb_2(Se^{4+}O_3)(Se^{6+}O_4)(OH)_4$

Origin: El Dragon mine, Potosi, Bolivia (type locality).
Experimental details: Raman scattering measurements have been performed on arbitrarily oriented crystals using a 633 nm He-Ne laser. The laser radiation power is not indicated. The Raman shifts have been determined for the maxima of individual peaks obtained as a result of the spectral curve analysis.
Raman shifts (cm^{-1}): 3428, (1919w), 1852w, 1604, 1576, 1457w, 1418, 1349w, 1095s, 934, 834, 764, 739w, 538w, 398s, 281, 247s, 178, 153, 139s.
Source: Frost and Keeffe (2008a).
Comments: No independent analytical data are provided for the sample used.

Schmitterite $(UO_2)(Te^{4+}O_3)$

Origin: Ozernyi district, Salla-Kuolajarvi, Kola Peninsula, Russia.
Experimental details: Raman scattering measurements have been performed on an arbitrarily oriented sample using 514.5 nm Ar$^+$ laser radiation. The nominal laser power was 50 mW.
Raman shifts (cm^{-1}): 1102, 1091, 882s, 849, 796, 406sh, 391, 323, 304w, 204w, 176w, 141, 122w, 111.
Source: Voloshin et al. (2015b).
Comments: The sample was characterized electron microprobe analyses. For the Raman spectrum of schmitterite see also Frost et al. (2006b).

Schneiderhöhnite $Fe^{2+}Fe^{3+}_3As^{3+}_5O_{13}$

Origin: Urucum mine, Doce River, Galileia, Minas Gerais, Brazil.
Experimental details: Raman scattering measurements have been performed on arbitrarily oriented crystals using a 633 nm He-Ne laser. The laser radiation power is not indicated. The Raman shifts

have been determined for the maxima of individual peaks obtained as a result of the spectral curve analysis.
Raman shifts (cm^{-1}): 382w, 370w, 353sh, 345, 318, 295s, 279, 259sh, 247s, 219sh, 214w, 198, (193), 184s, 164, 148, 141, 122w, 117.
Source: Bahfenne and Frost (2009).
Comments: No independent analytical data are provided for the sample used.

Schoenfliesite $MgSn(OH)_6$

Origin: Synthetic.
Experimental details: Raman scattering measurements have been performed on a poedered sample using 632.8 nm He-Ne laser radiation. The laserpower was varied between 1.94 and 0.07 mW.
Raman shifts (cm^{-1}): 980, 602s, ~460w, ~365w, 289.
Source: Barchiche et al. (2008).
Comments: The sample was characterized by powder X-ray diffraction data.

Schoepite $(UO_2)_8O_2(OH)_{12} \cdot 12H_2O$

Origin: Synthetic.
Experimental details: Raman scattering measurements have been performed on an arbitrarily oriented sample using 785 nm diode laser radiation. The laser radiation power is not indicated.
Raman shifts (cm^{-1}): 829s, 805sh, 746w, 551, 454, 438, 337, 260, 208.
Source: Stefaniak et al. (2008).
Comments: The sample was characterized by powder X-ray diffraction data. For the Raman spectra of schoepite see also Amme et al. (2002) and Frost et al. (2007h).

Scholzite $CaZn_2(PO_4)_2 \cdot 2H_2O$

Origin: Reaphook Hill, Martins Well, South Flinders Ranges, South Australia, Australia.
Experimental details: Raman scattering measurements have been performed on arbitrarily oriented crystals using a 633 nm He-Ne laser. The laser radiation power at the sample was 1 mW. The Raman shifts have been determined for the maxima of individual peaks obtained as a result of the spectral curve analysis.
Raman shifts (cm^{-1}): 3437, 3343, 3283, 3185sh, 1171, 1115, 1088w, 1053w, 1026w, 1000s, (935), 923.
Source: Frost (2004a).
Comments: No independent analytical data are provided for the sample used. Raman shifts below 900 cm^{-1} are not indicated.

Schorl $NaFe^{2+}_3Al_6(Si_6O_{18})(BO_3)_3(OH)_3(OH)$

Origin: Bonče, Prilep municipality, Republic of Macedonia.
Experimental details: Raman scattering measurements have been performed on an arbitrarily oriented sample using 532 nm Nd-YAG laser radiation. The laser radiation power is not indicated.
Raman shifts (cm^{-1}): 1066s, 980, 806, 784, 770, 705, 672, 536w, 369s, 314w, 239s, 217.
Source: Makreski and Jovanovski (2009).

Comments: The sample was characterized by powder X-ray diffraction data and electron microprobe analyses. For the Raman spectra of schorl see also Ertl et al. (2015) and Watenphul et al. (2016a, b).

Schorlomite $Ca_3Ti_2(SiFe^{3+}_2)O_{12}$

Origin: Wiluy River, Sakha-Yakutia, Russia.
Experimental details: Methods of sample preparation are not described. Raman scattering measurements have been performed using 514.5 nm Ar$^+$ laser radiation. The laser radiation power is not indicated.
Raman shifts (cm^{-1}): 946–952, 785–786, 728–739, 639–641w, 508–510s, 431–435, 340–350s, 295sh, 253–256, 212–219w, 157w.
Source: Galuskina et al. (2005).
Comments: A schorlomite variety enriched in Zr and Sc was used. The sample was characterized by electron microprobe analyses.

Schreibersite $(Fe,Ni)_3P$

Origin: Almahatta Sitta meteorite.
Experimental details: Raman scattering measurements have been performed on an arbitrarily oriented sample using 532 nm Nd-YAG laser radiation. The nominal laser radiation power was 22.5 mW. A 180°-scattering geometry was employed.
Raman shifts (cm^{-1}): ~650s, ~510, ~410s, ~395s, ~300s, ~220.
Source: Kaliwoda et al. (2013).
Comments: The sample was characterized by electron microprobe analyses. For the Raman spectrum of schreibersite see also La Cruz (2015).

Schreyerite $V^{3+}_2Ti^{4+}_3O_9$

Origin: Vihanti, Northern Ostrobothnia region, Finland.
Experimental details: Methods of sample preparation are not described. Raman scattering measurements have been performed using 633 nm He-Ne laser radiation. The nominal laser radiation power was 2 or 20 mW.
Raman shifts (cm^{-1}): 810s, 705s, 653s, 593, 532w, 468, 426, 358w, 308, 175s, 90.
Source: Voloshin et al. (2014).
Comments: The sample was characterized by powder X-ray diffraction data and electron microprobe analysis.

Schröckingerite $NaCa_3(UO_2)(SO_4)(CO3)_3F \cdot 10H_2O$

Origin: Wheal Edward, St. Just, Cornwall, UK.
Experimental details: Methods of sample preparation are not indicated. Raman scattering measurements have been performed using 785 nm laser radiation. The nominal laser power at the source was ~370 mW.
Raman shifts(cm^{-1}): 1093, 1009, ~980, 815s, ~745, ~620w, ~470w, ~305w, ~250w.
Source: Driscoll et al. (2014).
Comments: The sample was characterized by electron microprobe analysis. For the Raman spectrum of schröckingerite see also Frost et al. (2007d).

Schuetteite $Hg_3O_2(SO_4)$

Origin: Synthetic.
Experimental details: No data.
Ramanshifts (cm^{-1}): ~1090, ~1060, ~975s, ~620w, ~600w, ~520w, ~455, ~425, ~220s.
Source: Schofield (2004).
Comments: The sample was characterized by powder X-ray diffraction data.

Schultenite $Pb(AsO_3OH)$

Origin: Synthetic.
Experimental details: Raman scattering measurements have been performed on an oriented sample using 514.5 nm Ar$^+$ laser radiation. The laser radiation power is not indicated. A 180°-scattering geometry was employed.
Raman shifts (cm^{-1}): 827s, 799sh, 467, 357, 149, 93s, (73).
Source: Petzelt (1977).
Comments: No independent analytical data are provided for the sample used. The Raman shifts are given for the scattering geometry $y(p_1-)-y$. For the Raman spectrum of schultenite see also Młynarska et al. (2014).

Schumacherite $Bi_3O(VO_4)_2(OH)$

Origin: Wombat Hole Prospect, Morass Creek gorge, near Benambra, Victoria, Australia.
Experimental details: Raman scattering measurements have been performed on arbitrarily oriented crystals using a 633 nm He-Ne laser. The laser radiation power is not indicated. The Raman shifts have been determined for the maxima of individual peaks obtained as a result of the spectral curve analysis.
Raman shifts (cm^{-1}): 3616w, 3589w, 1557, 1041, 943, 809, 672s, 641, 498w, 410s, 341, 248, 197.
Source: Frost et al. (2006i).
Comments: No independent analytical data are provided for the sample used.

Schwertmannite $Fe^{3+}_{16}O_{16}(OH,SO_4)_{12-13} \cdot 10H_2O$ (?)

Origin: Synthetic.
Experimental details: Raman scattering measurements have been performed on a polycrystalline sample using He-Ne laser. The laser radiation power is not indicated.
Raman shifts (cm^{-1}): 1120w, 981, 715, 580sh, 544, 421s, 350, 318, 294sh.
Source: Mazzetti and Thistlethwaite (2002).
Comments: The sample was characterized by powder X-ray diffraction data.

Scolecite $Ca(Si_3Al_2O_{10}) \cdot 3H_2O$

Origin: Nasik, Maharashtra, India.
Experimental details: Micro-Raman scattering measurements have been performed on an arbitrarily oriented sample using 514.5 nm Ar$^+$ laser radiation. The laser power at the sample was 10 mW.
Raman shifts (cm^{-1}): 3536, 3472, 3405, 3212, 3182, 3087+3080, 1105, 1088, 1049+1044, 941, 718, 535s, 496, 480+472, 447s, 437, 426s, 354, 328+318, (301), 295s, 283+276+255, 245+241+224, 179, 171, 158, 146.

Source: Wopenka et al. (1998).
Comments: Bands whose intensities are significantly dependent upon polarization are indicated in the cited paper. For the Raman spectra of scolecite see also Pechar (1984) and Mozgawa (2001).

Scorodite $Fe^{3+}(AsO_4) \cdot 2H_2O$

Origin: Synthetic.
Experimental details: Raman scattering measurements have been performed on a powdered sample using 785 nm solid-state laser radiation. The laser radiation power at the sample was about 0.3 mW or some what more.
Raman shifts (cm^{-1}): 886s, 796s, 444, 416, 376, 333, 287, 254, 176s, 128.
Source: Das and Hendry (2011).
Comments: The sample was characterized by powder X-ray diffraction data. For the Raman spectra of scorodite see also Gomez et al. (2010a, 2011), Frost et al. (2015w), Culka et al. (2016b), and Kloprogge and Wood (2017).

Scotlandite $Pb(S^{4+}O_3)$

Origin: Leadhills, Scotland (type locality).
Experimental details: Raman scattering measurements have been performed on arbitrarily oriented crystals using a 633 nm He-Ne laser. Th elaser radiation power is not indicated.
Raman shifts (cm^{-1}): 975sh, 935s, 880sh, 622, 474s, 190, 144.
Source: Frost and Keeffe (2009d).
Comments: No independent analytical data are provided for the sample used.

Scottyite $BaCu_2Si_2O_7$

Origin: Wessels mine, Kalahari Manganese Fields, South Africa (type locality).
Experimental details: Raman scattering measurements have been performed on an arbitrarily oriented crystal using 532 nm laser radiation. The laser radiation power is not indicated.
Raman shifts (cm^{-1}): 1019w, 958w, 896s, 866w, 675s, 612, 578, 560, 459s.
Source: Yang et al. (2013b).
Comments: The sample was characterized by single-crystal X-ray diffraction data and electron microprobe analyses. The crystal structure is solved. For the Raman spectrum of scottyite see also Xia et al. (2014).

Scrutinyite PbO_2

Origin: Synthetic.
Experimental details: No data.
Raman shifts (cm^{-1}): 534w, 366w, 269, 133s, 80w,
Source: Inguanta et al. (2008).
Comments: The sample was characterized by powder X-ray diffraction data and electron microprobe analysis.

Sederholmite NiSe

Origin: Synthetic.
Experimental details: Raman scattering measurements have been performed on hollow nanospheres using 532 nm laser radiation. The laser radiation power is not indicated.
Raman shifts (cm^{-1}): 1075s, 539, 389, 207w.
Source: Shi et al. (2013).
Comments: The sample was characterized by powder X-ray diffraction data. The Raman shifts of a bulk sample are (cm^{-1}): 1060s, 524, 374, 192w.

Segnitite $PbFe^{3+}_3(AsO_4)(AsO_3OH)(OH)_6$

Origin: Broken Hill, New South Wales, Australia.
Experimental details: Raman scattering measurements have been performed on arbitrarily oriented crystals using a 633 nm He-Ne laser. The laser radiation power is not indicated. The Raman shifts have been determined for the maxima of individual peaks obtained as a result of the spectral curve analysis.
Raman shifts (cm^{-1}): (3467w), 3440sh, 3217, 2982sh, 1417w, 1231w, 1128w, 998w, 931sh, 860 +848s (unresolved doublet?), 811sh, 746sh, 689w, 572, 481s, (465), 441, 419sh, 370, (342), 318, 300, 250, 202, (195).
Source: Frost et al. (2005l).
Comments: The sample was characterized by powder X-ray diffraction data and electron microprobe analysis.

Seinäjokite $FeSb_2$

Origin: Synthetic.
Experimental details: Raman scattering measurements have been performed on an arbitrarily oriented sample using 514.5 nm Ar$^+$ laser radiation. The laser radiation power is not indicated.
Raman shifts (cm^{-1}): ~1060, ~700sh, ~660s, ~525.
Source: Xie et al. (2011).
Comments: The sample was characterized by powder X-ray diffraction data.

Sejkoraite-(Y) $Y_2[(UO_2)_8O_6(SO_4)_4(OH)_2] \cdot 26H_2O$

Origin: Červená vein, Jáchymov ore district, Western Bohemia, Czech Republic (type locality).
Experimental details: Methods of sample preparation are not described. Raman scattering measurements have been performed using 732 nm laser radiation. The laser radiation power is not indicated.
Raman shifts (cm^{-1}): 1537w, 1222w, 1157w, 1095w, 1014, 896w, 829s, 798sh, 670w, 546sh, 477sh, 461, 438sh, 404s, 369sh, 326w, 274, 262, 237sh, 211.
Source: Plášil et al. (2011a).
Comments: The sample was characterized by powder X-ray diffraction data and electron microprobe analyses. The crystal structure is solved.

Sekaninaite $Fe^{2+}_2Al_4Si_5O_{18}$

Origin: A miarolitic pegmatite at Zimnik, Strzegom-Sobótka massif, Sudetes, Poland.
Experimental details: Raman scattering measurements have been performed on an arbitrarily oriented sample using 532 nm Nd-YAG laser radiation. The nominal laser radiation power was 30 mW.
Raman shifts (cm^{-1}): 3630sh, 3597, 3579, 1158w, 1049w, 1004w, 989, 917w, 667, 599, 569s, 552s, 477w, 420, 296, 253, 152.
Source: Gadas et al. (2016).
Comments: The sample was characterized by electron microprobe analyses and LA-ICP-MS. For the Raman spectrum of sekaninaite see also Radica et al. (2013).

Selenium Se

Origin: Synthetic.
Experimental details: No data.
Raman shifts (cm^{-1}): 237 + 234s (unresolved doublet?), 140.
Source: Campos et al. (2004a).
Comments: The sample was characterized by powder X-ray diffraction data. For the Raman spectra of selenium see also Campos et al. (2004b).

Seligmannite $CuPbAsS_3$

Origin: Binntal, Switzerland.
Experimental details: Methods of sample preparation are not described. Raman scattering measurements have been performed in the spectral region from 50 to 600 cm^{-1} using 632.8 nm He-Ne laser radiation. The laser radiation power is not indicated. A 180°-scattering geometry was employed. The Raman shifts have been determined for the maxima of individual peaks obtained as a result of the spectral curve analysis.
Raman shifts (cm^{-1}): (363), 354s, 344sh, 334, 324, 311sh, 288, 231, 215sh, 202, 189sh, 172h, 155w, 121w, 106sh, 97s, 90s, 70s.
Source: Kharbish (2016).
Comments: The sample was characterized by powder X-ray diffraction data and electron microprobe analysis.

Sellaite MgF_2

Origin: Synthetic.
Experimental details: Raman scattering measurements have been performed on a single crystal in different scattering geometries, using 488 nm Ar$^+$ laser radiation. The nominal laser radiation power was ~75 mW. A 90°-scattering geometry was employed.
Raman shifts (cm^{-1}): 515w, 410s, 295, 92.
Source: Porto et al. (1967).
Comments: The band intensities are indicated for the sum of spectra obtained in different scattering geometries. For the Raman spectrum of sellaite see also Krishnan and Katiyar (1965).

Sénarmontite Sb_2O_3

Origin: Synthetic.
Experimental details: Methods of sample preparation are not described. Raman scattering measurements have been performed using 532 nm Nd-YAG laser radiation. The laser radiation power was below 1 mW.
Raman shifts (cm^{-1}): 451, 373w, 253s, 189, 115w.
Source: Makreski et al. (2013b).
Comments: The sample was characterized by thermal analysis. For the Raman spectra of Sénarmontite see also Cody et al. (1979), Voit et al. (2009), and Orman (2010).

Senegalite $Al_2(PO_4)(OH)_3 \cdot H_2O$

Origin: Jangada mine, Quadrilátero Ferrífero, municipality of Brumadinho, Minas Gerais, Brazil.
Experimental details: Raman scattering measurements have been performed on arbitrarily oriented crystals using a 633 nm He-Ne laser. The laser radiation power is not indicated. The Raman shifts have been determined for the maxima of individual peaks obtained as a result of the spectral curve analysis.
Raman shifts (cm^{-1}): 3614sh, 3610s, 3606sh, 3507–3505s, 3429sh, 3374sh, 3339, 3270w, 3206sh, 3099w, 2975w, 1753w, 1679w, (1587w), (1425w), 1377w, 1206w, 1179w, 1154, 1110w, 1071, 1029, (1026), 892w, 829w, 708, 677w, 635, 616sh, 581w, 559sh, 545, 501sh, 480, 462sh, 444, 417, 375, 364sh, 329sh, 318sh, (312w), 303, 237, 202sh, 193, 178, 166, (154w), 136 + 133 (unresolved doublet?), 102.
Source: Frost et al. (2013g).
Comments: The sample was characterized by qualitative electron microprobe analysis.

Sepiolite $Mg_4Si_6O_{15}(OH)_2 \cdot 6H_2O$

Origin: Durango, Mexico.
Experimental details: Methods of sample preparation are not described. Raman scattering measurements have been performed on randomly oriented crystals in back-scattering geometry, using 1064 nm Nd-YAG laser radiation. The nominal laser radiation power was 900 mW.
Raman shifts (cm^{-1}): ~1080, ~780, ~675s, ~380, ~335, ~290, ~265, ~230w, ~200s, ~170.
Source: McKeown et al. (2002).
Comments: The sample was characterized by powder X-ray diffraction data.

Sérandite $NaMn^{2+}{}_2Si_3O_8(OH)$

Origin: Poudrette quarry, Saint-Hilaire Mt., Montérégie (Rouville) Co., Québec, Canada (type locality).
Experimental details: Methods of sample preparation are not described. Raman scattering measurements have been performed using 638 nm laser radiation. The laser radiation power is not indicated.
Raman shifts (cm^{-1}): 1026s, 903, 666s, 422, 304, 168.
Source: Haring and McDonald (2014a).
Comments: No independent analytical data are provided for the sample used. For the Raman spectrum of sérandite see also Origlieri et al. (2017).

Serendibite $Ca_4[Mg_6Al_6]O_4[Si_6B_3Al_3O_{36}]$

Origin: Ratnapura area, Sri Lanka.
Experimental details: Raman scattering measurements have been performed on an arbitrarily oriented crystal using 325 or 514.5 nm laser radiation. The laser radiation power is not indicated.
Raman shifts (cm^{-1}): 997, 895, 756, 681, 635, 572, 530, 470, 405, 364, 310.
Source: Schmetzer et al. (2002).
Comments: The sample was characterized by electron microprobe analyses.

Serpierite $Ca(Cu,Zn)_4(SO_4)_2(OH)_6 \cdot 3H_2O$

Origin: Corchia, NW Italy.
Experimental details: Methods of sample preparation are not described. Raman scattering measurements have been performed on an arbitrarily oriented sample using 532 nm laser radiation. The laser radiation power at the sample was 4 mW.
Raman shifts (cm^{-1}): 3616w, 3570w, 1168w, 1132, 1115sh, 1085w, 991s, 651w, 605w, 474sh, 445, 426, 415, 338w, 244w, 218w.
Source: Coccato et al. (2016).
Comments: No independent analytical data are provided for the sample used.

Shattuckite $Cu_5(SiO_3)_4(OH)_2$

Origin: Navojoa, Sonora, Mexico.
Experimental details: Raman scattering measurements have been performed on arbitrarily oriented crystals using a 633 nm He-Ne laser. The laser radiation power is not indicated. The Raman shifts have been determined for the maxima of individual peaks obtained as a result of the spectral curve analysis.
Raman shifts (cm^{-1}): 3602, (3567w), 3424sh, 3367w, (3153w), 1564w, 1107w, 1054, (1000w), 981, 953sh, 887, 865w, 832w, 781, 739w, 667s, 549, 498, 439, 394, 349sh, 340, 326, 306sh, 255, 238, 211s, 191sh, (160w), 152, 135–139 sh, (117w), 113, (107w).
Source: Frost and Xi (2012b).
Comments: No independent analytical data are provided for the sample used.

Shcherbinaite V_2O_5

Origin: Synthetic.
Experimental details: Raman scattering measurements have been performed on a sample in rotated tube using 488 and 514.5 nm Ar$^+$, as well as 647.1 nm Kr$^+$ laser radiations. The laser radiation power is not indicated.
Raman shifts (cm^{-1}): 993, 703, 528.5, 483, 476, 405, 305, 285s, 198, 147s, 105.
Source: Sanchez et al. (1982).
Comments: No independent analytical data are provided for the sample used. For the Raman spectrum of shcherbinaite see also Menezes et al. (2009).

Shchurovskyite $K_2CaCu_6O_2(AsO_4)_4$

Origin: Arsenatnaya fumarole, Tolbachik volcano, Kamchatka, Russia (type locality).
Experimental details: Raman scattering measurements have been performed on an arbitrarily oriented crystal using 532 nm laser radiation. The laser radiation power is not indicated.
Raman shifts (cm^{-1}): 840s, 630, 486, 304, 135, 100.
Source: Pekov et al. (2015d).
Comments: The sample was characterized by powder X-ray diffraction data and electron microprobe analyses. The crystal structure is solved.

Shortite $Na_2Ca_2(CO_3)_3$

Origin: No data.
Experimental details: No data.
Raman shifts (cm^{-1}): 1523, 1470, 1440, 1407, 1387, 1091s, 1071s, 730, 719, 715, 695, 265s, 201, 171, 141.
Source: Shatskiy et al. (2015).
Comments: For the Raman spectra of shortite see also Frost and Dickfos (2007b, 2008).

Shuangfengite $IrTe_2$

Origin: Synthetic.
Experimental details: Raman scattering measurements have been performed in a nearly backscattering (xx) geometry on an oriented crystal, from the surface parallel to the (ab) plane using 532 nm solid-state laser radiation. The nominal laser radiation power was 5 mW.
Raman shifts (cm^{-1}): 166, 126.
Source: Glamazda et al. (2014).
Comments: No independent analytical data are provided for the sample used. For the Raman spectrum of shuangfengite see also Lazarević et al. (2014).

Shulamitite $Ca_3TiFe^{3+}AlO_8$

Origin: Central part of the Hatrurim Basin, Israel (type locality).
Experimental details: Raman scattering measurements have been performed on an arbitrarily oriented sample in a polished section using 514.5 nm Ar^+ laser radiation. The laser radiation output power was between 30 and 50 mW.
Raman shifts (cm^{-1}): 1501 (broad), 802sh, 742s, 561, 498sh, 388, 290, 238, 145w, 110w.
Source: Sharygin et al. (2013a).
Comments: The sample was characterized by powder X-ray diffraction data and electron microprobe analyses. The crystal structure is solved.

Siderite $Fe(CO_3)$

Origin: Minas Gerais, Brazil.
Experimental details: Methods of sample preparation are not described. Raman scattering measurements have been performed using 488 and 514.5 nm Ar^+ laser radiations. The nominal laser radiation power was in the range from 100 to 500 mW.

Raman shifts (cm^{-1}): 1738, 1088s, 731, 299, 194w.

Source: Rutt and Nicola (1974).

Comments: The sample was characterized by powder X-ray diffraction data. For the Raman spectra of siderite see also Buzgar and Apopei (2009), Das and Hendry (2011), Saheb et al. (2011), Frezzotti et al. (2012), Zhao and Guo (2014), and Andò and Garzanti (2014).

Sideronatrite $Na_2Fe^{3+}(SO_4)_2(OH)\cdot3H_2O$

Origin: Sierra Gorda, Chile.

Experimental details: Raman scattering measurements have been performed on an arbitrarily oriented sample using 532 nm laser radiation. The nominal laser radiation power was 2.5 mW. The Raman shifts have been determined for the maxima of individual peaks obtained as a result of the spectral curve analysis.

Raman shifts (cm^{-1}): 1646, 1223, 1189, 1159, 1117sh, 1106, 1024, 1013s, 996s, 624, 614sh, 600w, 536, 469, 458sh, 391, 259sh, 246s, 216sh, 203, 170, 115.

Source: Rouchon et al. (2012).

Comments: The sample was characterized by powder X-ray diffraction data and electron microprobe analyses.

Sidorenkite $Na_3Mn(PO_4)(CO_3)$

Origin: Alluaiv Mt., Lovozero massif, Kola Peninsula, Russia (type locality).

Experimental details: Raman scattering measurements have been performed on arbitrarily oriented crystals using a 633 nm He-Ne laser. The laser radiation power is not indicated. The Raman shifts have been determined for the maxima of individual peaks obtained as a result of the spectral curve analysis.

Raman shifts (cm^{-1}): 1074, 1044s, 1035sh, 1012s, 1004sh, 966sh, 959s, (953), 625w, 579, 469w, 414w, 297w, 252w, 202, 179w, 159, 129w.

Source: Frost et al. (2015c).

Comments: The sample was characterized by quqlitative electron microprobe analyses.

Sidwillite $MoO_3\cdot2H_2O$

Origin: Synthetic.

Experimental details: Raman scattering measurements have been performed on a polycrystalline sample using 488 and 514.5 nm Ar$^+$ laser radiations. The laser radiation power at the sample was 10–50 mW.

Raman shifts (cm^{-1}): 934, 771sh, 729s, 627, 418w, 386w, 353w, 331, 272s, 247, 216, 184, 168, 119, 95, 65.

Source: Seguin et al. (1995).

Comments: No independent analytical data are provided for the sample used. For the Raman spectrum of sidwillite see also Philip et al. (1988).

Siegenite $CoNi_2S_4$

Origin: Synthetic.

Experimental details: Raman spectrum was measured in argon atmosphere. Characteristics of laser radiation are not indicated.

Raman shifts (cm^{-1}): 373s, 342s, 301, 239, 150.
Source: Xia et al. (2015).
Comments: The sample was characterized by powder X-ray diffraction data.

Sigloite $Fe^{3+}Al_2(PO_4)_2(OH)_3 \cdot 7H_2O$

Origin: Siglo XX (Llallagua) mine, Andes Mts., Bustillo province, Potosi, Bolivia (type locality).
Experimental details: Raman scattering measurements have been performed on arbitrarily oriented crystals using a 633 nm He-Ne laser. The laser radiation power is not indicated. The Raman shifts have been determined for the maxima of individual peaks obtained as a result of the spectral curve analysis.
Raman shifts (cm^{-1}): 3615, 3552, 3493s, 3449s, 3422sh, 3356sh, 3118s, (2988), 1631w, (1532w), 1235–1231sh, 1228, 1167, 1099 + 1086 (unresolved doublet?), 1009s, 993sh, 888, 816, 693w, 659sh, 619sh, 596, 571, 550sh, 528, (506), (489), 453, 427, 401, 338sh, 308, 277, (267), 253, 241, 191, 177, 169sh, 140sh, 129, 112.
Source: Frost et al. (2013aa).
Comments: The sample was characterized by qualitative electron microprobe analysis.

Siidraite $Pb_2Cu(OH)_2I_3$

Origin: Broken Hill Cu-Zn-Pb ore deposit, Yancowinna Co., New South Wales, Australia (type locality).
Experimental details: Raman scattering measurements have been performed on an arbitrarily oriented crystal using 532 nm laser radiation. The laser radiation power at the sample was 1.8 mW. A 180°-scattering geometry was employed.
Raman shifts (cm^{-1}): 3455, 3443, 743w, 370, 332, 313sh, 277, ~248sh, ~217w, ~140sh, 128s, 116, 97.
Source: Welch et al. (2016).
Comments: The sample was characterized by single-crystal X-ray diffraction data and electron microprobe analyses. The crystal structure is solved.

Silicocarnotite $Ca_5[(PO_4)(SiO_4)](PO_4)$

Origin: Hatrurim basin, Negev Desert, Israel (type locality).
Experimental details: The laser radiation wavelength is not indicated. The laser radiation power at the sample was 44 mW.
Raman shifts (cm^{-1}): 1085, 1056, 1014 + 1004 (unresolved doublet?), 967s, (939w), 904sh, 893, 878, 850s, 788w, 734w, 711w, 693w, 671w, 640, 626, 584, 557, 474, 448, (418), 397, 318sh, 302, 275, 258, 234, 190sh, 153w.
Source: Galuskin et al. (2015a).
Comments: The sample was characterized by single-crystal X-ray diffraction data and electron microprobe analyses. The crystal structure is solved. The Raman shifts have been determined for the maxima of individual peaks obtained as a result of the spectral curve analysis. For the Raman spectra of silicocarnotite see also Serena et al. (2014, 2015).

Silicon Si

Origin: Dhofar 280 lunar highland meteorite.
Experimental details: Raman scattering measurements have been performed on arbitrarily oriented samples using 532 nm laser radiation. The laser radiation power at the samples was no more than 5 mW.
Raman shifts (cm^{-1}): ~950w, 504–515s, ~280w.
Source: Nazarov et al. (2012).
Comments: The samples were characterized by electron microprobe analyses. The strongest band of synthetic crystalline Si is observed at 520 cm^{-1}.

Sillénite $Bi_{12}SiO_{20}$

Origin: Synthetic.
Description: Synthesized from the stoichiometric mixture of oxides at 700 °C for 48 h. Cubic, space group $I23$.
Source: Betsch and White (1978).
Raman shifts (cm^{-1}): 621w, 538s, 458w, 328, 276s, 249sh, 205, 165, 143, 129s, 95sh, 87s, 66, 57s, 43w.

Sillimanite Al_2SiO_5

Origin: Premosello Chiovenda, Ossola valley, Verbano-Cusio-Ossola province, Piedmont, Italy.
Experimental details: Methods of sample preparation are not described. Raman scattering measurements have been performed on an arbitrarily oriented sample using 785 nm laser radiation. The laser radiation power is not indicated.
Raman shifts (cm^{-1}): ~1125, ~970, ~955, 870, ~707, ~596, ~480, 456, ~412, ~395, 309s, 235s, 142.
Source: Andò and Garzanti (2014).
Comments: No independent analytical data are provided for the sample used. For the Raman spectra of sillimanite see also Mernagh and Liu (1991) and Frezzotti et al. (2012).

Simonkolleite $Zn_5(OH)_8Cl_2·H_2O$

Origin: Artificial (a product of Zn corrosion in NaCl solution).
Experimental details: Raman scattering measurements have been performed on an arbitrarily oriented sample using 514.5 nm Ar$^+$ laser radiation. The laser radiation power at the sample was 40 mW.
Raman shifts (cm^{-1}): 3580, 3480s, 3450, 1030, 910, 730, 390s, 260s, 210.
Source: Bernard et al. (1993b).
Comments: No independent analytical data are provided for the sample used. For the Raman spectrum of simonkolleite see also Khamlich et al. (2013).

Sinhalite $MgAl(BO_4)$

Origin: No data.
Experimental details: Raman scattering measurements have been performed on an arbitrarily oriented sample using 632.8 nm He-Ne laser radiation. The laser radiation power is not indicated.
Raman shifts (cm^{-1}): 864, 554w, 488, 376.
Source: Ross (1972).
Comments: No independent analytical data are provided for the sample used. For the Raman spectrum of sinhalite see also Hayward et al. (1994).

Sinjarite $CaCl_2 \cdot 2H_2O$

Origin: Synthetic.
Experimental details: Methods of sample preparation are not described. Raman scattering measurements have been performed using 514.5 nm Ar^+ laser radiation. The laser radiation power is not indicated.
Raman shifts (cm^{-1}): 3486, 3452s, 3216w, 1638s, 1620w (at 21 °C); 3545w, 3491, 3475, 3437s, 3388, 3215w, 3211w, 1635s, 1630, 1616 (at −172 °C).
Source: Uriarte et al. (2015).
Comments: The sample was characterized by powder X-ray diffraction data. For the Raman spectrum of sinjarite see also Baumgartner and Bakker (2010).

Sinoite Si_2N_2O

Origin: Zakłodzie meteorite.
Experimental details: Raman scattering measurements have been performed on an arbitrarily oriented sample using 514.5 nm Ar^+ laser radiation. The laser radiation power at the sample was 1.2 mW.
Raman shifts (cm^{-1}): 1142w, 983w, 941w, 891w, 730w, 544, 496w, 455w, 373w, 328w, 217w, 185s.
Source: Ma et al. (2012a).
Comments: The mineral was identified by electron back-scatter diffraction. For the Raman spectrum of sinoite see also Sekine et al. (2006).

Skinnerite Cu_3SbS_3

Origin: Synthetic.
Experimental details: Raman scattering measurements have been performed on nanocrystals using 514 nm laser radiation. The laser radiation power is not indicated.
Raman shifts (cm^{-1}): 354.
Source: Qiu et al. (2013).
Comments: The sample was characterized by powder X-ray diffraction data and electron microprobe analyses.

Skippenite Bi_2Se_2Te

Origin: Synthetic.
Experimental details: Raman scattering measurements have been performed on a thin film. Characteristics of laser radiation are not indicated.
Raman shifts (cm^{-1}): 165, 147, 117s.
Source: Gopal et al. (2015).
Comments: The sample was characterized by powder X-ray diffraction data and electron microprobe analyses. For the Raman spectrum of skippenite see also Voloshin et al. (2015a).

Sklodowskite $Mg(UO_2)_2(SiO_3OH)_2 \cdot 6H_2O$

Origin: Eva mine, Northern Territory, Australia.
Experimental details: Raman scattering measurements have been performed on arbitrarily oriented crystals using a 633 nm He-Ne laser. The laser radiation power is not indicated. The Raman shifts have been determined for the maxima of individual peaks obtained as a result of the spectral curve analysis.
Raman shifts (cm^{-1}): 3506, 3420sh, 3316w, 1640w, 1528w, 1413w, 1312w, 1244w, 1150w, (986), 970, (957), 934w, 897w, 853w, 827w, 801, 777s, (756w), 549, 474, 414, 393sh, 318sh, 305, 282, 264, 217sh, 200 + 197s (unresolved doublet?), 156, 137, (127), 113.
Source: Frost et al. (2006e).
Comments: No independent analytical data are provided for the sample used.

Skorpionite $Ca_3Zn_2(PO_4)_2(CO_3)(OH)_2 \cdot H_2O$

Origin: Skorpion Zn mine, Lüderitz district, Karas region, Namibia (type locality).
Experimental details: Methods of sample preparation are not described. Raman scattering measurements have been performed using 488, 514.5 or 632.8 nm laser radiation. The nominal laser radiation power was 20 mW. A 180°-scattering geometry was employed.
Raman shifts (cm^{-1}): 3566s, 1633w, 1505, 1398, 1102, 1075s, 1054, 1016, 972s, 702, 639, 575, 468, 423, 384, 322, 276, 237.
Source: Krause et al. (2008).
Comments: The sample was characterized by powder X-ray diffraction data and electron microprobe analyses. The crystal structure is solved.

Smirnite $Bi^{3+}_2Te^{4+}O_5$

Origin: Synthetic.
Experimental details: Raman scattering measurements have been performed on an oriented single crystal using 632.8 nm He-Ne laser radiation. The nominal laser radiation power was 50 mW. A 90°-scattering geometry was employed.
Raman shifts (for the $y(xx)z$ scattering geometry, cm^{-1}): 768s, 740, 721sh, 625, 382w, 347, 282w, 254w, 239, 206w, 193w, 164w, 153w, 114, 102s, 89s, 64, 57s, 44w.
Source: Domoratskii et al. (2000).
Comments: No independent analytical data are provided for the sample used. For the Raman spectra of smirnite see also Klein et al. (1998).

Smithite $AgAsS_2$

Origin: Synthetic.
Experimental details: Methods of sample preparation are not described. Raman scattering measurements have been performed using 532 nm Nd-YAG laser radiation. The laser radiation power is not indicated.
Raman shifts (cm^{-1}): 362s, 324s, 301, 279w, 239, 207w, 176, 141, 120w.
Source: Minceva-Sukarova et al. (2003).
Comments: No independent analytical data are provided for the sample used.

Smithsonite $Zn(CO_3)$

Origin: Lavrion, Greece.
Experimental details: Methods of sample preparation are not described. Raman scattering measurements have been performed using 488 and 514.5 nm Ar$^+$ laser radiations. The nominal laser radiation power was in the range from 100 to 500 mW.
Raman shifts (cm^{-1}): 1735, 1406, 1090s, 726, 302, 194.
Source: Rutt and Nicola (1974).
Comments: The sample was characterized by powder X-ray diffraction data. For the Raman spectra of smithsonite see also Bouchard and Smith (2003) and Frezzotti et al. (2012).

Smythite $(Fe,Ni)_{3+x}S_4$ ($x \approx 0$–0.3)

Origin: Harrodsburg, Bloomington, Indiana, USA.
Experimental details: No data.
Raman shifts (cm^{-1}): 394, 358, 329, 326, 267, 262.
Source: Bon and Rakovan (2012).
Comments: The sample was characterized by single-crystal X-ray diffraction data. The crystal structure is solved.

Sobolevskite PdBi

Origin: Southern Sopchinskoe deposit, Monchegorsk district, Kola Peninsula, Russia.
Experimental details: Raman scattering measurements have been performed on an arbitrarily oriented sample using 514.5 nm Ar$^+$ laser radiation. The nominal laser power was 50 mW.
Raman shifts (cm^{-1}): 236w, 106w, 82, 63s.
Source: Voloshin et al. (2015a).
Comments: The sample was characterized by electron microprobe analyses. For the Raman spectrum of sobolevskite see also Bakker (2014).

Sodalite $Na_4(Si_3Al_3)O_{12}Cl$

Origin: Mogok, Myanmar.
Experimental details: No data.
Raman shifts (cm^{-1}): 1057–1062, 986–987, 973, 914, 463–464s, 451s, 263s.
Source: Culka et al. (2016a).
Comments: No independent analytical data are provided for the sample used. For the Raman spectra of sodalite see also Balassone et al. (2012), Hettmann et al. (2012), and Zahoransky et al. (2016).

Soddyite $(UO_2)_2(SiO_4) \cdot 2H_2O$

Origin: Sierra Albarrana, Córdoba, Spain.
Experimental details: Methods of sample preparation are not described. Raman scattering measurements have been performed using 632.8 nm He-Ne laser radiation. The laser radiation output power was 20 mW.
Raman shifts (cm^{-1}): 832s, 463, 404, 312, 293, 225, 195, 107.
Source: Bonales et al. (2015).
Comments: The sample was characterized by electron microprobe analyses. For the Raman spectra of soddyite see also Biwer et al. (1990), Giammar and Hering (2002), Frost et al. (2006d, h), and Amme et al. (2002).

Söhngeite $Ga(OH)_3$

Origin: Tsumeb mine, Namibia (type locality).
Experimental details: Methods of sample preparation are not described. Raman scattering measurements have been performed using 473 nm laser radiation. The nominal laser radiation power was 50 mW. A 180°-scattering geometry was employed.
Raman shifts (cm^{-1}): 3334, 3240, 3189sh, 3100, 3000, 923, 455w, 327s, 273w, 256, 185w.
Source: Welch and Kleppe (2016).
Comments: The sample was characterized by powder X-ray diffraction data and electron microprobe analyses. The crystal structure is solved.

Sonolite $Mn^{2+}_9(SiO_4)_4(OH)_2$

Origin: Franklin, Sussex Co., New Jersey, USA.
Experimental details: Raman scattering measurements have been performed on arbitrarily oriented crystals using a 633 nm He-Ne laser. The laser radiation power is not indicated. The Raman shifts have been determined for the maxima of individual peaks obtained as a result of the spectral curve analysis.
Raman shifts (cm^{-1}): 3555sh, 3544w, 3532sh, 948, 906, 848s, (838), 832sh, 814, 668, 638.
Source: Frost et al. (2007k).
Comments: The sample was characterized by electron microprobe analysis.

Sonoraite $Fe^{3+}(Te^{4+}O_3)(OH) \cdot H_2O$

Origin: Tombstone, Tombstone district, Cochise Co., Arizona, USA.
Experimental details: Raman scattering measurements have been performed on arbitrarily oriented crystals using a 633 nm He-Ne laser. The laser radiation power is not indicated. The Raman shifts have been determined for the maxima of individual peaks obtained as a result of the spectral curve analysis.
Raman shifts (cm^{-1}): 3450 + 3423 (unresolved doublet?), 3350sh, 3223sh, 3000sh, 994w, 911w, (804w), 779s, (714), 666s, 638sh, 521sh, 468, 425, 387, (374), 312, 267s, 253s, 234, 209, 159.
Source: Frost and Keeffe (2009c).
Comments: No independent analytical data are provided for the sample used. For the Raman spectrum of sonoraite see also Frost et al. (2015b).

Spangolite $Cu_6Al(SO_4)(OH)_{12}Cl \cdot 3H_2O$

Origin: Monte Fucinaia, central Western Italy.
Experimental details: Methods of sample preparation are not described. Raman scattering measurements have been performed on an arbitrarily oriented sample using 532 nm laser radiation. The laser radiation power at the sample was 4 mW.
Raman shifts (cm^{-1}): 968s, 615w, 520s, 410, 168.
Source: Coccato et al. (2016).
Comments: No independent analytical data are provided for the sample used.

Spencerite $Zn_4(PO_4)_2(OH)_2 \cdot 3H_2O$

Origin: Salmo, British Columbia, Canada.
Experimental details: Raman scattering measurements have been performed on arbitrarily oriented crystals using a 633 nm He-Ne laser. The laser radiation power at the sample was 1 mW. The Raman shifts have been determined for the maxima of individual peaks obtained as a result of the spectral curve analysis.
Raman shifts (cm^{-1}): 3516, 1095w, 1019w, 999s, 989w, 952w.
Source: Frost (2004a).
Comments: No independent analytical data are provided for the sample used.

Sperrylite $PtAs_2$

Origin: Synthetic.
Experimental details: Methods of sample preparation are not described. Raman scattering measurements have been performed using 18794.59 cm^{-1} laser radiation. The laser radiation power at the sample was between 1 and 2 mW.
Raman shifts (cm^{-1}): 293w, 279, 226, 216s.
Source: Bakker (2014).
Comments: No independent analytical data are provided for the sample used. For the Raman spectra of sperrylite see also Müller and Lutz (1991) and Mernagh and Hoatson (1995).

Spertiniite $Cu(OH)_2$

Origin: Artificial (a product of brass corrosion).
Experimental details: Micro-Raman scattering measurements have been performed on an arbitrarily oriented sample using 532 nm laser radiation. The laser radiation power at the sample was 74 µW.
Raman shifts (cm^{-1}): ~3555s, ~3305, ~950w, ~840w, ~495s, ~450, ~293.
Source: Schmutzler et al. (2016).
Comments: No independent analytical data are provided for the sample used.

Spessartine $Mn^{2+}_3Al_2(SiO_4)_3$

Origin: Lojane, municipality of Lipkovo, Republic of Macedonia.
Experimental details: Raman scattering measurements have been performed on an arbitrarily oriented sample using 1064 nm Nd-YAG laser radiation. The laser radiation power is not indicated.
Raman shifts (cm^{-1}): 1025sh, 976, 901s, 838, 632w, 607w, 557, 496, 454w, 413w, 377sh, 354, 293, 228, 183.
Source: Makreski et al. (2005b).
Comments: No independent analytical data are provided for the sample used. For the Raman spectra of spessartine see also Mingsheng et al. (1994), Kolesov and Geiger (1998), Bersani et al. (2009), Jovanovski et al. (2009), Frezzotti et al. (2012), and Andò and Garzanti (2014).

Sphalerite ZnS

Origin: Rio Tinto, Spain.
Experimental details: Methods of sample preparation are not described. Raman scattering measurements have been performed on an arbitrarily oriented sample using 514.5 nm Ar^+ laser radiation. The laser radiation power at the sample was between 1 and 10 mW. A 180°-scattering geometry was employed.

Raman shifts (cm^{-1}): 447, 420, 407, 397, 350s, 336, 287, 275s, 237, 219, 208, 178, 156, 144, 117.
Source: Mernagh and Trudu (1993).
Comments: The Raman shifts are given for Fe-bearing sphalerite. No independent quantitative analytical data are provided for the sample used. For the Raman spectra of sphalerite see also White (2009), Frezzotti et al. (2012), and Andò and Garzanti (2014).

Spherocobaltite $Co(CO_3)$

Origin: Synthetic.
Experimental details: Methods of sample preparation are not described. Raman scattering measurements have been performed using 514.5 nm Ar$^+$ laser radiation. The laser radiation power is not indicated.
Raman shifts (cm^{-1}): 1090s, 725w, 302, ~194.
Source: Chariton et al. (2017).
Comments: The sample was characterized by single-crystal X-ray diffraction data. For the Raman spectrum of spherocobaltite see also Rutt and Nicola (1974).

Spinel $MgAl_2O_4$

Origin: Synthetic.
Experimental details: Raman scattering measurements have been performed on an arbitrarily oriented crystal using 632.8 nm He-Ne and 473.1 nm Nd-YAG laser radiations. The laser radiation power at the sample was <1 mW.
Raman shifts (cm^{-1}): 768, 720w, 670, 562w, 493w, 408s, 375sh, 308.
Source: D'Ippolito et al. (2015).
Comments: The sample was characterized by electron microprobe analyses. For the Raman spectra of spinel see also Shoval et al. (2001), Jasinevicius (2009), Kojitani et al. (2013), Culka et al. (2016a, b), and Dongre et al. (2016).

Spionkopite $Cu_{39}S_{28}$

Origin: Synthetic.
Experimental details: No data.
Raman shifts (cm^{-1}): 474s.
Source: Parker et al. (2008).
Comments: The sharp band at 474 cm^{-1} corresponds to S–S pairs.

Spiroffite $Mn^{2+}{}_2Te^{4+}{}_3O_8$

Origin: Moctezuma mine, Sonora, Mexico (type locality).
Experimental details: Raman scattering measurements have been performed on arbitrarily oriented crystals using a 633 nm He-Ne laser. The laser radiation power is not indicated. The Raman shifts have been determined for the maxima of individual peaks obtained as a result of the spectral curve analysis.
Raman shifts (cm^{-1}): (773w), 743s, 721, 650sh, 466s, 394, 346s, 226, 148.
Source: Frost et al. (2009g).
Comments: No independent analytical data are provided for the sample used. The IR spectrum of presumed spiroffite given in the cited paper corresponds to quartz.

Spodumene $LiAlSi_2O_6$

Origin: Conțu-Negovanu pegmatite field, Lotru-Cibin Mts., Sibiu Co., Romania.
Experimental details: Raman scattering measurements have been performed on an arbitrarily oriented single crystal. Experimental details are not described.
Raman shifts (cm^{-1}): 1098, 1070s, 1017, 783, 705s, 582, 522, 438, 393, 355s, 296, 249.
Source: Buzatu and Buzgar (2010).
Comments: No independent analytical data are provided for the sample used. For the Raman spectra of spodumene see also Anderson et al. (2001) and Jasinevicius (2009).

Spurrite $Ca_5(SiO_4)_2(CO_3)$

Origin: Synthetic.
Experimental details: Methods of sample preparation are not described. Raman scattering measurements have been performed using 632.8 nm He-Ne laser radiation. The nominal laser radiation power was 20 mW.
Raman shifts (cm^{-1}): 1080s, 948w, 932w, 864, 852, 704, 547w, 520w, 404w, 389w.
Source: Gastaldi et al. (2008).
Comments: The sample was characterized by powder X-ray diffraction data.

Šreinite $Pb(UO_2)_4(BiO)_3(PO_4)_2(OH)_7 \cdot 4H_2O$

Origin: Horní Halže, Krušné Hory (Ore Mts.), Czech Republic (type locality).
Experimental details: Raman scattering measurements have been performed on an arbitrarily oriented sample using 785 nm laser radiation. The laser radiation power at the sample was 4 mW.
Raman shifts (cm^{-1}): 1361w, 1060w, 1023w, 975w, 871sh, 839sh, 797s, 678w, 595w, 507w, 457w, 449w, 334.
Source: Sejkora and Čejka (2007).
Comments: The sample was characterized by powder X-ray diffraction data and electron microprobe analyses.

Srilankite $(Ti,Zr)O_2$

Origin: Xiuyan meteorite crater, Xiuyan Co., Liaoning province, NE China.
Experimental details: Raman scattering measurements have been performed on an arbitrarily oriented Zr-free sample using 514.5 nm Ar$^+$ laser radiation. The laser radiation power is not indicated.
Raman shifts (cm^{-1}): 825w, 610, 572, 533, 442sh, 428s, 412sh, 357, 340w, 315, 287, 175s, 151.
Source: Chen et al. (2013a).
Comments: The sample was characterized by powder X-ray diffraction data and electron microprobe analyses. For the Raman spectrum of srilankite see also Mammone et al. (1981).

Stanfieldite $Ca_4Mg_5(PO_4)_6$

Origin: Artificial (a product of pyrometamorphic substitution of apatite in slag).
Experimental details: Micro-Raman scattering measurements have been performed on an arbitrarily oriented sample using 532 nm laser radiation. The laser radiation power at the sample was between 1 and 5 mW.

Raman shifts (cm^{-1}): 1293w, 1228w, 1158w, 1124w, 1075, 983sh, 974s, 968s, 753, 620, 611sh, 533, 514, 468, 402, 348, 293w, 176w.
Source: Schneider et al. (2013).
Comments: The sample was characterized by electron microprobe analyses.

Stanleyite $V^{4+}O(SO_4) \cdot 6H_2O$

Origin: Synthetic.
Experimental details: Raman scattering measurements have been performed on an arbitrarily oriented sample using 514.5 nm Ar$^+$ laser radiation. The laser radiation power at the sample was between 10 and 40 mW.
Raman shifts (cm^{-1}): 1078, 1028, 1006, 630, 450, 310.
Source: Hardcastle and Wachs (1991).
Comments: No independent analytical data are provided for the sample used. Band intensities are not indicated.

Stannite Cu_2FeSnS_4

Origin: Synthetic.
Experimental details: Raman scattering measurements have been performed on an arbitrarily oriented sample using 514.5 nm Ar$^+$ laser radiation. The laser radiation power is not indicated. A nearly 180°-scattering geometry was employed.
Raman shifts (cm^{-1}): 350w, 318s, 286.
Source: Himmrich and Haeuseler (1991).
Comments: For the Raman spectra of stannite see also Fontané et al. (2012) and Evrard et al. (2015).

Starkeyite $Mg(SO_4) \cdot 4H_2O$

Origin: Calingasta, San Juan province, Argentina.
Experimental details: Raman scattering measurements have been performed on an arbitrarily oriented sample using 632 nm He-Ne laser radiation. The laser radiation power is not indicated.
Raman shifts (cm^{-1}): ~1156, ~1120, ~1102s, ~1085w, ~615, ~560w, ~475.
Source: Peterson (2011).
Comments: The sample was characterized by powder X-ray diffraction data. For the Raman spectra of starkeyite see also Wang et al. (2006a) and Frezzotti et al. (2012).

Starovaite $KCu_5O(VO_4)_3$

Origin: Synthetic.
Experimental details: Raman scattering measurements have been performed at 250 K in a quasi-back-scattering geometry, from the surface parallel to the a-axis of a microtwinned crystal using 514.5 nm Ar$^+$ laser radiation. The laser radiation power at the sample was 10 mW.
Raman shifts (cm^{-1}): ~953s, ~908, ~860, ~834w.
Source: Choi et al. (2004).

Staurolite $Fe^{2+}_2Al_9Si_4O_{23}(OH)$

Origin: Štavica, municipality of Prilep, Republic of Macedonia.
Experimental details: Raman scattering measurements have been performed on an arbitrarily oriented sample using 1064 nm Nd-YAG laser radiation. The laser radiation power is not indicated.
Raman shifts (cm^{-1}): 3675, 3622w, 3571w, 3520sh, 3451sh, 3426s, 1022, 970s, 938sh, 898sh, 847sh, 684w, 640sh, 593, 543, 525sh, 487, 432, 399w.
Source: Makreski et al. (2005b).
Comments: No independent analytical data are provided for the sample used. For the Raman spectrum of staurolite see also Andò and Garzanti (2014).

Steedeite $NaMn_2[Si_3BO_9](OH)_2$

Origin: Poudrette quarry, Montérégie (Rouville) Co., Québec, Canada (type locality).
Experimental details: Raman scattering measurements have been performed on an arbitrarily oriented sample using 532 nm laser radiation. The laser radiation power is not indicated.
Raman shifts (cm^{-1}): 3443, 3317w, 1700, 1368, 1030, 1000s, 874, 836, 696, 636s, 431, 330, 264, 197w, 120.
Source: Haring and McDonald (2014a).
Comments: The sample was characterized by powder X-ray diffraction data and electron microprobe analyses. The crystal structure is solved.

Steenstrupine-(Ce) $Na_{14}Ce_6Mn^{2+}_2Fe^{3+}_2Zr(PO_4)_7Si_{12}O_{36}(OH)_2 \cdot 3H_2O$

Origin: Karnasurt Mt., Lovozero alkaline massif, Kola Peninsula, Russia.
Experimental details: Raman scattering measurements have been performed on a partly metamict sample annealed at 500 °C using 632.8 nm laser radiation. The laser radiation power is not indicated.
Raman shifts (cm^{-1}): ~957s, ~880sh, ~756, ~600 (broad).
Source: Kusz et al. (2010).
Comments: The sample was characterized by powder X-ray diffraction data and electron microprobe analyses.

Stephanite Ag_5SbS_4

Origin: Bohemia, Czech Republic.
Experimental details: Raman scattering measurements have been performed on an oriented crystal with the laser polarization parallel to the a-, b- and c-axes. 785 nm solid-state laser radiation was used. The nominal laser radiation power was 1.7 mW. A 180°-scattering geometry was employed.
Raman shifts (cm^{-1}): 335s, 317w, 306w, 301, 233, 204w, 178w.
Source: Kharbish et al. (2009).
Comments: The sample was characterized by electron microprobe analyses. The Raman shifts are given as the sum of the spectra of all scattering geometries.

Štěpite $U(AsO_3OH)_2 \cdot 4H_2O$

Origin: Geschieber vein, Jáchymov ore district, Western Bohemia, Czech Republic (type locality).
Experimental details: No data.
Raman shifts (cm^{-1}): 3552, 3484, 1641w (broad), 896s, 844s, 811s, 760s, 420, 401, 377, 368, 351, 322, 312, 287, 262, 235, 180, 160, 139, 115, 107.
Source: Plášil et al. (2013b).
Comments: The sample was characterized by powder X-ray diffraction data and electron microprobe analyses. The crystal structure is solved.

Stercorite $(NH_4)Na(PO_3OH) \cdot 4H_2O$

Origin: Petrogale Cave, Madura, Western Australia.
Experimental details: Raman scattering measurements have been performed on arbitrarily oriented crystals using a 633 nm He-Ne laser. The laser radiation power is not indicated. The Raman shifts have been determined for the maxima of individual peaks obtained as a result of the spectral curve analysis.
Raman shifts (cm^{-1}): 3158, 3024sh, 2900sh, 920s, 577, 476, 450sh, 396w, 345, 326sh, 216sh, 197 + 185 (unresolved doublet), (155), 143s, 110.
Source: Frost et al. (2011u).
Comments: The sample was characterized by powder X-ray diffraction data.

Steropesite Tl_3BiCl_6

Origin: Synthetic.
Experimental details: Raman scattering measurements have been performed on a powdered sample using 1036 nm Nd-YAG laser radiation. The nominal laser radiation power was between 30 and 300 mW. A 180°-scattering geometry was employed.
Raman shifts (cm^{-1}): 261s, 218, 117s.
Source: Beck and Benz (2010).
Comments: The sample was characterized by single-crystal X-ray diffraction data and X-ray absorption spectrum.

Stetindite $Ce(SiO_4)$

Origin: Stetind pegmatite, Tysfjord, Nordland, Norway (type locality).
Experimental details: No data.
Raman shifts (cm^{-1}): See comment below.
Source: Schlüter et al. (2009).
Comments: Raman micro-spectroscopy shows weak OH bands in the frequency range between 3200 and 3700 cm^{-1} corresponding to vibrations of OH groups. No other data on the Raman spectrum of stetindite are given in the cited paper.

Stibarsen SbAs

Origin: Synthetic.
Experimental details: No data.
Raman shifts (cm^{-1}): 216 (calculated).
Source: Zhang et al. (2016a).

Stibiconite $Sb^{3+}Sb^{5+}_2O_6(OH)$

Origin: Yucunani Mine, Tejocotes, Oaxaca, Mexico.
Experimental details: Raman scattering measurements have been performed on arbitrarily oriented crystals using a 633 nm He-Ne laser. The laser radiation power is not indicated. The Raman shifts have been determined for the maxima of individual peaks obtained as a result of the spectral curve analysis.
Raman shifts (cm^{-1}): 3603sh, 3424 + 3225 (unresolved doublet?), 3018sh, 2755w, 855sh, 827w, 736w, 609, 564w, (537w), 522, (508), 461, 409 + 400 (unresolved doublet?), 261, (250), 220sh, 199s, 146, 109w.
Source: Bahfenne and Frost (2010e).
Comments: No independent analytical data are provided for the sample used.

Stibioclaudetite $AsSbO_3$

Origin: Tsumeb mine, Tsumeb, Namibia (type locality).
Experimental details: Methods of sample preparation are not described. Raman scattering measurements have been performed using 514.5 nm Ar^+ laser radiation. The nominal laser radiation power was 100 mW.
Raman shifts (cm^{-1}): 817w, 766w, 726w, 631w, 620w, 517w, 477, 468, 430sh, 414s, 342, 323w, 298w, 273, 232, 210, 202, 183, 171s, 155s, 125w, 115w.
Source: Origlieri et al. (2009).
Comments: The sample was characterized by single-crystal X-ray diffraction data and electron microprobe analyses. For the Raman spectrum of Stibioclaudetite see also Origlieri (2005).

Stibiocolumbite $SbNbO_4$
Origin: Synthetic.
Experimental details: Raman scattering measurements have been performed on a polycrystalline sample using 514.5 nm Ar^+ laser radiation. The nominal laser radiation power was 200 mW.
Raman shifts (cm^{-1}): 913, 740w, 718, 620s (broad), 542, 448, 397, 377s, 350, 292, 269s, 239, 232, 192s, 168s, 128s, 87s, 75s, 54w, 37.
Source: Ayyub et al. (1986).
Comments: No independent analytical data are provided for the sample used. For the Raman spectrum of stibiocolumbite see also Ayyub et al. (1987).

Stibiopalladinite Pd_5Sb_2

Origin: Synthetic.
Experimental details: Methods of sample preparation are not described. Raman scattering measurements have been performed using 18794.59 cm^{-1} laser radiation. The laser radiation power at the sample was between 1 and 2 mW.
Raman shifts (cm^{-1}): 187w, 169, 108s.
Source: Bakker (2014).
Comments: No independent analytical data are provided for the sample used.

Stibnite Sb_2S_3

Origin: Schlaining, Oberwart, Burgenland, Austria.

Experimental details: Raman scattering measurements have been performed on an oriented crystal with the laser polarization parallel and perpendicular to the cleavage and elongation of kermesite. 632.8 nm He-Ne laser radiations were used. The nominal laser radiation power was 1.7 mW. A 180°-scattering geometry was employed.

Raman shifts (cm^{-1}): 310, 300, 281s, (254w), 237s, (225w), 207, 189, (180w), 156w, 125, 99, 71, 59s, 50, 39.

Source: Kharbish et al. (2009).

Comments: The sample was characterized by electron microprobe analysis. The Raman shifts are given as the sum of the spectra of all scattering geometries. For the Raman spectra of stibnite see also Mernagh and Trudu (1993), Minceva-Sukarova et al. (2003), Roy et al. (2008), Frost et al. (2010c), and Makreski et al. (2013b).

Stichtite $Mg_6Cr_2(CO_3)(OH)_{16} \cdot 4H_2O$

Origin: Synthetic.

Experimental details: Raman scattering measurements have been performed on arbitrarily oriented crystals using a 633 nm He-Ne laser. The laser radiation power is not indicated. The Raman shifts have been determined for the maxima of individual peaks obtained as a result of the spectral curve analysis.

Raman shifts (cm^{-1}): 1087s, 1067sh, 539s, (531), 458, (446), 366w, 328w, 317w, 292w, 248w, 215w, 153.

Source: Frost and Erickson (2004).

Comments: No independent analytical data are provided for the sample used. For the Raman spectrum of stichtite see also Mills et al. (2011b).

Stilbite-Ca $NaCa_4(Si_{27}Al_9)O_{72} \cdot 28H_2O$

Origin: Berufjördur, Sudur-Múlasýsla, Eastern Region, Iceland.

Experimental details: Methods of sample preparation are not described. Raman scattering measurements have been performed using 785 nm diode laser radiation. The maximum output powder of 300 mW could be filtered to diminish the power at the sample.

Raman shifts (cm^{-1}): 794, 644w, 497s, 459w, 410s, 320w.

Source: Jehlička et al. (2012).

Comments: No independent analytical data are provided for the sample used. For the Raman spectra of stilbite-Ca see also Mozgawa (2001), Makreski et al. (2009), Jehlička and Vandenabeele (2015), and Ma et al. (2016b).

Stilbite-Na $Na_9(Si_{27}Al_9)O_{72} \cdot 28H_2O$

Origin: Synthetic.

Experimental details: Raman scattering measurements have been performed on an arbitrarily oriented sample using 514.5 nm Ar^+ laser radiation. The laser radiation power at the sample was 1.5 mW. A 180°-scattering geometry was employed.

Raman shifts (cm^{-1}): 1133, 801, 618, 499s, 458, 411s, 152.

Source: Ma et al. (2016b).

Comments: The sample was characterized by powder X-ray diffraction data. The crystal structure is solved by the Rietveld method.

Stilleite ZnSe

Origin: Synthetic.
Experimental details: Raman scattering measurements have been performed on a thin film using 532 Nd-YAG laser radiation. The laser radiation power at the sample was 0.1 mW.
Raman shifts (cm^{-1}): 500, 252s, 202w.
Source: Perna et al. (2006).
Comments: The sample was characterized by powder X-ray diffraction data. For the Raman spectrum of stilleite see also Yang et al. (1999).

Stilpnomelane (K,Ca,Na)(Fe,Mg,Al)$_8$(Si,Al)$_{12}$(O,OH)$_{36} \cdot n$H$_2$O

Origin: Martian meteorite MIL 03346.
Experimental details: Raman scattering measurements have been performed on an arbitrarily oriented sample using 532.3 nm Nd-YAG laser radiation. The nominal laser radiation power was 14.5 mW.
Raman shifts (cm^{-1}): 3579w, 3568w, 1156, 897, 820w, 588s, 501, 379, 291.
Source: Kuebler (2013a).
Comments: The sample was characterized by electron microprobe analyses.

Stishovite SiO$_2$

Origin: Synthetic.
Experimental details: Raman scattering measurements have been performed on an arbitrarily oriented sample using 514.5 nm Ar$^+$ laser radiation. The laser radiation power at the sample was 16 mW. A 180°-scattering geometry was employed.
Raman shifts (cm^{-1}): 960, 750s, 584, 228.
Source: Liu and El Gorsey (2007). For the Raman spectra of stishovite see also Hemley (1987a, b), Von Czarnowski and Hübner (1987), Holtstam et al. (2003), Miyahara et al. (2013), and Spektor et al. (2016).

Stoiberite Cu$_5$O$_2$(VO$_4$)$_2$

Origin: Synthetic.
Experimental details: Methods of sample preparation are not described. Raman scattering measurements have been performed using 532.1 nm laser radiation. The laser radiation power is not indicated.
Raman shifts (cm^{-1}): ~950, ~900s, ~800s, ~560w, ~505w, ~410, ~330w.
Source: Kawada et al. (2015).
Comments: The sample was characterized by powder X-ray diffraction data. For the Raman spectrum of stoiberite see also Newhouse et al. (2016).

Stolzite Pb(WO$_4$)

Origin: Vysoká hill, near Havlíčkův Brod, Czech Republic.
Experimental details: Raman scattering measurements have been performed on an arbitrarily oriented sample using 532 nm laser radiation. The laser radiation output power was 4 mW.
Raman shifts (cm^{-1}): 905s, 766, 752, 357, 328, 324s, 192w, 178, 90, 77w, 71w, 64, 56.
Source: Pauliš et al. (2016).
Comments: The sample was characterized by powder X-ray diffraction data. For the Raman spectra of stolzite see also Frost et al. (2004d), Kloprogge et al. (2004b), and Andrade et al. (2014).

Stoppaniite $Fe^{3+}_2Be_3Si_6O_{18} \cdot H_2O$

Origin: Capranica, Vico volcanic complex, Latium, Italy (type locality).
Experimental details: No data in the cited paper.
Raman shifts (cm^{-1}): 3595–3588.
Source: Della Ventura et al. (2000).

Stottite $Fe^{2+}Ge(OH)_6$

Origin: Tsumeb mine, Tsumeb, Namibia (type locality).
Experimental details: Raman scattering measurements have been performed on an arbitrarily oriented single crystal using 514.5 nm Ar$^+$ laser radiation. The nominal laser radiation power was below 80 mW. A 135°-scattering geometry was employed.
Raman shifts (cm^{-1}): 3352, 3240, 3159, 3064, 636s, 418, 297, 266.
Source: Kleppe et al. (2012).
Comments: The sample was characterized by single-crystal X-ray diffraction data.

Strashimirite $Cu_4(AsO_4)_2(OH)_2 \cdot 2.5H_2O$

Origin: Zálesí deposit, Rychlebské Hory Mts., northern Moravia, Czech Republic.
Experimental details: Raman scattering measurements have been performed on arbitrarily oriented crystals using a 633 nm He-Ne laser. The laser radiation power is not indicated. The Raman shifts have been determined for the maxima of individual peaks obtained as a result of the spectral curve analysis.
Raman shifts (cm^{-1}): 3585w, 3488w, 3450w, 852s, 831sh, 554sh, 526sh, 497, 467sh, 393w, 337, 294s, 239sh, 220, 172sh, 152s.
Source: Frost et al. (2009i).
Comments: The sample was characterized by powder X-ray diffraction data and electron microprobe analyses.

Strengite $Fe^{3+}(PO_4) \cdot 2H_2O$

Origin: Iron Monarch, Middleback Ranges, South Australia, Australia.
Experimental details: Raman scattering measurements have been performed on arbitrarily oriented crystals using a 633 nm He-Ne laser. The laser radiation power is not indicated. The Raman shifts have been determined for the maxima of individual peaks obtained as a result of the spectral curve analysis.
Raman shifts (cm^{-1}): 1357, 1250, 1158, 1137, 1005, 985s, 744, 694, 560, 487, 447, 434, 398, 317, 303, 249, 204, 172s, 153, 135.
Source: Frost et al. (2004l).
Comments: No independent analytical data are provided for the sample used. For the Raman spectrum of strengite see also Kloprogge and Wood (2017).

Stringhamite $CaCu(SiO_4) \cdot H_2O$

Origin: Christmas mine, Gila Co., Arizona, USA.
Experimental details: Raman scattering measurements have been performed on arbitrarily oriented crystals using a 633 nm He-Ne laser. The laser radiation power at the sample was 0.1 mW. The Raman shifts have been determined for the maxima of individual peaks obtained as a result of the spectral curve analysis.
Raman shifts (cm^{-1}): 3239sh, 3193, 1371w, 1147w, 1061, (1027w), (997w), 980sh, 956s, 908, 848s, 825, 799, 764w, 693w, 626w, 570s, 519, 505, 431w, 396, 369, 341w, 326w, 303.
Source: Frost and Xi (2012f).
Comments: No independent analytical data are provided for the sample used.

Stromeyerite $CuAgS$

Origin: Artificial (a product of Ag-Cu alloy corrosion).
Experimental details: Raman scattering measurements have been performed on a polycrystalline sample using 514.5 nm Ar^+ laser radiation. The laser radiation power is not indicated.
Raman shifts (cm^{-1}): 266–280 (broad).
Source: De Caro et al. (2016).
Comments: The sample was characterized by powder X-ray diffraction data and electron microprobe analyses.

Stronadelphite $Sr_5(PO_4)_3F$

Origin: Synthetic.
Experimental details: Raman scattering measurements have been performed on a powder sample using 514.5 nm Ar^+ laser radiation. The laser radiation power is not indicated.
Raman shifts (cm^{-1}): 1055, 1042, 1029, 952s, 606sh, 595, 582, 575, 445, 423, 305w, 241, 208w, 196w, 186w, 174w.
Source: Zhai et al. (2015).
Comments: The sample was characterized by powder X-ray diffraction data.

Strontianite $Sr(CO_3)$

Origin: Drensteinfurt, Münsterland, North Rhine-Westphalia, Germany.
Experimental details: Methods of sample preparation are not described. Raman scattering measurements have been performed using 532 nm Nd-YAG laser radiation. The nominal laser radiation power was 100 mW.
Raman shifts (cm^{-1}): 1543w, 1445, 1069s, 700, 242.
Source: Buzgar and Apopei (2009).
Comments: No independent analytical data are provided for the sample used. For the Raman spectrum of strontianite see also Frezzotti et al. (2012).

Strontiofluorite SrF_2

Origin: Synthetic.
Experimental details: Raman scattering measurements have been performed on a single crystal using 253.65 nm Hg radiation.
Raman shifts (cm^{-1}): 285.
Source: Warrier and Krishnan (1964).

Strontiohurlbutite $SrBe_2(PO_4)_2$

Origin: Nanping No. 31 pegmatite, Fujian province, SE China (type locality).
Experimental details: Raman scattering measurements have been performed on an arbitrarily oriented single-crystal sample using 514.5 nm Ar^+ laser radiation. The laser radiation power at the sample was 5 mW.
Raman shifts (cm^{-1}): 1178, 1135, 1022s, 587, 575, 550, 494, 442, 421, 343, 204, 176.
Source: Rao et al. (2014).
Comments: The sample was characterized by powder X-ray diffraction data and electron microprobe analyses. The crystal structure is solved.

Strontiojoaquinite $(Na,Fe)_2Ba_2Sr_2Ti_2(SiO_3)_8(O,OH)_2·H_2O$

Origin: Junilla Claim, San Benito Co., California, USA.
Experimental details: Raman scattering measurements have been performed on arbitrarily oriented crystals using a 633 nm He-Ne laser. The laser radiation power is not indicated. The Raman shifts have been determined for the maxima of individual peaks obtained as a result of the spectral curve analysis.
Raman shifts (cm^{-1}): 3599w, 3587sh, 3519 + 3511 (unresolved doublet?), 3485sh, 1123s, 1062, 1031, 971s, 912, 892, 738w, 682 + 679 (unresolved doublet?), 621s, 602sh, 511, 468+437 (unresolved doublet?), 387–386, 357sh, 339, 284, 267, 193sh, 173, 159, 145.
Source: Frost and Pinto (2007).
Comments: The sample was characterized by electron microprobe analysis.

Strunzite $Mn^{2+}Fe^{3+}_2(PO_4)_2(OH)_2·6H_2O$

Origin: No data.
Experimental details: Raman scattering measurements have been performed on arbitrarily oriented crystals using a 633 nm He-Ne laser. The laser radiation power is not indicated. The Raman shifts have been determined for the maxima of individual peaks obtained as a result of the spectral curve analysis.
Raman shifts (cm^{-1}): 3483, 3410, 3340, 3120, 1120, 1048, 1000s, 975, 639, 567, 513, 469s, 433, 406, 329, 301, 281sh, 248, 199, 183, 168.
Source: Frost et al. (2002c).
Comments: The sample was characterized by powder X-ray diffraction data and quqlitative electron microprobe analysis.

Struvite-(K) $KMg(PO_4)·6H_2O$

Origin: Synthetic.
Experimental details: Methods of sample preparation are not described. Raman scattering measurements have been performed using 1064 nm Nd-YAG laser radiation. The laser radiation power is not indicated.
Raman shifts (cm^{-1}): 1075w, 1015w, 1005w, 985w, 946s, 569s, 470w, 430, and a series of bands in the range from 250 to 400 cm^{-1}.
Source: Stefov et al. (2004).
Comments: No independent analytical data are provided for the sample used.

Struvite $(NH_4)Mg(PO_4) \cdot 6H_2O$

Origin: No data.
Experimental details: Raman scattering measurements have been performed on arbitrarily oriented crystals using a 633 nm He-Ne laser. The laser radiation power is not indicated. The Raman shifts have been determined for the maxima of individual peaks obtained as a result of the spectral curve analysis.
Raman shifts (cm^{-1}): 3239sh, 3115, 2921sh, (2903sh), 2368w, 1077, 1013w, 950s, (942), 890, 564, 463w, 428, 300, 242, 229, 206.
Source: Frost et al. (2005j).
Comments: No independent analytical data are provided for the sample used. For the Raman spectrum of struvite see also García et al. (2013).

Studtite $(UO_2)(O_2)(H_2O)_2 \cdot 2H_2O$

Origin: Menzenschwand, Schwarzwald (Black Forest Mts.), Germany.
Experimental details: Methods of sample preparation are not described. Raman scattering measurements have been performed on an arbitrarily oriented single crystal using 632.8 nm He-Ne laser radiation. The laser radiation power at the sample was 6 mW. A 180°-scattering geometry was employed.
Raman shifts (cm^{-1}): 3473w, 3145, 1712w, 1685, 865s, 838sh, 819s, 810sh, 408w, 352w, 294w, 266w, 230w.
Source: Bastians et al. (2004).
Comments: No independent analytical data are provided for the sample used. For the Raman spectra of studtite see also Amme et al. (2002) and Colmenero et al. (2017).

Sturmanite $Ca_6Fe^{3+}_2(SO_4)_2[B(OH)_4](OH)_{11}O \cdot 25H_2O$

Origin: Black Rock mine, Kuruman Manganese Fields, Kalahari, South Africa (type locality).
Experimental details: Raman scattering measurements have been performed on arbitrarily oriented crystals using a 633 nm He-Ne laser. The laser radiation power is not indicated. The Raman shifts have been determined for the maxima of individual peaks obtained as a result of the spectral curve analysis.
Raman shifts (cm^{-1}): 3677sh, 3622 + 3600 (unresolved doublet?), 3479, (3401), 3276sh, 1776w, 1697, 1636sh, 1117w, 1069, 995sh, 990s, 981sh, 959w, 760w, 623w, 579s, 530, 501, 455, 383sh, 355, 268sh, (232), 205.
Source: Frost et al. (2014ag).
Comments: The sample was characterized by electron microprobe analysis.

Stützite $Ag_{5-x}Te_3$ ($x = 0.24-0.36$)

Origin: Coranda-Hondol open pit, Certej Au-Ag deposit, South Apuseni Mts., Romania.
Experimental details: Raman scattering measurements have been performed on an arbitrarily oriented grain in a polished section using 632.8 nm He-Ne laser radiation. The laser radiation power is not indicated.
Raman shifts (cm^{-1}): 147s, 80sh, 64.
Source: Apopei et al. (2014b).
Comments: The sample was characterized by electron microprobe analyses.

Sudoite $Mg_2Al_3(Si_3Al)O_{10}(OH)_8$

Origin: Semail ophiolite, Oman.
Experimental details: Methods of sample preparation are not described. Raman scattering measurements have been performed using 514.5 nm Ar^+ or 532 nm solid-state laser radiation. The laser radiation power is not indicated.
Raman shifts (cm^{-1}): 3696, 3692, 3666, 3646, 1109, 1080, 1045, 1005, 733, 702, 665, 616, 559s, 549, 477, 403, 388, 365, 259s, 211, 190, 94.
Source: Reynard et al. (2015).
Comments: The empirical formula of the sample used is $(Mg_{1.7}Fe_{0.3}Al_4)(Si_3Al)O_{10}(OH)_8$.

Sudovikovite $PtSe_2$

Origin: Synthetic.
Experimental details: Methods of sample preparation are not described. Raman scattering measurements have been performed using 532 nm laser radiation. The nominal laser radiation power was 10 mW.
Raman shifts (cm^{-1}): 206, 177.
Source: Altamura et al. (2014).
Comments: The sample was characterized by powder X-ray diffraction data.

Sulphohalite $Na_6(SO_4)_2ClF$

Origin: Searles Lake, San Bernardino Co., California, USA (type locality).
Experimental details: Raman scattering measurements have been performed on arbitrarily oriented crystals using a 633 nm He-Ne laser. The laser radiation power is not indicated. The Raman shifts have been determined for the maxima of individual peaks obtained as a result of the spectral curve analysis.
Raman shifts (cm^{-1}): (1132sh), 1128, (1120sh), 1021w, (1010sh), 1003s, (997sh), 986sh, 635, 624sh, (481sh), 472, 467sh, 159 + 146 (unresolved doublet?), 117 + 109 (unresolved doublet?).
Source: Frost et al. (2014z).
Comments: No independent analytical data are provided for the sample used. The sample was characterized by qualitative electron microprobe analysis.

Sulfur S

Origin: Mariana Arc.
Experimental details: Raman scattering measurements have been performed on an arbitrarily oriented sample using 532 nm Nd-YAG laser radiation. The laser radiation output power was 100 mW.
Raman shifts (cm^{-1}): 472s, 437, 246, 219s, 186w, 153.
Source: White (2009).
Comments: No independent analytical data are provided for the sample used. For the Raman spectra of sulfur see also Venkateswarlu (1940), Mycroft et al. (1990), Turcotte and Benner (1993), Munce et al. (2007), and Frezzotti et al. (2012).

Sulvanite Cu_3VS_4

Origin: Synthetic.
Experimental details: Raman scattering measurements have been performed on an arbitrarily oriented sample using Ar^+ or Kr^+ laser radiation. The laser radiation power at the sample was below 50 mW.
Raman shifts (cm^{-1}): 448sh, 440s, 376s, 301w, 201s, 147.
Source: Petritis et al. (1981).
Comments: The sample was characterized by powder X-ray diffraction data.

Suredaite $PbSnS_3$

Origin: Synthetic.
Experimental details: Raman scattering measurements have been performed on nanorods using 514.5 nm Ar^+ laser radiation. The laser radiation power is not indicated.
Raman shifts (cm^{-1}): 625, 442, 309s, 201, (143s).
Source: Wang et al. (2001b).
Comments: The sample was characterized by powder X-ray diffraction data.

Sursassite $Mn^{2+}{}_2Al_3(SiO_4)(Si_2O_7)(OH)_3$

Origin: Strategic Manganese Mine, near Woodstock, New Brunswick, Canada.
Experimental details: Raman scattering measurements have been performed on an arbitrarily oriented single crystal using 633 nm He-Ne laser radiation. The laser radiation power is not indicated.
Raman shifts (cm^{-1}): 3452, 3335, 3230, 2998sh, 1086w, 1026w, 925s, 866, 822, 705, 618, 554s, 491sh, 353, 283, (213), 151.
Source: Reddy and Frost (2007).
Comments: The sample was characterized by powder X-ray diffraction data. The Raman shifts have been determined for the maxima of individual peaks obtained as a result of the spectral curve analysis.

Susannite $Pb_4(SO_4)(CO_3)_2(OH)_2$

Origin: Herzog Julius Shaft, Astfeld, Schlackental, Harz Mts., Germany.
Experimental details: Raman scattering measurements have been performed on arbitrarily oriented crystals using a 633 nm He-Ne laser. The laser radiation power is not indicated. The Raman shifts have been determined for the maxima of individual peaks obtained as a result of the spectral curve analysis.
Raman shifts (cm^{-1}): 3630, 3550, 3513, 3447, 3377, 3307, 3241, 3179, 1154, 1105, 1048, 1026, 1011, 964s, 628, 602, 497, 470, 450, 427, 393, 363, 239, 203.
Source: Frost et al. (2003e).
Comments: Questionable data: bands of symmetric stretching vibrations of carbonate groups are unusually weak. No independent analytical data are provided for the sample used.

Suseinarguite $(Na_{0.5}Bi_{0.5})(MoO_4)$

Origin: Su Seinargiu, Sardinia, Italy (type locality).
Experimental details: Raman scattering measurements have been performed on an arbitrarily oriented sample using 514.5 nm Ar^+ laser radiation. The laser radiation power at the sample was 1.5 mW.
Raman shifts (cm^{-1}): 876s, 772, 376, 319s, 188, 131w.
Source: Orlandi et al. (2015).
Comments: The sample was characterized by powder X-ray diffraction data and electron microprobe analyses.

Svanbergite $SrAl_3(SO_4)(PO_4)(OH)_6$

Origin: Mt. Brussilof mine, Radium, British Columbia, Canada.
Experimental details: Raman scattering measurements have been performed on arbitrarily oriented crystals using a 633 nm He-Ne laser. The laser radiation power is not indicated. The Raman shifts have been determined for the maxima of individual peaks obtained as a result of the spectral curve analysis.
Raman shifts (cm^{-1}): 3518sh, 3467, 3415sh, 3319sh, (3215sh), 3151w, 3064sh, 1098, 1034sh, 1022s, 998, 981sh, 896, 654, 633, 616s, 602sh, 588sh, 572sh, 522, 486, 474sh, 392, 369sh, 280w, 246, 179.
Source: Frost and Palmer (2011b).
Comments: No independent analytical data are provided for the sample used.

Švenekite $Ca[AsO_2(OH)_2]_2$

Origin: Geschieber vein, Jáchymov ore district, Western Bohemia, Czech Republic (type locality).
Experimental details: Raman scattering measurements have been performed on an arbitrarily oriented sample using 532 nm laser radiation. The nominal laser radiation power was 3 mW.
Raman shifts (cm^{-1}): 3368, 2917w, 2385w, 929, 901, 871s, 840, 753s, 726, 541w, 498w, 417, 393, 358, 330, 289, 268, 223, 172w.
Source: Ondruš et al. (2013).
Comments: The sample was characterized by powder X-ray diffraction data and electron microprobe analyses. The crystal structure is solved.

Svornostite $K_2Mg[(UO_2)(SO_4)_2]_2 \cdot 8H_2O$

Origin: Geschieber vein, Jáchymov ore district, Western Bohemia, Czech Republic (type locality).
Experimental details: Raman scattering measurements have been performed on an arbitrarily oriented sample using 532 nm solid-state laser radiation. The nominal laser radiation power was 2.5 mW.
Raman shifts (cm^{-1}): 3622w, 3545, 3496, 1220w, 1200, 1155, 1110w, 1028, 989, 951w, 854s, 725w, 643, 610w, 458, 438, 322w, 268w, 207sh, 186, 132, 75.
Source: Plášil et al. (2015b).
Comments: The sample was characterized by powder X-ray diffraction data and electron microprobe analyses. The crystal structure is solved.

Swedenborgite $NaBe_4Sb^{5+}O_7$

Origin: Långban, near Pajsberg and Filipstad, Värmland, Sweden (type locality).
Experimental details: Raman scattering measurements have been performed on an arbitrarily oriented sample using 633 nm laser radiation.
Raman shifts (cm^{-1}): 906w, 797s, 767, ~740, 575, 427s.
Source: Gaft et al. (2013).
Comments: The Raman spectrum agrees well to that from the RRUFF database.

Symplesite $Fe^{2+}_3(AsO_4)_2 \cdot 8H_2O$

Origin: Laubach mine, Laufdorf, Wetzlar, Hesse, Germany.
Experimental details: Raman scattering measurements have been performed on an arbitrarily oriented sample using 532 nm Nd-YAG laser radiation. The laser radiation power is not indicated.
Raman shifts (cm^{-1}): 3448w, 892, 841s, 803, 768s, 520sh, 498s, 442, 373, 319w, 281, 249, 225sh, 207, 189w, 172w, 137w, 113w.
Source: Makreski et al. (2015b).
Comments: The sample was characterized by powder X-ray diffraction data.

Synchysite-(Ce) $CaCe(CO_3)_2F$

Origin: Soultz-sous-Forêts, Rhine Graben, France.
Experimental details: Raman scattering measurements have been performed on an arbitrarily oriented sample using 488 nm Ar$^+$ laser radiation. The laser radiation power at the sample was about 14 mW.
Raman shifts (cm^{-1}): 1101s, 1083s, 758w, 742, 515, 477, 454w, 276.
Source: Middleton et al. (2013).
Comments: The sample was characterized by electron microprobe analyses.

Syngenite $K_2Ca(SO_4)_2 \cdot H_2O$

Origin: Kalush mine, western Ukraine.
Experimental details: Raman scattering measurements have been performed on an arbitrarily oriented sample using 532 nm Nd-YAG laser radiation. The nominal laser radiation power was 100 mW.
Raman shifts (cm^{-1}): 3306, 1167, 1143, 1120w, 1084w, 1007s, 982s, 663w, 641, 495, 472sh, 442, 240w.
Source: Buzgar et al. (2009).
Comments: No independent analytical data are provided for the sample used. For the Raman spectra of syngenite see also Frezzotti et al. (2012) and Jentzsch et al. (2012a, 2013).

Szaibélyite $MgBO_2(OH)$

Origin: Vysoká-Zlatno Cu-Au porphyry-skarn deposit, Štiavnica Neogene strato volcano, Western Carpathians, Slovakia.
Experimental details: Raman scattering measurements have been performed on an arbitrarily oriented sample using 532 nm Nd-YAG laser radiation. The laser radiation power is not indicated. A 180°-scattering geometry was employed.

Raman shifts (cm^{-1}): 3567, 3559s, 1516w, 1463, 1284, 1186, 988, 915w, 836s, 661, 627, 611, 529, 493, 345, 321, 296, 184, 161.
Source: Bilohuščin et al. (2017).
Comments: The sample was characterized by electron microprobe analyses. For the Raman spectra of szaibélyite see also Frost et al. (2015aa) and Galuskina et al. (2008).

Szenicsite $Cu_3(MoO_4)(OH)_4$

Origin: Jardinera No 1 Mine, Inca de Oro, Chile (type locality).
Experimental details: Raman scattering measurements have been performed on arbitrarily oriented crystals using a 785 nm Nd-YAG laser. The laser radiation power is not indicated. The Raman shifts have been determined for the maxima of individual peaks obtained as a result of the spectral curve analysis.
Raman shifts (cm^{-1}): 3559, 3518, 3506w, 3503w, 3500w, 928, (903sh), 898s, (895sh), 827, 801w, 687w, 476, 408s, 349, 308, 280, 211, 147, 105.
Source: Frost et al. (2007a).
Comments: No independent analytical data are provided for the sample used. For the Raman spectrum of szenicsite see also Yang et al. (2012).

Szmikite $Mn(SO_4) \cdot H_2O$

Origin: Synthetic.
Experimental details: Raman scattering measurements have been performed on an arbitrarily oriented sample using 532 nm Nd-YAG laser radiation. The nominal laser radiation power was 100 mW.
Raman shifts (cm^{-1}): 1188, 1089, 1021s, 654w, 623, 493, 426, 263w.
Source: Buzgar et al. (2009).
Comments: No independent analytical data are provided for the sample used.

Szomolnokite $Fe(SO_4) \cdot H_2O$

Origin: Baia Sprie mining area, Romania.
Experimental details: Raman scattering measurements have been performed on an arbitrarily oriented sample using 632 nm Nd-YAG laser radiation. The nominal laser radiation power was 100 mW.
Raman shifts (cm^{-1}): 1191, 1091, 1020s, 667w, 623, 492, 427.
Source: Buzatu et al. (2016).
Comments: The sample was characterized by powder X-ray diffraction data. For the Raman spectra of szomolnokite see also Chio et al. (2007), Jentzsch et al. (2013), Rull et al. (2014), and Apopei et al. (2015).

Takedaite $Ca_3B_2O_6$

Origin: Fuka mine, Okayama prefecture, Japan (type locality).
Experimental details: Raman scattering measurements have been performed on arbitrarily oriented crystals using a 633 nm He-Ne laser. The laser radiation power is not indicated. The Raman shifts have been determined for the maxima of individual peaks obtained as a result of the spectral curve analysis.

Raman shifts (cm^{-1}): (1087s) (with shoulders), 929w, 911w, (715 + 712) (unresolved doublet?), 585w, 330w, (~282), 217w, (159 + 154) (unresolved doublet?).
Source: Frost et al. (2014n).
Comments: Questionable data. The sample was characterized by qualitative electron microprobe analysis (only Ca, O, and C have been found). The bands at 1087, 715 + 712, ~282, and 159 + 154 cm^{-1} correspond to calcite that is the main component in the sample used.

Takovite $Ni_6Al_2(CO_3)(OH)_{16} \cdot 4H_2O$

Origin: Kambalda, Western Australia, Australia.
Experimental details: Raman scattering measurements have been performed on arbitrarily oriented crystals using a 633 nm He-Ne laser. The laser radiation power is not indicated. The Raman shifts have been determined for the maxima of individual peaks obtained as a result of the spectral curve analysis.
Raman shifts (cm^{-1}): 3461, 2855, 2615, 1543, 1060s, 1042, 992, 697, 558s, 533, 492, 403, 324, 253, 225, 218.
Source: Frost et al. (2003h).
Comments: No independent analytical data are provided for the sample used.

Talc $Mg_3Si_4O_{10}(OH)_2$

Origin: Greiner, Zillertal, Austria.
Experimental details: Raman scattering measurements have been performed on an arbitrarily oriented sample using 257 nm Ar$^+$ laser radiation. The laser radiation power at the sample was below 3 mW.
Raman shifts (cm^{-1}): 3675s, 3660, 1051w, 793, 707, 676s, 469, 434, 363, 333w.
Source: Petry et al. (2006).
Comments: The sample was characterized by electron microprobe analyses. For Raman spectra of talc see also Blaha and Rosasco (1978), Rosasco and Blaha (1980), Wada and Kamitakahara (1991), and Frezzotti et al. (2012).

Talmessite $Ca_2Mg(AsO_4)_2 \cdot 2H_2O$

Origin: No data.
Experimental details: Raman scattering measurements have been performed on arbitrarily oriented crystals using a 633 nm He-Ne laser. The laser radiation power is not indicated. The Raman shifts have been determined for the maxima of individual peaks obtained as a result of the spectral curve analysis.
Raman shifts (cm^{-1}): 2882, 2376, 931, 905, 877, 836s, 814, 783, 455s, 445, 388, 363, 357, 305, 276, 212, 196, 173.
Source: Frost and Kloprogge (2003).
Comments: No independent analytical data are provided for the sample used. For the Raman spectrum of talmessite see also Frost (2009a).

Tangdanite $Ca_2Cu_9(AsO_4)_4(SO_4)_{0.5}(OH)_9 \cdot 9H_2O$

Origin: No data.

Experimental details: Raman scattering measurements have been performed on arbitrarily oriented crystals using a 633 nm He-Ne laser. The laser radiation power is not indicated. The Raman shifts have been determined for the maxima of individual peaks obtained as a result of the spectral curve analysis.

Raman shifts (cm^{-1}): 3489, 3403, 1136, 1007, 883, 841s, 802, 669, 618, 505, 493, 462, 414, 385, 359, 314, 260, 210, 179.

Source: Frost and Kloprogge (2003).

Comments: No independent analytical data are provided for the sample used. In the cited paper tangdanite was described with the old name clinotyrolite. For the Raman spectrum of tangdanite see also Frost et al. (2012n, 2015z).

Tangeite $CaCu(AsO_4)(OH)$

Origin: Tange gorge, Tyuya-Mayun, Kyrgyzstan (type locality).

Experimental details: Raman scattering measurements have been performed on arbitrarily oriented crystals using a 633 nm He-Ne laser. The laser radiation power is not indicated. The Raman shifts have been determined for the maxima of individual peaks obtained as a result of the spectral curve analysis.

Raman shifts (cm^{-1}): (3242), 3118, (868w), 842s, 823, 798, 768w, (715w), 507w, 482, 463, 392w, 367, 321s, 284, 264, 208, 184, 165, 145, 135.

Source: Martens et al. (2003c).

Comments: No independent analytical data are provided for the sample used.

Tantalite-(Fe) $Fe^{2+}Ta_2O_6$

Origin: Suzhou granite, Suzhou City, southern Jiangsu, China.

Experimental details: Methods of sample preparation are not described. Raman scattering measurements have been performed using 514.5 nm Ar$^+$ laser radiation. The nominal laser radiation power was 700 mW.

Raman shifts (cm^{-1}): 880.

Source: Wang et al. (1997).

Comments: The sample was characterized by powder X-ray diffraction data and electron microprobe analyses.

Tantalite-(Mg) $MgTa_2O_6$

Origin: Synthetic.
Experimental details: No data.
Raman shifts (cm^{-1}): 740sh, 712s, 662, 568w, 540w, 473, 425, 356w, 334, 253s, 185s.
Source: Husson et al. (1979).
Comments: No independent analytical data are provided for the sample used.

Tantalite-(Mn) $Mn^{2+}Ta_2O_6$

Origin: Alto do Giz pegmatite, Borborema pegmatite province, northeastern Brazil.
Experimental details: Raman scattering measurements have been performed on an arbitrarily oriented sample using 488 nm Ar⁺ laser radiation. The laser radiation power at the sample was 14 mW.
Raman shifts (cm^{-1}): 887s, 621, 547, 526, 324, 282, 238, 202, 121.
Source: Thomas et al. (2011a).
Comments: The sample was characterized by qualitative electron microprobe analyses.

Tantite orthorhombic polymorph Ta_2O_5

Origin: Synthetic.
Experimental details: Raman scattering measurements have been performed on an arbitrarily oriented sample using 514.5 nm Ar⁺ laser radiation. The laser radiation power is not indicated.
Raman shifts (cm^{-1}): 981, 948, 903, 844, 762, 711, 642, 612s, 562, 494, 458, 377, 338, 269, 245s, 196, 139, 106s, 78.
Source: Joseph et al. (2012).
Comments: The sample was characterized by powder X-ray diffraction data. For the Raman spectra of Ta_2O_5 see also Dobal et al. (2000) and Meng et al. (1997).

Taranakite $K_3Al_5(PO_3OH)_6(PO_4)_2 \cdot 18H_2O$

Origin: Jenolan Caves, New South Wales, Australia.
Experimental details: Raman scattering measurements have been performed on arbitrarily oriented crystals using a 633 nm He-Ne laser. The laser radiation power is not indicated. The Raman shifts have been determined for the maxima of individual peaks obtained as a result of the spectral curve analysis.
Raman shifts (cm^{-1}): 1149, 1126s, 1116sh, 1100sh, 1064w, 1026, 1010sh, (991), 962s, (946sh), (922sh), 811, 656, 648, 636, 615sh, 595, (580), 572sh, 560s, 547, (537sh), 529sh, 505sh, 489, 464, 444, 416s, (404), 396s, (388sh), 348w, 328, 304, 271, 260sh, 248, (237sh), 223, 202, 188s, 165, 155.
Source: Frost et al. (2011v).
Comments: The sample was characterized by powder X-ray diffraction data.

Tarapacáite $K_2(CrO_4)$

Origin: Synthetic.
Experimental details: Raman scattering measurements have been performed on an arbitrarily oriented sample using 457.9 nm Ar⁺ laser radiation. The nominal laser radiation power was between 5 and 30 mW.
Raman shifts (cm^{-1}): 906, 886, 883, 873, 859s, 395, 390, 388, 354s, 351s.
Source: Serghiou and Guillaume (2004).
Comments: For the Raman spectra of tarapacáite see also Kiefer and Bernstein (1972) and Huang and Butler (1990).

Tarbuttite $Zn_2(PO_4)(OH)$

Origin: Broken Hill, Zambia.
Experimental details: Raman scattering measurements have been performed on arbitrarily oriented crystals using a 633 nm He-Ne laser. The laser radiation power at the sample was 1 mW. The Raman shifts have been determined for the maxima of individual peaks obtained as a result of the spectral curve analysis.
Raman shifts (cm^{-1}): 3446, 1069, 1051, 1011, 965s.
Source: Frost (2004a).
Comments: No independent analytical data are provided for the sample used.

Tausonite $SrTiO_3$

Origin: Synthetic.
Experimental details: Raman scattering measurements have been performed on an arbitrarily oriented sample using the conventional 4358-Å mercury e line. The radiation power is not indicated.
Raman shifts (cm^{-1}): 1030, 720, 675, 620..., 360s, 310s, 250, 80.
Source: Perry et al. (1967).
Comments: No independent analytical data are provided for the sample used.

Tazheranite $(Zr,Ti,Ca)(O,\square)_2$

Origin: Synthetic (stabilized cubic ZrO_2).
Experimental details: Raman scattering measurements have been performed on an arbitrarily oriented sample using 632.8 nm He-Ne. The laser radiation power is not indicated.
Raman shifts (cm^{-1}): 625s, 480, 360, 250, 150.
Source: Phillippi and Mazdiyasni (1971).
Comments: For the Raman spectrum of tazheranite see also Galuskina et al. (2013a).

Tazzoliite $Ba_2CaSr_{0.5}Na_{0.5}Ti_2Nb_3SiO_{17}[PO_2(OH)_2]_{0.5}$

Origin: Euganei Hills, Padova, Italy (type locality).
Experimental details: Methods of sample preparation are not described. Raman scattering measurements have been performed using 514.5 nm Ar^+ laser radiation. The nominal laser radiation power was between 10 and 50 mW.
Raman shifts (cm^{-1}): 3516w, 1062w, 981w, 961w, 869w, 754s, 563s, 540, 328, 261s, 229s.
Source: Cámara et al. (2012a).
Comments: The sample was characterized by powder X-ray diffraction data and electron microprobe analyses. The crystal structure is solved.

Teepleite $Na_2B(OH)_4Cl$

Origin: No data.
Experimental details: Polarized Raman scattering measurements have been performed on a single crystal using Ar^+ laser radiation.
Raman shifts (cm^{-1}): 3555, 3535, 3525, 1195, 1185, 946, 854, 770, 743, 660, 505, 499, 429, 373, 185, 143, 135, 113.
Source: Devarajan et al. (1974).
Comments: No independent analytical data are provided for the sample used.

Teineite $Cu^{2+}(Te^{4+}O_3)\cdot 2H_2O$

Origin: Moctezuma Mine, Mexico.
Experimental details: Raman scattering measurements have been performed on arbitrarily oriented crystals using a 633 nm He-Ne laser. The laser radiation power is not indicated. The Raman shifts have been determined for the maxima of individual peaks obtained as a result of the spectral curve analysis.
Raman shifts (cm^{-1}): 3495, (3139), 3040, 2854sh, (2641w), 2286w, 778sh, 739s, 701, 667, 509s, 458sh, 384sh, 347, 319, 250sh, 235, 175, 131.
Source: Frost and Keeffe (2009b).
Comments: No independent analytical data are provided for the sample used.

Tellurantimony Sb_2Te_3

Origin: Synthetic.
Experimental details: Methods of sample preparation are not described. Raman scattering measurements have been performed using 514.5 nm Ar^+ laser radiation. The nominal laser radiation power was below 3 mW. A 180°-scattering geometry was employed.
Raman shifts (cm^{-1}): 164.5, 111, 68s.
Source: Chis et al. (2012).
Comments: No independent analytical data are provided for the sample used.

Tellurium Te

Origin: Synthetic.
Experimental details: Raman scattering measurements have been performed on a thin film using 532 nm Nd-YAG laser radiation. The laser radiation power is not indicated.
Raman shifts (cm^{-1}): 136, 116s.
Source: Russo et al. (2008).
Comments: For the Raman spectrum of tellurium see also Pine and Dresselhaus (1971).

Tellurobismuthite Bi_2Te_3

Origin: Synthetic.
Experimental details: Raman scattering measurements have been performed on a thin film using 532 nm Nd-YAG laser radiation. The laser radiation power at the sample was $<<2$ mW.
Raman shifts (cm^{-1}): 133, 101s, 60, 38.
Source: Xu et al. (2015a).
Comments: For the Raman spectra of tellurobismuthite see also Richter et al. (1977) and Chis et al. (2012).

Tengerite-(Y) $Y_2(CO_3)_3\cdot 2-3H_2O$

Origin: Paratoo copper mine, Yunta, Olary Province, South Australia, Australia.
Experimental details: Raman scattering measurements have been performed on arbitrarily oriented crystals using a 633 nm He-Ne laser. The laser radiation power is not indicated. The Raman shifts have been determined for the maxima of individual peaks obtained as a result of the spectral curve analysis.

Raman shifts (cm^{-1}): 3621w, 3367sh, 3281, 3241sh, 3047, 2920, 2789sh, 2657, 1689w, (1637), 1618, 1592sh, 1392w, 1334, 1114sh, 1100s, 1091sh, 1067, (1062), 1038w, 1006w, 775 + 765 (unresolved doublet?), 689 + 674 (unresolved doublet?), 611sh, 589, 553+544 (unresolved doublet?), 508, 479, 474, 417w, 408, 398, 355w.

Source: Frost et al. (2015s).

Comments: No independent analytical data are provided for the sample used. IR spectra of presumed tengerite presented in Figs. 1b, 2b, and 4b of the cited paper are wrong. Actually, they are IR spectra of a silicate with minor admixture of quartz. IR bands of the silicate and quartz are erroneously assigned to vibrations of carbonate groups.

Tennantite $Cu_6[Cu_4(Fe,Zn)_2]As_4S_{13}$

Origin: Tsumeb mine, Tsumeb, Namibia.

Experimental details: Raman scattering measurements have been performed on an arbitrarily oriented sample using 514.5 nm Ar$^+$ laser radiation. The laser radiation power at the sample was between 1 and 10 mW. A 180°-scattering geometry was employed.

Raman shifts (cm^{-1}): 377s, 344.

Source: Mernagh and Trudu (1993).

Comments: The sample was characterized by electron microprobe analyses. For the IR spectrum of tennantite see also Kharbish et al. (2009).

Tenorite CuO

Origin: Synthetic.

Experimental details: Raman scattering measurements have been performed on a thin film using 514.5 nm Ar$^+$ laser radiation. The laser radiation power is not indicated.

Raman shifts (cm^{-1}): 631, 346, 296s.

Source: Debbichi et al. (2012).

Comments: The sample was characterized by X-ray diffraction data.

Tephroite $Mn^{2+}_2(SiO_4)$

Origin: Franklin, New Jersey, USA.

Experimental details: No data.

Raman shifts (cm^{-1}): 933, 892, 840s, 806s, 516w, 387, 306w, 278w, 243.

Source: Welsh (2008).

Comments: The sample was characterized by electron microprobe analyses. For the Raman spectra of tephroite see also Stidham et al. (1976), Piriou and McMillan (1983), and Mouri and Enami (2008).

Tetrahedrite $Cu_6[Cu_4(Fe,Zn)_2]Sb_4S_{13}$

Origin: Kremnice, Slovakia.

Experimental details: Raman scattering measurements have been performed on an arbitrarily oriented sample using 632.8 nm He-Ne laser radiation. The laser radiation power at the sample was 4.25 mW. A 180°-scattering geometry was employed.

Raman shifts (cm^{-1}): 379sh, 366s, 356s, 331, 298, 258w.

Source: Kharbish et al. (2009).

Comments: The sample was characterized by electron microprobe analyses. The atomic ratio Sb:As is 3:1. For the Raman spectra of tetrahedrite see also Mernagh and Trudu (1993) and Rath et al. (2015).

Tetrawickmanite $Mn^{2+}Sn^{4+}(OH)_6$

Origin: Långban, near Pajsberg and Filipstad, Värmland, Sweden.
Experimental details: No data.
Raman shifts (cm^{-1}): 3374w, 3253, 3145, 3062sh.
Source: Lafuente et al. (2015).
Comments: The Raman spectrum was obtained only in the range of O–H-stretching vibrations.

Thaumasite $Ca_3Si(OH)_6(CO_3)(SO_4)\cdot 12H_2O$

Origin: Black Rock mine, Kuruman, Kalahari, Northen Cape Province, South Africa.
Experimental details: Raman scattering measurements have been performed on a single crystal using 458 nm solid-state diode source laser radiation, with the laser beam parallel to [100] and at the 180° polarization counterclockwise from [001]. The laser radiation power at the sample was 40 mW.
Raman shifts (cm^{-1}): ~3500, ~3440, ~3370, ~3100sh, 1685, 1112, 1066s, 983s, 887w, 658, 588w, 455w, 418w, 250w, 193w, 143w, 120w, 92.
Source: Gatta et al. (2012b).
Comments: In the cited paper, Raman spectra of thaumasite have been obtained in different scattering geometries. For the Raman spectrum of thaumasite see also Goryainov (2016).

Thecotrichite $Ca_3(CH_3COO)_3Cl(NO_3)_2\cdot 6H_2O$

Origin: Artificial (efflorescent salt occurringon surfaces of porous calcareous objects stored in wooden cabinets).
Experimental details: Raman scattering measurements have been performed on an arbitrarily oriented sample using 632.8 nm He-Ne laser radiation. The laser radiation power at the sample was 0.9 mW.
Raman shifts (cm^{-1}): 3470w, 3372w, 3011w, 2986w, 2956w, 2928, 1472s, 1431, 1349, 1058s, 1046s, 968s, 961s, 749, 710w, 667.
Source: Wahlberg et al. (2015).
Comments: The sample was characterized by powder X-ray diffraction data. The crystal structure is solved.

Theophrastite $Ni(OH)_2$

Origin: Synthetic.
Experimental details: Methods of sample preparation are not described. Raman scattering measurements have been performed using 514.5 nm Ar$^+$ laser radiation. The laser radiation power at the sample was 5 mW.
Raman shifts (cm^{-1}): 3601w, 3571, 450, 315.
Source: Gourrier et al. (2011).
Comments: The sample was characterized by powder X-ray diffraction data.

Thermonatrite $Na_2(CO_3)\cdot H_2O$

Origin: Synthetic.
Experimental details: Raman scattering measurements have been performed on an arbitrarily oriented particle using 532 nm Nd-YAG laser radiation. The laser radiation power at the sample was 2 mW.
Raman shifts (cm^{-1}): 3404w, 3278w, 2977w, 1535w, 1433w, 1394w, 1067s, 700w, 683w, 652w.

Source: Bouchard and Smith (2003).
Comments: The sample was characterized by powder X-ray diffraction data. For the Raman spectrum of thermonatrite see also Frezzotti et al. (2012).

Thometzekite $PbCu^{2+}_2(AsO_4)_2 \cdot 2H_2O$

Origin: Tsumeb mine, Tsumeb, Namibia (type locality).
Experimental details: Raman scattering measurements have been performed on arbitrarily oriented crystals using a 633 nm He-Ne laser. The laser radiation power is not indicated. The Raman shifts have been determined for the maxima of individual peaks obtained as a result of the spectral curve analysis.
Raman shifts (cm^{-1}): 3519w, 3282w, 2821w, 2345sh, 984w, 841s, 790sh, 728w, 499s, 428, 401, 356s, 322, 239w.
Source: Frost and Weier (2004e).
Comments: No independent analytical data are provided for the sample used.

Thomsonite-Ca $NaCa_2(Al_5Si_5)O_{20} \cdot 6H_2O$

Origin: Dobrna, Děčín, Bohemia, Czech Republic.
Experimental details: Raman scattering measurements have been performed on an arbitrarily oriented sample using 532 Nd-YAG or 785 nm diode laser radiation.
Raman shifts (cm^{-1}): 1071, 990, 968, 930w, 608, 536s, 494w, 474w, 443, 391w, 341w, 310w, 268w, 258w, 220w, 197w, 180, 167w, 156w, 120w.
Source: Jehlička et al. (2012).
Comments: No independent analytical data are provided for the sample used. For the Raman spectra of thomsonite-Ca see also Wopenka et al. (1998), Mozgawa (2001), and Jehlička and Vandenabeele (2015).

Thorianite ThO_2

Origin: Synthetic.
Experimental details: Raman scattering measurements have been performed on an arbitrarily oriented sample using 488 nm Ar$^+$ laser radiation. The nominal laser radiation power was between 30 and 150 mW.
Raman shifts (cm^{-1}): 1033w, 885w, 467s.
Source: Jayaraman et al. (1988).
Comments: The spectrum was obtained at 0.8 GPa.

Thorikosite $Pb_3O_3Sb^{3+}(OH)Cl_2$

Origin: Lavrion, Greece (type locality).
Experimental details: Raman scattering measurements have been performed on arbitrarily oriented crystals using a 633 nm He-Ne laser. The laser radiation power is not indicated. The Raman shifts have been determined for the maxima of individual peaks obtained as a result of the spectral curve analysis.

Raman shifts (cm^{-1}): 3602w, 3541sh, 3508+3504s (unresolved doublet?), (3488sh), 1085w, 730, (657), 596, 325s, 275 + 269s (unresolved doublet?), (155), 133s, 112, 105w.
Source: Frost and Bahfenne (2011b).
Comments: No independent analytical data are provided for the sample used.

Thorite Th(SiO$_4$)

Origin: Synthetic.
Experimental details: Polarized Raman scattering measurements have been performed on a single crystal, using 514.5 nm Ar$^+$ laser radiation. The nominal laser radiation power was 250 mW. A 90°-scattering geometry was employed.
Raman shifts (cm^{-1}): 920, 894, 855, 596, 517, 439, 312, 293, 264, 194, 129, 126.
Source: Syme et al. (1977).
Comments: For Raman spectra of thorite see also Lahalle et al. (1986) and Costin et al. (2012).

Thorneite Pb$_6$(Te$_2$O$_{10}$)(CO$_3$)Cl$_2$(H$_2$O)

Origin: Otto Mt., near Baker, San Bernardino Co., California, USA (type locality).
Experimental details: Raman scattering measurements have been performed using 514.5 nm Ar$^+$ laser radiation, with the light propagating parallel to the c axis of a single crystal. The nominal laser radiation power was 5 mW.
Raman shifts (cm^{-1}): ~3300 (broad), 1630w, 1056 (sharp), and a series of strong peaks below 900 cm^{-1}.
Source: Kampf et al. (2010a).
Comments: The sample was characterized by powder X-ray diffraction data and electron microprobe analyses. The crystal structure is solved.

Thortveitite Sc$_2$Si$_2$O$_7$

Origin: Synthetic.
Experimental details: Raman scattering measurements have been performed on a pulverized crystal using 514.5 nm Ar$^+$ laser radiation. The nominal laser radiation power was 4 W.
Raman shifts (cm^{-1}): 949, 932s, 688w, 543, 510, 445, 435s, 392s, 347, 280w, 253, 205, 194.
Source: Bretheau-Raynal et al. (1979).
Comments: The sample was characterized by electron microprobe analyses.

Thorutite (Th,U,Ca)Ti$_2$(O,OH)$_6$

Origin: Synthetic.
Experimental details: Raman scattering measurements have been performed on a powder sample using 514.5 nm Ar$^+$ laser radiation. The laser radiation power is not indicated.
Raman shifts (cm^{-1}): The strongest Raman peaks are observed at 760, 620, and 195 cm^{-1}.
Source: Zhang et al. (2011).
Comments: The sample was characterized by powder X-ray diffraction data.

Threadgoldite $Al(UO_2)_2(PO_4)_2(OH) \cdot 8H_2O$

Origin: South Alligator River, West Arnhem region, Northern Territory, Australia.
Experimental details: Raman scattering measurements have been performed on arbitrarily oriented crystals using a 633 nm He-Ne laser. The laser radiation power is not indicated. The Raman shifts have been determined for the maxima of individual peaks obtained as a result of the spectral curve analysis.
Raman shifts (cm^{-1}): 3576w, 3411w, 3158w, 1655w, 1107w, 1057w, 1026 + 1019s (unresolved doublet?), 999, 974s, 953w, (840w), 827s, (817), 612, 533, 451, 419, 398s, 391, 329, 292, 201s, 188, 146, 114.
Source: Frost et al. (2006a).
Comments: The sample was characterized by powder X-ray diffraction data and electron microprobe analyses.

Tiemannite $HgSe$

Origin: Synthetic.
Experimental details: Raman scattering measurements have been performed at 90 K on a cleavaged (111) plane, in a—$z(x,x)z$ polarization using 514.5 nm Ar^+ laser radiation.
Raman shifts (cm^{-1}): 133, 43.
Source: Szuszkiewicz et al. (1999).

Tilasite $CaMg(AsO_4)F$

Origin: No data.
Experimental details: Raman scattering measurements have been performed on arbitrarily oriented crystals using a 633 nm He-Ne laser. The laser radiation power is not indicated. The Raman shifts have been determined for the maxima of individual peaks obtained as a result of the spectral curve analysis.
Raman shifts (cm^{-1}): 1518w, 1318w, 1107w, 1056w, 820, 659w, 611, 493, 410s, 297, 245.
Source: Frost and Kloprogge (2003).
Comments: No independent analytical data are provided for the sample used. For the Raman spectrum of tilasite see also Downs et al. (2012).

Tilleyite $Ca_5Si_2O_7(CO_3)_2$

Origin: Kushiro, Hiba-gun, Hiroshima prefecture, Honshu Island, Japan.
Experimental details: Raman scattering measurements have been performed on arbitrarily oriented crystals using a 633 nm He-Ne laser. The laser radiation power is not indicated. The Raman shifts have been determined for the maxima of individual peaks obtained as a result of the spectral curve analysis.
Raman shifts (cm^{-1}): (3638sh), 3635w, (3629sh), 3594w, (3578w), 3574w, 1748w, 1738w, 1501w, 1436w, 1412w, 1093, 1086s, 1080sh, 1067w, 1047, (1018w), 1010, (999w), 986, 966w, 572, 546, 528, 505sh, 493, 483sh, 456, 450sh, 424sh, 413, 396, 380sh, 366, 353, 329, 321sh, 302sh, 282s, (273), 260sh, 242sh, 234, 204sh, 197, 176, 155, 143.
Source: Frost et al. (2015e).
Comments: The sample was characterized by qualitative electron microprobe analysis. Bands between 600 and 900 cm^{-1} are not indicated. The bands in the range from 3500 to 3700 cm^{-1} may correspond to an impurity.

Tiragalloite $Mn^{2+}_4As^{5+}Si_3O_{12}(OH)$

Origin: Valletta mine, Maira Valley, Cuneo province, Piedmont, Italy.
Experimental details: Raman scattering measurements have been performed on an arbitrarily oriented sample using 632.8 nm He-Ne laser radiation. The laser radiation output power of 80 mW was attenuated by means of a series of density filters.
Raman shifts (cm^{-1}): 1004, 975, 960s, 902s, 869s, 863, 803, 785, 661s, 647s, 549, 508, 481, 398, 364, 320, 286w, 218w, 181w, 153w.
Source: Cámara et al. (2015).
Comments: The sample was characterized by electron microprobe analyses.

Tissintite $(Ca,Na,\square)AlSi_2O_6$

Origin: Tissint Martian meteorite (type locality).
Experimental details: Raman scattering measurements have been performed on an arbitrarily oriented grain in a polished section using 514.5 nm Ar$^+$ laser radiation. The laser radiation power is not indicated.
Raman shifts (cm^{-1}): 997 (broad), 693s, 573, 523, 417sh, 377s, 203.
Source: Ma et al. (2015).
Comments: The sample was characterized by powder X-ray diffraction data and electron microprobe analyses.

Tistarite Ti_2O_3

Origin: Allende meteorite (type locality).
Experimental details: Raman scattering measurements have been performed on an arbitrarily oriented grain in a polished section using 514.5 nm Ar$^+$ laser radiation. The laser radiation power is not indicated.
Raman shifts (cm^{-1}): Only a figure of the Raman spectrum of tistarite is given in the cited paper. The strongest peak is observed at ~250 cm^{-1}.
Source: Ma and Rossman (2009a).
Comments: The sample was characterized by powder X-ray diffraction data and electron microprobe analyses.

Titanite $CaTi(SiO_4)O$

Origin: Village of Dunje, Municipality of Prilep, Republic of Macedonia.
Experimental details: Raman scattering measurements have been performed on a powdered sample using 514.5 nm Ar$^+$ or 1064 nm Nd-YAG laser radiation. The laser radiation power is not indicated.
Raman shifts (cm^{-1}): 912, 872, 856, 608s, 548, 467s, 425, 333w, 316, 286sh, 253, 233w, 208sh, 164, 146w.
Source: Makreski et al. (2005b).
Comments: For the Raman spectra of titanite see also Meyer et al. (1996), Jasinevicius (2009), Andò and Garzanti (2014), and Gaft et al. (2015).

Titanoholtite $(Ti_{0.75}\square_{0.25})Al_6BSi_3O_{18}$

Origin: Szklary pegmatite, Lower Silesia, Poland (type locality).
Experimental details: Raman scattering measurements have been performed on an arbitrarily oriented sample using 514.5 nm Ar^+ laser radiation. The laser radiation power at the sample was ~5.5 mW.
Raman shifts (cm^{-1}): 1055, 935s, 885, 624, 561s, 507s, 466s, 407s, 362s, 286, 211.
Source: Pieczka et al. (2013).
Comments: The sample was characterized by electron diffraction data and electron microprobe analyses.

Tlapallite $H_6(Ca,Pb)_2(Cu,Zn)_3O_2(SO_4)(Te^{4+}O_3)_4(Te^{6+}O_4)$

Origin: Mina Bambollita, Moctezuma, Sonora, Mexico (type locality).
Experimental details: Raman scattering measurements have been performed on arbitrarily oriented crystals using a 633 nm He-Ne laser. The laser radiation power is not indicated. The Raman shifts have been determined for the maxima of individual peaks obtained as a result of the spectral curve analysis.
Raman shifts (cm^{-1}): 2926w, 2867w, 2754w, 2594w, 2320w, 2206w, 1957w, 1571w, 1474sh, 1104, 1062w, 973s, 796 + 788s (unresolved doublet?), 744sh, 708, 691sh, 610sh, (523), 509, 474, 438s, 419sh, 383sh, 353sh, 314, 291, 258, 229, 189, 168, 146, 121s.
Source: Frost (2009b).
Comments: No independent analytical data are provided for the sample used.

Tobermorite $Ca_4Si_6O_{17}(H_2O)_2 \cdot (Ca \cdot 3H_2O)$

Origin: N'Chwaning II mine, Kalahari Manganese Fields, Republic of South Africa.
Experimental details: Methods of sample preparation are not described. Raman scattering measurements have been performed in nearly back-scattered geometry using 632.8 nm He-Ne laser radiation. The laser radiation power is not indicated.
Raman shifts (cm^{-1}): 3525, 1145w, 1013, 682s, 619, 530w, 475w, 447w, 425w, 366w, 321w.
Source: Biagioni et al. (2012).
Comments: The sample was characterized by single-crystal X-ray diffraction data. The crystal structure is solved. For the Raman spectrum of tobermorite see also Biagioni et al. (2013b).

Todorokite $(Na,Ca,K,Ba,Sr)_{1-x}(Mn,Mg,Al)_6O_{12} \cdot 3-4H_2O$

Origin: Synthetic.
Experimental details: Raman scattering measurements have been performed on a polycrystalline sample (pressed as a pellet) using 514.5 nm Ar^+ laser radiation. The laser radiation power at the sample was 2.5 mW.
Raman shifts (cm^{-1}): 643s, 359, 295.
Source: Feng et al. (2004).
Comments: The sample was characterized by powder X-ray diffraction data. For the Raman spectrum of todorokite see also Julien et al. (2004).

Tokyoite $Ba_2Mn^{3+}(VO4)_2(OH)$

Origin: Postmasburg Manganese Field, South Africa.
Experimental details: Raman scattering measurements have been performed on an arbitrarily oriented sample using 514.5 nm Ar⁺ laser radiation. The laser radiation power is not indicated.
Raman shifts (cm^{-1}): 935, 906s, 846, 830s, 720, 464, 420, 406, 390, 340, 305, 241.
Source: Costin et al. (2014).
Comments: As-rich variety. The sample was characterized by electron microprobe analyses.

Tolbachite $CuCl_2$

Origin: Synthetic.
Experimental details: Raman scattering measurements have been performed on an arbitrarily oriented sample using 632.8 or 785.5 nm laser radiation.
Raman shifts (cm^{-1}): 562, 290s, 234.
Source: Aceto et al. (2006).
Comments: For the Raman spectrum of tolbachite see also Burgio and Clark (2001).

Tondiite $Cu_3MgCl_2(OH)_6$

Origin: Vesuvius volcano, Italy (type locality).
Experimental details: Methods of sample preparation are not described. Raman scattering measurements have been performed using 647 nm Kr⁺ laser radiation. The laser radiation power at the sample was 6 mW. A 180°-scattering geometry was employed.
Raman shifts (cm^{-1}): 942, 695, 503s, 395, 363s.
Source: Malcherek et al. (2014).
Comments: The sample was characterized by electron microprobe analyses. The crystal structure is solved.

Tooeleite $Fe^{3+}_6(AsO_3)_4(SO_4)(OH)_4 \cdot 4H_2O$

Origin: Synthetic.
Experimental details: Raman scattering measurements have been performed on an arbitrarily oriented sample using a 514 nm diode laser. The laser radiation power is not indicated. The Raman shifts have been determined for the maxima of individual peaks obtained as a result of the spectral curve analysis.
Raman shifts (cm^{-1}): 1597sh, 1554, 1522, 1422, 1287s, 1085, 983, 870sh, 803, (758), 661sh, 604s, 508, 438sh, 464, 284.
Source: Liu et al. (2013).
Comments: The sample was characterized by powder X-ray diffraction data. The assignment of the strong band at 1287 cm^{-1} to asymmetric stretching vibrations of SO_4^{2-} given in the cited paper is questionable.

Topaz $Al_2SiO_4F_2$

Origin: Topaz Mountain, Thomas Range, Utah, USA.
Experimental details: Raman scattering measurements have been performed on arbitrarily oriented crystals using a 633 nm He-Ne laser. The laser radiation power is not indicated. The Raman shifts have been determined for the maxima of individual peaks obtained as a result of the spectral curve analysis.
Raman shifts (cm^{-1}): 1167, 1079, 1008, 985, 927s, 854, 643, 559, 545, 454, 400, 370, 325, 314, 284s, 265s, 237s.
Source: Kloprogge and Frost (2000a).
Comments: No independent analytical data are provided for the sample used. For the Raman spectra of topaz see also Bradbury and Williams (2003), Jasinevicius (2009), and Andò and Garzanti (2014).

Torbernite $Cu(UO_2)_2(PO_4)_2 \cdot 12H_2O$

Origin: Mount Painter, 9 km N of Arkaroola, South Australia.
Experimental details: Raman scattering measurements have been performed on arbitrarily oriented crystals using a 633 nm He-Ne laser. The laser radiation power at the sample was 1 mW. The Raman shifts have been determined for the maxima of individual peaks obtained as a result of the spectral curve analysis.
Raman shifts (cm^{-1}): 3359, 3197, 3032sh, 1004sh, (995), 988, 957, 900w, 826s, 808sh, 629, 464, 439, 406, 399, 290, 222.
Source: Frost (2004b).
Comments: No independent analytical data are provided for the sample used. For the Raman spectra of torbernite see also Frost and Weier (2004c, d) and Driscoll et al. (2014).

Toturite $Ca_3Sn_2(SiFe^{3+}_2)O_{12}$

Origin: Upper Chegem structure, Northern Caucasus, Kabardino-Balkaria, Russia (type locality).
Experimental details: Raman scattering measurements have been performed on an arbitrarily oriented grain in polished section using 514.5 nm Ar^+ laser radiation. The laser radiation output power was 40–60 mW. A 0°-scattering geometry was employed.
Raman shifts (cm^{-1}): 930w, 879w, 810w, 784w, 734, 678sh, 575, 527sh, 497 + 494s (unresolved doublet?), 413w, 345sh, 301s, (266), 244, 185w, 156w, 148.
Source: Galuskina et al. (2010c).
Comments: The Raman shifts have been determined for the maxima of individual peaks obtained as a result of the spectral curve analysis. The sample was characterized by backscatter electron diffraction data and electron microprobe analyses.

Trabzonite $Ca_4[Si_3O_9(OH)]OH$

Origin: Upper Chegem caldera, Northern Caucasus, Kabardino-Balkaria, Russia.
Experimental details: Raman scattering measurements have been performed on an arbitrarily oriented grain in polished section using 514.5 nm Ar^+ laser radiation. The laser radiation output power was 30–50 mW. A 180°-scattering geometry was employed.
Raman shifts (cm^{-1}): 3602, 3576, 1020sh, 1006, 957s, 902, 872, 660s.
Source: Armbruster et al. (2012).
Comments: The sample was characterized by electron microprobe analyses. The crystal structure is solved.

Tremolite $\square Ca_2(Mg_{5.0-4.5}Fe^{2+}_{0.0-0.5})Si_8O_{22}(OH)_2$

Origin: No data.
Experimental details: Raman scattering measurements have been performed on an arbitrarily oriented sample using 632.8 nm He-Ne laser radiation. The nominal laser radiation power was 20 mW.
Raman shifts (cm^{-1}): 1062, 1031, 950w, 932, 751w, 676s, 531w, 516w, 438w, 418, 396, 373, 355, 254, 234s, 225, 180, 162.
Source: Rinaudo et al. (2004).
Comments: The sample was characterized by selected areaelectron diffraction and electron microprobe analyses. For the Raman spectra of tremolite see also Blaha and Rosasco (1978), Petry et al. (2006), Makreski et al. (2006a), Apopei and Buzgar (2010), Apopei et al. (2011), Weber et al. (2012), Andò and Garzanti (2014), Bersani et al. (2014), and Leissner et al. (2015).

Trevorite $NiFe^{3+}_2O_4$

Origin: Synthetic.
Experimental details: Raman scattering measurements have been performed on an arbitrarily oriented sample using 647 nm Kr$^+$ laser radiation. The laser radiation output power was 5 mW (0.5 mW at the sample).
Raman shifts (cm^{-1}): 704s, 663sh, 590, 568, 487s, 456sh, 333, 211, 189w.
Source: Hosterman (2011).
Comments: The sample was characterized by powder X-ray diffraction data.

Tridymite SiO_2

Origin: Synthetic.
Experimental details: Raman scattering measurements have been performed on a poedered sample using 476.5 nm laser radiation. The nominal laser radiation power was 400 mW.
Raman shifts (cm^{-1}): 1086w, 786, 456, 426s, 407s, 468s, 336w, 293, 205, 152, 100w.
Source: Etchepare et al. (1978).
Comments: The sample was characterized by powder X-ray diffraction data. For the Raman spectra of tridymite see also Ilieva et al. (2007), Knyazev et al. (2012), and Wilson (2014).

Trilithionite $KLi_{1.5}Al_{1.5}(Si_3Al)O_{10}F_2$

Origin: Erajanir, Finland.
Experimental details: Raman scattering measurements have been performed with the electric field polarized perpendicular to the cleavage plane using 514.5 or 488 nm Ar$^+$ laser radiation. The laser radiation power is not indicated.
Raman shifts (cm^{-1}): 3691w, 1128, 1094, (750), 707s, (650), 561, (405), (300), 260, 244, 182s, 94s.
Source: Tlili et al. (1989).
Comments: A Fe- and Mn-rich variety. The sample was characterized by electron microprobe analyses.

Trinepheline $NaAlSiO_4$

Origin: Jadeite deposit of Tawmaw-Hpakant, Myanmar (type locality).
Experimental details: Methods of sample preparation are not described. Raman scattering measurements have been performed using 532 nm laser radiation. The nominal laser radiation power was 50 mW.
Raman shifts (cm^{-1}): 1031w, 676, 572w, 518w, 494s, 487sh, 453s, 406, 375w, 359w, 347w, 311, 304w, 264w, 223, 203, 153w.
Source: Ferraris et al. (2014).
Comments: The sample was characterized by electron microprobe analyses. The crystal structure is solved.

Triplite $(Mn^{2+},Fe^{2+})_2(PO_4)F$

Origin: Codera valley, Sondrio province, Central Alps, Italy.
Experimental details: Methods of sample preparation are not described. Raman scattering measurements have been performed using 632.8 nm Ar^+ laser radiation. The laser radiation power is not indicated.
Raman shifts (cm^{-1}): 3498w, 1120w, 1072w, 1036, 980.5s, 808w, 680w, 610.5, 605, 598, 573w, 468.5w, 450w, 429.5sh, 421, 398.5sh, 277.5w, 242.5w, 218.5w, 192.5w, 179.5w, 161w, 137.5w.
Source: Vignola et al. (2014).
Comments: The sample was characterized by electron microprobe analyses. The crystal structure is solved. For the Raman spectra of triplite see also Frezzotti et al. (2012) and Frost et al. (2014aj).

Trippkeite $Cu^{2+}As^{3+}_2O_4$

Origin: Synthetic.
Experimental details: Methods of sample preparation are not described. Raman scattering measurements have been performed using 633 nm He-Ne laser radiation. The laser radiation power is not indicated.
Raman shifts (cm^{-1}): 810sh, 780s, 657, 539, 496, (463sh), 446, 421w, 371s, 306w, 285, 235, 203, 193, 182, 141sh, 134s.
Source: Bahfenne (2011).
Comments: The sample was characterized by powder X-ray diffraction data. For Raman spectra of trippkeite see also Bahfenne et al. (2011a) and Kharbish (2012).

Tripuhyite $Fe^{3+}Sb^{5+}O_4$

Origin: Synthetic.
Experimental details: Micro-Raman scattering measurements have been performed on an arbitrarily oriented sample using 532 nm Ar^+ laser radiation. The laser radiation power at the sample was 20 μW.
Raman shifts (cm^{-1}): 740, 652s, 500, 420 (Sample 1); 767w, 652s, 465 (Sample 2).
Source: Bolanz (2014).
Comments: The samples were characterized by powder X-ray diffraction data.

Trogtalite $CoSe_2$

Origin: Synthetic.
Experimental details: Raman scattering measurements have been performed on a powder sample using 632.8 nm He-Ne laser radiation. The laser radiation power at the sample was below 0.5 mW.
Raman shifts (cm^{-1}): 188.
Source: Zhu et al. (2010).
Comments: The sample was characterized by powder X-ray diffraction data. For the Raman spectrum of trogtalite see also Zhang et al. (2014).

Troilite FeS

Origin: Synthetic.
Experimental details: Raman scattering measurements have been performed on a powdered sample using 514.5 nm Ar^+ laser radiation. The laser radiation power at the sample was 2 mW.
Raman shifts (cm^{-1}): 360, 310s, 160.
Source: Avril et al. (2013).
Comments: For the Raman spectra of troilite see also Ma et al. (2012a) and Kaliwoda et al. (2013).

Trona $Na_3(HCO_3)(CO_3) \cdot 2H_2O$

Origin: Synthetic.
Experimental details: Raman scattering measurements have been performed on arbitrarily oriented individual particles using 532 nm Nd-YAG laser radiation. The laser radiation power at the sample was 2 mW.
Raman shifts (cm^{-1}): 3440, 3059w, 2436w, 1720w, 1561w, 1430, 1058s, 846w, 697w, 639w.
Source: Jentzsch et al. (2013).
Comments: The sample was characterized by powder X-ray diffraction data. For the Raman spectrum of trona see also Frezzotti et al. (2012).

Tschermigite $(NH_4)Al(SO_4)_2 \cdot 12H_2O$

Origin: No data.
Experimental details: Raman scattering measurements have been performed on arbitrarily oriented crystals using a 633 nm He-Ne laser. The laser radiation power is not indicated. The Raman shifts have been determined for the maxima of individual peaks obtained as a result of the spectral curve analysis.
Raman shifts (cm^{-1}): 3364w, 3114w, 2982w, 2892w, 1680w, 1630w, 1132, 1102w, 991 + 983s (unresolved doublet?), 619s, 542sh, 507 + 499 (unresolved doublet?), 456s, 387, 330.
Source: Frost and Kloprogge (2001).
Comments: No independent analytical data are provided for the sample used.

Tsumcorite $PbZn_2(AsO_4)_2 \cdot 2H_2O$

Origin: Tsumeb mine, Tsumeb, Namibia (type locality).
Experimental details: Raman scattering measurements have been performed on arbitrarily oriented crystals using a 633 nm He-Ne laser. The laser radiation power is not indicated. The Raman shifts have been determined for the maxima of individual peaks obtained as a result of the spectral curve analysis.

Raman shifts (cm^{-1}): 3503w, 3245w, 2925sh, 927, (868sh), 834s, 746, 521sh, 493s, 439, 400, 361, (340sh), 293, 242w, 197.
Source: Frost and Xi (2012a).
Comments: No independent analytical data are provided for the sample used. For the Raman spectrum of tsumcorite see also Frost and Weier (2004e).

Tsumebite $Pb_2Cu(PO_4)(SO_4)(OH)$

Origin: Blue Bell mine, near Baker, San Bernardino Co., California, USA.
Experimental details: Raman scattering measurements have been performed on arbitrarily oriented crystals using a 633 nm He-Ne laser. The laser radiation power is not indicated. The Raman shifts have been determined for the maxima of individual peaks obtained as a result of the spectral curve analysis.
Raman shifts (cm^{-1}): 3648w, (3446sh), 3397, 3335, 3239sh, 1097w, 1061w, 971s, 935, (923sh), (852sh), 827, 606, 598sh, 554, 540w, 482sh, 468, 442s, 389, 358sh, 339, 321.
Source: Frost and Palmer (2011e).
Comments: No independent analytical data are provided for the sample used.

Tsumoite BiTe

Origin: Synthetic.
Experimental details: Raman scattering measurements have been performed on a thin film using 532 nm Nd-YAG laser radiation. The laser radiation power is not indicated.
Raman shifts (cm^{-1}): 117, (91), 88s, ~56.
Source: Russo et al. (2008).
Comments: The sample was characterized by powder X-ray diffraction data.

Tugarinovite MoO_2

Origin: Synthetic.
Experimental details: Raman scattering measurements have been performed on an arbitrarily oriented sample using 532 nm Nd-YAG laser radiation. The nominal laser radiation power was between 0.7 and 7 mW.
Raman shifts (cm^{-1}): 744s, 571, 496s, 464w, 363s, 230, 204s.
Source: Solferino and Anderson (2012).
Comments: The sample was characterized by powder X-ray diffraction data. For the Raman spectrum of tugarinovite see also Srivastava and Chase (1972).

Tuite $Ca_3(PO_4)_2$

Origin: Suizhou chondrite, China (type locality).
Experimental details: No data.
Raman shifts (cm^{-1}): 1095w, 997, 975s, 640w, 578, 411, 192.
Source: Xie et al. (2003).
Comments: The sample was characterized by powder X-ray diffraction data and electron microprobe analyses. For the Raman spectra of tuite see also Zhai et al. (2010, 2014) and Xie et al. (2016).

Tululite $Ca_{14}(Fe^{3+},Al)(Al,Zn,Fe^{3+},Si,P,Mn,Mg)_{15}O_{36}$

Origin: Tulul Al Hammam area, Siwaqa complex, Mottled Zone Formation, Dead Sea region, Jordan (type locality).
Experimental details: Raman scattering measurements have been performed on an arbitrarily oriented grain using 514.5 nm Ar^+ laser radiation. The laser radiation power at the sample was 17 mW. A 180°-scattering geometry was employed.
Raman shifts (cm^{-1}): 934w, 831w, 817w, 754sh, 636s, 550sh, 522, 456, 295sh, 260, 199w.
Source: Khoury et al. (2016a).
Comments: The sample was characterized by powder X-ray diffraction data and electron microprobe analyses. The crystal structure is solved.

Tunellite $SrB_6O_9(OH)_2 \cdot 3H_2O$

Origin: No data.
Experimental details: Raman scattering measurements have been performed on arbitrarily oriented crystals using a 633 nm He-Ne laser. The laser radiation power is not indicated. The Raman shifts have been determined for the maxima of individual peaks obtained as a result of the spectral curve analysis.
Raman shifts (cm^{-1}): 3614s, 3567s, (3526sh), 3430sh, 3383 + 3369s (unresolved doublet?), (3324), 3282, 3243sh, 1113sh, 1082, (1063), 1043, 994s, (979sh), 954sh, 901, 879w, 861w, 819, 790w, 737, 715sh, 677sh, 664sh, 639s, 601w, 568, 523, 480sh, 464s, 445sh, 426sh, 371, (350), 332s, 317sh, 297w, 289w, 270sh, 256, 243sh, 210sh, 192, (159), 150s, (141), 114 + 109 (unresolved doublet?).
Source: Frost et al. (2014e).
Comments: No independent analytical data are provided for the sample used.

Tungstenite WS_2

Origin: Synthetic.
Experimental details: Raman scattering measurements have been performed on a powder sample using 532 laser radiation. The nominal laser radiation power was 5 mW.
Raman shifts (cm^{-1}): 419, 350.
Source: Nuvoli et al. (2014).
Comments: For the Raman spectrum of tungstenite see also Štengl et al. (2015).

Tungstite $WO_3 \cdot H_2O$

Origin: No data.
Experimental details: Raman scattering measurements have been performed on an arbitrarily oriented sample using 488 nm Ar^+ laser radiation. The laser radiation power at the sample was between 0.06 and 0.15 mW.
Raman shifts (cm^{-1}): ~949s, ~650s, ~377w, ~237.
Source: Tarassov et al. (2002).
Comments: No independent analytical data are provided for the sample used.

Tunisite $NaCa_2Al_4(CO_3)_4(OH)_8Cl$

Origin: Condorcet, Nyons, Drôme, Rhône-Alpes, France.

Experimental details: Raman scattering measurements have been performed on arbitrarily oriented crystals using a 633 nm He-Ne laser. The laser radiation power at the sample was about 0.1 mW. The Raman shifts have been determined for the maxima of individual peaks obtained as a result of the spectral curve analysis.

Raman shifts (cm^{-1}): 3561sh, 3482s, 3419, 1945w, 1703w, 1683w, 1542sh, 1522, 1499sh, 1127s, 854, 842 + 838w (unresolved doublet?), 731, 676w, 534, 441sh, (425), 417, (408), 387sh, 350w, 325w, 293, 279w, 234sh, 221, 200sh, 188sh, 177, 164, 151sh, 132w, 115.

Source: Frost et al. (2015d).

Comments: The sample was characterized by qualitative electron microprobe analysis. For the Raman spectrum of tunisite see also Frost and Dickfos (2007b).

Turquoise $CuAl_6(PO_4)_4(OH)_8 \cdot 4H_2O$

Origin: Kouroudaiko mine, Falemeriver, Senegal.

Experimental details: Raman scattering measurements have been performed on an arbitrarily oriented sample using a 633 nm He-Ne laser. The laser radiation power is not indicated. The Raman shifts have been determined for the maxima of individual peaks obtained as a result of the spectral curve analysis.

Raman shifts (cm^{-1}): 3506, 3471s, 3453sh, 3290, 3092sh, 1614w, 1184sh, 1161, 1104, 1064sh, 1041s, (1031), 991sh, 935w, (836sh), 815, 642, 592, 571, 548, 511, 483, 469, 437sh, 423, (417sh), 385, 335, (320sh), 301sh, 277sh, 259w, 244sh, 231, (218sh), 210, 196sh, 175, 152w.

Source: Čejka et al. (2015).

Comments: The sample was characterized by electron microprobe analyses. For the Raman spectra of turquoise see also Guo et al. (2010) and Bernardino et al. (2016).

Tuzlaite $NaCaB_5O_8(OH)_2 \cdot 3H_2O$

Origin: Tuzla evaporite deposit, Bosnia and Herzegovina (type locality).

Experimental details: Raman scattering measurements have been performed on an arbitrarily oriented crystal using 514.5 nm Ar^+ laser radiation. The laser radiation power at the sample was 100 mW. A 90°-scattering geometry was employed.

Raman shifts (cm^{-1}): 3615, 3475s, 3434s, 3328w, 3228w, 3165w, 1247w, 1072w, 1027sh, 866, 827, 761, 704w, 663, 589, 546, 468, 447w, 366w, 340, 323w, 282w.

Source: Bermanec et al. (2003).

Comments: The sample was characterized by powder X-ray diffraction data.

Tychite $Na_6Mg_2(CO_3)_4(SO_4)$

Origin: Searles Lake, San Bernardino Co., California, USA (type locality).

Experimental details: Raman scattering measurements have been performed on an arbitrarily oriented sample using 488 nm Ar^+ laser radiation. The laser radiation power is not indicated.

Raman shifts (cm^{-1}): 1137s, 1103s, 1049, 995s, 970.

Source: Palaich et al. (2013).

Comments: The sample was characterized by powder X-ray diffraction data. For the Raman spectrum of tychite see also Schmidt et al. (2006).

Tyrolite $Ca_2Cu_9(AsO_4)_4(CO_3)(OH)_8 \cdot 11H_2O$

Origin: Brixlegg, Tyrol, Austria.
Experimental details: Raman scattering measurements have been performed on arbitrarily oriented crystals using a 633 nm He-Ne laser. The laser radiation power is not indicated. The Raman shifts have been determined for the maxima of individual peaks obtained as a result of the spectral curve analysis.
Raman shifts (cm^{-1}): 3545, 3438, 3379, 3303, 3204, 1667, 1635, 1498, 1463, 1431, 1370, 1088, 1058, 855, 830 (broad), 795, 755s, 717s, 598, 570, 534, 524, 503, 480, 433s, 355, 301, 262, 217, 202, 179.
Source: Kloprogge and Frost (2000b).
Comments: No independent analytical data are provided for the sample used.

Tyuyamunite $Ca(UO_2)_2(VO_4)_2 \cdot 5-8H_2O$

Origin: Chihuahua, Mexico.
Experimental details: Raman scattering measurements have been performed on arbitrarily oriented crystals using a 633 nm He-Ne laser. The laser radiation power is not indicated. The Raman shifts have been determined for the maxima of individual peaks obtained as a result of the spectral curve analysis.
Raman shifts (cm^{-1}): 975, 827, 747s, 644, 608, 582, 525, 470, 404, 369s, 345, 304, 239, 186, 155.
Source: Frost et al. (2005c).
Comments: No independent analytical data are provided for the sample used.

Ulexite $NaCaB_5O_6(OH)_6 \cdot 5H_2O$

Origin: An unknown locality in Morocco.
Experimental details: Raman scattering measurements have been performed on arbitrarily oriented crystals using a 633 nm He-Ne laser. The laser radiation power is not indicated. The Raman shifts have been determined for the maxima of individual peaks obtained as a result of the spectral curve analysis.
Raman shifts (cm^{-1}): 3490, (3433w), 3401, 1133, 1111sh, 1005s, 745w, 668w, 639w, 617w, 602w, 551w, 491, (475), 421sh, 412, 367, 305, 250, 200, 166.
Source: Kloprogge and Frost (1999a).
Comments: Questionable data: the strong band at 1005 cm^{-1} may correspond to a sulfate. No independent analytical data are provided for the sample used.

Ulrichite $CaCu(UO_2)(PO_4)_2 \cdot 4H_2O$

Origin: Lake Boga granite quarry, Northwest Victoria, Australia (type locality).
Experimental details: Raman scattering measurements have been performed on arbitrarily oriented crystals using 785 nm laser radiation. The laser radiation power is not indicated.
Raman shifts (cm^{-1}): (1077), 1028, (1009), 975, 812s, (458w).
Source: Faulques et al. (2015a, b).

Ulvöspinel $Fe^{2+}_2TiO_4$

Origin: Synthetic.
Experimental details: Raman scattering measurements have been performed at 1 GPa on an arbitrarily oriented sample using 514.5 nm Ar$^+$ laser radiation. The nominal laser radiation power was 10 mW.
Raman shifts (cm^{-1}): 681s, 493.
Source: Kyono et al. (2011).
Comments: The sample was characterized by electron microprobe analyses.

Umangite Cu_3Se_2

Origin: Synthetic.
Experimental details: Methods of sample preparation are not described. Raman scattering measurements have been performed using 514.5 nm Ar$^+$ laser radiation. The laser radiation power at the sample was 10 mW.
Raman shifts (cm^{-1}): 58sh, 49.
Source: Izquierdo-Roca et al. (2009).
Comments: No independent analytical data are provided for the sample used.

Umbite $K_2ZrSi_3O_9 \cdot H_2O$

Origin: Synthetic.
Experimental details: Raman scattering measurements have been performed using 1064 nm Nd-YAG laser radiation with a resolution of 1 cm^{-1}. No other data are provided.
Raman shifts (cm^{-1}): ~970, ~930, ~910.
Source: Lin and Rocha (2005).
Comments: In the cited paper, Raman spectrum of umbite below 900 cm^{-1} is given as a figure, without indication of Raman shifts. The sample was characterized by powder X-ray diffraction data.

Umbrianite $K_7Na_2Ca_2[Al_3Si_{10}O_{29}]F_2Cl_2$

Origin: Pian di Celle volcano, Umbria, Italy (type locality).
Experimental details: Raman scattering measurements have been performed on an arbitrarily oriented grain in a polished section using 514.5 nm Ar$^+$ laser radiation. The nominal laser radiation power was 50 mW. A 180°-scattering geometry was employed.
Raman shifts (cm^{-1}): 3120, 2970, 2810, 1142, 1036s, 735s, 646, 593s, 525s, 491, 400, 324, 256.
Source: Sharygin et al. (2013b).
Comments: The sample was characterized by powder X-ray diffraction data and electron microprobe analyses. The crystal structure is solved.

Ungemachite $K_3Na_8Fe^{3+}(SO_4)_6(NO_3)_2 \cdot 6H_2O$

Origin: Synthetic.
Experimental details: Raman scattering measurements have been performed on an arbitrarily oriented sample using 532 nm Nd-YAG laser radiation. The laser radiation power at the sample was 2 mW.
Raman shifts (cm^{-1}): 3420, 3370sh, 1711w, 1663w, 1383, 1192, 1163, 1144, 1047s, 1035, 1011, 952, 721, 655, 645sh, 619, 600w, 534w, 472, 464w, 446.
Source: Jentzsch et al. (2012b).
Comments: The sample was characterized by powder X-ray diffraction data. For the Raman spectrum of ungemachite see also Jentzsch et al. (2013).

Uraninite UO_2

Origin: Synthetic.
Experimental details: Raman scattering measurements have been performed on a powder sample using 514.5 nm Ar$^+$ laser and 785 nm diode laser radiations. The laser radiation power is not indicated.
Raman shifts (cm^{-1}): 1149s, 598 (broad), 445 (514.5 nm); 1343w, 1149w, 618w, 445s, 230w (785 nm).
Source: Stefaniak et al. (2008).
Comments: The sample was characterized by powder X-ray diffraction data.

Uranophane-α $Ca(UO_2)_2(SiO_3OH)_2 \cdot 5H_2O$

Origin: Dieresis uranium mine, Sierra Albarrana, Córdoba, Spain.
Experimental details: Raman scattering measurements have been performed on an arbitrarily oriented sample using 632.8 nm He-Ne laser radiation. The nominal laser radiation power was 20 mW.
Raman shifts (cm^{-1}): 967, 798s, ~548w, ~400w, ~292w, ~210.
Source: Bonales et al. (2015).
Comments: The sample was characterized by qualitative electron microprobe analysis. For the Raman spectra of uranophane see also Biwer et al. (1990), Amme et al. (2002), Frost et al. (2006e), and Driscoll et al. (2014).

Uranopilite $(UO_2)_6(SO_4)O_2(OH)_6 \cdot 14H_2O$

Origin: South Alligator River, Northern Territory, Australia.
Experimental details: Raman scattering measurements have been performed on arbitrarily oriented crystals using a 633 nm He-Ne laser. The laser radiation power is not indicated. The Raman shifts have been determined for the maxima of individual peaks obtained as a result of the spectral curve analysis.
Raman shifts (cm^{-1}): 3547, 3470, 1143w, 1117w, 1098w, 1011, 842s, (832sh), 547, 406, 320, 294, 253.
Source: Frost et al. (2005b).
Comments: No independent analytical data are provided for the sample used. For the Raman spectrum of uranopilite see also Frost et al. (2007i).

Uranosphaerite $Bi(UO_2)O_2(OH)$

Origin: Horní Halže, Krušné Hory (Czech Ore Mts.), Czech Republic.
Experimental details: Raman scattering measurements have been performed on an arbitrarily oriented single crystal using 633 nm laser radiation. The laser radiation power is not indicated. A 180-°-scattering geometry was employed.
Raman shifts (cm^{-1}): 3404, 884w, 794s, 600s, 524, 475, 387, 378, 348, 305, 278, 263, 245, 227, 184, 138.
Source: Sejkora et al. (2008).
Comments: The sample was characterized by powder X-ray diffraction data and electron microprobe analyses.

Urea solution $CO(NH_2)_2 \cdot aq$

Origin: Synthetic.
Experimental details: Raman scattering measurements have been performed on arbitrarily oriented crystals using a 633 nm He-Ne laser. The laser radiation power is not indicated. The Raman shifts have been determined for the maxima of individual peaks obtained as a result of the spectral curve analysis.
Raman shifts (cm^{-1}): 3435, 3357, 3323, 3242, 1649, 1581, 1047, 1012, ~548, ~379w.
Source: Frost et al. (2000).
Comments: No independent analytical data are provided for the sample used. For the Raman spectrum of urea in aqueous solution see also Spinner (1959).

Ushkovite $MgFe^{3+}_2(PO_4)_2(OH)_2 \cdot 8H_2O$

Origin: Linópolis, Divino das Laranjeiras, Minas Gerais, Brazil.
Experimental details: Raman scattering measurements have been performed on arbitrarily oriented crystals using a 633 nm He-Ne laser. The laser radiation power is not indicated. The Raman shifts have been determined for the maxima of individual peaks obtained as a result of the spectral curve analysis.
Raman shifts (cm^{-1}): 3517, 3495, (3449), 3343, 3286, (3225), 1611w, 1140sh, 1121, 1097, 1068, 1041, 1012sh, 991 + 984s (unresolved doublet?), 959sh, 835sh, 810w, 780w, (650), 637, 610 + 606w (unresolved doublet?), 583, 563, 548sh, 506, 492w, 469w, 442, (421sh), 410, 386, 321 + 313w (unresolved doublet?), 283sh, 274w, 261, 239sh, 216, 206, 183, 173sh, 156sh, 144, 132sh, 115sh, 109w.
Source: López et al. (2015b).
Comments: The sample was characterized by qualitative electron microprobe analysis.

Usturite $Ca_3(SbZr)(FeO_4)_3$

Origin: Upper Chegem volcanic structure, Kabardino-Balkaria, Northern Caucasus, Russia (type locality).
Experimental details: Raman scattering measurements have been performed on an arbitrarily oriented grain in a polished section using 514.5 nm Ar$^+$ laser radiation. The laser radiation output power was between 40 and 60 mW. A 0°-scattering geometry was employed.

Raman shifts (cm^{-1}): 815, 789, 751, 733sh, (615), 591, 565sh, 498s, 411w, 303 + 294s (unresolved doublet?), 262 + 244 (unresolved doublet?), 218sh, 187w, 161w, 149.
Source: Galuskina et al. (2010a).
Comments: The sample was characterized by single-crystal electron back-scatter diffraction, powder X-ray diffraction data and electron microprobe analyses.

Uvarovite $Ca_3Cr_2(SiO_4)_3$

Origin: Sweden (?).
Experimental details: Polarized Raman scattering measurements have been performed on a single crystal in different scattering geometries using 488 nm Ar$^+$ laser radiation. The nominal laser radiation power was 100 mW.
Raman shifts (cm^{-1}): 894, 828, ~618w, ~590w, 526, ~510w, 370s, ~272, ~242, 176.
Source: Kolesov and Geiger (1998).
Comments: No independent analytical data are provided for the sample used. For the Raman spectra of uvarovite see also Mingsheng et al. (1994), Bersani et al. (2009), Makreski et al. (2011), Frezzotti et al. (2012), and Andò and Garzanti (2014).

Uvite $CaMg_3(Al_5Mg)(Si_6O_{18})(BO_3)_3(OH)_3(OH)$

Origin: Brumado district, Bahia, Brazil.
Experimental details: Polarized micro-Raman scattering measurements have been performed using 488 and 514.5 nm Ar$^+$ laser radiation. The laser radiation power is not indicated.
Raman shifts (cm^{-1}): 3636w, 3592, (3554w), (3518w), ~705, ~670, ~372s, ~244, ~215.
Source: Fantini et al. (2014).
Comments: The sample was characterized electron microprobe analyses. For the Raman spectrum of uvite see also Hoang et al. (2011).

Vaesite NiS_2

Origin: Synthetic.
Experimental details: Methods of sample preparation are not described. Raman scattering measurements have been performed using 514.5 nm Ar$^+$ laser radiation. The laser radiation power at the sample was 15 mW. A 180°-scattering geometry was employed.
Raman shifts (cm^{-1}): 515w, 468s, 414w, 274, 263, 235w.
Source: Bishop et al. (2000).
Comments: The sample was characterized by powder X-ray diffraction data.

Vajdakite $(Mo^{6+}O_2)_2As^{3+}{}_2O_5 \cdot 3H_2O$

Origin: Jáchymov uranium deposit, Krušné Hory (Ore Mts.), Western Bohemia, Czech Republic (type locality).
Experimental details: Raman scattering measurements have been performed on arbitrarily oriented crystals using a 633 nm He-Ne laser. The laser radiation power is not indicated. The Raman shifts have been determined for the maxima of individual peaks obtained as a result of the spectral curve analysis.

Raman shifts (cm^{-1}): 3481, 3417, (3144sh), 3112w, 953-951s, 910sh, 898, (876w), 804, (799), 760, 720w, 604w, 560, 549, 521, 480, (473), 386s, 369, 333, 282w, 238sh, 227 + 224 (unresolved doublet?), 179, 157, 138w, 128.
Source: Čejka et al. (2010a).
Comments: Holotype sample was used.

Valentinite Sb_2O_3

Origin: Synthetic.
Experimental details: Raman scattering measurements have been performed on an arbitrarily oriented sample using 488 nm Ar$^+$ laser radiation. The laser radiation power at the sample was 300 mW. A 90°-scattering geometry was employed.
Raman shifts (cm^{-1}): 690w, 602w, 502, 449w, 294s, 269sh, 223s, 194, 100s, 103w, 71.
Source: Cody et al. (1979).
Comments: The sample was characterized by powder X-ray diffraction data. For the Raman spectrum of valentinite see also Orman (2010).

Vanackerite $Pb_4Cd(AsO_4)_3(Cl,OH)$

Origin: Tsumeb mine, Tsumeb, Namibia (type locality).
Experimental details: No data.
Raman shifts (cm^{-1}): ~830s, ~792, ~770, ~457, ~353.
Source: Schlüter et al. (2016).
Comments: The sample was characterized by powder X-ray diffraction data and electron microprobe analyses. The crystal structure is solved.

Vanadinite $Pb_5(VO_4)_3Cl$

Origin: Mibladen, Morocco.
Experimental details: Raman scattering measurements have been performed on arbitrarily oriented crystals using a 633 nm He-Ne laser. The laser radiation power is not indicated. The Raman shifts have been determined for the maxima of individual peaks obtained as a result of the spectral curve analysis.
Raman shifts (cm^{-1}): 827s, 811, 365, 323, 291.
Source: Frost et al. (2003a).
Comments: No independent analytical data are provided for the sample used. For the Raman spectra of vanadinite see also Levitt and Condrate, Sr (1970), Adams and Gardner (1974), and Bartholomäi and Klee (1978).

Vandendriesscheite $Pb_{1.6}(UO_2)_{10}O_6(OH)_{11} \cdot 11H_2O$

Origin: No data.
Experimental details: Raman scattering measurements have been performed on arbitrarily oriented crystals using a 633 nm He-Ne laser. The laser radiation power is not indicated. The Raman shifts have been determined for the maxima of individual peaks obtained as a result of the spectral curve analysis.

Raman shifts (cm^{-1}): 3530sh, 3475, 3262sh, (2904w), (2853w), 1624w, 1395, 854 + 852 (unresolved doublet?), 841s, (832), 819sh, 779, 703, 548, (503), 456, (425w), 404, (355), 332, 303, 273sh, 248, 218, 193.
Source: Frost et al. (2007h).
Comments: No independent analytical data are provided for the sample used.

Vanmeersscheite U(UO$_2$)$_3$(PO$_4$)$_2$(OH)$_6$·4H$_2$O

Origin: Kobokobo, Kivu, Democratic Republic of Congo (type locality).
Experimental details: Raman scattering measurements have been performed on arbitrarily oriented crystals using a 633 nm He-Ne laser. The laser radiation power is not indicated. The Raman shifts have been determined for the maxima of individual peaks obtained as a result of the spectral curve analysis.
Raman shifts (cm^{-1}): ~3390, 1208sh, 1153, 1110s, (1086sh), 1017 + 1013s (unresolved doublet?), 939sh, 860w, 650w, 624, 571, 453sh, 442s, 363, 290, 226s.
Source: Frost et al. (2009d).
Comments: No independent analytical data are provided for the sample used.

Vantasselite Al$_4$(PO$_4$)$_3$(OH)$_3$·9H$_2$O

Origin: Bihain, Vielsalm, Stavelot massif, Luxembourg province, Belgium (type locality).
Experimental details: Raman scattering measurements have been performed on arbitrarily oriented crystals using a 633 nm He-Ne laser. The laser radiation power is not indicated. The Raman shifts have been determined for the maxima of individual peaks obtained as a result of the spectral curve analysis.
Raman shifts (cm^{-1}): 3608sh, 3570sh, 3502s, (3436), 3399s, 3369sh, 3327sh, 3211sh, 2943sh, 1703sh, 1656sh, 1622sh, 1595, 1456w, 1299w, 1232w, 1200w, 1146, 1128sh, 1106sh, 1090s, (1076), 1051s, 1027sh, 1013s, 949, 930, 833sh, 813w, 715w, 649, 593, 557, 522 + 515, 494, 451 + 437 + 423 (unresolved triplet?), 374, 334sh, 317s.
Source: Frost et al. (2015v).
Comments: The sample was characterized by qualitative electron microprobe analysis.

Vanthoffite Na$_6$Mg(SO$_4$)$_4$

Origin: Synthetic (type locality).
Experimental details: Methods of sample preparation are not described. Raman scattering measurements have been performed 532 nm Nd-YAG laser radiation. The laser radiation power at the sample was about 2 mW.
Raman shifts (cm^{-1}): 1195w, 1178w, 1154w, 1129w, 1096w, 1076w, 1012s, 1002s, 643, 637, 629, 622, 613w, 603w, 473, 458, 452.
Source: Jentzsch et al. (2011).
Comments: No independent analytical data are provided for the sample used.

Vapnikite Ca_3UO_6

Origin: Jabel Harmun, Palestinian Autonomy, Israel (type locality).
Experimental details: Raman scattering measurements have been performed on an arbitrarily oriented grain in a polished section using 488 nm Ar^+ laser radiation. The laser radiation power at the sample was below 5 mW.
Raman shifts (cm^{-1}): 1446w, 725s, 474w, 391, 248w.
Source: Galuskin et al. (2014).
Comments: The sample was characterized by single-crystal X-ray diffraction data and electron microprobe analyses. The crystal structure is solved.

Variscite $Al(PO_4) \cdot 2H_2O$

Origin: Cioclovina cave, 40 km SE of Hunedoara, Şureanu Mts., Romania.
Experimental details: Raman scattering measurements have been performed on an arbitrarily oriented sample using 1064 nm Nd-YAG laser radiation. The laser radiation output power was 350 mW.
Raman shifts (cm^{-1}): 3400–3100 (broad), 1634, 1079sh, 1055, 1026, 605w, 562w, 434s, 225, 168, 144.
Source: Onac et al. (2012).
Comments: The sample was characterized by powder X-ray diffraction data and electron microprobe analyses. For the Raman spectra of variscite see also Frost et al. (2004l) and Litvinenko et al. (2016).

Västmanlandite-(Ce) $Ce_3CaMg_2Al_2Si_5O_{19}(OH)_2F$

Origin: Västmanl and Co., Bergslagen region, Sweden (type locality).
Experimental details: Raman scattering measurements have been performed on an arbitrarily oriented single crystal using 633 nm laser radiation. The laser radiation power is not indicated.
Raman shifts (cm^{-1}): 3671sh, ~3586, 3517s, ~3446, ~3317s, ~3201, ~2545w, ~2142w, ~1058, 1034, 1004, 968, ~944, 920, 900s, ~690, ~675, ~633, 574, 555, 501, 464, ~436, 412, 387, 361, 341, 329, 287, 234, 224, ~200.
Source: Holtstam et al. (2005).
Comments: The sample was characterized by powder X-ray diffraction data and electron microprobe analyses. The crystal structure is solved.

Vaterite $Ca(CO_3)$

Origin: Synthetic.
Experimental details: Methods of sample preparation are not described. Raman scattering measurements have been performed using 532 nm Nd-YAG laser radiation. The laser radiation power is not indicated.
Raman shifts (cm^{-1}): 1555w, 1542w, 1480w, 1460w, 1441w, 1421w, 1091s, 1085.5sh, 1081, 1075s, 881 + 878 + 874w, 751w, 743.5w, 738w, 685, 674, 667, 333sh, 302s, 268, 210, 175, 151, 120s, 106s.
Source: Wehrmeister et al. (2010).
Comments: No independent analytical data are provided for the sample used. For the Raman spectra of vaterite see also Behrens et al. (1995), Frezzotti et al. (2012), Kristova et al. (2014), and Sánchez-Pastor et al. (2016).

Vauquelinite $CuPb_2(CrO_4)(PO_4)(OH)$

Origin: Kintore Open Cut, Broken Hill, New South Wales, Australia.
Experimental details: Raman scattering measurements have been performed on arbitrarily oriented crystals using a 785 nm Nd-YAG laser. The laser radiation power at the sample was 1 mW. The Raman shifts have been determined for the maxima of individual peaks obtained as a result of the spectral curve analysis.
Raman shifts (cm^{-1}): (843), 827s, 375, 348s, 332sh.
Source: Frost (2004c).
Comments: No independent analytical data are provided for the sample used.

Vauxite $Fe^{2+}Al_2(PO_4)_2(OH)_2 \cdot 6H_2O$

Origin: Siglo XX mine (Llallagua), Bustillo province, Potosí department, Bolivia (type locality).
Experimental details: Raman scattering measurements have been performed on arbitrarily oriented crystals using a 633 nm He-Ne laser. The laser radiation power is not indicated. The Raman shifts have been determined for the maxima of individual peaks obtained as a result of the spectral curve analysis.
Raman shifts (cm^{-1}): 3648s, 3555, 3402, 3328 (broad), 3103s, 2918sh, 1696sh, 1633w, 1601w, 1567sh, 1370w, 1309sh, 1150sh, 1134 + 1122, 1105sh, 1075sh, 1059sh, 1046sh, 1027s, 1009s, 1000sh, 978s, (954sh), 918sh, 910w, 900sh, 535, 502 + 498w (unresolved doublet?), 478sh, 470, 461sh, 451sh, 418, 412sh, 399sh, 393, 370 + 364 (unresolved doublet?), (341sh), 332, 320sh, 284, 273sh, 267sh, 238, 230sh, 208, 181, 154 + 148 (unresolved doublet?), 132 + 127 (unresolved doublet?), 112 + 109 (unresolved doublet?).
Source: Scholz et al. (2015).
Comments: The sample was characterized by electron microprobe analysis.

Väyrynenite $BeMn^{2+}(PO_4)(OH)$

Origin: Viitaniemi pegmatite, Eräjärvi area, Orivesi, Finland (type locality).
Experimental details: Raman scattering measurements have been performed on arbitrarily oriented crystals using a 633 nm He-Ne laser. The laser radiation power is not indicated. The Raman shifts have been determined for the maxima of individual peaks obtained as a result of the spectral curve analysis.
Raman shifts (cm^{-1}): 3473sh, 3388sh, 3315w, (3249sh), 3219s, 3154sh, 1802sh, 1768w, 1660w, 1186sh, 1139sh, 1126, 1074, 1044, 1009 + 1004s (unresolved doublet?), 986, 936w, 898w, 800, 769w, 741, 707w, 642w, 619, 599, 573w, 538w, 518 + 506 (unresolved doublet?), 463, 404, 381, 353, 334, 287w, 266w, 238sh, 232, 220, 189 + 184 (unresolved doublet?), (171), 163, 129 + 123 (unresolved doublet?), 114sh.
Source: Frost et al. (2014m).
Comments: The sample was characterized by electron microprobe analyses.

Velikite Cu_2HgSnS_4

Origin: Synthetic.
Experimental details: Methods of sample preparation are not described. Raman scattering measurements have been performed using 514.5 nm Ar^+ laser radiation. The laser radiation power is not indicated. A 180°-scattering geometry was employed.
Raman shifts (cm^{-1}): 318s, 283.
Source: Himmrich and Haeuseler (1991).
Comments: The sample was characterized by powder X-ray diffraction data.

Versiliaite $(Fe^{2+}{}_2Fe^{3+}{}_2)(Fe^{3+}{}_2Sb^{3+}{}_6)O_{16}S$

Origin: An abandoned mine in the Karrantza valley, westerner area of the Basque Co., Spain.
Experimental details: Raman scattering measurements have been performed on an arbitrarily oriented sample using 514.5 nm Ar^+ laser radiation. The laser radiation power at the sample was 20 mW.
Raman shifts (cm^{-1}): 612, 505, 410s, 293s, 246, 226s.
Source: Goienaga et al. (2011).
Comments: The sample was characterized by X-ray fluorescence spectroscopy.

Vésigniéite $Cu_3Ba(VO_4)_2(OH)_2$

Origin: Vrančice deposit, central Bohemia, Czech Republic.
Experimental details: Raman scattering measurements have been performed on arbitrarily oriented crystals using a 633 nm He-Ne laser. The laser radiation power is not indicated. The Raman shifts have been determined for the maxima of individual peaks obtained as a result of the spectral curve analysis.
Raman shifts (cm^{-1}): 3463sh, 2896w, 2609sh, 1960w, 1636sh, 1559w, 1052w, 856s, 821s, 750, 511w, 466, (371sh), 355sh, 332, 307s, 185sh, 175, 162sh, 112w.
Source: Frost et al. (2011e).
Comments: The sample was characterized by powder X-ray diffraction data and electron microprobe analyses. For the Raman spectrum of Vésigniéite see also Wulferding et al. (2012).

Vesuvianite $(Ca,Na)_{19}(Al,Mg,Fe)_{13}(SiO_4)_{10}(Si_2O_7)_4(OH,F,O)_{10}$

Origin: Dosso degli Areti, Italy.
Experimental details: No data.
Raman shifts (cm^{-1}): 930s, 868, 696, 640, 410, 226.
Source: Andò and Garzanti (2014).
Comments: No independent analytical data are provided for the sample used. For the Raman spectra of vesuvianitein the OH region see Galuskin et al. (2007a).

Veszelyite $(Cu,Zn)_2Zn(PO_4)(OH)_3 \cdot 2H_2O$

Origin: Zdravo Vrelo, near Kreševo, Bosnia and Herzegovina.
Experimental details: Methods of sample preparation are not described. Raman scattering measurements have been performed using 514.5 nm Ar^+ laser radiation. The laser radiation power at the sample was 5 mW.

Raman shifts (cm^{-1}): 3566 (sharp), 3555sh, (3497sh), 3425, 2290sh, (3302sh), 3286, (3184sh), 2852w, 2662sh (weak), 2233, 2042, 1970, 1805, 1632, 1587, 1379, 1108w, 1045w, 1025w, 967, 951, 929, 882, 833, 624, 607, 556, 539, 486, 470, 439.
Source: Danisi et al. (2013).
Comments: The sample was characterized by single-crystal X-ray diffraction data. The crystal structure is solved. No data on band intensities below 2662 cm^{-1} are provided in the cited paper. The Raman shifts have been determined for the maxima of individual peaks obtained as a result of the spectral curve analysis.

Villamanínite CuS_2

Origin: Synthetic.
Experimental details: Raman scattering measurements have been performed on a polycrystalline aggregate using 514.5 nm Ar$^+$ laser radiation. The nominal laser radiation power was 100 mW.
Raman shifts (cm^{-1}): 512s, 264, 207.
Source: Anastassakis and Perry (1976).
Comments: No independent analytical data are provided for the sample used.

Vivianite $Fe^{2+}_3(PO_4)_2 \cdot 8H_2O$

Origin: Catavi Mine, Llallagua, Bolivia.
Experimental details: Raman scattering measurements have been performed on a cleavage plain using 514.5 nm Ar$^+$ laser radiation. The laser radiation power at the sampling objective was 50 mW.
Raman shifts (cm^{-1}): 1050, 986, 947s, 867m, 828w, 568, 532, 453, 422, 342w, 303w, 270w, 235, 227, 196, 162w, 126.
Source: Rodgers et al. (1993).
Comments: The sample was characterized Mössbauer spectroscopy and electron microprobe analyses. For the Raman spectra of vivianite see also Piriou and Poullen (1984), Frost et al. (2002f), and Hsu et al. (2014).

Vladimirivanovite $Na_6Ca_2[Al_6Si_6O_{24}](SO_4,S_3,S_2,Cl)_2 \cdot H_2O$

Origin: Tultuilazurite deposit, Baikal Lake region, Russia (type locality).
Experimental details: No data.
Raman shifts (cm^{-1}): 3124 (broad), 1189, (1183), 992, 799s, 726sh, 544 (sharp), 428, 353.
Source: Sapozhnikov et al. (2012).
Comments: The sample was characterized by powder X-ray diffraction, thermal data, and electron microprobe analyses. The crystal structure is solved.

Vladykinite $Na_3Sr_4(Fe^{2+}Fe^{3+})Si_8O_{24}$

Origin: Murun complex, eastern Siberia, Russia (type locality).
Experimental details: Raman scattering measurements have been performed on an arbitrarily oriented sample using 532 nm laser radiation. The laser radiation power is not indicated.
Raman shifts (cm^{-1}): 1039w, 991, 968, 915, 681w, 465s, 401 (very strong), 348, 264w, 203s, 167w, 129w.
Source: Chakhmouradian et al. (2014).
Comments: The sample was characterized by powder X-ray diffraction data, Mössbauer spectroscopy and electron microprobe analyses. The crystal structure is solved.

Voglite $Ca_2Cu(UO_2)(CO_3)_4 \cdot 6H_2O$

Origin: White Canyon No. 1 mine, Frey Point, San Juan Co., Utah, USA.

Experimental details: Raman scattering measurements have been performed on arbitrarily oriented crystals using a 633 nm He-Ne laser. The laser radiation power is not indicated. The Raman shifts have been determined for the maxima of individual peaks obtained as a result of the spectral curve analysis.

Raman shifts (cm^{-1}): 3535 + 3391w (broad; unresolved doublet?), 2939w, 1369, 1094, 836s, 749w, (261), 223, 148.

Source: Frost et al. (2008b).

Comments: No independent analytical data are provided for the sample used.

Volaschioite $Fe_4(SO_4)O_2(OH)_6 \cdot 2H_2O$

Origin: Fornovolasco, Apuan Alps, Tuscany, Italy (type locality).

Experimental details: Raman scattering measurements have been performed on an arbitrarily oriented sample using 632.8 nm He-Ne laser radiation. The laser radiation power is not indicated. A nearly 180°-scattering geometry was employed.

Raman shifts (cm^{-1}): 1530w, 1178, 1055, 1005, 941w, 527, 453s, 408s, 319s, 250.

Source: Biagioni et al. (2011b).

Comments: The sample was characterized by powder X-ray diffraction data and electron microprobe analyses. The crystal structure is solved.

Volborthite $Cu_3V_2O_7(OH)_2 \cdot 2H_2O$

Origin: Synthetic.

Experimental details: Raman scattering measurements have been performed on a polycrystalline sample using 532 nm laser radiation. The laser radiation power is not indicated.

Raman shifts (cm^{-1}): 894s, 820, 758, 476, 438w, 342, 236, 164w.

Source: Ni et al. (2010a).

Comments: The sample was characterized by powder X-ray diffraction data and electron microprobe analyses. For the Raman spectra of volborthite see also Frost et al. (2011e) and Wulferding et al. (2012).

Vonsenite $Fe^{2+}_2Fe^{3+}O_2(BO_3)$

Origin: Brosso mine, Torino, Italy.

Experimental details: Raman scattering measurements have been performed on arbitrarily oriented crystals using a 633 nm He-Ne laser. The laser radiation power is not indicated. The Raman shifts have been determined for the maxima of individual peaks obtained as a result of the spectral curve analysis.

Raman shifts (cm^{-1}): 1605w, 1462 + 1443 (unresolved doublet?), 1347sh, 1304, 1284sh, 1059, 997, 728sh, (687), 642, 529, 381s, 315, 324s, (315), 249s, (232), 158+145 (unresolved doublet?), 114.

Source: Frost et al. (2014ac).

Comments: The sample was characterized by electron microprobe analysis.

Vorlanite $CaUO_4$

Origin: Upper Chegem caldera, Kabardino-Balkaria, Northern Caucasus, Russia (type locality).
Experimental details: Raman scattering measurements have been performed from crystal cross-section approximately perpendicular to basal pinacoid in thin section using 514.5 nm Ar^+ laser radiation. The laser radiation output power was between 40 and 60 mW. A 0°-scattering geometry was employed.
Raman shifts (cm^{-1}): 1370, 683s, 524, 450w, 371w, 226.
Source: Galuskin et al. (2011a).
Comments: The sample was characterized by powder X-ray diffraction data and electron microprobe analyses. The crystal structure is solved. For the Raman spectra of vorlanite see also Galuskin et al. (2012a, 2013b, 2014).

Vrbaite $Hg_3Tl_4As_8Sb_2S_{20}$

Origin: Allchar, Republic of Macedonia.
Experimental details: Raman scattering measurements have been performed on an arbitrarily oriented sample using 632.8 nm He-Ne laser radiation. The laser radiation output power was 0.46 mW.
Raman shifts (cm^{-1}): 397, 383sh, 378, 370, 357, 346w, 322s, 306s, 244, 236, 195, 187w, 171w, 162w, 151, 130, 122w, 111, 103.
Source: Makreski et al. (2014).
Comments: The sample was characterized by electron microprobe analyses.

Vuorelainenite $Mn^{2+}V^{3+}_2O_4$

Origin: Synthetic.
Experimental details: Raman scattering measurements have been performed on a single crystal in different scattering geometries using 514.5 nm Ar^+ laser radiation. The nominal laser radiation power was about 10 mW.
Raman shifts (cm^{-1}): 585s, 479s, ~300, 178.
Source: Takubo et al. (2011).
Comments: No independent analytical data are provided for the sample used.

Vysokýite $U^{4+}[AsO_2(OH)_2]_4 \cdot 4H_2O$

Origin: Geschieber vein, Jáchymov ore district, Western Bohemia, Czech Republic (type locality).
Experimental details: Methods of sample preparation are not described. Raman scattering measurements have been performed using a 780 nm diode-pumped solid-state laser. The nominal laser radiation power was 5 mW.
Raman shifts (cm^{-1}): 2750w, 2230w (broad), 1545, 1425, 902s, 816s, 769sh, 595, 559, 427, 368, 324, 200 + 184 (unresolved doublet?), 99w, 61w.
Source: Plášil et al. (2013d).
Comments: The sample was characterized by powder X-ray diffraction data and electron microprobe analyses. The crystal structure is solved.

Vysotskite (Pd,Ni)S

Origin: Synthetic (Ni-free).
Experimental details: Raman scattering measurements have been performed on arbitrarily oriented grains in a polished section using 514.5 nm Ar$^+$ laser radiation. The laser radiation output power was 500 mW.
Raman shifts (cm^{-1}): 392, 368, 353w, 348, 326s.
Source: Pikl et al. (1999).
Comments: No independent analytical data are provided for the sample used.

Wadeite $K_2ZrSi_3O_9$

Origin: Saima alkaline complex, Liaodong Peninsula, northeastern China.
Experimental details: Raman scattering measurements have been performed on an arbitrarily oriented sample using 514.5 nm Ar$^+$ laser radiation. The laser radiation power is not indicated.
Raman shifts (cm^{-1}): 1057w, 990s, 930s, 734w, 629, 561s, 490s, 433, 370, 342, 191s, 153.
Source: Wu et al. (2015).
Comments: The sample was characterized by powder X-ray diffraction data and electron microprobe analyses. For the Raman spectrum of wadeite see also Geisinger et al. (1987).

Wadsleyite $Mg_2(SiO_4)$

Origin: Synthetic.
Experimental details: Raman scattering measurements have been performed on a powdered sample using 488 or 514.5 nm Ar$^+$ laser radiation. The laser radiation power at the sample was between 100 and 200 mW.
Raman shifts (cm^{-1}): 940, 919, 898, 850s, 836s, 588, 570, 528, 460, 408, 307, 280, 213.
Source: Akaogi et al. (1984).
Comments: The sample was characterized by powder X-ray diffraction and optical data. For Raman spectra of wadsleyite see also Kleppe et al. (2001, 2006) and Mao et al. (2011).

Wagnerite-*Ma5bc* $Mg_2(PO_4)F$

Origin: Larsemann Hills, Prydz Bay, East Antarctica.
Experimental details: Methods of sample preparation are not described. Raman scattering measurements have been performed using 514.5 nm Ar$^+$ laser radiation. The nominal laser radiation power was 5 mW.
Raman shifts (cm^{-1}): 3570 (other values are not indicated: only a figure is given in the cited paper).
Source: Ren et al. (2003).
Comments: The sample was characterized by powder X-ray diffraction data and electron microprobe analyses. The crystal structure is solved.

Waimirite-(Y) YF_3

Origin: Synthetic.
Experimental details: Raman scattering measurements have been performed on a polycrystalline sample using 514.5 nm Ar^+ laser radiation. The laser radiation power is not indicated. A 90°-scattering geometry was employed.
Raman shifts (cm^{-1}): 533 (broad), 444, 389, 366s, 349s, 342sh, 293, 262, 244, 220w, 189s, 171, 147w, 111, 75.
Source: Wilmarth et al. (1988).
Comments: For the Raman spectrum of waimirite-(Y) see also Lage et al. (2004).

Wakabayashilite $(As,Sb)_6As_4S_{14}$

Origin: Jas Roux, Hautes-Alpes, France.
Experimental details: Raman scattering measurements have been performed on a single crystal with the laser beam perpendicular to the fiber axis (c axis) using 632.8 nm He-Ne laser radiation. The laser radiation power at the sample was 2 mW.
Raman shifts (cm^{-1}): 398s, 382, 356s, 337s, 328, 315, 304, 227–229, 205s, 191, 167, 137, 131sh, 108, 87, 67sh, 60s.
Source: Bindi et al. (2014).
Comments: The sample was characterized by single-crystal X-ray diffraction data and electron microprobe analyses.

Wakefieldite-(Ce) $CeVO_4$

Origin: Synthetic.
Experimental details: No data.
Raman shifts (cm^{-1}): 1012w, 482, 244, 202, 148s, 120s.
Source: Au et al. (1996).
Comments: The sample was characterized by powder X-ray diffraction data.

Wakefieldite-(La) $LaVO_4$

Origin: Synthetic.
Experimental details: No data.
Raman shifts (cm^{-1}): 1000, 862s, 705w, 528w, 410w, 290w, 150s, 122.
Source: Au et al. (1996).
Comments: The sample was characterized by powder X-ray diffraction data. For the Raman spectra of wakefieldite-(La) see also Sun et al. (2010b) and Xie et al. (2012).

Wakefieldite-(Nd) $NdVO_4$

Origin: Synthetic.
Experimental details: No data.
Ramanshifts (cm^{-1}): 1000, 874s, 812, 700, 528, 482, 412w, 310, 290, 150s, 122s.
Source: Au et al. (1996).
Comments: The sample was characterized by powder X-ray diffraction data. For the Raman spectra of wakefieldite-(Nd) see also Au and Zhang (1997) and Moriyama et al. (2011).

Wakefieldite-(Y) YVO_4

Origin: Synthetic.
Experimental details: No data.
Raman shifts (cm^{-1}): 1090, 896, 842w, 820w, 662w, 494, 422, 398s, 320s, 280s, 150w, 122.
Source: Au et al. (1996).
Comments: The sample was characterized by powder X-ray diffraction data.

Walpurgite $Bi_4O_4(UO_2)(AsO_4)_2 \cdot 2H_2O$

Origin: Weisser Hirsch Mine, Schneeberg, Saxony, Germany (type locality).
Experimental details: Raman scattering measurements have been performed on arbitrarily oriented crystals using a 633 nm He-Ne laser. The laser radiation power is not indicated. The Raman shifts have been determined for the maxima of individual peaks obtained as a result of the spectral curve analysis.
Raman shifts (cm^{-1}): 3515w, 3375w, 1597w, 1323w, 1003w, 948w, 892w, 836s, (823), (795), 790s, 771sh, 615, 546, 513s, 491sh, 432, 398, 365, 331, 296, (278), 242, 208, (199), 154.
Source: Frost et al. (2006l).
Comments: No independent analytical data are provided for the sample used.

Walstromite $BaCa_2Si_3O_9$

Origin: Big Creek deposit, Fresno Co., California, USA (type locality).
Experimental details: Methods of sample preparation are not described. Raman scattering measurements have been performed using 633 nm laser radiation. The laser radiation power is not indicated.
Raman shifts (cm^{-1}): 1071, 1037, 988s, 650s, 501, 473w, 378, 291, 153, 124.
Source: Gaft et al. (2013).
Comments: No independent analytical data are provided for the sample used. For the Raman spectrum of walstromite see also Zedgenizov et al. (2014).

Wardite $NaAl_3(PO_4)_2(OH)_4 \cdot 2H_2O$

Origin: Lavrada Ilha, Minas Gerais, Brazil.
Experimental details: Raman scattering measurements have been performed on arbitrarily oriented crystals using a 785 nm laser. The laser radiation power is not indicated. The Raman shifts have been determined for the maxima of individual peaks obtained as a result of the spectral curve analysis.
Raman shifts (cm^{-1}): 3607s, 3588, 3542s, 3383, 3282w, 1579w, 1319w, 1178w, 1133, (1103), 1079, 1047s, 1029sh, 1013sh, 992s, 681, 609, 580sh, 486, 403sh, 392, 340w, 334, 323sh.
Source: Frost et al. (2014l).
Comments: The sample was characterized by qualitative electron microprobe analysis. For the Raman spectrum of wardite see also Kampf et al. (2014a).

Waterhouseite $Mn_7(PO_4)_2(OH)_8$

Origin: Iron Monarch deposit, Middleback Ranges, South Australia, Australia (type locality).
Experimental details: Raman scattering measurements have been performed on an arbitrarily oriented single crystal using 518 nm laser radiation. The laser radiation power is not indicated.
Raman shifts (cm^{-1}): 3555sh, 3510, 3439s, 3411sh, 1612w, 1076s, ~1050, ~1018, 984, 929s, 809s, 667, ~595sh, 571s, ~540sh, ~513s, 433, ~342, ~321, ~267, 174, ~137.
Source: Pring et al. (2005).
Comments: The sample was characterized by powder X-ray diffraction data and electron microprobe analyses. The crystal structure is solved.

Wavellite $Al_3(PO_4)_2(OH)_3 \cdot 5H_2O$

Origin: Zbirow, Bohemia, Czech Republic.
Experimental details: Raman scattering measurements have been performed on a powdered sample using 632.8 nm He-Ne laser radiation. The laser radiation output power was 18 mW.
Raman shifts (cm^{-1}): 3490, 3406w, 3198sh, 3078, 1145w, 1061sh, 1017s, 950w, 920w, 633, 559sh, 540, 408s, 311, 274, 213w.
Source: Capitelli et al. (2014).
Comments: The sample was characterized by single-crystal X-ray diffraction data and electron microprobe analyses. The crystal structure is solved.

Waylandite $BiAl_3(PO_4)_2(OH)_6$

Origin: Leucamp, Montsalvy, Cantal, Auvergne-Rhône-Alpes, France.
Experimental details: Raman scattering measurements have been performed on an arbitrarily oriented sample using 514.5 nm Ar^+ laser radiation. The nominal laser radiation power was 25 mW.
Raman shifts (cm^{-1}): 3170w, 1112w, 1021sh, 1012 (broad), 791s, 725, 612s, 706, 602w, 523, 470w, 450w, 415, 400sh, 306s, 280, 257, 226, 191w, 156, 139, 119, 95sh.
Source: Gama (2000).
Comments: The sample was characterized by electron microprobe analyses.

Weddellite $Ca(C_2O_4) \cdot 2H_2O$

Origin: Synthetic.
Experimental details: Methods of sample preparation are not described. Raman scattering measurements have been performed using 532 nm Nd-YAG laser radiation. The laser radiation power at the sample was between 50 and 100 mW.
Raman shifts (cm^{-1}): 3500–3200 (broad), 2941, 2855w, 1640, 1476s, 911s, 870w, 597w (broad), 506s, 188.
Source: Conti et al. (2015).
Comments: The sample was characterized by powder X-ray diffraction data. For the Raman spectra of weddellite see also Frost and Weier (2003) and Frost (2004d).

Weeksite $K_2(UO_2)_2(Si_5O_{13})\cdot 4H_2O$

Origin: Anderson's Mine, Yavapai Co., Arizona, USA.
Experimental details: Raman scattering measurements have been performed on arbitrarily oriented crystals using a 633 nm He-Ne laser. The laser radiation power is not indicated. The Raman shifts have been determined for the maxima of individual peaks obtained as a result of the spectral curve analysis.
Raman shifts (cm^{-1}): 3610sh, 3548, 3497sh, 3356sh, 1637, 1154, 1008, 962, 939, 814 + 810s (unresolved doublet?), 800, 765, 744, 574, 521, 480, 349, 333, 301, 266, 210s, 167.5, 113.
Source: Frost et al. (2006d, h).
Comments: No independent analytical data are provided for the sample used. For the Raman spectrum of weeksite see also Biwer et al. (1990).

Wegscheiderite $Na_5H_3(CO_3)_4$

Origin: Synthetic.
Experimental details: Raman scattering measurements have been performed on an arbitrarily oriented sample using 488 nm Ar^+ laser radiation. The laser radiation power is not indicated.
Raman shifts (cm^{-1}): 3120w (broad), 2930w (broad), 2500w (broad), 1910w (broad), 1700w, 1675sh, 1560w, 1450sh, 1429, 1391, 1355w (broad), 1270w (broad), 1057s, 1038s, 1022s, 841w, 699, 686, 654, 240, 224, 186, 153s, 118, 105s, 96w, 89, 72s, 55.
Source: Bertoluzza et al. (1981).
Comments: No independent analytical data are provided for the sample used.

Weissbergite $TlSbS_2$

Origin: Synthetic.
Experimental details: Polarized Raman scattering measurements have been performed on a single crystal in different configurations using 632.8 nm He-Ne laser radiation. The nominal laser radiation power was 17 mW.
Raman shifts (cm^{-1}): 334sh, 321s, 310s, 293sh, 275sh, ~250w, 178w, 162w, 145w, 127, 106, 94, 81, 62, 51, 41 (for E parallel to the a–c plane).
Source: Kharbish (2011).
Comments: The sample was characterized by single-crystal X-ray diffraction data and electron microprobe analyses. For the Raman spectra of weissbergite see also Minceva-Sukarova et al. (2003) and Makreski et al. (2013b, 2014).

Weloganite $Na_2Sr_3Zr(CO_3)_6\cdot 3H_2O$

Origin: Francon quarry, Québec province, Canada (type locality).
Experimental details: Raman scattering measurements have been performed on arbitrarily oriented crystals using a 633 nm He-Ne laser. The laser radiation power is not indicated. The Raman shifts have been determined for the maxima of individual peaks obtained as a result of the spectral curve analysis.

Raman shifts (cm^{-1}): (3444), 3403, 3376sh, 3329, (3325), 1740w, 1732sh, 1712w, 1700w, 1681, 1622, 1563w, 1548w, 1526w, 1417, 1385w, 1371w, 1350, 1082s, 1073sh, 1061sh, 870w, 762, 749, 736, 728, 696w, 682sh, 679, (657w), 550w, 424w, 372sh, 354, 326w, 312sh
Source: Frost et al. (2013ab).
Comments: No independent analytical data are provided for the sample used. For the Raman spectrum of weloganite see also Vard and Williams-Jones (1993).

Wendwilsonite $Ca_2Mg(AsO_4)_2 \cdot 2H_2O$

Origin: Bou Azzer district, Morocco.
Experimental details: Raman scattering measurements have been performed on arbitrarily oriented crystals using a 633 nm He-Ne laser. The laser radiation power is not indicated. The Raman shifts have been determined for the maxima of individual peaks obtained as a result of the spectral curve analysis.
Raman shifts (cm^{-1}): 3332w, 3119w, (3001), 1724w, 1624sh, 1098w, 970w, 871, 832s, 800, 714w, 669w, (626w), 478sh, 454, 425, 361, 341, 306, 286w, 244w, 212sh, 191, 164, (140w), 127w.
Source: Frost et al. (2014v).
Comments: The sample was characterized by qualitative electron microprobe analysis.

Wernerkrauseite $Ca(Fe^{3+},Mn^{3+})_2Mn^{4+}O_6$

Origin: Bellerberg volcano, Eifel, Germany (type locality).
Experimental details: Raman scattering measurements have been performed on a grain in a polished section using 488 nm Ar$^+$ laser radiation. The laser radiation power at the sample was between 10 and 20 mW.
Raman shifts (cm^{-1}): 1239w, 670sh, 622s, (558), 495, 408sh, (332w), 294, 169, 117sh.
Source: Galuskin et al. (2016b).
Comments: The sample was characterized by single-crystal X-ray diffraction data and electron microprobe analyses. The crystal structure is solved. The Raman shifts have been determined for the maxima of individual peaks obtained as a result of the spectral curve analysis.

Wetherillite $Na_2Mg(UO_2)_2(SO_4)_4 \cdot 18H_2O$

Origin: Blue Lizard mine, San Juan Co., Utah, USA (type locality).
Experimental details: Methods of sample preparation are not described. Raman scattering measurements have been performed using 532 nm laser radiation. The laser radiation power is not indicated.
Raman shifts (cm^{-1}): ~3600–3000, 1610, 1230, 1180, 1120, 1105, 1080, 1010s, 995, 922w, 890sh, ~830s, 815sh, 700w, 640, 615sh, 580sh, 506, 445, 385, ~240.
Source: Kampf et al. (2015b).
Comments: The sample was characterized by powder X-ray diffraction data and electron microprobe analyses. The crystal structure is solved.

Wheatleyite $Na_2Cu(C_2O_4)_2 \cdot 2H_2O$

Origin: Synthetic.
Experimental details: Raman scattering measurements have been performed on arbitrarily oriented crystals using a 633 nm He-Ne laser. The laser radiation power is not indicated. The Raman shifts have been determined for the maxima of individual peaks obtained as a result of the spectral curve analysis.
Raman shifts (cm^{-1}): 3519, 3448, 3359w, 1733s, 1714w, 1674, 1651sh, 1470, 1434s, 1262, 1066, 904, 860, 798, 585, 565+560s (ubresolved doublet?), 387, 277, 243, 210, 173s, 139sh, 127.
Source: Frost et al. (2008k).
Comments: The sample was characterized by powder X-ray diffraction data. For the Raman spectrum of wheatleyite see also Palacios et al. (2011).

Whelanite $Cu_2Ca_6[Si_6O_{17}(OH)](CO_3)(OH)_3(H_2O)_2$

Origin: Bawanamine, Milford, Utah, USA (type locality).
Experimental details: Raman scattering measurements have been performed on arbitrarily oriented single crystals using 532 nm laser radiation. The laser radiation power is not indicated. A 180°-scattering geometry was employed.
Raman shifts (cm^{-1}): 3599sh, 3558s, 3222w, 2954, 2917sh, 1600w, 1542w, 1471, 1085, 1012, 850, 715sh, 671s, 530w, 481, 458, 400+381 (unresolved doublet?), 254w, 217 + 201 (unresolved doublet?), 165 + 151 (unresolved doublet?), 106.
Source: Kampf et al. (2012).
Comments: The sample was characterized by powder X-ray diffraction data and electron microprobe analyses. The crystal structure is solved. For the Raman spectrum of whelanite see also Frost and Xi (2012e).

Whewellite $Ca(C_2O_4) \cdot H_2O$

Origin: Synthetic.
Experimental details: Methods of sample preparation are not described. Raman scattering measurements have been performed using 532 nm Nd-YAG laser radiation. The laser radiation power at the sample was between 50 and 100 mW.
Raman shifts (cm^{-1}): 3486, 3426, 3340, 3256, 3056, 2972w, 2919w, 1629, 1490s, 1463s, 896, 503, 140, and several bands between 160 and 250 cm^{-1}.
Source: Conti et al. (2015).
Comments: The sample was characterized by powder X-ray diffraction data. For the Raman spectra of whewellite see also Frost and Weier (2003), Frost et al. (2003k), Frost (2004d), Jehlička and Edwards (2008), and Conti et al. (2014).

Whitecapsite $H_{16}Fe^{2+}_5Fe^{3+}_{14}Sb^{3+}_6(AsO_4)_{18}O_{16} \cdot 120H_2O$

Origin: White Caps mine, Manhattan district, Nye Co., Nevada, USA (type locality).
Experimental details: Raman scattering measurements have been performed on an arbitrarily oriented grain using 632.8 nm laser radiation. The laser radiation power is not indicated.
Raman shifts (cm^{-1}): 3400 (broad), 3250, 2930sh, 2350w, 1655w, 1380w, 1165w, 1095w, 870s, 790sh, 585, 531s, 466, 318, 268, 202s, 175s, 120s.
Source: Pekov et al. (2014b).
Comments: The sample was characterized by powder X-ray diffraction data and electron microprobe analyses. The crystal structure is solved.

Whiteite [possibly, whiteite-(CaMnMg)] $CaMn^{2+}Mg_2Al_2(PO_4)_4(OH)_2 \cdot 8H_2O$ (?)

Origin: No data.
Experimental details: Raman scattering measurements have been performed on arbitrarily oriented crystals using a 633 nm He-Ne laser. The laser radiation power is not indicated. The Raman shifts have been determined for the maxima of individual peaks obtained as a result of the spectral curve analysis.
Raman shifts (cm^{-1}): 3552sh, 3496s, 3426, 3220, 2939, 1692w, 1607 + 1586 (unresolved doublet?), 1368 + 1334 (unresolved doublet?), 1266sh, 1173, 1076, 978 + 972 + 960s (unresolved triplet?), 630, 586 + 571 (unresolved doublet?), 553sh, 500sh, 479, (457), 432, 363, 303sh, 282, 238, 176sh, 150, 109.
Source: Frost et al. (2014ab).
Comments: No independent analytical data are provided for the sample used.

Whitlockite $Ca_9Mg(PO_3OH)(PO_4)_6$

Origin: Sixiangkot chondrite.
Experimental details: No data.
Raman shifts (cm^{-1}): 1107w, 1084, 1030, 1015w, 976s, 959s, 855w, 820w, 668, 622, 605, 595, 553, 480, 450, 410, 328, 236, 179.
Source: Chen et al. (1995).
Comments: No independent analytical data are provided for the sample used. For the Raman spectra of whitlockite see also Jolliff et al. (2006) and Tait et al. (2011).

Whitmoreite $Fe^{2+}Fe^{3+}_2(PO_4)_2(OH)_2 \cdot 4H_2O$

Origin: Hagendorf-South (Hagendorf-Süd) pegmatite, Bavaria, Germany.
Experimental details: Raman scattering measurements have been performed on arbitrarily oriented crystals using a 633 nm He-Ne laser. The laser radiation power is not indicated. The Raman shifts have been determined for the maxima of individual peaks obtained as a result of the spectral curve analysis.
Raman shifts (cm^{-1}): 1910w, 1157sh, 1144, 1032, 973s, 937s, 915, (617), 593, 565, 546sh, 474sh, 433, 305, 276, 243s, 190, 152 + 150 (unresolved doublet?).
Source: Frost et al. (2003b).
Comments: The sample was characterized by qualitative electron microprobe analysis.

Widenmannite $Pb_2(OH)_2[(UO_2)(CO_3)_2]$

Origin: Synthetic (type locality).
Experimental details: Methods of sample preparation are not described. Raman scattering measurements have been performed using 514.5 nm Ar$^+$ laser radiation. The nominal laser radiation power was 10 mW.
Raman shifts (cm^{-1}): 3592sh, 3568, 3078, 1509w, 1470w, 1381, 1348w, 1122, 1068, 1058w, 849s, 736w, 725, 355, 268, 246, 225, 211w, 191, 128w, 115w.
Source: Plášil et al. (2010b).
Comments: The sample was characterized by powder X-ray diffraction data and electron microprobe analyses.

Willemite Zn_2SiO_4

Origin: Synthetic.
Experimental details: Methods of sample preparation are not described. Raman scattering measurements have been performed using 514.5 nm Ar^+ laser radiation. The laser radiation power is not indicated.
Raman shifts (cm^{-1}): 951w, 911, 875s.
Source: Lin and Shen (1994).
Comments: The sample was characterized by powder X-ray diffraction data. For the Raman spectrum of willemite see also Annen and Davis (1993).

Willemseite $Ni_3Si_4O_{10}(OH)_2$

Origin: Scotia talc mine, Bon Accord area, Barberton district, South Africa.
Experimental details: Raman scattering measurements have been performed on an arbitrarily oriented sample using 532 nm laser radiation. The nominal laser radiation power was 0.5 mW.
Raman shifts (cm^{-1}): 3660s, 3645s, 3622s, 1043, 789, 671s, 410, 383s, 296s, 185s, 109.
Source: Villanova-de-Benavent et al. (2014).
Comments: The sample was characterized by powder X-ray diffraction data and electron microprobe analyses.

Winstanleyite $TiTe^{4+}_3O_8$

Origin: Synthetic.
Experimental details: Raman scattering measurements have been performed on a powdery sample using 632.8 nm He-Ne laser radiation. The laser radiation power is not indicated.
Raman shifts (cm^{-1}): 859w, 650, 475s, 385.
Source: Ghribi et al. (2015).
Comments: The sample was characterized by powder X-ray diffraction data.

Witherite $Ba(CO_3)_2$

Origin: Alston Moor, England (type locality).
Experimental details: Raman scattering measurements have been performed on a polycrystalline sample using 488 or 514.5 nm Ar^+ laser radiation. The nominal laser radiation power was 200 mW.
Raman shifts (cm^{-1}): 1423w, 1060s, 700w, 690, 222, 178w, 151s, 133s, 99w, 90, 78.
Source: Scheetz and White (1977).
Comments: No independent analytical data are provided for the sample used. For the Raman spectra of witherite see also Buzgar and Apopei (2009) and Frezzotti et al. (2012).

Wittichenite Cu_3BiS_3

Origin: Synthetic.
Experimental details: Raman scattering measurements have been performed on nanocrystals using Ar^+ laser radiation. The laser radiation power is not indicated.
Raman shifts (cm^{-1}): 459, 355w, 153s, 116s.
Source: Zhong et al. (2012).
Comments: The sample was characterized by powder X-ray diffraction data and qualitative electron microprobe analysis. For the Raman spectrum of wittichenite see also Yan et al. (2013).

Wollastonite $CaSiO_3$

Origin: Willsboro mine, Willsboro, Essex Co., New York, USA.

Experimental details: Raman scattering measurements have been performed on an arbitrarily oriented sample using 488 or 514.5 nm Ar^+ laser radiation. The nominal laser radiation power was 500 mW.

Raman shifts (cm^{-1}): 1153w, 1063sh, 1046, 1023w, 999w, 972s, 852, 668sh, 637, 622sh, 583w, 469w, 414, 339, 323, 306w, 282w, 251, 239, 229, 217w, 193, 163.

Source: Richet et al. (1998).

Comments: No independent analytical data are provided for the sample used. For the Raman spectrum of wollastonite see also Buzatu and Buzgar (2010).

Woodhouseite $CaAl_3(SO_4)(PO_4)(OH)_6$

Origin: Champion mine, White Mountains, Mono Co., California, USA.

Experimental details: Raman scattering measurements have been performed on arbitrarily oriented crystals using a 633 nm He-Ne laser. The laser radiation power is not indicated. The Raman shifts have been determined for the maxima of individual peaks obtained as a result of the spectral curve analysis.

Raman shifts (cm^{-1}): 3460, 3401sh, 3122, 3001sh, 1778w, 1168w, 1151w, 1096, 1032 + 1028s (unresolved doublet?), 1004, 988, 974sh, 666sh, 653, 618, 590, 534, 485+475 (unresolved doublet?), 408s, 364w, (258), 249, 230sh, 181, 142, 118.

Source: Frost et al. (2011s).

Comments: No independent analytical data are provided for the sample used. For the Raman spectrum of woodhouseite see also Maubec et al. (2012).

Wopmayite $Ca_6Na_3\square Mn(PO_4)_3(PO_3OH)_4$

Origin: Tanco Mine, Bernic Lake, Manitoba, Canada (type locality).

Experimental details: Raman scattering measurements have been performed on an arbitrarily oriented sample using 532 nm laser radiation. The nominal laser radiation power was between 5 and 12.5 mW. A 180°-scattering geometry was employed.

Raman shifts (cm^{-1}): No data: only a figure of the Raman spectrum of wopmanite is presented in the cited paper. The strongest band is observed at ~960 cm^{-1}.

Source: Cooper et al. (2013b).

Wulfenite $PbMoO_4$

Origin: Synthetic (commercial reactant).

Experimental details: Raman scattering measurements have been performed on an arbitrarily oriented sample using 785 nm laser radiation. The laser radiation power at the sample was 0.3 mW.

Raman shifts (cm^{-1}): 868s, 765, 742, 347w, 315s, 189w, 168w.

Source: Bayne and Butler (2014).

Comments: No independent analytical data are provided for the sample used. For the Raman spectra of wulfenite see also Frost et al. (2004c), Nitta et al. (2006), and Rotondo et al. (2012).

Wupatkiite $CoAl_2(SO_4)_4 \cdot 22H_2O$

Origin: Cloncurry, Queensland, Australia.
Experimental details: Raman scattering measurements have been performed on arbitrarily oriented crystals using a 633 nm He-Ne laser. The laser radiation power at the sample was 1 mW. The Raman shifts have been determined for the maxima of individual peaks obtained as a result of the spectral curve analysis.
Raman shifts (cm^{-1}): 3479sh, 3296, 2987sh, 1134, 1069, 1009, 995s, 976, 882, 779, 622, 601, 517, 468, 426, 390w, 287sh, 213.
Source: Locke et al. (2007).
Comments: The sample was characterized by electron microprobe analysis which corresponds to an intermediate member of the wupatkiite–halotrichite solid-solution series.

Wurtzite ZnS

Origin: Manus Spreading Centre, Bismarck Sea.
Experimental details: Raman scattering measurements have been performed on an arbitrarily oriented sample using 514.5 nm Ar^+ laser radiation. The laser radiation power at the sample was between 1 and 10 mW. A 180°-scattering geometry was employed.
Raman shifts (cm^{-1}): 402 (broad), 350s, 327sh, 297s, 245w, 233w, 222, 173, 158.
Source: Mernagh and Trudu (1993).
Comments: The sample was characterized by powder X-ray diffraction data. For the Raman spectra of wurtzite see also White (2009) and Ma et al. (2012a).

Wüstite FeO

Origin: Synthetic (type locality).
Experimental details: Raman scattering measurements have been performed on an arbitrarily oriented sample using 632.8 nm He-Ne laser radiation. The laser radiation power at the sample was 0.7 mW.
Raman shifts (cm^{-1}): 652.
Source: De Faria et al. (1997).
Comments: The sample was characterized by powder X-ray diffraction data.

Xenotime-(Y) $Y(PO_4)$

Origin: Synthetic.
Experimental details: Raman scattering measurements have been performed on arbitrarily oriented acicular crystals using 632.8 nm He-Ne laser radiation. The nominal laser radiation power was 25 mW.
Raman shifts (cm^{-1}): 1056s, 1023, 997s, 578w, 481, 330, 292.
Source: Richman (1966).
Comments: No independent analytical data are provided for the sample used. For the Raman spectra of xenotime-(Y) see also Liu et al. (2008), Qiong et al. (2008), Bracco et al. (2012), Frezzotti et al. (2012), Andò and Garzanti (2014), and Švecová et al. (2016).

Xieite $FeCr_2O_4$

Origin: Suizhou meteorite (type locality).
Experimental details: No data.
Raman shifts (cm^{-1}): 665sh, 605.
Source: Chen et al. (2008a).
Comments: The sample was characterized by powder X-ray diffraction data and electron microprobe analyses.

Xocolatlite $Ca_2Mn^{4+}{}_2Te^{6+}{}_2O_{12} \cdot H_2O$

Origin: Moctezuma deposit, Sonora, Mexico (type locality).
Experimental details: Raman scattering measurements have been performed on an arbitrarily oriented sample using 488 nm Ar$^+$ laser radiation. The laser radiation power is not indicated.
Raman shifts (cm^{-1}): 699, 630s, 520, 390, 246w.
Source: Grundler et al. (2008).
Comments: The sample was characterized by powder X-ray diffraction data and electron microprobe analyses.

Xocomecatlite $Cu_3(Te^{6+}O_4)(OH)_4$

Origin: Mina Bambollita, Moctezuma, Sonora, Mexico.
Experimental details: Raman scattering measurements have been performed on arbitrarily oriented crystals using a 633 nm He-Ne laser. The laser radiation power is not indicated. The Raman shifts have been determined for the maxima of individual peaks obtained as a result of the spectral curve analysis.
Raman shifts (cm^{-1}): 2926s, 2867, 2754, 2594, 2326, 2206, 1957, 1602, (1544), 1368w, 1314w, 1121w, 974, 796, 763, 710s, 680sh, 600sh, 509, 470, 438, 407sh, 291, 259, 231, 189, 161, 149.
Source: Frost and Keeffe (2009g).
Comments: The sample was characterized by chemical data.

Xonotlite $Ca_6Si_6O_{17}(OH)_2$

Origin: Point Sal, near Vandenberg Air Force Base, Santa Barbara Co., California, USA.
Experimental details: Raman scattering measurements have been performed on arbitrarily oriented crystals using a 633 nm He-Ne laser. The laser radiation power is not indicated. The Raman shifts have been determined for the maxima of individual peaks obtained as a result of the spectral curve analysis.
Raman shifts (cm^{-1}): (3665), 3627, 3611w, (3578), (3528), 3303sh, 2909w, (1660w), (1603w), (1488 + 1423w) (unresolved doublet?), 1070, 1042s, 1015w, 980, 961s, (953), 862s, 816sh, 777, 695s, 626, 593s, 524, 505, 445, 421, 393, 369, 335, 304, 271w, 259w, 234, 205, 158, 135, 105.
Source: Frost et al. (2012b).
Comments: No independent analytical data are provided for the sample used.

Yarrowite Cu_9S_8

Origin: Synthetic.
Experimental details: No data.
Raman shifts (cm^{-1}): 470s, 263.
Source: Kumar and Nagarajan (2011).
Comments: The sample was characterized by powder X-ray diffraction data.

Ye'elimite $Ca_4Al_6O_{12}(SO_4)$

Origin: Synthetic.
Experimental details: Methods of sample preparation are not described. Raman scattering measurements have been performed using 632.8 nm He-Ne laser radiation. The nominal laser radiation power was 20 mW.
Raman shifts (cm^{-1}): 991s, 616w, 521.
Source: Gastaldi et al. (2008).
Comments: The sample was characterized by powder X-ray diffraction data.

Yecoraite $Fe^{3+}{}_3Bi_5O_9(Te^{4+}O_3)(Te^{6+}O_4)_2 \cdot 9H_2O$

Origin: Marie Elena mine, Yecora, Sonora, Mexico (type locality).
Experimental details: Raman scattering measurements have been performed on arbitrarily oriented crystals using a 633 nm He-Ne laser. The laser radiation power is not indicated. The Raman shifts have been determined for the maxima of individual peaks obtained as a result of the spectral curve analysis.
Raman shifts (cm^{-1}): 3400sh, 3180w, 2936w, 2878sh, 979w, 808, 796w, 699 + 690 (unresolved doublet?), 640sh, 578, 470 + 465s (unresolved doublet?), (396), 390, 355, 301, 265, 206, 145, 128.
Source: Frost and Keeffe (2009h).
Comments: No independent analytical data are provided for the sample used.

Yimengite $K(Cr,Ti,Fe,Mg)_{12}O_{19}$

Origin: Synthetic.
Experimental details: Methods of sample preparation are not described. Raman scattering measurements have been performed using 632.8 nm He-Ne laser radiation. The nominal laser radiation power was between 4 and 8 mW.
Raman shifts (cm^{-1}): 695s, 629s, 545, 471, 285w.
Source: Konzett et al. (2005).
Comments: The sample was characterized by X-ray diffraction data and electron microprobe analyses.

Yingjiangite $K_2Ca(UO_2)_7(PO_4)_4(OH)_6 \cdot 6H_2O$

Origin: Xiazhuang U deposit, China.
Experimental details: Raman scattering measurements have been performed on arbitrarily oriented crystals using a 633 nm He-Ne laser. The laser radiation power is not indicated. The Raman shifts have been determined for the maxima of individual peaks obtained as a result of the spectral curve analysis.

Raman shifts (cm^{-1}): 3510w, 3375sh, 3180w, 1047, 1004, 841 + 836s (unresolved doublet?), 817sh, 567, 531, 437, 393, 269sh, 204s, 147s.
Source: Frost et al. (2008a).
Comments: No independent analytical data are provided for the sample used.

Yttriaite-(Y) Y_2O_3

Origin: Synthetic.
Experimental details: Raman scattering measurements have been performed on an arbitrarily oriented crystal using 514.5 nm Ar$^+$ laser radiation. The laser radiation power at the sample was 2.5 mW.
Raman shifts (cm^{-1}): The strongest peak is observed at 378 cm^{-1}. Raman shifts of other bands are not indicated.
Source: Mills et al. (2011c).
Comments: Raman spectrum of natural yttrialite-(Y) given in the cited paper as a figure differs significantly from that of synthetic Y_2O_3.

Yukonite $Ca_2Fe^{3+}{}_3(AsO_4)_3(OH)4 \cdot 4H_2O$

Origin: No data.
Experimental details: Methods of sample preparation are not described. Raman scattering measurements have been performed using 785 nm diode laser radiation. The laser radiation power at the sample was >0.3 mW.
Raman shifts (cm^{-1}): 1059s, 992s, 929, 854s, 633, 527, 449, 387s, 237, 137.
Source: Das and Hendry (2011).
Comments: The sample was characterized by powder X-ray diffraction data. The stromg bands at 1059 and 992 cm^{-1} may correspond to impurities (a carbonate and a sulfate). For the Raman spectra of yukonite see also Gomez et al. (2010a, b) and Gómez and Lee (2012).

Yuksporite $K_4(Ca,Na)_{14}(Sr,Ba)_2(\square,Mn,Fe)(Ti,Nb)_4(O,OH)_4(Si_6O_{17})_2(Si_2O_7)_3(H_2O,OH)_3$

Origin: Hackman valley, Yukspor Mt., Khibiny massif, Kola Peninsula, Russia (type locality).
Experimental details: Raman scattering measurements have been performed on arbitrarily oriented crystals using a 633 nm He-Ne laser. The laser radiation power is not indicated. The Raman shifts have been determined for the maxima of individual peaks obtained as a result of the spectral curve analysis.
Raman shifts (cm^{-1}): 3668w (sharp), 3628sh, 3562w, 3460w, 3298sh, 2908w, 1103sh, 1078s, 1074sh, 1045, 1008w, 979, 954, 929sh, 891sh, 870s, 845sh, 815 + 803 (unresolved doublet?), 764, 723, 670s, 656sh, 641s, 588, 542, 525, 473+463 (unresolved doublet?), 437sh, 426, 395, 370, 348sh, 307, 288, 262sh, 241, 211, 141.
Source: Frost et al. (2015j).
Comments: The sample was characterized by qualitative electron microprobe analysis.

Yurmarinite $Na_7(Fe^{3+},Mg,Cu)_4(AsO_4)_6$

Origin: Arsenatnaya fumarole, Tolbachik volcano, Kamchatka, Russia (type locality).
Experimental details: Raman scattering measurements have been performed on an arbitrarily oriented sample using 532 nm laser radiation. The nominal radiation power at the sample was about 30 mW.

Raman shifts (cm^{-1}): 931sh, 859s, 831, 794s, 481, 409s, 331, 288, 187, 162, 111sh.
Source: Pekov et al. (2014d).
Comments: The sample was characterized by powder X-ray diffraction data and electron microprobe analyses. The crystal structure is solved.

Yushkinite (Mg,Al)(OH)$_2$VS$_2$

Origin: Silova-Yakha River, Pai-Khoi Anticlinorium (type locality).
Experimental details: No data.
Raman shifts (cm^{-1}): 762, 570, 373s, 345, 301w.
Source: Koval'chuk, and Makeev (2007).
Comments: The sample was characterized by electron microprobe analyses.

Yvonite Cu(AsO$_3$OH)·2H$_2$O

Origin: Salsigne mine, north of Carcassonne, Aude, France (type locality).
Experimental details: Raman scattering measurements have been performed on arbitrarily oriented crystals using a 633 nm He-Ne laser. The laser radiation power is not indicated. The Raman shifts have been determined for the maxima of individual peaks obtained as a result of the spectral curve analysis.
Raman shifts (cm^{-1}): 3485, 3314, 3061, 2831, 953, 897, 863, 842, 824, 795, 756, 637, 559, 546, 490, 473, 360, 342.
Source: Frost et al. (2015w).
Comments: No independent analytical data are provided for the sample used. Band intensities are not indicated.

Żabińskiite Ca[Al$_{0.5}$(Ta,Nb)$_{0.5}$)](SiO$_4$)O

Origin: Piława Górna pegmatite, Góry Sowie Block, Poland (type locality).
Experimental details: Methods of sample preparation are not described. Raman scattering measurements have been performed using 514.5 nm Ar$^+$ laser radiation. The laser radiation power at the sample was 5 mW.
Raman shifts (cm^{-1}): 997, 835, 642w, 581, 487, 431w, 341s.
Source: Pieczka et al. (2016a).
Comments: The sample was characterized by single-crystal X-ray diffraction data and electron microprobe analyses. The crystal structure is solved.

Zadovite BaCa$_6$[(SiO$_4$)(PO$_4$)](PO$_4$)$_2$F

Origin: Hatrurim Basin, Negev Desert, Israel (type locality).
Experimental details: No data.
Raman shifts (cm^{-1}): 1031, 992w, 969s, 881s, 839sh, 627w, 589, 520w, 430, 389, 342sh, 299w, 222.
Source: Galuskin et al. (2015e).
Comments: The sample was characterized by single-crystal X-ray diffraction data and electron microprobe analyses. The crystal structure is solved.

Zálesíite $CaCu_6(AsO_4)_2(AsO_3OH)(OH)_6 \cdot 3H_2O$

Origin: Zálesí U deposit, Rychlebské Hory Mts., northern Moravia, Czech Republic (type locality).
Experimental details: Raman scattering measurements have been performed on arbitrarily oriented crystals using a 633 nm He-Ne laser. The laser radiation power is not indicated. The Raman shifts have been determined for the maxima of individual peaks obtained as a result of the spectral curve analysis.
Raman shifts (cm^{-1}): 3457sh, 3361w, 3124w, (3106w), 1102, 912sh, 873s, 839s, (806), 623 + 520 (unresolved doublet?), 534s, 489, 433, 378sh, 354, 278, 239, 214sh, 170sh, 155 + 143s (unresolved doublet?).
Source: Čejka et al. (2011c).
Comments: The sample was characterized by powder X-ray diffraction data and electron microprobe analyses.

Zanazziite $Ca_2Be_4Mg_5(PO_4)_6(OH)_4 \cdot 6H_2O$

Origin: Ponte do Piauimine, Piaui valley, municipality of Itinga, Minas Gerais, Brazil.
Experimental details: Raman scattering measurements have been performed on arbitrarily oriented crystals using a 633 nm He-Ne laser. The laser radiation power is not indicated. The Raman shifts have been determined for the maxima of individual peaks obtained as a result of the spectral curve analysis.
Raman shifts (cm^{-1}): 3447 + 3437s (unresolved doublet?), 3256sh, (3098), 1644w, 1569w, 1466w, 1096, 1064, 1047, 1007, 970 + 964s (unresolved doublet?), 756, 589, 568, 559, 487, 457, 419, (404), 371, 294sh, 264, 236, 182, 166, 145, 132sh, 117.
Source: Frost et al. (2013ae).
Comments: The sample was characterized by electron microprobe analyses.

Zaratite $Ni_3(CO_3)(OH)_4 \cdot 4H_2O$

Origin: Cape Ortegal, Galicia, Spain (type locality).
Experimental details: Raman scattering measurements have been performed on a massive sample using 532 nm laser radiation. The nominal laser radiation power was 10 mW.
Raman shifts (cm^{-1}): 3604, 3428w, 3328w, 3217sh, 3110w, 2983w, 2935, 2867w, 2753w, 1609, 1366s, 1073s, 972, 941, 788w, 685w, 536w, 458.
Source: Garcia-Guinea et al. (2013).
Comments: The sample was characterized by powder X-ray diffraction data and electron microprobe analyses. For the Raman spectra of zaratite see also Frost et al. (2008l) and LaIglesia et al. (2014).

Zdeněkite $NaPbCu_5(AsO_4)_4Cl \cdot 5H_2O$

Origin: Cap Garonne Mine, near le Pradet, France.
Experimental details: Raman scattering measurements have been performed on arbitrarily oriented crystals using a 633 nm He-Ne laser. The laser radiation power is not indicated. The Raman shifts have been determined for the maxima of individual peaks obtained as a result of the spectral curve analysis.
Raman shifts (cm^{-1}): (1109w), (936w), 850s, 795s, 537s, 486s, 445, 339s, 278, 247.
Source: Frost et al. (2007m).
Comments: No independent analytical data are provided for the sample used.

Zellerite $Ca(UO_2)(CO_3)_2 \cdot 5H_2O$

Origin: White Canyon No. 1 Mine, Frey Point, Utah, USA.
Experimental details: Raman scattering measurements have been performed on arbitrarily oriented crystals using a 633 nm He-Ne laser. The laser radiation power is not indicated. The Raman shifts have been determined for the maxima of individual peaks obtained as a result of the spectral curve analysis.
Raman shifts (cm^{-1}): 3514w, 3375sh, 2945w, 1374w, 1091, 854s, 758, 363w, 233, 147.
Source: Frost et al. (2008f).
Comments: Questionable data. No independent analytical data are provided for the sample used. In the figures given in the cited paper the mineral is named "Zellerite/Liebegite." The IR spectrum of the sample used corresponds to a sulfate.

Zemannite $Mg_{0.5}ZnFe^{3+}(TeO_3)_3 \cdot 4.5H_2O$

Origin: Mina Bambollita, Moctezuma, Sonora, Mexico.
Experimental details: Raman scattering measurements have been performed on arbitrarily oriented crystals using a 633 nm He-Ne laser. The laser radiation power is not indicated.
Raman shifts (cm^{-1}): 740s, 650, 460s, 375, 213, 136.
Source: Frost et al. (2008i).
Comments: No independent analytical data are provided for the sample used.

Zemkorite $Na_2Ca(CO_3)_2$

Origin: Product of heating of natural nyerereite to 400 °C.
Experimental details: Raman scattering measurements have been performed on an arbitrarily oriented sample using 514.5 nm Ar$^+$ laser radiation. The nominal laser radiation power was 20 mW.
Raman shifts (cm^{-1}): 1078s, 993w, 710.
Source: Golovin et al. (2014).
Comments: The sample was characterized by electron microprobe analyses.

Zhangpeishanite BaFCl

Origin: Synthetic.
Experimental details: Raman scattering measurements have been performed on an arbitrarily oriented crystal using 488 nm Ar$^+$ laser radiation. The laser radiation power is not indicated.
Raman shifts (cm^{-1}): 255, 215s, 165, 145.
Source: Sundarakannan et al. (2002).
Comments: No independent analytical data are provided for the sample used. For the Raman spectra of zhangpeishanite see also Scott (1968) and Sundarakkannan and Kesavamoorthy (1998).

Ziesite $Cu_2V^{5+}_2O_7$

Origin: Synthetic.
Experimental details: Methods of sample preparation are not described. Raman scattering measurements have been performed using 514.5 nm Ar⁺ laser radiation. The laser radiation power is not indicated.
Raman shifts (cm^{-1}): 950sh, 912s, 855, 786w, 389, 259, 192w.
Source: De Waal and Hutter (1994).
Comments: No independent analytical data are provided for the sample used.

Zincite ZnO

Origin: Franklin or Sterling Hill, New Jersey, USA
Experimental details: No data.
Raman shifts (cm^{-1}): 1603 (broad), 1080, 1004sh, 569, 522s, 486, 478, 438, 378w, 331w, 252w.
Source: Welsh (2008).
Comments: The sample was characterized by EDS analyses. For the Raman spectra of zincite see also Bouchard and Smith (2003) and Kunert et al. (2006).

Zincochromite $ZnCr_2O_4$

Origin: Synthetic.
Experimental details: Raman scattering measurements have been performed from the (100) face of a single crystal using unpolarized 488 nm Ar⁺ laser radiation. The laser radiation power is not indicated.
Raman shifts (cm^{-1}): 692s, 610, 515, 457w, 166.
Source: Lutz et al. (1991).
Comments: In the cited paper also polarized Raman spectra of zincochromite are given. For the Raman spectrum of zincochromite see also D'Ippolito et al. (2015).

Zincocopiapite $ZnFe^{3+}_4(SO_4)_6(OH)_2 \cdot 20H_2O$

Origin: Les Valettes, Wallis, Switzerland.
Experimental details: Raman scattering measurements have been performed on arbitrarily oriented crystals using a 633 nm He-Ne laser. The laser radiation power is not indicated. The Raman shifts have been determined for the maxima of individual peaks obtained as a result of the spectral curve analysis.
Raman shifts (cm^{-1}): 1231, 1162, 1159sh, 1099, 1021s, 1005s, 987sh, 893w, 860w, 738w, 624, (613sh), 565, 485, 450, 424, 302, 267s, 218.
Source: Frost (2011c).
Comments: No independent analytical data are provided for the sample used.

Zincospiroffite $Zn_2Te_3O_8$

Origin: Zhongshangou Au deposit, Chongli Co., Hebei Province, China (type locality).
Experimental details: Raman scattering measurements have been performed on an arbitrarily oriented sample using 514.5 nm Ar^+ laser radiation. The nominal laser radiation power was 20 mW.
Raman shifts (cm^{-1}): 748, 725s, 646, 578w, 536w, 407w, 344, 304w, 233w, 181, 123.
Source: Zhang et al. (2004).
Comments: The sample was characterized by powder X-ray diffraction data and electron microprobe analyses.

Zinkenite $Pb_9Sb_{22}S_{42}$

Origin: Zlatá Baňa, Slanské Vrchy Mts., central Slovakia.
Experimental details: Raman scattering measurements have been performed on a polycrystalline sample using 532 nm Nd-YAG laser radiation. Exciting radiation with the power densitiy of 8.5×10^{-3} Å mW mm^{-2} was used. A 180°-scattering geometry was employed.
Raman shifts (cm^{-1}): (335sh), 312s, (302sh), 282s, (271sh), 238, 204sh, 192, 156, 130w, 119w, 103, 75sh, 69, 58.
Source: Kharbish and Jeleň (2016).
Comments: The sample was characterized by electron microprobe analyses. For the Raman spectrum of zinkenite see also Goienaga et al. (2011).

Zippeite $K_3(UO_2)_4(SO_4)_2O_3(OH) \cdot 3H_2O$

Origin: Abandoned uranium mine at Pecs, Hungary.
Experimental details: Raman scattering measurements have been performed on an arbitrarily oriented sample using 785 nm Ar^+ laser radiation. The laser radiation power is not indicated.
Raman shifts (cm^{-1}): 1233w, 1091, 1012, 842s, 740, 398, 192.
Source: Stefaniak et al. (2009).
Comments: The sample was characterized by qualitative electron microprobe analysis. For the Raman spectra of zippeite see also Frost et al. (2005i) and Plášil et al. (2010a).

Zircon $ZrSiO_4$

Origin: Kozjak Mt., Kozjačija, Macedonia.
Experimental details: Raman scattering measurements have been performed on a powdered sample using 1064 nm Nd-YAG laser radiation. The laser radiation power is not indicated.
Raman shifts (cm^{-1}): 1008s, 973, 769, 438, 394w, 356s, 224, 200w, 180w.
Source: Makreski et al. (2005b).
Comments: No independent analytical data are provided for the sample used. For the Raman spectra of zircon see also Nicola and Rutt (1974), Syme et al. (1977), Geisler et al. (2003), Gucsik et al. (2004), Jasinevicius (2009), Frezzotti et al. (2012), Nhlabathi (2012), Andò and Garzanti (2014), and Grüneberger et al. (2016).

Zoisite $Ca_2Al_3[Si_2O_7][SiO_4]O(OH)$

Origin: No data.

Experimental details: Raman scattering measurements have been performed on an arbitrarily oriented single-crystal platelet using 532 nm laser radiation. The nominal laser radiation power was 150 mW.

Raman shifts (cm^{-1}): 3150w, 1092s, 1018s, 983, 946w, 928, 909, 889, 872, 860, 778w, 727w, 678, 623, 597, 574, 530, 493s, 456, 435, 420, 395, 337, 312w, 287, 280, 261, 215, 192.

Source: Mao et al. (2007).

Comments: The sample was characterized by single-crystal X-ray diffraction data and electron microprobe analyses. For the Raman spectra of zoisite see also Jasinevicius (2009), Andò and Garzanti (2014), and Weis et al. (2016).

Zorite $Na_6Ti_5Si_{12}O_{34}(O,OH)_5 \cdot 11H_2O$

Origin: Synthetic.

Experimental details: Raman scattering measurements have been performed on a powdery sample using 1064 nm Nd-YAG laser radiation. The laser radiation power is not indicated.

Raman shifts (cm^{-1}): 1050w, 940, 905, 870sh, 755s, and a series of bands between 200 and 550 cm^{-1}.

Source: Craveiro and Lin (2012).

Comments: The sample was characterized by powder X-ray diffraction data. For the Raman spectrum of zorite see also Ferdov et al. (2008).

Zuktamrurite FeP_2

Origin: Synthetic.

Experimental details: Polarized Raman scattering measurements have been performed on a cluster of needle-like crystals by the back-scattering technique, in different scattering geometries, using 514.5 nm Ar^+ laser radiation. The laser radiation power is not indicated.

Raman shifts (cm^{-1}): 448, 386–388, 323–326s.

Source: Lutz and Müller (1991).

Comments: No independent analytical data are provided for the sample used.

Zunyite $Al_{13}Si_5O_{20}(OH,F)_{18}Cl$

Origin: Zuni mine, San Juan Co., Colorado, USA (type locality).

Experimental details: Raman scattering measurements have been performed on arbitrarily oriented crystals using a 633 nm He-Ne laser. The laser radiation power is not indicated. The Raman shifts have been determined for the maxima of individual peaks obtained as a result of the spectral curve analysis.

Raman shifts (cm^{-1}): 3635, 3431, 3369, 3352, 3335, 3317, 3304, 1295, 1239, 1207, 1176, 1141, 1126, 1101, 1067, 994, 950s, 930s, 701, 612, 467, 444, 397s, 374sh, 360s, 334, 313, 271, 250, 213, 202s.

Source: Kloprogge and Frost (1999d).

Comments: No independent analytical data are provided for the sample used.

Zýkaite $Fe^{3+}_4(AsO_4)_3(SO_4)(OH) \cdot 15H_2O$

Origin: Kaňk, near Kutná Hora, Czech Republic (?).

Experimental details: Raman scattering measurements have been performed on an arbitrarily oriented sample using 785 nm laser radiation. The nominal laser radiation power was berween 60 and 120 mW.

Raman shifts (cm^{-1}): 1113w, 1063w, 998, 895, 883, 835–815 (broad), 472sh, 442, 412sh, 306sh, 285s, 217.

Source: Culka et al. (2016b).

Comments: The sample was characterized by powder X-ray diffraction data. For the Raman spectrum of zýkaite see also Frost et al. (2011m).

References

Abdija Z, Najdoski M, Koleva V, Runčevski T, Dinnebier RE, Šoptrajanov B, Stefov V (2014) Preparation, structural, thermogravimetric and spectroscopic study of magnesium potassium arsenate hexahydrate. Z anorg allg Chem 640(15):3177–3183

Abelló S, Bolshak E, Gispert-Guirado F, Farriol X, Montané D (2014) Ternary Ni–Al–Fe catalysts for ethanol steam reforming. Catal Sci Technol 4(4):1111–1122

Abraham S, Aruldhas G (1994) Infrared and polarized Raman spectra of $Cd(HCOO)_2 \cdot 2H_2O$. Phys Status Solidi A 144(2):485–491

Aceto M, Agostino A, Boccaleri E, Crivello F, Garlanda AC (2006) Evidence for the degradation of an alloy pigment on an ancient Italian manuscript. J Raman Spectrosc 37(10):1160–1170

Achary SN, Errandonea D, Muñoz A, Rodríguez-Hernández P, Manjón FJ, Krishna PSR, Patwe SJ, Grover V, Tyagi AK (2013) Experimental and theoretical investigations on the polymorphism and metastability of $BiPO_4$. Dalton Trans 42(42):14999–15015

Adams DM, Gardner IR (1974) Single-crystal vibrational spectra of apatite, vanadinite, and mimetite. J Chem Soc Dalton Trans 14:1505–1509

Adar F (2014) Molecular spectroscopy workbench. Raman spectra of metal oxides. Spectroscopy 29(10):14–19

Agakhanov AA, Pautov LA, Sokolova E, Hawthorne FC, Karpenko VY, Siidra O, Garanin VK (2016a) Mendeleevite-(Nd), $(Cs,\square)_6(\square,Cs)_6(\square,K)_6(REE,Ca)_{30}(Si_{70}O_{175})(OH,H_2O,F)_{35}$, a new mineral from the Darai-Pioz alkaline massif, Tajikistan. Mineral Mag. https://doi.org/10.1180/minmag.2016.080.076

Agakhanov AA, Pautov LA, Sokolova E, Abdu YA, Karpenko VY (2016b) Two astrophyllite-supergroup minerals: bulgakite, a new mineral from the Darai-Pioz alkaline massif, Tajikistan and revision of the crystal structure and chemical formula of nalivkinite. Can Mineral 54:33–48

Agnihotri OP, Sehgal HK (1972) Fundamental infrared lattice vibration spectrum and the laser-excited Raman spectrum of $MoSe_2$. Philosophical Mag 26(3):753–756

Agulló-Rueda F, Calleja JM, Martini M, Spinolo G, Cariati F (1987) Raman and infrared spectra of transition metal halide hexahydrates. J Raman Spectrosc 18(7):485–491

Ahn HS, Lee EP, Chang HY, Lee DW, Ok KM (2015) $Sr_3Bi_2(SeO_3)_6 \cdot H_2O$: a novel anionic layer consisting of second-order Jahn–Teller (SOJT) distortive cations. J Solid State Chem 221:73–78

Akaogi M, Ross NL, McMillan P, Navrotsky A (1984) The Mg_2SiO_4 polymorphs (olivine, modified spinel and spinel) – thermodynamic properties from oxide melt solution calorimetry, phase relations, and models of lattice vibrations. Am Mineral 69:499–512

Akrap A, Tran M, Ubaldini A, Teyssier J, Giannini E, Van Der Marel D, Lerch P, Homes CC (2012) Optical properties of Bi_2Te_2Se at ambient and high pressures. Phys Rev B 86(23):235207. (9 pp)

Aksenov SM, Chukanov NV, Rusakov VS, Panikorovskii TL, Gainov RR, Vagizov FG, Rastsvetaeva RK, Lyssenko KA, Belakovskiy DI (2016) Towards a revisitation of vesuvianite-group nomenclature: the crystal structure of Ti-rich vesuvianite from Alchuri, Shigar Valley, Pakistan. Acta Cryst B72:744–752

Aktas O, Truong KD, Otani T, Balakrishnan G, Clouter MJ, Kimura T, Quirion G (2011) Raman scattering study of delafossite magnetoelectric multiferroic compounds: $CuFeO_2$ and $CuCrO_2$. J Phys Condens Matter 24(3):036003. (11 pp)

Alekseev EV, Felbinger O, Wu S, Malcherek T, Depmeier W, Modolo G, Gesing TM, Krivovichev SV, Suleimanov EV, Gavrilova TA, Pokrovsky LD, Pugachev AM, Surovtsev NV, Atuchin VV (2013) K$[AsW_2O_9]$, the first member of the arsenate–tungsten bronze family: synthesis, structure, spectroscopic and non-linear optical properties. J Solid State Chem 204:59–63

Alharbi ND, Salah N, Habib SS, Alarfaj E (2012) Synthesis and characterization of nano-and microcrystalline cubes of pure and Ag-doped LiF. J Phys D 46(3):035305. (6 pp)

Ali AB, Awaleh MO, Leblanc M, Smiri LS, Maisonneuve V, Houlbert S (2004a) Hydrothermal synthesis, crystal structure, thermal behaviour, IR and Raman spectroscopy of $Na_3Y(CO_3)_3 \cdot 6H_2O$. Compt Rend Chim 7(6):661–668

Ali AB, Maisonneuve V, Houlbert S, Silly G, Buzaré JY, Leblanc M (2004b) Cation and anion disorder in new cubic rare earth carbonates $Na_2LiLn(CO_3)_3$ (Ln = Eu–Er, Yb, Lu, Y); synthesis, crystal structures, IR, Raman and NMR characterizations. Solid State Sci 6(11):1237–1243

Alibakhshi E, Ghasemi E, Mahdavian M (2012) A comparison study on corrosion behavior of zinc phosphate and potassium zinc phosphate anticorrosive pigments. Prog Color Colorants Coat 5:91–99

Alici E, Schmidt T, Lutz HD (1992) Zur Kenntnis des Calciumbromats und -iodats, Kristallstruktur, röntgenographische, IR- und Raman-spektroskopische und thermoanalytische Untersuchungen. Z anorg allg Chem 608:135–144. (in German)

Allen FM, Burnham CW (1992) A comprehensive structure-model for vesuvianite: symmetry variations and crystal growth. Can Miner 30:1–18

Almeida MAP, Cavalcante LS, Li MS, Varela JA, Longo E (2012) Structural refinement and photoluminescence properties of $MnWO_4$ nanorods obtained by microwave-hydrothermal synthesis. J Inorg Organometallic Polym Mater 22(1):264–271

Almeida RM, Andreeta MRB, Hernandes AC, Dias A, Moreira RL (2014) Optical phonon characteristics of an orthorhombic-transformed polymorph of $CaTa_2O_6$ single crystal fibre. Mater Res Express 1(1):016304-1–016304-13

Altamura G, Grenet L, Roger C, Roux F, Reita V, Fillon R, Fournier H, Perraud S, Mariette H (2014) Alternative back contacts in kesterite $Cu_2ZnSn(S_{1-x}Se_x)_4$ thin film solar cells. J Renew Sustain Energy 6(1):011401. (7 pp)

Amdouni N, Zarrouk H, Julien CM (2003) Synthesis, structure and intercalation of brannerite $LiWVO_6$ wet-chemical products. J Mater Sci 38(22):4573–4579

Amme M, Renker B, Schmid B, Feth MP, Bertagnolli H, Döbelin W (2002) Raman microspectrometric identification of corrosion products formed on UO_2 nuclear fuel during leaching experiments. J Nucl Mater 306(2):202–212

Amri M, Zouari N, Mhiri T, Gravereau P (2009) Synthesis, structure determination and calorimetric study of new cesium hydrogen selenate arsenate $Cs_4(SeO_4)(HSeO_4)_2(H_3AsO_4)$. J Alloys Compd 477(1):68–75

An S, Liu X, Yang L, Zhang L (2015) Enhancement removal of crystal violet dye using magnetic calcium ferrite nanoparticle: study in single-and binary-solute systems. Chem Eng Res Design 94:726–735

Anandan S, Lee GJ, Yang CK, Ashokkumar M, Wu JJ (2012) Sonochemical synthesis of Bi_2CuO_4 nanoparticles for catalytic degradation of nonylphenolethoxylate. Chem Eng J 183:46–52

Ananthanarayanan V (1961) Raman spectra of crystalline double sulphates. Z Phys 163(2):144–157

Anastassakis E (1973) Light scattering in transition metal diselenides $CoSe_2$ and $CuSe_2$. Solid State Commun 13(9):1297–1301

Anastassakis E, Perry CH (1976) Light scattering and IR measurements in XS_2 pyrite-type compounds. J Chem Phys 64(9):3604–3609

Anbalagan G, Mukundakumari S, Murugesan KS, Gunasekaran S (2009) Infrared, optical absorption, and EPR spectroscopic studies on natural gypsum. Vib Spectrosc 50(2):226–230

Anbalagan G, Sivakumar G, Prabakaran AR, Gunasekaran S (2010) Spectroscopic characterization of natural chrysotile. Vib Spectrosc 52(2):122–127

Anderson A (ed) (1973) The Raman Effect, Applications, vol 2. Marcel Dekker, New York

Anderson A, Chieh C, Irish DE, Tong JPK (1980) An X-ray crystallographic, Raman, and infrared spectral study of crystalline potassium uranyl carbonate, $K_4UO_2(CO_3)_3$. Can J Chem 58(16):1651–1658

Anderson AJ, Clark AH, Gray S (2001) The occurrence and origin of zabuyelite (Li_2CO_3) in spodumene-hosted fluid inclusions: implications for the internal evolution of rare-element granitic pegmatites. Can Mineral 39(6):1513–1527

Andò S, Garzanti E (2014) Raman spectroscopy in heavy-mineral studies. Geol Soc, London, Spec Publ 386(1):395–412

Andrade MB, Morrison SM, Di Domizio AJ, Feinglos MN, Downs RT (2012) Robertsite, $Ca_2Mn^{III}_3O_2(PO_4)_3 \cdot 3H_2O$. Acta Crystallogr E. https://doi.org/10.1107/S160053681203735X

Andrade MB, Atencio D, Persiano AIC, Ellena J (2013a) Fluorcalciomicrolite, $(Ca,Na,\square)_2Ta_2O_6F$, a new microlite-group mineral from Volta Grande pegmatite, Nazareno, Minas Gerais, Brazil. Mineral Mag 77(7):2989–2996

Andrade MB, Doell D, Downs RT, Yang H (2013b) Redetermination of katayamalite, $KLi_3Ca_7Ti_2(SiO_3)_{12}(OH)_2$. Acta Crystallogr E. https://doi.org/10.1107/S1600536813016620

Andrade MB, Yang H, Downs RT, Jenkins RA, Fay I (2014) Te-rich raspite, $Pb(W_{0.56}Te_{0.44})O_4$, from Tombstone, Arizona, USA: the first natural example of Te^{6+} substitution for W^{6+}. Am Mineral 99(7):1507–1510

Andrade MB, Yang H, Atencio D, Downs RT, Chukanov NV, Lemée-Cailleau MH, Persiano AIC, Goeta AE, Ellena J (2016) Hydroxycalciomicrolite, $Ca_{1.5}Ta_2O_6(OH)$, a new member of the microlite group from Volta Grande pegmatite, Nazareno, Minas Gerais, Brazil. Mineral Mag. https://doi.org/10.1180/minmag.2016.080.116

Andrade MB, Yang H, Downs RT, Färber G, ContreiraFilho RR, Evans SH, Loehn CW, Schumer BM (2017) Fluorlamprophyllite, $Na_3(SrNa)Ti_3(Si_2O_7)_2O_2F_2$, a new mineral from Poços de Caldas alkaline massif, Morro do Serrote, Minas Gerais, Brazil. Mineral Mag. https://doi.org/10.1180/minmag.2017.081.027

Andreani M, Daniel I, Pollet-Villard M (2013) Aluminum speeds up the hydrothermal alteration of olivine. Am Mineral 98(10):1738–1744

Annen MJ, Davis ME (1993) Raman and ^{29}Si MAS NMR spectroscopy of framework materials containing three-membered rings. Microporous Mater 1(1):57–65

Antony CJ, Aatiq A, Panicker CY, Bushiri MJ, Varghese HT, Manojkumar TK (2011) FT-IR and FT-Raman study of Nasicon type phosphates, ASnFe(PO$_4$)$_3$ [A= Na$_2$, Ca, Cd]. Spectrochim Acta A 78(1):415–419

Apollonov VN (1998) Nepskoeite Mg$_4$Cl(OH)$_7$·6H$_2$O, a new mineral from the Nepskoe K salt deposit. Zapiski VMO (Proc Russ Miner Soc) 127(1):41–46. (in Russian)

Apopei AI, Buzgar N (2010) The Raman study of amphiboles. Analele Stiintifice de Universitatii AI Cuza din Iasi. Sect 2 Geologie 56(1):57–83

Apopei AI, Buzgar N, Buzatu A (2011) Raman and infrared spectroscopy of kaersutite and certain common amphiboles. Analele Stiintifice ale Universitatii "Al. I. Cuza" din Iasi Seria Geologie 57(2):35–58

Apopei AI, Damian G, Buzgar N (2012) A preliminary Raman and FT-IR spectroscopic study of secondary hydrated sulfate minerals from the Hondol open pit (Metaliferi Mts., Romania). Rom J Mineral Depos 85(2):1–6

Apopei AI, Buzgar N, Damian G, Buzatu A (2014a) The Raman study of weathering minerals from the Coranda-Hondol Open Pit (Certej Gold-Silver Deposit) and their photochemical degradation products under laser irradiation. Can Mineral 52:1027–1038

Apopei AI, Damian G, Buzgar N, Milovska S, Buzatu A (2014b) New occurrences of hessite, petzite and stützite at Coranda-Hondol open pit (Certej gold-silver deposit, Romania). Carpath J Earth Environ Sci 9(2):71–78

Apopei AI, Buzgar N, Damian G, Buzatu A (2015) The Raman study of weathering minerals from the Coranda-Hondol Open Pit (Certej Gold-Silver Deposit) and their photochemical degradation products under laser irradiation. Can Mineral 52:1027–1038

Aramendia J, Gómez-Nubla L, Castro K, Madariaga JM (2014) Spectroscopic speciation and thermodynamic modeling to explain the degradation of weathering steel surfaces in SO$_2$ rich urban atmospheres. Microchem J 115:138–145

Arlt T, Armbruster T, Miletich R, Ulmer P, Peters T (1998) High pressure single-crystal synthesis, structure and compressibility of the garnet Mn$^{2+}_3$Mn$^{3+}_2$[SiO$_4$]$_3$. Phys Chem Minerals 26(2):100–106

Armbruster T, Lazic B, Gfeller F, Galuskin EV, Galuskina IO, Savelyeva VB, Zadov AE, Pertsev NN, Dzierżanowski P (2011) Chlorine content and crystal chemistry of dellaite from the Birkhin gabbro massif, Eastern Siberia, Russia. Mineral Mag 75(2):379–394

Armbruster T, Lazic B, Galuskina IO, Galuskin EV, Gnos E, Marzec KM, Gazeev VM (2012) Trabzonite, Ca$_4$[Si$_3$O$_9$(OH)]OH: crystal structure, revised formula, new occurrence and relation to killalaite. Mineral Mag 76(3):455–472

Armstrong CR, Nash KL, Griffiths PR, Clark SB (2011) Synthesis and characterization of françoisite-(Nd): Nd[(UO$_2$)$_3$O(OH)(PO$_4$)$_2$]·6H$_2$O. Am Mineral 96(2–3):417–422

Aronne A, Esposito S, Ferone C, Pansini M, Pernice P (2002) FTIR study of the thermal transformation of barium-exchanged zeolite A to celsian. J Mater Chem 12(10):3039–3045

Asensio MO, Yildirim M, Senberber FT, Kipcak AS, Derun EM (2016) Thermal dehydration kinetics and characterization of synthesized potassium borates. Rese Chem Intermed 42(5):4859–4878

Assaaoudi H, Ennaciri A, Rulmont A (2001) Vibrational spectra of hydrated rare earth orthophosphates. Vib Spectrosc 25(1):81–90

Atencio D, ContreiraFilho RR, Mills SJ, Coutinho J, Honorato SB, Ayala AP, Ellena J, Andrade MBD (2011) Rankamaite from the Urubu pegmatite, Itinga, Minas Gerais, Brazil: crystal chemistry and Rietveld refinement. Am Mineral 96(10):1455–1460

Atencio D, Ciriotti ME, Andrade MB (2013) Fluorcalcioroméite, (Ca,Na)$_2$Sb$^{5+}_2$(O,OH)$_6$F, a new roméite-group mineral from Starlera mine, Ferrera, Grischun, Switzerland: description and crystal structure. Mineral Mag 77(4):467–473

Au CT, Zhang WD (1997) Oxidative dehydrogenation of propane over rare-earth orthovanadates. J Chem Soc Faraday Trans 93:1195–1204

Au CT, Zhang WD, Wan HL (1996) Preparation and characterization of rare earth orthovanadates for propane oxidative dehydrogenation. Catal Lett 37:241–246

Augsburger MS, Juri MA, Pedregosa JC, Mercader RC (1992) Crystal data and spectroscopic studies of NaNiFe$_2$(AsO$_4$)$_3$. J Solid State Chem 101(1):66–70

Auzende AL, Daniel I, Reynard B, Lemaire C, Guyot F (2004) High-pressure behaviour of serpentine minerals: a Raman spectroscopic study. Phys Chem Minerals 31(5):269–277

Avril C, Malavergne V, Caracas R, Zanda B, Reynard B, Charon E, Bobocioiu E, Brunet F, Borensztajn S, Pont S, Tarrida M, Guyot F (2013) Raman spectroscopic properties and Raman identification of CaS–MgS–MnS–FeS–Cr$_2$FeS$_4$ sulfides in meteorites and reduced sulfur-rich systems. Meteor Planet Sci 48(8):1415–1426

Ayta WEF, Dantas NO, Silva ACA, Cano NF (2010) First evidence of crystalline KHSO$_4$: Mn grown by an aqueous solution method and the investigation of the effect of ionizing radiation exposure. J Cryst Growth 312(4):563–567

Ayyub P, Multani MS, Palkar VR, Vijayaraghavan R (1986) Vibrational spectroscopic study of ferroelectric SbNbO$_4$, antiferroelectric BiNbO$_4$, and their solid solutions. Phys Rev B 34(11):8137–8140

Ayyub P, Palkar VR, Multani MS, Vijayaraghavan R (1987) Structural, dielectric and vibrational properties of the stibiotantalite (Sb$_{1-x}$Bi$_x$)NbO$_4$ system for $0 \leq x \leq 1$. Ferroelectrics 76(1):93–106

Azdouz M, Manoun B, Essehli R, Azrour M, Bih L, Benmokhtar S, Hou AA, Lazor P (2010) Crystal chemistry, Rietveld refinements and Raman spectroscopy studies of the new solid solution series: Ba$_{3-x}$Sr$_x$(VO$_4$)$_2$ ($0 \leq x \leq 3$). J Alloys Compd 498(1):42–51

Azrour M, Azdouz M, Manoun B, Essehli R, Benmokhtar S, Bih L, El Ammari L, Ezzahi A, Ider A, Hou AA (2011) Rietveld refinements and vibrational spectroscopic studies of $Na_{1-x}K_xPb_4(PO_4)_3$ lacunar apatites ($0 \leq x \leq 1$). J Phys Chem Solids 72(11):1199–1205

Babouri L, Belmokre K, Kabir A, Abdelouas A, El Mendili Y (2015) Structural and electrochemical study of binary copper alloys corrosion in 3% NaCl solution. J Chem Pharm Res 7(4):1175–1186

Bahfenne S (2011) Single crystal Raman spectroscopy of selected arsenite, antimonite and hydroxyantimonate minerals. Dissertation, Queensland University of Technology

Bahfenne S, Frost RL (2009) Raman spectroscopic study of the arsenite minerals ludlockite $PbFe^{3+}_4As_{10}O_{22}$ and schneiderhöhnite $Fe^{2+}Fe^{3+}_3As^{3+}_5O_6$. Spectrochim Acta A 74(3):625–628

Bahfenne S, Frost RL (2010a) Raman spectroscopic study of the antimony bearing mineral langbanite. Spectrochim Acta A 75(2):710–712

Bahfenne S, Frost RL (2010b) Raman spectroscopic study of the antimonate mineral lewisite $(Ca,Fe,Na)_2(Sb,Ti)_2O_6(O,OH)_7$. Radiat Effects Defects Solids Incorp Plasma Sci Plasma Technol 165(1):46–53

Bahfenne S, Frost RL (2010c) Raman spectroscopic study of the antimony containing mineral partzite $Cu_2Sb_2(O,OH)_7$. Spectrosc Lett 43(3):202–206

Bahfenne S, Frost RL (2010d) Raman spectroscopic study of the multi-anion mineral dixenite $CuMn^{2+}_{14}Fe^{3+}(AsO_3)_5(SiO_4)_2(AsO_4)(OH)_6$. J Raman Spectrosc 41(4):465–468

Bahfenne S, Frost RL (2010e) Vibrational spectroscopic study of the antimonate mineral stibiconite $Sb^{3+}Sb^{5+}_2O_6(OH)$. Spectrosc Lett 43(6):486–490

Bahfenne S, Rintoul L, Frost RL (2011a) Single-crystal Raman spectroscopy of natural schafarzikite $FeSb_2O_4$ from Pernek, Slovak Republic. Am Mineral 96(5–6):888–894

Bahfenne S, Rintoul L, Frost RL (2011b) Single-crystal Raman spectroscopy of natural leiteite ($ZnAs_2O_4$) and comparison with the synthesised mineral. J Raman Spectrosc 42(4):659–666

Bahfenne S, Rintoul L, Langhof J, Frost RL (2011c) Single-crystal Raman spectroscopy of natural finnemanite and comparison with its synthesised analogue. J Raman Spectrosc 42(12):2119–2125

Bahfenne S, Rintoul L, Langhof J, Frost RL (2012) Single-crystal Raman spectroscopy of natural paulmooreite $Pb_2As_2O_5$ in comparison with the synthesized analog. Am Mineral 97(1):143–149

Bahrami M, Taton G, Conédéra V, Salvagnac L, Tenailleau C, Alphonse P, Rossi C (2014) Magnetron sputtered Al-CuO nanolaminates: effect of stoichiometry and layers thickness on energy release and burning rate. Propellants Explos Pyrotech 39(3):365–373

Bai L, Xue Y, Zhang J, Pan B, Wu C (2013) Synthetic potassium vanadium oxide $K_2V_6O_{16} \cdot 1.5H_2O$ superlong nanobelts: a 1D room-temperature ferromagnetic semiconductor. Eur J Inorg Chem 2013(20):3497–3505

Bai C, Han S, Pan S, Bian Q, Yang Z, Zhang X, Lin X, Jing Q (2014) Reinvestigation and characterization of the magnesium borate fluoride $Mg_5(BO_3)F$. Z anorg allg Chemie 640(10):2013–2018

Baioumy HM, Khedr MZ, Ahmed AH (2013) Mineralogy, geochemistry and origin of Mn in the high-Mn iron ores, Bahariya oasis, Egypt. Ore Geol Rev 53:63–76

Bajda T (2010) Solubility of mimetite $Pb_5(AsO_4)_3Cl$ at 5-55°C. Environ Chem 7:268–278

Bajda T, Mozgawa W, Manecki M, Flis J (2011) Vibrational spectroscopic study of mimetite–pyromorphite solid solutions. Polyhedron 30(15):2479–2485

Bakker RJ (2004) Raman spectra of fluid and crystal mixtures in the systems H_2O, $H_2O-NaCl$ and $H_2O-MgCl_2$ at low temperatures: applications to fluid-inclusion research. Can Mineral 42(5):1283–1314

Bakker RJ (2010) Raman spectroscopy: application to platinum-group minerals. In: Mogessie B et al (eds) Short Course: Mineralogy, geochemistry, and ore deposits of platinum group elements. 20th General Meeting of IMA, Budapest. http://fluids.unileoben.ac.at/Publications.html

Bakker RJ (2014) Application of combined Raman spectroscopy and Electron Probe Microanalysis to identify platinum group minerals. 11th EMAS regional workshop on electron probe microanalysis of materials today. Practical Aspects, pp 215–233

Balachandran UGEN, Eror NG (1982) Raman spectra of titanium dioxide. J Solid State Chem 42(3):276–282

Balakrishnan T, Bhagavannarayana G, Ramamurthi K (2008) Growth, structural, optical, thermal and mechanical properties of ammonium pentaborate single crystal. Spectrochim Acta A 71(2):578–583

Balassone G, Bellatreccia F, Mormone A, Biagioni C, Pasero M, Petti C, Mondillo M, Fameli G (2012) Sodalite-group minerals from the Somma–Vesuvius volcanic complex, Italy: a case study of K-feldspar-rich xenoliths. Mineral Mag 76(1):191–212

Ballirano P (2012) Haüyne: mutual cations/anionic groups arrangement and thermal expansion mechanism. Phys Chem Minerals 39(9):733–747

Balraj V, Vidyasagar K (1999) Syntheses and characterization of novel three-dimensional tellurites, $Na_2MTe_4O_{12}$ (M = W, Mo), with intersecting tunnels. Inorg Chem 38(25):5809–5813

Bang SE, Lee DW, Ok KM (2014) Variable framework structures and centricities in alkali metal yttrium selenites, $AY(SeO_3)_2$ (A = Na, K, Rb, and Cs). Inorg Chem 53(9):4756–4762

Banks E, Greenblatt M, McGarvey BR (1967) ESR and optical spectroscopy of CrO_4^{3-} in chlorospodiosite, Ca_2PO_4Cl. J Chem Phys 47(10):3772–3780

Banks E, Chianelli R, Korenstein R (1975) Crystal chemistry of struvite analogs of the type $MgMPO_4 \cdot 6H_2O$ ($M^+ = K^+$, Rb^+, Cs^+, Tl^+, NH_4^+). Inorg Chem 14(7):1634–1639

Banno Y, Miyawaki R, Momma K, Bunno M (2016) A CO_3-bearing member of the hydroxylapatite–hydroxylellestadite series from Tadano, Fukushima Prefecture, Japan: CO_3-SO_4 substitution in the apatite–ellestadite series. Mineral Mag 80(2):363–370

Banwell CN (1983) Fundamentals of molecular spectroscopy. McGraw-Hill (UK), London

Baran EJ (1973) Infrarotspektrum und Kraftkonstanten des CoO_4^{4-}-Ions. Z anorg allg Chemie 399(1):57–64. (in German)

Baran EJ (1976) Die Schwingungsspektren von $Ca_3(VO_4)_2$ und $Ca_3(AsO_4)_2$. Z anorg allg Chemie 427(2):131–136. (in German)

Baran EJ (1978) Das Schwingungsspektrum des Ditellurit-Ions. Z anorg allg Chemie 442(1):112–118. (in German)

Baran EJ (1994) Vibrational and electronic spectra of copper(II) chromate. Spectrochim Acta A 50(14):2385–2389

Baran EJ (1996) Vibrational spectra of $Sr_2(VO)V_2O_8$. J Raman Spectrosc 27(7):555–557

Baran EJ (1997) Vibrational spectra of $Ba_2(VO)V_2O_8$. J Raman Spectrosc 28:289–291

Baran EJ (2014) Natural oxalates and their analogous synthetic complexes. J Coord Chem 67(23–24):3734–3768

Baran EJ (2016) Natural iron oxalates and their analogous synthetic counterparts: a review. Chemie Erde-Geochem 76(3):449–460

Baran EJ, Aymonino PJ (1971) Die Infrarotspektren der Orthovanadate der leichteren Lanthanide. Z anorg allg Chemie 383(2):226–229. (in German)

Baran EJ, Aymonino PJ (1972) Die Infrarotspektren einiger Orthovanadate mit Apatitstruktur. Z anorg allg Chemie 390(1):77–84. (in German)

Baran EJ, Botto IL (1976) Das Schwingungsspektrum des synthetischen Carnotits. Mh Chem (Chemical Monthly) 107(3):633–639. (in German)

Baran EJ, Botto IL (1978) Die IR-Spektren einiger Doppeloxide mit Ilmenit-Struktur. Z anorg allg Chemie 444(1):282–288. (in German)

Baran EJ, Botto IL (1980) Die Schwingungsspektren einiger tellurhaltiger 2,6-Spinelle. Z anorg allg Chemie 463(1):185–192. (in German)

Baran EJ, Lii KH (1992) Vibrational spectrum of $Zn_2VO(PO_4)_2$. J Raman Spectrosc 23(2):125–126

Baran EJ, Rabe S (1999) The infrared spectrum of α-$(NH_4)_2(VO)_3(P_2O_7)_2$. J Mater Sci Lett 18:1779–1780

Baran EJ, Weil M (2004) Vibrational spectra of $Cd_2As_2O_7$. J Raman Spectrosc 35(2):178–180

Baran EJ, Aymonino PJ, Müller A (1972) Die Schwingungsspektren von Strontium- und Barium-Orthovanadat. J Molec Struct 11:453–457. (in German)

Baran EJ, Gentil LA, Pedregosa JC, Aymonino PJ (1974) Die Divanadate des Thoriums. Z anorg allg Chemie 410(3):301–312. (in German)

Baran EJ, Botto IL, Fournier LL (1981) Das Schwingungsspektrum von α-Te_2MoO_7 und ein Vorschlag zur Struktur der Telluromolybdate zwei wertiger Kationen. Z anorg allg Chemie 476(5):214–220. (in German)

Baran EJ, Botto IL, Muto F, Kumada N, Kinomura N (1986) Vibrational spectra of the ilmenite modifications of $LiNbO_3$ and $NaNbO_3$. J Mater Sci Lett 5(6):671–672

Baran EJ, Muto F, Kumada N, Kinomura N (1989) The infrared spectra of $TiPO_4$ and VPO_4. J Mater Sci Lett 8(11):1305–1306

Baran EJ, Vassalo MB, Lii K-H (1994) Vibrational Spectra of β-$LiVOPO_4$ and $NaVOPO_4$. J Raman Spectrosc 25:199–202

Baran EJ, Lii KH, Wu LS (1995) The infrared spectrum of $[Ni(H_2O)_4][VOPO_4]_2$. J Mater Sci Lett 14(5):324326

Baran EJ, Vassallo MB, Lii KH (1996) Vibrational spectrum of $RbVOPO_4$. Vibr Spectrosc 10(2):331–334

Baran J, Ilczyszyn MM, Marchewka MK, Ratajczak H (1999a) Vibrational studies of different modifications of the sodium hydrogen sulphate crystals. Spectrosc Lett Int J Rapid Commun 32(1):83–102

Baran EJ, Mormann T, Jeitschko W (1999b) Infrared and Raman spectra of $(Hg_2)_3(AsO_4)_2$ and $Hg_3(AsO_4)_2$. J Raman Spectrosc 30:1049–1051

Baran EJ, Schwendtner K, Kolitsch U (2006) Vibrational spectra of $ScAsO_4 \cdot H_2O$. J Raman Spectrosc 37(12):1453–1455

Barashkov MV, Komyak AI, Shashkov SN (2004) Vibrational spectra and structure of potassium alum $KAl(SO_4)_2 \cdot 12[(H_2O)_x (D_2O)_{1-x}]$. J Appl Spectrosc 71(3):328–3330

Barchiche CE, Rocca E, Hazan J (2008) Corrosion behaviour of Sn-containing oxide layer on AZ91D alloy formed by plasma electrolytic oxidation. Surf Coat Technol 202(17):4145–4152

Barnes JH (1986) Nakauriite, a new blue mineral from Cedar Hill. Pennsylvania Geol 17(5):6–8

Barpanda P, Liu G, Mohamed Z, Ling CD, Yamada A (2014) Structural, magnetic and electrochemical investigation of novel binary $Na_{2-x}(Fe_{1-y}Mn_y)P_2O_7$ ($0 \leq y \leq 1$) pyrophosphate compounds for rechargeable sodium-ion batteries. Solid State Ionics 268:305–311

Barshilia HC, Rajam KS (2004) Raman spectroscopy studies on the thermal stability of TiN, CrN, TiAlN coatings and nanolayered TiN/CrN, TiAlN/CrN multilayer coatings. J Mater Res 19(11):3196–3205

Bartholomäi G, Klee WE (1978) The vibrational spectra of pyromorphite, vanadinite and mimetite. Spectrochim Acta A 34(7):831–843

Barton IF, Yang H, Barton MD (2014) The mineralogy, geochemistry, and metallurgy of cobalt in the rhombohedral carbonates. Can Mineral 52:653–669

Bastians S, Crump G, Griffith WP, Withnall R (2004) Raspite and studtite: Raman spectra of two unique minerals. J Raman Spectrosc 35(8–9):726–731

Bates JB, Quist AS (1974) Vibrational spectra of solid and molten phases of the alkali metal tetrafluoroborates. Spectrochim Acta 31A:1317–1327

Bates JB, Quist AS, Boyd GE (1971) Infrared and Raman spectra of polycrystalline $NaBF_4$. J Chem Phys 54(1):124–126

Bauman RP, Porto SPS (1967) Lattice vibrations and structure of rare-earth fluorides. Phys Rev 161(3):842–847

Baumgartner M, Bakker RJ (2010) Raman spectra of ice and salt hydrates in synthetic fluid inclusions. Chem Geol 275(1):58–66

Bayne JM, Butler IS (2014) Variable-temperature and high-pressure micro-Raman spectra of inorganic artists' pigments: crystalline wulfenite, lead (II) molybdate (VI), $PbMoO_4$. Spectrosc Lett 47(8):616–620

Bechibani I, Litaiem H, Ktari L, Zouari N, Garcia-Granda-S, Dammak M (2014) Structural, thermal behavior, dielectric and vibrational studies of the new compound, sodium hydrogen arsenate tellurate $Na_2H_4As_2O_5(H_2TeO_4)$. J Phys Chem Solids 75(7):911–920

Bechir MB, Rhaiem AB, Guidara K (2014) A.c. conductivity and dielectric study of $LiNiPO_4$ synthesized by solid-state method. Bull Mater Sci 37(3):473–480

Beck J, Benz S (2010) Crystalline and glassy phases in the ternary system Tl/Bi/Cl: synthesis and crystal structures of the thallium (I) chloridobismutates (III) Tl_3BiCl_6 and $TlBi_2Cl_7$. Z anorg allg Chem. https://doi.org/10.1002/zaac.200900567

Begun GM, Rutenberg AC (1967) Vibrational frequencies and force constants of some Group IVa and Group Va hexafluoride ions. Inorg Chem 6(12):2212–2216

Begun GM, Beall GW, Boatner LA, Gregor WJ (1981) Raman spectra of the rare earth orthophosphates. J Raman Spectrosc 11(4):273–278

Behrens G, Kuhn LT, Ubic R, Heuer AH (1995) Raman spectra of vateritic calcium carbonate. Spectrosc Lett 28(6):983–995

Behrens EA, Poojary DM, Clearfield A (1996) Syntheses, crystal structures, and ion-exchange properties of porous titanosilicates, $HM_3Ti_4O_4(SiO_4)_3 \cdot 4H_2O$ (M = H^+, K^+, Cs^+), structural analogues of the mineral pharmacosiderite. Chem Mater 8(6):1236–1244

Beigi H, Bindu VH, Hamoon HZR, Rao KV (2011) Low temperature synthesis and characterization of $ZnTiO_3$ by sol-gel method. J Nano-Electron Phys 3(1):47–52

Belik AA, Izumi F, Stefanovich SY, Malakho AP, Lazoryak BI, Leonidov IA, Leonidova IA, Davydov SA (2002) Polar and centrosymmetric phases in solid solutions $Ca_{3-x}Sr_x(PO_4)_2$ ($0 \leq x \leq 16/7$). Chem Mater 14(7):3197–3205

Belik AA, Izumi F, Azuma M, Kamiyama T, Oikawa K, Pokholok KV, Lazoryak BI, Takano M (2005) Redox reactions in strontium iron phosphates: synthesis, structures, and characterization of $Sr_9Fe(PO_4)_7$ and $Sr_9FeD(PO_4)_7$. Chem Mater 17(22):5455–5464

Belik AA, Koo HJ, Whangbo MH, Tsujii N, Naumov P, Takayama-Muromachi E (2007) Magnetic properties of synthetic libethenite Cu_2PO_4OH: a new spin-gap system. Inorg Chem 46(21):8684–8689

Belik AA, Naumov P, Kim J, Tsuda S (2011) Low-temperature structural phase transition in synthetic libethenite $Cu_2(PO_4)(OH)$. J Solid State Chem 184(11):3128–3133

Belitsky IA, Fursenko BA, Gabuda SP, Kholdeev OV, Seryotkin YV (1992) Structural transformations in natrolite and edingtonite. Phys Chem Minerals 18(8):497–505

Belkouch J, Monceaux L, Bordes E, Courtine P (1995) Comparative structural study of mixed metals pyrophosphates. Mater Res Bull 30(2):149–160

Bellatreccia F, Cámara F, Ottolini L, Della Ventura G, Cibin G, Mottana A (2005) Wiluite from Ariccia, Latium, Italy: occurrence and crystal structure. Can Miner 43:1457–1468

Belluso E, Fornero E, Cairo S, Albertazzi G, Rinaudo C (2007) The application of micro-Raman spectroscopy to distinguish carlosturanite from serpentine-group minerals. Can Mineral 45(6):1495–1500

Beneventi P, Bersani D, Lottici PP, Kovacs L (1995) A Raman study of $Bi_4(Ge_xSi_{1-x})_3O_{12}$ crystals. Solid State Commun 93(2):143–146

Benhammou A, Yaacoubi A, Nibou L, Bonnet JP, Tanouti B (2011) Synthesis and characterization of pillared stevensites: application to chromate adsorption. Environ Technol 32(4):363–372

Benmokhtar S, El Jazouli A, Chaminade JP, Gravereau P, Guillen F, De Waal D (2004) Synthesis, crystal structure and optical properties of $BiMgVO_5$. J Solid State Chem 177(11):4175–4182

Benmokhtar S, Belmal H, El Jazouli A, Chaminade JP, Gravereau P, Pechev S, Grenier JC, Villeneuve G, De Waal D (2007a) Synthesis, structure, and physicochemical investigations of the new $\alpha Cu_{0.50}TiO(PO_4)$ oxyphosphate. J Solid State Chem 180(2):772–779

Benmokhtar S, Chaminade JP, Gravereau P, Menetrier M, Bouree F (2007b) New process of preparation, structure, and physicochemical investigations of the new titanyl phosphate $Ti_2O(H_2O)(PO_4)_2$. J Solid State Chem 180(10):2713–2722

Benreguia N, Barnabé A, Trari M (2015) Sol-gel synthesis and characterization of the delafossite $CuAlO_2$. J Sol-Gel Sci Technol 75(3):670–679

Berger J (1976) Infrared and Raman spectra of $CuSO_4 \cdot 5H_2O$; $CuSO_4 \cdot 5D_2O$; and $CuSeO_4 \cdot 5H_2O$. J Raman Spectrosc 5(2):103–114

Bermanec V, Furić K, Rajić M, Kniewald G (2003) Thermal stability and vibrational spectra of the sheet borate tuzlaite, $NaCa[B_5O_8(OH)_2] \cdot 3H_2O$. Am Mineral 88(2–3):271–276

Bermanec V, Tomašić N, Žigovečki Ž, Linarić MR, Furić K (2010) Dehydration processes in borate minerals: pentahydroborite and nifontovite from Fuka Mine, Okayama Prefecture, Japan. Mineral Mag 74(6):1013–1025

Bernard M-C, Hugot-Le Goff A, Thi BV, Cordoba de Torresi S (1993a) Electrochromic reactions in manganese oxides. I. Raman analysis. J Electrochem Soc 140(11):3065–3070

Bernard MC, Hugot-Le Goff A, Massinon D, Phillips N (1993b) Underpaint corrosion of zinc-coated steel sheet studied by in situ Raman spectroscopy. Corros Sci 35(5):1339–1349

Bernardino NDE, Izumi CM, de Faria DL (2016) Fake turquoises investigated by Raman microscopy. Forensic Sci Intern 262:196–200

Bernert T, Ruiz-Fuertes J, Bayarjargal L, Winkler B (2015) Synthesis and high (pressure, temperature) stability of $ZnTiO_3$ polymorphs studied by Raman spectroscopy. Solid State Sci 43:53–58

Bernstein MP, Sandford SA (1999) Variations in the strength of the infrared forbidden 2328.2 cm^{-1} fundamental of solid N_2 in binary mixtures. Spectrochim Acta A 55:2455–2466

Berryman EJ, Wunder B, Ertl A, Koch-Müller M, Rhede D, Scheidl K, Giester G, Heinrich W (2016) Influence of the X-site composition on tourmaline's crystal structure: investigation of synthetic K-dravite, dravite, oxy-uvite, and magnesio-foitite using SREF and Raman spectroscopy. Phys Chem Minerals 43 (2):83–102

Bersani D, Lottici PP, Montenero A (2000) A micro-Raman study of iron-titanium oxides obtained by sol-gel synthesis. J Mater Sci 35(17):4301–4305

Bersani D, Andò S, Vignola P, Moltifiori G, Marino IG, Lottici PP, Diella V (2009) Micro-Raman spectroscopy as a routine tool for garnet analysis. Spectrochim Acta A 73(3):484–491

Bersani D, Andò S, Scrocco L, Gentile P, Salvioli-Mariani E, Lottici PP (2014) Study of the composition of amphiboles in the tremolite–ferro-actinolite series by micro-Raman and SEM-EDXS. LPI Contributions Abstr 11th Geo Raman Int Conf 1783:5063. (2 pp)

Berthereau A (1995) Les matériaux vitreux pour l'optique non linéaire: – étude des verres à base d'oxyde de tellure a fort effet Kerr optique – le phénomène de génération de seconde harmonique dans un verre. Thése, Université Sciences et Technologies – Bordeaux I. (in French)

Bertolotti G, Bersani D, Lottici PP, Alesiani M, Malcherek T, Schlüter J (2012) Micro-Raman study of copper hydroxychlorides and other corrosion products of bronze samples mimicking archaeological coins. Analyt Bioanalyt Chem 402(4):1451–1457

Bertoluzza A, Monti P, Battaglia MA, Bonora S (1980) Infrared and Raman spectra of orthorhombic, monoclinic and cubic metaboric acid and their relation to the "strength" of the hydrogen bond present. J Molec Struct 64:123–136

Bertoluzza A, Monti P, Morelli MA, Battaglia MA (1981) A Raman and infrared spectroscopic study of compounds characterized by strong hydrogen bonds. J Molec Struct 73(1):19–29

Besnardiere J, Petrissans X, Ribot F, Briois V, Surcin C, Morcrette M, Buissette V, Le Mercier T, Cassaignon S, Portehault D (2016) Nanoparticles of low-valence vanadium oxyhydroxides: reaction mechanisms and polymorphism control by low-temperature aqueous chemistry. Inorg Chem 55:11502–11512

Best SP, Clark RJ, Hayward CL, Withnall R (1994) Polarized single-crystal Raman spectroscopy of danburite, $CaB_2Si_2O_8$. J Raman Spectrosc 25(7–8):557–563

Betsch RJ, White WB (1978) Vibrational spectra of bismuth oxide and the sillenite-structure bismuth oxide derivatives. Spectrochim Acta A 34:505–514

Bette S, Dinnebier RE, Freyer D (2014) $Ni_3Cl_{2.1}(OH)_{3.9} \cdot 4H_2O$, the Ni analogue to $Mg_3Cl_2(OH)_4 \cdot 4H_2O$. Inorg Chem 53(9):4316–4324

Bette S, Dinnebier RE, Röder C, Freyer D (2015) A solid solution series of atacamite type $Ni_{2x}Mg_{2-2x}Cl(OH)_3$. J Solid State Chem 228:131–140

Bette S, Rincke C, Dinnebier RE, Voigt W (2016) Crystal structure and hydrate water content of synthetic hellyerite, $NiCO_3 \cdot 5.5H_2O$. Z anor gallg Chem 642 (9–10):652–659

Beukes GJ, Schoch AE, De Bruiyn H, Van der Westhuizen WA, Bok LDC (1984) A new occurrence of the hydrated aluminum sulphate zaherite, from Pofadder, South Africa. Mineral Mag 48:131–135

Beurlen H, Thomas R, Melgarejo JC, JMR DS, Rhede D, Soares DR, MRR DS (2013) Chrysoberyl-sillimanite association from the Roncadeira pegmatite, Borborema Province, Brazil: implications for gemstone exploration. J Geosci 58(2):79–90

Bezur A, Kavich G, Stenger J, Torok E, Snow C (2015) Discovery of challacolloite, an uncommon chloride, on a fifteenth-century polychrome terracotta relief by Michele da Firenze. Appl Phys A 121(1):83–93

Bhake AM, Nair GB, Zade GD, Dhoble SJ (2016) Synthesis and characterization of novel $Na_{15}(SO_4)_5ClF_4:Ce^{3+}$ halosulfate phosphors. Luminescence. https://doi.org/10.1002/bio.3131

Bhalerao GM, Hermet P, Haines J, Cambon O, Keen DA, Tucker MG, Buixaderas E, Simon P (2012) Dynamic disorder and the α-β phase transition in quartz-type $FePO_4$ at high temperature investigated by total neutron scattering, Raman spectroscopy, and density functional theory. Phys Rev B 86(13):134104-1–134104-12

Bharti C, Sinha TP (2010) Dielectric properties of rare earth double perovskite oxide Sr_2CeSbO_6. Solid State Sci 12(4):498–502

Bharti C, Sinha TP (2011) Synthesis, structure and dielectric properties of a rare earth double perovskite oxide Ba_2CeTaO_6. Mater Res Bull 46(9):1431–1436

Bhide V, Husson E, Gasperin M (1980) Etude de niobates de structure GTB par absorption infra-rouge et diffusion Raman. Mater Res Bull 15(9):1339–1344. (in French)

Biagioni C, Bindi L (2016) Ordered distribution of Cu and Ag in the crystal structure of balkanite, $Cu_9Ag_5HgS_8$. Eur J Mineral. https://doi.org/10.1127/ejm/2017/0029-2591

Biagioni C, Bonaccorsi E, Merlino S, Parodi GC, Perchiazzi N, Chevrier V, Bersani D (2010) Roumaite, $(Ca,Na,\square)_3(Ca,REE,Na)_4(Nb,Ti)[Si_2O_7]_2(OH)F_3$, from Rouma Island, Los Archipelago, Guinea: a new mineral species related to dovyrenite. Can Mineral 48 (1):17–28

Biagioni C, Bonaccorsi E, Pasero M, Moëlo Y, Ciriotti ME, Bersani D, Callegari AM, Boiocchi M (2011a) Ambrinoite, $(K,NH_4)_2(As,Sb)_8S_{13} \cdot H_2O$, a new mineral from Upper Susa Valley, Piedmont, Italy: the first natural (K,NH$_4$)-hydrated sulfosalt. Am Mineral 96 (5–6):878–887

Biagioni C, Bonaccorsi E, Orlandi P (2011b) Volaschioite, Fe$^{3+}_4$(SO$_4$)O$_2$(OH)$_6$·2H$_2$O, a new mineral species from Fornovolasco, Apuan Alps, Tuscany, Italy. Can Mineral 49(2):605–614

Biagioni C, Bonaccorsi E, Merlino S, Bersani D, Forte C (2012) Thermal behaviour of tobermorite from N'Chwaning II mine (Kalahari Manganese Field, Republic of South Africa). II. Crystallographic and spectroscopic study of tobermorite 10 Å. Eur J Mineral 24(6):991–1004

Biagioni C, Bonaccorsi E, Cámara F, Cadoni M, Ciriotti ME, Bersani D, Kolitsch U (2013a) Lusernaite-(Y), Y$_4$Al(CO$_3$)$_2$(OH,F)$_{11}$·6H$_2$O, a new mineral species from Luserna Valley, Piedmont, Italy: description and crystal structure. Am Mineral 98(7):1322–1329

Biagioni C, Bonaccorsi E, Merlino S, Bersani D (2013b) New data on the thermal behavior of 14 Å tobermorite. Cem Concr Res 49:48–54

Biagioni C, Orlandi P, Nestola F, Bianchin S (2013c) Oxycalcioroméite, Ca$_2$Sb$_2$O$_6$O, from Bucadella Vena mine, Apuan Alps, Tuscany, Italy: a new member of the pyrochloresupergroup. Mineral Mag 77 (7):3027–3038

Billhardt HW (1969) Synthesis of lead pyrosilicate and other barysilite-like compounds. Am Mineral 54:510–521

Bilohuščin V, Uher P, Koděra P, Milovská S, Mikuš T, Bačík P (2017) Evolution of borate minerals from contact metamorphic to hydrothermal stages: Ludwigite-group minerals and szaibélyite from the Vysoká–Zlatnoskarn, Slovakia. Mineral Petrol. https://doi.org/10.1007/s00710-017-0518-y

Bindi L, Bonazzi P, Dei L, Zoppi A (2005) Does the bazhenovite structure really contain a thiosulfate group? A structural and spectroscopic study of a sample from the type locality. Am Mineral 90 (10):1556–1562

Bindi L, Carbone C, Cabella R, Lucchetti G (2011a) Bassoite, SrV$_3$O$_7$·4H$_2$O, a new mineral from Molinello mine, Val Graveglia, eastern Liguria, Italy. Mineral Mag 75(5):2677–2686

Bindi L, Nestola F, Kolitsch U, Guastoni A, Zorzi F (2011b) Fassinaite, Pb$^{2+}_2$(S$_2$O$_3$)(CO$_3$), the first mineral with coexisting thiosulphate and carbonate groups: description and crystal structure. Mineral Mag 75 (6):2721–2732

Bindi L, Bonazzi P, Zoppi M, Spry PG (2014) Chemical variability in wakabayashilite: a real feature or an analytical artifact? Mineral Mag 78(3):693–702

Bindi L, Christy AG, Mills SJ, Ciriotti ME, Bittarello E (2015a) New compositional and structural data validate the status of jamborite. Can Mineral. https://doi.org/10.3749/canmin.1400050

Bindi L, Pratesi G, Muniz-Miranda M, Zoppi M, Chelazzi L, Lepore GO, Menchetti S (2015b) From ancient pigments to modern optoelectronic applications of arsenic sulfides: bonazziite, the natural analogue of β-As$_4$S$_4$ from Khaidarkan deposit, Kyrgyzstan. Mineral Mag 79(1):121–131

Bindi L, Chen M, Xie X (2017) Discovery of the Fe-analogue of akimotoite in the shocked Suizhou L6 chondrite. Sci Rep 7:42674. (8 pp). https://doi.org/10.1038/srep42674

Binon J, Bonaccorsi E, Bernhardt HJ, Fransolet A-M (2004) The mineralogical status of "cavolinite" from Vesuvius, Italy, and crystallochemical data on the davyne subgroup. Eur J Mineral 16(3):511–520

Birch WD, Kolitsch U, Witzke T, Nasdala L, Bottrill RS (2000) Petterdite, the Cr-dominant analogue of dundasite, a new mineral species from Dundas, Tasmania, Australia and Callenberg, Saxony, Germany. Can Mineral 38(6):1467–1476

Birsöz B, Baykal A (2008) X-ray powder diffraction, FTIR, and Raman study of strontium boroarsenate, SrBAsO$_5$. Rus J Inorg Chem 53(7):1009–1012

Bishop JL, Murad E (2004) Characterization of minerals and biogeochemical markers on Mars: a Raman and IR spectroscopic study of montmorillonite. J Raman Spectrosc 35(6):480–486

Bishop DW, Thomas PS, Ray AS (2000) Micro Raman characterization of nickel sulfide inclusions in toughened glass. Mater Res Bull 35(7):1123–1128

Bissengaliyeva M, Ogorodova L, Vigasina M, Mel'chakova L, Kosova D, Bryzgalov I, Ksenofontov D (2016) Enthalpy of formation of natural hydrous copper sulfate: chalcanthite. J Chem Thermodyn 95:142–148

Biswas S, Steudtner R, Schmidt M, McKenna C, Vintró LL, Twamley B, Baker RJ (2016) An investigation of the interactions of Eu^{3+} and Am^{3+} with uranyl minerals: implications for the storage of spent nuclear fuel. Dalton Trans 45(15):6383–6393

Bittarello E, Ciriotti ME, Costa E, Gallo LM (2014) "Mohsite" of Colomba: identification as dessauite-(Y). Int J Mineral. https://doi.org/10.1155/2014/287069

Bittarello E, Cámara F, Ciriotti ME, Marengo A (2015) Ottensite, brizziite and mopungite from Pereta mine (Tuscany, Italy): new occurrences and crystal structure refinement of mopungite. Mineral Petrol 109 (4):431–442

Biwer BM, Ebert WL, Bates JK (1990) The Raman spectra of several uranyl-containing minerals using a microprobe. J Nucl Mater 175(3):188–193

Black L, Brooker A (2007) SEM–SCA: combined SEM–Raman spectrometer for analysis of OPC clinker. Adv Appl Ceram 106(6):327–334

Blaha JJ, Rosasco GJ (1978) Raman microprobe spectra of individual microcrystals and fibers of talc, tremolite, and related silicate minerals. Anal Chem 50 (7):892–896

Blasse G (1973) Vibrational spectra of yttrium niobate and tantalate. J Solid State Chem 7(2):169–171

Blasse G, 'T Lam RUE (1978) Some optical properties of aluminum and gallium niobate. J Solid State Chem 25:11–83

Blasse G, Corsmit AF (1974) Vibrational spectra of 1:2 ordered perovskites. J Solid State Chem 10(1):39–45

Blasse G, van den Heuvel GPM (1973) Some optical properties of tantalum borate ($TaBO_4$), a compound with unusual coordinations. Phys Status Solidi A 19:111–117

Blasse G, Van Den Heuvel GPM (1974) Vibrational spectra and structural considerations of compounds $NaLnTiO_4$. J Solid State Chem 10(3):206–210

Blonska-Tabero A (2009) A new iron lead vanadate $Pb_2FeV_3O_{11}$: synthesis and some properties. Mater Res Bull 44(8):1621–1625

Boch R, Dietzel M, Reichl P, Leis A, Baldermann A, Mittermayr F, Pölt P (2015) Rapid ikaite ($CaCO_3 \cdot 6H_2O$) crystallization in a man-made river bed: hydrogeochemical monitoring of a rarely documented mineral formation. Appl Geochem 63:366–379

Bode JHG, Kuijt HR, Lahey MT, Blasse G (1973) Vibrational spectra of compounds Ln_2MoO_6 and Ln_2WO_6. J Solid State Chem 8(2):114–119

Boghosian S, Eriksen KM, Fehrmann R, Nielsen K (1995) Synthesis, crystal structure redetermination and vibrational spectra of beta-$VOSO_4$. Acta Chem Scand 49:703–708

Bolanz RM (2014) Arsenic and antimony in the environment: release and possible immobilization mechanisms. Dissertation, Friedrich-Schiller-Universität Jena

Bon CE, Rakovan J (2012) The morphology and structure of smythite, $(Fe,Ni)_3S_4$, from Bloomington, Indiana. Contributed Papers in Specimen Mineralogy: 38th Rochester Mineralogical Symposium: Part. Rocks Minerals 87(2):171–172

Bonadeo HA, Silberman E (1970) The vibrational spectra of sodium, potassium and ammonium fluoroborates. Spectrochim Acta A 25:2337–2343

Bonales LJ, Menor-Salván C, Cobos J (2015) Study of the alteration products of a natural uraninite by Raman spectroscopy. J Nucl Mater 462:296–303

Bonales LJ, Colmenero F, Cobos J, Timón V (2016) Spectroscopic Raman characterization of rutherfordine: a combined DFT and experimental study. Phys Chem Chem Phys 18:16575–16584

Bondi M, Griffin WL, Mattioli V, Mottana A (1983) Chiavennite, $CaMnBe_2Si_5O_{13}(OH)_2 \cdot 2H_2O$, a new mineral from Chiavenna (Italy). Am Mineral 68:623–627

Bontchev RP, Moore RC (2004) A series of open-framework tin (II) phosphates: $A[Sn_4(PO_4)_3]$ (A = Na, K, NH_4). Solid State Sci 6(8):867–873

Bordes E, Courtine P, Johnson JW (1984) On the topotactic dehydration of $VOHPO_4 \cdot 0.5H_2O$ into vanadyl pyrophosphate. J Solid State Chem 55(3):270–279

Borel MM, Leclaire A, Chardon J, Daturi M, Raveau B (2000) Dimorphism of the vanadium (V) monophosphate $PbVO_2PO_4$: α-layered and β-tunnel structures. J Solid State Chem 149(1):149–154

Borovikova EY, Kurazhkovskaya VS (2006) Influence of fluorine on the formation of ordered and disordered vesuvianite modifications: IR spectroscopic investigation. Zapiski RMO (Proc Russ Miner Soc) 135(2):89–95. (in Russian)

Borovikova EY, Kurazhkovskaya VS, Boldyrev KN, Sukhanov MV, Pet'kov VI, Kokarev SA (2014) Vibrational spectra and factor-group analysis of double arsenates of zirconium and alkali metal $MZr_2(AsO_4)_3$ (M = Li–Cs). Vibr Spectrosc 73:158–163

Bortun AI, Bortun LN, Poojary DM, Xiang O, Clearfield A (2000) Synthesis, characterization, and ion exchange behavior of a framework potassium titanium trisilicate $K_2TiSi_3O_9 \cdot H_2O$ and its protonated phases. Chem Mater 12(2):294–305

Boscardin M, Rocchetti I, Zordan A, Zorzi F (2009) Scarbroite e Felsöbányaite: primo ritrovamentonel-Vicentino. Studi e Ricerche – Associazione Amici del Museo – Museo Civico "G. Zannato", Montecchio Maggiore (Vicenza) 16:47–56. (in Italian)

Boschetti C, Corradi A, Baraldi P (2008) Raman characterization of painted mortar in Republican Roman mosaics. J Raman Spectrosc 39(8):1085–1090

Bossy A, Grosbois C, Beauchemin S, Courtin-Nomade A, Hendershot W, Bril H (2010) Alteration of As-bearing phases in a small watershed located on a high grade arsenic-geochemical anomaly (French Massif Central). Appl Geochem 25(12):1889–1901

Bost N, Duraipandian S, Guimbretière G, Poirier J (2016) Raman spectra of synthetic and natural mullite. Vib Spectrosc 82:50–52

Bottger GL, Damsgard CV (1971) Second order Raman spectra of AgCl and AgBr crystals. Solid State Commun 9(15):1277–1280

Botto IL, Baran EJ (1976) Über Ammonium-Uranyl-Vanadat und die Produkte seiner thermischen Zersetzung. Z anorg allg Chemie 426(3):321–332. (in German)

Botto IL, Baran EJ (1977) KristallographischeDaten, IR-Spektrum und thermisches Verhalten von Cer (IV)-Diphosphat. Z anorg allg Chemie 430(1):283–288. (in German)

Botto IL, Baran EJ (1980) Die IR-Spektren einiger Doppeloxide des Typs $M^{II}SnO_3$. Z anorg allg Chemie 465(1):186–192. (in German)

Botto IL, Baran EJ (1981) IR-Spektren einiger Doppeloxide des Typs $Te_3M^{IV}O_8$. Z anorg allg Chemie 480(9):220–224. (in German)

Botto IL, Baran EJ (1982) Darstellung und Eigenschaften von $CeTe_2O_6$ und $ThTe_2O_6$, Verbindungen mit einer neuen Überstruktur des Fluorit-Typs. Z anorg allg Chemie 484(1):215–220. (in German)

Botto IL, Garcia AC (1989) Crystallographic data and vibrational spectrum of K_2SbAsO_6. Mat Res Bull 24(12):1431–1439

Botto IL, Vassallo M (1989) The vibrational spectrum of the $NaZnPO_4$ ferroelectric phase. J Mater Sci Lett 8(11):1336–1337

Botto IL, Baran EJ, Deliens M (1989) Vibrational spectrum of natural and synthetic metatyuyamunite. N Jb Miner Mh 5:212–218

Botto IL, Cabello CI, Minelli G, Occhiuzzi M (1994) Reductibility and spectroscopic behaviour of the $(NH_4)_4[H_6CuMo_6O_{24}]\cdot 4H_2O$ Anderson phase. Mater Chem Phys 39(1):21–28

Botto IL, Ramis AM, Schalamuk IB, Sánchez MA (1995) Thermal decomposition of Bi_2STe_2 tetradymite. Thermochim Acta 249:325–333

Botto IL, Barone VL, Castiglioni JL, Schalamuk IB (1997) Characterization of a natural substituted pyromorphite. J Mater Sci 32(24):6549–6553

Botto IL, Barone VL, Sanchez MA (2002) Spectroscopic and thermal contribution to the structural characterization of vandenbrandeite. J Mater Sci 37(1):177–183

Bouchard M, Smith DC (2003) Catalogue of 45 reference Raman spectra of minerals concerning research in art history or archaeology, especially on corroded metals and coloured glass. Spectrochim Acta A 59 (10):2247–2266

Bouchard-Abouchacra M (2001) Evaluation des capacités de la microscopie Raman dans la caractérisation minéralogique et physico-chimique de matériaux archéologiques: métaux, vitraux & pigments. Dissertation, Museum National D'Histoire Naturelle. (in French)

Bouhifd MA, Gruener G, Mysen BO, Richet P (2002) Premelting and calcium mobility in gehlenite $(Ca_2Al_2SiO_7)$ and pseudowollastonite $(CaSiO_3)$. Phys Chem Minerals 29(10):655–662

Bourdoiseau JA, Jeannin M, Sabot R, Rémazeilles C, Refait P (2008) Characterisation of mackinawite by Raman spectroscopy: effects of crystallisation, drying and oxidation. Corrosion Sci 50(11):3247–3255

Bourdoiseau JA, Jeannin M, Rémazeilles C, Sabot R, Refait P (2011) The transformation of mackinawite into greigite studied by Raman spectroscopy. J Raman Spectrosc 42(3):496–504

Bourrié G, Trolard F (2010) Identification criteria for fougerite and nature of the interlayered anion. 19th World Congress of Soil Science, Soil Solutions for a Changing World, Brisbane, Australia, 1–6 August 2010, pp 74–78

Boyadzhieva T, Koleva V, Zhecheva E, Nihtianova D, Mihaylov L, Stoyanova R (2015) Competitive lithium and sodium intercalation into sodium manganese phospho-olivine $NaMnPO_4$ covered with carbon black. RSC Adv 5(106):87694–87705

Brabers VAM (1969) Infrared spectra of cubic and tetragonal manganese ferrites. Phys Status Solidi B 33 (2):563–572

Brabers VAM (1976) Infrared spectra and ionic ordering of the lithium ferrite – aluminate and chromite systems. Spectrochim. Acta A 32(11):1709–1711

Bracco R, Balestra C, Castellaro F, Mills SJ, Ma C, Callegari AM, Boiocchi M, Bersani D, Ciriotti ME (2012) Nuovi minerali di Terre Rare da Costa Balzi Rossi, Magliolo (SV), Liguria. Micro 10:66–77. (in Italian)

Bradbury SE, Williams Q (2003) Contrasting bonding behavior of two hydroxyl-bearing metamorphic minerals under pressure: clinozoisite and topaz. Am Mineral 88(10):1460–1470

Braithwaite RSW, Pritchard R (1983) Nakauriite from Unst, Shetland. Mineral Mag 47:84–85

Brambilla L, Zerbi G, Nascetti S, Piemontesi F, Morini G (2004) Experimenteal and calculated vibrational spectra and structure of Ziegler-Natta catalyst precursor: 50/1 comilled $MgCl_2$-$TiCl_4$. Macromol Symp 213:287–301

Brandel V, Dacheux N, Genet M, Podor R (2001) Hydrothermal synthesis and characterization of the thorium phosphate hydrogenphosphate, thorium hydroxide phosphate, and dithorium oxide phosphate. J Solid State Chem 159(1):139–148

Brandmüller J, Moser H (1962) Einführung in die Ramanspektroskopie. Steinkopff Verlag, Darmstadt. (in German)

Bréard Y, Michel C, Hervieu M, Nguyen N, Ducouret A, Hardy V, Maignan A, Raveau B, Bourée F, André G (2004) Spin reorientation associated with a structural transition in the iron oxycarbonate $Sr_4Fe_2O_6CO_3$. Chem Mater 16(15):2895–2905

Breitinger DK, Brehm G, Mohr J, Colognesi D, Parker SF, Stolle A, Pimpl TN, Schwab RG (2006) Vibrational spectra of synthetic crandallite-type minerals – optical and inelastic neutron scattering spectra. J Raman Spectrosc 37(1–3):208–216

Bremard C, Laureyns J, Abraham F (1986) Vibrational spectra and phase transitions in columnar $M^I_3[M'^{III}(SO_4)_3]$ compounds. J Raman Spectrosc 17 (5):397–405

Breternitz J, Farrugia LJ, Godula-Jopek A, Saremi-Yarahmadi S, Malka IE, Hoang TK, Gregory DH (2015) Reaction of $[Ni(H_2O)_6](NO_3)_2$ with gaseous NH_3; crystal growth via in-situ solvation. J Cryst Growth 412:1–6

Bretheau-Raynal F, Dalbiez JP, Drifford M, Blanzat B (1979) Raman spectroscopic study of thortveitite structure silicates. J Raman Spectrosc 8(1):39–42

Britvin SN, Antonov AA, Krivovichev SV, Armbruster T, Burns PC, Chukanov NV (2003) Fluorvesuvianite, $Ca_{19}(Al,Mg,Fe^{2+})_{13}[SiO_4]_{10}[Si_2O_7]_4O(F,OH)_9$, a new mineral species from Pitkäranta, Karelia, Russia: description and crystal structure. Can Miner 41:1371–1380

Britvin SN, Kashtanov SA, Krzhizhanovskaya MG, Gurinov AA, Glumov OV, Strekopytov S, Kretser YL, Zaitsev AN, Chukanov NV, Krivovichev SV (2015) Perovskites with the framework-forming xenon. Angew Chem Int Ed 54:14340–14344

Britvin SN, Kashtanov SA, Krivovichev SV, Chukanov NV (2016) Xenon in rigid oxide frameworks: structure, bonding and explosive properties of layered perovskite $K_4Xe_3O_{12}$. J Am Chem Soc. https://doi.org/10.1021/jacs.6b09056

Brockner W, Hoyer LP (2002) Synthesis and vibrational spectrum of antimony phosphate, $SbPO_4$. Spectrochim Acta A 58(9):1911–1914

Brooker MH, Bates JB (1971) Raman and infrared spectral studies of anhydrous Li_2CO_3 and Na_2CO_3. J Chem Phys 54:4788–4796

Brooker MH, Eysel HH (1990) Raman study of sulfate orientational dynamics in. alpha.-potassium alum and in the deuterated and oxygen-18 enriched forms. J Phys Chem 94(2):540–544

Brooker MH, Sunder S, Taylor P, Lopata VJ (1983) Infrared and Raman spectra and X-ray diffraction studies of solid lead (II) carbonates. Can J Chem 61:494–502

Brotton SJ, Kaiser RI (2013) In situ Raman spectroscopic study of gypsum ($CaSO_4·2H_2O$) and epsomite ($MgSO_4·7H_2O$) dehydration utilizing an ultrasonic levitator. J Phys Chem Lett 4(4):669–673

Brugger J, Berlepsch P (1997) Johninnesite, $Na_2(Mn^{2+})_9$ $(MgMn)_7(AsO_4)_2(Si_6O_{17})_2(OH)_8$: a new occurrence in Val Ferrera (Graubünden, Switzerland). Schweiz Mineral Petrogr Mitt 77:449–455

Brugger J, Bonin M, Schenk KJ, Meisser N, Berlepsch P, Ragu A (1999) Description and crystal structure of nabiasite, $BaMn_9[(V,As)O_4]_6(OH)_2$, a new mineral from the Central Pyrénées (France). Eur J Mineral 11 (5):879–890

Brugger J, Krivovichev SV, Kolitsch U, Meisser N, Andrut M, Ansermet S, Burns PC (2002) Description and crystal structure of manganlotharmeyerite, $Ca(Mn^{3+}, \square, Mg)_2\{AsO_4,[AsO_2(OH)_2]\}_2(OH,H_2O)_2$, from the Starlera Mn deposit, Swiss Alps, and a redefinition of lotharmeyerite. Can Mineral 40(6):1597–1608

Brugger J, Meisser N, Etschmann B, Ansermet S, Pring A (2011) Paulscherrerite from the Number 2 Workings, Mount Painter Inlier, Northern Flinders Ranges, South Australia: "dehydrated schoepite" is a mineral after all. Am Mineral 96(2–3):229–240

Buhl JC (1991) Synthesis and properties of nitrite-nitrate sodalite solid solutions $Na_8[AlSiO_4]_6(NO_2)_{2-x}(NO_3)_x$; $0.4 \leq x \leq 1.8$. J Solid State Chem 91(1):16–24

Bühler K, Bues W (1961) Schwingungsspektren von Fluorophosphatschmelzen und -kristallen. Z anor gallg Chemie 308(1–6):62–71. (in German)

Bühn B, Rankin AH, Radtke M, Haller M, Knöchel A (1999) Burbankite a (Sr,REE,Na,Ca)-carbonate in fluid inclusions from carbonatite-derived fluids: identification and characterization using laser Raman spectroscopy, SEM-EDX, and synchrotron micro-XRF analysis. Am Mineral 84(7–8):1117–1125

Bühn B, Rankin AH, Schneider J, Dulski P (2002) The nature of orthomagmatic, carbonatitic fluids precipitating REE, Sr-rich fluorite: fluid-inclusion evidence from the Okorusu fluorite deposit, Namibia. Chem Geol 186(1):75–98

Bujakiewicz-Korońska R, Hetmańczyk Ł, Garbarz-Glos B, Budziak A, Koroński J, Hetmańczyk J, Antonova M, Kalvane A, Nałęcz D (2011) Investigations of low temperature phase transitions in $BiFeO_3$ ceramic by infrared spectroscopy. Ferroelectrics 417(1):63–69

Bulanov EN, Wang J, Knyazev AV, White T, Manyakina ME, Baikie T, Lapshin AN, Dong Z (2015) Structure and thermal expansion of calcium–thorium apatite, $[Ca_4]^F[Ca_2Th_4]^T[(SiO_4)_6]O_2$. Inorg Chem 54(23):11356–11361

Burba CM (2006) Vibrational spectroscopy of phosphate-based electrodes for lithium rechargeable batteries. Dissertation, University of Oklahoma

Burba CM, Frech R (2004) Raman and FTIR spectroscopic study of Li_xFePO_4 ($0 \leq x \leq 1$). J Electrochem Soc 151(7):A1032–A1038

Burba CM, Frech R (2006) Vibrational spectroscopic investigation of structurally-related $LiFePO_4$, $NaFePO_4$, and $FePO_4$ compounds. Spectrochim Acta A 65(1):44–50

Bürger H, Kneipp K, Hobert H, Vogel W, Kozhukharov V, Neov S (1992) Glass formation, properties and structure of glasses in the TeO_2–ZnO system. J Non-Cryst Solids 151:134–142

Burgio L, Clark RJ (2001) Library of FT-Raman spectra of pigments, minerals, pigment media and varnishes, and supplement to existing library of Raman spectra of pigments with visible excitation. Spectrochim Acta A 57(7):1491–1521

Burgio L, Clark RJ, Firth S (2001) Raman spectroscopy as a means for the identification of plattnerite (PbO_2), of lead pigments and of their degradation products. Analyst 126(2):222–227

Burlet C, Vanbrabant Y (2015) Study of the spectrochemical signatures of cobalt–manganese layered oxides (asbolane–lithiophorite and their intermediates) by Raman spectroscopy. J Raman Spectrosc 46(10):941–952

Burlet C, Vanbrabant Y, Goethals H, Thys T, Dupin L (2011) Raman spectroscopy as a tool to characterize heterogenite (CoO·OH) (Katanga Province, Democratic Republic of Congo). Spectrochim Acta A 80(1):138–147

Burlet C, Vanbrabant Y, Decree S (2014) Raman microspectroscopy as a tool to characterise cobalt-manganese layered oxides (heterogenite–asbolane–lithiophorite), study on crystalline and amorphous phases from the DRC (Democratic Republic of the Congo). 11th Geo Raman International Conference, St. Louis, Missouri, 15–19 June 2014, 1783:5080

Burns PC, Alexopoulos CM, Hotchkiss PJ, Locock AJ (2004) An unprecedented uranyl phosphate framework in the structure of $[(UO_2)_3(PO_4)O(OH)(H_2O)_2](H_2O)$. Inorg Chem 43(6):1816–1818

Burshtein Z, Shimony Y, Morganau S, Henderson DO, Mu R, Silberman E (1993) Symmetry lowering due to site-occupation disorder in vibrational spectra of gehlenite, $Ca_2(AlSi)AlO_7$. J Phys Chem Solids 54(9):1043–1049

Buvaneswari G, Varadaraju UV (2000) Synthesis and characterization of new apatite-related phosphates. J Solid State Chem 149(1):133–136

Buzatu A, Buzgar N (2010) The Raman study of single-chain silicates. Analele Stiintifice de Universitatii Al Cuza din Iasi. Sect 2 Geol 56(1):107–125

Buzatu A, Damian G, Buzgar N (2012) Raman and Infrared studies of weathering products from BaiaSprie ore deposit (Romania). Rom J Mineral Depos 85(2):7–10

Buzatu A, Dill HG, Buzgar N, Damian G, Maftei AE, Apopei AI (2016) Efflorescent sulfates from Baia Sprie mining area (Romania) – acid mine drainage and climatological approach. Sci Total Environ 542:629–641

Buzgar N (2008) The Raman study of certain K-Na dioctahedral micas. Rom J Mineral Depos – Rom J Mineral 83:45–49

Buzgar N, Apopei AI (2009) The Raman study of certain carbonates. Analele Stiintifice de Universitatii AI Cuza din Iasi. Sect 2 Geol 55(2):97–112

Buzgar N, Buzatu A, Sanislav IV (2009) The Raman study on certain sulfates. Analele Stiintifice de Universitatii AI Cuza din Iasi. Sect 2 Geol 55(1):5–23

Byer HH, Bobb LC, Lefkowitz I, Deaver BS Jr (1973) Raman and far-infrared spectra of proustite (Ag_3AsS_3) and pyrargyrite (Ag_3SbS_3). Ferroelectrics 5(1):207–217

Bykova EA, Bobrov AV, Sirotkina EA, Bindi L, Ovsyannikov SV, Dubrovinsky LS, Litvin YA (2014) X-ray single-crystal and Raman study of knorringite, $Mg_3(Cr_{1.58}Mg_{0.21}Si_{0.21})Si_3O_{12}$, synthesized at 16 GPa and 1.600°C. Phys Chem Minerals 41(4):267–272

Cadoni M, Ferraris G (2008) Penkvilksite-2O: $Na_2TiSi_4O_{11} \cdot 2H_2O$. Acta Crystallogr C. https://doi.org/10.1107/S0108270108031806

Cadoni M, Ferraris G (2009) Two new members of the rhodesitemero-plesiotype series close to delhayelite and hydrodelhayelite: synthesis and crystal structure. Eur J Mineral 21(2):485–493

Cadoni M, Bloise A, Ferraris G, Merlino S (2008) Order–disorder character and twinning in the structure of a new synthetic titanosilicate: $(Ba,Sr)_4Ti_6Si_4O_{24} \cdot H_2O$. Acta Crystallogr B 64(6):669–675

Caggiani MC, Acquafredda P, Colomban P, Mangone A (2014) The source of blue colour of archaeological glass and glazes: the Raman spectroscopy /SEM-EDS answers. J Raman Spectrosc 45(11–12):1251–1259

Cahen HT, de Wit JHW, Honders A, Broers GHJ, van den Dungen JPM (1980) Thermogalvanic power and fast ion conduction in δ-Bi_2O_3 and δ-$(Bi_2O_3)_{1-x}(R_2O_3)_x$ with R = Y, Tb–Lu. Solid State Ionics 1(5):425–440

Caldow GL, Van Cleave AB, Eager RL (1960) The infrared spectra of some uranyl compounds. Can J Chem 38(6):772–782

Camacho-López MA, Escobar-Alarcón L, Picquart M, Arroyo R, Córdoba G, Haro-Poniatowski E (2011) Micro-Raman study of the m-MoO_2 to α-MoO_3 transformation induced by cw-laser irradiation. Opt Mater 33(3):480–484

Carey C, Boucher T, Mahadevan S, Bartholomew P, Dyar MD (2015) Machine learning tools for mineral recognition and classification from Raman spectroscopy. J Raman Spectrosc 46(10):894–903

Cámara F, Oberti R, Chopin C, Medenbach O (2006) The arrojadite enigma: I. A new formula and a new model for the arrojadite structure. Am Mineral 91(8–9):1249–1259

Cámara F, Nestola F, Bindi L, Guastoni A, Zorzi F, Peruzzo L, Pedron D (2012a) Tazzoliite: a new mineral with a pyrochlore-related structure from the Euganei Hills, Padova, Italy. Mineral Mag 76(4):827–838

Cámara F, Sokolova E, Hawthorne FC (2012b) Kazanskyite, $BaTiNbNa_3Ti(Si_2O_7)_2O_2(OH)_2(H_2O)_4$, a Group-III Ti-disilicate mineral from the Khibiny alkaline massif, Kola Peninsula, Russia: description and crystal structure. Mineral Mag 76(3):473–492

Cámara F, Sokolova E, Abdu Y, Hawthorne FC, Khomyakov AP (2013) Kolskyite, (Ca☐) $Na_2Ti_4(Si_2O_7)_2O_4(H_2O)_7$, a Group-IV Ti-disilicate mineral from the Khibiny alkaline massif, Kola Peninsula, Russia: description and crystal structure. Can Mineral 51(6):921–936

Cámara F, Ciriotti ME, Bittarello E, Nestola F, Massimi F, Radica F, Costa E, Benna P, Piccoli GC (2014a) Arsenic-bearing new mineral species from Valletta mine, Maira Valley, Piedmont, Italy: I. Grandaite, $Sr_2Al(AsO_4)_2(OH)$, description and crystal structure. Mineral Mag 78(3):757–774

Cámara F, Sokolova E, Abdu YA, Hawthorne FC (2014b) Nafertisite, $Na_3Fe^{2+}_{10}Ti_2(Si_6O_{17})_2O_2(OH)_6F(H_2O)_2$, from Mt. Kukisvumchorr, Khibiny alkaline massif, Kola peninsula, Russia: refinement of the crystal structure and revision of the chemical formula. Eur J Mineral 26:689–700

Cámara F, Bittarello E, Ciriotti ME, Nestola F, Radica F, Marchesini M (2015) As-bearing new mineral species from Valletta mine, Maira Valley, Piedmont, Italy: II. Braccoite, $NaMn^{2+}_5[Si_5AsO_{17}(OH)](OH)$, description and crystal structure. Mineral Mag 79(1):171–189

Cámara F, Bittarello E, Ciriotti ME, Nestola F, Radica F, Massimi F, Balestra C, Bracco R (2016a) As-bearing new mineral species from Valletta mine, Maira Valley, Piedmont, Italy: III. Canosioite, $Ba_2Fe^{3+}(AsO_4)_2(OH)$, description and crystal structure. Mineral Mag. https://doi.org/10.1180/minmag.2016.080.097

Cámara F, Sokolova E, Abdu YA, Hawthorne FC, Charrier T, Dorcet V, Carpentier J-F (2016b) Fogoite-(Y), $Na_3Ca_2Y_2Ti(Si_2O_7)_2OF_3$, a Group-I TS-block mineral from the Lagoa do Fogo, the Fogo volcano, the São Miguel Island, the Azores: description and crystal structure. Mineral Mag. https://doi.org/10.1180/minmag.2016.080.103

Cámara F, Sokolova E, Abdu YA, Pautov LA (2016c) From structure topology to chemical composition. XIX. Titanium silicates: revision of the crystal structure and chemical formula of bafertisite, $Ba_2Fe^{2+}_4Ti_2(Si_2O_7)_2O_2(OH)_2F_2$, a group-II TS-block mineral. Can Mineral 54:49–63

Campos CEM, de Lima JC, Grandi TA, Machado KD, Pizani PS (2002) Structural studies of cobalt selenides prepared by mechanical alloying. Phys B 324(1):409–418

Campos CEM, De Lima JC, Grandi TA, Machado KD, Drago V, Pizani PS (2004a) Hexagonal CoSe formation in mechanical alloyed $Co_{75}Se_{25}$ mixture. Solid State Commun 131(3):265–270

Campos CEM, De Lima JC, Grandi TA, Machado KD, Drago V, Pizani PS (2004b) XRD, DSC, MS and RS studies of $Fe_{75}Se_{25}$ iron selenide prepared by mechanosynthesis. J Magnet Magnet Mater 270(1):89–98

Campostrini I, Gramaccioli CM, Demartin F (1999) Orlandiite, $Pb_3Cl_4(SeO_3)\cdot H_2O$, a new mineral species, and an associated lead-copper selenite chloride from the Baccu Locci mine, Sardinia, Italy. Can Mineral 37:1493–1498

Can N, Garcia Guinea JJ, Kibar R, Cetin A (2011) Luminescence behavior and Raman characterization of rhodonite from Turkey. Spectrosc Lett 44(7–8):566–569

Cancela LSG, Ramos JG, Gualberto GM (1983) Raman scattering with multiphonon process in single crystal of α-$Ni(SO_4)\cdot 6H_2O$. J Phys Soc Jpn 52(1):295–303

Capitani GC, Catelani T, Gentile P, Lucotti A, Zema M (2013) Cannonite [$Bi_2O(SO_4)(OH)_2$] from Alfenza (Crodo, Italy): crystal structure and morphology. Mineral Mag 77(8):3067–3080

Capitani GC, Mugnaioli E, Rius J, Gentile P, Catelani T, Lucotti A, Kolb U (2014) The Bi sulfates from the Alfenza Mine, Crodo, Italy: an automatic electron diffraction tomography (ADT) study. Am Mineral 99(2–3):500–510

Capitelli F, Della Ventura G, Bellatreccia F, Sodo A, Saviano M, Ghiara MR, Rossi M (2014) Crystal-chemical study of wavellite from Zbirov, Czech Republic. Mineral Mag 78(4):1057–1070

Capobianco JA, Cormier G, Bettinelli M, Moncorgé R, Manaa H (1992) Near-infrared intraconfigurational luminescence spectroscopy of the Mn^{5+} ($3d^2$) ion in Ca_2PO_4Cl, $Sr_5(PO_4)_3Cl$, Ca_2VO_4Cl and Sr_2VO_4Cl. J Lumin 54(1):1–11

Carbone C, Basso R, Cabella R, Martinelli A, Grice JD, Lucchetti G (2013) Mcalpineite from the Gambatesa mine, Italy, and redefinition of the species. Am Mineral 98:1899–1905

Cariati F, Masserano F, Martini M, Spinolo G (1989) Raman studies of $NiX_2\cdot 6H_2O$ and $FeCl_2\cdot 4H_2O$. J Raman Spectrosc 20:773–777

Carter EA, Hargreaves MD, Kee TP, Pasek MA, Edwards HG (2010) A Raman spectroscopic study of a fulgurite. Phil Trans R SocA 368(1922):3087–3097

Casciola M, Donnadio A, Montanari F, Piaggio P, Valentini V (2007) Vibrational spectra and H-bondings in anhydrous and monohydrate α-Zr phosphates. J Solid State Chem 180(4):1198–1208

Caspers HH, Buchanan RA, Marlin HR (1964) Lattice vibrations of LaF_3. J Chem Phys 41(1):94–99

Castriota M, Cosco V, Barone T, De Santo G, Carafa P, Cazzanelli E (2008) Micro-Raman characterizations of Pompei' smortars. J Raman Spectrosc 39(2):295–301

Castro K, Sarmiento A, Martínez-Arkarazo I, Madariaga JM, Fernández LA (2008) Green copper pigments biodegradation in cultural heritage: from malachite to moolooite, thermodynamic modeling, X-ray fluorescence, and Raman evidence. Anal Chem 80(11):4103–4110

Cavalcante LS, Moraes E, Almeida MAP, Dalmaschio CJ, Batista NC, Varela JA, Longo E, Li MS, Beltrán A (2013) A combined theoretical and experimental study of electronic structure and optical properties of β-$ZnMoO_4$ microcrystals. Polyhedron 54:13–25

Čejka J Jr, Muck A, Čejka J (1984) To the infrared spectroscopy of natural uranyl phosphates. Phys Chem Minerals 11:172–177

Čejka J, Frost RL, Sejkora J, Keeffe EC (2009a) Raman spectroscopic study of the uranyl sulphate mineral jáchymovite $(UO_2)_8(SO_4)(OH)_{14}\cdot 13H_2O$. J Raman Spectrosc 40(11):1464–1468

Čejka J, Sejkora J, Frost RL, Keeffe EC (2009b) Raman spectroscopic study of the uranyl mineral metauranospinite $Ca[(UO_2)(AsO_4)]_2\cdot 8H_2O$. J Raman Spectrosc 40(12):1786–1179

Čejka J, Sejkora J, Frost RL, Keeffe EC (2009c) Raman spectroscopic study of the uranyl mineral natrouranospinite $(Na_2,Ca)[(UO_2)(AsO_4)]_2\cdot 5H_2O$. J Raman Spectrosc 40(11):1521–1526

Čejka J, Bahfenne S, Frost RL, Sejkora J (2010a) Raman spectroscopic study of the arsenite mineral vajdakite [$(Mo^{6+}O_2)_2(H_2O)_2As^{3+}_2O_5$]·$H_2O$. J Raman Spectrosc 41(1):74–77

Čejka J, Sejkora J, Plášil J, Bahfenne S, Palmer SJ, Frost RL (2010b) Raman spectroscopic study of the uranyl carbonate mineral čejkaite and its comparison with synthetic trigonal $Na_4[UO_2(CO_3)_3]$. J Raman Spectrosc 41(4):459–464

Čejka J, Sejkora J, Bahfenne S, Palmer SJ, Plášil J, Frost RL (2011a) Raman spectroscopy of hydrogen-arsenate group (AsO_3OH) in solid-state compounds: cobalt mineral phase burgessite $Co_2(H_2O)_4[AsO_3OH]_2\cdot H_2O$. J Raman Spectrosc 42(2):214–218

Čejka J, Sejkora J, Plášil J, Bahfenne S, Palmer SJ, Frost RL (2011b) A vibrational spectroscopic study of hydrated Fe^{3+} hydroxyl-sulfates; polymorphic minerals butlerite and parabutlerite. Spectrochim Acta A 79(5):1356–1363

Čejka J, Sejkora J, Plášil J, Keeffe EC, Bahfenne S, Palmer SJ, Frost RL (2011c) A Raman and infrared spectroscopic study of Ca^{2+} dominant members of the mixite group from the Czech Republic. J Raman Spectrosc 42(5):1154–1159

Čejka J, Sejkora J, Jebavá I, Xi Y, Couperthwaite SJ, Frost RL (2013) A Raman spectroscopic study of the basic carbonate mineral callaghanite $Cu_2Mg_2(CO_3)(OH)_6\cdot 2H_2O$. Spectrochim Acta A 108:171–176

Čejka J, Sejkora J, Macek I, Frost RL, López A, Scholz R, Xi Y (2014a) A vibrational spectroscopic study of a hydrated hydroxy-phosphate mineral fluellite, $Al_2(PO_4)F_2(OH)\cdot 7H_2O$. Spectrochim Acta A 126:157–163

Čejka J, Sejkora J, Scholz R, López A, Xi Y, Frost RL (2014b) Raman and infrared spectroscopic studies of phurcalite from Red Canyon, Utah, USA – implications for the molecular structure. J Molec Struct 1068:14–19

Čejka J, Sejkora J, Macek I, Malíková R, Wang L, Scholz R, Xi Y, Frost RL (2015) Raman and infrared spectroscopic study of turquoise minerals. Spectrochim Acta A 149:173–182

Cernea M (2005) Sol-gel synthesis and characterization of BaTiO$_3$ powder. J Optoelectron Adv Mater 7(6):3015–3022

Cha J-H, Jung D-Y (2014) CuGaS$_2$ hollow spheres from Ga-CuS core-shell nanoparticles. Ultrasonics Sonochem 21(3):1194–1199

Chaalia S, Ayed B, Haddad A (2012) K$_2$Mn$_3$(AsO$_4$)$_3$: synthesis, crystalline structure and ionic conductivity. J Chem Crystallogr 42(9):941–946

Chadwick KM, Breeding CM (2008) Color variations and properties of johachidolite from Myanmar. Gems Gemol 44(3):246–251

Chahboun H, Groult D, Raveau B (1988) TaVO$_5$, a novel derivative of the series of monophosphate tungsten bronzes (PO$_2$)$_4$(WO$_3$)$_{2m}$. Mater Res Bulletin 23(6):805–812

Chaix-Pluchery O, Lucazeau G (1998) Vibrational study of transition metal disilicides, MSi$_2$ (M = Nb, Ta, V, Cr). J Raman Spectrosc 29(2):159–164

Chakhmouradian AR, Hughes JM, Rakovan J (2005) Fluorcaphite, a second occurrence and detailed structural analysis: simultaneous accommodation of Ca, Sr, Na, and LREE in the apatite atomic arrangement. Can Mineral 43(2):735–746

Chakhmouradian AR, Cooper MA, Medici L, Hawthorne FC, Adar F (2008) Fluorine-rich hibschite from silicocarbonatite, Afrikanda complex, Russia: crystal chemistry and conditions of crystallization. Can Mineral 46(4):1033–1042

Chakhmouradian AR, Cooper MA, Ball N, Reguir EP, Medici L, Abdu YA, Antonov AA (2014) Vladykinite, Na$_3$Sr$_4$(Fe^{2+}Fe^{3+})Si$_8$O$_{24}$: a new complex sheet silicate from peralkaline rocks of the Murun complex, eastern Siberia, Russia. Am Mineral 99(1):235–241

Chakhmouradian AR, Cooper MA, Medici L, Abdu YA, Shelukhina YS (2015) Anzaite-(Ce), a new rare-earth mineral and structure type from the Afrikanda silicocarbonatite, Kola Peninsula, Russia. Mineral Mag 79:1231–1244

Chakhmouradian AR, Cooper MA, Reguir EP, Moore MA (2017) Carbocernaite from the Bear Lodge carbonatite, Wyoming: revised structure, zoning and rare-earth fractionation on a microscale. Am Mineral. https://doi.org/10.2138/am-2017-6046

Chakir M, Jazouli AE, De Waal D (2003) Structural and vibrational studies of NaZr$_2$(AsO$_4$)$_3$. Mater Res Bull 38(13):1773–1779

Chakrabarty A, Pruseth KL, Sen AK (2011) First report of eudialyte occurrence from the Sushina hill region, Purulia district, West Bengal. J Geol Soc India 77(1):12–16

Chandra U, Sharma P, Parthasarathy G (2011a) High-pressure electrical resistivity, Mössbauer, thermal analysis, and micro-Raman spectroscopic investigations on microwave synthesized orthorhombic cubanite (CuFe$_2$S$_3$). Chem Geol 284(3):211–216

Chandra U, Singh N, Sharma P, Parthasarathy G, Garg AB, Mittal R, Mukhopadhyay R (2011b) High-pressure studies on synthetic orthorhombic cubanite (CuFe$_2$S$_3$). AIP Conf Proc 1349(1):143–144

Chandrappa GT, Chithaiah P, Ashoka S, Livage J (2011) Morphological evolution of (NH$_4$)$_{0.5}$V$_2$O$_5$·m H$_2$O fibers into belts, triangles, and rings. InorgChem 50(16):7421–7428

Chandrasekhar HR, Humphreys RG, Zwick U, Cardona M (1977) Infrared and Raman spectra of the IV-VI compounds SnS and SnSe. Phys Rev B 15(4):2177–2183

Chang HY, Kim SH, Ok KM, Halasyamani PS (2009) Polar or nonpolar? A$^+$cation polarity control in A$_2$Ti(IO$_3$)$_6$ (A = Li, Na, K, Rb, Cs, Tl). J Am Chem Soc 131(19):6865–6873

Charalambous FA, Ram R, Pownceby MI, Tardio J, Bhargava SK (2012) Chemical and microstructural characterisation studies on natural and heat treated brannerite samples. Miner Eng 39:276–288

Chariton S, Cerantola V, Ismailova L, Bykova E, Bykov M, Kupenko I, McCammon C, Dubrovinsky L (2017) The high-pressure behavior of spherocobaltite (CoCO$_3$): a single crystal Raman spectroscopy and XRD study. Phys Chem Minerals. https://doi.org/10.1007/s00269-017-0902-5

Chater R, Gavarri JR, Genet F (1986) Composes isomorphes MeX$_2$O$_4$E$_2$: I. Etude vibrationnelle de MnSb$_2$O$_4$ entre 4 et 300 K: champ de force et tenseurélastique. J Solid State Chem 63(2):295–307. (in French)

Chattopadhyay T, Carlone C, Jayaraman A, Schnering H (1982) Effect of temperature and pressure on the Raman spectrum of As$_4$S$_3$. J Phys Chem Solids 43(3):277–284

Chemtob SM, Arvidson RE, Fernández-Remolar DC, Amils R, Morris RV, Ming DW, Prieto-Ballesteros O, Mustard JF, Hutchison L, Stein TC, Donovan CE, Fairchild GM, Friedlander LR, Karas NM, Klasen MN, Mendenhall MP, Robinson EM, Steinhardt SE, Weber LR (2006) Identification of hydrated sulfates collected in the northern Rio Tinto valley by reflectance and Raman spectroscopy. Lunar Planet Sci 37:1941. (2 pp)

Chen X, Grandbois M (2013) In situ Raman spectroscopic observation of sequential hydrolysis of stannous chloride to abhurite, hydroromarchite, and romarchite. J Raman Spectrosc 44(3):501–506

Chen D, Jiao X (2001) Hydrothermal synthesis and characterization of Bi$_4$Ti$_3$O$_{12}$ powders from different precursors. Mater Res Bull 36(1):355–363

Chen M, Xie X (2015) Shock-produced akimotoite in the Suizhou L6 chondrite. Sci China Earth Sci 58(6):876–880

Chen X, Pei Y (2016) Effects of sodium pentaboratepentahydrate exposure on Chlorella vulgaris growth, chlorophyll content, and enzyme activities. Ecotoxicol Environ Safety 132:353–359

Chen M, Wopenka B, Xie X, El Goresy A (1995) A new high-pressure polymorph of chlorapatite in the shocked

Sixiangkou (L6) chondrite. XXVI Lunar Planetary Sci Conf 26:237–238

Chen G, Wu Y, Fu P (2006a) Growth and characterization of a new nonlinear optical crystal $Ca_5(BO_3)_3F$. J Crystal Growth 292(2):449–453

Chen X, Zhao Y, Chang X, Zuo J, Zang H, Xiao W (2006b) Syntheses and crystal structures of two new hydrated borates, $Zn_8[(BO_3)_3O_2(OH)_3]$ and $Pb[B_5O_8(OH)]·1.5H_2O$. J Solid State Chem 179(12):3911–3918

Chen X, Li M, Chang X, Zang H, Xiao W (2007a) Synthesis and crystal structure of a novel pentaborate, $Na_3ZnB_5O_{10}$. J Solid State Chem 180(5):1658–1663

Chen X, Li M, Zuo J, Chang X, Zang H, Xiao W (2007b) Syntheses and crystal structures of two pentaborates, $Na_3CaB_5O_{10}$ and $Na_3MgB_5O_{10}$. Solid State Sci 9(8):678–685

Chen Y, Liu Z, Ringer SP, Tong Z, Cui X, Chen Y (2007c) Selective oxidation synthesis of $MnCr_2O_4$ spinel nanowires from commercial stainless steel foil. Cryst Growth Design 7(11):2279–2281

Chen M, Shu J, Mao H-K (2008a) Xieite, a new mineral of high-pressure $FeCr_2O_4$ polymorph. Chin Sci Bull 53(21):3341–3345

Chen X, Li M, Chang X, Zang H, Xiao W (2008b) Synthesis and crystal structure of a new calcium borate, CaB_6O_{10}. J Alloys Compd 464(1):332–336

Chen X, Song F, Chang X, Zang H, Xiao W (2009) Syntheses and characterization of two oxoborates, $(Pb_3O)_2(BO_3)_2MO_4$ (M = Cr,Mo). J Solid State Chem 182:3091–3097

Chen X, Yang C, Chang X, Zang H, Xiao W (2010) Synthesis, crystal structure, and optical properties of a novel pentaborate, $K_2NaZnB_5O_{10}$. J Alloys Compd 492(1):543–547

Chen X, Yang C, Chu Z, Chang X, Zang H, Xiao W (2011) Synthesis, spectrum properties, and crystal structure of a new pentaborate, $Na_{2.18}K_{0.82}SrB_5O_{10}$. J ChemCrystallogr 41(6):816–822

Chen M, Gu X, Xie X, Yin F (2013a) High-pressure polymorph of TiO_2-II from the Xiuyan crater of China. Chin Sci Bull 58(36):4655–4662

Chen XB, Hien NTM, Han K, Sur JC, Sung NH, Cho BK, Yang IS (2013b) Raman studies of spin-phonon coupling in hexagonal $BaFe_{12}O_{19}$. J Appl Phys 114(1):013912-1–013912-6

Chen X, Wu L, Chang X, Xiao W (2014a) Synthesis, crystal structure, and spectrum properties of a new quaternary borate $NaSr_7AlB_{18}O_{36}$ with the cyclic $B_{18}O_{36}^{18-}$ group, notation of 6×(3 [2Δ+ 1T]). J Chem Crystallogr 44(11–12):572–579

Chen Y, Zhang Y, Feng S (2014b) Hydrothermal synthesis and properties of pigments Chinese purple $BaCuSi_2O_6$ and dark blue $BaCu_2Si_2O_7$. Dyes Pigm 105:167–173

Chen P, Xu K, Li X, Guo Y, Zhou D, Zhao J, Wu X, Wu C, Xie Y (2014c) Ultrathin nanosheets of feroxyhyte: a new two-dimensional material with robust ferromagnetic behavior. Chem Sci 5(6):2251–2255

Chen F, Zhao J, Xu J, Wu Y (2015a) Synthesis, structure, and optical properties of $BiCu_2(TeO_3)(SO_4)(OH)_3$. Z anorg allg Chemie 641(3–4):568–572

Chen HY, Lin YC, Lee JS (2015b) Crednerite–$CuMnO_2$ thin films prepared using atmospheric pressure plasma annealing. Appl Surface Sci 338:113–119

Cheng Z, Liu M (2007) Characterization of sulfur poisoning of Ni–YSZ anodes for solid oxide fuel cells using in situ Raman microspectroscopy. Solid State Ionics 178(13):925–935

Cheng K-W, Tsai W-T, Wu Y-H (2016) Photo-enhanced salt-water splitting using orthorhombic Ag_8SnS_6 photoelectrodes in photoelectrochemical cells. J Power Sources 317:81–92

Chi TTK, Gouadec G, Colomban P, Wang G, Mazerolles L, Liem NQ (2011) Off-resonance Raman analysis of wurtzite CdS ground to the nanoscale: structural and size-related effects. J Raman Spectrosc 42(5):1007–1015

Chia X, Ambrosi A, Sofer Z, Luxa J, Sedmidubský D, Pumera M (2015) Anti-MoS_2 nanostructures: Tl_2S and its electrochemical and electronic properties. ACS Nano 10(1):112–123

Chiavari C, Martini C, Montalbani S, Franzoni E, Bignozzi MC, Passeri MC (2016) The bronze panel (paliotto) of San Moisè in Venice: materials and causes of deterioration. Mater Corros 67(2):141–151

Chio CH, Sharma SK, Muenow DW (2007) The hydrates and deuterates of ferrous sulfate ($FeSO_4$): a Raman spectroscopic study. J Raman Spectrosc 38(1):87–99

Chio CH, Sharma SK, Ming LC, Muenow DW (2010) Raman spectroscopic investigation on jarosite–yavapaiite stability. Spectrochim Acta A 75(1):162–171

Chis V, Sklyadneva IY, Kokh KA, Volodin VA, Tereshchenko OE, Chulkov EV (2012) Vibrations in binary and ternary topological insulators: first-principles calculations and Raman spectroscopy measurements. Phys Rev B 86(17):174304. (12 pp)

Chitra S, Kalyani P, Yebka B, Mohan T, Haro-Poniatowski E, Gangadharan R, Julien C (2000) Synthesis, characterization and electrochemical studies of $LiNiVO_4$ cathode material in rechargeable lithium batteries. Mater Chem Phys 65(1):32–37

Cho A, Ahn S, Yun JH, Gwak J, Ahn SK, Shin K, .Yoo J, Song H, Yoon K (2013) The growth of $Cu_{2-x}Se$ thin films using nanoparticles. Thin Solid Films 546:299–307

Choi B-K, Lockwood DJ (1989) Raman spectrum of Na_2SO_4 (phases I and II). Solid State Commun 72(6):863–866

Choi KY, Lemmens P, Pommer J, Ionescu A, Güntherodt G, Hiroya S, Sakurai H, Yoshimura K, Matsuo A, Kindo K (2004) Random magnetism in the frustrated triangular spin ladder $KCu_5V_3O_{13}$. Phys Rev B 70(17):174417-1–174417-6

Choi KY, Oosawa A, Tanaka H, Lemmens P (2005) Inelastic light scattering experiments on the coupled spin dimer system $Tl_{1-x}K_xCuCl_3$. Prog Theor Phys Suppl 159:195–199

Choisnet J, Deschanvres A, Tarte P (1975) Spectres vibrationnels des silicates et germinates renfermant des anneaux, M_3O_9 (M = Si, Ge) – I. Attribution des frequencies caractéristiques de l'anneau M_3O_9, dans les composés de type bénitoïte, wadéite et tétragermanate. Spectrochim Acta A 31(8):1023–1034. (in French)

Chon MP, Tan KB, Khaw CC, Zainal Z, Yap YT, Chen SK, Tan PY (2014) Investigation of the phase formation and dielectric properties of $Bi_7Ta_3O_{18}$. J Alloys Compounds 590:479–485

Chopelas A (1999) Estimates of mantle relevant Clapeyron slopes in the $MgSiO_3$ system from high-pressure spectroscopic data. Am Mineral 84:233–244

Chopin C, Ferraris G, Prencipe M, Brunet F, Medenbach O (2001) Raadeite, $Mg_7(PO_4)_2(OH)_8$: a new dense-packed phosphate from Modum (Norway). Eur J Mineral 13(2):319–327

Chouaib S, Rhaiem AB, Guidara K (2011) Dielectric relaxation and ionic conductivity studies of $Na_2ZnP_2O_7$. Bull Mater Sci 34(4):915–920

Christy AG, Lowe A, Otieno-Alego V, Stoll M, Webster RD (2004) Voltammetric and Raman microspectroscopic studies on artificial copper pits grown in simulated potable water. J Appl Electrochem 34(2):225–233

Christy AG, Kampf AR, Mills SJ, Housley RM, Thorne B (2014) Crystal structure and revised chemical formula for burckhardtite, $Pb_2(Fe^{3+}Te^{6+})[AlSi_3O_8]O_6$: a double-sheet silicate with intercalated phyllotellurate layers. Mineral Mag 78(7):1763–1773

Chu S, Müller P, Nocera DG, Lee YS (2011) Hydrothermal growth of single crystals of the quantum magnets: clinoatacamite, paratacamite, and herbertsmithite. Appl Phys Lett 98(9):092508-1–092508-3

Chukanov NV (2014) Infrared spectra of mineral species: extended library. Springer-Verlag GmbH, Dordrecht. (1716 pp)

Chukanov NV, Chervonnyi AD (2016) Infrared spectroscopy of minerals and related compounds. Springer, Cham. (1109 pp)

Chukanov NV, Zubkova NV, Buhl J-C, Pekov IV, Ksenofontov DA, Depmeier W, Pushcharovskii DY (2011) Crystal structure of nitrate cancrinite synthesized under low-temperature hydrothermal conditions. Doklady Earth Sci 438(1):669–672

Chukanov NV, Nedelko VV, Blinova LN, Korshunova LA, Olysych LV, Lykova IS, Pekov IV, Buhl J-C, Depmeier W (2012a) The role of additional anions in microporous aluminosilicates with cancrinite-type framework. Russ J Phys Chem B 6(5):15–23

Chukanov NV, Scholz R, Aksenov SM, Rastsvetaeva RK, Pekov IV, Belakovskiy DI, Krambrock K, Paniago RM, Righi A, Martins RF, Belotti F, Bermanec V (2012b) Metavivianite, $Fe^{2+}Fe^{3+}_2(PO_4)_2(OH)_2 \cdot 6H_2O$: new data and formula revision. Mineral Mag 76(3):725–741

Chukanov NV, Krivovichev SV, Pakhomova AS, Pekov IV, Schäfer C, Vigasina MF, Van KV (2014a) Laachite, $(Ca,Mn)_2Zr_2Nb_2TiFeO_{14}$, a new zirconolite-related mineral from the Eifel volcanic region, Germany. Eur J Mineral 26:103–111

Chukanov NV, Scholz R, Zubkova NV, Pekov IV, Belakovskiy DI, Van KV, Lagoeiro L, Graça LM, Krambrock K, de Oliveira LCA, Menezes Filho LAD, Chaves MLCS, Pushcharovsky DY (2014b) Correianevesite, $Fe^{2+}Mn^{2+}_2(PO_4)_2 \cdot 3H_2O$, a new reddingite-group mineral from the Cigana mine, Conselheiro Pena, Minas Gerais, Brazil. Am Mineral 99:811–816

Chukanov NV, Aksenov SM, Rastsvetaeva RK, Schäfer C, Pekov IV, Belakovskiy DI, Scholz R, de Oliveira LCA, Britvin SN (2017a) Eleonorite, $Fe^{3+}_6(PO_4)_4O(OH)_4 \cdot 6H_2O$: validation as a mineral species and new data. Mineral Mag 81:61–76

Chukanov NV, Jonsson E, Aksenov SM, Britvin SN, Rastsvetaeva RK, Belakovskiy DI, Van KV (2017b) Roymillerite, $Pb_{24}Mg_9(Si_9AlO_{28})(SiO_4)(BO_3)(CO_3)_{10}(OH)_{14}O_4$, a new mineral: mineralogical characterization and crystal chemistry. Phys Chem Minerals. https://doi.org/10.1007/s00269-017-0893-2

Chukanov NV, Panikorovskii TL, Chervonnyi AD (2018) On the relationships between crystal-chemical characteristics of vesuvianite-group minerals and their IR spectra. Zapiski RMO (Proc Russ Miner Soc) 147(1):112–128. (in Russian)

Ciesielczuk J, Janeczek J, Dulski M, Krzykawski T (2016) Pseudomalachite–cornwallite and kipushite–philipsburgite solid solutions: chemical composition and Raman spectroscopy. Eur J Mineral. https://doi.org/10.1127/ejm/2016/0028-2536

Ciobotă V, Salama W, Tarcea N, Rösch P, El Aref M, Gaupp R, Popp J (2012) Identification of minerals and organic materials in Middle Eocene iron stones from the Bahariya Depression in the Western Desert of Egypt by means of micro-Raman spectroscopy. J Raman Spectrosc 43(3):405–410

Ciomartan DA, Clark RJH, McDonald LJ, Oldyha M (1996) Studies on the thermal decomposition of basic lead (II) carbonate by Fourier-transform Raman spectroscopy, X-ray diffraction and thermal analysis. J Chem Soc Dalton Trans 18:3639–3645

Clark RJ, Wang Q, Correia A (2007) Can the Raman spectrum of anatase in artwork and archaeology be used for dating purposes? Identification by Raman microscopy of anatase in decorative coatings on Neolithic (Yangshao) pottery from Henan, China. J Archaeolog Sci 34(11):1787–1793

Clavier N, Szenknect S, Costin DT, Mesbah A, Poinssot C, Dacheux N (2014) From thorite to coffinite: a spectroscopic study of $Th_{1-x}U_xSiO_4$ solid solutions. Spectrochim Acta A 118:302–307

Clavier N, Crétaz F, Szenknect S, Mesbah A, Poinssot C, Descostes M, Dacheux N (2016) Vibrational spectroscopy of synthetic analogues of ankoleite, chernikovite and intermediate solid solution. Spectrochim Acta A 156:143–150

Clearfield A, Roberts BD, Subramanian MA (1984) Preparation of $(NH_4)Zr_2(PO_4)_3$ and $HZr_2(PO_4)_3$. Mater Res Bull 19:219–226

Clearfield A, Bortun AI, Bortun LN, Cahill RA (1997) Synthesis and characterization of a novel layered sodium titanium silicate $Na_2TiSi_2O_7 \cdot 2H_2O$. Solvent Extraction Ion Exchange 15(2):285–304

Coccato A, Bersani D, Coudray A, Sanyova J, Moens L, Vandenabeele P (2016) Raman spectroscopy of green minerals and reaction products with an application in Cultural Heritage research. J Raman Spectrosc. https://doi.org/10.1002/jrs.4956

Cody CA, Levitt RC, Viswanath RS, Miller PJ (1978) Vibrational spectra of alkali hydrogen selenites, selenous acid, and their deuterated analogs. J Solid State Chem 26(3):281–291

Cody CA, DiCarlo L, Darlington RK (1979) Vibrational and thermal study of antimony oxides. Inorg Chem 18 (6):1572–1576

Coleyshaw EE, Griffith WP, Bowell RJ (1994) Fourier-transform Raman spectroscopy of minerals. Spectrochim Acta A 50(11):1909–1918

Coleyshaw EE, Crump G, Griffith WP (2003) Vibrational spectra of the hydrated carbonate minerals ikaite, monohydrocalcite, lansfordite and nesquehonite. Spectrochim Acta A 59(10):2231–2239

Colmenero F, Bonales LJ, Cobos J, Timón V (2017) Study of the thermal stability of studtite by in situ Raman spectroscopy and DFT calculations. Spectrochim Acta A 174:245–253

Colomban P (1986) Orientational disorder, glass/crystal transition and superionic conductivity in nasicon. Solid State Ionics 21(2):97–115

Colomban P, Courret H, Romain F, Gouadec G, Michel D (2000) Sol-gel-prepared pure and lithium-doped hexacelsian polymorphs: an infrared, Raman, and thermal expansion study of the β-phase stabilization by frozen short-range disorder. J Am Ceram Soc 83 (12):2974–2982

Colombo F, Rius J, Vallcorba O, Pannunzio Miner EV (2014) The crystal structure of sarmientite, (AsO_4) (SO_4)(OH)·$5H_2O$, solved *ab initio* from laboratory powder diffraction data. Mineral Mag 78:347–360

Comodi P, Liu Y, Stoppa F, Wooley AR (1999) A multi-method analysis of Si-, S- and *REE*-rich apatite from a new find of kalsilite-bearing leucitite (Abruzzi, Italy). Mineral Mag 63(5):661–672

Conti C, Casati M, Colombo C, Realini M, Brambilla L, Zerbi G (2014) Phase transformation of calcium oxalate dihydrate–monohydrate: effects of relative humidity and new spectroscopic data. Spectrochim Acta A 128:413–419

Conti C, Casati M, Colombo C, Possenti E, Realini M, Gatta GD, Merlini M, Brambilla L, Zerbi G (2015) Synthesis of calcium oxalate trihydrate: new data by vibrational spectroscopy and synchrotron X-ray diffraction. Spectrochim Acta A 150:721–730

Cooney RPJ, Hall JR, Hooper MA (1968) Raman spectra of mercury(I) and mercury(II) iodides in the solid state. Aust J Chem 21(9):2145–2152

Cooney TF, Scott ER, Krot AN, Sharma SK, Yamaguchi A (1999) Vibrational spectroscopic study of minerals in the Martian meteorite ALH84001. Am Mineral 84 (10):1569–1576

Cooper MA, Abdu YA, Ball NA, Hawthorne FC, Back ME, Tait KT, Schlüter J, Malcherek T, Pohl D, Gebhard G (2012a) Ianbruceite, ideally [$Zn_2(OH)$ (H_2O)(AsO_4)](H_2O)$_2$, a new arsenate mineral from the Tsumeb mine, Otjikoto (Oshikoto) region, Namibia: description and crystal structure. Mineral Mag 76(5):1119–1131

Cooper MA, Abdu YA, Ball NA, Černý P, Hawthorne FC, Kristiansen R (2012b) Aspedamite, ideally $\square_{12}$(Fe^{3+}, Fe^{2+})$_3$Nb$_4$[Th(Nb,Fe^{3+})$_{12}$$O_{42}$]{($H_2O$),(OH)}$_{12}$, a new heteropolyniobate mineral species from the Herrebøkasa Quarry, Aspedammen, Østfold, Southern Norway: description and crystal structure. Can Mineral 50(4):793–894

Cooper MA, Abdu YA, Hawthorne FC, Kampf AR (2013a) The crystal structure of comancheite, Hg^{2+}_{55} N^{3-}_{24}(OH,NH_2)$_4$(Cl,Br)$_{34}$, and crystal-chemical and spectroscopic discrimination of N^{3-} and O^{2-} anions in Hg^{2+} compounds. Min Mag 77(8):3217–3237

Cooper MA, Hawthorne FC, Abdu YA, Ball NA, Ramik RA, Tait KT (2013b) Wopmayite, ideally $Ca_6Na_3\square Mn$ (PO_4)$_3$(PO_3OH)$_4$, a new phosphate mineral from the Tanco Mine, Bernic Lake, Manitoba: description and crystal structure. Can Mineral 51(1):93–106

Cooper MA, Abdu YA, Hawthorne FC, Kampf AR (2016a) The crystal structure of gianellaite, [(NHg_2)$_2$] (SO_4)(H_2O)$_x$, a framework of (NHg_4) tetrahedra with ordered (SO_4) groups in the interstices. Mineral Mag. https://doi.org/10.1180/minmag.2016.080.028

Cooper MA, Hawthorne FC, Garcia-Veígas J, Alcobé X, Helvaci C, Grew ES, Ball NA (2016b) Fontarnauite, (Na,K)$_2$(Sr,Ca)(SO_4)[B_5O_8(OH)](H_2O)$_2$, a new sulfate-borate mineral from Doğanlar (Emet), Kütahya Province, Western Anatolia, Turkey. Can Mineral 53 (3):1–20. https://doi.org/10.3749/canmin.1400088

Cooper M, Hawthorne F, Langhof J, Hålenius U, Holtstam D (2016c) Wiklundite, ideally $Pb_2^{[4]}$(Mn^{2+}, Zn)$_3$(Fe^{3+},Mn^{2+})$_2$(Mn^{2+},Mg)$_{19}$($As^{3+}O_3$)$_2$[(Si,As^{5+}) O_4]$_6$(OH)$_{18}Cl_6$, a new mineral from Långban, Filipstad, Värmland, Sweden: description and crystal structure. Mineral Mag. https://doi.org/10.1180/minmag.2016.080.136

Čopjaková R, Škoda R, Galiová MV, Novák M, Cempírek J (2015) Sc-and *REE*-rich tourmaline replaced by Sc-rich *REE*-bearing epidote-group mineral from the mixed (NYF+ LCT) Kracovice pegmatite (Moldanubian Zone, Czech Republic). Am Mineral 100(7):1434–1451

Cornette J, Merle-Méjean T, Mirgorodsky A, Colas M, Smirnov M, Masson O, Thomas P (2011) Vibrational spectra of rhombohedral TeO_3 compared to those of ReO_3-like proto-phase and α-TeO_2 (paratellurite): lattice dynamic and crystal chemistry aspects. J Raman Spectrosc 42(4):758–764

Correia AM, Clark RJ, Ribeiro MI, Duarte ML (2007) Pigment study by Raman microscopy of 23 paintings by the Portuguese artist Henrique Pousão (1859–1884). J Raman Spectrosc 38(11):1390–1405

Corsmit AF, Blasse G (1974) The infrared spectrum of Ba$_2$NiTeO$_6$. J Inorg Nucl Chem 36(5):1155–1156

Costin DT, Mesbah A, Clavier N, Szenknect S, Dacheux N, Poinssot C, Ravaux J, Brau HP (2012) Preparation and characterization of synthetic Th$_{0.5}$U$_{0.5}$SiO$_4$ uranothorite. Prog Nucl Energy 57:155–160

Costin G, Fairey B, Tsikos H (2014) Tokyoite and As-rich tokyoite: new occurrence in the manganese ore of the Postmasburg Manganese Field, South Africa. Conference: IMA 2014, Johannesburg. https://doi.org/10.13140/2.1.2792.3202

Costin G, Fairey B, Tsikos H, Gucsik A (2015) Tokyoite, As-rich tokyoite, and noélbensonite: new occurrences from the Postmasburg manganese field, Northern Cape Province, South Africa. Can Mineral 53:981–990

Coutinho ML, Veiga JP, Alves LC, Mirão J, Dias L, Lima AM, Muralha VS, Macedo MF (2016) Characterization of the glaze and in-glaze pigments of the nineteenth-century relief tiles from the Pena National Palace, Sintra, Portugal. Appl Phys A 122(7):1–10

Craveiro R, Lin Z (2012) The influence of Fe on the formation of titanosilicate ETS-4. J Solid State Chem 190:162–168

Creton B, Bougeard D, Smirnov KS, Guilment J, Poncelet O (2008) Molecular dynamics study of hydrated imogolite. 1. Vibrational dynamics of the nanotube. J Phys Chem C 112(27):10013–10020

Crimmins LG (2012) Structure and chemistry of minerals in the Ca-(As,P)-(OH,F,Cl) apatite system: johnbaumite, svabite, and turneaureite from Franklin and Sterling Hill, New Jersey, USA. Dissertation, University of Miami

Croce A, Musa M, Allegrina M, Rinaudo C, Baris YI, Dogan AU, Powers A, Rivera Z, Bertino P, Yang H, Gaudino G, Carbone M (2013) Micro-Raman spectroscopy identifies crocidolite and erionite fibers in tissue sections. J Raman Spectrosc 44(10):1440–1445

Cuenca-Gotor VP, Sans JA, Ibañez J, Popescu C, Gomis O, Vilaplana R, Manjón FJ, Leonardo A, Sagasta E, Suárez-Alcubilla A, Gurtubay IG, Mollar M, Bergara A (2016) Structural, vibrational, and electronic study of α-As$_2$Te$_3$ under compression. J Phys Chem C 120(34):19340–19352

Cui M, Wang Y, Liu X, Zhu J, Sun J, Lv N, Meng C (2014) Solvothermal conversion of magadiite into zeolite omega in a glycerol–water system. J Chem Technol Biotechnol 89(3):419–424

Culka A, Jehlička J, Němec I (2009) Raman and infrared spectroscopic study of boussingaultite and nickelboussingaultite. Spectrochim Acta A 73(3):420–423

Culka A, Hyršl J, Jehlička J (2016a) Gem and mineral identification using GL Gem Raman and comparison with other portable instruments. Appl Phys A 122:959. (9 pp). https://doi.org/10.1007/s00339-016-0500-2

Culka A, Kindlová H, Drahota P, Jehlička J (2016b) Raman spectroscopic identification of arsenate minerals in situ at outcrops with handheld (532 nm, 785 nm) instruments. Spectrochim Acta A 154:193–199

D'Antonio MC, Palacios D, Coggiola L, Baran EJ (2007) Vibrational and electronic spectra of synthetic moolooite. Spectrochim Acta A 68(3):424–426

D'Antonio MC, Wladimirsky A, Palacios D, Coggiola L, González-Baró AC, Baran EJ, Mercader RC (2009) Spectroscopic investigations of iron(II) and iron(III) oxalates. J Braz Chem Soc 20:445–450

D'Antonio MC, Mancilla N, Wladimirsky A, Palacios D, González-Baró AC, Baran EJ (2010) Vibrational spectra of magnesium oxalates. Vib Spectrosc 53:218–221

Dacheux N, Clavier N, Wallez G, Quarton M (2007) Crystal structures of Th(OH)PO$_4$, U(OH)PO$_4$ and Th$_2$O(PO$_4$)$_2$. Condensation mechanism of M^{IV}(OH)PO$_4$ (M = Th, U) into M$_2$O(PO$_4$)$_2$. Solid State Sci 9(7):619–627

Dacheux N, De Kerdanie EDF, Clavier N, Podor R, Aupiais J, Szenknect S (2010) Kinetics of dissolution of thorium and uranium doped britholite ceramics. J Nucl Mater 404(1):33–43

Dahm M, Adam A (2001) Ab-initio-Berechnung des Tetracarbonatoscandat-Ions in Na$_5$Sc(CO$_3$)$_3$·2H$_2$O. Ein Kristallstruktur Bestimmung, Schwingungsspektren und thermischer Abbau. Z anorg allg Chemie 627(8):2023–2031. (in German)

Dal Bo F, Hatert F, Baijot M (2014) Crystal chemistry of synthetic M^{2+}Be$_2$P$_2$O$_8$ (M^{2+} = Ca, Sr, Pb, Ba) beryllophosphates. Can Mineral 52(2):337–350

Damak M, Kamoun M, Daoud A, Romain F, Lautie A, Novak A (1985) Vibrational study of hydrogen bonding and structural disorder in Na$_3$H(SO$_4$)$_2$, K$_3$H(SO$_4$)$_2$ and (NH$_4$)$_3$H(SO$_4$)$_2$ crystals. J Molec Struct 130(3):245–254

Daniel MF, Desbat B, Lassegues JC, Gerand B, Figlarz M (1987) Infrared and Raman study of WO$_3$ tungsten trioxides and WO$_3$·xH$_2$O tungsten trioxide hydrates. J Solid State Chem 67:235–247

Daniel P, Bulou A, Rousseau M, Nouet J, Fourquet JL, Leblanc M, Burriel R (1990) A study of the structural phase transitions in AlF$_3$: X-ray powder diffraction, differential scanning calorimetry (DSC) and Raman scattering investigations of the lattice dynamics and phonon spectrum. J Phys Conden Matter 2(26):5663–5677

Daniel I, Fiquet G, Gillet P, Schmidt MW, Hanfland M (2000) High-pressure behaviour of lawsonite. Eur J Mineral 12(4):721–733

Danisi RM, Armbruster T, Lazic B, Vulić P, Kaindl R, Dimitrijević R, Kahlenberg V (2013) In situ dehydration behavior of veszelyite (Cu,Zn)$_2$Zn(PO$_4$)(OH)$_3$·2H$_2$O: a single-crystal X-ray study. Am Mineral 98(7):1261–1269

Dardenne K, Vivien D, Ribot F, Chottard G, Huguenin D (1998) Mn (V) polyhedron size in Ba$_{10}$((P,Mn)O$_4$)$_6$F$_2$: vibrational spectroscopy and EXAFS study. Eur J Solid State Inorg Chem 35(6):419–431

Das S, Hendry MJ (2011) Application of Raman spectroscopy to identify iron minerals commonly found in mine wastes. Chem Geol 290(3):101–108

Das B, Reddy MV, Krishnamoorthi C, Tripathy S, Mahendiran R, SubbaRao GV, Chowdari BVR

(2009) Carbothermal synthesis, spectral and magnetic characterization and Li-cyclability of the Mo-cluster compounds, LiYMo$_3$O$_8$ and Mn$_2$Mo$_3$O$_8$. Electrochim Acta 54:3360–3373

Daturi M, Busca G, Borel MM, Leclaire A, Piaggio P (1997) Vibrational and XRD study of the system CdWO$_4$-CdMoO$_4$. J Phys Chem B 101(22):4358–4369

Daub M, Kazmierczak K, Gross P, Höppe H, Hillebrecht H (2013) Exploring a new structure family: alkali borosulfates Na$_5$[B(SO$_4$)$_4$], A$_3$[B(SO4)$_3$] (A = K, Rb), Li[B(SO$_4$)$_2$], and Li[B(S$_2$O$_7$)$_2$]. Inorg Chem 52:6011–6020

Davies JED (1973) Solid state vibrational spectroscopy–III[l]. The infrared and Raman spectra of the bismuth (III) oxide halides. J Inorg Nucl Chem 35:1531–1534

De Beer WHJ, Heyns AM, Richter PW, Clark JB (1980) High-pressure/high-temperature phase relations and vibrational spectra of CsSbF$_6$. J Solid State Chem 33(3):283–288

De Caro T, Caschera D, Ingo GM, Calandra P (2016) Micro-Raman innovative methodology to identify Ag–Cu mixed sulphides as tarnishing corrosion products. J Raman Spectrosc. https://doi.org/10.1002/jrs.4900

De Faria DLA, Venâncio Silva S, De Oliveira MT (1997) Raman microspectroscopy of some iron oxides and oxyhydroxides. J Raman Spectrosc 28(11):873–878

De Ferri L, Bersani D, Colomban P, Lottici PP, Simon G, Vezzalini G (2012) Raman study of model glass with medieval compositions: artificial weathering and comparison with ancient samples. J Raman Spectrosc 43(11):1817–1823

De La Pierre M, Belmonte D (2016) Ab initio investigation of majorite and pyrope garnets: lattice dynamics and vibrational spectra. Am Mineral 101(1):162–174

De las Heras C, Agulló-Rueda F (2000) Raman spectroscopy of NiSe$_2$ and NiS$_{2-x}$Se$_x$ (0<x< 2) thin films. J Phys Condens Matter 12(24):5317–5324

De Lima LC (2016) Espinélios do sistema Mg$_2$TiO$_4$-Mg$_2$SnO$_4$ obtidos pelo método Pechini-modificado: propriedades fotocatalíticas e antiadesão microbiana. Dissertação, Universidade Federalda Paraíba. (in Portuguese)

De Waal D, Heyns AM (1992) A reinvestigation of the thermal decomposition products of (NH$_4$)$_2$CrO$_4$ and (NH$_4$)$_2$Cr$_2$O$_7$. J Alloys Compd 187(1):171–180

De Waal D, Hutter C (1994) Vibrational spectra of two phases of copper pyrovanadate and some solid solutions of copper and magnesium pyrovanadate. Mater Res Bull 29(8):843–849

Deb SK, Manghnani MH, Ross K, Livingston RA, Monteiro PJM (2003) Raman scattering and X-ray diffraction study of the thermal decomposition of an ettringite-group crystal. Phys Chem Minerals 30(1):31–38

Debbichi L, Marco de Lucas MC, Pierson JF, Krüger P (2012) Vibrational properties of CuO and Cu$_4$O$_3$ from first-principles calculations, and Raman and infrared spectroscopy. J Phys Chem C 116(18):10232–10237

Degtyareva O, Struzhkin VV, Hemley RJ (2007) High-pressure Raman spectroscopy of antimony: as-type, incommensurate host-guest, and bcc phases. Solid State Commun 141:164–167

Del Bosque IS, Martínez-Ramírez S, Blanco-Varela MT (2014) FTIR study of the effect of temperature and nanosilica on the nano structure of C–S–H gel formed by hydrating tricalcium silicate. Constr Build Mater 52:314–323

Delaney MJ, Ushioda S (1976) Comparison of the Raman spectra of superionic conductors AgI and RbAg$_4$I$_5$. Solid State Commun 19(4):297–301

Della Ventura G, Rossi P, Parodi GC, Mottana A, Raudsepp M, Prencipe M (2000) Stoppaniite, (Fe,Al,Mg)$_4$(Be$_6$Si$_{12}$O$_{36}$)·(H$_2$O)$_2$(Na,□) a new mineral of the beryl group from Latium (Italy). Eur J Mineral 12(1):121–127

Demartin F, Gramaccioli CM, Campostrini I (2010) Pyracmonite, (NH$_4$)$_3$Fe(SO$_4$)$_3$, a new ammonium iron sulfate from La Fossa crater, Vulcano, Aeolian Islands, Italy. Can Mineral 48:307–313

Demartin F, Castellano C, Gramaccioli CM (2015) Campostriniite, (Bi^{3+},Na)$_3$(NH$_4$,K)$_2$Na$_2$(SO$_4$)$_6$·H$_2$O, a new sulfate isostructural with görgeyite, from La Fossa Crater, Vulcano, Aeolian Islands, Italy. Mineral Mag 79(4):1007–1018

Derun EM, Tugce SF (2014) Characterization and thermal dehydration kinetics of highly crystalline mcallisterite, synthesized at low temperatures. Sci World J. https://doi.org/10.1155/2014/985185

Derun EM, Kipcak AS, Senberber FT, Yilmaz MS (2015) Characterization and thermal dehydration kinetics of admontite mineral hydrothermally synthesized from magnesium oxide and boric acid precursor. Res Chem Intermed 41(2):853–866

Devanarayanan S, Morell G, Katiyar RS (1991) Raman spectroscopy of BeO at low temperatures. J Raman Spectrosc 22(6):311–314

Devarajan V, Shurvell HF (1977) Vibrational spectra and normal coordinate analysis of crystalline lithium metasilicate. Can J Chem 55(13):2559–2563

Devarajan V, Gräfe E, Funck E (1974) Vibrational spectra of complex borates – II. B(OH)$_4^-$-ion in teepleite, Raman spectrum and normal coordinate analysis. Spectrochim Acta A 30(6):1235–1242

Devi SA, Philip D, Aruldhas G (1994) Infrared, polarized Raman, and SERS spectra of borax. J Solid State Chem 113(1):157–162

Dey B, Jain YS, Verma AL (1982) Infrared and Raman spectroscopic studies of KHSO$_4$ crystals. J Raman Spectrosc 13(3):209–212

Dhandapani M, Thyagu L, Prakash PA, Amirthaganesan G, Kandhaswamy MA, Srinivasan V (2006) Synthesis and characterization of potassium magnesium sulphatehexahydrate crystals. Cryst Res Technol 41(4):328–331

Dhankhar S, Bhalerao G, Ganesamoorthy S, Baskar K, Singh S (2016) Growth and comparison of single crystals and polycrystalline brownmillerite Ca$_2$Fe$_2$O$_5$. J Crys Growth. https://doi.org/10.1016/j.jcrysgro.2016.09.051

Dhas NA, Gedanken A (1997) Characterization of sonochemically prepared unsupported and silica-supported nanostructured pentavalent molybdenum oxide. J Phys Chem B 101(46):9495–9503

Dias LN, Pinheiro MVB, Moreira RL, Krambrock K, Guedes KJ, Menezes Filho LAD, Karfunkel J, Schnellrath J, Scholz R (2011) Spectroscopic characterization of transition metal impurities in natural montebrasite/amblygonite. Am Mineral 96(1):42–52

Diez RP, Baran EJ, Lavat AE, Grasselli MC (1995) Vibrational and electronic spectra of some mixed oxides belonging to the Sr_2PbO_4 structural type. J Phys Chem Solids 56(1):135–139

Dill HG, Weber B (2010) Accessory minerals of fluorite and their implication regarding the environment of formation (Nabburg–Wölsendorf fluorite district, SE Germany), with special reference to fetid fluorite ("Stinkspat"). Ore Geol Rev 37(2):65–86

D'Ippolito V, Andreozzi GB, Bosi F, Hålenius U, Mantovani L, Bersani D, Fregola RA (2013) Crystallographic and spectroscopic characterization of a natural Zn-rich spinel approaching the endmember gahnite ($ZnAl_2O_4$) composition. Mineral Mag 77(7):2941–2953

D'Ippolito V, Andreozzi GB, Bersani D, Lottici PP (2015) Raman fingerprint of chromate, aluminate and ferrite spinels. J Raman Spectrosc 46(12):1255–1264

Djurek D, Prester M, Drobac DJ, Ivanda M, Vojta D (2015) Magnetic properties of nanoscaled paramelaconite Cu_4O_{3-x} ($x = 0.0$ and 0.5). J Magnetism Magnetic Mater 373:183–187

Dobal PS, Katiyar RS, Jiang Y, Guo R, Bhalla AS (2000) Micro-Raman scattering and X-ray diffraction studies of $(Ta_2O_5)_{1-x}(TiO_2)_x$ ceramics. J Appl Phys 87:8688–8694

Domoratskii KV, Pastukhov VI, Kudzin AY, Sadovskaya LY, Rizak VM, Stefanovich VA (2000) Raman scattering in the Bi_2TeO_5 single crystal. Phys Solid State 42(8):1443–1446

Dondur V, Dimitrijević R, Kremenović A, Damjanović L, Kićanović M, Cheong HM, Macura S (2005) Phase transformation of hexacelsians doped with Li, Na and Ca. Mater Sci Forum 494:107–112

Dong G, Pollard PJ (1997) Identification of ferropyrosmalite by laser Raman microprobe in fluid inclusions from metalliferous deposits in the Cloncurry District, NW Queensland, Australia. Mineral Mag 61(2):291–293

Dong L, Pan S, Yang Z, Zhao W, Dong X, Wang Y, Huang Y (2012) Synthesis, crystal structure, and properties of a new lead aluminum fluoride borate, $Pb_6Al(BO_3)_2OF_7$. Z anorg allg Chemie 638(14):2280–2285

Dongre AN, Viljoen KS, Rao NC, Gucsik A (2016) Origin of Ti-rich garnets in the groundmass of Wajrakarur field kimberlites, southern India: insights from EPMA and Raman spectroscopy. Mineral Petrol 110(2–3):295–307

Đorđević T, Kolitsch U, Nasadala L (2016) A single-crystal X-ray and Raman spectroscopic study of hydrothermally synthesized arsenates and vanadates with the descloizite and adelite structure types. Am Mineral 101(5):1135–1149

Downs GW, Yang BN, Thompson RM, Wenz MD, Andrade MB (2012) Redetermination of durangite, $NaAl(AsO_4)F$. Acta Crystallogr E. https://doi.org/10.1107/S160053681204384X

Drábek M, Frýda J, Šarbach M, Skála R (2017) Hydroxycalciopyrochlore from a regionally metamorphic marble at Bližná, Southwestern Czech Republic. N Jb Miner Abh 194(1):49–59

Driscoll RJP, Wolverson D, Mitchels JM, Skelton JM, Parker SC, Molinari M, Khan I, Geeson D, Allen GC (2014) A Raman spectroscopic study of uranyl minerals from Cornwall, UK. RSC Adv 4(103):59137–59149

Du W, Han B, Clark SM, Wang Y, Liu X (2017) Raman spectroscopic study of synthetic pyrope–grossular garnets: structural implications. Phys Chem Minerals. https://doi.org/10.1007/s00269-017-0908-z

Dubinina EV, Valizer PM (2011) Gorceixite: the first find in the Ilmenogorskii complex (Southern Urals). Doklady Earth Sci 439(1):961–963

Dudik JM, Johnson CR, Asher SA (1985) UV Resonance Raman studies of acetone, acetamide, and N-methylacetamide: models for the peptide bond. J Phys Chem 89(18):3805–3814

Dul K, Koleżyński A, Sitarz M, Madej D (2015) Vibrational spectra of a baghdadite synthetic analogue. Vib Spectrosc 76:1–5

Dultz W, Quilichini M, Scott JF, Lehmann G (1975) Phonon spectra of quartz isomorphs. Phys Rev B 11(4):1648–1653

Durand O (2006) Propriétés structurales et vibrationnelles des phases désordonnées dans le système TeO_2–Bi_2O_3. Dissertation, Université de Limoges. (in French)

Durig JR, Lau KK, Nagarajan G, Walker M, Bragin J (1969) Vibrational spectra and molecular potential fields of mercurous chloride, bromide, and iodide. J Chem Phys 50(5):2130–2139

Dutková E, Sayagués MJ, Briančin J, Zorkovská A, Bujňáková Z, Kováč J, Kováč J Jr, Baláž P, Ficeriová J (2016) Synthesis and characterization of $CuInS_2$ nanocrystalline semiconductor prepared by high-energy milling. J Mater Sci 51(4):1978–1984

Dzhagan VM, Litvinchuk AP, Kruszynska M, Kolny-Olesiak J, Valakh MY, Zahn DR (2014) Raman scattering study of Cu_3SnS_4 colloidal nanocrystals. J Phys Chem C 118(47):27554–27558

Echigo T, Kimata M (2008) Single-crystal X-ray diffraction and spectroscopic studies on humboldtine and lindbergite: weak Jahn–Teller effect of Fe^{2+} ion. Phys Chem Minerals 35(8):467–475

Echigo T, Kimata M, Kyono A, Shimizu M, Hatta T (2005) Re-investigation of the crystal structure of whewellite [$Ca(C_2O_4) \cdot H_2O$] and the dehydration mechanism of caoxite [$Ca(C_2O_4) \cdot 3H_2O$]. Mineral Mag 69(1):77–88

Echigo T, Kimata M, Maruoka T (2007) Crystal-chemical and carbon-isotopic characteristics of karpatite

($C_{24}H_{12}$) from the Picacho Peak Area, San Benito County, California: evidences for the hydrothermal formation. Am Mineral 92:1262–1269

Eder SH, Gigler AM, Hanzlik M, Winklhofer M (2014) Sub-micrometer-scale mapping of magnetite crystals and sulfur globules in magnetotactic bacteria using confocal Raman micro-spectrometry. PLoS One 9(9): e107356. (12 pp)

Edge A, Taylor HFW (1971) Crystal structure of thaumasite. Acta Cryst B27:594–601

Edwards HG, Jorge Villar SE, Bishop JL, Bloomfield M (2004) Raman spectroscopy of sediments from the Antarctic dry valleys; an analogue study for exploration of potential paleolakes on Mars. J Raman Spectrosc 35(6):458–462

Edwards HG, Villar SEJ, Jehlicka J, Munshi T (2005) FT-Raman spectroscopic study of calcium-rich and magnesium-rich carbonate minerals. Spectrochim Acta A 61(10):2273–2280

Edwards HG, Currie KJ, Ali HR, Villar SEJ, David AR, Denton J (2007) Raman spectroscopy of natron: shedding light on ancient Egyptian mummification. Anal Bioanal Chem 388(3):683–689

Effenberger H, Pippinger T, Libowitzky E, Lengauer CL, Miletich R (2014) Synthetic norsethite, BaMg (CO_3)$_2$: revised crystal structure, thermal behaviour and displacive phase transition. Mineral Mag 78(7):1589–1612

Efthimiopoulos I, Kemichick J, Zhou X, Khare SV, Ikuta D, Wang Y (2014) High-pressure studies of Bi_2S_3. J Phys Chem A 118(9):1713–1720

El Mendili Y, Minisini B, Abdelouas A, Bardeau JF (2014) Assignment of Raman-active vibrational modes of tetragonal mackinawite: Raman investigations and ab initio calculations. RSC Adv 4 (49):25827–25834

Elkhouni T, Colin CV, Strobel P, Salah AB, Amami M (2013) Effect of Ga substitution on the magnetic state of delafossite $CuCrO_2$ with antiferromagnetic triangular sublattice. J Superconduct Novel Magn 26 (6):2125–2134

Elliott P (2010) Crystal chemistry of cadmium oxysalt and associated minerals from Broken Hill, New South Wales. Dissertation, University of Adelaide

Elliott P, Kolitsch U, Giester G, Libowitzky E, McCammon C, Pring A, Burcj WD, Brugger J (2009) Description and crystal structure of a new mineral – plimerite, $ZnFe^{3+}_4(PO_4)_3(OH)_5$ – the Zn-analogue of rockbridgeite and frondelite, from Broken Hill, New South Wales, Australia. Mineral Mag 73(1):131–148

Elliott P, Giester G, Libowitzky E, Kolitsch U (2010) Description and crystal structure of liversidgeite, $Zn_6(PO_4)_4 \cdot 7H_2O$, a new mineral from Broken Hill, New South Wales, Australia. Am Mineral 95 (2–3):397–404

Elliott P, Kolitsch U, Willis AC, Libowitzky E (2013) Description and crystal structure of domerockite, $Cu_4(AsO_4)(AsO_3OH)(OH)_3 \cdot H_2O$, a new mineral from the Dome Rock Mine, South Australia. Mineral Mag 77(4):509–522

El-Metwally N, Al Thani MJ (1989) Preparation and infrared spectra of ($Cu(O)(H_2O)_3$) and ($Cu_2(O)(Cl)_2(H_2O)_2$) complexes formed in the reactions of Cu(II) salts with urea. J Phys Chem Solids 50 (2):183–186

Engel G (1973) Infrarotspektroskopische und röntgenographische Untersuchungen von Bleihydroxylapatit, Bleioxyapatit und Bleialkaliapatiten. J Solid State Chem 6(2):286–292. (in German)

English RB, Heyns AM (1984) An infrared, Raman, and single-crystal X-ray study of cesium hexafluorophosphate. J Crystallogr Spectrosc Res 14 (6):531–540

Enholm Z (2016) Mineral chemistry and parageneses of oxyborates in metamorphosed Fe-Mn oxide deposits. Degree Project at the Department of Earth Sciences, Uppsala University

Equeenuddin SM (2015) Occurrence of alpersite at Malanjkhand copper mine, India. Environ Earth Sci 73(7):3849–3853

Errandonea D, Manjón FJ, Garro N, Rodríguez-Hernández P, Radescu S, Mujica A, Moñoz A, Tu CY (2008) Combined Raman scattering and ab initio investigation of pressure-induced structural phase transitions in the scintillator $ZnWO_4$. Phys Rev B 78 (5):054116-1–054116-16

Errandonea D, Muñoz A, Rodríguez-Hernández P, Gomis O, Achary SN, Popescu C, Patwe SJ, Tyagi AK (2016) High-pressure crystal structure, lattice vibrations, and band structure of $BiSbO_4$. Inorg Chem 55:4958–4969

Ertl A, Kolitsch U, Dyar MD, Meyer HP, Rossman GR, Henry DJ, Prem M, Ludwig T, Nasdala L, Lengauer CL, Tillmanns E, Niedermayr G (2015) Fluor-schorl, a new member of the tourmaline supergroup, and new data on schorl from the cotype localities. Eur J Mineral. https://doi.org/10.1127/ejm/2015/0027-2501

Escobar ME, Baran EJ (1982) Darstellung und Eigenschaften einiger neuer Arsenat-und Vanadat-Halogen-Apatite. Z anorg allg Chemie 489(1):139–146. (in German)

Essehli R, El Bali B, Benmokhtar S, Fejfarová K, Dusek M (2009) Hydrothermal synthesis, structural and physico-chemical characterizations of two Nasicon phosphates: $M_{0.50}^{II}Ti_2(PO_4)_3$ (M = Mn, Co). Mater Res Bull 44(7):1502–1510

Essehli R, El Bali B, Benmokhtar S, Fuess H, Svoboda I, Obbade S (2010) Synthesis, crystal structure and infrared spectroscopy of a new non-centrosymmetric mixed-anion phosphate $Na_4Mg_3(PO_4)_2(P_2O_7)$. J Alloys Compd 493(1):654–660

Etchepare J, Merían M, Kaplan P (1978) Vibrational normal modes of SiO_2. II. Cristobalite and tridymite. J Chem Phys 68(4):1531–1537

Evrard C, Fouquet Y, Moëlo Y, Rinnert E, Etoubleau J, Langlade JA (2015) Tin concentration in hydrothermal sulphides related to ultramafic rocks along the Mid-Atlantic Ridge: a mineralogical study. Eur J Mineral 27(5):627–638

Ezzaafrani M, Ennaciri A, Harcharras M, Capitelli F (2014) Spectroscopic and structural investigation of BaNaP$_3$O$_9$·3H$_2$O cyclotriphosphate. Phosphorus Sulfur Silicon Relat Elem 189(12):1841–1850. https://doi.org/10.1080/10426507.2014.906419

Fakhfakh M, Madani A, Jouini N (2003) A$_3$Nb$_5$O$_{11}$(PO$_4$)$_2$ (A = Tl, K, Na) compounds: synthesis, crystal and vibrational characterization, conductivity study. Mater Res Bull 38(7):1215–1226

Falgayrac G, Sobanska S, Brémard C (2013) Heterogeneous microchemistry between CdSO$_4$ and CaCO$_3$ particles under humidity and liquid water. J Hazard Mater 248:415–423

Falk M, Knop O (1977) Infrared studies of water in crystalline hydrates: K$_2$HgCl$_4$·H$_2$O. Can J Chem 55(10):1736–1744

Fan Y, Hua Li G, Yang L, Ming Zhang Z, Chen Y, You Song T, HuaFeng S (2005) Synthesis, crystal structure, and magnetic properties of a three-dimensional hydroxide sulfate: Mn$_5$(SO$_4$)(OH)$_8$. Eur J Inorg Chem 2005(16):3359–3364

Fan X, Pan S, Hou X, Tian X, Han J (2010) Flux growth and morphology analysis of Na$_3$VO$_2$B$_6$O$_{11}$ crystals. J Cryst Growth 312(15):2263–2266

Fan C, Wang L, Fan X, Zhang Y, Zhao L (2015) The mineralogical characterization of argentiancryptomelane from XiangguangMn-Ag deposit, North China. J Mineral Petrol Sci 110(5):214–223

Fang Y, Ritter C, White T (2011) The crystal chemistry of Ca$_{10-y}$(SiO$_4$)$_3$(SO$_4$)$_3$Cl$_{2-x-2y}$F$_x$ ellestadite. Inorg Chem 50(24):12641–12650

Fantini C, Tavares MC, Krambrock K, Moreira RL, Righi A (2014) Raman and infrared study of hydroxyl sites in natural uvite, fluor-uvite, magnesio-foitite, dravite and elbaite tourmalines. Phys Chem Minerals 41(4):247–254

Farmer VC, Fraser AR, Tait JM (1979) Characterization of the chemical structures of natural and synthetic aluminosilicate gels and sols by infrared spectroscopy. Geochim Cosmochim Acta 43(9):1417–1420

Faulques E, Kalashnyk N, Massuyeau F, Perry DL (2015a) Spectroscopic markers for uranium (vi) phosphates: a vibronic study. RSC Adv 5(87):71219–71227

Faulques E, Massuyeau F, Kalashnyk N, Perry DL (2015b) Application of Raman and photoluminescence spectroscopy for identification of uranium minerals in the environment. Spectrosc Eur 27(1):14–17

Faulstich FR, Schnellrath J, De Oliveira LF, Scholz R (2013) Rockbridgeite inclusion in rock crystal from Galileia region, Minas Gerais, Brazil. Eur J Mineral 25(5):817–823

Faulstich FRL, Ávila CA, Neumann R, Silveira VSL, Callegario LS (2016) Gahnite from the São João Del Reipegmatitic province, Minas Gerais, Brazil: chemical composition and genetic implications. Can Mineral 54(6):1385–1402

Fendrich KV, Downs RT, Origlieri MJ (2016) Redetermination of ruizite, Ca$_2$Mn^{3+}$_2$[Si$_4$O$_{11}$(OH)$_2$](OH)$_2$·2H$_2$O. Acta Crystallogr E72(7):959–963

Feng XH, Tan WF, Liu F, Wang JB, Ruan HD (2004) Synthesis of todorokite at atmospheric pressure. Chem Mater 16(22):4330–4336

Feng L, Lin Y, Hu S, Xu L, Miao B (2011) Estimating compositions of natural ringwoodite in the heavily shocked Grove Mountains 052049 meteorite from Raman spectra. Am Mineral 96(10):1480–1489

Feng J-H, Hu C-L, Xu X, Kong F, Mao J-G (2015a) Na$_2$RE$_2$TeO$_4$(BO$_3$)$_2$ (RE = Y, Dy–Lu): luminescent and structural studies on a series of mixed metal borotellurates. Inorg Chem 54(5):2447–2454

Feng L-L, Li G-D, Liu Y, Wu Y, Chen H, Wang Y, Zou Y-C, Wang D, Zou X (2015b) Carbon-armored Co$_9$S$_8$ nanoparticles as all-pH efficient and durable H$_2$-evolving electrocatalysts. ACS Appl Mater Interfaces 7(1):980–988

Feng Y, Zhou W, Wang Y, Zhou J, Liu E, Fu Y, Ni Z, Wu X, Yuan H, Miao F, Wang B, Wan X, Xing D (2015c) Raman vibrational spectra of bulk to monolayer ReS$_2$ with lower symmetry. Phys Rev B 92(5):054110-1–054110-6

Fenske F, Lange H, Oertel G, Reinsperger GU, Schumann J, Selle B (1996) Characterization of semiconducting silicide films by infrared vibrational spectroscopy. Mater Chem Phys 43(3):238–242

Ferdov S, Lin Z, Ferreira RAS, Correia MR (2008) Hydrothermal synthesis, structural, and spectroscopic studies of vanadium substituted ETS-4. Micropor Mesopor Mater 110(2):436–441

Férid M, Horchani-Naifer K (2004) Synthesis, crystal structure and vibrational spectra of a new form of diphosphate NaLaP$_2$O$_7$. Mater Res Bull 39(14):2209–2217

Fermo P, Padeletti G (2012) The use of nano-particles to produce iridescent metallic effects on ancient ceramic objects. J Nanosci Nanotechnol 12(11):8764–8769

Fernandes PA, Salomé PMP, Da Cunha AF (2010) Cu$_x$SnS$_{x+1}$ ($x = 2, 3$) thin films grown by sulfurization of metallic precursors deposited by dc magnetron sputtering. Phys Status Solidi C 7(3–4):901–904

Fernandez JRL, de Souza-Parise M, Morais PC (2015) Structural characterization and simulation of colloidal MnS. Mater Res Express 2(9):095019. (5 pp)

Ferrari AM, Valenzano L, Meyer A, Orlando R, Dovesi R (2009) Quantum-mechanical ab initio simulation of the Raman and IR spectra of Fe$_3$Al$_2$Si$_3$O$_{12}$ almandine. J Phys Chem A 113(42):11289–11294

Ferraris C, Parodi GC, Pont S, Rondeau B, Lorand JP (2014) Trinepheline and fabriesite: two new mineral species from the jadeite deposit of Tawmaw (Myanmar). Eur J Mineral 26(2):257–265

Ferras Y, Robertson J, Swedlund PJ (2016) The application of Raman spectroscopy to probe the association of H$_4$SiO$_4$ with iron oxides. Aquat Geochem. https://doi.org/10.1007/s10498-016-9294-2

Ferrero S, Ziemann MA, Angel RJ, O'Brien PJ, Wunder B (2016) Kumdykolite, kokchetavite, and cristobalite crystallized in nanogranites from felsic granulites, Orlica-Sniezńik Dome (Bohemian Massif): not an evidence for ultrahigh-pressure conditions. Contrib Mineral Petrol 171(1):1–12

Ferroir T, Beck P, Van de Moortèle B, Bohn M, Reynard B, Simionovici A, El Goresy A, Gillet P (2008) Akimotoite in the Tenham meteorite: crystal chemistry and high-pressure transformation mechanisms. Earth Planet Sci Lett 275(1):26–31

Filipek E, Walczak J, Tabero P (1998) Synthesis and some properties of the phase $Cr_2V_4O_{13}$. J Alloys Compd 265(1):121–124

Filippi M, Doušová B, Machovič V (2007) Mineralogical speciation of arsenic in soils above the Mokrsko-west gold deposit, Czech Republic. Geoderma 139(1):154–170

Fillaux F, Lautié A, Tomkinson J, Kearley GJ (1991) Proton transfer dynamics in the hydrogen bond. Inelastic neutron scattering, infrared and Raman spectra of $Na_3H(SO_4)_2$, $K_3H(SO_4)_2$ and $Rb_3H(SO_4)_2$. Chem Phys 154(1):135–144

Finger LW, Hazen RM, Hemley RJ (1989) $BaCuSi_2O_6$: a new cyclosilicate with four-membered tetrahedral rings. Am Mineral 74:952–955

Fintor K, Walter H, Nagy S (2013) Petrographic and micro-Raman analysis of chondrulesand (Ca,Al)-rich inclusions of NWA 2086 CV3 type carbonaceous chondrite. 44th Lunar Planetary Sci Conf 44:1152

Fintor K, Park C, Nagy S, Pál-Molnár E, Krot AN (2014) Hydrothermal origin of hexagonal $CaAl_2Si_2O_8$ (dmisteinbergite) in a compact type A CAI from the Northwest Africa 2086 CV3 chondrite. Meteorit Planet Sci 49(5):812–823

Fischer H-H (2007) Beitrage zur Kristallchemie der Carbonate und Isonicotinate. Dissertation, Universität zu Köln. (in German)

Fitouri I, Falah C, Boughzala H (2015) Synthesis, infrared (IR) spectroscopy and single crystal structural study of a new arsenate $Cs_7Fe_7O_2(AsO_4)_8$. J Chem Crystallogr 45(5):231–237

Fleet ME, Henderson GS (1997) Structure-composition relations and Raman spectroscopy of high-pressure sodium silicates. Phys Chem Minerals 24(5):345–355

Fomichev VV, Kondratov OI (1994) Vibrational spectra of compounds with the wolframite structure. Spectrochim Acta A 50(6):1113–1120

Fontana MD, Idrissi H, Kugel GE, Wojcik K (1991) Raman spectrum in $PbTiO_3$ re-examined: dynamics of the soft phonon and the central peak. J Phys Conden Matter 3(44):8695–8705

Fontané X, Izquierdo-Roca V, Saucedo E, Schorr S, Yukhymchuk VO, Valakh MY, Pérez-Rodríguez A, Morante JR (2012) Vibrational properties of stannite and kesterite type compounds: Raman scattering analysis of $Cu_2(Fe,Zn)SnS_4$. J Alloys Compd 539:190–194

Fontané X, Izquierdo-Roca V, Fairbrother A, Espíndola-Rodríguez M, López-Marino S, Placidi M, Jawhari T, Saucedo E, Pérez-Rodríguez A (2013) Selective detection of secondary phases in $Cu_2ZnSn(S,Se)_4$ based absorbers by pre-resonant Raman spectroscopy. Proceedings of the IEEE 39th Photovoltaic Specialists Conference. https://doi.org/10.1109/PVSC.2013.6745001

Forbes P, Dubessy J (1988) Characterization of fresh and altered montroseite [V,Fe]OOH. A discussion of alteration processes. Phys Chem Minerals 15(5):438–445

Forneris R (1969) Infrared and Raman spectra of realgar and orpiment. Am Mineral 54(7–8):1062–1074

Fornero E, Allegrina M, Rinaudo C, Mazziotti-Tagliani S, Gianfagna A (2008) Micro-Raman spectroscopy applied on oriented crystals of fluoro-edenite amphibole. Per Mineral 77(2):5–14

Forray FL, Smith AML, Navrotsky A, Wright K, Hudson-Edwards KA, Dubbin WE (2014) Synthesis, characterization and thermochemistry of synthetic Pb–As, Pb–Cu and Pb–Zn jarosites. Geochim Cosmochim Acta 127:107–119

Fowler-Gerace N (2014) Textural and geochemical investigation of springwater pallasite olivine. Dissertation, University of Toronto

Frech R, Wang EC, Bates JB (1980) The i.r. and Raman spectra of $CaCO_3$ (aragonite). Spectrochim Acta 36A:915–919

Fresno F, Jana P, Reñones P, Coronado JM, Serrano DP, de la Peña O'Shea VA (2016) CO_2 reduction over $NaNbO_3$ and $NaTaO_3$ perovskite photocatalysts. Photochem Photobiol Sci. https://doi.org/10.1039/C6PP00235H

Freund GA, Kirby RD, Sellmyer DJ (1977) Raman scattering from pyrite structure $AuSb_2$. Solid State Commun 22(1):5–7

Frezzotti ML, Tecce F, Casagli A (2012) Raman spectroscopy for fluid inclusion analysis. J Geochem Exploration 112:1–20

Friedrich A, Winkler B, Morgenroth W, Perlov A, Milman V (2015) Pressure-induced spin collapse of octahedrally coordinated Mn^{3+} in the tetragonal hydrogarnet henritermierite $Ca_3Mn_2[SiO_4]_2[O_4H_4]$. Phys Rev B 92(1):014117-01–014117-11

Friis H, Weller MT, Kampf AR (2016) Hansesmarkite, $Ca_2Mn_2Nb_6O_{19} \cdot 20H_2O$, a new hexaniobate from a syenite pegmatite in the Larvik Plutonic Complex, southern Norway. Mineral Mag. https://doi.org/10.1180/minmag.2016.080.109

Frost RL (1995) Fourier transform Raman spectroscopy of kaolinite, dickite and halloysite. Clays Clay Minerals 43(2):191–195

Frost RL (2004a) An infrared and Raman spectroscopic study of natural zinc phosphates. Spectrochim Acta A 60(7):1439–1445

Frost RL (2004b) An infrared and Raman spectroscopic study of the uranyl micas. Spectrochim Acta A 60(7):1469–1480

Frost RL (2004c) Raman microscopy of selected chromate minerals. J Raman Spectrosc 35(2):153–158

Frost RL (2004d) Raman spectroscopy of natural oxalates. Analyt Chim Acta 517(1):207–214

Frost RL (2006) A Raman spectroscopic study of selected minerals of the rosasite group. J Raman Spectrosc 37(9):910–921

Frost RL (2009a) Raman and infrared spectroscopy of arsenates of the roselite and fairfieldite mineral subgroups. Spectrochim Acta A 71(5):1788–1794

Frost RL (2009b) Tlapallite $H_6(Ca,Pb)_2(Cu, Zn)_3SO_4(TeO_3)_4TeO_6$, a multi-anion mineral: a Raman spectroscopic study. Spectrochim Acta A 72 (4):903–906

Frost RL (2011a) Raman spectroscopic study of the uranyl titanate mineral holfertite $CaxU_{2-x}Ti(O_{8-x}OH_{4x})\cdot 3H_2O$ and the lack of metamictization. Radiat Effects Defects Solids 166(1):24–29

Frost RL (2011b) Raman spectroscopy of selected borate minerals of the pinakiolite group. J Raman Spectrosc 42(3):540–543

Frost RL (2011c) A Raman spectroscopic study of copiapites $Fe^{2+}Fe^{3+}_4(SO_4)_6(OH)_2\cdot 20H_2O$: environmental implications. J Raman Spectrosc 42(5):1130–1134

Frost RL (2011d) Raman spectroscopic study of the magnesium carbonate mineral hydromagnesite $(Mg_5[(CO_3)_4(OH)_2]\cdot 4H_2O)$. J Raman Spectrosc 42 (8):1690–1694

Frost RL, Bahfenne S (2009) Raman and mid-IR spectroscopic study of the magnesium carbonate minerals – brugnatellite and coalingite. J Raman Spectrosc 40 (4):360–365

Frost RL, Bahfenne S (2010a) Raman spectroscopic study of the antimonate mineral brizziite $NaSbO_3$. Radiat Effects Defects Solids Incorp Plasma Sci Plasma Technol 165(3):206–210

Frost RL, Bahfenne S (2010b) Thermal analysis and Hot-stage Raman spectroscopy of the basic copper arsenate mineral euchroite. J Therm Anal Calorim 100(1):89–94

Frost RL, Bahfenne S (2010c) Raman spectroscopi study of the antimonate mineral bahianite $Al_5Sb^{5+}_3O_{14}(OH)_2$. J Raman Spectrosc 41(2):207–211

Frost RL, Bahfenne S (2010d) Raman spectroscopic study of the arsenite minerals leiteite $ZnAs_2O_4$, reinerite $Zn_3(AsO_3)_2$ and cafarsite $Ca_5(Ti,Fe,Mn)_7(AsO_3)_{12}\cdot 4H_2O$. J Raman Spectrosc 41(3):325–328

Frost RL, Bahfenne S (2010e) Raman spectroscopic study of the antimonate mineral bottinoite $Ni[Sb_2(OH)_{12}]\cdot 6H_2O$ and in comparison with brandholzite $Mg[Sb^{5+}_2(OH)_{12}]\cdot 6H_2O$. J Raman Spectrosc 41(10):1353–1356

Frost RL, Bahfenne S (2010f) A Raman spectroscopic study of the mineral coquandite $Sb_6O_8(SO_4)\cdot (H_2O)$. Spectrochim Acta A 75(2):852–854

Frost RL, Bahfenne S (2011a) The Mineral Nealite $Pb_4Fe^{2+}(AsO_3)_2Cl_4\cdot 2H_2O$ – a Raman spectroscopic study. Spectrosc Lett 44(1):22–26

Frost RL, Bahfenne S (2011b) Raman spectroscopic study of the mineral thorikosite $Pb_3(OH)(SbO_3,AsO_3)Cl_2$: a mineral of archaeological significance. Spectrosc Lett 44:63–66

Frost RL, Bahfenne S (2011c) A Raman spectroscopic study of the antimony mineral klebelsbergite $Sb_4O_4(OH)_2(SO_4)$. J Raman Spectrosc 42 (2):219–223

Frost RL, Bouzaid JM (2007) Raman spectroscopy of dawsonite $NaAl(CO_3)(OH)_2$. J Raman Spectrosc 38:873–879

Frost RL, Čejka J (2009a) Raman spectroscopic study of the uranyl phosphate mineral dumontite $Pb_2[(UO_2)_3O_2(PO_4)_2]\cdot 5H_2O$. J Raman Spectrosc 40 (6):591–594

Frost RL, Čejka J (2009b) A Raman spectroscopic study of the uranyl mineral rutherfordine – revisited. J Raman Spectrosc 40(9):1096–1103

Frost RL, Dickfos MJ (2007a) Raman spectroscopy of halogen-containing carbonates. J Raman Spectrosc 38 (11):1516–1522

Frost RL, Dickfos M (2007b) Hydrated double carbonates – a Raman and infrared spectroscopic study. Polyhedron 26(15):4503–4508

Frost RL, Dickfos MJ (2008) Raman and infrared spectroscopic study of the anhydrous carbonate minerals shortite and barytocalcite. Spectrochim Acta A 71 (1):143–146

Frost RL, Dickfos MJ (2009) Raman spectroscopic study of the tellurite minerals: mackayite and quetzalcoatlite. Spectrochim Acta A 72(2):445–448

Frost RL, Erickson KL (2004) Vibrational spectroscopy of stichtite. Spectrochim Acta A 60(13):3001–3005

Frost RL, Erickson KL (2005) Raman spectroscopic study of the hydrotalcite desautelsite $Mg_6Mn_2CO_3(OH)_{16}\cdot 4H_2O$. Spectrochim Acta A 61 (11):2697–2701

Frost RL, Keeffe EC (2008a) Raman spectroscopic study of the schmiederite $Pb_2Cu_2[(OH)_4|SeO_3|SeO_4]$. J Raman Spectrosc 39(10):1408–1412

Frost RL, Keeffe EC (2008b) Raman spectroscopic study of the selenite minerals – chalcomenite $CuSeO_3\cdot 2H_2O$, clinochalcomenite and cobaltomenite. J Raman Spectrosc 39(12):1789–1793

Frost RL, Keeffe EC (2008c) Raman spectroscopic study of the tellurite mineral: poughite $Fe^3+_2SO_4(TeO_3)_2\cdot 3H_2O$ – a multi-anion mineral. J Raman Spectrosc 39(12):1794–1798

Frost RL, Keeffe EC (2009a) Raman spectroscopic study of the selenite mineral mandarinoite $Fe_2Se_3O_9\cdot 6H_2O$. J Raman Spectrosc 40(1):42–45

Frost RL, Keeffe EC (2009b) Raman spectroscopic study of the tellurite minerals: graemite $CuTeO_3\cdot H_2O$ and teineite $CuTeO_3\cdot 2H_2O$. J Raman Spectrosc 40 (2):128–132

Frost RL, Keeffe EC (2009c) Raman spectroscopic study of the tellurite mineral: sonoraite $Fe^{3+}Te^{4+}O_3(OH)\cdot H_2O$. J Raman Spectrosc 40(2):133–136

Frost RL, Keeffe EC (2009d) Raman spectroscopic study of the sulfite-bearing minerals scotlandite, hannebachite and orschallite: implications for the desulfation of soils. J Raman Spectrosc 40 (3):244–248

Frost RL, Keeffe EC (2009e) Raman spectroscopic study of kuranakhite $PbMn^{4+}Te^{6+}O_6$ – a rare tellurate mineral. J Raman Spectrosc 40(3):249–252

Frost RL, Keeffe EC (2009f) Raman spectroscopic study of the selenite mineral: ahlfeldite, $NiSeO_3\cdot 2H_2O$. J Raman Spectrosc 40(5):509–512

Frost RL, Keeffe EC (2009g) Raman spectroscopic study of the metatellurate mineral: xocomecatlite $Cu_3TeO_4(OH)_4$. J Raman Spectrosc 40(8):866–869

Frost RL, Keeffe EC (2009h) Raman spectroscopic study of the mixed anion mineral yecoraite $Bi_5Fe_3O_9(Te^{4+}O_3)(Te^{6+}O_4)_2 \cdot 9H_2O$. J Raman Spectrosc 40(9):1117–1120

Frost RL, Keeffe EC (2009i) Raman spectroscopic study of the tellurite mineral: rodalquilarite $H_3Fe^{3+}{}_2(Te^{4+}O_3)_4Cl$. Spectrochim Acta A 73(1):146–149

Frost RL, Keeffe EC (2011) The mixed anion mineral parnauite $Cu_9[(OH)_{10}|SO_4|(AsO_4)_2] \cdot 7H_2O$ – a Raman spectroscopic study. Spectrochim Acta A 81(1):111–116

Frost RL, Kloprogge JT (2001) Raman microscopy study of kalinite, tschermigite and lonecreekite at 298 and 77 K. N Jb Miner Mh 1:27–40

Frost RL, Kloprogge JT (2003) Raman spectroscopy of some complex arsenate minerals – implications for soil remediation. Spectrochim Acta A 59(12):2797–2804

Frost RL, Palmer SJ (2011a) A vibrational spectroscopic study of the mineral corkite $PbFe^{3+}{}_3(PO_4,SO_4)_2(OH)_6$. J Molec Struct 988(1):47–51

Frost RL, Palmer SJ (2011b) Molecular structure of the mineral svanbergite $SrAl_3(PO_4,SO_4)_2(OH)_6$ – a vibrational spectroscopic study. J Molec Struct 994(1):232–237

Frost RL, Palmer SJ (2011c) Raman spectroscopic study of pascoite $Ca_3V^{5+}{}_{10}O_{28} \cdot 17H_2O$. Spectrochim Acta A 78(1):248–252

Frost RL, Palmer SJ (2011d) A Raman and infrared spectroscopic study of the mineral delvauxite $CaFe^{3+}{}_4(PO_4, SO_4)_2(OH)_8 \cdot 4\text{-}6H_2O$ – a 'colloidal' mineral. Spectrochim Acta A 78(4):1250–1254

Frost RL, Palmer SJ (2011e) Vibrational spectroscopic study of the mineral tsumebite $Pb_2Cu(PO_4,SO_4)(OH)$. Spectrochim Acta A 79(5):1794–1797

Frost RL, Palmer SJ (2011f) Raman spectroscopic study of the minerals diadochite and destinezite $Fe^{3+}{}_2(PO_4, SO_4)_2(OH) \cdot 6H_2O$: implications for soil science. J Raman Spectrosc 42(7):1589–1595

Frost RL, Palmer SJ (2011g) Raman spectrum of decrespignyite $[(Y,REE)_4Cu(CO_3)_4Cl(OH)_5 \cdot 2H_2O]$ and its relation with those of other halogenated carbonates including bastnasite, hydroxybastnasite, parisite and northupite. J Raman Spectrosc 42(11):2042–2048

Frost RL, Palmer SJ (2011h) A vibrational spectroscopic study of the mixed anion mineral sanjuanite $Al_2(PO_4)(SO_4)(OH) \cdot 9H_2O$. Spectrochim Acta A 79(5):1210–1214

Frost RL, Pinto C (2007) Raman spectroscopy of the joaquinite minerals. J Raman Spectrosc 38(7):841–845

Frost RL, Reddy BJ (2010) Raman spectroscopic study of the uranyl titanate mineral betafite $(Ca,U)_2(Ti, Nb)_2O_6(OH)$ (OH): effect of metamictization. Radiat Effects Defects Solids Incorp Plasma Sci Plasma Technol 165(11):868–875

Frost RL, Reddy BJ (2011a) Raman spectroscopic study of the uranyl titanate mineral brannerite $(U,Ca,Y,Ce)_2(Ti, Fe)_2O_6$: effect of metamictisation. J Raman Spectrosc 42(4):691–695

Frost RL, Reddy BJ (2011b) The effect of metamictization on the Raman spectroscopy of the uranyl titanate mineral davidite $(La,Ce)(Y,U,Fe^{2+})(Ti,Fe^{3+})_{20}(O,OH)_{38}$. Radiat Effects Defects Solids 166(2):131–136

Frost RL, Rintoul L (1996) Lattice vibrations of montmorillonite: an FT Raman and X-ray diffraction study. Appl Clay Sci 11(2):171–183

Frost RL, Weier ML (2003) Raman spectroscopy of natural oxalates at 298 and 77 K. J Raman Spectrosc 34(10):776–785

Frost RL, Weier ML (2004a) The 'cave' mineral oxammite – a high resolution thermogravimetry and Raman spectroscopic study. N Jb Miner Mh 1:27–48

Frost RL, Weier ML (2004b) Vibrational spectroscopy of natural augelite. J Molec Struct 697(1–3):207–211

Frost RL, Weier M (2004c) Raman microscopy of selected autunite minerals. N Jb Miner Mh 2004(12):575–594

Frost RL, Weier M (2004d) Raman microscopy of autunite minerals at liquid nitrogen temperature. Spectrochim Acta A 60(10):2399–2409

Frost RL, Weier M (2004e) Raman and infrared spectroscopy of tsumcorite mineral group. N Jb Miner Mh 2004(7):317–336

Frost RL, Weier M (2006) Raman and infrared spectroscopy of the manganese arsenate mineral allactite. Spectrochim Acta A 65:623–627

Frost RL, Williams PA (2004) Raman spectroscopy of some basic chloride containing minerals of lead and copper. Spectrochim Acta A 60(8):2071–2077

Frost RL, Xi Y (2012a) Raman spectroscopy of selected tsumcorite $Pb(Zn,Fe^{3+})_2(AsO_4)_2(OH,H_2O)$ minerals – implications for arsenate accumulation. Spectrochim Acta A 86:224–230

Frost RL, Xi Y (2012b) Raman spectroscopic study of the mineral shattuckite $Cu_5(SiO_3)_4(OH)_2$. Spectrochim Acta A 87:241–244

Frost RL, Xi Y (2012c) Vibrational spectroscopic study of the copper silicate mineral kinoite $Ca_2Cu_2Si_3O_{10}(OH)_4$. Spectrochim Acta A 89:88–92

Frost RL, Xi Y (2012d) A vibrational spectroscopic study of plancheite $Cu_8Si_8O_{22}(OH)_4 \cdot H_2O$. Spectrochim Acta A 91:314–318

Frost RL, Xi Y (2012e) Whelanite $Ca_5Cu_2(OH)_2CO_3, Si_6O_{17} \cdot 4H_2O$ – a vibrational spectroscopic study. Spectrochim Acta A 91:319–323

Frost RL, Xi Y (2012f) Vibrational spectroscopy and solubility study of the mineral stringhamite $CaCuSiO_4 \cdot H_2O$. Spectrochim Acta A 91:324–328

Frost RL, Xi Y (2012g) Vibrational spectroscopic study of the mineral creaseyite $Cu_2Pb_2(Fe,Al)_2(Si_5O_{17}) \cdot 6H_2O$ – a zeolite mineral. Spectrochim Acta A 94:6–11

Frost RL, Xi Y (2012h) Vibrational spectroscopic study of the minerals cavansite and pentagonite $Ca(V^{4+}O)Si_4O_{10} \cdot 4H_2O$. Spectrochim Acta A 95:263–269

Frost RL, Xi Y (2012i) Raman spectroscopy of the borate mineral ameghinite $NaB_3O_3(OH)_4$. Spectrochim Acta A 96:89–94

Frost RL, Xi Y (2012j) Assessment of the molecular structure of the borate mineral sakhaite

Ca$_{12}$Mg$_4$(BO$_3$)$_7$(CO$_3$)$_4$Cl(OH)$_2$·H$_2$O using vibrational spectroscopy. Spectrochim Acta A 96:611–616

Frost RL, Xi Y (2012k) Raman spectroscopic study of the copper silicate mineral apachite Cu$_9$Si$_{10}$O$_{29}$·11H$_2$O. Spectrosc Lett 45(8):575–580

Frost RL, Xi Y (2012l) Molecular structure of the phosphate mineral brazilianite NaAl$_3$(PO$_4$)$_2$(OH)$_4$ – a semi precious jewel. J Molec Struct 1010:179–183

Frost RL, Xi Y (2012m) Vibrational spectroscopic study of the copper silicate mineral ajoite (K,Na)Cu$_7$AlSi$_9$O$_{24}$(OH)$_6$·3H$_2$O. J Molec Struct 1018:72–77

Frost RL, Xi Y (2012n) Vibrational spectroscopic study of the minerals nekoite Ca$_3$Si$_6$O$_{15}$·7H$_2$O and okenite Ca$_{10}$Si$_{18}$O$_{46}$·18H$_2$O – implications for the molecular structure. J Molec Struct 1020:96–104

Frost RL, Xi Y (2012o) Raman spectroscopic study of the minerals apophyllite-(KF) KCa$_4$Si$_8$O$_{20}$F·8H$_2$O and apophyllite-(KOH) KCa$_4$Si$_8$O$_{20}$(F,OH)·8H$_2$O. J Molec Struct 1028:200–207

Frost RL, Xi Y (2013a) Is chrysocolla (Cu, Al)$_2$H$_2$Si$_2$O$_5$(OH)$_4$·nH$_2$O related to spertiniite Cu(OH)$_2$? – a vibrational spectroscopic study. Vib Spectrosc 64:33–38

Frost RL, Xi Y (2013b) Vibrational spectroscopic study of the mineral penkvilksite Na$_2$TiSi$_4$O$_{11}$·2H$_2$O – a mineral used for the uptake of radionuclides. Radiat Effects Defects Solids 168(1):72–79

Frost RL, Xi Y (2013c) Vibrational spectroscopy of the borate mineral kotoite Mg$_3$(BO$_3$)$_2$. Spectrochim Acta A 103:151–155

Frost RL, Xi Y (2013d) A vibrational spectroscopic study of the so-called healing mineral papagoite CaCuAlSi$_2$O$_6$(OH)$_3$. Spectrosc Lett 46(5):344–349

Frost RL, Xi Y (2013e) Vibrational spectroscopy of the borate mineral henmilite Ca$_2$Cu[B(OH)$_4$]$_2$(OH)$_4$. Spectrochim Acta A 103:356–360

Frost RL, Xi Y (2014) Vibrational spectroscopy of the borate mineral pinnoite MgB$_2$O(OH)$_6$. Spectrochim Acta A 117:428–433

Frost RL, Fredericks PM, Bartlett JR (1993) Fourier transform Raman spectroscopy of kandite clays. Spectrochim Acta A 49(5):667–674

Frost RL, Kristof J, Rintoul L, Kloprogge JT (2000) Raman spectroscopy of urea and urea-intercalated kaolinites at 77 K. Spectrochim Acta A 56(9):1681–1691

Frost RL, Williams PA, Kloprogge JT, Leverett P (2001) Raman spectroscopy of descloizite and mottramite at 298 and 77 K. J Raman Spectrosc 32(11):906–911

Frost RL, Kloprogge T, Williams PA, Martens W, Johnson TE, Leverett P (2002a) Vibrational spectroscopy of the basic copper phosphate minerals: pseudomalachite, ludjibaite and reichenbachite. Spectrochim Acta A 58(13):2861–2868

Frost RL, Martens W, Kloprogge JT, Williams PA (2002b) Raman spectroscopy of the basic copper chloride minerals atacamite and paratacamite: implications for the study of copper, brass and bronze objects of archaeological significance. J Raman Spectrosc 33(10):801–806

Frost RL, Martens WN, Kloprogge T, Williams PA (2002c) Vibrational spectroscopy of the basic manganese and ferric phosphate minerals: strunzite, ferrostrunzite and ferristrunzite. N Jb Miner Mh 2002(11):481–496

Frost RL, Martens WN, Rintoul L, Mahmutagic E, Kloprogge JT (2002d) Raman spectroscopic study of azurite and malachite at 298 and 77 K. J Raman Spectrosc 33(4):252–259

Frost RL, Martens WN, Williams PA (2002e) Raman spectroscopy of the phase-related basic copper arsenate minerals olivenite, cornwallite, cornubite and clinoclase. J Raman Spectrosc 33(6):475–484

Frost RL, Martens W, Williams PA, Kloprogge JT (2002f) Raman and infrared spectroscopic study of the vivianite-group phosphates vivianite, baricite and bobierrite. Mineral Mag 66(6):1063–1073

Frost RL, Williams PA, Martens W, Kloprogge JT, Leverett P (2002g) Raman spectroscopy of the basic copper phosphate minerals cornetite, libethenite, pseudomalachite, reichenbachite and ludjibaite. J Raman Spectrosc 33(4):260–263

Frost RL, Crane M, Williams PA, Kloprogge JT (2003a) Isomorphic substitution in vanadinite [Pb$_5$(VO$_4$)$_3$Cl] – a Raman spectroscopic study. J Raman Spectrosc 34(3):214–220

Frost RL, Duong L, Martens W (2003b) Molecular assembly in secondary minerals – Raman spectroscopy of the arthurite group species arthurite and whitmoreite. N Jb Miner Mh 2003(5):223–240

Frost RL, Erickson KL, Weier ML, Mills S (2003c) Raman spectroscopy of the phosphate minerals: cacoxenite and gormanite. Asian Chem Lett 7(4):197–203

Frost RL, Kloprogge JT, Williams PA (2003e) Raman spectroscopy of lead sulphate-carbonate minerals – implications for hydrogen bonding. N Jb Miner Mh 2003(12):529–542

Frost RL, Martens W, Kloprogge JT, Ding Z (2003f) Raman spectroscopy of selected lead minerals of environmental significance. Spectrochim Acta A 59(12):2705–2711

Frost RL, Martens W, Williams PA, Kloprogge JT (2003g) Raman spectroscopic study of the vivianite arsenate minerals. J Raman Spectrosc 34(10):751–759

Frost RL, Weier ML, Kloprogge JT (2003h) Raman spectroscopy of some natural hydrotalcites with sulphate and carbonate in the interlayer. J Raman Spectrosc 34(10):760–768

Frost RL, Williams PA, Kloprogge JT, Martens W (2003i) Raman spectroscopy of the copper chloride minerals nantokite, eriochalcite and claringbullite – implications for copper corrosion. N Jb Miner Mh 2003(10):433–445

Frost RL, Williams PA, Martens W (2003j) Raman spectroscopy of the minerals boleite, cumengeite, diaboleite and phosgenite – implications for the analysis of cosmetics of antiquity. Mineral Mag 67(1):103–111

Frost RL, Yang J, Ding Z (2003k) Raman and FTIR spectroscopy of natural oxalates: implications for the evidence of life on Mars. Chin Sci Bull 48(17):1844–1852

Frost RL, Adebajo M, Weier ML (2004a) A Raman spectroscopic study of thermally treated glushinskite – the natural magnesium oxalate dihydrate. Spectrochim Acta A 60(3):643–651

Frost RL, Carmody O, Erickson KL, Weier ML, Čejka J (2004b) Molecular structure of the uranyl mineral andersonite – a Raman spectroscopic study. J Molec Struct 703(1):47–54

Frost RL, Duong L, Weier M (2004c) Raman microscopy of the molybdate minerals koechlinite, iriginite and lindgrenite. N Jb Miner Abh 180(3):245–260

Frost RL, Duong L, Weier M (2004d) Raman microscopy of selected tungstate minerals. Spectrochim Acta A 60(8):1853–1859

Frost RL, Erickson KL, Weier ML (2004e) Hydrogen bonding in selected vanadates: a Raman and infrared spectroscopy study. Spectrochim Acta A 60(10):2419–2423

Frost RL, Erickson KL, Weier ML, McKinnon AR, Williams PA, Leverett P (2004f) Effect of the lanthanide ionic radius on the Raman spectroscopy of lanthanide agardite minerals. J Raman Spectrosc 35(11):961–966

Frost RL, Henry DA, Erickson K (2004g) Raman spectroscopic detection of wyartite in the presence of rabejacite. J Raman Spectrosc 35(4):255–260

Frost RL, Leverett P, Williams PA, Weier ML, Erickson KL (2004h) Raman spectroscopy of gerhardtite at 298 and 77 K. J Raman Spectrosc 35(11):991–996

Frost RL, Martens W, Kloprogge JT (2004i) Synthetic deuterated erythrite – a vibrational spectroscopic study. Spectrochim Acta A 60(1):343–349

Frost RL, Mills SJ, Erickson KL (2004j) Thermal decomposition of peisleyite: a thermogravimetry and hot stage Raman spectroscopic study. Thermochim Acta 419(1):109–114

Frost RL, Weier ML, Adebajo MO (2004k) Thermal decomposition of metazeunerite – a high-resolution thermogravimetric and hot-stage Raman spectroscopic study. Thermochim Acta 419(1):119–129

Frost RL, Weier ML, Erickson KL, Carmody O, Mills S (2004l) Raman spectroscopy of phosphates of the variscite mineral group. J Raman Spectrosc 35(12):1047–1055

Frost RL, Weier M, Lyon WG (2004m) Metavivianite an intermediate mineral phase between vivianite, and ferro/ferristrunzite – a Raman spectroscopic study. N Jb Miner Mh 2004(5):228–240

Frost RL, Williams PA, Martens W, Leverett P, Kloprogge JT (2004n) Raman spectroscopy of basic copper (II) and some complex copper (II) sulfate minerals: implications for hydrogen bonding. Am Mineral 89(7):1130–1137

Frost RL, Adebajo MO, Erickson KL (2005a) Raman spectroscopy of synthetic and natural iowaite. Spectrochim Acta A 61(4):613–620

Frost RL, Carmody O, Erickson KL, Weier ML, Henry DO, Čejka J (2005b) Molecular structure of the uranyl mineral uranopilite – a Raman spectroscopic study. J Molec Struct 733(1):203–210

Frost RL, Čejka J, Weier ML, Martens W, Henry DA (2005c) Vibrational spectroscopy of selected natural uranyl vanadates. Vib Spectrosc 39(2):131–138

Frost RL, Erickson KL, Weier ML, Carmody O (2005d) Raman and infrared spectroscopy of selected vanadates. Spectrochim Acta A 61(5):829–834

Frost RL, Erickson KL, Čejka J, Reddy BJ (2005e) A Raman spectroscopic study of the uranyl sulphate mineral johannite. Spectrochim Acta A 61(11):2702–2707

Frost R, Erickson K, Kloprogge T (2005f) Vibrational spectroscopic study of the nitrate containing hydrotalcite mbobomkulite. Spectrochim Acta A 61:2919–2925

Frost RL, Erickson KL, Weier ML, Carmody O, Čejka J (2005g) Raman spectroscopic study of the uranyl tricarbonate mineral liebigite. J Molec Struct 737(2):173–181

Frost RL, Erickson KL, Weier ML, Leverett P, Williams PA (2005h) Raman spectroscopy of likasite at 298 and 77 K. Spectrochim Acta A 61(4):607–612

Frost RL, Weier ML, Bostrom T, Čejka J, Martens W (2005i) Molecular structure of the uranyl mineral zippeite – an XRD, SEM and Raman spectroscopic study. N Jb Miner Abh 181(3):271–280

Frost RL, Weier ML, Martens WN, Henry DA, Mills SJ (2005j) Raman spectroscopy of newberyite, hannayite and struvite. Spectrochim Acta A 62(1):181–188

Frost RL, Weier ML, Martens WN, Kloprogge JT, Kristóf J (2005k) Thermo-Raman spectroscopic study of the uranium mineral sabugalite. J Raman Spectrosc 36(8):797–805

Frost RL, Weier ML, Martens W, Mills S (2005l) Molecular structure of segnitite: a Raman spectroscopic study. J Molec Struct 752(1):178–185

Frost RL, Wills R-A, Martens WN (2005m) Raman spectroscopy of beaverite and plumbojarosite. J Raman Spectrosc 36(12):1106–1112

Frost RL, Čejka J, Weier M (2006a) A Raman spectroscopic study of the uranyl phosphate mineral threadgoldite. Spectrochim Acta A 65(3):797–801

Frost RL, Čejka J, Weier M, Ayoko GA (2006b) A Raman spectroscopic study of the uranyl tellurite mineral schmitterite. Spectrochim Acta A 65(3):571–574

Frost RL, Čejka J, Weier M, Ayoko GA (2006c) Raman spectroscopic study of the uranyl phosphate mineral dewindtite. J Raman Spectrosc 37(12):1362–1367

Frost RL, Čejka J, Weier ML, Martens W (2006d) A Raman and infrared spectroscopic study of the uranyl silicates – weeksite, soddyite and haiweeite: Part 2. Spectrochim Acta A 63(2):305–312

Frost RL, Čejka J, Weier ML, Martens W (2006e) Molecular structure of the uranyl silicates – a Raman spectroscopic study. J Raman Spectrosc 37(4):538–551

Frost RL, Čejka J, Weier M, Martens WN (2006f) A Raman spectroscopic study of the uranyl phosphate mineral parsonsite. J Raman Spectrosc 37(9):879–891

Frost RL, Čejka J, Weier ML, Martens WN (2006g) Raman spectroscopy study of selected uranophanes. J Molec Struct 788(1):115–125

Frost RL, Čejka J, Weier ML, Martens W, Kloprogge JT (2006h) A Raman and infrared spectroscopic study of the uranyl silicates – weeksite, soddyite and haiweeite. Spectrochim Acta A 64(2):308–315

Frost RL, Henry DA, Weier ML, Martens W (2006i) Raman spectroscopy of three polymorphs of $BiVO_4$: clinobisvanite, dreyerite and pucherite, with comparisons to $(VO_4)^{3-}$-bearing minerals: namibite, pottsite and schumacherite. J Raman Spectrosc 37(7):722–732

Frost RL, Reddy BJ, Martens WN, Weier M (2006j) The molecular structure of the phosphate mineral turquoise – a Raman spectroscopic study. J Molec Struct 788(1):224–231

Frost RL, Weier ML, Čejka J, Ayoko GA (2006k) Raman spectroscopy of uranyl rare earth carbonate kamotoite-(Y). Spectrochim Acta A 65(3):529–534

Frost RL, Weier ML, Čejka J, Kloprogge JT (2006l) Raman spectroscopy of walpurgite. J Raman Spectrosc 37(5):585–590

Frost RL, Weier M, Martens W (2006m) Using Raman spectroscopy to identify mixite minerals. Spectrochim Acta A 63(1):60–65

Frost RL, Weier M, Martens WN (2006n) Raman microscopy of synthetic goudeyite ($YCu_6(AsO_4)_2(OH)_6 \cdot 3H_2O$). Spectrochim Acta A 63(3):685–689

Frost RL, Weier ML, Martens WN, Mills SJ (2006o) The hydroxylated nickel carbonates otwayite and paraotwayite – a SEM, EDX and vibrational spectroscopic study. N Jb Miner Abh 183(1):107–116

Frost RL, Weier ML, Martens WN, Mills SJ (2006p) Thermo Raman spectroscopic study of kintoreite. Spectrochim Acta A 63(2):282–288

Frost RL, Weier ML, Reddy BJ, Čejka J (2006q) A Raman spectroscopic study of the uranyl selenite mineral haynesite. J Raman Spectrosc 37(8):816–821

Frost RL, Wills RA, Weier ML, Martens W, Mills S (2006r) A Raman spectroscopic study of selected natural jarosites. Spectrochim Acta A 63(1):1–8

Frost RL, Bouzaid J, Butler IS (2007a) Raman spectroscopic study of the molybdate mineral szenicsite and comparison with other paragenetically related molybdate minerals. Spectrosc Lett 40(4):603–614

Frost RL, Bouzaid JM, Martens WN, Reddy BJ (2007b) Raman spectroscopy of the borosilicate mineral ferroaxinite. J Raman Spectrosc 38(2):135–141

Frost RL, Bouzaid JM, Palmer S (2007c) The structure of mimetite, arsenian pyromorphite and hedyphane – a Raman spectroscopic study. Polyhedron 26(13):2964–2970

Frost RL, Čejka J, Ayoko GA, Dickfos MJ (2007d) Raman spectroscopic study of the multi-anion uranyl mineral schroeckingerite. J Raman Spectrosc 38(12):1609–1614

Frost RL, Čejka J, Ayoko GA, Weier M (2007e) A Raman spectroscopic study of the uranyl phosphate mineral bergenite. Spectrochim Acta A 66(4):979–984

Frost RL, Čejka J, Ayoko GA, Weier ML (2007f) Raman spectroscopic and SEM analysis of sodium zippeite. J Raman Spectrosc 38(10):1311–1319

Frost RL, Čejka J, Ayoko GA, Weier ML (2007g) Vibrational spectroscopic study of hydrated uranyl oxide: curite. Polyhedron 26(14):3724–3730

Frost RL, Čejka J, Weier ML (2007h) Raman spectroscopic study of the uranyl oxyhydroxide hydrates: becquerelite, billietite, curite, schoepite and vandendriesscheite. J Raman Spectrosc 38(4):460–466

Frost RL, Čejka J, Weier ML, Martens WN, Ayoko GA (2007i) Raman spectroscopy of uranopilite of different origins – implications for molecular structure. J Raman Spectrosc 38(4):398–409

Frost RL, Hales MC, Reddy BJ (2007j) Aurichalcite – an SEM and Raman spectroscopic study. Polyhedron 26(13):3291–3300

Frost RL, Palmer SJ, Bouzaid JM, Reddy BJ (2007k) A Raman spectroscopic study of humite minerals. J Raman Spectrosc 38(1):68–77

Frost RL, Weier ML, Mills SJ (2007l) A vibrational spectroscopic study of perhamite, an unusual silicophosphate. Spectrochim Acta A 67(3):604–610

Frost RL, Weier ML, Williams PA, Leverett P, Kloprogge JT (2007m) Raman spectroscopy of the sampleite group of minerals. J Raman Spectrosc 38(5):574–583

Frost RL, Wills RA, Martens WN (2007n) A Raman spectroscopic study of synthetic giniite. Spectrochim Acta A 66(1):42–47

Frost RL, Čejka J, Ayoko G (2008a) Raman spectroscopic study of the uranyl phosphate minerals phosphuranylite and yingjiangite. J Raman Spectrosc 39(4):495–502

Frost RL, Čejka J, Ayoko GA, Dickfos MJ (2008b) Raman spectroscopic study of the uranyl carbonate mineral voglite. J Raman Spectrosc 39(3):374–379

Frost RL, Čejka J, Dickfos MJ (2008c) Raman and infrared spectroscopic study of the molybdate-containing uranyl mineral calcurmolite. J Raman Spectrosc 39(7):779–785

Frost RL, Čejka J, Dickfos MJ (2008d) Raman spectroscopic study of the uranyl selenite mineral demesmaekerite $Pb_2Cu_5(UO_2)_2(SeO_3)_6(OH)_6 \cdot 2H_2O$. J Raman Spectrosc 40:476–480

Frost RL, Čejka J, Keeffe EC, Dickfos MJ (2008e) Raman spectroscopic study of the uranyl selenite mineral marthozite $Cu[(UO_2)_3(SeO_3)_2O_2] \cdot 8H_2O$. J Raman Spectrosc 39(10):1413–1418

Frost RL, Dickfos MJ, Čejka J (2008f) Raman spectroscopic study of the uranyl carbonate mineral zellerite. J Raman Spectrosc 39(5):582–586

Frost RL, Dickfos MJ, Čejka J (2008g) Raman spectroscopic study of the uranyl mineral, compreignacite,

Frost RL, Dickfos MJ, Keeffe EC (2008g) Raman spectroscopic study of the uranyl mineral $K_2[(UO_2)_3O_2(OH)_3]_2 \cdot 7H_2O$. J Raman Spectrosc 39(9):1158–1161

Frost RL, Dickfos MJ, Keeffe EC (2008h) Raman spectroscopic study of the tellurite minerals: rajite and denningite. Spectrochim Acta A 71(4):1512–1515

Frost RL, Dickfos MJ, Keeffe EC (2008i) Raman spectroscopic study of the tellurite minerals: emmonsite $Fe^{3+}{}_2Te^{4+}{}_3O_9 \cdot 2H_2O$ and zemannite $Mg_{0.5}[Zn^{2+}Fe^{3+}(TeO_3)_3]4.5H_2O$. J Raman Spectrosc 39(12):1784–1788

Frost RL, Jagannadha Reddy B, Dickfos MJ (2008j) Raman spectroscopy of the nickel silicate mineral pecoraite – an analogue of chrysotile (asbestos). J Raman Spectrosc 39(7):909–913

Frost RL, Locke A, Martens WN (2008k) Synthesis and Raman spectroscopic characterisation of the oxalate mineral wheatleyite $Na_2Cu(C_2O_4)_2 \cdot 2H_2O$. J Raman Spectrosc 39(7):901–908

Frost RL, Dickfos MJ, Jagannadha Reddy B (2008l) Raman spectroscopy of hydroxy nickel carbonate minerals nullaginite and zaratite. J Raman Spectrosc 39(9):1250–1256

Frost RL, Bahfenne S, Graham J (2009a) Raman spectroscopic study of the magnesium carbonate minerals – artinite and dypingite. J Raman Spectrosc 40(8):855–860

Frost RL, Čejka J, Dickfos MJ (2009b) Raman spectroscopic study of the uranyl tellurite mineral moctezumite $PbUO_2(TeO_3)_2$. J Raman Spectrosc 40:38–41

Frost RL, Čejka J, Dickfos MJ (2009c) Raman spectroscopic study of the mineral guilleminite $Ba(UO_2)_3(SeO_3)_2(OH)_4 \cdot 3H_2O$. J Raman Spectrosc 40(4):355–359

Frost RL, Čejka J, Dickfos MJ (2009d) Raman spectroscopic study of the uranyl minerals vanmeersscheite $U(OH)_4[(UO_2)_3(PO_4)_2(OH)_2] \cdot 4H_2O$ and arsenouranylite $Ca(UO_2)[(UO_2)_3(AsO_4)_2(OH)_2] \cdot (OH)_2 \cdot 6H_2O$. Spectrochim Acta A 71(5):1799–1803

Frost RL, Čejka J, Sejkora J, Ozdín D, Bahfenne S, Keeffe EC (2009e) Raman spectroscopic study of the antimonate mineral brandholzite $Mg[Sb_2(OH)_{12}] \cdot 6H_2O$. J Raman Spectrosc 40(12):1907–1910

Frost RL, Čejka J, Sejkora J, Plášil J, Bahfenne S, Palmer SJ (2009f) Raman microscopy of the mixite mineral $BiCu_6(AsO_4)_3(OH)_6 \cdot 3H_2O$ from the Czech Republic. J Raman Spectrosc 41(5):566–570

Frost RL, Dickfos MJ, Keeffe EC (2009g) Raman spectroscopic study of the tellurite minerals: carlfriesite and spiroffite. Spectrochim Acta A 71(5):1663–1666

Frost RL, Plášil J, Čejka J, Sejkora J, Keeffe EC, Bahfenne S (2009h) Raman spectroscopic study of the uranyl mineral pseudojohannite $Cu_{6.5}[(UO_2)_4O_4(SO_4)_2]_2(OH)_5 \cdot 25H_2O$. J Raman Spectrosc 40(12):1816–1821

Frost RL, Sejkora J, Čejka J, Keeffe EC (2009i) Vibrational spectroscopic study of the arsenate mineral strashimirite $Cu_8(AsO_4)_4(OH)_4 \cdot 5H_2O$ – relationship to other basic copper arsenates. Vibrational Spectroscopy 50(2):289–297

Frost RL, Sejkora J, Čejka J, Keeffe EC (2009j) Raman spectroscopic study of the mixed anion sulphate–arsenate mineral parnauite $Cu_9[(OH)_{10}|SO_4|(AsO_4)_2] \cdot 7H_2O$. J Raman Spectrosc 40(11):1546–1550

Frost RL, Bahfenne S, Čejka J, Sejkora J, Plášil J, Palmer SJ (2010a) Raman and infrared study of phyllosilicates containing heavy metals (Sb, Bi): bismutoferrite and chapmanite. J Raman Spectrosc 41(7):814–819

Frost RL, Bahfenne S, Čejka J, Sejkora J, Plášil J, Palmer SJ (2010b) Raman spectroscopic study of the hydrogen-arsenate mineral pharmacolite $Ca(AsO_3OH) \cdot 2H_2O$ – implications for aquifer and sediment remediation. J Raman Spectrosc 41(10):1348–1352

Frost RL, Bahfenne S, Keeffe EC (2010c) Raman spectroscopic study of the mineral gerstleyite $Na_2(Sb,As)_8S_{13} \cdot 2H_2O$ and comparison with some heavy-metal sulfides. J Raman Spectrosc 41(12):1779–1783

Frost RL, Bakon KH, Palmer SJ (2010d) Raman spectroscopic study of synthetic reevesite and cobalt substituted reevesite $(Ni,Co)_6Fe_2(OH)_{16}(CO3) \cdot 4H_2O$. J Raman Spectrosc 41(1):78–83

Frost RL, Čejka J, Sejkora J, Plášil J, Bahfenne S, Palmer SJ (2010e) Raman spectroscopy of the basic copper arsenate mineral: euchroite. J Raman Spectrosc 41(5):571–575

Frost RL, Keeffe EC, Bahfenne S (2010f) A Raman spectroscopic study of the antimonite mineral peretaite $Ca(SbO)_4(OH)_2(SO_4)_2 \cdot 2H_2O$. Spectrochim Acta A 75(5):1476–1479

Frost RL, Palmer SJ, Keeffe EC (2010g) Raman spectroscopic study of the hydroxyl-arsenate-sulfate mineral chalcophyllite $Cu_{18}Al_2(AsO_4)_4(SO_4)_3(OH)_{24} \cdot 36H_2O$. J Raman Spectrosc 41(12):1769–1774

Frost RL, Bahfenne S, Čejka J, Sejkora J, Palmer SJ, Škoda R (2010h) Raman microscopy of haidingerite $Ca(AsO_3OH) \cdot H_2O$ and brassite $Mg(AsO_3OH) \cdot 4H_2O$. J Raman Spectrosc 41(6):690–693

Frost RL, Bahfenne S, Čejka J, Sejkora J, Plášil J, Palmer SJ, Keefe EC, Němec I (2011a) Dussertite $BaFe^{3+}{}_3(AsO_4)_2(OH)_5$ – a Raman spectroscopic study of a hydroxyl-arsenate mineral. J Raman Spectrosc 42(1):56–61

Frost RL, Čejka J, Sejkora J, Plášil J, Reddy BJ, Keeffe EC (2011b) Raman spectroscopic study of a hydroxy-arsenate mineral containing bismuth – atelestite $Bi_2O(OH)(AsO_4)$. Spectrochim Acta A 78(1):494–496

Frost RL, Palmer SJ, Bahfenne S (2011c) A near-infrared and Raman spectroscopic study of the mineral richelsdorfite $Ca_2Cu5Sb[Cl|(OH)_6|(AsO_4)_4] \cdot 6H_2O$. Spectrochim Acta A 78(4):1302–1304

Frost RL, Palmer SJ, Čejka J (2011d) The application of Raman spectroscopy to the study of the uranyl mineral coconinoite $Fe^{3+}{}_2Al_2(UO_2)_2(PO_4)_4(SO_4)(OH)_2 \cdot 20H_2O$. Spectrosc Lett 44(6):381–387

Frost RL, Palmer SJ, Čejka J, Sejkora J, Plášil J, Bahfenne S, Keeffe EC (2011e) A Raman spectroscopic study of the different vanadate groups in solid-state compounds – model case: mineral phases vésigniéite [BaCu$_3$(VO$_4$)$_2$(OH)$_2$] and volborthite [Cu$_3$V$_2$O$_7$(OH)$_2$·2H$_2$O]. J Raman Spectrosc 42(8):1701–1710

Frost RL, Palmer SJ, Čejka J, Sejkora J, Plášil J, Jebavá I, Keeffe EC (2011f) A Raman spectroscopic study of M^{2+}M^{3+} sulfate minerals, römerite Fe^{2+}Fe^{3+}$_2$(SO$_4$)$_4$·14H$_2$O and botryogen Mg^{2+}Fe^{3+}(SO$_4$)$_2$(OH)·7H$_2$O. J Raman Spectrosc 42(4):825–830

Frost RL, Palmer SJ, Henry DA, Pogson R (2011g) A Raman spectroscopic study of the 'cave' mineral ardealite Ca$_2$(HPO$_4$)(SO$_4$)·4H$_2$O. J Raman Spectrosc 42(6):1447–1454

Frost RL, Palmer SJ, Reddy BJ (2011h) Raman spectroscopic study of the uranyl titanate mineral euxenite (Y, Ca,U,Ce,Th)(Nb,Ta,Ti)$_2$O$_6$. J Raman Spectrosc 42(5):1160–1162

Frost RL, Palmer SJ, Spratt HJ, Martens WN (2011i) The molecular structure of the mineral beudantite PbFe$_3$(AsO$_4$,SO$_4$)$_2$(OH)$_6$ – implications for arsenic accumulation and removal. J Molec Struct 988(1):52–58

Frost RL, Palmer SJ, Theiss F (2011j) Synthesis and Raman spectroscopic characterisation of hydrotalcites based on the formula Ca$_6$Al$_2$(CO$_3$)(OH)$_{16}$·4H$_2$O. J Raman Spectrosc 42(5):1163–1167

Frost RL, Palmer SJ, Xi Y (2011k) A vibrational spectroscopic study of the mineral hinsdalite PbAl$_3$(SO$_4$)(PO$_4$)(OH)$_6$. J Molec Struct 1001(1):43–48

Frost RL, Palmer SJ, Xi Y (2011l) A Raman spectroscopic study of the mono-hydrogen phosphate mineral dorfmanite Na$_2$(PO$_3$OH)·2H$_2$O and in comparison with brushite. Spectrochim Acta A 82(1):132–136

Frost RL, Palmer SJ, Xi Y (2011m) Vibrational spectroscopy of the multi-anion mineral zykaite Fe$_4$(AsO$_4$)$_3$(SO$_4$)(OH)·15H$_2$O – implications for arsenate removal. Spectrochim Acta A 83(1):444–448

Frost RL, Palmer SJ, Xi Y (2011n) Raman spectroscopy of the multi anion mineral arsentsumebite Pb$_2$Cu(AsO$_4$)(SO$_4$)(OH) and in comparison with tsumebite Pb$_2$Cu(PO$_4$)(SO$_4$)(OH). Spectrochim Acta A 83(1):449–452

Frost RL, Palmer SJ, Xi Y (2011o) The molecular structure of the multianion mineral hidalgoite PbAl$_3$(SO$_4$)(AsO$_4$)(OH)$_6$ – implications for arsenic removal from soils. J Molec Struct 1005(1):214–219

Frost RL, Palmer SJ, Xi Y, Tan K (2011p) Raman spectroscopy of the multi-anion mineral mallestigite Pb$_3$Sb^{5+}(SO$_4$)(AsO$_4$)(OH)$_6$·3H$_2$O: a mineral of archaeological significance. Spectrochim Acta A 83(1):432–436

Frost RL, Sejkora J, Čejka J, Plášil J, Bahfenne S, Keeffe EC (2011q) Raman spectroscopy of hydrogen arsenate group (AsO$_3$OH)$^{2-}$ in solid-state compounds: cobalt-containing zinc arsenate mineral, koritnigite (Zn,Co)(AsO$_3$OH)·H$_2$O. J Raman Spectrosc 42(3):534–539

Frost RL, Xi Y, Palmer SJ (2011r) Are the 'cave' minerals archerite (K,NH$_4$)H$_2$PO$_4$ and biphosphammite (K,NH$_4$)H$_2$PO$_4$ identical? A molecular structural study. J Molec Struct 1001(1):49–55

Frost RL, Xi Y, Palmer SJ (2011s) Molecular structure of the mineral woodhouseite CaAl$_3$(PO$_4$,SO$_4$)$_2$(OH)$_6$. J Molec Struct 1001(1):56–61

Frost RL, Xi Y, Palmer SJ (2011t) The structure of the mineral leogangite Cu$_{10}$(OH)$_6$(SO$_4$)(AsO$_4$)$_4$·8H$_2$O – implications for arsenic accumulation and removal. Spectrochim Acta A 82(1):221–227

Frost RL, Xi Y, Palmer SJ, Millar GJ, Tan K, Pogson RE (2011u) Vibrational spectroscopy of synthetic stercorite H(NH$_4$)Na(PO$_4$)·4H$_2$O – a comparison with the natural cave mineral. Spectrochim Acta A 84(1):269–274

Frost RL, Xi Y, Palmer SJ, Pogson RE (2011v) Vibrational spectroscopic analysis of taranakite (K,NH$_4$)Al$_3$(PO$_4$)$_3$(OH)·9(H$_2$O) from the Jenolan Caves, Australia. Spectrochim Acta A 83(1):106–111

Frost RL, Xi Y, Palmer SJ, Pogson R (2011w) Vibrational spectroscopic analysis of the mineral crandallite CaAl$_3$(PO$_4$)$_2$(OH)$_5$·(H$_2$O) from the Jenolan Caves, Australia. Spectrochim Acta A 82(1):461–466

Frost RL, Couperthwaite SJ, Xi Y (2012a) Vibrational spectroscopy of the multianion mineral kemmlitzite (Sr,Ce)Al$_3$(AsO$_4$)(SO$_4$)(OH)$_6$. Spectrosc Lett 45(7):482–486

Frost RL, Mahendran M, Poologanathan K, Xi Y (2012b) Raman spectroscopic study of the mineral xonotlite Ca$_6$Si$_6$O$_{17}$(OH)$_2$ – a component of plaster boards. Mater Res Bull 47(11):3644–3649

Frost RL, Palmer SJ, Xi Y (2012c) Raman spectroscopy of the multi-anion mineral schlossmacherite (H$_3$O,Ca)Al$_3$(AsO$_4$,PO$_4$,SO$_4$)$_2$(OH)$_6$. Spectrochim Acta A 87:209–213

Frost RL, Palmer SJ, Xi Y (2012d) Is the mineral borickyite (Ca,Mg)(Fe^{3+},Al)$_4$(PO$_4$,SO$_4$,CO$_3$)(OH)$_8$·3-7.5H$_2$O the same as delvauxite CaFe^{3+}$_4$(PO$_4$,SO$_4$)$_2$(OH)$_8$·4-6H$_2$O. Spectrochim Acta A 92:377–381

Frost RL, Xi Y, Palmer SJ, Pogson RE (2012e) Identification of montgomeryite mineral [Ca$_4$MgAl$_4$(PO$_4$)$_6$·(OH)$_4$·12H$_2$O] found in the Jenolan Caves – Australia. Spectrochim Acta A 94:1–5

Frost RL, Xi Y, Palmer SJ, Tan K, Millar GJ (2012f) Vibrational spectroscopy of synthetic archerite (K, NH$_4$)H$_2$PO$_4$$^-$ and in comparison with the natural cave mineral. J Molec Struct 1011:128–133

Frost RL, Xi Y, Pogson RE (2012g) Raman spectroscopic study of the mineral arsenogorceixite BaAl$_3$AsO$_3$(OH)(AsO$_4$,PO$_4$)(OH,F)$_6$. Spectrochim Acta A 91:301–306

Frost RL, Xi Y, Pogson RE, Millar GJ, Tan K, Palmer SJ (2012h) Raman spectroscopy of synthetic CaHPO$_4$·2H$_2$O – and in comparison with the cave mineral brushite. J Raman Spectrosc 43(4):571–576

Frost RL, Xi Y, Scholz R, Belotti FM (2012i) Infrared and Raman spectroscopic characterization of the phosphate mineral kosnarite KZr$_2$(PO$_4$)$_3$ in comparison with other pegmatitic phosphates. Transit Metal Chem 37 (8):777–782

Frost RL, Xi Y, Scholz R, Belotti FM, Lagoeiro LE (2012j) Chemistry, Raman and infrared spectroscopic characterization of the phosphate mineral reddingite: $(MnFe)_3(PO_4)_2(H_2O,OH)_3$, a mineral found in lithium-bearing pegmatite. Phys Chem Minerals 39 (10):803–810

Frost RL, Xi Y, Scholz R, Belotti FM, Menezes Filho LAD (2012k) Raman and infrared spectroscopic characterization of beryllonite, a sodium and beryllium phosphate mineral – implications for mineral collectors. Spectrochim Acta A 97:1058–1062

Frost RL, Xi Y, Tan K, Millar GJ, Palmer SJ (2012l) Vibrational spectroscopic study of the mineral pitticite Fe, AsO_4, SO_4, H_2O. Spectrochim Acta A 85 (1):173–178

Frost RL, Xi Y, Scholz R (2012m) Assessment of the molecular structure of the borate mineral boracite $Mg_3B_7O_{13}Cl$ using vibrational spectroscopy. Spectrochim Acta A 96:946–951

Frost RL, Xi Y, Couperthwaite SJ (2012n) Vibrational spectroscopic study of multianion mineral clinotyrolite $Ca_2Cu_9[(As,S)O_4]_4(OH)_{10}\cdot 10(H_2O)$. Spectrochim Acta A 95:258–262

Frost RL, López A, Scholz R, Xi Y, da Silveira AJ, Lima RMF (2013a) Characterization of the sulphate mineral amarantite using infrared, Raman spectroscopy and thermogravimetry. Spectrochim Acta A 114:85–91

Frost RL, Lópes A, Scholz R, Xi Y (2013b) Infrared and Raman spectroscopic characterization of the arsenate mineral ceruleite $Cu_2Al_7(AsO_4)_4(OH)_{13}\cdot 11.5H_2O$. Spectrochim Acta A 116:518–523

Frost RL, López A, Scholz R, Xi Y, Belotti FM (2013c) Infrared and Raman spectroscopic characterization of the carbonate mineral huanghoite – and in comparison with selected rare earth carbonates. J Molec Struct 1051:221–225

Frost RL et al (2013d) Xxx

Frost RL, López A, Xi Y, Granja A, Scholz R, Lima RMF (2013e) Vibrational spectroscopy of the phosphate mineral kovdorskite – $Mg_2(PO_4)(OH)\cdot 3H_2O$. Spectrochim Acta A 114:309–315

Frost RL, López A, Xi Y, Lima RMF, Scholz R, Granja A (2013f) The molecular structure of the borate mineral inderite $Mg(H_4B_3O_7)(OH)\cdot 5H_2O$ – a vibrational spectroscopic study. Spectrochim Acta A 116:160–164

Frost RL, López A, Xi Y, Murta N, Scholz R (2013g) The molecular structure of the phosphate mineral senegalite $Al_2(PO_4)(OH)_3\cdot 3H_2O$ – a vibrational spectroscopic study. J Molec Struct 1048:420–425

Frost RL, López A, Xi Y, Scholz R, da Costa GM, Belotti FM, Lima RMF (2013h) Vibrational spectroscopy of the mineral meyerhofferite $CaB_3O_3(OH)_5\cdot H_2O$ – an assessment of the molecular structure. Spectrochim Acta A 114:27–32

Frost RL, López A, Xi Y, Scholz R, da Costa GM, Lima RMF, Granja A (2013i) The spectroscopic characterization of the sulphate mineral ettringite from Kuruman manganese deposits, South Africa. Vib Spectrosc 68:266–271

Frost RL, López A, Xi Y, Scholz R, Graça LM, Lagoeiro L (2013j) Vibrational spectroscopic characterization of the sulphate mineral leightonite $K_2Ca_2Cu(SO_4)_4\cdot 2H_2O$ – implications for the molecular structure. Spectrochim Acta A 112:90–94

Frost RL, López A, Xi Y, Granja A, Scholz R (2013k) Vibrational spectroscopic characterization of the phosphate mineral kulanite $Ba(Fe^{2+},Mn^{2+},Mg)_2(Al,Fe^{3+})_2(PO_4)_3(OH)_3$. Spectrochim Acta A 115:22–25

Frost RL, Palmer SJ, Xi Y, Čejka J, Sejkora J, Plášil J (2013l) Raman spectroscopic study of the hydroxyphosphate mineral plumbogummite $PbAl_3(PO_4)_2(OH,H_2O)_6$. Spectrochim Acta A 103:431–434

Frost RL, Scholz R, López A, Xi Y (2013m) Vibrational spectroscopic characterization of the sulphate mineral khademite $Al(SO_4)F\cdot 5H_2O$. Spectrochim Acta A 116:165–169

Frost RL, Scholz R, Lópes A, Xi Y, Gobac ŽŽ, Horta LFC (2013n) Raman and infrared spectroscopic characterization of the phosphate mineral paravauxite $Fe^{2+}Al_2(PO_4)_2(OH)_2\cdot 8H_2O$. Spectrochim Acta A 116:491–496

Frost RL, Scholz R, López A, Xi Y, Granja A, Gobac ŽŽ, Lima RMF (2013o) Infrared and Raman spectroscopic characterization of the silicate mineral olmiite $CaMn^{2+}[SiO_3(OH)](OH)$ – implications for the molecular structure. J Molec Struct 1053:22–26

Frost RL, Xi Y, Beganovic M, Belotti FM, Scholz R (2013p) Vibrational spectroscopy of the phosphate mineral lazulite – $(Mg,Fe)Al_2(PO_4)_2(OH)_2$ found in the Minas Gerais, Brazil. Spectrochim Acta A 107:241–247

Frost RL, Xi Y, López A, Scholz R, de Carvalho LC, e Souza FB (2013q) Vibrational spectroscopic characterization of the phosphate mineral barbosalite – implications for the molecular structure. J Molec Struct 1051:292–298

Frost RL, Xi Y, Millar G, Tan K, Palmer SJ (2013r) Vibrational spectroscopy of natural cave mineral monetite $CaHPO_4$ and the synthetic analog. Spectrosc Lett 46(1):54–59

Frost RL, Xi Y, Pogson RE, Scholz R (2013s) A vibrational spectroscopic study of philipsbornite $PbAl_3(AsO_4)_2(OH)_5\cdot H_2O$ – molecular structural implications and relationship to the crandallite subgroup arsenates. Spectrochim Acta A 104:257–261

Frost RL, Xi Y, Scholz R (2013t) Vibrational spectroscopic characterization of the phosphate mineral anapaite $Ca_2Fe^{2+}(PO_4)_2\cdot 4(H_2O)$. Spectrosc Lett 46(6):441–446

Frost RL, Xi Y, Scholz R (2013u) A vibrational spectroscopic study of the phosphate mineral cyrilovite $Na(Fe^{3+})_3(PO_4)_2(OH)_4\cdot 2(H_2O)$ and in comparison with wardite. Spectrochim Acta A 108:244–250

Frost RL, Xi Y, Scholz R (2013v) Vibrational spectroscopy of the copper (II) disodium sulphatedihydrate mineral kröhnkite $Na_2Cu(SO_4)_2\cdot 2H_2O$. Spectrosc Lett 46(6):447–452

Frost RL, Xi Y, Scholz R, Belott FM (2013w) Vibrational spectroscopic characterization of the phosphate mineral ludlamite $(Fe,Mn,Mg)_3(PO_4)_2 \cdot 4H_2O$ – a mineral found in lithium bearing pegmatites. Spectrochim Acta A 103:143–150

Frost RL, Xi Y, Scholz R, Belotti FM (2013x) Vibrational spectroscopic characterization of the phosphate mineral bermanite – $Mn^{2+}Mn^{3+}_2(PO_4)_2(OH)_2 \cdot 4H_2O$. Spectrochim Acta A 105:359–364

Frost RL, Xi Y, Scholz R, Belotti FM, Beganovic M (2013y) SEM–EDX, Raman and infrared spectroscopic characterization of the phosphate mineral frondelite $(Mn^{2+})(Fe^{3+})_4(PO_4)_3(OH)_5$. Spectrochim Acta A 110:7–13

Frost RL, Xi Y, Scholz R, Belotti FM, Cândido Filho M (2013z) Infrared and Raman spectroscopic characterization of the borate mineral colemanite – $CaB_3O_4(OH)_3 \cdot H_2O$ – implications for the molecular structure. J Molec Struct 1037:23–28

Frost RL, Xi Y, Scholz R, Belotti FM, Cândido Filho M (2013aa) The phosphate mineral sigloite $Fe^{3+}Al_2(PO_4)_2(OH)_3 \cdot 7(H_2O)$, an exception to the paragenesis rule – a vibrational spectroscopic study. J Molec Struct 1033:258–264

Frost RL, Xi Y, Scholz R, Belotti FM, Cândido Filho M (2013ab) Infrared and Raman spectroscopic characterization of the carbonate mineral weloganite – $Sr_3Na_2Zr(CO_3)_6 \cdot 3H_2O$ and in comparison with selected carbonates. J Molec Struct 1039:101–106

Frost RL, Xi Y, Scholz R, Belotti FM, Filho MC (2013ac) Infrared and Raman spectroscopic characterization of the phosphate mineral leucophosphite $KFe^{3+}_2(PO_4)_2(OH) \cdot 2H_2O$. Spectrosc Lett 46(6):415–420

Frost RL, Xi Y, Scholz R, Belotti FM, Lopez A (2013ad) Infrared and Raman spectroscopic characterization of the phosphate mineral fairfieldite – $Ca_2(Mn^{2+},Fe^{2+})_2(PO_4)_2 \cdot 2(H_2O)$. Spectrochim Acta A 106:216–223

Frost RL, Xi Y, Scholz R, Belotti FM, Menezes Filho LAD (2013ae) A vibrational spectroscopic study of the phosphate mineral zanazziite – $Ca_2(MgFe^{2+})(MgFe^{2+})_4Be_4(PO_4)_6 \cdot 6(H_2O)$. Spectrochim Acta A 104:250–256

Frost RL, Xi Y, Scholz R, de Brito Ribeiro CA (2013af) The molecular structure of the phosphate mineral chalcosiderite – a vibrational spectroscopic study. Spectrochim Acta A 111:24–30

Frost RL, Xi Y, Scholz R, Horta LFC (2013ag) The phosphate mineral arrojadite-(KFe) and its spectroscopic characterization. Spectrochim Acta A 109:138–145

Frost RL, Xi Y, Scholz R, Lima RMF, Horta LFC, Lopez A (2013ah) Thermal analysis and vibrational spectroscopic characterization of the boro silicate mineral datolite – $CaBSiO_4(OH)$. Spectrochim Acta A 115:376–381

Frost RL, Xi Y, Scholz R, López A, Belotti FM (2013ai) Infrared and Raman spectroscopic characterization of the silicate-carbonate mineral carletonite – $KNa_4Ca_4Si_8O_{18}(CO_3)_4(OH,F) \cdot H_2O$. J Molec Struct 1042:1–7

Frost RL, Xi Y, Scholz R, López A, Belotti FM (2013aj) Vibrational spectroscopic characterization of the phosphate mineral hureaulite – $(Mn,Fe)_5(PO_4)_2(HPO_4)_2 \cdot 4(H_2O)$. Vib Spectrosc 66:69–75

Frost RL, Xi Y, Scholz R, López A, Belotti FM, Chaves ML (2013ak) Raman and infrared spectroscopic characterization of the phosphate mineral lithiophilite – $LiMnPO_4$. Phosphorus Sulfur Silicon Related Elem 188(11):1526–1534

Frost RL, Xi Y, Scholz R, López A, Granja A (2013al) Infrared and Raman spectroscopic characterisation of the sulphate mineral creedite – $Ca_3Al_2SO_4(F,OH) \cdot 2H_2O$ – and in comparison with the alums. Spectrochim Acta A 109:201–205

Frost RL, Xi Y, Scholz R, López A, Lima RMF, Ferreira CM (2013am) Vibrational spectroscopic characterization of the phosphate mineral series eosphorite–childrenite – $(Mn,Fe)Al(PO_4)(OH)_2(H_2O)$. Vib Spectrosc 67:14–21

Frost RL, Xi Y, Scholz R, Tazava E (2013an) Spectroscopic characterization of the phosphate mineral florencite-La – $LaAl_3(PO_4)_2(OH,H_2O)_6$, a potential tool in the REE mineral prospection. J Molec Struct 1037:148–153

Frost RL, Čejka J, Scholz R, López A, Theiss FL, Xi Y (2014a) Vibrational spectroscopic study of the uranyl selenite mineral derriksite $Cu_4UO_2(SeO_3)_2(OH)_6 \cdot H_2O$. Spectrochim Acta A 117:473–477

Frost RL, Gobac ŽŽ, López A, Xi Y, Scholz R, Lana C, Lima RMF (2014b) Characterization of the sulphate mineral coquimbite, a secondary iron sulphate from Javier Ortega mine, Lucanas Province, Peru – using infrared, Raman spectroscopy and thermogravimetry. J Molec Struct 1063:251–258

Frost RL, López A, Belotti FM, Xi Y, Scholz R (2014c) A vibrational spectroscopic study of the phosphate mineral lulzacite $Sr_2Fe^{2+}(Fe^{2+},Mg)_2Al_4(PO_4)_4(OH)_{10}$. Spectrochim Acta A 127:243–247

Frost RL, López A, de Oliveira Gonçalves G, Scholz R, Xi Y (2014d) A vibrational spectroscopic study of the arsenate mineral bayldonite $Cu_3PbO(AsO_3OH)_2(OH)_2$ – a comparison with other basic arsenates. J Molec Struct 1056:267–272

Frost RL, López A, Scholz R, Xi Y (2014e) Vibrational spectroscopy of the borate mineral tunellite $SrB_6O_9(OH)_2(3H_2O)$ – implications for the molecular structure. J Molec Struct 1059:40–43

Frost RL, López A, Scholz R, Xi Y (2014f) Vibrational spectroscopy of the borate mineral chambersite $Mn_3B_7O_{13}Cl$ – implications for the molecular structure. Spectrochim Acta A 120:270–273

Frost RL, López A, Scholz R, Xi Y, CândidoFilho M (2014g) A vibrational spectroscopic study of the phosphate mineral churchite $(REE)(PO_4) \cdot 2H_2O$. Spectrochim Acta A 127:429–433

Frost RL, López A, Theiss FL, Aarão GM, Scholz R (2014h) A vibrational spectroscopic study of the phosphate mineral rimkorolgite $(Mg,Mn^{2+})_5(Ba,Sr)(PO_4)_4 \cdot 8H_2O$ from Kovdor massif, Kola Peninsula, Russia. Spectrochim Acta A 132:762–766

Frost RL, López A, Theiss FL, Romano AW, Scholz R (2014i) A vibrational spectroscopic study of the silicate mineral analcime – $Na_2(Al_4SiO_4O_{12}) \cdot 2H_2O$ – a natural zeolite. Spectrochim Acta A 133:521–525

Frost RL, López A, Theiss FL, Scholz R, Belotti FM (2014j) A vibrational spectroscopic study of the borate mineral ezcurrite $Na_4B_{10}O_{17} \cdot 7H_2O$ – implications for the molecular structure. J Molec Struct 1070:45–51

Frost RL, López A, Theiss FL, Scholz R, Souza L (2014k) The molecular structure of the phosphate mineral kidwellite $NaFe^{3+}_9(PO_4)_6(OH)_{11} \cdot 3H_2O$ – a vibrational spectroscopic study. J Molec Struct 1074:429–434

Frost RL, López A, Xi Y, Cardoso LH, Scholz R (2014l) A vibrational spectroscopic study of the phosphate mineral minyulite $KAl_2(OH,F)(PO_4)_2 \cdot 4(H_2O)$ and in comparison with wardite. Spectrochim Acta A 124:34–39

Frost RL, López A, Xi Y, Gobac ŽŽ, Scholz R (2014m) The molecular structure of the phosphate mineral vaeyrynenite: a vibrational spectroscopic study. Spectrosc Lett 47(4):253–260

Frost RL, López A, Xi Y, Graça LM, Scholz R (2014n) A vibrational spectroscopic study of the borate mineral takedaite $Ca_3(BO_3)_2$. Spectrochim Acta A 132:833–837

Frost RL, López A, Xi Y, Lana C, Souza L, Scholz R, Sejkora J, Čejka J (2014o) A vibrational spectroscopic study of the arsenate minerals cobaltkoritnigite and koritnigite. Spectrochim Acta A 125:313–318

Frost RL, López A, Xi Y, Scholz R (2014p) Vibrational spectroscopic characterization of the phosphate mineral althausite $Mg_2(PO_4)(OH,F,O)$ – implications for the molecular structure. Spectrochim Acta A 120:252–256

Frost RL, López A, Xi Y, Scholz R (2014q) A study of the phosphate mineral kapundaite $NaCa(Fe^{3+})_4(PO_4)_4(OH)_3 \cdot 5(H_2O)$ using SEM/EDX and vibrational spectroscopic methods. Spectrochim Acta A 122:400–404

Frost RL, López A, Xi Y, Scholz R (2014r) A vibrational spectroscopic study of the silicate mineral inesite $Ca_2(Mn,Fe)_7Si_{10}O_{28}(OH) \cdot 5H_2O$. Spectrochim Acta A 128:207–211

Frost RL, López A, Xi Y, Scholz R, Gandini AL (2014s) A vibrational spectroscopic study of the silicate mineral ardennite-(As). Spectrochim Acta A 118:987–991

Frost RL, López A, Xi Y, Scholz R, Lana C (2014t) A vibrational spectroscopic study of the silicate mineral plumbophyllite $Pb_2Si_4O_{10} \cdot H_2O$. Spectrochim Acta A 128:665–670

Frost RL, López A, Xi Y, Scholz R, Souza L, Lana C (2014u) The molecular structure of the borate mineral rhodizite $(K,Cs)Al_4Be_4(B,Be)_{12}O_{28}$ – a vibrational spectroscopic study. Spectrochim Acta A 128:291–294

Frost RL, Scholz R, López A, Belotti FM, Xi Y (2014v) Structural characterization and vibrational spectroscopy of the arsenate mineral wendwilsonite. Spectrochim Acta A 118:737–743

Frost RL, Scholz R, López A, Firmino BE, Lana C, Xi Y (2014w) A Raman and infrared spectroscopic characterisation of the phosphate mineral phosphohedyphane $Ca_2Pb_3(PO_4)_3Cl$ from the Roote mine, Nevada, USA. Spectrochim Acta A 127:237–242

Frost RL, Scholz R, López A, Lana C, Xi Y (2014x) A Raman and infrared spectroscopic analysis of the phosphate mineral wardite $NaAl_3(PO_4)_2(OH)_4 \cdot 2(H_2O)$ from Brazil. Spectrochim Acta A 126:164–169

Frost RL, Scholz R, López A, Theiss FL (2014y) Vibrational spectroscopic study of the natural layered double hydroxide manasseite now defined as hydrotalcite-$2H$ – $Mg_6Al_2(OH)_{16}[CO_3]$ $4H_2O$. Spectrochim Acta A 118:187–191

Frost RL, Scholz R, López A, Theiss FL (2014z) Vibrational spectroscopic characterization of the sulphate-halide mineral sulphohalite – implications for evaporites. Spectrochim Acta A 133:794–798

Frost RL, Scholz R, López A, Xi Y (2014aa) Raman spectroscopy of the arsenate minerals maxwellite and in comparison with tilasite. Spectrochim Acta A 123:416–420

Frost RL, Scholz R, López A, Xi Y (2014ab) A vibrational spectroscopic study of the phosphate mineral whiteite $CaMn^{++}Mg_2Al_2(PO_4)_4(OH)_2 \cdot 8(H_2O)$. Spectrochim Acta A 124:243–248

Frost RL, Scholz R, López A, Xi Y, Belotti FM (2014ac) Infrared and Raman spectroscopic characterization of the borate mineral vonsenite $Fe^{2+}_2Fe^{3+}BO_5$. Spectrosc Lett 47(7):512–517

Frost RL, Scholz R, López A, Xi Y, de Siqueira Queiroz C, Belotti FM, Cândido Filho M (2014ad) Raman, infrared and near-infrared spectroscopic characterization of the herderite – hydroxylherderite mineral series. Spectrochim Acta A 118:430–437

Frost RL, Scholz R, Lópes A, Xi Y, Gobac ŽŽ, de Carvalho Lana C (2014ae) Vibrational spectroscopy of the borate mineral gaudefroyite from N'Chwaning II mine, Kalahari, Republic of South Africa. Spectrochim Acta A 120:265–269

Frost RL, Scholz R, López A, Xi Y, Graça LM (2014af) Infrared and Raman spectroscopic characterization of the borate mineral hydroboracite $CaMg[B_3O_4(OH)_3]_2 \cdot 3H_2O$ – implications for the molecular structure. J Molec Struct 1059:20–26

Frost RL, Scholz R, López A, Xi Y, Lana C (2014ag) Vibrational spectroscopy of the sulphate mineral sturmanite from Kuruman manganese deposits, South Africa. Spectrochim Acta A 133:24–30

Frost RL, Theiss FL, López A, Scholz R (2014ah) Vibrational spectroscopic study of the sulphate mineral glaucocerinite $(Zn,Cu)_{10}Al_6(SO_4)_3(OH)_{32} \cdot 18H_2O$ – a natural layered double hydroxide. Spectrochim Acta A 127:349–354

Frost RL, Xi Y, López A, Corrêa L, Scholz R (2014ai) The molecular structure of the vanadate mineral mottramite $[PbCu(VO_4)(OH)]$ from Tsumeb, Namibia – a vibrational spectroscopic study. Spectrochim Acta A 122:252–256

Frost RL, Xi Y, López A, Moreira VA, Scholz R, Lima RMF, Gandini AL (2014aj) Assessment of the molecular structure of an intermediate member of the triplite-zwieselite mineral series: a Raman and infrared study. Spectrosc Lett 47(3):214–222

Frost RL, Xi Y, Scholz R, da Costa Alves Pereira M (2014ak) Vibrational spectroscopy of the borate mineral olshanskyite $Ca_3[B(OH)_4]_4(OH)_2$. Carbonates Evaporites 29(1):33–39

Frost RL, Lopez A, Scholz R, Xi Y, Lana C (2014al) The molecular structure of the phosphate mineral beraunite $Fe^{2+}Fe^{3+}_5(PO_4)_4(OH)_5 \cdot 4H_2O$ – A vibrational spectroscopic study. Spectrochim Acta A 128:408–412

Frost RL, López A, Scholz R (2015a) Raman and infrared spectroscopic study of kamphaugite-(Y). Spectrochim Acta A 143:67–71

Frost RL, López A, Scholz R (2015b) A SEM, EDS and vibrational spectroscopic study of the tellurite mineral: sonoraite $Fe^{3+}Te^{4+}O_3(OH) \cdot H_2O$. Spectrochim Acta A 147:225–229

Frost RL, López A, Scholz R, Belotti FM, Xi Y (2015c) A vibrational spectroscopic study of the anhydrous phosphate mineral sidorenkite $Na_3Mn(PO_4)(CO_3)$. Spectrochim Acta A 137:930–934

Frost RL, López A, Scholz R, de Oliveira FA (2015d) SEM, EDX and vibrational spectroscopic study of the mineral tunisite $NaCa_2Al_4(CO_3)_4Cl(OH)_8$. Spectrochim Acta A 136:911–917

Frost RL, López A, Scholz R, de Oliveira FA (2015e) Scanning electron microscopy with energy dispersive spectroscopy and Raman and infrared spectroscopic study of tilleyite $Ca_5Si_2O_7(CO_3)_2$-Y. Spectrochim Acta A 149:333–337

Frost RL, López A, Scholz R, Lana C, Xi Y (2015f) A Raman spectroscopic study of the arsenate mineral chenevixite $Cu_2Fe^{3+}_2(AsO_4)_2(OH)_4 \cdot H_2O$. Spectrochim Acta A 135:192–197

Frost RL, López A, Scholz R, Lima RMF (2015g) Vibrational spectroscopic study of poldervaartite CaCa $[SiO_3(OH)(OH)]$. Spectrochim Acta A 137:827–831

Frost RL, López A, Scholz R, Sampaio NP, de Oliveira FA (2015h) SEM, EDS and vibrational spectroscopic study of dawsonite $NaAl(CO_3)(OH)_2$. Spectrochim Acta A 136:918–923

Frost RL, López A, Scholz R, Theiss F, da Costa GM (2015i) Structural characterization of the borate mineral inyoite – $CaB_3O_3(OH)_5 \cdot 4H_2O$. J Molec Struct 1080:99–104

Frost RL, López A, Scholz R, Theiss FL, Romano AW (2015j) SEM, EDX, Infrared and Raman spectroscopic characterization of the silicate mineral yuksporite. Spectrochim Acta A 137:607–611

Frost RL, López A, Scholz R, Wang L (2015k) A Raman and infrared spectroscopic study of the sulphate mineral aluminite $Al_2(SO_4)(OH)_4 \cdot 7H_2O$. Spectrochima Acta A 148:232–236

Frost RL, López A, Scholz R, Xi Y (2015l) Vibrational spectroscopy of the borate mineral priceite – implications for the molecular structure. Spectrosc Lett 48(2):101–106

Frost RL, López A, Theiss FL, Graça LM, Scholz R (2015m) A vibrational spectroscopic study of the silicate mineral lomonosovite $Na_5Ti_2(Si_2O_7)(PO_4)O_2$. Spectrochim Acta A 134:53–57

Frost RL, López A, Theiss FL, Romano AW, Scholz R (2015n) A vibrational spectroscopic study of the silicate mineral pectolite – $NaCa_2Si_3O_8(OH)$. Spectrochim Acta A 134:58–62

Frost RL, López A, Theiss FL, Romano AW, Scholz R (2015o) An SEM, EDS and vibrational spectroscopic study of the silicate mineral meliphanite $(Ca,Na)_2Be$ $[(Si,Al)_2O_6(F,OH)]$. Spectrochim Acta A 136:216–220

Frost RL, López A, Theiss FL, Scholz R, Romano AW (2015p) A vibrational spectroscopic study of the silicate mineral normandite – $NaCa(Mn^{2+},Fe^{2+})(Ti,Nb,Zr)Si_2O_7(O,F)_2$. Spectrochim Acta A 135:801–804

Frost RL, López A, Wang L, Romano AW, Scholz R (2015q) A vibrational spectroscopic study of the silicate mineral harmotome – $(Ba,Na,K)_{1-2}(Si,Al)_8O_{16} \cdot 6H_2O$ – a natural zeolite. Spectrochim Acta A 137:70–74

Frost RL, López A, Wang L, Scholz R, Sampaio NP (2015r) SEM, EDX and Raman an infrared spectroscopic study of brianyoungite $Zn_3(CO_3,SO_4)(OH)_4$ from Esperanza Mine, Laurion District, Greece. Spectrochim Acta A 149:279–284

Frost RL, López A, Wang L, Scholz R, Sampaio NP, de Oliveira FA (2015s) A vibrational spectroscopic study of tengerite-(Y) $Y_2(CO_3)_3 \cdot 2-3H_2O$. Spectrochim Acta A 137:612–616

Frost RL, Lópes A, Xi Y, Scholz R (2015t) A vibrational spectroscopic study of the silicate mineral kornerupine. Spectrosc Lett 48(7):487–491

Frost RL, López A, Xi Y, Scholz R (2015u) A vibrational spectroscopic study of the copper bearing silicate mineral luddenite. Spectrochim Acta A 137:717–720

Frost RL, Scholz R, Belotti FM, López A, Theiss FL (2015v) A vibrational spectroscopic study of the phosphate mineral vantasselite $Al_4(PO_4)_3(OH)_3 \cdot 9H_2O$. Spectrochim Acta A 147:185–192

Frost RL, Scholz R, Jirásek J, Belott FM (2015w) An SEM–EDX and Raman spectroscopic study of the fibrous arsenate mineral liskeardite and in comparison with other arsenates kaňkite, scorodite and yvonite. Spectrochim Acta A 151:566–575

Frost RL, Scholz R, Lima RMF, López A (2015x) SEM, EDS and vibrational spectroscopic study of the sulphate mineral rostite $Al(SO_4)(OH,F) \cdot 5(H_2O)$. Spectrochim Acta A 151:616–620

Frost RL, Scholz R, López A (2015y) Infrared and Raman spectroscopic characterization of the carbonate bearing silicate mineral aerinite – implications for the molecular structure. J Molec Struct 1097:1–5

Frost RL, Scholz R, López A (2015z) Raman and infrared spectroscopic characterization of the arsenate-bearing mineral tangdanite – and in comparison with the discredited mineral clinotyrolite. J Raman Spectrosc 46(10):920–926

Frost RL, Scholz R, López A, Belotti FM (2015aa) The molecular structure of the borate mineral szaibelyite $MgBO_2(OH)$ – a vibrational spectroscopic study. J Molec Struct 1089:20–24

Frost RL, Scholz R, López A, Xi Y (2015ab) Raman and infrared spectroscopic characterization of the silicate mineral lamprophyllite. Spectrosc Lett 48(10):701–704

Frost RL, Scholz R, Wang L (2015ac) A Raman and infrared spectroscopic study of the phosphate mineral pseudolaueite and in comparison with strunzite and ferrostrunzite. J Chem Crystallogr 45(8–9):391–400

Frost RL, López A, Scholz R, Firmino B (2016a) SEM, EDX and vibrational spectroscopic study of the carbonate mineral donnayite-(Y) $NaCaSr_3Y(CO_3)_6 \cdot 3H_2O$. Carbonates Evaporites 31(1):1–8

Frost RL, Scholz R, López A (2016b) A Raman and infrared spectroscopic study of the phosphate mineral laueite. Vib Spectrosc 82:31–36

Frost RL, Scholz R, Ruan X, Lima RMF (2016c) A thermogravimetric, scanning electron microscope and vibrational spectroscopic study of the phosphate mineral santabarbaraite from Santa Barbara mine, Tuscany, Italy. J Therm Anal Calorim 124(2):639–644

Frost RL, Scholz R, Wang L (2016d) Vibrational spectroscopic study of the phosphate mineral kryzhanovskite and in comparison with reddingite – implications for the molecular structure. J Molec Struct 1118:203–211

Fu X, Yang G, Sun J, Zhou J (2012) Vibrational spectra of copper sulfate hydrates investigated with low-temperature Raman spectroscopy and terahertz time domain spectroscopy. J Phys Chem A 116 (27):7314–7318

Fuchs Y, Mellini M, Memmi I (2001) Crystal-chemistry of magnesiocarpholite: controversial X-ray diffraction, Mössbauer, FTIR and Raman results. Eur J Mineral 13(3):533–543

Fujimori H, Komatsu H, Ioku K, Goto S, Watanabe T (2005) Vibrational spectra of Ca_3SiO_5: ultraviolet laser Raman spectroscopy at high temperatures. J Am Ceram Soc 88(7):1995–1998

Fujita J, Martell AE, Nakamoto K (1962) Infrared spectra of metal chelate compounds. VI. A normal coordinate treatment of oxalate metal complexes. J Chem Phys 36:324–331

Furman E, Brafman O, Makovsky J (1976) Approximation to long-wavelength lattice dynamics of SbSI-type crystals. Phys Rev B 13(4):1703. (8 pp)

Furukawa T, Brawer SA, White WB (1979) Raman and infrared spectroscopic studies of the crystalline phases in the system Pb_2SiO_4-$PbSiO_3$. J Am Ceram Soc 62 (7–8):351–356

Gabal MA, Elroby SA, Obaid AY (2012) Synthesis and characterization of nano-sized ceria powder via oxalate decomposition route. Powder Technol 229:112–118

Gabelica-Robert M, Tarte P (1979) Vibrational spectrum of akermanite-like silicates and germinates. Spectrochim Acta A 35:649–654

Gabelica-Robert M, Tarte P (1981) Vibrational spectrum of fresnoite ($Ba_2TiOSi_2O_7$) and isostructural compounds. Phys Chem Minerals 7(1):26–30

Gadas P, Novák M, Szuszkiewicz A, Szełęg E, Vašinová Galiová M, Haifler J (2016) Magnesium-rich Na,Be, Li-rich sekaninaite from miarolitic pegmatite at Zimnik, Strzegom-Sobótka massif, Sudetes, Poland. Can Mineral 54(4):971–987

Gaft M, Nagli L, Waychunas G, Weiss D (2004) The nature of blue luminescence from natural benitoite $BaTiSi_3O_9$. Phys Chem Minerals 31(6):365–373

Gaft M, Yeates H, Nagli L (2013) Laser-induced time-resolved luminescence of natural margarosanite $Pb(Ca,Mn)_2Si_3O_9$, swedenborgite $NaBe_4SbO_7$ and walstromite $BaCa_2Si_3O_9$. Eur J Mineral 25:71–77

Gaft M, Reisfeld R, Panczer G (2015) Identification of minerals. In: Modern luminescence spectroscopy of minerals and materials. Springer, pp 315–322

Gaitán M, Jerez A, Pico C, Veiga ML (1985) Ditellurium (IV) trioxide selenate: a new solid phase in the system Te–Se–O_2. Mater Res Bull 20(9):1069–1074

Galera-Gomez PA, Sanz-Pinilla S, Otero-Aenlle E, Gonzáles-Díaz PF (1982) Infrared spectra of arsenate and vanadate strontium apatites. Spectrochim Acta A 38(2):253–259

Galuskin EV, Armbruster T, Malsy A, Galuskina IO, Sitarz M (2003) Morphology, composition and structure of low-temperature P4/nnc high-fluorine vesuvianite whiskers from Polar Yakutia, Russia. Can Miner 41:843–856

Galuskin E, Janeczek J, Kozanecki M, Sitarz M, Jastrzębski W, Wrzalik R, Stadnicka K (2007a) Single-crystal Raman investigation of vesuvianite in the OH region. Vib Spectrosc 44(1):36–41

Galuskin EV, Pertsev NN, Armbruster T, Kadiyski M, Zadov AE, Galuskina IO, Dzierżanowski P, Wrzalik R, Kislov EV (2007b) Dovyrenite $Ca_6Zr[Si_2O_7]_2(OH)_4$ – a new mineral from skarned carbonate xenoliths in basic-ultrabasic rocks of the Ioko-Dovyrenmassif, Northern Baikal region, Russia. MineralogiaPolonica 38(1):15–28

Galuskin EV, Gazeev VM, Lazic B, Armbruster T, Galuskina IO, Zadov AE, Pertsev NN, Wrzalik R, Dzierżanowski P, Gurbanov AG, Bzowska G (2009) Chegemite $Ca_7(SiO_4)_3(OH)_2$ – a new humite-group calcium mineral from the Northern Caucasus, Kabardino-Balkaria, Russia. Eur J Mineral 21 (5):1045–1059

Galuskin EV, Armbruster T, Galuskina IO, Lazic B, Winiarski A, Gazeev VM, Dzierżanowski P, Zadov AE, Pertsev NN, Wrzalik R, Gurbanov AG, Janeczek J (2011a) Vorlanite ($CaU^{6+})O_4$ – a new mineral from the Upper Chegem caldera, Kabardino-Balkaria, Northern Caucasus, Russia. Am Mineral 96 (1):188–196

Galuskin EV, Galuskina IO, Gazeev VM, Dzierzanowski P, Prusik K, Pertsev NN, Zadov AE, Bailau R, Gurbanov AG (2011b) Megawite, $CaSnO_3$: a new perovskite-group mineral from skarns of the Upper Chegem caldera, Kabardino-Balkaria, Northern Caucasus, Russia. Mineral Mag 75(5):2563–2572

Galuskin EV, Galuskina IO, Lazic B, Armbruster T, Zadov AE, Krzykawski T, Banasik K, Gazeev VM,

Pertsev NN (2011c) Rusinovite, $Ca_{10}(Si_2O_7)_3Cl_2$: a new skarn mineral from the Upper Chegem caldera, Kabardino-Balkaria, northern Caucasus, Russia. Eur J Mineral 23(5):837–844

Galuskin EV, Galuskina IO, Dubrovinsky LS, Janeczek J (2012a) Thermally induced transformation of vorlanite to "protovorlanite": restoration of cation ordering in self-irradiated $CaUO_4$. Am Mineral 97(5–6):1002–1004

Galuskin EV, Gfeller F, Savelyeva VB, Armbruster T, Lazic B, Galuskina IO, Többens DM, Zadov AE, Dzierżanowski P, Pertsev NN, Gazeev VM (2012b) Pavlovskyite $Ca_8(SiO_4)_2(Si_3O_{10})$: a new mineral of altered silicate-carbonate xenoliths from the two Russian type localities, Birkhin massif, Baikal Lake area and Upper Chegem caldera, North Caucasus. Am Mineral 97(4):503–512

Galuskin EV, Kusz J, Armbruster T, Bailau R, Galuskina IO, Ternes B, Murashko M (2012c) A reinvestigation of mayenite from the type locality, the Ettringer-Bellerberg volcano near Mayen, Eifel district, Germany. Mineral Mag 76(3):707–716

Galuskin EV, Lazic B, Armbruster T, Galuskina IO, Pertsev NN, Gazeev VM, Włodyka R, Dulski M, Dzierżanowski P, Zadov AE, Dubrovinsky LS (2012d) Edgrewite $Ca_9(SiO_4)_4F_2$–hydroxyledgrewite $Ca_9(SiO4)_4(OH)_2$, a new series of calcium humite-group minerals from altered xenoliths in the ignimbrite of Upper Chegem caldera, Northern Caucasus, Kabardino-Balkaria, Russia. Am Mineral 97(11–12):1998–2006

Galuskin EV, Galuskina IO, Bailau R, Prusik K, Gazeev VM, Zadov AE, Pertsev NN, Jeżak L, Gurbanov AG, Dubrovinsky L (2013a) Eltyubyuite, $Ca_{12}Fe^{3+}_{10}Si_4O_{32}Cl_6$ – the Fe^{3+} analogue of wadalite: a new mineral from the Northern Caucasus, Kabardino-Balkaria, Russia. Eur J Mineral 25(2):221–229

Galuskin EV, Kusz J, Armbruster T, Galuskina IO, Marzec K, Vapnik Y, Murashko M (2013b) Vorlanite, $(CaU^{6+})O_4$, from Jabel Harmun, Palestinian Autonomy, Israel. Am Mineral 98(11–12):1938–1942

Galuskin EV, Galuskina IO, Kusz J, Armbruster T, Marzec KM, Dzierżanowski P, Murashko M (2014) Vapnikite Ca_3UO_6–a new double-perovskite mineral from pyrometamorphiclarnite rocks of the JabelHarmun, Palestinian Autonomy, Israel. Mineral Mag 78(3):571–582

Galuskin EV, Galuskina IO, Gfeller F, Krüger B, Kusz J, Vapnik Y, Dulski M, Dzierżanowski P (2015a) Silicocarnotite, $Ca_5[(PO_4)(SiO_4)](PO_4)$, a new "old" mineral from the Negev Desert, Israel, and the ternesite-silicocarnotite solid solution: indicators of high-temperature alteration of pyrometamorphic rocks of the Hatrurim Complex, Southern Levant. Eur J Mineral. https://doi.org/10.1127/ejm/2015/0027-2494

Galuskin EV, Galuskina IO, Kusz J, Gfeller F, Armbruster T, Bailau R, Dulski M, Gazeev VM, Pertsev NN, Zadov AE, Dzierżanowski P (2015b) Mayenite supergroup, Part II: Chlorkyuygenite from Upper Chegem, Northern Caucasus, Kabardino-Balkaria, Russia, a new microporous mineral with "zeolitic" H_2O. Eur J Mineral 27(1):113–122

Galuskin EV, Gfeller F, Armbruster T, Galuskina IO, Vapnik Y, Dulski M, Murashko M, Dzierżanowski P, Sharygin VV, Krivovichev SV, Wirth R (2015c) Mayenite supergroup, Part III: Fluormayenite, $Ca_{12}Al_{14}O_{32}[\square_4F_2]$, and fluorkyuygenite, $Ca_{12}Al_{14}O_{32}[(H_2O)_4F_2]$, two new minerals from pyrometamorphic rocks of the Hatrurim Complex, South Levant. Eur J Mineral 27(1):123–136

Galuskin EV, Gfeller F, Armbruster T, Galuskina IO, Vapnik Y, Murashko M, Włodyka R, Dzierżanowski P (2015d) New minerals with a modular structure derived from hatrurite from the pyrometamorphic Hatrurim Complex. Part I. Nabimusaite, $KCa_{12}(SiO_4)_4(SO_4)_2O_2F$, from larnite rocks of Jabel Harmun, Palestinian Autonomy, Israel. Mineral Mag 79(5):1061–1072

Galuskin EV, Gfeller F, Galuskina IO, Pakhomova A, Armbruster T, Vapnik Y, Włodyka R, Dzierżanowski P, Murashko M (2015e) New minerals with a modular structure derived from hatrurite from the pyrometamorphic Hatrurim Complex. Part II. Zadovite, $BaCa_6[(SiO_4)(PO_4)](PO_4)_2F$ and aradite, $BaCa_6[(SiO_4)(VO_4)](VO_4)_2F$, from paralavas of the Hatrurim Basin, Negev Desert, Israel. Mineral Mag 79(5):1073–1087

Galuskin EV, Gfeller F, Galuskina IO, Armbruster T, Krzatala A, Vapnik Y, Kusz J, Dulski M, Gardocki M, Gurbanov AG, Dzierzanowski P (2016a) New minerals with modular structure derived from hatrurite from the pyrometamorphic rocks, Part III: Gazeevite, $BaCa_6(SiO_4)_2(SO_4)_2O$, from Israel and Palestine Autonomy, South Levant and from South Ossetia, Greater Caucasus. Mineral Mag. https://doi.org/10.1180/minmag.2016.080.105

Galuskin EV, Krüger B, Krüger H, Blass G, Widmer R, Galuskina IO (2016b) Wernerkrauseite, $CaFe^{3+}_2Mn^{4+}O_6$: the first nonstoichiometric post-spinel mineral, from Bellerberg volcano, Eifel, Germany. Eur J Mineral 28(2):485–493

Galuskina IO, Galuskin EV, Dzierżanowski P, Armbruster T, Kozanecki M (2005) A natural scandian garnet. Am Mineral 90(10):1688–1692

Galuskina IO, Kadiyski M, Armbruster T, Galuskin EV, Pertsev NN, Dzierżanowski P, Wrzalik R (2008) A new natural phase in the system Mg_2SiO_4–Mg_2BO_3F–$Mg_2BO_3(OH)$: composition, paragenesis and structure of OH-dominant pertsevite. Eur J Mineral 20(5):951–964

Galuskina IO, Lazic B, Armbruster T, Galuskin EV, Gazeev VM, Zadov AE, Pertsev NN, Jeżak L, Wrzalik R, Gurbanov AG (2009) Kumtyubeite $Ca_5(SiO_4)_2F_2$ – a new calcium mineral of the humite group from Northern Caucasus, Kabardino-Balkaria, Russia. Am Mineral 94(10):1361–1370

Galuskina IO, Galuskin EV, Armbruster T, Lazic B, Dzierżanowski P, Gazeev VM, Prusi KK, Pertsev NN, Winiarski A, Zadov AE, Wrzali KR, Gurbanov AG (2010a) Bitikleite-(SnAl) and bitikleite-(ZrFe):

new garnets from xenoliths of the Upper Chegem volcanic structure, Kabardino-Balkaria, Northern Caucasus, Russia. Am Mineral 95(7):959–967

Galuskina IO, Galuskin EV, Armbruster T, Lazic B, Kusz J, Dzierżanowski P, Gazeev VM, Pertsev NN, Prusik J, Zadov AE, Winiarski A, Wrzalik R, Gurbanov AG (2010b) Elbrusite-(Zr) – a new uraniangarnet from the Upper Chegem caldera, Kabardino-Balkaria, Northern Caucasus, Russia. Am Mineral 95(8–9):1172–1181

Galuskina IO, Galuskin EV, Dzierżanowski P, Gazeev VM, Prusik K, Pertsev NN, Winiarski A, Zadov AE, Wrzalik R (2010c) Toturite $Ca_3Sn_2Fe_2SiO_{12}$ – a new mineral species of the garnet group. Am Mineral 95 (8–9):1305–1311

Galuskina IO, Galuskin EV, Lazic B, Armbruster T, Dzierżanowski P, Prusik K, Wrzalik R (2010d) Eringaite, $Ca_3Sc_2(SiO_4)_3$, a new mineral of the garnet group. Mineral Mag 74(2):365–373

Galuskina IO, Galuskin EV, Prusik K, Gazeev VM, Pertsev NN, Dzierżanowski P (2013a) Irinarassite $Ca_3Sn_2SiAl_2O_{12}$ – new garnet from the Upper Chegem Caldera, Northern Caucasus, Kabardino-Balkaria, Russia. Mineral Mag 77(6):2857–2866

Galuskina IO, Galuskin EV, Kusz J, Dzierżanowski P, Prusik K, Gazeev VM, Pertsev NN, Dubrovinsky L (2013b) Dzhuluite, $Ca_3SbSnFe^{3+}{}_3O_{12}$, a new bitikleite-group garnet from the Upper Chegem Caldera, Northern Caucasus, Kabardino-Balkaria, Russia. Eur J Mineral 25(2):231–239

Galuskina IO, Vapnik Y, Lazic B, Armbruster T, Murashko M, Galuskin EV (2014) Harmunite $CaFe_2O_4$: a new mineral from the Jabel Harmun, West Bank, Palestinian Autonomy, Israel. Am Mineral 99(5–6):965–975

Galuskina IO, Krüger B, Galuskin EV, Armbruster T, Gazeev VM, Włodyka R, Dulski M, Dzierżanowski P (2015) Fluorchegemite, $Ca_7(SiO_4)_3F_2$, a new mineral from the edgrewite-bearing endoskarn zone of an altered xenolith in ignimbrites from upper Chegem caldera, Northern Caucasus, Kabardino-Balkaria, Russia: occurrence, crystal structure, and new data on the mineral assemblages. Can Mineral 53(2):325–344

Galuskina IO, Galuskin EV, Prusik K, Vapnik Y, Juroszek R, Jeżak L, Murashko M (2016a) Dzierżanowskite, $CaCu_2S_2$ – a new natural thiocuprate from Jabel Harmun, Judean Desert, Palestine Autonomy, Israel. Mineral Mag. https://doi.org/10.1180/minmag.2016.080.153

Galuskina IO, Galuskin EV, Vapnik Y, Prusik K, Stasiak M, Dzierżanowski P, Murashko M (2016b) Gurimite, $Ba_3(VO_4)_2$, and hexacelsian, $BaAl_2Si_2O_8$ – two new minerals from schorlomite-rich paralava of the Hatrurim Complex, Negev Desert, Israel. Mineral Mag. https://doi.org/10.1180/minmag.2016.080.147

Galuskina IO, Galuskin EV, Pakhomova AS, Widmer R, Armbruster T, Krüger B, Grew ES, Vapnik Y, Dzierażanowski P, Murashko M (2017) Khesinite, $Ca_4Mg_2Fe^{3+}{}_{10}O_4[(Fe^{3+}{}_{10}Si_2)O_{36}]$, a new rhönite-group (sapphirine supergroup) mineral from the Negev Desert, Israel – natural analogue of the SFCA phase. Eur J Mineral. https://doi.org/10.1127/ejm/2017/0029-2589

Gama S˙ (2000) Evénements métallogéniques à W-Bi (Au) à 305 Ma en Châtaigneraie du Cantal: apport d'une analyse multi-spectrométrique (micro PIXE-PIGE et Raman) des minéraux et de sfluides occlus à l'identification des sources de fluids hydrothermaux. Minéralogie. Thése, Université d'Orléans. (in French)

Gao J, Huang W, Wu X, Fan D, Wu Z, Xia D, Qin S (2015) Compressibility of carbonophosphate bradleyite $Na_3Mg(PO_4)(CO_3)$ by X-ray diffraction and Raman spectroscopy. Phys Chem Minerals 42 (3):191–201

Garavelli A, Pinto D, Mitolo D, Bindi L (2014) Leguernite, $Bi_{12.67}O_{14}(SO_4)_5$, a new Bi oxysulfate from the fumarole deposit of La Fossa crater, Vulcano, Aeolian Islands, Italy. Mineral Mag 78(7):1629–1646

Garbout A, Taazayet-Belgacem IB, Férid M (2013) Structural, FT-IR, XRD and Raman scattering of new rare-earth-titanatepyrochlore-type oxides $LnEuTi_2O_7$ (Ln = Gd, Dy). J Alloys Compd 573:43–52

Garcia CS, Abedin MN, Sharma SK, Misra AK, Ismail S, Singh UN, Refaat TF, Elsayed-Ali HE, Sandford SP (2006) Remote pulsed laser Raman spectroscopy system for detecting water, ice, and hydrous minerals. Proc SPIE. https://doi.org/10.1117/12.680879

García FP, Hernandez JC, Cruz VER, Santillan YM, Marzo MAM, Moreno FP (2013) Recovery and characterization of struvite from sediment and sludge resulting from the process of acid whey electrocoagulation. Asian J Chem 25(14):8005–8009

Garcia-Guinea J, La Iglesia A, Crespo-Feo E, Del Tánago JG, Correcher V (2013) The status of zaratite: investigation of the type specimen from Cape Ortegal, Galicia, Spain. Eur J Mineral 25(6):995–1004

Gasanly NM, Magomedov AZ, Melnik NN, Salamov BG (1993) Raman and infrared studies of $AgIn_5S_8$ and $CuIn_5S_8$ single crystals. Phys Status Solidi (b) 177(1): K31–K35

Gasharova B, Mihailova B, Konstantinov L (1997) Raman spectra of various types of tourmaline. Eur J Mineral 9:935–940

Gastaldi D, Boccaleri E, Canonico F (2008) In situ Raman study of mineral phases formed as by-products in a rotary kiln for clinker production. J Raman Spectrosc 39(7):806–812

Gatta G, Lotti P, Kahlenberg V, Haefeker U (2012a) The low-temperature behaviour of cancrinite: an in situ single-crystal X-ray diffraction study. Mineral Mag 76(4):933–948

Gatta GD, McIntyre GJ, Swanson JG, Jacobsen SD (2012b) Minerals in cement chemistry: a single-crystal neutron diffraction and Raman spectroscopic study of thaumasite, $Ca_3Si(OH)_6(CO_3)(SO_4)\cdot 12H_2O$. Am Mineral 97:1060–1069

Gatta GD, Jacobsen SD, Vignola P, McIntyre GJ, Guastella G, Abate LF (2014) Single-crystal neutron diffraction and Raman spectroscopic study of hydroxylherderite, CaBePO$_4$(OH,F). Mineral Mag 78(3):723–738

Gatta GD, Rotiroti N, Bersani D, Bellatreccia F, Della Ventura G, Rizzato S (2015a) A multi-methodological study of the (K,Ca)-variety of the zeolite merlinoite. Mineral Mag 79(7):1755–1767

Gatta GD, Scheidl KS, Pippinger T, Skála R, Lee Y, Miletich R (2015b) High-pressure behavior and crystal–fluid interaction under extreme conditions in paulingite [PAU-topology]. Micropor Mesopor Mater 206:34–41

Gatta G, Bosi F, Fernandez Diaz MT, Hålenius U (2016) H-bonding scheme in allactite: a combined single-crystal neutron/X-ray diffraction, EPMA-WDS, FTIR and OAS study. Mineral Mag. https://doi.org/10.1180/minmag.2016.080.020

Gavarri JR, Chater R, Ziółkowski J (1988) The chemical bonds in MeSb$_2$O$_4$ (Me = Mn, Ni, Fe, Zn) isomorphous compounds: thermal expansion, force constants, energies. J Solid State Chem 73(2):305–316

Ge J-P, Li Y-D (2003) Ultrasonic synthesis of nanocrystals of metal selenides and tellurides. J Mater Chem 13(4):911–915

Gee AR, O'Shea DC, Cummins HZ (1966) Raman scattering and fluorescence in calcium fluoride. Solid State Commun 4(1):43–46

Gehring P, Benia HM, Weng Y, Dinnebier R, Ast CR, Burghard M, Kern K (2013) A natural topological insulator. Nano Lett 13(3):1179–1184

Gehring P, Vaklinova K, Hoyer A, Benia HM, Skakalova V, Argentero G, Eder F, Burghard M, Kern K (2015) Dimensional crossover in the quantum transport behaviour of the natural topological insulator Aleksite. Sci Rep. https://doi.org/10.1038/srep11691

Geisinger KL, Ross NL, McMillan PF, Navrotsky A (1987) K$_2$Si$_4$O$_9$; energetics and vibrational spectra of glass, sheet silicate, and wadeite-type phases. Am Mineral 72(9–10):984–994

Geisler T, Zhang M, Salje EK (2003) Recrystallization of almost fully amorphous zircon under hydrothermal conditions: an infrared spectroscopic study. J Nucl Mater 320(3):280–291

Geisler T, Berndt J, Meyer HW, Pollok K, Putnis A (2004) Low-temperature aqueous alteration of crystalline pyrochlore: correspondence between nature and experiment. Mineral Mag 68(6):905–922

Genceli FE, Lutz M, Spek AL, Witkamp GJ (2007) Crystallization and characterization of a new magnesium sulfate hydrate MgSO$_4$·11H$_2$O. Cryst Growth Design 7(12):2460–2466

Genceli FE, Horikawa S, Iizuka Y, Sakurai T, Hondoh T, Kawamura T, Witkamp GJ (2009) Meridianiite detected in ice. J Glaciol 55(189):117–122

Genchev G, Erbe A (2016) Raman spectroscopy of mackinawite FeS in anodic iron sulfide corrosion products. J Electrochem Soc 163(6):C333–C338

Georgiev M, Wildner M, Stoilova D, Karadjova V (2007) Preparation, crystal structure and infrared spectroscopy of the new compound rubidium beryllium sulfate dihydrate, Rb$_2$Be(SO$_4$)$_2$·2H$_2$O. Vibr Spectrosc 44(2):266–272

Georgiev M, Bancheva T, Marinova D, Stoyanova R, Stoilova D (2016) On the formation of solid solutions with blödite-and kröhnkite-type structures. I. Synthesis, vibrational and EPR spectroscopic investigations of Na$_2$Zn$_{1-x}$Cu$_x$(SO$_4$)$_2$·4H$_2$O ($0<x<0.14$). IJSRST 2(5):279–292

Gerding H (1941) Das Ramanspektrum des Festen Selendioxyds. Recueil Travaux Chim Pays-Bas 60(10):728–731. (in German)

Gerlinger H, Schaack G (1986) Crystal-field states of the Ce^{3+} ion in CeF$_3$: a demonstration of vibronic interaction in ionic rare-earth compounds. Phys Rev B 33(11):7438–7450

Gfeller F, Armbruster T, Galuskin EV, Galuskina IO, Lazic B, Savelyeva VB, Zadov AE, Dzierżanowski P, Gazeev VM (2013) Crystal chemistry and hydrogen bonding of rustumite Ca$_{10}$(Si$_2$O$_7$)$_2$(SiO$_4$)(OH)$_2$Cl$_2$ with variable OH, Cl, F. Am Mineral 98(2–3):493–500

Gfeller F, Widmer R, Krüger B, Galuskin EV, Galuskina IO, Armbruster T (2015) The crystal structure of flamite and its relation to Ca$_2$SiO$_4$ polymorphs and nagelschmidtite. Eur J Mineral 27(6):755–769

Ghorbel K, Litaiem H, Ktari L, Garcia-Granda S, Dammak M (2015) X-ray single crystal, thermal analysis and vibrational study of (NH$_4$)$_2$(SO$_4$)$_{0.92}$H(AsO$_4$)$_{0.08}$·Te(OH)$_6$. J Molec Struct 1079:225–231

Ghribi N, Karray R, Laval JP, Kabadou A, Ben Salah A (2015) X-ray powder diffraction and Raman vibrational study of new doped compound TiTe$_3$O$_8$:Ce^{4+}. J Mater Environ Sci 6(4):989–996

Ghule A, Bhongale C, Chang H (2003) Monitoring dehydration and condensation processes of Na$_2$HPO$_4$·12H$_2$O using thermo-Raman spectroscopy. Spectrochim Acta A 59(7):1529–1539

Giammar DE, Hering JG (2002) Equilibrium and kinetic aspects of soddyite dissolution and secondary phase precipitation in aqueous suspension. Geochim Cosmochim Acta 66(18):3235–3245

Gianfagna A, Mazziotti-Tagliani S, Croce A, Allegrina M, Rinaudo C (2014) As-rich apatite from Mt. Calvario: characterization by micro-Raman spectroscopy. Can Mineral 52(5):799–808

Giarola M, Sanson A, Rahman A, Mariotto G (2011) Vibrational dynamics of YPO$_4$ and ScPO$_4$ single crystals: an integrated study by polarized Raman spectroscopy and first principles calculations. Phys Rev B 83:224302-1–224302-8

Gieré R, Williams CT, Wirth R, Ruschel K (2009) Metamict fergusonite-(Y) in a spessartine-bearing granitic pegmatite from Adamello, Italy. Chem Geol 261(3):333–345

Gies H (1983) Studies on clathrasils. III. Crystal structure of melanophlogite, a natural clathrate compound of silica. Z Kristallogr 164:247–257

Gies H, Gerke H, Liebau F (1982) Chemical composition and synthesis of melanophlogite, a clathrate compound of silica. N Jb Mineral Mh 3:119–124

Giese B, McNaughton D (2002) Density functional theoretical (DFT) and surface-enhanced Raman spectroscopic study of guanine and its alkylated derivatives Part 1. DFT calculations on neutral, protonated and deprotonated guanine. Phys Chem Chem Phys 4 (20):5161–5170

Giester G, Lengauer CL, Pristacz H, Rieck B, Topa D, Von Bezing KL (2016) Cairncrossite, a new Ca-Sr (-Na) phyllosilicate from the Wessels Mine, Kalahari Manganese Field, South Africa. Eur J Mineral. https://doi.org/10.1127/ejm/2016/0028-2519

Giguére PA, Harvey KB (1956) On the infrared absorption of water and heavy water in condensed states. Can J Chem 34:798–808

Gilg HA, Gast N (2015) Determination of titanium content in pyrope by Raman spectroscopy. J Raman Spectrosc. https://doi.org/10.1002/jrs.4838

Gillet P, Reynard B, Tequi C (1989) Thermodynamic properties of glaucophane new data from calorimetric and spectroscopic measurements. Phys Chem Minerals 16:659–667

Gillet P, Chen M, Dubrovinsky L, El Goresy A (2000) Natural $NaAlSi_3O_8$-hollandite in the shocked Sixiangkou meteorite. Science 287(5458):1633–1636

Giuliani G, Ohnenstetter D, Palhol F, Feneyrol J, Boutroy E, De Boissezon H, Lhomme T (2008) Karelianite and vanadian phlogopite from the Merelani Hills gem zoisite deposits, Tanzania. Can Mineral 46 (5):1183–1194

Glamazda A, Choi KY, Lemmens P, Yang JJ, Cheong SW (2014) Proximity to a commensurate charge modulation in $IrTe_{2-x}Se_x$ ($x=$ 0 and 0.45) revealed by Raman spectroscopy. New J Phys 16(9):093061. (12 pp)

Godec J, Colombo L (1976) Interpretation of the vibrational spectrum of crystalline phenanthrene. J Chem Phys 65:4693–4700

Godelitsas A, Astilleros JM, Hallam KR, Löns J, Putnis A (2003) Microscopic and spectrosopic investigation of the calcite surface interacted with Hg(II) in aqueous solutions. Mineral Mag 67(6):1193–1204

Goienaga N, Arrieta N, Carrero JA, Olivares M, Sarmiento A, Martinez-Arkarazo I, Fernández LA, Madariaga JM (2011) Micro-Raman spectroscopic identification of natural mineral phases and their weathering products inside an abandoned zinc/lead mine. Spectrochim Acta A 80(1):66–74

Golovin AV, Korsakov AV, Zaitsev AN (2014) In Situ high-temperature Raman spectroscopic study of zemkorite and $(Na,K)_2Ca(CO_3)_2$ γ-phase. 11th Int GeoRaman Conf 5046

Golovin AV, Korsakov AV, Zaitsev AN (2015) In situ ambient and high-temperature Raman spectroscopic studies of nyerereite $(Na,K)_2Ca(CO_3)_2$: can hexagonal zemkorite be stable at earth-surface conditions. J Raman Spectrosc 46(10):904–912

Golovin AV, Korsakov AV, Gavryushkin PN, Zaitsev AN, Thomas VG, Moine BN (2017) Raman spectra of nyerereite, gregoryite, and synthetic pure $Na_2Ca(CO_3)_2$: diversity and application for the study micro inclusions. J Raman Spectrosc. https://doi.org/10.1002/jrs.5143

Golubev YA, Martirosyan OV (2012) The structure of the natural fossil resins of North Eurasia according to IR-spectroscopy and microscopic data. Phys Chem Miner 39:247–258

Golubev YA, Isaenko SI, Prikhodko AS, Borgardt NI, Suvorova EI (2016) Raman spectroscopic study of natural nanostructured carbon materials: shungite vs. anthraxolite. Eur J Mineral. https://doi.org/10.1127/ejm/2016/0028-2537

Gómez MA, Lee K (2012) Vibrational spectroscopy of complex synthetic and industrial products. In: De Caro D (ed) Vibrational spectroscopy. INTECH. https://doi.org/10.5772/32982

Gomez MA, Assaaoudi H, Becze L, Cutler JN, Demopoulos GP (2010a) Vibrational spectroscopy study of hydrothermally produced scorodite $(FeAsO_4 \cdot 2H_2O)$, ferric arsenate sub-hydrate (FAsH; Fe $AsO_4 \cdot 0.75H_2O$) and basic ferric arsenate sulfate (BFAS; $Fe[(AsO_4)_{1-x}(SO4)_x(OH)_x] \cdot wH_2O$). J Raman Spectrosc 41(2):212–221

Gomez MA, Becze L, Blyth RIR, Cutler JN, Demopoulos GP (2010b) Molecular and structural investigation of yukonite (synthetic & natural) and its relation to arseniosiderite. Geochim Cosmochim Acta 74 (20):5835–5851

Gomez MA, Le Berre JF, Assaaoudi H, Demopoulos GP (2011) Raman spectroscopic study of the hydrogen and arsenate bonding environment in isostructural synthetic arsenates of the variscite group – M^{3+} $AsO_4 \cdot 2H_2O$ (M^{3+} = Fe, Al, In and Ga): implications for arsenic release in water. J Raman Spectrosc 42 (1):62–71

Gomez MA, Ventruti G, Celikin M, Assaaoudi H, Putz H, Becze L, Lee KE, Demopoulos GP (2013) The nature of synthetic basic ferric arsenate sulfate (Fe $(AsO_4)_{1-x}(SO_4)x(OH)x$) and basic ferric sulfate $(FeOHSO_4)$: their crystallographic, molecular and electronic structure with applications in the environment and energy. RSC Adv 3(37):16840–16849

Gómez-Nubla L, Aramendia J, Vallejuelo SFO, Castro K, Madariaga JM (2013) From portable to SCA Raman devices to characterize harmful compounds contained in used black slag produced in electric arc furnace of steel industry. J Raman Spectrosc 44(8):1163–1171

Gönen ZS, Kizilyalli M, Pamuk HÖ (2000) Synthesis and characterization of Na_2GdOPO_4 and Na_2LaOPO_4. J Alloys Compd 303:416–420

Gong WL, Ewing RC, Wang LM, Xie HS (1995) Aeschynite and euxenite structure-types as host phases for rare-earths and actinides from HLW. In: MRS Proceedings, vol 412. Cambridge University Press, p 377

Gopal NO, Narasimhulu KV, Rao JL (2004) EPR, optical, infrared and Raman spectral studies of actinolite mineral. Spectrochim Acta A 60(11):2441–2448

Gopal RK, Singh S, Chandra R, Mitra C (2015) Weak-antilocalization and surface dominated transport in topological insulator Bi_2Se_2Te. AIP Adv 5(4):047111-1–047111-10

Gopinath AB, Devanarayanan S, Castro A (1998) Vibrational spectra of three anhydrous rare-earth selenites $R_2Se_3O_9$ (R = La, Sm and Lu). Spectrochim Acta A 54(6):785–791

Goreva JS, Ma C, Rossman GR (2001) Fibrous nanoinclusions in massive rose quartz: the origin of rose coloration. Am Mineral 86(4):466–472

Goryainov SV (2016) Raman study of thaumasite $Ca_3Si(OH)_6(SO_4)(CO_3) \cdot 12H_2O$ at high pressure. J Raman Spectrosc. https://doi.org/10.1002/jrs.4936

Goryainov SV, Smirnov MB (2001) Raman spectra and lattice-dynamical calculations of natrolite. Eur J Mineral 13(3):507–519

Goryainov SV, Krylov AS, Pan Y, Madyukov IA, Smirnov MB, Vtyurin AN (2012) Raman investigation of hydrostatic and nonhydrostatic compressions of OH- and F-apophyllites up to 8 GPa. J Raman Spectrosc 43(3):439–447

Goryainov SV, Likhacheva AY, Rashchenko SV, Shubin AS, Afanas'ev VP, Pokhilenko NP (2014) Raman identification of lonsdaleite in Popigai impactites. J Raman Spectrosc 45(4):305–313

Goryainov SV, Krylov AS, Vtyurin AN, Pan Y (2015) Raman study of datolite $CaBSiO_4(OH)$ at simultaneously high pressure and high temperature. J Raman Spectrosc 46(1):177–181

Goryainov SV, Pan Y, Smirnov MB, Sun W, Mi JX (2017) Raman investigation on the behavior of parasibirskite $CaHBO_3$ at high pressure. Spectrochim Acta A 173:45–52

Gottschall R, Schöllhorn R, Muhler M, Jansen N, Walcher D, Gütlich P (1998) Electronic state of nickel in barium nickel oxide, $BaNiO_3$. Inorg Chem 37(7):1513–1518

Gourrier L, Deabate S, Michel T, Paillet M, Hermet P, Bantignies J-L, Henn F (2011) Characterization of unusually large "pseudo-single crystal" of β-nickel hydroxide. J Phys Chem C 115:15067–15074

Gow RN (2015) Spectroelectrochemistry and modeling of enargite (Cu_3AsS_4) under atmospheric conditions. Dissertation, University of Montana

Goypiron A, De Villepin J, Novak A (1980) Raman and infrared study of $KHSO_4$ crystal. J Raman Spectrosc 9(5):297–303

Graeser S, Schwander H, Demartin F, Gramaccioli CM, Pilati T, Reusser E (1994) Fetiasite $(Fe^{2+},Fe^{3+},Ti)_3O_2[As_2O_5]$, a new arsenite mineral: its description and structure determination. Am Mineral 79:996–1002

Graeser S, Hetherington CJ, Gieré R (2003) Ganterite, a new barium-dominant analogue of muscovite from the Berisal Complex, Simplon Region, Switzerland. Can Mineral 41(5):1271–1280

Graf L (1978) Phonons and electronic transitions in the Raman spectra of $FeCl_2 \cdot 2H_2O$ and isomorphic compounds. Solid State Commun 27(12):1361–1365

Graf L, Schaack G (1976) Magnon-phonon coupling in antiferromagnetic $FeCl_2 \cdot 2H_2O$. Z Phys B 24(1):83–89

Graham CM, Tareen JA, Mcmillan PF, Lowe BM (1992) An experimental and thermodynamic study of cymrite and celsian stability in the system $BaO-Al_2O_3-SiO_2-H_2O$. Eur J Mineral 4(2):251–269

Graham IT, Pogson RE, Colchester DM, Baines A (2003) Zeolite crystal habits, compositions, and paragenesis; Blackhead Quarry, Dunedin, New Zealand. Mineral Mag 67(4):625–637

Graham IT, Pogson RE, Colchester DM, Hergt J, Martin R, Williams PA (2007) Pink lanthanite-(Nd) from Whitianga Quarry, Coromandel Peninsula, New Zealand. Can Mineral 45(6):1389–1396

Grandhe BK, Bandi VR, Jang K, Lee HS, Shin DS, Yi SS, Jeong JH (2012) Effect of sintering atmosphere and lithium ion co-doping on photoluminescence properties of $NaCaPO_4$: Eu^{2+} phosphor. Ceram Int 38(8):6273–6279

Grasselli MC, Baran EJ (1984) IR spectroscopic characterization of tetrabasic lead sulphate. J Mater Sci Lett 3(11):949–950

Grey IE, Shanks FL, Wilson NC, Mumme WG, Birch WD (2011) Carbon incorporation in plumbogummite-group minerals. Mineralag 75(1):145–158

Grey IE, Keck E, Mumme WG, Pring A, Macrae CM, Glenn AM, Davidson CJ, Shamks FL, Mills SJ (2016a) Kummerite, $Mn^{2+}Fe^{3+}Al(PO_4)_2(OH)_2 \cdot 8H_2O$, a new laueite-group mineral from the HagendorfSüd pegmatite, Bavaria, with ordering of Al and Fe^{3+}. Mineral Mag. https://doi.org/10.1180/minmag.2016.080.061

Grey IE, Betterton J, Kampf AR, Macrae CM, Shanks FL, Price JR (2016b) Penberthycroftite, $[Al_6(AsO_4)_3(OH)_9(H_2O)_5] \cdot 8H_2O$, a second new hydrated aluminium arsenate mineral from the Penberthy Croft mine, St. Hilary, Cornwall. Mineral Mag. https://doi.org/10.1180/minmag.2016.080.069

Grey I, Keck E, Kampf AR, Macrae CM, Glenn AM, Price JR (2016c) Wilhelmgümbelite, $[ZnFe^{2+}Fe^{3+}_3(PO_4)_3(OH)_4(H_2O)_5] \cdot 2H_2O$, a new schoonerite-related mineral from the Hagendorf Süd pegmatite, Bavaria. Mineral Mag. https://doi.org/10.1180/minmag.2016.080.098

Grice JD, Van Velthuizen J, Dunn PJ, Newbury DE, Etz ES, Nielsen CH (1986) Moydite (Y,REE)[B(OH)$_4$](CO$_3$), a new mineral species from the Evans-Lou pegmatite, Quebec. Can Mineral 24:665–673

Grice JD, Rowe R, Poirier G (2015) Hydroterskite: a new mineral species from the Saint-Amable Sill, Quebec, and acomparison with terskite and elpidite. Can Mineral 53(5):821–832

Griffith WP (1970) Raman studies on rock-forming minerals. Part II. Minerals containing MO_3, MO_4, and MO_6 groups. J Chem Soc A Inorg Phys Theor 286–291

Grishchenko RO, Emelina AL, Makarov PY (2013) Thermodynamic properties and thermal behavior of Friedel's salt. Thermochim Acta 570:74–79

Grishina S, Dubessy J, Kontorovich A, Pironon J (1992) Inclusions in salt beds resulting from thermal metamorphism by dolerite sills (eastern Siberia, Russia). Eur J Mineral 4(5):1187–1202

Groat LA, Hawthorne FC, Erict ⊠S (1995) The chemistry of vesuvianite. Can Miner 33:19–48

Groat LA, Kek S, Bismayer U, Schmidt C, Krane HG, Meyer H, Nestor L, Tendeloo GV (1996) A synchrotron radiation, HRTEM, X-ray powder diffraction, and Raman spectroscopic study of malayaite, $CaSnSiO_5$. Am Mineral 81(5–6):595–602

Groß U, Rüdiger S, Kemnitz E, Brzezinka KW, Mukhopadhyay S, Bailey C, Wander A, Harrison N (2007) Vibrational analysis study of aluminum trifluoride phases. J Phys Chem A 111(26):5813–5819

Grosse P, Richter W (1970) Absorption spectra of tellurium in the spectral range from 18 to 460 cm^{-1}. Phys Status Solidi B 41(1):239–246

Grundler PV, Brugger J, Meisser N, Ansermet S, Borg S, Etschmann B, Testemale D, Bolin T (2008) Xocolatlite, $Ca_2Mn^{4+}{}_2Te_2O_{12} \cdot H_2O$, a new tellurate related to kuranakhite: description and measurement of Te oxidation state by XANES spectroscopy. Am Mineral 93(11–12):1911–1920

Grüneberger AM, Schmidt C, Jahn S, Rhede D, Loges A, Wilke M (2016) Interpretation of Raman spectra of the zircon-hafnon solid solution. Eur J Mineral. https://doi.org/10.1127/ejm/2016/0028-2551

Grzechnik A, McMillan PF (1997) In situ high pressure Raman spectra of $Ba_3(VO_4)_2$. Solid State Commun 102 (8):569–574

Gualtieri AF (2000) Study of NH_4^+ in the zeolite phillipsite by combined synchrotron powder diffraction and IR spectroscopy. Acta Crystallogr B 56 (4):584–593

Guang F, Xiangun G, Guowu L, Apeng Y, Ganfu S (2016) Oxynatromicrolite,$(Na,Ca,U)_2Ta_2O_6(O,F)$, a new member of the pyrochloresupergroup from Guanpo, Henan Province, China. Mineral Mag. https://doi.org/10.1180/minmag.2016.080.121

Guc M, Ursaki VV, Bodnar IV, Lozhkin DV, Arushanov E, Izquierdo-Roca V, Rodríguez AP (2012) Raman scattering investigation of $Mn_xFe_{1-x}In_2S_4$ solid solutions. Mater Chem Phys 136(2):883–888

Gucsik A, Zhang M, Koeberl C, Salje EKH, Redfern SAT, Pruneda JM (2004) Infrared and Raman spectra of $ZrSiO_4$ experimentally shocked at high pressures. Mineral Mag 68:801–811

Guilherme LR, Massabni AC, Dametto AC, de Souza Corrêa R, de Araujo AS (2010) Synthesis, infrared spectroscopy and crystal structure determination of a new decavanadate. J Chem Crystallogr 40(11):897–901

Guillaume F, Huang S, Harris KD, Couzi M, Talaga D (2008) Optical phonons in millerite (NiS) from single-crystal polarized Raman spectroscopy. J Raman Spectrosc 39(10):1419–1422

Guin SN, Banerjee S, Sanyal D, Pati SK, Biswas K (2016) Origin of the order–disorder transition and the associated anomalous change of thermopower in $AgBiS_2$ nanocrystals: a combined experimental and theoretical study. Inorg Chem. https://doi.org/10.1021/acs.inorgchem.6b00997

Guizzetti G, Marabelli F, Patrini M, Pellegrino P, Pivac B, Miglio L, Meregalli V, Lange H, Henrion W, Tomm V (1997) Measurement and simulation of anisotropy in the infrared and Raman spectra of β-$FeSi_2$ single crystals. Phys Rev B 55(21): 14290–14297

Güler H, Tekin B (2009) Synthesis and crystal structure $CoNi_2(BO_3)_2$. Inorg Mater 45(5):538–542

Güner FEG, Sakurai T, Hondoh T (2013) Ernstburkeite, $Mg(CH_3SO_3)_2 \cdot 12H_2O$, a new mineral from Antarctica. Eur J Mineral 25(1):79–84

Guńka PA, Dranka M, Piechota J, Żukowska GZ, Zalewska A, Zachara J (2012) As_2O_3 polymorphs: theoretical insight into their stability and ammonia template claudetite II crystallization. Cryst Growth Des 12(11):5663–5670

Günter JR, Amberg M (1989) "High-temperature" magnesium tungstate, $MgWO_4$, prepared at moderate temperature. Solid State Ionics 32:141–146

Guo F, Fu P, Wang J, Liu F, Yang Z, Wu Y (2000) Hydrothermal synthesis, characterization and nonlinear optical effect of orthorhombic phase $Ca_2B_6O_{11} \cdot H_2O$. Chinese Sci Bull 45(19):1756–1760

Guo K, Lu B, Xia Y, Qi L (2010) Thermal phase transformation of turquoise. J Chin Ceramic Soc 38 (4):694–699. (in Chinese, English abstr)

Guo X, Ushakov SV, Curtius H, Bosbach D, Navrotsky A (2014a) Energetics of metastudtite and implications for nuclear waste alteration. Proc Natl Acad Sci 111 (50):17737–17742

Guo X, Wu H, Pan S, Yang Z, Yu H, Zhang B, Han J, Zhang F (2014b) Synthesis, crystal structure, and characterization of a congruent melting compound magnesium strontium diborate $MgSrB_2O_5$. Z anor gallg Chemie 640(8–9):1805–1809

Guo H, Li Y, Fang X, Zhang K, Ding J, Yuan N (2016) Co-sputtering deposition and optical-electrical characteristic of Cu_2CdSnS_4 thin films for use in solar cells. Mater Lett 162:97–100

Gurzhiy VV, Tyumentseva OV, Kornyakov IV, Krivovichev SV, Tananaev IG (2014) The role of potassium atoms in the formation of uranyl selenates: the crystal structure and synthesis of two novel compounds. J Geo Sci 59(2):123–133

Gusain M, Rawat P, Nagarajan R (2015) Influence of reaction conditions for the fabrication of Cu_2SnS_3 and Cu_3SnS_4 in ethylene glycol. Mater Res Express 2 (5):055501. (5 pp)

Gutwirth J, Wagner T, Vlcek M, Drasar C, Benes L, Hrdlicka M, Frumar M, Schwarz J, Ticha H (2006) Ag-Sb-S thin films prepared by RF magnetron sputtering and their properties. Mater Res Soc Symp Proc 918:0918-H07-01-G08-01. (10 pp)

Hadrich A, Lautie A, Mhiri T (2000) Vibrational study of structural phase transitions in $(NH_4)_2HPO_4$ and $(ND_4)_2DPO_4$. J Raman Spectrosc 31(7):587–593

Hadrich A, Lautié A, Mhiri T, Romain F (2001) Vibrational behaviour of K_2HPO_4, $K_2HPO_4 \cdot 3H_2O$ and their deuterated derivatives with temperature. Vib Spectrosc 26(1):51–64

Haeuseler H, Haxhillazi G (2003) Vibrational spectra of the peroxochromates $(NH_4)_3[Cr(O_2)_4]$, $K_3[Cr(O_2)_4]$ and $Rb_3[Cr(O_2)_4]$. J Raman Spectrosc 34(5):339–344

Hagemann H, Lucken A, Bill H, Gysler-Sanz J, Stalder HA (1990) Polarized Raman spectra of beryl and bazzite. Phys Chem Minerals 17(5):395–401

Hahn RB (1951) Phosphates of niobium and tantalum. J Amer Chem Soc 73:5091–5093

Hålenius U, Westlund E (1998) Manganese valency and the colour of the $Mn_2AsO_4(OH)$ polymorphs eveite and sarkinite. Mineral Mag 62(1):113–119

Haley LV, Wylie IW, Koningstein JA (1982) An investigation of the lattice and interlayer water vibrational spectral regions of muscovite and vermiculite using Raman microscopy. A Raman microscopic study of layer silicates. J Raman Spectrosc 13(2):203–205

Halonen L, Carrington T Jr (1988) Fermi resonances and local modes in water, hydrogen sulfide, and hydrogen selenide. J Chem Phys 88:4171–4185

Hamdouni M, Walha S, Kabadou A, Duhayon C, Sutter JP (2013) Synthesis and crystal structures of various phases of the microporous three-dimensional coordination polymer $[Zr(OH)_2(C_2O_4)]_n$. Crystal Growth Design 13(11):5100–5106

Han S, Pan S, Yang Z, Wang Y, Zhang B, Zhang M, Huang Z, Dong L, Yu H (2013) Synthesis, structure characterization, and optical properties of the aluminosilicate $Li_2Na_3AlSi_2O_8$. Z anor gallg Chemie 639(5):779–783

Hänni HA, Kiefert L, Chalain JP (1997) A Raman microscope in the gemmological laboratory: first experiences of application. J Gemm 25:394–407

Hanson RC, Fjeldly TA, Hochheimer HD (1975) Raman scattering from five phases of silver iodide. Phys Status Solidi B 70(2):567–576

Hansteen TH, Burke EA (1994) Aphthitalite in high-temperature fluid illclusions in quartz from the Eikeren-Skrim granite complex, the Oslo paleorift. Norsk Geologisk Tidsskrift 74:238–240

Hanuza J, Haznar A, Mączka M, Pietraszko A, Lemiec A, Van der Maas JH, Lutz ETG (1997) Structure and vibrational properties of tetragonal scheelite $NaBi(MoO_4)_2$. J Raman Spectrosc 28(12):953–963

Hanuza J, Mączka M, Lorenc J, Kaminskii AA, Bohaty L, Becker P (2008a) Polarised IR and Raman spectra of non-centrosymmetric $Na_3Li(SeO_4)_2 \cdot 6H_2O$ crystal – a new Raman laser material. Spectrochim Acta A 71(1):68–72

Hanuza J, Mączka M, Lorenc J, Kaminskii AA, Becker P, Bohatý L (2008b) Polarized Raman and IR spectra of non-centrosymmetric PbB_4O_7 single crystal. J Raman Spectrosc 39(3):409–414

Hanuza J, Ptak M, Mączka M, Hermanowicz K, Lorenc J, Kaminskii AA (2012) Polarized IR and Raman spectra of $Ca_2MgSi_2O_7$, $Ca_2ZnSi_2O_7$ and $Sr_2MgSi_2O_7$ single crystals: temperature-dependent studies of commensurate to incommensurate and incommensurate to normal phase transitions. J Solid State Chem 191:90–101

Hao S-Y (2008) Ultrafine CeO_2: microwave-assisted heating prepartion and ploshing properties. Chin J Inorg Chem 24(6):1012–1016

Harcharras M, Capitelli F, Ennaciri A, Brouzi K, Moliterni AGG, Mattei G, Bertolasi V (2003) Synthesis, X-ray crystal structure and vibrational spectroscopy of the acidic pyrophosphate $KMg_{0.5}H_2P_2O_7 \cdot H_2O$. J Solid State Chem 176(1):27–32

Hardcastle FD, Wachs IE (1991) Determination of vanadium-oxygen bond distances and bond orders by Raman spectroscopy. J Phys Chem 95(13):5031–5041

Haring MM, McDonald AM (2014a) Steedeite, $NaMn_2[Si_3BO_9](OH)_2$: characterization, crystal-structure determination, and origin. Can Mineral 52(1):47–60

Haring MMM, McDonald AM (2014b) Franconite, $NaNb_2O_5(OH) \cdot 3H_2O$: structure determination and the role of H bonding, with comments on the crystal chemistry of franconite-related minerals. Mineral Mag 78(3):591–608

Haring MMM, McDonald AM (2016) Nolzeite, $Na(Mn,\square)_2[Si_3(B,Si)O_9(OH)_2] \cdot 2H_2O$, a new pyroxenoid mineral. Mineral Mag. https://doi.org/10.1180/minmag.2016.080.089

Haring MMM, McDonald AM (2017) Charleshatchettite, $CaNb_4O_{10}(OH)_2 \cdot 8H_2O$, a new mineral from Mont Saint-Hilaire, Québec, Canada: description, crystal-structure determination and origin. Am Mineral. https://doi.org/10.2138/am-2017-5926

Haring MM, McDonald AM, Cooper MA, Poirier GA (2012) Laurentianite, $[NbO(H_2O)]_3(Si_2O_7)_2[Na(H_2O)_2]_3$, a new mineral from Mont Saint-Hilaire, Québec: description, crystal-structure determination and paragenesis. Can Mineral 50(5):1265–1128

Harlov DE, Förster HJ, Schmidt C (2003) High PT experimental metasomatism of a fluorapatite with significant britholite and fluorellestadite components: implications for LREE mobility during granulite-facies metamorphism. Mineral Mag 67(1):61–72

Harrington JA, Harley RT, Walker CT (1971) Impurity induced vibrational spectra in BaF_2:H. Solid State Comm 9(10):683–687

Harrison KL, Manthiram A (2013) Microwave-assisted solvothermal synthesis and characterization of various polymorphs of $LiVOPO_4$. Mech Eng 25(9):1751–1760

Harvey KB, Morrow BA, Shurvell HF (1963) The infrared absorption of some crystalline inorganic formates. Can J Chem 41(5):1181–1187

Hatert F, Rebbouh L, Hermann RP, Fransolet A-M, Long GJ, Grandjean F (2005) Crystal chemistry of the hydrothermally synthesized $Na_2(Mn_{1-x}Fe^{2+}{}_x)_2Fe^{3+}(PO_4)_3$ alluaudite-type solid solution. Am Mineral 90(4):653–662

Hatipoglu M, Babalik H (2012) Micro-Raman characterization of the lapeyreite mineral, the Alpes-Maritimes Region, Nice, France. Asian J Chem 24(5):1941–1944

Hawthorne FC, Cooper MA, Abdu YA, Ball NA, Back ME, Tait KT (2012) Davidlloydite, ideally $Zn_3(AsO_4)_2(H_2O)_4$, a new arsenate mineral from the Tsumeb mine, Otjikoto (Oshikoto) region, Namibia: description and crystal structure. Mineral Mag 76(1):45–57

Hawthorne FC, Abdu YA, Ball NA, Pinch WW (2013) Carlfrancisite: $Mn^{2+}{}_3(Mn^{2+},Mg,Fe^{3+},Al)_{42}(As^{3+}O_3)_2(As^{5+}O_4)_4[(Si,As^{5+})O_4]_6[(As^{5+},Si)O_4]_2(OH)_{42}$, a new arseno-silicate mineral from the Kombat mine, Otavi Valley, Namibia. Am Mineral 98(10):1693–1696

Hawthorne FC, Abdu YA, Ball NA, Černý P, Kristiansen R (2014) Agakhanovite-(Y), ideally $(YCa)_2KBe_3Si_{12}O_{30}$, a new milarite-group mineral from the Heftetjern pegmatite, Tørdal, Southern Norway: description and crystal structure. Am Mineral 99(10):2084–2088

Hayward CL, Best SP, Clark RJ, Ross NL, Withnall R (1994) Polarised single crystal Raman spectroscopy of sinhalite, $MgAlBO_4$. Spectrochim Acta A 50(7):1287–1294

He M, Chen XL, Zhou T, Hu BQ, Xu YP, Xu T (2001) Crystal structure and infrared spectra of $Na_2Al_2B_2O_7$. J Alloys Compd 327(1):210–214

He M, Chen XL, Gramlich V, Baerlocher C, Zhou T, Hu BQ (2002) Synthesis, structure, and thermal stability of $Li_3AlB_2O_6$. J Solid State Chem 163(2):369–376

Helan M, Berchmans LJ (2011) Low-temperature synthesis of lithium manganese oxide using $LiCl$-Li_2CO_3 and manganese acetate eutectic mixture. Mater Manuf Process 26(11):1369–1373

Hemley RJ (1987a) Pressure dependence of Raman spectra of SiO_2 polymorphs: α-quartz, coesite, and stishovite. In: Manghnani MH, Syono Y (eds) High-pressure research of mineral physics, vol 39. Terra Scientific Publishing Company (TERRAPUB)/American Geophysical Union, Tokyo/Washington, DC, pp 347–359

Hemley RJ (1987b) Pressure dependence of Raman spectra of SiO_2 polymorphs: α-quartz, coesite, and stishovite. In: High-pressure research in mineral physics: a volume in honor of Syun-iti Akimoto. Terra Scientific Publishing Company (TERRAPUB), Tokyo, pp 347–359

Herman RG, Bogdan CE, Sommer AJ, Simpson DR (1987) Discrimination among carbonate minerals by Raman spectroscopy using the laser microprobe. Appl Spectrosc 41(3):437–440

Hettmann K, Wenzel T, Marks M, Markl G (2012) The sulfur speciation in S-bearing minerals: new constraints by a combination of electron microprobe analysis and DFT calculations with special reference to sodalite-group minerals. Am Mineral 97(10):1653–1661

Heuser J, Bukaemskiy AA, Neumeier S, Neumann A, Bosbach D (2014) Raman and infrared spectroscopy of monazite-type ceramics used for nuclear waste conditioning. Prog Nucl Energy 72:149–155

Heyns AM, Harden PM (1999) Evidence for the existence of Cr(IV) in chromium-doped malayaite Cr^{4+}: $CaSnOSiO_4$: a resonance Raman study. J Phys Chem Solids 60(2):277–284

Heyns AM, van den Berg MM (1995) $KSbF_6$ revisited. I – a Raman and infrared study of the tetragonal phase I. J Raman Spectrosc 26(8–9):847–854

Heyns AM, Richter PW, Clark JB (1981) The vibrational spectra and crystallographic properties of $CsPF_6$. J Solid State Chem 39(1):106–113

Heyns AM, Range K-J, Wildenauer M (1990) The vibrational spectra of $NbBO_4$, $TaBO_4$, $NaNb_3O_8$ and $NaTa_3O_8$. Spectrochim Acta A 46(11):1621–1628

Himmrich M, Haeuseler H (1991) Far infrared studies on stannite and wurtzstannite type compounds. Spectrochim Acta A 47(7):933–942

Hinteregger E, Wurst K, Niederwieser N, Heymann G, Huppertz H (2014) Pressure-supported crystal growth and single-crystal structure determination of Li_2SiF_6. Z Kristallogr 229(2):77–82

Hirano S-I, Hayashi T, Kageyama T (1987) Synthesis of $LiAlO_2$ powder by hydrolysis of metal alkoxides. J Am Ceram Soc 70(3):171–174

Ho JC, Batabyal SK, Pramana SS, Lum J, Pham VT, Li D, Xiong Q, Tok AIY, Wong LH (2012) Optical and electrical properties of wurtzite copper indium sulfide nanoflakes. Mater Express 2(4):344–350

Hoang LH, Hien NTM, Chen XB, Minh NV, Yang IS (2011) Raman spectroscopic study of various types of tourmalines. J Raman Spectrosc 42(6):1442–1446

Hoekstra HR, Siegel S (1971) Preparation and properties of Cr_2UO_6. J Inorg Nucl Chem 33(9):2867–2873

Hoekstra HR, Siegel S (1973) The uranium trioxide-water system. J Inorg Nucl Chem 35:761–779

Hofmeister AM, Ito E (1992) Thermodynamic properties of $MgSiO_3$ ilmenite from vibrational spectra. Phys Chem Minerals 18(7):423–432

Hofmeister AM, Hoering TC, Virgo D (1987) Vibrational spectroscopy of beryllium aluminosilicates: heat capacity calculations from band assignments. Phys Chem Minerals 14(3):205–224

Hofmeister AM, Wopenka B, Locock AJ (2004) Spectroscopy and structure of hibonite, grossite, and $CaAl_2O_4$: implications for astronomical environments. Geochim Cosmochim Acta 68(21):4485–4503

Hölsä J, Piriou B, Räsänen M (1993) IR- and Raman-active normal vibrations of rare earth oxyfluorides, REOF; RE = Y, La, and Gd. Spectrochim Acta A 49(4):465–470

Holtstam D (1997) Härneite, a new, Zr-Sb oxide mineral isostructural with calzirtite, from Långban, Sweden. Eur J Mineral 9:843–848

Holtstam D, Broman C, Söderhielm J, Zetterqvist A (2003) First discovery of stishovite in an iron meteorite. Meteor Planet Sci 38(11):1579–1583

Holtstam D, Kolitsch U, Andersson UB (2005) Västmanlandite-(Ce) – a new lanthanide-and F-bearing sorosilicate mineral from Västmanland, Sweden. Eur J Mineral 17(1):129–141

Honda K, Yokoyama S, Tanaka SI (1999) Assignment of the Raman active vibration modes of beta-Si_3N_4 using micro-Raman scattering. J Appl Phys 85:7380–7384

Hong W, He S, Huang S, Wang Y, Hou H, Zhu X (1999) A Raman spectral studies of rare-earth (REE) fluorocarbonate minerals. Guangpuxueyu Guangpu fen xi = Guangpu 19(4):546–549. (in Chinese, English Abstr)

Hood WC, Steidl PF (1973) Synthesis of benstonite at room temperature. Am Mineral 58:341–342

Hoshino K, Nagatomi A, Watanabe M, Okudaira T, Beppu Y (2006) Nahcolite in fluid inclusions from the Ryoke metamorphic rocks and its implication for fluid genesis. J Mineral Petrol Sci 101(5):254–259

Hosterman BD (2011) Raman spectroscopic study of solid solution spinel oxides. Dissertation, University of Nevada

Hosterman BD, Farley JW, Johnson AL (2013) Spectroscopic study of the vibrational modes of magnesium nickel chromite, $Mg_xNi_{1-x}Cr_2O_4$. J Phys Chem Solids 74(7):985–990

Houlbert S, Chaabane TB, Bardeau JF, Bulou A, Smiri L (2004) Vibrational study of $Li_6P_6O_{18} \cdot 3H_2O$ and ab initio calculations in P_6O_{18} and $P_6O_{18} \cdot 3H_2O$. Spectrochim Acta A 60(1):251–259

Hrazdil V, Houzar S, Sejkora J, Koníčková Š, Jarošová L (2016) Linarite from the Ag-Pb ore deposit at Kletné near Suchdol nad Odrou (Jeseníky Culm, Vítkov Highlands). Acta Musei Silesiae Sci Natur 65(1):88–96

Hsu T-W, Jiang W-T, Wang Y (2014) Authigenesis of vivianite as influenced by methane-induced sulfidization in cold-seep sediments off southwestern Taiwan. J Asian Earth Sci 89:88–97

Hu T, Lin JB, Kong F, Mao JG (2008) $Mg_7V_4O_{16}(OH)_2 \cdot H_2O$: a magnesium vanadate with a novel 3D magnesium oxide open framework. Inorg Chem Comm 11(9):1012–1014

Hu S, Johnsson M, Lemmens P, Schmid D, Menzel D, Tapp J, Möller A (2014) Acentric pseudo-kagome structures: the solid solution $(Co_{1-x}Ni_x)_3Sb_4O_6F_6$. Chem Mater 26(12):3631–3636

Huang Y, Butler IS (1990) High-pressure micro-Raman spectra of potassium chromate(VI) and potassium tungstate(VI). Appl Spectrosc 44(8):1326–1328

Huang J, Sleight AW (1992) Synthesis, crystal structure, and optical properties of a new bismuth magnesium vanadate: $BiMg_2VO_6$. J Solid State Chem 100(1):170–178

Huang Y, Jiang Z, Schwieger W (1998) A vibrational spectroscopic study of kanemite. Micropor Mesopor Mater 26(1):215–219

Huang Y, Jiang Z, Schwieger W (1999a) A structural investigation of the singly layered silicates, silinaite and makatite, by vibrational spectroscopy. Can J Chem 77(4):495–501

Huang Y, Jiang Z, Schwieger W (1999b) Vibrational spectroscopic studies of layered silicates. Chem Mater 11(5):1210–1217

Huang E, Chen CH, Huang T, Lin EH, Xu J-A (2000) Raman spectroscopic characteristics of Mg-Fe-Ca pyroxenes. Am Mineral 85:473–479

Huang Y, Wang X, Lee HS, Cho E, Jang K, Tao Y (2007) Synthesis, vacuum ultraviolet and ultraviolet spectroscopy of Ce^{3+} ion doped olgite $Na(Sr,Ba)PO_4$. J Phys D 40(24):7821–7825

Huang Y, Jiang C, Cao Y, Shi L, Seo HJ (2009) Luminescence and microstructures of Eu^{3+}-doped in triple phosphate $Ca_8MgR(PO_4)_7$ (R = La, Gd, Y) with whitlockite structure. Mater Res Bull 44(4):793–798

Huang P, Kong Y, Li Z, Gao F, Cui D (2010) Copper selenidenanosnakes: bovine serum albumin-assisted room temperature controllable synthesis and characterization. Nanoscale Res Lett 5(6):949–956

Huang H, Yao W, He R, Chen C, Wang X, Zhang Y (2013a) Synthesis, crystal structure and optical properties of a new beryllium borate, $CsBe_4(BO_3)_3$. Solid State Sci 18:105–109

Huang H, He R, Yao W, Lin Z, Chen C, Zhang Y (2013b) Noncentrosymmetric mixed-cation borate: crystal growth, structure and optical properties of $Cs_2Ca[B_4O_5(OH)_4]_2 \cdot 8H_2O$. J Crystal Growth 380:176–181

Huang Z, Pan S, Yang Z, Yu H, Dong X, Zhao W, Dong L, Su X (2013c) $Pb_8M(BO_3)_6$ (M = Zn, Cd): two new isostructural lead borates compounds with two-dimensional $\infty[Pb_8B_6O_{18}]^{2-}$ layer structure. Solid State Sci 15:73–78

Hudson-Edwards KA, Smith AM, Dubbin WE, Bennett AJ, Murphy PJ, Wright K (2008) Comparison of the structures of natural and synthetic Pb-Cu-jarosite-type compounds. Eur J Mineral 20(2):241–252

Hurai V, Wierzbicka-Wieczorek M, Pentrák M, Huraiová M, Thomas R, Swierczewska A, Luptáková J (2014) X-Ray diffraction and vibrational spectroscopic characteristics of hydroxylclinohumite from Ruby-Bearing Marbles (Luc Yen district, Vietnam). Int J Mineral. https://doi.org/10.1155/2014/648530

Husson E, Repelin Y, Dao NQ, Brusset H (1977a) Etude par spectrophotométries d'absorption infrarouge et de diffusion Raman des niobates de structure columbite. Spectrochim Acta A 33(11):995–1001. (in French)

Husson E, Repelin Y, Dao NQ, Brusset H (1977b) Characterization of different bondings in some divalent metal niobates of columbite structure. Mater Res Bull 12(12):1199–1206

Husson E, Repelin Y, Brusset H, Cerez A (1979) Spectres de vibration etcalcul du champ de force des antimoniates et des tantalates de structure trirutile. Spectrochim Acta A 35:1177–1187. (in French)

Husson E, Repelin Y, Vandenborre MT (1984) Spectres de vibration et champ de force de l'antimoniate et de l'arseniate de calcium $CaSb_2O_6$ et $CaAs_2O_6$. Spectrochim Acta A 40(11):1017–1020. (in French)

Husson E, Genet F, Lachgar A, Piffard Y (1988a) The vibrational spectra of some antimony phosphates. J Solid State Chem 75(2):305–312

Husson E, Lachgar A, Piffard Y (1988b) The vibrational spectra of the layered compounds $K_3Sb_3M_2O_{14} \cdot xH_2O$ (M = P, As): normal coordinate analysis of $K_3Sb_3P_2O_{14} \cdot xH_2O$. J Solid State Chem 74:138–146

Hwang SL, Shen P, Chu HT, Yui TF, Liou JG, Sobolev NV, Zhang R-Y, Shatsky VS, Zayachkovsky AA (2004) Kokchetavite: a new potassium-feldspar polymorph from the Kokchetav ultrahigh-pressure terrane. Contrib Mineral Petrol 148(3):380–389

Hwang SL, Shen P, Chu HT, Yui TF, Liou JG, Sobolev NV (2009) Kumdykolite, an orthorhombic polymorph of albite, from the Kokchetav ultrahigh-pressure massif, Kazakhstan. Eur J Mineral 21(6):1325–1334

Hwang SL, Shen P, Chu HT, Yui TF, Varela ME, Iizuka Y (2016) Kuratite, $Ca_4(Fe^{2+}{}_{10}Ti_2)O_4[Si_8Al_4O_{36}]$, the Fe^{2+}-analogue of rhönite, a new mineral from the D'Orbigny angrite meteorite. Mineral Mag. https://doi.org/10.1180/minmag.2016.080.043

Ibáñez-Insa J, Elvira JJ, Llovet X, Pérez-Cano J, Oriols N, Busquets-Masó M, Hernández S (2017) Abellaite, $NaPb_2(CO_3)_2(OH)$, a new supergene mineral from the Eureka mine, Lleida province, Catalonia, Spain. Eur J Mineral. https://doi.org/10.1127/ejm/2017/0029-2630

Ignat'eva LN, Merkulov EB, Stremousova EA, Plotnichenko VG, Koltashev VV, Buznik VM (2006) Effect of bismuth trifluoride on the characteristics of fluoroindate glasses: the InF_3-BiF_3-BaF_2 system. Russ J Inorg Chem 51(10):1641–1645

Iliev MN, Hadjiev VG, Litvinchuk AP (2013) Raman and infrared spectra of brookite (TiO_2): experiment and theory. Vib Spectrosc 64:148–152

Ilieva D, Kovacheva D, Petkov C, Bogachev G (2001) Vibrational spectra of $R(PO_3)_3$ metaphosphates (R = Ga, In, Y, Sm, Gd, Dy). J Raman Spectrosc 32(11):893–899

Ilieva A, Mihailova B, Tsintsov Z, Petrov O (2007) Structural state of microcrystalline opals: a Raman spectroscopic study. Am Mineral 92(8–9):1325–1333

Ingale A, Bansal ML, Roy AP (1989) Resonance Raman scattering in HgTe: TO-phonon and forbidden-LO-phonon cross section near the E_1 gap. Phys Rev B 40(18):12353. (6 pp)

Inguanta R, Piazza S, Sunseri C (2008) Growth and characterization of ordered PbO_2 nanowire arrays. J Electrochem Soc 155(12):K205–K210

Inoue H, Nanba T, Hagihara H, Kanazawa T, Yasui I (1988) Computer simulation of Raman spectra of fluoride glasses. Mater Sci Forum 32:403–408

Ishii M, Saeki M (1992) Raman and infrared spectroscopic studies of Ba_2TiS_5 and Ba_2TiS_4. Phys Stat Solidi B 169:K53–K58

Ishii M, Wada H (2000) Raman and infrared studies of a silver–tantalum sulfide with a layered structure. Mater Res Bull 35(8):1361–1368

Ishii M, Shibata K, Nozaki H (1993) Anion distributions and phase transitions in $CuS_{1-x}Se_x$ ($x = 0$–1) studied by Raman spectroscopy. J Solid State Chem 105(2):504–511

Ito M, Hori J, Kurisaki H, Okada H, Kuroki AJP, Ogita N, Udagawa M, Fujii H, Nakamura F, Fujita T, Suzuki T (2003) Pressure-induced superconductor-insulator transition in the spinel compound $CuRh_2S_4$. Phys Rev Lett 91(7):077001. (4 pp)

Ivanov VG, Hadjiev VG, Litvinchuk AP, Dimitrov DZ, Shivachev BL, Abrashev MV, Lorenz B, Iliev MN (2014) Lattice dynamics and spin-phonon coupling in $CaMn_2O_4$: a Raman study. Phys Rev B 89(18):184307-1–184307-8

Ivanyuk GY, Yakovenchuk VN, Pakhomovsky YA, Panikorovskii TL, Konoplyova NG, Bazai AV, Bocharov VN, Antonov AA, Selivanova EA (2017) Goryainovite, Ca_2PO_4Cl, a new mineral from the Stora Sahavaara iron ore deposit (Norrbotten, Sweden). GFF 139(1):75–82

Izquierdo-Roca V, Saucedo E, Ruiz CM, Fontané X, Calvo-Barrio L, Álvarez-Garcia J, Grand P-P, Jaime-Ferrer JS, Pérez-Rodríguez A, Morantr JR, Bermudez V (2009) Raman scattering and structural analysis of electrodeposited $CuInSe_2$ and S-rich quaternary CuIn$(S,Se)_2$ semiconductors for solar cells. Phys Status Solidi A 206(5):1001–1004

Jacco JC (1986) The infrared spectra of $KTiOPO_4$ and a K_2O–P_2O_5–TiO_2 glass. Mater Res Bull 21(10):1189–1194

Jain YS, Bist HD, Verma AL (1974) Optical phonons in α-$Ni(SO_4) \cdot 6H_2O$ single crystal. J Raman Spectrosc 2:327–339

Jakeš D, Sedláková LN, Moravec J, Germanič J (1968) The manganese, cobalt, nickel, copper, silver and mercury uranates. J Inorg Nucl Chem 30(2):525–533

Jamal A, Raahman MM, Khan SB, Abdullah MM, Faisaal M, Asiri AM, Khan AAP, Akhtar K (2013) Simple growth and characterization of α-Sb_2O_4: evaluation of their photo-catalytic and chemical sensing applications. J Chem Soc Pak 35(3):570–576

Jambor JL, Roberts AC, Grice JD, Birkett TC, Groat LA, Zajac S (1998) Gerenite-(Y), $(Ca,Na)_2(Y, REE)_3Si_6O_{18} \cdot 2H_2O$, a new mineral species, and an associated Y-bearing gadolinite-group mineral, from the Strange Lake Peralkaline Complex, Quebec-Labrador. Can Mineral 36:793–800

Jambor JL, Viñals J, Groat LA, Raudsepp M (2002) Cobaltarthurite, $Co^{2+}Fe^{3+}{}_2(AsO_4)_2(OH)_2 \cdot 4H_2O$, a new member of the arthurite group. Can Mineral 40(2):725–732

Jana YM, Halder P, Biswas AA, Roychowdhury A, Das D, Dey S, Kumar S (2016) Synthesis, X-ray Rietveld analysis, infrared and Mössbauer spectroscopy of R_2FeSbO_7 (R^{3+} = Y, Dy, Gd, Bi) pyrochlore solid solution. J Alloys Compd 656:226–236

Janáková S, Salavcová L, Renaudin G, Filinchuk Y, Boyer D, Boutinaud P (2007) Preparation and structural investigations of sol–gel derived Eu^{3+}-doped $CaAl_2O_4$. J Phys Chem Solids 68(5):1147–1151

Janeczek J, Ciesielczuk J, Dulski M, Krzykawski T (2016) Chemical composition and Raman spectroscopy of cornubite and its relation to cornwallite in Miedzianka, the Sudety Mts., Poland. N Jb Miner Abh 193(3):265–274

Jaquet R, Haeuseler H (2008) Vibrational analysis of the $H_4I_2O_{10}^{2-}$ ion in $CuH_4I_2O_{10} \cdot 6H_2O$. J Raman Spectrosc 39(5):599–606

Jasinevicius R (2009) Characterization of vibrational and electronic features in the Raman spectra of gem minerals. Dissertation, University of Arizona

Jaszczak JA, Rumsey MS, Bindi L, Hackney SA, Wise MA, Stanley CJ, Spratt J (2016) Merelaniite, $Mo_4Pb_4VSbS_{15}$, a new molybdenum-essential member of the cylindrite group, from the Merelani Tanzanite Deposit, Lelatema Mountains, Manyara Region, Tanzania. Minerals 6(4):115. (19 pp)

Javed QU, Wang F, Toufiq AM, Rafiq MY, Iqbal MZ, Kamran MA (2013) Preparation, characterizations and optical property of single crystalline $ZnMn_2O_4$ nanoflowers via template-free hydrothermal synthesis. J Nanosci Nanotechnol 13(4):2937–2942

Jay WH, Cashion JD (2013) Raman spectroscopy of Limehouse porcelain sherds supported by Mössbauer spectroscopy and scanning electron microscopy. J Raman Spectrosc 44(12):1718–1732

Jay WH, Cashion JD, Blenkinship B (2015) Lancaster delftware: a Raman spectroscopy, electron microscopy and Mössbauer spectroscopy compositional study. J Raman Spectrosc 46(12):1265–1282

Jayaraman A, Kourouklis GA, Van Uitert LG (1988) A high pressure Raman study of ThO_2 to 40 GPa and pressure-induced phase transition from fluorite structure. J Phys 30(3):225–231

Jehlička J, Edwards HGM (2008) Raman spectroscopy as a tool for the non-destructive identification of organic minerals in the geological record. Org Geochem 39(4):371–386

Jehlička J, Vandenabeele P (2015) Evaluation of portable Raman instruments with 532 and 785-nm excitation for identification of zeolites and beryllium containing silicates. J Raman Spectrosc 46(10):927–932

Jehlička J, Edwards HG, Villar SE, Pokorný J (2005) Raman spectroscopic study of amorphous and crystalline hydrocarbons from soils, peats and lignite. Spectrochim Acta A 61(10):2390–2398

Jehlička J, Edwards HGM, Villar SE, Frank O (2006) Raman spectroscopic study of the complex aromatic mineral idrialite. J Raman Spectrosc 37(7):771–776

Jehlička J, Edwards HG, Villar SEJ (2007a) Raman spectroscopy of natural accumulated paraffins from rocks: evenkite, ozokerite and hatchetine. Spectrochim Acta A 68(4):1143–1148

Jehlička J, Žáček V, Edwards HGM, Shcherbakova E, Moroz T (2007b) Raman spectra of organic compounds kladnoite ($C_6H_4(CO)_2NH$) and hoelite ($C_{14}H_8O_2$) – rare sublimation products crystallising on self-ignited coal heaps. Spectrochim Acta A 68(4):1053–1057

Jehlička J, Edwards HGM, Vítek P (2009a) Assessment of Raman spectroscopy as a tool for the non-destructive identification of organic minerals and biomolecules for Mars studies. Planet Space Sci 57(5):606–613

Jehlička J, Vitek P, Edwards HGM, Hargreaves MD, Čapoun T (2009b) Fast detection of sulphate minerals (gypsum, anglesite, baryte) by a portable Raman spectrometer. J Raman Spectrosc 40(8):1082–1086

Jehlička J, Vandenabeele P, Edwards HGM (2012) Discrimination of zeolites and beryllium containing silicates using portable Raman spectroscometric equipment with near-infrared excitation. Spectrochim Acta A 86:341–346

Jeitschko W, Sleight AW, McClellan WR, Weiher JF (1976) A comprehensive study of disordered and ordered scheelite-related $Bi_3(FeO_4)(MoO_4)_2$. Acta Crystallographica B 32(4):1163–1170

Jenkins TE (1986) Anharmonic interactions in solids: a Raman investigation of some vibrational states in ammonium fluorosilicate. J Phys C 19(7):1065–1070

Jentzsch PV, Kampe B, Rösch P, Popp J (2011) Raman spectroscopic study of crystallization from solutions containing $MgSO_4$ and Na_2SO_4: Raman spectra of double salts. J Phys Chem A 115(22):5540–5546

Jentzsch PV, Bolanz RM, Ciobotă V, Kampe B, Rösch P, Majzlan J, Popp J (2012a) Raman spectroscopic study of calcium mixed salts of atmospheric importance. Vib Spectrosc 61:206–213

Jentzsch PV, Ciobotă V, Bolanz RM, Kampe B, Rösch P, Majzlan J, Popp J (2012b) Raman and infrared spectroscopic study of synthetic ungemachite, $K_3Na_8Fe(SO_4)_6(NO_3)_2 \cdot 6H_2O$. J Molec Struct 1022:147–152

Jentzsch PV, Ciobotă V, Kampe B, Rösch P, Popp J (2012c) Origin of salt mixtures and mixed salts in atmospheric particulate matter. J Raman Spectrosc 43(4):514–519

Jentzsch PV, Kampe B, Ciobotă V, Rösch P, Popp J (2013) Inorganic salts in atmospheric particulate matter: Raman spectroscopy as an analytical tool. Spectrochim Acta A 115:697–708

Jia Y-Z, Gao S-Y, Xia S-P, Li J, Li W (2001) Raman spectroscopy of supersaturated aqueous solution of borate. Chem J Chin Univ 22(1):99–103. (in Chinese, English Abstr)

Jia G, Wu Z, Wang P, Yao J, Chang K (2016) Morphological evolution of self-deposition Bi2Se3 nanosheets by oxygen plasma treatment. Sci Rep. https://doi.org/10.1038/srep22191

Jiang T, Ozin GA (1997) Tin (IV) sulfide-alkylamine composite mesophase: a new class of thermotropic liquid crystals. J Mater Chem 7(11):2213–2222

Jiang XM, Xu ZN, Zhao ZY, Guo SP, Guo GC, Huang JS (2011) Syntheses, crystal structures, and optical properties of indium arsenic (III) oxide halides: $In_2(As_2O_5)Cl_2$ and $In_4(As_2O_5)(As_3O_7)Br_3$. Eur J Inorg Chem 2011(26):4069–4076

Jiang YR, Lee WW, Chen KT, Wang MC, Chang KH, Chen CC (2014) Hydrothermal synthesis of β-ZnMoO$_4$ crystals and their photocatalytic degradation of Victoria Blue R and phenol. J Taiwan Inst Chem Eng 45(1):207–218

Jin GB, Soderholm L (2015) Solid-state syntheses and single-crystal characterizations of three tetravalent thorium and uranium silicates. J Solid State Chem 221:405–410

Johnson JE (2015) Manganese: minerals, microbes, and the evolution of oxygenic photosynthesis. Dissertation. California Institute of Technology

Johnston MG, Harrison WT (2011) New BaM$_2$(SeO$_3$)$_3$·nH$_2$O (M= Co, Ni, Mn, Mg; n≈ 3) Zemannite-type frameworks: single-crystal structures of BaCo$_2$(SeO$_3$)$_3$·3H$_2$O, BaMn$_2$(SeO$_3$)$_3$·3H$_2$O and BaMg$_2$(SeO$_3$)$_3$·3H$_2$O. Eur J Inorg Chem 2011 (19):2967–2974

Johnston CT, Helsen J, Schoonheydt RA, Bish DL, Agnew SF (1998) Single-crystal Raman spectroscopic study of dickite. Am Mineral 83(1):75–84

Johnstone IW, Fletcher JR, Bates CA, Lockwood DJ, Mischler G (1978) Temperature dependent electron-phonon coupling in FeCl$_2$ observed by Raman scattering. J Phys C 11(21):4425–4438

Jolliff BL, Hughes JM, Freeman JJ, Zeigler RA (2006) Crystal chemistry of lunar merrillite and comparison to other meteoritic and planetary suites of whitlockite and merrillite. Am Mineral 91: 1583–1595

Jonsson E (2003) Mineralogy and parageneses of Pb oxychlorides in Långban-type deposits, Bergslagen, Sweden. GFF 125(2):87–98

Joseph C, Bourson P, Fontana MD (2012) Amorphous to crystalline transformation in Ta$_2$O$_5$ studied by Raman spectroscopy. J Raman Spectrosc 43(8):1146–1150

Jouini A, Férid M, Gâcon JC, Grosvalet L, Thozet A, Trabelsi-Ayadi M (2006) Crystal structure, vibrational spectra and optical properties of praseodymium cyclotriphosphate PrP$_3$O$_9$·H$_2$O. Mater Res Bull 41 (7):1370–1377

Jovanovski G, Makreski P, Kaitner B, Boev B (2009) Silicate minerals from Macedonia. Complementary use of vibrational spectroscopy and X-ray powdered diffraction for identification and detection purposes. Croat Chem Acta 82(2):363–386

Julien C, Rougier A, Haro-Poniatowski E, Nazri GA (1998) Vibrational spectroscopy of lithium manganese spinel oxides. Molec Cryst Liquid Cryst Sci Technol A 311(1):81–87. https://doi.org/10.1080/10587259808042370

Julien C, Barnier S, Ivanov I, Guittard M, Pardo MP, Chilouet A (1999) Vibrational studies of copper thiogallate solid solutions. Mater Sci Eng B 57 (2):102–109

Julien C, Massot M, Baddour-Hadjean R, Franger S, Bach S, Pereira-Ramos JP (2003) Raman spectra of birnessite manganese dioxides. Solid State Ionics 159 (3):345–356

Julien CM, Massot M, Poinsignon C (2004) Lattice vibrations of manganese oxides. Part I. Periodic structures. Spectrochim Acta A 60:689–700

Jun L, Shuping X, Shiyang G (1995) FT-IR and Raman spectroscopic study of hydrated borates. Spectrochim Acta A 51(4):519–532

Kaczmarski M, Eichner A, Mielcarek S, Olejniczak I, Mróz B (2000) IR temperature study of internal vibrations in K$_3$Na(SeO$_4$)$_2$. Vib Spectrosc 23(1):77–81

Kagi H, Nagai T, Loveday JS, Wada C, Parise JB (2003) Pressure-induced phase transformation of kalicinite (KHCO$_3$) at 2.8 GPa and local structural changes around hydrogen atoms. Am Mineral 88(10):1446–1451

Kagunya W, Baddor-Hadjean R, Kooli F, Jones W (1998) Vibrational modes in layered double hydroxides and their calcined derivatives. Chem Phys 236:225–234

Kahlenberg V, Girtler D, Arroyabe E, Kaindl R, Többens DM (2010) Devitrite (Na$_2$Ca$_3$Si$_6$O$_{16}$) – structural, spectroscopic and computational investigations on a crystalline impurity phase in industrial soda-lime glasses. Mineral Petrol 100(1–2):1–9

Kaliwoda M, Hochleitner R, Hoffmann VH, Mikouchi T, Gigler AM, Schmahl WW (2011) New Raman spectroscopic data of Almahata Sitta meteorite. In: Conference on Micro-Raman and Luminescence Studies in the Earth and Planetary Sciences (CORALS II). LPI Contribution No. 1616, Lunar and Planetary Institute, Houston, p XX

Kaliwoda M, Hochleitner R, Hoffmann VH, Mikouchi T, Gigler AM, Schmahl WW (2013) New Raman spectroscopic data of the Almahata Sitta meteorite. Spectrosc Lett 46(2):141–146

Kaminskii AA, Haussühl E, Haussühl S, Eichler HJ, Ueda K, Hanuza J, Takaichi K, Rhee H, Gad GMA (2004) α-Alums: K, Rb, Tl and NH$_4$Al(SO$_4$)$_2$·12H$_2$O – a new family of $\chi^{(3)}$ – active crystalline materials for Raman laser converters with large frequency shifts. Laser Phys Lett 1(4):205–211

Kaminskii AA, Rhee H, Lux O, Eichler HJ, Bohatý L, Becker P, Liebertz J, Ueda K, Shirakawa A, Koltashev VV, Hanuza J, Dong J, Stavrovskii DB (2011) Many-phonon stimulated Raman scattering and related cascaded and cross-cascaded $\chi^{(3)}$-nonlinear optical effects in melilite-type crystal Ca$_2$ZnSi$_2$O$_7$. Laser Phys Lett 8 (12):859–874

Kaminskii AA, Haussühl E, Eichler HJ, Hanuza J, Mączka M, Yoneda H, Shirakawa A (2015) Lithium silicate, LiAlSi$_4$O$_{10}$ (petalite) – a novel monoclinic SRS-active crystal. Laser Phys Lett 12(8):085002. (8 pp)

Kaminsky F, Wirth R, Matsyuk S, Schreiber A, Thomas R (2009) Nyerereite and nahcolite inclusions in diamond: evidence for lower-mantle carbonatitic magmas. Mineral Mag 73(5):797–816

Kamoun LA, Remain F, Novak A (1988) Etude par spectrométrie infrarouge et Raman des phases cristallines basses temperatures de $(NH_4)_3H(SO_4)_2$. J Raman Spectrosc 19:329–335. (in French)

Kampf AR, Moore PB, Jonsson EJ, Swihart GH (1998) Philolithite: a new mineral from Långban, Värmland, Sweden. Mineral Rec 29(3):201–206

Kampf AR, Adams PM, Kolitsch U, Steele IM (2009a) Meurigite-Na, a new species, and the relationship between phosphofibrite and meurigite. Am Mineral 94(5–6):720–727

Kampf AR, Rossman GR, Housley RM (2009b) Plumbophyllite, a new species from the Blue Bell claims near Baker, San Bernardino County, California. Am Mineral 94(8–9):1198–1204

Kampf AR, Housley RM, Marty J (2010a) Lead-tellurium oxysalts from Otto Mountain near Baker, California: III. Thorneite, $Pb_6(Te^{6+}{}_2O_{10})(CO_3)Cl_2(H_2O)$, the first mineral with edge-sharing octahedral tellurate dimers. Am Mineral 95(10):1548–1553

Kampf AR, Rossman GR, Steele IM, Pluth JJ, Dunning GE, Walstrom RE (2010b) Devitoite, a new heterophyllosilicate mineral with astrophyllite-like layers from eastern Fresno County, California. Can Min 48(1):29–40

Kampf AR, Downs RT, Housley RM, Jenkins RA, Hyršl J (2011a) Anorpiment, As_2S_3, the triclinic dimorph of orpiment. Mineral Mag 75(6):2857–2867

Kampf AR, Mills SJ, Rossman GR, Steele IM, Pluth JJ, Favreau G (2011b) Afmite, $Al_3(OH)_4(H_2O)_3(PO_4)(PO_3OH) \cdot H_2O$, a new mineral from Fumade, Tarn, France: description and crystal structure. Eur J Mineral 23(2):269–277

Kampf AR, Yang H, Downs RT, Pinch WW (2011c) The crystal structures and Raman spectra of aravaipaite and calcioaravaipaite. Am Mineral 96(2–3):402–407

Kampf AR, Mills SJ, Merlino S, Pasero M, McDonald AM, Wray WB, Hindman JR (2012) Whelanite, $Cu_2Ca_6[Si_6O_{17}(OH)](CO_3)(OH)_3(H_2O)_2$, an (old) new mineral from the Bawana mine, Milford, Utah. Am Mineral 97(11–12):2007–2015

Kampf AR, Mills SJ, Housley RM, Rossman GR, Marty J, Thorne B (2013a) Lead-tellurium oxysalts from Otto Mountain near Baker, California: X. Bairdite, $Pb_2Cu^{2+}{}_4Te^{6+}{}_2O_{10}(OH)_2(SO_4)(H_2O)$, a new mineral with thick HCP layers. Am Mineral 98(7):1315–1321

Kampf AR, Mills SJ, Housley RM, Rossman GR, Marty J, Thorne B (2013b) Lead-tellurium oxysalts from Otto Mountain near Baker, California: XI. Eckhardite, $(Ca,Pb)Cu^{2+}Te^{6+}O_5(H_2O)$, a new mineral with HCP stair-step layers. Am Mineral 98(8–9):1617–1623

Kampf AR, Mills SJ, Housley RM, Rossman GR, Nash BP, Dini M, Jenkins RA (2013c) Joteite, $Ca_2CuAl[AsO_4][AsO_3(OH)]_2(OH)_2 \cdot 5H_2O$, a new arsenate with a sheet structure and unconnected acid arsenate groups. Mineral Mag 77(6):2811–2823

Kampf AR, Mills SJ, Nash BP, Housley RM, Rossman GR, Dini M (2013d) Camaronesite, $[Fe^{3+}(H_2O)_2(PO_3OH)]_2(SO_4) \cdot 1\text{-}2H_2O$, a new phosphate-sulfate from the Camarones Valley, Chile, structurally related to taranakite. Mineral Mag 77(4):453–465

Kampf AR, Adams PM, Housley RM, Rossman GR (2014a) Fluorowardite, $NaAl_3(PO_4)_2(OH)_2F_2 \cdot 2H_2O$, the fluorine analog of wardite from the Silver Coin mine, Valmy, Nevada. Am Mineral 99(4):804–810

Kampf AR, Hughes JM, Nash BP, Wright SE, Rossman GR, Marty J (2014b) Ophirite, $Ca_2Mg_4[Zn_2Mn^{3+}{}_2(H_2O)_2(Fe^{3+}W_9O_{34})_2] \cdot 46H_2O$, a new mineral with a heteropolytungstate tri-lacunary Keggin anion. Am Mineral 99(5–6):1045–1051

Kampf AR, Kasatkin AV, Cejka J, Marty J (2015a) Plášilite, $Na(UO_2)(SO_4)(OH) \cdot 2H_2O$, a new uranyl sulfate mineral from the Blue Lizard mine, San Juan County, Utah, USA. J Geosci 60(1):1–10

Kampf AR, Plášil J, Kasatkin AV, Marty J (2015b) Bobcookite, $NaAl(UO_2)_2(SO_4)_4 \cdot 18H_2O$ and wetherillite, $Na_2Mg(UO_2)_2(SO_4)_4 \cdot 18H_2O$, two new uranyl sulfate minerals from the Blue Lizard mine, San Juan County, Utah, USA. Mineral Mag 79(3):695–714

Kampf AR, Plášil J, Kasatkin AV, Marty J, Čejka J (2015c) Fermiite, $Na_4(UO_2)(SO_4)_3 \cdot 3H_2O$ and oppenheimerite, $Na_2(UO_2)(SO_4)_2 \cdot 3H_2O$, two new uranyl sulfate minerals from the Blue Lizard mine, San Juan County, Utah, USA. Mineral Mag 79(5):1123–1142

Kampf AR, Cámara F, Ciriotti ME, Nash BP, Balestra C, Chiappino L (2016a) Castellaroite, $Mn^{2+}{}_3(AsO_4)_2 \cdot 4H_2O$, a new mineral from Italy related to metaswitzerite. Eur J Mineral. https://doi.org/10.1127/ejm/2016/0028-2535

Kampf AR, Cooper MA, Mills SJ, Housley RM, Rossman GR (2016b) Lead-tellurium oxysalts from Otto Mountain near Baker, California: XII. Andychristyite, $PbCu^{2+}Te^{6+}O_5(H_2O)$, a new mineral with HCP stair-step layers. Mineral Mag. https://doi.org/10.1180/minmag.2016.080.042

Kampf AR, Housley RM, Rossman GR (2016c) Wayneburnhamite, $Pb_9Ca_6(Si_2O_7)_3(SiO_4)_3$, an apatite polysome: the Mn-free analog of ganomalite from Crestmore, California. Am Mineral 101(11):2423–2429

Kampf AR, Mills SJ, Nash BP (2016d) Pauladamsite, $Cu_4(SeO_3)(SO_4)(OH)_4 \cdot 2H_2O$, a new mineral from the Santa Rosa mine, Darwin district, California, USA. Mineral Mag 80(6):949–958

Kampf AR, Plášil J, Čejka J, Marty J, Škoda R, Lapčák L (2016e) Alwilkinsite-(Y), a new rare-earth uranyl sulfate mineral from the Blue Lizard mine, San Juan County, Utah, USA. Mineral Mag. https://doi.org/10.1180/minmag.2016.080.139

Kampf AR, Plášil J, Kasatkin AV, Marty J, Čejka J, Lapčák L (2016f) Shumwayite, $[(UO_2)(SO_4)(H_2O)_2]_2 \cdot H_2O$, a new uranyl sulfate mineral from Red Canyon, San Juan County, Utah, USA. Mineral Mag. https://doi.org/10.1180/minmag.2016.080.091

Kampf AR, Plášil J, Kasatkin AV, Marty J, Čejka J (2016g) Klaprothite, péligotite and ottohahnite, three new sodium uranyl sulfate minerals with bidentate UO_7–SO_4 linkages from the Blue Lizard mine, San Juan County, Utah, USA. Mineral Mag. https://doi.org/10.1180/minmag.2016.080.120

Kampf AR, Richards RP, Nash BP, Murowchick JB, Rakovan JF (2016h) Carlsonite, $(NH_4)_5Fe^{3+}_3O(SO_4)_6 \cdot 7H_2O$, and huizingite-(Al), $(NH_4)_9Al_3(SO_4)_8(OH)_2 \cdot 4H_2O$, two new minerals from a natural fire in an oil-bearing shale near Milan, Ohio. Am Mineral 101(9):2095–2107

Kampf AR, Rossman GR, Ma C (2016i) Kyawthuite, $Bi^{3+}Sb^{5+}O_4$, a new gem mineral from Mogok, Burma (Myanmar). Mineral Mag. https://doi.org/10.1180/minmag.2016.080.102

Kampf AR, Adams PM, Barwood H, Nash BP (2017a) Fluorwavellite, $Al_3(PO_4)_2(OH)_2F \cdot 5H_2O$, the fluorine analogue of wavellite. Am Mineral. https://doi.org/10.2138/am-2017-5948

Kampf AR, Cooper MA, Nash BP, Cerling TE, Marty J, Hummer DR, Celestian AJ, Rose TP, Trebisky TJ (2017b) Rowleyite, $[Na(NH_4,K)_9Cl_4][V^{5+,4+}_2(P,As)O_8]_6 \cdot n[H_2O,Na,NH_4,K,Cl]$, a new mineral with a microporous framework structure. Am Mineral. https://doi.org/10.2138/am-2017-5977

Kanzaki M, Xue X, Amalberti J, Zhang Q (2012) Raman and NMR spectroscopic characterization of high-pressure K-cymrite $(KAlSi_3O_8 \cdot H_2O)$ and its anhydrous form (kokchetavite). J Mineral Petrol Sci 107(2):114–119

Kao LS, Peacor DR, Coveney RM Jr, Zhao G, Dungey KE, Curtis MD, Penner-Hahn JE (2001) AC/MoS_2 mixed-layer phase (MoSC) occurring in metalliferous black shales from southern China, and new data on jordisite. Am Mineral 86:852–861

Kaoua S, Krimi S, Péchev S, Gravereau P, Chaminade JP, Couzi M, El Jazouli A (2013) Synthesis, crystal structure, and vibrational spectroscopic and UV–visible studies of $Cs_2MnP_2O_7$. J Solid State Chem 198:379–385

Karimova OV, Burns PC (2007) Structural units in three uranyl perrhenates. Inorg Chem 46(24):10108–10113

Karwowski Ł, Helios K, Kryza R, Muszyński A, Drożdżewsk P (2013) Raman spectra of selected mineral phases of the Morasko iron meteorite. J Raman Spectrosc 44(8):1181–1186

Karwowski Ł, Kusz J, Muszyński A, Kryza R, Sitarz M, Galuskin EV (2015) Moraskoite, $Na_2Mg(PO_4)F$, a new mineral from the Morasko IAB-MG iron meteorite (Poland). Mineral Mag 79(2):387–398

Karwowski Ł, Kryza R, Muszyński A, Kusz J, Helios K, Drożdżewski P, Galuskin EV (2016) Czochralskiite, $Na_4Ca_3Mg(PO_4)_4$, a second new mineral from the Morasko IAB-MG iron meteorite (Poland). Eur J Minerals. https://doi.org/10.1127/ejm/2016/0028-2557

Karydis DA, Boghosian S, Nielsen K, Eriksen KM, Fehrmann R (2002) Crystal structure and spectroscopic properties of $Na_2K_6(VO)_2(SO_4)_7$. Inorg Chem 41(9):2417–2421

Kasatkin AV, Plášil J, Marty J, Agakhanov AA, Belakovskiy DI, Lykova IS (2014) Nestolaite, $CaSeO_3 \cdot H_2O$, a new mineral from the Little Eva mine, Grand County, Utah, USA. Mineral Mag 78:497–505

Kasatkin AV, Plášil J, Pekov IV, Belakovskiy DI, Nestola F, Čejka J, Vigasina MF, Zorzi F, Thorne B (2015) Karpenkoite, $Co_3(V_2O_7)(OH)_2 \cdot 2H_2O$, a cobalt analogue of martyite from the Little Eva mine, Grand County, Utah, USA. J Geosci 60(4):251–257

Kassandrov EG, Mazurov MP (2009) Magmatogenic manganese ores of the South MinusaIntermontane Trough. Geol Ore Depos 51(5):356–370

Katerinopoulou A, Katerinopoulos A, Voudouris P, Bieniok A, Musso M, Amthauer G (2009) A multi-analytical study of the crystal structure of unusual Ti–Zr–Cr–rich andradite from the Maroniaskarn, Rhodope massif, western Thrace, Greece. Mineral Petrol 95:113–124

Kavun VY, Slobodyuk AB, Voit EI, Sinebryukhov SL, Merkulov EB, Goncharuk VK (2010) Ionic mobility and structure of glasses in ZrF_4–BiF_3–MF_2 (M = Sr, Ba, Pb) systems according to NMR, IR, and Raman spectroscopic data. J Struct Chem 51(5):862–868

Kawada T, Hinokuma S, Machida M (2015) Structure and SO_3 decomposition activity of nCuO–V_2O_5/SiO_2 ($n = 0, 1, 2, 3$ and 5) catalysts for solar thermochemical water splitting cycles. Catal Today 242:268–273

Kawamoto Y, Kono A (1986) Raman spectroscopic study of AlF_3–CaF_2–BaF_2 glasses. J Non-crystal Solids 85(3):335–345

Kawasaki T, Nakano N, Osanai Y (2011) Osumilite and a spinel+quartz association in garnet–sillimanite gneiss from Rundvågshetta, Lützow-Holm Complex, East Antarctica. Gondwana Res 19(2):430–445

Kawasaki T, Adachi T, Nakano N, Osanai Y (2013) Possible armalcolitepseudomorph-bearing garnet–sillimanite gneiss from Skallevikshalsen, Lützow-Holm Complex, East Antarctica: implications for ultrahigh-temperature metamorphism. Geol Soc, London, Special Publ 383(1):135–167

Kendix E, Moscardi G, Mazzeo R, Baraldi P, Prati S, Joseph E, Capelli S (2008) Far infrared and Raman spectroscopy analysis of inorganic pigments. J Raman Spectrosc 39:1104–1112

Keramidas VG, Deangelis BA, White WB (1975) Vibrational spectra of spinels with cation ordering on the octahedral sites. J Solid State Chem 15(3):233–245

Kessler H, Olazcuaga R, Hatterer A, Hagenmuller P (1979) Etude des phases Na_4XO_4 (X = Sn, Pb) et de K_4SnO_4 par spectrophotométrie d'absorption infrarouge et diffusion Raman. Z anorg allg Chemie 458(1):195–201. (in French)

Ketani M, Abraham F, Mentré O (1999) Channel structure in the new $BiCoPO_5$. Comparison with $BiNiPO_5$. Crystal structure, lone pair localization and infrared characterization. Solid State Sci 1:449–460

Ketatni M, Abraham F, Mentre O (1999) Channel structure in the new $BiCoPO_5$. Comparison with $BiNiPO_5$. Crystal structure, lone pair localisation and infrared characterisation. Solid State Sci 1(6):449–460

Khadka DB, Kim J (2014) Structural transition and band gap tuning of $Cu_2(Zn,Fe)SnS_4$ chalcogenide for photovoltaic application. J Phys Chem C 118(26):14227–14237

Khamlich S, Bello A, Fabiane M, Ngom BD, Manyala N (2013) Hydrothermal synthesis of simonkolleite microplatelets on nickel foam-graphene for electrochemical supercapacitors. J Solid State Electrochem 17(11):2879–2886

Khanderi J, Shi L, Rothenberger A (2015) Hydrolysis of bis (dimethylamido) tin to tin(II) oxyhydroxide and its selective transformation into tin(II) or tin(IV) oxide. Inorg Chim Acta 427:27–32

Khaoulaf R, Ezzaafrani M, Ennaciri A, Harcharras M, Capitelli F (2012) Vibrational study of dipotassium zinc bis(dihydrogendiphosphate) dihydrate, $K_2Zn(H_2P_2O_7)_2·2H_2O$. Phosphorus, sulfur, and silicon and the related elements 187(11):1367–1376. https://doi.org/10.1080/10426507.2012.685669

Kharbish S (2011) Raman spectroscopic investigations of some Tl-sulfosalt minerals containing pyramidal (As, Sb)S_3 groups. Am Mineral 96(4):609–616

Kharbish S (2012) Raman spectra of minerals containing interconnected As(Sb)O_3 pyramids: trippkeite and schafarzikite. J Geosci 57(1):53–62

Kharbish S (2016) Micro-Raman spectroscopic investigations of extremely scarce Pb–As sulfosalt minerals: baumhauerite, dufrénoysite, gratonite, sartorite, and seligmannite. J Raman Spectrosc. https://doi.org/10.1002/jrs.4973

Kharbish S, András P (2014) Investigations of the Fe sulfosalts berthierite, garavellite, arsenopyrite and gudmundite by Raman spectroscopy. Mineral Mag 78(5):1287–1299

Kharbish S, Jeleň S (2016) Raman spectroscopy of the Pb-Sb sulfosalts minerals: boulangerite, jamesonite, robinsonite and zinkenite. Vib Spectrosc 85:157–166

Kharbish S, Libowitzky E, Beran A (2007) The effect of As-Sb substitution in the Raman spectra of tetrahedrite-tennantite and pyrargyrite-proustite solid solutions. Eur J Mineral 19(4):567–574

Kharbish S, Libowitzky E, Beran A (2009) Raman spectra of isolated and interconnected pyramidal XS$_3$ groups (X = Sb, Bi) in stibnite, bismuthinite, kermesite, stephanite and bournonite. Eur J Mineral 21(2):325–333

Kharbish S, András P, Luptáková J, Milovská S (2014) Raman spectra of oriented and non-oriented Cu hydroxyl-phosphate minerals: libethenite, cornetite, pseudomalachite, reichenbachite and ludjibaite. Spectrochim Acta A 130:152–163

Khomyakov AP, Kurova TA, Nechelyustov GN, Piloyan GO (1983) Barentsite, $Na_7AlH_2(CO_3)_4F_4$, a new mineral. Zapiski RMO (Proc Russ Miner Soc) 112(4):474–479. (in Russian)

Khomyakov AP, Nechelyustov GN, Rastsvetaeva RK, Rozenberg KA (2013) Davinciite, $Na_{12}K_3Ca_6Fe^{2+}{}_3Zr_3(Si_{26}O_{73}OH)Cl_2$, a new K,Na-ordered mineral of the eudialyte group from the Khibiny Alkaline Pluton, Kola Peninsula, Russia. Geol Ore Depos 55(7):532–540

Khorari S, Rulmont A, Cahay R, Tarte P (1995) Structure of the complex arsenates $NaCa_2M^{2+}{}_2 (AsO_4)_3$ (M^{2+} = Mg, Ni, Co): first experimental evidence of a garnet-alluaudite reversible polymorphism. J Solid State Chem 118:267–273

Khorari S, Rulmont A, Tarte P (1997) Alluaudite-like structure of the arsenate $Na_3In_2(AsO_4)_3$. J Solid State Chem 134(1):31–37

Khosravi I, Yazdanbakhsh M, Eftekhar M, Haddadi Z (2013) Fabrication of nanodelafossite $LiCo_{0.5}Fe_{0.5}O_2$ as the new adsorbent in efficient removal of reactive blue 5 from aqueous solutions. Mater Res Bull 48(6):2213–2219

Khoury HN, Sokol EV, Kokh SN, Seryotkin YV, Nigmatulina EN, Goryainov SV, Belogub EV, Clark ID (2016a) Tululite, $Ca_{14}(Fe^{3+},Al)(Al,Zn,Fe^{3+},Si,P,Mn,Mg)_{15}O_{36}$: a new Ca zincate-aluminate from combustion metamorphic marbles, central Jordan. Mineral Petrol 110(1):125–140

Khoury HN, Sokol EV, Kokh SN, Seryotkin YV, Kozmenko OA, Goryainov SV, Clark ID (2016b) Intermediate members of the lime-monteponite solid solutions ($Ca_{1-x}Cd_xO$, $x = 0.36$–0.55): discovery in natural occurrence. Am Mineral 101(1):146–161

Kiefer W, Bernstein HJ (1972) The resonance Raman effect of the permanganate and chromate ions. Molec Phys 23(5):835–851

Kiefert L, Karampelas S (2011) Use of the Raman spectrometer in gemmological laboratories: review. Spectrochim Acta A 80(1):119–124

Kiefert L, Schmetzer K (1998) Distinction of taaffeite and musgravite. J Gemmol 26:65–16

Kim CY, Condrate RA Sr (1984) The vibrational spectra of crystalline $W_2O_3(PO_4)_2$ and related tungsten phosphate glasses. J Phys Chem Solids 45(11):1213–1218

Kim HS, Stair PC (2004) Bacterially produced manganese oxide and todorokite: UV Raman spectroscopic comparison. J Phys Chem B 108(44):17019–17026

Kim K-W, Kim Y-H, Lee S-Y, Lee J-W, Joe K-S, Lee E-H, Kim J-S, Song K-C (2009) Precipitation characteristics of uranyl ions at different pHs depending on the presence of carbonate ions and hydrogen peroxide. Environ Sci Technol 43(7):2355–2361

Kim MK, Kim SH, Chang HY, Halasyamani PS, Ok KM (2010a) New noncentrosymmetric tellurite phosphate material: synthesis, characterization, and calculations of $Te_2O(PO_4)_2$. Inorg Chem 49(15):7028–7034

Kim SW, Chang HY, Halasyamani PS (2010b) Selective pure-phase synthesis of the multiferroic $BaMF_4$ (M = Mg, Mn, Co, Ni, and Zn) family. J Am Chem Soc 132(50):17684–17685

Kim YH, Lee DW, Ok KM (2013) α-$ScVSe_2O_8$, β-$ScVSe_2O_8$, and $ScVTe_2O_8$: new quaternary mixed metal oxides composed of only second-order Jahn–Teller distortive cations. Inorg Chem 52(19):11450–11456

Kim YH, Lee DW, Ok KM (2014) Noncentrosymmetric $YVSe_2O_8$ and centrosymmetric $YVTe_2O_8$: macroscopic centricities influenced by the size of lone pair cation linkers. Inorg Chem 53(2):1250–1256

Kimata M (1993) Crystal structure of $KBSi_3O_8$ isostructural with danburite. Mineral Mag 57:157–164

Kimura M, Mikouchi T, Suzuki A, Miyahara M, Ohtani E, Goresy AE (2009) Kushiroite, $CaAlAlSiO_6$: a new mineral of the pyroxene group from the ALH 85085 CH chondrite, and its genetic significance in refractory inclusions. Am Mineral 94(10):1479–1482

Kingma KJ, Hemley RJ (1994) Raman spectroscopic study of microcrystalline silica. Am Mineral 79 (3–4):269–273

Kinner T, Bhandari KP, Bastola E, Monahan BM, Haugen NO, Roland PJ, Begioni TP, Ellingson RJ (2016) Majority carrier type control of cobalt iron sulfide ($Co_xFe_{1-x}S_2$) pyrite nanocrystals. J Phys Chem C 120 (10):5706–5713

Kipcak AS, Baysoy DY, Derun EM, Piskin S (2013) Characterization and neutron shielding behavior of dehydrated magnesium borate minerals synthesized via solid-state method. Adv Mater Sci Eng. https://doi.org/10.1155/2013/747383

Kipcak AS, Derun EM, Pişkin S (2014) Synthesis and characterization of magnesium borate minerals of admontite and mcallisterite obtained via ultrasonic mixing of magnesium oxide and various sources of boron: a novel method. Turk J Chem 38(5):792–805

Kirkpatrick RJ, Yarger JL, McMillan PF, Ping Y, Cong X (1997) Raman spectroscopy of CSH, tobermorite, and jennite. Adv Cem Based Mater 5(3):93–99

Klein RS, Fortin W, Kugel GE (1998) Raman spectra in Bi_2TeO_5 as a function of the temperature and the polarization. J Phys Condens Matter 10(16):3659–3658

Kleppe AK, Jephcoat AP (2004) High-pressure Raman spectroscopic studies of FeS_2 pyrite. Mineral Mag 68 (3):433–441

Kleppe AK, Jephcoat AP, Olijnyk H, Slesinger AE, Kohn SC, Wood BJ (2001) Raman spectroscopic study of hydrous wadsleyite (β-Mg_2SiO_4) to 50 GPa. Phys Chem Minerals 28(4):232–241

Kleppe AK, Jephcoat AP, Welch MD (2003) The effect of pressure upon hydrogen bonding in chlorite: a Raman spectroscopic study of clinochlore to 26.5 GPa. Am Mineral 88(4):567–573

Kleppe AK, Jephcoat AP, Smyth JR (2006) High-pressure Raman spectroscopic study of Fo_{90} hydrous wadsleyite. Phys Chem Minerals 32(10):700–709

Kleppe AK, Welch MD, Crichton WA, Jephcoat AP (2012) Phase transitions in hydroxide perovskites: a Raman spectroscopic study of stottite, $FeGe(OH)_6$, to 21 GPa. Mineral Mag 76(4):949–962

Klingenberg B, Vannice MA (1996) Influence of pretreatment on lanthanum nitrate, carbonate, and oxide powders. Chem Mater 8(12):2755–2768

Kloprogge JT, Frost RL (1999a) Raman microscopic study of some borate minerals: ulexite, kernite, and inderite. Appl Spectrosc 53(3):356–364

Kloprogge JT, Frost RL (1999b) Raman microscopy study of cafarsite. Appl Spectrosc 53(7):874–880

Kloprogge JT, Frost RL (1999c) Raman microscopy study of basic aluminum sulfate. J Mater Sci 34 (17):4199–4202

Kloprogge JT, Frost RL (1999d) Raman and infrared microscopy study of zunyite, a natural Al_{13} silicate. Spectrochim Acta A 55(7):1505–1513

Kloprogge JT, Frost RL (1999e) Raman microprobe spectroscopy of hydrated halloysite from a neogene cryptokarst from Southern Belgium. J Raman Spectrosc 30(12):1079–1108

Kloprogge JT, Frost RL (2000a) Raman microscopic study at 300 and 77 K of some pegmatite minerals from the Iveland-Evje area, Aust-Agder, Southern Norway. Spectrochim Acta A 56(3):501–513

Kloprogge JT, Frost RL (2000b) Raman microscopy study of tyrolite: a multi-anion arsenate mineral. Appl Spectrosc 54(4):517–521

Kloprogge JT, Frost RL (2000c) The effect of synthesis temperature on the FT-Raman and FT-IR spectra of saponites. Vib Spectrosc 23(1):119–127

Kloprogge JT, Wood BJ (2017) X-ray photoelectron Spectroscopic and Raman microscopic investigation of the variscite group minerals: variscite, strengite, scorodite and mansfieldite. Spectrochim Acta A 185: 163–172

Kloprogge JT, Case MH, Frost RL (2001a) Raman microscopic study of the Li amphibole holmquistite, from the Martin Marietta Quarry, Bessemer City, NC, USA. Mineral Mag 65(6):775–785

Kloprogge JT, Ruan H, Duong LV, Frost RL (2001b) FT-IR and Raman microscopic study at 293 K and 77 K of celestine, $SrSO4$, from the middle triassic limestone (Muschelkalk) in Winterswijk, Netherlands. Geologie en Mijnbouw (Netherlands J Geosci) 80 (2):41–48

Kloprogge JT, Visser D, Ruan H, Frost RL (2001c) Infrared and Raman spectroscopy of holmquistite, $Li_2(Mg, Fe^{2+})_3(Al,Fe^{3+})_2(Si,Al)_8O_{22}(OH)_2$. J Mater Sci Lett 20 (16):1497–1499

Kloprogge JT, Wharton D, Hickey L, Frost RL (2002) Infrared and Raman study of interlayer anions CO_3^{2-}, NO_3^-, SO_4^{2-} and ClO_4^- in Mg/Al-hydrotalcite. Am Mineral 87(5–6):623–629

Kloprogge JT, Martens WN, Nothdurft L, Duong LV, Webb GE (2003) Low temperature synthesis and characterization of nesquehonite. J Mater Sci Lett 22 (11):825–829

Kloprogge JT, Hickey L, Duong LV, Martens WN, Frost RL (2004a) Synthesis and characterization of $K_2Ca_5(SO_4)_6\cdot H_2O$, the equivalent of görgeyite, a rare evaporite mineral. Am Mineral 89(2–3):266–272

Kloprogge JT, Weier ML, Duong LV, Frost RL (2004b) Microwave-assisted synthesis and characterisation of divalent metal tungstate nanocrystalline minerals: ferberite, hübnerite, sanmartinite, scheelite and stolzite. Mater Chem Phys 88(2):438–443

Kloprogge JT, Duong LV, Weier M, Martens WN (2006) Nondestructive identification of arsenic and cobalt minerals from Cobalt City, Ontario, Canada: arsenolite, erythrite, and spherocobaltite on pararammelsbergite. Appl Spectrosc 60(11):1293–1296

Knight DS, White WB (1989) Characterization of diamond films by Raman spectroscopy. J Mater Res 4(02):385–393

Knittle E, Kaner RB, Jeanloz R, Cohen ML (1995) High-pressure synthesis, characterization, and equation of state of cubic C-BN solid solutions. Phys Rev B 51(18):12149–12155

Knop O, Brisse F, Castelliz L (1969) Pyrochlores. V. Thermoanalytic, X-ray, neutron, infrared, and dielectric studies of $A_2Ti_2O_7$ titanates. Can J Chem 47(6):971–990

Knyazev AV, Mączka M, Kuznetsova NY (2010) Thermodynamic modeling, structural and spectroscopic studies of the $KNbWO_6$–$KSbWO_6$–$KTaWO_6$ system. Thermochim Acta 506(1):20–27

Knyazev AV, Mączka M, Ladenkov IV, Bulanov EN, Ptak M (2012) Crystal structure, spectroscopy, and thermal expansion of compounds in M^I_2O–Al_2O_3–TiO_2 system. J Solid State Chem 196:110–118

Knyrim JS, Schappacher FM, Pöttgen R, Schmedt auf der Guenne J, Johrendt D, Huppertz H (2007) Pressure-induced crystallization and characterization of the tin borate β-SnB_4O_7. Chem Mater 19(2):254–262

Kobsch A, Downs RT, Domanik KJ (2016) Redetermination of metarossite, $CaV^{5+}_2O_6·2H_2O$. Acta Crystallogr E 72(9):1280–1284

Koch-Müller M, Hofmeister AM, Fei Y, Liu Z (2002) High-pressure IR-spectra and the thermodynamic properties of chloritoid. Am Miner 87:609–622

Kohlrausch KWF (1943) Ramanspektren. Akademische Verlagsgesellschaft, Leipzig

Kojitani H, Többens DM, Akaogi M (2013) High-pressure Raman spectroscopy, vibrational mode calculation, and heat capacity calculation of calcium ferrite-type $MgAl_2O_4$ and $CaAl_2O_4$. Am Mineral 98(1):197–206

Kolesov BA (2018) Applied Raman spectroscopy. Publishing House of the Siberian Branch of the RAS, Novosibirsk. (in Russian)

Kolesov BA, Geiger CA (1998) Raman spectra of silicate garnets. Phys Chem Minerals 25(2):142–151

Kolesov BA, Geiger CA (2002) Raman spectroscopic study of H_2O in bikitaite: "One-dimensional ice". Am Mineral 87(10):1426–1431

Kolesov BA, Geiger CA (2003) Molecules in the SiO_2-clathrate melanophlogite: a single-crystal Raman study. Am Miner 88(8–9):1364–1368

Kolesov BA, Geiger CA (2005) The vibrational spectrum of synthetic hydrogrossular (katoite) $Ca_3Al_2(O_4H_4)_3$: a low-temperature IR and Raman spectroscopic study. Am Mineral 90(8–9):1335–1341

Koleva V, Effenberger H (2007) Crystal chemistry of M $[PO_2(OH)_2]_2·2H_2O$ compounds (M = Mg, Mn, Fe, Co, Ni, Zn, Cd): structural investigation of the Ni, Zn and Cd salts. J Solid State Chem 180(3):956–967

Koleva V, Stefov V, Najdoski M, Cahil A (2015) Thermal, spectral and microscopic studies of water-rich hydrate of the type $Mg_2KH(PO_4)_2·15H_2O$. Thermal transformations. Thermochim Acta 619:20–25

Kolitsch U, Bernhardt H-J, Lengauer CL, Blass G, Tillmanns E (2006) Allanpringite, $Fe_3(PO_4)_2(OH)_3·5H_2O$, a new ferric iron phosphate from Germany, and its close relation to wavellite. Eur J Mineral 18(6):793–801

Kolitsch U, Bernhardt HJ, Krause W, Blass G (2008) Pattersonite, $PbFe_3(PO_4)_2(OH)_4[(H_2O)_{0.5}(OH)_{0.5}]_2$, a new supergene phosphate mineral: description and crystal structure. Eur J Mineral 20(2):281–288

Kolitsch U, Atencio D, Chukanov NV, Zubkova NV, Menezes Filho LAD, Coutinho JMV, Birch WD, Sghlütter J, Pohl D, Kampf AR, Steele IM, Favreau G, Nasdala L, Mocke S, Giester G, Pushcharovsky DY (2010) Bendadaite, a new iron arsenate mineral of the arthurite group. Mineral Mag 74(3):469–486

Kolitsch U, Merlino S, Holtstam D (2012) Molybdophyllite: crystal chemistry, crystal structure, OD character and modular relationships with britvinite. Mineral Mag 76(3):493–516

Kompanchenko AA, Sidorov MY, Voloshin AV (2016) Raman spectroscopy as a method of diagnostic of minerals forming microscopic segregations. In: Voytekhovsky YL (ed) Regional geology, mineralogy and mineral resources of Kola Peninsula. K & M, Apatity. (in Russian)

Kong WG, Wang A, Chou I-M (2011a) Experimental determination of the phase boundary between kornelite and pentahydrated ferric sulfate at 0.1 MPa. Chem Geol 284(3):333–338

Kong WG, Wang A, Freeman JJ, Sobron P (2011b) A comprehensive spectroscopic study of synthetic Fe^{2+}, Fe^{3+}, Mg^{2+} and Al^{3+}copiapite by Raman, XRD, LIBS, MIR and vis–NIR. J Raman Spectrosc 42(5):1120–1129

Kononov OV, Klyuchareva SV, Guseva EV, Kabalov YK, Topor ND, Orlov RY (1990) Thaumasite from hydrothermally altered skarns from the Tyrnyauzskoe deposit (Northern Caucasus). Bull Moscow Univ 4(1):82–87. (in Russian)

Konstantinović MJ, Brink J, Popović ZV, Moshchalkov VV, Isobe M, Ueda Y (2002) Orbital dimerization and dynamic Jahn-Teller effect in $NaTiSi_2O_6$. arXiv preprint cond-mat/0210191. (6 pp)

Konzett J, Yang H, Frost DJ (2005) Phase relations and stability of magnetoplumbite-and crichtonite-series phases under upper-mantle PT conditions: an experimental study to 15 GPa with implications for LILE metasomatism in the lithospheric mantle. J Petrol 46(4):749–781

Korinevsky VG (2015) Spessartine-andradite in scapolite pegmatite, Ilmenymountains, Russia. Can Mineral 53(4):623–632

Korinevsky VG, Kotlyarov VA, Korinevsky EV, Mironov AB, Shtenberg MV (2016) Magnesiohögbomite (Mg, Fe^{2+},Zn)$_8$(Al,Ti,Fe^{3+})$_{20}$O$_{38}$(OH)$_2$ from Ilmenogorsky-Vishnevogorsky complex. Mineral 2:20–33. (in Russian)

Korsakov AV, Golovin AV, De Gussem K, Sharygin IS, Vandenabeele P (2009) First finding of burkeite in melt inclusions in olivine from sheared lherzolite xenoliths. Spectrochim Acta A 73(3):424–427

Korthuis VC, Hoffmann RD, Huang J, Sleight AW (1993) Synthesis and crystal structure of potassium and sodium vanadium phosphates. Chem Mater 5(2):206–209

Korybska-Sadło I, Sitarz M, Król M, Gunia P (2016) Vibrational spectroscopic characterization of the magnesium borate-phosphate mineral lüneburgite. Spectrosc Lett. https://doi.org/10.1080/00387010.2016.1236819

Koszowska E, Wesełucha-Birczyńska A, Borzęcka-Prokop B, Porębska E (2005) Micro and FT-Raman characterisation of destinezite. J Molec Struct 744:845–854

Kotková J, Škoda R, Machovič V (2014) Kumdykolite from the ultrahigh-pressure granulite of the Bohemian Massif. Am Mineral 99(8–9):1798–1801

Koutsopoulos S (2002) Synthesis and characterization of hydroxyapatite crystals: a review study on the analytical methods. J Biomed Mater Res 62(4):600–612

Koutsovitis P, Perraki M, Magganas A (2013) *REE*-rich allanites from a plagiogranite occurrence in South Othris, Greece. In: Anniversary Meeting of the Petrology Group of the Mineralogical Society of Poland, p 50

Kovacheva D, Petrov K (1998) Preparation of crystalline ZnSnO$_3$ from Li$_2$SnO$_3$ by low-temperature ion exchange. Solid State Ionics 109(3):327–332

Koval'chuk NS, Makeev AB (2007) Typomorphism and parasteresis of yushkinite (Pai-Khoi anticlinorium). Zapiski RMO (Proc Russ Miner Soc) 136(5):1–21. (in Russian)

Kowalczyk LN, Condrate RA Sr (1974) Vibrational spectra of spodiosite analogs. J Am Ceram Soc 57(2):102–105

Kowitz A, Güldemeister N, Reimold WU, Schmitt RT, Wünnemann K (2013) Diaplectic quartz glass and SiO$_2$ melt experimentally generated at only 5 GPa shock pressure in porous sandstone: laboratory observations and meso-scale numerical modeling. Earth Planet Sci Lett 384:17–26

Kozlova MN, Mironov YV, Grayfer ED, Smolentsev AI, Zaikovski VI, Nebogatikova NA, Podlipskaya TY, Fedorov VE (2015) Synthesis, crystal structure, and colloidal dispersions of vanadium tetrasulfide (VS$_4$). Chemistry 21(12):4639–4645

Kramar S, Tratnik V, Hrovatin IM, Mladenović A, Pristacz H, Rogan Šmuc N (2015) Mineralogical and chemical characterization of roman slag from the archaeological site of Castra (Ajdovščina, Slovenia). Archaeometry 57(4):704–719

Krause W, Bernhardt H-J, Effenberger H, Kolitsch U, Lengauer C (2003) Redefinition of arhbarite, Cu$_2$Mg(AsO$_4$)(OH)$_3$. Mineral Mag 67(5):1099–1107

Krause W, Bernhardt HJ, Braithwaite RSW, Kolitsch U, Pritchard R (2006) Kapellasite, Cu$_3$Zn(OH)$_6$Cl$_2$, a new mineral from Lavrion, Greece, and its crystal structure. Mineral Mag 70(3):329–340

Krause W, Effenberger H, Bernhardt HJ, Medenbach O (2008) Skorpionite, Ca$_3$Zn$_2$(PO$_4$)$_2$(CO$_3$)(OH)$_2$·H$_2$O, a new mineral from Namibia: description and crystal structure. Eur J Mineral 20(2):271–280

Kreisel J, Lucazeau G, Vincent H (1998a) Raman spectra and vibrational analysis of BaFe$_{12}$O$_{19}$ hexagonal ferrite. J Solid State Chem 137(1):127–137

Kreisel J, Pignard S, Vincent H, Senateur JP, Lucazeau G (1998b) Raman study of BaFe$_{12}$O$_{19}$ thin films. Appl Phys Lett 73:1194–1196

Kreisel J, Lucazeau G, Vincent H (1999) Raman study of substituted barium ferrite single crystals, BaFe$_{12-2x}$Me$_x$Co$_x$O$_{19}$ (Me = Ir, Ti). J Raman Spectrosc 30(2):115–120

Kremenović A, Colomban P, Piriou B, Massiot D, Florian P (2003) Structural and spectroscopic characterization of the quenched hexacelsian. J Phys Chem Solids 64(11):2253–2268

Krenn K, Kaindl R, Hoinkes G (2004) Pumpellyite in metapelites of the Schneeberg Complex (Eastern Alps, Austria). Eur J Mineral 16(4):661–669

Kreske S, Devarajan V (1982) Vibrational spectra and phase transitions in ferroelectric-ferroelastic langbeinites: K$_2$Mn$_2$(SO$_4$)$_3$, (NH$_4$)$_2$Cd$_2$(SO$_4$)$_3$ and Tl$_2$Cd$_2$(SO$_4$)$_3$. J Phys C 15(36):7333–7350

Krishnakumar T, Pinna N, Kumari KP, Perumal K, Jayaprakash R (2008) Microwave-assisted synthesis and characterization of tin oxide nanoparticles. Mater Lett 62(19):3437–3440

Krishnamurthy N, Soots V (1970) Raman spectra of CdF$_2$ and PbF$_2$. Can J Phys 48(9):1104–1107

Krishnamurti D (1955) Raman spectra of borax, kernite and colemanite. Proceed Ind Acad Sci A 41(1):7–11

Krishnamurti D (1958) The Raman spectra of crystalline sulphates of Ni and Mn. Proceed Ind Acad Sci A 48(6):355–363

Krishnan RS, Katiyar RS (1965) The Raman spectrum of magnesium fluoride. J Phys 26(11):627–629

Krishnan RS, Ramanujam PS (1973) Raman spectrum of calcium formate. J Raman Spectrosc 1(6):533–538

Kristiansen R (2016) Personal Commun

Kristova P, Hopkinson LJ, Rutt KJ, Hunter HM, Cressey G (2014) Carbonate mineral paragenesis and reaction kinetics in the system MgO–CaO–CO$_2$–H$_2$O in presence of chloride or nitrate ions at near surface ambient temperatures. Appl Geochem 50:6–24

Krivovichev SV, Burns P (2000) Crystal chemistry of basic lead carbonates. II. Crystal structure of synthetic 'plumbonacrite'. Mineral Mag 64:1969–1975

Krivovichev SV, Turner R, Rumsey M, Siidra OI, Kirk CA (2009) The crystal structure and chemistry of mereheadite. Mineral Mag 73:103–117

Kruger A, Heyns AM (1997) A Raman and infrared study of $(NH_4)_2ZrF_6$. Vib Spectrosc 14(2):171–181

Krüger H, Tropper P, Haefeker U, Kaindl R, Tribus M, Kahlenberg V, Wikete C, Fuchs MR, Olieric V (2014) Innsbruckite, $Mn_{33}(Si_2O_5)_{14}(OH)_{38}$ – a new mineral from the Tyrol, Austria. Mineral Mag 78(7):1613–1628

Krzemnicki MS, Reusser E (1998) Graeserite, $Fe_4Ti_3AsO_{13}(OH)$, a new mineral species of the derbylite group from the Monte Leone nappe, Binntal region, Western Alps, Switzerland. Can Mineral 36:1083–1088

Kučerová G, Ozdín D, Lalinská-Voleková B (2013) Primárny nízkotermálny delafossit ($CuFeO_2$) z odkaliska Slovinky (Slovensko). Bull Mineralpetrolog Odd Nár Muz (Praha) 21(1):78–83. ISSN 1211-0329. (in Czech)

Kuebler KE (2013a) A combined electron microprobe (EMP) and Raman spectroscopic study of the alteration products in Martian meteorite MIL 03346. J Geophys Res Planets 118(3):347–368

Kuebler KE (2013b) A comparison of the iddingsite alteration products in two terrestrial basalts and the Allan Hills 77005 martian meteorite using Raman spectroscopy and electron microprobe analyses. J Geophys Res Planets 118(4):803–830

Kuebler KE, Wang A, Jolliff BL (2011) Review of terrestrial laihunite and stilpnomelane analogs, identified as potential secondary alteration phases in MIL 03346. 42nd Lunar Planet Sci Conf 42:1022

Kulikova OV, Kulyuk LL, Radautsan SI, Ratseev SA, Strumban EE, Tezlevan VE, Tsitsanu VI (1988) Influence of defect generation processes in $CdIn_2S_4$ single crystals on the photoluminescence and Raman scattering spectra. Phys Status Solidi A 107(1):373–377

Kumar P, Nagarajan R (2011) An elegant room temperature procedure for the precise control of composition in the Cu-S system. Inorg Chem 50(19):9204–9206

Kunert HW, Barnas J, Brink DJ, Malherbe J (2006) Raman active modes of one-, two-, and three-phonon processes in the most important compounds and semiconductors with the rhombic, tetragonal, regular, trigonal, and hexagonal structures. J Phys IV France 132:329–336

Kurazhkovskaya VS, Borovikova EY (2003) IR spectra of high-symmetry and low-symmetry vesuvianites. Zapiski RMO (Proc Russ Miner Soc) 132(1):109–121. (in Russian)

Kurazhkovskaya VS, Borovikova EY, Alferova MS (2005) Infrared spectra, unit cell parameters and optical sign of boron-bearing vesuvianites and wiluites. Zapiski RMO (Proc Russ Miner Soc) 134:82–91. (in Russian)

Kurdakova SV, Grishchenko RO, Druzhinina AI, Ogorodova LP (2014) Thermodynamic properties of synthetic calcium-free carbonate cancrinite. Phys Chem Minerals 41(1):75–83

Kurzawa M, Blonska-Tabero A (2002) The synthesis and selected properties of new compounds: $Mg_3Fe_4(VO_4)6$ and $Zn_3Fe_4(VO_4)_6$. Mater Res Bull 37(5):849–858

Kushwaha AK (2008) Lattice dynamics at zone-center of sulphide and selenide spinels. Commun Theor Phys 50(6):1422–1426

Kustova GN, Burgina EB, Sadykov VA, Poryvaev SG (1992) Vibrational spectroscopic investigation of the goethite thermal decomposition products. Phys Chem Minerals 18(6):379–382

Kustova GN, Chesalov YA, Plyasova LM, Molina IY, Nizovskii AI (2011) Vibrational spectra of $WO_3 \cdot nH_2O$ and WO_3 polymorphs. Vibr Spectrosc 55(2):235–240

Kusz J, Malczewski D, Zubko M, Häger T, Hofmeister W (2010) High temperature study of metamictsteenstrupine. Solid State Phenom 163:253–255

Kutsay O, Yan C, Chong YM, Ye Q, Bello I, Zhang WJ, Zapien JA, Zhou ZF, Li YK, Garashchenko V, Gontar AG, Novikov NV, Lee ST (2010) Studying cubic boron nitride by Raman and infrared spectroscopies. Diamond Related Mater 19(7):968–971

Kyono A, Kimata M (2001) The crystal structure of synthetic $TlAlSi_3O_8$. Influence of the inert-pair effect of thallium on the feldspar structure. Eur J Mineral 13(5):849–856

Kyono A, Ahart M, Yamanaka T, Gramsch S, Mao HK, Hemley RJ (2011) High-pressure Raman spectroscopic studies of ulvöspinel Fe_2TiO_4. Am Mineral 96(8–9):1193–1198

La Cruz NL (2015) Schreibersite: synthesis, characterization and corrosion and possible implications for origin of life. Thesis. University of South Florida

La Iglesia A, Garcia-Guinea J, González del Tánago J (2014) La zaratita de Cabo Ortegal (A Coruña): historia de su descubrimiento y caracterización con técnicas analíticas no destructivas. Estudios Geológicos. (in Spanish). https://doi.org/10.3989/egeol.41353.275

Labajos FM, Rives V (1996) Thermal evolution of chromium (III) ions in hydrotalcite-like compounds. Inorg Chem 35(18):5313–5318

Lafuente B, Downs RT (2016) Redetermination of brackebuschite, $Pb_2Mn^{3+}(VO_4)_2(OH)$. Acta Crystallogr E 72(3):293–296

Lafuente B, Yang H, Downs RT (2015) Crystal structure of tetrawickmanite, $Mn^{2+}Sn^{4+}(OH)_6$. Acta Crystallogr E 71:234–237

Lafuente B, Downs RT, Origlieri MJ, Domanik KJ, Gibbs RB, Rumsey MS (2016) New data on hemihedrite from Arizona. Mineral Mag. https://doi.org/10.1180/minmag.2016.080.148

Lage MM, Righi A, Matinaga FM, Gesland JY, Moreira RL (2004) Raman-spectroscopic study of lanthanide trifluorides with the β-YF_3 structure. J Phys Condens Matter 16(18):3207–3218

Lager GA, Xie Q, Ross FK, Rossman GR, Armbruster T, Rotella FJ, Schultz AJ (1999) Hydrogen-atom position in $P4/nnc$ vesuvianite. Can Miner 37:763–768

Lahalle MP, Krupa JC, Lepostollec M, Forgerit JP (1986) Low-temperature Raman study on $ThSiO_4$ single crystal and related infrared spectra at room temperature. J Solid State Chem 64:181–187

Lahti SI, Saikkonen R (1985) Bityite $2M_1$ from Eräjärvi compared with related Li-Be brittle micas. Bull Geol Soc Finland 57:207–215

Laihunite Research Group, Guiyang Institute of Geochemistry, Academia Sinica and Geological Team 101 (1976) Laihunite – a new iron silicate mineral. Geochim 2:95–103. (in Chinese, English Abstr)

Lajzérowicz J (1966) Étude par diffraction des rayons X et absorption infra-rouge de la barysilite, $MnPb_8 \cdot 3Si_2O_7$, et de composes isomorphes. Acta Crystallogr 20 (3):357–363. (in French)

Lambruschi E, Gatta G, Adamo I, Bersani D, Salvioli-Mariani E, Lottici PP (2014) Raman and structural comparison between the new gemstone pezzottaite Cs $(Be_2Li)Al_2Si_6O_{18}$ and Cs-beryl. J Raman Spectrosc 45 (11–12):993–999

Lannin JS (1977) Raman scattering properties of amorphous As and Sb. Phys Rev B 15(8):3863. (15 pp)

Lanteigne S, Schindler M, McDonald AM, Skeries K, Abdu Y, Mantha NM, Murayama M, Hawthorne FC, Hochella MF Jr (2012) Mineralogy and weathering of smelter-derived spherical particles in soils: implications for the mobility of Ni and Cu in the surficial environment. Water Air Soil Pollut 223(7): 3619–3641

Larkin PJ (2011) Infrared and Raman spectroscopy: principles and spectral interpretation. Elsevier

Latturner SE, Sachleben J, Iversen BB, Hanson J, Stucky GD (1999) Covalent guest-framework interactions in heavy metal sodalites: structure and properties of thallium and silver sodalite. J Phys Chem B 103 (34):7135–7144

Laurs BM, Pezzotta F, Simmons WB, Falster AU, Muhlmeister S (2002) Rhodizite-londonite from the Antsongombato pegmatite, Central Madagascar. Gems Gemol 38(4):326–339

Lavat AE, Grasselli MC, Baran EJ (1989) The IR spectra of the $(Cr_xFe_{1-x})VO_4$ phases. J Solid State Chem 78 (2):206–208

Lavrova GV, Burgina EB, Matvienko AA, Ponomareva VG (2006) Bulk and surface properties of ionic salt $CsH_5(PO_4)_2$. Solid State Ionics 177(13):1117–1122

Lazarević N, Bozin ES, Šćepanović M, Opačić M, Lei H, Petrovic C, Popović ZV (2014) Probing Ir Te_2 crystal symmetry by polarized Raman scattering. Phys Rev B 89(22):224301-1–224301-6

Lazic B, Armbruster T, Savelyeva VB, Zadov AE, Pertsev NN, Dzierżanowski P (2011) Galuskinite, $Ca_7(SiO_4)_3(CO_3)$, a new skarn mineral from the Birkhin gabbro massif, Eastern Siberia, Russia. Mineral Mag 75(5):2631–2648

Lazic B, Armbruster T, Liebich BW, Perfler L (2012) Hydrogen-bond system and dehydration behavior of the natural zeolite parthéite. Am Mineral 97 (11–12):1866–1873

Le Cléac'h A, Gillet P (1990) IR and Raman spectroscopic study of natural lawsonite. Eur J Mineral 2(1): 43–53

Leander F, de Capitani C, Sun TT, Hanni HA, Atichat W (2014) A comparative study of jadeite, omphacite and kosmochlor jades from Myanmar, and suggestions for a practical nomenclature. J Gemm 34(3):210–229

Lecker A (2013) Synthese, Strukturchemie und physikalische Untersuchungen an Mangan-, Eisen- und Quecksilber-Chalkogenometallatverbindungen. Dissertation, Universität Regensburg, Institut für Anorganische Chemie. (in German)

Lee DW, Ok KM (2014) New polymorphs of ternary sodium tellurium oxides: hydrothermal synthesis, structure determination, and characterization of β-$Na_2Te_4O_9$ and $Na_2Te_2O_6 \cdot 1.5H_2O$. Inorg Chem 53 (19):10642–10648

Lee CY, Macquart R, Zhou Q, Kennedy BJ (2003) Structural and spectroscopic studies of $BiTa_{1-x}Nb_xO_4$. J Solid State Chem 174(2):310–318

Lee DW, Oh SJ, Halasyamani PS, Ok KM (2011) New quaternary tellurite and selenite: synthesis, structure, and characterization of centrosymmetric $InVTe_2O_8$ and noncentrosymmetric $InVSe_2O_8$. Inorg Chem 50 (10):4473–4480

Lee DW, Bak DB, Kim SB, Kim J, Ok KM (2012) Effect of the framework flexibility on the centricities in centrosymmetric $In_2Zn(SeO_3)_4$ and noncentrosymmetric $Ga_2Zn(TeO_3)_4$. Inorg Chem 51(14):7844–7850

Lee EP, Song SY, Lee DW, Ok KM (2013) New bismuth selenium oxides: syntheses, structures, and characterizations of centrosymmetric $Bi_2(SeO_3)_2(SeO_4)$ and $Bi_2(TeO_3)_2(SeO_4)$ and noncentrosymmetric $Bi(SeO_3)$ $(HSeO_3)$. Inorg Chem 52(7):4097–4103

Lee CW, Park HK, Park S, Han HS, Seo SW, Song HJ, Shin S, Kim D-W, Hong KS (2015) Ta-substituted $SnNb_{2-x}Ta_xO_6$ photocatalysts for hydrogen evolution under visible light irradiation. J Mater Chem A. https://doi.org/10.1039/c4ta05885b

Lehnen T, Valldor M, Nižňanský D, Mathur S (2014) Hydrothermally grown porous $FeVO_4$ nanorods and their integration as active material in gas-sensing devices. J Mater Chem A 2(6):1862–1868

Leissner L, Schlüter J, Horn I, Mihailova B (2015) Exploring the potential of Raman spectroscopy for crystallochemical analyses of complex hydrous silicates: I. Amphiboles. Am Mineral 100(11–12):2682–2694

Leite CAF, Guimaraes RB, Fernandes JC, Continentino MA, Paschoal CWA, Ayala AP, Guedes I (2002) Temperature-dependent Raman scattering study of $Fe_3O_2BO_3$ ludwigite. J Raman Spectrosc 33(1):1–5

Lenaz D, Lughi V (2013) Raman study of $MgCr_2O_4$–$Fe^{2+}Cr_2O_4$ and $MgCr_2O_4$–$MgFe^{3+}{}_2O_4$ synthetic series: the effects of Fe^{2+} and Fe^{3+} on Raman shifts. Phys Chem Minerals 40(6):491–498

Lengauer CL, Hrauda N, Kolitsch U, Krickl R, Tillmanns E (2009) Friedrichbeckeite, K $(\square_{0.5}Na_{0.5})_2(Mg_{0.8}Mn_{0.1}Fe_{0.1})_2(Be_{0.6}Mg_{0.4})_3[Si_{12}O_{30}]$, a new milarite-type mineral from the Bellerberg volcano, Eifel area, Germany. Mineral Petrol 96 (3–4):221–232

Lepore GO, Bindi L, Zanetti A, Ciriotti ME, Medenbach O, Bonazzi P (2015) Balestraite, $KLi_2VSi_4O_{10}O_2$, the first member of the mica group with octahedral V^{5+}. Am Mineral 100(2–3):608–614

Lepot L, Denoël S, Gilbert B (2006) The technique of the mural paintings of the Tournai Cathedral. J Raman Spectrosc 37(10):1098–1103

Levin D, Soled SL, Ying JY (1996) Crystal structure of an ammonium nickel molybdate prepared by chemical precipitation. Inorg Chem 35(14):4191–4197

Levitt SR, Condrate RA Sr (1970) Vibrational spectra of lead apatites. Am Mineral 55(9–10):1562–1575

Lewis DW, Salih D, Ruiz-Salvador AR, Emerich H, van Beek W, White CLIM, Green MA (2006) Computational and experimental studies of the structure and dynamics of water in natural zeolites. In: Zeolite '06–7th International Conference on the Occurrence, Properties, and Utilization of Natural Zeolites, Socorro, New Mexico, USA, 16–21 July 2006, pp 161–162

Li W, Chen G (1990) Lishizhenite – a new zinc sulphate mineral. Acta Mineral Sinica 10(4):299–305. (in Chinese, English abstr)

Li D, Peng N, Bancroft GM (1994) The vibrational spectra and structure of nordenskiöldine. Can Mineral 32:81–86

Li J, Xia S-P, Gao S-Y (1995) FT-IR and Raman spectroscopic study of hydrated borates. Spectrochim Acta A 51(4):519–532

Li G, Feng S, Li L, Li X, Jin W (1997) Mild hydrothermal syntheses and thermal behaviors of hydrogarnets $Sr_3Al_2(OH)_{12}$ (M = Cr, Fe, and Al). Chem Mater 9 (12):2894–2901

Li Z, Wu Y, Fu P, Pan S, Lin Z, Chen C (2003) Czochralski crystal growth and properties of $Na_5[B_2P_3O_{13}]$. J Cryst Growth 255(1):119–122

Li XZ, Wang C, Chen XL, Li H, Jia LS, Wu L, Du YX, Xu YP (2004a) Syntheses, thermal stability, and structure determination of the novel isostructural $RBa_3B_9O_{18}$ (R = Y, Pr, Nd, Sm, Eu, Gd, Tb, Dy, Ho, Er, Tm, Yb). Inorg Chem 43(26):8555–8560

Li YH, Ling YH, Bai XD (2004b) Preparation and characterization of anisotropic ammonium titanium phosphate crystals via hydrothermal route. Key Eng Mater 280:597–600

Li Y, Chen G, Zhang H, Li Z, Sun J (2008) Electronic structure and photocatalytic properties of $ABi_2Ta_2O_9$ (A = Ca, Sr, Ba). J Solid State Chem 181 (10):2653–2659

Li H, Pan S, Wu H, Yang Z (2011a) Growth, structure and properties of the non-centrosymmetric hydrated borate $(NH_4)_2CaB_8O_{26}H_{24}$. Mater Chem Phys 129 (1):176–179

Li J, Pan S, Zhao W, Tian X, Han J, Fan X (2011b) Synthesis and crystal structure of a novel boratotungstate: $Pb_6B_2WO_{12}$. Solid State Sci 13(5):966–969

Li H, Zhao Y, Pan S, Wu H, Yu H, Zhang F, Poeppelmeier KR (2013) Synthesis and Structure of $KPbBP_2O_8$ – a congruent melting borophosphate with nonlinear optical properties. Eur J Inorg Chem 2013(18): 3185–3190

Li G, Zhang B, Yu F, Novakova AA, Krivenkov MS, Kiseleva TY, Chang L, Rao J, Polyakov AO, Blake GR, de Groot RA, Palstra TTM (2014) High-purity Fe_3S_4 greigite microcrystals for magnetic and electrochemical performance. Chem Mater 26 (20):5821–5829

Li Y, Shen J, Hu Y, Qiu S, Min G, Song Z, Sun Z, Li C (2015) General flame approach to chainlike MFe_2O_4 Spinel (M = Cu, Ni, Co, Zn) nanoaggregates for reduction of nitroaromatic compounds. Ind Eng Chem Res 54(40):9750–9757

Liao S, Chen ZP, Tian XZ, Wu WW (2009) Synthesis and regulation of α-$LiZnPO_4 \cdot H_2O$ via a solid-state reaction at low-heating temperatures. Mater Res Bull 44 (6):1428–1431

Libowitzky E (1999) Correlation of O–H stretching frequencies and O–H$\cdots$O hydrogen bond lengths in minerals. Monatshefte für Chemie 130:1047–1059

Libowitzky E (2006) Crystal structure dynamics: evidence by diffraction and spectroscopy. Croat Chem Acta 79 (2):299–309

Liegeois-Duyckaerts M (1975) Vibrational studies of molybdates, tungstates and related compounds – IV. Hexagonal perovskites: $Ba_2B^{II}TeO_6$ (B^{II} = Ni, Co, Zn). Spectrochim Acta A 31(11):1585–1588

Liegeois-Duyckaerts M (1985) Spectroscopic and structural studies of the hexagonal perovskite Ba_2CoTeO_6. Spectrochim Acta A 41(4):523–529

Liermann HP, Downs RT, Yang H (2006) Site disorder revealed through Raman spectra from oriented single crystals: a case study on karooite ($MgTi_2O_5$). Am Mineral 91(5–6):790–793

Likhacheva AY, Goryainov SV, Seryotkin YV, Litasov KD, Momma K (2016) Raman spectroscopy of chibaite, natural MTN silica clathrate, at high pressure up to 8 GPa. Micropor Mesopor Mater 224:100–106

Lin CC (2001) High-pressure Raman spectroscopic study of Co-and Ni-olivines. Phys Chem Minerals 28 (4):249–257

Lin C-C (2004) Pressure-induced polymorphism in enstatite ($MgSiO_3$) at room temperature: clinoenstatite and orthoenstatite. J Phys Chem Solids 65(5):913–921

Lin Z, Rocha J (2005) ^{29}Si MAS NMR and Raman spectroscopy studies of synthetic microporous (Zr, Hf)-umbite. Stud Surf Sci Catal 158:861–868

Lin C-C, Shen P (1994) Sol-gel synthesis of zinc orthosilicate. J Non-Cryst Solids 171(3):281–289

Lin CC, Liu LG, Irifune T (1999) High-pressure Raman spectroscopic study of chondrodite. Phys Chem Minerals 26(3):226–233

Lin CC, Liu LG, Mernagh TP, Irifune T (2000) Raman spectroscopic study of hydroxyl-clinohumite at various pressures and temperatures. Phys Chem Minerals 27(5):320–331

Lin YH, Adebajo MO, Kloprogge JT, Martens WN, Frost RL (2006) X-ray diffraction and Raman spectroscopic studies of Zn-substituted carrboydite-like compounds. Mater Chem Phys 100(1):174–186

Lin CH, Chen CS, Shiryaev AA, Zubavichus YV, Lii KH (2008) $K_3(U_3O_6)(Si_2O_7)$ and $Rb_3(U_3O_6)(Ge_2O_7)$: a pentavalent-uranium silicate and germanate. Inorg Chem 47(11):4445–4447

Lin C-C, Leung KS, Shen P, Chen S-F (2015) Elasticity and structure of the compounds in the wollastonite $(CaSiO_3)$–Na_2SiO_3 system: from amorphous to crystalline state. J Mater Sci Mater Med. https://doi.org/10.1007/s10856-014-5361-7

Lindsay JW, Robinson HN, Bramlet HL, Johnson AJ (1970) The thermal decomposition of neptunium (IV) oxalate. J Inorg Nucl Chem 32(5):1559–1567

Ling ZC, Wang A (2010) A systematic spectroscopic study of eight hydrous ferric sulfates relevant to Mars. Icarus 209(2):422–433

Ling Z, Wang A (2015) Spatial distributions of secondary minerals in the Martian meteorite MIL 03346, 168 determined by Raman spectroscopic imaging. J Geophys Res Planets 120(6):1141–1159

Ling ZC, Xia HR, Ran DG, Liu FQ, Sun SQ, Fan JD, Zang HJ, Wang JY, Yu LL (2006) Lattice vibration spectra and thermal properties of $SrWO_4$ single crystal. Chem Phys Lett 426(1):85–90

Ling ZC, Wang A, Li C (2009) Comparative spectroscopic study of three ferric sulfates: kornelite, lausenite and pentahydrate. 40th Lunar Planetary Sci Conf 40:1867

Ling ZC, Wang A, Jolliff BL (2011) Mineralogy and geochemistry of four lunar soils by laser-Raman study. Icarus 211(1):101–113

Linnow K, Steiger M, Lemster C, De Clercq H, Jovanović M (2013) In situ Raman observation of the crystallization in $NaNO_3$–Na_2SO_4–H_2O solution droplets. Environ Earth Sci 69(5):1609–1620

Litvinenko AK, Sorokina ES, Karampelas S, Krivoschekov NN, Serov R (2016) Variscite from central Tajikistan: preliminary results. Gems Gemol 52(1):60–65

Liu LG, El Gorsey A (2007) High-pressure phase transitions of the feldspars, and further characterization of lingunite. Int Geol Rev 49(9):854–860

Liu H, Tang D (2009) Synthesis of ZnV_2O_6 powder and its cathodic performance for lithium secondary battery. Mater Chem Phys 114(2):656–659

Liu J, Wang Y, Lan G, Zheng J (2001) Vibrational spectra of barium formate crystal. J Raman Spectrosc 32(12):1000–1003

Liu Y, Cao J, Wang Y, Zeng J, Qian Y (2002) Aqueous ammonia route to $Cu_{1.8}S$ with triangular and rod-like shapes. Inorg Chem Commun 5(6):407–410

Liu Q, Su Y, Yu H, Han W (2008) YPO_4 nanocrystals: preparation and size-induced lattice symmetry enhancement. J Rare Earths 26(4):495–500

Liu L-G, Lin C-C, Yung YJ, Mernagh TP, Irifune T (2009) Raman spectroscopic study of K-lingunite at various pressures and temperatures. Phys Chem Minerals 36(3):143–149

Liu W, Liu Y, Qiu X (2012) The first discovery of hemusite in China and its mineralogical features. Acta Mineral Sinica 4:5. (in Chinese, English abstr)

Liu J, Cheng H, Frost RL, Dong F (2013) The mineral tooeleite $Fe_6(AsO_3)_4SO_4(OH)_4 \cdot 4H_2O$ – an infrared and Raman spectroscopic study-environmental implications for arsenic remediation. Spectrochim Acta A 103:272–275

Liu J, Ming D, Cheng H, Xu Z, Frost RL (2015a) Spectroscopic vibrations of austinite ($CaZnAsO_4 \cdot OH$) and its mineral structure: implications for identification of secondary arsenic-containing mineral. Spectrochim Acta A 135:351–355

Liu L, Zhang F, Pan S, Lei C, Zhang R, Dong X, Wang Z, Yu H, Yang Z (2015b) Synthesis, crystal structure and properties of a new barium calcium borate, $Ba_2Ca_2(B_2O_5)_2$. Solid State Sci 39:105–109

Liu Y, Mei D, Xu J, Wu Y (2015c) Hydrothermal synthesis, structures and optical properties of $A_2Zn_3(SeO_3)_4 \cdot XH_2O$ (A = Li, Na, K; X = 2 or 0). J Solid State Chem 232:193–199

Liu Y, Xiao C, Lyu M, Lin Y, Cai W, Huang P, Tong W, Zou Y, Xie Y (2015d) Ultrathin Co_3S_4 nanosheets that synergistically engineer spin states and exposed polyhedra that promote water oxidation under neutral conditions. Angew Chem Int 127(38):11383–11387

Liu J, Sun J, Huang X, Li G, Liu B (2015e) Goldindec: a novel algorithm for Raman spectrum baseline correction. Appl Spectrosc 69(7):834–842

Liu J, He L, Dong F, Frost RL (2017) Infrared and Raman spectroscopic characterizations on new Fe sulphoarsenatehilarionite ($Fe^{(III)}_2(SO_4)(AsO_4)(OH) \cdot 6H_2O$): implications for arsenic mineralogy in supergene environment of mine area. Spectrochim Acta A 170:9–13

Locke AJ, Martens WN, Frost RL (2007) Natural halotrichites – an EDX and Raman spectroscopic study. J Raman Spectrosc 38(11):1429–1435

Łodziński M, Sitarz M (2009) Chemical and spectroscopic characterization of some phosphate accessory minerals from pegmatites of the Sowie Góry Mts, SW Poland. J Molec Struct 924:442–447

López A, Frost RL, Scholz R, Gobac ŽŽ, Xi Y (2013a) Vibrational spectroscopy of the silicate mineral plumbotsumite $Pb_5(OH)_{10}Si_4O_8$ – an assessment of the molecular structure. J Molec Struct 1054:228–233

López A, Frost RL, Xi Y, Scholz R, Belotti FM, Ribeiro É (2013b) Assessment of the molecular structure of natrodufrénite – $NaFe^{2+}Fe^{3+}_5(PO_4)_4(OH)_6 \cdot 2(H_2O)$, a secondary pegmatite phosphate mineral from Minas Gerais, Brazil. J Molec Struct 1051:265–270

López A, Xi Y, Frost RL (2013c) Raman and infrared spectroscopic study of the mineral goyazite $SrAl_3(PO_4)_2(OH)_5 \cdot H_2O$. Spectrochim Acta A 116:204–208

López MC, Ortiz GF, Arroyo-de Dompablo EM, Tirado JL (2014a) An unnoticed inorganic solid electrolyte: dilithium sodium phosphate with the nalipoite structure. Inorg Chem 53(4):2310–2316

López A, Frost RL, Scholz R, Xi Y, Amaral A (2014b) Infrared and Ramans Spectroscopic characterization of the silicate mineral gilalite $Cu_5Si_6O_{17} \cdot 7H_2O$. Spectrosc Lett 47(6):488–493

López A, Frost RL, Xi Y (2014c) Vibrational spectroscopy of the multianion mineral gartrellite from the Anticline Deposit, Ashburton Downs, Western Australia. Spectrochim Acta A 123:54–58

López A, Frost RL, Xi Y, Scholz R (2014d) Vibrational spectroscopic characterization of the sulphate-carbonate mineral burkeite: implications for evaporites. Spectrosc Lett 47(7):564–570

López A, Frost RL, Xi Y, Scholz R (2014e) Vibrational spectroscopic characterization of the arsenate mineral barahonaite: implications for the molecular structure. Spectrosc Lett 47(7):571–578

López A, Frost RL, Xi Y, Scholz R (2014f) A vibrational spectroscopic study of the sulphate mineral glauberite. Spectrosc Lett 47(10):740–745

López A, Scholz R, Frost RL (2015a) Raman and infrared spectroscopic study of the borate mineral kaliborite. Spectrosc Lett 48(10):712–716

López A, Scholz R, Frost RL, Belotti FM (2015b) SEM, EDX and vibrational spectroscopic study of the phosphate mineral ushkovite $MgFe^{3+}_2(PO_4)_2(OH)_2 \cdot 8H_2O$ – implications of the molecular structure. J MolecStruct 1081:329–333

López-Sánchez J, Serrano A, Del Campo A, Abuín M, Rodríguez de la Fuente O, Carmona N (2016) Sol–gel synthesis and micro-Raman characterization of ε-Fe_2O_3 micro- and nanoparticles. Chem Mater 28 (2):511–518

Losertová L, Buřival Z, Losos Z (2014) Waylandite and petitjeanite, two new phosphates for locality Cetoraz near Pacov (Czech Republic). Bull Mineral-petrolog Odd Nár Muz (Praha) 22(2):269–274. ISSN 1211-0329. (in Czech)

Lotti P (2014) Cancrinite-group minerals at non-ambient conditions: a model of the elastic behavior and structure evolution. Dissertation, University of Milan

Loubbidi L, Chagraoui A, Yakine I, Orayech B, Naji M, Igartua JM, Tairi A (2014) Crystal structural and Raman vibrational studies of $Bi_{1-x}Sb_{1-x}Te_{2x}O_4$ solid solution with $0 \le x \le 0.1$. Open Access Libr J 1 (09):1–10

Loun J, Čejka J, Sejkora J, Plášil J, Novák M, Frost RL, Palmer SJ, Keeffe EC (2011) A Raman spectroscopic study of bukovskýite $Fe_2(AsO_4)(SO_4)(OH) \cdot 7H_2O$, a mineral phase with a significant role in arsenic migration. J Raman Spectrosc 42(7):1596–1600

Lucovsky G, Sladek RJ, Allen JW (1977) IR reflectance spectra of Ti_2O_3: infrared-active phonons and Ti 3d electronic effects. Phys Rev B 16(12):5452–5459

Lugo GJ, Mazón P, Baudin C, De Aza PN (2015) Nurse's A-phase: synthesis and characterization in the binary system Ca_2SiO_4–$Ca_3(PO_4)_2$. J Am Ceram Soc 98 (10):3042–3046

Lussier A, Ball NA, Hawthorne FC, Henry DJ, Shimizu R, Ogasawara Y, Ota T (2016) Maruyamaite, $K(MgAl_2)$ $(Al_5Mg)(BO_3)_3(Si_6O_{18})(OH)_3O$, a potassium-dominant tourmaline from the ultrahigh-pressure Kokchetav massif, northern Kazakhstan: description and crystal structure. Am Mineral 101(2):355–361

Lutz HD, Müller B (1991) Lattice vibration spectra. LXVIII. Single-crystal Raman spectra of marcasite-type iron chalcogenides and pnictides, FeX_2 (X = S, Se, Te; P, As, Sb). Phys Chem Minerals 18(4):265–268

Lutz HD, Pobitschka W, Frischemeier B, Becker RA (1978) Gitterschwingungsspektren. XIX – Infrarot- und Ramanspektren von $BaBr_2 \cdot 2H_2O$ und $BaBr_2 \cdot 2D_2O$. J Raman Spectrosc 7(3):130–136. (in German)

Lutz HD, Wa G, Kliche G, Haeuseler H (1983) Lattice vibration spectra, XXXIII: far-infrared reflection spectra, TO and LO phonon frequencies, optical and dielectric constants, and effective charges of the spinel-type compounds MCr_2S_4 (M = Mn, Fe, Co, Zn, Cd, Hg), MCr_2Se_4 (M = Zn, Cd, Hg), and MIn_2S_4 (M = Mn, Fe, Co, Ni, Cd, Hg). J Solid State Chem 48(2):196–208

Lutz HD, Becker W, Müller B, Jung M (1989) Raman single crystal studies of spinel type MCr_2S_4 (M = Mn, Fe, Co, Zn, Cd), MIn_2S_4 (M = Mn, Fe, Co, Ni), $MnCr_{2-2x}In_{2x}S_4$ and $Co_{1-x}Cd_xCr_2S_4$. J Raman Spectrosc 20(2):99–103

Lutz HD, Müller B, Steiner HJ (1991) Lattice vibration spectra. LIX. Single crystal infrared and Raman studies of spinel type oxides. J Solid State Chem 90(1):54–60

Lutz HD, Schmidt M, Weckler B (1993) Infrared and Raman studies on calcium, zinc and cadmium hydroxide halides $Ca\{O(H,D)\}Cl$, $Cd\{O(H, D)\}Cl$, $Zn\{O(H, D)\}F$ and β-$Zn\{O(H,D)\}Cl$. J Raman Spectrosc 24 (11):797–804

Lutz HD, Möller H, Schmidt M (1994) Lattice vibration spectra. Part LXXXII. Brucite-type hydroxides M $(OH)_2$ (M= Ca, Mn, Co, Fe, Cd) – IR and Raman spectra, neutron diffraction of $Fe(OH)_2$. J Molec Struct 328:121–132

Lutz HD, Beckenkamp K, Peter ST (1995) Laurionite-type M(OH)X (M = Ba, Pb; X = Cl, Br, I) and Sr(OH)I. An IR and Raman spectroscopic study. Spectrochim Acta A 51(5):755–767

Lutz HD, Jung C, Mörtel R, Jacobs H, Stahl R (1998) Hydrogen bonding in solid hydroxides with strongly polarising metal ions, β-$Be(OH)_2$ and ϵ-$Zn(OH)_2$. Spectrochim Acta A 54(7):893–901

Lyalina L, Zolotarev A Jr, Selivanova E, Savchenko Y, Krivovichev S, Mikhailova Y, Kadyrova G, Zozulya D (2016) Batievaite-(Y) $Y_2Ca_2Ti[Si_2O_7]_2(OH)_2(H_2O)_4$, a new mineral from nepheline syenite pegmatite in the Sakharjok massif, Kola Peninsula, Russia. Mineral Petrol 110(6):895–904

Ma C (2010) Hibonite-(Fe), $(Fe,Mg)Al_{12}O_{19}$, a new alteration mineral from the Allende meteorite. Am Mineral 95(1):188–191

Ma C, Rossman GR (2008) Barioperovskite, BaTiO$_3$, a new mineral from the Benitoite Mine, California. Am Mineral 93(1):154–157

Ma C, Rossman GR (2009a) Tistarite, Ti$_2$O$_3$, a new refractory mineral from the Allende meteorite. Am Mineral 94(5–6):841–844

Ma C, Rossman GR (2009b) Davisite, CaScAlSiO$_6$, a new pyroxene from the Allende meteorite. Am Mineral 94(5–6):845–848

Ma J, Wu Q, Ding Y (2007) Assembly and deagglomeration of lanthanum orthoborate nanobundles. J Amer Ceram Soc 90(12):3890–3895

Ma C, Simon SB, Rossman GR, Grossman L (2009) Calcium Tschermak's pyroxene, CaAlAlSiO$_6$, from the Allende and Murray meteorites: EBSD and micro-Raman characterizations. Am Mineral 94(10):1483–1486

Ma C, Connolly HC Jr, Beckett JR, Tschauner O, Rossman GR, Kampf AR, Zega TJ, Smith SAS, Schrader DL (2011a) Brearleyite, Ca$_{12}$Al$_{14}$O$_{32}$Cl$_2$, a new alteration mineral from the NWA 1934 meteorite. Am Mineral 96(8–9):1199–1206

Ma C, Kampf AR, Connolly HC, Beckett JR, Rossman GR, Smith SAS, Schrader DL (2011b) Krotite, CaAl$_2$O$_4$, a new refractory mineral from the NWA 1934 meteorite. Am Mineral 96(5–6):709–715

Ma C, Beckett JR, Rossman GR (2012a) Buseckite, (Fe, Zn,Mn)S, a new mineral from the Zakłodzie meteorite. Am Mineral 97(7):1226–1233

Ma C, Beckett JR, Rossman GR (2012b) Browneite, MnS, a new sphalerite-group mineral from the Zakłodzie meteorite. Am Mineral 97(11–12):2056–2059

Ma C, Tschauner O, Beckett JR, Rossman GR, Liu W (2012c) Panguite, (Ti^{4+},Sc,Al,Mg,Zr,Ca)$_{1.8}$O$_3$, a new ultra-refractory titania mineral from the Allende meteorite: synchrotron micro-diffraction and EBSD. Am Mineral 97(7):1219–1225

Ma C, Tschauner O, Beckett JR, Rossman GR, Liu W (2013) Kangite, (Sc,Ti,Al,Zr,Mg,Ca,□)$_2$O$_3$, a new ultra-refractory scandia mineral from the Allende meteorite: synchrotron micro-Laue diffraction and electron backscatter diffraction. Am Mineral 98(5–6): 870–878

Ma C, Beckett JR, Rossman GR (2014a) Monipite, MoNiP, a new phosphide mineral in a Ca-Al-rich inclusion from the Allende meteorite. Am Mineral 99(1):198–205

Ma C, Beckett JR, Rossman GR (2014b) Allendeite (Sc$_4$Zr$_3$O$_{12}$) and hexamolybdenum (Mo, Ru, Fe), two new minerals from an ultrarefractory inclusion from the Allende meteorite. Am Mineral 99(4):654–666

Ma C, Tschauner O, Beckett JR, Liu Y, Rossman GR, Zhuravlev K, Prakapenka V, Dera P, Taylor LA (2015) Tissintite, (Ca,Na,□)AlSi$_2$O$_6$, a highly-defective, shock-induced, high-pressure clinopyroxene in the Tissint martian meteorite. Earth Planet Sci Lett 422:194–205

Ma C, Tschauner O, Beckett JR, Liu Y, Rossman GR, Sinogeikin SV, Smith JS, Taylor LA (2016a) Ahrensite, γ-Fe$_2$SiO$_4$, a new shock-metamorphic mineral from the Tissint meteorite: implications for the Tissint shock event on Mars. Geochim Cosmochim Acta 184:240–256

Ma Y, Liu Z, Geng A, Vogt T, Lee Y (2016b) Structural and spectroscopic studies of alkali-metal exchanged stilbites. Micropor Mesopor Mater 224:339–348

Maczka M, Hanuza J, Lutz ETG, Van der Maas JH (1999) Infrared activity of KAl(MoO$_4$)$_2$ and NaAl(MoO$_4$)$_2$. J Solid State Chem 145(2):751–756

Mączka M, Hanuza J, Fuentes AF, Amador U (2002) Vibrational characteristics of new double tungstates Li$_2$M^{II}(WO$_4$)$_2$ (M = Co, Ni and Cu). J Raman Spectrosc 33(1):56–61

Mączka M, Pietraszko A, Hanuza J, Majchrowski A (2010) Raman and IR spectra of noncentrosymmetric Bi$_{0.21}$La$_{0.91}$Sc$_{2.88}$(BO$_3$)$_4$ single crystal with the huntite-type structure. J Raman Spectrosc 41(10):1297–1301

Mączka M, Ptak M, Kurnatowska M, Hanuza J (2013) Synthesis, phonon and optical properties of nanosized CoCr$_2$O$_4$. Mater Chem Phys 138(2):682–688

Mączka M, Szymborska-Małek K, Gągor A, Majchrowski A (2015) Growth and characterization of acentric BaHf(BO$_3$)$_2$ and BaZr(BO$_3$)$_2$. J Solid State Chem 225:330–334

Madon M, Price GD (1989) Infrared spectroscopy of the polymorphic series (enstatite, ilmenite, and perovskite) of MgSiO$_3$, MgGeO$_3$, and MnGeO$_3$. J Geophys Res Solid Earth 94(B11):15687–15701

Madon M, Gillet P, Julien C, Price GD (1991) A vibrational study of phase transitions among the GeO$_2$ polymorphs. Phys Chem Minerals 18:7–18

Madsen LD, Weaver L (1998) Characterization of lead oxide thin films produced by chemical vapor deposition. J Am Ceram Soc 81(4):988–996

Maguregui M, Martinez-Arkarazo I, Angulo M, Castro K, Fernández LA, Madariaga JM (2008) Portable spectroscopic analysis of nitrates affecting to cultural heritage materials. In: Castillejo M, Moreno P, Oujja M, Radvan R, Ruiz J (eds) Lasers in the conservation of artworks, pp 177–182

Majumdar AS, Mathew G (2015) Raman-infrared (IR) spectroscopy study of natural cordierites from Kalahandi, Odisha. J Geol Soc India 86(1):80–92

Majzlan J, Michallik R (2007) The crystal structures, solid solutions and infrared spectra of copiapite-group minerals. Miner Mag 71(5):553–569

Majzlan J, Zittlau AH, Grevel K-D, Schliesser J, Woodfield BF, Dachs E, Števko M, Chovan M, Plášil J, Sejkora J, Milovská S (2015) Thermodynamic properties and phase equilibria of the secondary copper minerals libethenite, olivenite, pseudomalachite, kröhnkite, cyanochroite, and devilline. Can Mineral 53(5):937–960

Makreski P, Jovanovski G (2009) Minerals from Macedonia: XXIII. Spectroscopic and structural characterization of schorl and beryl cyclosilicates. Spectrochim Acta A 73(3):460–467

Makreski P, Jovanovski G, Minceva-Sukarova B, Soptrajanov B, Green A, Engelen B, Grzetic I (2004) Vibrational spectra of $M^I_3M^{III}S_3$ type synthetic minerals (M^I = Tl or Ag and M^{III} = As or Sb). Vib Spectrosc 35(1):59–65

Makreski P, Jovanovski G, Dimitrovska S (2005a) Minerals from Macedonia: XIV. Identification of some sulfate minerals by vibrational (infrared and Raman) spectroscopy. Vib Spectrosc 39(2):229–239

Makreski P, Jovanovski G, Stojančeska S (2005b) Minerals from Macedonia: XIII. Vibrational spectra of some commonly appearing nesosilicate minerals. J Molec Struct 744:79–92

Makreski P, Jovanovski G, Gajović A (2006a) Minerals from Macedonia: XVII. Vibrational spectra of some common appearing amphiboles. Vib Spectrosc 40(1):98–109

Makreski P, Jovanovski G, Gajović A, Biljan T, Angelovski D, Jaćimović R (2006b) Minerals from Macedonia: XVI. Vibrational spectra of some common appearing pyroxenes and pyroxenoids. J Molec Struct 788(1):102–114

Makreski P, Jovanovski G, Kaitner B, Gajović A, Biljan T (2007) Minerals from Macedonia: XVIII. Vibrational spectra of some sorosilicates. Vib Spectrosc 44(1):162–170

Makreski P, Jovanovski G, Kaitner B (2009) Minerals from Macedonia: XXIV. Spectra-structure characterization of tectosilicates. J Molec Struct 924:413–419

Makreski P, Runčevski T, Jovanovski G (2011) Minerals from Macedonia: XXVI. Characterization and spectra–structure correlations for grossular and uvarovite. Raman study supported by IR spectroscopy. J Raman Spectrosc 42(1):72–77

Makreski P, Jovanovski S, Pejov L, Kloess G, Hoebler HJ, Jovanovski G (2013a) Theoretical and experimental study of the vibrational spectra of sarkinite $Mn_2(AsO_4)(OH)$ and adamite $Zn_2(AsO_4)(OH)$. Spectrochim Acta A 113:37–42

Makreski P, Petruševski G, Ugarković S, Jovanovski G (2013b) Laser-induced transformation of stibnite (Sb_2S_3) and other structurally related salts. Vib Spectrosc 68:177–182

Makreski P, Jovanovski G, Boev B (2014) Micro-Raman spectra of extremely rare and endemic Tl-sulfosalts from Allchar deposit. J Raman Spectrosc 45(7):610–617

Makreski P, Jovanovski S, Pejov L, Petruševski G, Ugarković S, Jovanovski G (2015a) Theoretical and experimental study of the vibrational spectra of liroconite, $Cu_2Al(AsO_4)(OH)_4\cdot 4H_2O$ and bayldonite, $Cu_3Pb[O(AsO_3OH)_2(OH)_2]$. Vib Spectrosc 79:36–43

Makreski P, Stefov S, Pejov L, Jovanovski G (2015b) Theoretical and experimental study of the vibrational spectra of (para)symplesite and hörnesite. Spectrochim Acta A 144:155–162

Makreski P, Stefov S, Pejov L, Jovanovski G (2017) Minerals from Macedonia: XXIX. Experimental and theoretical study of the vibrational spectra of extremely rare Tl-sulfate mineral from Allchar – dorallcharite. Vib Spectrosc 89:85–91

Malakho AP, Morozov VA, Pokholok KV, Lazoryak BI, Van Tendeloo G (2005) Layered ordering of vacancies of lead iron phosphate $Pb_3Fe_2(PO_4)_4$. Solid State Sci 7(4):397–404

Malcherek T, Mihailova B, Schlüter J, Husdal T (2012) Atelisite-(Y), a new rare earth defect silicate of the KDP structure type. Eur J Mineral 24(6):1053–1060

Malcherek T, Bindi L, Dini M, Ghiara MR, Molina Donoso A, Nestola F, Rossi M, Schlüter J (2014) Tondiite, $Cu_3Mg(OH)_6Cl_2$, the Mg-analogue of herbertsmithite. Mineral Mag 78:583–590

Malik V, Pokhriyal M, Uma S (2016) Single step hydrothermal synthesis of beyerite, $CaBi_2O_2(CO_3)_2$ for the fabrication of UV-visible light photocatalyst BiOI/$CaBi_2O_2(CO_3)_2$. RSC Adv 6(44):38252–38262

Malinina GA, Stefanovsky SV (2014) Structure and vibrational spectra of slags produced from radioactive waste. J Appl Spectrosc 81(2):200–204

Mammone JF, Nicol M, Sharma SK (1981) Raman spectra of TiO_2-II, TiO_2-III, SnO_2, and GeO_2 at high pressure. J Phys Chem Solids 42(5):379–384

Manara D, Grandjean A, Neuville DR (2009) Advances in understanding the structure of borosilicate glasses: a Raman spectroscopy study. Am Mineral 94(5–6):777–784

Manca SG, Baran EJ (1981) Characterization of the monoclinic form of praseodymium chromate (V). J Phys Chem Solids 42(10):923–925

Mancilla N, Caliva V, D'Antonio MC, González-Baró AC, Baran EJ (2009a) Vibrational spectroscopic investigation of the hydrates of manganese (II) oxalate. J Raman Spectrosc 40:915–920

Mancilla N, D'Antonio MC, González-Baró AC, Baran EJ (2009b) Vibrational spectra of lead (II) oxalate. J Raman Spectrosc 40(12):2050–2052

Manghnani MH, Vijayakumar V, Bass JD (1998) High-pressure Raman scattering study of majorite-garnet solid solutions in the system $Mg_4Si_4O_{12}$–$Mg_3Al_2Si_3O_{12}$. In: Manghnani MH, Yagi T (eds) Properties of earth and planetary materials at high pressure and temperature. Geophys Monogr 101:129–138

Manonmoni JV, Bhagavannarayana G, Ramasamy G, Meenakshisundaram S, Amutha M (2014) Growth, structure and spectral studies of a novel mixed crystal potassium zinc manganese sulphate. Spectrochim Acta A 117:9–12

Manoun B, Downs RT, Saxena SK (2006) A high-pressure Raman spectroscopic study of hafnon, $HfSiO_4$. Am Mineral 91(11–12):1888–1892

Manuella FC, Carbone S, Ottolini L, Gibilisco S (2012) Micro-Raman spectroscopy and SIMS characterization of oxykinoshitalite in an olivine nephelinite from the Hyblean Plateau (Sicily, Italy). Eur J Mineral 24(3):527–533

Mao Z, Jiang F, Duffy TS (2007) Single-crystal elasticity of zoisite $Ca_2Al_3Si_3O_{12}(OH)$ by Brillouin scattering. Am Mineral 92(4):570–576

Mao Z, Jacobsen SD, Frost DJ, McCammon CA, Hauri EH, Duffy TS (2011) Effect of hydration on the single-crystal elasticity of Fe-bearing wadsleyite to 12 GPa. Am Mineral 96(10):1606–1612

Mariappan CR, Govindaraj G, Ramya L, Hariharan S (2005) Synthesis, characterization and electrical conductivity studies on $A_3Bi_2P_3O_{12}$ (A = Na, K) materials. Mater Res Bull 40(4):610–618

Marinova D, Kostov V, Nikolova R, Kukeva R, Zhecheva E, Sendova-Vasileva M, Stoyanova R (2015) Fromkröhnkite – to alluaudite-type of structure: novel method of synthesis of sodium manganese sulfates with electrochemical properties in alkali-metal ion batteries. J Mater Chem A 3 (44):22287–22299

Markov YF, Roginskii EM (2011) Nanoclusters in mixed crystals Hg_2Hal_2. Bull Russ Acad Sci (Phys) 75 (10):1317–1323

Marshukova NK, Pavlovskii AB, Sidorenko GA (1984) Mushistonite, $(Cu,Zn,Fe)Sn(OH)_6$, a new tin mineral. Zapiski RMO (Proc Russ Miner Soc) 113(5):612–617. (in Russian)

Martens W, Frost RL, Kloprogge JT, Williams PA (2003a) Raman spectroscopic study of the basic copper sulphates – implications for copper corrosion and 'bronze disease'. J Raman Spectrosc 34(2):145–151

Martens WN, Frost RL, Kloprogge JT, Williams PA (2003b) The basic copper arsenate minerals olivenite, cornubite, cornwallite, and clinoclase: an infrared emission and Raman spectroscopic study. Am Mineral 88(4):501–508

Martens W, Frost RL, Williams PA (2003c) Molecular structure of the adelite group of minerals – a Raman spectroscopic study. J Raman Spectrosc 34:104–111

Martin SW, Bloyer DR (1991) Preparation and infrared characterization of thioborate compounds and polycrysts. J Am Ceram Soc 74(5):1003–1010

Martina I, Wiesinger R, Jembrih-Simbürger D, Schreiner M (2012) Micro-Raman characterization of silver corrosion products: instrumental set up and reference database. Raman spectrosc e-Preserv Sci 9:1–8

Martínez-Arkarazo I, Angulo M, Zuloaga O, Usobiaga A, Madariaga JM (2007) Spectroscopic characterisation of moonmilk deposits in Pozalagua tourist cave (Karrantza, Basque Country, north of Spain). Spectrochim Acta A 68(4):1058–1064

Martínez-Ramírez S, Fernández-Carrasco L (2011) Raman spectroscopy: application to cementitious systems. In: Doyle SG (ed) Construction and building: design, materials, and techniques. Nova Science. ISBN: 978-1-61761-371-5

Martins T, Chakhmouradian AR, Medici L (2014) Perovskite alteration in kimberlites and carbonatites: the perole of kassite, $CaTi_2O_4(OH)_2$. Phys Chem Minerals 41 (6):473–484

Martins T, Kressall R, Medici L, Chakhmouradian AR (2016) Cancrinite-vishnevite solid solution from Cinder Lake (Manitoba, Canada): crystal chemistry and implications for alkaline igneous rocks. Mineral Mag. https://doi.org/10.1180/minmag.2016.080.165

Mary SS, Kirupavathy SS, Mythili P, Srinivasan P, Kanagasekaran T, Gopalakrishnan R (2008) Studies on the growth, optical, electrical and spectral properties of potassium pentaborate (KB5) single crystals. Spectrochim Acta A 71(1):10–16

Masingboon C, Thongbai P, Maensiri S, Yamwong T, Seraphin S (2008) Synthesis and giant dielectric behavior of $CaCu_3Ti_4O_{12}$ ceramics prepared by polymerized complex method. Mater Chem Phys 109 (2):262–270

Masingboon C, Thongbai P, Maensiri S, Yamwong T (2009) Nanocrystalline $CaCu_3Ti_4O_{12}$ powder by PVA sol-gel route: synthesis, characterization and its giant dielectric constant. Appl Phys A 96(3): 595–602

Maslar JE, Hurst WS, Bowers WJ, Hendricks JH, Aquino MI, Levin I (2001) In situ Raman spectroscopic investigation of chromium surfaces under hydrothermal conditions. Appl Surf Sci 180(1):102–118

Massaferro A, Kremer E, Wagner CC, Baran EJ (1999) Vibrational spectra of $Pb_4(PO_4)_2SO_4$. J Raman Spectrosc 30(3):225–226

Mathlouthi M, Seuvre AM, Koenig JL (1986) FT-IR and laser-Raman spectra of guanine and guanosine. Carbohydrate Res 146(1):15–27

Mathur MS, Frenzel CA, Bradley EB (1968) New measurements of the Raman and the IR spectrum of $K_2Cr_2O_7$. J Molec Struct 2(6):429–435

Matović V, Erić S, Srećković-Batoćanin D, Colomban P, Kremenović A (2014) The influence of building materials on salt formation in rural environments. Environ Earth Sci 72(6):1939–1951

Matson DW, Sharma SK, Philpotts JA (1986) Raman spectra of some tectosilicates and of glasses along the orthoclase-anorthite and nepheline-anorthite joins. Am Mineral 71(5–6):694–704

Mattes R, Müller G, Becher HJ (1972) Schwingungsspektren und Struktur von Dioxotrifluoromolybdaten und -wolframaten. Z anorg allg Chemie 389 (2):177–187. (in German)

Maubec N, Lahfid A, Lerouge C, Wille G, Michel K (2012) Characterization of alunite supergroup minerals by Raman spectroscopy. Spectrochim Acta A 96:925–939

Mayerhöfer TG, Dunken HH (2001) Single-crystal IR spectroscopic investigation on fresnoite, Sr-fresnoite and Ge-fresnoite. Vib Spectrosc 25(2):185–195

Mazzetti L, Thistlethwaite PJ (2002) Raman spectra and thermal transformations of ferrihydrite and schwertmannite. J Raman Spectrosc 33(2):104–111

McBride JR, Hass KC, Weber WH (1991) Resonance-Raman and lattice-dynamics studies of single-crystal PdO. Phys Rev B 44(10):5016–5028

McDonald AM, Proenza JA, Zaccarini F, Rudashevsky NS, Cabri LJ, Stanley CJ, Rudashevsky VN, Melgarejo JC, Lewis JF, Longo F, Bakker RJ (2010) Garutiite, (Ni,Fe,Ir), a new hexagonal polymorph of native Ni from Loma Peguera, Dominican Republic. Eur J Mineral 22(2):293–304

McDonald AM, Back ME, Gault RA, Horváth L (2013) Peatite-(Y) and ramikite-(Y), two new Na-Li-Y±Zr phosphate-carbonate minerals from the Poudrette Pegmatite, Mont Saint-Hilaire, Quebec. Can Mineral 51(4):569–596

McDonald AM, Tarassoff P, Chao GY (2015) Hogarthite, $(Na,K)_2CaTi_2Si_{10}O_{26} \cdot 8H_2O$, a new member of the lemoynite group from Mont Saint-Hilaire, Quebec: characterization, crystal-structure determination, and origin. Can Mineral 53(1):13–30

McKendry IG, Thenuwara AC, Sun J, Peng H, Perdew JP, Strongin DR, Zdilla MJ (2016) Water oxidation catalyzed by cobalt oxide supported on the mattagamite phase of $CoTe_2$. ACS Catal 6:7393–7397

McKeown DA (2005) Raman spectroscopy and vibrational analyses of albite: from 25 °C through the melting temperature. Am Mineral 90:1506–1517

McKeown DA (2008) Raman spectroscopy, vibrational analysis, and heating of buergerite tourmaline. Phys Chem Minerals 35(5):259–270

McKeown DA, Bell MI (1998) Linked four-membered silicate rings: vibrational analysis of gillespite. Phys Chem Minerals 25(4):273–281

McKeown DA, Kim CC, Bell MI (1995) Vibrational analysis of dioptase $Cu_6Si_6O_{18} \cdot 6(H_2O)$ and its puckered six-membered ring. Phys Chem Minerals 22(3):137–144

McKeown DA, Bell MI, Etz ES (1999) Raman spectra and vibrational analysis of the trioctahedral mica phlogopite. Am Mineral 84(5–6):970–976

McKeown DA, Post JE, Etz ES (2002) Vibrational analysis of palygorskite and sepiolite. Clays Clay Minerals 50(5):666–679

McMaster S, Ram R, Charalambous F, Tardio J, Bhargava S (2013) Characterisation studies on natural and heat treated betafite. Chemeca 2013: Challenging Tomorrow, pp 529–534

McMaster SA, Ram R, Charalambous F, Pownceby MI, Tardio J, Bhargava SK (2014) Synthesis and characterisation of the uranium pyrochlore betafite $[(Ca,U)_2(Ti,Nb,Ta)_2O_7]$. J Hazard Mater 280:478–486

McMaster SA, Ram R, Pownceby MI, Tardio J, Bhargava S (2015) Characterisation and leaching studies on the uranium mineral betafite $[(U,Ca)_2(Nb,Ti,Ta)_2O_7]$. Minerals Eng 81:58–70

McMillan P, Akaogi M, Ohtani E, Williams Q, Nieman R, Sato R (1989) Cation disorder in garnets along the $Mg_3Al_2Si_3O_{12}$-$Mg_4Si_4O_{12}$ join: an infrared, Raman and NMR study. Phys Chem Minerals 16:428–435

McPherson GL, Chang JR (1973) Infrared and structural studies of $M^IM^{III}X_3$ type transition metal halides. Inorg Chem 12(5):1196–1198

Mees F, Hatert F, Rowe R (2008) Omongwaite, $Na_2Ca_5(SO_4)_6 \cdot 3H_2O$, a new mineral from recent salt lake deposits, Namibia. Mineral Mag 72(6):1307–1318

Melendres CA, Tani BS (1986) Evidence for the presence of sulfate in the passive film on nickel in concentrated sulfuric acid solution. J Electrochem Soc 133(5):1059–1060

Melghit K, Al-Mungi AS (2007) New form of iron orthovanadate $FeVO_4 \cdot 1.5H_2O$ prepared at normal pressure and low temperature. Mater Sci Eng B 136(2):177–181

Melghit K, Al-Belushi AK, Al-Amri I (2007) Short reaction time preparation of zinc pyrovanadate at normal pressure. Ceram Int 33(2):285–288

Menezes Filho LAD, Chaves MLDSC, Dias CH, Atencio D (2016) Recent mineral discoveries in the Coronel Murta, Taquaral, and Medina pegmatite fields, northeastern Minas Gerais, Brazil. REM-Int Eng J 69(3):301–307

Menezes Filho LAD, Chaves MLSC, Chukanov NV, Atencio D, Scholz R, Pekov I, da Costa GM, Morrison SM, Andrade MB, Freitas ETF, Downs RT, Belakovskiy DI (2017) Parisite-(La), $CaLa_2(CO_3)_3F_2$, a new mineral from Novo Horizonte, Bahia, Brazil. Mineral Mag 82(1):133–144

Menezes WG, Reis DM, Benedetti TM, Oliveira MM, Soares JF, Torresi RM, Zarbin AJ (2009) V_2O_5 nanoparticles obtained from a synthetic barianditelike vanadium oxide: synthesis, characterization and electrochemical behavior in an ionic liquid. J Colloid Interface Scie 337(2):586–593

Meng JF, Rai BK, Katiyar RS, Bhalla AS (1997) Raman investigation on $(Ta_2O_5)_{1-x}(TiO_2)_x$ system at different temperatures and pressures. J Phys Chem Solids 58(10):1503–1506

Meng L, Burris S, Bui H, Pan WP (2005) Development of an analytical method for distinguishing ammonium bicarbonate from the products of an aqueous ammonia CO_2 scrubber. Anal Chem 77(18):5947–5952

Mer A, Obbade S, Rivenet M, Renard C, Abraham F (2012) $[La(UO_2)V_2O_7][(UO_2)(VO_4)]$ the first lanthanum uranyl-vanadate with structure built from two types of sheets based upon the uranophane aniontopology. J Solid State Chem 185:180–186

Merkel S, Goncharov AF, Mao HK, Gillet P, Hemley RJ (2000) Raman spectroscopy of iron to 152 gigapascals: implications for Earth's inner core. Science 288(5471):1626–1629

Merkle RKW, Pikl R, Verryn SMC, Waal DD (1999) Raman spectra of synthetic 'braggite', (Pd,Pt,Ni)S. Mineral Mag 63(3):363–363

Mernagh TP, Hoatson DM (1995) A laser-Raman microprobe study of platinum-group minerals from the Munni Munni layered intrusion, West Pilbara Block, Western Australia. Can Mineral 33:409–417

Mernagh TP, Liu LG (1991) Raman spectra from the Al_2SiO_5 polymorphs at high pressures and room temperature. Phys Chem Minerals 18(2):126–130

Mernagh TP, Trudu AG (1993) A laser Raman microprobe study of some geologically important sulphide minerals. Chem Geol 103(1–4):113–127

Mernagh TP, Liu LG, Lin CC (1999) Raman spectra of chondrodite at various temperatures. J Raman Spectrosc 30(10):963–969

Mernagh TP, Kamenetsky VS, Kamenetsky MB (2011) A Raman microprobe study of melt inclusions in kimberlites from Siberia, Canada, SW Greenland and South Africa. Spectrochim Acta A 80(1):82–87

Mesbah A, Szenknect S, Clavier N, Lozano-Rodriguez J, Poinssot C, Den Auwer C, Ewing RC, Dacheux N (2015) Coffinite, $USiO_4$, is abundant in nature: so why is it so difficult to synthesize? Inorg Chem 54(14):6687–6696

Meyer HW, Zhang M, Bismayer U, Salje EKH, Schmidt C, Kek S, Morgenroth W, Bleser T (1996) Phase transformation of natural titanite: an infrared, Raman spectroscopic, optical birefringence and X-ray diffraction study. Phase Transit 59(1–3):39–60

Michel JP, Beauverger M, Museur L, Kanaev A, Petitet JP (2007) High-pressure cell for studies of ultra-short laser impulses action on materials. High Press Res 27(3):353–359

Michiba K, Miyawaki R, Minakawa T, Terada Y, Nakai I, Matsubara S (2013) Crystal structure of hydroxyl-bastnäsite-(Ce) from Kamihouri, Miyazaki Prefecture, Japan. J Mineral Petrol Sci 108(6):326–334

Middleton AW, Förster H-J, Uysal IT, Golding SD, Rhede D (2013) Accessory phases from the Soultzmonzo granite, Soultz-sous-Forêts, France: implications for titanite destabilisation and differential *REE*, Y and Th mobility in hydrothermal systems. Chem Geol 335:105–117

Mikkelsen A, Andersen AB, Engelsen SB, Hansen HCB, Larsen O, Skibsted LH (1999) Presence and dehydration of ikaite, calcium carbonate hexahydrate, in frozen shrimp shell. J Agric Food Chem 47(3):911–917

Mikuli E, Hetmańczyk Ł, Medycki W, Kowalska A (2007) Phase transitions and molecular motions in $[Zn(NH_3)_4](BF_4)_2$ studied by nuclear magnetic resonance, infrared and Raman spectroscopy. J Phys Chem Solids 68(1):96–103

Milekhin A, Sveshnikova L, Duda T, Surovtsev N, Adichtchev S, Zahn DRT (2011) Optical phonons in nanoclusters formed by the Langmuir-Blodgett technique. Chin J Phys 49(1):63–70

Milenov TI, Tenev T, Miloushev I, Avdeev GV, Luo CW, Chou WC (2014) Preliminary studies of the Raman spectra of Ag_2Te and Ag_5Te_3. Opt Quant Electron 46:573–580

Miller FA, Wilkins CH (1952) Infrared spectra and characteristic frequencies of inorganic ions, their use in qualitative analysis. Analyt Chem 24(8):1253–1294

Miller KH, Stephens PW, Martin C, Constable E, Lewis RA, Berger H, Carr GL, Tanner DB (2012) Infrared phonon anomaly and magnetic excitations in single-crystal $Cu_3Bi(SeO_3)_2O_2Cl$. Phys Rev B 86(17):174104. (12 pp)

Mills SJ, Frost RL, Kloprogge JT, Weier ML (2005) Raman spectroscopy of the mineral rhodonite. Spectrochim Acta A 62(1):171–175

Mills SJ, Groat LA, Wilson SA, Birch WD, Whitfield PS, Raudsepp M (2008) Angastonite, $CaMgAl_2(PO_4)_2(OH)_4 \cdot 7H_2O$: a new phosphate mineral from Angaston, South Australia. Mineral Mag 72(5):1011–1020

Mills SJ, Kolitsch U, Miyawaki R, Groat LA, Poirier G (2009) Joëlbruggerite, $Pb_3Zn_3(Sb^{5+},Te^{6+})As_2O_{13}(OH, O)$, the Sb^{5+} analog of dugganite, from the Black Pine mine, Montana. Am Mineral 94(7):1012–1017

Mills SJ, Kampf AR, Kolitsch U, Housley RM, Raudsepp M (2010) The crystal chemistry and crystal structure of kuksite, $Pb_3Zn_3Te^{6+}P_2O_{14}$, and a note on the crystal structure of yafsoanite, $(Ca,Pb)_3Zn(TeO_6)_2$. Am Mineral 95(7):933–938

Mills SJ, Kampf AR, Sejkora J, Adams PM, Birch WD, Plášil J (2011a) Iangreyite: a new secondary phosphate mineral closely related to perhamite. Mineral Mag 75(2):327–336

Mills SJ, Whitfield PS, Wilson SA, Woodhouse JN, Dipple GM, Raudsepp M, Francis CA (2011b) The crystal structure of stichtite, re-examination of barbertonite, and the nature of polytypism in MgCr hydrotalcites. Am Mineral 96(1):179–187

Mills SJ, Kartashov PM, Ma C, Rossman GR, Novgorodova MI, Kampf AR, Raudsepp M (2011c) Yttriaite-(Y): the natural occurrence of Y_2O_3 from the Bol'shayaPol'yariver, Subpolar Urals, Russia. Am Mineral 96(7):1166–1170

Mills SJ, Kampf AR, McDonald AM, Favreau G, Chiappero PJ (2012a) Forêtite, a new secondary arsenate mineral from the Cap Garonne mine, France. Mineral Mag 76(3):769–775

Mills SJ, Sejkora J, Kampf AR, Grey IE, Bastow TJ, Ball NA, Adams PM, Raudsepp M, Cooper MA (2012b) Krásnoite, the fluorophosphate analogue of perhamite, from the Huber open pit, Czech Republic and the Silver Coin mine, Nevada, USA. Miner Mag 76(3):625–634

Mills SJ, Kampf AR, McDonald AM, Bindi L, Christy AG, Kolitsch U, Favreau G (2013) The crystal structure of parnauite: a copper arsenate–sulphate with translational disorder of structural rods. Eur J Mineral 25(4):693–704

Mills SJ, Christy AG, Schnyder C, Favreau G, Price JR (2014a) The crystal structure of cameroraite and structural variation in the cyanotrichite family of merotypes. Mineral Mag 78(7):1527–1552

Mills SJ, Kampf AR, Christy AG, Housley RM, Rossman GR, Reynolds RE, Marty J (2014b) Bluebellite and mojaveite, two new minerals from the central Mojave Desert, California, USA. Mineral Mag 78(5):1325–1340

Mills SJ, Kampf AR, Christy AG, Housley RM, Thorne B, Chen YS, Steele IM (2014c) Favreauite, a new selenite mineral from the El Dragón mine, Bolivia. Eur J Mineral 26(6):771–781

Mills SJ, Christy AG, Kampf AR, Birch WD, Kasatkin A (2017a) Hydroxykenoelsmoreite, the first new mineral from the Republic of Burundi. Eur J Minerals. https://doi.org/10.1127/ejm/2017/0029-2618

Mills SJ, Christy AG, Rumsey MS, Spratt J, Bittarello E, Favreau G, Ciriotti ME, Berbain C (2017b) Hydroxyferroroméite, a new secondary weathering mineral from Oms, France. Eur J Mineral. https://doi.org/10.1127/ejm/2017/0029-2594

Minceva-Sukarova B, Jovanovski G, Makreski P, Soptrajanov B, Griffith W, Willis R, Grzetic I (2003) Vibrational spectra of $M^IM^{III}S_2$ type synthetic minerals (M^I = Tl or Ag and M^{III} = As or Sb). J Molec Struct 651–653:181–189

Minch R, Seoung DH, Ehm L, Winkler B, Knorr K, Peters L, Borkowski LA, Parise JB, Lee Y, Dubrovinsky L, Depmeier W (2010) High-pressure behavior of otavite ($CdCO_3$). J Alloys Compd 508 (2):251–257

Mingsheng P, Mao HK, Dien L, Chao ECT (1994) Raman spectroscopy of garnet-group minerals. Chin J Geochem 13(2):176–183

Mirgorodsky AP, Baraton MI, Quintard P (1989) Lattice dynamics of silicon oxynitride, Si_2N_2O: vibrational spectrum, elastic and piezoelectric properties. J Phys Condens Matter 1(50):10053–10066

Mirgorodsky AP, Merle-Méjean T, Champarnaud JC, Thomas P, Frit B (2000) Dynamics and structure of TeO_2 polymorphs: model treatment of paratellurite and tellurite; Raman scattering evidence for new γ-and δ-phases. J Phys Chem Solids 61(4):501–509

Mironova-Ulmane N, Kuzmin A, Grube M (2009) Raman and infrared spectromicroscopy of manganese oxides. J Alloys Compd 480(1):97–99

Mishra B, Bernhardt H-J (2009) Metamorphism, graphite crystallinity, and sulfide anatexis of the Rampura–Agucha massive sulfide deposit, northwestern India. Mineral Depos 44(2):183–204

Mitchell RH, Welch MD, Kampf AR, Chakhmouradian AK, Spratt J (2015) Barrydawsonite-(Y), $Na_{1.5}Y_{0.5}CaSi_3O_9H$: a new pyroxenoid of the pectolite–serandite group. Mineral Mag 79(3):671–686

Miyahara M, Kaneko S, Ohtani E, Sakai T, Nagase T, Kayama M, Nishido H, Hirao N (2013) Discovery of seifertite in a shocked lunar meteorite. Nat Commun 4:1737. (7 pp)

Młynarska M, Manecki M, Bajda T (2014) Structural and Raman spectroscopy studies of schultenite – phosphoschultenite isomorphic series. Geol Geophys Environ 40(1):110–111

Moëlo Y, Lasnier B, Palvadeau P, Léone P, Fontan F (2000) La lulzacite, $Sr_2Fe^{2+}(Fe^{2+},Mg)_2Al_4(PO_4)_4(OH)_{10}$, un nouveau phosphate de strontium (Saint-Aubin-des-Châteaux, Loire-Atlantique, France). Compt Rend l'Académie Sci IIA 330(5):317–324. (in French)

Moëlo Y, Lulzac Y, Rouer O, Palvadeau P, Gloaguen É, Léone P (2002) Scandium mineralogy: pretulite with scandian zircon and xenotime-(Y) within an apatite-rich oolitic ironstone from Saint-Aubin-Des-Châteaux, Armorican Massif, France. Can Mineral 40 (6):1657–1673

Moenke H (1962) Mineralspektren I. Akademie Verlag, Berlin. (in German)

Mohanan K (1993) Vibrational spectroscopic studies of olivines, pyroxenes, and amphiboles at hightemperature and pressures. Dissertation, University of Hawaii

Mohanan K, Sharma SK, Bishop FC (1993) A Raman spectral study of forsterite-monticellite solid solutions. Am Mineral 78:42–48

Molina-Mendoza AJ, Giovanelli E, Paz WS, Niño MA, Island JO, Evangeli C, Island JO, Evangeli C, Aballe L, Foerster M, van der Zant HSJ, Rubio-Bollinger G, Agraït N, Palacios JJ, Pérez EM, Castellanos-Gomez A (2016) Franckeite: a naturally occurring van der Waals heterostructure. arXiv preprint arXiv:1606.06651

Momma K, Ikeda T, Nishikubo K, Takahashi N, Honma C, Takada M, Furukawa Y, Nagase T, Kudoh Y (2011) New silica clathrate minerals that are isostructural with natural gas hydrates. Nat Commun 2:196. (7 pp)

Moore RK, White WB (1970) Study of order-disorder in rock-salt-related structures by infrared spectroscopy. J Amer Ceram Soc 53(12):679–682

Moreira RL, Rubinger CPL, Krambrock K, Dias A (2010a) Polarized Raman scattering and infrared spectroscopy of a natural manganocolumbite single crystal. J Raman Spectrosc 41(9):1044–1049

Moreira RL, Teixeira NG, Andreeta MRB, Hernandes AC, Dias A (2010b) Polarized micro-Raman scattering of $CaNb_2O_6$ single crystal fibers obtained by laser heated pedestal growth. Cryst Growth Des 10 (4):1569–1573

Morillas H, Maguregui M, Paris C, Bellot-Gurlet L, Colomban P, Madariaga JM (2015) The role of marine aerosol in the formation of (double) sulfate/nitrate salts in plasters. Microchem J 123:148–157

Morillas H, Maguregui M, García-Florentino C, Marcaida I, Madariaga JM (2016) Study of particulate matter from Primary/Secondary Marine Aerosol and anthropogenic sources collected by a self-made passive sampler for the evaluation of the dry deposition impact on built heritage. Sci Total Environ 550:285–296

Moriyama T, Miyawaki R, Yokoyama K, Matsubara S, Hirano H, Murakami H, Watanabe Y (2011) Wakefieldite-(Nd), a new neodymium vanadate mineral in the Arase stratiform ferromanganese deposit, Kochi Prefecture, Japan. Resour Geol 61(1):101–110

Moroz T, Shcherbakova E, Kostrovsky V (2004) Vibrational spectra of kladnoite, natural analogue of phthalimide $C_6H_4(CO)_2NH$. Mitt Österr Miner Ges 149:73

Morris RE, Harrison WT, Stucky GD, Cheetham AK (1991) The syntheses and crystal structures of two novel aluminum selenites, $Al_2(SeO_3)_3 \cdot 6H_2O$ and $AlH(SeO_3)_2 \cdot 2H_2O$. J Solid State Chem 94(2):227–235

Morrison SM, Downs RT, Yang H (2012) Redetermination of kovdorskite, $Mg_2(PO_4)(OH)\cdot 3H_2O$. Acta Crystallogr E. https://doi.org/10.1107/S1600536812000256

Morrison SM, Domanik KJ, Origlieri MJ, Downs RT (2013) Agardite-(Y), $Cu^{2+}{}_6Y(AsO_4)_3(OH)_6\cdot 3H_2O$. Acta Crystallogr E 69(9):i61–i62

Morrison SM, Domanik KJ, Yang H, Downs RT (2016) Petersite-(Ce), $Cu^{2+}{}_6Ce(PO_4)_3(OH)_6\cdot 3H_2O$, a new mixite group mineral from Yavapai County, Arizona, USA. Can Mineral 54(6):1505–1511

Morss LR (1974) Crystal structure of dipotassium sodium fluoroaluminate (elpasolite). J Inorg Nucl Chem 36(12):3876–3878

Mouri T, Enami M (2008) Raman spectroscopic study of olivine-group minerals. J Mineral Petrol Sci 103(2):100–104

Mozgawa W (2001) The relation between structure and vibrational spectra of natural zeolites. J Molec Struct 596(1):129–137

Müller B, Lutz HD (1991) Single crystal Raman studies of pyrite-type RuS_2, $RuSe_2$, OsS_2, $OsSe_2$, PtP_2, and $PtAs_2$. Phys Chem Minerals 17(8):716–719

Muller O, White WB, Roy R (1969a) Infrared spectra of the chromates of magnesium, nickel and cadmium. Spectrochim Acta A 25(8):1491–1499

Muller O, White WB, Roy R (1969b) X-ray diffraction study of the chromates of nickel, magnesium and cadmium. Z Krist 130:112–120

Müller K, Ciminelli VS, Dantas MSS, Willscher S (2010) A comparative study of As (III) and As (V) in aqueous solutions and adsorbed on iron oxy-hydroxides by Raman spectroscopy. Water Res 44(19):5660–5672

Müller D, Hochleitner R, Fehr KT (2015) Raman spectroscopic investigations of natural jennite from Maroldsweisach, Bavaria, Germany. J Raman Spectrosc. https://doi.org/10.1002/jrs.4865

Munce CG, Parker GK, Holt SA, Hope GA (2007) A Raman spectroelectrochemical investigation of chemical bath deposited Cu_xS thin films and their modification. Colloids Surf A 295(1):152–158

Muniz-Miranda M, Sbrana G, Bonazzi P, Menchetti S, Pratesi G (1996) Spectroscopic investigation and normal mode analysis of As_4S_4 polymorphs. Spectrochim Acta A 52(11):1391–1401

Muraki N, Katagiri G, Sergo V, Pezzotti G, Nishida T (1997) Mapping of residual stresses around an indentation in β-Si_3N_4 using Raman spectroscopy. J Mater Sci 32(20):5419–5423

Murli C, Sharma SM, Kulshreshtha SK, Sikka SK (1997) High pressure study of phase transitions in α-$FePO_4$. Pramana – J Phys 49(3):285–291

Murphy PJ, Smith AM, Hudson-Edwards KA, Dubbin WE, Wright K (2009) Raman and IR spectroscopic studies of alunite-supergroup compounds containing Al, Cr^{3+}, Fe^{3+} and V^{3+} at the B site. Can Mineral 47(3):663–681

Murshed MM, Mendive CB, Curti M, Šehović M, Friedrich A, Fischer M, Gesing TM (2015) Thermal expansion of mullite-type $Bi_2Al_4O_9$: a study by X-ray diffraction, vibrational spectroscopy and density functional theory. J Solid State Chem 229:87–96

Murthy TSN, Srinivas V, Saibabu G, Salagram M (1992) Structural distortions in $CrO_4{}^{2-}$ ion in $3CdSO_4\cdot 8H_2O$ crystals from IR studies. J Solid State Chem 97(2):358–365

Murugan R, Ghule A, Chang H (2000) Thermo-Raman spectroscopic studies on polymorphism in Na_2SO_4. J Phys Condens Matter 12(5):677–700

Musić S, Dragčević Đ, Maljković M, Popović S (2003) Influence of chemical synthesis on the crystallization and properties of zinc oxide. Mater Chem Phys 77(2):521–530

Musumeci A, Frost RL (2007) A spectroscopic and thermoanalytical study of the mineral hoganite. Spectrochim Acta A 67(1):48–57

Mutschke H, Min M, Tamanai A (2013) Laboratory-based grain-shape models for simulating dust infrared spectra. Astron Astrophys manuscr No. 12267, 8 pp

Mycroft JR, Bancroft GM, McIntyre NS, Lorimer JW, Hill IR (1990) Detection of sulphur and polysulphides on electrochemically oxidized pyrite surfaces by X-ray photoelectron spectroscopy and Raman spectroscopy. J Electroanal Chem Interfacial Electrochem 292(1–2):139–152

Nabar MA, Mhatre BG (2001) Studies on triple orthovanadates: VIII. Synthesis and spectrostructural characterization of triple orthovanadates $BaLnTh(VO_4)_3$ (Ln = La or Pr) and $BaLnCe(VO_4)3$ (Ln = La, Pr, Nd or Sm). J Alloys Compd 323:83–85

Nabar MA, Sakhardande RR (1985) Synthesis and crystal chemistry of triple orthoarsenates, $CaLnTh(AsO_4)_3$. J Crystallographic Spectroscopic Res 15(3):263–269

Naddari T, El Feki H, Savariault JM, Salles P, Salah AB (2003) Structure and ionic conductivity of the lacunary apatite $Pb_6Ca_2Na_2(PO_4)_6$. Solid State Ionics 158(1):157–166

Naemura K, Shimizu I, Svojtka M, Hirajima T (2015) Accessory priderite and burbankite in multiphase solid inclusions in the orogenic garnet peridotite from the Bohemian Massif, Czech Republic. J Mineral Petrol Sci 110:20–28

Naïli H, Mhiri T, Jaud J (2001) Crystal structure and characterization of $CsH_5(AsO_4)_2$: a new cesium pentahydrogen arsenate, and comparison with $CsH_5(PO_4)_2$ and $RbH_5(AsO_4)_2$. J Solid State Chem 161(1):9–16

Nakagawa T, Kihara K, Harada K (2001) The crystal structure of low melanophlogite. Am Miner 86:1506–1512

Nakajima A, Yoshihara A, Ishigame M (1994) Defect-induced Raman spectra in doped CeO_2. Phys Rev B 50(18):13297–13307

Nakamoto K (2009) Infrared and Raman spectra of inorganic and coordination compounds, 6th edn. Wiley, New York

Nakashima S, Mishima H, Mitsuishi A (1973) Raman scattering in yellow and red mecury iodides. J Raman Spectrosc 1(4):325–340

Nanba T, Kawashima I, Ikezawa M (1987) Far-infrared and Raman scattering spectra due to the two-photon processes in thallous halide crystals. J Phys Soc Jpn 50(9):3063–3070

Narang SN, Patel ND, Kartha VB (1994) Infrared and Raman spectral studies and normal modes of a-B_2O_3. J Molec Struct 327:221–235

Narasimham KV, Girija M (1967) Absorption spectrum of potassium uranyl sulphate. Defence Sci J 17(2):95–106

Narasimhulu KV, Sunandana CS, LakshmanaRao J (2000) Spectroscopic studies of Cu^{2+} ions doped in $KZnClSO_4 \cdot 3H_2O$ crystals. Phys Status Solidi B 217(2):991–997

Narsimhulu M, Saritha A, Raju B, Hussain KA (2015) Synthesis, structure, optical, thermal, dielectric and magnetic properties of cation deficient $K_{2.72}[Fe(C_2O_4)_3]\cdot 3.17H_2O$ crystals. Int J Innov Res Sci Eng Technol 4:7548–7555

Nasdala L, Pekov IV (1993) Ravatite, $C_{14}H_{10}$, a new organic mineral species from Ravat, Tadzhikistan. Eur J Mineral 5:699–705

Nasdala L, Witzke T, Ullrich B, Brett R (1998) Gordaite $NaZn_4(SO_4)(OH)_6Cl \cdot 6H_2O$: second occurrence in the Juan de Fuca Ridge, and new data. Am Mineral 83(9–10):1111–1116

Nasdala L, Smith DC, Kaindl R, Ziemann MA (2004) Raman spectroscopy: analytical perspectives in mineralogical research (Chapter 7). In: Beran A, Libowitsky E (eds) Spectroscopic methods in mineralogy, vol 6. EMU notes in mineralogy, pp 281–343

Nasdala L, Wildner M, Wirth R, Groschopf N, Pal DC, Möller A (2006) Alpha particle haloes in chlorite and cordierite. Mineral Petrol 86(1–2):1–27

Nasdala L, Blaimauer D, Chanmuang C, Corfu F, Lengauer CL, Ruschel K, Škoda R, Wirth R, Zeug M, Zoysa G (2016) Ekanite ($Ca_2ThSi_8O_{20}$) from Sri Lanka: concordant U–Th–Pb isotope system in spite of metamictization. In: Gadas P, Plášil J, Laufek F (eds) New minerals and mineralogy in the 21st century. International Scientific Symposium Jáchymov 2016, Book of abstracts and Fieldtrip guidebook, pp 59–62

Nassau K, Cooper AS, Shiever JW, Prescott BE (1973) Transition metal iodates. III. Gel growth and characterization of six cupric iodates. J Solid State Chem 8(3):260–273

Natkaniec-Nowak L, Dumańska-Słowik M, Ertl A (2009) "Watermelon" tourmaline from the Paprok mine (Nuristan, Afghanistan). N Jb Miner Abh 186(2):185–193

Nayak M, Kutty TRN (1998) Luminescence of Fe^{3+} doped $NaAlSiO_4$ prepared by gel to crystallite conversion. Mater Chem Phys 57(2):138–146

Nazarov M (2010) Raman spectroscopy of double activated $YNbO_4:Eu^{3+}, Tb^{3+}$ and $YTaO4:Eu^{3+}, Tb^{3+}$. Mold J Phys Sci 9(1):5–15

Nazarov MA, Demidova SI, Anosova MO, Kostitsyn YA, Ntaflos T, Brandstaetter F (2012) Native silicon and iron silicides in the Dhofar 280 lunar meteorite. Petrology 20(6):506–519

Nefedov EI, Griffin WL, Kristiansen R (1977) Minerals of the schoenfliesite – wickmanite series from Pitkäranta, Karelia, U.S.S.R. Can Mineral 15:437–445

Nemanich RJ (1986) Raman spectroscopy for semiconductor thin film analysis. Proc Mater Res Soc 69:23–37

Nestola F, Guastoni A, Cámara F, Secco L, Dal Negro A, Pedron D, Beran A (2009) Aluminocerite-Ce: a new species from Baveno, Italy: description and crystal-structure determination. Am Mineral 94(4):487–493

Nestola F, Mittempergher S, Di Toro G, Zorzi F, Pedron D (2010) Evidence of dmisteinbergite (hexagonal form of $CaAl_2Si_2O_8$) in pseudotachylyte: a tool to constrain the thermal history of a seismic event. Am Mineral 95(2–3):405–409

Nestola F, Burnham AD, Peruzzo L, Tauro L, Alvaro M, Walter MJ, Gunter M, Anzolini C, Kohn S (2016) Tetragonal Almandine-Pyrope Phase, TAPP: finally a name for it, the new mineral jeffbenite. Mineral Mag. https://doi.org/10.1180/minmag.2016.080.059

Neuville DR, Cormier L, Flank AM, Massiot D (2002) XANES and Raman spectrometry on glasses and crystals in the CAS system. Rep EMPG IX, Zurich, Switzerland, 24–27 March 2002

Newhouse P, Boyd DA, Shinde A, Guevarra D, Zhou L, Soedarmadji E, Li G, Neaton J, Gregoire J (2016) Solar fuel photoanodes prepared by inkjet printing of copper vanadates. J Mater Chem A. https://doi.org/10.1039/C6TA01252C

Nguyen SD, Halasyamani PS (2012) Synthesis, structure, and characterization of new Li^+-d^0-lone-pair-oxides: noncentrosymmetric polar $Li_6(Mo_2O_5)_3(SeO_3)_6$ and centrosymmetric $Li_2(MO_3)(TeO_3)(M = Mo^{6+}$ or $W^{6+})$. Inorg Chem 51(17):9529–9538

Nguyen N, Choisnet J, Raveau B (1980) Silicates synthétiques à structure milarite. J Solid State Chem 34:1–9

Nhlabathi TN (2012) An investigation into the fluorination capabilities of ammonium acid fluoride under microwave radiation with respect to zircon. Dissertation, University of Pretoria

Ni S, Wang X, Zhou G, Yang F, Wang J, He D (2010a) Hydrothermal synthesis and magnetic property of $Cu_3(OH)_2V_2O_7 \cdot nH_2O$. Mater Lett 64(4):516–519

Ni S, Wang X, Zhou G, Yang F, Wang J, He D (2010b) Crystallized $Zn_3(VO_4)_2$: synthesis, characterization and optical property. J Alloys Compd 491(1):378–381

Nicola JH, Rutt HN (1974) A comparative study of zircon ($ZrSiO_4$) and hafnon ($HfSiO_4$) Raman spectra. J Phys C 7(7):1381–1386

Nie L, Wang H, Chai Y, Liu S, Yuan R (2016) In situ formation of flower-like $CuCo_2S_4$ nanosheets/graphene composites with enhanced lithium storage properties. RSC Adv 6(44):38321–38327

Nien YT, Zaman B, Ouyang J, Chen IG, Hwang CS, Yu K (2008) Raman scattering for the size of CdSe and CdSnanocrystals and comparison with other techniques. Mater Lett 62(30):4522–4524

Nieuwoudt MK, Comins JD, Cukrowski I (2011) The growth of the passive film on iron in 0.05 M NaOH studied in situ by Raman micro-spectroscopy and electrochemical polarisation. Part I: Near-resonance enhancement of the Raman spectra of iron oxide and oxyhydroxide compounds. J Raman Spectrosc 42(6):1335–1339

Nilges T, Reiser S, Hong JH, Gaudin E, Pfitzner A (2002) Preparation, structural, Raman and impedance spectroscopic characterisation of the silver ion conductor $(AgI)_2Ag_3SbS_3$. Phys Chem Chem Phys 4(23):5888–5894

Nishio-Hamane D, Minakawa T, Okada H (2014) Iwateite, $Na_2BaMn(PO_4)_2$, a new mineral from the Tanohata mine, Iwate Prefecture, Japan. J Mineral Petrol Sci 109(1):34–37

Nishio-Hamane D, Momma K, Miyawaki R, Minakawa T (2016a) Bunnoite, a new hydrous manganese aluminosilicate from Kamo Mountain, Kochi prefecture, Japan. Mineral Petrol. https://doi.org/10.1007/s00710-016-0454-2

Nishio-Hamane D, Momma K, Ohnishi M, Shimobayashi N, Miyawaki R, Tomita N, Okuma R, Kampf AR, Minakawa T (2016b) Iyoite, $MnCuCl(OH)_3$, and misakiite, $Cu_3Mn(OH)_6Cl_2$: new members of the atacamite family from Sadamisaki Peninsula, Ehime Prefecture, Japan. Mineral Mag. https://doi.org/10.1180/minmag.2016.080.104

Nitta E, Kimata M, Hoshino M, Echigo T, Hamasaki S, Shinohara H, Nishida N, Hatta T, Shimizu M (2006) High-temperature volcanic sublimates from Iwodake volcano, Satsuma-Iwojima, Kyushu, Southwestern Japan. Jpn Mag Mineral Petrol Sci 35(6):270–281. (in Japanese)

Noguchi T, Nakamura T, Misawa K, Imae N, Aoki T, Toh S (2009) Laihunite and jarosite in the Yamato 00 nakhlites: alteration products on Mars? J Geophys Res 114:E10004. (13 pp)

Noguera O, Merle-Méjean T, Mirgorodsky AP, Smirnov MB, Thomas P, Champarnaud-Mesjard J-C (2003) Vibrational and structural properties of glass and crystalline phases of TeO_2. J Non-Cryst Solids 330:50–60

Nolas GS, Slack GA, Caillat T, Meisner GP (1996) Raman scattering study of antimony-based skutterudites. J Appl Phys 79(5):2622–2626

Noureldine D, Anjum DH, Takanae K (2014) Flux-assisted synthesis of $SnNb_2O_6$ for tuning photocatalytic properties. Phys Chem Chem Phys 16(22):10762–10769

Nuvoli D, Rassu M, Alzari V, Sanna R, Malucelli G, Mariani A (2014) Preparation and characterization of polymeric nanocomposites containing exfoliated tungstenite at high concentrations. Compos Sci Technol 96:97–102

Nyhus P, Cooper SL, Fisk Z (1995) Electronic Raman scattering across the unconventional charge gap in FeSi. Phys Rev B 51(21):15626–15629

Nytko EA (2008) Synthesis, structure, and magnetic properties of spin-1/2 kagomé antiferromagnets. Dissertation, Massachusetts Institute of Technology

O'Neill AE, Uy D, Jagner M (2006) Characterization of phosphates found in vehicle-aged exhaust gas catalysts: a Raman study. SAE Techn Paper No 2006-01-0410. (15 pp)

Obbade S, Dion C, Saad M, Yagoubi S, Abraham F (2004) $Pb(UO_2)(V_2O_7)$, a novel lead uranyl divanadate. J Solid State Chem 177(11):3909–3917

Oçafrain A, Chaminade JP, Viraphong O, Cavagnat R, Couzi M, Pouchard M (1996) Growth by the heat exchanger method and characterization of neighborite, $NaMgF_3$. J Cryst Growth 166(1):414–418

Ogawara T, Akai J (2014) Graphite-3R in a fault fracture zone associated with black jadeitite from Kanayamadani, Itoigawa, central Japan. J Mineral Petrol Sci 109(3):125–137

Ogawara T, Mashima H, Akai J (2010) Raman analysis of rengeite and orthorhombic polymorph of rengeite from Itoigawa region, central Japan. Ann Meet Jpn Assoc Mineral Sci. (in Japanese). https://doi.org/10.14824/jakoka.2010.0.99.0

Ohnishi M, Kobayashi S, Kusachi I (2002) Ktenasite from the Hirao mine at Minoo, Osaka, Japan. J Mineral Petrol Sci 97(4):185–189

Ohsaka T, Fujiki Y (1982) Raman spectra in hollandite type compounds $K_{1.6}Mg_{0.8}Ti_{7.2}O_{16}$ and $K_{1.6}Al_{1.6}Ti_{6.4}O_{16}$. Solid State Commun 44(8):1325–1327

Ohwada K (1972) Far-infrared spectrum of uranyl fluoride. J Inorg Nucl Chem 34(7):2357–2358

Ohwada K, Soga T, Iwasaki M (1972) The infrared spectrum of tripotassium uranyl fluoride. Spectrochim Acta A 28(5):933–938

Okada T, Narita T, Nagai T, Yamanaka T (2008) Comparative Raman spectroscopic study on ilmenite-type $MgSiO_3$ (akimotoite), $MgGeO_3$, and $MgTiO_3$ (geikielite) at high temperatures and high pressures. Am Mineral 93(1):39–47

Okotrub KA, Surovtsev NV (2013) Raman scattering evidence of hydrohalite formation on frozen yeast cells. Cryobiology 66(1):47–51

Olds TA, Plášil J, Kampf AR, Škoda R, Burns PC, Čejka J, Bourgoin V, Boulliard J-C (2016a) Gauthierite, $KPb[(UO_2)_7O_5(OH)_7]\cdot 8H_2O$, a new uranyl-oxide hydroxyhydrate mineral from Shinkolobwe with a novel uranyl-anion sheet-topology. Eur J Mineral. https://doi.org/10.1127/ejm/2017/0029-2586

Olds TA, Sadergaski LR, Plášil J, Kampf AR, Burns PC, Steele IM, Marty J, Carlson S, Mills OP (2016b) Leószilárdite, the first Na,Mg-containing uranyl carbonate from the Markey Mine, San Juan County, Utah, USA. Mineral Mag. https://doi.org/10.1180/minmag.2016.080.149

Onac BP, Effenberger H, Ettinger K, Cinta Panzaru S (2006) Hydroxylellestadite from Cioclovina Cave (Romania): microanalytical, structural, and vibrational spectroscopy data. Am Mineral 91(11–12):1927–1931

Onac BP, Kearns J, Breban R, Cîntǎ Pânzaru S (2012) Variscite ($AlPO_4\cdot 2H_2O$) from Cioclovina Cave (Sureanu Mountains, Romania): a tale of a missing phosphate. Studia UBB Geologia 49(1):3–14

Ondruš P, Veselovský F, Skála R, Sejkora J, Pažout R, Frýda J, Gabašová A, Vajdak J (2006) Lemanskiite, NaCaCu$_5$(AsO$_4$)$_4$Cl·5H$_2$O, a new mineral species from the Abundancia mine, Chile. Can Mineral 44:523–531

Ondruš P, Skála R, Plášil J, Sejkora J, Veselovský F, Čejka J, Kallistová A, Hloušek J, Fejfarová K, Škoda R, Dušek M, Gabasova A, Machovič V, Lapčák L (2013) Švenekite, Ca[AsO$_2$(OH)$_2$]$_2$, a new mineral from Jáchymov, Czech Republic. Mineral Mag 77(6):2711–2724

Ono A (1985) Phase relations in the system NH$_4$Zr$_2$(PO$_4$)$_3$ – (NH$_4$)$_3$M$_2$(PO$_4$)$_3$: M = Y, Al or In. J Mater Sci Lett 4(8):936–939

Ono H, Hosokawa Y, Shinoda K, Koyanagi K, Yamaguchi H (2001) Ta–O phonon peaks in tantalum oxide films on Si. Thin Solid Films 381(1):57–61

Onodera A, Liu X, Kyokane D, Kura K, Machida K, Adachi GY, Su W (1999) Pressure-induced amorphization of SrB$_2$O$_4$. J Phys Chem Solids 60(10):1737–1743

Orberger B, Vymazalova A, Wagner C, Fialin M, Gallien JP, Wirth R, Pasava J, Montagnac G (2007) Biogenic origin of intergrown Mo-sulphide-and carbonaceous matter in Lower Cambrian black shales (Zunyi Formation, southern China). Chem Geol 238(3):213–231

Origlieri MJ (2005) Crystal chemistry of selected Sb, As and P minerals. Dissertation, University of Arizona

Origlieri MJ, Downs RT, Pinch WW, Zito GL (2009) Stibioclaudetite AsSbO$_3$ a new mineral from Tsumeb, Namibia. Mineral Rec 40(3):209–214

Origlieri MJ, Yang H, Downs RT, Posner ES, Domanik KJ, Pinch WW (2012) The crystal structure of bartelkeite, with a revised chemical formula, PbFeGe^{VI}(Ge^{IV}$_2$O$_7$)(OH)$_2$·H$_2$O, isotypic with high-pressure P2$_1$/m lawsonite. Am Mineral 97(10):1812–1815

Origlieri MJJ, Downs RT, Yang H, Hoffman DR, Ducea MN, Post JE (2017) Marshallsussmanite, NaCaMnSi$_3$O$_8$(OH), a new pectolite group mineral providing insight into hydrogen bonding in pyroxenoids. Mineral Mag. https://doi.org/10.1180/minmag.2017.081.049

Orlandi P, Biagioni C, Moëlo Y, Langlade J, Faulques E (2015) Suseinargiuite, (Na$_{0.5}$Bi$_{0.5}$)(MoO$_4$), the Na-Bi analogue of wulfenite, from Su Seinargiu, Sardinia, Italy. Eur J Mineral 27(5):695–699

Orlandi P, Biagioni C, Zaccarini F (2017) Cabvinite, Th$_2$F$_7$(OH)·3H$_2$O, the first natural actinide halide. Am Mineral. https://doi.org/10.2138/am-2017-6013

Orman RG (2010) Characterisation of novel antimony (III) oxide-containing glasses. Dissertation, University of Warwick

Orman RG, Holland D, Hannon AC (2008) Antimony oxychloride glass and its relation to crystalline onoratoite, Sb$_8$O$_{11}$Cl$_2$. Phys Chem Glasses – Eur J Glass Sci Technol B 49(1):15–18

Orozbaev RT, Yoshida K, Bakirov AB, Hirajima T, Takasu A, Sakiev KS, Tagiri M (2011) Preiswerkite and högbomite within garnets of Aktyuzeclogite, Northern Tien Shan, Kyrgyzstan. J Mineral Petrol Sci 106(6):320–325

Ōsaka T (1971) Far-infrared absorption spectra of mercurous halides. J Chem Phys 54(3):863–867

Osaka A, Takahashi K, Ikeda M (1984) Infrared study of trivalent cations B and Fe in amorphous and crystalline phosphates. J Mater Sci Lett 3(1):36–38

Ospitali F, Bersani D, Di Lonardo G, Lottici PP (2008) 'Green earths': vibrational and elemental characterization of glauconites, celadonites and historical pigments. J Raman Spectrosc 39(8):1066–1073

Ossa FO, Hofmann A, Vidal O, Kramers JD, Belyanin G, Cavalazzi B (2016) Unusual manganese enrichment in the Mesoarchean Mozaan Group, Pongola Supergroup, South Africa. Precambrian Res 281:414–433

Ostrooumov M, Taran Y, Arellano-Jiménez M, Ponce A, Reyes-Gasga J (2009) Colimaite, K$_3$VS$_4$ – a new potassium-vanadium sulfide mineral from the Colima volcano, State of Colima (Mexico). Rev Mex Cienc Geol 26:600–608

Ostroumov M, Fritsch E, Faulques E, Chauvet O (2002) Etude spectrometrique de la lazurite du Pamir, Tajikistan. Can Mineral 40(3):885–893. (in French)

Ouerfelli N, Guesmi A, Molinié P, Mazza D, Zid MF, Driss A (2007) The iron potassium diarsenate KFe(As$_2$O$_7$) structural, electric and magnetic behaviors. J Solid State Chem 180(10):2942–2949

Ouerfelli N, Smida YB, Zid MF (2015) Synthesis, crystal structure and electrical properties of a new iron arsenate Na$_{2.77}$K$_{1.52}$Fe$_{2.57}$(AsO$_4$)$_4$. J Alloys Compd 651:616–622

Owen OS, Kung MC, Kung HH (1992) The effect of oxide structure and cation reduction potential of vanadates on the selective oxidative dehydrogenation of butane and propane. Catal Lett 12(1–3):45–50

Oyamada R (1974) Raman spectra of the fused PbCl$_2$-KCl system. J Phys Soc Jpn 36(3):903–905

Pachoud E, Zhang W, Tapp J, Liang KC, Lorenz B, Chu PC, Halasyamani PS (2013) Top-seeded single-crystal growth, structure, and physical properties of polar LiCrP$_2$O$_7$. Cryst Growth Design 13(12):5473–5480

Pagès-Camagna S, Colinart S, Coupry C (1999) Fabrication processes of archaeological Egyptian blue and green pigments enlightened by Raman microscopy and scanning electron microscopy. J Raman Spectrosc 30(4):313–317

Pagliai M, Bonazzi P, Bindi L, Muniz-Miranda M, Cardini G (2011) Structural and vibrational properties of arsenic sulfides: alacránite (As$_8$S$_9$). J Phys Chem A 115(17):4558–4562

Palacios D, Wladimirsky A, D'Antonio MC, González-Baró AC, Baran EJ (2011) Vibrational spectra of double oxalates of the type M^I$_2$Cu(C$_2$O$_4$)$_2$·2H$_2$O (M^I = Na$^+$, K$^+$, NH$_4^+$). Spectrochim Acta A 79:1145–1148

Palaich SE, Manning CE, Schauble E, Kavner A (2013) Spectroscopic and X-ray diffraction investigation of the behavior of hanksite and tychite at high pressures, and a model for the compressibility of sulfate minerals. Am Mineral 98(8–9):1543–1549

Palenzona A, Martinelli A (2007) La nakauriite del Monte Ramazzo, Genova. Rivista Mineral Ital 31(1):48–51. (in Italian)

Palmer SJ, Frost RL (2011) The structure of the mineral arthurite $(AsO_4,PO_4,SO_4)_2(O,OH)_2 \cdot 4H_2O$ – a Raman spectroscopic study. J Mol Struct 994(1): 283–288

Palmer SJ, Grand LM, Frost RL (2011) The synthesis and spectroscopic characterisation of hydrotalcite formed from aluminate solutions. Spectrochim Acta A 79(1):156–160

Palmeri R, Frezzotti ML, Godard G, Davies RJ (2009) Pressure-induced incipient amorphization of α-quartz and transition to coesite in an eclogite from Antarctica: a first record and some consequences. J Metamorphic Geol 27(9):685–705

Paluszkiewicz C, Żabiński W (1992) Far infrared spectra of vesuvianite: preliminary report. Mineralogia Polonica 23(1):13–16

Palvadeau P, Euzen P, Queignec M, Venien JP (1991) Characterization of Mn_7SiO_{12}, a synthetic equivalent of "braunite" a natural mineral with various manganese sites. Mater Res Bull 26(9):841–848

Palve BM, Jadkar SR, Pathan HM (2016) A simple chemical route to synthesis the CuS nanocrystal powder at room temperature and phase transition. J Mater Sci Mater Electron. https://doi.org/10.1007/s10854-016-5318-3

Pan S, Smit JP, Marvel MR, Stern CL, Watkins B, Poeppelmeier KR (2006) Synthesis, structure and properties of $Pb_2CuB_2O_6$. Mater Res Bull 41(5):916–924

Panikorovskii TL, Krivovichev SV, Galuskin EV, Shilovskikh VV, Mazur AS, Bazai AV (2016a) Si-deficient, OH-substituted, boron-bearing vesuvianite from Sakha-Yakutia, Russia: a combined single-crystal, ^1H MAS-NMR and IR spectroscopic study. Eur J Mineral 28:931–941

Panikorovskii TL, Zolotarev AA Jr, Krivovichev SV, Shilovskikh VV, Bazay AV (2016b) Crystal chemistry of Cu-bearing vesuvianites ("cyprines") from Kleppan (Norway). Zapiski RMO 145(1):131–142. (in Russian)

Panikorovskii TL, Krivovichev SV, Zolotarev AA, Antonov AA (2016c) Crystal chemistry of low-symmetry (P4nc) vesuvianite from the Kharmankul' Cordon (South Urals, Russia). Zapiski RMO 145(3):94–104. (in Russian)

Panikorovskii TL, Krivovichev SV, Yakovenchuk VN, Shilovskikh VV, Mazur AS (2016d) Crystal chemistry of Na-bearing vesuvianite from fenitized gabbroid of the Western Keivy (Kola peninsula, Russia). Zapiski RMO 145(5):83–95. (in Russian)

Panikorovskii TL, Chukanov NV, Aksenov SM, Mazur AS, Avdontseva EY, Shilovskikh VV, Krivovichev SV (2017a) Alumovesuvianite, $Ca_{19}Al(Al,Mg)_{12}Si_{18}O_{69}(OH)_9$, a new vesuvianite-group member from the Jeffrey mine, Asbestos, Estrie Region, Québec, Canada. Miner Petrol 111(6):833–842

Panikorovskii TL, Mazur AS, Bazai AV, Shilovskikh VV, Galuskin EV, Chukanov NV, Rusakov VS, Zhukov YM, Avdontseva EY, Aksenov SM, Krivovichev SV (2017b) X-ray diffraction and spectroscopic study of wiluite: implications for the vesuvianite-group nomenclature. Phys Chem Mineral 44(8):577–593

Panikorovskii TL, Shilovskikh VV, Avdontseva EY, Zolotarev AA, Pekov IV, Britvin SN, Hålenius U, Krivovichev SV (2017c) Cyprine, $Ca_{19}Cu^{2+}(Al,Mg)_{12}Si_{18}O_{69}(OH)_9$, a new vesuvianite-group mineral from the Wessels mine, South Africa. Eur J Mineral 29:295–306

Panikorovskii TL, Shilovskikh VV, Avdontseva EY, Zolotarev AA, Karpenko VY, Mazur AS, Yakovenchuk VN, Krivovichev SV, Pekov IV (2017d) Magnesiovesuvianite, $Ca_{19}Mg(Al,Mg)_{12}Si_{18}O_{69}(OH)_9$, a new vesuvianite-group mineral. J Geosc (Czech Republic) 62:25–36

Paques-Ledent MT (1978) Vibrational spectra and structure of $LiB^{2+}PO_4$ compounds with B = Sr, Ba, Pb. J Solid State Chem 23(1):147–154

Paques-Ledent MT, Tarte P (1974) Vibrational studies of olivine-type compounds – II orthophosphates, -arsenates and -vanadates $A^I B^{II} X^V O_4$. Spectrochim Acta A 30(3):673–689

Parajón-Costa BS, Mercader RC, Baran EJ (2013) Spectroscopic characterization of mixed cation diphosphates of the type $M^I Fe^{III} P_2O_7$ (with M^I = Li, Na, K, Rb, Cs, Ag). J Phys Chem Solids 74(2):354–359

Parker G, Hope GA, Woods R (2008) Raman spectroscopic identification of surface species in the leaching of chalcopyrite. Colloids Surf A 318:160–168

Pasteris JD, Yoder CH, Sternlieb MP, Liu S (2012) Effect of carbonate incorporation on the hydroxyl content of hydroxylapatite. Mineral Mag 76(7):2741–2759

Pauliš P, Kopecký S, Vrtiška L, Čejka J, Pour O, Laufek F (2016) Stolzit z Vysoké u Havlíčkova Brodu (Česká Republika). Bull Mineral-petrolog Odd Nár Muz (Praha) 24(1):95–99. ISSN 1211-0329. (in Czech)

Pautov LA, Agakhanov AA, Sokolova E, Hawthorne FC, Karpenko VY, Siidra OI, Garanin VK, Abdu YA (2015) Khvorovite, $Pb^{2+}_4Ca_2[Si_8B_2(SiB)O_{28}]F$, a new hyalotekite-group mineral from the Darai-Pioz alkaline massif, Tajikistan: description and crystal structure. Mineral Mag 79(4):949–963

Pavunny SP, Kumar A, Katiyar RS (2010) Raman spectroscopy and field emission characterization of delafossite $CuFeO_2$. J Appl Phys 107(1):013522-1–013522-7

Payen J-L, Durand J, Cot L, Galigne J-L (1979) Etude structurale du monofluorophosphate de potassium K_2PO_3F. Can J Chem 57(8):886–889. (in French). https://doi.org/10.1139/v79-146

Pažout R, Sejkora J, Maixner J, Dušek M, Tvrdý J (2015) Refikite from Krásno, Czech Republic: a crystal-and molecular-structure study. Mineral Mag 79(1):59–70

Peacor DR, Simmons WB Jr, Essene EJ, Heinrich EW (1982) New data on and discreditation of "texasite", "albrittonite", "cuproartiniter", "cuprohydromagnesite"and "yttromicrolite" with corrected data on nickelbischofite, rowlandite, and yttrocrasite. Am Mineral 67:156–169

Pechar F (1983) Study of the complex vibrational spectra of natural mesolite. Cryst Res Technol 18(8):1045–1052

Pechar F (1984) Study of the vibrational spectra of natural skolecite. Cryst Res Technol 19(4):541–548

Pechar F, Rykl D (1983) Study of the complex vibrational spectra of natural zeolite chabazite. Zeolites 3(4):333–336

Pedregosa JC, Baran EJ, Aymonino PJ (1974) Kristallchemisches Verhalten und IR-Spektren einiger Divanadate des Thortveitit-Typs und verwandter Strukturen. Z anorg allg Chemie 404(3):308–320. (in German)

Pedro M, Trombe JC, Castro A (1995) On the rare-earth selenites $Pr_2Se_4O_{11}$, $R_2Se_3O_9$ and R_2SeO_5. J Mater Sci Lett 14(14):994–997

Pekov IV (2016) Personal Commun

Pekov IV, Yapaskurt VO, Polekhovsky YS, Vigasina MF, Siidra OI (2014a) Ekplexite $(Nb,Mo)S2·(Mg_{1-x}Al_x)(OH)_{2+x}$, kaskasite $(Mo,Nb)S_2·(Mg_{1-x}Al_x)(OH)_{2+x}$ and manganokaskasite $(Mo,Nb)S2·(Mn_{1-x}Al_x)(OH)_{2+x}$, three new valleriite-group mineral species from the Khibiny alkaline complex, Kola Peninsula, Russia. Mineral Mag 78(3):663–680

Pekov IV, Zubkova NV, Göttlicher J, Yapaskurt VO, Chukanov NV, Lykova IS, Belakovskiy DI, Jensen MC, Leising JF, Nikischer AJ, Pushcharovsky DY (2014b) Whitecapsite, a new hydrous iron and trivalent antimony arsenate mineral from the White Caps mine, Nevada, USA. Eur J Mineral 26(4):577–587

Pekov IV, Zubkova NV, Yapaskurt VO, Belakovskiy DI, Vigasina MF, Sidorov EG, Pushcharovsky DY (2014c) New arsenate minerals from the Arsenatnaya fumarole, Tolbachik volcano, Kamchatka, Russia. II. Ericlaxmanite and kozyrevskite, two natural modifications of $Cu_4O(AsO_4)_2$. Mineral Mag 78(7):1553–1569

Pekov IV, Zubkova NV, Yapaskurt VO, Belakovskiy DI, Lykova IS, Vigasina MF, Sidorov EG, Pushcharovsky DY (2014d) New arsenate minerals from the Arsenatnaya fumarole, Tolbachik volcano, Kamchatka, Russia. I. Yurmarinite, $Na_7(Fe^{3+},Mg,Cu)_4(AsO_4)_6$. Mineral Mag 78(4):905–918

Pekov IV, Zubkova NV, Belakovskiy DI, Lykova IS, Yapaskurt VO, Vigasina MF, Sidorov EG, Pushcharovsky DY (2015a) Sanguite, $KCuCl_3$, anewmineralfromthe Tolbachik volcano, Kamchatka, Russia. Can Mineral 53(4):633–641

Pekov IV, Zubkova NV, Yapaskurt VO, Belakovskiy DI, Vigasina MF, Pushcharovsky DY (2015b) New arsenate minerals from the Arsenatnaya fumarole, Tolbachik volcano, Kamchatka, Russia. III. Popovite, $Cu_5O_2(AsO_4)_2$. Mineral Mag 79(1):133–144

Pekov IV, Zubkova NV, Belakovskiy DI, Yapaskurt VO, Vigasina MF, Lykova IS, Sidorov EG, Pushcharovsky DY (2015c) Chrysothallite $K_6Cu_6Tl^{3+}Cl_{17}(OH)_4·H_2O$, a new mineral species from the Tolbachik volcano, Kamchatka, Russia. Mineral Mag 79(2):365–376

Pekov IV, Zubkova NV, Belakovskiy DI, Yapaskurt VO, Vigasina MF, Sidorov EG, Pushcharovsky DY (2015d) New arsenate minerals from the Arsenatnaya fumarole, Tolbachik volcano, Kamchatka, Russia. IV. Shchurovskyite, $K_2CaCu_6O_2(AsO_4)_4$ and dmisokolovite, $K_3Cu_5AlO_2(AsO_4)_4$. Mineral Mag 79(7):1737–1753

Pekov IV, Zubkova NV, Yapaskurt VO, Britvin SN, Vigasina MF, Sidorov EG, Pushcharovsky DY (2015e) New zinc and potassium chlorides from fumaroles of the Tolbachik volcano, Kamchatka, Russia: mineral data and crystal chemistry. II. Flinteite, K_2ZnCl_4. Eur J Mineral. https://doi.org/10.1127/ejm/2015/0027-2459

Pekov IV, Zubkova NV, Yapaskurt VO, Lykova IS, Belakovskiy DI, Vigasina MF, Sidorov EG, Britvin SN, Pushcharovsky DY (2015f) New zinc and potassium chlorides from fumaroles of the Tolbachik volcano, Kamchatka, Russia: mineral data and crystal chemistry. I. Mellizinkalite, $K_3Zn_2Cl_7$. Eur J Mineral 27(2):247–253

Pekov IV, Yapaskurt VO, Belakovskiy DI, Vigasina MF, Zubkova NV, Sidorov EG (2016a) New arsenate minerals from the Arsenatnaya fumarole, Tolbachik volcano, Kamchatka, Russia. VII. Pharmazincite, $KZnAsO_4$. Mineral Mag. https://doi.org/10.1180/minmag.2016.080.146

Pekov IV, Yapaskurt VO, Britvin SN, Vigasina MF, Lykova IS, Zubkova NV, Krivovichev SV, Sidorov EG (2016b) Romanorlovite, a new copper and potassium hydroxychloride from the Tolbachik volcano, Kamchatka, Russia. Zapiski RMO (Proc Russ Miner Soc) 145(4):36–46. (in Russian)

Pekov IV, Yapaskurt VO, Britvin SN, Zubkova NV, Vigasina MF, Sidorov EG (2016c) New arsenate minerals from the Arsenatnaya fumarole, Tolbachik volcano, Kamchatka, Russia. V. Katiarsite, $KTiO(AsO_4)$. Mineral Mag. https://doi.org/10.1180/minmag.2016.080.007

Pekov IV, Zubkova NV, Yapaskurt VO, Polekhovsky YS, Vigasina MF, Belakovskiy DI, Britvin SN, Sidorov EG, Pushcharovsky DY (2016d) New arsenateminerals from the Arsenatnaya fumarole, Tolbachik volcano, Kamchatka, Russia. VI. Melanarsite, $K_3Cu_7Fe^{3+}O_4(AsO_4)_4$. Mineral Mag 80(5):855–867

Penel G, Leroy G, Rey C, Bres E (1998) Micro Raman spectral study of the PO_4 and CO_3 vibrational modes in synthetic and biological apatites. Calcif Tissue Int 63(6):475–481

Peng GW, Chen SK, Liu HS (1995) Infrared absorption spectra and their correlation with the Ti-O bond length variations for TiO_2 (rutile), Na-titanates, and Na-titanosilicate (natisite, $Na_2TiOSiO_4$). Appl Spectrosc 49(11):1646–1651

Peretyazhko IS, Prokofiev VY, Zagorskii VE, Smirnov SZ (2000) Role of boric acids in the formation of pegmatite and hydrothermal minerals: petrologic consequences of sassolite (H_3BO_3) discovery in fluid inclusions. Petrol 8(3):214–237

Peretyazhko IS, Savina EA, Khromova EA (2016) Minerals of the rhönite-kuratite series in Paralavas from a new combustion metamorphic complex of Choir–Nyalga basin (Central Mongolia): chemistry, mineral assemblages, and formation conditions. https://doi.org/10.1180/minmag.2016.080.143

Perna G, Lastella M, Ambrico M, Capozzi V (2006) Temperature dependence of the optical properties of ZnSe films deposited on quartz substrate. Appl Phys A 83 (1):127–130

Perraki M, Faryad SW (2014) First finding of microdiamond, coesite and other UHP phases in felsic granulites in the Moldanubian Zone: implications for deep subduction and a revised geodynamic model for Variscan Orogeny in the Bohemian Massif. Lithos 202:157–166

Perrin J, Vielzeuf D, Laporte D, Ricolleau A, Rossman GR, Floquet N (2016) Raman characterization of synthetic magnesian calcites. Am Mineral 101 (11):2525–2538

Perry CH, Fertel JH, McNelly TF (1967) Temperature dependence of the Raman spectrum of $SrTiO_3$ and $KTaO_3$. J Chem Phys 47(5):1619–1625

Pertsev NN, Malinko SV, Vakhrushev VA, Fitsev BP, Sokolova EV, Nikitina IB (1980) Shabynite, a new hydrous magnesium borate chloride. Zapiski VMO (Proc Russ Miner Soc) 109(5):569–573. (in Russian)

Peterson RC (2011) Cranswickite $MgSO_4 \cdot 4H_2O$, a new mineral from Calingasta, Argentina. Am Mineral 96 (5–6):869–877

Petit S, Decarreau A, Gates W, Andrieux P, Grauby O (2015) Hydrothermal synthesis of dioctahedral smectites: the Al–Fe^{3+} chemical series. Part II: Crystal-chemistry. App Clay Sci 104:96–105

Petritis D, Martinez G, Levy-Clement C, Gorochov O (1981) Investigation of the vibronic properties of Cu_3VS_4, Cu_3NbS_4, and Cu_3TaS_4 compounds. Phys Rev B 23(12):6773–6786

Petrov D (2014) Laser "Raman" spectroscopy of anglesite and cubanite from deposit "Chelopech". Annu Univ Mining Geol "St. Ivan Rilski", Geol Geophys 57 (I):77–82

Petrova EV, Vorontsova MA, Manomenova VL, Rashkovich LN (2012) Some properties of aqueous solutions of α-$Ni(SO_4) \cdot 6H_2O$. Crystallogr Rep 57 (4):579–584

Petry R, Mastalerz R, Zahn S, Mayerhöfer TG, Völksch G, Viereck-Götte L, Kreher-Hartman B, Holz L, Lankers M, Popp J (2006) Asbestos mineral nnalysis by UV Raman and energy-dispersive X-ray spectroscopy. Chem Phys Chem 7(2):414–420

Petzelt J (1977) Raman study of the ferroelectric phase transition in $PbHPO_4$ and $PbHAsO_4$. J Chem Phys 67 (9):3890–3896

Peytavin S, Brun G, Cot L, Maurin M (1972a) Etude vibrationnelle des sulfates et séléniates doubles dihydratés de sodium et de métal divalent: $Na_2M^{II}(SO_4)_2 \cdot 2H_2O$ (M^{II} = Cr, Cd, Mn, Cu) $Na_2M^{II}(SeO_4)_2 \cdot 2H_2O$ (M^{II} = Cd, Mn). Spectrochim Acta A 28(10):1995–2003. (in French)

Peytavin S, Brun G, Guillermet J, Cot L, Maurin M (1972b) Spectres de vibration des sulfates doubles de sodium et d'un métal divalent $Na_2M^{II}(SO_4)_2 \cdot 4H_2O$ (M^{II} = Mg, Fe, Co, Ni, Zn). Spectrochim Acta A 28 (10):2005–2011. (in French)

Pham-Thi M, Colomban P, Novak A (1985) Vibrational study of $H_3OUO_2PO_4 \cdot 3H_2O$ (HUP) and related compounds. Phase transitions and conductivity mechanisms: Part I, $KUO_2PO_4 \cdot 3H_2O$ (KUP). J Phys Chem Solids 46(4):493–504

Philip D, Aruldhas G, Ramakrishnan V (1988) Infrared and Raman spectra of aquamolybdenum(VI) oxide hydrate ($MoO_3 \cdot 2H_2O$). J Phys 30(2):129–133

Phillippi CM, Mazdiyasni KS (1971) Infrared and Raman spectra of zirconia polymorphs. J Am Ceram Soc 54 (5):254–258

Pieczka A, Evans RJ, Grew ES, Groat LA, Ma C, Rossman GR (2013) The dumortierite supergroup. II. Three new minerals from the Szklary pegmatite, SW Poland: niboholtite, $(Nb_{06}\square_{04})Al_6BSi_3O_{18}$, titanoholtite, $(Ti_{0.75}\square_{0.25})Al_6BSi_3O_{18}$, and szklaryite, $\square Al_6BAs^{3+}_3O_{15}$. Mineral Mag 77(6):2841–2856

Pieczka A, Hawthorne FC, Cooper MA, Szełęg E, Szuszkiewicz A, Turniak K, Nejbert K, Ilnicki S (2015) Pilawite-(Y), $Ca_2(Y,Yb)_2[Al_4(SiO_4)_4O_2(OH)_2]$, a new mineral from the Piława Górna granitic pegmatite, southwestern Poland: mineralogical data, crystal structure and association. Mineral Mag 79 (5):1143–1157

Pieczka A, Hawthorne FC, Ma C, Rossman GR, Szełęg E, Szuszkiewicz A, Turniak K, Nejbert K, Ilnicki SS, Buffat P, Rutkowski B (2016a) Żabińskiite, ideally $Ca(Al_{0.5}(Ta,Nb)_{0.5})(SiO_4)O$, a new mineral of the titanite group from the Piława Górna pegmatite, the Góry Sowie Block, southwestern Poland. Mineral Mag. https://doi.org/10.1180/minmag.2016.080.110

Pieczka A, Szełęg E, Szuszkiewicz A, Gołębiowska B, Zelek S, Ilnicki S, Nejbert K, Turniak K (2016b) Cs-Bearing beryl evolving to pezzottaite from the Juliannapegmatitic system, SW Poland. Can Mineral 54:115–124

Pignatelli I, Mugnaioli E, Mosser-Ruck R, Barres O, Kolb U, Michau N (2014) A multi-technique, micrometer- to atomic-scale description of a synthetic analogue of chukanovite, $Fe_2(CO_3)(OH)_2$. Eur J Mineral 26:221–229

Piilonen PC, McDonald AM, Grice JD, Rowe R, Gault RA, Poirier G, Cooper MA, Kolitsch U, Roberts AC, Lechner W, Palfi AG (2010) Arisite-(Ce), a new rare-earth fluorcarbonate from the Arisphonolite, Namibia, Mont Saint-Hilaire and the Saint-Amable sill, Quebec, Canada. Can Mineral 48(3):661–671

Pikl R, De Waal D, Aatiq A, El Jazouli A (1998) Vibrational spectra and factor group analysis of $Li_{2x}Mn_{0.5-x}Ti_2(PO_4)$ {x = 0, 0.25, 0.50}. Mater Res Bull 33(6):955–961

Pikl R, De Waal D, Merkle RKW, Verryn SMC (1999) Raman spectroscopic identification of synthetic "braggite" (Pt, Pd, Ni)S samples in comparison with synthetic "cooperite". Appl Spectrosc 53(8):927–930

Pilati T, Gramaccioli CM, Pezzotta F, Fermo P, Bruni S (1998) Single-crystal vibrational spectrum of phenakite, Be_2SiO_4, and its interpretation using a transferable empirical force field. J Phys Chem A 102(26):4990–4996

Pillai VM, Nayar VU, Jordanovska VB (1997) Infrared and Raman spectra of $Na_2Cu(SO_4)_2 \cdot 2H_2O$ and $(CH_3NH_3)_2M(II)(SO_4)_2 \cdot 6H_2O$ with $M(II)$ = Cu, Zn, and Ni. J Solid State Chem 133(2):407–415

Pine AS, Dresselhaus G (1971) Raman spectra and lattice dynamics of tellurium. Phys Rev B 4(2):356–371

Pînzaru SC, Onac BP (2009) Raman study of natural berlinite from a geological phosphate deposit. Vib Spectrosc 49(2):97–100

Pirayesh H, Nychka JA (2013) Sol-gel synthesis of bioactive glass-ceramic 45S5 and its *in vitro* dissolution and mineralization behavior. J Am Ceram Soc 96(5):1643–1650

Piriou B, McMillan P (1983) The high-frequency vibrational spectra of vitreous and crystalline orthosilicates. Am Mineral 68:426–443

Piriou B, Poullen JF (1984) Raman study of vivianite. J Raman Spectrosc 15(5):343–346

Piro OE, Echeverría GA, González-Baró AC, Baran EJ (2016) Crystal and molecular structure and spectroscopic behavior of isotypic synthetic analogs of the oxalate minerals stepanovite and zhemchuzhnikovite. Phys Chem Minerals 43(4):287–300

Piro OE, Echeverría GA, González-Baró AC, Baran EJ (2017) Crystal structure and spectroscopic behavior of synthetic novgorodovaite $Ca_2(C_2O_4)Cl_2 \cdot 2H_2O$ mineral and its perfect twin triclinic $Ca_2(C_2O_4)Cl_2 \cdot 7H_2O$ analogue. Phys Chem Minerals. https://doi.org/10.1007/s00269-017-0907-0

Piseri L, Pollini I (1984) Vibrational structure of d-excitons in layered manganese dihalides. J Phys C 17(25):4519–4527

Piszczek P, Grodzicki A, Engelen B (2003) Infrared and Raman studies of water molecule normal vibrations in crystalline hydrates which form the chain structures. J Molec Struct 646(1):45–54

Plášil J, Čejka J, Sejkora J, Hloušek J, Goliáš V (2009) New data for metakirchheimerite from Jáchymov (St. Joachimsthal), Czech Republic. J Geosci 54:373–384

Plášil J, Buixaderas E, Čejka J, Sejkora J, Jehlička J, Novák M (2010a) Raman spectroscopic study of the uranyl sulphate mineral zippeite: low wavenumber and U–O stretching regions. Anal Bioanal Chem 397(7):2703–2715

Plášil J, Čejka J, Sejkora J, Škácha P, Goliáš V, Jarka P, Laufek F, Jehlička J, Němec I, Strnad L (2010b) Widenmannite, a rare uranyl lead carbonate: occurrence, formation and characterization. Mineral Mag 74(1):97–110

Plášil J, Sejkora J, Čejka J, Novák M, Viñals J, Ondruš P, Veselovský F, Škácha P, Jehlička J, Goliáš V, Hloušek J (2010c) Metarauchite, $Ni(UO_2)_2(AsO_4)_2 \cdot 8H_2O$, from Jáchymov, Czech Republic, and Schneeberg, Germany: a new member of the autunite group. Can Mineral 48(2):335–350

Plášil J, Sejkora J, Čejka J, Škácha P, Goliáš V, Ederová J (2010d) Characterization of phosphate-rich metalodèvite from Příbram, Czech Republic. Can Mineral 48(1):113–122

Plášil J, Dušek M, Novák M, Čejka J, Císařová I, Škoda R (2011a) Sejkoraite-(Y), a new member of the zippeite group containing trivalent cations from Jáchymov (St. Joachimsthal), Czech Republic: description and crystal structure refinement. Am Mineral 96(7):983–991

Plášil J, Fejfarová K, Novák M, Dušek M, Škoda R, Hloušek J, Čejka J, Majzlan J, Sejkora J, Machovič V, Talla D (2011b) Běhounekite, $U(SO_4)_2(H_2O)_4$, from Jáchymov (St Joachimsthal), Czech Republic: the first natural U^{4+}sulphate. Miner Mag 75(6):2739–2753

Plášil J, Hauser J, Petříček V, Meisser N, Mills SJ, Škoda R, Fejarova K, Čejka J, Sejkora J, Hloušek J, Johannet J-M, Machovič V, Lapčak L (2012a) Crystal structure and formula revision of deliensite, $Fe[(UO_2)_2(SO_4)_2(OH)_2](H2O)_7$. Mineral Mag 76(7):2837–2860

Plášil J, Hloušek J, Veselovský F, Fejfarová K, Dušek M, Škoda R, Novák M, Čejka J, Ondruš P (2012b) Adolfpateraite, $K(UO_2)(SO_4)(OH)(H_2O)$, a new uranyl sulphate mineral from Jáchymov, Czech Republic. Am Mineral 97(2–3):447–454

Plášil J, Kasatkin AV, Škoda R, Novák M, Kallistová A, Dušek M, Skála R, Fejfarová K, Čejka J, Meisser N, Goethals H, Machovič V, Lapčak L (2013a) Leydetite, $Fe(UO_2)(SO_4)_2(H_2O)_{11}$, a new uranyl sulfate mineral from Mas d'Alary, Lodève, France. Mineral Mag 77(4):429–441

Plášil J, Fejfarová K, Hloušek J, Škoda R, Novák M, Sejkora J, Čejka J, Dušek M, Veselovský F, Ondruš P, Majzlan J, Mrázek Z (2013b) Štěpite, $U(AsO_3OH)_2 \cdot 4H_2O$, from Jáchymov, Czech Republic: the first natural arsenate of tetravalent uranium. Mineral Mag 77(1):137–152

Plášil J, Kampf AR, Kasatkin AV, Marty J, Škoda R, Silva S, Čejka J (2013c) Meisserite, $Na_5(UO_2)(SO_4)_3(SO_3OH)(H_2O)$ a new uranyl sulfate mineral from the Blue Lizard mine, San Juan County, Utah, USA. Mineral Mag 77(7):2975–2988

Plášil J, Hloušek J, Škoda R, Sejkora J, Čejka J, Veselovský F, Majzlan J (2013d) Vysokýite, $U^{4+}[AsO_2(OH)_2]_4 \cdot 4H_2O$, a new mineral from Jáchymov, Czech Republic. Mineral Mag 77(8):3055–3066

Plášil J, Kampf AR, Kasatkin AV, Marty J (2014a) Bluelizardite, $Na_7(UO_2)(SO_4)_4Cl(H_2O)_2$, a new uranyl sulfate mineral from the Blue Lizard mine, San Juan County, Utah, USA. J Geosci 59(2):145–158

Plášil J, Škoda R, Fejfarová K, Čejka J, Kasatkin AV, Dušek M, Talla D, Lapčák L, Machovič V, Dini M (2014b) Hydroniumjarosite, $(H_3O)^+Fe_3(SO_4)_2(OH)_6$, from Cerros Pintados, Chile: single-crystal X-ray diffraction and vibrational spectroscopic study. Mineral Mag 78(3):535–547

Plášil J, Veselovský F, Hloušek J, Škoda R, Novák M, Sejkora J, Čejka J, Škácha P, Kasatkin AV (2014c) Mathesiusite, $K_5(UO_2)_4(SO_4)_4(VO_5)(H_2O)_4$, a new uranyl vanadate-sulfate from Jáchymov, Czech Republic. Am Mineral 99(4):625–632

Plášil J, Čejka J, Škoda R (2014d) Chalcoalumite, $CuAl_4(SO_4)(OH)_{12}\cdot 3H_2O$, from the Červená vein, Jáchymov (Czech Republic). Bull Mineral-petrolog Odd Nár Muz (Praha) 22(2):227–232. ISSN 1211-0329. (in Czech)

Plášil J, Hloušek J, Kasatkin AV, Belakovskiy DI, Čejka J, Chernyshov D (2015a) Ježekite, $Na_8[(UO_2)(CO_3)_3](SO_4)_2\cdot 3H_2O$, a new uranyl mineral from Jáchymov, Czech Republic. J Geosci 60(4):259–267

Plášil J, Hloušek J, Kasatkin AV, Novák M, Čejka J, Lapčák L (2015b) Svornostite, $K_2Mg[(UO_2)(SO_4)_2]_2\cdot 8H_2O$, a new uranyl sulfate mineral from Jáchymov, Czech Republic. J Geosci 60(2):113–121

Plášil J, Hloušek J, Kasatkin AV, Škoda R, Novák M, Čejka J (2015c) Geschieberite, $K_2(UO_2)(SO_4)_2(H_2O)_2$, a new uranyl sulfate mineral from Jáchymov. Mineral Mag 79(1):205–216

Plášil J, Meisser N, Čejka J (2015d) The crystal structure of $Na_6[(UO_2)(SO_4)_4](H_2O)_4$: X-ray and Raman spectroscopy study. Can Mineral. https://doi.org/10.3749/canmin.1400095

Plášil J, Sejkora J, Čejka J, Pavlíček R, Babka K, Škoda R (2016a) Výskyt boltwooditu na uranovém ložisku Kladská (Česká republika). Bull Mineral-petrolog Odd Nár Muz (Praha) 24(2):298–303. ISSN 1211-0329. (in Czech)

Plášil J, Škoda R, Čejka J, Bourgoin V, Boulliard JC (2016b) Crystal structure of the uranyl-oxide mineral rameauite. Eur J Mineral. https://doi.org/10.1127/ejm/2016/0028-2568

Plášil J, Škácha P, Sejkora J, Kampf AR, Škoda R, Čejka J, Hloušek J, Kasatkin AV, Pavlíček R, Babka K (2017) Plavnoite, a new K–Mn member of the zippeite group from Jáchymov, Czech Republic. Eur J Mineral 29:117–128

Podsiadlo S, Weisbrod G, Bialoglowski M, Jastrzebski D, Fadaghi M, Ostrowski A (2015) Synthesis and crystal growth of microcrystals of the cubic and new orthorhombic polymorphs of $(NH_4)_2SnCl_6$. Cryst Res Technol 50(9–10):764–768

Pookmanee P, Kojinok S, Puntharod R, Sangsrichan S, Phanichphant S (2013) Preparation and characterization of $BiVO_4$ powder by the sol-gel method. Ferroelectrics 456(1):45–54

Popov VA, Kolisnichenko SV, Blinov IA (2016) Nickelean copper and nakauriite from the Blue vein in ultramafites of the Verkhniy Ufaley Region, Southern Urals. Preprint of the Institute of Mineralogy, Uralian Branch of the Russian Academy of Sciences, Miass. (in Russian)

Popović L, Manoun B, De Waal D, Nieuwoudt MK, Comins JD (2003) Raman spectroscopic study of phase transitions in Li_3PO_4. J Raman Spectrosc 34(1):77–83

Popović ZV, Konstantinović MJ, Dohčević-Mitrović Z, Isobe M, Ueda Y (2006) Orbital and lattice dynamics in pyroxenes. Phys B Condens Matter 378:1072–1074

Porta P, Minelli G, Botto IL, Baran EJ (1991) Structural, magnetic, and optical investigation of Ni_6MnO_8. J Solid State Chem 92(1):202–207

Porter Y, Halasyamani PS (2003) Syntheses, structures, and characterization of new lead (II)-tellurium (IV)-oxide halides: $Pb_3Te_2O_6X_2$ and $Pb_3TeO_4X_2$ (X = Cl or Br). Inorg Chem 42(1):205–209

Porter Y, Bhuvanesh NSP, Halasyamani PS (2001) Synthesis and characterization of non-centrosymmetric $TeSeO_4$. Inorg Chem 40(6):1172–1175

Porto SPS, Fleury PA, Damen TC (1967) Raman Spectra of TiO_2, MgF_2, ZnF_2, FeF_2, and MnF_2. Phys Rev 154(2):522–526

Postl W, Moser B (1988) Nakauriite from Lobminggraben. Mineralogische Notizen aus der Steiermark. Mitt Abt Miner Landesmuseum Joanneum 56:5–47. (in German)

Potgieter-Vermaak S, Rotondo G, Novakovic V, Rollins S, Van Grieken R (2012) Component-specific toxic concerns of the inhalable fraction of urban road dust. Environ Geochem Health 34(6):689–696

Poudret L, Prior TJ, McIntyre LJ, Fogg AM (2008) Synthesis and crystal structures of new lanthanide hydroxyhalide anion exchange materials, $Ln_2(OH)_5X\cdot 1.5H_2O$ (X = Cl, Br; Ln = Y, Dy, Er, Yb). Chem Mater 20(24):7447–7453

Poupon M, Barrier N, Petit S, Clevers S, Dupray V (2015) Hydrothermal synthesis and dehydration of $CaTeO_3(H_2O)$: an original route to generate new $CaTeO_3$ polymorphs. Inorg Chem 54(12):5660–5670

Povarennykh AS (1970) Spectres infrarouges de certains minéraux de Madagascar. Bull Soc Fr Minéral Cristallogr 93:224–234. (in French)

Powers DA, Rossman GR, Schugar HJ, Gray HB (1975) Magnetic behavior and infrared spectra of jarosite, basic iron sulfate, and their chromate analogs. J Solid State Chem 13(1–2):1–13

Pradhan AK, Choudhary RNP (1987) Vibrational spectra of rare earth orthoniobates. Phys Status Solidi B 143: K161–K166

Pradhan AK, Choudhary RNP, Wanklyn BM (1987) Raman and infrared spectra of $YAsO_4$. Phys Status Solidi B 139:337–334

Prasad PSR, Prasad KS (2007) Dehydration behavior of natural cavansite: an in-situ FTIR and Raman spectroscopic study. Mater Chem Phys 105(2):395–400

Prekajski M, Kremenović A, Babić B, Rosić M, Matović B, Radosavljević-Mihajlović A, Radović M (2010) Room-temperature synthesis of nanometric α-Bi_2O_3. Mater Lett 64(20):2247–2250

Price LS, Parkin IP, Hardy AM, Clark RJ, Hibbert TG, Molloy KC (1999) Atmospheric pressure chemical vapor deposition of tin sulfides (SnS, Sn_2S_3, and SnS_2) on glass. Chem Mater 11(7):1792–1799

Pring A, Kolitsch U, Birch WD (2005) Description and unique crystal-structure of waterhouseite, a new hydroxy manganese phosphate species from the Iron Monarch deposit, Middleback Ranges, South Australia. Can Mineral 43(4):1401–1410

Pring A, Wallwork K, Brugger J, Kolitsch U (2006) Paratooite-(La), a new lanthanum-dominant rare-earth copper carbonate from Paratoo, South Australia. Mineral Mag 70(1):131–138

Prinsloo LC, Colomban P, Brink JD, Meiklejohn I (2011) A Raman spectroscopic study of the igneous rocks on Marion Island: a possible terrestrial analogue for the geology on Mars. J Raman Spectrosc 42(4):626–632

Prodjosantoso AK, Laksono EW, Utomo MP (2015) Sintesis dan karaterisasi SnO_2 sebagai upaya pengembangan produk hilir timah putih untuk meningkatkan devisa nasional. J Penelitian Saintek 16(2):99–110. (in Indonesian)

Psycharis V, Kapoutsis IA, Chryssikos GD (1999) Crystal structure and vibrational spectra of Li_2BAlO_4. J Solid State Chem 142(1):214–219

Ptak M, Mączka M, Pikul A, Tomaszewski PE, Hanuza J (2014) Magnetic and low temperature phonon studies of $CoCr_2O_4$ powders doped with Fe(III) and Ni (II) ions. J Solid State Chem 212:218–226

Pucka G, Dobosz SM, Paluszkiewicz CZ (2000) Mechanizm wiązania mas fosforanowych w świetle badań spektroskopowych w podczerwieni. Krzepnięcie Metalii Stopów 2(44):521–526. (in Polish)

Qader M (2011) Identification and dissolution behavior of the secondary uranium minerals in the corrosion products of Depleted Uranium (DU) ammunition formed in soils. Dissertation. Johannes Gutenberg-Universität Mainz

Qiong LIU, Yiguo SU, Hongsheng YU, Wei HAN (2008) YPO_4 nanocrystals: preparation and size-induced lattice symmetry enhancement. J Rare Earths 26(4):495–500

Qiu X, Ji S, Chen C, Liu G, Ye C (2013) Synthesis, characterization, and surface-enhanced Raman scattering of near infrared absorbing Cu_3SbS_3 nanocrystals. Cryst Eng Comm 15(48):10431–10434

Raade G, Mladeck MH, Kristiansen R, Din VK (1984) Kaatialaite, a new ferric arsenate mineral from Finland. Amer Mineral 69:383–387

Raade G, Grice JD, Erambert M, Kristiansson P, Witzke T (2008) Proshchenkoite-(Y) from Russia – a new mineral species in the vicanite group: descriptive data and crystal structure. Mineral Mag 72:1071–1082

Raade G, Grice JD, Rowe R (2016) Ferrivauxite, a new phosphate mineral from Llallagua, Bolivia. Mineral Mag 80(2):311–324

Radepont M (2013) Understanding of chemical reactions involved in pigment discoloration, inparticular in mercury sulfide (HgS) blackening. Analytical chemistry. Dissertation, Université Pierre et Marie Curie, Universiteit Antwerpen. (in English)

Radica F, Capitelli F, Bellatreccia F, Della Ventura G, Cavallo A, Piccinini M, Hawthorne FC (2013) Spectroscopy and X-ray structure refinement of sekaninaite from Dolní Bory (Czech Republic). Mineral Mag 77(4):485–498

Rafique MY, Li-Qing P, Iqbal MZ, Hong-Mei Q, Farooq MH, Zhen-Gang G, Tanveer M (2013) Growth of monodisperse nanospheres of $MnFe_2O_4$ with enhanced magnetic and optical properties. Chin Phys B 22(10):107101-1–107101-7

Raison PE, Jardin R, Bouëxière D, Konings RJM, Geisler T, Pavel CC, Rebizant J, Popa K (2008) Structural investigation of the synthetic $CaAn(PO_4)_2$ ($An =$ Th and Np) cheralite-like phosphates. Phys Chem Minerals 35(10):603–609

Raj AME, Jayanthi DD, Jothy VB (2008) Optimized growth and characterization of cadmium oxalate single crystals in silica gel. Solid State Sci 10(5):557–562

Rajaji V, Malavi PS, Yamijala SSRKC, Sorb YA, Dutta U, Guin SN, Joseph B, Pati SK, Biswas K, Narayana C (2016) Pressure induced structural, electronic topological, and semiconductor to metal transition in $AgBiSe_2$. Appl Phys Lett 109(17):171903-1–171903-5

Rakovan J, Schmidt GR, Gunter ME, Nash B, Marty J, Kampf AR, Wise WS (2011) Hughesite, $Na_3AlV_{10}O_{28}·22H_2O$, a new member of the pascoite family of minerals from the Sunday mine, San Miguel County, Colorado. Can Mineral 49(5):1253–1265

Ramakrishnan V, Aruldhas G (1987) Raman and infrared studies of $Na_2HPO_4·2H_2O$. J Raman Spectrosc 18(2):145–146

Ramanaiah MV, Ravikumar RVSSN, Srinivasulu G, Reddy BJ, Rao PS (1996) Detailed spectroscopic studies on cornetite from Southern Shaba, Zaire. Ferroelectrics 175(1):175–182

Ramaraghavulu R, Sivaiah K, Buddhudu S (2012) Structural and dielectric properties of LiV_3O_8 ceramic powders. Ferroelectrics 432(1):55–64

Rancourt DG, Mercier PHJ, Cherniak DJ, Desgreniers S, Kodama H, Robert JL, Murad E (2001) Mechanisms and crystal chemistry of oxidation in annite: resolving the hydrogen-loss and vacancy reactions. Clays Clay Minerals 49(6):455–491

Rangan KK, Prasad BR, Subramanian CK, Gopalakrishnan J (1993) Coupled substitution of niobium and silicon in $KTiOPO_4$ and $KTiOAsO_4$. Synthesis and nonlinear optical properties of $KTi_{1-x}Nb_xOX_{1-x}Si_xO4$ (X = P, As). Inorg Chem 32(20):4291–4293

Ranieri V, Bourgogne D, Darracq S, Cambon M, Haines J, Cambon O, Leparc R, Levelut C, Largeteau A, Demazeau G (2009) Raman scattering study of α-quartz and $Si_{1-x}Ge_xO_2$ solid solutions. Phys Rev B 79:1–9

Rao BL (1941) Raman spectra of some crystalline nitrates and sulphates. Proc Indian Acad Sci 14:48–55. https://doi.org/10.1007/BF03049125

Rao KS, Buddhudu S (2010) Structural, thermal and dielectric properties of BiNbO$_4$ ceramic powder. Ferroelectrics Lett 37(6):101–109

Rao C, Wang RC, Hu H (2011) Paragenetic assemblages of beryllium silicates and phosphates from the Nanping no. 31 granitic pegmatite dyke, Fujian province, southeastern China. Can Mineral 49(5):1175–1187

Rao C, Wang R, Hatert F, Gu X, Ottolini L, Hu H, Dong C, Dal Bo F, Baijot M (2014) Strontiohurlbutite, SrBe$_2$(PO$_4$)$_2$, a new mineral from Nanping No. 31 pegmatite, Fujian Province, Southeastern China. Am Mineral 99(2–3):494–499

Rao C, Hatert F, Wang RC, Gu XP, Dal Bo F, Dong CW (2015) Minjiangite, BaBe$_2$(PO$_4$)$_2$, a new mineral from Nanping No. 31 pegmatite, Fujian Province, southeastern China. Mineral Mag 79(5):1195–1202

Raschke MB, Anderson EJ, Allaz J, Friis H, Smyth JR, Tschernich R, Becker R (2016) Crystal chemistry of brannockite, KLi$_3$Sn$_2$Si$_{12}$O$_{30}$, from a new occurrence in the Golden Horn Batholith, Washington State, USA. Eur J Mineral 28(1):153–161

Rath T, MacLachlan AJ, Brown MD, Haque SA (2015) Structural, optical and charge generation properties of chalcostibite and tetrahedrite copper antimony sulfide thin films prepared from metal xanthates. J Mater Chem A 3(47):24155–24162

Ratheesh R, Suresh G, Nayar VU, Morris RE (1997) Vibrational spectra of three aluminium selenities Al$_2$(SeO$_3$)$_3$·3H$_2$O, Al$_2$(SeO$_3$)$_3$·6H$_2$O and AlH(SeO$_3$)$_2$·H$_2$O. Spectrochim Acta A 53(12):1975–1979

Ratnam BV, Jayasimhadri M, Jang K (2014) Luminescent properties of orange emissive Sm^{3+}–activated thermally stable phosphate phosphor for optical devices. Spectrochim Acta A 132:563–567

Rauch M, Keppler H, Häfner W, Poe B, Wokaun A (1996) A pressure-induced phase transition in MgSiO$_3$-rich garnet revealed by Raman spectroscopy. Am Mineral 81:1289–1292

Ravi V, Adyanthaya S, Aslam M, Pethkar S, Choube VD (1999) Synthesis of bismuth tin pyrochlore. Mater Lett 40(1):11–13

Ravikumar RVSSN, Chandrasekhar AV, Reddy BJ, Reddy YP, Ikeda K (2002) X-Ray powder diffraction, DTA and vibrational studies of CdNH$_4$PO$_4$·6H$_2$O crystals. Crystal Res Technol 37(10):1127–1132

Ravindran TR, Arora AK, Parthasarathy G (2012) Raman spectroscopic study of pressure induced amorphization in cavansite. J Phys. https://doi.org/10.1088/1742-6596/377/1/012004

Reddy BJ, Frost RL (2005) Spectroscopic characterization of chromite from the Moa-Baracoa Ophiolitic Massif, Cuba. Spectrochim Acta A 61(8):1721–1728

Reddy BJ, Frost RL (2007) Electronic and vibrational spectra of Mn rich sursassite. Spectrochim Acta A 66(2):312–317

Reddy KM, Moorthy LR, Reddy BJ (1987) Electronic and vibrational absorption spectra in falcondoite. Solid State Commun 64(7):1085–1088

Reddy BJ, Frost RL, Martens WN (2005) Characterization of conichalcite by SEM, FTIR, Raman and electronic reflectance spectroscopy. Mineral Mag 69(2):155–167

Reddy SL, Reddy GS, Wain DL, Martens WN, Reddy BJ, Frost RL (2006) Electron paramagnetic resonance, optical absorption and IR spectroscopic studies of the sulphate mineral apjohnite. Spectrochim Acta A 65(5):1227–1233

Reddy SL, Maheswaramma KS, Reddy GS, Reddy BJ, Endo T, Frost RL (2010) Optical absorption, infrared, Raman, and EPR spectral studies on natural iowaite mineral. Transition Met Chem 35(3):331–336

Reich S, Ferrari AC, Arenal R, Loiseau A, Bello I, Robertson J (2005) Resonant Raman scattering in cubic and hexagonal boron nitride. Phys Rev B 71(20):205201-1–205201-12

Rémazeilles C, Refait P (2009) Fe(II) hydroxycarbonate Fe$_2$(OH)$_2$CO$_3$ (chukanovite) as iron corrosion product: synthesis and study by Fourier Transform Infrared Spectroscopy. Polyhedron 28(4):749–756

Rémazeilles C, Saheb M, Neff D, Guilminot E, Tran K, Bourdoiseau J-A, Sabot R, Jeannin M, Matthiesen H, Dillmann P, Refait P (2010) Microbiologically influenced corrosion of archaeological artefacts: characterisation of iron(II) sulfides by Raman spectroscopy. J Raman Spectrosc 41(11):1425–1433

Remy C, Reynard B, Madon M (1997) Raman spectroscopic investigations of dicalcium silicate: polymorphs and high-temperature phase transformations. J Am Ceram Soc 80(2):413–423

Ren Y, Ximen L, Peng Z (1983) Daqingshanite – a new mineral recently discovered in China. Geochem 2:180–184

Ren L, Grew ES, Xiong M, Ma Z (2003) Wagnerite-Ma5bc, a new polytype of Mg$_2$(PO$_4$)(F,OH), from granulite-faciesparagneiss, Larsemann Hills, Prydz Bay, East Antarctica. Can Mineral 41:393–411

Renaudin G, Segni R, Mentel D, Nedelec JM, Leroux F, Taviot-Gueho C (2007) A Raman study of the sulfated cement hydrates: ettringite and monosulfoaluminate. J Adv Concr Technol 5(3):299–312

Repelin Y, Husson E, Brusset H (1979) Etude par spectroscopies d'absorptioni.r. et de diffusion Raman des composés A^{II}B$^V_2$O$_6$ de structure de type "blocs 1×2" – I. Etude du niobate de Baryum BaNb$_2$O$_6$. Spectrochim Acta A 35:937–948. (in French)

Repelin Y, Husson E, Dao NQ, Brusset H (1980) Etude par spectroscopies d'absorption i.r. et de diffusion Raman des composés A^{II}B$^V_2$O$_6$ de structure de type "blocs 1×2" – III. Etude des niobate et tantalite de plomb rhomboèdriques PbNb$_2$O$_6$ et PbTa$_2$O$_6$. Spectrochim Acta A 36:253–258. (in French)

Reshak AH, Chen X, Auluck S, Kamarudin H (2012a) Structural, electronic properties and charge density distribution of the LiNaB$_4$O$_7$: theory and experiment. Mater Chem Phys 137(1):346–352

Reshak AH, Chen X, Auluck S, Kamarudin H (2012b) Single-crystal oxoborate $(Pb_3O)_2(BO_3)_2WO_4$: growth and characterization. Mater Res Bull 47:2552–2560

Reshetnyak NB, Bukanov VV (1991) A method of non-destructive diagnosis of gems (laser spectroscopy of Raman scattering of light). Geoinformmark, Moscow. (in Russian)

Reynard B, Guyot F (1994) High-temperature properties of geikielite ($MgTiO_3$-ilmenite) from high-temperature high-pressure Raman spectroscopy – some implications for $MgSiO_3$-ilmenite. Phys Chem Minerals 21(7):441–450

Reynard B, Bezacier L, Caracas R (2015) Serpentines, talc, chlorites, and their high-pressure phase transitions: a Raman spectroscopic study. Phys Chem Miner 42(8):641–649

Ribeiro J, Flores D, Ward CR, Silva LF (2010) Identification of nanominerals and nanoparticles in burning coal waste piles from Portugal. Scie Total Environ 408 (23):6032–6041

Richet P (1990) GeO_2 vs SiO_2: glass transition and thermodynamic properties of polymorphs. Phys Chem Minerals 17:79–88

Richet P, Mysen BO, Ingrin J (1998) High-temperature X-ray diffraction and Raman spectroscopy of diopside and pseudowollastonite. Phys Chem Mineral 25 (6):401–414

Richman I (1966) Raman spectra of YPO_4 and $YbPO_4$. JOSA 56(11):1589–1590

Richter W, Köhler H, Becker CR (1977) A Raman and far-infrared investigation of phonons in the rhombohedral V_2–VI_3 compounds Bi_2Te_3, Bi_2Se_3, Sb_2Te_3 and $Bi_2(Te_{1-x}Se_x)_3$ ($0 < x < 1$), $(Bi_{1-y}Sb_y)_2Te_3$ ($0 < y < 1$). Phys Status Solidi B 84(2):619–628

Rickers K, Thomas R, Heinrich W (2006) The behavior of trace elements during the chemical evolution of the H_2O-, B-, and F-rich granite-pegmatite-hydrothermal system at Ehrenfriedersdorf, Germany: a SXRF study of melt and fluid inclusions. Mineral Deposita 41:229–245

Rieger F, Mudring A-V (2007) Phase transition in Tl_2TeO_3: influence and origin of the thallium lone pair distortion. Inorg Chem 46(2):446–452

Rigotti G, Punte G, Rivero BE, Escobar ME, Baran EJ (1981) Crystal data and vibrational spectra of the rare earth decavanadates. J Inorg Nucl Chem 43 (11):2811–2814

Rigotti G, Lavat AE, Escobar ME, Baran EJ (1983) KristallographischeDaten, IR-Spektrum und thermisches Verhalten von Aluminium-Dekavanadat. Z anorg allg Chemie 501(6):184–190. (in German)

Rinaudo C, Gastaldi D, Belluso E (2003) Characterization of chrysotile, antigorite and lizardite by FT-Raman spectroscopy. Can Mineral 41(4):883–890

Rinaudo C, Belluso E, Gastaldi D (2004) Assessment of the use of Raman spectroscopy for the determination of amphibole asbestos. Mineral Mag 68(3):455–465

Rinaudo C, Cairo S, Gastaldi D, Gianfagna A, Mazziotti Tagliani S, Tosi G, Conti C (2006) Characterization of fluoro-edenite by μ-Raman and μ-FTIR spectroscopy. Mineral Mag 70(3):291–298

Rincón C, Quintero M, Power C, Moreno E, Quintero E, Henao JA, Macías MA, Morocoima M (2015) Raman spectra of $Cu_2B^{II}C^{IV}X^{VI}_4$ magnetic quaternary semiconductor compounds with tetragonal stannite type structure. J Appl Phys 117(20):205701-1–205701-6

Rintoul L, Bahfenne S, Frost RL (2011) Single-crystal Raman spectroscopy of brandholzite Mg[Sb $(OH)_6]_2 \cdot 6H_2O$ and bottinoite Ni[Sb$(OH)_6]_2 \cdot 6H_2O$ and the polycrystalline Raman spectrum of mopungite Na[Sb$(OH)_6$]. J Raman Spectrosc 42(5):1147–1153

Riquelme F, Ruvalcaba-Sil JL, Alvarado-Ortega J, Estrada-Ruiz E, Galicia-Chávez M, Porras-Múzquiz H, Stojanoff V, Siddons DP, Miller L (2014) Amber from México: coahuilite, simojovelite and bacalite. In: MRS Proceedings, vol 1618. Cambridge University Press, pp 169–180. https://doi.org/10.1557/opl.2014.466

Rissom C, Schmidt H, Voigt W (2008) Crystal structure and thermal properties of a new double salt: $K_2SiF_6 \cdot KNO_3$. Cryst Res Technol 43(1):74–82

Ristić M, Popović S, Musić S (2005) Application of sol–gel method in the synthesis of gallium (III)-oxide. Mater Lett 59(10):1227–1233

Rius J, Allmann R (1984) The superstructure of the double layer mineral wermlandite $[Mg_7(Al_{0.57}Fe^{3+}_{0.43}) (OH)_{18}]^{2+} \cdot [(Ca_{0.6},Mg_{0.4})(SO_4)_2(H_2O)_{12}]^{2-}$. Z Kristallogr 168:133–144

Rocchiccioli-Deltcheff C (1973) Comparaison des spectres d'absorption infrarouge de niobates et tantalates de mètaux monovalents. Spectrochim Acta A 29(1):93–106. (in French)

Rocha AL, Baran EJ (1988) Spektroskopisches und thermisches Verhalten vom $VOSeO_3 \cdot 3H_2O$ und $VOSeO_3$. Z anorg allg Chemie 564(1):141–147. (in German)

Rocha J, Brandão P, Lin Z, Kharlamov A, Anderson MW (1996) Novel microporous titanium–niobium–silicates with the structure of nenadkevichite. Chem Commun 5:669–670

Rocquet P, Couzi M, Tressaud A, Chaminade JP, Hauw C (1985) Structural phase transition in chiolite $Na_5Al_3F_{14}$. I. Raman scattering and X-ray diffraction study. J Phys C 18(36):6555–6569

Rodgers KA (1992) The laser Raman identity of gibbsite pseudomorphous after crocoite from Dundas, Tasmania. Papers Proc Royal Soc Tasmania 126:95–99

Rodgers KA (1993) Routine identification of aluminium hydroxide polymorphs with the laser Raman microprobe. Clay Minerals 28(1):85–99

Rodgers K, Gregory MR, Cooney RP (1989) Bayerite, Al $(OH)_3$ from Raoul Island, Kermadec Group, South Pacific. Clay Minerals 24(3):531–538

Rodgers KA, Kobe HW, Childs CW (1993) Characterization of vivianite from Catavi, Llallagua, Bolivia. Mineral Petrol 47(2–4):193–208

Rodríguez-Clemente R, Serna CJ, Ocaña M, Matijević E (1994) The relationship of particle morphology and structure of basic copper (II) compounds obtained by homogeneous precipitation. J Cryst Growth 143 (3):277–286

Romann J, Chevallier V, Merlen A, Valmalette JC (2009) Self-assembly and Raman spectroscopy of additive coated nanocrystals. MRS Proc. https://doi.org/10.1557/PROC-1176-Y06-20

Romero M, Gómez RW, Marquina V, Pérez-Mazariego JL, Escamilla R (2014) Synthesis by molten salt method of the AFeO$_3$ system (A = La, Gd) and its structural, vibrational and internal hyperfine magnetic field characterization. Phys B 443

Roncaglia DI, Botto IL, Baran EJ (1985) Vibrational spectrum of Pb$_2$CrO$_5$. J Mater Sci Lett 4:1427–1428

Roncaglia DI, Botto IL, Baran EJ (1993) Vibrational spectrum of synthetic rooseveltite. N Jb Miner Mh 6:249–253

Ronchi L-H (2003) Contribution à la connaissance de la ceinture à fluorine du Vale do Ribeira (Parana-Sao Paulo), Brésil. Etude géologique, minéralogique et géochimique. Thèse, Université d'Orléans. (in French)

Rondeau B, Fritsch E, Lefevre P, Guiraud M, Fransolet A-M, Lulzac Y (2006) A Raman investigation of the amblygonite–montebrasite series. Can Mineral 44(5):1109–1117

Roqué-Rosell J, Mosselmans JFW, Proenza JA, Labrador M, Galí S, Atkinson KD, Quinn PD (2010) Sorption of Ni by "lithiophorite-asbolane" intermediates in Moa Bay lateritic deposits, eastern Cuba. Chem Geol 275(1):9–18

Rosasco GJ, Blaha JJ (1980) Raman microprobe spectra and vibrational mode assignments of talc. Appl Spectrosc 34(2):140–144

Rosi F, Manuali V, Miliani C, Brunetti BG, Sgamellotti A, Grygar T, Hradil D (2009) Raman scattering features of lead pyroantimonate compounds. Part I: XRD and Raman characterization of Pb$_2$Sb$_2$O$_7$ doped with tin and zinc. J Raman Spectrosc 40(1):107–111

Rosi F, Grazia C, Gabrieli F, Romani A, Paolantoni M, Vivani R, Brunetti BG, Colomban P, Miliani C (2016) UV–Vis-NIR and micro Raman spectroscopies for the non destructive identification of Cd$_{1-x}$Zn$_x$S solid solutions in cadmium yellow pigments. Microchem J 124:856–867

Ross SD (1972) The vibrational spectra of some minerals containing tetrahedrally co-ordinated boron. Spectrochim Acta A 28(8):1555–1561

Rotondo GG, Darchuk L, Swaenen M, Van Grieken R (2012) Micro-Raman and SEM analysis of minerals from the Darhib mine, Egypt. J Anal Sci Methods Instrum 2:42–47

Rouchon V, Badet H, Belhadj O, Bonnerot O, Lavédrine B, Michard JG, Miska S (2012) Raman and FTIR spectroscopy applied to the conservation report of paleontological collections: identification of Raman and FTIR signatures of several iron sulfate species such as ferrinatrite and sideronatrite. J Raman Spectrosc 43(9):1265–1274

Roy BN (1987) Spectroscopic analysis of the structure of silicate glasses along the joint xMAlO$_2$–$(1-x)$SiO$_2$ (M = Li, Na, K, Rb, Cs). J Amer Ceram Soc 70(3):183–192

Roy P, Srivastava SK, Nayak BB, Saxena AK (2008) Morphology evolution of Sb$_2$S$_3$ under hydrothermal conditions: flowerlike structure to nanorods. Cryst Growth Design 8(6):2019–2023

Rozov K, Berner U, Taviot-Gueho C, Leroux F, Renaudin G, Kulik D, Diamond LW (2010) Synthesis and characterization of the LDH hydrotalcite-pyroaurite solid solution series. Cem Concr Res 40:1248–1254

Rozov KB, Berner U, Kulik DA, Diamond LW (2011) Solubility and thermodynamic properties of carbonate-bearing hydrotalcite–pyroaurite solid solutions with a 3:1 Mg/(Al+ Fe) mole ratio. Clays Clay Minerals 59(3):215–232

RRUFF (2007) Infrared spectrum for heklaite. http://rruff.info/heklaite/display=default/. Accessed 5 Apr 2007

Ruan HD, Frost RL, Kloprogge JT (2001) Comparison of Raman spectra in characterizing gibbsite, bayerite, diaspore and boehmite. J Raman Spectrosc 32(9):745–750

Ruiz-Fuertes J, López-Moreno S, Errandonea D, Pellicer-Porres J, Lacomba-Perales R, Segura A, Rodríguez-Hernández P, Muñoz A, Romero AH, González J (2010) High-pressure phase transitions and compressibility of wolframite-type tungstates. J Appl Phys 107(8):083506-01–083506-10

Rull F, Guerrero J, Venegas G, Gázquez F, Medina J (2014) Spectroscopic Raman study of sulphate precipitation sequence in Rio Tinto mining district (SW Spain). Environ Sci Pollut Res 21(11):6783–6792

Rulmont A (1972) Spectre infra-rouge de quelques oxyhalogénures de bismuth et de terres rares. Spectrochim Acta A 28:1287–1296. (in French)

Rulmont A, Almou M (1989) Vibrational spectra of metaborates with infinite chain structure: LiBO$_2$, CaB$_2$O$_4$, SrB$_2$O$_4$. Spectrochim Acta A 45(5):603–610

Rulmont A, Tarte P (1987) Infrared spectrum of crystalline and glassy borosilicates M^{I}BSi$_2$O$_6$. J Mater Sci Lett 6(1):38–40

Rulmont A, Tarte P, Winand JM (1987) Vibrational spectrum of crystalline and glassy LiBGeO$_4$: structural analogies with BAsO$_4$. J Mater Sci Lett 6(6):659–662

Rulmont A, Tarte P, Choisnet J (1992) Vibrational spectra of tellurates with the garnet structure. Spectrochim Acta A 48(7):921–930

Runčevski T, Wu C, Yu H, Yang B, Dinnebier RE (2013) Structural characterization of a new magnesium oxysulfate hydrate cement phase and its surface reactions with atmospheric carbon dioxide. J Am Ceram Soc 96(11):3609–3616

Ruschel K, Nasdala L, Kronz A, Hanchar JM, Többens DM, Škoda R, Finger F, Möller A (2012) A Raman spectroscopic study on the structural disorder of monazite-(Ce). Mineral Petrol 105(1–2):41–55

Russell JD, Milodowski AE, Fraser AR, Clark DR (1983) New IR and XRD data for leadhillite of ideal composition. Mineral Mag 47:371–375

Russo V, Bailini A, Zamboni M, Passoni M, Conti C, Casari CS, Li Bassi A, Bottani CE (2008) Raman spectroscopy of Bi-Te thin films. J Raman Spectrosc 39(2):205–210

Rutt HN, Nicola JH (1974) Raman spectra of carbonates of calcite structure. J Phys C 7(24):4522–4528

Ryskin YI, Stavitskaya GP (1990) Asymmetry of the water molecule in crystal hydrates: IR spectra of dioptase and apophyllite. Bull Acad Sci URRS Chem Sci 39:1610–1614

Saad S, Obbade S, Yagoubi S, Renard C, Abraham F (2008) A new uranyl niobate sheet in the cesium uranyl niobate $Cs_9[(UO_2)_8O_4(NbO_5)(Nb_2O_8)_2]$. J Solid State Chem 181(4):741–750

Sabadel JC, Armand P, Cachau-Herreillat D, Baldeck P, Doclot O, Ibanez A, Philippot E (1997) Structural and nonlinear optical characterizations of tellurium oxide-based glasses: TeO_2-BaO-TiO_2. J Solid State Chem 132(2):411–419

Sagatowska I (2010) Chromite in serpentine mud volcanoes of the Mariana forearc: implications for abiotic organic reactions. Dissertation, Stockholm Unisersity

Saheb M, Neff D, Bellot-Gurlet L, Dillmann P (2011) Raman study of a deuterated iron hydroxycarbonate to assess long-term corrosion mechanisms in anoxic soils. J Raman Spectrosc 42(5):1100–1108

Sahoo PP, Sumithra S, Madras G, Row TG (2009) Synthesis, structure and photocatalytic properties of β-$ZrMo_2O_8$. Bull Mater Sci 32(3):337–342

Sakai A, Negishi M, Fujiwara E, Moriyoshi C, Itoh K (2001) Micro-Raman apectra of langbeinite-type $K_2Mn_2(SO_4)_3$ and $(NH_4)_2Cd_2(SO_4)_3$ near the phase transition temperature. J Phys Soc Jpn 70(11):3452–3456

Sakurai T, Ohno H, Horikawa S, Iizuka Y, Uchida T, Hondoh T (2010) A technique for measuring microparticles in polar ice using micro-Raman spectroscopy. Int J Spectrosc. https://doi.org/10.1155/2010/384956

Salentine CG (1987) Synthesis, characterization, and crystal structure of a new potassium borate, $KB_3O_5 \cdot 3H_2O$. Inorg Chem 26(1):128–132

Salvadó MA, Pertierra P, Bortun AI, Trobajo C, García JR (2005) New hydrothermal synthesis and structure of $Th_2(PO_4)_2(HPO_4) \cdot H_2O$: the first structurally characterized thorium hydrogen phosphate. Inorg Chem 44(10):3512–3517

Sanchez C, Livage J, Lucazeau G (1982) Infrared and Raman study of amorphous V_2O_5. J Raman Spectrosc 12(1):68–72

Sanchez-Moral S, Fernandez-Cortes A, Cuezva S, Cañaveras JC, Correcher V, Miller AZ, Dionisio A, Marques JM, Saiz-Jimenez C, Afonso MJ, Chaminé HI, Furio M, Garcia-Guinea J (2011) Uranyl-evansites from Porto (Northwest Portugal) and Galicia (Northwest Spain): structure and assignment of spectra catholuminescence and Raman bands. Spectrosc Lett 44(7–8):511–515

Sánchez-Pastor N, Cruz UA, Gigler AM, Park S, Jordan G, Schmahl W, Fernández-Díaz L (2010) Microprobe and Raman investigation of the zoning in synthetic Ca (CO_3,CrO_4) crystals. Revista Soc Esp Mineral 13:197–198

Sánchez-Pastor N, Pinto AJ, Astilleros JM, Fernández-Díaz L, Gonçalves MA (2013) Raman spectroscopic characterization of a synthetic, non-stoichiometric Cu–Ba uranyl phosphate. Spectrochim Acta A 113:196–202

Sánchez-Pastor N, Oehlerich M, Astilleros JM, Kaliwoda M, Mayr CC, Fernández-Díaz L, Schmahl WW (2016) Crystallization of ikaite and its pseudomorphic transformation into calcite: Raman spectroscopy evidence. Geochim Cosmochim Acta 175:271–281

Saniger JM (1995) Al-O infrared vibrational frequencies of γ-alumina. Mater Lett 22(1):109–113

Sanjeewa LD, McGuire MA, Garlea VO, Hu L, Chumanov G, McMillen CD, Kolis JW (2015) Hydrothermal synthesis and characterization of novel brackebuschite-type transition metal vanadates: $Ba_2M(VO_4)_2OH$, $M = V^{3+}$, Mn^{3+}, and Fe^{3+}, with interesting Jahn–Teller and spin-liquid behavior. Inorg Chem 54(14):7014–7020

Sanjeewa LD, McGuire MA, McMillen CD, Willett D, Chumanov G, Kolis JW (2016) Honeycomb-like S= 5/2 spin-lattices in manganese (II) vanadates. Inorg Chem 55:9240–9249

Santos VP, Soares OSGP, Bakker JJW, Pereira MFR, Órfão JJM, Gascon J, Kapteijn F, Figueiredo JL (2012) Structural and chemical disorder of cryptomelane promoted by alkali doping: influence on catalytic properties. J Catal 293:165–174

Sapozhnikov AN, Kaneva EV, Cherepanov DI, Suvorova LF, Levitsky VI, Ivanova LA, Reznitsky LZ (2012) Vladimirivanovite, $Na_6Ca_2[Al_6Si_6O_{24}](SO_4,S_3,S_2, Cl)_2 \cdot H_2O$, a new mineral of sodalite group. Geol Ore Depos 54(7):557–564

Sapozhnikov AN, Kaneva EP, Surorova LP, Levitsky VI, Ivanova LA (2016) Sulfhydrylbystrite, $Na_5K_2Ca[Al_6Si_6O_{24}](S_5)(SH)$, a new mineral with the LOS framework, and re-interpretation of bystrite: cancrinite-group minerals with novel extra-framework anions. Mineral Mag. https://doi.org/10.1180/minmag.2016.080.106

Sarma LP, Prasad PSR, Ravikumar N (1998) Raman spectroscopic study of phase transitions in natural gypsum. J Raman Spectrosc 29(9):851–856

Sarr O, Diop L (1984) The vibrational spectra of the crystalline tripotassium hydrogen pyrophosphates $K_3HP_2O_7 \cdot 3H_2O$ and $K_3HP_2O_7$. Spectrochim Acta A 40(11):1011–1015

Sarr O, Diop L (1987) Ir and Raman spectra of $M_3HP_2O_7 \cdot nH_2O$ (M = Na, Cs; $n = 0, 1, 9$). Correlation between the POP bridge vibrational frequencies and the POP angle value. Spectrochim Acta A 43(8):999–1005

Sasaki K, Tanaike O, Konno H (1998) Distinction of jarosite-group compounds by Raman spectroscopy. Can Mineral 36:1225–1235

Sasaki K, Nakamuta Y, Hirajima T, Tuovinen OH (2009) Raman characterization of secondary minerals formed during chalcopyrite leaching with Acidithiobacillus ferrooxidans. Hydrometallurgy 95(1):153–158

Sato T, Aoki K, Kino R, Kuroe H, Sekine T, Hase M, Ito T, Eisaki H (2014) Raman scattering in $(Cu, Zn)_3(Mo,W)_2O_9$. JPS Conf Proc. https://doi.org/10.7566/JPSCP.3.014035

Scheetz BE, White WB (1977) Vibrational spectra of the alkaline earth double carbonates. Am Mineral 62:36–50

Scheuermann W, Schutte CJH (1973a) Raman and infrared spectra of $BaCrO_4$ and $BaSeO_4$. J Raman Spectrosc 1(6):605–618

Scheuermann W, Schutte CJH (1973b) Raman and infrared spectra of $SrSeO_4$ and $PbSeO_4$. J Raman Spectrosc 1(6):619–627

Schiffer J, Hornig DF (1961) On a reported new form of ice. J Chem Phys 35:1136–1137

Schingaro E, Kullerud K, Lacalamita M, Mesto E, Scordari F, Zozulya D, Erambert R, Ravna EJ (2014) Yangzhumingite and phlogopite from the Kvaløya lamproite (North Norway): structure, composition and origin. Lithos 210:1–13

Schlüter J, Pohl D, Golla-Schindler U (2008) Santarosaite, CuB_2O_4, a new mineral with disordered structure from the Santa Rosa mine, Atacama desert, Chile. N Jb Miner Abh 185(1):27–32

Schlüter J, Malcherek T, Husdal TA (2009) The new mineral stetindite, $CeSiO_4$, a cerium end-member of the zircon group. N Jb Miner Abh 186(2):195–200

Schlüter J, Geisler T, Pohl D, Stephan T (2010) Krieselite, $Al_2GeO_4(F,OH)_2$: a new mineral from the Tsumeb mine, Namibia, representing the Ge analogue of topaz. N Jb Miner Abh 187(1):33–40

Schlüter J, Malcherek T, Mihailova B, Gebhard G (2013) The new mineral erikapohlite, $Cu_3(Zn,Cu,Mg)_4Ca_2(AsO_4)_6\cdot 2H_2O$, the Ca-dominant analogue of keyite, from Tsumeb, Namibia. N Jb Miner Mh 190(3):319–325

Schlüter J, Malcherek T, Mihailova B (2014) Galloplumbogummite from Tsumeb, Namibia, a new member of the alunite group with tetravalent charge balance. N Jb Miner Abh 191(3):301–309

Schlüter J, Malcherek T, Gebhard G (2016) Vanackerite, a new lead cadmium arsenate of the apatite supergroup from Tsumeb, Namibia. N Jb Miner Abh 193(1):79–86

Schmetzer K, Bosshart G, Bernhardt H-J, Gübelin EJ, Smith CP (2002) Serendibite from Sri Lanka. Gems Gemology 38:73–79

Schmetzer K, Burford M, Kiefert L, Bernhardt H-J (2003) The first transparent faceted grandidierite, from Sri Lanka. Gems Gemol 39:32–37

Schmid T, Dariz P (2015) Shedding light onto the spectra of lime: Raman and luminescence bands of CaO, Ca(OH)$_2$ and $CaCO_3$. J Raman Spectrosc 46(1):141–146

Schmidt M, Lutz HD (1993) Hydrogen bonding in basic copper salts: a spectroscopic study of malachite, $Cu_2(OH)_2CO_3$, and brochantite, $Cu_4(OH)_6SO_4$. Phys Chem Minerals 20(1):27–32

Schmidt GR, Reynard J, Yang H, Downs RT (2006) Tychite, $Na_6Mg_2(CO_3)_4(SO_4)$: structure analysis and Raman spectroscopic data. Acta Crystallogr E. https://doi.org/10.1107/S160053680603491X

Schmidt BC, Gehlken P-L, Böttcher ME (2013) Vibrational spectra of $BaMn(CO_3)_2$ and a re-analysis of the Raman spectrum of $BaMg(CO_3)_2$. Eur J Mineral 25:137–144

Schmidt A, Lerch M, Eufinger J-P, Janek J, Tranca I, Islam MM, Bredow T, Dolle R, Wiemhöfer H-D, Boysen H, Hölzel M (2014) Chlorine ion mobility in Cl-mayenite $(Ca_{12}Al_{14}O_{32}Cl_2)$: an investigation combining high-temperature neutron powder diffraction, impedance spectroscopy and quantum-chemical calculations. Solid State Ionics 254:48–58

Schmutzler B, Eggert G, Kuhn-Wawrzinek CF (2016) Copper (II) hydroxide on artefacts: corrosion, conservation, colourants. Stud Conserv:1–7. https://doi.org/10.1080/00393630.2016.1215591

Schneider P, Tropper P, Kaindl R (2013) The formation of phosphoran olivine and stanfieldite from the pyrometamorphic breakdown of apatite in slags from a prehistoric ritual immolation site (Goldbichl, Igls, Tyrol, Austria). Mineral Petrol 107(2):327–340

Schofield K (2004) Elegant elusive oxidation mechanism of mercury. Prepr Pap Am Chem Soc Div Fuel Chem 49(1):294–295

Scholz R, Frost RL, Xi Y, Graça LM, Lagoeiro L, López A (2013a) Vibrational spectroscopic characterization of the phosphate mineral phosphophyllite – $Zn_2Fe(PO_4)_2\cdot 4H_2O$, from Hagendorf Süd, Germany and in comparison with other zinc phosphates. J Molec Struct 1039:22–27

Scholz R, Xi Y, Frost RL (2013b) The molecular structure of matioliite – $NaMgAl_5(PO_4)_4(OH)_6\cdot 2(H_2O)$ – a pegmatite mineral from Minas Gerais, Brazil. J Molec Struct 1033:265–271

Scholz R, Frost RL, Frota L, Belotti FM, López A (2015) SEM, EDX and vibrational spectroscopy of the phosphate mineral vauxite from Llallagua, Bolívia. Spectrochim Acta A 151:149–155

Schönenberger UW, Günter JR, Oswald HR (1971) Polymorphism of copper (II) hydroxide. J Solid State Chem 3(2):190–193

Schumer BN, Yang H, Downs RT (2016) Natropalermoite, $Na_2SrAl_4(PO_4)_4(OH)_4$, a new mineral isostructural with palermoite, from the Palermo No. 1 mine, Groton, New Hampshire, USA. Mineral Mag. https://doi.org/10.1180/minmag.2016.080.133

Schutte CJH, Van Rensburg DJJ (1971) Low-temperature infrared and Raman studies. X. The vibrational behaviour of ammonium fluoro-borate – its phase changes and the rotational freedom of its ions. J Molec Struct 10:484–489

Schwalowsky L, Bismayer U, Lippmann T (1996) The improper ferroelastic phase transition of letovicite, $(NH_4)_3H(SO_4)_2$: an optical birefringence, X-ray diffraction and Raman spectroscopic study. Phase Transit 59(1–3):61–76

Schwendt P, Joniaková D (1975) Vibrational spectra of vanadium(V) compounds. II. Vibrational spectra of divanadates with nonlinear bridge VOV. Chemzvesti 29(3):381–386

Scott JF (1968) Raman spectra of BaClF, BaBrF, and SrClF. J Chem Phys 49(6):2766–2769

Seetharaman S, Bhat HL, Narayanan PS (1983) Raman spectroscopic studies on sodium metavanadate. J Raman Spectrosc 14(6):401–405

Seguin L, Figlarz M, Cavagnat R, Lassègues JC (1995) Infrared and Raman spectra of MoO_3 molybdenum trioxides and $MoO_3 \cdot xH_2O$ molybdenum trioxide hydrates. Spectrochim Acta A 51(8):1323–1344

Seidel H, Ehrhardt H, Viswanathan K, Johannes W (1974) Darstellung, Struktur und Eigenschaften von Kupfer (II)-Carbonat. Z anorg allg Chemie 410(2):138–148. (in German)

Sejkora J, Čejka J (2007) Šreinite from Horní Halže, the Krušné hory Mountains, Czech Republic, a new mineral species, its comparison with asselbornite from Schneeberg, and new data for asselbornite. N Jb Miner Abh 184(2):197–206

Sejkora J, Ozdín D, Vitáloš J, Tuček P, Čejka J, Ďuďa R (2007) Schafarzikite from the type locality Pernek (Malé Karpaty Mountains, Slovak Republic) revisited. Eur J Mineral 19:419–427

Sejkora J, Čejka J, Kolitsch U (2008) Uranosphaerite from Horní Halže near Měděnec (Krušné hory Mountains, Czech Republic): description and vibrational characteristics. N Jb Miner Abh 185(1):91–98

Sejkora J, Čejka J, Frost RL, Bahfenne S, Plášil J, Keeffe EC (2010) Raman spectroscopy of hydrogen-arsenate group (AsO_3OH) in solid-state compounds: copper mineral phase geminite $Cu(AsO_3OH) \cdot H_2O$ from different geological environments. J Raman Spectrosc 41(9):1038–1043

Sejkora J, Plášil J, Filip J (2011) Plimerite from Krásno near Horní Slavkov ore district, Czech Republic. J Geosci 56:215–229

Sejkora J, Čejka J, Malíková R, López A, Xi Y, Frost RL (2014) A Raman spectroscopic study of a hydrated molybdate mineral ferrimolybdite, $Fe_2(MoO_4)_3 \cdot 7-8H_2O$. Spectrochim Acta A 130:83–89

Sejkora J, Grey IE, Kampf AR, Price JR, Čejka J (2016) Tvrdýite, $Fe^{2+}Fe^{3+}_2Al_3(PO_4)_4(OH)_5(OH_2)_4 \cdot 2H_2O$, a new phosphate mineral from Krásno near Horní Slavkov, Czech Republic. Mineral Mag. https://doi.org/10.1180/minmag.2016.080.045

Sekine T, Izumi M, Nakashizu T, Uchinokura K, Matsuura E (1980) Raman scattering and infrared reflectance in $2H$-$MoSe_2$. J Phys Soc Jpn 49(3):1069–1077

Sekine T, He H, Kobayashi T, Shibata K (2006) Sinoite (Si_2N_2O) shocked at pressures of 28 to 64 GPa. Am Mineral 91(2–3):463–466

Serena S, Sainz MA, Caballero A (2014) Single-phase silicocarnotite synthesis in the subsystem $Ca_3(PO_4)_2$–Ca_2SiO_4. Ceramics Int 40:8245–8252

Serena S, Caballero A, De Aza PN, Sainz MA (2015) New evaluation of the in vitro response of silicocarnotite monophasic material. Ceram Int 41(8):9411–9419

Serghiou G, Guillaume C (2004) Stability of K_2CrO_4 to 50 GPa using Raman spectroscopy measurements. J Solid State Chem 177(12):4672–4679

Serna CJ, García-Ramos JV, Peña MJ (1985) Vibrational study of dawsonite type compounds $MAl(OH)_2CO_3$ (M= Na, K, NH_4). Spectrochim Acta A 41(5):697–702

Shahar A, Bassett WA, Mao HK, Chou IM, Mao W (2005) The stability and Raman spectra of ikaite, $CaCO_3 \cdot 6H_2O$, at high pressure and temperature. Am Mineral 90(11–12):1835–1839

Shantakumari C (1953) Raman spectra of crystalline sulphates of zinc, magnesium and sodium. Proc Indian Acad Sci 37:393–401. https://doi.org/10.1007/BF03052656

Shao H, Padmanathan N, McNulty D, O'Dwyer C, Razeeb KM (2016) Supercapattery based on binder-free $Co_3(PO_4)_2 \cdot 8H_2O$ multilayer nano/microflakes on nickel foam. ACS Appl Mater Interf. https://doi.org/10.1021/acsami.6b08354

Shao-Long T, Qiang S, Ya-Qin Y, Jian-Fang M (1996) A new phosphate, $KMgY(PO4)_2$, isostructural with xenotime. Chin J Chem 14(1):25–30

Sharma SK, Pandya DK (1974) Laser-Raman spectra of crystalline $(NH_4)_2FeCl_5 \cdot H_2O$, $K_2FeCl_5 \cdot H_2O$ and $K_2FeCl_5 \cdot D_2O$. J Inorg Nucl Chem 36(5):1165–1166

Sharma SK, Simons B, Yoder HS Jr (1983) Raman study of anorthite, calcium Tschermak's pyroxene, and gehlenite in crystalline and glassy states. Am Mineral 68(11–12):1113–1125

Sharma SK, Yoder HS, Matson DW (1988) Raman study of some melilites in crystalline and glassy states. Geochim Cosmochim Acta 52(8):1961–1967

Sharma SK, Parthasarathy G, Ravindran TR, Arora AK, Kumar B (2001) Laser-Raman spectroscopic on graphite from East Antarctica. Polar Geosci 14:157–164

Sharygin VV, Stoppa F, Kolesov BA (1996a) Cuspidine in melilitolites of San Venanzo, Italy. Transact (Doklady) Russ Acad Sci Earth Sci Sect 349(5):747–751

Sharygin VV, Stoppa F, Kolesov BA (1996b) Zr-Ti disilicates from the Pian di Celle volcano, Umbria, Italy. Eur J Mineral 8:1199–1212

Sharygin VV, Lazic B, Armbruster TM, Murashko MN, Wirth R, Galuskina IO, Galuskin EV, Vapnik Y, Britvin SN, Logvinova AM (2013a) Shulamitite $Ca_3TiFe^3+AlO_8$ – a new perovskite-related mineral from Hatrurim Basin, Israel. Eur J Mineral 25(1):97–111

Sharygin VV, Pekov IV, Zubkova NV, Khomyakov AP, Stoppa F, Pushcharovsky DY (2013b) Umbrianite, $K_7Na_2Ca_2[Al_3Si_{10}O_{29}]F_2Cl_2$, a new mineral species from melilitolite of the Pian di Celle volcano, Umbria, Italy. Eur J Mineral 25(4):655–669

Sharygin IS, Golovin AV, Korsakov AV, Pokhilenko NP (2013c) Eitelite in sheared peridotite xenoliths from Udachnaya-East kimberlite pipe (Russia) – a new locality and host rock type. Eur J Mineral 25(5):825–834

Sharygin VV, Kryvdik SG, Karmanov NS, Nigmatulina EN (2014) Chlorine-bearing annite from Khlebodarovka enderbites, Azov Sea region, Ukrainian shield. 30th International Conference on "Ore potential of alkaline, kimberlite and carbonatite magmatism", Antalya, Turkey, 29 September–02 October 2014, p 162

Shatskiy A, Gavryushkin PN, Sharygin IS, Litasov KD, Kupriyanov IN, Higo Y, Borzdov YM, Funakoshi K, Palyanov YN, Ohtani E (2013) Melting and subsolidus phase relations in the system Na_2CO_3–$MgCO_3 \pm H_2O$ at 6 GPa and the stability of $Na_2Mg(CO_3)_2$ in the upper mantle. Am Mineral 98(11–12):2172–2182

Shatskiy A, Gavryushkin PN, Litasov KD, Koroleva ON, Kupriyanov IN, Borzdov YM, Funakoshi K, Palyanov YN, Ohtani E (2015) Raman spectroscopic and X-ray diffraction studies of the Na-Ca carbonates synthesized under upper mantle conditions. Eur J Mineral. https://doi.org/10.1127/ejm/2015/0027-2426

Sherwin R, Clark RJ, Lauck R, Cardona M (2005) Effect of isotope substitution and doping on the Raman spectrum of galena (PbS). Solid State Commun 134(8):565–570

Shi F, Rocha J, Trindade T (2005) Synthetic $NaMnPO_4$ microtubules. Mater Lett 59(6):652–655

Shi W, Zhang X, Che G (2013) Hydrothermal synthesis and electrochemical hydrogen storage performance of porous hollow NiSe nanospheres. Int J Hydrogen Energy 38(17):7037–7045

Shimura T, Akai J, Lazic B, Armbruster T, Shimizu M, Kamei A, Tsukada K, Owada M, Yuhara M (2012) Magnesiohögbomite-2N4S: a new polysome from the central Sør Rondane Mountains, East Antarctica. Am Mineral 97(2–3):268–280

Shin Y, Lee DW, Choi KY, Koo HJ, Ok KM (2013) VSb$(SeO_3)_4$, first selenite containing V^{3+} cation: synthesis, structure, characterization, magnetic properties, and calculations. Inorg Chem 52(24):14224–14230

Shinde VV, Dhoble SJ (2015) Wet chemical synthesis of $Na_{21}(SO_4)_7F_6Cl$:Ce optoelectronics materials. IOP Conf Series Mater Sci Eng 73(1):012038. (7 pp). https://doi.org/10.1088/1757-899X/73/1/012038

Shinde VV, Shinde SV, Dhoble NS, Dhoble SJ (2015) Luminescence investigations of RE^{3+} (RE^{3+} = Dy^{3+}, Eu^{3+}, Tb^{3+}) activated $Na_{21}(SO_4)_7F_6Cl$ optoelectronic nanophosphors under near UV excitation for LED. J Inorg Organometal Polym Mater 25(3):593–600

Shirozu H, Ishida K (1982) Infrared study of some 7 Å and 14 Å layer silicates by deuteration. Mineral J (Jpn) 11(4):161–171

Shoval S, Boudeulle M, Yariv S, Lapides I, Panczer G (2001) Micro-Raman and FT-IR spectroscopy study of the thermal transformations of St. Claire dickite. Opt Mater 16(1):319–327

Shoval S, Gaft M, Panczer G (2003) Luminescence of Cr^{3+} in natural and calcined diaspore. J Therm Anal Calorim 71(3):699–706

Shu J, Zhu X, Yi T (2009) RETRACTED: $CuCrO_2$ as anode material for lithium ion batteries. Electrochim Acta 54(10):2795–2799

Siebert H (1959) Ultrarotspektren von Tellursäuren, Telluraten und Antimonaten. Z anorg allg Chemie 301(3–4):161–170. (in German)

Silva EN, Ayala AP, Guedes I, Paschoal CWA, Moreira RL, Loong CK, Boatner LA (2006) Vibrational spectra of monazite-type rare-earth orthophosphates. Opt Mater 29(2):224–230

Silva MDP, Silva FC, Sinfrônio FSM, Paschoal AR, Silva EN, Paschoal CWA (2014) The effect of cobalt substitution in crystal structure and vibrational modes of $CuFe_2O_4$ powders obtained by polymeric precursor method. J Alloys Compd 584:573–580

Simakin AG, Eremyashev VE, Kucherinenko YV (2010) New data on dmisteinbergite. Zapiski RMO (Proc Russ Miner Soc) 139(3):102–108. (in Russian)

Singh N, Hiller J, Metiu H, McFarland E (2014) Investigation of the electrocatalytic activity of rhodium sulfide for hydrogen evolution and hydrogen oxidation. Electrochim Acta 145:224–230

Siqueira KP, Moreira RL, Dias A (2010) Synthesis and crystal structure of lanthanide orthoniobates studied by vibrational spectroscopy. Chem Mater 22(8):2668–2674

Sirbescu M-LC, Nabelek PI (2003) Dawsonite: an inclusion mineral in quartz from the Tin Mountain pegmatite, Black Hills, South Dakota. Am Mineral 88(7):1055–1059

Sithole J, Ngom BD, Khamlich S, Manikanadan E, Manyala N, Saboungi ML, Knoessen D, Nemutudi R, Maaza M (2012) Simonkolleite nano-platelets: synthesis and temperature effect on hydrogen gas sensing properties. Appl Surface Sci 258(20):7839–7843

Sivakumar T, Chang HY, Baek J, Halasyamani PS (2007) Two new noncentrosymmetric polar oxides: synthesis, characterization, second-harmonic generating, and pyroelectric measurements on $TlSeVO_5$ and $TlTeVO_5$. Chem Mater 19(19):4710–4715

Škácha P, Buixaderas E, Plášil J, Sejkora J, Goliáš V, Vlček V (2014a) Permingeatite, Cu_3SbSe_4, from Příbram (Czech Republic): description and Raman spectroscopy investigations of the luzonite-subgroup of minerals. Can Mineral 52(3):501–511

Škácha P, Plášil J, Sejkora J, Čejka J, Škoda R, Meisser N (2014b) Unique occurrence of bayleyite, $Mg_2(UO_2)(CO_3)_3 \cdot 18H_2O$, from Jáchymov. Bull Mineral-petrolog Odd NárMuz (Praha) 22(2):240–247. ISSN 1211-0329. (in Czech)

Škácha P, Sejkora J, Plášil J (2017) Selenide mineralization in the Příbram uranium and base-metal district (Czech Republic). Fortschr Mineral. https://doi.org/10.3390/min7060091

Škoda R, Plášil J, Čopjaková R, Novák M, Jonsson E, Vašinová Galiová M, Holtstam D (2017) Gadolinite-(Nd), a new member of the gadolinite supergroup from Fe-REE deposits of Bastnäs-type, Sweden. Mineral Mag. https://doi.org/10.1180/minmag.2017.081.047

Slade RC, Knowles JA, Jones DJ, Rozière J (1997) The isomorphous acid salts α-$Zr(HPO_4)_2 \cdot H_2O$, α-$Ti(HPO_4)_2 \cdot H_2O$ and α-$Zr(HAsO_4)_2 \cdot H_2O$. Comparative thermochemistry and vibrational spectroscopy. Solid State Ionics 96(1):9–19

Smirnov M, Sukhomlinov S, Mirgorodsky A, Masson O, Béchade E, Colas M, Merle-Méjean T, Julien I, Thomas P (2010) Raman and infrared spectra of doped $La_{8+x}Sr_{2-y}(SiO_4)_6O_{2+\delta}$ compounds compared to the ab initio-obtained spectroscopic characteristics of fully stoichiometric $La_8Sr_2(SiO_4)_6O_2$. J Raman Spectrosc 41(12):1700–1707

Smit WMA, Dirksen GJ, Stufkens DJ (1990) Infrared and Raman spectra of the elpasolites $Cs_2NaSbCl_6$ and $Cs_2NaBiCl_6$: evidence for a pseudo Jahn-Teller distorted ground state. J Phys Chem Solids 51(2):189–196

Smith AJ, Meek PE, Liang WY (1977) Raman scattering studies of SnS_2 and $SnSe_2$. J Phys C 10(8):1321–1333

Smith CP, Bosshart G, Graeser S, Hänni H, Günther D, Hametner K, Gübelin EJ (2003) Poudretteite: a rare gem species from the Mogok valley. Gems Gemol 39:24–31

Snyder MQ, McCool BA, DiCarlo J, Tripp CP, DeSisto WJ (2006) An infrared study of the surface chemistry of titanium nitride atomic layer deposition on silica from $TiCl_4$ and NH_3. Thin Solid Films 514(1):97–102

Sobkowiak A, Roberts MR, Younesi R, Ericsson T, Häggström L, Tai CW, Andersson AM, Edström K, Gustafsson T, Björefors F (2013) Understanding and controlling the surface chemistry of $LiFeSO_4F$ for an enhanced cathode functionality. Chem Mater 25(15):3020–3029

Sobron P, Alpers CN (2013) Raman spectroscopy of efflorescent sulfate salts from Iron Mountain Mine Superfund site, California. Astrobiol 13(3):270–278

Sohr G, Wilhelm D, Vitzthum D, Schmitt MK, Huppertz H (2014) The high-pressure borate $HP-(NH_4)B_3O_5$. Z anorg allg Chemie 640(14):2753–2758

Sokol EV, Seryotkin YV, Kokh SN, Vapnik Y, Nigmatulina EN, Goryainov SV, Belogub EV, Sharygin VV (2015) Flamite, $(Ca,Na,K)_2(Si,P)O_4$, a new mineral from ultra high temperature combustion metamorphic rocks, Hatrurim Basin, Negev Desert, Israel. Mineralog Mag 79(3):583–596

Sokol EV, Kokh SN, Khoury HN, Seryotkin YV, Goryainov SV (2016) Long-term immobilisation of Cd^{2+} at the Tulul Al Hammam natural analogue site, central Jordan. Appl Geochem 70:43–60

Sokolova E, Hawthorne FC, Abdu YA (2013) From structure topology to chemical composition. XV. Titanium silicates: revision of the crystal structure and chemical formula of schüllerite, $Na_2Ba_2Mg_2Ti_2(Si_2O_7)_2O_2F_2$, from the Eifel volcanic region, Germany. Can Mineral 51(5):715–725

Sokolova E, Abdu Y, Hawthorne FC, Genovese A, Cámara F, Khomyakov AP (2015a) From structure topology to chemical composition. XVIII. Titanium silicates: revision of the crystal structure and chemical formula of betalomonosovite, a Group-IV TS-block mineral from the Lovozero alkaline massif, Kola Peninsula, Russia. Can Mineral 53(3):401–428

Sokolova E, Cámara F, Abdu YA, Hawthorne FC, Horváth L, Pfenninger-Horváth E (2015b) Bobshannonite, $Na_2KBa(Mn,Na)_8(Nb,Ti)_4(Si_2O_7)_4O_4(OH)_4(O,F)_2$, a new TS-block mineral from Mont Saint-Hilaire, Québec, Canada: description and crystal structure. Mineral Mag 79(7):1791–1811

Solferino G, Anderson AJ (2012) Thermal reduction of molybdite and hematite in water and hydrogen peroxide-bearing solutions: insights on redox conditions in Hydrothermal Diamond Anvil Cell (HDAC) experiments. ChemGeol 322:215–222

Souag R, Kamel N, Hammadi M, Kamel Z, Moudir D, Aouchiche F, Mouheb Y, Kamariz S (2015) Study of leaching of a $2M$-Zirconolite $(Ca_{0.83}Ce_{0.17}ZrTi_{1.66}Al_{0.34}O_7)$ in acidic and basic media. J Ceram Proc Res 16(1):150–155

Sourisseau C, Cavagnat R, Fouassier M, Maraval P (1990) Electronic, vibrational and resonance Raman spectra of the layered semiconducting compound NbS_3. J Raman Spectrosc 21(6):337–349

Souza de Araujo A, Carlos Diniz J, da Silva AOS, Alves de Melo RA (1997) Hydrothermal synthesis of cerium aluminophosphate. J Alloys Compd 250(1):532–535

Spandler C, Morris C (2016) Geology and genesis of the Toongi rare metal (Zr, Hf, Nb, Ta, Y and REE) deposit, NSW, Australia, and implications for rare metal mineralization in peralkaline igneous rocks. Contrib Mineral Petrol 171(12):104. (24 pp)

Spektor K, Nylen J, Mathew R, Edén M, Stoyanov E, Navrotsky A, Leinenweber K, Häussermann U (2016) Formation of hydrous stishovite from coesite in high-pressure hydrothermal environments. Am Mineral 101(11):2514–2524

Spengler W, Kaiser R, Christensen AN, Müller-Vogt G (1978) Raman scattering, superconductivity, and phonon density of states of stoichiometric and nonstoichiometric TiN. Phys Rev B 17(3):1095–1101

Spinner E (1959) The vibration spectra and structures of the hydrochlorides of urea, thiourea and acetamide. The basic properties of amides and thioamides. Spectrochim Acta 15:95–109

Spratt HJ, Rintoul L, Avdeev M, Martens WN (2013) The crystal structure and vibrational spectroscopy of jarosite and alunite minerals. Am Mineral 98(10):1633–1643

Srinivasan TT, Srivastava CM, Venkataramani N, Patni MJ (1984) Infrared absorption in spinel ferrites. Bull Mater Sci 6(6):1063–1067

Srivastava R, Chase LL (1972) Raman spectra of CrO_2 and MoO_2 single crystals. Solid State Commun 11:349–353

Sronsri C, Noisong P, Danvirutai C (2014) Synthesis and properties of $LiM^{II}PO_4$ (M^{II} = Mg, $Mn_{0.5}Mg_{0.5}$, $Co_{0.5}Mg_{0.5}$) affected by isodivalent doping and Li-sources. Solid State Sci 36:80–88

Stalder R, Kronz A, Schmidt BC (2009) Raman spectroscopy of synthetic $(Mg,Fe)SiO_3$ single crystals. An analytical tool for natural orthopyroxenes. Eur J Mineral 21(1):27–32

Stamatovska N, Makreski P, Pejov L, Jovanovski G (2011) Minerals from Macedonia. XXVII: Theoretical and experimental study of the vibrational spectra of endemic nežilovite. J Molec Struct 993(1):104–108

Stanila D, Smith W, Anderson A (2000) Raman spectra of selenium dioxide at high pressures. Spectrosc Lett 33(4):555–567

Stanimirova T, Kerestedjian T, Kirov G (2016) Dehydration and rehydration of Zn-hydroxy sulfate minerals with interrupted decorated hydroxide sheets. Appl Clay Sci. https://doi.org/10.1016/j.clay.2016.08.032

Stavrakieva D, Ivanova Y, Pyrov J (1988) On the composition of the crystal phases in the PbO TeO$_2$ system. J Mater Sci 23(5):1871–1876

Stefaniak EA, Alsecz A, Sajó IE, Worobiec A, Máthé Z, Török S, Van Grieken R (2008) Recognition of uranium oxides in soil particulate matter by means of μ-Raman spectrometry. J Nucl Mater 381(3):278–283

Stefaniak EA, Alsecz A, Frost R, Máthé Z, Sajó IE, Török S, Worobiec A, Van Grieken R (2009) Combined SEM/EDX and micro-Raman spectroscopy analysis of uranium minerals from a former uranium mine. J Hazard Mater 168(1):416–423

Stefov V, Šoptrajanov B, Petruševski V (1992) Vibrational spectra of hexaaqua complexes. II. External motions of water molecules in the spectra of AlCl$_3$·6H$_2$O. J Molec Struct 267:203–208

Stefov V, Šoptrajanov B, Spirovski F, Kuzmanovski I, Lutz HD, Engelen B (2004) Infrared and Raman spectra of magnesium ammonium phosphate hexahydrate (struvite) and its isomorphous analogues. I. Spectra of protiated and partially deuterated magnesium potassium phosphate hexahydrate. J Molec Struct 689(1):1–10

Steiner M, Tropper P, Vavtar F, Kaindl R, Krismer M (2010) Balkanite from the Cu ore deposit Röhrerbühel, Kitzbühel (N-Tyrol, Austria). N Jb Miner Abh 187(2):207–215

Štengl V, Henych J (2013) Strongly luminescent monolayered MoS$_2$ prepared by effective ultrasound exfoliation. Nanoscale 5(8):3387–3394

Štengl V, Tolasz J, Popelková D (2015) Ultrasonic preparation of tungsten disulfide single-layers and quantum dots. RSC Adv 5(109):89612–89620

Stepakova LV, Skripkin MY, Chernykh LV, Starova GL, Hajba L, Mink J, Sandström M (2008) Vibrational spectroscopic and force field studies of copper (II) chloride and bromide compounds, and crystal structure of KCuBr$_3$. J Raman Spectrosc 39(1):16–31

Števko M, Sejkora J, Peterec D (2015) Grumiplucite from the Rudňany deposit, Slovakia: a second world-occurrence and new data. J Geo Sci 60(4):269–281

Stidham HD, Bates JB, Finch CB (1976) Vibrational spectra of synthetic single crystal tephroite, Mn$_2$SiO$_4$. J Phys Chem 80(11):1226–1234

Stodolski R, Heidemann D, Wieker W, Pilz W (1985) Thermische Zersetzung von Afwillit Ca$_3$(SiO$_3$OH)$_2$·2H$_2$O. Z anorg allg Chemie 527(8):150–160. (in German)

Stoilova D, Koleva V (2000) IR study of solid phases formed in the Mg(HCOO)$_2$–Cu(HCOO)$_2$–H$_2$O system. J Molec Struct 553(1):131–139

Stoilova D, Vassileva V (1999) X-ray powder diffraction study and vibrational spectra of calcium cadmium formate. Crystal Res Technol 34(3):397–401

Stoilova D, Vassileva V (2002) Infrared spectroscopic study of solids in the Cu$_2$(OH)$_3$Cl (paratacamite) – Zn$_5$(OH)$_8$Cl$_2$·H$_2$O (simonkolleite) series. Compt Rend de l'Academie Bulgare Sci 55(7):51–54

Stoilova D, Marinova D, Georgiev M (2009) Hydrogen bond strength in chromates with kröhnkite-type chains, K$_2$Me(CrO$_4$)$_2$·2H$_2$O (Me = Mg, Co, Ni, Zn, Cd). Vib Spectrosc 50(2):245–249

Stranford GT, Condrate RA Sr (1984a) The vibrational spectra and normal coordinate analyses of VSO$_5$, VPO$_5$ and VMoO$_5$ phases. Spectrosc Lett 17(2):85–113

Stranford GT, Condrate RA Sr (1984b) The vibrational spectra of α-MoPO$_5$ and α-NbPO$_5$. J Solid State Chem 52(3):248–253

Stranford GT, Condrate RA Sr (1984c) The infrared and Raman spectra of β-TaPO$_5$ and β-NbPO$_5$. J Mater Sci Lett 3(4):303–306

Stranford GT, Condrate RA Sr (1990) A vibrational spectral study of hydrated tantalum phosphate (TaPO$_5$) phases. J Solid State Chem 85(2):326–331

Strobel P, Ibarra-Palos A, Anne M, Poinsignon C, Crisci A (2003) Cation ordering in Li$_2$Mn$_3$MO$_8$ spinels: structural and vibration spectroscopy studies. Solid State Sci 5(7):1009–1018

Su Y, Balmer ML, Bunker BC (2000) Raman spectroscopic studies of silicotitanates. J Phys Chem B 104(34):8160–8169

Suchanek W, Yashima M, Kakihana M, Yoshimura M (1998) β-Rhenanite (β-NaCaPO$_4$) as weak interphase for hydroxyapatite ceramics. J Eur Ceram Soc 18(13):1923–1929

Sumathi S, Gopal B (2015) A new insight into biomedical applications of an apatite like oxyphosphate – BiCa$_4$(PO$_4$)$_3$O. Ceram Intern 41(3):4852–4860

Sun Z, Luo K, Tan F, Zhang J (1994) Pingguite – a new bismuth tellurite mineral. Acta Mineral Sin 14:315–321

Sun HY, Sun W, Huang YX, Mi JX (2010a) Low temperature flux synthesis and characterizations of a new layered barium borate BaB$_8$O$_{11}$(OH)$_4$. Z anorg allg Chemie 636(6):977–981

Sun L, Zhao X, Li Y, Li P, Sun H, Cheng X, Fan W (2010b) First-principles studies of electronic, optical, and vibrational properties of LaVO$_4$ polymorph. J Appl Phys 108(9):093519-1–093519-10

Sun W, Huang YX, Pan Y, Mi JX (2015) Synthesis and magnetic properties of centennialite: a new S = ½ Kagomé antiferromagnet and comparison with herbertsmithite and kapellasite. Phys Chem Minerals 43(2):1–10

Sundarakannan B, Ravindran TR, Kesavamoorthy R, Satyanarayana SVM (2002) High pressure Raman spectroscopic study of BaFCl. Solid State Commun 124(10):385–389

Sundarakkannan B, Kesavamoorthy R (1998) Anharmonic behaviour of BaFCl using Raman scattering. Eur Phys J B 3(2):179–183

Sushchinsky MM (1981) Raman scattering of light and the structure of matter. Nauka Publishing House, Moscow. (in Russian)

Suzuki J, Ito M, Sugiura T (1976) A new copper sulfate-carbonate hydroxide hydrate mineral, $(Mn,Ni,Cu)_8(SO_4)_4(CO_3)(OH)_6 \cdot 48H_2O$, from Nakauri, Aichi Prefecture, Japan. J Mineral Petrol Econ Geol 71:183–192

Suzuki Y, Wakihara T, Itabashi K, Ogura M, Okubo T (2009) Cooperative effect of sodium and potassium cations on synthesis of ferrierite. Top Catal 52 (1–2):67–74

Švecová E, Čopjaková R, Losos Z, Škoda R, Nasdala L, Cícha J (2016) Multi-stage evolution of xenotime–(Y) from Písek pegmatites, Czech Republic: an electron probe micro-analysis and Raman spectroscopy study. Mineral Petrol. https://doi.org/10.1007/s00710-016-0442-6

Sweet LE, Buck EC, Henager CH, Hu S, Meier DE, Peper SM, Schwantes JM, Su Y-F, Sams RL, Blake TA, Johnson TJ (2012) Investigations into the polymorphs and hydration products of UO_3. Proc SPIE 8358:83581R-1. (7 pp)

Syme RWG, Lockwood DJ, Kerr HJ (1977) Raman spectrum of synthetic zircon ($ZrSiO_4$) and thorite ($ThSiO_4$). J Phys C 10(8):1335–1348

Syrbu NN, Bogdanash M, Moldovyan NA (1996a) Vibrational modes in $ZnAl_2S_4$ and $CdIn_2S_4$ crystals. Infrared Phys Technol 37(7):763–768

Syrbu NN, Radautsan SI, Cretu RV, Tezlevan VE, Moldoveanu NA (1996b) Far infrared and Raman optical study of $CdInGaS_4$, $CdIn_2S_4$, $HgInGaS_4$ and $CdIn_2S_2Se_2$ crystals. Cryst Res Technol 31(3):307–314

Szilágyi IM, Madarász J, Hange F, Pokol G (2004) Online evolved gas analyses (EGA by TG-FTIR and TG/DTA-MS) and solid state (FTIR, XRD) studies on thermal decomposition and partial reduction of ammonium paratungstate tetrahydrate. Solid State Ionics 172 (1):583–586

Szuszkiewicz W, Dynowska E, Gorecka J, Witkowska B, Jouanne M, Morhange JF, Julien C, Hennion B (1999) Peculiarities of the lattice dynamics of cubic mercury chalcogenides. Phys Status Solidi B 215 (1):93–98

Tait KT, Yang H, Downs RT, Li C, Pinch WW (2010) The crystal structure of esperite, with a revised chemical formula, $PbCa_2(ZnSiO_4)_3$, isostructural with beryllonite. Am Mineral 95(5–6):699–705

Tait KT, Barkley MC, Thompson RM, Origlieri MJ, Evans SH, Prewitt CT, Yang H (2011) Bobdownsite, a new mineral species from Big Fish River, Yukon, Canada, and its structural relationship with whitlockite-type compounds. Can Mineral 49(4):1065–1078

Tait KT, Dicecco V, Ball NA, Hawthorne FC, Kampf AR (2014) Backite, $Pb_2Al(TeO_6)Cl$, a new tellurate mineral from the Grand Central mine, Tombstone Hills, Cochise County, Arizona: description and crystal structure. Can Mineral 52:935–942

Tait K, Ball NA, Hawthorne FC (2015) Pieczkaite, ideally $Mn_5(PO_4)_3Cl$, a new apatite-supergroup mineral from Cross Lake, Manitoba, Canada: description and crystal structure. Am Mineral 100: 1047–1052

Takahashi Y, Iwasaki K, Masai H, Fujiwara T (2008) Raman spectroscopic study of benitoite-type compounds. J Ceram Soc Jpn 116(1358):1139–1142

Takubo K, Kubota R, Suzuki T, Kanzaki T, Miyahara S, Furukawa N, Katsufuji T (2011) Evolution of phonon Raman spectra with orbital ordering in spinel MnV_2O_4. Phys Rev B 84(9):094406-1–094406-8

Tang ZH, Chen X, Li M (2008) Synthesis and crystal structure of a new strontium borate, $Sr_2B_{16}O_{26}$. Solid State Sci 10(7):894–900

Tang Y, Cui M, Guo W, Zhang S, Yang M, He Z (2015) Syntheses, structure, and magnetic properties of new 3d–4f heterometallic hydroxysulfates $Ln_2Cu(SO_4)_2(OH)_4$ (Ln = Sm, Eu, Tb, or Dy) with a two-dimensional triangle network. Cryst Growth Design 15(6):2742–2747

Tao Y, Shiyang G, Lixia Z, Shuping X, Kaibei Y (2002) Crystal growth and crystal structure of magnesium oxysulfate $2MgSO_4(Mg(OH)_2) \cdot 2H_2O$. J Molec Struct 616(1):247–252

Tao Z, Zhang W, Huang Y, Wei D, Seo HJ (2014) A novel pyrophosphate $BaCr_2(P_2O_7)_2$ as green pigment with high NIR solar reflectance and durable chemical stability. Solid State Sci 34:78–84

Tarassov M, Mihailova B, Tarassova E, Konstantinov L (2002) Chemical composition and vibrational spectra of tungsten-bearing goethite and hematite from Western Rhodopes, Bulgaria. Eur J Mineral 14 (5):977–986

Tarte P (1963) Etude infra-rouge des orthosilicates et des orthogermanates – III: Structures du type spinelle. Spectrochim Acta 19(1):49–71. (in French)

Tarte P, Cahay R, Rulmont A, Werding G (1985) Infrared spectrum of synthetic isotopic species of sinhalite $MgAlBO_4$. Spectrochim Acta A 41(10):1215–1219

Tarte P, Rulmont A, Merckaert-Ansay C (1986) Vibrational spectrum of nasicon-like, rhombohedral orthophosphates $M^IM^{IV}_2(PO_4)_3$. Spectrochim Acta A 42(9):1009–1016

Tarte P, Rulmont A, Sbai K, Simonot-Grange MH (1987) Vibrational spectrum of $KM^{II}P_3O_9$ tricyclophosphates with the benitoite structure. Spectrochim Acta A 43 (3):337–343

Taylor JCW, Weichman FL (1971) Raman effect in cuprous oxide compared with infrared absorption. Can J Phys 49(5):601–605

Taylor MD, Cheung TT, Hussein MA (1972) Variations of the infrared spectra with the nature and structure of the rare earth metal halides. J Inorg Nucl Chem 34 (10):3073–3079

Taylor P, Sunder S, Lopata VJ (1984) Structure, spectra, and stability of solid bismuth carbonates. Can J Chem 62(12):2863–2873

Teng MK, Massot M, Balkanski M, Ziolkiewicz S (1978) Atomic substitution and ferroelectric phase transition in $Bi_xSb_{1-x}SI$. Phys Rev B 17(9):3695. (6 pp)

Ternane R, Ferid M, Kbir-Ariguib N, Trabelsi-Ayedi M (2000) The silver lead apatite $Pb_8Ag_2(PO_4)_6$: hydrothermal preparation. J Alloys Compd 308(1):83–86

Ternane R, Ferid M, Panczer G, Trabelsi-Ayadi M (2008) Structural, optical and scintillation properties of cerium cyclotriphosphates and polyphosphates. J Phys Chem Solids 69(7):1684–1690

Terra O, Dacheux N, Clavier N, Podor R, Audubert F (2008) Preparation of optimized uranium and thorium bearing brabantite or monazite/brabantite solid solutions. J Am Ceram Soc 91(11):3673–3682

Theiss F, López A, Frost RL, Scholz R (2015a) Spectroscopic characterisation of the LDH mineral quintinite $Mg_4Al_2(OH)_{12}CO_3 \cdot 3H_2O$. Spectrochim ActaA 150:758–764

Theiss FL, López A, Scholz R, Frost RL (2015b) A SEM, EDS and vibrational spectroscopic study of the clay mineral fraipontite. Spectrochim Acta A 147:230–234

Thema FT, Beukes P, Gurib-Fakim A, Maaza M (2015) Green synthesis of MonteponiteCdO nanoparticles by *Agathosmabetulina* natural extract. J Alloys Compd 646:1043–1048

Thema FT, Manikandan E, Gurib-Fakim A, Maaza M (2016) Single phase bunsenite NiO nanoparticles green synthesis by *Agathosmabetulina* natural extract. J Alloys Compd 657:655–661

Thomas R (2002) Determination of the H_3BO_3 concentration in fluid and melt inclusions in granite pegmatites by laser Raman microprobe spectroscopy. Am Mineral 87(1):56–68

Thomas R, Davidson P (2010) Hambergite-rich melt inclusions in morganite crystals from the Muiane pegmatite, Mozambique and some remarks on the paragenesis of hambergite. Mineral Petrol 100(3–4):227–239

Thomas R, Davidson P (2017) Hingganite-(Y) from a small aplite vein in granodiorite from Oppach, Lusatian Mts., E-Germany. Mineral Petrol. https://doi.org/10.1007/s00710-016-0489-4

Thomas R, Webster JD (1999) Characteristics of berlinite from the Ehrenfriedersdorf pegmatite, Erzgebirge, Germany. Z geol Wiss 27(5/6):443–454

Thomas GH, Falk M, Knop O (1974) Infrared studies of water in crystalline hydrates: $K_2CuCl_4 \cdot 2H_2O$. Can J Chem 52(7):1029–1041

Thomas R, Webster JD, Rhede D (1998) Strong phosphorus enrichment in a pegmatite-forming melt. Acta Univ Carolinae Geol 42(1):150–164

Thomas R, Davidson P, Hahn A (2008) Ramanite-(Cs) and ramanite-(Rb): new cesium and rubidium pentaborate tetrahydrate minerals identified with Raman spectroscopy. Am Mineral 93(7):1034–1042

Thomas R, Davidson P, Beurlen H (2011a) Tantalite-(Mn) from the Borborema Pegmatite Province, northeastern Brazil: conditions of formation and melt-and fluid-inclusion constraints on experimental studies. Miner Deposita 46(7):749–759

Thomas R, Webster JD, Davidson P (2011b) Be-daughter minerals in fluid and melt inclusions: implications for the enrichment of Be in granite-pegmatite systems. Contrib Mineral Petrol 161(3):483–495

Tite T, Shu GJ, Chou FC, Chang YM (2010) Structural and thermal properties of MnSi single crystal. Appl Phys Lett 97(3):031909-1–031909-3

Tlili A, Smith DC (2007) Raman spectroscopic study of synthetic Na-Mg-Al-Si trioctahedral micas compared with their Ge-and Ga-equivalents. In: Rull-Pérez F, Edwards H, Smith D, Vandenabeele P (eds) Selected topics in Raman spectroscopic applications: geology-bio-materials-art. Edition: Universidad Valladolid. ISBN 978-84-690-9239-2

Tlili A, Smith DC, Beny JM, Boyer H (1989) A Raman microprobe study of natural micas. Mineral Mag 53(2):165–179

Többens DM, Kahlenberg V, Kaindl R (2005) Characterization and *ab initio* XRPD structure determination of a novel silicate with *vierer* single chains: the crystal structure of $NaYSi_2O_6$. Inorg Chem 44(25):554–9560

Tomašić N, Gajović A, Bermanec V, Rajić M (2004) Recrystallization of metamict Nb–Ta–Ti–*REE* complex oxides: a coupled X-ray-diffraction and Raman spectroscopy study of aeschynite-(Y) and polycrase-(Y). Can Mineral 42(6):1847–1857

Tomašić N, Bermanec V, Gajović A, Rajić Linarić M (2008) Metamict minerals: an insight into a relic crystal structure using XRD, Raman spectroscopy, SAED and HRTEM. Croat Chem Acta 81(2):391–400

Tomašić N, Gajović A, Bermanec V, Linarić MR, Su D, Škoda R (2010) Preservation of the samarskite structure in a metamict ABO_4 mineral: a key to crystal structure identification. Eur J Mineral 22(3):435–442

Tomasini EP, Dubois CMF, Little NC, Centeno SA, Maier MS (2015) Identification of pyroxene minerals used as black pigments in painted human bones excavated in Northern Patagonia by Raman spectroscopy and XRD. Microchem J 121:157–162

Tomaszewski PE, Mączka M, Majchrowski A, Waśkowska A, Hanuza J (2005) Crystal structure and vibrational properties of $KMg_4(PO_4)_3$. Solid State Sci 7(10):1201–1208

Tong H-J, Qian Z-G, Reid GP, Zhang Y-H (2011) High temporal and spatial resolution measurements of the rapid efflorescence of sea salt droplets. Acta Phys-Chim Sin 27(11):2521–2527

Tong Y, Ma J, Zhao S, Huo H, Zhang H (2015) A Salt-assisted combustion method to prepare well-dispersed octahedral $MnCr_2O_4$ spinel nanocrystals. J Nanomater. https://doi.org/10.1155/2015/214978

Tongay S, Sahin H, Ko C, Luce A, Fan W, Liu K, Zhou J, Huang Y-S, Ho C-Y, Yan J, Ogletree DF, Aloni S, Ji J, Li S, Li J, Peeters FM, Wu J (2014) Monolayer behaviour in bulk ReS_2 due to electronic and vibrational decoupling. Nat Commun. https://doi.org/10.1038/ncomms4252

Toumi M, Tlili A (2008) Rietveld refinement and vibrational spectroscopic study of alunite from El Gnater, Central Tunisia. Russ J Inorg Chem 53(12):1845–1853

Toumi M, Hlel F, Chaabane TB, Smiri L, Laligant Y, Emery J (1998) X-ray powder structure determination of $Li_6P_6O_{18}\cdot 3H_2O$. Eur J Solid State Inorg Chem 35(10):689–697

Tran TT, Halasyamani PS (2013) New fluoride carbonates: centrosymmetric $KPb(CO_3)_2F$ and noncentrosymmetric $K_{2.70}Pb_{5.15}(CO_3)_5F_3$. Inorg Chem 52(5):2466–2473

Tran TH, Tran KA, Pham TH, Le TV, Le QM (2012) Effect of the soft-template agents on size, shape and optical properties of YVO_4: Eu^{3+} nanomaterials. Adv Natur Sci Nanosci Nanotechnol 3(3):035012. (4 pp)

Trefler M, Wilkinson GR (1969) Motion of ammonium ions in non-cubic crystal sites. Discus Faraday Soc 48:108–115

Treiman AH (2008) Rhönite in Luna 24 pyroxenes: first find from the Moon, and implications for volatiles in planetary magmas. Am Mineral 93(2–3):488–491

Trentelman K, Stodulski L, Pavlosky M (1996) Characterization of pararealgar and other light-induced transformation products from realgar by Raman microspectroscopy. Anal Chem 68(10):1755–1761

Tribaudino M, Artoni A, Mavris C, Bersani D, Lottici PP, Belletti D (2008) Single-crystal X-ray and Raman investigation on melanophlogite from Varano Marchesi (Parma, Italy). Am Mineral 93(1):88–94

Tribaudino M, Mantovani L, Bersani D, Lottici PP (2011) Raman investigation on pigeonite in ureilite. Spectrosc Lett 44(7–8):480–485

Trolard F, Bourrié G, Abdelmoula M, Refait P, Feder F (2007) Fougerite, a new mineral of the pyroauriteiowaite group: description and crystal structure. Clays Clay Minerals 55(3):323–334

Tsuda H, Jongebloed WL, Stokroos I, Arends J (1993) Combined Raman and SEM study on CaF_2 formed on/in enamel by APF treatments. Caries Res 27(6):445–454

Tsunogae T, Santosh M (2005) Ti-free högbomite in spinel-and sapphirine-bearing Mg-Al rock from the Palghat-Cauvery shear zone system, southern India. Mineral Mag 69(6):937–949

Tuinstra F, Koenig JL (1970) Raman spectrum of graphite. J Chem Phys 53(3):1126–1130

Tumiati S, Godard G, Martin S, Nimis P, Mair V, Boyer B (2005) Dissakisite-(La) from the Ulten zone peridotite (Italian Eastern Alps): a new end-member of the epidote group. Am Mineral 90(7):1177–1185

Turcotte SB, Benner RE (1993) Surface analysis of electrochemically oxidized metal sulfides using Raman spectroscopy. J Electroanal Chem 347:195–205

Tyutyunnik AP, Zubkov VG, Tarakina NV, Krasil'nikov VN, Perelyaeva LA, Baklanova IV, Svensson G (2006) Synthesis, crystal structure and vibrational spectra of $KCrV_2O_7$ and $RbCrV_2O_7$. Solid State Sci 8(11):1344–1352

Uchida H, Downs RT, Yang H (2006) Crystal-chemical investigation of kalsilite from San Venanzo, Italy, using single-crystal X-ray diffraction and Raman spectroscopy. Geochim Cosmochim Acta 70(18):A677

Uchida H, Downs RT, Thompson RM (2007) Reinvestigation of eakerite, $Ca_2SnAl_2Si_6O_{18}(OH)_2\cdot 2H_2O$: H-atom positions by single-crystal X-ray diffraction and correlation with Raman spectroscopic data. Acta Crystallogr E. https://doi.org/10.1107/S1600536807002000

Uher P, Milovská S, Milovský R, Koděra P, Bačík P, Bilohuščin V (2015) Kerimasite,$\{Ca_3\}[Zr_2](SiFe^{3+}{}_2)O_{12}$ garnet from the Vysoká-Zlatnoskarn, Štiavnica stratovolcano, Slovakia. Mineral Mag 79(3):715–734

Ullrich A, Miletich R, Nestola F, Weikusat C, Ohashi H (2009) Lattice compression and structural behavior of $NaVSi_2O_6$ clinopyroxene to 11 GPa. Am Mineral 94(4):557–564

Unger WK, Farnworth B, Irwin JC, Pink H (1978) Raman and infrared spectra of $CdIn_2S_4$ and $ZnIn_2S_4$. Solid State Commun 25(11):913–915

Unger WK, Karecki D, Clayman BP, Irwin JC, Pink H (1979) Raman and far–infrared spectra of $NaCrS_2$. Solid State Commun 29(3):149–151

Unnikrishnan NV, Ittyachen MA (2016) Growth and characterization of Sm^{3+} doped cerium oxalate single crystals. J Mater Res Technol. https://doi.org/10.1016/j.jmrt.2016.01.001

Uriarte LM, Dubessy J, Boulet P, Baonza VG, Bihannic I, Robert P (2015) Reference Raman spectra of synthesized $CaCl_2 \cdot nH_2O$ solids ($n = 0, 2, 4, 6$). J Raman Spectrosc 46(10):822–828

Ursaki VV, Manjón FJ, Tiginyanu IM, Tezlevan VE (2002) Raman scattering study of pressure-induced phase transitions in MIn_2S_4spinels. J Phys Condens Matter 14(27):6801–6813

Utyuzh AN, Timofeev YA, Stepanov GN (2010) Effect of pressure on Raman spectra of SnS_2 single crystals. Phys Solid State 52(2):352–356

Uztetik-Amour A, Kizilyalli M (1995) Solid-state synthesis, X-ray powder diffraction, and IR data of Na_2GdOPO_4. J Solid State Chem 120(2):275–278

Vallette Campos MM, Alvarado Aguayo T (2015) Vibrational spectroscopy for the study of Chilean cultural heritage. Heritage Sci 3(1):1–10

Van Loosdrecht PHM, van Bentum PJM, Balzuweit K (1992) Raman study of incommensurately modulated calaverite ($AuTe_2$). Ferroelectrics 125(1):517–522

Vandenabeele P (2013) Practical Raman Spectroscopy – an introduction. Wiley, London

Vandenborre MT, Husson E, Brusset H, Cerez A (1980) Spectres de vibration et calcul du champ de force des antimoniates de structure 'type $PbSb_2O_6$'. Spectrochim Acta A 36:1045–1052. (in French)

Vandenborre MT, Husson E, Fourquet JL (1982) Spectres vibrationnels et champ de force de divers composés de formule $A_2B_2O_7$ et $A_2B_2O_6$ de structure pyrochlore. Spectrochim Acta A 38(9):997–1003. (in French)

Vard E, Williams-Jones AE (1993) A fluid inclusion study of vug minerals in dawsonite-altered phonolite sills, Montreal, Quebec: implications for HFSE mobility. Contrib Mineral Petrol 113(3):410–423

Vardeny Z, Brafman O (1979) Phonon anomalies in Cu halides. Phys Rev B 19(6):3276–3288

Vassileva M, Vergilov I, Dobrev S, Kolkovski B (2005) Mineralogical features of Zn-Mn spinel (hetaerolite) from the Au-base-metal Madjarovo deposit, Eastern Rrhodopes. Annu Univ Mining Geol "St. Ivan Rilski", Geol Geophys 48(I):23–28

Velchuri R, Kumar BV, Devi VR, Prasad G, Prakash DJ, Vithal M (2011a) Preparation and characterization of rare earth orthoborates, $LnBO_3$ (Ln = Tb, La, Pr, Nd, Sm, Eu, Gd, Dy, Y) and $LaBO_3$:Gd, Tb, Eu by metathesis reaction: ESR of $LaBO_3$:Gd and luminescence of $LaBO_3$:Tb, Eu. Mater Res Bull 46(8):1219–1226

Velchuri R, Kumar BV, Devi VR, Prakash DJ, Vithal M (2011b) Solid-state syntheses of rare-earth–doped $Sr_{1-x}Ln_{2x/3}MgP_2O_7$ (Ln = Gd, Eu, Dy, Sm, Pr, and Nd; $x = 0.05$) by metathesis reactions and their spectroscopic characterization. Spectrosc Lett 44(4):258–266

Venkateswaran CS (1936) The Raman spectra of selenious acid and its sodium salts. Proc Indian Acad Sci A 3(6):533–543

Venkateswarlu K (1940) Raman spectrum of sulphur. Proc Indian Acad Sci 12(14):453–461

Ventruti G, Della Ventura G, Orlando R, Scordari F (2015) Structure refinement, hydrogen-bond system and vibrational spectroscopy of hohmannite, $Fe^{3+}_2[O(SO_4)_2] \cdot 8H_2O$. Mineral Mag 79:11–24

Ventruti G, Della Ventura G, Bellatreccia F, Lacalamita M, Schingaro E (2016) Hydrogen bond system and vibrational spectroscopy of the iron sulfate fibroferrite, $Fe(OH)SO_4 \cdot 5H_2O$. Eur J Mineral. https://doi.org/10.1127/ejm/2016/0028-2571

Verble JL, Humphrey FM (1974) Infrared and Raman spectra of MnS_2. Solid State Commun 15(10):1693–1697

Vermaak MKG (2005) Fundamentals of the flotation behaviour of palladium bismuth tellurides. Dissertation, University of Pretoria

Verwoerd WJ (2008) Kamphaugite-(Y) from the Goudini carbonatite, South Africa. Can Mineral 46:1007–1022

Vieira Ferreira LF, Conceição DS, Ferreira DP, Santos LF, Casimiro TM, Ferreira Machado I (2014) Portuguese 16th century tiles from Santo António da Charneca's kiln: a spectroscopic characterization of pigments, glazes and pastes. J Raman Spectrosc 45(9):838–847

Vignola P, Hatert F, Bersani D, Diella V, Gentile P, Risplendente A (2012) Chukhrovite-(Ca), $Ca_{4.5}Al_2(SO_4)F_{13} \cdot 12H_2O$, a new mineral species from the Val Cavallizza Pb-Zn-(Ag) mine, Cuasso al Monte, Varese province, Italy. Eur J Mineral 24:1069–1076

Vignola P, Gatta GD, Hatert F, Guastoni A, Bersani D (2014) On the crystal-chemistry of a near-endmember triplite, $Mn^{2+}_2(PO_4)F$, from the Codera Valley (Sondrio Province, Central Alps, Italy). Can Mineral 52(2):235–245

Vignola P, Gatta GD, Rotiroti N, Gentile P, Hatert F, Baijot M, Risplendente A, Pavese A (2016) Albertiniite, $Fe^{2+}(SO_3) \cdot 3H_2O$, a new sulfite mineral species from the Monte Falò Pb-Zn mine, Coiromonte municipality, Verbano Cusio Ossola province, Piedmont, Italy. Mineral Mag. https://doi.org/10.1180/minmag.2016.080.033

Vigouroux JP, Husson E, Calvarin G, Dao NQ (1982) Etude par spectroscopié vibrationnelle des oxydes Pb_3O_4, $SnPb_2O_4$ et $SnPb(Pb_2O_4)_2$. Spectrochim Acta A 38(4):393–398. (in French)

Villanova-de-Benavent C, Aiglsperger T, Jawhari T, Proenza JA (2012) Micro-Raman spectroscopy of garnierite minerals: a useful method for phase identification. Revista Soc Española Mineral 16(12):180–181

Villanova-de-Benavent C, Tredoux M, Aiglsperger T, Proenza JA (2014) Ni-Mg-phyllosilicates from Bon Accord, Barberton, South Africa: new data on willemseite and nimite. Proceedings of the 21st General Meeting of the International Mineralogical Association "IMA2014", p 657

Vilminot S, Richard-Plouet M, André G, Swierczynski D, Bourée-Vigneron F, Kurmoo M (2003) Hydrothermal synthesis in the system $Ni(OH)_2$–$NiSO_4$: nuclear and magnetic structures and magnetic properties of $Ni_3(SO_4)_2(OH)_2 \cdot 2H_2O$. Inorg Chem 42(21):6859–6867

Vimal G, Mani KP, Jose G, Biju PR, Joseph C, Unnikrishnan NV, Ittyachen MA (2014) Growth and spectroscopic properties of samarium oxalate single crystals. J Cryst Growth 404:20–25

Viñals J, Jambor JL, Raudsepp M, Roberts AC, Grice JD, Kokinos M, Wise W (2008) Barahonaite-(Al) and barahonaite-(Fe), new Ca-Cu arsenate mineral species, from Murcia province, southeastern Spain, and Gold hill, Utah. Can Mineral 46:205–217

Vivekanandan K, Selvasekarapandian S, Kolandaivel P (1995) Raman and FT-IR studies of $Pb_4(NO_3)_2(PO_4)_2 \cdot 2H_2O$. Mater Chem Phys 39:284–289

Vlaev LT, Genieva SD, Gospodinov GG (2005) Study of the crystallization fields of cobalt (II) selenites in the system $CoSeO_3$-SeO_2-H_2O. J Thermal Anal Calorimetry 81(2):469–475

Voit EI, Panasenko AE, Zemnukhova LA (2009) Vibrational spectroscopic and quantum chemical study of antimony (III) oxide. J Struct Chem 50(1):60–66

Volkovich VA, Griffiths TR, Fray DJ, Fields M (1998) Vibrational spectra of alkali metal (Li, Na and K) uranates and consequent assignment of uranate ion site symmetry. Vib Spectrosc 17(1):83–91

Voll D, Beran A, Schneider H (2006) Variation of infrared absorption spectra in the system $Bi_2Al_{4-x}Fe_xO_9$ ($x = 0$–4), structurally related to mullite. Phys Chem Minerals 33(8–9):623–628

Voloshin AV, Karpov SM, Isaenko SI, Chernyavsky AV, Sergeeva NE (2014) Raman spectroscopy of vanadium association minerals in massive sulfide deposits of the pyrrhotite gorge (Kola region, Russia) and Vihanti (Finland). Zapiski RMO (Proc Russ Miner Soc) 143(4):55–60. (in Russian)

Voloshin AV, Chernyavsky AV, Bocharov VN, Vasil'ev EA (2015a) Raman spectroscopy of minerals from tetradymite and aleksite groups. In: Voytekhovsky YL (ed) Geology and base minerals of Kola region. K & M, Apatity. (in Russian)

Voloshin AV, Karpov SM, Chernyavsky AV (2015b) New mineral data. First finds in Russia and Kola region. In: Voytekhovsky YL (ed) Geology and base minerals of Kola region. K & M, Apatity. (in Russian)

Von Czarnowski A, Hübner K (1987) Raman and infrared investigations of stishovite and their interpretation. Phys Status Solidi B 142(1):K91–K96

Vtyurin AN, Isaenko LI, Krylova SN, Yelisseyev A, Shebanin AP, Turchin PP, Zamkova NG, Zinenko VI (2004) Raman spectra and elastic properties of KPb_2Cl_5 crystals. Phys Status Solidi C 1(11):3142–3145

Vuk AŠ, Orel B, Dražič G (2001) IR spectroelectrochemical studies of $Fe_2V_4O_{13}$, $FeVO_4$ and $InVO_4$ thin films obtained via sol-gel synthesis. J Solid State Electrochem 5(7–8):437–449

Vymazalová A, Zaccarini F, Bakker RJ (2014) Raman spectroscopy characterisation of synthetic platinum-group minerals (PGM) in the Pd–Sn–Te and Pd–Pb–Te ternary systems. Eur J Mineral 26(6):711–716

Wada N, Kamitakahara WA (1991) Inelastic neutron-and Raman-scattering studies of muscovite and vermiculite layered silicates. Phys Rev B 43(3):2391–2397

Wahlberg N, Runčevski T, Dinnebier RE, Fischer A, Eggert G, Iversen BB (2015) Crystal structure of thecotrichite, an efflorescent salt on calcareous objects stored in wooden cabinets. Cryst Growth Design 15(6):2795–2800

Walenta K, Theye T (2012) Heisenbergite, a new uranium mineral from the uranium deposit of Menzenschwand in the Southern Black Forest, Germany. N Jb Miner Abh 189(2):117–123

Wallwork K, Kolitsch U, Pring A, Nasdala L (2002) Decrespignyite-(Y), a new copper yttrium rare earth carbonate chloride hydrate from Paratoo, South Australia. Mineral Mag 66(1):181–188

Wang A, Zhou Y (2014) Experimental comparison of the pathways and rates of the dehydration of Al-, Fe-, Mg-and Ca-sulfates under Mars relevant conditions. Icarus 234:162–173

Wang XW, Noble RJ, Finnigan DJ, Spitulnik F, McGinnis P, Fu LP, Petrou AP (1993) Laser deposition of cubic and hexagonal structured CdS thin films. Proc SPIE 1793, Integr Opt Microstruct. https://doi.org/10.1117/12.141211

Wang R, Xu S, Francois F, Lin C (1997) Tantalite in Suzhou granite: mineralogy and geological implications. Chin J Geochem 16(4):353–358

Wang A, Haskin LA, Jolliff BL (1998a) Characterization of mineral products of oxidation and hydration by laser Raman spectroscopy – implications for *in situ* petrologic investigation on the surface of Mars. 29th Annual Lunar and Planetary Science Conference

Wang R-P, Zhou Y-L, Pan S-H, Zhang H, Guo X-X, Xiong X-M, Lu H-B, Zhen Z-H, Yang G-Z (1998b) Structural characteristics of CeO_2 films grown on biaxially textured nickel (001). J Appl Phys 84(4):1994–1997

Wang A, Jolliff BL, Haskin LA (1999) Raman spectroscopic characterization of a highly weathered basalt: igneous mineralogy, alteration products, and a microorganism. J Geophys Res 104(11):27067–27077

Wang A, Jolliff BL, Haskin LA, Kuebler KE, Viskupic KM (2001a) Characterization and comparison of structural and compositional features of planetary quadrilateral pyroxenes by Raman spectroscopy. Am Mineral 86(7–8):790–806

Wang C, Tang K, Yang Q, Shen G, Hai B, An C, Zuo J, Qian Y (2001b) Characterization of $PbSnS_3$ nanorods prepared via an iodine transport hydrothermal method. J Solid State Chem 160(1):50–53

Wang G, Wu Y, Fu P, Liang X, Xu Z, Chen C (2002) Crystal growth and properties of β-Zn_3BPO_7. Chem Mater 14(5):2044–2047

Wang Z, Saxena SK, Neumeier JJ (2003) Raman scattering study on pressure-induced phase transformation of marokite ($CaMn_2O_4$). J Solid State Chem 170(2):382–389

Wang A, Freeman JJ, Jolliff BL, Chou IM (2006a) Sulfates on Mars: a systematic Raman spectroscopic study of hydration states of magnesium sulfates. Geochim Cosmochim Acta 70(24):6118–6135

Wang CM, Liao CH, Chen PL, Lii KH (2006b) $UF_3(H_2O)(C_2O_4)_{0.5}$: a fluorooxalate of tetravalent uranium with a three-dimensional framework structure. Inorg Chem 45(4):1436–1438

Wang Y, Yi Z, Li Y, Yang Q, Wang D (2007) Hydrothermal synthesis of potassium niobate powders. Ceram Intern 33(8):1611–1615

Wang A, Freeman JJ, Arvidson R (2008) Study of two structural polymorphs of $MgSO_4 \cdot H_2O$ by Raman, IR, XRD, and humidity buffer experiments – implication for martian kieserite. 39th Lunar and Planetary Science Conference (1391):2172

Wang C, Xue S, Hu J, Tang K (2009a) Raman, far infrared, and Mössbauer spectroscopy of $CuFeS_2$ nanocrystallites. Jpn J Appl Phys 48:023003-1–023003-3

Wang Y, Pan S, Tian X, Zhou Z, Liu G, Wang J, Jia D (2009b) Synthesis, structure, and properties of the noncentrosymmetric hydrated borate $Na_2B_5O_8(OH) \cdot 2H_2O$. Inorg Chem 48(16):7800–7804

Wang X, Ming-Chou I, Hu W, Burruss RC, Sun Q, Song Y (2011) Raman spectroscopic measurements of CO_2 density: experimental calibration with high-pressure optical cell (HPOC) and fused silica capillary capsule (FSCC) with application to fluid inclusion observations. Geochim Cosmochim Acta 75:4080–4093

Wang Y, Liang J, Wen T, Li K, Wang Y, Li G, Lin J (2012) Syntheses and properties of a series of chromium vanadates $ACrV_2O_7$ (A = Na, K, Rb, Cs) with layered structure. J Solid State Chem 192:1–6

Wang Y, Lu X, Wang T, Pan F, Yan Y, Zhang Z (2013) Hydrothermal synthesis of flower-like ammonium illite constructed by nanosheets from coal series kaolin. Mater Lett 96:233–236

Wang M, Zhang P, Li J, Jiang C (2014) The effects of Mn loading on the structure and ozone decomposition activity of MnO$_x$ supported on activated carbon. Chin J Catal 35(3):335–341

Wang A, Freeman JJ, Jolliff BL (2015) Understanding the Raman spectral features of phyllosilicates. J Raman Spectrosc 46(10):829–845

Wang T, Sanchez C, Groenen J, Sciau P (2016) Raman spectroscopy analysis of terra sigillata: the yellow pigment of marbled sigillata. J Raman Spectrosc. https://doi.org/10.1002/jrs.4906

Warrier AVR, Krishnan RS (1964) Raman spectrum of strontium fluoride (SrF$_2$). Naturwissenschaften 51(1):8–9

Watanabe I, Noguchi S, Shimizu T (1983) Study on local structure in amorphous Sb-S films by Raman scattering. J Non-Cryst Solids 58(1):35–40

Watenphul A, Burgdorf M, Schlüter J, Horn I, Malcherek T, Mihailova B (2016a) Exploring the potential of Raman spectroscopy for crystallochemical analyses of complex hydrous silicates: II. Tourmalines. Am Mineral 101(4):970–985

Watenphul A, Schlüter J, Bosi F, Skogby H, Malcherek T, Mihailova B (2016b) Influence of the octahedral cationic-site occupancies on the framework vibrations of Li-free tourmalines, with implications for estimating temperature and oxygen fugacity in host rocks. Am Mineral 101(11):2554–2563

Watté J, Wuyts K, Silverans RE, Van Hove M, Van Rossum M (1994) A combined x-ray diffraction and Raman analysis of Ni/Au/Te-ohmic contacts to n-GaAs. J Appl Phys 75(4):2055–2060

Weber I, Böttger U, Jessberger EK, Hübers HW, Pavlov SG, Schröder S, Tarcea N, Dörfer T (2012) Raman spectroscopy of Mars relevant minerals for planetary exploration. 43rd Lunar Planetary Sci Conf 43:1793–1794

Wehrmeister U, Soldati AL, Jacob DE, Häger T, Hofmeister W (2010) Raman spectroscopy of synthetic, geological and biological vaterite: a Raman spectroscopic study. J Raman Spectrosc 41(2):193–201

Wei Q, Cheng JW, He C, Yang GY (2014) An acentric calcium borate Ca$_2$[B$_5$O$_9$](OH)·H$_2$O: ynthesis, structure, and nonliner optical property. Inorg Chem 53(21):11757–11763

Wei C, Bai Y, Deng A, Bao Y (2016) Universal synthesis of air stable, phase pure, controllable FeSe$_2$ nanocrystals. Nanotechnol 27(16):165702. (9 pp)

Weil M (2000) Preparation, single crystal structure analysis, and thermal behaviour of the acidic mercury (I) phosphate (Hg$_2$)$_2$(H$_2$PO$_4$)(PO$_4$). Z anorg allg Chemie 626(8):1752–1756. (in English)

Weis FA, Lazor P, Skogby H, Stalder R, Eriksson L (2016) Polarized IR and Raman spectra of zoisite: insights into OH-dipole orientation and the luminescence. Eur J Mineral. https://doi.org/10.1127/ejm/2016/0028-2528

Welch MD, Kleppe AK (2016) Polymorphism of the hydroxide perovskite Ga(OH)$_3$ and possible proton-driven transformational behavior. Phys Chem Minerals 43:515–526

Welch MD, Mitchell RH, Kampf AR, Chakhmouradian AR, Smith D, Carter M (2014) Crystal structure and topological affinities of magbasite, KBaFe^3$^+$Mg$_7$Si$_8$O$_{22}$(OH)$_2$F$_6$: a trellis structure related to amphibole and carpholite. Mineral Mag 78:29–45

Welch MD, Rumsey MS, Kleppe AK (2016) A naturally-occurring new lead-based halocuprate (I). J Solid State Chem 238:9–14

Welsh FS (2008) Identification of 1850s brown zinc paint made with franklinite and zincite at the US Capitol. APT Bull J Preserv Technol 39(1):17–30

Wesełucha-Birczyńska A, Tobola T, Natkaniec-Nowak L (2008) Raman microscopy of inclusions in blue halites. Vib Spectrosc 48(2):302–307

White SN (2009) Laser Raman spectroscopy as a technique for identification of seafloor hydrothermal and cold seep minerals. Chem Geol 259(3):240–252

White WB, Keramidas VG (1972) Vibrational spectra of oxides with the C-type rare earth oxide structure. Spectrochim Acta A 28(3):501–509

Wickleder M, Logemann C, Schwarzer S (2016) Oxidizing rhodium with sulfuric acid: the sulfates Rh$_2$(SO$_4$)$_3$ and Rh$_2$(SO$_4$)$_3$·2H$_2$O. Eur J Inorg Chem. https://doi.org/10.1002/ejic.201601247

Widulle F, Kramp S, Pyka NM, Göbel A, Ruf T, Debernardi A, Lauck R, Cardona M (1999) The phonon dispersion of wurtzite CdSe. Phys B 263–264:448–451

Wieghardt K, Siebert H (1971) Zur Kenntnis der Hexafluoromanganate (III). Z anorg allg Chemie 381(1):12–20. (in German)

Wienold J, Jentoft RE, Ressler T (2003) Structural investigation of the thermal decomposition of ammonium heptamolybdate by in situ XAFS and XRD. Eur J Inorg Chem 2003(6):1058–1071

Wiewióra A, Hida T (1996) X-Ray determination of superstructure of pyrophyllite from Yano-Shokozan mine, Hiroshima, Japan. Clay Sci 10(1):15–35

Wijzen F, Rulmont A, Tarte P (1994) Origin of spurious bands in the infrared spectrum of Ba$_2$TiO$_4$. Spectrochim Acta A 50(4):677–681

Wild S, Elliott H, Thompson DP (1978) Combined infrared and X-ray studies of β-silicon nitride and β'-sialons. J Mater Sci 13(8):1769–1775

Wildner M, Tillmanns E, Andrut M, Lorenz J (2003) Sailaufite, (Ca,Na,□)$_2$Mn$_3$O$_2$(AsO$_4$)$_2$(CO$_3$)·3H$_2$O, a new mineral from Hartkoppe hill, Ober-Sailauf (Spessart mountains, Germany), and its relationship to mitridatite–group minerals and pararobertsite. Eur J Mineral 15:555–564

Wilkins RWT (1971) Infrared spectroscopy in the mineralogical analysis of uranium ores. N Jb Miner Mh 11:441–450

Wilmarth WR, Begun GM, Nave SE, Peterson JR (1988) The Raman spectra of polycrystalline orthorhombic YF$_3$, SmF$_3$, HoF$_3$, YbF$_3$, and single crystal TbF$_3$. J Chem Phys 89(2):711–715

Wilson MJ (2014) The structure of opal-CT revisited. J Non-Cryst Solids 405:68–75

Windom BC, Sawyer WG, Hahn DW (2011) A Raman spectroscopic study of MoS_2 and MoO_3: applications to tribological systems. Tribol Lett 42(3):301–310

Witt K, Mecke R (1967) Grundschwingungen von Phenanthren aus experimentellen Ergebnissen. Z Naturforsch A 22(8):1247–1254. (in German)

Witzke T, Zhen S, Seff K, Doering T, Nasdala L, Kolitsch U (2001) Ronneburgite, $K_2MnV_4O_{12}$, a new mineral from Ronneburg, Thuringia, Germany: description and crystal structure. Am Mineral 86(9):1081–1086

Witzke T, Kolitsch U, Krause W, Wiechowski A, Medenbach O, Kampf AR, Steele IM, Favreau G (2006) Guanacoite, $Cu_2Mg_2(Mg_{0.5}Cu_{0.5})(OH)_4(H_2O)_4(AsO_4)_2$, a new arsenate mineral species from the El Guanaco Mine, near Taltal, Chile: description and crystal structure. Eur J Mineral 18(6):813–821

Włodyka R, Wrzalik R (2004) Apophyllite from the Międzyrzeczesill near Bielsko-Biała, the type of area of the teschenite-picrite association. Mineral Polonica 35(1):19–32

Wopenka B, Freeman JJ, Nikischer T (1998) Raman spectroscopic identification of fibrous natural zeolites. Appl Spectrosc 52(1):54–63

Wopenka B, Freeman JJ, Grew E (1999) Raman spectroscopic identification of B-free and B-rich kornerupine (prismatine). Am Mineral 84(4):550–554

Wu L, Chen XL, Li H, He M, Xu YP, Li XZ (2005) Structure determination and relative properties of novel cubic borates MM'4 (BO3) 3 (M = Li, M' = Sr; M = Na, M' = Sr, Ba). Inorg Chem 44(18):6409–6414

Wu L, Chen XL, Xu YP, Sun YP (2006a) Structure determination and relative properties of novel noncentrosymmetric borates $MM'_4(BO_3)_3$ (M = Na, M' = Ca and M = K, M' = Ca, Sr). Inorg Chem 45(7):3042–3047

Wu L, Chen XL, Zhang Y, Kong YF, Xu JJ, Xu YP (2006b) Ab initio structure determination of novel borate $NaSrBO_3$. J Solid State Chem 179(4):1219–1224

Wu L, Sun JC, Zhang Y, Jin SF, Kong YF, Xu JJ (2010a) Structure determination and relative properties of novel chiral orthoborate $KMgBO_3$. Inorg Chem 49(6):2715–2720

Wu SY, Liu XQ, Chen XM (2010b) Hydrothermal synthesis of $NaNbO_3$ with low NaOH concentration. Ceram Int 36(3):871–877

Wu H, Pan S, Poeppelmeier KR, Li H, Jia D, Chen Z, Fan X, Yang Y, Rondinelli JM, Luo H (2011) $K_3B_6O_{10}Cl$: a new structure analogous to perovskite with a large second harmonic generation response and deep UV absorption edge. J Am Chem Soc 133(20):7786–7790

Wu B, Wang R-C, Yang J-H, Wu F-Y, Zhang W-L, Gu X-P, Zhang A-C (2015) Wadeite ($K_2ZrSi_3O_9$), an alkali-zirconosilicate from the Saimaagpaitic rocks in northeastern China: its origin and response to multi-stage activities of alkaline fluids. Lithos 224:126–142

Wulferding D, Lemmens P, Scheib P, Röder J, Mendels P, Chu S, Han T, Lee YS (2010) Interplay of thermal and quantum spin fluctuations in the kagome lattice compound herbertsmithite. Phys Rev B 82(14):144412-1–144412-6

Wulferding D, Lemmens P, Yoshida H, Okamoto Y, Hiroi Z (2012) The spin dynamics in distorted kagome lattices: a comparative Raman study. J Phys Condens Matter 24(18):185602. (4 pp)

Xhaxhiu K, Saraçi E, Bente K (2013) Sequestration of supercritical CO_2 by mercury oxide. Chem Papers 67(6):594–600

Xia Y, Ma Q, Zhang Z, Liu Z, Feng J, Shao A, Wang W, Fu Q (2014) Development of Chinese barium copper silicate pigments during the Qin Empire based on Raman and polarized light microscopy studies. J Archaeolog Sci 49:500–509

Xia C, Li P, Gandi AN, Schwingenschlögl U, Alshareef HN (2015) Is $NiCo_2S_4$ really a semiconductor? Chem Mater 27(19):6482–6485

Xiao B, Dellen J, Schlenz H, Bosbach D, Suleimanov EV, Alekseev EV (2014) Unexpected structural complexity in cesium thorium molybdates. Cryst Growth Design 14(5):2677–2684

Xiao B, Schlenz H, Dellen J, Bosbach D, Suleimanov EV, Alekseev EV (2015) From two-dimensional layers to three-dimensional frameworks: expanding the structural diversity of uranyl compounds by cation-cation interactions. Cryst Growth Design 15(8):3775–3784

Xie X, Minitti ME, Chen M, Mao HK, Wang D, Shu J, Fei Y (2002) Natural high-pressure polymorph of merrillite in the shock veins of the Suizhou meteorite. Geochim Cosmochim Acta 66(13):2439–2444

Xie X, Minitti ME, Chen M, Mao HK, Wang D, Shu J, Fei Y (2003) Tuite, γ-$Ca_3(PO_4)_2$: a new mineral from the Suizhou L6 chondrite. Eur J Mineral 15(6):1001–1005

Xie HD, Shen DZ, Wang XQ, Shen GQ (2007) Growth and characterization of $KBi(WO_4)_2$ single crystals. Cryst Res Technol 42(1):18–22

Xie J, Zheng Y-X, Pan R-J, Liu S-Y, Song W-T, Cao G-S, Zhu T-J, Zhao X-B (2011) Sb-based alloy (NiSb, $FeSb_2$) nanoparticles decorated graphene prepared by one-step solvothermal route as anode for Li-ion batteries. Int. J Electrochem Sci 6:4811–4821

Xie B, Lu G, Wang Y, Guo Y, Guo Y (2012) Selective synthesis of tetragonal $LaVO_4$ with different vanadium sources and its luminescence performance. J Alloys Compd 544:173–180

Xie X, Chen M, Zhai S, Wang F (2014) Eutectic metal + troilite + Fe-Mn-Na phosphate + Al-free chromite assemblage in shock-produced chondritic melt of the Yanzhuang chondrite. Meteor Planet Sci 49(12):2290–2304

Xie X, Yang H, Gu X, Downs RT (2015) Chemical composition and crystal structure of merrillite from the Suizhou meteorite. Am Mineral 100(11–12):2753–2756

Xie X, Gu X, Chen M (2016) An occurrence of tuite, γ-$Ca_3(PO_4)_2$, partly transformed from Ca-phosphates in the Suizhou meteorite. Meteorit Planet Sci 51(1):195–202

Xiong Z-H, Zhao M-Z, He J-G, Li Y-P, Li H-Z (2016) Raman spectra of bredigite at high temperature and high pressure. Spectrosc Spectral Anal 36(10):3404–3409

X-ray Laboratory, Peking Institute of Uranium Geology, Wuhan Geological College (1978) Orthobrannerite – a new mineral of the brannerite group. Acta Geol Sin 52:241–251. (in Chinese, English abstr)

Xu G (1998) A fluid inclusion study of syntectonic Zn-Pb-Ag mineralization at Dugald River, Northwest Queensland, Australia. Econc Geol 93(8):1165–1179

Xu J, Gilson DF, Butler IS (1998) FT-Raman and high-pressure FT-infrared spectroscopic investigation of monocalcium phosphate monohydrate, Ca$(H_2PO_4)_2 \cdot H_2O$. Spectrochima Acta A 54(12):1869–1878

Xu J, Butler IS, Gilson DF (1999) FT-Raman and high-pressure infrared spectroscopic studies of dicalcium phosphate dihydrate ($CaHPO_4 \cdot 2H_2O$) and anhydrous dicalcium phosphate ($CaHPO_4$). Spectrochim Acta A 55(14):2801–2809

Xu S, Wu W, Xiao W, Yang J, Chen J, Ji S, Liu Y (2008) Moissanite in serpentinite from the Dabie Mountains in China. Mineral Mag 72(4):899–908

Xu H, Song Y, Pan W, Chen Q, Wu X, Lu P, Gong Q, Wang S (2015a) Vibrational properties of epitaxial Bi_4Te_3 films as studied by Raman spectroscopy. AIP Adv 5(8):087103-1–087103-7

Xu X, Yang J, Robinson PT, Xiong F, Ba D, Guo G (2015b) Origin of ultrahigh pressure and highly reduced minerals in podiform chromitites and associated mantle peridotites of the Luobusa ophiolite, Tibet. Gondwana Res 27(2):686–700

Xue D-J, Yang B, Yuan Z-K, Wang G, Liu X, Zhou Y, Hi L, Pan D, Chen S, Tang J (2015) $CuSbSe_2$ as a potential photovoltaic absorber material: studies from theory to experiment. Adv Energy Mater 5(23):1501203. (9 pp)

Yagoubi S, Obbade S, Benseghir M, Abraham F, Saadi M (2007) Synthesis, crystal structure, cationic mobility, thermal evolution and spectroscopic study of $Cs_8(UO_2)_4(WO_4)_4(WO_5)_2$ containing infinite uranyl tungstate chains. Solid State Sci 9(10):933–943

Yagoubi S, Renard C, Abraham F, Obbade S (2013) Molten salt flux synthesis and crystal structure of a new open-framework uranyl phosphate $Cs_3(UO_2)_2(PO_4)O_2$: spectroscopic characterization and cationic mobility studies. J Solid State Chem 200:13–21

Yakovenchuk VN, Ivanyuk GY, Pakhomovsky YA, Panikorovskii TL, Britvin SN, Krivovichev SV, Shilovskikh VV, Bocharov VN (2017) Kampelite, $Ba_3Mg_{1.5}Sc_4(PO_4)_6(OH)_3 \cdot 4H_2O$, a new very complex Ba-Sc phosphate mineral from the Kovdor phoscorite-carbonatite complex (Kola Peninsula, Russia). Mineral Petrol. https://doi.org/10.1007/s00710-017-0515-1

Yakubovich OV, Massa W, Chukanov NV (2008) Crystal structure of britvinite $[Pb_7(OH)_3F(BO_3)_2(CO_3)][Mg_{4.5}(OH)_3(Si_5O_{14})]$: a new layered silicate with an original type of silicon-oxygen networks. Crystallogr Repts 53(2):206–215

Yan G, Zhang S, Zhao M, Ding J, Li D (1992) Jianshuiite – a new magnesic mineral of chalcophanite group. Acta Mineral Sin 12:69–77. (in Chinese, English Abstr)

Yan T, Wang X, Long J, Lin H, Yuan R, Dai W, Li Z, Fu X (2008) Controlled preparation of In_2O_3, InOOH and $In(OH)_3$ via a one-pot aqueous solvothermal route. New J Chem 32(11):1843–1846

Yan C, Gu E, Liu F, Lai Y, Li J, Liu Y (2013) Colloidal synthesis and characterizations of wittichenite copper bismuth sulphide nanocrystals. Nanoscale 5(5):1789–1792

Yang Z, Giester G (2016) Hydrogen-bonding system in amarillite, $NaFe(SO_4)_2(H_2O)_6$: the structure refinement. Eur J Mineral. https://doi.org/10.1127/ejm/2016/0028-2567

Yang TR, Lu CC, Chou WC, Feng ZC, Chua SJ (1999) Infrared and Raman spectroscopic study of $Zn_{1-x}Mn_xSe$ materials grown by molecular-beam epitaxy. Phys Rev B 60(23):16058–16064

Yang J, Cheng GH, Zeng JH, Yu SH, Liu XM, Qian YT (2001) Shape control and characterization of transition metal diselenides MSe_2 (M= Ni, Co, Fe) prepared by a solvothermal-reduction process. Chem Mater 13(3):848–853

Yang M, Yu J, Shi L, Chen P, Li G, Chen Y, Xu R, Gao S (2006) Synthesis, dtructure, and magnetic property of a new open-framework manganese borophosphate, $[NH_4]_4[Mn_9B_2(OH)_2(HPO_4)_4(PO_4)_6]$. Chem Mater 18(2):476–481

Yang H, Costin G, Keogh J, Lu R, Downs RT (2007a) Cobaltaustinite, $CaCo(AsO_4)(OH)$. Acta Crystallogr E 63(2):i53–i55

Yang JS, Dobrzhinetskaya L, Bai WJ, Fang QS, Robinson PT, Zhang J, Green HW (2007b) Diamond- and coesite-bearing chromitites from the Luobusa ophiolite, Tibet. Geol 35(10):875–878

Yang N, Yang H, Jia J, Pang X (2007c) Formation and magnetic properties of nanosized $PbFe_{12}O_{19}$ particles synthesized by citrate precursor technique. J Alloys Compd 438:263–267

Yang H, Dembowski RF, Conrad PG, Downs RT (2008a) Crystal structure and Raman spectrum of hydroxyl-bästnasite-(Ce), $CeCO_3(OH)$. Am Mineral 93(4):698–701

Yang Y, Fang H, Zheng J, Li L, Li G, Yan G (2008b) Towards the understanding of poor electrochemical activity of triclinic $LiVOPO_4$: experimental characterization and theoretical investigations. Solid State Sci 10(10):1292–1298

Yang H, Jenkins RA, Downs RT, Evans SH, Tait KT (2011a) Rruffite, $Ca_2Cu(AsO_4)_2 \cdot 2H_2O$, a new member of the roselite group, from Tierra Amarilla, Chile. Can Mineral 49(3):877–884

Yang H, Sun HJ, Downs RT (2011b) Hazenite, $KNaMg_2(PO_4)_2 \cdot 14H_2O$, a new biologically related phosphate mineral, from Mono Lake, California, USA. Am Mineral 96(4):675–681

Yang JJ, Martens WN, Frost RL (2011c) Transition of chromium oxyhydroxide nanomaterials to chromium oxide: a hot-stage Raman spectroscopic study. J Raman Spectrosc 42(5):1142–1146

Yang YW, Stevenson RA, Siegel AM, Downs GW (2011d) Redetermination of eveite, $Mn_2AsO_4(OH)$, based on single-crystal X-ray diffraction data. Acta Crystallogr E. https://doi.org/10.1107/S1600536811044266

Yang Y, Pan S, Li H, Han J, Chen Z, Zhao W, Zhou Z (2011e) $Li_4Cs_3B_7O_{14}$: synthesis, crystal structure, and optical properties. Inorg Chem 50(6): 2415–2419

Yang Y, Wu J, Wang Y, Zhu J, Liu R, Meng C (2011f) Synthesis, crystal structure and characterization of a new protonated magnesium borophosphate: $(H_3O)Mg(H_2O)_2[BP_2O_8]\cdot H_2O$. Z anorg allg Chemie 637 (1):137–141. (in English)

Yang H, Jenkins RA, Thompson RM, Downs RT, Evans SH, Bloch EM (2012) Markascherite, $Cu_3(MoO_4)(OH)_4$, a new mineral species polymorphic with szenicsite, from Copper Creek, Pinal County, Arizona, USA. Am Mineral 97(1):197–202

Yang H, Downs RT, Evans SH, Jenkins RA, Bloch EM (2013a) Rongibbsite, $Pb_2(Si_4Al)O_{11}(OH)$, a new zeolitic aluminosilicate mineral with an interrupted framework from Maricopa County, Arizona, USA. Am Mineral 98(1):236–241

Yang H, Downs RT, Evans SH, Pinch WW (2013b) Scottyite, the natural analog of synthetic $BaCu_2Si_2O_7$, a new mineral from the Wessels mine, Kalahari Manganese Fields, South Africa. Am Mineral 98 (2–3):478–484

Yang H, Downs RT, Evans SH, Pinch WW (2014) Lavinskyite, $K(LiCu)Cu_6(Si_4O_{11})_2(OH)_4$, isotypic with plancheite, a new mineral from the Wessels mine, Kalahari Manganese Fields, South Africa. Am Mineral 99(2–3):525–530

Yang H, Andrade MB, Downs RT, Gibbs RB, Jenkins RA (2016a) Raygrantite, $Pb_{10}Zn(SO_4)_6(SiO_4)_2(OH)_2$, a new mineral isostructural with iranite, from the Big Horn Mountains, Maricopa County, Arizona, USA. Can Mineral 54:625–634

Yang Z, Giester G, Mao Q, Ma Y, Zhang D, Li H (2016b) Zincobotryogen, $ZnFe^{3+}(SO_4)_2(OH)\cdot 7H_2O$: validation as a mineral species and new data. Mineral Petrol. https://doi.org/10.1007/s00710-016-0484-9

Yanqing Z, Erwei S, Suxian C, Wenjun L, Xingfang H (2000) Hydrothermal preparation and characterization of brookite-type TiO_2 nanocrystallites. J Mater Sci Lett 19(16):1445–1448

Yeon J, Kim SH, Hayward MA, Halasyamani PS (2011) "A" cation polarity control in $ACuTe_2O_7$ (A = Sr^{2+}, Ba^{2+}, or Pb^{2+}). Inorg Chem 50(17):8663–8670

Yin P-F, Sun L-L, Gao Y-L, Wang S-Y (2008) Preparation and characterization of Co_9S_8 nanocrystalline and nanorods. Bull Mater Sci 31(4):593–596

Yin J, Li G, Yang G, Ge X, Xu H, Wang J (2015) Fluornatropyrochlore, a new pyrochlore supergroup mineral from the Boziguoer rare earth element deposit, Baicheng County, Akesu, Xinjiang, China. Can Mineral 53:455–460

Yoder CH, Pasteris JD, Krol KA, Weidner VL, Schaeffer RW (2012) Synthesis, structure, and solubility of carbonated barium chlor- and hydroxylapatites. Polyedron 44:143–149

Yong W, Botis S, Shieh SR, Shi W, Wither AC (2012) Pressure-induced phase transition study of magnesiochromite ($MgCr_2O_4$) by Raman spectroscopy and X-ray diffraction. Phys Earth Planet Interiors 196:75–82

You J, Jiang G, Xu K (2001) High temperature Raman spectra of sodium disilicate crystal, glass and its liquid. J Non-Cryst Solids 282(1):125–131

Yu YS, Lee HC, Kim HK, Han SG, Lee JH, Song JY, Lee GI, Jang MS (1990) The crystal growth of $BiNbO_4$. Ferroelectrics 107(1):225–228

Yu Z-T, Shi Z, Jiang Y-S, Yuan H-M, Chen J-S (2002) A chiral lead borate containing infinite and finite chains built up from BO_4 and BO_3 units. Chem Mater 14 (3):1314–1318

Yu H, Pan S, Wu H, Su X, Yang Z (2014) Synthesis, structures, optical properties and electronic structures of two mixed metal borates $MBaB_5O_9$ (M = Na, K). J Alloys Compd 585:602–607

Yu H, Young J, Wu H, Zhang W, Rondinelli JM, Halasyamani PS (2016) Electronic, crystal chemistry, and nonlinear optical property relationships in the dugganite $A_3B_3CD_2O_{14}$ family. J Am Chem Soc 138 (14):4984–4989

Yuan AQ, Wu J, Huang ZY, Wu K, Liao S, Tong ZF (2008) Synthesis of $NH_4FePO_4\cdot H_2O$ nano-plates via solid-state reaction at low temperature and its thermochemistry properties. Mater Res Bull 43(6):1339–1345

Yun-Fang W, Ya-Han C, Yung-Tang N, In-Gann C (2013) Crystal structure and optical performance of Al^{3+} and Ce^{3+} codoped $Ca_3Sc_2Si_3O_{12}$ green phosphors for white LEDs. J Am Ceram Soc 96(1):234–240

Yuran D, Li W (1998) Study on Raman spectrum characteristics of hsianghualite, baotite, huanghoite and jixianite. Rock Mineral Anal 3:181–184. (in Chinese, English Abstr)

Yuvaraj S, Karthikeyan K, Kalpana D, Lee YS, Selvan RK (2016) Surfactant-free hydrothermal synthesis of hierarchically structured spherical $CuBi_2O_4$ as negative electrodes for Li-ion hybrid capacitors. J Colloid Interface Sci 469:47–56

Yuzyuk YI, Simon P, Gagarina E, Hennet L, Thiaudiere D, Torgashev VI, Raevskaya SI, Raevskii IP, Sauvajol JL (2005) Modulated phases in $NaNbO_3$: Raman scattering, synchrotron x-ray diffraction, and dielectric investigations. J Phys Condens Matter 17 (33):4977–4990

Zaccarini F, Thalhammer OA, Princivalle F, Lenaz D, Stanley CJ, Garuti G (2007) Djerfisherite in the GuliDunite Complex, Polar Siberia: a primary or metasomatic phase? Can Mineral 45(5):1201–1211

Zaccarini F, Bakker RJ, Garuti G, Aiglsperger T, Thalhammer OA, Campos L, Proenza JA, Lewis JF (2010) Platinum group minerals in chromitite bodies of the Santa Elena Nappe, Costa Rica: mineralogical characterization by electron microprobe and Raman spectroscopy. Boletín Soc Geol Mexicana 62 (1):161–171

Zaccarini F, Bindi L, Pushkarev E, Garuti G, Bakker RJ (2016) Multi-analytical characterization of minerals of the bowieite–kashinite series from the Svetly Bor complex, Urals, Russia, and comparison with worldwide occurrences. Can Mineral 54:461–473

Zadov AE, Pekov IV, Zubkova NV, Gazeev VM, Chukanov NV, Yapaskurt VO, Kartasheov PM, Galuskin EV, Galuskina IO, Pertsev NN, Gurbanov AG, Pusharovsky DY (2013) Aklimaite, $Ca_4[Si_2O_5(OH)_2](OH)_4·5H_2O$, a new natural hydrosilicate from Mount Lakargi, the Northern Caucasus, Russia. Geol Ore Depos 55(7):541–548

Zaghib K, Julien CM (2005) Structure and electrochemistry of $FePO_4·2H_2O$ hydrate. J Power Sources 142(1):279–284

Zahoransky T, Friis H, Marks MA (2016) Luminescence and tenebrescence of natural sodalites: a chemical and structural study. Phys Chem Minerals 43:459–480

Zaitsev AN, Keller J, Spratt J, Jeffries TE, Sharygin VV (2009) Chemical composition of nyerereite and gregoryite from natrocarbonatites of Oldoinyo Lengai volcano, Tanzania. Geol Ore Depos 51(7):608–616

Zaitsev AN, Williams CT, Britvin SN, Kuznetsova IV, Spratt J, Petrov SV, Keller J (2010) Kerimasite, $Ca_3Zr_2(Fe^{3+}_2Si)O_{12}$, a new garnet from carbonatites of Kerimasi volcano and surrounding explosion craters, northern Tanzania. Mineral Mag 74(5):803–820

Zaitsev AN, Chakhmouradian AR, Siidra OI, Spratt J, Williams CT, Stanley CJ, Petrov SV, Britvin SN, Polyakova EA (2011) Fluorine-, yttrium-and lanthanide-rich cerianite-(Ce) from carbonatitic rocks of the Kerimasi volcano and surrounding explosion craters, Gregory Rift, northern Tanzania. Mineral Mag 75(6):2813–2822

Zaitsev AN, Avdontseva EY, Britvin SN, Demény A, Homonnay Z, Jeffries TE, Keller J, Krivovichev VG, Markl G, Platonova NV, Siidra OI, Spratt J, Vennemann T (2013) Oxo-magnesio-hastingsite, $NaCa_2(Mg_2Fe^{3+}_3)(Al_2Si_6)O_{22}O_2$, a new anhydrous amphibole from the Deeti volcanic cone, Gregory rift, northern Tanzania. Mineral Mag 77(6):2773–2792

Zaitsev AN, Britvin SN, Kearsley A, Wenzel T, Kirk C (2017) Jörgkellerite, $Na_3Mn^{3+}_3(PO_4)_2(CO_3)O_2·5H_2O$, a new layered phosphate-carbonate mineral from the Oldoinyo Lengai volcano, Gregory rift, northern Tanzania. Mineral Petrol 111:373–381

Zajzon N, Váczi T, Fehér B, Takács Á, Szakáll S, Weiszburg TG (2013) Pyrophanite pseudomorphs after perovskite in Perkupa serpentinites (Hungary): a microtextural study and geological implications. Phys Chem Minerals 40(8):611–623

Zákutná D, Repko A, Matulková I, Nižňanský D, Ardu A, Cannas C, Mantlíková A, Vejpravová J (2014) Hydrothermal synthesis, characterization, and magnetic properties of cobalt chromite nanoparticles. J Nanoparticle Res 16(2):1–14

Zavorotynska O, Corno M, Damin A, Spoto G, Ugliengo P, Baricco M (2011) Vibrational properties of MBH_4 and MBF_4 crystals (M = Li, Na, K): a combined DFT, infrared and Raman study. J Phys Chem C 115:18890–18900

Zedgenizov DA, Shatskiy A, Ragozin AL, Kagi H, Shatsky VS (2014) Merwinite in diamond from São Luiz, Brazil: a new mineral of the Ca-rich mantle environment. Am Mineral 99(2–3):547–550

Zhai S, Wu X, Ito E (2010) High-pressure Raman spectra of tuite, γ-$Ca_3(PO_4)_2$. J Raman Spectrosc 41(9):1011–1013

Zhai S, Akaogi M, Kojitani H, Xue W, Ito E (2014) Thermodynamic investigation on β- and γ-$Ca_3(PO_4)_2$ and the phase equilibria. Phys Earth Planet Interiors 228:144–149

Zhai S, Shieh SR, Xue W, Xie T (2015) Raman spectra of stronadelphite $Sr_5(PO_4)_3F$ at high pressures. Phys Chem Minerals 42(7):579–585

Zhai S, Yin Y, Shieh SR, Chang YY, Xie T, Xue W (2016) Raman spectroscopic study of $MnAl_2O_4$ galaxite at various pressures and temperatures. Phys Chem Minerals. https://doi.org/10.1007/s00269-016-0845-2

Zhan P (2009) Large scale hydrothermal synthesis of β-Co$(OH)_2$ hexagonal nanoplates and their conversion into porous Co_3O_4 nanoplates. J Alloys Compd 478(1):823–826

Zhang G, Redhammer J, Salje EKH, Mookherjee MM (2002a) $LiFeSi_2O_6$ and $NaFeSi_2O_6$ at low temperatures: an infrared spectroscopic study. Phys Chem Minerals 29:609–616

Zhang G, Wu Y, Fu P, Wang G, Liu H, Fan G, Chen C (2002b) A new sodium samarium borate $Na_3Sm_2(BO_3)_3$. J Phys Chem Solids 63(1):145–149

Zhang M, Xu H, Salje EKH, Heaney PJ (2003) Vibrational spectroscopy of beta-eucryptite ($LiAlSiO_4$): optical phonons and phase transition(s). Phys Chem Minerals 30:457–462

Zhang PH, Zhu JC, Zhao ZH, Gu XP, Lin JF (2004) Zincospiroffite, a new tellurite mineral species from the Zhongshangou gold deposit, Hebei Province, People's Republic of China. Can Mineral 42(3):763–768

Zhang G, Wu Y, Li Y, Chang F, Pan S, Fu P, Chen C (2005) Flux growth and characterization of a new oxyborate crystal $Na_3La_9O_3(BO_3)_8$. J Cryst Growth 275(1):e1997–e2001

Zhang D, Yoshioka F, Ikeue K, Machida M (2008) Synthesis and oxygen release/storage properties of Ce-substituted La-oxysulfates, $(La_{1-x}Ce_x)_2O_2SO_4$. Chem Mater 20(21):6697–6703

Zhang G, Liu Z, Zhang J, Fan F, Liu Y, Fu P (2009a) Crystal growth, structure, and properties of a non-centrosymmetric fluoride borate, $Ba_3Sr_4(BO_3)_3F_5$. Cryst Growth Des 9(7):3137–3141

Zhang H, Chen G, Li X, Wang Q (2009b) Electronic structure and water splitting under visible light irradiation of $BiTa_{1-x}Cu_xO_4$ ($x = 0.00$–0.04) photocatalysts. Int J Hydrogen Energy 34(9):3631–3638

Zhang SY, Jiang HL, Sun CF, Mao JG (2009c) Syntheses, crystal structures, and properties of five new transition metal molybdenum(VI) selenites and tellurites. Inorg Chem 48(24):11809–11820

Zhang A, Hsu W, Li Q, Liu Y, Jiang Y, Tang G (2010a) SIMS Pb/Pb dating of Zr-rich minerals in lunar meteorites Miller Range 05035 and LaPazIcefield 02224: implications for the petrogenesis of mare basalt. Sci China Earth Sci 53(3):327–334

Zhang L, Ling LS, Qu Z, Tong W, Tan S, Zhang YH (2010b) Study of lattice dynamics in the $CuIr_2S_4$ system. Eur Phys J B 77(1):83–86

Zhang FX, Lang M, Liu Z, Ewing RC (2011) Phase stability of some actinides with brannerite structure at high pressures. J Solid State Chem 184(11):2834–2839

Zhang M, Pan S, Han J, Yang Z, Su X, Zhao W (2012a) $Li_2Sr_4B_{12}O_{23}$: a new alkali and alkaline-earth metal mixed borate with $[B_{10}O_{18}]^{6-}$ network and isolated $[B_2O_5]^{4-}$ unit. J Solid State Chem 190:92–97

Zhang W, Wang X, Shen G, Shen D (2012b) Top-seeded growth, optical properties and theoretical studies of noncentrosymmetric $Te_2V_2O_9$. Cryst Res Technol 47(2):163–168

Zhang Y, Karatchevtseva I, Qin M, Middleburgh SC, Lumpkin GR (2013) Raman spectroscopic study of natural and synthetic brannerite. J Nucl Mater 437(1):149–153

Zhang H, Lei L, Zhang X (2014) One-step synthesis of cubic pyrite-type $CoSe_2$ at low temperature for efficient hydrogen evolution reaction. RSC Adv 4(97):54344–54348

Zhang Z, Fu Y, Zhou C, Lai Y (2015) Facile synthesis of $CuSbS_2$ blocks, and their lithium ion storage performance. J Electron Mater 44(1):252–257

Zhang S, Xie M, Cai B, Zhang H, Ma Y, Chen Z, Zhu Z, Hu Z, Zeng H (2016a) Semiconductor-topological insulator transition of two-dimensional SbAs induced by biaxial tensile strain. Phys Rev B 93(24):245303-1–245303-7

Zhang Y, Zhang Y, Zhao X, Zhang Y (2016b) Sol–gel synthesis and properties of europium–strontium copper silicates blue pigments with high near-infrared reflectance. Dyes Pigments 131:154–159

Zhao K, Guo H (2014) Behavior and mechanism of arsenate adsorption on activated natural siderite: evidences from FTIR and XANES analysis. Environ Sci Pollut Res 21(3):1944–1953

Zhao H, Wang J, Li J, Xu G, Zhang H, Yu L, Gao W, Xia H, Boughton RI (2008a) Lattice vibrations and thermal properties of stoichiometric $KYb(WO_4)_2$ crystal. Cryst Growth Design 8(11):3978–3983

Zhao WY, Wei P, Wu XY, Wang W, Zhang QJ (2008b) Lattice vibration characterization and magnetic properties of M-type barium hexaferrite with excessive iron. J Appl Phys 103(6):063902. (5 pp)

Zhao W, Pan S, Han J, Zhou Z (2011) Synthesis, crystal structure and optical properties of a new lead bismuth borate. Inorg Chim Acta 379(1):130–134

Zhao W, Pan S, Wang Y, Yang Z, Wang X, Han J (2012) Structure, growth and properties of a novel polar material, $KSr_4(BO_3)_3$. J Solid State Chem 195:73–78

Zhao C, Wang F, Sun Y, Zhou Y (2013a) Synthesis and characterization of β-$Yb_2Si_2O_7$ powders. Ceram Int 39(5):5805–5811

Zhao X-M, Zhang J, Berlie A, Qin Z-X, Huang Q-W, Jiang S, Zhang J-B, Tang L-Y, Liu J, Zhang C, Zhong G-H, Lin H-Q, Chen X-J (2013b) Phase transformations and vibrational properties of coronene under pressure. J Chem Phys 139(14):144308-1–144308-7

Zhao L, Liu W, Cao L, Su G, Gao R, Yang H (2015) A new member of ferrous sulfates, $FeSO_4 \cdot 2H_2O$ with PtS topology showing spin-canted long-range antiferromagnetic ordering. J Solid State Chem 231:58–63

Zheng HL, Zhang ZC, Zhou JG, Yang SS, Zhao J (2012) Vibrational spectra of $CaGa_2O_4$, Ca_2GeO_4, $CaIn_2O_4$ and $CaSnO_3$ prepared by electrospinning. Appl Phys A 108(2):465–473

Zhigadlo ND, Zhang M, Salje EKH (2001) An infrared spectroscopic study of $Li_2B_4O_7$. J Phys Condens Matter 13(30):6551–6561

Zhizhin GN, Mavrin BN, Shabanov VF (1984) Optical vibrational spectra of crystals. Nauka Publishing House, Moscow. (in Russian)

Zhong J, Xiang W, Cai Q, Liang X (2012) Synthesis, characterization and optical properties of flower-like Cu3BiS3 nanorods. Mater Lett 70:63–66

Zhou T, Schwarz U, Hanfland M, Liu ZX, Syassen K, Cardona M (1998) Effect of pressure on the crystal structure, vibrational modes, and electronic excitations of HgO. Phys Rev B 57(1):153–160

Zhou D, Huang G, Chen X, Xu J, Gong S (2004) Synthesis of $LaAlO_3$ via ethylenediaminetetraacetic acid precursor. Mater Chem Phys 84(1):33–36

Zhou Y, Hoffmann S, Huang YX, Prots Y, Schnelle W, Menezes PW, Carrillo-Cabrera W, Sichelschmidt J, Mi J-X, Kniep R (2011) $K_3Ln[OB(OH)_2]_2[HOPO_3]_2$ (Ln = Yb, Lu): layered rare-earth dihydrogen borate monohydrogen phosphates. J Solid State Chem 184(6):1517–1522

Zhou Y, Leng M, Xia Z, Zhong J, Song H, Liu X, Yang B, Zhang J, Chen J, Zhou K, Han J, Cheng Y, Tang J (2014) Solution-processed antimony selenide heterojunction solar cells. Adv Energy Mater. https://doi.org/10.1002/aenm.201301846

Zhu L, Teo M, Wong PC, Wong KC, Narita I, Ernst F, Mitchell KAR, Campbell SA (2010) Synthesis, characterization of a $CoSe_2$ catalyst for the oxygen reduction reaction. Appl Catal A 386(1):157–165

Zhu D, Yun S, Nai X, Zhao D, Liu X, Li W (2013) Synthesis and characterization of strontium chloroborate whiskers. Cryst Res Technol 48(1):6–10

Zhukova ES, Mikheykin AS, Torgashev VI, Bush AA, Yuzyuk YI, Sashin AE, Prokhorov AS, Dressel M, Gorshunov BP (2016) Crucial influence of crystal site disorder on dynamical spectral response in artificial magnetoplumbites. Solid State Sci 62:13–21

Zhuo JQ, Caban-Acevedo M, Liang HF, Samad L, Ding Q, Fu YP, Li MX, Jin S (2015) High-performance electrocatalysis forhydrogen evolution reaction using se-doped pyrite-phase nickeldiphosphide nanostructures. ACS Catal 5:6355–6361

Zubko M, Kusz J, Malczewski D, Häger T, Hofmeister W (2013) High temperature study of metamict rinkite. Solid State Phenom 203–204:331–334

Index

A

Abellaite, 95, 741
Abelsonite, 741
Abhurite, 742
Acanthite, 742
Acetamide solution, 742
Actinolite, 742, 743
Adachiite, 743
Adamite, 743
Adelite, 743
Admontite, 71, 744
Adolfpateraite, 744
Aegirine, 744
Aegirine-augite, 267
Aegirine Li analogue, 744
Aerinite, 745
Aeschynite-(Y), 745
Afmite, 745
Afwillite, 745
Agakhanovite-(Y), 746
Agardite-(Ce), 629, 746
Agardite-(La), 746
Agardite-(Nd), 630
Agardite-(Y), 746
Agricolaite, 747
Ahlfeldite, 747
Ahrensite, 747
Aikinite, 747
Ajoite, 748
Akaganeite, 748
Åkermanite, 245, 748
Åkermanite Sr analogue, 748
Akimotoite, 321, 749
Aklimaite, 749
Alabandite, 749
Alacránite, 749
Alamosite, 749
Alamosite polymorph, 266
Alarsite, 750
Albertiniite, 750
Albite, 750
Aleksite, 750
Alforsite, 431
Alforsite F-analogue, 431

Alforsite OH-analogue, 751
Alforsite vanadate analogue, 578
Alfredopetrovite, 661
Allactite, 617, 751
Allanite (Ce), 751
Allanite-(Nd), 751
Allanpringite, 752
Allendeite, 752
Allophane, 752
Almandine, 752
Almarudite, 753
Alpersite, 501
Alstonite, 753
Altaite, 753
Althausite, 458, 753
Alum-(K), 754
Aluminite, 754
Aluminium acid selenite hydrate, 639
Aluminium decavanadate hydrate, 556
Aluminium niobate, 125
Aluminium phosphate hydrate, 355
Aluminium selenite hydrate, 640
Aluminocerite-(Ce), 754
Aluminocopiapite, 501, 755
Alumohydrocalcite, 755
Alumovesuvianite, 251
Alunite, 755
Alunogen, 517, 755
Alwilkinsite-(Y), 755
Amakinite, 210
Amarantite, 756
Amarillite, 502
Amblygonite, 756
Ambrinoite, 756
Ameghinite, 756
Amesite, 757
Ammoniojarosite, 757
Ammoniozippeite, 532
Ammonium bicarbonate, 80
Ammonium calcium borate, 43
Ammonium cuprooxopolymolybdate $(NH_4)_4[H_6CuMo_6O_{24}]\cdot 4H_2O$, 665
Ammonium dichromate, 590
Ammonium heptamolybdate, 666

Ammonium iron(II) phosphate hydrate, 357
Ammonium magnesium phosphate, 357
Ammonium manganese(II) borophosphate, 393
Ammonium nickel molybdate $(NH_4)Ni_2(HMoO_4)(MoO_4)(OH)_2$, 666
Ammonium paratungstate tetrahydrate, 707
Ammonium pentaborate, 41
Ammonium sulfate tellurate, 536
Ammonium titanophosphate, 358
Ammonium uranyl vanadate hydrate, 556
Ammonium vanadyl compound $(NH_4)_{0.5}V_2O_5 \cdot nH_2O$, 557
Ammonium vanadyl pyrophosphate, 355
Ammonium zirconofluoride, 226
Analcime, 757
Anapaite, 758
Anatase, 758
Ancylite-(Ce), 758
Andalusite, 759
Andersonite, 759
Andradite, 759
Andychristyite, 760
Angastonite, 760
Anglesite, 760
Anhydrite, 8, 760
Anilite, 761
Ankerite, 761
Ankoleite, 432
Annabergite, 761
Annite, 761
Annite Cl-analogue, 761
Anorpiment, 762
Anorthite, 762
Antarcticite, 762
Anthophyllite, 763
Anthraxolite, 763
Antigorite, 763, 764
Antimonselite, 764
Antimony, 764
Antimony(III) phosphate, 359
Antimony(V) oxophosphate, 359
Antlerite, 764
Antofagastaite, 474
Apachite, 765
Aphthitalite, 765
Apjohnite, 765
Apuanite, 765
Aradite, 766
Aragonite, 8, 731, 766
Arangasite, 535
Arapovite-related silicate, 766
Aravaipaite, 767
Arcanite, 767
Archerite, 767
Ardealite, 535, 767
Ardennite-(As), 768
Arfvedsonite, 768
Argentojarosite, 768
Argutite, 768
Arhbarite, 617

Arisite (Ce), 769
Armalcolite, 769
Arrojadite-(KFe), 769
Arsenbrackebuschite, 769
Arsendescloisite Sr-analogue, 770
Arseniosiderite, 770
Arsenogorceixite, 770
Arsenolamprite, 770
Arsenolite, 770
Arsenopyrite, 771
Arsentsumebite, 771
Arsenuranylite, 771
Arthurite, 772
Artinite, 772
Arzakite, 772
Asbolane, 772, 773
Aspedamite, 773
Asselbornite, 773
Atacamite, 773
Atelestite, 774
Atelisite (Y), 774
Athabascaite, 774
Atokite, 774
Augelite, 775
Augite, 775
Aurichalcite, 775
Aurostibite, 776
Austinite, 776
Autunite, 776
Avicennite, 776
Avogadrite, 776
Awaruite, 777
Axinite-(Fe), 777
Azurite, 777

B

Bacalite, 104
Backite, 777
Badalovite, 632
Baddeleyite, 778
Bafertisite, 778
Baghdadite, 778
Bahianite, 778
Bairdite, 779
Balestraite, 779
Balkanite, 779
Bambollaite Te analogue, 779
Baotite, 780
Barahonaite-(Al), 780
Barahonaite-(Fe), 780
Bararite, 780
Barberiite, 781
Barbosalite, 781
Barentsite, 84
Bariandite Al-free analogue, 781
Barićite, 781
Barioferrite, 782
Barioperovskite, 782
Barium borate, 42

Index

Barium bromide dihydrate, 664
Barium calcium diborate, 43
Barium calcium tellurate, 680
Barium cerium tantalite, 126
Barium chromium pyrophosphate, 361
Barium cobalt antimonate, 127
Barium cobaltate, 127
Barium cobalt selenite hydrate, 640
Barium cobalt tellurate, 680
Barium copper tellurate tellurite, 681
Barium formate, 98
Barium lanthanum thorium orthovanadate, 558
Barium magnesium fluoride, 227
Barium manganese fluoride, 227
Barium nickel oxide $BaNiO_2$, 128
Barium nickel oxide $BaNiO_3$, 129
Barium nickel tellurate, 682
Barium niobate, 129
Barium sodium cyclotriphosphate hydrate, 361
Barium strontium orthoborate fluoride, 21
Barium titanate, 130
Barium titanium sulfide, 481
Barium vanadyl phosphate, 362
Barium vanadyl vanadate, 558
Barium zinc tellurate, 682
Barium zirconium orthoborate, 22
Barnesite, 782
Barringtonite, 782
Barrydawsonite-(Y), 782
Bartelkeite, 783
Barylite, 783
Barysilite, 246
Baryte, 783
Baryte selenate analogue, 641
Barytocalcite, 784
Bassanite, 784
Bassoite, 784
Bastnäsite-(Ce), 784
Batievaite-(Y), 340
Batiferrite co-bearing, 785
Baumhauerite, 785
Bavenite, 785
Bayerite, 785
Bayldonite, 786
Bayleyite, 90, 786
Baylissite NH_4-analogue, 786
Bazhenovite, 786
Bazirite, 341, 787
Bazzite, 787
Mereheadite, 77
Beaverite-(Cu), 787
Beaverite-(Zn), 529
Becquerelite, 787
Behierite, 788
Behoite, 788
Běhounekite, 788
Belakovskiite, 788
Bellidoite, 789
Bendadaite, 789

Benitoite, 789
Benstonite, 789
Beraunite, 351, 353, 789
Berborite, 40
Berdesinskiite, 790
Bergenite, 790
Berlinite, 790
Berlinite tetragonal polymorph, 791
Bermanite, 791
Bernalite, 791
Berndtite, 791
Berthierite, 791
Bertrandite, 792
Beryl, 792
Beryl Cs-bearing, 792
Beryllonite, 793
Berzeliite, 793
Berzeliite polymorph alluaudite-type, 793
Betalomonosovite, 344, 345, 793
Bettertonite, 618
Beudantite, 794
Beusite, 794
Beyerite, 794
Bianchite/Goslarite, 794
Bikitaite, 795
Billietite, 795
Biotite, 795
Biphosphammite, 795
Birnessite, 795, 796
Bischofite, 796
Bismite, 174, 796
Bismoclite, 796
Bismuth, 796
Bismuth(III) aluminate $Bi_2Al_4O_9$, 131
Bismuth(III) aluminoferrite $Bi_2Fe_3AlO_9$, 132
Bismuth(III) calcium oxophosphate, 363
Bismuth copper sulfate tellurite, 482
Bismuth ferrite, 135
Bismuth(III) ferrite dimolybdate, 667
Bismuthinite, 797
Bismuth(III) magnesium oxovanadate, 560
Bismuth(III) magnesium oxovanadate $BiMg(VO_4)O$, 559
Bismuth molybdate, 668
Bismuth(III) nickel oxophosphate $BiNi(PO_4)O$, 364
Bismuth(III) stannate pyrochlore-type, 133
Bismuth sulfate, 483
Bismuth(III) tantalate $Bi_7Ta_3O_{18}$, 133
Bismuth(III) tellurite selenate, 642
Bismuth(III) titanate $Bi_4Ti_3O_{12}$, 134
Bismuth tungstate, 708
Bismuthyl sulfate, 483
Bismutite, 797
Bismutocolumbite, 175, 797
Bismutoferrite, 797
Bismutotantalite, 798
Bismutotantalite triclinic dimorph, 176
Bitikleite, 798
Bityite, 327
Bixbyite, 798

Blatterite, 798
Blödite, 799
Bluebellite, 799
Bluelizardite, 799
Bobcookite, 471, 799
Bobdownsite, 800
Bobierrite, 800
Bobshannonite, 800
Bohdanowiczite, 800
Böhmite, 801
Boleite, 801
Boltwoodite, 801
Bonaccordite, 801
Bonattite, 802
Bonazziite, 802
Boracite, 802
Borax, 803
Bornite, 803
Boron arsenate, 599
Boron phosphate, 365
Bosiite, 803
Botallackite, 803
Botryogen, 803
Bottinoite, 804
Boulangerite, 804
Bournonite, 804
Boussingaultite, 804, 805
Bowieite, 805
Braccoite, 805
Bracewellite, 805
Brackebuschite, 806
Bradaczekite, 626
Bradleyite, 806
Braggite, 806
Brandholzite, 806
Brannerite, 807
Brannockite, 807
Brassite, 807
Braunite, 176, 807
Brazilianite, 808
Bredigite, 808
Breithauptite, 808
Brewsterite-Sr, 808
Brianyoungite, 809
Briartite, 809
Bridgmanite, 321
Bridgmanite trigonal polymorph, 322
Britvinite, 15, 16, 809
Brizziite, 809
Brizzite polymorph, 177
Brochantite, 810
Bromargyrite, 810
Bromellite, 178, 810
Brookite, 178, 811
Browneite, 811
Brownleeite, 811
Brownmillerite, 811
Brucite, 2, 12, 812
Brucite Co-analogue, 179

Brüggenite, 812
Brugnatellite, 812
Brunogeierite, 597
Brushite, 812
Buchwaldite, 813
Buchwaldite dimorph, 450
Bukovskýite, 813
Bulgakite, 345
Bunnoite, 813
Bunsenite, 813
Burangaite, 814
Burbankite, 814
Burckhardtite, 814
Burgessite, 815
Burkeite, 815
Buseckite, 815
Bustamite, 815
Butlerite, 816
Buttgenbachite, 816
Byströmite, 816

C
Cabalzarite, 816
Cabvinite, 817
Cacoxenite, 817
Cadmium formatedihydrate, 98
Cadmium molybdate, 668
Cadmium oxalate trihydrate, 99
Cadmium stannate, 136
Cadmium tungstate, 708
Cadmoindite, 817
Cadmoselite, 817
Cafarsite, 818
Cafetite, 818
Cahnite, 818
Cairncrossite, 818
Calamaite, 476
Calaverite, 819
Calciborite, 819
Calcioaravaipaite, 819
Calciojohillerite, 634
Calciolangbeinite, 820
Calcio-olivine, 819
Calciopetersite, 820
Calcite, 8, 731, 820
Calcium antimonite, 138
Calcium arsenate $CaAs_2O_6$, 602
Calcium borate, 45
Calcium chlorarsenate, 601
Calcium chlorophosphate ("chlor-spodiosite"), 368
Calcium copper titanate $CaCu_3Ti_4O_{12}$, 139
Calcium dihydrophosphate monohydrate, 369
Calcium hydroxychloride, 540
Calcium indium oxide Ca_2InO_4, 135
Calcium iron(III) tin orthophosphate, 432
Calcium magnesium lanthanum phosphate, 370
Calcium magnesium yttrium phosphate, 370
Calcium niobate columbite-type, 139
Calcium orthoarsenate trigonal polymorph, 603

Calcium orthoborate fluoride, 23
Calcium orthophosphate orthosilicate, 347
Calcium orthovanadate trigonal polymorph, 560
Calcium plumbate, 140
Calcium samarium thorium arsenate, 603
Calcium strontium orthophosphate whitlockite-type, 422
Calcium tellurite monohydrate, 683
Calcium tetraborate, 45
Calcurmolite, 821
Calderite, 821
Calderónite, 821
Caledonite, 821
Callaghanite, 822
Calomel, 822
Calumetite, 822
Camaronesite, 822
Camerolaite, 823
Campostriniite, 503
Canavesite, 823
Cancrinite, 727, 823
Cancrinite Ca-free analogue, 306
Cancrinite CO_3-deficient, 301
Cancrinite NO_3-analogue, 304
Cancrinite NO_3-analogue low-hydrous, 305
Cancrinite SO_4-rich, 823
Canfieldite, 824
Cannonite, 824
Canosioite, 619
Canutite, 637
Caoxite, 104, 105
Carbocernaite, 824
Carbonatecyanotrichite, 824
Carletonite, 825
Carlfrancisite, 825
Carlfriesite, 825
Carlhintzeite, 226
Carlinite, 825
Carlosturanite, 826
Carlsbergite, 826
Carlsonite, 515
Carminite, 826
Carnallite, 826
Carnegieite, 826
Carnegieite (high), 306
Carnegieite (low), 307
Carnotite, 827
Carpathite, 827
Carpholite, 827
Carrboydite, 827
Carrollite, 828
Caryopilite, 828
Cassiterite, 828
Castellaroite, 597, 828
Caswellsilverite, 829
Catalanoite, 829
Catapleiite, 829
Catapleiite heating product, 346
Cattierite, 829
Cavansite, 829

Cebaite (Ce), 830
Čejkaite, 830
Celadonite, 830
Celestine, 830
Celsian, 831
Centennialite, 545
Cerianite-(Ce), 831
Cerium metaphosphate trihydrate, 372
Cerium(III) polyphosphate, 372
Cerium(IV) pyrophosphate, 371
Černýite, 831
Ceruleite, 831
Cerussite, 16, 832
Cervantite, 832
Cesanite, 832
Cesàrolite, 118
Cesium acid (pentahydrogen) arsenate, 600
Cesium acid arsenate selenate, 642
Cesium acid (pentahydrogen) phosphate, 366
Cesium antimony chloride, 539
Cesium beryllium orthoborate, 22
Cesium borosilicate pollucite-type, 330
Cesium calcium borate, 44
Cesium copper chloride, 537
Cesium fluormolybdate $CsMoO_2F_3$, 669
Cesium hexafluorphosphate, 228, 229
Cesium iron arsenate $Cs_7Fe_7O_2(AsO_4)_8$, 601
Cesium iron sulfate, 484
Cesium magnesium chloride, 538
Cesium manganese(II) pyrophosphate, 367
Cesium sodium stibiochloride, 539
Cesium stibiofluoride, 229
Cesium thorium molybdate, 670
Cesium uranyl niobate $Cs_2(UO_2)_2(Nb_2O_8)$, 137
Cesium uranyl niobate
  $Cs_9[(UO_2)_8O_4(NbO_5)(Nb_2O_8)_2]$, 137
Cesium uranyl oxophosphate, 368
Cesium uranyl tungstate
  $Cs_4[(UO_2)_4(WO_5)(W_2O_8)O_2]$, 709
Cesium uranyl tungstate $Cs_4[(UO_2)_7(WO_5)_3O_3]$, 710
Cesium uranyl tungstate $Cs_8(UO_2)_4(WO_4)_4(WO_5)_2$, 710
Chabazite-Ca, 832
Chalcanthite, 832
Chalcoalumite, 833
Chalcocite, 833
Chalcocyanite, 833
Chalcomenite, 833
Chalconatronite, 834
Chalcophanite, 834
Chalcophyllite, 834
Chalcopyrite, 834
Chalcosiderite, 835
Chalcostibite, 835
Challacolloite, 835
Chambersite, 835
Chamosite, 836
Changbaiite, 836
Changoite, 522, 836
Changoite (slightly deuterated), 503

Chapmanite, 836
Charlesite, 521
Charoite, 836
Chegemite, 837
Chekhovichite, 837
Chenevixite, 837
Cheralite, 837
Cheralite La-bearing, 620
Cheremnykhite trigonal dimorph, 585
Chernikovite, 838
Chernovite-(Y), 838
Chervetite, 838
Chiavennite, 327, 838
Chibaite, 207, 839
Childrenite, 839
Chiolite, 839
Chloraluminite, 839
Chlorapatite, 840
Chlorargyrite, 840
Chlorellestadite, 335
Chloritoid, 840
Chlorkyuygenite, 840
Chlormayenite, 180, 840
Chlorocalcite, 841
Chloromagnesite, 841
Chlorophoenicite, 630
Chloroxiphite, 841
Chondrodite, 841
Chromatite, 842
Chromite, 842
Chromium disilicide, 320
Chromium iron(III) orthovanadate, 561
Chromium uranium oxide Cr_2UO_6, 141
Chromium vanadate $Cr_2V_4O_{13}$, 562
Chrysoberyl, 181, 842
Chrysocolla, 843
Chrysothallite, 843
Chrysotile, 9, 843
Chudobaite, 599
Chukanovite, 85, 843
Chukhrovite (Ca), 843
Churchite-(Nd), 844
Churchite-(Y), 844
Cinnabar, 844
Claringbullite, 844
Clarkeite, 210
Claudetite, 845
Clausthalite, 845
Clinoatacamite, 845
Clinobisvanite, 579, 846
Clinocervantite, 224, 846
Clinochalcomenite, 846
Clinochlore, 846
Clinoclase, 847
Clinoenstatite, 847
Clinohedrite, 847
Clinometaborite, 847
Clinoptilolite (Na), 848
Clinotobermorite-like mineral, 848

Clinozoisite, 848
Clintonite, 848
Coahuilite, 106
Coalingite, 94, 849
Cobaltarthurite, 849
Cobaltaustinite, 849
Cobalt dinickel orthoborate, 39
Cobalt ferrite spinel-type, 142
Cobaltkoritnigite, 849
Cobaltomenite, 661, 850
Cobaltpentlandite, 850
Cobalt selenite, 662
Cobalt selenite hydrate, 663
Cobalt zinc tellurium oxide, 141
Coccinite, 850
Cochromite, 850
Cochromite Ni-bearing, 182
Coconinoite, 851
Coesite, 851
Coffinite, 851
Colemanite, 851
Colimaite, 852
Colinowensite, 279, 852
Coloradoite, 852
Columbite-(Mg), 852
Columbite-(Mn), 183, 853
Comancheite, 545, 853
Combeite, 853
Compreignacite, 853
Conichalcite, 853
Connellite, 854
Cookeite, 292, 297
Cooperite, 854
Copiapite, 504, 854
Copper acid diperiodate hydrate, 702
Copper(II) carbonate, 81
Copper chromate, 590
Copper divanadate hydroxide hydrate, 562
Copper(II) hydroxide, 142
Copper iodate, 699, 700
Copper molybdate selenite, 644
Copper oxychloride hydrate, 541
Copper strontium formate, 100
Copper tinanium oxyphosphate, 373
Coquandite, 855
Coquimbite, 855
Corderoite, 855
Cordierite, 855
Cordylite (Ce), 856
Corkite, 856
Cornetite, 856
Cornubite, 856
Cornwallite, 857
Coronadite, 857
Correianevesite, 857
Corundum, 183, 858
Cosalite, 858
Cotunnite, 858
Coulsonite, 858

Covellite, 858
Crandallite, 859
Cranswickite, 859
Creaseyite, 859
Crednerite, 859
Creedite, 860
Cristobalite, 860
Crocoite, 860
Cryptohalite, 861
Cryptomelane, 861
Cu,Al-hydroxyphosphate, 459
Cubanite, 861
Cuboargyrite, 861
Cumengeite, 549, 861
Cummingtonite, 862
Cuprite, 862
Cuprocopiapite, 862
Cuproiridsite, 862
Cupromolybdite, 863
Cuprorhodsite, 863
Cuprorivaite, 863
Cuprosklodowskite, 863
Cuprospinel, 864
Cuprotungstite, 864
Curienite, 864
Curite, 864
Cuspidine, 244, 865
Cyanochroite, 865
Cyanotrichite, 865
Cymrite, 865
Cyprine, 250
Cyrilovite, 866
Czochralskiite, 866

D

Dachiardite (Na), 866
Dalnegorskite, 269
Danalite, 329
Danburite, 866
Daqingshanite-(Ce), 456
Darapskite, 867
Darrellhenryite, 332, 333, 867
Dashkovaite, 867
Datolite, 867
Daubréelite, 868
Davidite-(La), 868
Davidlloydite, 868
Davinciite, 282
Davisite, 868
Dawsonite, 869
Decrespignyite-(Y), 869
Delafossite Al analogue, 184
Delafossite, 869
Delhayelite, 869
Deliensite, 870
Dellaite, 870
Deltalumite, 122
Delvauxite, 870
Demartinite, 871

Demesmaekerite, 871
Demicheleite-(Br), 871
Demicheleite-(Cl), 871
Demicheleite-(I), 871
Denningite, 696, 872
Depmeierite, 302
Derriksite, 872
Desautelsite, 872
Dessauite-(Y), 872
Destinezite, 873
Deveroite-(Ce), 106, 110
Devilline, 873
Devitoite, 873
Devitrite, 873
Dewindtite, 874
Diaboleite, 874
Diadochite, 874
Diamond, 874
Diaoyudaoite, 220
Diaspore, 875
Dicalcium hexaborate monohydrate, 46
Dickinsonite-(KMnNa), 875
Dickite, 875
Digenite, 875
Dimorphite, 876
Diomignite, 876
Diopside, 876
Dioptase, 876
Dipotassium sodium zinc pentaborate, 47
Dissakisite-(La), 877
Dixenite, 877
Djerfisherite, 877
Dmisokolovite, 877
Dmisteinbergite, 313, 878
Dolomite, 878
Domerockite, 878
Donnayite-(Y), 878
Dorallcharite, 879
Dorfmanite, 879
Dorrite, 267
Double perovskite $KBa(XeNaO_6)$, 704
Double perovskite $KCa(XeNaO_6)$, 704
Double perovskite $KSr(XeNaO_6)$, 705
Double-ring borate $(Na,K)_3Sr(B_5O_{10})$, 47
Dovyrenite, 879
Downeyite, 879
Doyleite, 880
Dravite, 880
Dreyerite, 880
Drugmanite, 467
Drysdallite, 880
Dualite, 283
Dufrénoysite, 881
Duftite, 881
Dumontite, 881
Dumortierite, 881
Dundasite, 882
Durangite, 882
Dussertite, 882

Duttonite, 211
Dwornikite, 882
Dypingite, 883
Dysprosium copper hydroxysulfate
  $Dy_2Cu(SO_4)_2(OH)_4$, 485
Dysprosium decavanadate hydrate, 563
Dzhalindite, 883
Dzhuluite, 883
Dzierżanowskite, 883

E

Eakerite, 884
Eastonite, 884
Ecandrewsite, 884
Ecdemite, 885
Eckhardite, 885
Edgrewite, 885
Edingtonite, 885
Edoylerite, 886
Effenbergerite, 886
Eitelite, 886
Ekanite, 886
Ekplexite, 887
Elbaite, 887
Elbrusite, 888
Eleomelanite, 470
Eleonorite, 888
Elpasolite, 888
Eltyubyuite, 888
Embreyite, 595
Emmonsite, 889
Enargite, 889
Enstatite, 889
Eosphorite, 890
Epidote, 890
Epidote-(Sr), 257
Epistilbite, 890
Epsomite, 890, 891
Ericlaxmanite, 891
Erikapohlite, 891
Eringaite, 891
Eriochalcite, 892
Erionite-Ca, 892
Erionite-K, 313
Erlichmanite, 892
Ernstburkeite, 892
Erythrite, 893
Erythrosiderite, 893
Eskebornite, 893
Eskolaite, 894
Esperite, 894
Ettringite, 731, 894
Euchroite, 894
Euclase, 895
Eucryptite-β, 895
Eugsterite, 895
Eulytine, 896
Euxenite-(Y), 896
Evansite, 896

Eveite, 896
Evenkite, 897
Ezcurrite, 897

F

Fabriesite, 897
Fairfieldite, 897
Falcondoite, 286, 898
Falottaite, 898
Fangite, 898
Farringtonite, 898
Fassinaite, 899
Faujasite-Ca, 316
Faujasite-Na, 899
Favreauite, 899
Fayalite, 899
Feitknechtite, 900
Felsőbányaite, 900
Ferberite, 900
Fergusonite-(Ce)-β, 900
Fergusonite-(La)-β, 901
Fergusonite-(Nd)-β, 901
Fergusonite-(Y), 901, 902
Fergusonite-(Y)-β, 901
Fermiite, 902
Feroxyhyte, 902
Ferriakasakaite-(La), 258
Ferriallanite-(Ce), 902
Ferricopiapite, 505, 903
Ferricoronadite, 117
Ferri-eckermannite, 903
Ferrierite-K, 903
Ferrierite-Na, 903
Ferrierite-NH_4, 319
Ferri-fluoro-katophorite, 277
Ferri-fluoro-leakeite, 274
Ferrihollandite, 118
Ferrihydrite, 904
Ferri-kaersutite, 904
Ferri-leakeite, 274
Ferrilotharmeyerite, 904
Ferrimolybdite, 905
Ferrinatrite, 473, 905
Ferrisaponite, 300
Ferrisepiolite, 297
Ferristrunzite, 905
Ferrivauxite, 434
Ferro-actinolite, 905
Ferrocarpholite, 906
Ferroceladonite, 906
Ferro-ferri-fluoro-leakeite, 273
Ferro-ferri-katophorite, 271
Ferro-ferri-nybøite, 272
Ferro-glaucophane, 906
Ferrohögbomite, 907
Ferro-hornblende, 906
Ferrokësterite, 907
Ferronigerite-2$N1S$, 215
Ferro-pargasite, 271

Ferrorhodonite, 265
Ferroselite, 907
Ferrosilite, 907
Ferrostrunzite, 908
Ferrovesuvianite, 251, 256
Ferruccite, 908
Fersmite, 908
Feruvite, 908
Fervanite (?), 586
Fetiasite, 620
Fibroferrite, 519
Fichtelite, 909
Fiedlerite, 909
Fiedlerite-1A, 550
Finnemanite, 909
Flamite, 909
Flinteite, 910
Florencite-(La), 910
Florencite-(Nd), 452
Fluellite, 910
Fluocerite-(Ce), 911
Fluocerite-(La), 911
Fluorapatite As-rich, 911
Fluorapatite, 911
Fluorapophyllite-(K), 735, 736, 912
Fluorbarytolamprophyllite, 336
Fluor-buergerite, 912
Fluorcalciobritholite, 912
Fluorcalciomicrolite, 185, 912
Fluorcalciopyrochlore, 215
Fluorcalcioroméite, 913
Fluorcaphite, 913
Fluorcarmoite-(BaNa), 451
Fluorchegemite, 238
Fluor-dravite, 334
Fluor-elbaite, 913
Fluorite, 914
Fluorkyuygenite, 914
Fluorlamprophyllite, 914
Fluormayenite, 915
Fluormayenite-related garnet, 914
Fluornatropyrochlore, 185
Fluorocronite, 915
Fluoro-edenite, 915
Fluoro-pargasite, 278
Fluorowardite, 916
Fluor-schorl, 913
Fluor-uvite, 916
Fluorvesuvianite, 257
Fluorwavellite, 439, 916
Fogoite-(Y), 341
Foitite, 916
Fontarnauite, 71
Foordite, 917
Forêtite, 917
Formanite-(Y), 917
Formicaite, 917
Fornacite, 918
Forsterite, 918

Fougèrite, 918
Fowlerite, 264
Fraipontite, 918
Francevillite, 919
Francisite, 919
Franckeite, 919
Françoisite-(Nd), 919
Franconite, 920
Frankdicksonite, 920
Franklinite, 173, 920
Freboldite, 920
Fredrikssonite, 921
Fresnoite, 921
Friedrichbeckeite, 921
Frohbergite, 921
Frolovite, 922
Frondelite, 922
Fulgurite (a high-silicon glass), 922
Fupingqiuite, 465

G
Gadolinite-(Nd), 922
Gadolinite-(Y), 326
Gahnite, 923
Gaidonnayite, 923
Galaxite, 923
Galeite, 523
Galena, 923
Galileiite, 924
Gallite, 924
Gallium hydroxyde hydrate Ga(OH)$_3 \cdot n$H$_2$O, 186
Gallium(III) oxide, 124
β-Gallium(III)-oxide, 131
Gallium sulfide, 924
Gallium(III) oxyhydroxide, 124
Galloplumbogummite, 924
Galuskinite, 925
Gamagarite, 925
Gamma-alumina, 180
Gananite, 235, 925
Ganomalite, 253, 926
Ganterite, 926
Garavellite, 926
Garnet, 926
Gartrellite, 927
Garutiite, 927
Gaspéite, 927
Gaudefroyite, 927
Gauthierite, 209
Gaylussite, 928
Gazeevite, 928
Gearksutite, 928
Geffroyite, 928
Gehlenite, 929
Geikielite, 929
Geminite, 929
Gerdtremmelite, 631
Gerenite-(Y), 279
Gerhardtite, 929

Gerstleyite, 930
Geschieberite (?), 505
Geschieberite, 930
Ghiaraite, 930
Gianellaite, 506
Gibbsite, 930
Gilalite, 931
Gillardite, 931
Gillespite, 931
Giniite, 932
Gismondine, 932
Glass, 300
Glauberite, 932
Glaucocerinite, 932
Glauconite, 932
Glaucophane, 933
Glaukosphaerite, 933
Glushinskite, 933
Gmelinite-Na, 934
Goethite, 934
Goldfieldite, 934
Goldmanite, 934
Gonnardite, 934
Goosecreekite, 935
Gorceixite, 935
Gordaite, 507
Görgeyite, 935
Gormanite, 935
Goryainovite, 441, 936
Goslarite, 936
Götzenite, 936
Goudeyite, 936
Gowerite, 937
Goyazite, 937
Graemite, 937
Graeserite, 937
Graftonite, 938
Gramaccioliite-(Y), 938
Grandaite, 938
Grandidierite, 938
Graphite, 729, 939
Gratonite, 939
Greenockite, 939
Gregoryite, 939
Greigite, 940
Griceite, 940
Grimaldiite, 940
Grimselite, 941
Grossite, 941
Grossular, 941
Groutite, 941
Grumiplucite, 942
Grunerite, 757, 942
Guanacoite, 942
Guanine, 943
Gudmundite, 943
Gugiaite, 328
Guilleminite, 943
Guimarãesite, 466

Gunningite, 943
Gurimite, 944
Gwihabaite, 944
Gypsum, 944
Gyrolite, 944

H
Hafnon, 945
Häggite, 212
Haidingerite, 945
Haiweeite, 945
Hakite, 945
Halite, 732, 946
Halloysite-10Å, 946
Halotrichite, 946
Hambergite, 946
Hanjiangite, 947
Hanksite, 947
Hannayite, 947
Hannebachite, 947
Hansesmarkite, 948
Haradaite, 269
Hardystonite, 246, 948
Harmotome, 948
Harmunite, 948
Harmunite cubic polymorph, 187
Harmunite Mn^{4+}-bearing, 949
Hartite, 949
Hashemite, 949
Hatchetine, 949
Hatrurite triclinic polymorph, 239
Hatrurite, 950
Hauerite, 950
Hausmannite, 187, 950
Haüyne, 951
Hawleyite, 951
Hawthorneite, 951
Haynesite, 951
Hazenite, 952
Heazlewoodite, 952
Hectorite, 952
Hedenbergite, 952
Hedyphane, 953
Heisenbergite, 953
Heklaite, 236
Heliophyllite, 953
Hellyerite, 953
Hematite, 954
Hemihedrite, 954
Hemimorphite, 954
Hemleyite, 955
Hemusite, 955
Hendricksite, 955
Henmilite, 955
Henritermierite, 956
Henryite, 956
Herbertsmithite, 956
Hercynite, 956
Herderite, 957

Herzenbergite, 957
Hessite, 957
Hetaerolite, 957
Heterogenite, 958
Heterosite, 958
Heulandite, 958
Hexaamminenickel(II) nitrate, 111
Hexacelsian, 308, 958
Hexaferrum, 959
Hexahydrite, 959
Hexahydroborite, 959
Hiärneite, 959
Hibonite, 960
Hibonite-(Fe), 960
Hidalgoite, 960
Hieratite, 960
Hilarionite, 961
Hingganite-(Y), 326, 961
Hinsdalite, 961
Hiortdahlite, 961
Hisingerite, 961
Hochelagaite, 962
Hoelite, 962
Hoganite, 962
Hogarthite, 962
Hohmannite, 963
Holdawayite, 963
Holfertite, 963
Hollandite, 963
Hollingworthite, 964
Holmquistite, 964
Honessite, 964
Hopeite, 964, 965
Hörnesite, 965
Hsianghualite, 965
Huanghoite-(Ce), 965
Huanzalaite, 717, 966
Hubeite, 966
Hübnerite, 966
Hughesite, 966
Huizingite-(Al), 516
Humberstonite, 967
Humboldtine, 967
Humite, 967, 968
Hummerite, 968
Hungchaoite, 968
Huntite, 968
Hureaulite, 969
Hurlbutite, 969
Hyalite, 223
Hydrobiotite, 296
Hydroboracite, 969
Hydrocalumite, 222, 546, 969
Hydrocerussite, 15–17, 970
Hydrochamosite-1M, 299
Hydrodelhayelite-related compound, 970
Hydrogarnet $Sr_3Al_2(OH)_{12}$, 164
Hydrohalite, 553, 971
Hydrohonessite, 971

Hydrokenomicrolite, 188
Hydromagnesite, 971
Hydronaujakasite, 298
Hydroniumjarosite, 971, 972
Hydronium jarosite Pb,As-bearing, 512
Hydronium jarosite Pb,Cu-bearing, 513
Hydronium jarosite Pb,Zn-bearing, 514
Hydroromarchite, 216, 972
Hydrotalcite, 972
Hydrotalcite-2H, 972
Hydroterskite, 343
Hydrotungstite, 706, 973
Hydrovesuvianite, 261
Hydroxyapophyllite-(K), 973
Hydroxycalciobetafite, 973
Hydroxycalciomicrolite, 974
Hydroxycalciopyrochlore, 974
Hydroxycalcioroméite, 974
Hydroxycancrinite (?), 301
Hydroxyferroroméite, 974
Hydroxykenoelsmoreite, 975
Hydroxylapatite, 975
Hydroxylbastnäsite-(Ce), 975
Hydroxylchondrodite, 976
Hydroxylclinohumite, 976
Hydroxyledgrewite, 976
Hydroxylellestadite, 977
Hydroxylgadolinite-(Y), 325
Hydroxylgugiaite, 324
Hydroxylherderite, 977
Hydrozincite, 977
Hydrucerussite-like mineral 9-40, 87
Hypersthene, 978

I
Ianbruceite, 978
Iangreyite, 978
Ice, 978
Idaite, 979
Idrialite, 979
Ikaite, 95, 979
Ilesite, 980
Ilmenite, 980
Ilsemannite, 189
Ilvaite, 980
Imogolite, 287, 980
Inderite, 981
Indite, 981
Indium oxide, 143
Indium vanadate selenite $In(VSe_2O_8)$, 644
Indium vanadate tellurite $In(VTe_2O_8)$, 684
Indium zinc selenite $In_2Zn(SeO_3)_4$, 645
Inesite, 981
Innsbruckite, 981
Insizwaite, 982
Inyoite, 982
Iodargyrite, 982
Iowaite, 983
Iranite, 596, 983

Irarsite, 983
Iriginite, 984
Irinarassite, 984
Iron, 984
Iron(II) acid phosphate hydrate, 376
Iron(III) basic sulfate, 485, 486
Iron dimolybdate selenite hydrate
    $Fe_2(Mo_2O_7)(SeO_3)_2 \cdot H_2O$, 646
Iseite, 984
Ishikawaite, 224
Isocubanite, 985
Isokite, 985
Isomertieite, 985
Ivanyukite-Cs, 342
Ivsite, 507, 985
Iwakiite-hausmannite intermediate member, 189
Iwateite, 985
Iyoite, 986

J
Jáchymovite, 986
Jacobsite, 986
Jadeite, 986
Jakobssonite, 987
Jalpaite, 987
Jamborite, 987
Jamesonite, 987
Janchevite, 589
Jarosite, 988
Jeffbenite, 988
Jennite, 988
Ježekite, 989
Jianshuiite, 212
Jixianite, 989
Joaquinite-(Ce), 989
Joegoldsteinite, 508, 989
Joëlbruggerite, 990
Johachidolite, 990
Johannite, 990
Johnbaumite, 991
Johnbaumite Sr-analogue, 621
Johninnesite, 991
Jordisite, 991
Jörgkellerite, 433
Joteite, 992
Jouravskite, 530
Juansilvaite, 638
Junitoite, 249
Jurbanite, 529

K
Kaatialaite, 622
Kaersutite, 992
Kainite, 992
Kalgoorlieite, 992
Kaliborite, 993
Kalicinite, 993
Kalininite, 508, 993
Kalinite, 993

Kaliophilite, 994
Kalsilite, 312, 994
Kamacite, 994
Kamarizaite, 628
Kamiokite, 679
Kamotoite-(Y), 994
Kampelite, 994
Kamphaugite-(Y), 995
Kanemite, 288
Kangite, 995
Kaňkite, 995
Kanoite, 996
Kaolinite, 2, 996
Kapellasite, 996
Kapundaite, 996
Karelianite, 997
Karpenkoite, 997
Karrooite, 997
Kashinite, 997
Kasolite, 998
Kassite, 998
Katayamalite, 998
Katerinopoulosite, 520
Katiarsite, 622
Katoite, 998
Kawazulite, 999
Kazanskyite, 999
Keilite, 999
Keiviite-(Yb), 247, 1000
Kemmlitzite, 1000
Kempite, 1000
Kentbrooksite, 895
Kentrolite, 1000
Kenyaite, 1001
Kerimasite, 1001
Kermesite, 1001
Keyite, 1001
Khademite, 1002
Khatyrkite, 1002
Khesinite, 263, 1002
Khvorovite, 1002
Kiddcreekite, 1003
Kidwellite, 1003
Kieftite, 1003
Kieserite, 1003
Kilchoanite, 1003
Killalaite, 1004
Kimzeyite, 1004
Kinoite, 1004
Kinoshitalite, 1005
Kintoreite, 1005
Kipushite, 1005
Kirschsteinite, 241, 1005
Kladnoite, 1006
Klaprothite, 1006
Klebelsbergite, 1006
Klockmannite, 1006
Knorringite, 1007
Kobokoboite, 1007

Koechlinite, 1007
Kojonenite, 1007
Kokchetavite, 1007
Koktaite, 1008
Kolskyite, 1008
Kolwezite, 1008
Konyaite, 1008
Koritnigite, 1009
Kornelite, 1009
Kornerupine, 1009
Kosmochlor, 1010
Kosnarite NH_4-analogue, 435
Kosnarite NH_4-analogue cubic polymorph, 435
Kosnarite, 1010
Kotoite, 1010
Kottenheimite dimorph, 471
Köttigite, 1010
Kotulskite, 1011
Kovdorskite, 1011
Kozyrevskite, 1011
Krásnoite, 437
Kremersite, 1011
Krieselite, 1012
Kröhnkite, 1012
KröhnkiteMn analogue, 508
Krotite, 1012
Krut'aite, 1012
Kryzhanovskite, 1013
Ktenasite, 523, 1013
Kuksite, 1013
Kuksite trigonal dimorph, 459
Kuksite trigonal Mg analogue, 460
Kulanite, 1013
Kuliginite, 551
Kullerudite, 1014
Kumdykolite, 1014
Kummerite, 436
Kumtyubeite, 1014
Kuramite, 1014
Kuranakhite, 1015
Kuratite, 1015
Kurnakovite, 1015
Kusachiite, 217, 1015
Kushiroite, 1016
Kutnohorite, 1016
Kuzminite, 1016
Kyanite, 1016
Kyawthuite, 190, 1017

L

Laachite, 733, 1017
Lacroixite, 1017
Lafossaite, 1017
Laihunite, 242, 1018
Lakargiite, 1018
Lamprophyllite, 1018
Lanarkite, 1019
Långbanite, 1019
Langbeinite, 1019

Langite, 1019
Lanmuchangite, 1020
Lansfordite, 1020
Lantanum nitrate hexahydrate, 112
Lanthanite-(Nd), 1020
Lanthanum aluminum oxide, 144
Lanthanum calcium oxophosphate, 374
Lanthanum iron(III) oxide, 144
Lanthanum molybdate, 670
Lanthanum orthoborate, 24
Lanthanum orthosilicate, 238
Lanthanum oxosulfate, 487
Lanthanum selenite, 646, 647
Lanthanum strontium oxophosphate, 375
Lanthanum uranyl orthovanadate divanadate, 564
Lapeyreite, 1020
Larnite, 1021
Laueite, 1021
Laumontite, 1021
Laurentianite, 1021
Laurionite, 1022
Laurionite Ba-analogue, 547
Laurionite I-analogue, 703
Laurite, 1022
Lausenite, 1022
Lautarite, 1023
Lavendulan, 1023
Lavenite Fe-analogue, 338
Lavinskyite, 1023
Lawrencite, 1023
Lawsonite, 247, 1024
Layered perovskite $BaBi_2Ta_2O_9$, 191
Layered perovskite $CaBi_2Ta_2O_9$, 191
Layered perovskite $K_4Xe_3O_{12}$, 706
Layered perovskite $SrBi_2Ta_2O_9$, 192
Lazaridisite, 524, 1024
Lazulite, 1024
Lazurite, 1024
Lead, 1025
Lead aluminium orthoborate fluoride $Pb_6Al(BO_3)_2OF_7$, 24
Lead beryllium phosphate hurlbutite-type, 376
Lead bismuth orthoborate, 25
Lead borate $Pb_6B_{11}O_{18}(OH)_9$, 48
Lead borate PbB_4O_7, 49
Lead cadmium orthoborate, 26
Lead copper orthoborate, 26
Lead copper tellurate tellurite, 684
Lead iron(III) phosphate, 377
Lead iron(III) trivanadate, 564
Lead orthoborate chromate, 591
Lead orthoborate molybdate, 671
Lead orthoborate tungstate, 27, 29
Lead(II) oxalate, 100
Lead(II) oxysulfate, 487
Lead phosphate nitrate hydrate, 378
Lead phosphate sulfate, 378
Lead selenate, 643
Lead silver phosphate apatite-type, 379
Lead sodium calcium phosphate apatite-type, 380

Lead(II) stannate Pb_2SnO_4, 145
Lead tin oxide $Pb^{2+}_4Pb^{4+}Sn^{4+}O_8$, 146
Lead uranyl divanadate, 565
Leadhillite, 533, 534, 1025
Lechatelierite, 1025
Lecoqite-(Y), 85
Leguernite, 1025
Leifite, 329
Leightonite, 1025
Leiteite, 1026
Lemanskiite, 623, 624, 1026
Lemoynite, 1026
Leogangite, 1027
Leószilárdite, 1027
Lepidocrocite, 200, 1027
Lesukite Cu-bearing variety, 121
Letovicite, 1027
Leucite, 1028
Leucophosphite, 1028
Lévyne-Ca, 1028
Leydetite, 1028
Libethenite, 1029
Liebenbergite, 1029
Liebigite, 1029
Likasite, 1030
Lime, 1030
Linarite, 1030
Lindbergite, 1030, 1031
Lindgrenite, 1031
Lindsleyite, 1031
Línekite, 90
Lingunite, 1031
Lingunite K-analogue, 1031
Linnaeite, 1032
Linzhiite, 1032
Liroconite, 1032
Lishizhenite, 509
Liskeardite, 1032
Litharge, 1033
Lithiophilite, 1033
Lithiophorite, 1033
Lithiophosphate, 1033
Lithiotantite, 1034
Lithium aluminate $LiAl_5O_8$, 146
Lithium aluminate $LiAlO_2$-beta, 147
Lithium aluminate $LiAlO_2$-gamma, 148
Lithium aluminium orthoborate, 28
Lithium aluminium oxide-alpha, 148
Lithium aluminium oxide-gamma, 149
Lithium aluminoborate, 49
Lithium cesium borate $Li_4Cs_3B_7O_{14}$, 50
Lithium chromium pyrophosphate, 380
Lithium cobalt(III) iron(III) oxide delafossite-type, 149
Lithium copper tungstate, 712
Lithium cyclo-hexaphosphate trihydrate, 383
Lithium dimolybdate selenite, 672
Lithium ferrite $LiFe^{3+}_5O_8$, 150
Lithium hexafluorosilicate, 230
Lithium iron(III) oxide, 151

Lithiumiron(III) pyrophosphate, 381
Lithium iron(II) sulfate fluoride tavorite-type, 488
Lithium iron(III) tungstate wolframite-type, 711
Lithium magnesium manganese(IV) oxide spinel-type, 151
Lithium magnesium phosphate olivine-type, 382
Lithium manganese oxide spinel-type, 152
Lithium metasilicate, 265
Lithium molybdate tellurite, 672
Lithium nickel phosphate triphylite-type, 382
Lithium nickel tungstate, 713
Lithium nickel vanadate, 566
Lithium niobateilmenite-type, 153
Lithium sodium borate $LiNaB_4O_7$, 51
Lithium strontium borate $Li_2Sr_4B_{12}O_{23}$, 52
Lithium strontium orthoborate, 29
Lithium strontium orthophosphate, 384
Lithium tetraborate, 53
Lithium trivanadate, 566
Lithium tungstate tellurite, 685
Lithium tungstate vanadate brannerite-type, 567
Lithium vanadyl phosphate, 384–386
Lithium zinc niobium oxide spinel-type, 153
Lithium zinc phosphate monohydrate, 386
Lithium zinc selenite $Li_2Zn_3(SeO_3)_4 \cdot 2H_2O$, 648
Lithium zirconium arsenate, 604, 605
Liversidgeite, 1034
Livingstonite, 1034
Lizardite, 1034
Löllingite, 1035
Lomonosovite, 1035
Lonecreekite, 1035
Lonsdaleite, 1035
Lópezite, 1036
Lorándite, 1036
Lorenzenite, 1036
Löweite, 1036
Luanshiweiite, 284
Luddenite, 1037
Ludjibaite, 1037
Ludlamite, 1037
Ludlockite, 627, 1037
Ludwigite, 1038
Lueshite, 193, 1038
Luinaite-(OH), 334
Lulzacite, 453, 1038
Lüneburgite, 41, 1038
Luobusaite, 322
Luogufengite, 1039
Lusernaite-(Y), 1039

M
Macedonite, 1039
Mackayite, 1039
Mackinawite, 1040
Macquartite, 1040
Magadiite, 1040
Magbasite, 1040
Maghemite, 1041

Magnesio-arfvedsonite, 1041
Magnesioaubertite, 474
Magnesiocanutite, 636
Magnesiocarpholite, 734, 1041
Magnesiochloritoid, 236, 1041
Magnesiochromite, 1042
Magnesiocopiapite, 509, 1042
Magnesio-ferri-hornblende, 270
Magnesioferrite, 1042
Magnesio-foitite, 1042
Magnesiohögbomite-2N3S, 122, 219
Magnesiohögbomite-2N4S, 222, 1043
Magnesio-hornblende, 1043
Magnesiokoritnigite, 598
Magnesiopascoite, 580
Magnesiotaaffeite-2N'2S, 1043
Magnesiotaaffeite-6N'3S, 1043
Magnesiovesuvianite, 253
Magnesiovoltaite, 480
Magnesite, 1043
Magnesium acid phosphate hydrate, 387
Magnesium borophosphate, 365
Magnesium chromate, 592
Magnesium hydroxychlorite atacamite-type, 551
Magnesium hydroxysulfate hydrate, 489
Magnesium orthoborate fluoride, 30
Magnesium oxychloride hydrate $Mg_3Cl_2(OH)_4 \cdot 4H_2O$, 541
Magnesium strontium diorthoborate, 54
Magnesium sulfate hydroxide $Mg_6(SO_4)(OH)_{10} \cdot 7H_2O$, 490
Magnesium tellurite $MgTe_2O_5$, 686
Magnesium vanadate $Mg_7V_4O_{16}(OH)_2 \cdot H_2O$, 568
Magnetite, 1044
Magnetoplumbite, 1044
Majorite, 1044
Makatite, 1044
Malachite, 1045
Malayaite, 1045
Malladrite, 1045
Mallardite, 1045
Mallestigite, 1045
Mandarinoite, 1046
Manganese acid phosphate hydrate, 388
Manganese(II) antimony(III) oxide, 154
Manganese hydroxysulfate $Mn_5(SO_4)(OH)_8$, 489
Manganese(II) titanium orthophosphate, 388
Manganese(II) titanium phosphate $MnTi_4(PO_4)_6$, 389
Mangangordonite, 457
Manganite, 1046
Manganlotharmeyerite, 1046
Manganochromite, 1046
Manganolangbeinite, 1047
Manganosite, 1047
Manitobaite, 354
Manjiroite, 1047
Marcasite, 1047
Margarite, 289, 1047
Margarosanite, 1048
Marićite, 1048

Markascherite, 1048
Marokite, 1048
Marthozite, 1048
Martinandresite, 318
Martyite, 735, 1049
Maruyamaite, 1049
Mascagnite, 1049
Maskelynite, 1049
Massicot, 1050
Masuyite, 206
Mathesiusite, 1050
Mathiasite, 1050
Matildite, 1050
Matioliite, 1051
Matlockite, 1051
Mattagamite, 1051
Matteuccite, 1051
Maxwellite, 1051
Mazzite-Na, 317
Mbobomkulite, 1052
Mcallisterite, 1052
Mcalpineite, 1052
Mcconnellite, 1052
Mcguinnessite, 1053
Megawite, 1053
Meisserite, 1053
Meixnerite, 1053
Melanarsite, 738, 739, 1054
Melanophlogite, 5–11, 1054
Melanterite, 1054
Meliphanite, 1054
Mellite, 1054
Mellizinkalite, 1055
Mendeleevite-(Nd), 322
Mendipite, 1055
Mercallite, 510, 1055
Mercury(I) acid phosphate, 390
Mercury(I) orthoarsenate, 605
Mercury(II) orthoarsenate, 606
Merelaniite, 1055
Merenskyite, 1056
Meridianiite, 1056
Merlinoite, 314
Merrillite, 1056
Merrillite Na-free analogue, 1056
Merwinite, 1056
Mesolite, 1057
Meta-ankoleite, 1057
Meta-autunite, 1057
Metacinnabar, 1057
Metahewettite, 1058
Metakirchheimerite, 1058
Metalodèvite, 1058
Metamunirite, 1058
Metarauchite, 1058
Metarossite, 1059
Metasideronatrite, 476
Metastibnite, 1059
Metastudtite, 193, 1059

Metathénardite, 479, 1059
Metatorbernite, 1060
Metatyuyamunite, 1060
Metauranocircite-I, 1060
Metauranospinite, 1060
Metavariscite, 1061
Metavivianite, 1061
Metazeunerite, 1061
Meurigite-Na, 1061
Meyerhofferite, 1062
Meymacite monoclinic analogue, 1062
Miargyrite, 1062
Microcline, 724, 1062
Miersite, 1062
Mikasaite, 1063
Milarite, 1063
Millerite, 1063
Millosevichite, 1063
Mimetite, 1064
Minguzzite, 1064
Minium, 194, 1064
Minjiangite, 360, 438, 469
Minnesotaite, 1064
Minyulite, 352, 1064
Mirabilite, 1065
Misakiite, 1065
Mitscherlichite, 1065
Mixite, 1065
Moctezumite, 1066
Mogánite, 1066
Mohite, 1066
Mohrite, 1066
Moissanite, 1066
Mojaveite, 1067
Molybdenite, 1067
Molybdite, 1067
Molybdofornacite, 1067
Molybdophyllite, 14, 15, 1068
Molybdyl phosphate, 390
Molysite, 1068
Monazite-(Ce), 1068
Monazite-(La), 1068
Monazite-(Nd), 1068
Monazite-(Sm), 1069
Moncheite, 1069
Monetite, 1069
Monipite, 1069
Monohydrocalcite, 1070
Montebrasite, 1070
Monteponite, 1070
Montgomeryite, 1070
Monticellite, 1070
Montmorillonite, 296, 1071
Montroseite, 1071
Montroydite, 217, 1071
Moolooite, 1071
Mopungite, 195, 1072
Moraskoite, 1072
Mordenite, 1072

Morenosite, 1072
Morimotoite, 237
Moschelite, 1073
Mosesite, 1073
Mottramite, 1073
Motukoreaite-related mineral, 533
Mountkeithite, 1073
Moydite-(Y), 1073
Mukhinite, 1074
Mukhinite V-rich analogue, 1074
Mullite, 1074
Muscovite, 734, 1074
Mushistonite, 213

N
Nabiasite, 1075
Nabimusaite, 1075
Nacrite, 1075
Nadorite, 1075
Nafertisite, 1076
Nagelschmidtite, 1076
Nahcolite, 1076
Nahpoite, 1076
Nakauriite, 92, 93
Nalipoite, 451
Namibite, 1077
Nantokite, 1077
Naquite, 1077
Narsarsukite, 1077
Nasonite, 245
Natalyite, 1077
Natisite, 1078
Natrite, 1078
Natroalunite, 1078
Natrochalcite, 1078
Natrodufrénite, 351, 1079
Natrojarosite, 1079
Natrolemoynite, 1079
Natrolite, 1079
Natron, 1080
Natroniobite, 1080
Natropalermoite, 1080
Natrophilite, 461, 1080
Natrosilite, 1081
Natrouranospinite, 1081
Natroxalate, 1081
Natrozippeite, 1081
Naumannite, 1082
Nealite, 1082
Negevite, 1082
Neighborite, 1082
Nekoite, 1082
Nenadkevichite, 1083
Nepheline, 1083
Neptunium(IV) oxalate hexahydrate, 101
Nesquehonite, 1083
Nestolaite, 1083
Newberyite, 1084
Nežilovite, 1084

Nichromite, 195
Nickel antimonate fluoride, 231
Nickelaustinite, 1084
Nickelbischofite, 1084
Nickelboussingaultite, 1085
Nickelhexahydrite, 1085
Nickel hydroxychlorite atacamite-type, 552
Nickel hydroxysulfate hydrate
    $Ni_3(SO_4)_2(OH)_2 \cdot 2H_2O$, 491
Nickeline, 1085
Nickel manganese(IV) oxide, 155
Nickel oxychloride hydrate $Ni_3Cl_{2.1}(OH)_{3.9} \cdot 4H_2O$, 542
Nickelpicromerite, 1085
Nickel vanadyl phosphate hydrate, 391
Nickenichite, 635
Nierite hexagonal polyborph, 1086
Nierite, 113
Nifontovite, 1086
Nimite, 1086
Niningerite, 1086
Niobium sulfide NbS_3, 491
Nioboholtite, 1086
Niobylphosphate, 392
Niter, 1087
Nitratine, 114, 1087
Nitrobarite, 1087
Nitrocalcite, 1087
Nitromagnesite, 1088
Nobleite, 1088
Noelbensonite, 1088
Nolanite, 1088
Nolzeite, 331
Nontronite, 1088
Norbergite, 1089
Nordenskiöldine, 1089
Nordstrandite, 1089
Normandite, 1089
Norsethite, 1090
Northupite, 1090
Nosean, 1090
Nováčekite-II, 1090
Novgorodovaite, 1091
Nsutite, 1091
Nullaginite, 1091
Nyerereite, 1091

O
Offretite, 1092
Okenite, 1092
Okhotskite, 254
Oldhamite, 1092
Olgite, 1092
Olivenite, 1093
Olivine P-rich variety, 1093
Olmiite, 1093
Olshanskyite, 1093
Omongwaite, 1094
Omphacite, 1094
Onoratoite, 1094

Opal-A, 1094
Opal-CT, 1094
Ophirite, 1095
Oppenheimerite, 1095
Ordoñezite, 1095
Orlandiite, 663
Orpiment, 1095
Orschallite, 1096
Orthobrannerite, 214
Orthoclase, 1096
Orthojoaquinite-(Ce), 1096
Osakaite, 521, 1096
Osbornite, 114, 1097
Oskarssonite, 1097
Osumilite, 1097
Otavite, 1097
Ottemannite, 1098
Ottensite, 1098
Ottohahnite, 1098
Otwayite, 1098
Oxammite, 1099
Oxo-magnesio-hastingsite, 276
Oxybetafite-(Gd), 196
Oxybetafite-(Sm), 196
Oxybismuthobetafite, 197
Oxybritholite thorium analogue, 243
Oxycalcioroméite, 1099
Oxy-dravite, 1099
Oxykinoshitalite, 1099
Oxynatromicrolite, 1100
Oxyplumboroméite, 1100
Oxypyromorphite, 439
Oyelite, 1100
Ozokerite, 1100

P
Pabstite, 280, 1100
Padmaite, 1101
Paganoite phosphate analogue, 461
Pakhomovskyite, 1101
Palermoite, 1101
Palladinite, 1101
Palladosilicide, 1102
Palygorskite, 1102
Panguite, 1102
Panichiite, 1102
Papagoite, 1102
Parabutlerite, 477, 1103
Paracoquimbite, 1103
Paragonite, 1103
Paraguanajuatite, 1103
Parahopeite, 1104
Paramelaconite, 197
Paramontroseite, 1104
Paranatrolite, 1104
Paraotwayite, 1104
Parapierrotite, 1105
Pararealgar, 1105
Pararobertsite, 1105

Parascholzite, 1105
Parascorodite, 1105
Parasibirskite, 1106
Parasymplesite, 1106
Paratacamite, 1106
Paratellurite, 1107
Paratooite-(La), 94, 1107
Paravauxite, 1107
Pargasite, 1107
Parisite-(Ce), 79, 1108
Parisite-(La), 78, 1108
Parnauite, 1108
Parsonsite, 1108
Parthéite, 1109
Partzite, 1109
Pašavaite, 1109
Pascoite, 1110
Patrónite, 1110
Pattersonite, 1110
Pauflerite, 1110
Pauflerite tetragonal dimorph, 517
Pauladamsite, 519
Paulingite-K, 1111
Paulmooreite, 1111
Pauloabibite, 1111
Paulscherrerite, 1111
Pavlovskyite, 1112
Pearlite, 323
Peatite-(Y), 1112
Pecoraite, 1112
Pectolite, 1112
Peisleyite, 1113
Péligotite, 1113
Penberthycroftite, 624
Penikisite, 453
Penkvilksite-$2O$, 1113
Penroseite, 1113
Pentagonite, 1114
Pentahydrite, 1114
Pentahydroborite, 1114
Pentlandite, 1114
Peretaite, 1115
Perhamite, 1115
Perite, 1115
Permingeatite, 1115
Perovskite, 1116
Pertsevite-(OH), 1116
Petalite, 1116
Petersite-(Ce), 1116
Petewilliamsite-related Cd diarsenate, 625
Petitjeanite, 455, 1117
Petrukite, 1117
Petterdite, 1117
Petzite, 1117
Pezzottaite, 1117
Pharmacolite, 1118
Pharmacosiderite, 1118
Pharmazincite, 1118
Phenakite, 1118

Philipsbornite, 1119
Philipsburgite, 1119
Phillipsite-K, 1119
Phillipsite-NH_4, 315
Philolithite, 1119
Phlogopite, 734, 1120
Phoenicochroite, 1120
Phosgenite, 1120
Phosphammite, 1120
Phosphohedyphane, 1121
Phosphophyllite, 1121
Phosphorrösslerite, 462
Phosphosiderite, 1121
Phosphuranylite, 1121
Phurcalite, 1122
Pickeringite, 1122
Picromerite, 1122
Picromerite dimorph (?), 512
Picropharmacolite, 1122
Pieczkaite, 1123
Piemontite, 1123
Pigeonite, 1123
Pilawite-(Y), 240
Pilsenite, 1123, 1124
Pimelite, 1124
Pinakiolite, 1124
Pingguite, 696
Pinnoite, 1124
Pirssonite, 1125
Pitticite, 1125
Plancheite, 1125
Plášilite, 511, 1125
Platarsite, 1126
Platinum, 1126
Plattnerite, 1126
Plavnoite, 1126
Plimerite, 1127
Plombièrite, 1127
Plumbogummite, 1127
Plumbojarosite, 1127
Plumbophyllite, 289, 1128
Plumbotellurite, 698
Plumbotsumite, 1128
Poitevinite, 1128
Pokrovskite, 1128
Poldervaartite, 1129
Pollucite, 1129
Polyakovite-(Ce), 336
Polycrase-(Y), 1129
Polydymite, 1129
Polyhalite, 1130
Popovite, 1130
Portlandite, 1130
Posnjakite, 1130
Potarite, 1131
Potassic-ferri-leakeite, 275
Potassic-ferro-pargasite, 276
Potassic-magnesio-fluoro-arfvedsonite, 273
Potassium acid phosphate, 394

Potassium acid pyrophosphate hydrate, 394
Potassium acid selenite, 648
Potassium acid tellurate hydrate
  $K_2[TeO_2(OH)_4]\cdot 3H_2O$, 686
Potassium aluminium molybdate, 673
Potassium antimonate tungstate, 713
Potassium antimony fluoride, 231
Potassium antimony oxoarsenate, 607
Potassium antimony oxophosphate, 396
Potassium antimony(V) oxophosphate, 395
Potassium arsenate tungstate $K(AsW_2O_9)$, 714
Potassium barium borate $KBaB_5O_9$, 54
Potassium bismuth(III) phosphate, 396
Potassium bismuth(III) tungstate, 715
Potassium borate $KB_3O_3(OH)_4\cdot H_2O$, 73
Potassium borate $KB_3O_5\cdot H_2O$, 55
Potassium borosilicate pollucite-type, 330
Potassium borosulfate $K_5[B(SO_4)_4]$, 492
Potassium calcium orthoborate, 31
Potassium chloride borate perovskite-related
  $K_3B_6O_{10}Cl$, 56
Potassium chromium divanadate, 568, 569
Potassium decavanadate decahydrate, 570
Potassium difluorphosphate, 397
Potassium diuranate, 155
Potassium hexavanadate hydrate, 570
Potassium hydronium uranyl selenate hydrate, 649
Potassium iron diarsenate (pyroarsenate), 608
Potassium iron pyrophosphate, 398
Potassium lead borophosphate, 398
Potassium lead carbonate fluoride, 81
Potassium lead phosphate, 399
Potassium magnesium acid phosphate hydrate, 400
Potassium magnesium acid pyrophosphate hydrate, 400
Potassium magnesium arsenate hexahydrate, 608
Potassium magnesium chromate hydrate, 592
Potassium magnesium orthoborate, 31
Potassium magnesium orthophosphate $KMg_4(PO_4)_3$, 401
Potassium magnesium yttrium phosphate
  (xenotime-type), 402
Potassium manganese arsenate, 609
Potassium manganese(III) fluoride, 232
Potassium mercury chloride hydrate, 543
Potassium monofluorphosphate, 402
Potassium nickel chromate hydrate, 593
Potassium niobate, 156
Potassium niobate KNb_7O_{18}, 157
Potassium niobate perovskite-type, 157
Potassium niobate tungstate, 158
Potassium niobium oxophosphate, 403
Potassium pentaborate, 56
Potassium peroxochromate, 593
Potassium sodium iron arsenate
  $Na_{2.77}K_{1.52}Fe_{2.57}(AsO_4)_4$, 610
Potassium sodium selenate, 650
Potassium sodium vanadyl sulfate, 493
Potassium sodium zinc borate $K_2NaZnB_5O_{10}$, 57
Potassium strontium orthoborate, 32
Potassium tantalate tungstate, 159

Potassium tantalite perovskite-type, 159
Potassium tin orthophosphate, 404
Potassium titanium iodate, 701
Potassium titanium oxophosphate, 404
Potassium titanium silicate $K_2TiSi_3O_9\cdot H_2O$, 338
Potassium triborate, 58
Potassium uranium(V) sorosilicate, 349
Potassium uranyl fluoride, 232
Potassium urinate, 160
Potassiumvanadyl phosphate, 405
Potassium ytterbium acid orthoborate acid
  orthophosphate, 406
Potassium ytterbium tungstate, 715
Potassium yttrium selenite, 650
Potassium zinc acid pyrophosphate hydrate, 406
Potassium zinc cyclotriphosphate benitoite-type, 407
Potassium zinc hydrogen phosphate, 463
Potassium zinc selenite $K_2Zn_3(SeO_3)_4$, 651
Potassium zinc sulfate chloride trihydrate, 493
Potassium zinc sulfate hexahydrate, 494
Potassium zirconium arsenate, 610
Pottsite, 1131
Poubaite, 1131
Poudretteite, 1131
Poughite, 697, 1131
Povondraite, 1132
Powellite, 1132
Praseodymium chromate(V), 594
Praseodymium cyclotriphosphatetrihydrate, 408
Prehnite, 1132
Preiswerkite, 1132
Pretulite, 1133
Příbramite, 1133
Priceite, 76, 1133
Priderite Al-analogue, 1133
Priderite Cr-analogue, 1134
Priderite Mg-analogue, 1133
Prismatine, 1134
Probertite, 76
Proshchenkoite-(Y), 332
Protoimogolite, 290
Proustite, 1134
Pseudoboleite, 1134
Pseudobrookite, 1135
Pseudocotunnite, 1135
Pseudojohannite, 1135
Pseudolaueite, 1135
Pseudomalachite, 1136
Pseudowollastonite, 1136
Pucherite, 585, 1136
Pumpellyite-(Al), 1136
Pyracmonite, 525
Pyrargyrite, 1137
Pyrite, 1137
Pyroaurite, 1137
Pyrochroite, 214, 1137
Pyrolusite, 1138
Pyromorphite As-rich, 1138
Pyromorphite, 1138

Pyrope, 1139
Pyrophanite, 198, 1139
Pyrophyllite, 291, 1139
Pyrosmalite-(Fe), 1139
Pyrosmalite-(Mn), 1139
Pyroxmangite, 1140
Pyrrhotite, 1140

Q
Qandilite, 1140
Qingheiite, 1140
Qingsongite, 1141
Qingsongite (C-bearing), 115, 116
Quadridavyne, 1141
Quartz, 1141
Quenstedtite, 1141
Quetzalcoatlite, 1141
Quintinite, 1142
Quintinite-related hydroxyde carbonate $Mg_4Cr_2(OH)_{12}(CO_3) \cdot nH_2O$, 87
Quintinite-related hydroxyde carbonate $Ni_4Cr_2(OH)_{12}(CO_3) \cdot nH_2O$, 88

R
Raadeite, 463
Rabejacite, 1142
Raguinite, 1142
Rajite, 1143
Ramanite-(Cs), 1143
Ramanite-(Rb), 1143
Rambergite, 1143
Rameauite, 1144
Ramikite-(Y), 1144
Ramsdellite, 1144
Ranciéite, 1144
Rankamaite, 1145
Rankinite, 1145
Rapidcreekite, 1145
Raspite, 1145
Rasvumite, 1146
Ravatite, 1146
Raygrantite, 1146
Realgar, 1146
Rebulite, 1147
Reedmergnerite, 1147
Reevesite, 1147
Refikite, 107
Reichenbachite, 1147
Reinerite, 1148
Reinhardbraunsite, 1148
Rengeite, 1148
Reppiaite, 582
Retgersite, 1148
Reyerite, 1148
Rhabdophane-(Ce), 1149
Rhabdophane-(Nd), 1149
Rheniite, 1149
Rhodium sulfate, 525

Rhodium sulfate hydrate, 526
Rhodizite, 1149
Rhodochrosite, 1150
Rhodonite, 1150
Rhomboclase, 473, 1150
Rhönite, 1150
Richelsdorfite, 1151
Richterite, 1151
Riebeckite, 1151
Riebeckite (Crocidolite), 1151
Rimkorolgite, 1152
Ringwoodite, 1152
Rinkite, 1152
Riomarinaite, 1152
Riotintoite, 472
Rippite, 282
Robertsite, 1152
Robinsonite, 1153
Rockbridgeite, 444, 1153
Rodalquilarite, 1153
Rodolicoite, 1153
Roedderite Na-free analogue, 284
Roküknite, 1154
Romanèchite, 1154
Romanorlovite, 737, 1154
Romarchite, 218, 1154
Römerite, 475, 1155
Rondorfite, 1155
Rongibbsite, 319, 1155
Ronneburgite, 587, 1155
Rooseveltite, 1156
Roquesite, 1156
Rosasite, 1156
Roscherite, 467
Roselite, 1156
Rosiaite, 1157
Rostite, 1157
Rouaite, 1157
Roumaite, 1157
Rowlandite-like mineral, 249
Rowleyite, 1158
Roxbyite, 1158
Roymillerite, 97
Rozenite, 1158
Rruffite, 1158
Rubicline, 308
Rubidium beryllium sulfate hydrate, 495
Rubidium iron(III) pyrophosphate, 408
Rubidium vanadyl phosphate, 409
Rucklidgeite, 1159
Rudashevskyite, 1159
Ruizite, 1159
Rusinovite, 1159
Russellite, 1159
Rustumite, 1160
Rutherfordine, 1160
Rutile, 198, 1160
Rynersonite, 1160

S

Sabugalite, 1161
Sahlinite, 1161
Sailaufite, 1161
Sakhaite, 1161
Salammoniac, 1162
Saléeite, 1162
Samarium metaphosphate, 410
Samarium orthoborate, 33
Samarium oxalate decahydrate, 102
Samarskite-(Y), 208, 225, 1162
Samarskite-(Yb), 208
Sampleite, 1162
Sanderite Fe^{2+} analogue, 526
Sanderite, 1162
Sanguite, 553, 1163
Sanidine, 1163
Sanjuanite, 1163
Sanmartinite, 718, 1163
Santabarbaraite, 1164
Santarosaite, 1164
Santite, 1164
Saponite, 1164
Sapphirine, 323
Sarcopside, 1165
Sarkinite, 1165
Sarmientite, 1165
Sartorite, 1165
Sassolite, 1166
Satimolite, 75
Sborgite, 73
Scacchite, 1166
Scandium arsenate monohydrate, 611
Scandium lanthanum orthoborate, 33
Scandium vanadate tellurite, 687
Scandium vanadyl selenite, 652
Scarbroite, 89
Schafarzikite, 1166
Schäferite, 555
Schäferite Ni analogue, 584
Schairerite, 527
Scheelite, 1166
Schiavinatoite, 1166
Schlossmacherite, 1167
Schmiederite, 1167
Schmitterite, 1167
Schneiderhöhnite, 1167
Schoenfliesite, 199, 1168
Schoepite, 1168
Scholzite, 1168
Schorl, 1168
Schorlomite, 1169
Schreibersite, 1169
Schreyerite, 1169
Schröckingerite, 1169
Schuetteite, 1170
Schülerite-type mineral, 343
Schultenite, 1170
Schumacherite, 1170
Schwartzembergite, 537
Schwertmannite, 1170
Scolecite, 1170
Scorodite, 1171
Scotlandite, 1171
Scottyite, 248, 1171
Scrutinyite, 1171
Sederholmite, 1172
Segnitite, 1172
Seinäjokite, 1172
Sejkoraite-(Y), 1172
Sekaninaite, 1173
Selenium, 1173
Seligmannite, 1173
Sellaite, 1173
Sénarmontite, 1174
Senegalite, 1174
Sepiolite, 1174
Sérandite, 1174
Serendibite, 1175
Serpierite, 1175
Shannonite, 96
Shattuckite, 1175
Shcherbinaite, 1175
Shchurovskyite, 1176
Shortite, 1176
Shuangfengite, 1176
Shulamitite, 1176
Shumwayite, 514
Siderite, 1176
Sideronatrite, 1177
Sidorenkite, 1177
Sidwillite, 1177
Siegenite, 1177
Sigloite, 1178
Siidraite, 1178
Silicocarnotite, 1178
Silicon, 8, 1179
Silinaite, 291
Sillénite, 1179
Sillimanite, 1179
Silver indium sulfide $AgIn_5S_8$, 495
Silver iron(III) pyrophosphate, 410
Silver tantalum sulfide, 496
Simojovelite, 108
Simonkolleite, 547, 548, 1179
Sinhalite, 72, 1179
Sinjarite, 1180
Sinoite, 116, 1180
Skinnerite, 1180
Skippenite, 1180
Sklodowskite, 1181
Skorpionite, 1181
Smirnite, 1181
Smirnovskite, 440
Smithite, 1181
Smithsonite, 1182
Smythite, 1182
Sobolevskite, 1182

Sodalite, 1182
Sodalite Ca-Al-Mo-analogue, 120
Sodalite Ca-Al-Mo-W-analogue, 119
Sodalite Ca-Al-Mo-W-analogue, 120
Sodalite nitrite analogue, 309
Soddyite, 1182
Sodium acid diarsenite tellurite, 688
Sodium acid pyrophosphate hydrate, 411
Sodium acid selenite, 653
Sodium acid tellurate $Na_2[TeO_2(OH)_4]$, 688
Sodium aluminium molybdate, 674
Sodium aluminum borate $Na_2Al_2B_2O_7$, 58
Sodium barium borate $NaBaB_5O_9$, 59
Sodium bismuth molybdate scheelite-type, 674
Sodium borate $Na_2B_5O_8(OH)·2H_2O$, 60
Sodium borophosphate $Na_5(B_2P_3O_{13})$, 60
Sodium borosulfate $Na_5[B(SO_4)_4]$, 61
Sodium cadmium selenate hydrate, 654
Sodium cadmium sulfate hydrate, 497
Sodium calcium orthoborate, 34, 35
Sodium calcium pentaborate $Na_3Ca(B_5O_{10})$, 62
Sodium calcium silicophosphate $Na_2Ca_4(PO_4)_2SiO_4$ (apatite-type), 457
Sodium diuranate, 161
Sodium gadolinium oxophosphate, 412
Sodium indium arsenate (alluaudite-type), 612
Sodium iron(II) iron(III) phosphate alluaudite-type, 412
Sodium iron(II) pyrophosphate, 413
Sodium iron(III) pyrophosphate, 414
Sodium iron(III) tin orthophosphate, 441
Sodium lanthanum orthoborate, 35
Sodium lanthanum pyrophosphate, 414
Sodium lead neodymium arsenate chloride (apatite-type), 613
Sodium lithium aluminosilicate $Na_3Li_2(AlSi_2O_8)$, 285
Sodium lithium gadolinium carbonate $Na_2LiGd(CO_3)_3$, 82
Sodium lithium selenate hydrate, 654
Sodium magnesium orthophosphate pyrophosphate $Na_4Mg_3(PO_4)_2(P_2O_7)$, 415
Sodium magnesium pentaborate $Na_3MgB_5O_{10}$, 63
Sodium manganese(II) iron(III) phosphate alluaudite-type, 416
Sodium manganese(II) sulfate alluaudite-type, 497
Sodium molybdenum(VI) tellurite, 689
Sodium nickel iron(III) arsenate, 613
Sodium niobium oxophosphate, 416
Sodium samarium orthoborate, 36
Sodium scandium carbonate $Na_5Sc(CO_3)_3·2H_2O$, 83
Sodium stannate, 161
Sodium strontium aluminum borate $NaSr_7AlB_{18}O_{36}$, 64
Sodium strontium orthoborate, 37
Sodium tantalite perovskite-type, 162
Sodium tellurate tellurite hydrate $Na_2Te_2O_6·1.5H_2O$, 690
Sodium tellurite β-$Na_2Te_4O_9$, 690
Sodium thioborate $Na_3B_3S_6$, 498
Sodium thioborate Na_3BS_3, 499
Sodium tin orthophosphate, 417
Sodium tin phosphate, 442
Sodium titanate $Na_2Ti_3O_7$, 200
Sodium titanate $Na_2Ti_6O_{13}$, 207
Sodium titanium iodate, 702
Sodium titanium phosphate, 445
Sodium titanium silicate $Na_2TiSi_2O_7·2H_2O$, 339
Sodium tungsten tellurite, 691
Sodium uranate, 163
Sodium vanadyl borate $Na_3(VO_2)B_6O_{11}$, 64
Sodium vanadyl phosphate $Na(VO)PO_4$, 418
Sodium vanadyl phosphate $Na_2(VO_2)(PO_4)$, 418
Sodium yttrium selenite, 655
Sodium yttrium tellurate borate $Na_2Y_2(Te^{6+}B_2O_{10})$, 65
Sodium yttrium titanate, 163
Sodium zinc orthophosphate, 419
Sodium zinc pentaborate $Na_3ZnB_5O_{10}$, 66
Sodium zinc pyrophosphate, 420
Sodium zinc selenite $Na_2Zn_3(SeO_3)_4·2H_2O$, 656
Sodium zirconium arsenate, 614, 615
Söhngeite, 1183
Somersetite, 86
Sonolite, 1183
Sonoraite, 1183
Spangolite, 1183
Spencerite, 1184
Sperrylite, 1184
Spertiniite, 201, 1184
Spessartine, 1184
Spessartine Ca-rich, 241
Sphaerobertrandite, 324
Sphaerobismoite, 202
Sphalerite, 1184
Spherocobaltite, 1185
Spinel, 1185
Spionkopite, 1185
Spiroffite, 1185
Spodumene, 1186
Spurrite, 1186
Šreinite, 1186
Srilankite, 1186
Stanfieldite, 1186
Stanleyite, 1187
Stannite, 1187
Starkeyite, 1187
Starovaite, 1187
Staurolite, 1188
Steedeite, 1188
Steenstrupine-(Ce), 1188
Stepanovite, 108
Stephanite, 522, 1188
Štěpite, 1189
Stercorite, 1189
Steropesite, 1189
Stetindite, 1189
Stevensite, 293
Stibarsen, 1189
Stibiconite, 1190
Stibioclaudetite, 1190
Stibiocolumbite, 1190
Stibiopalladinite, 1190
Stibnite, 1191

Stichtite, 79, 1191
Stilbite-Ca, 1191
Stilbite-Na, 1191
Stilleite, 1192
Stilpnomelane, 1192
Stishovite, 1192
Stoiberite, 1192
Stolzite, 1192
Stoppaniite, 1193
Stottite, 1193
Strashimirite, 1193
Strengite, 1193
Stringhamite, 1194
Stromeyerite, 1194
Stronadelphite, 1194
Strontianite, 1194
Strontiofluorite, 1194
Strontiohurlbutite, 445, 1195
Strontiojoaquinite, 1195
Strontium aluminum hydroxide, 164
Strontium bismuth(III) selenite hydrate, 656
Strontium borate chloride $Sr_2B_5O_9Cl$, 67
Strontium borate SrB_2O_4, 67
Strontium borate SrB_8O_{13}, 68
Strontium boroarsenate $Sr(BAsO_5)$, 69
Strontium cerium antimonate perovskite-type, 164
Strontium copper tellurate tellurite, 692
Strontium iron(III) oxycarbonate, 83
Strontium iron phosphate whitlockite-related, 420
Strontium magnesium niobate, 165
Strontium magnesium pyrophosphate, 421
Strontium selenate, 657
Strontium tungstate, 716
Strontium vanadyl vanadate, 571
Strunzite, 1195
Struvite, 1196
Struvite Cd analogue, 356
Struvite-(K), 1195
Studtite, 1196
Sturmanite, 1196
Stützite, 1196
Sudoite, 294, 1197
Sudovikovite, 1197
Sulfhydrylbystrite, 310
Sulfur, 1197
Sulphohalite, 1197
Sulvanite, 1198
Suredaite, 1198
Sursassite, 1198
Susannite, 1198
Suseinargiuite, 1199
Svabite, 633
Svanbergite, 1199
Švenekite, 1199
Svornostite, 1199
Swamboite-(Nd), 349
Swedenborgite, 1200
Symplesite, 628, 1200
Synchysite-(Ce), 1200

Syngenite, 1200
Szaibélyite, 1200
Szenicsite, 1201
Szmikite, 1201
Szomolnokite, 478, 1201

T

Takedaite, 1201
Takovite, 1202
Talc, 1202
Talmessite, 1202
Tangdanite, 1203
Tangeite, 1203
Tantalite-(Fe), 1203
Tantalite-(Mg), 1203
Tantalite-(Mn), 1204
Tantalum oxyarsenate, 615
Tantalum oxyphosphate, 422
Tantalum oxyvanadate, 572
Tantite, 202
Tantite orthorhombic polymorph, 1204
Taranakite, 1204
Tarapacáite, 1204
Tarbuttite, 1205
Tausonite, 1205
Tazheranite, 1205
Tazzoliite, 1205
Teepleite, 1205
Teineite, 1206
Tellurantimony, 1206
Tellurite rhombohedral polymorph, 166
Tellurium, 694, 1206
Tellurium(IV) oxosulfate, 499
Tellurium(IV) oxovanadate, 572
Tellurium oxumolybdate, 675
Tellurium(IV) oxyphosphate, 423
Tellurium(IV) oxyselenate, 658
Tellurium oxyselenite, 658
Tellurium(IV) tin oxide Te_3SnO_8, 166
Tellurium(IV) titanium oxide Te_3TiO_8, 167
Tellurium(IV) zirconium oxide Te_3ZrO_8, 168
Tellurobismuthite, 1206
Telluroperite, 555
Tengerite-(Y), 1206
Tennantite, 1207
Tenorite, 1207
Tephroite, 1207
Terlinguacreekite, 549
Tetraferrinontronite, 295
Tetrahedrite, 1207
Tetrammine zinc borofluoride, 233
Tetrawickmanite, 1208
Thadeuite, 468
Thallium feldspar, 311
Thallium sodalite, 311
Thallium tellurite, 692
Thallium(I) selenite vanadate $TlSeVO_5$, 573
Thallium(I) tellurite vanadate $TlTeVO_5$, 574
Thaumasite, 732, 734, 1208

Thecotrichite, 1208
Theophrastite, 1208
Thermonatrite, 1208
Thometzekite, 1209
Thomsonite-Ca, 1209
Thorianite, 1209
Thorikosite, 1209
Thorite, 1210
Thorium divanadate cubic polymorph, 574
Thorium divanadate orthorhombic polymorph, 575
Thorium hydrogenphosphate, 424
Thorium tellurite, 693
Thorneite, 1210
Thortveitite, 1210
Thorutite, 1210
Threadgoldite, 1211
Tiemannite, 1211
Tilasite, 1211
Tilleyite, 1211
Tin(IV) hydroxide, 168
Tin tetraborate, 69
Tiragalloite, 1212
Tissintite, 1212
Tistarite, 203, 1212
Titanite, 1212
Titanium acid phosphate monohydrate, 425
Titanium(III) orthophosphate, 424
Titanium oxophosphate hydrate, 426
Titanoholtite, 1213
Tlapallite, 1213
Tobelite hydrated variety, 286
Tobermorite, 278, 1213
Todorokite, 1213
Tokyoite, 1214
Tolbachite, 554, 1214
Tondiite, 1214
Tooeleite, 1214
Topaz, 723, 725, 1215
Torbernite, 1215
Toturite, 1215
Trabzonite, 1215
Tremolite, 1216
Trevorite, 1216
Trevorite Co-analogue, 203
Triammoniun hydrogen disulfate, 500
Triazolite, 110
Tridymite, 1216
Trilithionite, 1216
Trinepheline, 1217
Triphylite Mg-analogue, 446
Triplite, 1217
Trippkeite, 1217
Tripuhyite, 1217
Trogtalite, 1218
Troilite, 1218
Trolleite, 454
Trona, 1218
Tschermigite, 1218
Tschernichite, 317

Tsumcorite, 1218
Tsumebite, 1219
Tsumoite, 1219
Tugarinovite, 1219
Tuite, 1219
Tululite, 1220
Tunellite, 1220
Tungstenite, 1220
Tungsten(VI) oxyphosphate, 427
Tungsten trioxide monoclinic, 169
Tungsten trioxide orthorhombic, 170
Tungsten trioxide triclinic, 170
Tungstite, 1220
Tunisite, 1221
Turquoise, 1221
Tuzlaite, 1221
Tvrdýite, 352, 447
Tychite, 1221
Tyretskite (monoclinic polytype), 74
Tyrolite, 1222
Tyuyamunite, 1222

U
Uedaite-(Ce), 254
Ulexite, 1222
Ulfanderssonite-(Ce), 243
Ulrichite, 1222
Ulvospinel Zn-analogue, 204
Ulvöspinel, 1223
Umangite, 1223
Umbite, 1223
Umbrianite, 1223
Ungemachite, 1224
Uraninite, 1224
Uranium(IV) oxalate fluoride hydrate, 102
Uranophane-α, 1224
Uranopilite, 1224
Uranosphaerite, 1225
Uranospinite, 638
Uranyl fluoride, 234
Uranyl nitrate hexahydrate, 113
Uranyl oxy-hydroxyphosphate, 427
Uranyl perrhenate hydrate, 719
Urea solution, 1225
Ushkovite, 1225
Usturite, 1225
Uvarovite, 1226
Uvite, 1226

V
Vaesite, 1226
Vajdakite, 1226
Valentinite, 1227
Vanackerite, 1227
Vanadinite Sr,OH-analogue, 580
Vanadinite, 1227
Vanadium(III) antimony(V) selenite, 659
Vanadium oxide bariandite-type, 171
Vanadyl molybdate, 678

Index

β-Vanadyl pyrophosphate, 363
γ-Vanadyl pyrophosphate, 374
Vanadyl selenite, 660
Vanadyl sulfate, 517
Vandenbrandeite, 204
Vandenbrandeite hydrogen-free analogue, 205
Vandendriesscheite, 1227
Vanmeersscheite, 1228
Vantasselite, 1228
Vanthoffite, 1228
Vapnikite, 1229
Variscite, 1229
Variscite-4O, 353
Varulite, 455
Västmanlandite-(Ce), 1229
Vaterite, 1229
Vauquelinite, 1230
Vauxite, 1230
Väyrynenite, 443, 1230
Velikite, 1231
Versiliaite, 1231
Vésigniéite, 1231
Vesuvianite, 2, 5, 259, 1231
Vesuvianite B-bearing, 262, 263
Vesuvianite Cr-bearing, 260
Vesuvianite S-bearing, 260
Veszelyite, 1231
Villamanínite, 1232
Vishnevite, 303
Vishnevite CO_3-bearing, 304
Vishnevite potassium analogue, 302
Vittingeite, 268
Vivianite, 731, 1232
Vladimirivanovite, 1232
Vladykinite, 1232
Voglite, 1233
Volaschioite, 1233
Volborthite, 1233
Vonsenite, 1233
Vorlanite, 1234
Vrbaite, 1234
Vuorelainenite, 1234
Vyacheslavite, 464
Vyacheslavite anhydrous Th analogue, 464
Vysokýite, 626, 1234
Vysotskite, 1235

W
Wadalite, 237
Wadeite, 1235
Wadeite dimorph, 280
Wadeite Rb analogue, 281
Wadsleyite, 1235
Wagnerite-$Ma5bc$, 1235
Waimirite-(Y), 1236
Waimirite-(Yb), 235
Wakabayashilite, 1236
Wakefieldite-(Ce), 1236
Wakefieldite-(La), 1236
Wakefieldite-(Nd), 1236
Wakefieldite-(Pr), 581

Wakefieldite-(Y), 588, 1237
Walpurgite, 1237
Walstromite, 1237
Wardite, 1237
Waterhouseite, 1238
Wavellite, 1238
Wavellite-(OH), 443
Waylandite, 1238
Weddellite, 1238
Weeksite, 1239
Wegscheiderite, 1239
Weissbergite, 1239
Weloganite, 1239
Wendwilsonite, 1240
Wermlandite carbonate analogue, 91
Wernerbaurite, 584
Wernerkrauseite, 1240
Wesselsite, 298
Wetherillite, 1240
Wheatleyite, 1241
Whelanite, 1241
Whewellite, 1241
Whitecapsite, 1241
Whiteite [possibly, whiteite-(CaMnMg)], 1242
Whitlockite, 1242
Whitmoreite, 1242
Widenmannite, 1242
Wiklundite, 348
Wilhelmgümbelite, 448
Willemite, 1243
Willemseite, 1243
Wiluite, 252, 255
Windhoekite Na-bearing variety, 293
Winstanleyite, 1243
Witherite, 1243
Wittichenite, 1243
Wollastonite, 1244
Wölsendorfite, 221
Woodallite, 123
Woodhouseite, 1244
Wopmayite, 1244
Wulfenite, 1244
Wupatkiite, 1245
Wurtzite, 1245
Wüstite, 1245

X
Xanthoxenite, 438
Xenotime-(Y), 1245
Xieite, 1246
Ximengite polymorph, 448, 449
Xocolatlite, 1246
Xocomecatlite, 1246
Xonotlite, 1246

Y
Yafsoanite, 695
Yangzhumingite, 295
Yarrowite, 1247
Yavapaiite, 478
Yecoraite, 1247

Ye'elimite, 1247
Yimengite, 1247
Yingjiangite, 1247
Yoshimuraite, 337
Yttriaite-(Y), 1248
Yttrium barium borate YBa$_3$B$_9$O$_{18}$, 70
Yttrium hydroxychloride hydrate, 543
Yttrium iron antimony(V) oxide pyrochlore-type, 172
Yttrium metaphosphate, 428
Yttrium oxide, 172
Yttrium tungstate, 717
Yttrium vanadyl oxyselenite, 660
Yttrium vanadyl oxytellurite, 694
Yukonite, 1248
Yuksporite, 1248
Yurmarinite, 1248
Yushkinite, 1249
Yvonite, 1249

Z

Żabińskiite, 1249
Zadovite, 1249
Zaherite, 518
Zálesíite, 1250
Zanazziite, 1250
Zaratite, 1250
Zdeněkite, 1250
Zellerite, 1251
Zemannite, 698, 1251
Zemkorite, 1251
Zhangpeishanite, 1251
Zhemchuzhnikovite, 109
Ziesite, 1252
Ziesite and blossite polymorph, 582
Ziminaite, 588
Ziminaite monoclinic polymorph, 583
Zinc basic pyrovanadate hydrate, 577
Zinc hydroxychloride, 544
Zinc hydroxyfluoride, 234
Zinc iron(III) orthovanadate, 576
Zincite, 206, 1252
Zinc molybdate, 676
Zincoberaunite, 350
Zincobotryogen, 528
Zincochromite, 1252
Zincocopiapite, 1252
Zinc orthoborate hydroxide, 38
Zinc orthoborate orthophosphate, 39
Zinc orthovanadate, 576
Zincospiroffite, 1253
Zincovelesite-6N6S, 220
Zincovoltaite, 531
Zinc stannate, 174
Zinc telluromolybdate, 676
Zinc vanadyl oxide Zn(VO$_2$)$_2$O$_2$, 578
Zinc vanadyl phosphate, 429
Zinkenite, 1253
Zippeite, 1253
Zircon, 1253
Zirconium acid arsenate monohydrate, 616
Zirconium acid phosphate monohydrate, 429, 430
Zirconium basic oxalate, 103
Zirconium molybdenum oxide (monoclinic), 677
Zirconium molybdenum oxide (trigonal), 678
Zirconolite-2M, 218
Zoisite, 1254
Zorite, 1254
Zuktamrurite, 1254
Zunyite, 1254
Zýkaite, 1255